Advanced Organic Chemistry

THIRD EDITION

Part A: Structure and Mechanisms

Advanced Organic Chemistry

PART A: Structure and Mechanisms
PART B: Reactions and Synthesis

Advanced Organic Chemistry

THIRD EDITION

Part A: Structure and Mechanisms

FRANCIS A. CAREY
and RICHARD J. SUNDBERG

University of Virginia
Charlottesville, Virginia

PLENUM PRESS • NEW YORK AND LONDON

Library of Congress Cataloging in Publication Data

(Revised for 3rd ed.)
Carey, Francis A., 1937-
Advanced organic chemistry.
Includes bibliographical references.
Contents: pt. A. Structure and mechanisms—pt. B. Reactions and synthesis.
1. Chemistry, Organic. I. Sundberg, Richard J., 1938- . I. Title.
QD251.2.C36 1990 547 90-6851
ISBN 0-306-43440-7 (Part A)
ISBN 0-306-43447-4 (pbk.: Part A)
ISBN 0-306-43456-3 (Part B)
ISBN 0-306-43457-1 (pbk.: Part B)

10 9 8 7 6

A Division of Plenum Publishing Corporation
233 Spring Street, New York, N.Y. 10013

Printed in the United States of America

Preface to the Third Edition

The purpose of this edition, like that of the earlier ones, is to provide the basis for a deeper understanding of the *structures* of organic compounds and the *mechanisms* of organic reactions. The level is aimed at advanced undergraduates and beginning graduate students. Our goals are to solidify the student's understanding of basic concepts provided by an introduction to organic chemistry and to present more information and detail, including quantitative information, than can be presented in the first course in organic chemistry.

The first three chapters consider the fundamental topics of bonding theory, stereochemistry, and conformation. Chapter 4 discusses the techniques that are used to study and characterize reaction mechanisms. Chapter 9 focuses on aromaticity and the structural basis of aromatic stabilization. The remaining chapters consider basic reaction types, including substituent effects and stereochemistry. As compared to the earlier editions, there has been a modest degree of reorganization. The emergence of free-radical reactions in synthesis has led to the inclusion of certain aspects of free-radical chemistry in Part B. The revised chapter, Chapter 12, emphasizes the distinctive mechanistic and kinetic aspects of free-radical reactions. The synthetic applications will be considered in Part B. We have also split the topics of aromaticity and the reactions of aromatic compounds into two separate chapters, Chapters 9 and 10. This may facilitate use of Chapter 9, which deals with the nature of aromaticity, at an earlier stage if an instructor so desires.

Both the language of valence bond theory and resonance and that of molecular orbital theory are used in the discussion of structural effects on reactivity. Our intention is to illustrate the use of both types of interpretation, with the goal of facilitating the student's ability to understand and apply both of these viewpoints of structure. Nearly all reaction types and concepts are illustrated by specific examples from the chemical literature. Such examples, of course, cannot provide breadth of coverage, and those that are cited have been selected merely to illustrate the mechanism or interpretation. Such illustrations are not meant to suggest any priority of the specific example which has been selected. Whenever possible, references to reviews which can provide a broader coverage of the topic are given.

Some new problems have been added. The general level is similar to that of the earlier editions, and we expect many of the problems will present a considerable challenge to the student. Most represent applications of concepts to new systems and circumstances, rather than review of material explicitly presented in the text. References to the literature material upon which the problems are based are given at the end of the book for nearly all problems.

The companion volume, Part B, has been revised to reflect the continuing development and evolution of synthetic practice. Part B emphasizes the synthetic applications of organic reactions. We believe that the material in Parts A and B can provide a level of preparation which will permit the student to assimilate and apply the primary and review literature of organic chemistry.

During the preparation of the third edition, F.A.C. has been involved in other writing endeavors, and the primary responsibility for new errors or omission of new results rests with R.J.S. We hope that this text will continue to serve students in fostering an understanding of organic chemistry. We continue to welcome comments and suggestions from colleagues which can improve the treatment of material in this text.

F. A. Carey
R. J. Sundberg

Charlottesville, Virginia

Contents of Part A

Contents of Part B

Chapter 7. Organometallic Compounds of Group I and II Metals

Chapter 8. Reactions Involving the Transition Metals

Chapter 9. Carbon–Carbon Bond-Forming Reactions of Compounds of Boron, Silicon, and Tin

List of Tables

List of Figures

List of Schemes

1

Chemical Bonding and Structure

Introduction

Organic chemistry is a broad field which intersects with such diverse areas as biology, medicine and pharmacology, polymer technology, agriculture, and petroleum engineering. At the core of organic chemistry are fundamental concepts of molecular structure and reactivity of carbon-containing compounds. The purpose of this text is to cover the central core of organic chemistry. This knowledge can be used within organic chemistry or applied to other fields, such as those named above, which require significant contributions from organic chemistry. One organizational approach to organic chemistry divides it into three main areas—*structure, dynamics,* and *synthesis. Structure* includes the description of bonding in organic molecules and the methods for determining, analyzing, and predicting molecular structure. *Dynamics* refers to study of the physical properties and chemical transformations of molecules. *Synthesis* includes those activities which are directed toward finding methods which convert existing substances into different compounds. These three areas are all interrelated, but synthesis is built on knowledge of both structure and reactions (chemical dynamics), while understanding dynamic processes ultimately rests on detailed knowledge about molecular structure. A firm grounding in the principles of structure and chemical bonding is therefore an essential starting point for fuller appreciation of dynamics and synthesis. In this first chapter, we will discuss the ideas that have proven most useful to organic chemists for describing and correlating facts, concepts, and theories about the structure of organic molecules.

Structural formulas serve as key devices for communication of chemical information, and it is important to recognize the symbolic relationship between structural formulas and molecular structure. The current system of structural formulas arose

Fig. 1.1. Idealized view of σ-bond formation by overlap of (a) an s and a p orbital and (b) two p orbitals.

of a p orbital of one atom with an s or p orbital of another is illustrated in Fig. 1.1. The electron distribution that results is cylindrically symmetric with respect to the internuclear axis and defines a σ bond.

The electronic configuration of a carbon atom in its ground state, established unambiguously by spectroscopic measurements as $1s^2 2s^2 2p^2$, precludes a simple rationalization of the tetrahedral bonding at carbon. Pauling suggested that the four atomic orbitals ($2s$, $2p_x$, $2p_y$, $2p_z$) are replaced by a set of four equivalent *hybrid orbitals*, designated sp^3. The approximate shapes of these orbitals are shown in Fig. 1.2. Notice particularly that the probability distribution is highly directional for the sp^3 orbitals, with the region of greatest probability concentrated to one side of the nucleus.

Orbital hybridization has two important consequences: First, four bonds, rather than two, may be formed to carbon. Second, the highly directional sp^3 orbitals provide for more effective overlap and stronger bonds. Thus, although an isolated carbon atom with one electron in each of four equivalent sp^3-hybridized orbitals would be of higher energy than the spectroscopic ground state, the energy required in a formal sense to promote two electrons from a $2s$ orbital to sp^3 orbitals is more than compensated for by the formation of four bonds rather than two. In addition, each of the bonds is stronger due to the directional properties of the hybrid orbitals. Tetrahedral geometry is predicted by the mathematical description of hybridization. Methane is found experimentally to be a perfect tetrahedron, with each H—C—H bond angle equal to 109.5°.

The descriptive valence bond approach to the bonding in ethylene and acetylene and their congeners is analogous to that for methane. In ethylene (Fig. 1.3) each carbon bears three ligands and possesses sp^2 hybridization. Three sp^2 orbitals are generated from the $2s$ and two of the $2p$ orbitals. The three sp^2 orbitals are coplanar and orthogonal to the remaining $2p$ orbital. A bond is formed between the two carbon atoms by overlap of an sp^2 orbital of each. The four hydrogens are bonded by σ bonds involving hydrogen $1s$ orbitals and the remaining two sp^2 hybrid orbitals. Additional bonding between the two carbon atoms is portrayed as resulting from overlap of the unhybridized p orbitals on each carbon atom, each of which contains

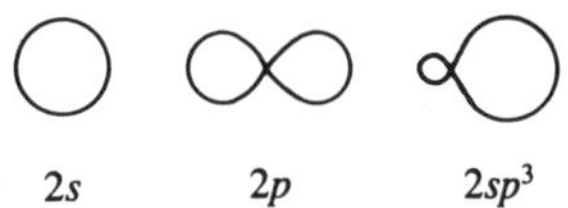

Fig. 1.2. Cross section of angular dependence of orbitals.

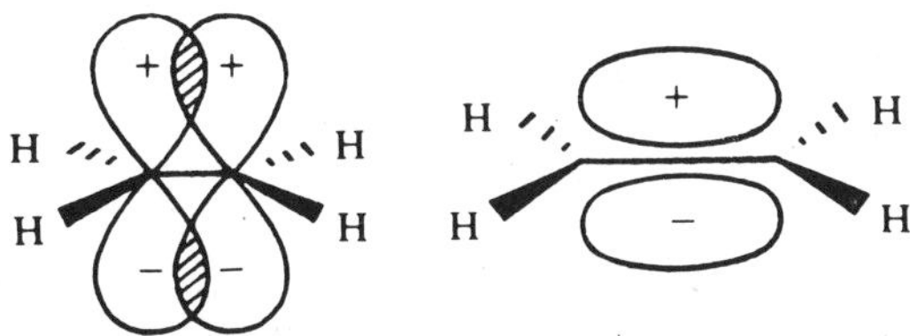

Fig. 1.3. The π bond in ethylene.

one electron. This overlap is somewhat less effective than that of a σ bond and defines a π bond. The electron distribution in a π bond is concentrated above and below the plane of the σ framework. The molecule is planar, and the plane defined by the nuclei represents a nodal plane for the π system. The probability of finding an electron occupying the π orbital in this plane is zero.

The hybridization at each carbon atom of acetylene is *sp*, and the two carbon atoms are considered as bonded by a σ bond and two π bonds, as shown in Fig. 1.4.

The relation between the number of ligands on carbon (its coordination number), hybridization, and molecular geometry is summarized in Table 1.1. Unless all the ligands on a particular carbon atom are identical, there will be deviations from the perfectly symmetrical structures implied by the hybridization schemes. For example, in contrast to methane and carbon tetrachloride, which are tetrahedral with bond angles of 109.5°, cyclohexane has a C—C—C angle of 111.5°. The H—C—H angle in formaldehyde is 118°, rather than 120°. Benzene, however, is a regular hexagon with 120° bond angles.

Large deviations in bond angles from the normal values are found in cyclopropane, cyclobutane, and other molecules containing three- and four-membered rings. These molecules are less stable than molecules with larger rings, and the difference in energy is referred to as *angle strain.* Since the three carbon atoms of a cyclopropane ring are required by symmetry to be at the vertices of an equilateral triangle, the internuclear angles are 60°. This arrangement represents a serious distortion of the normal tetrahedral bond angle and engenders unique chemical and physical properties. To develop a valence bond model of the bonding in cyclopropane, it is assumed that the carbon atoms will adopt the hybridization that produces the most stable

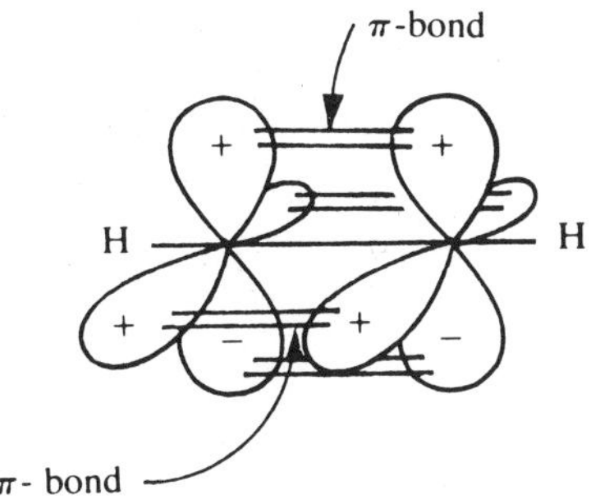

Fig. 1.4. π-Bonding in acetylene.

Table 1.1. Dependence of Structure on Hybridization of Carbon

Number of ligands	Hybridization	Geometry	Examples
4	sp^3	Tetrahedral	Methane, cyclohexane, methanol, carbon tetrachloride
3	sp^2	Trigonal	Ethylene, formaldehyde, benzene methyl cation, carbonate ion
2	sp	Linear	Acetylene, carbon dioxide, hydrogen cyanide, allene

bonding arrangement.[4] The orbitals used for forming the carbon-carbon bonds in cyclopropane can overlap more effectively if they have more *p* character than normal sp^3 bonds, since additional *p* character corresponds to a reduced bond angle. Consequently, the orbitals used for bonding to hydrogen must have increased *s* character. This adjustment in hybridization can be described quantitatively by assignment of numerical values to the "percent *s* character" in the C—H bonds. The values of 33% and 17%, respectively, have been suggested for the C—H and the C—C bond of cyclopropane on the basis of nuclear magnetic resonance (NMR) measurements.[5] The picture of the bonding in cyclopropane indicates that the region of maximum orbital overlap would not correspond to the internuclear axis. The C—C bonds are described as "bent bonds" (Fig. 1.5).

The change in hybridization is associated with a change in electronegativity. The greater the *s* character, the greater is the electronegativity of a particular carbon atom. As a result, strained carbon atoms are more electronegative than unstrained ones.[6] Figure 1.6 shows some calculated charges for cyclopropane and other strained hydrocarbons in comparison with the unstrained reference cyclohexane. Notice that the greater the distortion from the normal tetrahedral angle, the greater is the negative charge on carbon.

Even more drastic distortions from ideal geometry are found when several small rings are assembled into bicyclic, tricyclic, and more complex molecules. The synthesis of such highly strained molecules is not only a challenge to the imagination and skill of chemists, but also provides the opportunity to test bonding theories by probing the effects of unusual bonding on the properties of molecules. One series of such molecules are the *propellanes.*[7] The structure of some specific propellanes and the strain energies of the molecules are shown in Fig. 1.7.

Each of the molecules in Fig. 1.7 has been synthesized, and some of their physical and structural properties have been analyzed. In the propellanes with small rings, the bridgehead carbon must be severely flattened to permit bonding. In order to attain this geometry, the hybridization at the bridgehead carbons must change as the size of the bridges decreases. While the hybridization at the bridgehead carbon

4. For a review of various descriptions of the bonding in cyclopropane, see A. de Meijer, *Angew. Chem. Int. Ed. Eng.* **18**, 809 (1979).
5. F. J. Weigert and J. D. Roberts, *J. Am. Chem. Soc.* **89**, 5962 (1967).
6. K. B. Wiberg, R. F. W. Bader, and C. D. H. Lau, *J. Am. Chem. Soc.* **109**, 1001 (1987).
7. K. B. Wiberg, *Acc. Chem. Res.* **17**, 379 (1984).

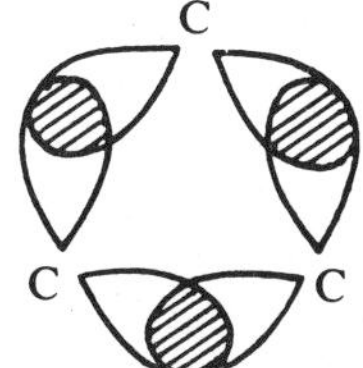

Fig. 1.5. Bent bonds in cyclopropane.

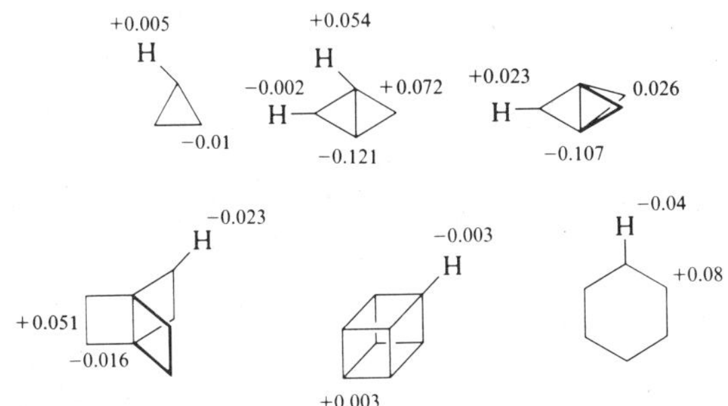

Fig. 1.6. Charge distributions in strained cyclic hydrocarbons in comparison with cyclohexane. Data are from K. B. Wiberg, R. F. W. Bader, and C. D. H. Lau, *J. Am. Chem. Soc.* **109**, 1001 (1987).

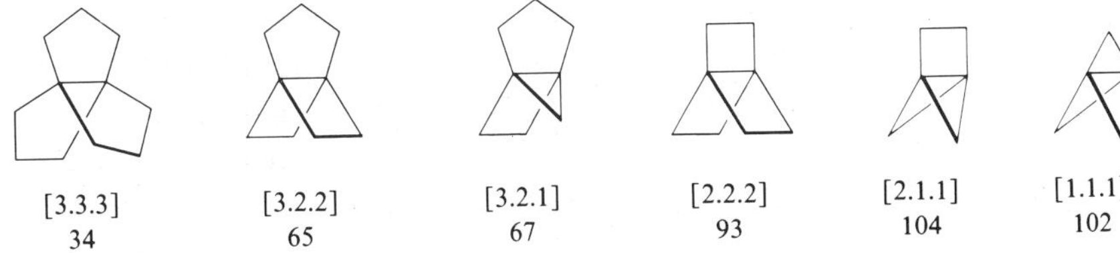

Fig. 1.7. Strain energies of some propellanes in kcal/mol.

in [4.4.4]propellane can be approximately the normal sp^3, in [2.2.2]propellane the flattening of the bridgehead must result in a change to approximately sp^2 hybridization, with the central bond between the two bridgehead carbons being a σ bond formed by overlap of two p orbitals. The distortion is still more extreme in [1.1.1]propellane, where each bridgehead carbon is an "inverted carbon" with all four bonds to one side. The resulting bond is quite special in its characteristics and is not adequately described as a localized bond.[8]

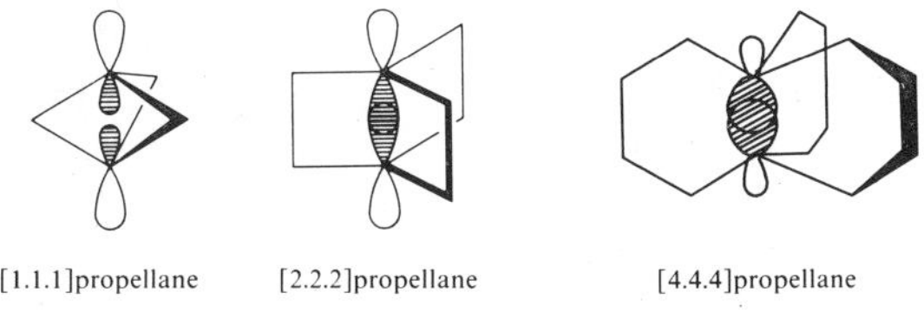

8. J. E. Jackson and L. C. Allen, *J. Am. Chem. Soc.* **106**, 591 (1984).

The distortion in the bond angles in propellanes leads both to strain and unusual chemical reactivity. [3.2.1]Propellane, for example, is found to have a strain energy of 67 kcal/mol, as compared to 27 kcal/mol for cyclopropane. The molecule is exceptionally reactive and undergoes a variety of reactions involving the cleavage of the central bond under mild conditions. For example, it undergoes bromination instantaneously at −50°C.[9] The strain should be less in [3.3.3]propellane, where a smaller distortion at the bridgehead carbons is required to permit bonding. This is reflected by the lower strain energy of 34 kcal/mol. On the other hand, the smaller bridges lead to increased strain. [2.2.1]Propellane conforms to the expectation that it would be highly reactive. It can be observed when isolated in solid argon at 45°K but decomposes at temperatures higher than this and cannot be isolated as a pure substance.[10]

A second concept that makes valence bond theory useful for the structural description of complex molecules is *resonance theory*. Resonance theory is an extension of valence bond theory which recognizes that for many molecules more than one Lewis structure can be written. Its usefulness in organic chemistry lies in its being a convenient way of depicting electron delocalization. Resonance theory is particularly useful in describing conjugated systems and reactive intermediates. Arguments based on resonance theory are usually made in a qualitative way, although a full mathematical treatment can be applied.[11] The elements of resonance theory that are necessary for qualitative applications can be summarized as follows:

a. Whenever alternative Lewis structures can be written for a molecule differing only in assignment of electrons among the nuclei, with the nuclear positions being constant for all the structures, then the molecule is not adequately represented by a single Lewis structure, but has properties of all of them.
b. All structures are restricted to the maximum number of valence electrons which is appropriate for each atom, that is, two for hydrogen and eight for the first-row elements.
c. Some individual Lewis structures are more stable than others. The structures that approximate the actual molecule most closely are those that incorporate the following features: maximum number of covalent bonds, minimum separation of unlike charges, and placement of any negative charges on the most electronegative atom (or any positive charge on the most electropositive atom). Stated in another way, the most favorable (lowest-energy) resonance structure makes the greatest contribution to the true (hybrid) structure.

9. K. B. Wiberg and G. J. Burgmaier, *J. Am. Chem. Soc.* **94**, 7396 (1972).
10. F. H. Walker, K. B. Wiberg, and J. Michl, *J. Am. Chem. Soc.* **104**, 2056 (1982).
11. For a classical presentation of resonance theory, see G. W. Wheland, *Resonance Theory in Organic Chemistry*, Wiley, New York, 1955. Models of molecular structure based on mathematical descriptions of valence bond theory have been developed: F. W. Bodrowicz and W. A. Goddard III, in *Modern Theoretical Chemistry, Methods of Electronic Structure Theory*, H. F. Schaefer III (ed.), Plenum Press, New York, 1977, Vol. 3, Chapter 4; A. Voter and W. A. Goddard III, *Chem. Phys.* **57**, 253 (1981); N. D. Epiotis, *Unified Valence Bond Theory of Electronic Structure*, Springer-Verlag, Berlin, 1983.

d. In most cases, the delocalization of electrons, as represented by the writing of alternative Lewis structures, is associated with enhanced stability relative to a single localized structure. This is not always true, however, since molecules and ions are known in which electron delocalization produces an increase in energy relative to a localized model.

The use of the resonance concept can be illustrated by considering the relative acidity of 2-methylpropene (isobutene) and 2-propanone (acetone). The relative acidity is indicated by the pK values, which are, respectively, ~45 and ~25. The difference of ~20 pK units (20 powers of 10!) shows that it is much easier for a proton to be removed from acetone than from isobutene. The main reason for the difference in acidity is the difference in stability of the two conjugate bases. A resonance-stabilized anion is generated in each case, but one of the contributing structures for the acetone anion has a negative charge on oxygen. For the anion of isobutene, both structures show the negative charge on carbon. Since oxygen is a more electronegative element than carbon, application of resonance theory leads to the conclusion that the acetone anion will be more stable than the isobutene anion and that acetone will therefore be more acidic.

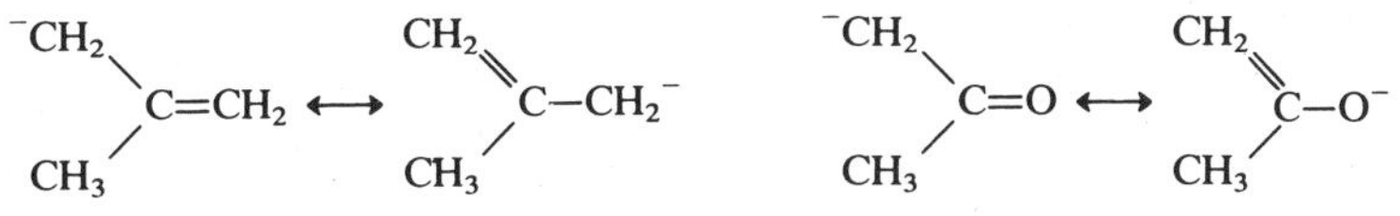

$$(^-CH_2)(CH_3)C{=}CH_2 \longleftrightarrow (CH_2{=})(CH_3)C{-}CH_2^- \qquad (^-CH_2)(CH_3)C{=}O \longleftrightarrow (CH_2{=})(CH_3)C{-}O^-$$

Another simple illustration of the resonance concept is the allyl cation, which is known to be a particularly stable carbocation. This stability can be understood by recognizing that the cationic charge is delocalized over two carbon atoms, as represented by the two equivalent resonance structures. The delocalization imposes a structural requirement. The p orbitals on the three contiguous carbon atoms must all be aligned in the same direction to permit electron delocalization. As a result, there is an energy barrier to rotation about the carbon–carbon bonds in the allyl cation. The most stable geometry is planar, and it is estimated that the barrier for rotation is 25–28 kcal/mol.[12]

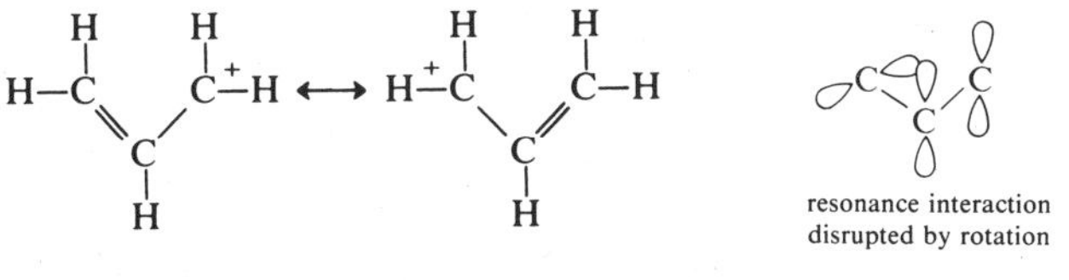

$$H_2C{=}CH{-}\overset{+}{C}H_2 \longleftrightarrow H_2\overset{+}{C}{-}CH{=}CH_2$$

resonance interaction disrupted by rotation

Carbonyl compounds having a carbon–carbon double bond adjacent to the carbonyl group provide a good example of how structural features can be related to resonance interactions. While only a single uncharged structure can be drawn, two structures with charges can be drawn. Of these, the structure with a negative

12. H. Mayr, W. Forner, and P. v. R. Schleyer, *J. Am. Chem. Soc.* **97**, 752 (1975).

charge on oxygen is far more important because of the higher electronegativity of oxygen relative to carbon. The structure with a positive charge on oxygen is very unfavorable and would make only a minor contribution.

major contributor ⟷ significant contributor ⟷ minor contributor

Some of the structural features of this class of compounds which are in accord with the resonance picture are as follows. (1) The C=O bond is not as strong as in saturated carbonyl compounds. This is revealed by the infrared stretching frequency, which comes at lower energy (typically 1690 cm^{-1} versus 1730 cm^{-1} for saturated compounds). (2) Carbon-13 NMR spectroscopy also reveals that the β-carbon is less shielded (lower electron density) than is the case for a simple alkene. This results from the delocalization of π electrons from this carbon to the carbonyl oxygen. (3) The chemical reactivity of the double bond is also affected by the presence of the conjugated carbonyl group. Simple alkenes are not very reactive toward nucleophiles. In contrast, double bonds adjacent to carbonyl groups do react with nucleophiles. The partial positive charge depicted by the resonance structure makes the β-carbon subject to nucleophilic attack. Also, the carbonyl group stabilizes the negative charge that develops at the α-carbon as a result of nucleophilic attack.

Judgment must be exercised in cases where two of the general resonance criteria are in conflict. For example, the structure **B** is an important contributor to the resonance hybrid for the intermediate in the nitration of methoxybenzene (anisole), even though the structure has a positive charge on oxygen. This unfavorable feature is compensated for by the additional covalent bond present in structure **B** compared with the other major contributor, structure **A**, in which the positive charge is on carbon.

A ⟷ **B**

Benzene is a very familiar example of a molecule that is not adequately represented by any single valence bond structure. Either single structure with alternating single and double bonds erroneously suggests alternating "short" and "long" bonds around the ring. Actually, benzene is a perfectly hexagonal molecule.

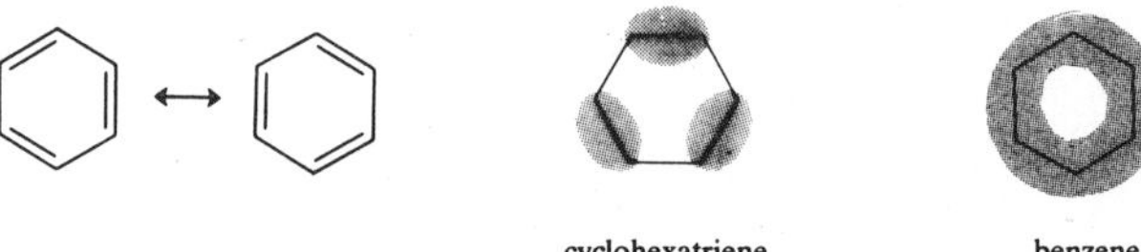

cyclohexatriene　　benzene

The second structure with the alternate disposition of double bonds is necessary to depict the symmetrical nature of benzene. Benzene is also an excellent example of the stabilization that is usually associated with delocalized (resonance stabilized) molecules. Benzene is considerably more stable than the hypothetical localized cyclohexatriene structure would be. The principal contribution to the greater stability of benzene is the decreased electron–electron repulsion. If the double bonds were localized as in cyclohexatriene, they would on the average be restricted to a smaller region of space. In the delocalized structure, benzene, the π electrons have equal distribution between all six carbon atoms. The stabilization resulting from the electron delocalization that occurs in conjugated systems is referred to as *resonance energy*. Unfortunately, this energy is not a measurable quantity, since it is a difference between a real molecule (benzene) and a hypothetical one (cyclohexatriene) for which no experimental data can be obtained. Calculation of resonance stabilization therefore depends upon assumptions about the reference system. Estimated values for the resonance energy of benzene range from 20 to 40 kcal/mol. The value 36 kcal/mol is frequently adopted. It is certainly clear that benzene is stabilized as a result of the electron delocalization implied by the resonance structures, but by just how much is experimentally indeterminate.

It must be emphasized that resonance structures do not represent separate molecules. A single structure exists. Resonance structures are alternative descriptions that, *taken together*, describe the real molecule. To use resonance theory in a qualitative way to predict features of structure and reactivity, one must make judgments about the qualitative weighting of all the possible contributing structures. The student's objective should be to develop both a capacity for making such judgments and a recognition of the limitations to the reliability of analyses based on qualitative resonance theory. Much more will be said about resonance theory and resonance structures, particularly in Chapters 9 and 10 where aromaticity and aromatic substitution are discussed.

1.2. Bond Energies, Lengths, and Dipoles

Of the various geometric parameters associated with molecular shape, the one most nearly constant from molecule to molecule and most nearly independent of remote substituent effects is bond length. Bond lengths to carbon depend strongly on the hybridization of the carbon involved but are little influenced by other factors. Table 1.2 lists the interatomic distances for some of the most common bonds in organic molecules. The near constancy of bond lengths from molecule to molecule reflects the fact that the properties of individual bonds are, to a good approximation, independent of the remainder of the molecule. Other features of molecular structure that are closely related to bond strength, such as force constants for bond stretching, are also very similar from molecule to molecule.

Table 1.3 gives some bond energy data. Part A includes bond energies for some simple diatomic molecules and generalized values for some of the types of bonds

Table 1.2. Bond Lengths (Å)[a]

sp^3	C–H	1.09	sp^3–sp^3	C–C	1.54	sp^2–sp^2	C–C	1.46	C–O	1.42
sp^2	C–H	1.086	sp^3–sp^2	C–C	1.50	sp^2–sp^2	C=C	1.34	C=O	1.22
sp	C–H	1.06	sp^3–sp	C–C	1.47	sp–sp	C≡C	1.20		

a. From experimental values tabulated for simple molecules by M. J. S. Dewar and W. Thiel, *J. Am. Chem. Soc.* **99**, 4907 (1977).

found most often in organic molecules. The assumption that bond energies are independent of the remainder of the molecule is a rather rough one. Part B of Table 1.3 lists some specific C—H, C—C, and other bond energies. It is apparent that some are substantially different from the generalized values. For example, the CH_2—H bond dissociation energies listed for propene and toluene are 85 kcal/mol, which is significantly less than for a C—H bond in methane (104 kcal/mol). The reason for the relative weakness of these particular bonds is that the allyl and benzyl radicals that are produced by the bond dissociations are stabilized by resonance. A similar explanation lies behind the diminished strength of the sp^3–sp^3 carbon–carbon bond in ethylbenzene. The general trend toward weaker C—C bonds with increased substitution that can be recognized in Table 1.3 reflects the increased stability of substituted radicals relative to primary radicals.

$$H_2C{=}CH{-}CH_2\cdot \longleftrightarrow \cdot H_2C{-}CH{=}CH_2$$

C_6H_5—$CH_2\cdot$ ⟷ ·C_6H_5=CH_2 ⟷ C_6H_5=CH_2 ⟷ C_6H_5=CH_2

Table 1.3. Bond Energies (kcal/mol)

A. Some common bond energies[a]

H—H	103	C—H	98	C=C	145
C—C	81	N—H	92	C≡C	198
O—O	34	O—H	109	N≡N	225
Cl—Cl	57	Cl—H	102	C=O	173
Br—Br	45	Br—H	87	C—O	79
I—I	36	I—H	71	C—N	66

B. Some specific bond dissociation energies[b]

H_3C—H	104	H_3C—CH_3	88	H_3C—F	108
CH_3CH_2—H	98	H_5C_2—CH_3	85	H_3C—Cl	84
$H_2C{=}CH$—H	104	$(CH_3)_2CH$—CH_3	83	H_3C—Br	70
$H_2C{=}CHCH_2$—H	85	$PhCH_2$—CH_3	70	H_3C—I	56
$PhCH_2$—H	85	H_5C_2—C_2H_5	82	H_3C—OH	91
H_2N—H	103	$(CH_3)_2CH$—$CH(CH_3)_2$	78		
CH_3NH—H	92				
CH_3O—H	102	$H_2C{=}CH_2$	163[c]		
		HC≡CH	230[c]		

a. From Table 1, G. J. Janz, *Thermodynamic Properties of Organic Compounds*, Academic Press, New York, 1967.
b. Except where noted, from J. A. Kerr, *Chem. Rev.* **66**, 465 (1966).
c. From S. W. Benson, *J. Chem. Educ.* **42**, 502 (1965).

The bond energies in Table 1.3 refer to *homolytic* bond dissociation to uncharged radical fragments. Many reactions involve heterolytic bond cleavages. The energy for heterolytic cleavage of C—H or C—C bonds in the gas phase is very high, largely because of the energy required for charge separation. However, in solution, where stabilization of the ions by solvation becomes possible, heterolytic bond dissociation can become energetically feasible. Heterolytic bond dissociations are even more sensitive to structural changes than homolytic ones. Table 1.4 gives a series of comparable ionization energies for cleavage of H^- from hydrocarbons and Cl^- from chlorides in the gas phase. Besides noting the higher energies for heterolytic bond dissociations in comparison with the homolytic bond dissociation energies, it can be seen that branching decreases the energy requirement for heterolytic bond cleavage more dramatically than for homolytic cleavage.

Smaller, but nevertheless significant, differences in energies of organic molecules also result from less apparent differences in structure. Table 1.5 gives the heats of formation of some hydrocarbons. These energy values represent the heat evolved on formation of the compound from its constituent elements under standard conditions. The heats of formation therefore permit precise comparison of the stabilities of *isomeric compounds.* The more negative the heat of formation, the greater is the stability. Direct comparison of compounds having different elemental composition is not meaningful, since the total number of bonds formed is then different.

Part A of Table 1.5 shows all the acyclic C_4-C_6 and some of the C_8 hydrocarbons. A general trend is discernible in the heats of formation data. Branched-chain hydrocarbons are more stable than straight-chain hydrocarbons. For example, ΔH_f for *n*-octane is −49.82 kcal/mol, whereas the most highly branched isomer possible, 2,2,3,3-tetramethylbutane, is the most stable of the octanes, with ΔH_f of −53.99 kcal/mol. Similar trends are observed in the other series.

Part B of Table 1.5 gives heats of formation for the C_4, C_5, and some of the C_6 alkenes. A general relationship is also observed for the alkenes. The more highly substituted the double bond, the more stable is the compound. There are also other factors that enter into alkene stability. *trans*-Alkenes are usually more stable than

Table 1.4. Heterolytic Bond Dissociation Energies for Some C—H and C—Cl Bonds[a]

	$R{-}H \rightarrow R^+ + H^-$	$R{-}Cl \rightarrow R^+ + Cl$
R	ΔE_{C-H} (kcal/mol)	ΔE_{C-Cl} (kcal/mol)
CH_3	312.2	227.1
CH_3CH_2	272.6	190.3
$(CH_3)_2CH$	249.9	171.0
$CH_2{=}CH$	290.2	
$CH_2{=}CHCH_2$	255.3	

a. Data from C. G. Screttas, *J. Org. Chem.* **45**, 333 (1980).

Table 1.5. Standard Heats of Formation of Some Hydrocarbons (kcal/mol)[a]

A. Saturated hydrocarbons			
C_4		C_8	
n-Butane	−30.15	*n*-Octane	−49.82
i-Butane	−32.15	2-Methylheptane	−51.50
C_5		3-Methylheptane	−50.82
n-Pentane	−35.00	4-Methylheptane	−50.69
i-Pentane	−36.90	2,2-Dimethylhexane	−53.71
Neopentane	−36.97	2,3-Dimethylhexane	−51.13
		2,4-Dimethylhexane	−52.44
C_6		3,3-Dimethylhexane	−52.61
n-Hexane	−39.96	2,2,3-Trimethylpentane	−52.61
2-Methylpentane	−41.66	2,2,4-Trimethylpentane	−53.57
3-Methylpentane	−41.02	2,2,3,3-Tetramethylbutane	−53.99
2,3-Dimethylbutane	−42.49		
2,2-Dimethylbutane	−44.35		
B. Alkenes			
C_4		C_6	
1-Butene	−.03	1-Hexene	−9.96
trans-2-Butene	−2.67	*trans*-2-Hexene	−12.56
cis-2-Butene	−1.67	*cis*-2-Hexene	−11.56
2-Methylpropene	−4.04	*trans*-3-Hexene	−12.56
C_5		*cis*-3-Hexene	−11.56
1-Pentene	−5.00	2-Methyl-1-pentene	−13.56
trans-2-Pentene	−7.59	3-Methyl-1-pentene	−11.02
cis-2-Pentene	−6.71	4-Methyl-1-pentene	−11.66
2-Methyl-1-butene	−8.68	2-Methyl-2-pentene	−14.96
3-Methyl-1-butene	−6.92	3-Methyl-2-pentene	−14.32
2-Methyl-2-butene	−10.17	2,3-Dimethyl-1-butene	−14.78
		3,3-Dimethyl-1-butene	−14.25
		2,3-Dimethyl-2-butene	−15.91

a. From F. D. Rossini, K. S. Pitzer, R. L. Arnett, R. M. Braun, and G. C. Pimentel, *Selected Values of Physical and Thermodynamic Properties of Hydrocarbons and Related Compounds*, Carnegie Press, Pittsburgh, 1953.

cis-alkenes, probably largely because of increased nonbonded repulsion in the *cis* isomers.[13]

Table 1.6 gives the heats of hydrogenation for some pairs of *cis*- and *trans*-alkenes. Since hydrogenation leads to the same saturated product, the difference in the two heats of hydrogenation corresponds to the energy difference between the two compounds. This difference is seen to increase from 1.0 kcal/mol to nearly 10 kcal/mol as the groups on the double bond increase in size from methyl to *t*-butyl.[14]

13. For a theoretical discussion of this point, see N. D. Epiotis, R. L. Yates, and F. Bernardi, *J. Am. Chem. Soc.* **97**, 5961 (1975).
14. A review of the use of heats of hydrogenation for evaluation of the enthalpy of organic molecules is given by J. L. Jensen, *Prog. Phys. Org. Chem.* **12**, 189 (1976).

Table 1.6. Heats of Hydrogenation of Some Alkenes (kcal/mol)[a]

$CH_3CH{=}CHCH_3$	*cis*	28.6
	trans	27.6
$CH_3CH{=}CHC(CH_3)_3$	*cis*	30.8
	trans	26.5
$(CH_3)_3CCH{=}CHC(CH_3)_3$	*cis*	36.2
	trans	26.9
$(CH_3)_3CCH_2CH{=}CHCH_2C(CH_3)_3$	*cis*	26.9
	trans	26.0

a. In acetic acid; from R. B. Turner, A. D. Jarrett, P. Goebel, and B. J. Mallon, *J. Am. Chem. Soc.* **95**, 790 (1973).

Table 1.7. Atomic and Group Electronegativities

A. Atomic electronegativities[a]

H 2.1	C 2.5; **2.35**	N 3.0; **3.16**	O 3.5; **3.52**	F 4.0; **4.00**
	Si 1.8; **1.64**	P 2.1; **2.11**	S 2.5; **2.52**	Cl 3.0; **2.84**
		As 2.0; **1.99**	Se 2.4; **2.40**	Br 2.8; **2.52**
				I 2.5

B. Empirical electronegativities for some organic functional groups[b]

CH_3	2.3	H	2.28	F	3.95
CH_2Cl	2.75	NH_2	3.35	Cl	3.03
$CHCl_2$	2.8	$^+NH_3$	3.8	Br	2.80
CCl_3	3.0	NO_2	3.4	I	2.28
CF_3	3.35	OH	3.7		
Ph	3.0				
$CH{=}CH_2$	3.0				
$C{\equiv}CH$	3.3				
$C{\equiv}N$	3.3				

a. From L. Pauling, *The Nature of the Chemical Bond*, third edition, Cornell University Press, Ithaca, New York, 1960. Boldface values from G. Simons, M. E. Zandler, and E. R. Talaty, *J. Am. Chem. Soc.* **98**, 7869 (1976).
b. From P. R. Wells, *Prog. Phys. Org. Chem.* **6**, 111 (1968).

Another important property of chemical bonds is their *polarity.* In general, it is to be expected that the pair of electrons in a covalent bond will be subject to a probability distribution that favors one of the two atoms. The tendency of an atom to attract electrons is called *electronegativity.* There are a number of different approaches to assigning electronegativity, and most are numerically scaled to a definition originally proposed by Pauling.[15] Table 1.7, part A, gives the original Pauling values and also a more recent set based on theoretical calculation of electron distributions. The concept of electronegativity can also be expanded to include functional groups. Part B of Table 1.7 gives some numerical values which are scaled

15. For leading references, see G. Simons, M. E. Zandler, and E. R. Talaty, *J. Am. Chem. Soc.* **98**, 7869 (1976).

Table 1.8. Bond and Group Dipoles for Some Organic Functional Groups[a]

Bond moments[b]		Bond moments[b]		Group moments[b]	
C—H	0.4	C—N	0.22	MeO	1.3
C—F	1.41	C—O	0.74	NH_2	1.2
C—Cl	1.46	C=O	2.3	CO_2H	1.7
C—Br	1.38	C≡N	3.5	COMe	2.7
C—I	1.19			NO_2	3.1
				CN	4.0

a. From C. P. Smyth, *Dielectric Behavior and Structure*, McGraw-Hill Book Company, New York, 1955, pp. 244, 253.
b. In e.s. units × 10^{18}

to be numerically consistent with elemental electronegativities. These electronegativity values can serve to convey a qualitative impression of the electron-attracting capacity of these groups.

The unequal distribution of electron density in covalent bonds produces a bond dipole having the units of charge times distance.[16] Bonds with significant bond dipoles are described as being polar. The bond and group dipole moments of some typical substituents are shown in Table 1.8. It is possible to estimate with a fair degree of precision the dipole moment of a molecule as the vector sum of the component bond dipoles. A qualitative judgment of bond polarity can be made by comparing the difference in electronegativity of the bound atoms or groups. The larger the difference, the greater will be the bond dipole.

For most purposes, hydrocarbon groups can be considered to be nonpolar. There are however small dipoles associated with C—H bonds and with bonds between carbons of different hybridization or substitution pattern. Normal sp^3 carbon is found to be slightly negatively charged relative to hydrogen.[17] The electronegativity order for carbon is $sp > sp^2 > sp^3$. Scheme 1.1 lists some dipole moments for some hydrocarbons and some other organic molecules.

The polarity of covalent bonds is considered to be the basis of a number of structure–reactivity relationships in organic chemistry. The pK_a values of some derivatives of acetic acid are presented in Table 1.9. These data illustrate that substitution of a more electronegative atom or group for hydrogen increases the equilibrium constant for ionization, that is, makes the derivative a stronger acid. The highly electronegative fluorine atom causes a larger increase in acidity than the somewhat less electronegative chlorine atom. A slight acid-weakening effect of a methyl substituent is observed when propionic acid is compared with acetic acid. The data refer to measurements in solution, and some care must be taken in interpreting the changes in acidity solely on the basis of electronegativity. In fact,

16. For more detailed discussion of dipole moments, see L. E. Sutton, in *Determination of Organic Structures by Physical Methods*. Vol 1, E. A. Braude and F. C. Nachod (eds.), Academic Press, New York, 1955, Chapter 9; V. I. Minkin, O. A. Osipov, and Y. A. Zhdanov, *Dipole Moments in Organic Chemistry*, Plenum Press, New York, 1970.
17. K. B. Wiberg, R. F. W. Bader, and C. D. H. Lau, *J. Am. Chem. Soc.* **109**, 1001 (1987).

Scheme 1.1. Dipole Moments for Some Organic Compounds[a]

A. Hydrocarbons

Compound	Dipole moment
$H_2C{=}CHCH_2CH_3$	0.34
$HC{\equiv}CCH_2CH_3$	0.800
cis-$CH_3CH{=}CHCH_3$	0.253
trans-$CH_3CH{=}CHCH_3$	0 (symmetry)
$H_2C{=}C(CH_3)_2$	0.503
$CH_3C{\equiv}CCH_3$	0 (symmetry)
$H_2C{=}CHCH_3$	0.366
$HC{\equiv}CCH_3$	0.781

B. Substituted Molecules

Compound	Dipole moment
cis-$ClCH{=}CHCl$	1.90
trans-$ClCH{=}CHCl$	0 (symmetry)
$H_2C{=}CCl_2$	1.34
Tetrahydrofuran	1.63
CH_3CN	3.92
CH_3NO_2	3.46
CH_3OCH_3	1.30
CH_3OH	1.70
CH_3CO_2H	1.74
CH_3COCH_3	2.88

a. Units are in debye. Data are from *Handbook of Chemistry and Physics*, CRC Press, Inc., Boca Raton, Florida (1979).

Table 1.9. Acidity of Substituted Acetic Acids

	H_2O (pK_a)[a]	Gas phase (ΔG)[b]
$(CH_3)_3CCO_2H$	5.0	
$(CH_3)_2CHCO_2H$	4.9	
$CH_3CH_2CO_2H$	4.9	340.3
CH_3CO_2H	4.8	341.5
FCH_2CO_2H		331.0
$ClCH_2CO_2H$	2.7	328.8
F_3CCO_2H		316.3
Cl_3CCO_2H	0.7	
$NCCH_2CO_2H$	2.6	
$O_2NCH_2CO_2H$	1.3	

a. From *Stability Constants* and *Stability Constants, Supplement No. 1*, Special Publications 17 (1964) and 25 (1971), The Chemical Society, London.
b. ΔG for $AH \rightarrow H^+ + A^-$ at 300 K; from J. B. Cummings and P. Kebarle, *Can. J. Chem.* **56**, 1 (1978).

measurements in the gas phase show that propionic acid is a slightly stronger acid than acetic acid.[18] The second column in Table 1.9 shows the free energy for dissociation of some of the same acids in the gas phase. The effect of strongly electron-withdrawing groups is still evident. There is frequently a large difference in the energy of chemical processes in the gas and solution phases because of the importance of solvation. In the case of acid dissociation, both the proton and carboxylate anions will be strongly solvated and this greatly favors the dissociation process in solution relative to the gas phase. We can discuss trichloroacetic acid as an example. The polar C—Cl bonds have replaced the slightly polar C—H

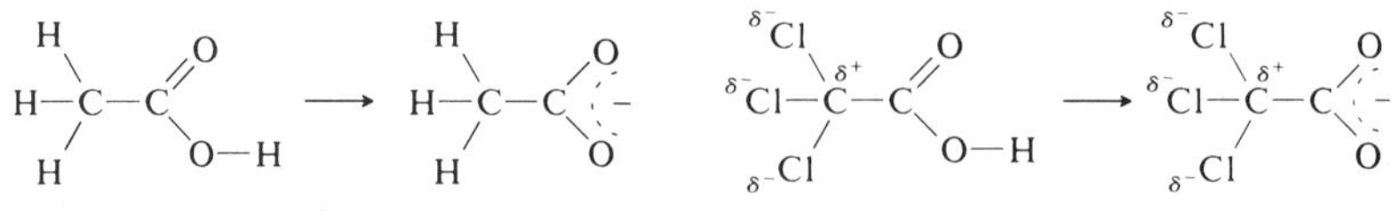

$$H_3C{-}CO_2H \longrightarrow H_3C{-}CO_2^- \qquad \overset{\delta^-}{Cl_3}\overset{\delta^+}{C}{-}CO_2H \longrightarrow \overset{\delta^-}{Cl_3}\overset{\delta^+}{C}{-}CO_2^-$$

very little dipolar stabilization of the anion

dipolar stabilization of the anion

For the acid dissociation equilibrium

$$RCO_2H + H_2O \rightleftharpoons RCO_2^- + H_3O^+$$

dissociation places a negative charge on the carboxylate residue and increases the electron density at the carboxyl group carbon and oxygen atoms. For acetic acid, where $R = CH_3$, this increase occurs adjacent to the carbon on the methyl group, which bears a very small negative charge. In the case of trichloracetic acid, where $R = CCl_3$, the corresponding carbon is somewhat positive as a result of the C—Cl bond dipoles. The development of negative charge is more favorable in the second case because of the favorable electrostatic effect. As a result, trichloracetic acid is a stronger acid than acetic acid. Substituent effects such as these which result from nearby bond dipoles are called *inductive effects.* Inductive effects are primarily electrostatic in nature and are effective only at short range.

It is always important to keep in mind the *relative* aspect of substituent effects. Thus, the effect of the chlorine atoms in the case of trichloroacetic acid is primarily to stabilize the dissociated anion. The acid is more highly dissociated than in the unsubstituted case because there is a more favorable energy *difference* between the parent acid and the anion. It is the energy differences, not the absolute energies, that determine the extent of ionization. As we will discuss more fully in Chapter 4, there are other mechanisms by which substituents can affect the energy of reactants and products. The detailed understanding of substituent effects will require that we separate inductive effects from these other factors.

A fundamental atomic property which is closely related to electronegativity is *polarizability.* Polarizability describes the response of the electrons of atoms to nearby

18. R. Yamdagni and P. Kebarle, *J. Am. Chem. Soc.* **95**, 4050 (1973).

Table 1.10. Hardness of Some Atoms, Acids, and Bases[a]

Atom	η	Acid	η_A	Base	η_B
H	6.4	H^+	∞	H^-	6.8
Li	2.4	Li^+	35.1	F^-	7.0
C	5.0	Mg^{2+}	32.5	Cl^-	4.7
N	7.3	Na^+	21.1	Br^-	4.2
O	6.1	Ca^{2+}	19.7	I^-	3.7
F	7.0	Al^{3+}	45.8	CH_3^-	4.0
Na	2.3	Cu^+	6.3	NH_2^-	5.3
Si	3.4	Cu^{2+}	8.3	OH^-	5.6
P	4.9	Fe^{2+}	7.3	SH^-	4.1
S	4.1	Fe^{3+}	13.1	CN^-	5.3
Cl	4.7	Hg^{2+}	7.7	H_2O	7.0
		Pb^{2+}	8.5	NH_3	6.9
		Pd^{2+}	6.8	H_2S	5.3

a. From R. G. Parr and R. G. Pearson, *J. Am. Chem. Soc.* **105**, 7512 (1983).

charges. The qualitative terms *softness* and *hardness* are used to describe, respectively, the ease or difficulty of distortion. Highly electronegative atoms tend to be hard whereas less electronegative atoms are softer. Polarizability is also a function of atomic number. Larger atoms are softer than smaller atoms of similar electronegativity. The charge on an atom also influences polarizability. Metal cations, for example, become harder as the oxidation number increases. Table 1.10 gives numerical values for the hardness of some elements that are frequently of interest in organic chemistry.

A useful precept for understanding Lewis acid–base interactions is that hard acids prefer hard bases and soft acids prefer soft bases. The hard–hard interactions are dominated by electrostatic attraction while soft–soft interactions are dominated by mutual polarization.[19]

The ideas about bond length, bond energy, polarity, and polarizability discussed in this section are very useful because of the relative constancy of these properties from molecule to molecule. Thus, data obtained from simple well-studied molecules can provide a good guide to the properties of substances whose structures are known but which have yet to be studied in detail. Organic chemists have usually discussed this transferability of properties in terms of valence bond theory. Thus, the properties are thought of as characteristic of the various types of atoms and bonds. The properties of the molecule are thought of as the sum of the properties of the bonds. This has been a highly fruitful conceptual approach in organic chemistry. As we shall see in the next section, there is an alternative description of molecules which is also highly informative and useful.

19. R. G. Pearson, *J. Am. Chem. Soc.* **85**, 3533 (1963); T. L. Ho, *Hard and Soft Acids and Bases in Organic Chemistry*, Academic Press, New York, 1977; W. B. Jensen, *The Lewis Acid–Base Concept*, Wiley-Interscience, New York, 1980, Chapter 8.

1.3. Molecular Orbital Theory and Methods

The second broad approach to the description of molecular structure that is of importance in organic chemistry is molecular orbital theory. Molecular orbital (MO) theory discards the idea that bonding electron pairs are localized between specific atoms in a molecule and instead pictures electrons as being distributed among a set of molecular orbitals of discrete energies. In contrast to the orbitals described by valence bond theory, which are usually concentrated between two specific atoms, these orbitals can extend over the entire molecule. Molecular orbital theory is based on the Schrödinger equation,

$$H\psi = E\psi$$

in which ψ is a wave function describing an orbital, H is the Hamiltonian operator, and E is the energy of an electron in a particular orbital. The wave function describes the interaction of the electron with the other electrons and nuclei of the molecule. The total electronic energy is the sum of the individual electron energies:

$$E = \int \psi H \psi \, d\tau \qquad \text{when} \qquad \int \psi^2 \, d\tau = 1$$

In order to make the mathematics tractable, a number of approximations are made. The choice of approximations has produced a variety of molecular orbital methods, the judicious application of which can provide valuable insight into questions of bonding, structure, and dynamics. The discussion that follows will not be sufficiently detailed or complete for the reader to understand how the calculations are performed or the details of the approximations. Instead, the nature of the information which is obtained will be described, and the ways in which organic chemists have applied the results of MO theory will be illustrated. Several excellent books are available which provide detailed treatment ofvarious aspects of molecular orbital theory.[20]

All but the simplest molecular orbital calculations are done on computers, and there is a necessary trade-off between the accuracy of the calculations and the expense in terms of computer time. In general, the more severe the approximations, the more limited is the range of applicability of the particular calculation. The organic chemist who wishes to make use of the results of molecular orbital calculations must therefore make a judgment about the applicability of the various methods to the particular problem. In general, the programs which are used for the calculations

20. M. J. S. Dewar, *The Molecular Orbital Theory of Organic Chemistry*, McGraw-Hill, New York, 1969; W. T. Borden, *Modern Molecular Orbital Theory for Organic Chemists*, Prentice-Hall, Englewood Cliffs, New Jersey, 1975; H. E. Zimmerman, *Quantum Mechanics for Organic Chemists*, Academic Press, New York, 1975; I. G. Cszimadia, *Theory and Practice of MO Calculations on Organic Molecules*, Elsevier, Amsterdam, 1976; M. J. S. Dewar and R. C. Dougherty, *The PMO Theory of Organic Chemistry*, Plenum Press, New York, 1975; T. A. Albright, J. K. Burdett, and M.-H. Whangbo, *Orbital Interactions in Chemistry*, Wiley, New York, 1985; W. G. Richards and D. L. Cooper, *An Initio Molecular Orbital Calculations for Chemists*, Second Edition, Clarendon Press, Oxford, 1983; W. J. Hehre, L. Radom, P. v. R. Schleyer, and J. Pople, *Ab Initio Molecular Orbital Theory*, Wiley-Interscience, New York, 1986.

are available, and the complexity of use tends to increase with the sophistication of the calculation.[21]

Mathematically, the molecular orbitals are treated as linear combinations of atomic orbitals, so that the wave function, ψ, is expressed as a sum of individual atomic orbitals multiplied by appropriate weighting factors (coefficients):

$$\psi = c_1\phi_1 + c_2\phi_2 + \cdots c_n\phi_n$$

The coefficients indicate the contribution of each atomic orbital to the molecular orbital. This method of representing molecular orbital wave functions in terms of combinations of atomic orbital wave functions is known as the *linear combination of atomic orbitals (LCAO) approximation.* The combination of atomic orbitals chosen is called the *basis set.* A minimum basis set for molecules containing C, H, O, and N would consist of $2s$, $2p_x$, $2p_y$, and $2p_z$ orbitals for each C, N, and O and a $1s$ orbital for each hydrogen. The basis sets are mathematical expressions describing the properties of the atomic orbitals.

Two main streams of computational techniques branch out from this point. These are referred to as *ab initio* and *semiempirical* calculations. In both *ab initio* and semiempirical treatments, mathematical formulations of the wave functions which describe hydrogen-like orbitals are used. Examples of wave functions that are commonly used are Slater-type orbitals (abbreviated STO) and Guassian-type orbitals (GTO). There are additional variations which are designated by additions to the abbreviations. Both *ab initio* and semiempirical calculations treat the linear combination of orbitals by iterative computations which establish a self-consistent electrical field (SCF) and minimize the energy of the system. The minimum-energy combination is taken to describe the molecule.

The various semiempirical methods differ in the approximations which are made concerning repulsions between electrons in different orbitals. The approximations are then corrected for by "parameterization," whereby parameters are included in the protocol to adjust the results to match *ab initio* calculations or experimental data. The reliability and accuracy of the methods has evolved, and the increasing power of computers has permitted wider application of the more accurate methods. The earliest semiempirical methods to be applied extensively to organic molecules included the Extended Hückel Theory[22] and the CNDO (complete neglect of differential overlap) methods.[23] These methods give correct representations of the shapes and trends in charge distribution in the various molecular orbitals but are only roughly reliable in describing molecular geometry. These methods tend to make large errors in calculation of total energies of molecules and can be used only to compare energies in closely related structures where errors can be assumed to cancel. Improved semiempirical calculations give more satisfactory representations of

21. Many computation programs have been made available through the Quantum Chemistry Program Exchange, Chemistry Department, Indiana University, Bloomington, Indiana.
22. R. Hoffmann, *J. Chem. Phys.* **39**, 1397 (1963).
23. J. A. Pople and G. A. Segal, *J. Chem. Phys.* **44**, 3289 (1966).

charge distributions, molecular geometry, and ground state total energies. Among these methods are MINDO-3,[24] MNDO,[25] and AM1.[26] (The acronyms refer to semidescriptive titles of the methods.) There are differences among the methods in the ranges of compounds for which the results are satisfactory.

Ab initio calculations are iterative procedures based on *self-consistent field* (SCF) methods. Electron–electron repulsion is specifically taken into account. Normally, calculations are approached by the Hartree–Fock closed-shell approximation, which treats a single electron at a time interacting with an aggregate of all the other electrons. Self-consistency is achieved by a procedure in which a set of orbitals is assumed, and the electron–electron repulsion is calculated; this energy is then used to calculate a new set of orbitals, and these in turn are used to calculate a new repulsive energy. The process is continued until convergence occurs and self-consistency is achieved.[27]

The individual *ab initio* calculations are further identified by abbreviations for the basis set orbitals which are used. These abbreviations include, for example, STO-3G,[28] 4-31G,[29] and 6-31G.[30] A fundamental difference between *ab initio* methods and the semiempirical methods is the absence of adjustable parameters in the former. In general, the *ab initio* calculations make fewer assumptions and therefore the computations are more complex.

Another distinguishing aspect of MO methods is the extent to which they deal with *electron correlation.* The Hartree–Fock approximation does not deal with correlation between individual electrons, and the results are expected to be in error because of this, giving energies above the exact energy. Molecular orbital methods that include electron correlation have been developed and can be used to assess the importance of electron correlation when necessary.

Present MO methods give excellent results on ground state molecular geometry and charge distribution. They can also give excellent agreement with experimental data in the calculation of relative molecular energies. The energy changes associated with dynamic processes can be studied by calculation of molecular energy as a function of molecular distortion. Much effort is also being devoted to the description of reaction processes. This is a very formidable task because information about solvent participation and exact separation of reacting molecules is imprecise. In cases where good assumptions about such variables can be made, *ab initio* MO calculations can give good estimates of the energy changes associated with chemical reactions.

24. R. C. Bingham, M. J. S. Dewar, and D. H. Lo, *J. Am. Chem. Soc.* **97**, 1285, 1294, 1302 (1975).
25. M. J. S. Dewar and W. Thiel, *J. Am. Chem. Soc.* **99**, 4907 (1977).
26. M. J. S. Dewar, E. G. Zoebisch, E. F. Healy, and J. J. P. Stewart, *J. Am. Chem. Soc.* **107**, 3902 (1985).
27. C. C. J. Roothaan, *Rev. Mod. Phys.* **23**, 69 (1951); R. Pariser and R. G. Parr, *J. Chem. Phys.* **21**, 767 (1953); J. A. Pople, *J. Phys. Chem.* **61**, 6, (1957).
28. W. J. Hehre, R. F. Stewart, and J. A. Pople, *J. Chem. Phys.* **51**, 2657 (1971).
29. R. Ditchfield, W. J. Hehre, and J. A. Pople, *J. Chem. Phys.* **54**, 724 (1971).
30. W. J. Hehre, R. Ditchfield, and J. A. Pople, *J. Chem. Phys.* **56**, 2257 (1972).

The relative merits of various methods have been discussed somewhat in the literature.[31] In general, it can be stated that the *ab initio* type of calculations will be more reliable but the semiempirical calculations are faster in terms of computer time. The complexity of calculation also increases rapidly as the number of atoms in the molecule increases. A choice of methods is normally made on the basis of evidence that the method is adequate for the problem at hand and the availability of appropriate computational facilities. Results should be subjected to critical evaluation by comparison with experimental data or checked by representative calculations using higher-level methods.

The results of all types of MO calculations include the energy of each MO, the total electronic energy of the molecule relative to the separated atoms, and the coefficients of the AOs contributing to each MO. Such information may be related directly to a number of physical and chemical properties. The total electronic energy obtained by summing the energies of the occupied orbitals gives the calculated molecular energy. Comparison of isomeric molecules permits conclusions about the relative stabilities of the compounds. Conclusions about molecular stability can be checked by comparison with thermodynamic data when it is available. Conformational effects can be probed by calculating the total energy as a function of molecular geometry. The minimum energy should correspond to the most favorable molecular structure. Most calculations are done on single molecules, not an interacting array of molecules. Thus, the results are most comparable to the situation in the gas phase, where intermolecular forces are weak. The types of data which are readily obtained by MO calculations are illustrated in the following paragraphs.

Table 1.11 lists some reported deviations of calculated ΔH_f values from experimental ones for some small hydrocarbons. The extent of deviation gives an indication of the accuracy of the various types of MO calculations in this application.

The coefficients for the AOs that comprise each MO may be related to the electron density at each atom by the equation

$$q_r = \sum_j n_j c_{jr}^2$$

which gives the electron density at atom r as the sum over all the occupied molecular orbitals of the product of the number of electrons in each orbital and the *square* of the coefficient at atom r for each orbital. To illustrate, the coefficients for the methyl cation according to the CNDO/2 method are given in Table 1.12. There are seven MOs generated from the three hydrogen $1s$ and carbon $2s$, $2p_x$, $2p_y$, and $2p_z$ atomic orbitals.

31. J. A. Pople, *J. Am. Chem. Soc.* **97**, 5306 (1975); W. J. Hehre, *J. Am. Chem. Soc.* **97**, 5308 (1975); T. A. Halgren, D. A. Kleier, J. H. Hall, Jr., L. D. Brown, and W. N. Lipscomb, *J. Am. Chem. Soc.* **100**, 6595 (1978); M. J. S. Dewar and G. P. Ford, *J. Am. Chem. Soc.* **101**, 5558 (1979); W. J. Hehre, *Acc. Chem. Res.* **9**, 399 (1976); M. J. S. Dewar, E. G. Zoebisch, E. F. Healy, and J. J. P. Stewart, *J. Am. Chem. Soc.* **107**, 3902 (1985); J. N. Levine, *Quantum Chemistry*, Third Edition, Allyn and Bacon, Boston, 1983, pp. 507–512; W. Hehre, L. Radom, P. v. R. Schleyer, and J. A. Pople, *Ab Initio Molecular Orbital Calculations*, Wiley-Interscience, New York, 1986, Chapter 6.

Table 1.11. Deviations of Calculated Values from Experimental ΔH_f Data for Various MO Methods

	Deviation (kcal/mol)			
Hydrocarbon	MNDO	AM1	3-21G	6-31G
Methane	5.9	9.0	−0.9	−0.5
Ethane	0.3	2.6	0.2	1.9
Ethene	3.1	4.0	−1.6	−2.4
Allene	−1.6	0.6	−2.6	−6.8
1,3-Butadiene	2.7	3.6	−4.7	
Cyclopropane	−1.5	5.1		−2.4
Cyclobutane	−18.7	0.2		
Cyclopentane	−12.0	−10.5		
Cyclohexane	−5.3	−9.0		
Benzene	1.5	2.2		

a. Energy comparisons are from M. J. S. Dewar, E. G. Zoebisch, E. F. Healy, and J. J. P. Stewart, *J. Am. Chem. Soc.* **107**, 3902 (1985).

The electron densities are calculated from the coefficients of ψ_1, ψ_2, and ψ_3 only because these are the occupied orbitals for the six-electron system. The carbon atom is calculated to have 3.565 electrons (exclusive of the $1s$ electrons), and each hydrogen atom is calculated to have 0.812 electrons. Since neutral carbon has four valence electrons, its net charge in the methyl cation is +0.435 electrons. Each hydrogen atom has a charge of +0.188. The total charge $0.435 + 3(0.188) = 1.000$ electron. A sample calculation of the hydrogen electron density from the orbital coefficients follows:

$$q_H = 2(0.3528)^2 + 2(0.0999)^2 + 2(0.5210)^2$$
$$= 0.812$$

Further examination of Table 1.12 reveals that the lowest unoccupied molecular orbital, ψ_4, is a pure p orbital, localized on carbon. This is revealed by the coefficients,

Table 1.12. Coefficients of Wave Functions Calculated for Methyl Cation by the CNDO/2 Approximation[a]

Orbital	C_{2s}	C_{2p_x}	C_{2p_y}	C_{2p_z}	H	H	H
ψ_1	0.7915	0.0000	0.0000	0.0000	0.3528	0.3528	0.3528
ψ_2	0.0000	0.1431	0.7466	0.0000	0.0999	0.4012	−0.5011
ψ_3	0.0000	0.7466	−0.1431	0.0000	0.5210	−0.3470	−0.1740
ψ_4	0.0000	0.0000	0.0000	1.0000	0.0000	0.0000	0.0000
ψ_5	−0.6111	0.0000	0.0000	0.0000	0.4570	0.4570	0.4570
ψ_6	0.0000	0.5625	−0.3251	0.0000	−0.5374	0.5377	−0.0003
ψ_7	0.0000	0.3251	0.5625	0.0000	−0.3106	−0.3101	0.6207

a. The orbital energies (eigenvalues) are not given. The lowest-energy orbital is ψ_1; the highest-energy orbital, ψ_7.

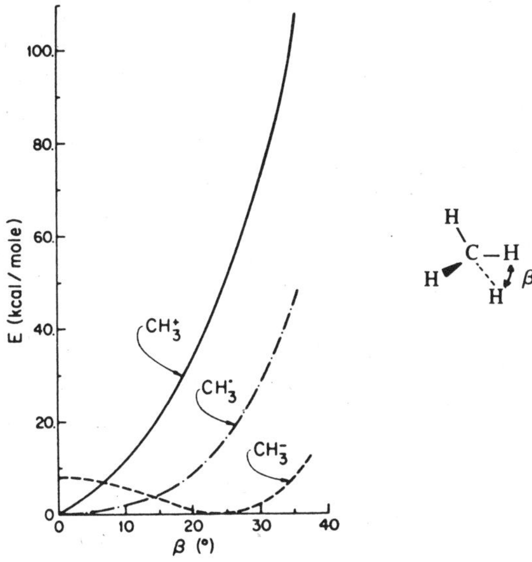

Fig. 1.8. Total energy as a function of distortion from planarity for methyl cation, methyl radical, and methyl anion. [Reproduced from *J. Am. Chem. Soc.* **98**, 6483 (1976).]

which are 0.00 for all but the $2p_z$ orbital. The molecular orbital picture is in agreement with the usual qualitative hybridization picture for the methyl cation.

The use of molecular orbital methods to probe the relationship between structure and energy can be illustrated by a study of CH_3^+, $CH_3\cdot$, and CH_3^-. The study employed *ab initio* calculations and the 4-31G basis set and was aimed at exploring the optimum geometry and resistance to deformation in each of these reaction intermediates.[32] Figure 1.8 is a plot of the calculated energy as a function of deformation from planarity for the three species. While CH_3^+ and $CH_3\cdot$ are found to have minimum energy at $\beta = 0°$, that is, when the molecule is planar, CH_3^- is calculated to have a nonplanar equilibrium geometry. This calculated result is in good agreement with a variety of experimental observations which will be discussed in later chapters where these intermediates are considered in more detail.

Substituent effects on intermediates such as these can also be analyzed by MO methods. Take, for example, methyl cations where adjacent substituents with lone pairs of electrons can form strong π bonds. The π bonding can be expressed in either valence bond or MO terminology.

$$:X{-}\overset{+}{C}H_2 \leftrightarrow {}^{+}X{=}CH_2$$

δ+ δ+
X—C(H)(H)

32. E. D. Jemmiss, V. Buss, P. v. R. Schleyer, and L. C. Allen, *J. Am. Chem. Soc.* **98**, 6483 (1976).

Table 1.13. Calculated Charge Transfer and Stabilization Resulting from Substituents on the Methyl Cation

Substituent	Electron density in C $2p_z$ orbital (4-31G)	Stabilization (kcal/mol) 4-31G[a]	Stabilization (kcal/mol) 6-31G*[b]
F	0.35	2.1	14.16
OH	0.49	48	53.77
NH_2	0.58	93	87.33
CH_3	0.31	30	

a. Y. Apeloig, P. v. R. Schleyer, and J. A. Pople, *J. Am. Chem. Soc.* **99**, 1291 (1977).
b. F. Bernardi, A. Bottini, and A. Venturini, *J. Am. Chem. Soc.* **108**, 5395 (1986).

Table 1.14. Calculated Stabilization of Methyl Anion by Substituents

Substituent	Stabilization (kcal/mol) 4-31G[a]	Stabilization (kcal/mol) 6-31G*[b]	Substituent	Stabilization (kcal/mol) 4-31G[a]
BH_2	68	61.4	C≡N	61
CH_3	2	1.4	NO_2	98
NH_2	5	3.3	$CH{=}CH_2$	38
OH	15	7.9	CF_3	57
F	25	14.6	CH=O	72

a. A. Pross, D. J. DeFrees, B. A. Levi, S. K. Pollack, L. Radom, and W. J. Hehre, *J. Org. Chem.* **46**, 1693 (1981).
b. G. W. Spitznagel, T. Clark, J. Chandrasekhar, and P. v. R. Schleyer, *J. Comput. Chem.* **3**, 363 (1982).

An *ab initio* study using 4-31G basis set orbitals gave the charge densities shown in Table 1.13.[33] The table also shows the calculated stabilization relative to methyl cation resulting from electron release by the substituent.

The π-donor effects of the fluoro, oxygen, and nitrogen groups are partially counterbalanced by the inductive electron withdrawal through the σ bond. In the case of the oxygen and nitrogen substituents, the π-donor effect is dominant and these substituents strongly stabilize the carbocation. For the fluorine substituent, the balance is much closer and the overall stabilization is calculated to be quite small. We will return to the case of the methyl group and its stabilization effect on the methyl cation a little later.

In the case of the methyl anion, stabilization will result from electron-accepting substituents. Table 1.14 gives some stabilization energies calculated for a range of substituents.[34] Those substituents which have a low-lying π orbital capable of accepting electrons from the carbon $2p_z$ orbital (i.e., BH_2, C≡N, NO_2, CH=O) are strongly stabilizing. For electronegative substituents without π-acceptor capacity, a smaller stabilization of the methyl anion is calculated. The order is

33. Y. Apeloig, P. v. R. Schleyer, and J. A. Pople, *J. Am. Chem. Soc.* **99**, 1291 (1977).
34. A. Pross, D. J. DeFrees, B. A. Levi, S. K. Pollack, L. Radom, and W. J. Hehre, *J. Org. Chem.* **46**, 1693 (1981).

$F > OH > NH_2$, which parallels the ability of these substituents to act as σ-electron acceptors. The strong effect of the trifluoromethyl group is a combination of both σ- and π-bond effects.

1.4. Qualitative Application of Molecular Orbital Theory

As with valence bond theory, the full mathematical treatment of MO theory is too elaborate to apply to all situations. It is important to be able to develop qualitative approaches based on the fundamental concepts of molecular orbital theory which can be applied without the need for detailed calculations. A key tool for this type of analysis is a qualitative molecular orbital energy diagram. The construction of a qualitative energy level diagram may be accomplished without recourse to detailed calculations by keeping some basic principles in mind. These principles can be illustrated by referring to some simple examples. Consider first diatomic species formed from atoms in which only the 1*s* orbitals are involved in the bonding scheme. The two 1*s* orbitals can combine in either a bonding or an antibonding manner to give two molecular orbitals, as indicated in Fig. 1.9.

The number of molecular orbitals (bonding + nonbonding + antibonding) is equal to the sum of the atomic orbitals in the basis set from which they are generated. The bonding combination is characterized by a positive overlap in which the coefficients are of like sign, while the antibonding combination is characterized by a negative overlap with coefficients of opposite sign.

Orbitals are occupied by electrons, beginning with the orbital of lowest energy and filling each orbital with a maximum of two electrons (the Aufbau principle). The number of electrons is determined by the number of electrons present on the interacting atoms. The orbitals in Fig. 1.9 could be applied to systems such as H_2^+ (one electron), H_2 (two electrons), He_2^+ (three electrons), or He_2 (four electrons). A reasonable conclusion would be that H_2 would be the most stable of these diatomic species because it has the largest net number of electrons in the bonding orbital

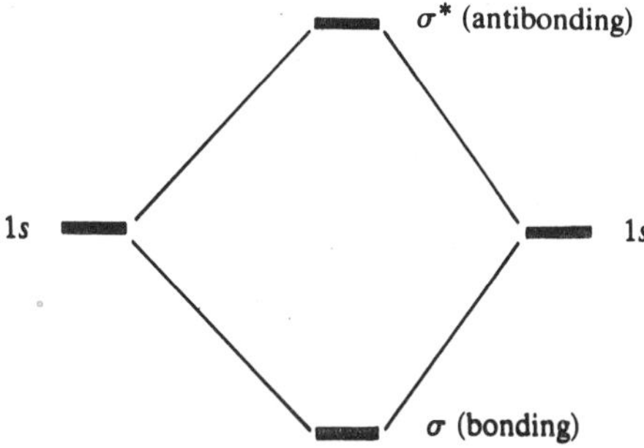

Fig. 1.9. Graphic description of combination of two 1*s* orbitals to give two molecular orbitals.

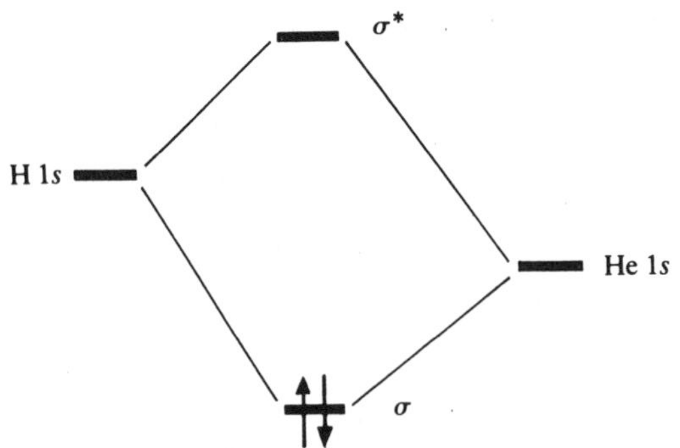

Fig. 1.10. Energy level diagram for HHe^+.

(two). The He_2 molecule has no net bonding because the anitbonding orbital contains two electrons and cancels the bonding contribution of the occupied bonding orbital. Both H_2^+ and He_2^+ have one more electron in bonding orbitals than in antibonding orbitals. They have been determined to have bond energies of 61 and 60 kcal/mol, respectively. The bond energy of H_2, for comparison, is 103 kcal/mol.

A slight adjustment in the energy level diagram allows it to be applied to heteronuclear diatomic species such as HHe^+. Rather than being a symmetrical diagram, the He 1*s* level is lower than the H 1*s* level due to the increased nuclear charge on helium. The diagram that results from this slight modification is shown in Fig. 1.10. Calculations for the HHe^+ ion indicate a bond energy of 43 kcal/mol.[35]

The second-row elements including carbon, oxygen and nitrogen involve *p* atomic orbitals as well as 2*s* orbitals. An example of a heteronuclear diatomic molecule involving these elements is carbon monoxide, C≡O. The carbon monoxide molecule has 14 electrons, and the orbitals for each atom are 1*s*, 2*s*, $2p_x$, $2p_y$, and $2p_z$. For most chemical purposes, the carbon 1*s* and oxygen 1*s* electrons are ignored. This simplification is valid because the energy gap between the 1*s* and 2*s* levels is large and the effect of the 1*s* levels on the valence electrons is very small. The ten valence electrons are distributed among eight molecular orbitals generated by combining the four valence atomic orbitals from carbon with the four from oxygen, as illustrated in Fig. 1.11. Figure 1.12 illustrates in a qualitative way the interactions between the atomic orbitals that gives rise to the molecular orbitals.

Figure 1.12 shows the various combinations of atomic orbitals which interact with one another. Each pair of atomic orbitals leads to a bonding and an antibonding combination. The 2*s* orbitals give the σ and σ^* orbitals. The $2p_x$ and $2p_y$ combinations form molecular orbitals that are π in character. The $2p_z$ combination gives a σ-type orbital labeled σ' as well as the corresponding antibonding orbital. The lower five orbitals are each doubly occupied, accounting for the ten valence shell electrons in the molecule. Of these five occupied orbitals, one is antibonding, with the net result of six bonding electrons, in agreement with the triple bond found in the Lewis structure for carbon monoxide. The shapes of the molecular orbitals can also be

35. H. H. Michels, *J. Chem. Phys.* **44**, 3834 (1966).

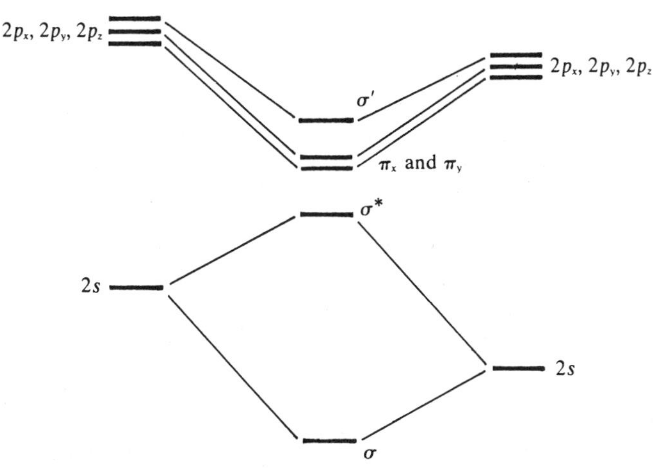

Fig. 1.11. Energy levels in the carbon monoxide molecule. (Adapted from H. B. Gray and G. P. Haight, *Basic Principles of Chemistry*, W. A. Benjamin, New York, 1967, p. 289.)

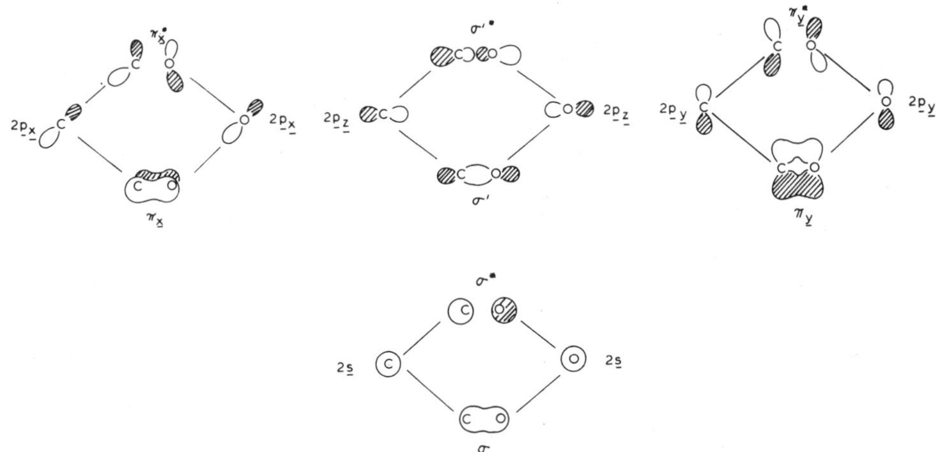

Fig. 1.12. Interaction of atomic orbitals of carbon and oxygen leading to molecular orbitals of carbon monoxide.

depicted as in Fig. 1.13. Here the nodes in the molecular orbitals are represented by a change from full to dashed lines, and the size of the lobes are scaled to represent atomic coefficients. One gains from these pictures an impression of the distortion of the bonding π orbital toward oxygen as a result of the greater electronegativity of the oxygen atom.

Just as we were able to state some guiding rules for qualitative application of resonance theory, it is possible to state some conditions by which to test the correctness of an MO energy level diagram derived by qualitative considerations.

a. The total number of molecular orbitals must equal the number of atomic orbitals from which they were constructed.
b. The symmetry of the molecular orbitals must conform to the symmetry of the molecule. That is, if a molecule possesses a plane of symmetry, for example, *all* the molecular orbitals must be either symmetric (unchanged) or antisymmetric (unchanged except for sign) with respect to that plane.
c. Atomic orbitals that are orthogonal to one another do not interact. Thus, two different carbon $2p$ orbitals will not contribute to the same molecular orbital.
d. The energies of similar atomic orbitals (s or p) are lower for elements of higher electronegativity.
e. The relative energy of molecular orbitals in a molecule increases with the number of nodes in the orbital.

By applying these rules and recognizing the elements of symmetry present in the molecule, it is possible to construct molecular orbital diagrams for more complex molecules. In the succeeding paragraphs, the MO diagrams of methane and ethylene are constructed from these kinds of considerations.

To provide a basis for comparison, Fig. 1.14 gives the results of an *ab initio* calculation on the methane molecule.[36] This particular calculation used as a basis set the $1s$, $2s$, and three $2p$ orbitals of carbon and the $1s$ orbitals of the four hydrogens. The lowest molecular orbital is principally $1s$ in character. A significant feature of this and other MO calculations of methane is that, unlike a picture involving localized bonds derived from sp^3 hybrid carbon orbitals, there are not four equivalent orbitals. We can obtain an understanding of this feature of the MO picture by a qualitative analysis of the origin of the methane molecular orbitals. For simplicity, we will consider the orbitals to be derived from the carbon $2s$, $2p_x$, $2p_y$, and $2p_z$ orbitals and ignore the carbon $1s$ orbital. The most convenient frame of reference for the tetrahedral methane molecule is a cube with hydrogen atoms at alternate corners and the carbon atom centered in the cube, as shown in Fig. 1.15.

This orientation of the molecule reveals that methane possesses three twofold symmetry axes, one each along the x, y, and z axes. Because of this molecular symmetry, the proper molecular orbitals of methane must possess symmetry with respect to these same axes. There are two possibilities: the orbital may be unchanged by 180° rotation about the axis (symmetric), or it may be transformed into an orbital

36. W. E. Palke and W. N. Lipscomb, *J. Am. Chem. Soc.* **88**, 2384 (1966).

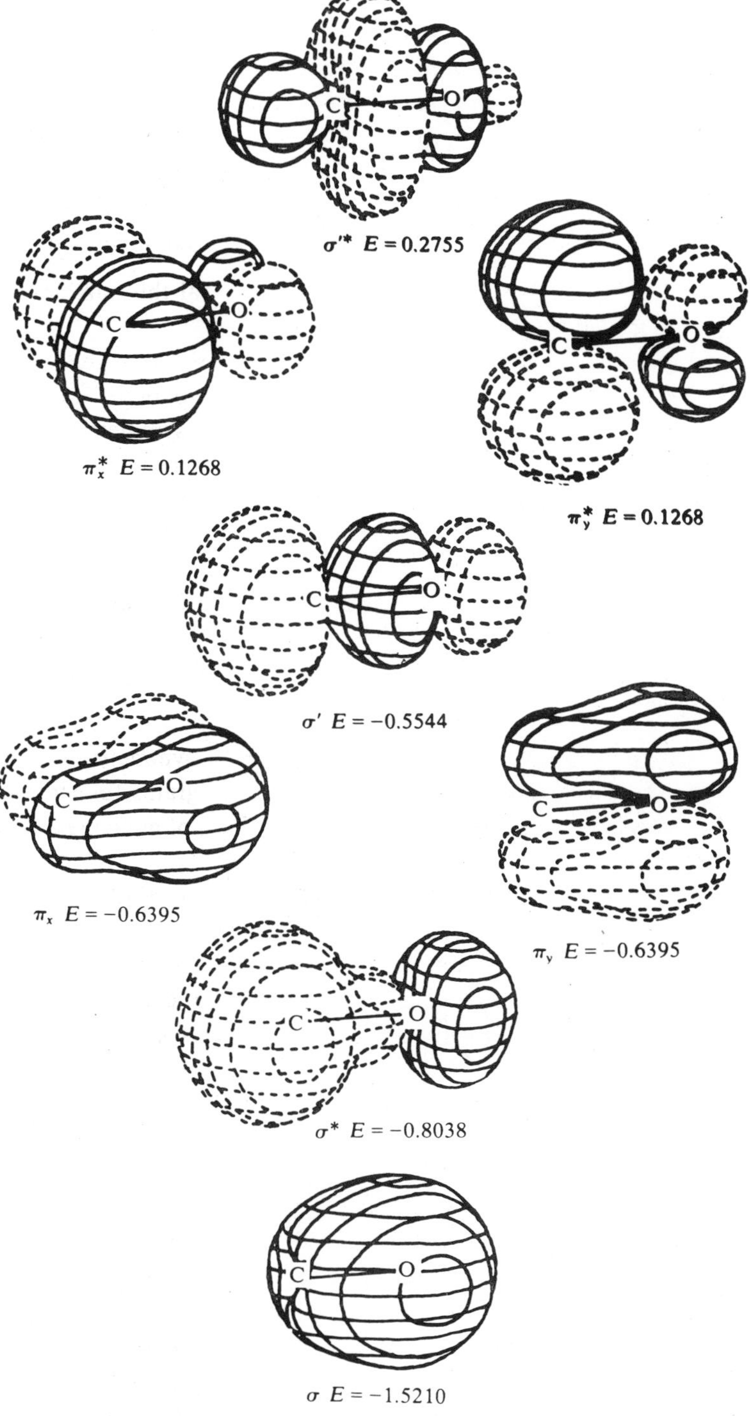

Fig. 1.13. Representation of the molecular orbitals of carbon monoxide. Energies are given in atomic units = 27.21 eV. (From W. L. Jorgensen and L. Salem, *The Organic Chemist's Book of Orbitals*, Academic Press, New York, 1973, Reproduced by permission.)

of identical shape but opposite sign by the symmetry operation (antisymmetric). The carbon $2s$ orbital is symmetric with respect to each axis, but the three $2p$ orbitals are each antisymmetric with respect to two of the axes and symmetric with respect to one. The combinations which give rise to molecular orbitals that meet these symmetry requirements are shown in Fig. 1.16.

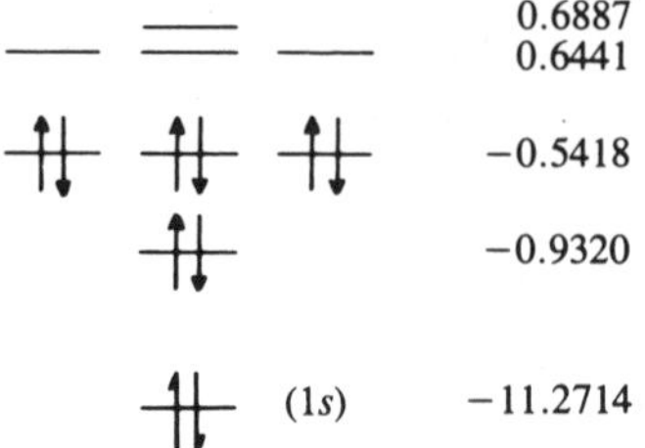

Fig. 1.14. Molecular orbital energy diagram for methane. Energies are in atomic units.

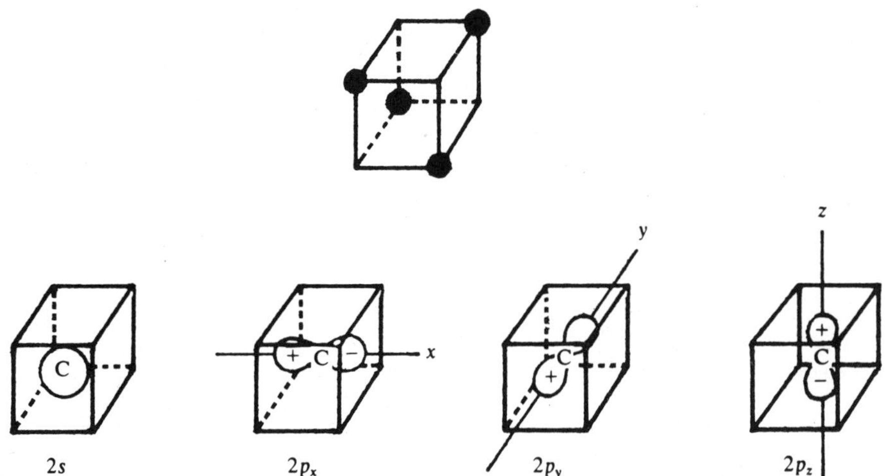

Fig. 1.15. Atomic orbitals of carbon relative to methane in a cubic frame of reference.

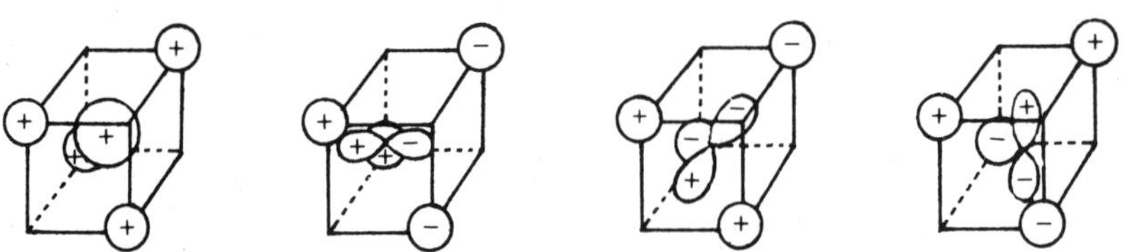

Fig. 1.16. Atomic orbital combinations giving rise to bonding molecular orbitals for methane.

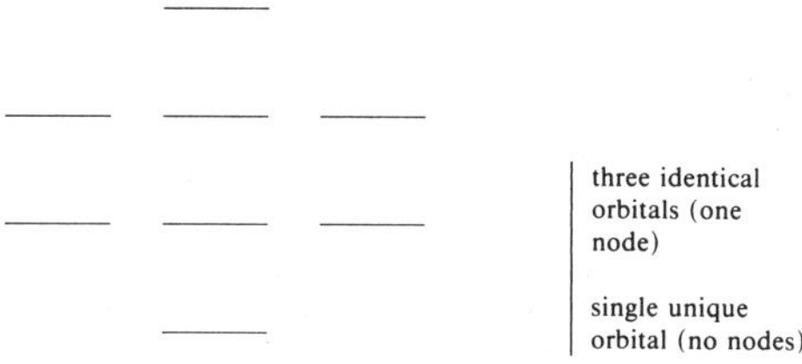

Fig. 1.17. Qualitative molecular molecular orbital diagram for methane.

The bonding combination of the carbon $2s$ orbital with the four $1s$ hydrogen orbitals leads to a molecular orbital that encompasses the entire molecule and has no nodes. Each of the MOs derived from a carbon $2p$ orbital has a node at carbon. The three combinations are equivalent, but higher in energy than the MO with no nodes. The four antibonding orbitals arise from similar combinations, but with the carbon and hydrogen orbitals having opposite signs in the region of overlap. Thus, the molecular orbital diagram arising from these considerations shows one bonding MO with no nodes and three degenerate (same energy) MOs with one node. The diagram is given in Fig. 1.17. Note that except for inclusion of the $1s$ orbital, this qualitative picture corresponds to the calculated orbital diagram in Fig. 1.14.

A qualitative approach cannot assign energies to the orbitals. In many cases, it is, however, possible to assign an ordering of energies. The relationship between relative energy and number of nodes has already been mentioned. In general, σ-type orbitals are lower in energy than π-type orbitals because of this factor. Conversely, antibonding σ orbitals are higher in energy than antibonding π orbitals. Orbitals derived from more electronegative atoms are lower in energy than those derived from less electronegative atoms.

We might ask if the prediction of MO theory that methane has one molecular orbital which is more stable than any other can be tested. The *ionization potential* is the energy required to remove an electron from a molecule and is quite high for most organic molecules, being on the order of 200 kcal/mol. The methods of choice for determining ionization potentials are photoelectron spectroscopy[37] and ESCA (electron spectroscopy for chemical analysis).[38] These techniques are complementary in that ionization potentials up to about 20 eV (1 eV = 23.06 kcal) are determined by photoelectron spectroscopy, corresponding to the binding energy of the valence electrons, while ESCA measures binding energies of core electrons. An ultraviolet source (photoelectron spectroscopy) or X-ray source (ESCA) emits photons, which are absorbed by the sample, resulting in ejection of an electron and formation of

37. For a review, see C. R. Brundle and M. B. Robin, in *Determination of Organic Structures by Physical Methods*, Vol. 3, F. C. Nachod and J. J. Zuckerman (eds.), Academic Press, New York, 1971, Chapter 1.
38. D. W. Turner, *Annu. Rev. Phys. Chem.* **21**, 107 (1970).

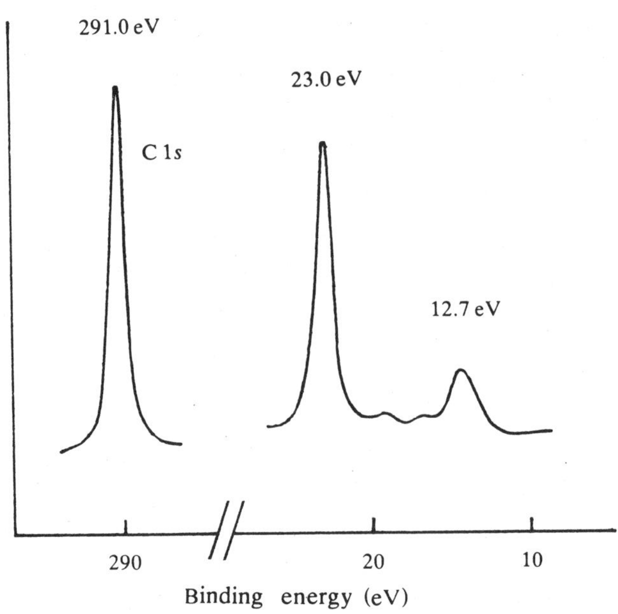

Fig. 1.18. ESCA spectrum of methane.

a positive ion. The kinetic energy of the emitted electron is determined and related to its binding energy by the equation

$$\text{binding energy} = \text{photon energy} - \text{kinetic energy of emitted electron}$$

This equation is the same as for the photoelectric effect observed for electron emission from metallic surfaces. These measurements allow the construction of orbital energy diagrams directly from experimental data and provide a way of critically examining bonding theories.

The ESCA spectrum of methane is presented in Fig. 1.18, where it can clearly be seen to be consistent with molecular orbital theory. There are two bands for the valence electrons at 12.7 and 23.0 eV, in addition to the band for the core electrons at 291.0 eV.[39] It should be emphasized that these values are the binding energies of electrons in the three orbitals of differing energy, and are *not* the energies required for successive ejection of first one, then a second, and then a third electron. The intensities bear no relation to the number of orbitals or number of electrons and differ from each other because the efficiency for ionizations are different for different orbitals.

The construction of the molecular orbitals of ethylene is similar to the process used for carbon monoxide, but the total number of atomic orbitals is greater, twelve instead of eight, because of the additional atomic orbitals from hydrogen. We must

39. U. Gelius, in *Electron Spectroscopy*, D. A. Shirley (ed.), American Elsevier, New York, 1972, pp. 311–344.

first define the geometry of ethylene. Ethylene is known from experiment to be a planar molecule.

This geometry possesses three important elements of symmetry, the molecular plane and two planes which bisect the molecule. All MOs must be either symmetric or antisymmetric with respect to each of these symmetry planes. With the axes defined as in the diagram above, the orbitals arising from carbon $2p_z$ have a node in the molecular plane. These are the familiar π and π^* orbitals.

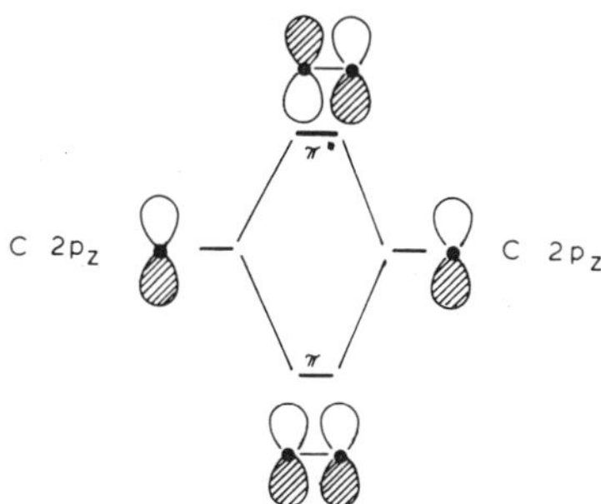

The π orbital is symmetric with respect to both the x–z plane and the y–z plane. It is antisymmetric with respect to the molecular (x–y) plane. On the other hand, π^* is antisymmetric with respect to the y–z plane.

The orbitals which remain are the four H 1s, two C 1s, and four C 2p orbitals. All lie in the molecular plane. The combinations using the C 2s and H 1s orbitals can take only two forms that meet the molecular symmetry requirements. One, σ, is bonding between all atoms, whereas the other, σ^*, is antibonding between all nearest-neighbor atoms. No other combination corresponds to the symmetry of the ethylene molecule.

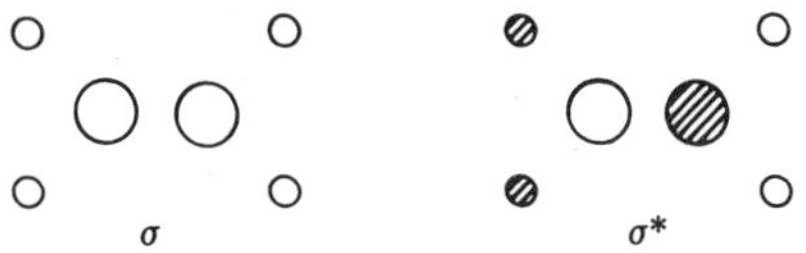

Let us next consider the interaction of $2p_y$ with the four hydrogen 1s orbitals. There are four possibilities which conform to the molecular symmetry:

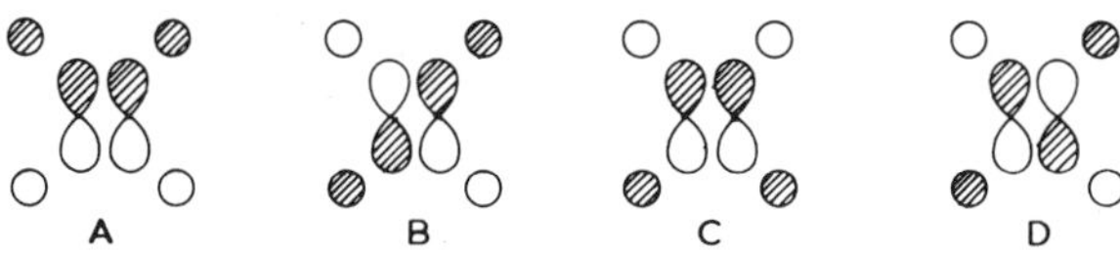

D	0.89
σ^*	0.84
H	0.63
G	0.62
C	0.59
π^*	0.24
π	−0.37
B	−0.51
F	−0.56
A	−0.64
E	−0.78
σ	−1.0

Fig. 1.19. Ethylene molecular orbital energy levels. Energies are given in atomic units.

Orbital **A** is bonding between all nearest-neighbor atoms, whereas **B** is bonding within the CH_2 units but antibonding with respect to the two carbons. The orbital labeled **C** is C—C bonding but antibonding with respect to the hydrogens. Finally, orbital **D** is antibonding with respect to all nearest-neighbor atoms. Similarly, the $2p_x$ orbitals must be considered. Again, four possible combinations arise. Notice that the nature of the overlap of the $2p_x$ orbitals is different from that in the $2p_y$ case, so that the two sets of MOs should have different energies.

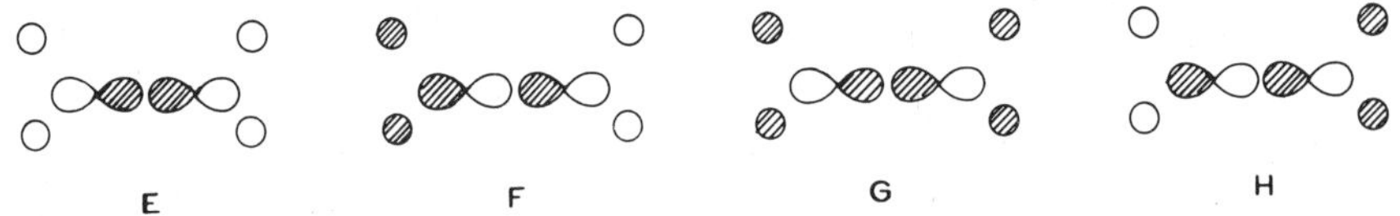

The final problem in construction of a qualitative MO diagram for ethylene is the relative placement of the orbitals. There are some guidelines whicich are useful. First, since π-type interactions are normally weaker than σ-type, we expect the separation between σ and σ^* to be greater than that between π and π^*. Within the sets **ABCD** and **EFGH**, we can predict the ordering **A** < **B** < **C** < **D** and **E** < **F** < **G** < **H** on the basis that C—H bonding interactions will outweigh C—C antibonding interactions arising from relatively weak *p*-*p* overlaps. Placement of the set **ABCD** in relation to **EFGH** is not qualitatively obvious. Calculations give the results shown in Fig. 1.19.[40] Pictorial representations of the orbitals are given in Fig. 1.20.

The kind of qualitative considerations that have been used to construct the ethylene MO diagram do not give an indication of how much each atomic orbital contributes to the individual MOs. This information is obtained from the coefficients provided by an MO calculation. Without these coefficients we cannot specify the

40. W. L. Jorgensen and L. Salem, *The Organic Chemist's Book of Orbitals*, Academic Press, New York, 1973.

shapes of the MOs very precisely. However, the qualitative ideas do permit conclusions about the *symmetry* of the orbitals. As will be seen in Chapter 11, just knowing the symmetry of the MOs provides very useful insight into many chemical reactions.

Figures 1.21 and 1.22 are another type of representation of the ethylene molecule derived from MO calculations. Figure 1.21 is a log scale plot of the σ-electron density. It shows the highest density around the nuclear positions as indicated by the pronounced peaks corresponding to the atomic positions and also indicates the continuous nature of the σ-electron distribution. A representation of the π-electron density is given in Fig. 1.22. This depicts the density in a plane bisecting the molecule and perpendicular to the plane of the molecule. The diagram shows that the π-electron density drops to zero in the nodal plane of the π system.

1.5. Hückel Molecular Orbital Theory

Before the advent of computers enabled detailed molecular orbital calculations to be performed routinely, it was essential that greatly simplifying approximations be applied to the molecular systems of interest to organic chemists. The most useful of these approximations was that incorporated in the Hückel molecular orbital (HMO) theory for treatment of conjugated systems. HMO theory is based on the approximation that the π system can be treated independently of the σ framework in conjugated planar molecules and that it is the π system that is of paramount importance in determining the chemical and spectroscopic properties of conjugated polyenes and aromatic compounds. The rationalization for treating the σ and π systems as independent of each other is based on their orthogonality. The σ skeleton of a planar conjugated system lies in the nodal plane of the π system and does not interact with it. Because of its simplicity, HMO theory has been extremely valuable in the application of molecular orbital concepts to organic chemistry. It provides a good qualitative description of the π-molecular orbitals in both cyclic and acyclic conjugated systems. In favorable cases such as aromatic ring systems, it provides a quite thorough analysis of the relative stability of related structures.

In the HMO approximation, the π-electron wave function is expressed as a linear combination of the p_z atomic orbitals (for the case in which the plane of the molecule coincides with the x–y plane). Minimizing the total π-electron energy with respect to the coefficients leads to a series of equations from which the atomic coefficients can be extracted. Although the mathematical operations involved in solving the equations are not difficult, we will not describe them in detail but will instead concentrate on the interpretation of the results of the calculations. For many systems, the Hückel MO energies and atomic coefficients have been tabulated.[41]

41. C. A. Coulson and A. Streitwieser, Jr., *Dictionary of π-Electron Calculations*, W. H. Freeman, San Francisco, 1965; E. Heilbronner and P. A. Straub, *Hückel Molecular Orbitals*, Springer-Verlag, Berlin, 1966.

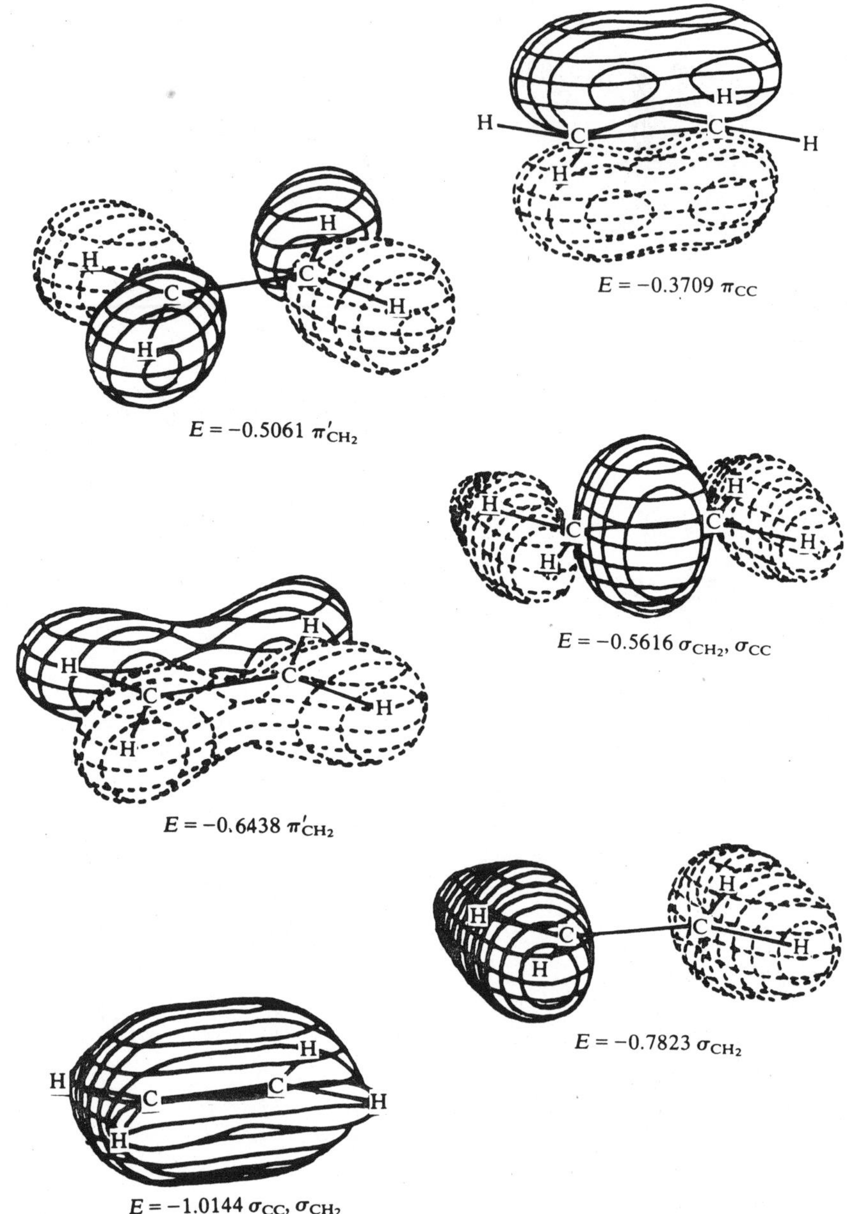

Fig. 1.20. Representation of the molecular orbitals of ethylene. (From W. L. Jorgensen and L. Salem, *The Organic Chemist's Book of Orbitals*, Academic Press, New York, 1973. Reproduced by permission.)

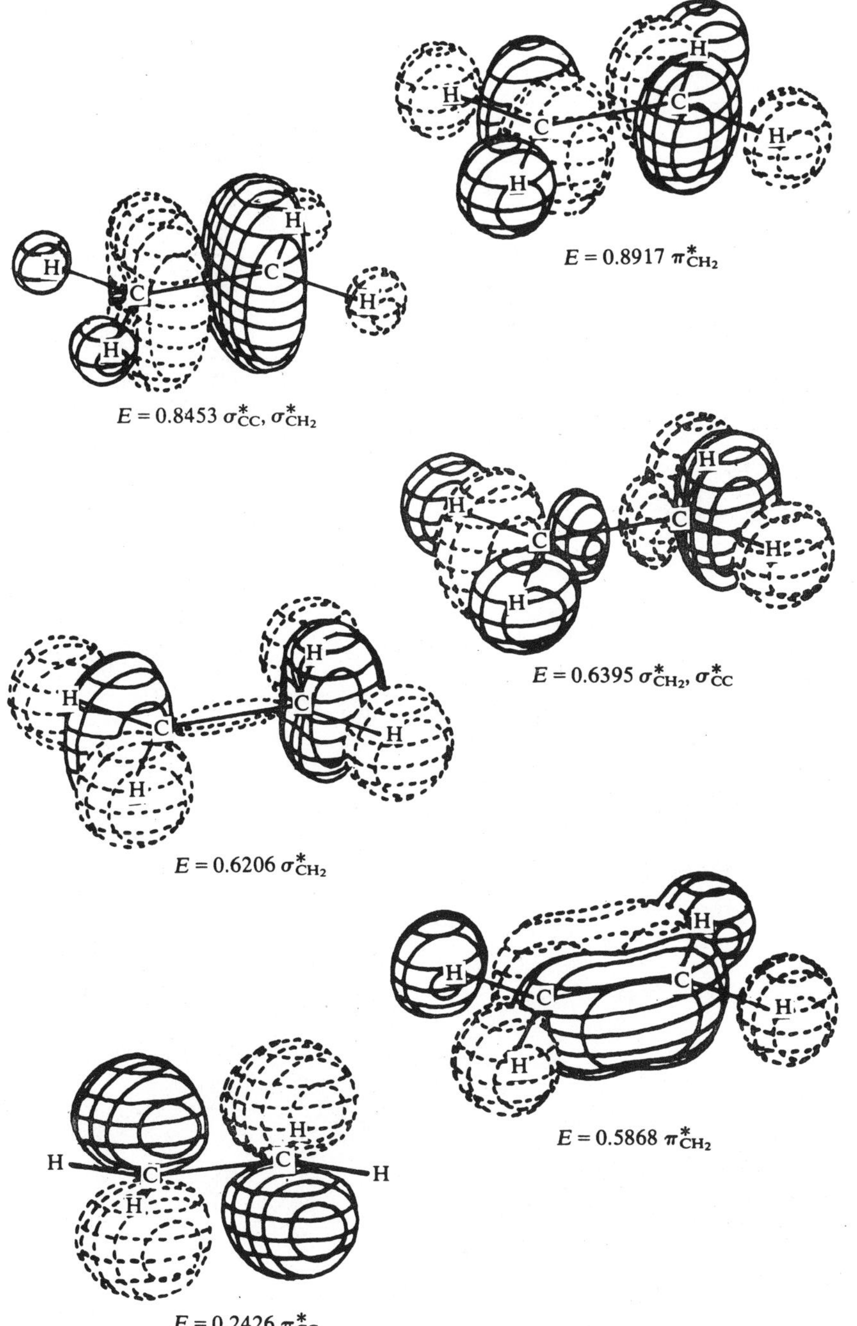

$E = 0.8917\ \pi^*_{CH_2}$

$E = 0.8453\ \sigma^*_{CC}, \sigma^*_{CH_2}$

$E = 0.6395\ \sigma^*_{CH_2}, \sigma^*_{CC}$

$E = 0.6206\ \sigma^*_{CH_2}$

$E = 0.5868\ \pi^*_{CH_2}$

$E = 0.2426\ \pi^*_{CC}$

The most easily obtained information from such calculations is the relative orderings of the energy levels and the atomic coefficients. Solutions are readily available for a number of general cases of frequently encountered delocalized systems, which we will illustrate by referring to some typical examples. Consider first *linear polyenes* of formula C_nH_{n+2} such as 1,3-butadiene, 1,3,5-hexatriene, and so forth. The energy levels for such compounds are given by the expression

$$E = \alpha + m_j\beta$$

where

$$m_j = 2\cos\frac{j\pi}{n+1} \qquad \text{for } j = 1, 2, \ldots, n$$

and n is the number of carbon atoms in the conjugated chain. This calculation generates a series of molecular orbitals with energies expressed in terms of the quantities α and β, which symbolize the *Coulomb integral* and *resonance integral*, respectively. The Coulomb integral, α, is related to the binding energy of an electron

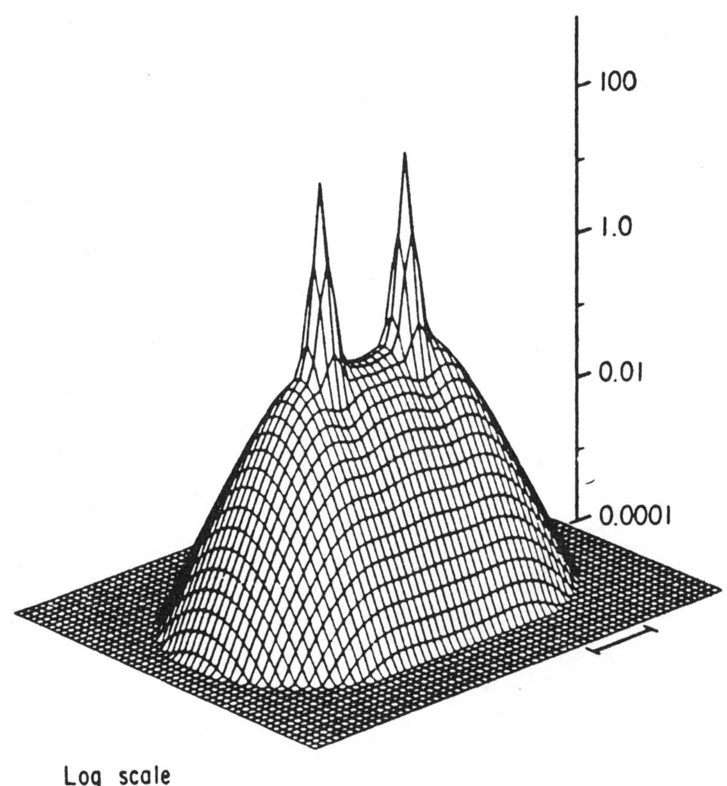

Fig. 1.21. Log scale plot of σ-electron density in a plane bisecting the carbon atoms and perpendicular to the plane of the molecule. (From A Streitwieser, Jr., and P. H. Owens, *Orbital and Electron Density Diagrams*, Macmillan, New York, 1973. Reproduced with permission.)

in a $2p$ orbital, and this is taken to be a constant for all carbon atoms but will vary for heteroatoms as a result of the difference in electronegativity. The resonance integral, β, is related to the energy of an electron in the field of two or more nuclei. In the Hückel method, β is assumed to be zero when nuclei are separated by distances greater than the normal bonding distance. The approximation essentially assumes that the electron is affected only by nearest-neighbor nuclei. Both α and β are negative numbers and represent unspecified units of energy.

The coefficient corresponding to the contribution of the $2p$ AO of atom r to the jth MO is given by

$$c_{rj} = \left(\frac{2}{n+1}\right)^{1/2}\left(\sin\frac{rj\pi}{n+1}\right)$$

Carrying out the numerical operations for 1,3,5-hexatriene gives the results shown in Table 1.15. Since the molecule has six-π-electrons, ψ_1, ψ_2, and ψ_3 are all doubly occupied, giving a total π-electron energy of $6\alpha + 6.988\beta$. The general solution for this system is based on the assumption that the electrons are delocalized. If this

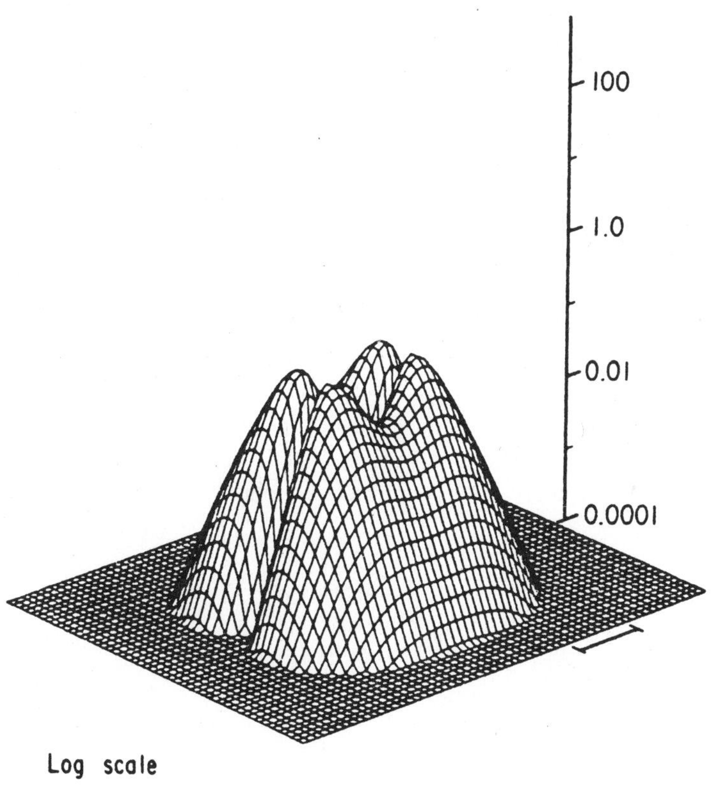

Fig. 1.22. Log scale plot of π-electron density in a plane bisecting the carbon atoms and perpendicular to the plane of the molecule. (From A. Streitwieser, Jr., and P. H. Owens, *Orbital and Electron Density Diagrams*, Macmillan, New York, 1973. Reproduced with permission.)

assumption were not made and the molecule were considered to be composed of alternating single and double bonds, the total π-electron energy would have been $6\alpha + 6\beta$, or identical to that for three ethylene units. The differences between the electron energy calculated for a system of delocalized electrons and that calculated for alternating single and double bonds is referred to as the *delocalization energy* and is a measure of the extra stability afforded a molecule containing delocalized electrons compared to a molecule containing localized "bonds." The calculated delocalization energy (DE) for 1,3,5-hexatriene is 0.988β. The value of β (as expressed in conventional energy units) is not precisely defined. One of the frequently used values is 18 kcal/mol, which is based on the value of 36 kcal/mol for the resonance energy of benzene, for which the calculated π-DE is 2β.

Inspection of the coefficients and a familiarity with the way they translate into symmetry properties of orbitals can be used in an extremely powerful way to aid in understanding a number of aspects of the properties of conjugated unsaturated compounds. Such considerations apply particularly well to the class of reactions classified as *concerted*, which will be described in detail in Chapter 11. It can be seen in Table 1.15 that the coefficients are all of like sign in the lowest-energy orbital, ψ_1, and that the number of times that a sign change occurs in the wave function increases with the energy of the orbital. A change in sign of the coefficients of the AOs on adjacent atoms corresponds to an antibonding interaction between the two atoms, and a node exists between them. Thus, ψ_1 has no nodes, ψ_2 has one, ψ_3 has two, and so on up to ψ_6, which has five nodes and no bonding interactions in its combination of AOs. A diagrammatic view of the bonding and antibonding interactions among the AOs of 1,3,5-hexatriene is presented in Fig. 1.23. Notice that for the bonding orbitals ψ_1, ψ_2, and ψ_3, there are more bonding interactions (positive overlap) than antibonding interactions (negative overlap), while the opposite is true of the antibonding orbitals.

The success of simple HMO theory in dealing with the relative stabilities of cyclic conjugated polyenes is impressive. Simple resonance arguments lead to confusion when one tries to compare the unique stability of benzene with the elusive and unstable nature of cyclobutadiene. (Two apparently analogous resonance structures can be drawn in each case.) This contrast is readily explained by Hückel's rule, which states that a species composed of a planar monocyclic array of atoms,

Table 1.15. Energy Levels and Coefficients for HMOs of 1,3,5-Hexatriene

π-Orbital: ψ_j	m_j	c_1	c_2	c_3	c_4	c_5	c_6
ψ_1	1.802	0.2319	0.4179	0.5211	0.5211	0.4179	0.2319
ψ_2	1.247	0.4179	0.5211	0.2319	−0.2319	−0.5211	−0.4179
ψ_3	0.445	0.5211	0.2319	−0.4179	−0.4179	0.2319	0.5211
ψ_4	−0.445	0.5211	−0.2319	−0.4179	0.4179	0.2319	−0.5211
ψ_5	−1.247	0.4179	−0.5211	0.2319	0.2319	−0.5211	0.4179
ψ_6	−1.802	0.2319	−0.4179	0.5211	−0.5211	0.4179	−0.2319

each of which contributes a p orbital to the π system, will be strongly stabilized (aromatic) if the number of electrons in the π system is $4n + 2$, where n is an integer. By this criterion, benzene, with six π electrons, is aromatic, while cyclobutadiene, with four, is not. An understanding of the theoretical basis for Hückel's rule can be gained by examining the results of HMO calculations. For cyclic polyenes, the general solution for the energy levels is

$$E = \alpha + m_j\beta$$

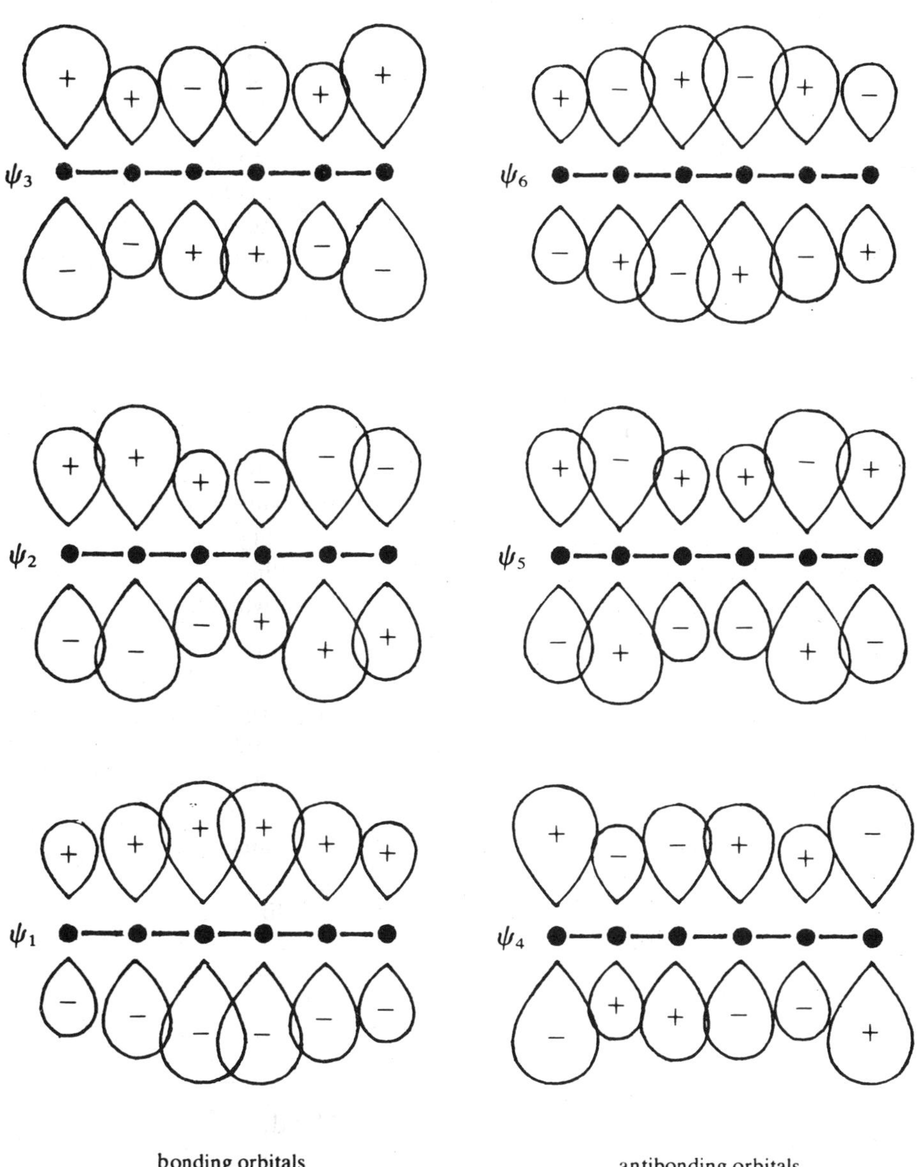

Fig. 1.23. Graphic representation of π-molecular orbitals of 1,3,5-hexatriene as combinations of $2p$ AOs. The sizes of the orbitals are roughly proportional to the coefficients of the Hückel wave functions.

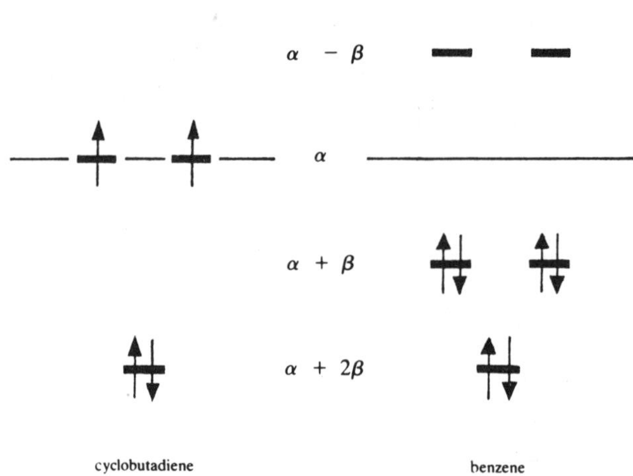

Fig. 1.24. Energy level diagrams for cyclobutadiene and benzene.

where

$$m_j = 2\cos\frac{2j\pi}{n} \qquad \text{for } j = 0, \pm1, \pm2, \ldots, \begin{cases} \pm(n-1)/2 \text{ for } n \text{ odd} \\ \pm n/2 \text{ for } n \text{ even} \end{cases}$$

and n is the number of carbon atoms in the ring. This solution gives the energy level diagrams for cyclobutadiene and benzene shown in Fig. 1.24.

The total π-electron energy of benzene is $6\alpha + 8\beta$, corresponding to a DE of 2β. Cyclobutadiene is predicted to have a triplet ground state (for a square geometry) and zero DE, since the π-electron energy is $4\alpha + 4\beta$, the same as that for two independent double bonds. Thus, at this level of approximation, HMO theory predicts no stabilization for cyclobutadiene from delocalization and furthermore predicts that the molecule will have unpaired electrons, which would lead to very high reactivity. The extreme instability of cyclobutadiene is then understandable. More elaborate MO calculations modify this picture somewhat. They predict that cyclobutadiene will be a rectangular molecule, as will be discussed in Chapter 9. These calculations, nevertheless, agree with simple HMO theory in concluding that there will be no stabilization of cyclobutadiene resulting from the conjugated double bonds.

A useful mnemonic device for quickly setting down the HMOs for cyclic systems is *Frost's circle.*[42] If a regular polygon of n sides is inscribed in a circle of diameter 4β with one corner at the lowest point, the points at which the corners of the polygon touch the circle define the energy levels. The energy levels obtained with Frost's circle for benzene and cyclobutadiene are shown in Fig. 1.25.

42. A. A. Frost and B. Musulin, *J. Chem. Phys.* **21**, 572 (1953).

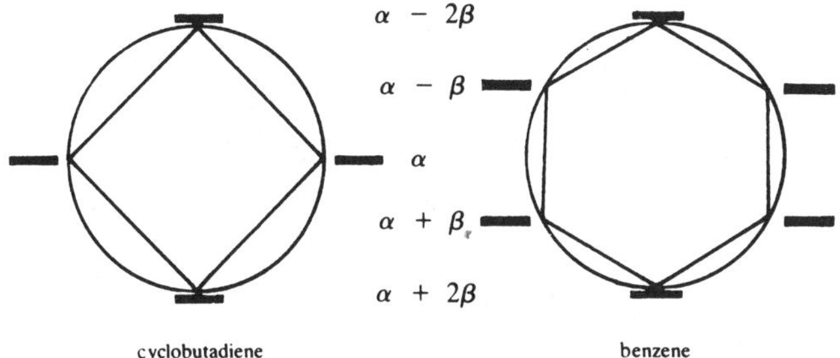

Fig. 1.25. Energy level diagrams for cyclobutadiene and benzene, illustrating the application of Frost's circle.

The energy level diagrams for charged C_3H_3 and C_5H_5 systems are readily constructed and are presented in Fig. 1.26. Cyclopropenyl cation has a total of two π electrons, which occupy the bonding HMO, and a total π-electron energy of $2\alpha + 4\beta$. This gives a DE of 2β and is indicative of a stabilized species. Addition of two more π electrons to the system to give cyclopropenide anion requires population of higher-energy antibonding orbitals and results in a net destabilization of the molecule. The opposite is true for the C_5H_5 case, where the anion is more stabilized than the cation.

Monocyclic conjugated systems are referred to as *annulenes*, and there exists ample experimental evidence to support the conclusions based on application of HMO theory to neutral and charged annulenes. The relationship between stability and structure in cyclic conjugated systems will be explored more fully in Chapter 9.

While Hückel's $4n + 2$ rule applies only to monocyclic systems, HMO theory is applicable to many other systems. HMO calculations of fused-ring systems are carried out in much the same way as for monocyclic species and provide energy levels and atomic coefficients for the systems. The incorporation of heteroatoms is also possible. Because of the underlying assumption of orthogonality of the σ and π systems of electrons, HMO theory is restricted to planar molecules.

While the Hückel method has now been supplanted by more complete treatments for theoretical analysis of organic reactions, the pictures of the π orbitals of both linear and cyclic conjugated polyene systems are correct as to symmetry and the relative energy of the orbitals. In many reactions where the π system is the

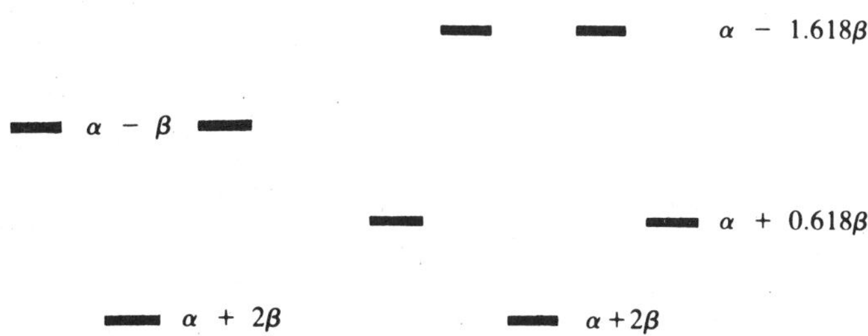

Fig. 1.26. Energy level diagrams for C_3H_3 and C_5H_5 systems.

primary site of reactivity, these orbitals correctly describe the behavior of the systems. For that reason, the student should develop a familiarity with the qualitative description of the π orbitals of typical linear polyenes and conjugated cyclic hydrocarbons. These orbitals will be the basis for further discussion in Chapters 9 and 11.

1.6. Perturbation Molecular Orbital Theory

The construction of molecular orbital diagrams under the guidance of the general principles and symmetry restrictions which have been outlined can lead to useful insights into molecular structure. Now we need to consider how these structural concepts can be related to reactivity. In valence bond terminology, structure is related to reactivity in terms of substituent effects. The impact of inductive effects and resonance effects on the electron distribution and stability of reactants, transition states, and intermediates is assessed. In molecular orbital theory, reactivity is related to the relative energies and shapes of the orbitals which are involved as the reactants are transformed to products. Reactions which can take place through relatively stable intermediates and transition states are more favorable and more rapid than reactions which proceed through less stable ones. The *symmetry* of the molecular orbitals is a particularly important feature of many analyses of reactivity based on molecular orbital theory. The shapes of orbitals also affect the energy of reaction processes. Orbital shapes are quantified by the atomic coefficients. The strongest overlap (bonding when the overlapping orbitals have the same sign) occurs when the interacting orbitals on two reaction centers have high coefficients on the two atoms which undergo bond formation.

The qualitative description of reactivity in molecular orbital terms must begin with a basic understanding of the molecular orbitals of the reacting systems. At this point, we have developed a familiarity with the MOs of ethylene and conjugated unsaturated systems from the discussion of Hückel MO theory and the construction of the ethylene MOs from atomic orbitals. To apply these ideas to new systems, we need to be able to understand how a change in structure will affect the MOs. One approach to this problem is called *perturbation molecular orbital theory* or PMO for short.[43] In this approach, a system under analysis is compared to another related system for which the MO pattern is known. In PMO theory the MO characteristics of the new system are deduced by analyzing how the change in structure, the perturbation, affects the MO pattern. The types of changes which can be handled in a qualitative way are substitution of atoms by other elements with the resulting change in electronegativity and changes in connectivity which revise the pattern of

43. C. A. Coulson and H. C. Longuet-Higgins, *Proc. R. Soc. London Ser. A*, **192**, 16 (1947); L. Salem, *J. Am. Chem. Soc.* **90**, 543 (1968); M. J. S. Dewar and R. C. Dougherty, *The PMO Theory of Organic Chemistry*, Plenum Press, New York, 1975; G. Klopman, *Chemical Reactivity and Reaction Paths*, Wiley-Interscience, New York, 1974, Chapter 4.

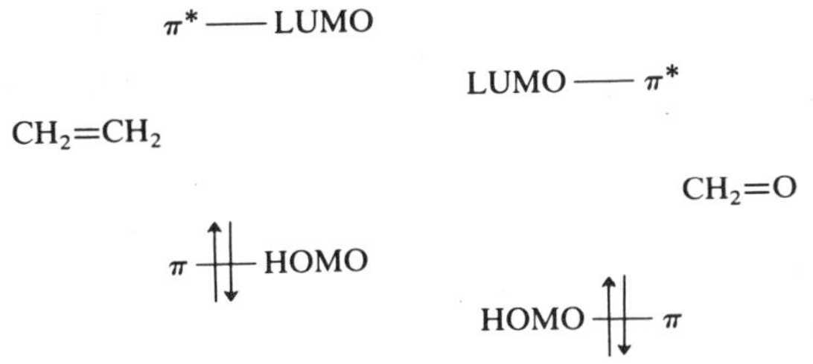

π^* —— LUMO

LUMO —— π^*

$CH_2{=}CH_2$

$CH_2{=}O$

π ⇅ HOMO

HOMO ⇅ π

Fig. 1.27. Relative energy of the π and π^* orbitals in ethylene and formaldehyde.

direct orbital overlap. The fundamental thesis of PMO theory is that the resulting changes in the MO energies are relatively small and can be treated as adjustments on the original MO system.

Another aspect of qualitative application of MO theory is the analysis of interactions of the orbitals in reacting molecules. As molecules approach one another and reaction proceeds, there is a mutual perturbation of the orbitals. This process continues until the reaction is complete and the new product (or intermediate in a multistep reaction) is formed. PMO theory incorporates the concept of *frontier orbital control.* This concept proposes that the most important interactions will be between a particular pair of orbitals.[44] These orbitals are the highest filled orbital (the HOMO, highest occupied molecular orbital) of one reactant and the lowest unfilled orbital (the LUMO, lowest unoccupied molecular orbital) of the other reactant. The basis for concentrating attention on these two orbitals is that they will usually be the closest in energy of the interacting orbitals. A basic postulate of PMO theory is that interactions are strongest between orbitals that are close in energy. Frontier orbital theory proposes that these strong initial interactions can then guide the course of the reaction as it proceeds to completion. A further general feature of PMO theory is that only MOs of matching symmetry can interact so as to lead to bond formation. Thus, analysis of a prospective reaction path by PMO will direct attention to the relative energy and symmetry of the interacting orbitals.

These ideas can be illustrated here by considering some very simple cases. We will return to frontier orbital theory in more detail in Chapter 11. Let us consider the fact that the double bonds of ethylene and formaldehyde have quite different chemical reactivities. Formaldehyde reacts readily with nucleophiles whereas ethylene does not. The π bond in ethylene is more reactive than the formaldehyde $C{=}O$ π bond toward electrophiles. We have already described the ethylene MOs in Figs. 1.19 and 1.20. How will those of formaldehyde differ? In the first place, the higher atomic number of oxygen provides two additional electrons so that in place of the CH_2 group of ethylene, the oxygen of formaldehyde has two pairs of nonbonding electrons. The key change, however, has to do with the frontier orbitals, the π (HOMO) and π^* (LUMO) orbitals. These are illustrated in Fig. 1.27. One

44. K. Fukui, *Acc. Chem. Res.* **4**, 57 (1971); I. Fleming, *Frontier Orbitals and Organic Chemical Reactions*, Wiley, New York, 1976; L. Salem, *Electrons in Chemical Reactions*, Wiley, New York, 1982, Chapter 6.

significant difference between the two molecules is the lower energy of the π and π^* orbitals in formaldehyde. These are lower in energy than the corresponding ethylene orbitals because they are derived in part from the lower-lying (more electronegative) $2p$ orbital of oxygen. Because of its lower energy, the π^* orbital is a better acceptor of electrons from the HOMO of any attacking nucleophile than is the LUMO of ethylene. On the other hand, we also can see why ethylene is more reactive toward electrophiles than formaldehyde. In electrophilic attack it is the HOMO that is involved as an electron donor to the approaching electrophile. In this case, the fact that the HOMO of ethylene lies higher in energy than the HOMO of formaldehyde will mean that electrons can be more easily attracted by the approaching electrophile. The unequal electronegativities of the oxygen and carbon atoms also distort the π molecular orbital. Relative to the symmetrical distribution in ethylene, the formaldehyde π MO has a higher atomic coefficient at oxygen. This results in a net positive charge in the vicinity of the carbon atom, which is a favorable circumstance for approach by a nucleophilic reactant.

Perturbation theory can be expressed in a quantitative way. We will not consider the quantitative theory here but do want to point out two features of the quantitative theory which underlie the qualitative application of PMO theory. One principle is that the degree of perturbation is a function of the degree of overlap of the orbitals. Thus, in the qualitative application of PMO theory, it is important to consider the shape of the orbitals (as indicated quantitatively by their coefficients) and the proximity which can be achieved within the limits of the geometry of the reacting molecules. Secondly, the strength of a perturbation depends on the relative energy of the interacting orbitals. The closer the orbitals are in energy, the greater will be their mutual interaction. This principle, if used in conjunction with reliable estimates of relative orbital energies, can be of value in predicting the relative importance of various possible interactions.

Let us illustrate these ideas by returning to the comparisons of the reactivities of ethylene and formaldehyde toward a nucleophilic species and an electrophilic species. The interactions (perturbations) that arise as both a nucleophile and an electrophile approach are sketched in Fig. 1.28.

The electrophilic species $\mathbf{E}^+$ must have a low-lying empty orbital. The strongest interaction will be with the ethylene π orbital, and this leads to a strong perturbation which has a stabilizing effect on the complex since the electrons are located in an orbital that is stabilized. The same electrophilic species would lie further from the π orbital of formaldehyde since the formaldehyde orbitals are shifted to lower energy relative to the ethylene orbitals. As a result, the interaction between the electrophilic species and the formaldehyde HOMO will be weaker than the corresponding interaction in the case of ethylene. The conclusion is that such an electrophile will undergo a greater stabilizing attraction on approaching within bonding distance of ethylene than for formaldehyde. In the case of $\mathbf{Nu}^-$, a strong bonding interaction with π^* of formaldehyde is possible [(D), Fig. 1.28]. In the case of ethylene, there may be a stronger interaction with the HOMO, but this is a destabilizing interaction since both orbitals are filled and the lowering of one orbital is canceled by the raising of the other. Thus, we conclude that a nucleophile with a high-lying HOMO will interact more favorably with formaldehyde than with ethylene.

The ideas of PMO theory can be used in a slightly different way to describe substituent effects. Let us consider, for example, the effect of a π-donor substituent or a π-acceptor substituent on the MO levels of ethylene and upon the reactivity of substituted ethylenes. We can take the amino group as an example of a π-donor substituent. The nitrogen atom provides an additional $2p_z$ orbital and two electrons to the π system. The overall shape of the π orbitals for aminoethylene will be very similar to those of an allyl anion but with some distortion since the system is no longer symmetrical. The highest charge density should be on the terminal atoms, that is, the nitrogen atom and the β-carbon, since the HOMO has a node at the center carbon. Furthermore, the HOMO should be considerably higher in energy than the HOMO in ethylene. The HOMO in aminoethylene will resemble ψ_2 of the allyl anion. It will not be quite as high in energy as the allyl ψ_2 because of the higher electronegativity of the nitrogen atom, but it will be substantially higher than the HOMO of ethylene. Thus, we expect aminoethylene, with its high-lying HOMO, to be more reactive toward electrophiles than ethylene. Furthermore, the HOMO has the highest coefficients on the terminal atoms so we expect an electrophile to become bonded to the β-carbon or nitrogen, but not the α-carbon. On the other hand, the LUMO will now correspond to the higher-energy ψ_3 of the allyl anion so we expect aminoethylene to be even less reactive toward nucleophiles than is ethylene.

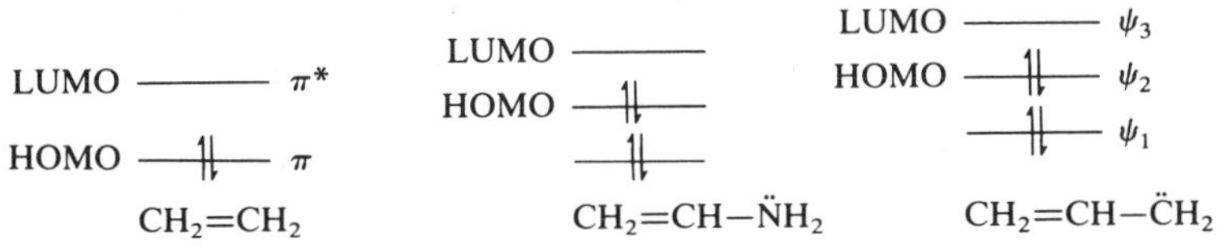

LUMO —— π^* HOMO —⥮— π $CH_2{=}CH_2$ | LUMO —— HOMO —⥮— —⥮— $CH_2{=}CH{-}\ddot{N}H_2$ | LUMO —— ψ_3 HOMO —⥮— ψ_2 —⥮— ψ_1 $CH_2{=}CH{-}\ddot{C}H_2$

π MO energy levels for ethylene with a π-donor substituent.

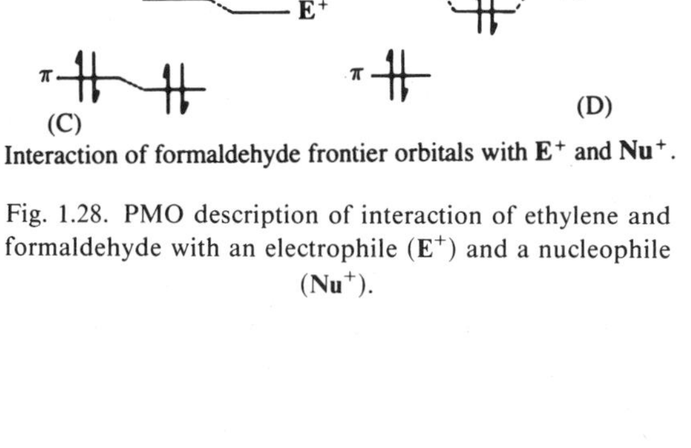

π^*, π, E^+ (A) π^*, π, Nu^- (B)

Interaction of ethylene frontier orbitals with $\mathbf{E}^+$ and $\mathbf{Nu}^-$

π^*, π, E^+ (C) π^*, π, Nu^- (D)

Interaction of formaldehyde frontier orbitals with $\mathbf{E}^+$ and $\mathbf{Nu}^+$.

Fig. 1.28. PMO description of interaction of ethylene and formaldehyde with an electrophile ($\mathbf{E}^+$) and a nucleophile ($\mathbf{Nu}^+$).

An example of a π-acceptor group would be the formyl group, as in acrolein:

$$CH_2{=}CHCH{=}O$$

In this case, the π-molecular orbitals should resemble those of butadiene. Relative to butadiene, however, the acrolein orbitals will lie somewhat lower in energy because of the effect of the more electronegative oxygen atom. This factor will also increase the electron density at oxygen relative to carbon.

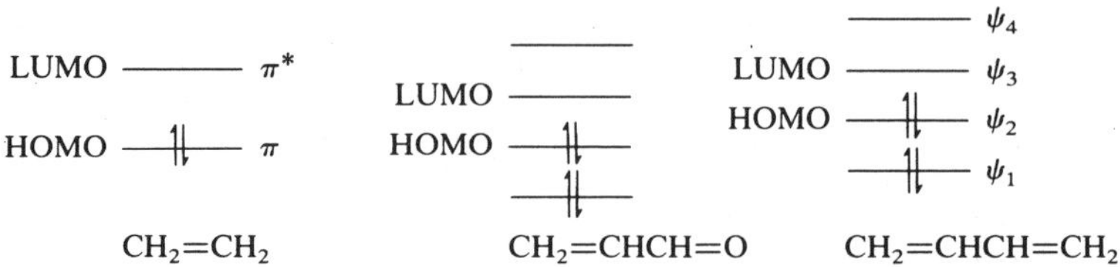

π MO energy levels for ethylene with a π-acceptor substituent.

The LUMO, which will be the frontier orbital in reactions with nucleophiles, has a large coefficient on the β-carbon atom, whereas the two occupied orbitals are distorted in such a way as to have larger coefficients on oxygen. The overall effect is that the LUMO is relatively low-lying and has a high coefficient on the β-carbon atom. The frontier orbital theory therefore predicts nucleophiles will react preferentially at the β-carbon atom. A numerical value can be assigned to the coefficient by an MO calculation. Figure 1.29 gives the orbital coefficients as derived from CNDO calculations.

Both MO theory and experimental measurements provide a basis for evaluation of the energetic effects of conjugation between a double bond and adjacent substituents. Table 1.16 gives some representative values. The theoretical values ΔE are for the reaction

$$CH_2{=}CH{-}X + CH_4 \rightarrow CH_3{-}X + CH_2{=}CH_2$$

and are calculated at the *ab initio* level of theory. These values refer to the gas phase. The ΔH values are based on the experimentally determined thermodynamic ΔH_f values of the compounds. Notice that both electron-withdrawing and electron-accepting substituents result in a net stabilization of the conjugated system. This stabilization results from the lowering in energy of the lowest-lying MO in each

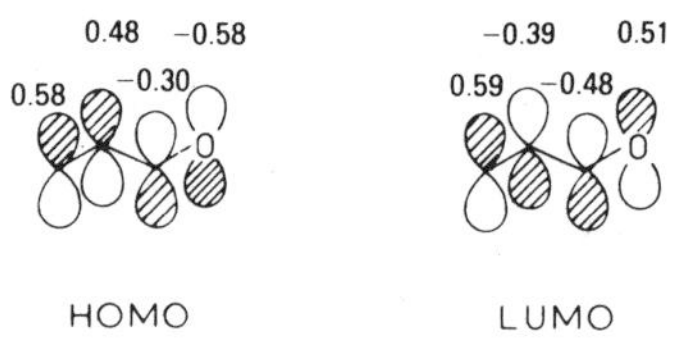

Fig. 1.29. Orbital coefficients for the HOMO and LUMO of acrolein. [From K. N. Houk and P. Strozier, *J. Am. Chem. Soc.* **95**, 4094 (1973).]

case. The effect on the HOMO is different for electron-withdrawing as compared with electron-accepting substituents. For donor substituents. the HOMO is raised in energy, relative to ethylene. For electron-accepting substituents, it is lowered relative to ethylene.

Notice that the MO picture gives the same qualitative picture of the substituent effect as described by resonance structures. The amino group is pictured by resonance as an electron donor which causes a buildup of electron density at the β carbon, whereas the formyl group is an electron acceptor which diminishes electron density at the β carbon.

$$CH_2{=}CH{-}NH_2 \leftrightarrow {}^{-}CH_2{-}CH{=}NH_2^{+} \qquad CH_2{=}CH{-}CH{=}O \leftrightarrow {}^{+}CH_2{-}CH{=}CH{-}O^{-}$$

The chemical reactivity of these two substituted ethylenes is in agreement with the ideas encompassed by both the MO and resonance descriptions. Enamines, as amino-substituted alkenes are called, are very reactive toward electrophilic species, and it is the β-carbon that is the site of attack. For example, enamines are protonated on the β-carbon. Acrolein is an electrophilic alkene, as predicted, and nucleophiles attack the β-carbon.

Frontier orbital theory also provides the basic framework for analysis of the effect that the symmetry of orbitals has upon reactivity. One of the basic tenets of PMO theory is that the symmetries of two orbitals must match to permit a strong interaction between them. This symmetry requirement, used in the context of frontier orbital theory, can be a very powerful tool for predicting reactivity. As an example, let us examine the approach of an allyl cation and an ethylene molecule and ask whether the following reaction is likely to occur.

H
$H_2C{\cdots}C(H){\cdots}CH_2^{+}$
$H_2C{=}CH_2$
$\xrightarrow{?}$ H, +

The positively charged allyl cation would be expected to be the electron acceptor in any initial interaction with ethylene. Therefore, to consider this reaction in terms of frontier orbital theory, the question we need to answer is "do the ethylene HOMO

Table 1.16. Stabilization Resulting from Conjugation of Ethylene with Substituents

Substituent	ΔE (kcal/mol)	ΔH (kcal/mol)	Substituent	ΔE (kcal/mol)	ΔH (kcal/mol)
H	0	0	NH_2	13.3	13.3
F	6.4	6.7 (3.4)	CN	3.3	4.8
CH_3	4.3	5.4	$COCH_3$	4.0	10.5 (3.8)
OCH_3	10.9	12.3 (4.9)	CO_2CH_3	8.0	11.9 (3.4)

a. From A. Greenberg and T. A. Stevenson, *J. Am. Chem. Soc.* **107**, 3488 (1985). Values in parentheses are from J. Hine and M. J. Skoglund, *J. Org. Chem.* **47**, 4766 (1982) and are based on experimental equilibrium measurement values. These measurements are in solution, and the difference between the two sets of values may reflect the effect of solvent.

and allyl cation LUMO interact favorably as the reactants approach one another?" The orbitals which are involved are shown in Fig. 1.30. If we analyze a symmetrical approach, which would be necessary to simultaneously form the two new bonds, we see that the symmetries of the two orbitals do not match. Any bonding interaction developing at one end would be canceled by an antibonding interaction at the other end. The conclusion that is drawn from this analysis in terms of PMO theory is that this particular reaction process is not favorable. We would need to consider other modes of approach to analyze the problem more thoroughly, but this analysis indicates that simultaneous (concerted) bond formation between ethylene and the allyl cation to form a cyclopentyl cation is not possible.

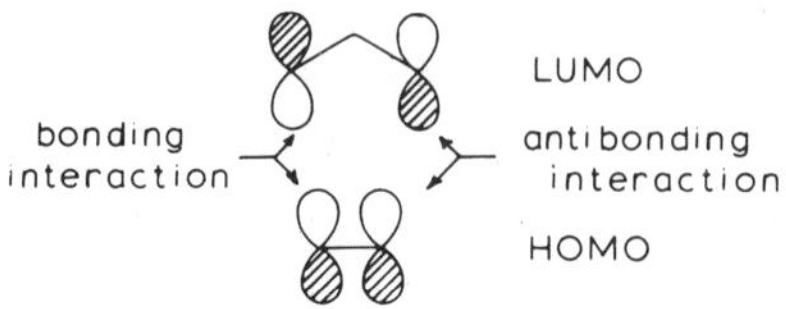

Let us now consider another hypothetical reaction, this time between the allyl cation and butadiene. Again the assumption will be made that it is the π electrons that will govern the course of the reaction. We will also be slightly more formal about the issue of symmetry. This can be done by recognizing the elements of symmetry which would be maintained as the reaction proceeded. If the reaction is to proceed in a single step, the geometry must permit simultaneous overlap of the orbitals on the carbons where new bonds are being formed. A geometry of approach which permits such a simultaneous overlap is shown below.

The allyl cation could approach from the top (or bottom) of the *cis* conformation of butadiene, and the new σ bonds would be formed from the π orbitals. This

Fig. 1.30. MOs for ethylene and allyl cation.

arrangement would maintain a plane of symmetry during the course of the reaction. The plane bisects butadiene between C-2 and C-3 and the allyl cation at C-2. The orbitals can be classified as symmetric (S) or antisymmetric (A) with respect to this plane. This gives rise to the MO diagram shown in Fig. 1.31. Since strong interactions will occur only between orbitals of the same symmetry, the mutual perturbation of the approaching reactants will affect the orbital energy levels as shown in the diagram. As in all such perturbations, one orbital of the interacting pair will be stabilized and the other will move to higher energy. The energy levels of the perturbed orbitals at some point on the way to the transition state are shown in the diagram. Eventually, when the reaction has proceeded to completion, a new set of orbitals belonging to the product will have been formed. The energy levels of the product are shown in the center of the diagram but we will be considering only the initial perturbed set. The lowest-lying orbitals of both butadiene and the allyl cation are filled. These will interact, with one moving down in energy and the other up. These two changes in energy are partially compensating, with the total energy change being a net increase in the energy of the system. Both the HOMO and LUMO are antisymmetric and will interact strongly, but in this case, since only two electrons are involved, only the energy of the stabilized orbital will affect the total energy since the destabilized orbital is empty. This HOMO–LUMO interaction then contributes a net bonding contribution as the transition state is approached. From this analysis we conclude that there is the possibility of a favorable bonding interaction

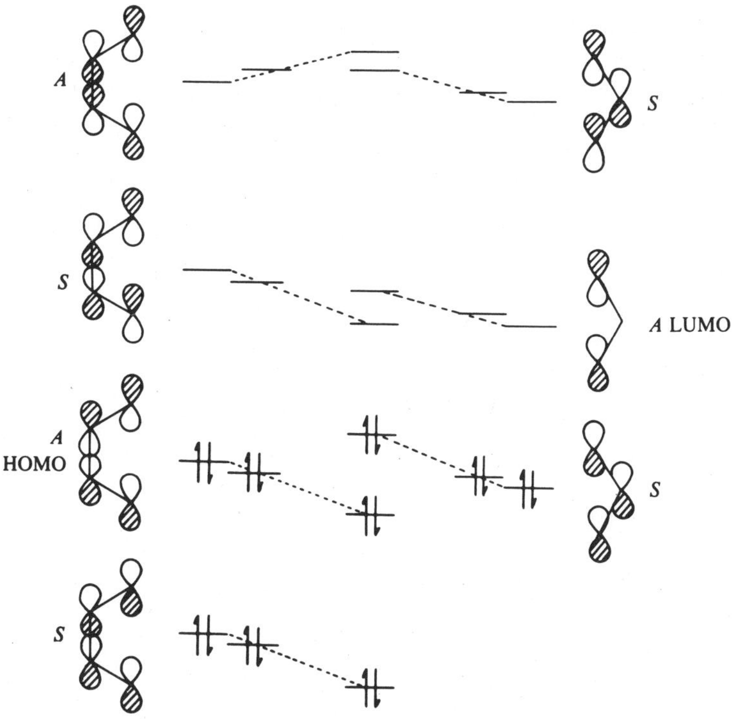

Fig. 1.31. MO diagram showing mutual perturbation of MOs of butadiene and allyl cation.

between the two reactant species. Notice that the reaction is only *permitted*, and nothing can be said about its actual efficiency or energy requirement on the basis of the analysis given. Such matters as steric hindrance to approach of the reactants and the geometric requirements for satisfactory overlap of the orbitals could still cause the reaction to proceed with difficulty. The analysis does establish, however, that there is a pathway by which the orbitals of the reactants can interact in a way that is favorable for reaction.

A more complete analysis of interacting molecules would examine all of the involved MOs in a similar way. A *correlation diagram* would be constructed to determine which reactant orbital is transformed into which product orbital. Reactions which permit smooth transformation of the reactant orbitals to product orbitals without intervention of high-energy transition states or intermediates can be identified in this way. If no such transformation is possible, a much higher activation energy is likely since the absence of a smooth transformation implies that bonds must be broken before they can be reformed. This treatment is more complete than the frontier orbital treatment since it focuses attention not only on the reactants but also on the products. We will describe this method of analysis in detail in Chapter 11. The PMO approach which has been described here is a useful and simple way to apply MO theory to reactivity problems in a qualitative way, and we will employ it in subsequent chapters to problems in reactivity which are best described in MO terms.

It is worth noting that in the case of the reaction of ethylene and butadiene with the allyl cation, the MO description has provided a prediction which would not have been recognized by a pictorial application of valence bond terminology. Thus, we can write an apparently satisfactory description of both reactions:

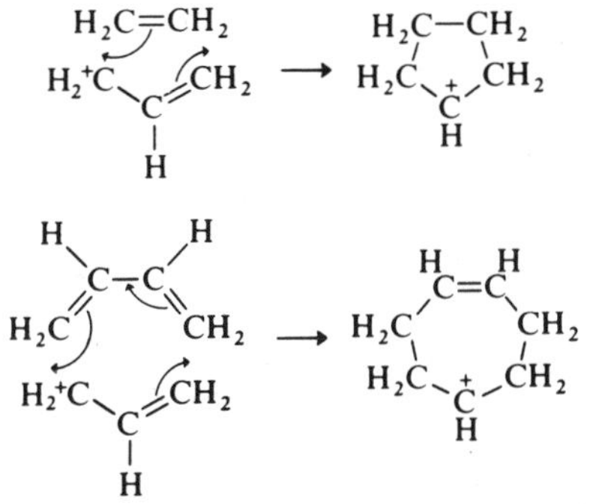

$H_2C{=}CH_2$ + $H_2\overset{+}{C}{-}CH{=}CH_2$ → $H_2C{-}CH_2$ / $H_2C{-}\overset{+}{C}H{-}CH_2$ (cyclopentyl cation)

$H_2C{=}CH{-}CH{=}CH_2$ + $H_2\overset{+}{C}{-}CH{=}CH_2$ → $H_2C{-}CH{=}CH{-}CH_2$ / $H_2C{-}\overset{+}{C}H{-}CH_2$ (cycloheptenyl cation)

It is only on considering the symmetry of the interacting orbitals that we find reason to suspect that only the second of the two reactions is possible.

1.7. Interactions between σ and π Systems—Hyperconjugation

One of the key assumptions of the Hückel approximation is the noninteraction of the π-orbital system with the σ-molecular framework. This is a good approximation for completely planar π molecules where the σ framework is in the nodal plane of the π system. For other molecules, as for example when an sp^3 carbon is added as a substituent group, this approximation is no longer entirely valid. Qualitative application of molecular orbital theory can be enlightening in describing interactions between the π system and substituent groups. In valence bond theory, a special type of resonance called hyperconjugation is used to describe such interactions. For example, much chemical and structural evidence indicates that alkyl groups substituted on a carbon–carbon double bond act as electron donors to the π system. In valence bond language, "no bond" resonance structures are introduced to indicate this electronic interaction.

```
     H                  H
     |                  |
  H—C       H       ⁺H  C       H
     |\    /            |\\    /
    H C=C    ⟷         H C—C:⁻
     /    \              /    \
    H      H            H      H
```

The molecular orbital picture of such interactions flows from the idea that individual orbitals encompass the entire molecule. Thus, while the MO description of ethylene involved no interaction between the C $2p_z$ orbitals and the H $1s$ orbitals (see p. 35 to recall this discussion), this strict separation would not exist in propene since the hydrogens of the methyl group are not in the nodal plane of the π bond. The origin of interactions of these hydrogens with the π orbital can be indicated as in Fig. 1.32, which shows propene in a geometry in which two of the hydrogen $1s$ atomic orbitals are in a position to interact with the $2p_z$ orbital of carbon-2.

An *ab initio* calculation using a STO-3G basis set was carried out on propene in two distinct geometries, eclipsed and staggered.

```
  H,,  /H               H /H
  H—C                 H—C
      \    /H              \    /H
       C=C                  C=C
      /    \               /    \
     H      H             H      H
     eclipsed             staggered
```

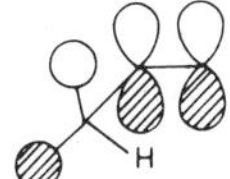

Fig. 1.32. Interactions between two hydrogen $1s$ orbitals and carbon $2p_z$ orbitals stabilize the eclipsed conformation of propene.

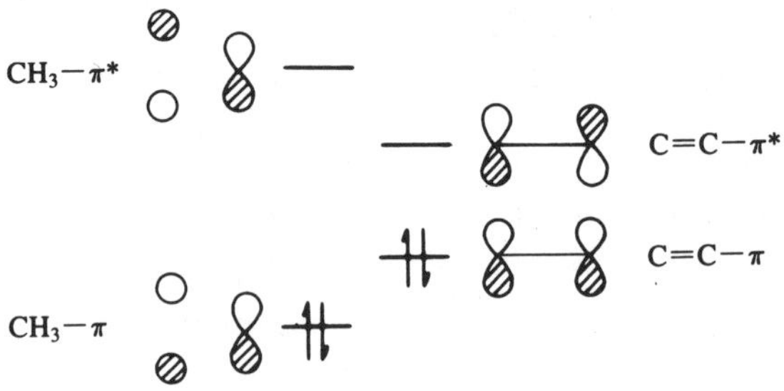

$CH_3-\pi^*$

$C{=}C-\pi^*$

$C{=}C-\pi$

$CH_3-\pi$

Fig. 1.33. Interactions between CH_3-π and CH_3-π^* orbitals and carbon $2p_z$ orbitals.

The calculations of the optimum geometry show a slight lengthening of the C—H bonds because of the electron release to the π system. These calculations also reveal a barrier to rotation of the methyl group of about 1.5–2.0 kcal/mol. Interaction between the hydrogens and the π system favors the eclipsed conformation to this extent.[45] Let us examine the reason for the preference for the eclipsed conformation. This issue can be approached by analyzing the interactions between the carbon $2p_z$ orbitals and the CH_3 fragment in a little more detail. The bonding and antibonding combinations that arise from interaction of the appropriate CH_3-π and CH_3-π^* orbitals with the $2p_z$ orbitals are shown in Fig. 1.33. The strongest interaction is a repulsive one between the filled CH_3-π and C=C-π orbitals. It is this interaction which is primarily responsible for the favored eclipsed conformation. The eclipsed structure minimizes the repulsion by maximizing the separation between the hydrogens and the π bond. The second interaction is the stabilizing hyperconjugative one between CH_3-π and C=C-π^*. This is a bonding interaction since π^* is an empty orbital and can accept electron density from CH_3-π. It is this bonding interaction which transfers electron density from the methyl group to the terminal carbon of the double bond. There is the possibility of a corresponding interaction between C=C-π and CH_3-π^*, which can become important in systems where the π orbital of the double bond is of higher energy. Notice that there is a correspondence between the MO picture and the valence bond resonance structure in that both specify a net transfer of electron density from C—H bonds to the π bond with a net strengthening of the bond between C-2 and C-3 but a weakening of the C-1—C-2 π bond.

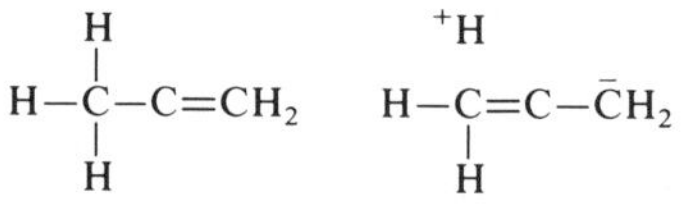

H–C(H)(H)–C=CH₂ ⟷ H–C(⁺H)(H)=C–C̄H₂

hyperconjugation

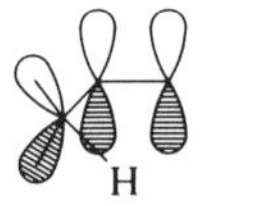

repulsive interaction between CH_3-π and π

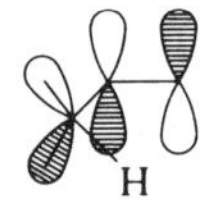

attractive interaction between CH_3-π and π^*

45. W. J. Hehre, J. A. Pople, and A. J. P. Devaquet, *J. Am. Chem. Soc.* **98**, 664 (1976); A. Pross, L. Radom, and N. V. Riggs, *J. Am. Chem. Soc.* **102**, 2253 (1980); K. B. Wiberg and E. Martin, *J. Am. Chem. Soc.* **107**, 5035 (1985); A. E. Dorigo, D. W. Pratt, and K. N. Houk, *J. Am. Chem. Soc.* **109**, 6589 (1987).

One of the fundamental structural facets of organic chemistry, which has been explained most satisfactorily in MO terms, is the existence of a small barrier to rotation about single bonds. In ethane, for example, it is known that the staggered conformation is about 3 kcal/mol more stable than the eclipsed conformation so that the eclipsed conformation represents a barrier for transformation of one staggered conformation into another by rotation.

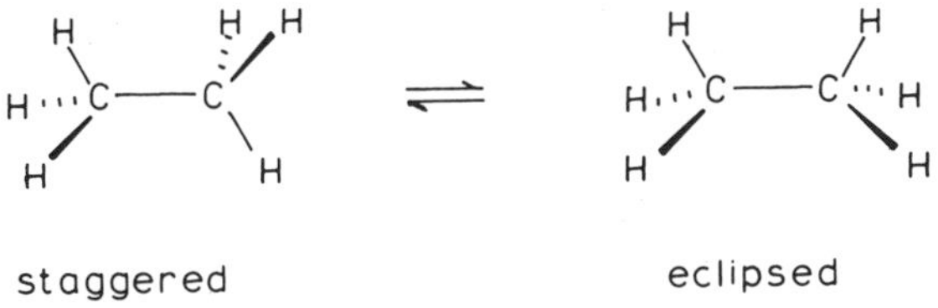

staggered ⇌ eclipsed

Valence bond theory offers no immediate qualitative explanation since the σ bond which is involved is cylindrically symmetric. A steric argument based on repulsions between hydrogens also fails since, on detailed examination of this hypothesis, it is found that the hydrogens are too small and too distant from one another to account for the observed energy. Molecular orbital ideas and calculations, however, succeed in correctly predicting and calculating the magnitude of the ethane rotational barrier.[46] The origin of the barrier is a repulsive interaction between the filled C—H orbitals which is maximal in the eclipsed geometry.

The interaction can be further examined by consideration of the ethane MOs.[47] Since ethane contains two carbon atoms and six hydrogens, the molecular orbitals are constructed from six H 1*s*, two C 2*s*, and six C 2*p* orbitals. Figure 1.34 depicts the seven bonding MOs, assuming the staggered geometry. The σ, σ', and σ_x orbitals are not affected much by the rotation of the two CH_3 groups with respect to one another because the H 1*s* orbitals all have the same sign within each CH_3 group. The other MOs, however, are of a π type, having a nodal plane derived from the nodal plane of the C $2p_z$ orbitals. The extent of the overlap in these orbitals clearly changes as the two CH_3 groups are rotated with respect to one another. Analysis of the relative magnitude of the bonding and antibonding interactions that take place as rotation occurs indicates that the change in energy of these two pairs of MOs is the source of the ethane rotational barrier.

The interaction of the lone pair electrons on an amine nitrogen with adjacent C—H bonds is another example of a hyperconjugative effect which can be described in MO language. The lone pair electrons, when properly aligned with the C—H bond, lead to a donation of electron density from the lone pair orbital to the

46. R. M. Pitzer, *Acc. Chem. Res.* **16**, 207 (1983); R. Hoffmann, *J. Chem. Phys.* **39**, 1397 (1963); R. M. Pitzer and W. N. Lipscomb, *J. Chem. Phys.* **39**, 1995 (1963); J. A. Pople and G. A. Segal, *J. Chem. Phys.* **43**, 5136 (1966).
47. J. P. Lowe, *Prog. Phys. Org. Chem.* **6**, 1 (1968); J. P. Lowe, *J. Am. Chem. Soc.* **92**, 3799 (1970); J. P. Lowe, *Science* **179**, 527 (1973).

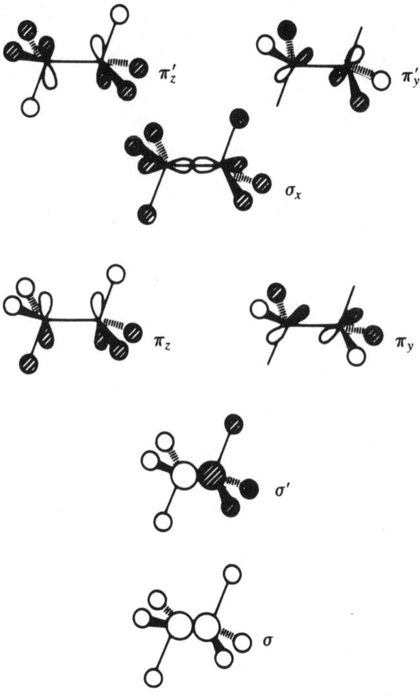

Fig. 1.34. Molecular orbitals of ethane revealing π character of π_z, π_y, π_z', and π_y' orbitals. Only the filled orbitals are shown.

antibonding C—H orbital. The overall effect is to weaken the C—H bond.

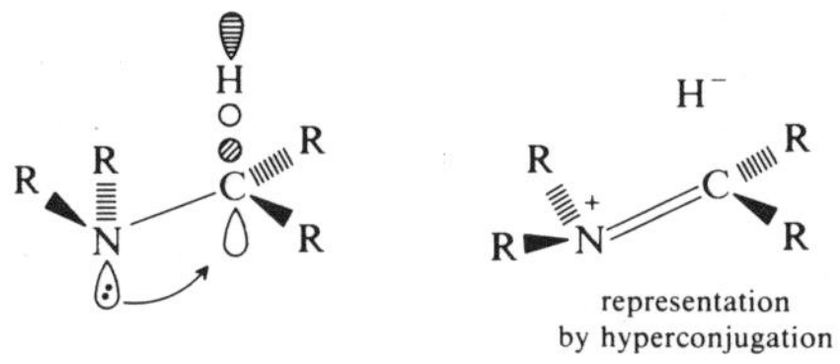

Electron donation from N n orbital to C—H σ^* orbital.

In acyclic structures, such effects are averaged by rotation but in cyclic structures differences in C—H bond strengths, based on the different alignments, can be recognized.[48] The C—H bonds that are in an *anti* orientation to the lone pair are weaker than the C—H bonds in other orientations.

The ideas that have been presented in this section illustrate the approach that is used to express structure and reactivity effects within the framework of a molecular orbital description of structure. In the chapters which follow, both valence bond theory and molecular orbital theory will be used in the discussion of structure and

48. A. Pross, L. Radom, and N. V. Riggs, *J. Am. Chem. Soc.* **102**, 2253 (1980).

reactivity. Qualitative valence bond terminology is normally most straightforward for saturated systems. Molecular orbital theory provides useful insights into conjugated systems and into effects which depend upon the symmetry of the molecules under discussion.

General References

T. A. Albright, J. K. Burdett, and M.-H. Whangbo, *Orbital Interactions in Chemistry*, Wiley, New York, 1985.
W. T. Borden, *Modern Molecular Orbital Theory for Organic Chemists*, Prentice-Hall, Englewood Cliffs, New Jersey, 1975.
I. G. Cszimadia, *Theory and Practice of MO Calculations on Organic Molecules*, Elsevier, Amsterdam, 1976.
M. J. S. Dewar, *The Molecular Orbital Theory of Organic Chemistry*, McGraw-Hill, New York, 1969.
M. J. S. Dewar and R. C. Dougherty, *The PMO Theory of Organic Chemistry*, Plenum Press, New York, 1975.
L. N. Ferguson, *Organic Molecular Structure*, Willard Grant Press, Boston, 1975.
I. Fleming, *Frontier Orbitals and Organic Chemical Reactions*, Wiley, New York, 1976.
W. J. Hehre, L. Radom, P. V. R. Schleyer, and J. Pople, *Ab Initio Molecular Orbital Theory*, Wiley-Interscience, New York, 1986.
C. K. Ingold, *Structure and Mechanism in Organic Chemistry*, Second Edition, Cornell University Press, Ithaca, New York, 1969.
A. Liberles, *Introduction to Theoretical Organic Chemistry*, Macmillan, New York, 1968.
M. Orchin, *The Importance of Antibonding Orbitals*, Houghton Mifflin, Boston, 1967.
W. G. Richards and D. L. Cooper, *Ab Initio Molecular Orbital Calculations for Chemists*, Second Edition, Clarondon Press, Oxford, 1983.
L. Salem, *Electrons in Chemical Reactions*, Wiley, New York, 1982.
A. Streitwieser, Jr., *Molecular Orbital Theory for Organic Chemists*, Wiley, New York, 1961.
K. B. Wiberg, *Physical Organic Chemistry*, Wiley, New York, 1964.
R. B. Woodward and R. Hoffmann, *The Conservation of Orbital Symmetry*, Verlag Chemie, Weinheim, 1970.
H. E. Zimmerman, *Quantum Mechanics for Organic Chemists*, Academic Press, New York, 1975.

Problems

(*References for these problems will be found on page* 773.)

1. Use thermochemical relationships to obtain the required information.

   (a) The heats of formation of cyclohexane, cyclohexene, and benzene are, respectively, −29.5, −1.1, and +19.8 kcal/mol. Estimate the resonance energy of benzene using these data.
   (b) Calculate ΔH for the air oxidation of benzaldehyde to benzoic acid given that the heats of formation of benzaldehyde and benzoic acid are −8.8 and −70.1 kcal/mol, respectively.
   (c) Using the appropriate heats of formation in Table 1.5, calculate the heat of hydrogenation of 2-methyl-1-pentene.

2. Suggest an explanation for the following observations:

(a) The dipole moment of the hydrocarbon calicene has been estimated to be as large as 5.6 D.

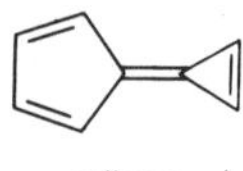

calicene

(b) The measured dipole moment of *p*-nitroaniline (6.2 D) is larger than the value calculated using empirical group moments (5.2 D).

(c) The dipole moment of furan is smaller than and in the opposite direction from that of pyrrole.

O 0.71 D   N H 1.80 D

3. Predict the energetically preferred site of protonation for each of the following molecules and explain the basis of your prediction.

(a) $C_6H_5CH{=}N{-}C_6H_5$

(b) $CH_3{-}C(=O){-}NHCH_3$

(c) N H (pyrrole)

(d) N, $-NH_2$ (2-aminopyridine)

4. What physical properties such as absorption spectra, bond length, dipole moment, etc., could be examined to obtain evidence of resonance interactions in the following molecules? What deviations from "normal" physical properties would you expect to find?

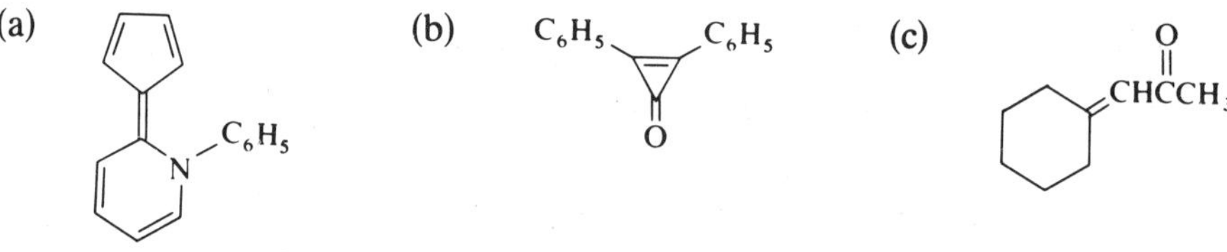

5. Certain C—H bonds have significantly lower bond dissociation energies than do the "normal" C—H bonds in saturated hydrocarbons. Offer a structural rationalization of the lowered bond energy in each of the following compounds, relative to the saturated hydrocarbon C—H bond taken as a reference. (The bond dissociation energies are given in kcal/mol.)

(a) $C_6H_5{-}CH_2{-}H$ (85) versus $CH_3{-}H$ (103)

(b) $HOCH_2-H-H$ (92) versus CH_3-H (103)

(c) $CH_3\underset{\|}{\overset{}{C}}-H$ (88) versus CH_3CH_2-H (98)
O

6. (a) Carboxamides have substantial rotational barriers on the order of 20 kcal/mol for the process

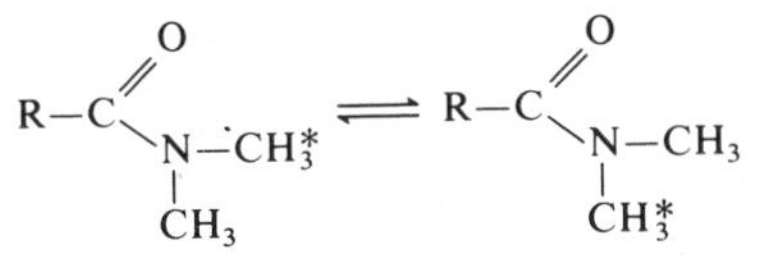

$$R-C(=O)-N(CH_3)-CH_3^* \rightleftharpoons R-C(=O)-N(CH_3^*)-CH_3$$

Develop a structural explanation for the existence of this barrier in both resonance and molecular orbital terminology.

(b) In the gas phase the rotational barrier of *N,N*-dimethylformamide is about 19.4 kcal/mol, which is about 1.5 kcal/mol less than in solution. Is this change consistent with the ideas you presented in (a)? Explain.

(c) Explain the relative rates of alkaline hydrolysis of the following pairs of carboxamides.

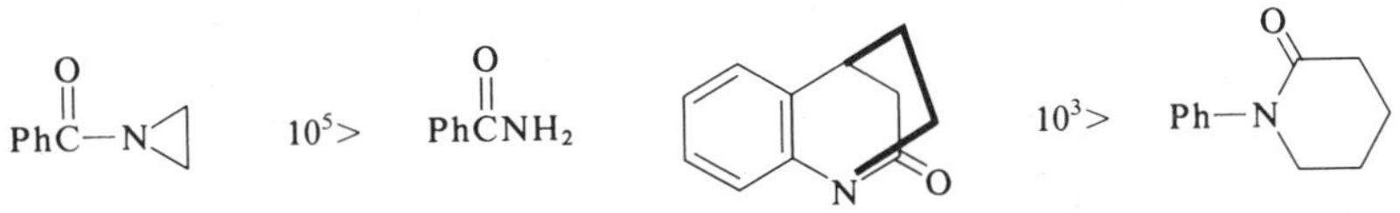

PhC(=O)—N(aziridine) $10^5>$ PhC(=O)NH_2 (bridged benzolactam, N, O) $10^3>$ Ph—N(piperidinone, =O)

7. Construct a qualitative MO diagram showing how the π-molecular orbitals in the following molecules are modified by the addition of the substituent:

(a) vinyl fluoride, compared to ethylene
(b) acrolein, compared to ethylene
(c) acrylonitrile, compared to ethylene
(d) benzyl cation, compared to benzene
(e) propene, compared to ethylene
(f) fluorobenzene, compared to benzene

8. The data below give the stabilization calculated in kcal/mol by MO methods for the reaction:

$$CH_3-X + CH_2=CH_2 \rightarrow CH_4 + CH_2=CH-X$$

X	ΔH
F	−6.4
OCH_3	−10.9
NH_2	−13.3

(a) Demonstrate that this indicates that there is a stabilizing interaction between the substituent and a carbon-carbon double bond.

(b) Draw resonance structures showing the nature of the interaction.

(c) Construct a qualitative MO diagram which rationalizes the existence of a stabilizing interaction.

(d) Both in resonance and molecular orbital terminology, explain the order of the stabilization N > O > F.

9. Construct a qualitative MO diagram for the H-bridged ethyl cation by analyzing the interaction of the ethylene MOs given in Fig. 1.20 with a proton approaching the center of the ethylene molecule from a direction perpendicular to the molecular plane. Indicate which ethylene orbitals will be lowered by this interaction and which will be raised or left relatively unchanged. Assume that the hydrogens of ethylene are slightly displaced away from the direction of approach of the proton.

10. In the Hückel treatment, atomic orbitals on nonadjacent atoms are assumed to have no interaction. They are neither bonding nor antibonding. The concept of *homoconjugation* suggests that such orbitals may interact, especially in rigid structures which direct orbitals toward one another. Consider, for example, bicyclo[2.2.1]hepta-2,5-diene:

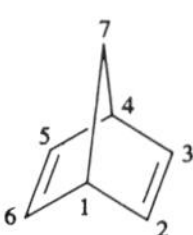

(a) Construct the MO diagram according to simple Hückel theory and assign energies to the orbitals.

(b) Extend the MO description by allowing a significant interaction between the C-2 and C-6 and between the C-3 and C-5 orbitals. Construct a qualitative MO diagram by treating the interaction as a perturbation on the orbitals shown for (a).

11. (a) Sketch the nodal properties of the highest occupied molecular orbital of pentadienyl cation ($CH_2{=}CHCH{=}CHCH_2{}^+$).

(b) Two of the π-MOs of pentadienyl are given below. Specify which one is of lower energy, and classify each as to whether it is bonding, nonbonding, or antibonding. Explain your reasoning.

```
1  2  3  4  5
o—o—o—o—o
```

$$\psi_x = 0.50\phi_1 + 0.50\phi_2 - 0.50\phi_4 - 0.50\phi_5$$

$$\psi_y = 0.58\phi_1 - 0.58\phi_3 + 0.58\phi_5$$

12. Charge densities for a series of fluorobenzenes calculated by the CNDO/2 MO method are as shown:

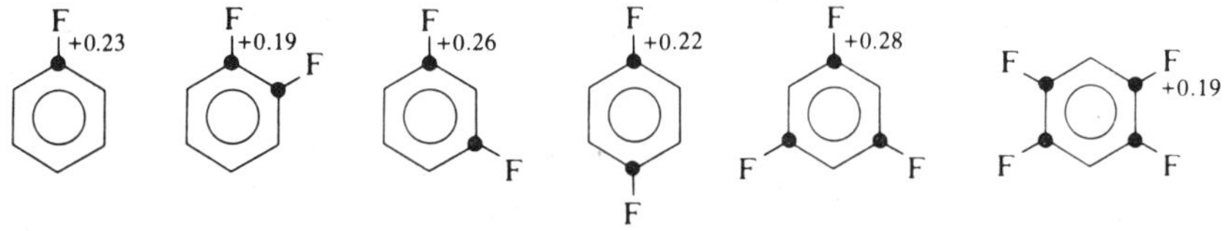

Can the relative magnitudes of the charges at the fluorine-substituted carbon atoms be rationalized on the basis of inductive effects? Resonance effects? What relationship do you believe would be observed between the X-ray photoelectron spectra (ESCA) of the compounds and the charges as calculated by CNDO?

13. Calculate the energy levels and coefficients for 1,3-butadiene using Hückel MO theory.

14. (a) Estimate from HMO theory the delocalization energy, expressed in units of β, of cyclobutadienyl dication ($C_4H_4^{2+}$).
    (b) Estimate, in units of β, the energy associated with the long-wavelength UV-VIS absorption of 1,3,5,7-octatetraene. Does it appear at longer or shorter wavelengths than the corresponding absorption for 1,3,5-hexatriene?

15. Addition of methylmagnesium bromide to 2-methylcyclohexanone followed by iodine-catalyzed dehydration of the resulting alcohol gave three alkenes in the ratio A:B:C = 3:31:66. Each isomer gave a mixture of *cis*- and *trans*-1,2-dimethylcyclohexane on catalytic hydrogenation. When the alkene mixture is heated with a small amount of sulfuric acid, the ratio A:B:C is changed to 0.0:15:85. Assign structures to A, B, and C.

16. The propellanes are highly reactive substances which readily undergo reactions involving rupture of the central bond. It has been suggested that the polymerization of propellanes occurs by a dissociation of the central bond:

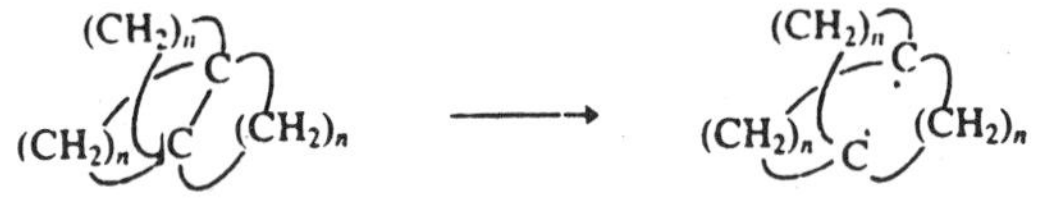

Somewhat surprisingly perhaps, it has been found that [1.1.1]propellane is considerably *less reactive* than [2.2.1]propellane. Use the theoretically calculated enthalpy data below to estimate the bond dissociation energy of the central bond in each of the three propellanes shown. How might this explain the relative reactivity of the [1.1.1] and [2.2.1] propellanes?

Enthalpy for addition of hydrogen to give the corresponding bicycloalkane

[2.2.1]propellane + H_2 → bicyclo[2.2.1]heptane, $\Delta H = -99$ kcal/mol

[2.1.1]propellane + H_2 → bicyclo[2.1.1]hexane, $\Delta H = -73$ kcal/mol

[1.1.1]propellane + H_2 → bicyclo[1.1.1]pentane, $\Delta H = -39$ kcal/mol

Assume that the bond dissociation energy of the bridghead hydrogens in each bicycloalkane is 104 kcal/mol. Indicate and discuss any other assumptions you have made.

17. Examine the following thermochemical data pertaining to hydrogenation of unsaturated eight-membered ring hydrocarbons to give cyclooctane:

Unsaturated ring hydrocarbon	$-\Delta H$ (kcal/mol)
cis,cis,cis,cis-1,3,5,7-Cyclooctatetraene	97.96
cis,cis,cis-1,3,5-Cyclooctatriene	76.39
cis,cis,cis-1,3,6-Cyclooctatriene	79.91
cis,cis-1,5-Cyclooctadiene	53.68
cis,cis-1,4-Cyclooctadiene	52.09
cis,cis-1,3-Cyclooctadiene	48.96
trans-Cyclooctene	32.24
cis-Cyclooctene	22.98

(a) Discuss the differences observed in each isomeric series of compounds, and offer an explanation for these differences.

(b) Comment on whether the conjugation present in cyclooctatetraene has a stabilizing or destabilizing effect on the C=C bonds.

18. Cyclic amines such as piperidine and its derivatives show substantial differences in the properties of the axial C-2 and C-6 versus the equatorial C-2 and C-6 C—H bond.

The axial C—H bonds are *weaker* than the equatorial C—H bonds as can be demonstrated by a strongly shifted C—H stretching frequency in the IR spectrum. Axial C-2 and C-6 methyl groups *lower* the ionization potential of the lone pair electrons on nitrogen substantially more than do equatorial C-2 or C-6 methyl groups. Discuss the relationship between these observations and provide a rationalization in terms of qualitative MO theory.

19. (a) The strain energy of spiropentane (62.5 kcal/mol) is considerably greater than twice that of cyclopropane (27.5 kcal/mol). Suggest an explanation.

(b) The fractional s character in bonds to carbon in organic molecules may be estimated by its relation to ^{13}C—^{13}C coupling constants, as determined by NMR. Estimate the fractional s character of C-1 in its bond to C-3 of spiropentane, given the following information

$$s_{1(3)} = \frac{J_{^{13}C-^{13}C}}{K s_{3(1)}}$$

where K is a constant equal to 550 Hz, the $^{13}C—^{13}C$ coupling constant J between C-1 and C-3 is observed to be 20.2 Hz, and $s_{3(1)}$ is the s character at C-3 in its bond to C-1.

20. The ionization potential of ethylene is 10.52 eV. How will the ionization potential change as a result of the introduction of the substituents in acrylonitrile and vinyl acetate? Explain your reasoning.

21. Predict which compound would give the faster (k) or more complete (K) reaction. Explain the basis for your prediction.

(a) 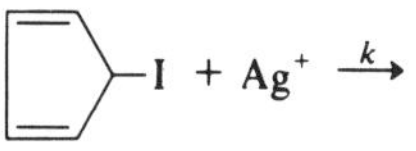 3-iodocyclopentene $+ Ag^+ \xrightarrow{k}$

or

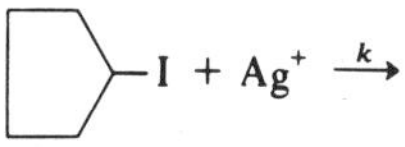

iodocyclopentane $+ Ag^+ \xrightarrow{k}$

(b) $Ph_3CH + {}^-NH_2 \overset{K}{\rightleftharpoons} Ph_3C^- + NH_3$

or

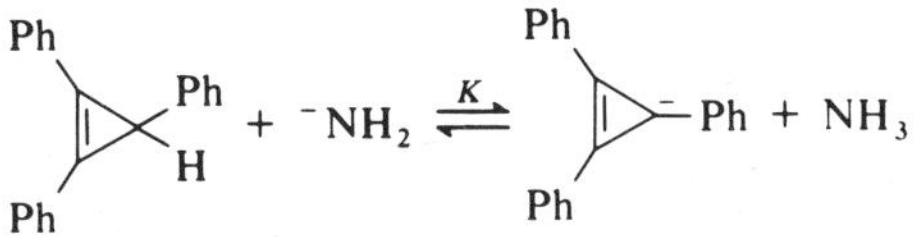

1,2,3-triphenylcyclopropene (3-H) $+ {}^-NH_2 \overset{K}{\rightleftharpoons}$ 1,2,3-triphenylcyclopropenyl anion $+ NH_3$

(c) 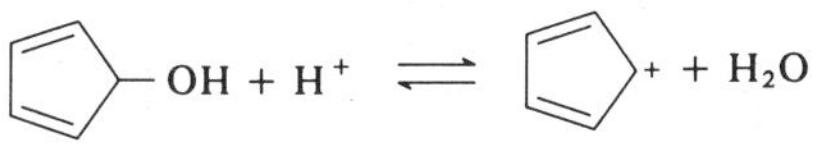 cyclopentadienol (C–OH) $+ H^+ \rightleftharpoons$ cyclopentadienyl cation $+ H_2O$

allyl alcohol (OH) $+ H^+ \rightleftharpoons$ allyl cation $+ H_2O$

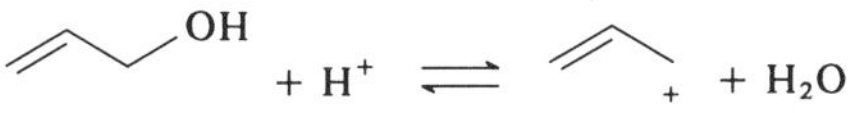

22. Computational comparison of structures of the benzyl cation (**A**) and singlet phenylcarbene (**B**) indicates a much greater double-bond character for the exocyclic bond in **A** than in **B**.

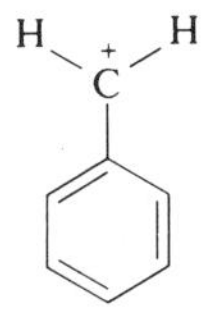

A ($C_6H_5CH_2^+$) **B** ($C_6H_5\ddot{C}H$)

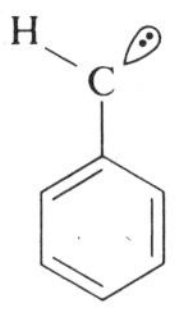

Can you provide a rationalization of this difference in terms of both valence bond-resonance and PMO considerations? Explain.

23. The ionization potentials of some substituted norbornadienes have been measured by photoelectron spectroscopy. The values which pertain to the π orbitals are shown:

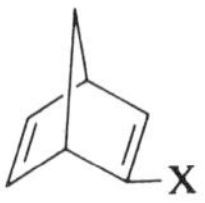

X	IP (eV) 1	IP (eV) 2
H	8.69	9.55
CH_3O	8.05	9.27
CN	9.26	10.12

Use PMO theory to describe the effect of the substituents on the photoelectron spectra. Use an MO diagram to explain the interaction of the substituents with the π bonds. Explicitly take into account the fact that the two orbitals interact and therefore cannot be treated as separate entities (see Problem 10).

24. *Ab initio* MO calculations using 4-31G orbitals indicate that the eclipsed conformation of acetaldehyde is more stable than the staggered conformation.

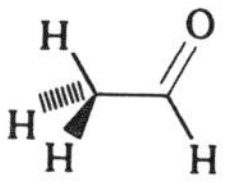

$E = -152.685$ au
H/C = O eclipsed

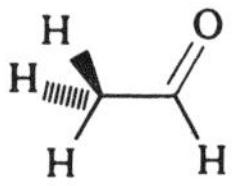

$E = -152.683$ au
H/C = O staggered

Provide a rationalization of this structural effect in terms of PMO theory. Construct a qualitative MO diagram for each conformation, and point out the significant differences that can account for the preference for the eclipsed conformation.

2

Stereochemical Principles

Introduction

For most combinations of atoms, a number of molecular structures that differ from each other in the sequence of bonding of the atoms are possible. Each individual molecular assembly is called an *isomer*, and the *constitution* of a compound is the particular combination of bonds and sequence of atoms (molecular connectivity) which is characteristic of that structure. Propanal, allyl alcohol, acetone, 2-methyloxirane, and cyclopropanol each correspond to the molecular formula C_3H_6O but differ in constitution.

$CH_3CH_2CH{=}O$	$CH_2{=}CHCH_2OH$	$CH_3C(=O)CH_3$	$CH_3CH\text{—}CH_2$ (O bridging)	cyclopropyl–OH
propanal	allyl alcohol	acetone	methyloxirane	cyclopropanol

When structures of the same constitution differ in spatial arrangement, they are *stereoisomers*. Stereoisomers are described by specifying their topology and the nature of their relationship to other stereoisomers of the same constitution. Stereoisomers differ in *configuration*, and in order to distinguish between stereoisomeric compounds, it is necessary to specify the configuration.[1,2] If two stereoisomers are related by being nonsuperimposable mirror images, the molecules are *enantiomeric*. Structures that have nonsuperimposable mirror images are called *chiral*. *Chirality* is the property of any molecule (or other object) of being nonsuperimposable on its mirror image. It is possible to separate the enantiomers of chiral

1. The IUPAC rules and definitions for fundamental stereochemistry are given with examples in *J. Org. Chem.* **35**, 2849 (1970); see also G. Krow, *Top. Stereochem.* **5**, 31 (1969).
2. K. Mislow and M. Raban, *Top. Stereochem.* **1**, 1 (1967); J. K. O'Loane, *Chem. Rev.* **80**, 41 (1980).

compounds. Samples containing only one enantiomer are called enantiomerically pure or *homochiral*. Stereoisomers that are not enantiomers are *diastereomers*.

One characteristic of chiral molecules is that the separated enantiomers cause the plane of polarized light to rotate by opposite but equal amounts. Samples that have an excess of one enantiomer over the other show a net rotation and are said to be *optically active*. Samples that contain only one of the enantiomers are said to be *optically pure*. Samples that have equal amounts of two enantiomers show zero net rotation and are called *racemic mixtures*.

In addition to constitution and configuration, there is a third significant level of structure, that of *conformation*. Conformations are discrete molecular arrangements that differ in spatial arrangement as a result of facile rotations about single bonds. The subject of conformational interconversion will be discussed in detail in Chapter 3. A special case arises when rotation about single bonds is restricted by steric or other factors so that the different conformations can be separated. The term *atropisomer* is applied to this type of stereoisomerism which depends upon restricted bond rotation.[3]

In this chapter, configurational relationships will be emphasized. Both structural and dynamic aspects of stereochemical relationships will be considered. We will be concerned both with the fundamental principles of stereochemistry and the conventions that have been adopted to describe the spatial arrangements of molecules. We will consider the stereochemical consequences of chemical reactions so as to provide a basis for understanding the relationship between stereochemistry and reaction mechanism that will be encountered later in the book.

2.1. Enantiomeric Relationships

The relationship between chirality and optical activity is historically such a close one that chemists sometimes use the terms improperly. It is important to recognize that optical activity refers to just one property of chiral molecules, namely, the ability to rotate plane-polarized light. Measurement of optical activity is useful both for determining the configuration of chiral molecules and for investigating the stereochemical relationship between reactants and products. The mechanics of measuring optical rotation will not be discussed here since the basic method is well described in most introductory texts. Both the sign and magnitude of optical rotation are dependent on the conditions of the measurement, including temperature, solvent, and the wavelength of the light. By convention, single-wavelength measurements are usually made at the 589-nm emission line of sodium arc lamps. This wavelength is known as the sodium D line, and optical rotations measured at this wavelength are designated $[\alpha]_D$.

Pure enantiomeric substances show rotations that are equal in magnitude but opposite in direction. Unequal mixtures of enantiomers rotate light in proportion

3. M. Oki, *Top. Stereochem.* **14**, 1 (1983).

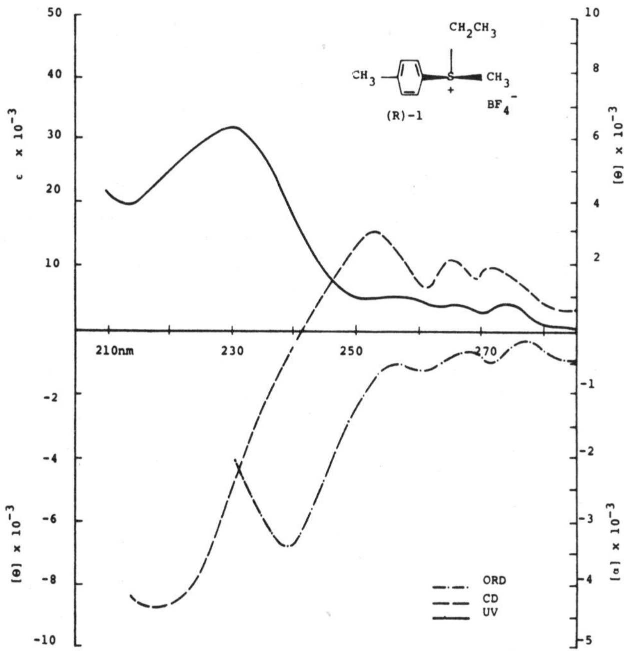

Fig. 2.1. UV absorption, ORD, and CD curves of ethyl methyl *p*-tolyl sulfonium tetrafluoroborate. [Reproduced with permission from *J. Org. Chem.* **41**, 3099 (1976).]

to the composition. The relationship between optical purity and measured rotation is

$$\text{optical purity (\%)} = \frac{[\alpha]_{\text{mixture of enantiomers}}}{[\alpha]_{\text{pure enantiomer}}} \times 100$$

The optical purity is numerically equivalent to the *enantiomeric excess*, which is defined as

Enantiomeric excess (%)
= [mole fraction (major enantiomer) − mole fraction (minor enantiomer)] × 100

Measurement of rotation as a function of wavelength is useful in structural studies aimed at determining the chirality of a molecule. This technique is called *optical rotatory dispersion.*[4] The resulting plot of rotation against wavelength is called an ORD curve. The shape of the ORD curve is determined by the configuration of the molecule and its absorption spectrum. In many cases, the ORD curve can be used to specify the configuration of a molecule by relating it to similar molecules of known configuration.

4. P. Crabbe, *Top. Stereochem.* **1**, 93 (1967); C. Djerassi, *Optical Rotatory Dispersion*, McGraw-Hill, New York, 1960; P. Crabbe, *Optical Rotatory Dispersion and Circular Dichroism in Organic Chemistry*, Holden Day, San Francisco, 1965; E. Charney, *The Molecular Basis of Optical Activity. Optical Rotatory Dispersion and Circular Dichroism*, Wiley, New York, 1979.

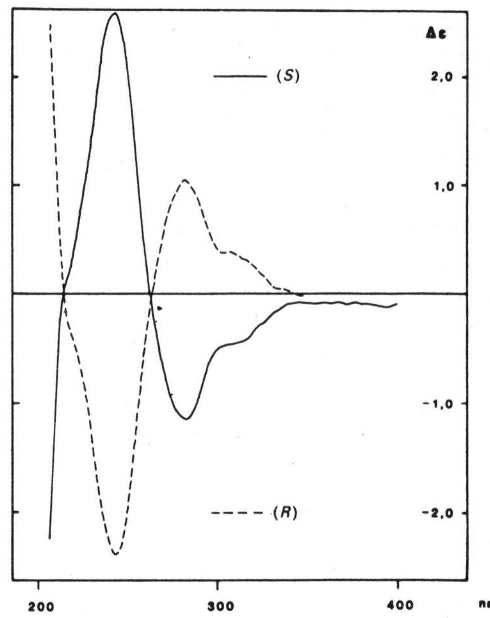

Fig. 2.2. CD spectra of (S)- and (R)-2-amino-l-phenyl-l-propanone hydrochloride. [Reproduced from *Helv. Chim. Acta* **69**, 1498 (1986).]

Enantiomeric substances also show differential absorption of circularly polarized light. This is called *circular dichroism* and is quantitatively expressed as the molecular ellipticity,

$$\theta = 3330(\varepsilon_L - \varepsilon_R)$$

where ε_L and ε_R are the extinction coefficients of left and right circularly polarized light. Figure 2.1 shows the UV, ORD, and CD spectra of an enantiomerically pure sulfonium ion salt.[5]

The molecular ellipticity is analogous to specific rotation in that two enantiomers have exactly opposite values at each wavelength. The two enantiomers will thus show circular dichroism (CD) spectra having opposite signs. A compound with several absorption bands may show both positive and negative bands. Figure 2.2 shows the CD curves for both enantiomers of 2-amino-1-phenyl-1-propane.[6]

Although measurements of optical rotation and ORD and CD spectra have historically been the main methods for determining optical purity and assigning configuration, other analytical techniques are also available. High-performance liquid chromatography (HPLC) using chiral column packing material can resolve enantiomers on both an analytical and a preparative scale. Chiral packing materials for gas–liquid chromatography (GLC) have also been developed. Several other

5. K. K. Andersen, R. L. Caret, and D. L. Ladd, *J. Org. Chem.* **41**, 3096 (1976).
6. J.-P. Wolf and H. Pfander, *Helv. Chim. Acta* **69**, 1498 (1986).

approaches to determining optical purity depend upon formation of diastereomers, and these will be discussed in Section 2.2.

Compounds in which chirality is the result of one or more carbon atoms having four nonidentical substituents represent the largest class of chiral molecules. A molecule having a single carbon atom with four nonidentical ligands is chiral. Carbon atoms with four nonidentical ligands are referred to as asymmetric carbon atoms since the molecular environment at such a carbon atom possesses no element of symmetry. 2-Butanol is an example of a chiral molecule and exists as two nonsuperimposable mirror images. Carbon-2 is an asymmetric carbon.

H CH_2CH_3 C H_3C OH   H CH_2CH_3 C HO CH_3 H H C H_3C OH   H H C HO CH_3

Ethanol is an achiral molecule. The plane defined by atoms C-1, C-2, and O is a plane of symmetry. Any carbon atom with two identical ligands is located in a plane of symmetry that includes the two nonidentical ligands. Any molecule, no matter how complex, which possesses a plane of symmetry is achiral.

The necessary criterion that an object not be superimposable on its mirror image can be met by a number of types of compounds in addition to those containing asymmetric carbon atoms, an important example being sulfoxides containing nonidentical substituents on sulfur.[7] Sulfoxides are nonplanar, and there is a sufficient barrier to inversion at sulfur that the pyramidal sulfoxides maintain their configuration at room temperature. Unsymmetrical sulfoxides are therefore chiral and exist as enantiomers. Sulfonium salts with three nonidentical ligands are also chiral as a result of their pyramidal shape. Some examples of chiral derivatives of sulfur are given in Scheme 2.1.

Although unsymmetrically substituted amines are chiral, the configuration is not stable because of rapid inversion at nitrogen. The activation energy for pyramidal inversion at phosphorus is much higher than at nitrogen, and many optically active phosphines have been prepared.[8] The barrier to inversion is usually in the range of 30–35 kcal/mol so that enantiomerically pure phosphines are stable at room temperature but racemize by inversion at elevated temperatures. Asymmetrically substituted tetracoordinate phosphorus compounds such as phosphonium salts and phosphine oxides are also chiral. Scheme 2.1 includes some examples of chiral phosphorus compounds.

The chirality, or handedness, of a molecule is described by specifying its configuration. The system that has received general acceptance is the *Cahn–Ingold–Prelog convention*, which uses the descriptors *R* and *S*. The *Fischer convention*,

7. For reviews of chiral sulfoxides, see M. Cinquini, F. Cozzi, and F. Montanari, *Studies in Organic Chemistry* **19**, 355 (1985); M. R. Barbachy and C. R. Johnson, in *Asymmetric Synthesis*, Vol. 4, J. D. Morrison and J. W. Scott (eds.), Academic Press, New York, 1984, Chapter 2.
8. D. Valentine, Jr., in *Asymmetric Synthesis*, Vol. 4, J. D. Morrison and J. W. Scott (eds.), Academic Press, New York, 1984, Chapter 3.

Scheme 2.1. Chiral Compounds of Sulfur and Phosphorus

1[a]

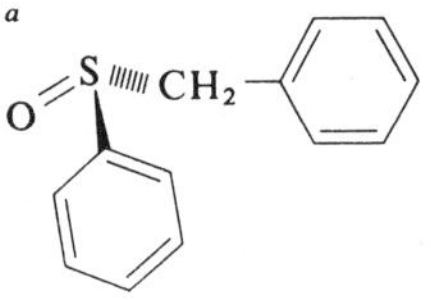

$[\alpha]_D = +92.4°$
R-enantiomer

2[b]

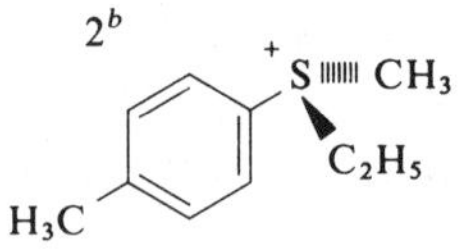

$[\alpha]_D = -15.8°$
R-enantiomer

3[c]

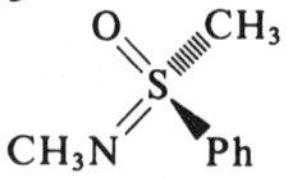

$[\alpha]_D = +172.4°$
S-enantiomer

4[d]

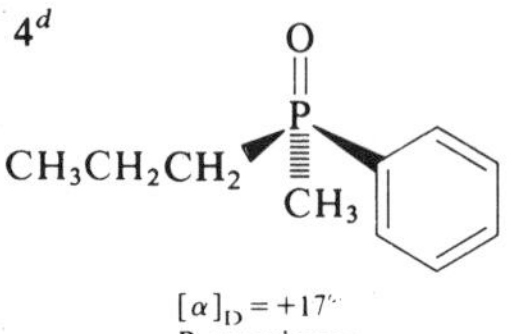

$[\alpha]_D = +17°$
R-enantiomer

5[e]

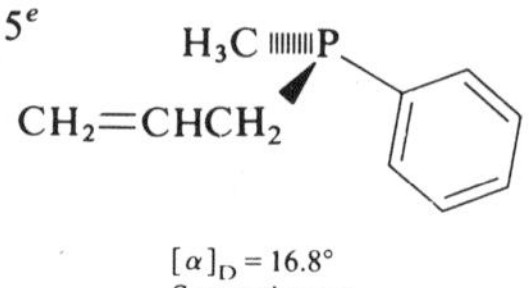

$[\alpha]_D = 16.8°$
S-enantiomer

6[f]

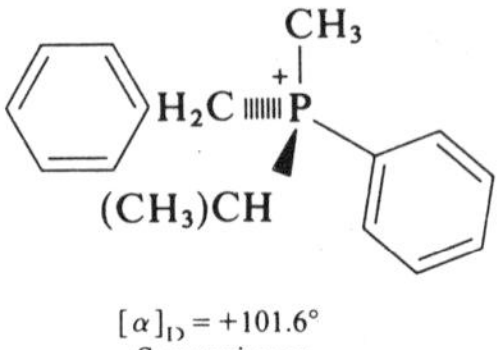

$[\alpha]_D = +101.6°$
S-enantiomer

7[g]

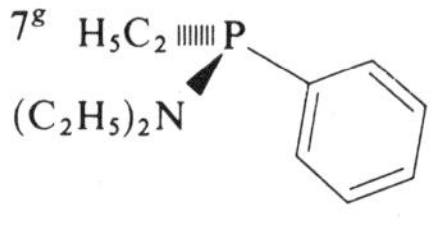

$[\alpha]_D = -86.9°$
R-enantiomer

a. C. R. Johnson and D. McCants, Jr., *J. Am. Chem. Soc.* **87**, 5404 (1965).
b. K. K. Andersen, R. L. Caret, and D. L. Ladd, *J. Org. Chem.* **41**, 3096 (1976).
c. C. R. Johnson and C. W. Schroeck, *J. Am. Chem. Soc.* **95**, 7418 (1973); C. R. Johnson, C. W. Schroeck, and J. R. Shanklin, *J. Am. Chem. Soc.* **95**, 7424 (1973).
d. O. Korpiun, R. A. Lewis, J. Chickos, and K. Mislow, *J. Am. Chem. Soc.* **90**, 4842 (1968).
e. L. Horner, H. Winkler, A. Rapp, A. Mentrup, H. Hoffman, and P. Beck, *Tetrahedron Lett.*, 161 (1961).
f. W.-D. Balzer, *Chem. Ber.* **102**, 3546 (1969).
g. L. Horner and M. Jordan, *Phosphorus and Sulfur* **8**, 225 (1980).

employing the descriptors D and L, is historically important and is still used with certain types of molecules.

The Cahn–Ingold–Prelog descriptors *R* and *S* are assigned by using the *sequence rule* to assign a priority order to the substituents on the atom to which a configuration is being assigned. The substituent atoms are assigned decreasing priority in the order of decreasing atomic number. When two or more of the substituent atoms are the same element (e.g., carbon), the next attached atoms in those substituents are compared. This process of substituent comparison is continued until the order of priority of all substituents has been established. An atom that is multiply bonded is counted once for each formal bond. When the substituent group priority has been established, the molecule is viewed in an orientation which places the lowest-priority substituent behind the chiral center. The three remaining substituents project toward the viewer. The remaining substituents have one of two possible arrangements. The substituents decrease in priority in either a clockwise manner or in a counterclockwise manner. In the former case, the configuration *R* (for Latin *rectus*, right-handed) is assigned. If the priority decreases in the counterclockwise sense, the atom is of *S* (for Latin *sinister*, left-handed) configuration.

The configuration of the 2-butanol enantiomer shown below is established as *S* as follows. The highest-priority atom bonded to the asymmetric carbon is O; the lowest is H. The remaining two atoms are both C, and the choice as to which of these is of higher priority is made by comparing their ligands. The methyl group has (H, H, H), while the ethyl group has (C, H, H); therefore, the ethyl group is of higher priority than the methyl group. The complete priority list is $OH > C_2H_5 > CH_3 > H$. When viewed from the side opposite the lowest-priority ligand, the remaining groups appear in order of decreasing priority in counterclockwise fashion, and the configuration is *S*:

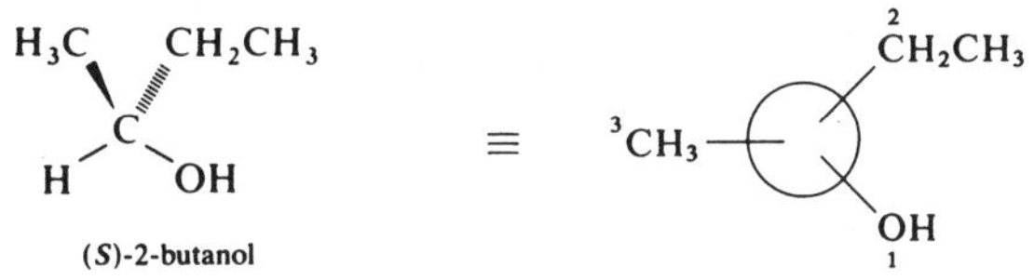

(*S*)-2-butanol

Some other examples of assignment of configuration are illustrated below.

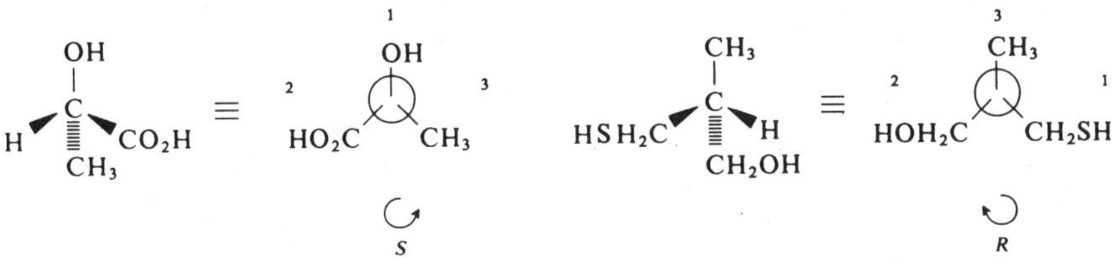

When the chiral center is tricoordinate, as is the case for sulfoxides, sulfonium salts, and phosphines, then a "phantom atom" of atomic number zero is taken to occupy the lowest-priority site of a presumed tetrahedral atom. Application of the sequence rule in the usual manner allows the assignment of the *R* configuration to the benzyl phenyl sulfoxide enantiomer and the *S* configuration to the methylallylphenylphosphine enantiomer shown in Scheme 2.1.

Glyceraldehyde is the point of reference for describing the configuration of carbohydrates and other natural substances following the *Fischer convention.* The two enantiomers of glyceraldehyde were originally arbitrarily assigned the configurations D and L as shown below. Subsequently, a determination of the configuration of sodium rubidium tartrate by X-ray crystallography and the relationship of this material to D-glyceraldehyde established that the original arbitrary assignments were the correct ones.

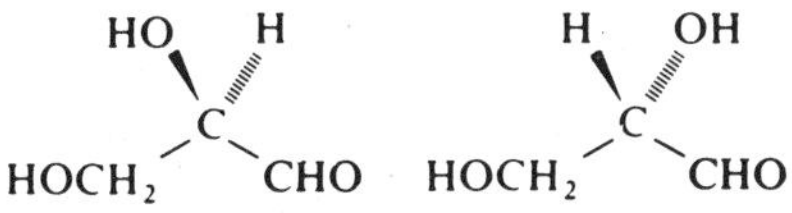

D-(+)-glyceraldehyde L-(−)-glyceraldehyde

In the Fischer convention, the configuration of a molecule is described by the descriptors D and L, which are assigned by comparison with the reference molecule glyceraldehyde. It is convenient in employing the Fischer convention to use *projection formulas.* These are planar representations defined in such a way as to convey three-dimensional structural information. The molecule is oriented with the major carbon chain aligned vertically with the most oxidized terminal carbon at the top. The vertical bonds at each carbon are directed back, away from the viewer, and the horizontal bonds are directed forward to the viewer. The D and L forms of glyceraldehyde are shown below with the equivalent Fischer projection formulas.

CHO, H–C–OH, CH_2OH ≡ CHO, H–+–OH, CH_2OH
D

CHO, HO–C–H, CH_2OH ≡ CHO, HO–+–H, CH_2OH
L

The configuration of any other chiral molecule to be assigned in the Fischer convention is done by comparison with D- and L-glyceraldehyde. The molecule is aligned with the chain vertical and the most oxidized carbon at the top, as specified by the Fischer convention. The chiral atom with the highest number (at the lowest position in the Fischer projection) is compared with C-2 of glyceraldehyde. If the configuration is that of D-glyceraldehyde, the molecule is assigned the D-configuration, whereas if it is that of L-glyceraldehyde, it is assigned the L-configuration. This is illustrated below with several carbohydrates.

D-ribose: CHO; H–OH; H–OH; H–OH; CH_2OH

L-fucose: CHO; HO–H; H–OH; H–OH; HO–H; CH_3

D-mannose: CHO; HO–H; HO–H; H–OH; H–OH; CH_2OH

D-fructose: CH_2OH; C=O; HO–H; H–OH; H–OH; CH_2OH

All of the amino acids found in proteins have the L-configuration as illustrated for alanine, serine, and leucine.

L-alanine: CO_2H; H_2N–H; CH_3

L-serine: CO_2H; H_2N–H; CH_2OH

L-leucine: CO_2H; H_2N–H; $CH_2CH(CH_3)_2$

At the present time, use of the Fischer convention is almost entirely restricted to carbohydrates, amino acids, and biologically important molecules of closely related structural types. The problem with more general use is that there are no adequate rules for deciding whether a chiral atom is "like" D-glyceraldehyde or L-glyceraldehyde when the structures are not closely similar to the reference molecules. This relationship is clear for carbohydrates and amino acids.

The property of chirality is determined by molecular topology, and there are many molecules that are chiral even though they do not possess an asymmetrically substituted atom. Examples include certain allenes, spiranes, alkylidenecycloalkanes, and biaryls as well as other specific examples. Some specific molecules that have been isolated in optically active form are given in Scheme 2.2. The configuration of these molecules is established by subrules in the Cahn–Ingold–Prelog convention. We will not describe these here. Discussion of these rules can be found in Ref. 1.

There is no direct relationship between the configurational descriptors *R* and *S* or D and L and the sign of optical rotation of the molecule. *R* or *S* molecules can have either + or − signs for rotation as can D or L molecules. Thus, even though a structure can be specified on the basis of these conventions, additional information is necessary to establish which molecule of an enantiomeric pair possesses the specified configuration. Determination of the ***absolute configuration*** establishes the configuration of a specific enantiomer. There are several approaches to this problem. One is to establish a direct structural relationship to a molecule of known configuration by chemical transformation.[9] This is the way in which most of the reference molecules whose absolute configurations are known were initially assigned. The existence of a base of molecules whose absolute configurations are known has permitted the development of correlations based on the CD and ORD curves of cerain types of chromophores. When chromophores are located close to chiral centers, the spectroscopic properties are affected in a predictable way so that the sign and shape of the ORD and CD curve can be a reliable basis for configurational assignment.[10] While routine X-ray crystal structure determination does not provide the absolute configuration of the molecule, special analysis of the diffraction data does allow assignment of absolute configuration.[11] These methods are important, for example, in assigning the absolute configuration of new natural products.

2.2. Diastereomeric Relationships

Diastereomers are defined as stereoisomers that are not related as an object and its mirror image. Consider the four structures in Fig. 2.3. These structures exemplify the four stereoisomers of 2,3,4-trihydroxybutanal. The configurations of C-2 and C-3 are indicated. Each of the four structures is stereoisomeric with respect to any of the others. The 2*R*,3*R* and 2*S*,3*S* isomers are enantiomeric, as are the 2*R*,3*S* and 2*S*,3*R* pair. The 2*R*,3*S* isomer is diastereomeric with the 2*S*,3*S* and 2*R*,3*R* isomers since they are stereoisomers but not enantiomers. Any given structure

9. For a review of chemical methods for determining absolute configuration, see *Stereochemistry, Fundamentals and Methods*, Vol. 3, H. B. Kagan (ed.), G. Theime, Stuttgart, 1977.
10. K. Nakanishi and N. Harada, *Circular Dichroism Spectroscopy*: *Exiton Coupling in Organic Stereochemistry*, University Science Books, Mill Valley, California, 1983; D. N. Kirk, *Tetrahedron* **42**, 777 (1986).
11. D. Rogers, *Acta Crystallogr., Sect.* 1 **A37**, 734 (1981).

Scheme 2.2. Examples of Chiral Molecules Lacking Asymmetric Atoms

1[a] *R*-(−)-Glutinic acid

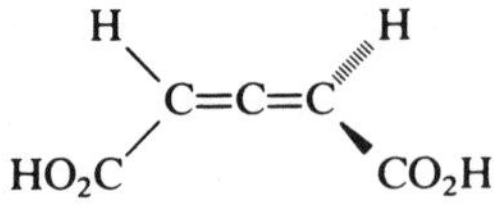

2[b] *R*-(−)-1,3-Dimethylallene

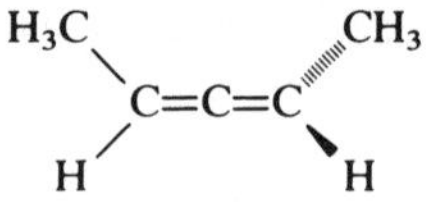

3[c] *S*-(+)-1,1′-Binaphthyl

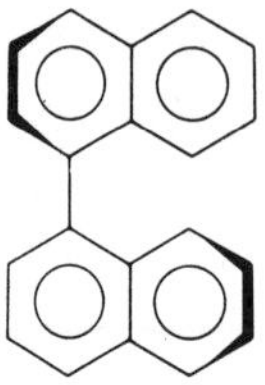

4[d] *R*-(+)-2,2′-Diamino-6,6′-dimethylbiphenyl

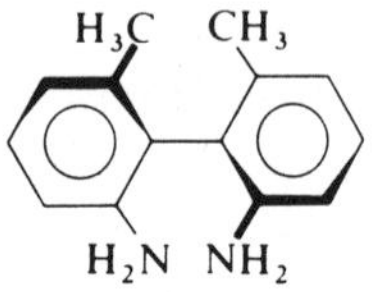

5[e] *R*-(−)-*trans*-Cyclooctene

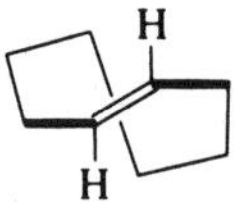

6[f] *cis,trans*-1,3-cyclooctadiene

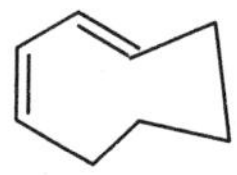

7[g] *S*-(+)-Spiro[3.3]-hepta-1,5-diene

8[h] *R*-(+)-1,1′-Spirobiindan

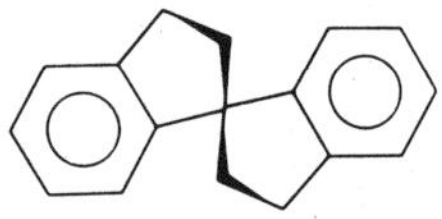

9[i]

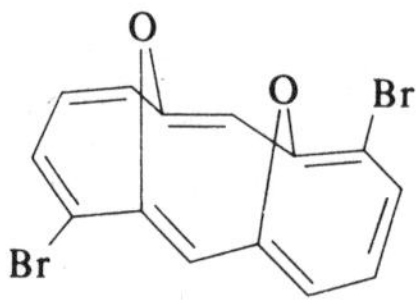

10[j]

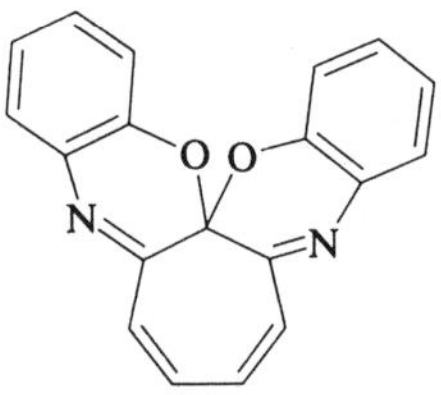

a. W. C. Agosta, *J. Am. Chem. Soc.* **86**, 2638 (1964).
b. W. L. Waters, W. S. Linn, and M. C. Caserio, *J. Am. Chem. Soc.* **90**, 6741 (1968).
c. P. A. Browne, M. M. Harris, R. Z. Mazengo, and S. Singh, *J. Chem. Soc., C*, 3990 (1971).
d. L. H. Pignolet, R. P. Taylor, and W. DeW. Horrocks, Jr., *J. Chem. Soc. Chem. Commun.*, 1443 (1968).
e. A. C. Cope and A. S. Mehta, *J. Am. Chem. Soc.* **86**, 1268 (1964).
f. R. Isaksson, J. Roschester, J. Sandström, and L.-G. Wirstrand, *J. Am. Chem. Soc.* **107**, 4074 (1985).
g. L. A. Hulshof, M. A. McKervey, and H. Wynberg, *J. Am. Chem. Soc.* **96**, 3906 (1974).
h. J. H. Brewster and R. T. Prudence, *J. Am. Chem. Soc.* **95**, 1217 (1973); R. K. Hill and D. A. Cullison, *J. Am. Chem. Soc.* **95**, 1229 (1973).
i. E. Vogel, W. Tückmantel, K. Schlögl, M. Widhalm, E. Kraka, and D. Cremer, *Tetrahedron Lett.* **25**, 4925 (1984).
j. N. Harada, H. Uda, T. Nozoe, Y. Okamoto, H. Wakabayashi, and S. Ishikawa, *J. Am. Chem. Soc.* **109**, 1661 (1987).

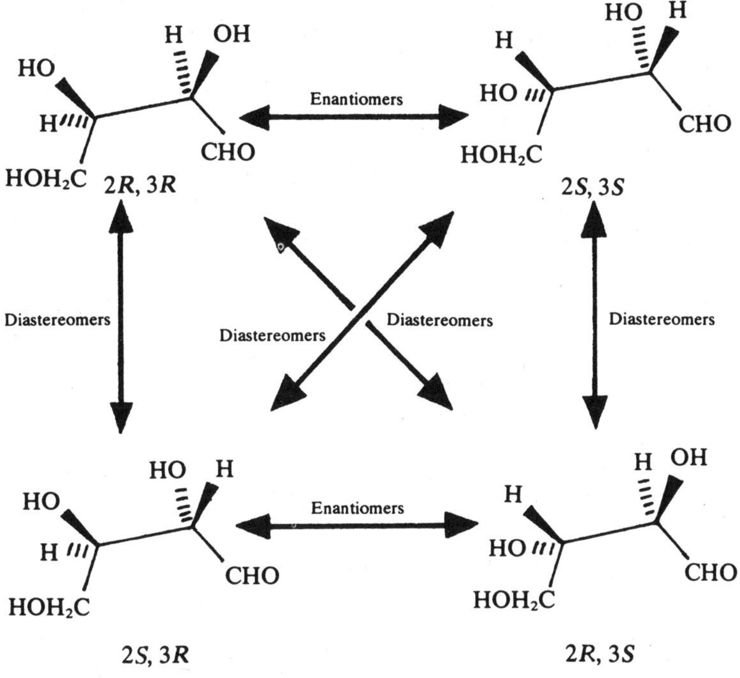

Fig. 2.3. Stereoisomeric relationships in 2,3,4-trihydroxybutanal.

can have only one enantiomer. All other stereoisomers of that molecule are diastereomeric.

Diastereosomers differ in both physical properties and chemical reactivity. They generally have different melting points, boiling points, solubility characteristics, etc. The specific rotations of diastereomeric molecules can differ both in magnitude and sign. The difference in chemical reactivity can be a slight difference in rate or two diastereomers can lead to entirely different products, depending on the nature and mechanism of the particular reaction.

The specification of the configuration of diastereomers is done by a direct extension of the sequence rule. Each chiral center is designated *R* or *S* by application of the sequence rule. The structures in Fig. 2.3 can serve as examples for study.

Some use is still made of the terms *erythro* and *threo* to specify the relative configuration of two adjacent chiral atoms. The terms are derived from the sugars erythrose and threose. The terms were originally defined such that a Fischer projection formula in which the two adjacent substituents were on the same side was the *erythro* isomer and that with the substituents on opposite sides was the *threo* isomer.

D-erythrose	L-erythrose	D-threose	L-threose
CHO	CHO	CHO	CHO
H—OH	HO—H	HO—H	H—OH
H—OH	HO—H	H—OH	HO—H
CH_2OH	CH_2OH	CH_2OH	CH_2OH

Unfortunately, as with the Fischer system itself, assignment of molecules which are not closely related to the reference molecules becomes a subjective matter of assigning which substituents are "similar." The application of the terminology to cases where the chiral centers are not adjacent is also ambiguous. As a result, the system is not a general method of specifying stereochemical relationships.

Fischer projection formulas can be used to represent molecules with several stereoisomeric centers, and they are commonly used for carbohydrate molecules. For other types of structures, a common practice is to draw the molecule in an extended conformation with the main chain horizontal. In this arrangement, each tetrahedral carbon has two additional substituents, one facing out and one in. The orientation is specified with solid wedged bonds for substituents facing out and with dashed bonds for substituents that point in.

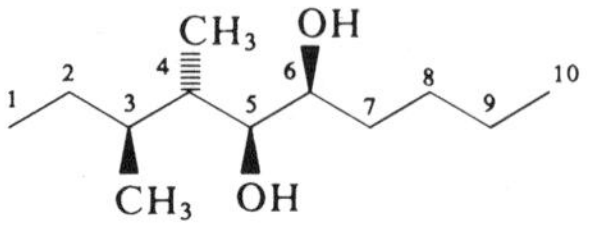

representation of 3,4-dimethyldecane-5,6-diol.

Sometimes the words *syn* and *anti* are then used to specify the relationship between the substituents. For instance, in the structure above, the 5- and 6-hydroxyls are in a *syn* relationship while the 3- and 4-methyls are *anti* with respect to one another.

Since chirality is a property of a molecule as a whole, the juxtaposition of two or more asymmetric centers in a molecule may result in an achiral molecule. For example, there are three stereoisomers of tartaric acid (2,3-dihydroxybutanedioic acid). Two of these are chiral and optically active but the third is not.

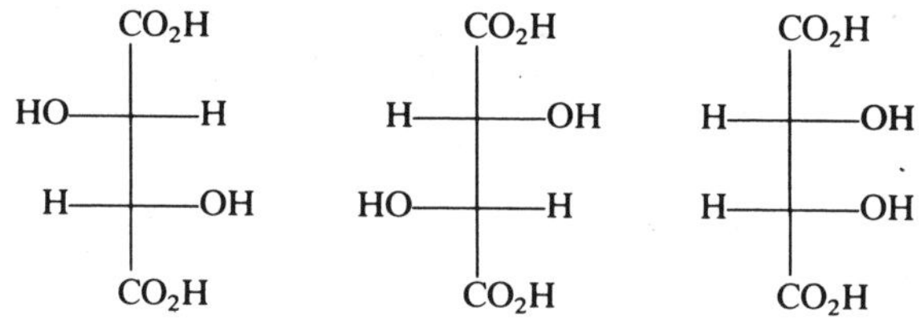

D-tartaric acid	L-tartaric acid	*meso*-tartaric acid
CO_2H	CO_2H	CO_2H
HO—C—H	H—C—OH	H—C—OH
H—C—OH	HO—C—H	H—C—OH
CO_2H	CO_2H	CO_2H

The reason for the lack of chirality of the third stereoisomer is that the two asymmetric carbons are located with respect to each other in such a way that a molecular plane of symmetry exists. Compounds that contain asymmetric carbons but are nevertheless achiral are called *meso* forms. This situation occurs whenever pairs of asymmetric centers are arranged in the molecule in such a way that a plane of symmetry exists.

Incorporation of chiral centers into cyclic structures produces interesting examples of *meso* forms. Certain dimethylcycloalkanes are achiral. All *cis*-dimethylcycloalkanes are achiral since all contain a plane of symmetry. *trans*-Dimethylcycloalkanes whose ring size is an odd number are chiral. With even-membered rings, the chirality depends upon the substitution pattern. Ring sizes 3–6 are classified in Scheme 2.3. Three stereoisomers are possible for both 1,2-dimethylcyclopropane and 1,2-dimethylcyclobutane. There is a *meso* form (the *cis* isomer) and a pair of enantiomers (the *trans* isomer). Both stereoisomers of 1,3-dimethylcyclobutane are achiral. The *cis*-1,3-isomer possesses two planes of symmetry, one passing through C-1 and C-3 and the other through C-2 and C-4. The *trans* isomer possesses a symmetry plane through C-1 and C-3.

Since the presence of a plane of symmetry in a molecule ensures that it will be achiral, one approach to classification of stereoisomers as chiral or achiral is to examine the molecule for symmetry elements. There are other elements of symmetry in addition to planes of symmetry which ensure that a molecule will be superimposable on its mirror image. The case of *trans*-1,3-dibromo-*trans*-2,4-dimethylcyclobutane is illustrative. This molecule does not possess a plane of symmetry but the mirror images are superimposable as illustrated below. This molecule possesses a *center of symmetry*. The center of the molecule is a point from which any line drawn through the molecule encounters an identical environment in either direction from the center of symmetry.

Scheme 2.3. Chiral and Achiral Disubstituted Cycloalkanes

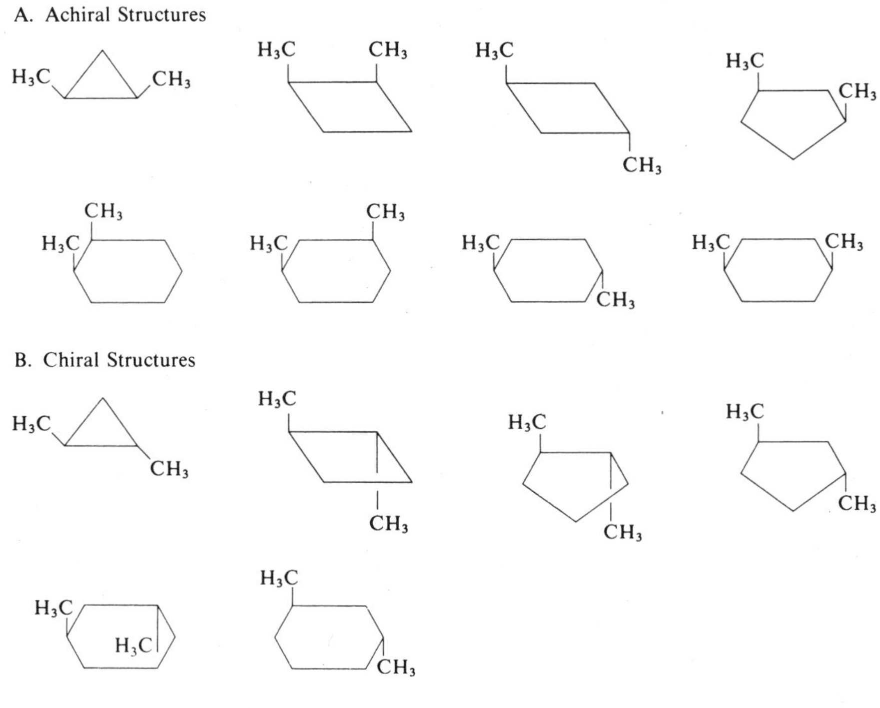

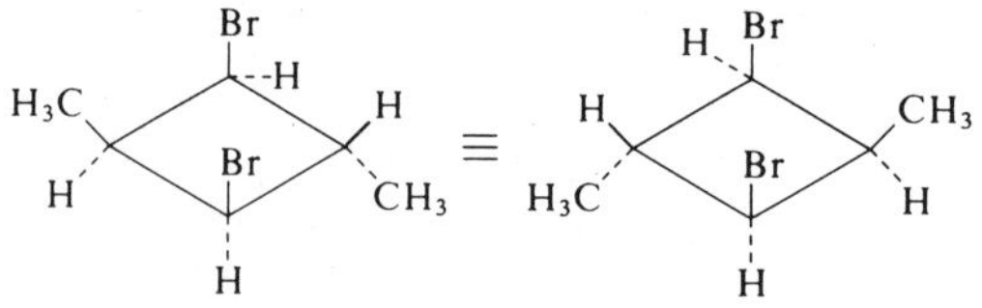

trans,cis,cis-1,3-Dibromo-2,4-dimethylcyclobutane has a plane of symmetry. The mirror images are superimposable.

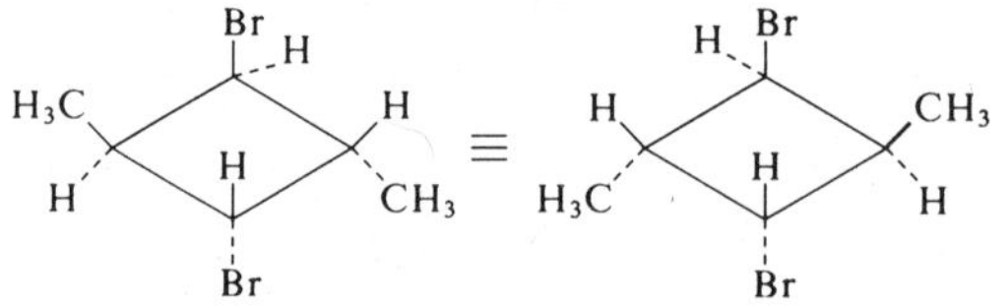

trans,trans,cis-1,3-Dibromo-2,4-dimethylcyclobutane has a center of symmetry. The mirror images are superimposable.

A particularly striking example of a *meso* form is the antibiotic nonactin.[12] (Work Problem 2.24 to convince yourself that nonactin is a *meso* form.)

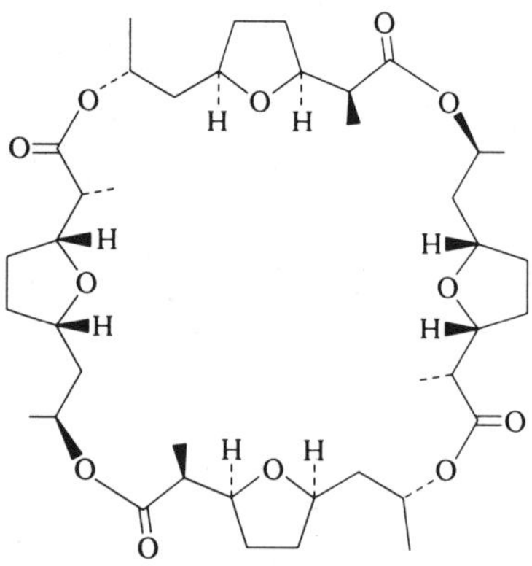

nonactin

As is evident from the examples in Scheme 2.2, chirality is not uniquely associated with asymmetrically substituted atoms. There are examples of chiral molecules lacking asymmetric atoms and chiral molecules containing two or more asymmetrically substituted carbons. Nevertheless, the most important cases of stereoisomerism in organic chemistry are those which involve asymmetrically substituted atoms. In such cases, the maximum number of stereoisomers is given by 2^n, where n is the number of asymmetric centers in the molecule. The actual number of stereoisomers is reduced in those cases where elements of symmetry render particular stereoisomers achiral.

Diastereomeric relationships provide the basis on which a range of chemical and physical separation processes occur. The process of *resolution* is the separation of a mixture containing equal amounts of a pair of enantiomers (racemic mixture). Separation is frequently effected by converting the mixture of enantiomers into a

12. J. Dominguez, J. D. Dunitz, H. Gerlach, and V. Prelog, *Helv. Chim. Acta* **45** 129 (1962); H. Gerlach and V. Prelog, *Liebigs Ann. Chem.* **669**, 121 (1963); B. T. Kilbourn, J. D. Dunitz, L. A. R. Pioda, and W. Simon, *J. Mol. Biol.* **30**, 559 (1967).

Scheme 2.4. Resolution of 2-Phenyl-3-methylbutanoic Acid[a]

Ph, H, CO_2H, $(CH_3)_2CH$ — R-(−)

H, Ph, CO_2H, $(CH_3)_2CH$ — S-(+)

racemic mixture (461 g)

CH_3, Ph, NH_2, H — R-(+)

mixture 353 g of diastereomeric ammonium carboxylate salts (*R*-acid, *R*-amine and *S*-acid, *R*-amine). recrystallized from ethanol–water

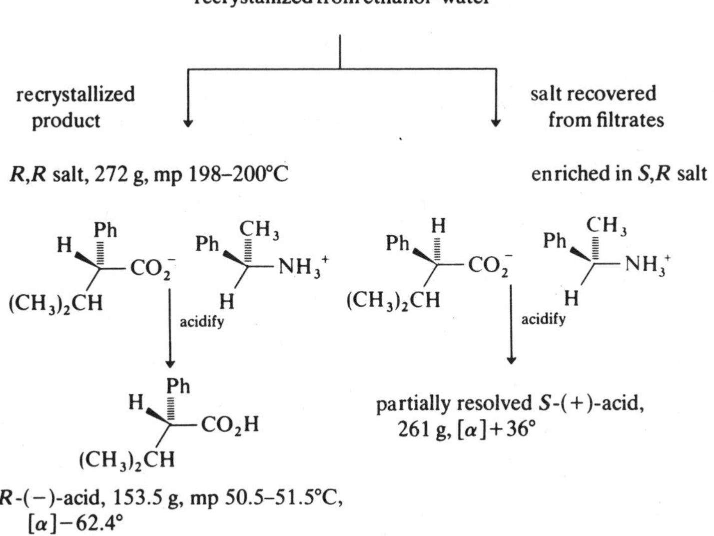

a. C. Aaron, D. Dull, J. L. Schmiegel, D. Jaeger, Y. Ohashi, and H. S. Mosher, *J. Org. Chem.* **32**, 2797 (1967).

mixture of diastereomers by reaction with a pure enantiomer of a second reagent, the *resolving agent.*[13] Since the two resulting products will be diastereomeric, they can be separated. The separated diastereomers can then be reconverted to the pure enantiomers by reversing the initial chemical transformation. An example of this method is shown in Scheme 2.4 for the resolution of a racemic carboxylic acid by way of a diastereomeric salt resulting from reaction with an enantiomerically pure amine. The *R*-acid–*R*-amine and *S*-acid–*R*-amine salts are separated by fractional recrystallization. The resolved acids are regenerated by reaction with a strong acid, which liberates the carboxylic acid from the amine salt.

13. For reviews of resolution methods, see S. H. Wilen, *Top. Stereochem.* **6**, 107 (1971); S. H. Wilen, A. Collet, and J. Jacques, *Tetrahedron* **33**, 2725 (1977); A. Collet, M. J. Brienne, and J. Jacques, *Chem. Rev.* **80**, 215 (1980); J. Jacques, A. Collet, and S. H. Wilen, *Enantiomers, Racemates and Resolutions*, Wiley-Interscience, New York, 1981.

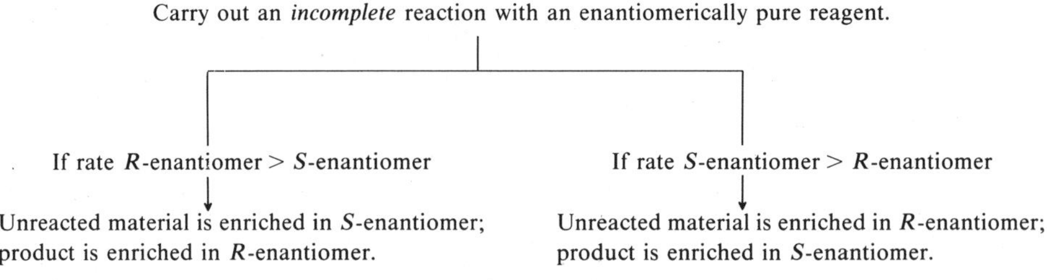
S-enantiomer + R-enantiomer
Carry out an incomplete reaction with an enantiomerically pure reagent.
If rate R-enantiomer > S-enantiomer
Unreacted material is enriched in S-enantiomer;
product is enriched in R-enantiomer.
If rate S-enantiomer > R-enantiomer
Unreacted material is enriched in R-enantiomer;
product is enriched in S-enantiomer.

Fig. 2.4. Basis of kinetic resolution.

Although the traditional method of separating the diastereomeric compounds generated in a resolution procedure is fractional crystallization, chromatographic procedures are sometimes more convenient. Diastereomeric compounds exhibit differential absorption on achiral materials and can be separated by standard column chromatography or, if necessary, by taking advantage of the greater separation powers of high-performance liquid chromatography (HPLC).

An alternative means of resolution depends on the difference in rates of reaction of two enantiomers with a chiral reagent. The transition state energies for reaction of each enantiomer with one enantiomer of a chiral reagent will be different. This is because the two transition states (*R*-substrate···*R*-reactant) and (*S*-substrate···*R*-reactant) are diastereomeric. *Kinetic resolution* is the term used to describe the separation of enantiomers by selective reaction with an optically pure reagent. Figure 2.4 summarizes the basis of kinetic resolution. Because the separation is based on differential rates of reaction, the degree of resolution which can be achieved depends on both the magnitude of the rate difference and on the extent of reaction. The greater the difference in the two rates, the higher will be the enantiomeric purity of both the reacted and the unreacted enantiomer. The extent of enantiomeric purity can be controlled by controlling the degree of conversion. As the extent of conversion increases, the enantiomeric purity of the *unreacted* enantiomer becomes very high.[14] The relationship between the relative rate of reaction, extent of conversion, and enantiometric purity of the unreacted enantiomer is shown in Fig. 2.5. Scheme 2.5 gives some specific examples of kinetic resolution procedures.

Preparation of optically active materials by use of chiral catalysts is also based on differences in transition state energies. While the reactant is part of a complex or intermediate containing a chiral catalyst, it is in a chiral environment. The intermediates and complexes containing enantiomeric reactant and a homochiral catalyst are diastereomeric and differ in energy. This energy difference can then control selection between the stereoisomeric products of the reaction. If the reaction creates a new chiral center in the reactant molecule, there can be a preference for formation of one enantiomer over the other.

14. V. S. Martin, S. S. Woodard, T. Katsuki, Y. Yamada, M. Ikeda, and K. B. Sharpless, *J. Am. Chem. Soc.* **103**, 6237 (1981).

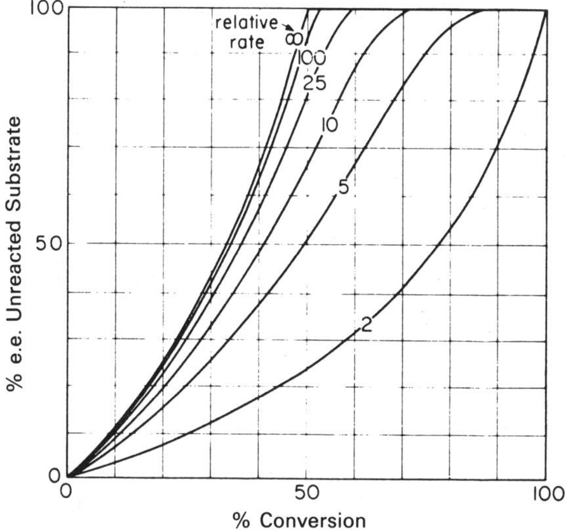

Fig. 2.5. Dependence of enantiomeric excess on relative rate of reaction with a chiral reagent in a kinetic resolution. [Reproduced from V. S. Martin, S. S. Woodward, T. Katsuki, Y. Yamada, M. Ikeda, and K. B. Sharpless, *J. Am. Chem. Soc.* **103**, 6237 (1981) by permission of the American Chemical Society.]

Enzymes constitute a particularly important group of chiral catalysts.[15] Enzymes are highly efficient and selective catalysts and can carry out a variety of transformations. Because the enzymes are derived from optically pure L-amino acids, they are homochiral, and usually one enantiomer of a reactant is significantly more reactive than the other. The reason is that the interaction of the enzyme with one enantiomer bears a diastereomeric relationship to the interaction of the enzyme with the other. Since enzyme catalysis is usually based on a specific fit to an "active site," the degree of selection between the two enantiomers is often very high. Enzyme-catalyzed reactions can therefore be used to resolve organic compounds. The most completely characterized enzymes available are those which catalyze hydrolysis of esters and amides (esterases, lipases, peptidases, acylases) and those which oxidize alcohols to ketones or aldehydes (dehydrogenases). Purified enzymes can be used, or the reaction can be done by incubating the reactant with a microorganism (yeast, for example) that produces an appropriate enzyme during fermentation. Scheme 2.6 gives some specific examples of enzymatic resolutions.

Other methods of resolution depend upon the difference between two enantiomers in *noncovalent* binding to a chiral substance. This is the basis for resolution by chromatography on homochiral adsorbents. The noncovalent binding between enantiomers and the chromatographic adsorbent establishes diastereomeric com-

15. J. B. Jones, *Tetrahedron* **42**, 3351 (1986); J. B. Jones, in *Asymmetric Synthesis*, Vol. 5, J. D. Morrison (ed.), Academic Press, New York, 1985, Chapter 9; G. M. Whitesides and C.-H. Wong, *Angew. Chem. Int. Ed. Engl.*, **24**, 617 (1985).

Scheme 2.5. Examples of Kinetic Resolutions

1[a] $C_6H_5CH(NH_2)CH(CH_3)_2 + (C_6H_5CH(CH_2CH_3)C(=O))_2O \longrightarrow C_6H_5CH(CH(CH_3)_2)NHC(=O)CH(CH_2CH_3)C_6H_5$

ratio of $R,S : S,S = 1.75$ with 0.25 equiv anhydride

2[b] 1-(2-hydroxy-2-phenylethyl)pyrrolidine, $N\text{-}CH_2CH(OH)C_6H_5$ $\xrightarrow[\text{(+) diisopropyl tartrate}]{Ti(OR)_4,\ t\text{-BuOOH}}$ $N\text{-}CH_2CH(OH)C_6H_5$

37% yield recovered, 95% e.e.

3[c] $(PhOCH(CH_3)CO)_2$ (1 equiv) + $CH_3CH(OH)\text{-}C_5H_4N$ (S-isomer, 1 equiv) $\longrightarrow$ $PhOCH(CH_3)C(=O)OCH(CH_3)\text{-}C_5H_4N$ + $PhOCH(CH_3)CO_2H$

92% R,S ester
8% S,S ester

84% e.e. S

4[d] $CH_3O_2CC(=CH_2)CH(OH)CH_3 \xrightarrow[\text{S-2,3-}bis\text{(diphenylphosphino)butane complex of rhodium}]{H_2} CH_3O_2CCH(CH_3)CH(OH)CH_3$

6.5 : 1 enantioselectivity

5[e] 3-methylcyclohex-2-en-1-ol $\xrightarrow[H_2]{S\text{-2,2'-}bis\text{(diphenylphosphino)-1,1'-binaphthyl complex of Ru(OAc)}_2}$ 3-methylcyclohex-2-en-1-ol

48% recovery, 96% e.e.

a. Y. Hiraki and A. Tai, *Bull. Chem. Soc. Jpn.* **57**, 1570 (1984).
b. S. Miyano, L. D. L. Lu, S. M. Viti, and K. B. Sharpless, *J. Org. Chem.* **48**, 3608 (1983).
c. U. Salz and C. Rüchardt, *Chem. Ber.* **117**, 3457 (1984).
d. J. M. Brown and I. Cutting, *J. Chem. Soc., Chem. Commun.*, 578 (1985).
e. M. Kitamura, I. Kasahara, K. Manabe, and R. Noyori, *J. Org. Chem.* **53**, 708 (1988).

plexes, and these have differing binding affinities. The positions of the equilibrium between the bound and the unbound states are therefore different for the two enantiomers. This means that the two enantiomers will move through the column at different rates and can be separated. Although this principle has long been recognized, it is only fairly recently, with the development of techniques for preparing enantiomerically pure adsorbents and improvements in chromatographic methods, that this method of resolution has become very practical.[16] A study which demon-

16. W. H. Pirkle and J. Finn, *Asymmetric Synthesis*, Vol. 1, J. D. Morrison (ed.), Academic Press, New York, 1983, Chapter 6; D. W. Armstrong, *J. Liq. Chromatogr.* **7**(Suppl. 2), 353 (1984); S. Allenmark, *J. Biochem. Biophys. Methods* **9**, 1 (1984).

1[a] $CH_3CH(OH)(CH_2)_5CH_3 + C_3H_7CO_2CH_2CH(O_2CC_3H_7)CH_2O_2CC_3H_7 \xrightarrow{\text{Candida cylindracea lipase}} CH_3CH(OH)(CH_2)_5CH_3 + CH_3CH(O_2CC_3H_7)(CH_2)_5CH_3$

S-enantiomer 97% e.e.; *R*-enantiomer 92% e.e.

2[b] $CH_3O{-}C_6H_4{-}CH_2CH_2CH(NHCOCH_3)CO_2H \xrightarrow{\text{Aspergillus acylase}} CH_3O{-}C_6H_4{-}CH_2CH_2CH(NH_2)CO_2H$

L-enantiomer, 82% yield

3[c] $CH_3CH(Br)CO_2H + C_4H_9OH \xrightarrow{\text{yeast lipase}} CH_3CH(Br)CO_2C_4H_9$

R-enantiomer 96% e.e. at 45% conversion

4[d] $C_6H_5{-}CH_2CH(NHCOCH_3)CO_2CH_3 \xrightarrow{\text{Subtilisin Carlsberg (Alcalase®)}} C_6H_5{-}CH_2CH(NHCOCH_3)CO_2H$

S-enantiomer, 98% e.e.

5[e] $C_6H_5{-}CH(O_2CCH_3)CH_3 \xrightarrow{\text{Rhizopus nigricans}} C_6H_5{-}CH(OH)CH_3$

R-enantiomer, 98% e.e. at 28% conversion

a. B. Cambou and A. M. Klibanov, *J. Am. Chem. Soc.* **106**, 2687 (1984).
b. N. Kosui, M. Waki, T. Kato, and N. Izumiya, *Bull. Chem. Soc. Jpn.* **55**, 918 (1982).
c. G. Kirchner, M. P. Scollar, and A. M. Klibanov, *J. Am. Chem. Soc.* **107**, 7072 (1985).
d. J. M. Roper and D. P. Bauer, *Synthesis*, 1041 (1983).
e. H. Ziffer, K. Kawai, M. Kasai, M. Imuta, and C. Froussios, *J. Org. Chem.* **48**, 3017 (1983).

strates the present capability of the technique reports the resolution of a number of aromatic compounds on a 1–8-g scale. The adsorbent is a silica that has been derivatized with a chiral reagent. Specifically, hydroxyl groups on the silica surface are covalently bound to a derivative of *R*-phenylglycine. Using medium-pressure chromatography apparatus, the racemic mixture is passed through the column and, when resolution is successful, the separated enantiomers are isolated as completely resolved fractions.[17] The combination of the excellent separation capacity of HPLC columns with the use of chiral column packing materials has provided a powerful technique for analysis and separation of enantiomeric materials. Gas–liquid

17. W. H. Pirkle and J. M. Finn, *J. Org. Chem.* **47**, 4037 (1982). For other examples of chiral HPLC adsorbents, see W. H. Pirkle and M. H. Hyun, *J. Org. Chem.* **49**, 3043 (1984); W. H. Pirkle, T. C. Pochapsky, G. S. Mahler, D. E. Corey, D. S. Reno, and D. M. Alessi, *J. Org. Chem.* **51**, 4991 (1986).

chromatography (GLC) done using chiral stationary phases is also a means of separation of enantiomeric compounds.[18]

The difference in the physical properties of diastereomers is also the basis for a particularly sensitive method for assessing the enantiomeric purity of compounds. Although, in principle, enantiomeric purity can be determined by measuring the optical rotation, this value is reliable only if the rotation of the pure compound is accurately known. This is never the case for a newly prepared material and is also often not the case for previously prepared compounds. If a derivative of a chiral compound is prepared in which a new chiral center is introduced, the two enantiomers will give difference diastereomers. Since these will have different physical properties, the relative amounts can be determined. NMR spectroscopy provides one of the most convenient ways of detecting and quantitating the two diastereomeric products. A pure enantiomer will give only a single spectrum, but a partially resolved material will show the spectra having intensities corresponding to the amounts of the two diastereomeric derivatives. The most widely used derivatizing reagent for the NMR method is a compound known as *Mosher's reagent.*[19] One reason that this compound is particularly useful is that the aromatic ring usually induces markedly different chemical shifts in the two diastereomeric products that are formed.

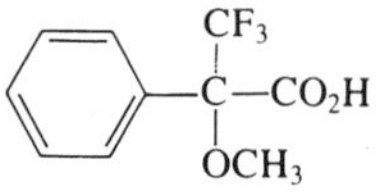

CF_3
C_6H_5—C—CO_2H
OCH_3

Mosher's reagent

Changes in NMR spectra can also be observed as the result of formation of noncovalent complexes between chiral molecules and another chiral reagent. This is the basis of the use of *chiral shift reagents* to determine the enantiomeric purity of chiral substances.[20] Several of the lanthanide elements have the property of forming strong complexes with alcohols, ketones, and other types of organic functional groups having Lewis base character. If the lanthanide ion is in a chiral environment as the result of an enantiomerically pure ligand, two diastereomeric complexes are formed. The lanthanide elements induce large NMR shifts, and, as a result, two strongly shifted spectra are seen for the two complexed enantiomers. The relative intensities of the two spectra correspond to the ratio of enantiomers present in the sample. Figure 2.6 shows the NMR spectrum of an unequal mixture of the two enantiomers of 1-phenylethylamine in the presence of a chiral shift reagent.[21]

18. V. Schurig, in *Asymmetric Synthesis*, Vol. 1, J. D. Morrison (ed.), Academic Press, New York, 1983, Chapter 5.
19. J. A. Dale, D. L. Dull, and H. S. Mosher, *J. Org. Chem.* **34**, 2543 (1969).
20. G. R. Sullivan, *Top. Stereochem.* **10**, 287 (1978); R. R. Fraser, in *Asymmetric Synthesis*, Vol. 1, J. D. Morrison (ed.), Academic Press, New York, 1983, Chapter 9.
21. G. M. Whitesides and D. W. Lewis, *J. Am. Chem. Soc.* **93**, 5914 (1971); M. D. McCreary, D. W. Lewis, D. L. Wernick, and G. M. Whitesides, *J. Am. Chem. Soc.* **96**, 1038 (1974).

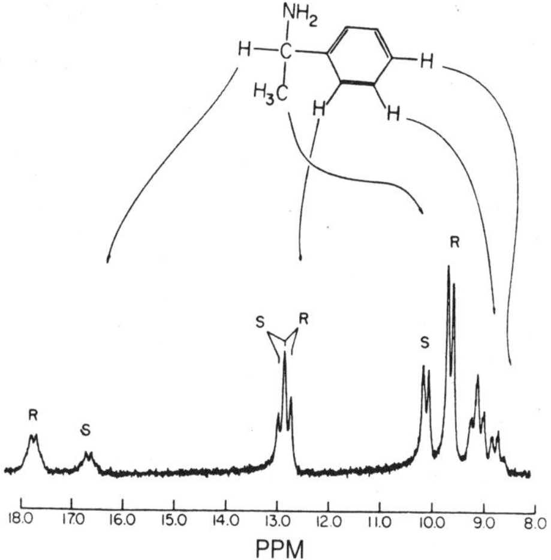

Fig. 2.6. NMR spectrum of l-phenylethylamine in presence of chiral shift reagent, showing differential chemical shift of methine and methyl signals and indicating ratio of *R*- to *S*-enantiomers. [Reproduced with permission from *J. Am. Chem. Soc.* **93**, 5914 (1971).]

As stated earlier, geometric isomers of alkenes are broadly classified as being diastereomeric, since they are stereoisomeric but not enantiomeric. The specification of the geometry of double bonds as *cis* and *trans* suffers from the same amibiguity as the Fischer convention; that is, it requires a judgment about the "similarity" of groups. The sequence rule is the basis for an unambiguous method for assignment of alkene geometry.[22] The four substituents on the double bond are taken in pairs. The sequence rules are used to determine if the higher-priority groups on the atoms forming the double bond are on the same or the opposite side of the double bond. If the higher-priority groups are on the same side, the descriptor is Z (from German *zusammen*, together); if they are on opposite sides, the descriptor is E (from German *entgegen*, opposite). As in applying the sequence rule to chiral centers, if the atoms directly attached to the double bond have the same atomic number, the priorities are assigned by comparing subsequent atoms in the substituent until priority can be established. The system can also be applied to multiple bonds involving elements other than carbon, such as, for example, C=N. The Z and E descriptors have replaced *syn* and *anti* for describing the stereochemistry of oximes, for example. As in the case of chiral centers, if an atom at a double bond does not have two substituents (as is the case for oximes), then a "phantom ligand" with atomic number zero is assumed and assigned the lower priority. Scheme 2.7 shows some stereoisomeric compounds named according to the sequence rule convention.

22. J. E. Blackwood, C. L. Gladys, K. L. Loening, A. E. Petrarca, and J. E. Rush, *J. Am. Chem. Soc.* **90**, 509 (1968).

2.3. Stereochemistry of Dynamic Processes

Up to this point, we have emphasized the stereochemical properties of molecules as objects, without concern for processes which affect the molecular shape. The term *dynamic stereochemistry* applies to the topological features of such processes. The cases that are of most importance in organic chemistry are chemical reactions, conformational processes, and noncovalent complex formation. In order to understand the stereochemical aspects of a dynamic process, it is essential not only that the stereochemical relationship between starting and product states be established,

Scheme 2.7. Stereoisomeric Alkenes and Related Molecules with the Double-Bond Geometry Named According to the Sequence Rule

1[a] (*Z*)-3-Decenoic acid (the sex pheromone of the furniture carpet beetle)

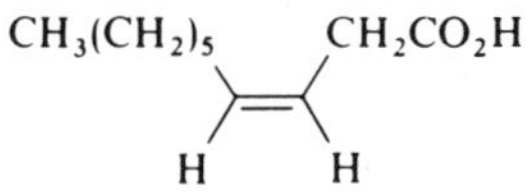

$CH_3(CH_2)_5$ CH_2CO_2H H H

2[b] Methyl (2*E*,6*E*,10*Z*)-10,11-epoxy-3,7,11-trimethyltridecadienoate (the juvenile hormone of the tobacco hornworm)

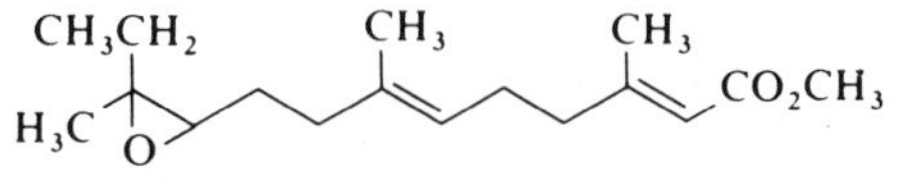

CH_3CH_2 CH_3 CH_3 CO_2CH_3 H_3C O

3[c] Nitrones and oxime ethers

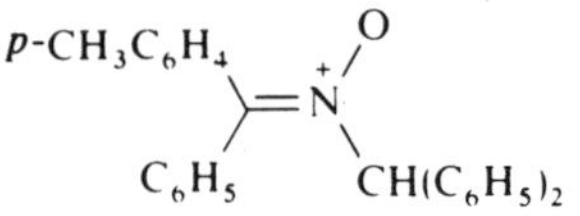

p-$CH_3C_6H_4$ O N^+ C_6H_5 $CH(C_6H_5)_2$ *Z*

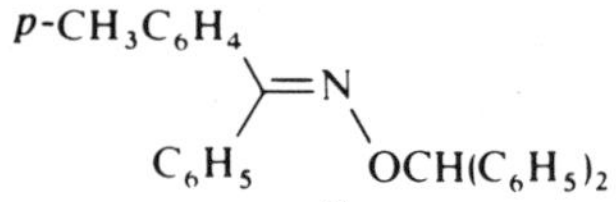

p-$CH_3C_6H_4$ N C_6H_5 $OCH(C_6H_5)_2$ *E*

4[d] (2*Z*,4*Z*,6*E*,8*E*)-9-(3′-Furyl)-2,6-dimethylnona-2,4,6,8-tetraen-4-olide (dihydrofreelingyne)

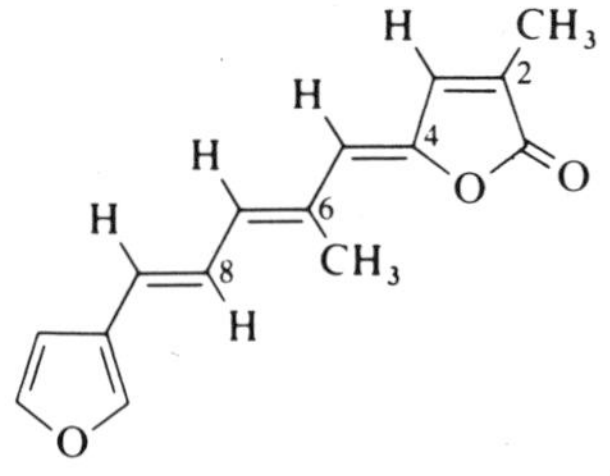

H CH_3 2 H 4 H O O 6 H CH_3 8 H O

a. H. Fukui, F. Matsumara, M. C. Ma, and W. E. Burkholder, *Tetrahedron Lett.* 3536 (1974).
b. R. C. Jennings, K. J. Judy, and D. A. Schooley, *J. Chem. Soc. Chem. Commun.*, 21 (1975).
c. T. S. Dobashi and E. J. Grubbs, *J. Am. Chem. Soc.* **95**, 5070 (1973).
d. C. F. Ingham and R. A. Massy-Westropp, *Aust. J. Chem.* **27**, 1491 (1974).

but also that the spatial features of proposed intermediates and transition states be consistent with the experimental observations.

In describing the stereochemical features of chemical reactions, we can distinguish between two types: stereospecific reactions and stereoselective reactions.[23] A *stereospecific reaction* is one in which stereoisomeric starting materials afford stereoisomerically different products under the same reaction conditions. A *stereoselective reaction* is one in which a single reactant has the capacity of forming two or more stereoisomeric products in a particular reaction but one is formed preferentially. A stereospecific reaction is a special, more restrictive, case of a stereoselective reaction.

The stereochemistry of the most familiar reaction types such as addition, substitution, and elimination are described by terms which specify the stereochemical relationship between the reactants and products. Addition and elimination reactions are classified as *syn* or *anti*, depending on whether the covalent bonds that are made or broken are on the same or opposite faces of the plane of the double bond.

syn elimination / *syn* addition; *anti* elimination / *anti* addition

Substitution reactions at tetrahedral centers are classified as proceeding with retention or inversion of configuration or with racemization. The term *retention of configuration* applies to a process in which the relative spatial arrangement at the reaction center is the same in the reactant and the product. *Inversion of configuration* describes a process in which the topology of the substitution site in the product has a mirror image relationship to that of the reactant. A substitution which generates both possible enantiomers of a product from a single enantiomer of the reactant occurs with *racemization.* Such a process can result in complete or partial racemization depending on whether the product is a racemic mixture or if an excess of one enantiomer is formed. The term *epimerization* is used to describe the special case of racemization of a single stereogenic center in a diastereomer, while maintaining the configuration of the other centers.

While it is convenient to use optically active substrates to probe the stereochemistry of substitution reactions, it should be emphasized that the stereochemistry of a reaction is a property of the mechanism, not of the means of determining it. Thus, it is proper to speak of the hydrolysis of methyl iodide as proceeding with inversion, even though the stereochemistry is not directly discernible because of the archiral nature of the reactant and product.

$$H_2O + H_3C{-}I \longrightarrow H^+ \; HO{-}CH_3 \; I^-$$

Inversion of configuration at carbon in hydrolysis of methyl iodide.

23. E. L. Eliel, *Stereochemistry of Carbon Compounds*, McGraw-Hill, New York, 1962, p. 436.

Some stereospecific reactions are listed in Scheme 2.8. Examples of stereoselective reactions are presented in Scheme 2.9. As can be seen in Scheme 2.8, the starting materials in these stereospecific processes are stereoisomeric pairs and the products are stereoisomeric with respect to each other. Each reaction proceeds to give a single stereoisomer without contamination by the alternative stereoisomer. The stereochemical relationships are determined by the reaction mechanism. Detailed discussion of the mechanisms of these reactions will be deferred until later sections, but some comments can be made here to illustrate the concept of stereospecificity.

Entries 1 and 2 in Scheme 2.8 are typical of concerted *syn* addition to alkene double bonds. On treatment with peroxyacetic acid, the Z-alkene affords the *cis*-oxirane, while the E-alkene affords only the *trans*-oxirane. Similarly, addition of dibromocarbene to Z-2-butene yields exclusively 1,1-dibromo-*cis*-2,3-dimethylcyclopropane, while only 1,1-dibromo-*trans*-2,3-dimethylcyclopropane is formed from E-2-butene. There are also numerous stereospecific *anti* additions. Entry 3 shows the *anti* stereochemistry typical of bromination of simple alkenes.

Nucleophilic substitution reactions at sp^3 carbon by direct displacement proceed with inversion of configuration at the carbon atom bearing the leaving group. Thus, *cis*-4-*t*-butylcyclohexyl *p*-toluenesulfonate is converted by thiophenoxide ion to *trans*-4-*t*-butylcyclohexyl phenyl thioether. The stereoisomeric *trans* *p*-toluenesulfonate gives the *cis* phenyl thioether (entry 4). 2-Octyl *p*-tolenesulfonate esters react with acetate ion to give the substitution product of inverted configuration, as can be demonstrated by the use of optically active reactant (entry 5).

Entry 6 is an example of a stereospecific elimination reaction of an alkyl halide in which the transition state requires the proton and bromide ion being lost to be in an *anti* orientation with respect to each other. The diastereomeric *threo*- and *erythro*-1,2-diphenyl-1-bromopropanes undergo β elimination to produce stereoisomeric products. Entry 7 is an example of a pyrolytic elimination requiring a *syn* orientation of the proton being removed and the nitrogen atom of the amine oxide group. The elimination proceeds through a cyclic transition state in which the proton is transferred to the oxygen of the amine oxide group.

The stereoselective reactions in Scheme 2.9 include examples that are completely stereoselective (e.g., entries 2 and 3), one that is highly stereoselective (entry 6), and others in which the stereoselectivity is modest to low (entries 1, 4, 5, and 7). In the highly stereoselective acid-catalyzed opening of methylcyclopropyl carbinol (entry 2) and the addition of formic acid to norbornene (entry 3), only a single stereoisomer is produced from each reactant. Reduction of 4-*t*-butylcyclohexanone is typical of the reduction of unhindered cyclohexanones in that the major diastereomer produced has an equatorial hydroxyl group. Certain other reducing agents, particularly sterically bulky ones, exhibit the opposite stereoselectivity and favor the formation of the axial hydroxyl group.

The alkylation of 1-methyl-4-*t*-butylpiperidine with benzyl chloride (entry 7) provides only a slight excess of one diastereomer over the other. It is also observed that the ratio can be reversed by changing solvents.

We have previously seen (Scheme 2.8, entry 6) that the dehydrohalogenation of alkyl halides is a stereospecific reaction requiring an *anti* orientation of the proton and the halide leaving group in the transition state. The elimination reaction is also moderately stereoselective (Scheme 2.9, entry 1) in the sense that the more stable of the two alkene isomers is formed preferentially. Both isomers are formed by *anti* elimination processes but involve topologically distinct hydrogens. The base-catalyzed elimination reaction of 2-iodobutane affords three times as much *trans*-2-butene as *cis*-2-butene.

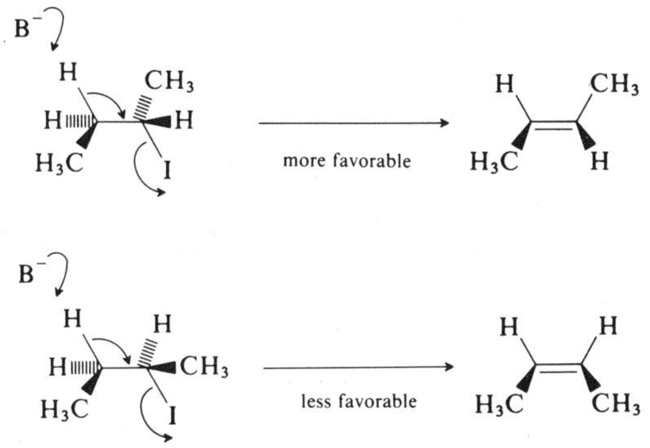

more favorable

less favorable

Moderate stereoselectivity is also seen in the addition of phenoxycarbene to cyclohexene, in which the product ratio is apparently influenced by steric considerations that favor introduction of the larger group (PhO versus H) in the less crowded *exo* position.

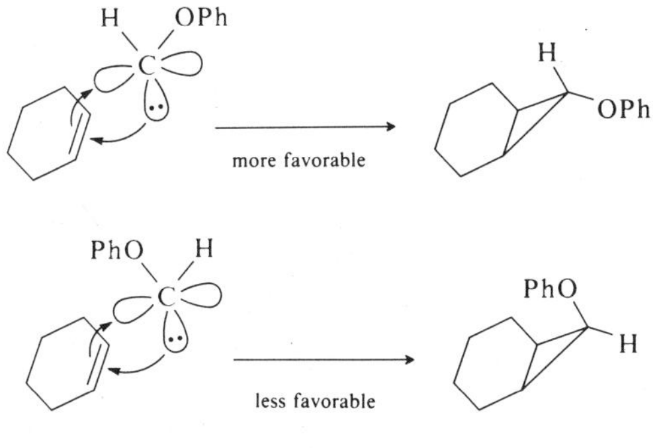

more favorable

less favorable

The addition of methylmagnesium iodide to 2-phenylpropanal is stereoselective in producing twice as much *erythro*-3-phenyl-2-butanol as *threo* isomer (Scheme 2.9, entry 5) The stereoselective formation of a particular configuration at a new chiral center in a reaction of a chiral reactant is called *asymmetric induction.* This particular case is one where the stereochemistry can be predicted on the basis of an empirical correlation called *Cram's rule.* The structural and mechanistic basis of Cram's rule will be discussed in Chapter 3.

Standing in contrast to stereospecific and stereoselective processes are the racemization processes which result in formation of products of both configurations. The most common mechanistic course by which organic reactions lead to racemic products is by cleavage of one of the ligands from an asymmetric carbon to give a planar or rapidly inverting tricoordinate intermediate, such as a carbocation or free

Scheme 2.8.

A. Stereospecific addition to alkenes

1[a] Epoxidation

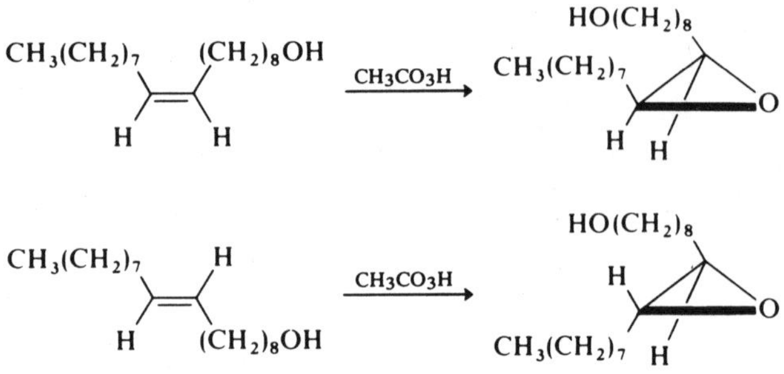

$CH_3(CH_2)_7$ / H, $(CH_2)_8OH$ / H (cis) $\xrightarrow{CH_3CO_3H}$ epoxide with $HO(CH_2)_8$, $CH_3(CH_2)_7$, H, H, O

$CH_3(CH_2)_7$ / H, H / $(CH_2)_8OH$ (trans) $\xrightarrow{CH_3CO_3H}$ epoxide with $HO(CH_2)_8$, H, $CH_3(CH_2)_7$, H, O

2[b] Addition of dibromocarbene

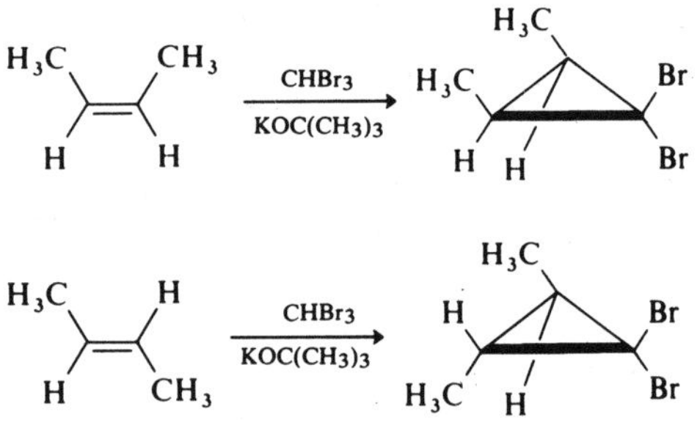

H_3C / H, CH_3 / H (cis) $\xrightarrow[KOC(CH_3)_3]{CHBr_3}$ cyclopropane with H_3C, H_3C, H, H, Br, Br

H_3C / H, H / CH_3 (trans) $\xrightarrow[KOC(CH_3)_3]{CHBr_3}$ cyclopropane with H_3C, H, H_3C, H, Br, Br

3[c] Bromination

$H_3C(H)C{=}C(H)CH_3$ (trans) $\xrightarrow{Br_2}$ $H_3C(H)(Br)C{-}C(H)(CH_3)(Br)$ *meso*

$H_3C(H)C{=}C(H)CH_3$ (cis) $\xrightarrow{Br_2}$ $H_3C(H)(Br)C{-}C(CH_3)(H)(Br)$ *d,l*

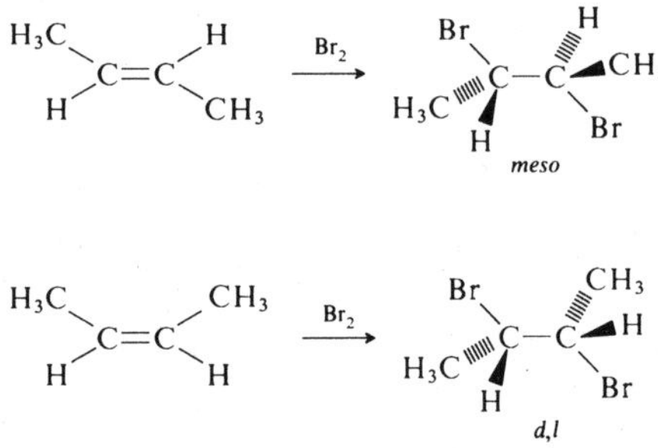

a. L. P. Witnauer and D. Swern, *J. Am. Chem. Soc.* **72**, 3364 (1950).
b. P. S. Skell and A. Y. Garner, *J. Am. Chem. Soc.* **78**, 3409 (1956).
c. A. Modro, G. H. Schmid, and K. Yates, *J. Org. Chem.* **42**, 3673 (1977).

B. Nucleophilic substitution

4[d] *cis*- and *trans*-4-*t*-Butylcyclohexyl *p*-toluenesulfonate

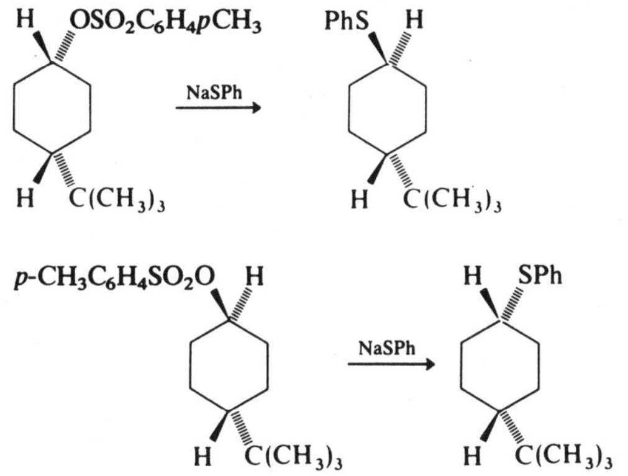

H $OSO_2C_6H_4pCH_3$ → NaSPh → PhS H

H $C(CH_3)_3$ H $C(CH_3)_3$

p-$CH_3C_6H_4SO_2O$ H → NaSPh → H SPh

H $C(CH_3)_3$ H $C(CH_3)_3$

5[e,f] *S*-(+)- and *R*-(−)-2-Octyl *p*-toluenesulfonate

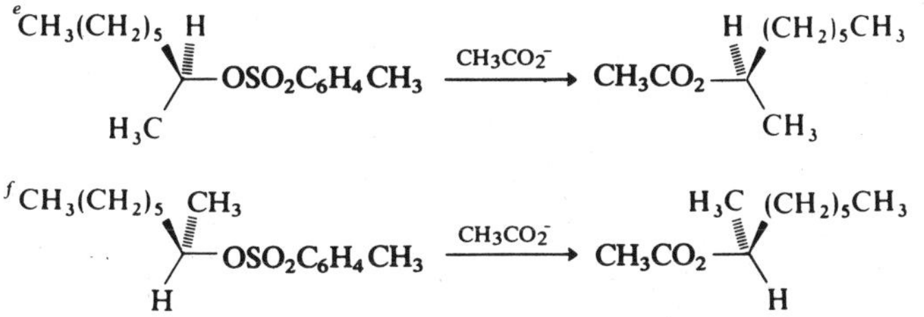

[e] $CH_3(CH_2)_5$ H, H_3C —$OSO_2C_6H_4CH_3$ → $CH_3CO_2^-$ → CH_3CO_2— H $(CH_2)_5CH_3$, CH_3

[f] $CH_3(CH_2)_5$ CH_3, H —$OSO_2C_6H_4CH_3$ → $CH_3CO_2^-$ → CH_3CO_2— H_3C $(CH_2)_5CH_3$, H

C. Elimination

6[g] Dehydrohalogenation

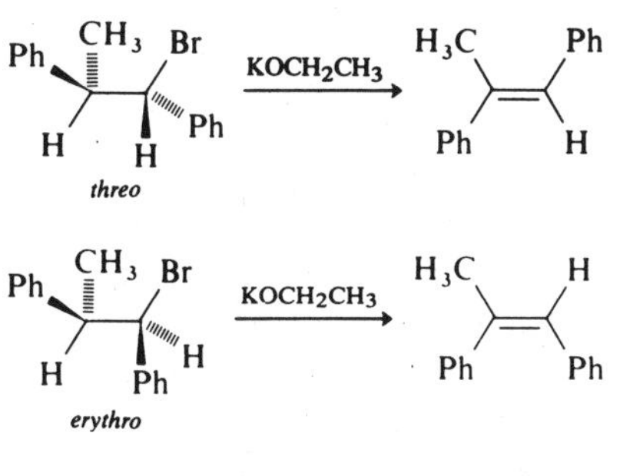

Ph, CH_3, Br, H, H, Ph (*threo*) → $KOCH_2CH_3$ → H_3C, Ph, Ph, H

Ph, CH_3, Br, H, Ph, H (*erythro*) → $KOCH_2CH_3$ → H_3C, H, Ph, Ph

7[h] Pyrolysis of amine oxides

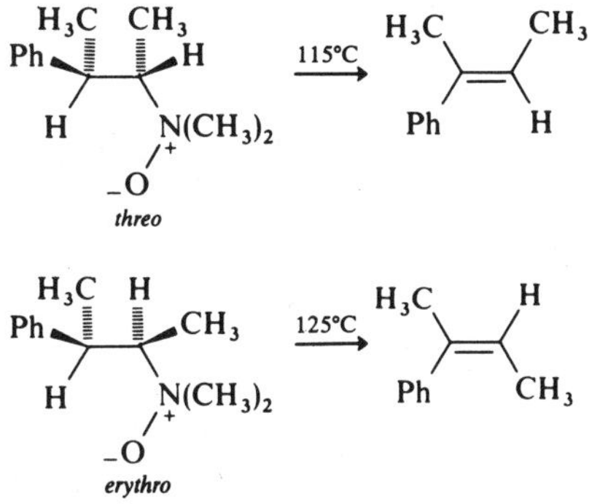

H_3C CH_3, Ph, H, H, $\overset{+}{N}(CH_3)_2$, ^-O (*threo*) → 115°C → H_3C, CH_3, Ph, H

H_3C H, Ph, CH_3, H, $\overset{+}{N}(CH_3)_2$, ^-O (*erythro*) → 125°C → H_3C, H, Ph, CH_3

d. E. L. Eliel and R. S. Ro. *J. Am. Chem. Soc.* **79**, 5995 (1957).
e. A. Streitwieser, Jr., and A. C. Waiss, Jr., *J. Org. Chem.* **27**, 290 (1962).
f. H. Philips, *J. Chem. Soc.*, 2552 (1925).
g. D. J. Cram, F. D. Greene, and C. H. DePuy, *J. Am. Chem. Soc.* **78**, 790 (1956).
h. D. J. Cram and J. E. McCarty, *J. Am. Chem. Soc.* **76**, 5740 (1954).

Scheme 2.9.

A. Formation of alkenes

1[a] Dehydrohalogenation

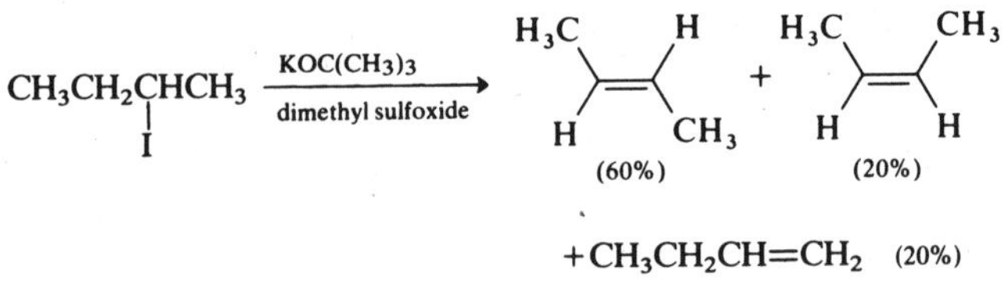

$CH_3CH_2CHICH_3 \xrightarrow[\text{dimethyl sulfoxide}]{KOC(CH_3)_3}$ cis-2-butene (60%) + trans-2-butene (20%) + $CH_3CH_2CH{=}CH_2$ (20%)

2[b] Acid-catalyzed ring-opening of cyclopropylcarbinols

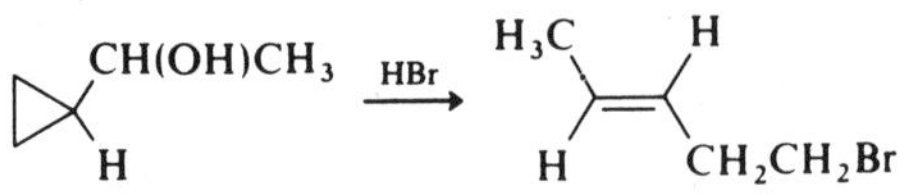

cyclopropyl-CH(OH)CH_3 $\xrightarrow{HBr}$ H_3C(H)C=C(H)CH_2CH_2Br

B. Addition to alkenes

3[c] Addition of formic acid to norbornene

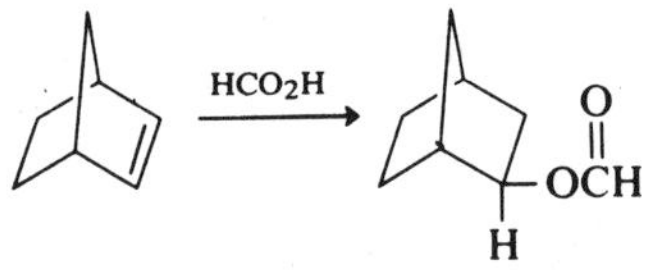

norbornene $\xrightarrow{HCO_2H}$ exo-norbornyl formate (OCH=O, H)

4[d] Addition of phenoxycarbene to cyclohexene

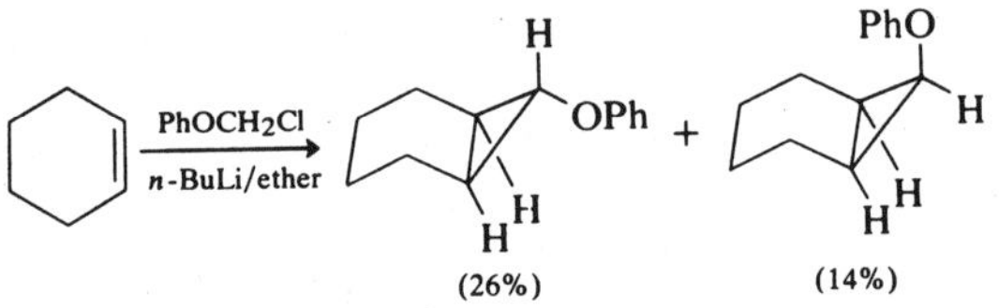

cyclohexene $\xrightarrow[n\text{-BuLi/ether}]{PhOCH_2Cl}$ 7-phenoxynorcarane (26%) + isomer (14%)

a. R. A. Bartsch, G. M. Pruss, B. A. Bushaw, and K. E. Wiegers, *J. Am. Chem. Soc.* **95**, 3405 (1973).
b. M. Julia, S. Julia, and S.-Y. Tchen, *Bull. Soc. Chim. Fr.*, 1849 (1961).
c. D. C. Kleinfelter and P. von R. Schleyer, *Org. Synth.* **V**, 852 (1973).
d. U. Schöllkopf, A. Lerch, and W. Pitteroff, *Tetrahedron Lett.*, 241 (1962).

radical. In the absence of any special solvation effects, such intermediates will produce equal quantities of the two possible enantiomeric products. Nucleophilic substitution proceeding through a carbocation intermediate is a familiar example of such a process. This case will be discussed in detail in Chapter 5.

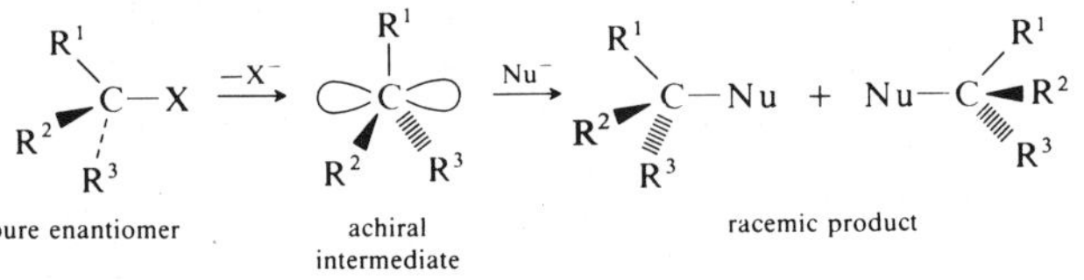

$R^1R^2R^3C{-}X \xrightarrow{-X^-} [R^1R^2R^3C^+] \xrightarrow{Nu^-} R^1R^2R^3C{-}Nu + Nu{-}CR^1R^2R^3$

pure enantiomer — achiral intermediate — racemic product

C. Addition to carbonyl groups

5[e]

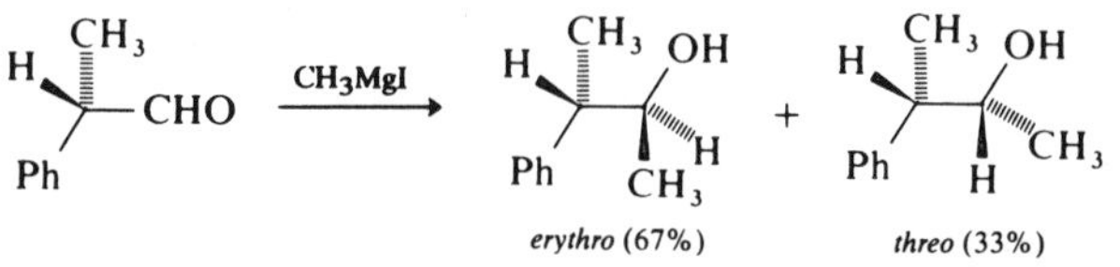

CH_3MgI

erythro (67%) threo (33%)

6[f]

$LiAlH_4$

(90%)

(10%)

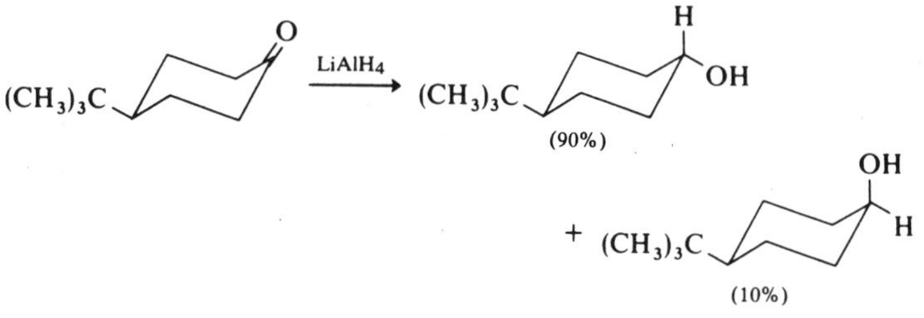

D. Formation of quaternary ammonium salts

7[g]

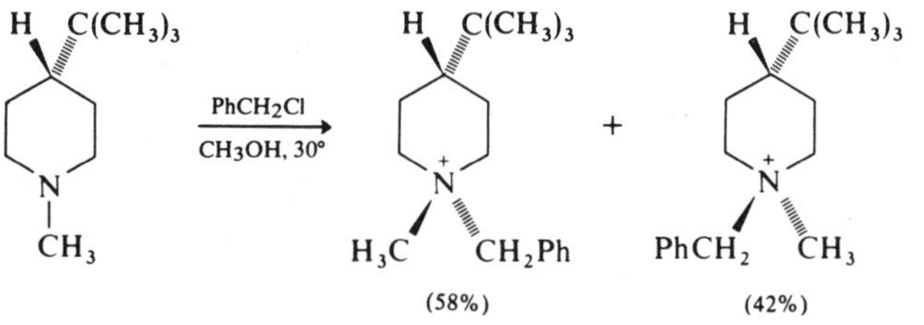

$PhCH_2Cl$, CH_3OH, 30°

(58%) (42%)

e. D. J. Cram and F. A. Abd Elhafez, *J. Am. Chem. Soc.* **74**, 5828 (1952).
f. E. L. Eliel and M. N. Rerick, *J. Am. Chem. Soc.* **82**, 1367 (1960).
g. A. T. Bottini and M. K. O'Rell, *Tetrahedron Lett.*, 423 (1967).

The term racemization can be used to describe any process which leads to formation of both configurations at a chiral center and is not restricted to processes which involve bond cleavage. Examples would be pyramidal inversion at trivalent nitrogen, sulfur, or phosphorus. The rate of racemization of such compounds depends upon the barrier to the inversion process. For ammonia and simple amines, the barrier is very low and inversion of configuration at nitrogen is rapid at room temperature. Thus, although unsymmetrically substituted amines are chiral, the process of racemization is too rapid to allow separation of the enantiomers. Incorporation of the nitrogen into a three-membered ring serves to raise the barrier for inversion, due to the additional strain in the planar transition state. For aziridine the energy barrier to pyramidal inversion is 12 kcal/mol. While this is too low for separation of enantiomers, separate NMR spectra for stereoisomeric aziridines can be observed.[24]

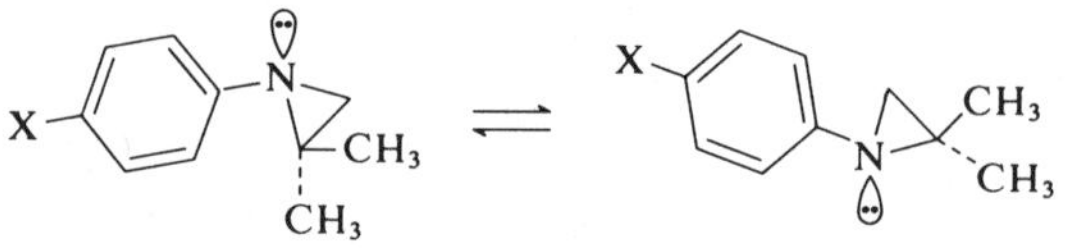

inversion barrier 8–12 kcal/mol depending on X

Certain substituted aziridines can be isolated as enantiomers as the result of still higher barriers. Most of these compounds are *N*-chloro or *N*-alkoxy aziridines.[25]

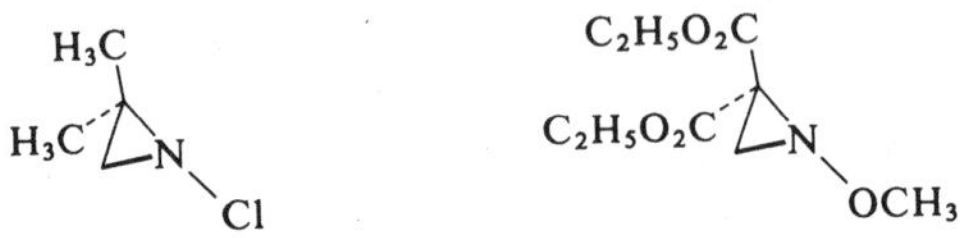

While the barriers for pyramidal inversion are low for first-row elements, the heavier elements have much higher barriers to inversion. The preferred bonding angle at trivalent phosphorus and sulfur is about 100°. A greater distortion is required to reach a planar transition state. Typical barriers for trisubstituted phosphines are 30–35 kcal/mol, while for sulfoxides the barriers are about 35–45 kcal/mol. Many phosphines and sulfoxides have been isolated in optically active form, and they undergo racemization by pyramidal inversion only at high temperature.[26]

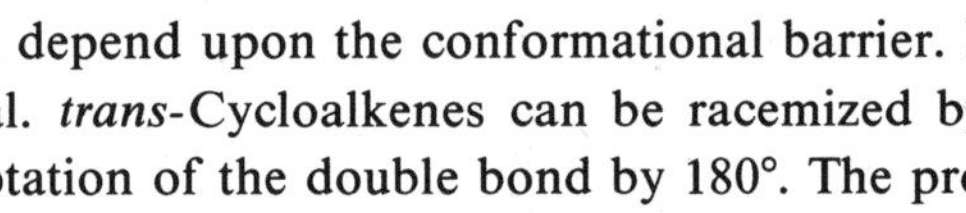

Ref. 27

inversion barrier 32.7 kcal/mol at 130°

Molecules that are chiral as a result of barriers to conformational interconversion can be racemized if the enantiomeric conformers are interconverted. The rate of racemization will depend upon the conformational barrier. For example, *trans*-cyclooctene is chiral. *trans*-Cycloalkenes can be racemized by a conformational process involving rotation of the double bond by 180°. The process is represented

24. J. D. Andose, J. M. Lehn, K. Mislow, and J. Wagner, *J. Am. Chem. Soc.* **92**, 4050 (1970).
25. S. J. Brois, *J. Am. Chem. Soc.* **90**, 506, 508 (1968); S. J. Brois, *J. Am. Chem. Soc.* **92**, 1079 (1970); V. F. Rudchenko, O. A. D'yachenko, A. B. Zolotoi, L. O. Atovmyan, I. I. Chervin, and R. G. Kostyanovsky, *Tetrahedron* **38**, 961 (1982).
26. For a review of racemization via vibrational inversion at chiral atoms, see J. B. Lambert, *Top Stereochem.* **6**, 19 (1971).
27. R. D. Baechler and K. Mislow, *J. Am. Chem. Soc.* **93**, 773 (1971).

below but is more easily seen by working with a molecular model.

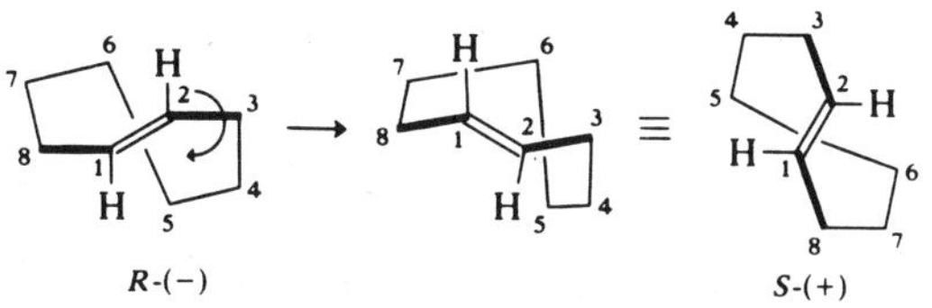

R-(−) → ≡ S-(+)

Since one of the vinyl hydrogens must "slip through" the ring, the ease of rotation depends upon the ring size. *trans*-Cyclooctene is quite stable to thermal racemization and can be recovered with no loss of enantiomeric purity after 7 days at 61°C.[28] When the ring size is larger, rotation of the double bond through the ring can occur more easily and racemization takes place more rapidly. The half-life for racemization of *trans*-cyclononene is 5 min at 0°C.[29] The rate of racemization of *trans*-cyclodecene is so fast that racemization occurs immediately on its release from the platinum complex employed for its resolution.[29]

The dynamic stereochemistry of biaryls is similar. The energy barrier for racemization of optically active 1,1-binaphthyl (Scheme 2.2, entry 3, p. 76) is 21–23 kcal/mol.[30] The two rings are not coplanar in the ground state, and the racemization takes place by rotation about the 1,1′ bond.

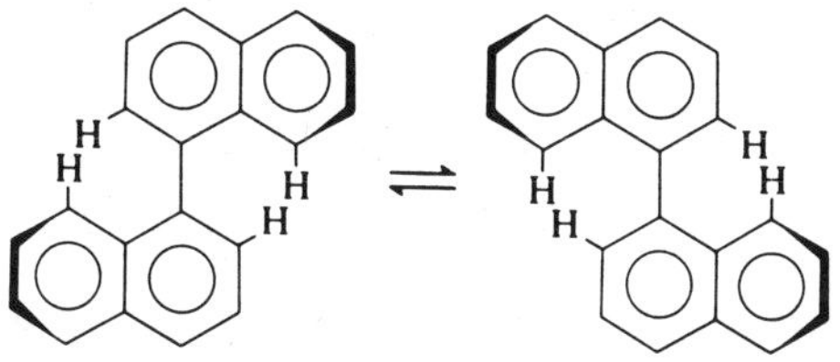

Rotation about the 1,1′ bond is resisted by van der Waals interactions between the hydrogens shown in the structures. These hydrogens crowd each other when the two naphthyl groups are coplanar, and the racemization process requires the hydrogens to move past each other. The existence of enantiomeric substituted biphenyls also depends on steric interactions between substituents. The relationship between the rate of racemization and the size of the substituents has been investigated.[31] There is a correlation between the barrier to rotation and the extent of steric

28. A. C. Cope, C. R. Ganellin, H. W. Johnson, Jr., T. V. VanAuken, and H. J. S. Winkler, *J. Am. Chem. Soc.* **85**, 3276 (1963). The activation energy is 35.6 kcal/mol: A. C. Cope and B. A. Pawson, *J. Am. Chem. Soc.* **87**, 3649 (1965).
29. A. C. Cope, K. Banholzer, H. Keller, B. A. Pawson, J. J. Whang, and H. J. S. Winkler, *J. Am. Chem. Soc.* **87**, 3644 (1965).
30. A. K. Colter and L. M. Clemens, *J. Phys. Chem.* **68**, 651 (1964).
31. F. H. Westheimer, in *Steric Effects in Organic Chemistry*, M. S. Newman (ed.), Wiley, New York, 1956, Chapter 12.

repulsion between substituents.[32]

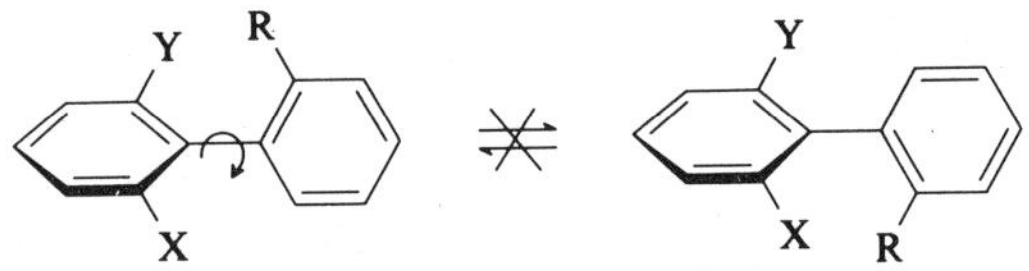

There are also many cases where conformational barriers are large enough to permit observation of stereoisomers by spectroscopic methods even though the enantiomers cannot be isolated at room temperature. NMR is a particularly powerful tool in this regard since the dependence of the appearance of the spectrum on rates of conformational interconversion is easily interpreted.

2.4. Prochiral Relationships

It is frequently necessary to distinguish between identical ligands, that, even though bonded to the same atom, may be topologically nonequivalent. Let us consider 1,3-propanediol as an example. If a process occurs in which a proton at C-2 is substituted by another ligand, say, deuterium, the two possible substitution modes generate identical products. The two protons at C-2 are therefore topologically equivalent and are termed *homotopic* ligands.

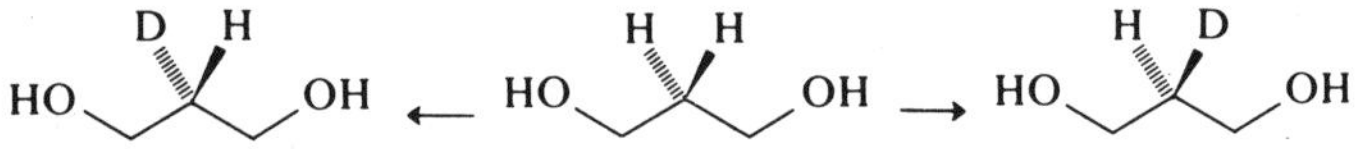

Substitution products are superimposable. There is a plane of symmetry defined by the atoms H–C(2)–D.

If a similar process occurred involving the two protons at C-1, a stereochemically different situation would result. Substitution at C-1 produces a chiral product, 1-*deuterio*-1,3-propanediol:

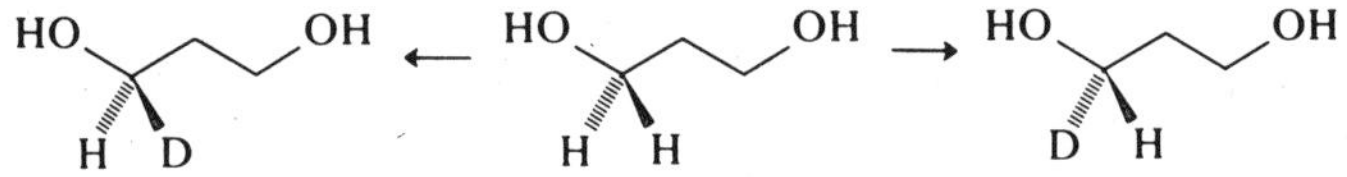

(*R*)-1-*deuterio*-1,3-propanediol

(*S*)-1-*deuterio*-1,3-propanediol

The two protons at C-1 are topologically nonequivalent since substitution of one produces a product that is stereochemically distinct from that produced by substitution of the other. Ligands of this type are termed *heterotopic*, and, because the products of substitution are enantiomers, the more precise term *enantiotopic* also

32. G. Bott, L. D. Field, and S. Sternhell, *J. Am. Chem. Soc.* **102**, 5618 (1980).

applies.[33] If a chiral assembly is generated when a point ligand is replaced by a new point ligand, the original assembly is *prochiral.* Both C-1 and C-3 of 1,3-propanediol are *prochiral centers.*

The sequence rule may be applied directly to the specification of heterotopic ligands in prochiral molecules using the descriptors *pro-R* and *pro-S.* The assignment is done by selecting one of the heterotopic ligands at the prochiral center and arbitrarily assigning it a higher priority than the other, without disturbing the priorities of the remaining ligands. If application of the sequence rule results in assignment of *R* as the configuration of the prochiral center, then the selected ligand is *pro-R.* If the prochiral center is *S*, then the selected ligand is *pro-S.* It is customary to designate prochirality in structures by subscripts *R* or *S* at the appropriate atoms. For 1,3-propanediol, the prochiral hydrogens are as indicated below:

HO OH
H_SH_R H_RH_S

Enantiotopic atoms or groups are equivalent in all chemical and physical respects except toward a chiral reagent. Many examples of discrimination between enantiotopic groups are found among enzyme-catalyzed reactions. The enzymes *liver alcohol dehydrogenase* and *yeast alcohol dehydrogenase,* for example, distinguish between the enantiotopic C(1) hydrogens of ethanol. Ethanol is a prochiral molecule, and it has been shown that oxidation to acetaldehyde by either enzyme results in the loss of the *pro-R* hydrogen. Both enzymes require nicotinamide adenine dinucleotide (NAD^+) as a coenzyme, which serves as the immediate hydrogen acceptor. Incubation of (*S*)-1-*deuterio*-ethanol with the enzyme–coenzyme system produces exclusively acetaldehyde-1-*d*, while the same treatment of (*R*)-1-*deuterio*-ethanol affords acetaldehyde containing no deuterium.

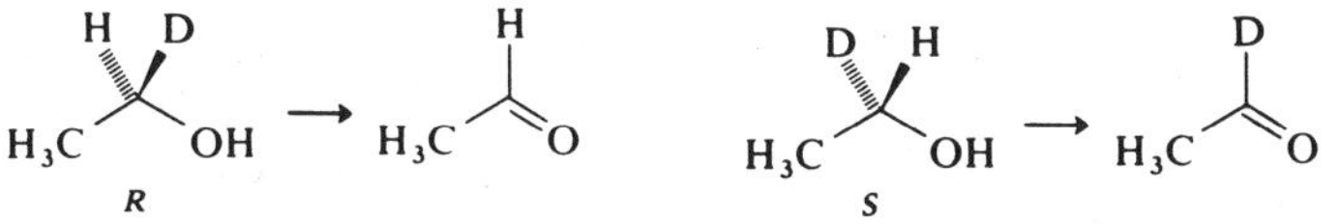

H D, H_3C OH, *R* → H, H_3C O   D H, H_3C OH, *S* → D, H_3C O

The enzyme-catalyzed interconversion of acetaldehyde and ethanol serves to illustrate a second important feature of prochiral relationships, that of *prochiral faces.* Addition of a fourth ligand, different from the three already present, to the carbonyl carbon of acetaldehyde will produce a chiral molecule. The original molecule presents to the approaching reagent two faces which bear a mirror image relationship to one another and are therefore enantiotopic. The two faces may be classified as *re* (from *rectus*) or *si* (from *sinister*) according to the sequence rule. If the substituents viewed from a particular face appear clockwise in order of decreasing

33. For a more complete discussion of definitions and terminology, see E. L. Eliel, *J. Chem. Educ.* **57**, 52 (1980); E. L. Eliel, *Top. Curr. Chem.* **105**, 1 (1982); K. R. Hanson, *J. Am. Chem. Soc.* **88**, 2731 (1966).

priority, then that face is *re*; if counterclockwise, then *si*. The *re* and *si* faces of acetaldehyde are shown below.

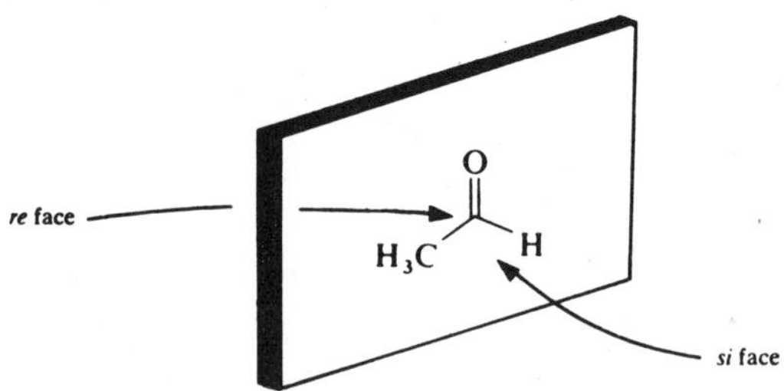

Reaction of an achiral reagent with a molecule exhibiting enantiotopic faces will produce equal quantities of enantiomers, and a racemic mixture will result. The achiral reagent sodium borodeuteride, for example, will produce racemic 1-*deuterio*-ethanol. Chiral reagents can discriminate between the prochiral faces, and an optically active product can result. Enzymatic reduction of acetaldehyde-1-*d* produces *R*-1-*deuterio*-ethanol that is optically pure.[34]

Fumaric acid is converted to L-malic acid by hydration in the presence of the enzyme *fumarase*. From the structure of the substrate and the configuration of the product, it is apparent that the hydroxyl group has been added to the *si* face of one of the carbon atoms of the double bond. Each of the trigonal carbon atoms of an alkene has its face specified separately. The molecule of fumaric acid shown below is viewed from the *re*–*re* face.

HO_2C H / H CO_2H (*re* *re*) → CO_2H, HO—H, H—H, CO_2H ≡ HO H, HO_2C CO_2H

As was the case for kinetic resolution of enantiomers, enzymes typically exhibit a high degree of selectivity toward enantiotopic functional groups. Selective reactions of enantiotopic groups provide enantiomerically enriched products. Thus, the treatment of a racemic material containing two enantiotopic functional groups is a means of effecting resolution. Most successful examples reported to date have involved hydrolysis. Several examples are outlined in Scheme 2.10.

Processes which create an excess of one enantiomer of a pair are called *enantioselective*. Most enzyme-catalyzed processes, such as the examples just discussed, are very enantioselective, leading to products of high enantiomeric purity. Reactions with smaller chiral reagents exhibit a wide range of enantioselectivity. A frequent objective of the study of such reactions is finding the best reagent and conditions to optimize the enantioselectivity of the reaction.

34. H. R. Levy, F. A. Loewus, and B. Vennesland, *J. Am. Chem. Soc.* **79**, 2949 (1957).

Scheme 2.10. Enantioselective Transformations Based on Enzyme-Catalyzed Reactions Which Differentiate between Enantiomers or Enantiotopic Substituents

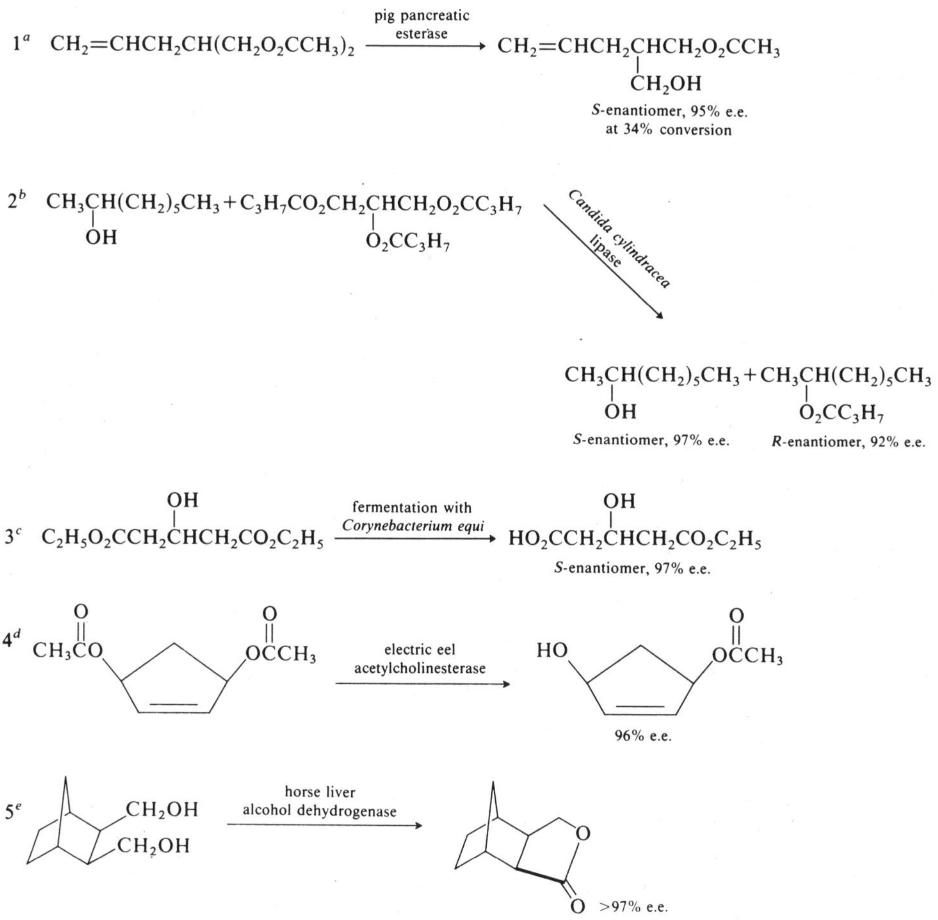

1[a] $CH_2{=}CHCH_2CH(CH_2O_2CCH_3)_2 \xrightarrow{\text{pig pancreatic esterase}} CH_2{=}CHCH_2CH(CH_2OH)CH_2O_2CCH_3$

S-enantiomer, 95% e.e. at 34% conversion

2[b] $CH_3CH(OH)(CH_2)_5CH_3 + C_3H_7CO_2CH_2CH(O_2CC_3H_7)CH_2O_2CC_3H_7 \xrightarrow{\textit{Candida cylindracea}\text{ lipase}} CH_3CH(OH)(CH_2)_5CH_3 + CH_3CH(O_2CC_3H_7)(CH_2)_5CH_3$

S-enantiomer, 97% e.e. R-enantiomer, 92% e.e.

3[c] $C_2H_5O_2CCH_2CH(OH)CH_2CO_2C_2H_5 \xrightarrow{\text{fermentation with } \textit{Corynebacterium equi}} HO_2CCH_2CH(OH)CH_2CO_2C_2H_5$

S-enantiomer, 97% e.e.

4[d] cis-3,5-bis(acetoxy)cyclopentene (CH_3CO_2–, –O_2CCH_3) $\xrightarrow{\text{electric eel acetylcholinesterase}}$ HO–cyclopentene–O_2CCH_3

96% e.e.

5[e] bicyclic diol (CH_2OH, CH_2OH) $\xrightarrow{\text{horse liver alcohol dehydrogenase}}$ bicyclic lactone

>97% e.e.

a. Y.-F. Wang and C. J. Sih, *Tetrahedron Lett.* **25**, 4999 (1984).
b. B. Cambou and A. M. Klibanov, *J. Am. Chem. Soc.* **106**, 2687 (1984).
c. A. S. Gopalan and C. J. Sih, *Tetrahedron Lett.* **25**, 5235 (1984).
d. D. R. Deardorff, A. J. Matthews, D. S. McMeekin, and C. L. Craney, *Tetrahedron Lett.* **27**, 1255 (1986).
e. K. P. Lok, I. J. Jakovac, and J. B. Jones, *J. Am. Chem. Soc.* **107**, 2521 (1985).

Chiral chemical reagents can react with prochiral centers in achiral substances to give partially or completely enantiomerically pure product. An example of such processes is the preparation of optically active sulfoxides from achiral sulfides using a chiral oxidant. To convert a sulfide to an optically active sulfoxide, the reagent must preferentially react with one of the two prochiral faces of the sulfide.

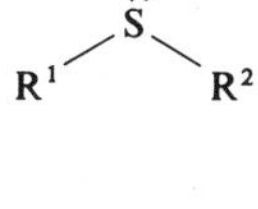

R^1–S–R^2

An achiral reagent cannot distinguish between these two faces. In a complex with a chiral reagent, however, the two (phantom ligand) electron pairs are in different environments (enantiotopic). The two complexes are therefore diastereomeric and are formed and react at different rates. Two reaction systems that have been used successfully for enantioselective formation of sulfoxides are illustrated below. In the first example, the $Ti(O\text{-}i\text{-}Pr)_4$–$t$-BuOOH–diethyl tartrate reagent is chiral by virtue of the presence of the chiral tartrate ester in the reactive complex. With simple aryl methyl sulfides, up to 90% enantiomeric excess in the product is obtained.

CH_3S–Ph $\xrightarrow[t\text{-BuOOH}]{R,R\text{-diethyl tartrate},\ Ti(O\text{-}i\text{-}Pr)_4}$ H_3C–$S^+(O^-)$–Ph 89% e.e. Ref. 35

The second method uses sodium periodate ($NaIO_4$) as the oxidant in the presence of the readily available protein bovine serum albumin. In this procedure, the sulfide is complexed in the chiral environment of the protein. Although the oxidant is achiral, it encounters the sulfide in a chiral environment in which the two faces of the sulfide are differentiated.

$(CH_3)_2CHS$–Ph $\xrightarrow[\text{bovine serum albumin}]{NaIO_4}$ $(CH_3)_2CH$–$S^+(O^-)$–Ph 81% e.e. Ref. 36

Another important example of an enantioselective reaction mediated by a chiral catalyst is the hydrogenation of 3-substituted 2-acetamido acrylic acid derivatives.

$RCH{=}C(NHCOCH_3)CO_2H \xrightarrow[\text{catalyst}]{H_2} RCH_2\text{–}CH(NHCOCH_3)CO_2H$

Depending on the stereoselectivity of the reaction, either the *R* or the *S* configuration can be generated at C-2 in the product. This corresponds to enantioselective synthesis of the D and L enantiomers of α-amino acids. The hydrogenation using stereoselective chiral catalysts has been carefully investigated.[37] The most effective catalysts for the reaction are rhodium complexes with chiral phosphine ligands. Table 2.1 records some illustrative results. The details of the catalytic mechanism need not be considered here. The fundamental point is that the chiral environment at the catalytic

35. P. Pitchen, E. Dunack, M. N. Desmukh, and H. B. Kagan, *J. Am. Chem. Soc.* **106**, 8188 (1984); F. DiFuria, G. Modena, and R. Seraglia, *Synthesis*, 325 (1984).
36. T. Sugimoto, T. Kokubu, J. Miyazaki, S. Tanimoto, and M. Okano, *J. Chem. Soc., Chem. Commun.*, 402 (1979); see also S. Colonna, S. Banfi, F. Fontana, and M. Sommaruga, *J. Org. Chem.* **50**, 769 (1985).
37. W. S. Knowles, *Acc. Chem. Res.* **16**, 106 (1983); D. Valentine, Jr., and J. W. Scott, *Synthesis*, 329 (1978).

Table 2.1. Enantioselective Reduction of 2-Acetamido Acrylic Acids by Chiral Phosphine Complexes of Rhodium

$$\underset{H}{\overset{R}{}}\!\!>C=C<\!\!\underset{CO_2H}{\overset{NHCOCH_3}{}} \xrightarrow[H_2]{\text{catalyst}} RCH_2\underset{\displaystyle CO_2H}{\underset{|}{C}}HNHCOCH_3$$

	R	Chiral ligand	Configuration of product	Enantiomeric excess (%)
1[a]		CH_3O; $P{-}CH_2CH_2{-}P$; OCH_3	*S*	94
2[b]	$(CH_3)_2CH-$	H_3C, CH_3; $H{-}C{-}C{-}H$; Ph_2P, PPh_2	*R*	100
3[c]	H_3C; N H	H; CH_2PPh_2; CH_2PPh_2; H	*R*	86
4[d]		Ph_2P; N; CH_2PPh_2; $(CH_3)_3CO_2C$	*R*	91
5[e]	H_3C; N H	H; Ph_2P; O; CH_3; CH_3; O; Ph_2P; H	*R*	86

a. B. D. Vineyard, W. S. Knowles, M. J. Sabacky, G. L. Bachman, and D. J. Weinkauf, *J. Am. Chem. Soc.* **99**, 5946 (1977).
b. M. B. Fryzuk and B. Bosnich, *J. Am. Chem. Soc.* **99**, 6262 (1977).
c. U. Hengartner, D. Valentine, Jr., K. K. Johnson, M. E. Larschied, F. Pigott, F. Scheidl, J. W. Scott, R. C. Sun, J. M. Townsend, and T. H. Williams, *J. Org. Chem.* **44**, 3741 (1979).
d. K. Achiwa, *J. Am. Chem. Soc.* **98**, 8265 (1976).
e. J. M. Townsend, J. F. Blount, R. C. Sun, S. Zawoiski, and D. Valentine, Jr., *J. Org. Chem.* **45**, 2995 (1980).

rhodium atoms causes a preference for reaction at either the *re* or the *si* face of the reactant. The hydrogen delivered from the catalyst then establishes the configuration at the new chiral center at C-2.

Another basic reaction that has been studied extensively with the objective of developing enantioselective reagents is the reduction of ketones. All unsymmetrical ketones possess prochiral faces, and chiral reductants should be enantioselective. One class of such reductants are derivatives of the common reducing agents lithium aluminum hydride and sodium borohydride that have been modified with chiral ligands.[38] A number of other types of chemical reagents[39] and enzymes[40] are also employed for enantioselective reduction of ketones. Table 2.2 provides some examples of enantioselective reduction of acetophenone.

The concept of heterotopic atoms, groups, and faces can be extended from enantiotopic to diastereotopic types. If each of two nominally equivalent ligands in a molecule is replaced by a test group and the molecules that are generated are diastereomeric, then the ligands are *diastereotopic.* Similarly, if reaction at one face of a trigonal atom generates a molecule diastereomeric with that produced at the alternate face, the faces are diastereotopic.

As an example of a molecule with diastereotopic ligands, consider the amino acid L-phenylalanine. The two protons at C-3 are diastereotopic, since substitution of either of them would generate a molecule with two chiral centers. Because the chiral center already present is *S*, the two diastereomers would be the 2*S*,3*R* and the 2*S*,3*S* stereoisomers. As in the case of enantiotopic protons, diastereotopic protons are designated *pro-R* or *pro-S.* The enzyme *phenylalanine ammonia lyase* catalyzes the conversion of phenylalanine to *trans*-cinnamic acid by a process involving *anti* elimination of the amino group and the 3-*pro-S* hydrogen. This stereochemical course has been demonstrated using deuterium-labeled L-phenylalanine as shown[41]:

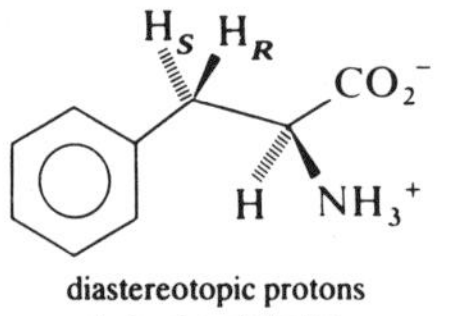

H_S H_R CO_2^- H NH_3^+

diastereotopic protons in L-phenylalanine

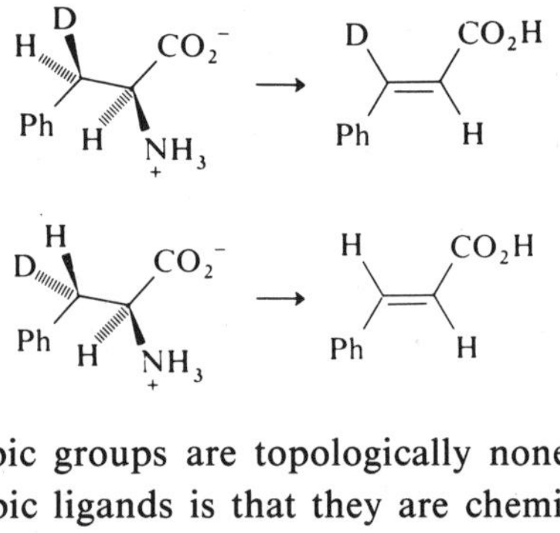

D H CO_2^- Ph H $\overset{+}{N}H_3$ → D CO_2H Ph H

H D CO_2^- Ph H $\overset{+}{N}H_3$ → H CO_2H Ph H

The environments of diastereotopic groups are topologically nonequivalent. An important property of diastereotopic ligands is that they are chemically nonequivalent toward achiral as well as chiral reagents, and they can also be distinguished by physical probes, especially NMR spectroscopy. As a consequence of their nonequivalence, they experience different shielding effects and have different chemical shifts in the NMR spectrum. (Enantiotopic groups have identical chemical shifts and are not distinguishable in NMR spectra.) A clear example of this effect can be seen in the proton NMR spectra of the *cis* and *trans* isomers of 1-benzyl-2,6-dimethylpiperidine shown in Fig. 2.7.[42] The methylene protons of the benzyl group

38. E. F. Grandbois, S. J. Howard, and J. D. Morrison, in *Asymmetric Synthesis*, J. D. Morrison (ed.), Vol. 2, Academic Press, New York, 1983, Chapter 3.
39. H. C. Brown, W. S. Park, B. T. Cho, and P. V. Ramachandran, *J. Org. Chem.* **52**, 5406 (1987).
40. C. J. Sih and C. S. Chen, *Angew. Chem. Int. Ed. Engl.* **23**, 570 (1984).
41. R. H. Wightman, J. Staunton, A. R. Battersby, and K. R. Hanson, *J. Chem. Soc., Perkin Trans. 1*, 2355 (1972).
42. R. K. Hill and T. H. Chan, *Tetrahedron* **21**, 2015 (1965); for additional discussion of chemical shift nonequivalence in diastereotopic groups, see W. B. Jennings, *Chem. Rev.* **75**, 307 (1975).

Table 2.2. Examples of Enantioselective Reduction of Acetophenone

Enantiomeric excess (%)	Reductant	Ref.
20–30	$LiAlH_4$, 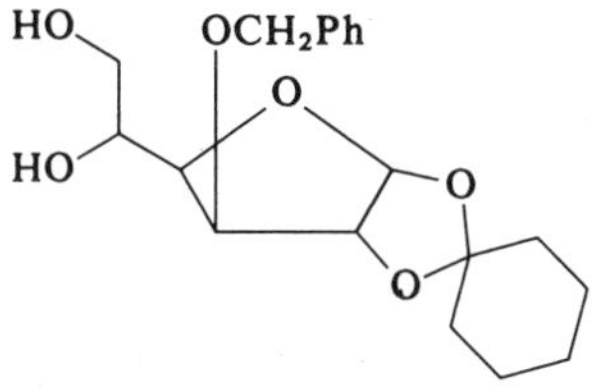 [structure: HO, HO, OCH_2Ph, O, O, O]	a
95	$LiAlH_4$, 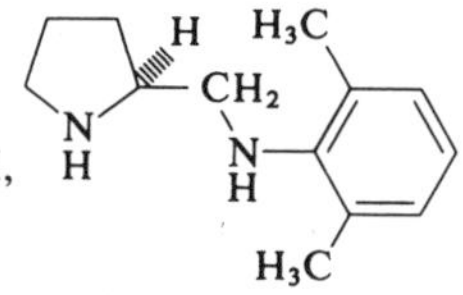 [structure: N H, H, H_3C, CH_2, N H, H_3C]	b
65–75	$LiAlH_4$, 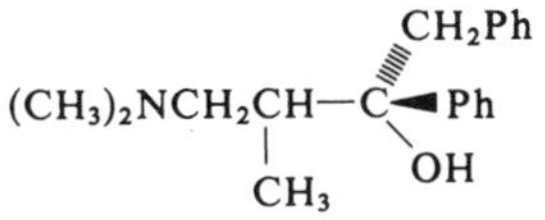 $(CH_3)_2NCH_2CH(CH_3)—C(CH_2Ph)(Ph)OH$	c
64	$NaBH_4$, $(CH_3)_2CHCO_2H$ 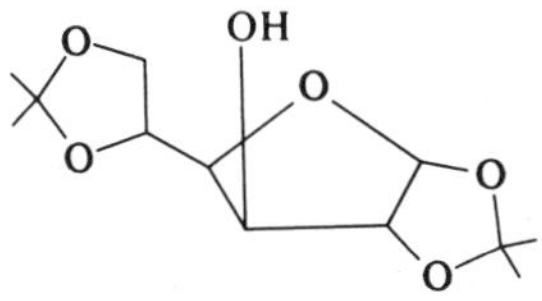 [structure: O, O, OH, O, O, O]	d
70–95	$LiAlH_3OC_2H_5$, 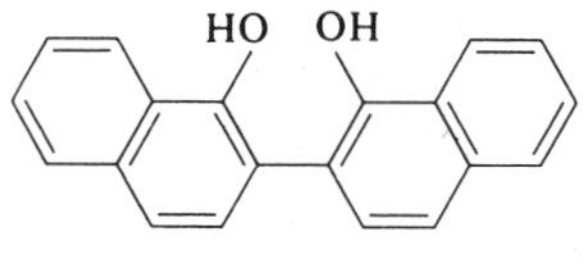 [structure: HO OH]	e
70	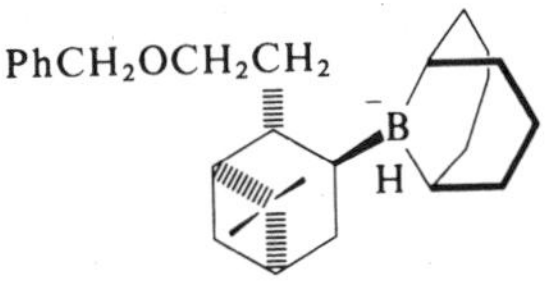 [structure: $PhCH_2OCH_2CH_2$, B, H]	f
96	BH_3, [structure: Ph Ph, H, O, B, N, CH_3] 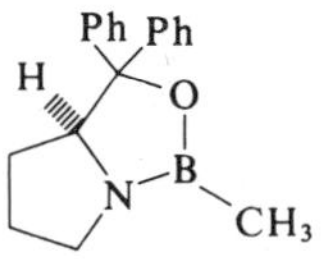	g

a. S. R. Landor, B. J. Miller, and A. P. Tatchell, *J. Chem. Soc., C*, 2280 (1966).
b. M. Asami and T. Mukaiyama, *Heterocycles* **12**, 499 (1979).
c. S. Yamaguchi and H. S. Mosher, *J. Org. Chem.* **38**, 1870 (1973); C. J. Reich, G. R. Sullivan, and H. S. Mosher, *Tetrahedron L.*, 1505 (1973).
d. J. D. Morrison, E. R. Grandbois, and S. I. Howard, *J. Org. Chem.* **45**, 4229 (1980): A. Hirao, S. Nakahama, H. Mochizuki, S. Itsuno, and N. Yamazaki, *J. Org. Chem.* **45**, 4231 (1980).
e. R. Noyori, I. Tomino, Y. Tanimoto, and M. Nishizawa, *J. Am. Chem. Soc.* **106**, 6709 (1984).
f. M. M. Midland and A. Kazubski, *J. Org. Chem.* **47**, 2495 (1982).
g. E. J. Corey, R. K. Bakshi, S. Shibata, C.-P. Chen, and V. K. Singh, *J. Am. Chem. Soc.* **109**, 7925 (1987).

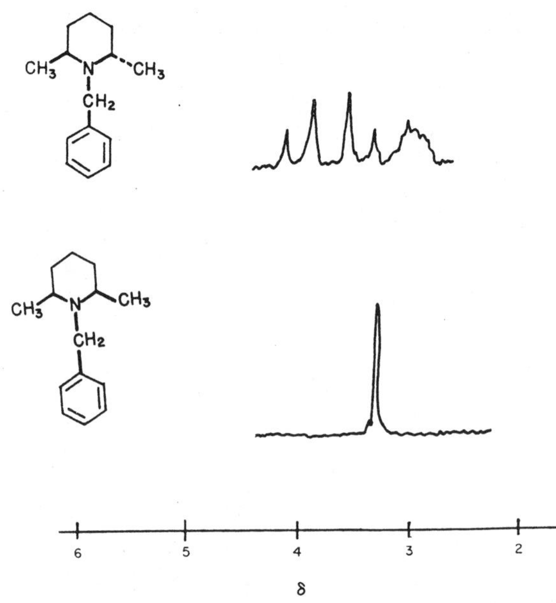

Fig. 2.7. Equivalent benzyl CH_2 protons in 1-benzyl-*cis*-2,6-dimethylpiperidine compared with nonequivalent protons in *trans* isomer. [Reproduced from *Tetrahedron* **21**, 2015 (1965).]

of the *cis* isomer are enantiotopic and appear as a sharp singlet. The methylene protons of the *trans*-isomer are diastereotopic and appear as a four-line *AB* pattern.

cis-isomer-methylene protons of benzyl group enantiotopic   *trans*-isomer-methylene protons of benzyl group diastereotopic

In general, any pair of hydrogens in a methylene group in a chiral molecule are diastereotopic. Whether the topological difference is detectable in the NMR spectrum is determined by the proximity to the chiral center and the particular shielding effects of the molecule.

Similarly, the two faces at a trigonal carbon in a molecule containing a chiral center are diastereotopic. Both chiral and achiral reactants can distinguish between these diastereotopic faces. Many examples of diastereoselective transformations of such compounds are known. One of the cases which has been examined closely is addition reactions at a trigonal center adjacent to a chiral carbon. Particular attention has been given to the case of nucleophilic addition to carbonyl centers.

1) Nu^- 2) H^+ *or*

Such reactions are usually *diastereoselective*; that is, one of the diastereomeric products is formed in excess. For the case of the nucleophile being hydride from a metal hydride reductant or alkyl or aryl groups from organometallic reagents, the preferred stereochemistry can frequently be predicted on the basis of *Cram's rule.* This empirically based prediction is done by considering a conformation in which the sterically most demanding of the three substituents at the adjacent chiral center is *anti* to the carbonyl group. The preferred stereoisomeric product is that in which the nucleophile is added from the face of the carbonyl group occupied by the smaller of the remaining substituents.[43]

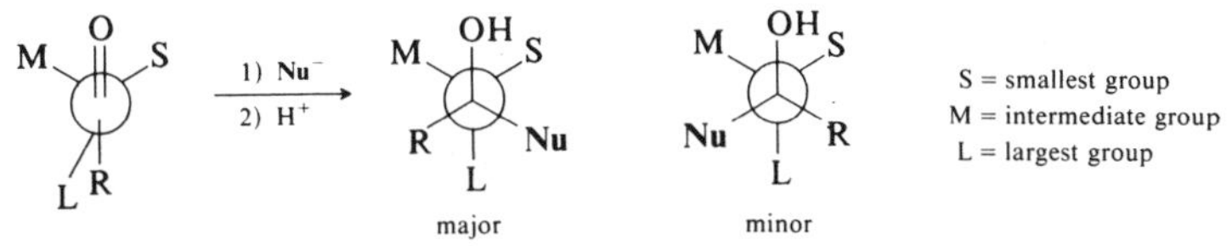

S = smallest group
M = intermediate group
L = largest group

major minor

We will discuss the structural and mechanistic basis of Cram's rule in Chapter 3. As would probably be expected, the influence of a chiral center on the diastereoselectivity of reaction is diminished when the chiral center is more remote from the reaction site.

A method for stereoselective synthesis of α-amino acids can be used to illustrate a diastereoselective chemical reaction.[44] The key is the stereoselective reduction of a carbon–nitrogen double bond in which the hydrogen atom addition is highly preferred from one diastereotopic face. The sequence is shown below for the synthesis of D-alanine. The optical purity observed is 96% in this instance, with optical yields of 92–97% reported for other amino acids. The chirality which is present in the reactant molecule directs the course of the addition of hydrogen in the step in which the new chiral center is created. This occurs as the result of a steric effect. It is easier for the reactant to approach the surface of the hydrogenation catalyst from the less bulky side of the molecule. After the reduction, the newly created chiral product is obtained by a reaction which also releases the original chiral reactant in a form in which it can be reconverted to the chiral hydrazine.

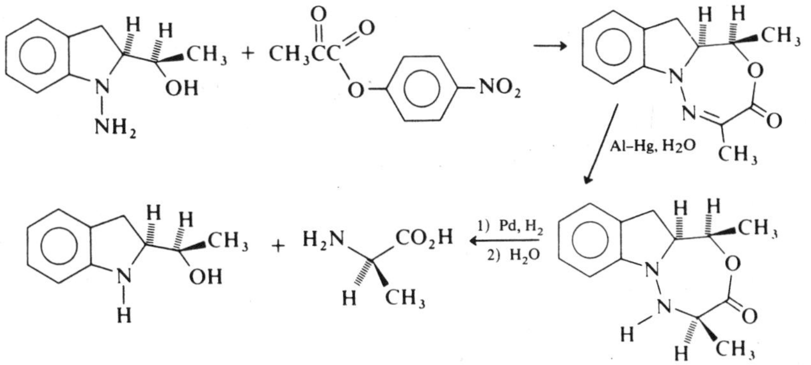

43. D. J. Cram and F. A. Abd Elhafez, *J. Am. Chem. Soc.* **74**, 5828 (1952); D. J. Cram and K. R. Kopecky, *J. Am. Chem. Soc.* **81**, 2748 (1959); E. L. Eliel, in *Asymmetric Synthesis*, Vol. 2, J.D. Morrison (ed.), Academic Press, New York, 1983, Chapter 5.
44. E. J. Corey, H. S. Sachdev, J. Z. Gougoutas, and W. Saenger, *J. Am. Chem. Soc.* **92**, 2488 (1970).

General References

General

E. L. Eliel, *Stereochemistry of Carbon Compounds*, McGraw-Hill, New York, 1962.
J. Jacques, A. Collet, and S. H. Wilen, *Enantiomers, Racemates and Resolutions*, Wiley, New York, 1981.
H. B. Kagan, *Stereochemistry, Fundamentals and Methods*, G. Thieme, Stuttgart, 1977.
K. Mislow, *Introduction to Stereochemistry*, W. A. Benjamin, New York, 1966.
G. Natta and M. Farina, *Stereochemistry*, Harper and Row, New York, 1972.

Stereoselective and Stereospecific Reactions

Y. Izumi and A. Tai, *Stereodifferentiating Reactions*, Academic Press, New York, 1977.
J. D. Morrison (ed.), *Asymmetric Synthesis*, Vols. 1-4, Academic Press, New York, 1983-1984.
J. D. Morrison and H. S. Mosher, *Asymmetric Organic Reactions*, 2nd Printing, American Chemical Society, Washington, DC., 1976.

Stereochemistry in Biological Processes

W. L. Alworth, *Stereochemistry and Its Application in Biochemistry*, Wiley-Interscience, New York, 1972.
R. Bentley, *Molecular Asymmetry in Biology*, Vols. I and II, Academic Press, New York, 1969, 1970.

Problems

(*References for these problems will be found on page* 774.)

1. Indicate whether the relationship in each of the following pairs of compounds is identical, enantiomeric, or diastereomeric:

(a) CHO; H—NH$_2$; HO—H; CH$_2$OH and CH$_2$OH; H—OH; H—NH$_2$; CHO

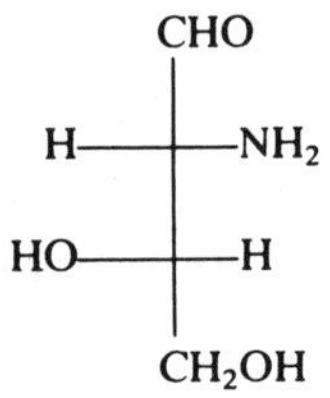

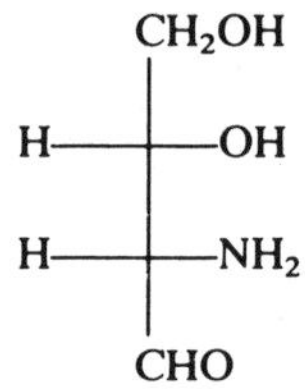

(b) H CH$_3$ and H$_3$C H

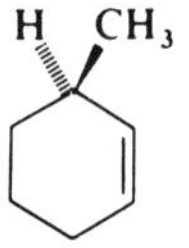

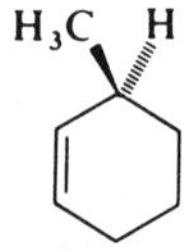

(c) and O

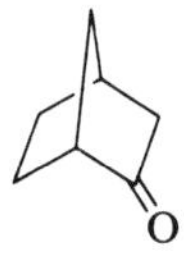

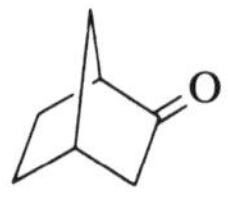

(d) 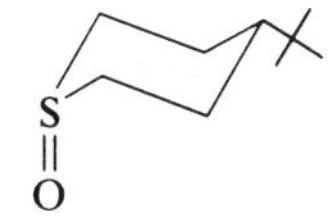 and

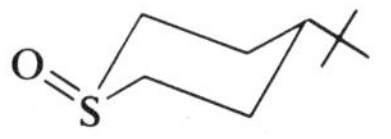

(e) 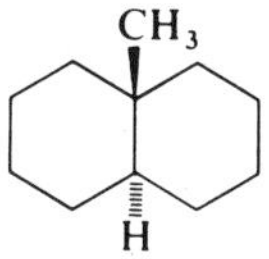 CH_3 H and CH_3 H

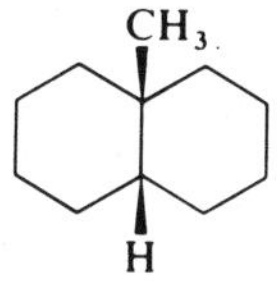

(f) 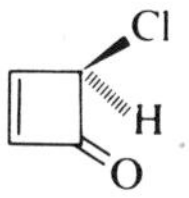 Cl H O and H Cl O

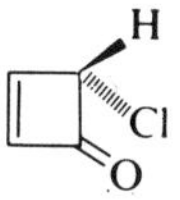

(g) 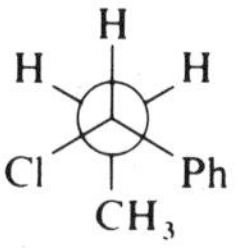 H H H Cl Ph CH_3 and Cl H CH_3 H Ph H

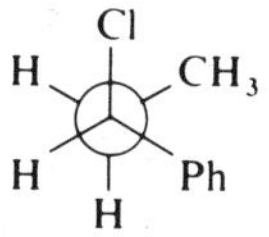

2. The structure originally proposed for cordycepic acid, $[\alpha]_D = +40.3°$, has been shown to be incorrect. Suggest a reason to be skeptical about the original structure, which is given below:

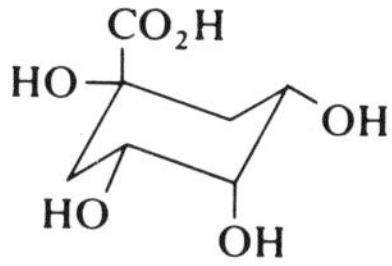

CO_2H HO OH HO OH

3. Each reaction in the sequence shown is reported to proceed with retention of configuration; yet the starting material has the *R* configuration, and the product has the *S* configuration. Reconcile this apparent contradiction.

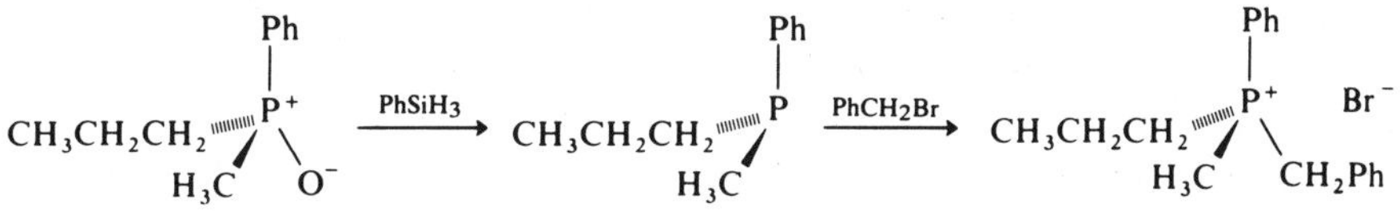

$CH_3CH_2CH_2$ Ph P^+ H_3C O^- $\xrightarrow{PhSiH_3}$ $CH_3CH_2CH_2$ Ph P H_3C $\xrightarrow{PhCH_2Br}$ $CH_3CH_2CH_2$ Ph P^+ Br^- H_3C CH_2Ph

4. Using the sequence rule, specify the configuration at each chiral center in the following molecules:

(a) 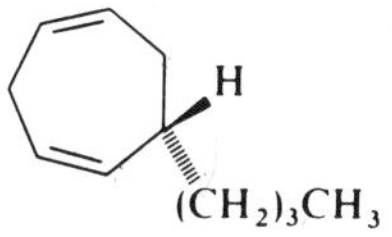 H $(CH_2)_3CH_3$

(b) 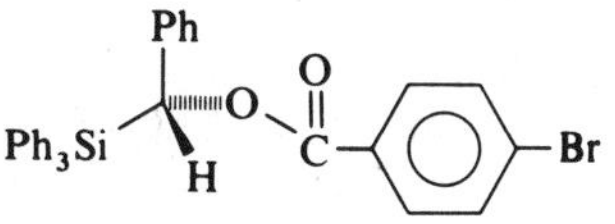 Ph Ph$_3$Si H O O C Br

(c) HO CH_3 H CH_3 O O

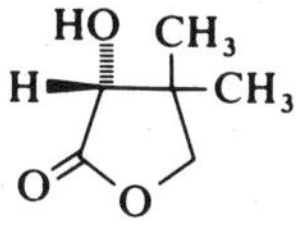

(d) 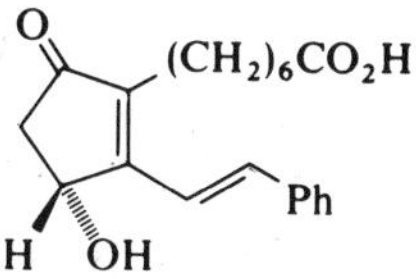 O $(CH_2)_6CO_2H$ H OH Ph

(e) 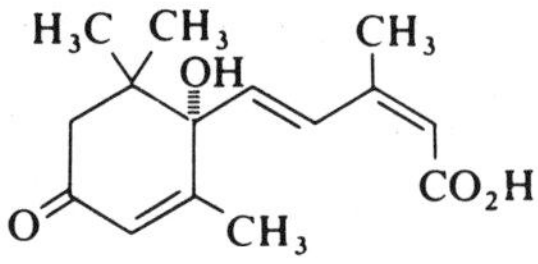 (f) 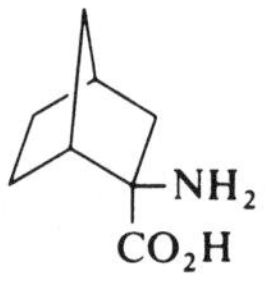 (g)

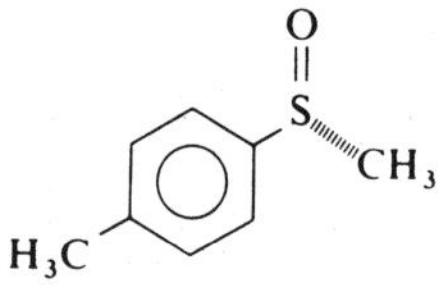

5. Draw structural formulas for each of the following compounds, clearly showing stereochemistry:
   (a) (E)-3,7-dimethyl-2,6-octadien-1-ol (geraniol)
   (b) (R)-4-methyl-4-phenyl-2-cyclohexenone
   (c) L-*erythro*-2-(methylamino)-1-phenylpropan-1-ol [(−)-ephedrine]
   (d) (7R,8S)-7,8-epoxy-2-methyloctadecane (the sex attractant of the female gypsy moth)
   (e) methyl (1S)-cyano-(2R)-phenylcyclopropanecarboxylate
   (f) (Z)-2-methyl-2-butenol
   (g) (E)-(3-methyl-2-pentenylidene)triphenylphosphorane

6. The racemization of medium-ring *trans*-cycloalkenes depends upon ring size and substitution, as indicated by the data below. Discuss these relative reactivities in terms of the structures of the cycloalkenes and the mechanism of racemization.

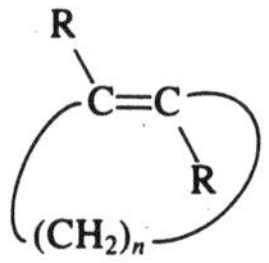

Ring size	n	R	$t_{1/2}$ for racemization
8	6	H	10^5 years at 25°
9	7	H	10 sec at 25°
10	8	CH_3	3 days at 100°
12	10	CH_3	1 day at 25°

7. Compound **A** can be prepared in enantiomerically pure form. It is racemized by heating to 120°C with an E_a of about 30 kcal/mol. Suggest a mechanism for the racemization process.

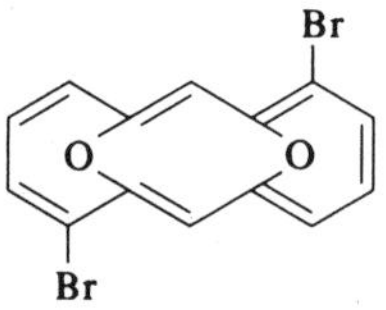

A

8. Reaction of 1,3,5-*tris*(thiomethyl)benzene with potassium hydroxide and then with *tris*(2-bromoethyl)methane gives a product of formula $C_{16}H_{22}S_3$ having the following NMR signals: δ −1.7, septet, 1H; 1.0, multiplet, 6H; 2.1, multiplet, 6H; 3.0, singlet, 6H; 7.1, singlet, 3H. Propose a structure which is consistent

with the observed properties of this material and explain the basis of your proposal.

9. The substance chaetochromin A, structure **A**, has been shown by X-ray diffraction to have the absolute configuration indicated in the structure. The CD spectra of **A** and the related compounds cephalochromin (**B**) and ustilaginoidin (**C**) are shown in the figure. Deduce the absolute stereochemistry of cephalochromin and ustilaginoidin from these data, and draw perspective structures indicating the absolute configuration.

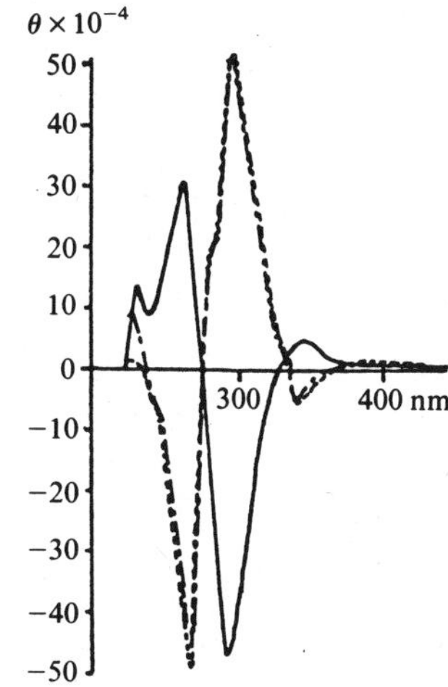

CD Spectra of chaetochromin A, cephalochromin, and ustilaginoidin A (in dioxane). –·–·–, Chaetochromin A (**A**); - - -, cephalochromin (**B**); ——, ustilaginoidin (**C**).

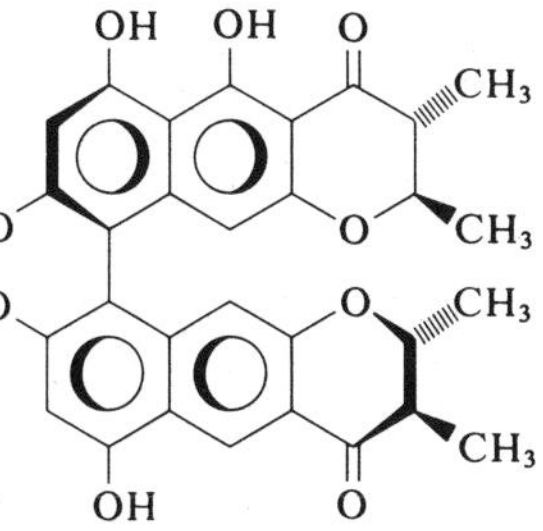

A
chaetochromin A

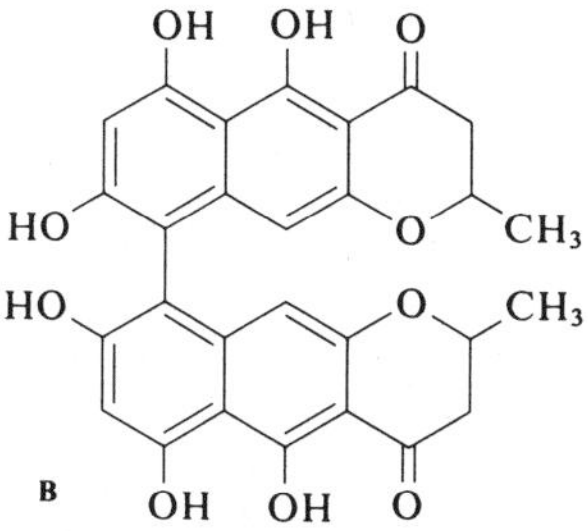

B
cephalochromin

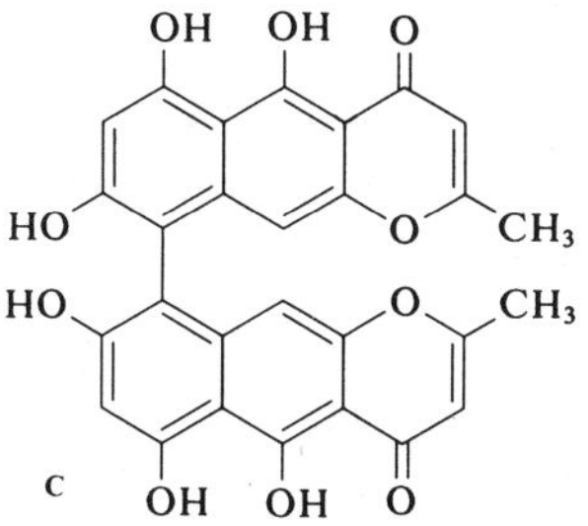

C
ustilaginoidin

10. When partially resolved samples of 5-hydroxymethylpyrrolidin-2-one are allowed to react with benzaldehyde in the presence of an acid catalyst, two products, **A** ($C_{12}H_{13}NO_2$) and **B** ($C_{24}H_{26}N_2O_4$), are formed. The ratio **A**:**B** depends on the enantiomeric purity of the starting material. When the starting material is optically pure, only **A** is formed. When it is racemic, only **B** is formed. Partially resolved material gives both **A** and **B**. The more nearly it is enantiomerically pure, the less **B** is formed. The product **A** is optically active but **B** is achiral. Develop an explanation for these observations, including structures for **A** and **B**.

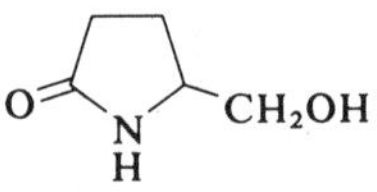

O N H CH_2OH

11. Give the product(s) described for each reaction. Specify all aspects of stereochemistry.

    (a) stereospecific *anti* addition of bromine to *cis*- and *trans*-cinnamic acid
    (b) solvolysis of (*S*)-3-bromooctane in methanol with 6% racemization
    (c) stereospecific *syn* elimination of acetic acid from (*R*,*S*)-1,2-diphenylpropyl acetate
    (d) stereoselective epoxidation of bicyclo[2.2.1]hept-2-ene proceeding 94% from the *exo* direction

12. Compound **A** can be resolved to give an optically pure substance, $[\alpha]_D^{25} = -124°$. Oxidation gives the pure ketone **B**, which is optically active, $[\alpha]_D^{25} \simeq -439°$. Heating the alcohol **A** gives partial conversion (an equilibrium is established) to an isomer with $[\alpha]_D^{25} = +22°$. Oxidation of this isomer gives the enantiomer of the ketone **B**. Heating either enantiomer of the ketone leads to the racemic mixture. Explain the stereochemical relationships between these compounds.

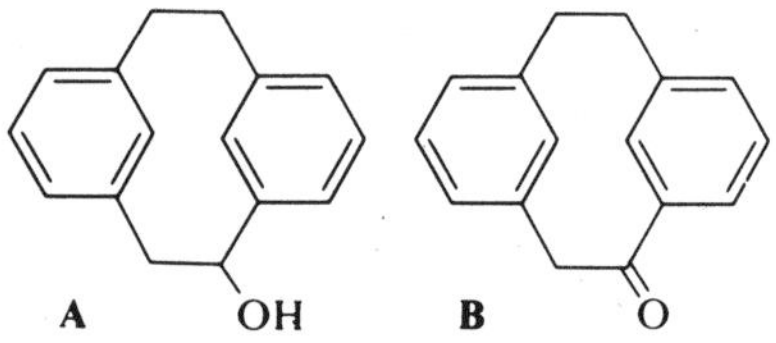

A OH **B** O

13. Some of the compounds shown contain diastereotopic atoms or groups. Which possess this characteristic? For those that do, indicate the atoms or groups that are diastereotopic and indicate which atom or group is *pro-R* and which is *pro-S*.

(a) 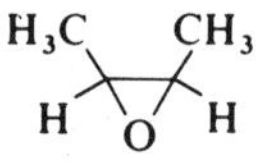 H_3C CH_3 H O H

(b) 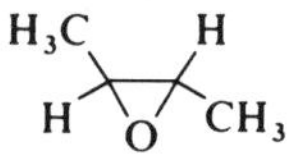 H_3C H H O CH_3

(c) $C_6H_5CHCO_2CH(CH_3)_2$ with OCH_3 on the CH

(d) $(CH_3)_2CHCHCO_2H$ with NH_2 on the second CH (i.e., $(CH_3)_2CHCH(NH_2)CO_2H$)

(e) $C_6H_5CH_2CH(NH_2)C(=O)NHCH_2CO_2H$

(f) $BrCH_2CH(OC_2H_5)_2$

14. Indicate which of the following molecules are chiral and which are achiral. For each molecule that is achiral, indicate the element of symmetry that is present in the molecule.

(a)

(b) CH$_3$ CH$_3$ O O CH$_3$ CH$_3$

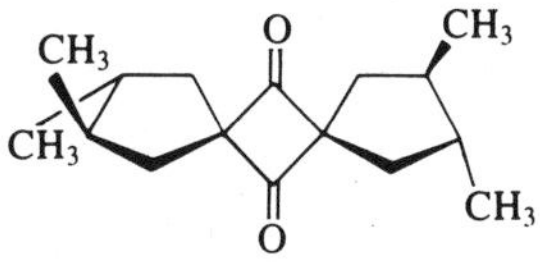

(c)

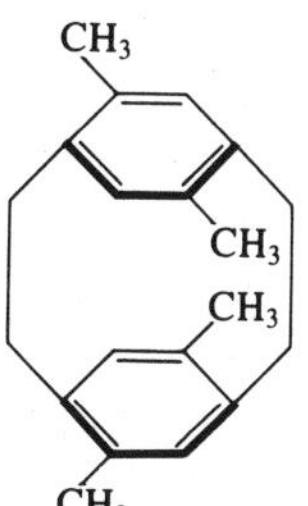

(d) H H H H

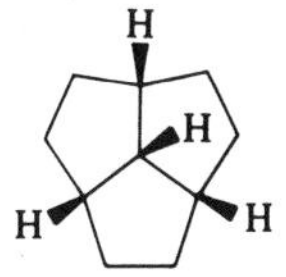

(e)

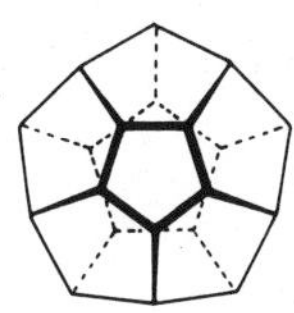

$C_{20}H_{20}$ (dodecahedrane)

(f) O O

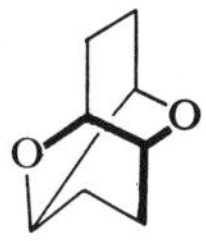

(g)

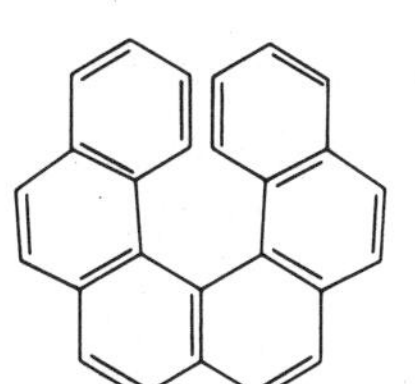

(h)

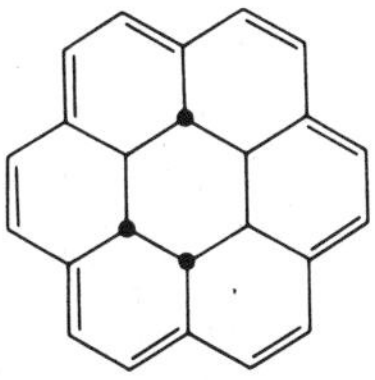

(i) CH$_3$O $\overset{+}{N}$ OCH$_3$

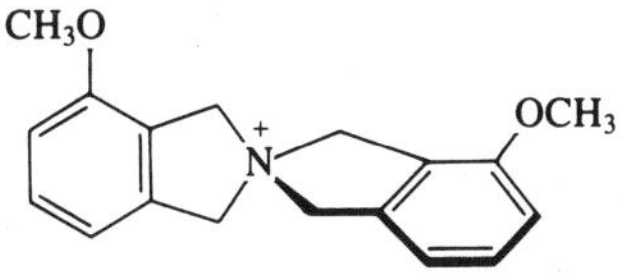

(j) O O

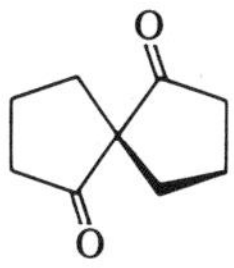

(k) OH HO

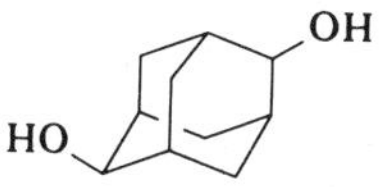

(l) O O

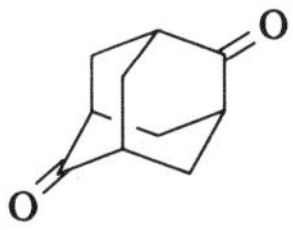

(m) O O

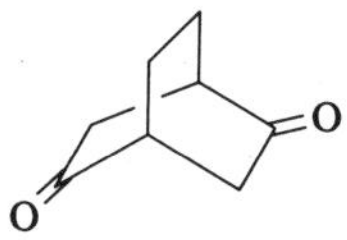

15. Assign configurations, using the sequence rule, to each chiral center of the stereoisomeric isocitric acids and allocitric acids:

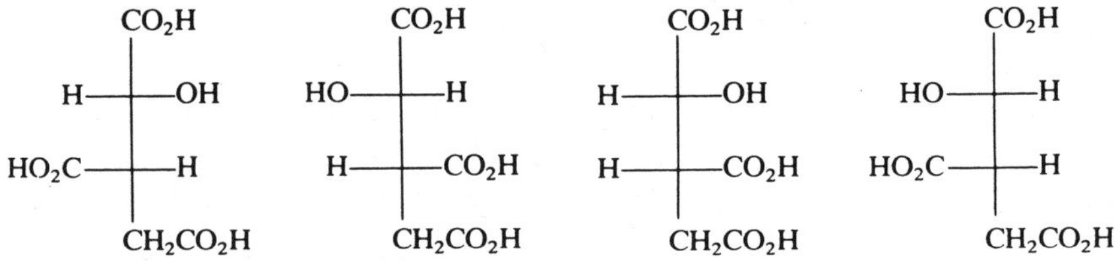

$$\begin{array}{c} CO_2H \\ H-\!\!\!+\!\!\!-OH \\ HO_2C-\!\!\!+\!\!\!-H \\ CH_2CO_2H \end{array} \quad \begin{array}{c} CO_2H \\ HO-\!\!\!+\!\!\!-H \\ H-\!\!\!+\!\!\!-CO_2H \\ CH_2CO_2H \end{array} \quad \begin{array}{c} CO_2H \\ H-\!\!\!+\!\!\!-OH \\ H-\!\!\!+\!\!\!-CO_2H \\ CH_2CO_2H \end{array} \quad \begin{array}{c} CO_2H \\ HO-\!\!\!+\!\!\!-H \\ HO_2C-\!\!\!+\!\!\!-H \\ CH_2CO_2H \end{array}$$

isocitric acids — alloisocitric acids

16. The enzyme enolase catalyzes the following reaction:

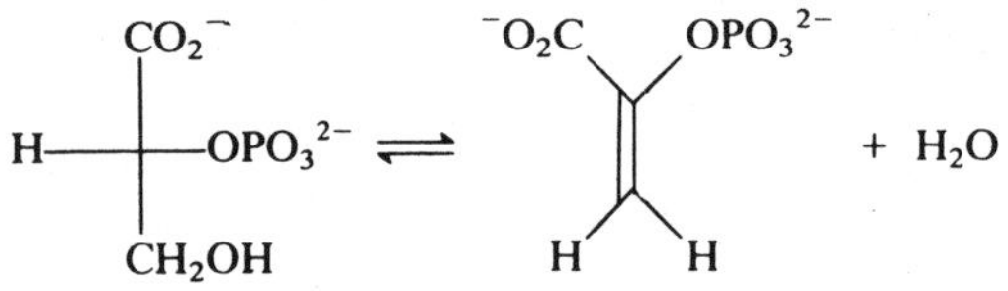

$$\begin{array}{c} CO_2^- \\ H-\!\!\!+\!\!\!-OPO_3^{2-} \\ CH_2OH \end{array} \rightleftharpoons \begin{array}{c} ^-O_2C \quad OPO_3^{2-} \\ C{=}C \\ H \quad H \end{array} + H_2O$$

When (2*R*,3*R*)-2-phosphoglycerate-3-*d* was used as the substrate, the *E*-isomer of phosphoenolpyruvate-3-*d* was produced. Is the stereochemistry of elimination *syn* or *anti*?

17. An important sequence in valine biosynthesis in bacteria is

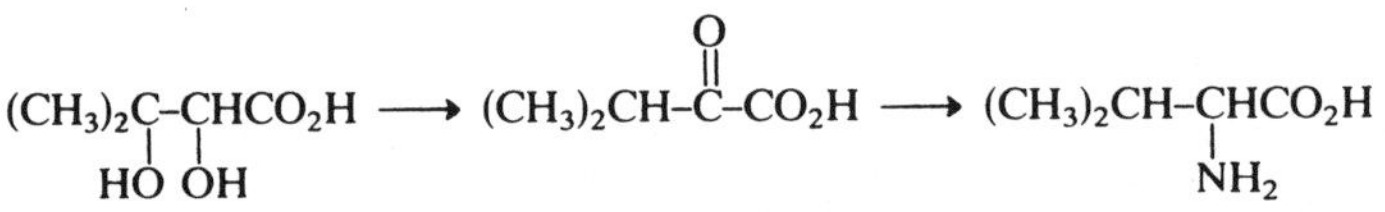

$$(CH_3)_2\underset{HO}{C}-\underset{OH}{C}HCO_2H \longrightarrow (CH_3)_2CH-\overset{O}{\overset{\|}{C}}-CO_2H \longrightarrow (CH_3)_2CH-\underset{NH_2}{C}HCO_2H$$

valine

The stereochemical aspects of this sequence have been examined, using a diol substrate in which one of the methyl groups has been replaced by CD_3. Given the information that labeled starting diol of configuration 2*R*,3*R* produces labeled valine of configuration 2*S*,3*S* deduce whether the C-3 hydroxyl group is replaced with overall retention or inversion of configuration.

18. 1,2-Diphenyl-1-propanol may be prepared in either of two ways:
(a) lithium aluminum hydride reduction of 1,2-diphenyl-1-propanone
(b) reaction of 2-phenylpropanal with phenylmagnesium bromide.
Which method would you choose to prepare the *threo* isomer? Explain.

19. A mixture of tritium (^3H = T) labeled **A** and **B** was carried through the reaction sequence shown:

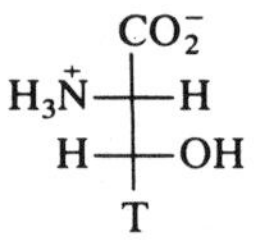

$$\begin{array}{c} CO_2^- \\ H_3\overset{+}{N}-\!\!\!+\!\!\!-H \\ H-\!\!\!+\!\!\!-OH \\ T \end{array} \ \textbf{A} \qquad \begin{array}{c} CO_2^- \\ H-\!\!\!+\!\!\!-\overset{+}{N}H_3 \\ HO-\!\!\!+\!\!\!-H \\ T \end{array} \ \textbf{B}$$

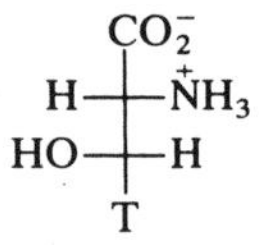

$$HOCH_2\underset{^+NH_3}{C}HCO_2^- \xrightarrow[\text{oxidase}]{\text{D-amino acid}} HOCH_2\overset{O}{\overset{\|}{C}}CO_2H \xrightarrow{H_2O_2} HOCH_2CO_2H \xrightarrow[\text{oxidase}]{\text{glycolate}} H\overset{O}{\overset{\|}{C}}CO_2H$$

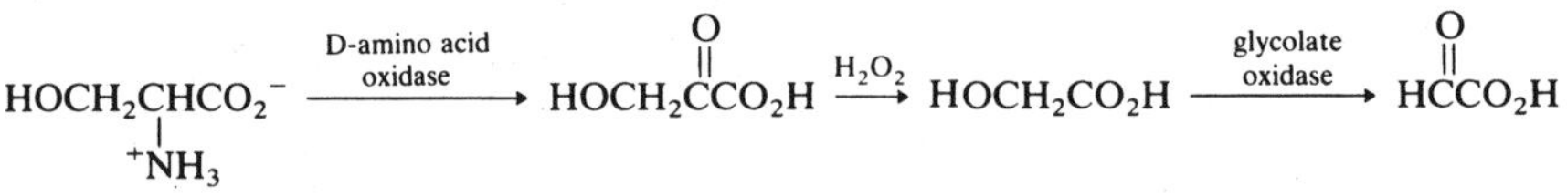

D-Amino acid oxidase will oxidase only serine having the *R* configuration at C-2. Glycolate oxidase will remove only the *pro-R* hydrogen of glycolic acid. Does the product ($O{=}CHCO_2H$) contain tritium? Explain your reasoning.

20. Enzymatic oxidation of naphthalene by bacteria proceeds by way of the intermediate *cis*-diol shown. Which prochiral faces of C-1 and C-2 of naphthalene are hydroxylated in this process?

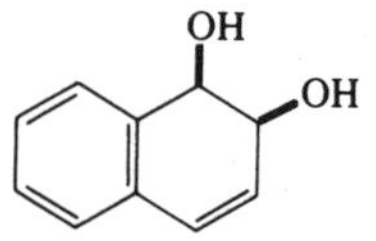

OH
OH

21. An amino acid having the constitution shown has been isolated from horse chestnuts. It is configurationally related to L-proline and has the *R* configuration at C-3. Write a stereochemically correct representation for this compound.

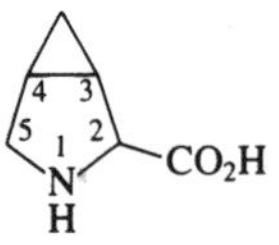

4 3
5 1 2
N CO_2H
H

(horse-chestnut amino acid)

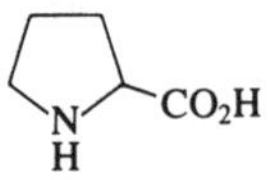

N CO_2H
H

(proline—no stereochemistry implied)

22. (a) One of the diastereomers of 2,6-dimethylcyclohexyl benzyl ether exhibits two doublets for the benzylic protons in its NMR spectrum. Deduce the stereochemistry of this isomer.
    (b) The NMR spectrum of the highly hindered molecule trimesitylmethane indicates that there are two enantiomeric species present in solution, the interconversion of which is separated by a barrier of 22 kcal/mol. Discuss the source of the observed chirality of this molecule.

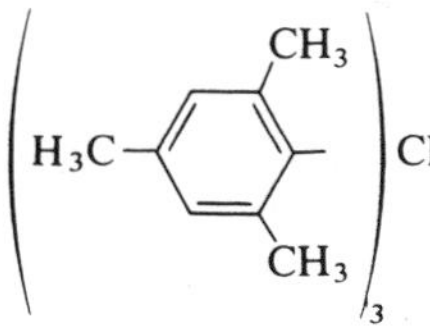

CH_3
H_3C— —CH
CH_3
3

23. A synthesis of the important biosynthetic intermediate mevalonic acid starts with the enzymatic hydrolysis of the diester **A** by pig liver esterase. The *pro-R* group is selectively hydrolyzed. Draw a three-dimensional structure of the product.

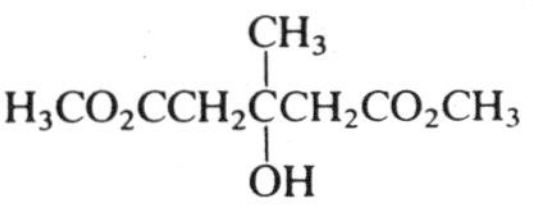

CH_3
$H_3CO_2CCH_2CCH_2CO_2CH_3$
OH

A

24. The structure of a natural product is shown without any specification of stereochemistry. It is a pure substance which gives no indication of being a mixture of stereoisomers and has zero rotation. It is not a racemic mixture because it does not yield separate peaks on a chiral HPLC column. When the material is completely hydrolyzed, it gives a racemic sample of the product shown. Deduce the complete stereochemical structure of the natural product from this information.

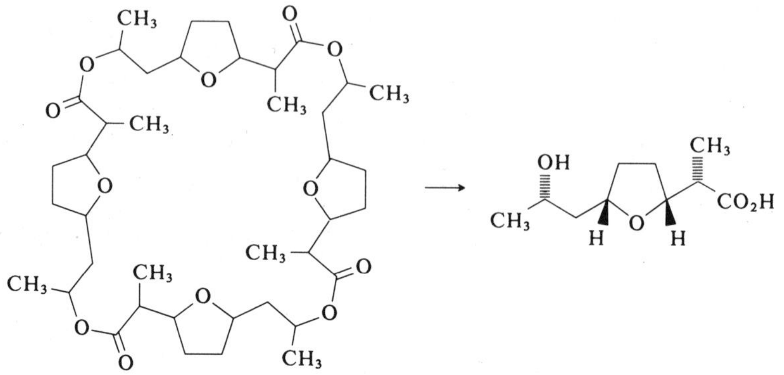

CH_3 O O CH_3 CH_3 O CH_3 O O O CH_3 CH_3 O CH_3 O O O CH_3 → OH CH_3 H O H CH_3 CO_2H

3

Conformational, Steric, and Stereoelectronic Effects

The total energy of a molecule is directly related to its geometry. Several aspects of molecular geometry can be recognized, and, to some extent, components of the total energy can be dissected and attributed to specific structural features. Among the factors that contribute to total energy and have a recognizable connection with molecular geometry are nonbonded repulsions, ring strain in cyclic systems, torsional strain arising from nonoptimum rotational alignment, and destabilization resulting from distortion of bond lengths or bond angles from optimal values. Conversely, there are stabilizing interactions which have geometric constraints. Most of these can be classed as stereoelectronic effects; that is, *a particular geometric relationship is required to maximize the stabilizing interaction.* In addition, there are other molecular interactions, such as hydrogen bond formation and dipole–dipole interactions, for which the strength of the interaction will be strongly dependent on the geometry of the molecule. A molecule will adopt the minimum-energy geometry that is available by rotations about single bonds and adjustment of bond angles and bond lengths. Since bond angles and bond lengths vary relatively little from molecule to molecule, molecular shape is primarily determined by the rotational processes. The various shapes that a given molecule can attain are called *conformations.* The principles on which analysis of conformational equilibria is based have largely been developed using a classical-mechanical framework. The interpretation of molecular geometry can also be approached from the molecular orbital viewpoint.

Many molecules exhibit *strain* caused by nonideal geometry. Any molecule will adopt a minimum-energy conformation at equilibrium, but the structural adjustments may not compensate entirely for nonideal bonding arrangements. *Strain energy* is the excess energy relative to an unstrained reference molecule. This chapter will focus on the following interrelated topics: the sources of strain; the structural evidence of strain; and the energetic and reactivity consequences of strain.[1]

From a molecular orbital viewpoint, the energy of a molecule is the sum of the energies of the occupied molecular orbitals. Calculation of total energy in different spatial arrangements reveals the energy as a function of geometry. The physical interpretation is given in terms of the effectiveness of orbital overlap. Maximum overlap between orbitals that have a bonding interaction lowers the molecular energy, whereas overlap of antibonding orbitals raises the energy. The term *stereoelectronic effect* is used to encompass relationships between structure, conformation, energy, and reactivity that can be traced to geometry-dependent orbital interactions.

3.1. Steric Strain and Molecular Mechanics

A system of analyzing the energy differences among molecules and among various geometries of a particular molecule has been developed and is based on some fundamental concepts formalized by Westheimer.[2] The method is now known by the term *molecular mechanics*, although the expressions *empirical force field calculations* or the *Westheimer method* are sometimes applied.[3]

A molecule will adopt that geometry which minimizes its total energy. The minimum-energy geometry will be strained to a degree that depends on the extent to which its structural parameters deviate from their ideal values. The energy for a particular kind of distortion is a function of the amount of distortion and the opposing force. The total strain energy, originally called *steric energy*, is the sum of several contributions:

$$E_{\text{steric}} = E(r) + E(\theta) + E(\phi) + E(d)$$

where $E(r)$ is the energy component associated with stretching or compression of single bonds, $E(\theta)$ is the energy of bond-angle distortion, $E(\phi)$ is the torsional strain, and $E(d)$ are the energy increments that result from nonbonded interactions between atoms.

The mathematical expressions for the force fields are derived from classical-mechanical potential energy functions. The energy required to stretch a bond or to bend a bond angle increases as the square of the distortion.

Bond stretching: $\quad E(r) = 0.5k_r(r - r_0)^2$

1. For a review of the concept of strain, see K. B. Wiberg, *Angew. Chem. Int. Ed. Engl.* **25**, 312 (1986).
2. F. H. Westheimer, in *Steric Effects in Organic Chemistry*, M. S. Newman (ed.), Wiley, New York, 1956, Chapter 12.
3. For reviews, see J. E. Williams, P. J. Stang, and P. v. R. Schleyer, *Annu. Rev. Phys. Chem.* **19**, 531 (1968); D. B. Boyd and K. P. Lipkowitz, *J. Chem. Educ.* **59**, 269 (1982); P. J. Cox, *J. Chem. Educ.* **59**, 275 (1982); N. L. Allinger, *Adv. Phys. Org. Chem.* **13**, 1 (1976); E. Osawa and H. Musso, *Top. Stereochem.* **13**, 117 (1982); U. Burkett and N. L. Allinger, *Molecular Mechanics*, ACS Monograph No. 177, American Chemical Society, Washington, D.C., 1982.

where k_r is the stretching force constant, r the bond length, and r_0 the normal bond length.

Bond-angle bending: $E(\theta) = 0.5k_\theta(\Delta\theta)^2$

where k_θ is the bending force constant and $\Delta\theta$ is the deviation of the bond angle from its normal value.

The torsional strain is a sinusoidal function of the torsion angle. Torsional strain results from the barrier to rotation about single bonds as described for ethane on page 57. For molecules with a threefold barrier, such as ethane, the form of the torsional barrier is

$$E(\phi) = 0.5\ V_0(1 + \cos 3\phi)$$

where V_0 is the rotational energy barrier and ϕ is the torsional angle. For hydrocarbons, V_0 can be taken as being equal to the ethane barrier (2.8–2.9 kcal/mol). The potential energy diagram for rotation about the C—C bond of ethane is given in Fig. 3.1. The stereoelectronic origin of the ethane barrier was discussed in Chapter 1.

Nonbonded interaction energies are the most difficult contributions to evaluate and may be attractive or repulsive. When two uncharged atoms approach each other,

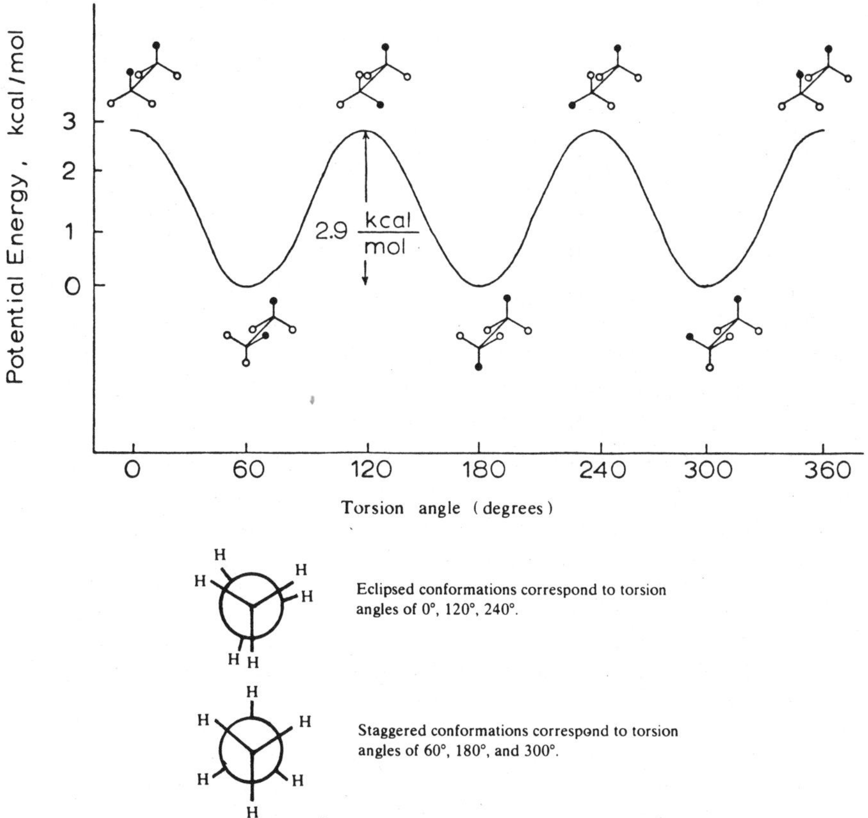

Fig. 3.1. Potential energy as a function of torsion angle for ethane.

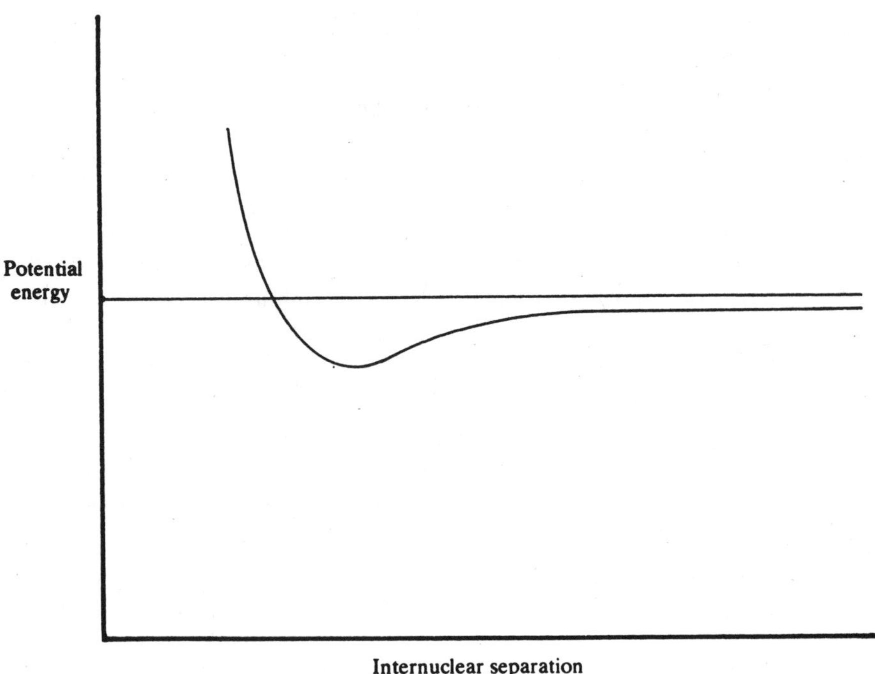

Fig. 3.2. Energy as a function of internuclear distance for nonbonded atoms.

the interaction between them is very small at large distances, becomes increasingly attractive as the separation approaches the sum of their van der Waals radii, but then becomes strongly repulsive as the separation becomes less than the sum of their van der Waal radii. This behavior is represented graphically by a Morse potential diagram such as in Fig. 3.2. The attractive interaction results from a polarization of the electrons of each atom by the other. Such attractive forces are called *London forces* or *dispersion forces* and are relatively weak interactions. London forces vary inversely with the sixth power of internuclear distance and become negligible as internuclear separation increases. At distances smaller than the sum of the van der Waals radii, the much stronger repulsive forces are dominant. Table 3.1 lists van der Waals radii for atoms and groups most commonly encountered in organic molecules.

The interplay between torsional strain and nonbonded interactions can be illustrated by examining the conformations of *n*-butane. The relationship between

Table 3.1. Van der Waals Radii of Several Atoms and Groups (Å)[a]

H	1.20			CH_3	2.0		
N	1.55	P	1.80				
O	1.52	S	1.80				
F	1.47	Cl	1.75	Br	1.85	I	1.98

a. From A. Bondi, *J. Phys. Chem.* **68**, 441 (1964).

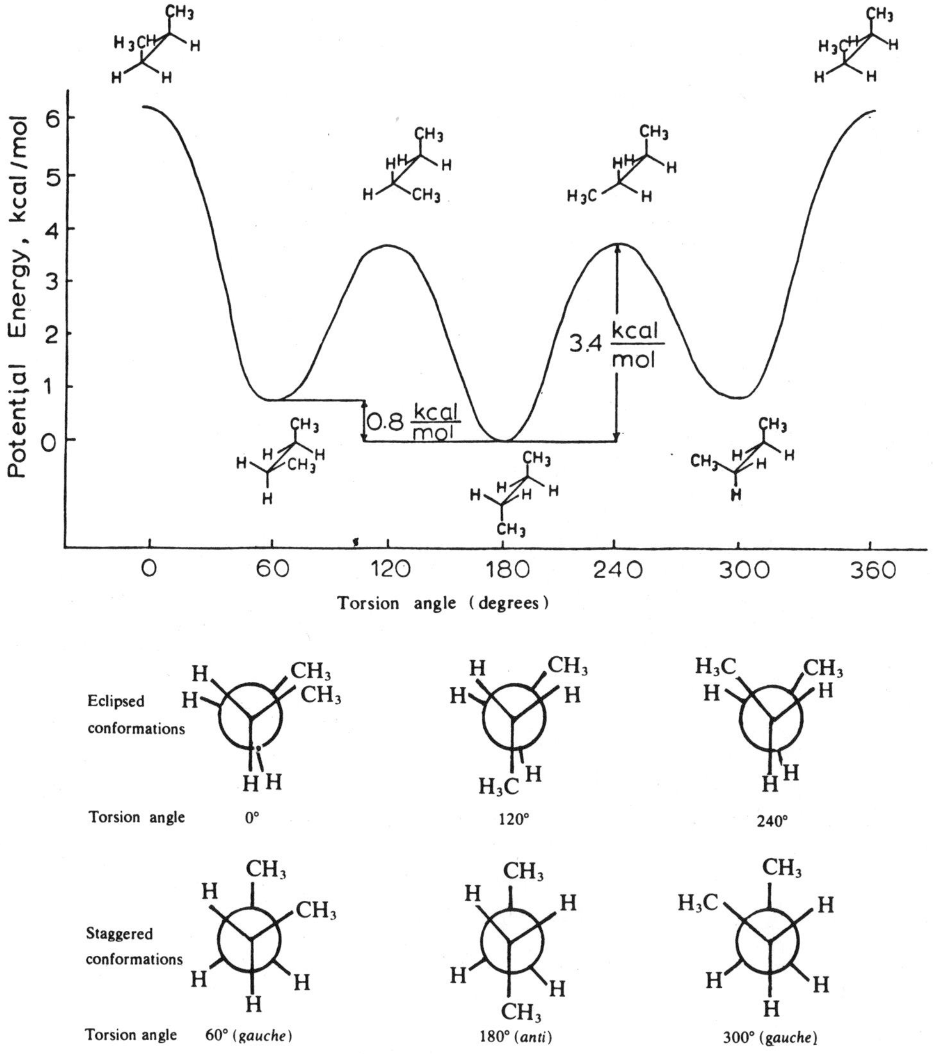

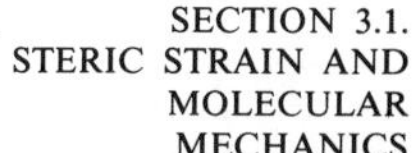

Fig. 3.3. Potential energy diagram for rotation about C-2—C-3 bond of *n*-butane.

potential energy and the torsion angle for rotation about the C-2—C-3 bond is presented in Fig. 3.3.

The potential energy diagram of *n*-butane resembles that of ethane in having three maxima and three minima but differs in that one of the minima is of lower energy than the other two, and one of the maxima is of higher energy than the other two. The minima correspond to staggered conformations. Of these, the *anti* conformation is lower in energy than the two *gauche* conformations. The energy difference between the *anti* and *gauche* conformations in *n*-butane is about 0.8 kcal/mol.[4] The maxima correspond to eclipsed conformations, with the highest-energy conformation

4. G. J. Szasz, N. Sheppard, and D. H. Rank, *J. Chem. Phys.* **16**, 704 (1948); P. B. Woller and E. W. Garbisch, Jr., *J. Am. Chem. Soc.* **94**, 5310 (1972).

being the one with the two methyl groups eclipsed with each other. The methyl-methyl eclipsed conformation is about 2.6 kcal/mol higher in energy than the methyl-hydrogen eclipsed conformation and 6 kcal/mol higher in energy than the staggered *anti* conformation.

The rotational profile of *n*-butane can be understood as a superimposition of van der Waals forces on the ethane rotational energy profile. The two *gauche* conformations are raised in energy relative to the *anti* conformation by an energy increment resulting from the van der Waals repulsion between the two methyl groups of 0.8 kcal/mol. The eclipsed conformations all incorporate 2.9 kcal/mol of torsional strain relative to the staggered conformations, just as is true in ethane. The methyl-methyl eclipsed conformation is further strained by the van der Waals repulsion between the methyl groups. The van der Waals repulsion between methyl and hydrogen in the other eclipsed conformations is smaller. If we subtract the torsional-strain contribution of 2.9 kcal/mol, we conclude that the methyl-methyl eclipsing interaction destabilizes the 0° conformation by an additional 3.2 kcal/mol of van der Waals strain. The 120° and 240° eclipsed conformations are strained by 0.6 kcal/mol over and above the torsional strain, or by 0.3 kcal/mol for each of the two methyl-hydrogen repulsions.

The populations of the various conformations are related to the energy between them by the equation

$$\Delta G^\circ = -RT \ln K$$

For the case of *n*-butane, the equilibrium

$$gauche \rightleftharpoons anti$$

has $\Delta H^\circ = -0.8$ kcal/mol. Since there are two enantiomeric *gauche* conformers, the free energy also reflects an entropy contribution:

$$\Delta S^\circ = -R \ln 2$$

and

$$\Delta G^\circ = \Delta H^\circ - T\,\Delta S^\circ$$

$$\Delta G^\circ = -0.8 \text{ kcal/mol} - (-RT \ln 2)$$

At 298 K

$$\Delta G^\circ = -0.8 \text{ kcal/mol} + 0.41 \text{ kcal/mol} = -0.39 \text{ kcal/mol}$$

and

$$K = (anti)/(gauche) = 1.9$$

This corresponds to a distribution of 66% *anti* and 34% *gauche.* Table 3.2 gives the relationship between free-energy difference, equilibrium constant, and percent composition.

Examples of attractive nonbonded interactions can be found in certain halogenated hydrocarbons. In 1-chloropropane, for example, the *gauche* conformation is slightly preferred over the *anti* conformation at equilibrium. This is not solely the

Table 3.2. Composition–Equilibrium–Free-Energy Relationships[a]

More stable isomer (%)	Equilibrium constant (K)	Free energy ΔG°_{25} (kcal/mol)
50	1	0.0
55	1.22	−0.119
60	1.50	−0.240
65	1.86	−0.367
70	2.33	−0.502
75	3.00	−0.651
80	4.00	−0.821
85	5.67	−1.028
90	9.00	−1.302
95	19.00	−1.744
98	49.00	−2.306
99	99.00	−2.722
99.9	999.00	−4.092

a. From E. L. Eliel, *Stereochemistry of Carbon Compounds*, McGraw-Hill, New York, 1962.

result of the statistical (entropy) factor of 2:1 but also reflects a ΔH value of 0.3 ± 0.3 kcal/mol, which is attributed to a London attractive force. The chlorine atom and methyl group are separated by about the sum of their van der Waal radii in the *gauche* conformation.[5]

The separation of the total strain energy into component elements of bond-length strain, bond-angle strain, torsional strain, and nonbonded repulsion in a qualitative way is useful for analysis and rationalization of structural and steric effects on equilibria and reactivity. The quantitative application of the principles of molecular mechanics for calculation of ground state geometries, heats of formation, and strain energies has been developed to a high level. Minimization of the total strain energy of a molecule, expressed by a multiparameter equation, can be accomplished by iterative computation. The calculational methods have been refined to the point that geometries of hydrocarbons of moderate size can be calculated to an accuracy of 0.01 Å in bond length and 1–2° in bond angle.[6] A similar degree of accuracy can be expected from calculations on a wide variety of molecules including both the usual functional groups and certain intermediates such as carbocations.[7] In these types of systems, terms for dipole–dipole interactions and mutual polarization must be included in the parameterized equations.

Several systems of parameters and equations for carrying out the calculations have been developed.[8] The most frequently used in organic chemistry is that

5. W. E. Steinmetz, F. H. Hickernell, I. K. Mun, and L. H. Scharpen, *J. Mol. Spectrosc.* **68**, 173 (1977); N. Y. Morino and K. Kuchitsu, *J. Chem. Phys.* **28**, 175 (1958).
6. N. L. Allinger, M. A. Miller, F. A. VanCatledge, and J. A. Hirsch, *J. Am. Chem. Soc.* **89**, 4345 (1967); N. L. Allinger, *J. Am. Chem. Soc.* **99**, 8127 (1977).
7. For a summary, see N. L. Allinger, *Adv. Phys. Org. Chem.* **13**, 1 (1976).
8. E. M. Engler, J. D. Andose, and P. v. R. Schleyer, *J. Am. Chem. Soc.* **95**, 8005 (1973); D. N. White and M. J. Bovill, *J. Chem. Soc., Perkin Trans. 2,* 1610 (1977); S. R. Niketic and K. Rasmussen, *The Consistent Force Field*, Springer-Verlag, Berlin, 1977.

Table 3.3. Correlation between Intramolecular Strain and Activation Energies for Dissociation of C—C Bonds in Substituted Ethanes

$$R^1-C(R^2)(R^3)-C(R^2)(R^3)-R^1 \rightarrow 2R^1-\dot{C}(R^2)(R^3)$$

R^1	R^2	R^3	$\Delta G^{\ddagger}$ (kcal/mol)	MM strain (kcal/mol)	R^1	R^2	R^3	$\Delta G^{\ddagger}$ (kcal/mol)	MM strain (kcal/mol)
H	H	H	79	0	Ph	H	CH_3	50	3.6
CH_3	H	H	69	0	Ph	H	C_2H_5	49.7	4.8
CH_3	CH_3	H	68	2.7	Ph	H	$CH(CH_3)_2$	47.4	8.4
CH_3	CH_3	CH_3	60	6.9	Ph	H	$C(CH_3)_3$	42.1	22.2
CH_3	CH_3	C_2H_5	55.3	12.3	Ph	CH_3	CH_3	37.9	18.4
CH_3	CH_3	$CH(CH_3)_2$	47.3	22.4	Ph	CH_3	C_2H_5	34.9	23.7
CH_3	CH_3	$C(CH)_3$	33.7	45.3	Ph	CH_3	$CH(CH_3)_2$	26.7	40.9

a. C. Rüchardt and H.-D. Beckhaus, *Angew. Chem. Int. Ed. Engl.* **19**, 429 (1980).

developed by Allinger and co-workers, frequently referred to as MM (molecular mechanics) calculations. These calculations can be run readily on moderate-size computers or on microcomputers.[9] The computations involve iterations to locate an energy minimum. Precautions must be taken to establish that a true ("global") minimum, as opposed to a local energy minimum, has been achieved. This can be accomplished by using a number of different initial geometries and comparing the structures of the minima that are located.

An example of the application of molecular mechanics in the investigation of chemical reactions is a study of the correlation between steric strain in a molecule and the ease of rupture of carbon–carbon bonds. For a series of hexasubstituted ethanes, it was found that there is a good correlation between the strain calculated by the molecular mechanics method and the rate of thermolysis.[10] Some of the data are shown in Table 3.3.

3.2. Conformations of Acyclic Molecules

Simple hydrocarbons represent rather well-behaved extensions of the conformational principles illustrated previously in the analysis of rotational equilibria in ethane and *n*-butane. The staggered conformations correspond to potential energy minima, and the eclipsed conformations to potential energy maxima. Of the staggered conformations, *anti* forms are more stable than *gauche*. The magnitudes of the

9. N. L. Allinger and Y. H. Yuh, MM2, No. 395, Quantum Chemistry Exchange, Chemistry Department, University of Indiana, Bloomington, Indiana; A variety of computer programs for energy minimization are also available from commercial sources.
10. C. Rüchardt and H.-D. Beckhaus, *Angew. Chem. Int. Ed. Engl.* **19**, 429 (1980).

Table 3.4. Rotational Energy Barriers of Compounds of the Type CH_3–X[a]

Compound	Barrier height (kcal/mol)
Alkanes	
1. CH_3–CH_3	2.88
2. CH_3–CH_2CH_3	3.4
3. CH_3–$CH(CH_3)_2$	3.9
4. CH_3–$C(CH_3)_3$	4.7
5. CH_3–SiH_3	1.7
Haloethanes	
6. CH_3–CH_2F	3.3
7. CH_3–CH_2Cl	3.7
8. CH_3–CH_2Br	3.7
9. CH_3–CH_2I	3.2
Heteroatom substitution	
10. CH_3–NH_2	1.98
11. CH_3–$NHCH_3$	3.62
12. CH_3–OH	1.07
13. CH_3–OCH_3	2.7

a. Taken from the compilation of J. P. Lowe, *Prog. Phys. Org. Chem.* **6**, 1 (1968); barriers are those for rotation about the bond indicated in the formula.

barriers to rotation of many small organic molecules have been measured.[11] Some representative examples are listed in Table 3.4. The experimental techniques used to study rotational processes include microwave spectroscopy, electron diffraction, ultrasonic absorption, and infrared spectroscopy.[12]

Substitution of methyl groups for hydrogen atoms on one of the carbon atoms produces a regular increase of about 0.6 kcal/mol in the height of the rotational energy barrier. The barrier in ethane is 2.88 kcal/mol. In propane, the barrier is 3.4 kcal/mol, corresponding to an increase of 0.5 kcal/mol for methyl-hydrogen eclipsing. When two methyl-hydrogen eclipsing interactions occur, as in 2-methylpropane, the barrier is raised to 3.9 kcal/mol. The increase in going to 2,2-dimethylpropane, in which the barrier is 4.7 kcal/mol, is 1.8 kcal/mol for the total of three methyl-hydrogen eclipsing interactions.

H H H H H H 2.88

H H CH3 H H H 3.4

H H3C CH3 H H H 3.9

H H3C CH3 H H CH3 4.7

Rotational barrier increases with the number of CH_3/H eclipsing interaction.

The rotational barrier in methylsilane (Table 3.4, entry 5) is significantly smaller than that in ethane (1.7 versus 2.88 kcal/mol). This probably reflects the decreased electron–electron repulsions in the eclipsed conformation resulting from the longer

11. For a review, see J. P. Lowe, *Prog. Phys. Org. Chem.* **6**, 1 (1968).
12. Methods for determination of rotational barriers are discussed in Ref. 11 and by E. Wyn-Jones and R. A. Pethrick, *Top. Stereochem.* **5**, 205 (1969).

carbon–silicon bond length (1.87 Å) compared to the carbon–carbon bond (1.54 Å) in ethane.

The haloethanes all have similar rotational barriers of 3.2–3.7 kcal/mol. The increase in the barrier height relative to ethane is probably due to a van der Waals repulsive effect. The heavier halogens have larger van der Waals radii, but this is offset by the longer bond lengths, so that the net effect is a relatively constant rotational barrier for each of the ethyl halides.

Changing the atom bound to a methyl group from carbon to nitrogen to oxygen, as in going from ethane to methylamine to methanol, produces a regular decrease in the rotational barrier from 2.88 to 1.98 to 1.07 kcal/mol. This closely approximates the 3:2:1 ratio of the number of H–H eclipsing interactions in these three molecules.

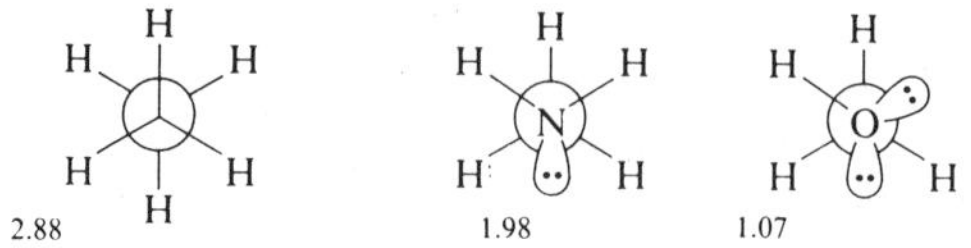

2.88 1.98 1.07

Rotational barrier decreases with the number of H/H eclipsing interaction.

Entries 11 and 13 in Table 3.4 present data relating the effect of methyl substitution on methanol and methylamine. The data show an increased response to methyl substitution. While the propane barrier is 3.4 kcal/mol (compared to 2.88 in ethane), the dimethylamine barrier is 3.62 kcal/mol (compared to 1.98 in methylamine), and in dimethyl ether the barrier is 2.7 kcal/mol (compared to 1.07 in methanol). Thus, while the increase in the barrier due to methyl–hydrogen eclipsing is 0.5 kcal/mol for propane, the increase for both dimethylamine and dimethyl ether is 1.6 kcal/mol. This increase in the barrier is attributed to greater van der Waals repulsions resulting from the shorter C—N and C—O bonds, relative to the C—C bond.

There are two families of conformations available to terminal alkenes. These are the eclipsed and bisected conformations shown below for propene. The eclipsed conformation is preferred by about 2 kcal/mol.[13]

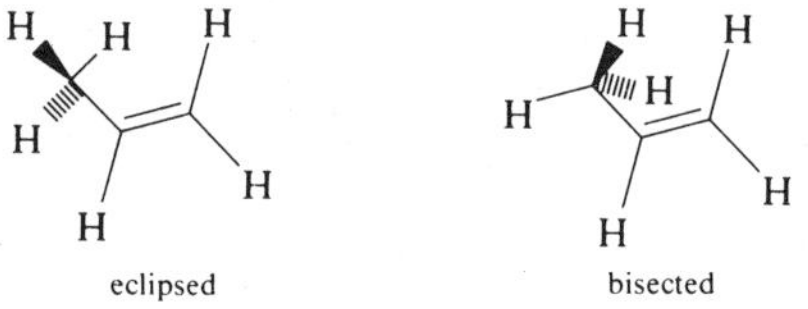

eclipsed bisected

The origin of the preference for the eclipsed conformation of propene can be explained in MO terms by focusing attention on the interaction between the double bond and the π component of the orbitals associated with the methyl group. The dominant interaction is a repulsive one between the filled methyl group orbitals and the filled π orbital of the double bond. This repulsive interaction is greater in the

13. K. B. Wiberg and E. Martin, *J. Am. Chem. Soc.* **107**, 5035 (1985); A. E. Dorigo, D. W. Pratt, and K. N. Houk, *J. Am. Chem. Soc.* **109**, 6591 (1987).

bisected conformation than in the eclipsed conformation.[14]

major repulsive interaction

reduced repulsive interaction

With more substituted terminal alkenes, additional conformations are available as indicated for 1-butene.

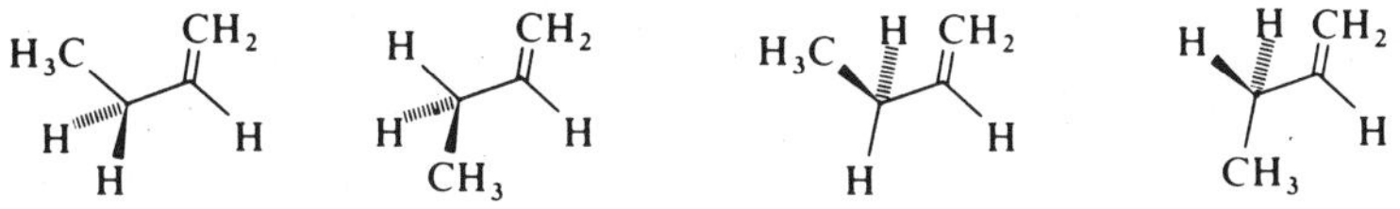

A **B** **C** **D**

Conformations **A** and **B** are of the eclipsed type whereas **C** and **D** are bisected. It has been determined by microwave spectroscopy that the eclipsed conformations are more stable than the bisected ones and that **B** is about 0.15 kcal more stable than **A**.[15] MO calculations at the 6-31G* level have found relative energies of 0.00, −0.25, 1.75, and 1.74 kcal/mol, respectively, for **A–D**.[13]

Further substitution can introduce additional factors, especially nonbonded repulsions, which influence conformational equilibria. For example, methyl substitution at C-2, as in 2-methyl-1-butene, introduces a methyl–methyl *gauche* interaction in the conformation analogous to **B** with the result that in 2-methyl-1-butene the two eclipsed conformations are of approximately equal energy.[16] Increasing the size of the group at C-3 increases the preference for the eclipsed conformation analogous to **B** at the expense of that analogous to **A**. 4,4-Dimethyl-1-pentene, for example, exists mainly in the hydrogen-eclipsed conformation.

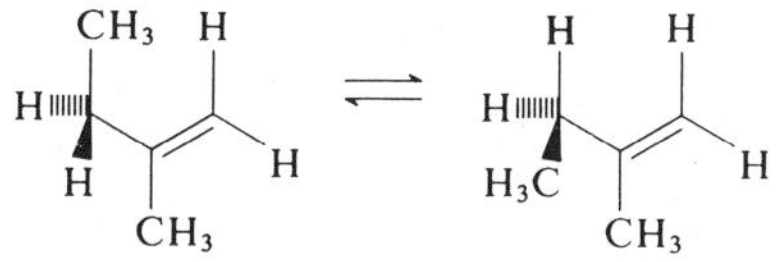

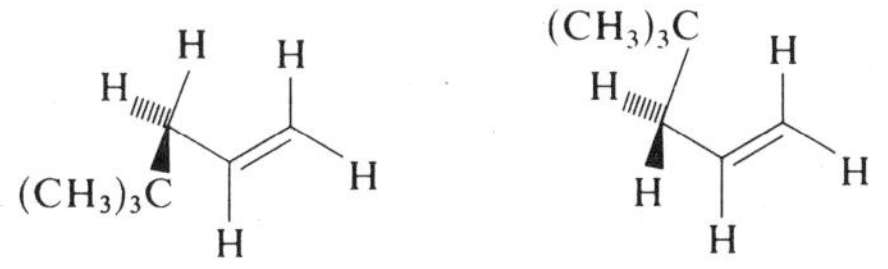

favored

disfavored

The preferred conformations of carbonyl compounds, like 1-alkenes, are eclipsed rather than bisected, as shown below for acetaldehyde and propionaldehyde. The barrier for acetaldehyde is 1.1 kcal/mol.[17] This is about one-third of the barrier in ethane, and molecular orbital calculations indicate that the origin of the barrier is largely the hydrogen–hydrogen repulsion in the conformation in which the aldehyde hydrogen is eclipsed with a hydrogen of the methyl group.[13] In propion-

14. W. J. Hehre, J. A. Pople, and A. J. P. Devaquet, *J. Am. Chem. Soc.* **98**, 664 (1976).
15. S. Kondo, E. Hirota, and Y. Morino, *J. Mol. Spectrosc.* **28**, 471 (1968).
16. T. Shimanouchi, Y. Abe, and K. Kuchitsu, *J. Mol. Struct.* **2**, 82 (1968).
17. R. W. Kilb, C. C. Lin, and E. B. Wilson, Jr., *J. Chem. Phys.* **26**, 1695 (1957).

aldehyde, it is the methyl group, rather than the hydrogen, that is eclipsed with the carbonyl group in the most stable conformation. The energy difference between the two eclipsed conformations has been determined to be 0.9 kcal/mol by microwave spectroscopy.[18] A number of other aldehydes have been studied by NMR spectroscopy and found to have analogous rotameric compositions.[19] When the alkyl substituent becomes very sterically demanding, the hydrogen-eclipsed conformation becomes more stable. This is the case with $(CH_3)_3CCH_2CH{=}O$.

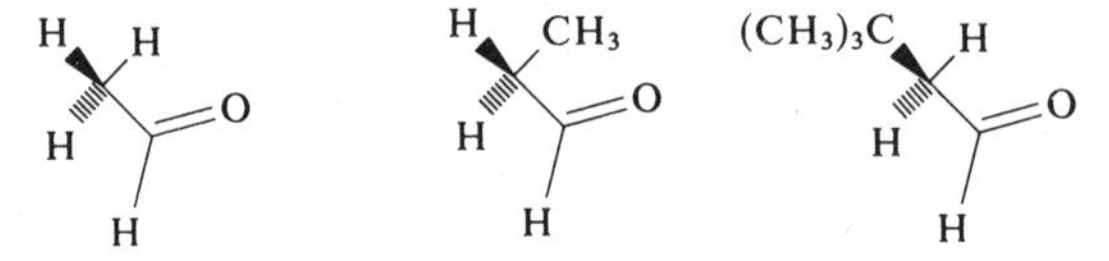

Preferred conformations for acetaldehyde, propionaldehyde, and 3,3-dimethylbutyraldehyde.

Ketones also favor eclipsed conformations. The preferred rotamer is that in which the alkyl group, rather than a hydrogen, is eclipsed with the carbonyl group because this conformation allows the two alkyl groups to be in an *anti* orientation. The alternative rotamer would have the substituents *gauche.*

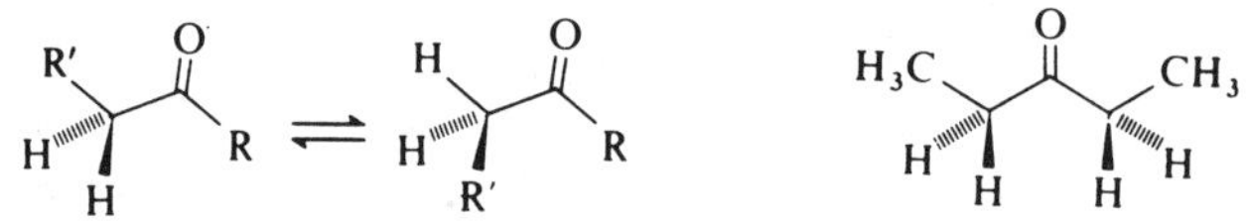

more stable

less stable

Preferred conformation for 3-pentanone

Electron diffraction studies of 3-pentanone indicate the conformation shown above to be the most stable rotamer, in accord with this generalization.[20]

1,3-Dienes would be expected to adopt conformations in which the double bonds are coplanar, so as to permit effective orbital overlap and electron delocalization. The two alternative planar conformations for 1,3-butadiene are referred to as *s-trans* and *s-cis.* In addition to the two planar conformations, there is a third conformation, referred to as the *skew* conformation, which is cisoid but not planar. Various types of studies have shown that the *s-trans* conformation is the most stable one for 1,3-butadiene.[21] A small amount of one of the cisoid conformations is also present in equilibrium with the major conformer. The planar *s-cis* conformation involves a van der Waals repulsion between the hydrogens on C-1 and C-4. This is

18. S. S. Butcher and E. B. Wilson, Jr., *J. Chem. Phys.* **40**, 1671 (1964).
19. G. J. Karabatsos and N. Hsi, *J. Am. Chem. Soc.* **87**, 2864 (1965).
20. C. Romers and J. E. G. Creutzberg, *Rec. Trav. Chim.* **75**, 331 (1956).
21. A. Almenningen, O. Bastiansen, and M. Traetteburg, *Acta Chem. Scand.* **12**, 1221 (1958); K. K. Kuchitsu, T. Fukuyama, and Y. Morino, *J. Mol. Struc.* **1**, 643 (1967); R. L. Lipnick and E. W. Garbisch, Jr., *J. Am. Chem. Soc.* **95**, 6370 (1973).

relieved in the skew conformation.

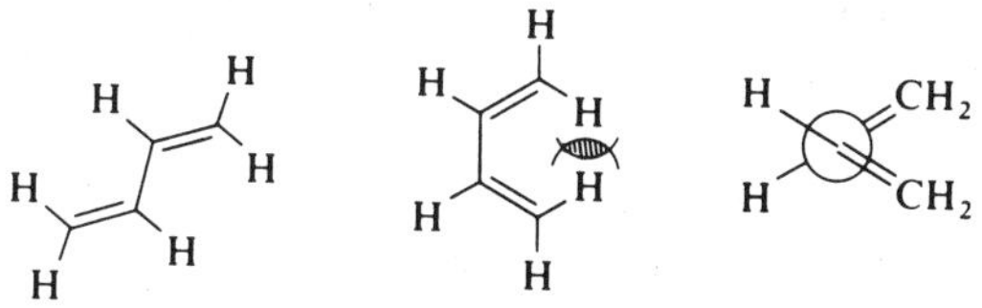

s-trans s-cis skew

The IR and UV spectra of samples of 1,3-butadiene that have an increased content of the higher-energy conformer have been recorded under matrix isolation conditions. These data favor a nearly planar *s-cis* conformation for the less stable conformer.[22] The barrier for conversion of the less stable conformation to the *s-trans* is 3.9 kcal/mol. This energy maximum presumably refers to the conformation (transition state) in which the two π bonds are mutually perpendicular. Various MO calculation find the *s-trans* conformation to be 2–5 kcal/mol lower in energy than either the planar or the skew cisoid conformation.[23] Most high-level calculations favor the *skew* conformation over the planar *s-cis*, but the energy differences found are quite small.[24]

The case of α,β-unsaturated carbonyl compounds is analogous to that of 1,3-dienes in that stereoelectronic factors favor coplanarity of the C=C—C=O system. The rotamers that are important are the *s-trans* and *s-cis* conformations. Microwave data indicate that the *s-trans* form is the only conformation present in detectable amounts in acrolein (2-propen-1-one).[25] The equilibrium distribution of *s-trans* and *s-cis* conformations of α,β-unsaturated ketones depends on the extent of van der Waals interaction between substituents.[26] Methyl vinyl ketone has no unfavorable van der Waals repulsions between substituents and exists predominantly as the *s-trans* conformer:

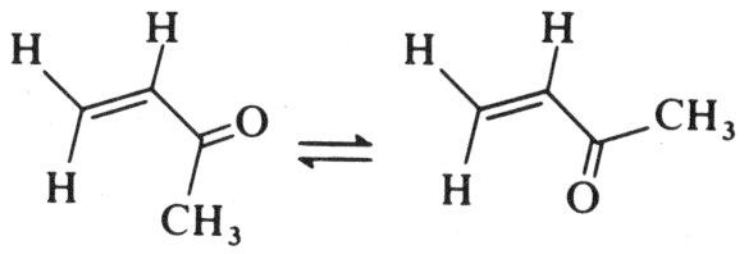

s-trans (73%) ⇌ *s-cis* (27%)

When larger alkyl groups are substituted for methyl, the mole fraction of the *s-cis* form progressively increases as the size of the alkyl group increases.[27]

22. M. E. Squillacote, R. S. Sheridan, O. L. Chapman, and F. A. L. Anet, *J. Am. Chem. Soc.* **101**, 3657 (1979); J. J. Fischer and J. Michl, *J. Am. Chem. Soc.* **109**, 1056 (1987).
23. A. J. P. Devaquet, R. E. Townshend, and W. J. Hehre, *J. Am. Chem. Soc.* **98**, 4068 (1976).
24. J. Breulet, T. J. Lee, and H. F. Schaefer III, *J. Am. Chem. Soc.* **106**, 6250 (1984); D. Feller and E. R. Davidson, *Theor. Chim. Acta* **68**, 57 (1985).
25. E. A. Cherniak and C. C. Costain, *J. Chem. Phys.* **45**, 104 (1966).
26. G. Montaudo, V. Librando, S. Caccamese, and P. Maravigna, *J. Am. Chem. Soc.* **95**, 6365 (1973).
27. A. Bienvenue, *J. Am. Chem. Soc.* **95**, 7345 (1973).

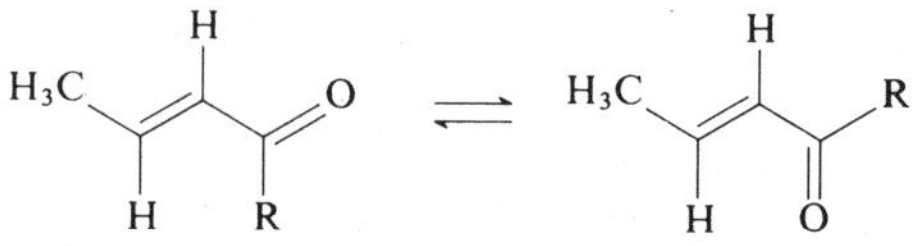

R =	s-trans	s-cis
CH_3	0.7	0.3
CH_3CH_2	0.55	0.45
$(CH_3)_2CH$	0.3	0.7
$(CH_3)_3C$	0.0	1.0

An unfavorable methyl–methyl interaction destabilizes the *s-trans* conformation of 4-methyl-3-penten-2-one (mesityl oxide) relative to the *s-cis* conformation, and the equilibrium favors the *s-cis* form.

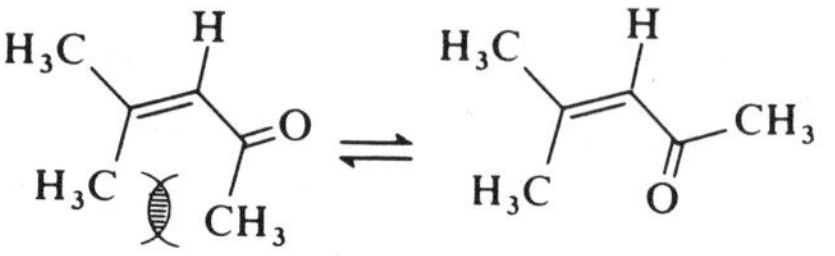

s-trans (28%) ⇌ *s-cis* (72%)

3.3. Conformations of Cyclohexane Derivatives

A particularly well understood aspect of conformational analysis is concerned with compounds containing six-membered rings. A major reason for the depth of study and resulting detailed knowledge has to do with the nature of the system itself. Cyclohexane and its derivatives lend themselves well to thorough analysis, since they are characterized by a small number of energy minima. The most stable conformations are separated by rotational energy barriers that are somewhat higher, and more easily measured, than rotational barriers in noncyclic compounds or in other ring systems.

The most stable conformation of cyclohexane is the chair. Electron diffraction studies in the gas phase reveal a slight flattening of the chair compared with the geometry obtained when using tetrahedral molecular models. The torsion angles are 55.9°, compared with 60° for the "ideal" chair conformation, and the axial C—H bonds are not perfectly parallel, but are oriented outward by about 7°. The C—C bonds are 1.528 Å, the C—H bonds are 1.119 Å, and the C—C—C angles are 111.05°.[28]

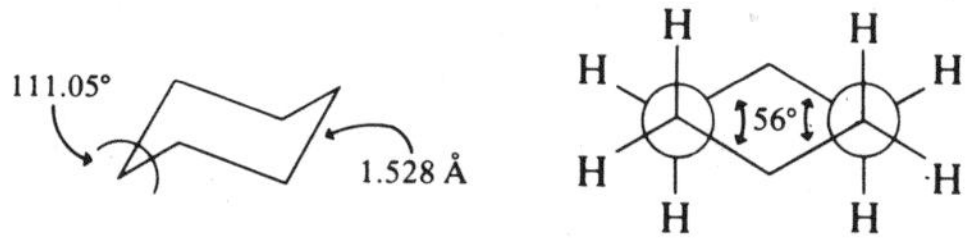

111.05° 1.528 Å 56°

structural features of cyclohexane chair conformation

28. H. J. Geise, H. R. Buys, and F. C. Mijlhoff, *J. Mol. Struct.* 9, 447 (1971).

Two other nonchair conformations of cyclohexane that have normal bond angles and bond lengths are the *twist* and the *boat*.[29] Both the twist and boat conformations are less stable than the chair. Molecular mechanics calculations indicate that the twist conformation is about 5 kcal/mol, and the boat about 6.4 kcal/mol, higher in energy than the chair.[6] A direct measurement of the chair-twist energy difference has been made using low-temperature IR spectroscopy.[30] The chair was determined to be 5.5 kcal/mol lower in enthalpy than the twist. The twist and boat conformations are more flexible than the chair but are destabilized by torsional strain. In addition, the boat conformation is further destabilized by a van der Waals repulsion between the "flagpole" hydrogens, which are separated from each other by about 1.83 Å, a distance considerably less than the sum of their van der Waals radii of 2.4 Å.

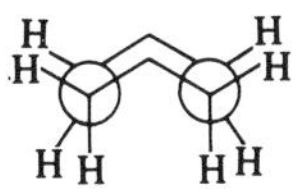

eclipsing in boat conformation

flagpole interaction in boat conformation

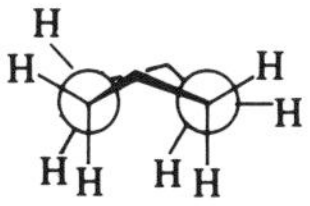

partial relief of eclipsing in twist conformation

Interconversion of chair forms is known as *conformational inversion* and occurs by rotation about carbon–carbon bonds. For cyclohexane, the first-order rate constant for ring inversion is 10^4–$10^5\ s^{-1}$ at 300 K. The enthalpy of activation is 10.8 kcal/mol.[31] Calculation of the geometry of the transition state by molecular mechanics suggests a half-twist form lying 12.0 kcal/mol above the chair. The transition state incorporates 0.2 kcal/mol of compression energy from bond deformation, 2.0 kcal/mol of bond-angle strain, 4.4 kcal/mol of van der Waals stain, and 5.4 kcal/mol of torsional strain.[6] Figure 3.4 presents an energy diagram illustrating the process of conformational inversion in cyclohexane. The boat form is not shown in the diagram because the chair forms can interconvert without passing through the boat. The boat lies 1–2 kcal/mol above the twist conformation and is a transition state for interconversion of twist forms.

Substitution on a cyclohexane ring does not significantly affect the rate of conformational inversion, but does affect the equilibrium distribution between alternative chair forms. All substituents that were axial in one chair conformation become equatorial on ring inversion, and vice versa. For methylcyclohexane, $\Delta G°$ for the equilibrium

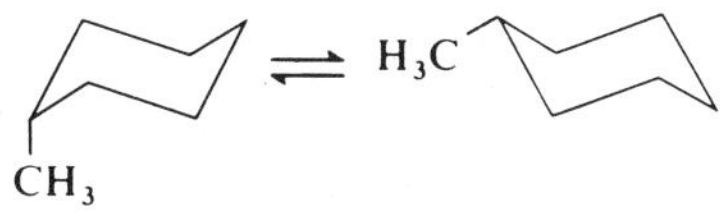

29. For a review of nonchair conformations of six-membered rings, see G. M. Kellie and F. G. Riddell, *Top. Stereochem.* **8**, 225 (1974).
30. M. Squillacote, R. S. Sheridan, O. L. Chapman, and F. A. L. Anet, *J. Am. Chem. Soc.* **97**, 3244 (1975).
31. F. A. L. Anet and A. J. R. Bourn, *J. Am. Chem. Soc.* **89**, 760 (1967).

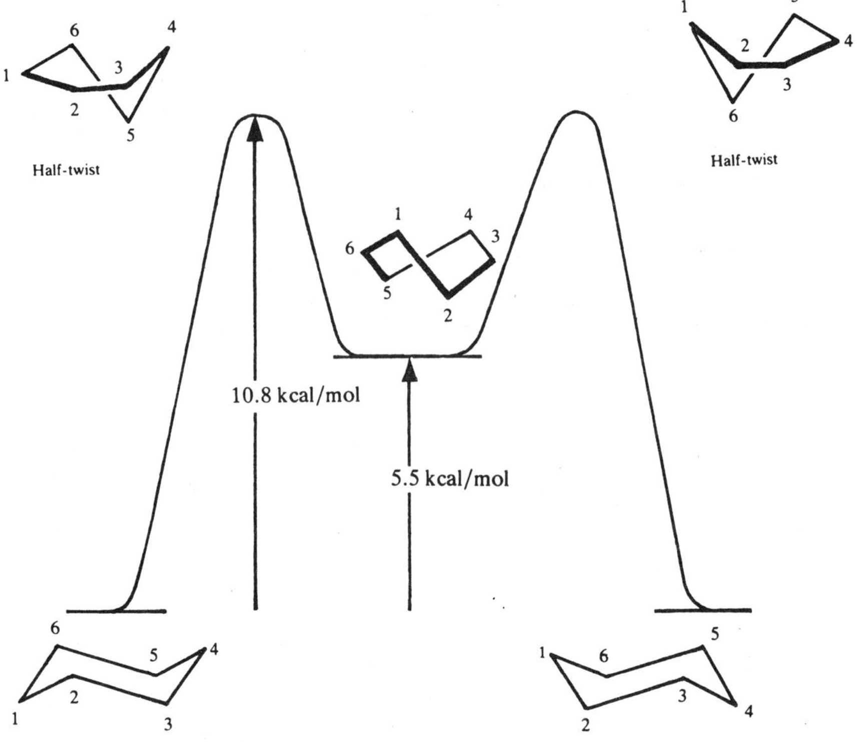

Fig. 3.4. Energy diagram for ring inversion of cyclohexane. [For a rigorous analysis of ring inversion in cyclohexane, see H. M. Pickett and H. L. Strauss, *J. Am. Chem. Soc.* **92**, 7281 (1970).]

is −1.8 kcal/mol, corresponding to a composition with 95% of the equatorial methyl conformation. Two factors contribute to the preference for the equatorial conformation. The equatorial methyl conformation corresponds to an *anti* arrangement with respect to the C-2—C-3 and C-6—C-5 bonds, whereas the axial methyl group is in a *gauche* relationship to these bonds. We have seen earlier that the *gauche* conformation of *n*-butane is 0.8 kcal/mol higher in energy than the *anti* conformation. In addition, there is a van der Waals repulsion between the axial methyl group and the axial hydrogens at C-3 and C-5. Interactions of this type are called *1,3-diaxial interactions.*

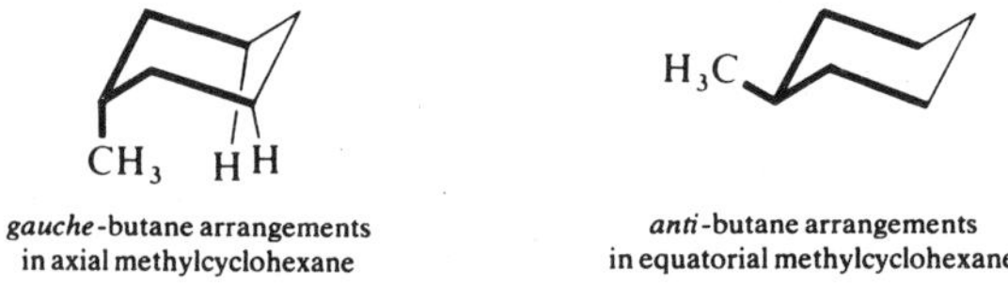

gauche-butane arrangements in axial methylcyclohexane

anti-butane arrangements in equatorial methylcyclohexane

Energy differences between conformations of substituted cyclohexanes can be measured by several physical methods, as can the kinetics of the ring inversion processes. Nuclear magnetic resonance spectroscopy has been especially valuable

Table 3.5. Half-Life for Conformational Inversion of Cyclohexyl Chloride at Various Temperatures[a]

Temperature (°C)	Half-life
25	1.3×10^{-5} s
−60	2.5×10^{-2} s
−120	23 min
−160	22 yr

a. F. R. Jensen and C. H. Bushweller, *J. Am. Chem. Soc.* **91**, 3223 (1969).

for both thermodynamic and kinetic studies.[32] In NMR terminology, the transformation of an equatorial substituent to an axial one and vice versa is called a *site exchange process.* Depending on the rate of the process, the difference in the chemical shift of the nucleus between the two sites, and the field strength of the spectrometer, the spectrum will be either a weighted average spectrum (rapid site exchange, $k > 10^5\ s^{-1}$), a superposition of the spectra of the two conformers reflecting the conformational composition (slow site exchange, $k < 10^2\ s^{-1}$), or, at intermediate rates of exchange, a broadened spectrum. Analysis of the temperature dependence of the spectra can lead to the activation parameters for the conformational process. Figure 3.5 illustrates the change in appearance of a simple spectrum.

For substituted cyclohexanes, the slow-exchange condition is met at temperatures below about −50°C. Table 3.5 presents data for the half-life for conformational equilibrium of cyclohexyl chloride as a function of temperature.

From these data it can be seen that conformationally pure solutions of equatorial cyclohexyl chloride could be maintained at low temperature. This has been accomplished experimentally.[33] Crystallization of cyclohexyl chloride at low temperature affords crystals containing only the equatorial conformer. When the solid is dissolved at −150°C, the NMR spectrum of the solution exhibits only the signal characteristic of the equatorial conformer. When the solution is warmed, the conformation equilibrium is reestablished.

The free-energy difference between conformers is referred to as the *conformational free energy,* or sometimes as the *A-value.* For substituted cyclohexanes it is conventional to specify the value of $-\Delta G°$ for the equilibrium

$$\text{axial} \rightleftharpoons \text{equatorial}$$

Since $\Delta G°$ will be negative when the equatorial conformation is more stable than the axial, the value of $-\Delta G°$ is positive for substituents that favor the equatorial

32. G. Binsch, *Top. Stereochem.* **3**, 97 (1968); F. G. Riddell, *Nucl. Magn. Reson.* **12**, 246 (1983); J. Sandstrom, *Dynamic NMR Spectroscopy*, Academic Press, New York, 1982; J. L. Marshall, *Nuclear Magnetic Resonance*, Verlag Chemie, Deerfield Beach, Florida, 1983; M. Oki, *Applications of Dynamic NMR to Organic Chemistry*, VCH Publishers, Inc., Deerfield Beach, Florida, 1985; Y. Takeuchi and A. P. Marchand (eds.), *Applications of NMR Spectroscopy in Stereochemistry and Conformational Analysis*, VCH Publishers, Deerfield Beach, Florida, 1986.
33. F. R. Jensen and C. H. Bushweller, *J. Am. Chem. Soc.* **91**, 3223 (1969).

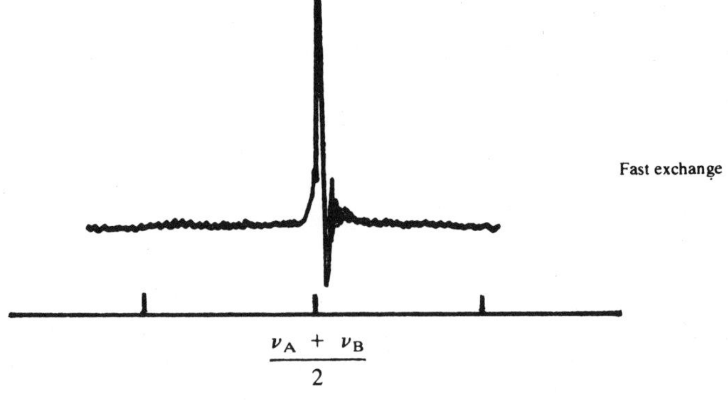

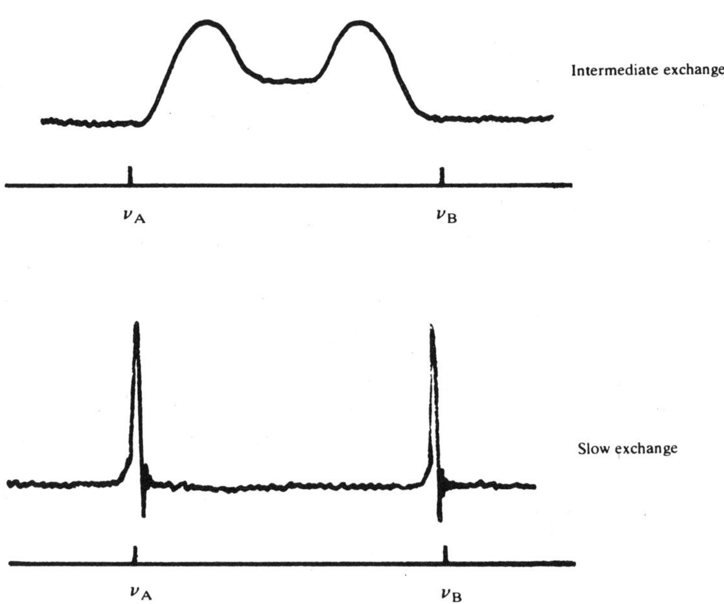

Fig. 3.5. Appearance of NMR spectra for system undergoing two-site exchange ($A \rightleftharpoons B$).

position. The larger the $-\Delta G°$, the greater is the preference for the equatorial position.

The case of cyclohexyl iodide provides an example of the use of NMR spectroscopy to determine the conformational equilibrium constant and the value of $-\Delta G°$. At −80°C, the NMR spectrum shows two distinct peaks in the area of the CHI signal, as shown in Fig. 3.6.[34] The multiplet at higher field is a triplet of triplets with coupling constants of 3.5 and 12 Hz. This pattern is characteristic of a hydrogen in an axial position with two axial–axial couplings and two axial–equatorial couplings. The broader peak at lower field is characteristic of a proton at an equatorial

34. F. R. Jensen, C. H. Bushweller, and B. H. Beck, *J. Am. Chem. Soc.* **91**, 334 (1969).

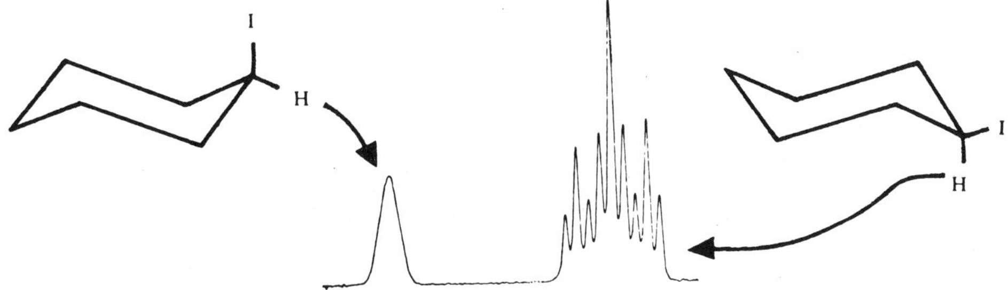

Fig. 3.6. NMR spectrum of cyclohexyl iodide at −80°C. Only the lowest-field signals are shown (100-MHz spectrum). [Reproduced from *J. Am. Chem. Soc.* **91**, 344 (1969), by permission of the American Chemical Society.]

position and reflects the four equatorial-equatorial couplings of such a proton. The relative area of the two peaks is 3.4:1 in favor of the conformer with the axial hydrogen. This corresponds to a $-\Delta G^\circ$ value of 0.47 kcal/mol for the iodo substituent.

Conformational free-energy values for many substituent groups on cyclohexane have been determined by NMR methods; some are recorded in Table 3.6.

A second important method for measuring conformational free energies involves establishing an equilibrium between *diastereomers* differing only in the orientation

Table 3.6. Conformational Free Energies ($-\Delta G^\circ$) for Substituent Groups

Substituent	$-\Delta G^\circ$ (kcal/mol)	Ref.
—F	0.24–0.28	a
—Cl	0.53	a
—Br	0.48	a
—I	0.47	a
$-CH_3$	1.8	b
$-CH_2CH_3$	1.8	b
$-CH(CH_3)_2$	2.1	b
$-C(CH_3)_3$	>4.5	c
$-CH{=}CH_2$	1.7	d
$-C{\equiv}CH$	0.5	e
$-C_6H_5$	2.9	d
—CN	0.15–0.25	a
$-O_2CCH_3$	0.71	a
$-CO_2H$	1.35	c
$-CO_2C_2H_5$	1.1–1.2	c
—OH (aprotic solvents)	0.52	c
—OH (protic solvents)	0.87	c
$-OCH_3$	0.60	c
$-NO_2$	1.16	a
—HgBr	0	a

a. F. R. Jensen and C. H. Bushweller, *Adv. Alicyclic Chem.* **3**, 140 (1971).
b. N. L. Allinger and L. A. Freiberg, *J. Org. Chem.* **31**, 804 (1966).
c. J. A. Hirsch, *Top Stereochem.* **1**, 199 (1967).
d. E. L. Eliel and M. Manoharan, *J. Org. Chem.* **46**, 1959 (1981).
e. H. J. Schneider and V. Hoppen, *J. Org. Chem.* **43**, 3866 (1978).

of the designated substituent group. The equilibrium constant can then be determined and used to calculate the free-energy difference between the isomers. For example, *cis*- and *trans*-4-t-butylcyclohexanol can be equilibrated using a nickel catalyst in refluxing benzene to give a mixture containing 28% *cis*-4-*t*-butylclohexanol and 72% *trans*-*t*-butylcyclohexanol.[35]

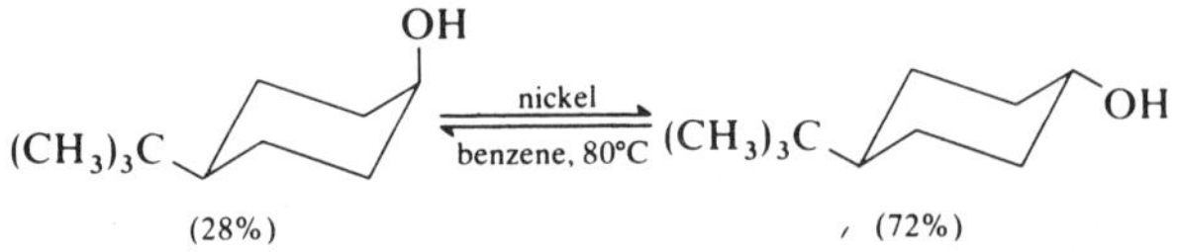

$(CH_3)_3C$ ... OH (28%) $\xrightarrow[\text{benzene, 80°C}]{\text{nickel}}$ $(CH_3)_3C$ ... OH (72%)

If it is assumed that only conformations that have the *t*-butyl group equatorial are significant, then the free-energy change for the equilibration is equal to the free-energy difference between an axial and an equatorial hydroxyl group. The equilibrium constant leads to a $-\Delta G°$ value of 0.7 kcal/mol for the hydroxyl substituent. This approach also assumes that the *t*-butyl group does not distort the ring in any way or interact directly with the hydroxyl group.

There are several other methods available for determining conformational free energies.[36] Values for many substituents in addition to those listed in Table 3.6 have been compiled.[37]

Some insight into the factors which determine the $-\Delta G°$ values for various substituents can be gained by considering some representative groups. Among the halogens, fluorine has the smallest preference for an equatorial conformation. The other halogens have very similar conformational free energies. This is the result of compensating trends in van der Waals radii and bond lengths. Although the van der Waals radii increase with atomic number, the bond length also increases so the net effect is very small. There may also be a contribution from attractive London forces, which would increase with the size of the halogen atom.

The alkyl groups methyl, ethyl, and isopropyl have similar conformational energies, with $-\Delta G°$ for isopropyl being only slightly larger than for methyl and ethyl. The similar values for the three substituents reflect the fact that rotation about the bond between the substituent and the ring allows the ethyl and isopropyl groups to adopt a conformation that minimizes the effect of the additional methyl substituents.

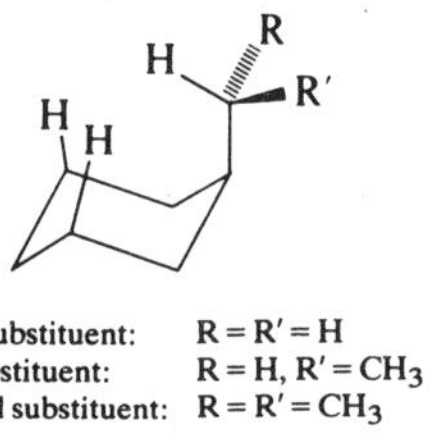

methyl substituent: R = R′ = H
ethyl substituent: R = H, R′ = CH_3
isopropyl substituent: R = R′ = CH_3

35. E. L. Eliel and S. H. Schroeter, *J. Am. Chem. Soc.* **87**, 5031 (1965).
36. F. R. Jensen and C. H. Bushweller, *Adv. Alicyclic Chem.* **3**, 139 (1971).
37. J. A. Hirsch, *Top. Stereochem.* **1**, 199 (1967); E. L. Eliel, N. L. Allinger, S. J. Angyal, and G. A. Morrison, *Conformational Analysis*, Interscience, New York, 1965, pp. 436–443; M. Charton and B. I. Charton, *J. Chem. Soc. B*, 43 (1967).

A *t*-butyl substituent in the axial orientation experiences a strong van der Waals repulsion with the *syn*-axial hydrogens which cannot be relieved by rotation about the bond to the ring. As a result, the $-\Delta G°$ value for *t*-butyl group is much larger than for the other alkyl groups. The value of about 5 kcal/mol has been calculated by molecular mechanics.[38] Experimental attempts to measure the $-\Delta G°$ value for *t*-butyl have provided only a lower limit since very little of the axial conformation is present and the energy difference is very similar to that between the chair and twist forms of the cyclohexane ring.

The strong preference for a *t*-butyl group to occupy the equatorial position has made it a useful group for the study of conformationally biased systems. A *t*-butyl group will ensure that the equilibrium lies heavily to the side having the *t*-butyl group equatorial but does not stop the process of conformational inversion. It should be emphasized that "conformationally biased" is not synonymous with "conformationally locked." Since ring inversion can still occur, it is inappropriate to think of the systems as being "locked" in a single conformation.

When two or more substituents are present on a cyclohexane ring, the interactions between the substituents must be included in the analysis. The dimethylcyclohexanes provide an example in which a straightforward interpretation is in complete agreement with the experimental data. For 1,2-, 1,3-, and 1,4-dimethylcyclohexane, the free-energy change for the *cis* ⇌ *trans* isomerization is given below.[6]

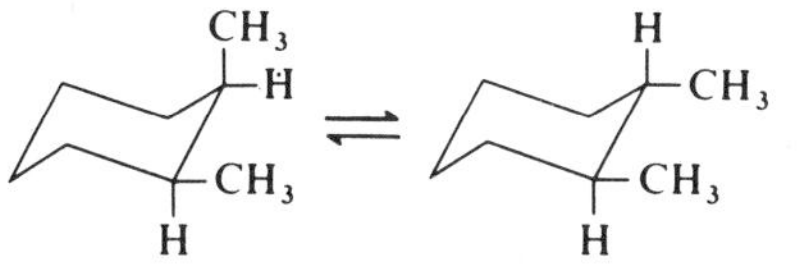

cis $\Delta G° = -1.87$ kcal/mol *trans*

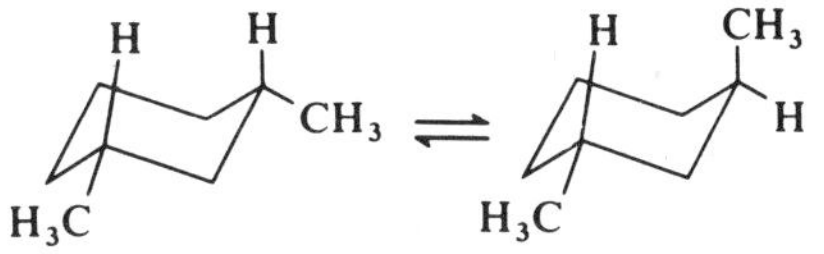

cis $\Delta G° = +1.96$ kcal/mol *trans*

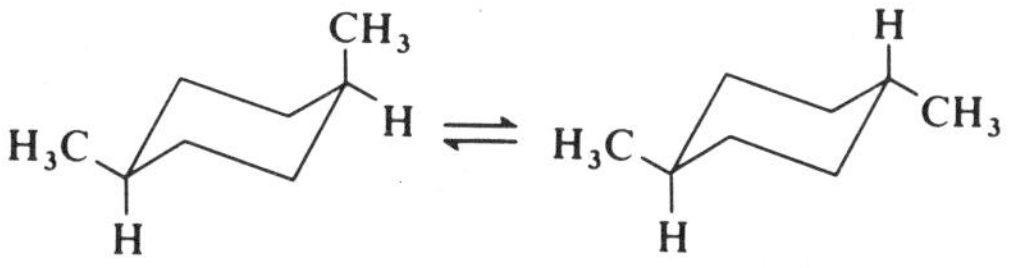

cis $\Delta G° = -1.90$ kcal/mol *trans*

The more stable diastereomer in each case is the one having both methyl groups equatorial. The free-energy difference favoring the diequatorial isomer is about the same for each case (about 1.9 kcal/mol) and is close to the $-\Delta G°$ value of the methyl group (1.8 kcal/mol). This implies that there are no important interactions present that are not also present in methylcyclohexane. This is reasonable since in each case the axial methyl group interacts only with the 3,5-diaxial hydrogens as in methylcyclohexane.

Conformations in which there is a 1,3-diaxial interaction between substituent groups larger than hydrogen are destabilized by van der Waals repulsion. Equilibra-

38. N. L. Allinger, J. A. Hirsch, M. A. Miller, I. J. Tyminski, and F. A. VanCatledge, *J. Am. Chem. Soc.* **90**, 1199 (1968); B. van de Graf, J. M. A. Baas, and B. M. Wepster, *Recl. Trav. Chim. Pays-Bas* **97**, 268 (1978); J. M. A. Baas, A. van Veen, and B. M. Wepster, *Recl. Trav. Chim. Pays-Bas* **99**, 228 (1980).

tion of *cis*- and *trans*-1,1,3,5-tetramethylcyclohexane establishes that the *cis* isomer is favored by 3.7 kcal/mol.[39] This provides a value for a 1,3-diaxial methyl interaction that is 1.9 kcal/mol higher than that for the 1,3-methyl–hydrogen interaction.

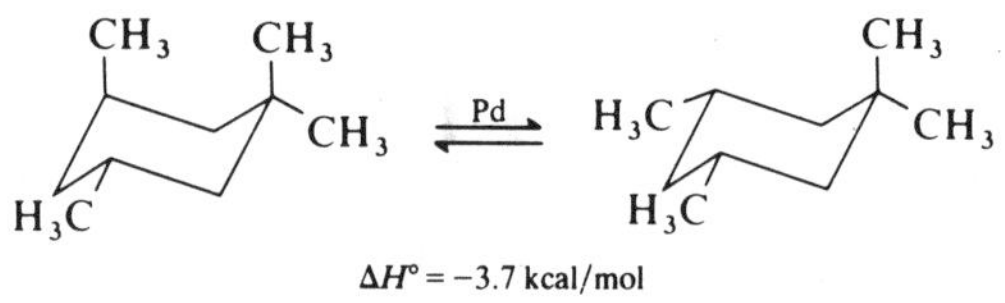

$\Delta H° = -3.7$ kcal/mol

The decalin (bicyclo[4.4.0]decane) ring system provides another important system for study of conformational effects in cyclohexane rings. The energy difference can be analyzed by noting that the *cis* isomer has three more *gauche*-butane interactions than does the *trans* isomer. Assigning a value of 0.8 kcal/mol to the *gauche* interaction would predict an enthalpy difference of 2.4 kcal/mol between the two isomers. Equilibration of the *cis* and *trans* isomers shows that the *trans* isomer is favored by about 2.7–3.0 kcal/mol.

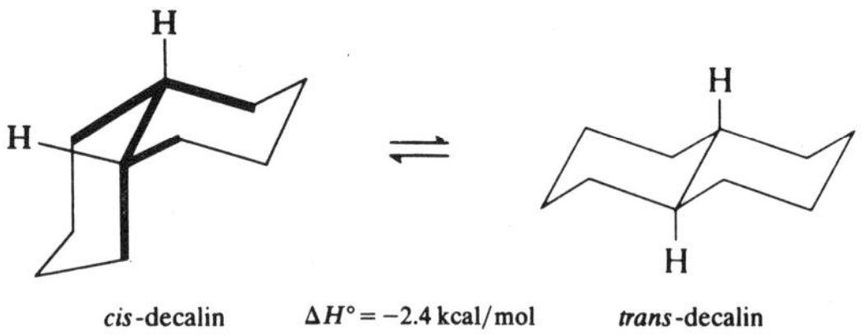

cis-decalin $\Delta H° = -2.4$ kcal/mol *trans*-decalin

There is an important difference between the *cis*- and *trans*-decalin systems with respect to their conformational flexibility. *trans*-Decalin, because of the nature of the ring fusion, is incapable of ring inversion. *cis*-Decalin is conformationally mobile and undergoes ring inversion at a rate only slightly slower than cyclohexane ($\Delta G^{\ddagger} = 12.3$–$12.4$ kcal/mol).[40] The *trans*-decalin system than represents a "conformationally locked" system and can be used to determine the difference in stability and reactivity of groups in axial and equatorial environments.

The effect of introducing sp^2-hybridized atoms into open-chain molecules was discussed earlier, and it was noted that torsional barriers in 1-alkenes and aldehydes are somewhat smaller than in alkanes. Similar effects are noted when sp^2 centers are incorporated into six-membered rings. Whereas the free-energy barrier for ring inversion in cyclohexane is 10.3 kcal/mol, it is reduced to 7.7 kcal/mol in methylenecyclohexane[41] and to 4.9 kcal/mol in cyclohexanone.[42]

By analogy with acyclic aldehydes and ketones, an alkyl group at C-2 of a cyclohexanone ring would be expected to be more stable in the equatorial than in

39. N. L. Allinger and M. A. Miller, *J. Am. Chem. Soc.* **83**, 2145 (1961).
40. F. R. Jensen and B. H. Beck, *Tetrahedron Lett.*, 4523 (1966); D. K. Dalling, D. M. Grant, and L. F. Johnson, *J. Am. Chem. Soc.* **93**, 367 (1971); B. E. Mann, *J. Magn. Reson.* **21**, 17 (1976).
41. J. T. Gerig, *J. Am. Chem. Soc.* **90**, 1065 (1968).
42. F. R. Jensen and B. H. Beck, *J. Am. Chem. Soc.* **90**, 1066 (1968).

the axial orientation. The equatorial orientation is eclipsed with the carbonyl group and corresponds to the more stable conformation of open-chain ketones. This conformation also avoids 3,5-diaxial interactions with *syn*-diaxial hydrogens as in cyclohexane. Conformational free energies for 2-alkyl substituents in cyclohexanones have been determined by equilibrium studies. The value for the methyl group is similar to that in cyclohexane while the values for ethyl and isopropyl are somewhat smaller than in cyclohexane. This has been attributed to a repulsive steric interaction with the carbonyl oxygen for the larger substituents.[43]

The conformational energy of an alkyl group at C-3 of cyclohexanone is substantially less than that of an alkyl group in cyclohexane because of reduced 1,3-diaxial interactions. A C-3 methyl group in cyclohexanone has a $-\Delta G°$ of 1.3–1.4 kcal/mol.[35]

The preferred conformation of 2-bromo- and 2-chlorocyclohexanones depends upon the polarity of the solvent. In solvents of low dielectric constant, the halogen substituent is more stable in the axial orientation. For example, in carbon tetrachloride the Br-axial conformation of 2-bromocyclohexanone is favored by 3:1.[44] Optical rotatory dispersion measurements on optically active *trans*-2-chloro-5-methylcyclohexanone demonstrate that the preferred conformation in octane is the one in which both the chlorine and the methyl group are axial.[45] The conformational equilibrium is reversed in the more polar solvent methanol.

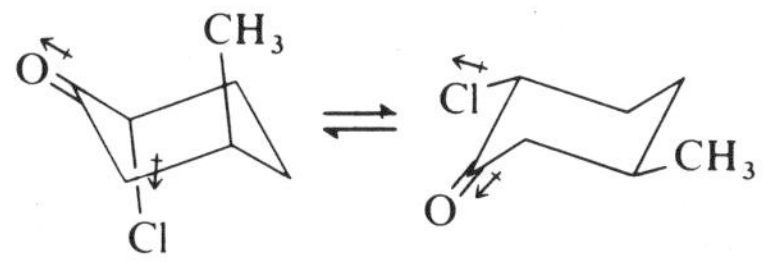

CH₃ O Cl ⇌ Cl CH₃ O

more stable conformer in octane — more stable conformer in methanol

The *α-haloketone effect*, as this phenomenon is known, is believed to be the result of dipolar interactions between the carbonyl group and the carbon–halogen bond. The bond dipoles largely cancel in the conformation with an axial chlorine but are additive for that with an equatorial chlorine. The conformation with the smaller dipole moment will be favored in solvents of low dielectric constant.

The relative preference for the axial orientation for α-chloroketones can also be explained in molecular orbital terms. The axial arrangement places the C—Cl bond nearly perpendicular to the plane of the carbonyl group. This permits interaction between the π orbitals of the carbonyl group and the σ orbitals associated with the C—Cl bond. There are two interactions possible: $\sigma \rightarrow \pi^*$ donation and $\pi \rightarrow \sigma^*$ donation. This σ–π delocalization would not be possible with the equatorial orientation of the chlorine since the C—Cl bond then lies approximately in the

43. N. L. Allinger and H. M. Blatter, *J. Am. Chem. Soc.* **83**, 994 (1961); B. Rickborn, *J. Am. Chem. Soc.* **84**, 2414 (1962); E. L. Eliel, N. L. Allinger, S. J. Angyal, and G. A. Morrison, *Conformational Analysis*, Interscience, New York, 1965, pp. 113–114.
44. J. Allinger and N. L. Allinger, *Tetrahedron* **2**, 64 (1958).
45. C. Djerassi, *Optical Rotatory Dispersion*, McGraw-Hill, New York, 1960, pp. 125–126.

nodal plane of the carbonyl group.

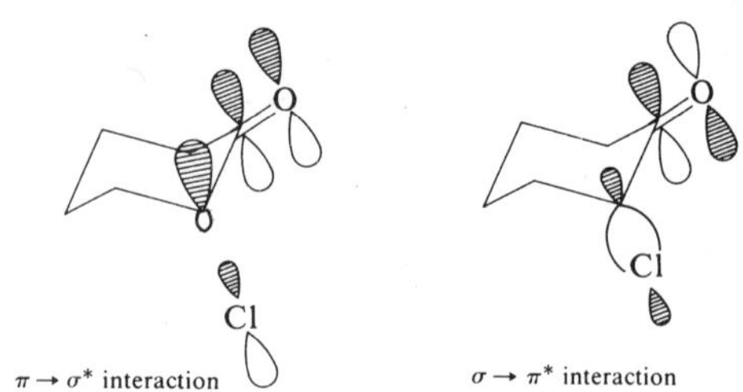

It has been found that alkylidenecyclohexanes bearing alkyl groups of moderate size at C-2 tend to adopt the conformation with the alkyl group axial in order to relieve unfavorable van der Waals interactions with the alkylidene group.[46] This results from van der Waals repulsion between the alkyl group in the equatorial position and *cis* substituents on the exocyclic double bond. The term *allylic strain* is used to designate this steric effect. The repulsive energy is minimal for methylenecyclohexanes, but molecular mechanics calculations indicate that the axial conformation **B** is 2.6 kcal/mol more stable than **A** with an exocyclic isopropylidene group.[47]

CH_3 CH_3 CH_3 H ⇌ CH_3 H CH_3 CH_3

A **B**

The conformation of cyclohexene is described as a half-chair. Structural parameters determined on the basis of electron diffraction and microwave spectroscopy reveal that the double bond can be accommodated into the ring without serious distortion.[48]

half-chair conformation of cyclohexene

The C-1—C-2 bond length is 1.335 Å, and the C-1—C-2—C-3 bond angle is 123°. The substituents at C-3 and C-6 are tilted from the usual axial and equatorial directions and are referred to as *pseudoaxial* and *pseudoequatorial*. The activation energy for ring inversion is 5.3 kcal/mol.[49] The preference for equatorial orientation

46. F. Johnson, *Chem. Rev.* **68**, 375 (1968).
47. N. L. Allinger, J. A. Hirsch, M. A. Miller, and I. J. Tyminski, *J. Am. Chem. Soc.* **90**, 5773 (1968).
48. J. F. Chiang and S. H. Bauer, *J. Am. Chem. Soc.* **91**, 1898 (1969); L. H. Scharpen, J. E. Wollrab, and D. P. Ames, *J. Chem. Phys.* **49**, 2368 (1968).
49. F. A. L. Anet and M. Z. Haq, *J. Am. Chem. Soc.* **87**, 3147 (1965).

Table 3.7. Strain Energies of Cycloalkanes

Cycloalkane	Strain energy (kcal/mol)[a]
Cyclopropane	28.1[b]
Cyclobutane	26.3
Cyclopentane	7.3
Cyclohexane	1.4
Cycloheptane	7.6
Cyclooctane	11.9
Cyclononane	15.5
Cyclodecane	16.4
Cyclododecane	11.8

a. Estimated values taken from E. M. Engler, J. D. Andose, and P. v. R. Schleyer, *J. Am. Chem. Soc.* **95**, 8005 (1973).
b. Estimated value taken from P. v. R. Schleyer, J. E. Williams, and K. R. Blanchard, *J. Am. Chem. Soc.* **92**, 2377 (1970).

of a methyl group in cyclohexene is less than in cyclohexane because of the ring distortion and the removal of one 1,3-diaxial interaction. A value of 1 kcal/mol has been suggested for the $-\Delta G°$ value for a methyl group in 4-methylcyclohexene.[50]

3.4. Carbocyclic Rings Other Than Six-Membered

The most important structural features that influence the conformation and reactivity of cycloalkanes differ depending on whether small (cyclopropane and cyclobutane), common (cyclopentane, cyclohexane, and cycloheptane), medium (cyclooctane through cycloundecane), or large (cyclododecane and up) rings are considered. The small rings are dominated by angle strain and torsional strain. The common rings are relatively low in strain and their conformational preferences are most influenced by torsional factors. Medium rings exhibit conformational preferences and chemical properties indicating that cross-ring van der Waals interactions play an important role. Large rings become increasingly flexible and possess a large number of low-energy conformations. Table 3.7 presents data on the strain energies of cycloalkanes up to cyclododecane (with the exclusion of cycloundecane).

The cyclopropane ring is necessarily planar, and the question of conformation does not arise. The C—C bond lengths are slightly shorter than normal at 1.5 Å, and the H-C-H angle of 115° is opened from the tetrahedral angle.[51]

Cyclobutane adopts a puckered conformation.[52] Substituents then occupy axial-like or equatorial-like positions. 1,3-Disubstituted cyclobutanes show small energy

50. B. Rickborn and S.-Y. Lwo, *J. Org. Chem.* **30**, 2212 (1965).
51. O. Bastiansen, F. N. Fritsch, and K. Hedberg, *Acta Crystallogr.* **17**, 538 (1964).
52. A. Almenningen, O. Bastiansen, and P. N. Skancke, *Acta Chem. Scand.* **15**, 711 (1961).

preferences for the *cis* isomer since this places both substituents in equatorial-like positions.[53]

R, R′ ⇌ R, R′

trans *cis*

R = R′ = Br $\Delta G^\circ = -0.4$ kcal/mol (Ref. 53a)
R = CH_3; R′ = CO_2CH_3 $\Delta G^\circ = -0.3$ kcal/mol (Ref. 53b)

The energy differences and the barrier to inversion are both smaller than in cyclohexane.

Cyclopentane is nonplanar, and the two most easily described geometries are the envelope and the half-chair.[54] In the envelope conformation, one carbon atom is displaced from the plane of the other four. In the half-chair conformation, three carbons are coplanar, with one of the remaining two being above the plane and the other below. The energy differences between the conformers are very small, and rapid interconversion of conformers occurs.[55] All of the carbon atoms in effect rapidly move through planar and nonplanar positions. The process is called *pseudorotation.*

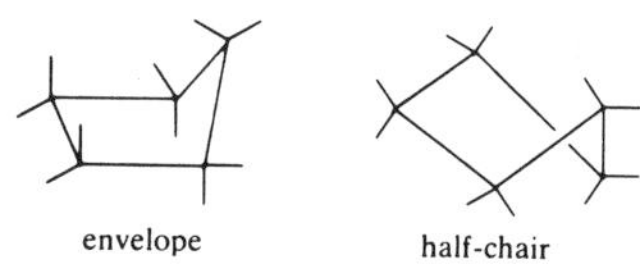

As ring size increases, there are progressively more conformations that need to be considered. For cycloheptane, four conformations have been calculated to be particularly stable.[56] NMR investigations indicate the twist–chair is the most stable.[57]

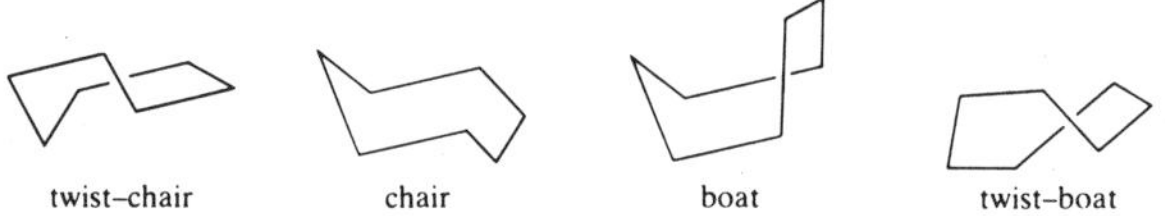

The total spread in energies calculated for the four conformations is only 2.7 kcal/mol. The individual twist–chair conformations interconvert rapidly by pseudorotation.[58]

53. (a) K. B. Wiberg and G. M. Lampman, *J. Am. Chem. Soc.* **88**, 4429 (1966); (b) N. L. Allinger and L. A. Tushaus, *J. Org. Chem.* **30**, 1945 (1965).
54. A. C. Legon, *Chem. Rev.* **80**, 231 (1980); B. Fuchs, *Top. Stereochem.* **10**, 1 (1978).
55. W. J. Adams, H. J. Geise, and L. S. Bartell, *J. Am. Chem. Soc.* **92**, 5013 (1970); J. B. Lambert, J. J. Papay, S. A. Kahn, K. A. Kappauf, and E. S. Magyar, *J. Am. Chem. Soc.* **96**, 6112 (1974).
56. J. B. Hendrickson, *J. Am. Chem. Soc.* **89**, 7036 (1967).
57. J. B. Hendrickson, R. K. Boeckman, Jr., J. D. Glickson, and E. Grunwald, *J. Am. Chem. Soc.* **95**, 494 (1973).
58. D. F. Bocian, H. M. Pickett, T. C. Rounds, and H. L. Strauss, *J. Am. Chem. Soc.* **97**, 687 (1975).

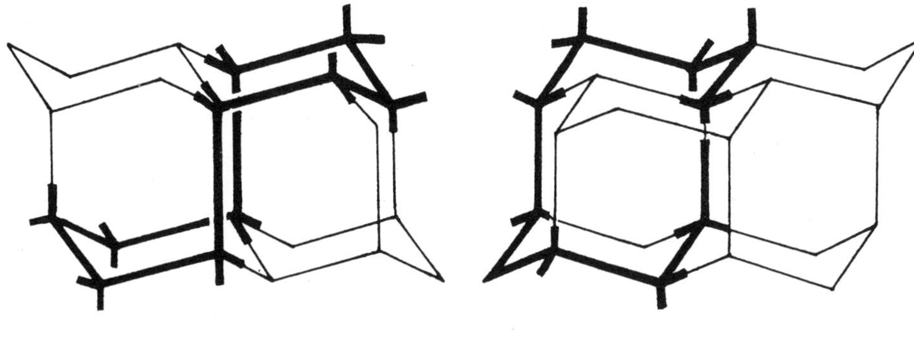

Fig. 3.7. Equivalent diamond-lattice conformations of cyclodecane (boat–chair–boat).

In the case of cyclooctane, a total of 11 conformations have been suggested for consideration and their relative energies calculated. The boat–chair conformation was calculated to be the most stable conformation.[56] This prediction was confirmed by analysis of the temperature dependence of the ^{19}F NMR spectra of fluorocyclooctanes.[59] The energy for interconversion of conformers is 5–8 kcal/mol. A few of the most stable conformations are shown below.

The conformational possibilities for larger rings quickly become very large. One interesting and simplifying concept has emerged. The diamond lattice is the most stable arrangement for a large array of sp^3 carbon atoms. There are also both theoretical and experimental results that show that complex polycyclic saturated hydrocarbons are most stable in diamond-type structures. Adamantane is the most familiar example of this type of structure.

It might be anticipated that large flexible rings would adopt similar structures incorporating the chair cyclohexane conformation. Conformations for C_{10} through C_{24} cycloalkanes corresponding to diamond-lattice sections have been identified by systematic topological analysis using models or computation.[60] This type of relationship is illustrated in Fig. 3.7 for cyclodecane.

59. J. E. Anderson, E. S. Glazer, D. L. Griffin, R. Knorr, and J. D. Roberts, *J. Am. Chem. Soc.* **91**, 1386 (1969); see also F. A. L. Anet and M. St. Jacques, *J. Am. Chem. Soc.* **88**, 2585, 2586 (1966).
60. J. Dale, *J. Chem. Soc.*, 93 (1963); M. Saunders, *Tetrahedron* **23**, 2105 (1967); J. Dale, *Top. Stereochem.* **9**, 199 (1976).

Studies of cyclodecane derivatives by X-ray crystallographic methods have demonstrated that the boat–chair–boat conformation is adopted in the solid state.[61] (Notice that "boat" is used here in a different sense than for cyclohexane.) As was indicated in Table 3.7 (p. 141), cyclodecane is significantly more strained than cyclohexane. Examination of the boat–chair–boat conformation reveals that the source of most of this strain is the close van der Waals contacts between two sets of three hydrogens on either side of the molecule, as indicated in the drawing below. Distortion of the molecule to twist forms relieves this interaction but introduces torsional strain.

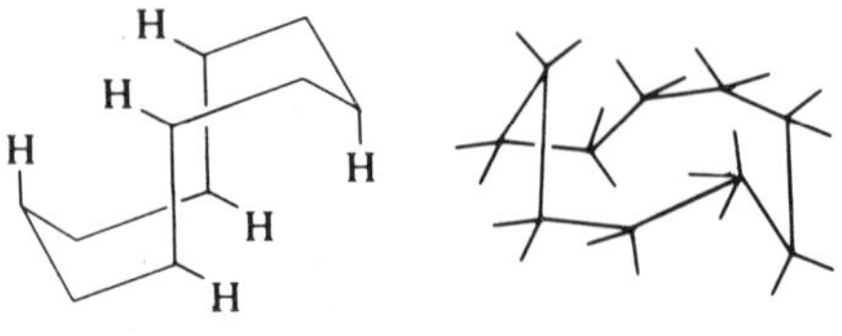

H H H H H H

boat–chair–boat

twist–boat–chair

The X-ray diffraction study of 1,1,9,9-tetramethylcyclododecane reveals a diamond-like conformation is adopted.[62] Both molecular mechanics calculations and physical properties indicate that the 16-membered ring in cyclohexadecane also adopts a diamond-lattice conformation.[63]

3.5. The Effect of Heteroatoms on Conformational Equilibria

The replacement of carbon by other elements produces changes in several structural parameters, and consequently affects the conformational characteristics of the molecule. In this section, we will first describe some stereochemical features of heterocyclic analogs of cycloalkanes.[64] For the purpose of elaborating conformational principles, the discussion will focus on six-membered rings, so that the properties may be considered in the context of a ring system possessing a limited number of low-energy conformations.

The most obvious changes that occur on introduction of a heteroatom into a six-membered ring have to do with bond lengths and angles. Both the carbon–oxygen and carbon–nitrogen bond lengths (1.43 and 1.47 Å, respectively) are shorter than the carbon–carbon bond length of 1.54 Å, while the carbon–sulfur bond length (1.82 Å) is considerably longer. The normal valence angles are somewhat smaller than tetrahedral at oxygen and nitrogen, and significantly so for sulfur, for which

61. J. D. Dunitz, in *Perspectives in Structural Chemistry*, Vol II, J. D. Dunitz and J. A. Ibers (eds.), John Wiley, New York, 1968, pp. 1–70.
62. P. Groth, *Acta Chem. Scand.* **A28**, 808 (1974).
63. N. L. Allinger, B. Gorden, and S. Profeta, Jr., *Tetrahedron* **36**, 859 (1980).
64. For reviews, see J. B. Lambert and S. I. Featherman, *Chem. Rev.* **75**, 611 (1975); F. G. Ridell, *The Conformational Analysis of Heterocyclic Compounds*, Academic Press, New York, 1980; E. L. Eliel, *Acc. Chem. Res.* **3**, 1 (1970).

the normal C—S—C angle is about 100°. The six-membered heterocycles containing oxygen (tetrahydropyran), nitrogen (piperidine), and sulfur (thiane) all resemble the chair conformation of cyclohexane but are modified so as to accommodate the bond lengths and bond angles characteristeric of the heteroatom. The rings are all somewhat more puckered than cyclohexane. Because of the shorter C—O bond distances, 2-substituents in tetrahydropyran and 1,3-dioxane rings have larger conformational free energies than in cyclohexane rings. The shorter bonds lead to stronger repulsive interaction with the axial hydrogens at C-4 and C-6.

tetrahydropyran piperidine thiane 1,3-dioxane 1,3-dithiane

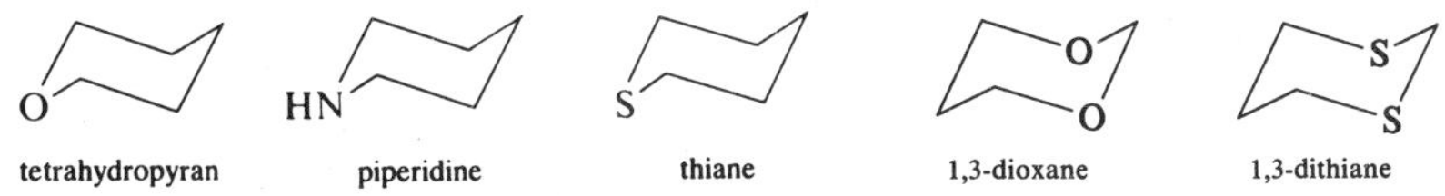

An important feature associated with heterocyclic rings is the reduction in steric repulsions for axial substituents that results from replacement of a methylene group in cyclohexane by oxygen, nitrogen, or sulfur. This effect is readily apparent in *cis*-2-methyl-5-*t*-butyl-1,3-dioxane, in which the preferred conformation has the *t*-butyl group axial and the methyl group equatorial.[65] Divalent oxygen has no exocyclic substituents so the 1,3-diaxial interaction, which is the main unfavorable interaction for axial substituents in cyclohexanes, is not present.

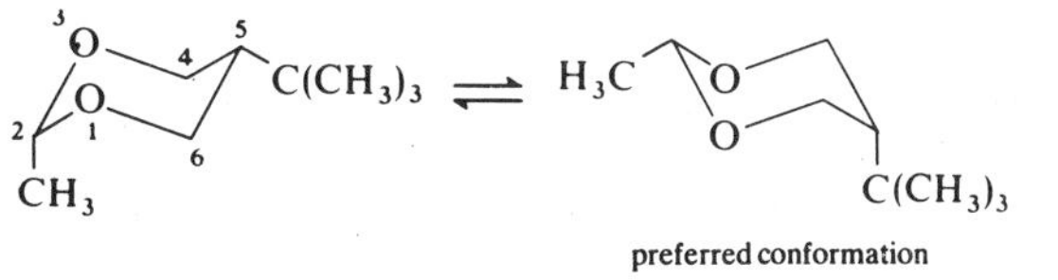

$C(CH_3)_3 \rightleftharpoons H_3C$ ... $C(CH_3)_3$

preferred conformation

It is consistently found that 5-alkyl substituents in 1,3-dioxane exhibit a smaller equatorial preference than they do in cyclohexane. This decreased preference is due to decreased van der Waals repulsions in the axial orientation, since there are no hydrogens that are *syn*-axial to the 5-alkyl substituent. As mentioned above, a 2-alkyl substituent, on the other hand, has a greater preference for the equatorial orientation in 1,3-dioxane than in cyclohexane, presumably because the decreased C—O bond length (relative to C—C) brings an axial 2-alkyl group into closer contact with the *syn*-axial hydrogens at C-4 and C-6, resulting in an increased van der Waals repulsion. Similarly, an axial 4-alkyl substituent in a 1,3-dioxane suffers a greater van der Waals repulsion with the axial hydrogen at C-2 than it does in cyclohexane. Table 3.8 presents $-\Delta G°$ values for several groups in tetrahydropyrans, 1,3-dioxanes, and 1,3-dithianes, along with their comparative $-\Delta G°$ values in cyclohexane. The general point to be recognized is that the conformational free energy is a function not only of the size of the group but also of the molecular environment that it encounters.[66]

65. E. L. Eliel and M. C. Knoeber, *J. Am. Chem. Soc.* **90**, 3444 (1968).
66. For a review of conformational analysis of dioxanes, see M. J. O. Anteunis, D. Tavernier, and F. Borremans, *Heterocycles* **4**, 293 (1976).

Table 3.8. Comparison of Conformational Free-Energy Values for Substituents on Tetrahydropyran, 1,3-Dioxane, and 1,3-Dithiane Rings with Those for Cyclohexane

	$-\Delta G°$ (kcal/mol)					
		Tetrahydro-pyran[a]	1,3-dioxane[b]		1,3-dithiane[c]	
Group	Cyclohexane	2-Position	2-Position	5-Position	2-Position	5-Position
CH_3-	1.8	2.9	4.0	0.8	1.8	1.0
CH_3CH_2-	1.8		4.0	0.7	1.5	0.8
$(CH_3)_2CH-$	2.1		4.2	1.0	1.5	0.8
$(CH_3)_3C-$	>4.5			1.4	>2.7	
$CH_2{=}CH-$	1.7	2.3				
$CH{\equiv}C-$	0.5	0.3				

a. E. L. Eliel, K. D. Hargrave, K. M. Pietrusiewicz, and M. Manoharan, *J. Am. Chem. Soc.* **104**, 3635 (1982).
b. E. L. Eliel and M. C. Knoeber, *J. Am. Chem. Soc.* **90**, 3444 (1968); F. W. Nader and E. L. Eliel, *J. Am. Chem. Soc.* **92**, 3050 (1970).
c. E. L. Eliel and R. O. Hutchins, *J. Am. Chem. Soc.* **91**, 2703 (1969).

The decreased preference for the equatorial orientation of a 5-alkyl group in 1,3-dioxanes and 1,3-dithianes is evident from these data. It is also interesting that the increased preference for the equatorial orientation of a 2-methyl group in 1,3-dioxane disappears in going to 1,3-dithiane. The conformational free energies of 2-alkyl substituents in 1,3-dithiane are very similar to those in cyclohexane (actually, slightly smaller) because of the longer length of the C—S bond compared to C—O.

When a polar substituent is present, interactions between the substituent and the ring heteroatom can become an important factor in the position of the conformational equilibrium. In some cases, the interactions are straightforward and readily assessed. For example, the preferred conformation of 5-hydroxy-1,3-dioxane has the hydroxyl group in the axial position.[67] This conformation is favored because hydrogen bonding of the hydroxyl group with the ring oxygen, which is possible only when the hydroxyl group is axial, serves as a stabilizing force for this conformation.

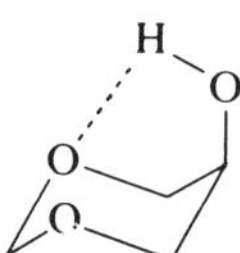

The incorporation of heteroatoms can result in stereoelectronic effects that have a pronounced effect on conformation and, ultimately, on reactivity. It is known from numerous examples in carbohydrate chemistry that pyranose sugars substituted

67. J. L. Alonso and E. B. Wilson, *J. Am. Chem. Soc.* **102**, 1248 (1980); N. Baggett, M. A. Bukhari, A. B. Foster, J. Lehmann, and J. M. Webber, *J. Chem. Soc.*, 4157 (1963).

with an electron-withdrawing group such as halogen or alkoxy at C-1 are often more stable when the substituent has an axial orientation, rather than an equatorial one. This tendency is not limited to carbohydrates but carries over to simpler ring system such as 2-substituted tetrahydropyrans. The phenomenon is known as the *anomeric effect*, since it involves a substituent at the anomeric position in carbohydrate pyranose rings.[68] Scheme 3.1 lists several compounds that exhibit the anomeric effect, along with some measured equilibrium distributions. In entries 1-3, the equilibria are between diastereomers, while entries 4-6 illustrate the anomeric effect in conformationally mobile systems. In all cases, the more stable isomer is written on the right.

The magnitude of the anomeric effect depends on the nature of the substituent and decreases with increasing dielectric constant of the medium. The effect of the substituent can be seen by comparing the related 2-chloro- and 2-methoxy-substituted tetrahydropyrans in entries 2 and 3. The 2-chloro compound exhibits a significantly greater preference for the axial orientation than the 2-methoxy. Entry 3 also provides data regarding the effect of solvent polarity, where it is observed that the equilibrium constant is larger in carbon tetrachloride ($\varepsilon = 2.2$) than in acetonitrile ($\varepsilon = 37.5$).

Compounds in which conformational, rather than configurational, equilibria are influenced by the anomeric effect are depicted in entries 4–6. Single-crystal X-ray diffraction studies have unambiguously established that all the chlorine atoms of *trans,syn,trans*-2,3,5,6-tetrachloro-1,4-dioxane occupy axial sites in the crystal. Each chlorine in the molecule is bonded to an anomeric carbon and is subject to the anomeric effect. Equally striking is the observation that all the substituents of the tri-*O*-acetyl-β-D-xylopyranosyl chloride shown in entry 5 are in the axial orientation *in solution*. Here, no special crystal packing forces can be invoked to rationalize the preferred conformation. The anomeric effect of a single chlorine is sufficient to drive the equilibrium in favor of the conformation that puts the three acetoxy groups in axial positions.

Several structural factors have been considered as possible cause of the anomeric effect. In localized bond terminology, it can be recognized that there will be a dipole–dipole interaction between the polar bonds at the anomeric carbon. This dipole–dipole interaction is reduced in the axial conformation, and this factor might contribute to the solvent dependence of the anomeric effect. From the molecular orbital viewpoint, the anomeric effect is expressed as resulting from an interaction between the lone pair electrons on the pyran oxygen and the σ^* orbital associated with the bond to the C-2 substituent.[69] When the C—X bond is axial, an interaction

68. For reviews, see P. L. Durrette and D. Horton, *Adv. Carbohydr. Chem. Biochem.* **26**, 49 (1971); R. U. Lemieux, *Pure Appl. Chem.* **25**, 527 (1971); W. A. Szarek and D. Horton (eds.), *Anomeric Effects*, ACS Symposium Series, No. 87, American Chemical Society, Washington, D.C., 1979; A. J. Kirby, *The Anomeric Effect and Related Stereoelectronic Effects at Oxygen*, Springer-Verlag, Berlin, 1983; P. Deslongchamps, *Stereoelectronic Effects in Organic Chemistry*, Pergamon Press, Oxford, 1983.
69. S. Wolfe, A. Rauk, L. M. Tel, and I. G. Csizmaida, *J. Chem. Soc. B*, 136 (1971); S. O. David, O. Eisenstein, W. J. Hehre, L. Salem, and R. Hoffmann, *J. Am. Chem. Soc.* **95**, 306 (1973); F. A Van-Catledge, *J. Am. Chem. Soc.* **96**, 5693 (1974).

Scheme 3.1. Equilibria in Compounds That Exhibit the Anomeric Effect

1[a] Glucose pentaacetate

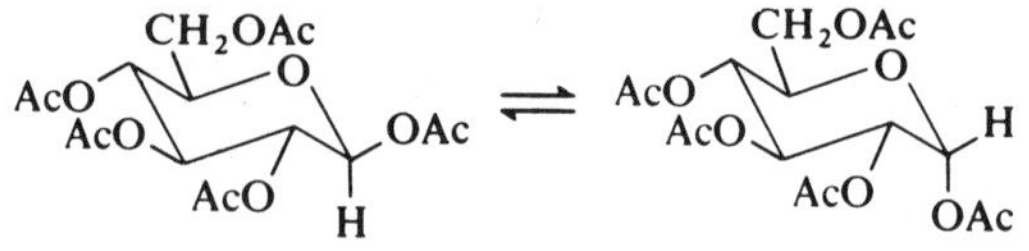

CH$_2$OAc, AcO, AcO, AcO, O, OAc, H ⇌ CH$_2$OAc, AcO, AcO, AcO, O, H, OAc

β α

$K = 5$ (in 50% acetic acid : acetic anhydride, 0.1 M H_2SO_4 at 25°C)
$\Delta H° = -1.4$ kcal/mol

2[b] 2-Chloro-4-methyltetrahydropyran

H_3C, O, Cl ⇌ H_3C, O, Cl

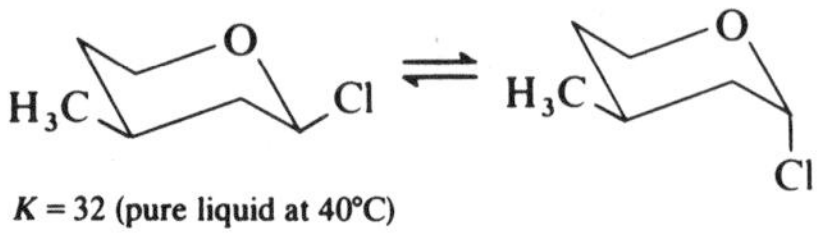

$K = 32$ (pure liquid at 40°C)

3[c] 2-Methoxy-6-methyltetrahydropyran

CH_3, O, OCH_3 ⇌ CH_3, O, OCH_3

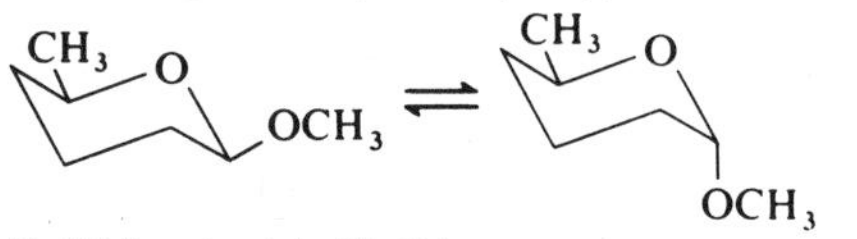

$K = 3.4$ (in carbon tetrachloride)
$K = 1.8$ (in acetonitrile)

4[d] *trans*, *syn*, *trans*-2,3,5,6-Tetrachloro-1,4-dioxane

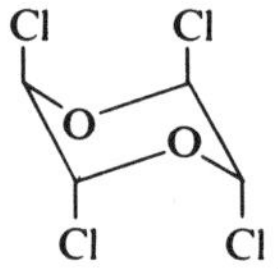

Cl, Cl, O, O, Cl, Cl

Equilibrium constant not known in solution; crystalline form has all chlorines axial.

5[e] Tri-*O*-acetyl-β-D-xylopyranosyl chloride

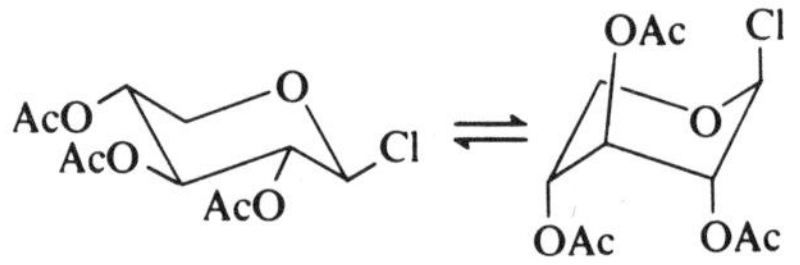

AcO, AcO, AcO, O, Cl ⇌ OAc, Cl, O, OAc, OAc

The NMR spectrum in $CDCl_3$ indicates that the all-axial form is strongly favored. The equilibrium constant is not known.

6[f] α-D-Altropyranose pentaacetate

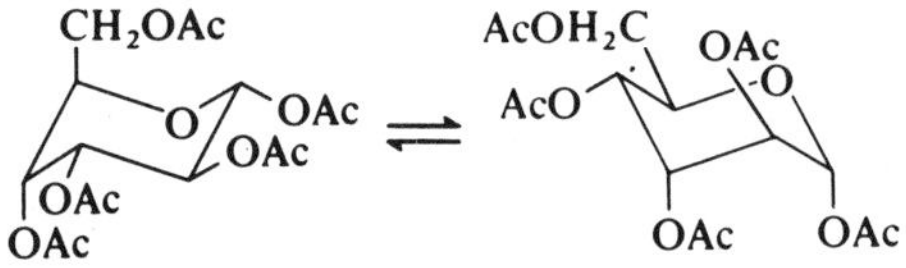

CH$_2$OAc, O, OAc, OAc, OAc, OAc ⇌ AcOH$_2$C, OAc, O, AcO, OAc, OAc

only form present at equilibrium by NMR analysis

a. W. A. Bonner, *J. Am. Chem. Soc.* **73**, 2659 (1951).
b. C. B. Anderson and D. T. Sepp, *J. Org. Chem.* **32**, 607 (1967).
c. E. L. Eliel and C. A. Giza, *J. Org. Chem.* **33**, 3754 (1968).
d. E. W. M. Rutten, N. Nibbering, C. H. MacGillavry, and C. Romers, *Rec. Trav. Chim.* **87**, 888 (1968).
e. C. V. Holland, D. Horton, and J. S. Jewell, *J. Org. Chem.* **32**, 1818 (1967).
f. B. Coxon, *Carbohydr. Res.* **1**, 357 (1966).

between an occupied p-type orbital on oxygen (lone pair electrons) with the antibonding σ^* orbital of the C—X combination is possible. This permits delocalization of the lone pair electrons and would be expected to shorten the C—O bond while lengthening the C—X bond. In 2-alkoxy tetrahydropyran derivatives, there is a correlation between the length of the exocyclic C—O bond and the nature of the oxygen substituent. The more electron withdrawing the group, the longer is the bond to the oxygen. This indicates that the extent of the anomeric effect increases with the electron-accepting capacity of the exocyclic oxygen.[70]

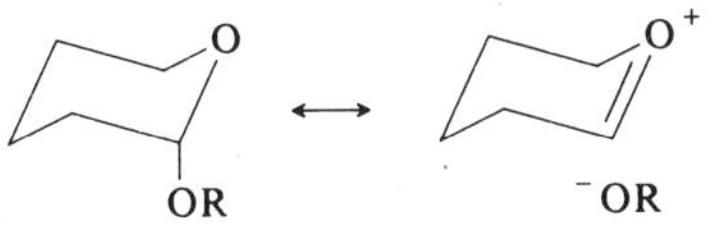

O ⟷ O^+
OR $^-$OR

Extent of bond lengthening increases with electron-accepting capacity of OR.

The anomeric effect is also present in acyclic systems and stabilizes conformations that allow antiperiplanar alignment of the C—X bond with a lone pair orbital of the heteroatom. In addition to α-haloethers, anomeric effects are prominent in determining the conformation of acetals and α-alkoxyamines.

Molecular orbital calculations (4-31G) have found 4 kcal as the difference between the two conformations shown below for methoxymethyl chloride.[71]

H H H
O—C O—C H
H_3C Cl H_3C Cl

sc *ap*
favored

The conformation marked *sc* is appropriate for overlap of an oxygen nonbonded electron pair with the σ^* C—Cl orbital. Because of the donor-acceptor nature of the interaction, it is enhanced in the order $F < O < N$ for the donor atom and $N < O < F$ for the acceptor atom.

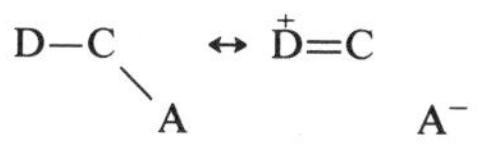

$$\text{D—C(—A)} \leftrightarrow \overset{+}{\text{D}}\text{=C} \quad \text{A}^-$$

An isodesmic reaction series

$$CH_4 + X{-}CH_2{-}Y \rightarrow CH_3{-}X + CH_3{-}Y$$

70. A. J. Briggs, R. Glenn, P. G. Jones, A. J. Kirby, and P. Ramaswamy, *J. Am. Chem. Soc.* **106**, 6200 (1984).
71. G. A. Jeffrey and J. H. Yates, *J. Am. Chem. Soc.* **101**, 820 (1979).

should measure the stabilizing effect. The following results were obtained using 3-21G level calculations.[72]

Stabilization Calculated for Anomeric Effect (kcal/mol)

	Y (acceptor)		
X (donor)	NH_2	OH	F
NH_2	10.6	12.7	17.6
OH		17.4	16.2
F			13.9

Simple acyclic acetals can possess six distinct conformations.

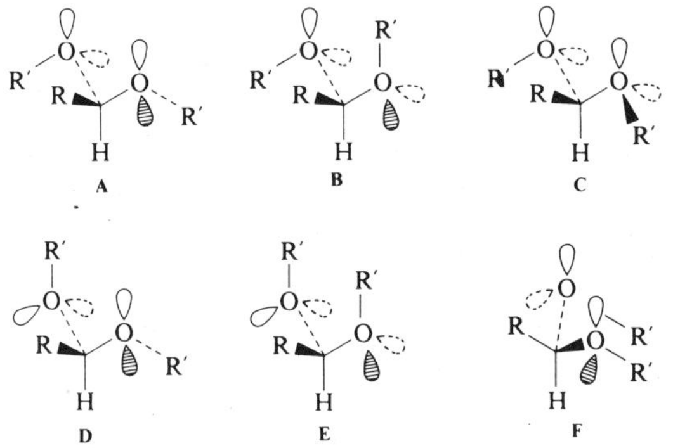

A B C D E F

The preferred conformation is **D** because it maximizes the number of antiperiplanar relationships between nonbonded electron pairs and C—O bonds while avoiding the R′-R′ repulsions in conformations **E** and **F**.

In cyclic systems such as **1**, the dominant conformation is the one with the maximum anomeric effect. In the case of **1**, only conformation **1A** corresponds to the preferred geometry.[73] Other effects, such as torsional strain and nonbonded repulsion, contribute to the conformational equilibrium, of course. Normally, a value of about 1.5 kcal/mol is assigned to the stabilization due to an optimum anomeric interaction in an acetal. This is substantially less than the value suggested by the theoretical results.

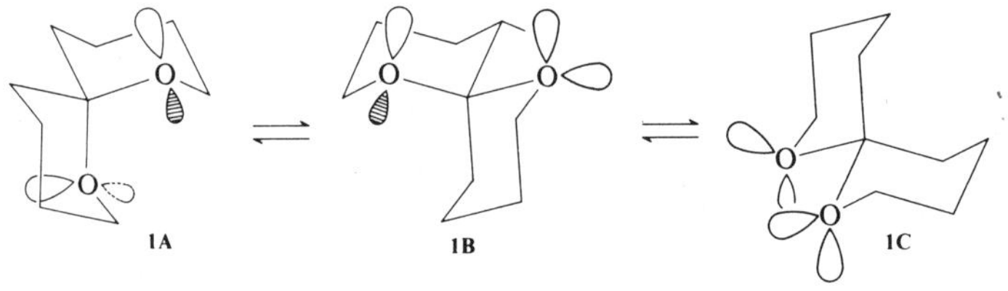

1A 1B 1C

72. P. v. R. Schleyer, E. D. Jemmis, and G. W. Spitznagel, *J. Am. Chem. Soc.* **107**, 6393 (1985).
73. P. Deslongchamps, D. D. Rowan, N. Pothier, G. Sauve, and J. K. Saunders, *Can. J. Chem.* **59**, 1132 (1981).

The molecular mechanics approach to conformational analysis has the virtue of analyzing conformational factors in terms of molecular properties that are easy to describe in physical terms. Contributions from bond-length distortions, bond-angle strain, van der Waals repulsions, and dipole–dipole interactions can be calculated. The use of carefully chosen potential functions can give highly precise information as to the relative energies of various conformations. The accuracy and reliability of the calculation depend on the potential functions and parameters that describe the various interactions.

The analysis of conformational equilibria can also be approached from the molecular orbital viewpoint.[74] The most stable conformation can be identified by searching for the energy minimum as a function of molecular geometry. Conformational barriers can be evaluated by calculation of the energy of the presumed transition state for the conformational change.[75]

The interpretation of molecular orbital calculations on conformational isomers is not as straightforward as for molecular mechanics methods. Because MO calculations treat all of the bonding forces of the molecule, the difference between two conformations represents only a small part of the total energy. Furthermore, unlike the molecular mechanics model in which energies are assigned to specific interatomic interactions, the energy of a specific molecular orbital may encompass contributions from a number of intermolecular interactions. Thus, the identification of the structural features responsible for the energy difference between two conformers may be very difficult.

Molecular orbital methods have been useful in identifying contributions to rotational barriers that may not be recognized as one of the usual effects included in molecular mechanics approaches. The case of fluoromethanol is illustrative. There is a substantial barrier to rotation of the hydroxyl hydrogen with respect to the fluoromethyl group, with the preferred orientation having the hydroxyl hydrogen *gauche* to the fluorine.[76] This conformation is 12.6 kcal/mol more stable than that having the fluorine *anti* to the hydroxyl hydrogen.

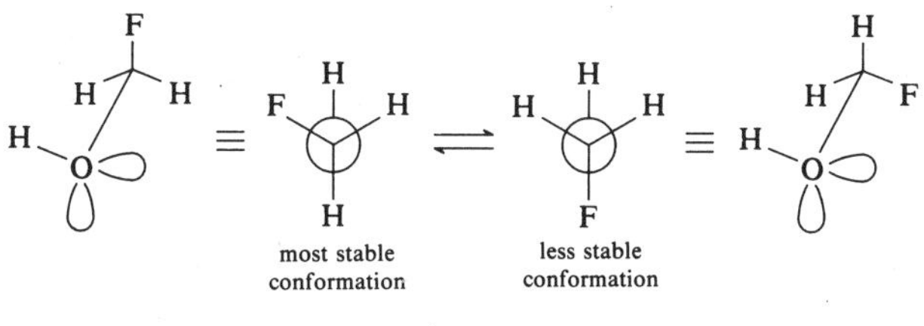

most stable conformation

less stable conformation

The preference for the *gauche* arrangement is an example of the anomeric effect. An oxygen lone pair is *anti* to fluorine in the stable conformation but not in the unstable conformation.

74. A. Golebiewski and A. Parczewski, *Chem. Rev.* **74**, 519 (1974).
75. W. J. Hehre, L. Radom, P. v. R. Schleyer, and J. A. Pople, *Ab Initio Molecular Orbital Theory*, Wiley, New York, 1986, pp. 261–270.
76. S. Wolfe, M.-H. Whangbo, and D. J. Mitchell, *Carbohydr. Res.* **69**, 1 (1979).

3.7. Conformational Effects on Reactivity

Conformational effects on reactivity have been particularly thoroughly studied in cyclohexane systems. The difference between an axial and an equatorial environment of a functional group can lead to significant differences in reaction rates. One of the most common ways of studying the effect of orientation on reactivity is to use an appropriately placed *t*-butyl or other large substituent to ensure that the reacting group is overwhelmingly in the equatorial or the axial position. The conformationally rigid *trans*-decalin system can also be used to assess reactivity differences between functional groups in axial versus equatorial positions.

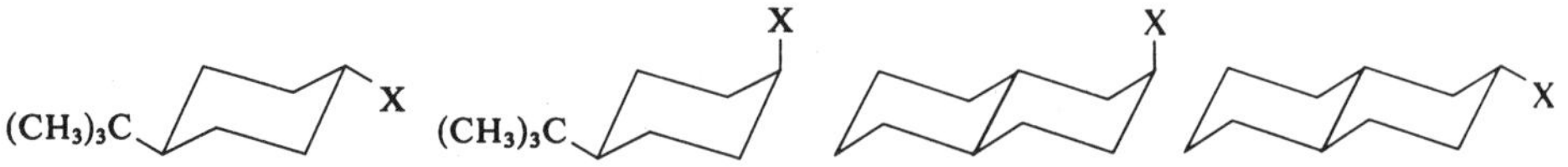

$(CH_3)_3C$ — X $(CH_3)_3C$ — X

Scheme 3.2 gives some data that illustrate the difference in reactivity between groups in axial and equatorial positions. It should be noted that a group can be either more reactive or less reactive in an axial position than in the corresponding equatorial position.

Scheme 3.2. Effects of Functional-Group Orientation on Rates and Equilibria

cis-4-*t*-Butylcyclohexanol		*trans*-4-*t*-Butylcyclohexanol
$(CH_3)_3C$ — OH	versus	$(CH_3)_3C$ — OH
Relative rate of oxidation (CrO_3) = 3.23[a] Relative rate of acetylation = 1.00[b]		Relative rate of oxidation (CrO_3) = 1.00[a] Relative rate of acetylation = 3.70[b]
$(CH_3)_3C$ — $CO_2C_2H_5$	versus	$(CH_3)_3C$ — $CO_2C_2H_5$
Relative rate of saponification = 1.00[c]		Relative rate of saponification = 19.8[c]
$(CH_3)_3C$ — CO_2H	versus	$(CH_3)_3C$ — CO_2H
$pK_a = 8.23$[d]		$pK_a = 7.79$[d]

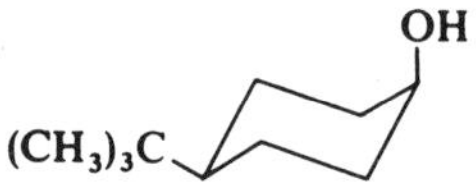

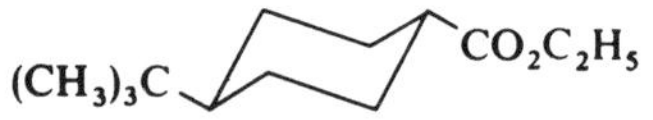

a. E. L. Eliel, S. H. Schroeter, T. J. Brett, F. J. Biros, and J.-C. Richer, *J. Am. Chem. Soc.* **88**, 3327 (1966).
b. E. L. Eliel and F. J. Biros, *J. Am. Chem. Soc.* **88**, 3334 (1966).
c. E. L. Eliel, H. Haubenstock, and R. V. Acharya, *J. Am. Chem. Soc.* **83**, 2351 (1961).
d. R. D. Stolow, *J. Am. Chem. Soc.* **81**, 5806 (1959).

The effect of conformation on reactivity is intimately associated with the details of the mechanism of a reaction. The examples of Scheme 3.2 can illustrate some of the ways in which substituent orientation can affect reactivity. It has been shown that oxidation of *cis*-4-*t*-butylcyclohexanol is faster than oxidation of the *trans* isomer, but the rates of acetylation are in the opposite order. Let us consider the acetylation first. The rate of the reaction will depend on the free energy of activation for the rate-determining step. For acetylation, this step involves nucleophilic attack by the hydroxyl group on the acetic anhydride carbonyl group to form the tetrahedral intermediates **I** and **J**. An approximate energy diagram is given in Fig. 3.8.

HO–C_6H_{10}–$C(CH_3)_3$ + $(CH_3\overset{\displaystyle O}{\overset{\|}{C}})_2O$ $\longrightarrow$ HO–$\underset{\displaystyle CH_3CO(=O)}{\overset{\displaystyle}{CH_3C}}$–O–$C_6H_{10}$–$C(CH_3)_3$ $\longrightarrow$ product

1 OH axial (*cis*)
2 OH equatorial (*trans*)

I *cis*
J *trans*

Because its hydroxyl group occupies an equatorial position, the *trans* isomer **2** is more stable than the *cis* isomer **1** by an amount equal to $-\Delta G°$ for the hydroxyl group. It can be assumed that the transition state for the rate-determining step will resemble the tetrahedral intermediates **I** and **J**. Since the substituent group has become larger as the acetylating reagent is bonded to the hydroxyl group, the value of $-\Delta G°$ for the substituent at the transition state should be *greater than* that for the hydroxyl group, 0.7 kcal/mol. Intermediate **I**, then, must be higher in energy

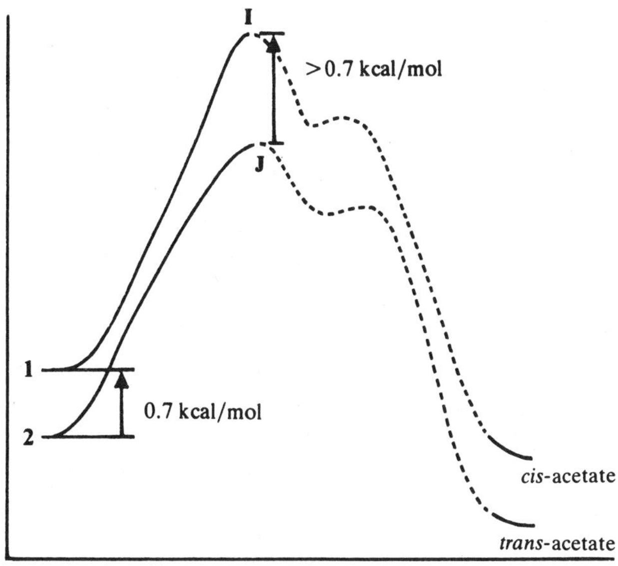

Fig. 3.8. Approximate energy diagram for acetylation of *cis*- and *trans*-4-*t*-butylcyclohexanol.

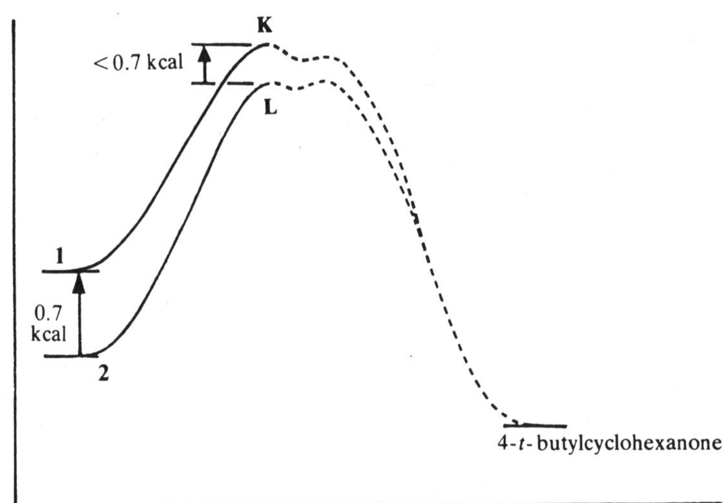

Fig. 3.9. Approximate energy diagram for oxidation of *cis*- and *trans*-4-*t*-butylcyclohexanol.

than **J** by more than 0.7 kcal/mol. From this information it can be predicted that **1** will acetylate more slowly than **2**, because a larger free energy of activation will be required. This is illustrated in Fig. 3.8. As shown by the data in Scheme 3.2, the prediction is correct.

Extensive research has established that axial cyclohexanols are more reactive than equatorial alcohols toward chromic acid oxidation.[77] The basis for this effect can be seen by analyzing the free energies of activation for the reactions. The available evidence indicates that the rate-determining step is breakdown of a chromate ester intermediate. The transition state involves partial cleavage of the C—H bond proceeding toward loss of chromium. An approximate energy diagram is given in Fig. 3.9.

K **L**

The diaxial interactions that are responsible for a large portion of the conformational free energy of the hydroxyl group are relieved in the transition state as the reaction proceeds toward sp^2 hybridization at the carbon atom undergoing oxidation. Because the substituent is effectively becoming smaller as reaction proceeds, the energy difference between the diastereomeric transition states is less than that

77. E. L. Eliel, S. H. Schroeter, T. J. Brett, F. J. Biros, and J.-C. Richer, *J. Am. Chem. Soc.* **88**, 3327 (1966); P. Mueller and J.-C. Perlberger, *J. Am. Chem. Soc.* **98**, 8407 (1976).

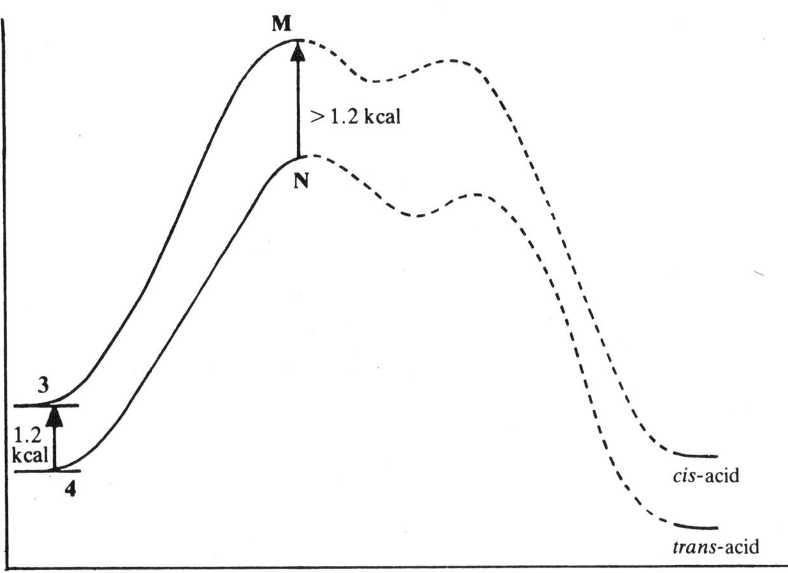

Fig. 3.10. Approximate energy diagram for saponification of ethyl esters of *cis*- and *trans*-4-*t*-butylcyclohexanecarboxylic acid.

between the reactant alcohols. Under these circumstances, the higher-energy *cis* isomer is the more reactive of the two alcohols.

A similar analysis of the hydrolysis of the esters **3** and **4** is possible. From Table 3.6 (p. 135), we see that the conformational free energy of the carboethoxy group is 1.2 kcal/mol. The *cis* isomer is this much higher in energy than the *trans* isomer. The transition states resemble intermediates **M** and **N**. The substituent group increases in size as the transition state is reached. As a result, the difference in energy between **M** and **N** must be greater than 1.2 kcal/mol. As can be concluded by examining the approximate energy diagram in Fig. 3.10, the *trans* isomer, with the equatorial carboethoxy group, hydrolyzes significantly faster than the *cis* isomer.

$(CH_3)_3C$–(cyclohexane)–$CO_2C_2H_5$ (**3**) ⟶ $(CH_3)_3C$–(cyclohexane)–$C(O^-)(OH)(OC_2H_5)$ (**M**)

$(CH_3)_3C$–(cyclohexane)–$CO_2C_2H_5$ (**4**) ⟶ $(CH_3)_3C$–(cyclohexane)–$C(OH)(O^-)(OC_2H_5)$ (**N**)

Many examples of reactivity effects that are due to the anomeric effect have been identified. For example, CrO_3 can oxidize some pyranose acetals, leading eventually to δ-ketoesters.

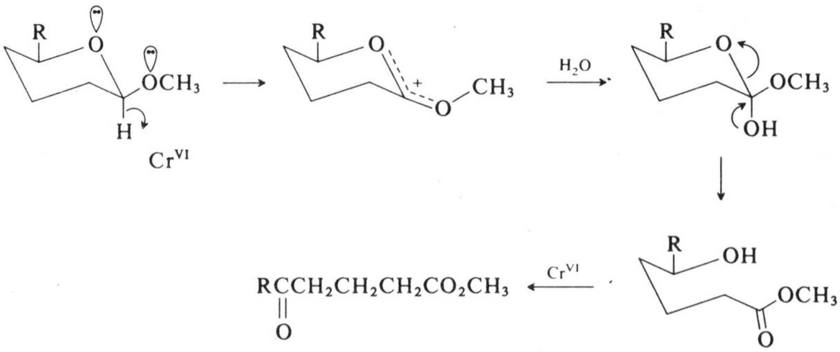

$RCCH_2CH_2CH_2CO_2CH_3$ (C=O)

Isomers with equatorial 2-alkoxy groups are more reactive than those with axial 2-alkoxy groups.[78] The greater reactivity of the equatorial isomers is the result of the alignment of the lone pairs on both the endocyclic and the exocyclic oxygen to assist in abstraction of the hydrogen.

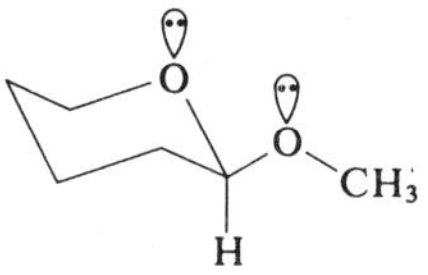

two lone pair orbitals in antiperiplanar arrangement

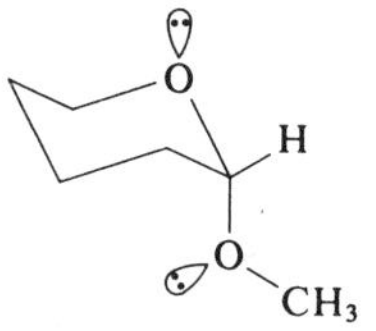

only one lone pair orbital in antiperiplanar arrangement

Other reagents which oxidize acetals such as ozone and *N*-bromosuccinimide show similar reactivity trends.[79]

Another example is the lack of oxygen exchange with solvent in the hydrolysis of gluconolactone. Simple acyclic esters usually undergo oxygen exchange at a rate that is competitive with hydrolysis. This occurs through the intermediacy of the addition intermediate.

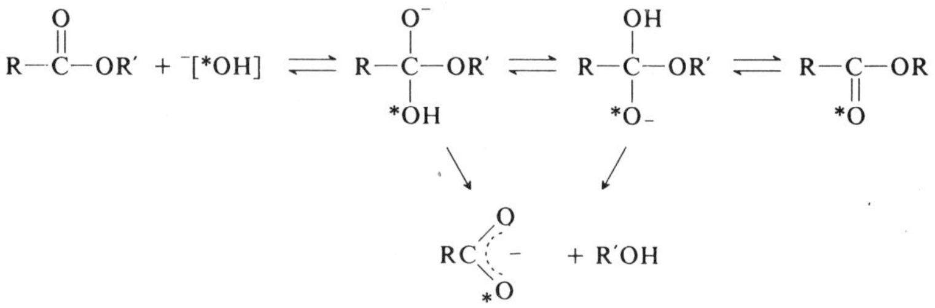

$$R{-}\overset{O}{\overset{\|}{C}}{-}OR' + {}^{-}[{}^{*}OH] \rightleftharpoons R{-}\overset{O^-}{\overset{|}{\underset{{}^{*}OH}{\underset{|}{C}}}}{-}OR' \rightleftharpoons R{-}\overset{OH}{\overset{|}{\underset{{}^{*}O^-}{\underset{|}{C}}}}{-}OR' \rightleftharpoons R{-}\underset{{}^{*}O}{\underset{\|}{C}}{-}OR$$

$$RCO_2^{-}\,(\text{one O labelled } {}^{*}O) + R'OH$$

78. S. J. Angyal and K. James, *Aust. J. Chem.* **23**, 1209 (1970).
79. P. Deslongchamps, C. Moreau, D. Frehel, and R. Chevenert, *Can. J. Chem.* **53**, 1204 (1975) and preceding papers; C. W. McClelland, *J. Chem. Soc., Chem. Commun.*, 751 (1979).

Gluconolactone shows no exchange.[80] The reason is that the tetrahedral intermediate is formed and breaks down stereoselectively.

HOCH2 O HO HO HO O [*OH] HOCH2 *OH HO HO HO O− O HOCH2 *O− HO HO HO OH O

gluronolactone

reverse reaction favored by two antiperiplanar lone pair orbitals

expulsion of equatorial OH favored by only one lone pair orbital

HOCH2 O* HO HO HO O

Even though proton exchange can occur in the tetrahedral intermediate, the anomeric effect leads to preferential loss of the axial oxygen.

3.8. Angle Strain and Its Effect on Reactivity

Another important factor in reactivity is angle strain. Table 3.9 gives some data on the total angle strain for some cyclic, bicyclic, and tricyclic hydrocarbons. Six-membered rings are nearly strain-free, while the strain energy in smaller rings increases from 6–7 kcal/mol for cyclopentane to about 30 kcal/mol for cyclopropane. In more complex structures, total strain increases as molecular geometry requires greater distortion from optimal bond angles.

Because of the increased ground state energy resulting from angle strain, reactions which lead to ring opening often proceed much more readily than do similar reactions in unstrained systems. For example, while normal saturated hydrocarbons are inert to bromine in the dark, cyclopropane reacts rapidly, giving ring-opened products.[81] The products arise from ring opening to yield a carbocation, followed by capture by bromide ion. Two minor products arise from rearrangement

80. Y. Pocker and E. Green, *J. Am. Chem. Soc.* **95**, 113 (1973).
81. J. B. Lambert and B. A. Iwanetz, *J. Org. Chem.* **37**, 4082 (1972); J. B. Lambert and K. Kobayashi, *J. Org. Chem.* **41**, 671 (1976); P. S. Skell, J. C. Day, and K. H. Shea, *J. Am. Chem. Soc.* **94**, 1126 (1972); J. B. Lambert, W. J. Schulz, Jr., P. H. Mueller, and K. Kobayashi, *J. Am. Chem. Soc.* **106**, 792 (1984).

Table 3.9. Strain Energies in Some Alicyclic Compounds (kcal/mol)[a]

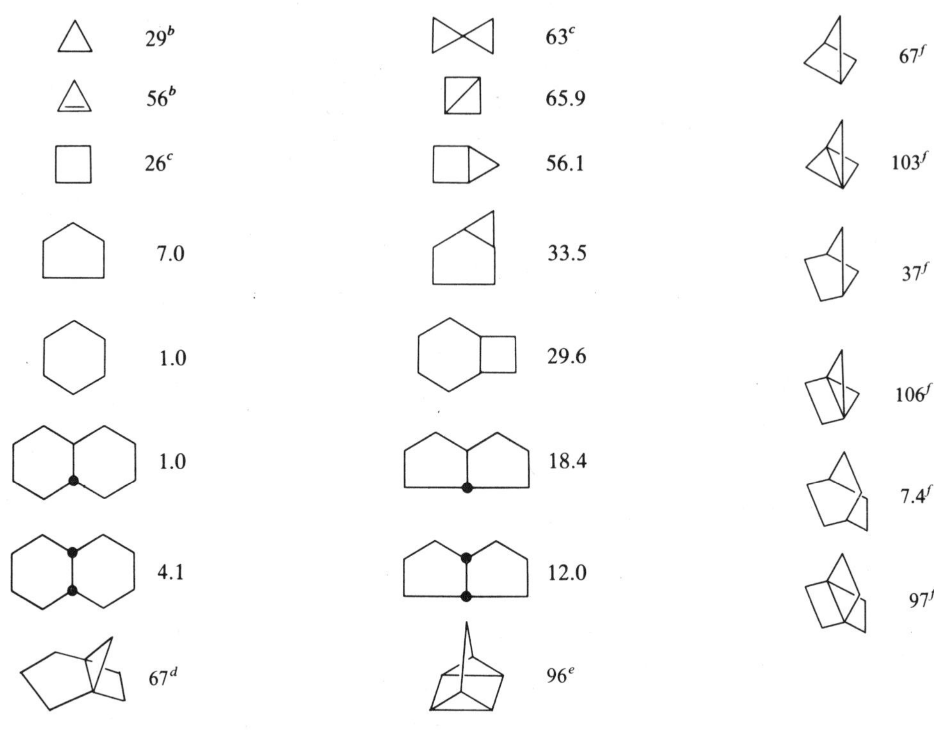

a. Data from S. Chang, D. McNally, S. Shary-Tehrany, M. J. Hickey, and R. H. Boyd, *J. Am. Chem. Soc.* **92**, 3109 (1970), except where noted otherwise.
b. G. L. Closs, in *Advances in Alicyclic Chemistry*, Vol. 1, H. Hart and G. J. Karabatsos (eds.), Academic Press, New York, 1966, p. 67.
c. P. v. R. Schleyer, J. E. Williams, and K. R. Blanchard, *J. Am. Chem. Soc.* **92**, 2377 (1970).
d. K. B. Wiberg, H. A. Connon, and W. E. Pratt, *J. Am. Chem. Soc.* **101**, 6970 (1979).
e. D. S. Kabakoff, J.-C. G. Bünzli, J. F. M. Oth, W. B. Hammon, and J. A. Berson, *J. Am. Chem. Soc.* **97**, 1510 (1975).
f. K. B. Wiberg, *J. Am. Chem. Soc.* **105**, 1227 (1983).

of the cationic intermediate.

$$\text{trans-1,2-dimethylcyclopropane} + Br_2 \rightarrow \underset{\sim 60\%}{CH_3CH(Br)CH(CH_3)CH_2Br} + \underset{\sim 20\%}{BrCH_2CH(Br)C(CH_3)_2} + \underset{\sim 20\%}{BrCH_2C(CH_3)(Br)CH_2CH_3} + \text{minor products}$$

Kinetic and product structure studies of the reaction of cyclopropanes and bicyclic compounds containing three-membered rings have shown that the protonation of the cyclopropane ring is followed by addition of the nucleophile at the most substituted carbon. The product composition is determined by the ability of the more highly substituted carbons to sustain more of the positive charge. The substitution at the incipient carbocation is the most important factor in determining the degree of reactivity. The relative rates of cyclopropane and its methyl, 1,1-

dimethyl, 1,1,2-trimethyl, and 1,1,2,2-tetramethyl derivatives, as given below, demonstrate the accelerating effect of electron-donating methyl groups.[82]

$\xrightarrow{CH_3CO_2H}$ solvolysis products

substituents	relative rate
none	1
1-Methyl	91
1,1-Dimethyl	639
1,1,2-Trimethyl	668
1,1,2,2-Tetramethyl	2135

Bicyclo[1.1.0]butane is an example of a molecule in which severe angle strain results in decreased stability and greatly enhanced reactivity. The central bond in bicyclo[1.1.0]butane is formed from nearly pure *p* orbitals of the two bridgehead carbons.[83] Bicyclo[1.1.0]butane and its derivatives undergo ring opening very readily with the halogens, weak acids, and a variety of other reagents. The products are derived from cleavage of both the central and the peripheral bonds.

$\xrightarrow{Cl_2}$ $H_2C{=}CHCH_2CHCl_2$ (45%) + Cl—◇—Cl (4%) + Cl△CH_2Cl (41%) + △CH_2Cl Cl (6%)

Ref. 84a

$\xrightarrow{CH_3CO_2H}$ $H_2C{=}CH(CH_2)_2OCCH_3$ (O) 10% + ◇—$OCCH_3$ (O) 23% + ▷—CH_2OCCH_3 (O) 67% Ref. 84a

CH_3, H_3C, H_3C $\xrightarrow{I_2}$ I CH_3 H_3C H_3C I H + $(CH_3)_2C{=}\overset{CH_3}{C}CH_2CHI_2$ Ref. 84b

The propellanes (see pp. 7–8) represent another interesting group of molecules whose reactivity reflects the effects of distorted bond angles. As the bridges in the propellanes are made smaller, the molecules become more reactive. [2.2.1]Propellane can be isolated in solid argon at 45 K but decomposes at higher temperatures and has not been obtained as a pure substance.[85] The trend toward decreasing stability with increasing strain make a sharp reversal at [1.1.1]propellane. This compound

82. K. B. Wiberg and S. R. Kass, *J. Am. Chem. Soc.* **107**, 988 (1985); K. B. Wiberg, S. R. Kass, and K. C. Bishop III, *J. Am. Chem. Soc.* **107**, 996 (1985).
83. J. M. Schulman and G. J. Fisanick, *J. Am. Chem. Soc.* **92**, 6653 (1970); R. D. Bertrand, D. M. Grant, E. L. Allred, J. C. Hinshaw, and A. B. Strong, *J. Am. Chem. Soc.* **94**, 997 (1972); D. R. Whitman and J. F. Chiang, *J. Am. Chem. Soc.* **94**, 1126 (1972).
84. (a) K. B. Wiberg and G. Szeimies, *J. Am. Chem. Soc.* **92**, 571 (1970); (b) W. R. Moore, K. G. Taylor, P. Muller, S. S. Hall, and Z. L. F. Gaibel, *Tetrahedron Lett.*, 2365 (1970).
85. F. H. Walker, K. B. Wiberg, and J. Michl, *J. Am. Chem. Soc.* **104**, 2056 (1982); K. B. Wiberg and F. H. Walker, *J. Am. Chem. Soc.* **104**, 5239 (1982).

is observed to decompose much more slowly than [2.2.1]propellane. To understand this situation, it must be recognized that a major factor in determining stability is the strength of the central bond toward homolytic cleavage, which provides a path for decomposition. This energy is strongly influenced by the *difference* in the strain energy of the reactant and the resulting diradical. The more strain relieved by the bond rupture, the more reactive is the molecule. The substantially increased stability of [1.1.1]propellane is due to the fact that not much strain is relieved at the diradical stage because the diradical remains highly strained. As the structural formula implies, [1.1.1]propellane has a very unusual shape. All four bonds to each bridgehead carbon are directed to the same side of the atom.[86]

The molecule tetrahedrane represents still another increment of angle strain.

tetrahedrane

$C(CH_3)_3$
$(CH_3)_3C$ — — $C(CH_3)_3$
$C(CH_3)_3$

The parent compound has not yet been synthesized but the tetra-*t*-butyl derivative is known. Molecular orbital calculations estimate that the breaking of a C—C bond in tetrahedrane would require only about 10 kcal/mol, indicating that the molecule would have only a short existence even at quite low temperatures.[87] The tetra-*t*-butyl derivative, however, is stable up to 100°C.[88]

Alkenes exhibit large strain energy when molecular geometry does not permit all the bonds to the two sp^2-hybridized carbons to be coplanar. An example that illustrates this point is *trans*-cycloheptene:

H
H

With only five methylene units available to bridge the *trans* positions, the molecule is highly strained and very reactive. Isolation of *trans*-cycloheptene has not been possible, but evidence for its formation has been obtained by trapping experiments.[89] The alkene is generated in the presence of a reagent with which it is expected to react rapidly, in this case, the very reactive Diels–Alder diene 2,5-diphenyl-3,4-isobenzofuran. The adduct that is isolated has the structure anticipated for that derived from *trans*-cycloheptene. The lifetime of *trans*-cycloheptene has been measured after generation by photoisomerization of the *cis* isomer. The activation energy for isomerization to *cis*-cycloheptene is about 17 kcal/mol. The lifetime in pentane is on the order of minutes at 0°.[90]

86. L. Hedberg and K. Hedberg, *J. Am. Chem. Soc.* **107**, 7257 (1985).
87. H. Kollmar, *J. Am. Chem. Soc.* **102**, 2617 (1980).
88. G. Maier, S. Pfriem, U. Schafer, and R. Matusch, *Angew. Chem. Int. Ed. Engl.* **17**, 520 (1978).
89. E. J. Corey, F. A. Carey, and R. A. E. Winter, *J. Am. Chem. Soc.* **87**, 934 (1965).
90. Y. Inoue, S. Takamuku, and H. Sakurai, *J. Chem. Soc., Perkin Trans. 2,* 1635 (1977); Y. Inoue, T. Ueoka, T. Kuroda, and T. Hakushi, *J. Chem. Soc., Perkin Trans. 2,* 983 (1983).

Table 3.10. Relative Stabilities of *cis*- and *trans*-Cycloalkenes

Cycloalkene	$\Delta H°$(*trans* ⇌ *cis*) (kcal/mol)	Ref.
Cycloheptene	−20.3	a
Cyclooctene	−9.7	b
Cyclononene	−2.8	b
Cyclodecene	−3.5	b
Cycloundecene	+0.1	b
Cyclododecene	+0.4	b

a. Calculated value, from N. L. Allinger and J. T. Sprague, *J. Am. Chem. Soc.* **94**, 5734 (1972).
b. From R. B. Turner and W. R. Meador, *J. Am. Chem. Soc.* **79**, 4133 (1957); A. C. Cope, P. T. Moore, and W. R. Moore, *J. Am. Chem. Soc.* **82**, 1744 (1960).

While *trans*-cyclohexene has been postulated as a reactive intermediate, it has not been observed directly. MO calculations at the 6-31G* level predict it to be 56 kcal/mol less stable than the *cis* isomer and the barrier to isomerization to be about 15 kcal/mol.[91]

trans-Cyclooctene is also significantly strained, but less so than *trans*-cycloheptene. As the ring size is increased, the amount of strain decreases. The *trans* isomers of both cyclononene and cyclodecene are less stable than the corresponding *cis* isomers, but for cycloundecene and cyclododecene, the *trans* isomers are the more stable.[92] Table 3.10 gives data concerning the relative stabilities of the *cis* and *trans* isomers of the C_7 through C_{12} cycloalkenes.

The geometry of bicyclic rings can also cause distortion of the alkene bond from coplanarity. An example is bicylo[2.2.1]hept-1-ene:

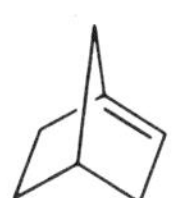

Attempts to construct a model of this molecule will show that the geometry of the bicyclic system does not permit coplanarity of the atoms bound to the sp^2 carbons. As a result of the strain, the molecule has, at most, transitory existence.[93] The absence of such "bridgehead double bonds" was noted long ago and formulated as *Bredt's rule.* As the structural basis for Bredt's rule became clear, it was evident that the prohibition against bridgehead double bonds would not be absolute.[94] When the bridges of the bicyclic system are large enough to permit planarity of the double bond, bridgehead alkenes are capable of existence. It has been proposed that the limit for unstable but isolable bridgehead alkenes is reached when the size of the

91. J. Verbeek, J. H. van Lenthe, P. J. J. A. Timmermans, A. Mackor, and P. H. M. Budzelaar, *J. Org. Chem.* **52**, 2955 (1987).
92. A. C. Cope, P. T. Moore, and W. R. Moore, *J. Am. Chem. Soc.* **82**, 1744 (1960).
93. R. Keese and E.-P. Krebs, *Angew. Chem. Int. Ed. Engl.* **11**, 518 (1972).
94. G. Kobrich, *Angew. Chem. Int. Ed. Engl.* **12**, 464 (1973).

Scheme 3.3. Bridgehead Alkenes[a]

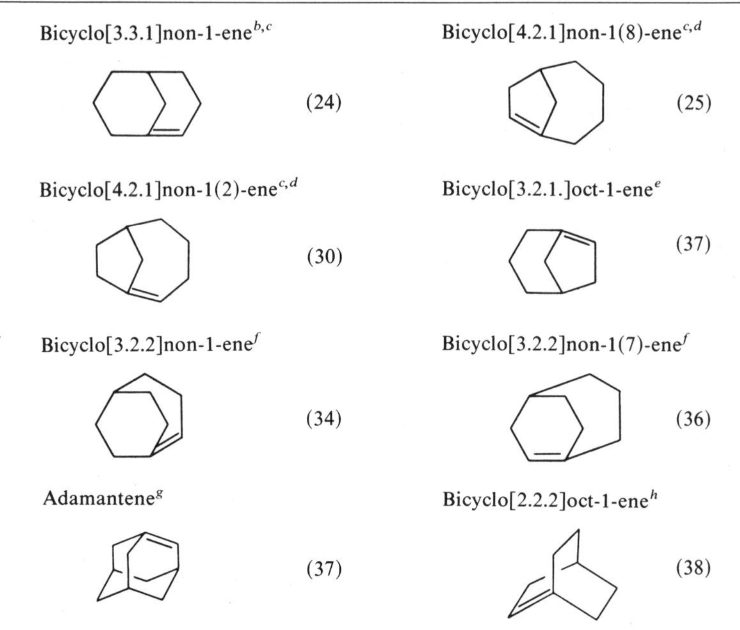

a. Strain energies calculated by molecular mechanics (Ref. 97) are given in parentheses in kcal/mol.
b. J. R. Wiseman and W. A. Pletcher, *J. Am. Chem. Soc.* **92**, 956 (1970); J. A. Marshall and H. Faubl, *J. Am. Chem. Soc.* **89**, 5965 (1967); M. Kim and J. D. White, *J. Am. Chem. Soc.* **99**, 1172 (1977).
c. K. B. Becker, *Helv. Chim. Acta* **60**, 81 (1977).
d. J. R. Wiseman, H.-F. Chan, and C. J. Ahola, *J. Am. Chem. Soc.* **91**, 2812 (1969).
e. W. G. Dauben and J. D. Robbins, *Tetrahedron Lett.* 151 (1975).
f. Transitory existence only; J. R. Wiseman and J. A. Chong, *J. Am. Chem. Soc.* **91**, 7775 (1969).
g. Transitory existence only; A. H. Alberts, J. Strating, and H. Wynberg, *Tetrahedron Lett.* 3047 (1973); J. E. Gano and L. Eizenberg, *J. Am. Chem. Soc.* **95**, 972 (1973); D. J. Martella, M. Jones, Jr., and P. v. R. Schleyer, *J. Am. Chem. Soc.* **100**, 2896 (1978); R. T. Conlin, R. D. Miller and J. Michl, *J. Am. Chem. Soc.* **101**, 7637 (1979).
h. A. D. Wolf and M. Jones, Jr., *J. Am. Chem. Soc.* **95**, 8209 (1973); H. H. Grootveld, C. Blomberg, and F. Bickelhaupt, *J. Chem. Soc., Chem. Commun.*, 542 (1973).

largest ring containing the double bond is at least eight atoms. Bridgehead alkenes in which the largest ring is made up of seven atoms are expected to be capable only of short existence.[95] These proposals have subsequently been tested and verified by the development of successful syntheses of bridgehead alkenes, such as those shown in Scheme 3.3.[96] The strained double bonds in these molecules are exceptionally reactive and undergo a variety of addition reactions. The total strain in the bridgehead alkenes can be computed by molecular mechanics methods. Some of the calculated strain energies are included in Scheme 3.3. This total strain energy can be dissected to indicate that fraction of the total strain which is due to the twist of the carbon–

95. J. R. Wiseman, *J. Am. Chem. Soc.* **89**, 5966 (1967); J. R. Wiseman and W. A. Pletcher, *J. Am. Chem. Soc.* **92**, 956 (1970).
96. For reviews of the synthesis and properties of bridgehead alkenes, see G. L. Buchanan, *Chem. Soc. Rev.* **3**, 41 (1974); K. J. Shea, *Tetrahedron* **36**, 1683 (1980); R. Keese, *Angew. Chem. Int. Ed. Engl.* **14**, 528 (1975); G. Szeimies, in *Reactive Intermediates*, Vol. 3, R. A. Abramovitch (ed.), Plenum Press, New York, 1983, Chapter 5.

	Reaction	Ring size =	Relative rate 3	4	5	6	7	8
1.[a]	$Br(CH_2)_xCO_2^-$ → lactone		8.3×10^{-4}	0.31	90	1	0.0052	6×10^{-5}
2.[b]	$Br(CH_2)_xNH_2$ → cyclic amine		0.07	0.001	100	1	0.002	—
3.[c]	$PhC(=O)(CH_2)_xCl$ → nucleophilic participation in solvolysis		—	0.37	36	1	0.13	—
4.[d]	o-^-O-C_6H_4-$O(CH_2)_xBr$ → cyclic ether formation		—	—	—	1	0.01	4×10^{-4}
5.[e]	$ArSO_2\bar{N}(CH_2)_xCl$ → cyclization		17	33	—	1	—	—

a. C. Galli, G. Illuminati, L. Mandolini, and P. Tamborra, *J. Am. Chem. Soc.* **99**, 2591 (1977); L. Mandolini, *J. Am. Chem. Soc.* **100**, 550 (1978).
b. D. F. DeTar and W. Brooks, Jr., *J. Org. Chem.* **43**, 2245 (1978); D. F. DeTar and N. P. Luthra, *J. Am. Chem. Soc.* **102**, 4505 (1980).
c. D. J. Pasto and M. P. Serve, *J. Am. Chem. Soc.* **87**, 1515 (1965).
d. G. Illuminati, L. Mandolini, and B. Masci, *J. Am. Chem. Soc.* **96**, 1422 (1974).
e. R. Bird, A. C. Knipe, and C. J. M. Stirling, *J. Chem. Soc., Perkin Trans. 2*, 1215 (1973).

carbon double bond. This strain proves to be a quite reliable predictor of the stability of bridgehead alkenes.[97]

3.9. Relationships between Ring Size and Facility of Ring Closure

Many examples of intramolecular reactions have served to establish a rough correlation between the rate of a reaction and the size of the ring being formed.[98] Although different reaction types exhibit large quantitative differences, and there are exceptions, the order $5 > 6 > 3 > 7 > 4 > 8\text{–}10$ is a rough guide of relative reactivity for many systems. Some quantitative data on typical reactions involving nucleophilic substitution or participation are shown in Scheme 3.4.

Table 3.11 gives rate data for ring closure of a series of ω-bromoalkyl malonate anions for ring sizes 4–13, 17, and 21. The rates range from a maximum of $6 \times 10^2\ s^{-1}$ for the five-membered ring to $1.4 \times 10^{-6}\ s^{-1}$ for the ten-membered ring.[99]

$$Br{-}(CH_2)_n\bar{C}(CO_2C_2H_5) \longrightarrow (CH_2)_n\,C(CO_2C_2H_5)_2$$

97. W. F. Maier and P. v. R. Schleyer, *J. Am. Chem. Soc.* **103**, 1891 (1981).
98. G. Illuminati and L. Mandolini, *Acc. Chem. Res.* **14**, 95 (1981); L. Mandolini, *Adv. Phys. Org. Chem.* **22**, 1 (1986).
99. M. A. Casadei, C. Galli, and L. Mandolini, *J. Am. Chem. Soc.* **106**, 1051 (1984).

Table 3.11. Relative Rates of Cyclization of Diethyl (ω-Bromoalkyl)-malonate Ester Anions as a Function of Ring Size[a]

Ring size	Relative rate
4	0.58
5	833
6	1.0
7	8.7×10^{-3}
8	1.5×10^{-4}
9	1.7×10^{-5}
10	1.4×10^{-6}
11	2.9×10^{-6}
12	4.0×10^{-4}
13	7.4×10^{-4}
17	2.9×10^{-3}
21	4.3×10^{-3}

a. M. A. Casadei, G. Galli, and L. Mandolini, *J. Am. Chem. Soc.* **106**, 1051 (1984).

Figure 3.11 shows the relative reactivity as a function of ring size for two intramolecular displacement reactions, namely, conversion of ω-bromoalkanecarboxylates to lactones and formation of ethers from ω-bromoalkyl monoethers of 1,2-dihydroxybenzene.

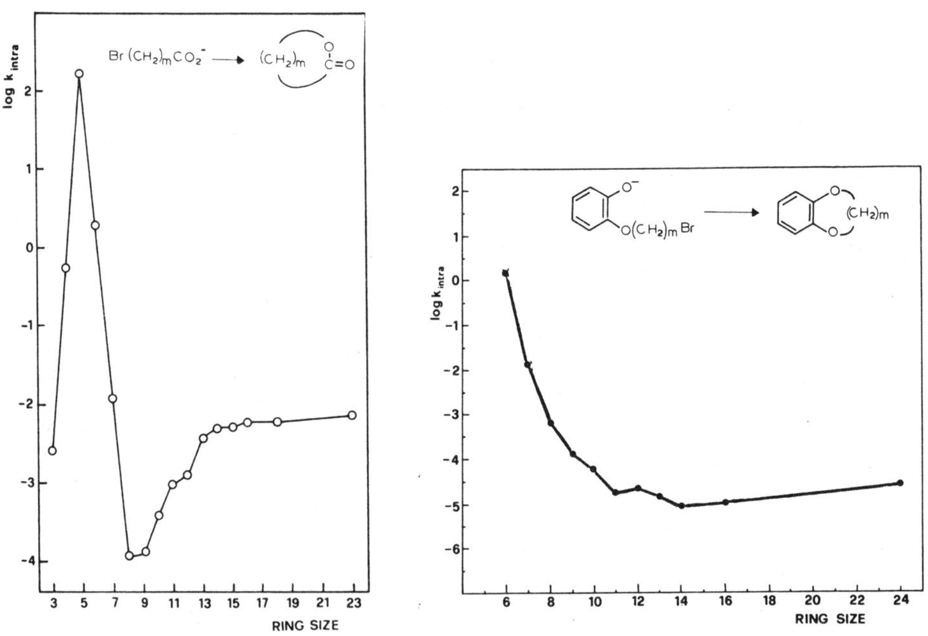

Fig. 3.11. Rates of ring closure for ω-bromoalkanecarboxylates (left) and ω-bromoalkyloxyphenolates (right). [Reproduced with permission from *Acc. Chem. Res.* **14**, 95 (1981).]

Table 3.12. Classification of Ring-Closure Types[a]

	Exocyclic bonds			Endocyclic bonds[b]	
Ring size	*sp* (*dig*)	sp^2 (*trig*)	sp^3 (*tet*)	*sp* (*dig*)	sp^2 (*trig*)
3	unfav	fav	fav	fav	unfav
4	unfav	fav	fav	fav	unfav
5	fav	fav	fav	fav	unfav
6	fav	fav	fav	fav	fav
7	fav	fav	fav	fav	fav

a. J. E. Baldwin, *J. Chem. Soc., Chem. Commun.*, 734 (1976).
b. The category *endo–tet* also exists but is somewhat rare and is not discussed here.

The disssection of the energy of activation of typical ring closure reactions usually shows some consistent features. The ΔH^{+} for formation of three- and four-membered rings is normally higher than ΔH^{+} for the corresponding five- and six-membered rings, while ΔS^{+} is least negative for the three-membered rings, of comparable magnitude for four-, five-, and six-membered rings, and then becomes more negative as the ring size increases above seven. The ΔH^{+} term reflects the strain which develops in the closure of three-membered rings, while the large negative entropy associated with eight-membered and larger rings reflects the relative improbability of achieving the required molecular orientation. Because the combination of the two factors is most favorable for five- and six-membered rings, the maximum rate is observed for these ring sizes.

Superimposed on this broad relationship between enthalpy and entropy are more variable and individualized structural features, including changes in solvation and the effect of branching on the intervening chain. Most important, however, are geometric (stereoelectronic) constraints on the transition state for ring closure.[100] There will be a preferred direction of approach which will vary depending on the type of reaction that is involved. While the reactions shown in Scheme 3.4, which are all intramolecular nucleophilic substitutions, reveal a general trend $5 > 6 > 3 > 7 > 8$, reactions with other mechanisms may exhibit a different relationship. For example, the formation of cyclopropyl ketones from 4-halo enolates is often faster than formation of the cyclopentyl analogs from 6-halo enolates.[101]

A systematic effort to correlate ease of ring closure with the stereoelectronic requirements of the transition state has been developed by Baldwin and co-workers. They classify ring closures with respect to three factors: (a) ring size, (b) the hybridization of the carbon at the reaction site, and (c) the relationship (endocyclic or exocyclic) of the reacting bond to the forming ring. Certain types are found to be favorable whereas others are unfavorable for stereoelectronic reasons. The relationships are summarized in Table 3.12.

The classification can be made clear with a few examples. All of the nucleophilic substitutions shown in Scheme 3.4 are of the *exo-tet* classification. The reacting

100. G. Illuminati and L. Mandolini, *Acc. Chem. Res.* **14**, 95 (1981).
101. C. J. M. Stirling, *J. Chem. Educ.* **50**, 844 (1973).

atom is of sp^3 hybridization (tetrahedral = tet) and the reacting bond, that is, the bond to the leaving group, is exocyclic to the forming ring:

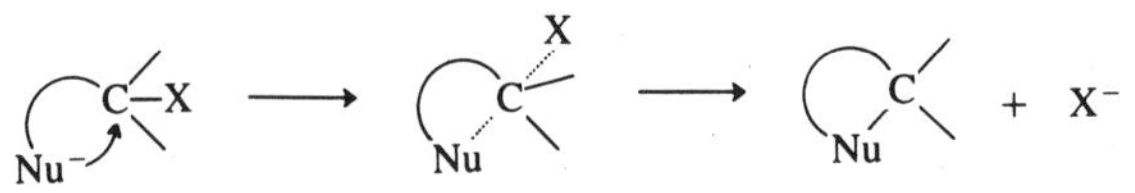

An example of an *exo-trig* process would be lactonization of a ω-hydroxycarboxylic acid:

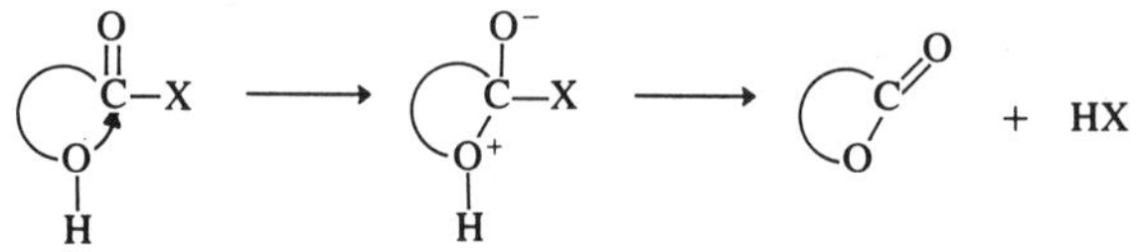

An example of an *exo-dig* process would be the base-catalyzed cyclization of an ε-hydroxy-α,β-ynone:

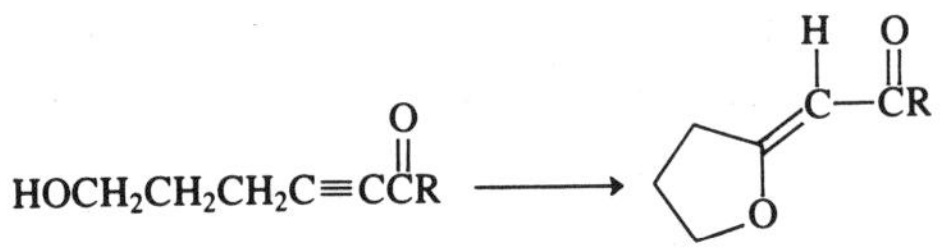

$HOCH_2CH_2CH_2C{\equiv}CCR$ (with C=O) $\longrightarrow$

Let us focus attention on the unfavorable ring closures. Why, for example, should formation of a five-membered ring by an *endo-trig* process be difficult? The answer is provided by a consideration of the trajectory of approach of the nucleophile.[102] If Z is an electron-attracting conjugating group of the type necessary to activate the double bond to nucleophilic attack, the reaction would involve the LUMO of the conjugated system, a π^* orbital. The nucleophile cannot approach in the nodal plane of the π system, and thus it must attack from above or below.

This stereoelectronic requirement would lead to a large distortion of the normal geometry of a five-membered ring and introduce strain. It is this distortion and strain which disfavors the 5-*endo-trig* cyclization. In contrast, 5-*endo-dig* cyclization is feasible since the acetylenic system provides an orbital which is available for a nearly planar mode of approach.

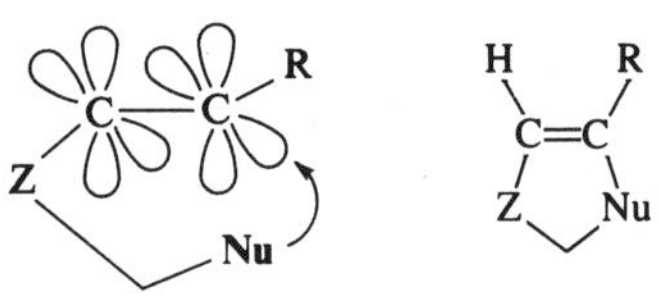

102. J. E. Baldwin, *J. Chem. Soc., Chem. Commun.*, 738 (1976).

In agreement with these analyses, it was found that compound **5** was unreactive toward base-catalyzed cyclization to **6**, even though the compound would appear to be reactive toward intramolecular conjugate addition. On the other hand, the acetylene **7** is readily cyclized to **8**[103]:

$$(CH_3)_2C(OH)C(=O)CH{=}CHPh \ (\mathbf{5}) \not\longrightarrow \mathbf{6}$$

$$(CH_3)_2C(OH)C(=O)C{\equiv}CPh \ (\mathbf{7}) \xrightarrow{CH_3ONa} \mathbf{8}$$

The terms favored and disfavored imply just that. Other factors will determine the absolute ease of a given ring closure, but these relationships point out the need to recognize the specific stereoelectronic requirements which may be imposed on the transition state in ring closure reactions.

3.10. Torsional and Stereoelectronic Effects on Reactivity

Torsional strain refers to the component of total molecular energy which results from nonoptimal arrangement of vicinal bonds, as in the eclipsed conformation of ethane. The origin and stereoelectronic nature of torsional strain was discussed in Section 3.1. The preference for staggered arrangements around single bonds is general for all alkanes, and when geometric constraints enforce an eclipsed arrangement, the molecule suffers torsional strain. Torsional strain that develops in a transition state raises the energy requirement of the reaction.

One case in which torsional effects play a major role is in reactions that involve hybridization changes at a ring atom. A general relationship concerning the relative ease of conversion of carbon atoms in a ring from sp^3 to sp^2 or vice versa has been developed. It is useful in comparing the reactivity of cyclohexanones with that of cyclopentanones. It is observed that reactions which convert an sp^2 carbon to an sp^3 carbon in a six-membered ring are more favorable than the corresponding reactions in a five-membered ring.

For example, cyclohexanone is reduced by sodium borohydride 23 times faster than cyclopentanone.[104] The explanation for this difference lies in the relative torsional strain in the two systems. Converting an sp^2 atom in a five-membered ring to sp^3 increases the torsional strain because of the increase in the number of eclipsing

103. J. E. Baldwin, R. C. Thomas, L. I. Kruse, and L. Silberman, *J. Org. Chem.* **42**, 3846 (1977).
104. H. C. Brown and K. Ichikawa, *Tetrahedron* **1**, 221 (1957).

interactions. A similar change in a six-membered ring leads to a completely staggered (chair) arrangement and reduces torsional strain.

=O + HY ⟶ OH Y more favorable

=O + HY ⟶ OH Y less favorable

Conversely, processes that convert sp^3 carbons to sp^2 carbons are more favorable for five-membered than for six-membered rings. This can be illustrated by the data for acetolysis of cyclopentyl versus cyclohexyl tosylate. The former proceeds with an enthalpy of activation about 3 kcal/mol less than for the cyclohexyl compound. A molecular mechanics analysis of the difference found that it was largely accounted for by the relief of torsional strain in the cyclopentyl case.[105] Notice that there is an angle strain effect operating in the opposite direction, since there will be some resistance to the expansion of the bond angle at the reaction center to 120° in the cyclopentyl ring.

H OSO$_2$Ar $CH_3CO_2^-$ H OSO$_2$Ar CH_3CO_2 H

H OSO$_2$Ar CH_3CO_2 H OSO$_2$Ar CH_3CO_2 H

There is another stereoelectronic aspect to the reactivity of the carbonyl group in cyclohexanone. This has to do with the preference for approach of reactants from the axial or the equatorial direction. The chair conformation of cyclohexanone places the carbonyl group in an unsymmetrical environment. It is observed that small nucleophiles prefer to approach the carbonyl group of cyclohexanone from the axial direction. How do the differences in the C—C bonds (on the axial side) as opposed to the C—H bonds (on the equatorial side) influence the reactivity of cyclohexanone?

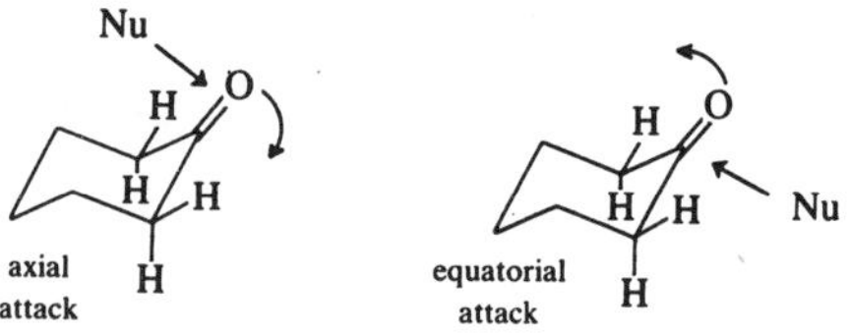

105. H.-J. Schneider and F. Thomas, *J. Am. Chem. Soc.* **102**, 1424 (1980); H. C. Brown and G. Ham, *J. Am. Chem. Soc.* **78**, 2735 (1956).

Several possible effects have been considered. One is the interaction between the σ bonds and the π^* orbital of the carbonyl group. This interaction could distort the shape of the carbonyl LUMO. One proposal is that the (C—C σ)-(C=O π^*) interaction distorts the C=O LUMO so that it has greater density on the axial (C—C) side.[106] An alternative view is that the axial C—H bonds (at C-2 and C-6) can preferentially stabilize the transition state for axial attack by electron donation into the σ^* orbital of the developing bond to the nucleophile. This proposal views C—H bonds as better hyperconjugative electron donors than C—C bonds.[107] A third view emphasizes flattening of the carbonyl group, which makes the C-2 and C-6 axial C—H bonds more nearly antiperiplanar with the approaching nucleophile. The axial trajectory would then be favored by the nucleophile.[108] These three viewpoints are not necessarily mutually exclusive. They differ in the relative importance assigned to specific components, but all focus on differential interaction of the carbonyl group with adjacent σ bonds.

Torsional effects also play a major role in the preference for axial approach. In the initial ketone, the carbonyl group is almost eclipsed by the equatorial C-2 and C-6 C—H bonds. This torsional strain is relieved by axial attack, but equatorial approach increases it somewhat since the oxygen atom must move through a fully eclipsed arrangement.[109]

More bulky nucleophiles usually approach the cyclohexanone carbonyl from the equatorial direction. This is called *steric approach control* and is the result of van der Waals-type repulsions. Larger nucleophiles encounter the 3,5-axial hydrogens on the axial approach trajectory.[110]

A very important relationship between stereochemistry and reactivity arises in the case of reaction at an sp^2 carbon adjacent to a chiral center. Using nucleophilic addition to the carbonyl group as an example, it can be seen that two diastereomeric products are possible. The stereoselectivity and predictability of such reactions are important in controlling stereochemistry in synthesis, for example.

106. J. Klein, *Tetrahedron Lett.*, 4037 (1973); *Tetrahedron* **30**, 3349 (1974).
107. A. S. Cieplak, *J. Am. Chem. Soc.* **103**, 4540 (1981).
108. N. T. Ahn, *Top Curr. Chem.* **88**, 195 (1980).
109. M. Cherest, H. Felkin, and N. Prudent, *Tetrahedron Lett.*, 2199 (1968); M. Cherest and H. Felkin, *Tetrahedron Lett.*, 2205 (1968).
110. W. G. Dauben, G. Fonken, and D. S. Noyce, *J. Am. Chem. Soc.* **92**, 709 (1970); H. C. Brown and W. C. Dickason, *J. Am. Chem. Soc.* **92**, 709 (1970); D. C. Wigfield, *Tetrahedron* **35**, 449 (1979); T. Wipke and P. Gund, *J. Am. Chem. Soc.* **98**, 8107 (1976).

A number of years ago an empirical relationship, now called *Cram's rule*, was recognized. When R^1, R^2, and R^3 differed in size, and there were not other perturbing factors, the major product was that arising from the following transformation.[111]

R_s = smallest group
R_m = intermediate group
R_l = largest group

major   minor

Various aspects have been considered in rationalizing this result in structural terms. The most generally satisfactory approach has been based on a transition state model advanced by Felkin and co-workers, in which the larger group is oriented perpendicularly to the carbonyl group. Nucleophilic addition to the carbonyl group occurs from the opposite side.[112] The preference for this transition state has been supported by molecular orbital calculations.[113]

STO-3G calculations find the corresponding transition state to be several kcal/mol more stable than other possible conformations.[114] The origin for the preference for this transition state conformation is believed to be a stabilization of the C=O* LUMO by the σ orbital of the perpendicularly oriented substituent.

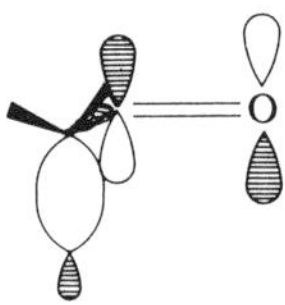

The more stable the LUMO, the stronger is the interaction with the HOMO of the approaching nucleophile. The observed (Cram's rule) stereoselectivity is then a combination of stereoelectronic effects which establish a preference for a perpendicular substituent and a steric effect which establishes a preference for the nucleophile to approach from the direction occupied by the smallest substituent.

Another system in which the factors controlling the direction of reagent approach has been studied systematically is the bicyclo[2.2.1]heptene ring system.

111. D. J. Cram and F. A. Abd Elhafez, *J. Am. Chem. Soc.* **74**, 5828 (1952).
112. M. Cherest, H. Felkin, and N. Prudent, *Tetrahedron Lett.*, 4199 (1968).
113. Y.-D. Wu and K. N. Houk, *J. Am. Chem. Soc.* **109**, 908 (1987).
114. N. T. Ahn, *Top. Curr. Chem.* **88**, 145 (1980).

Table 3.13. Comparison of the Stereochemistry of Reactions with Bicyclo[2.2.1]heptene and 7,7-Dimethylbicyclo[2.2.1]heptene[a]

Reagent	Bicyclo[2.2.1]heptene		7,7-Dimethylbicyclo[2.2.1]heptene	
	exo	*endo*	*exo*	*endo*
B_2H_6 (hydroboration)	99.5	0.5	22	78
RCO_3H (epoxidation)	99.5	0.5	12	88
H_2, Pd (hydrogenation)	90	10	10	90

a. H. C. Brown, J. H. Kawakami, and K. T. Liu, *J. Am. Chem. Soc.* **95**, 2209 (1973).

The stereochemistry of a number of reactions of the parent system and the 7,7-dimethyl derivative has been examined.[115] Some of the results are given in Table 3.13. These reactions reveal a reversal of the preferred direction of attack with the introduction of the methyl substituents. The methyl groups have a similar effect in controlling the stereochemistry of reduction of the related ketones.[116]

NaBH4 → OH (major) + OH (minor)

H_3C CH_3 → NaBH4 → H_3C CH_3 OH (minor) + H_3C CH_3 OH (major)

The preference for *endo* attack in 7,7-dimethylnorbornene is certainly steric in origin, with the 7-methyl substituents shielding the *exo* direction of approach. The origin of the preferred *exo* attack in norbornene is more subject to discussion. A purely steric explanation views the *endo* hydrogens at C-5 and C-6 as sterically shielding the *endo* approach. There probably is also a major torsional effect.

115. H. C. Brown, J. Kawakami, and K.-T. Liu, *J. Am. Chem. Soc.* **95**, 2209 (1973).
116. H. C. Brown and J. Muzzio, *J. Am. Chem. Soc.* **88**, 211 (1966).

Comparison of the *exo* and *endo* modes of approach shows that greater torsional strain develops in the *endo* mode of approach.[117]

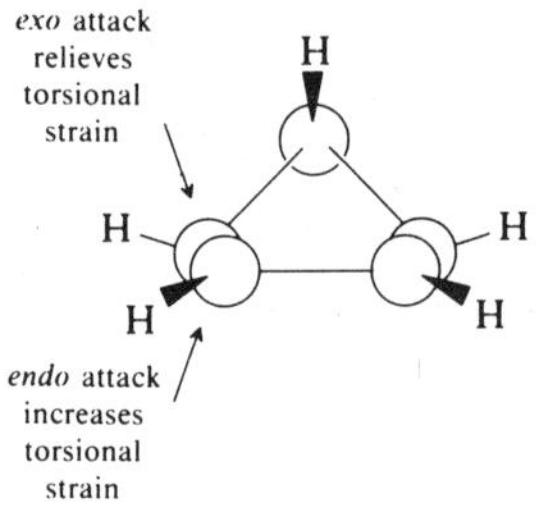

exo attack relieves torsional strain

endo attack increases torsional strain

General References

G. Chiurdoglu (ed.), *Conformational Analysis, Scope and Present Limitations*, Academic Press, New York, 1971.

J. Dale, *Stereochemistry and Conformational Analysis*, Verlag Chemie, New York, 1978.

E. L. Eliel, *Stereochemistry of Carbon Compounds*, McGraw-Hill, New York, 1962.

E. L. Eliel, N. L. Allinger, S. J. Angyal, and G. A. Morrison, *Conformational Analysis*, Wiley-Interscience, New York, 1965.

A. Greenberg and J. F. Liebman, *Strained Organic Molecules*, Academic Press, New York, 1978.

M. Hanack, *Conformation Theory*, Academic Press, New York, 1965.

L. M. Jackman and F. A. Cotton (eds.), *Dynamic Nuclear Magnetic Resonance Spectroscopy*, Academic Press, New York, 1975, Chapters 3, 6, 7, 14.

M. S. Newman (ed.), *Steric Effects in Organic Chemistry*, Wiley, New York, 1956.

M. Oki, *Application of Dynamic NMR Spectroscopy to Organic Chemistry*, VCH Publishers, Deerfield Beach, Florida, 1985.

Problems

(*References for these problems will be found on page* 775.)

1. Estimate $\Delta H°$ for each of the following conformational equilibria:

(a)

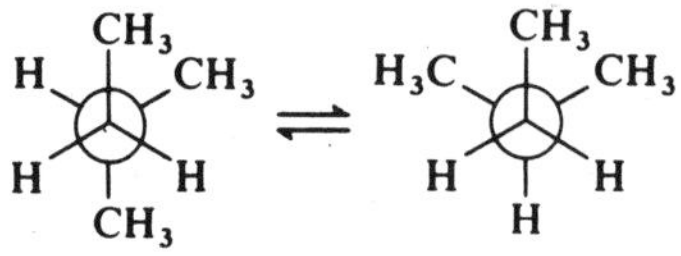

(b)

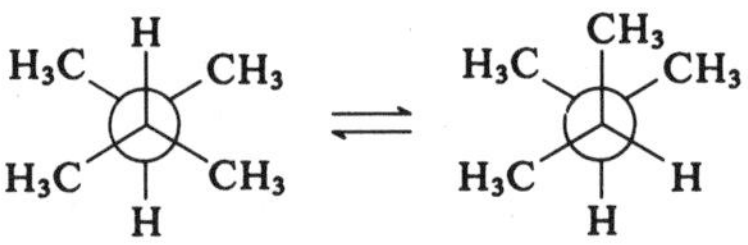

117. N. G. Rondan, M. N. Paddon-Row, P. Caramella, J. Mareda, P. Mueller, and K. N. Houk, *J. Am. Chem. Soc.* **104**, 4974 (1982); M. N. Paddon-Row, N. G. Rondan, and K. N. Houk, *J. Am. Chem. Soc.* **104**, 7162 (1982); K. N. Houk, N. G. Rondan, F. K. Brown, W. L. Jorgensen, J. D. Madura, and D. G. Spellmayer, *J. Am. Chem. Soc.* **105**, 5980 (1983).

(c) [Structures: conformational equilibrium of a dimethylcyclohexanone bearing two CH_3 groups]

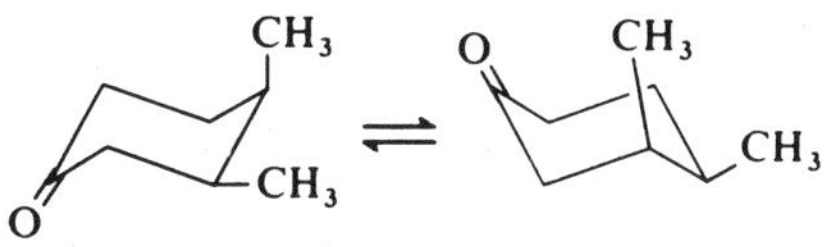

2. Draw a clear three-dimensional representation showing the preferred conformation of *cis,cis,trans*-perhydro-9b-phenalenol (**A**):

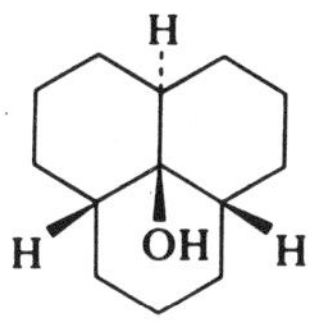

[Structure **A**: perhydrophenalene ring system with H, OH, H, H substituents]

A

3. The *trans* : *cis* ratio at equilibrium for 4-*t*-butylcyclohexanol has been established for several solvents near 80°C:

Solvent	*trans* (%)	*cis* (%)
Cyclohexane	70.0	30.0
Benzene	72.5	27.5
1,2-Dimethoxyethane	71.0	29.0
Tetrahydrofuran	72.5	27.5
t-Butyl alcohol	77.5	22.5
Isopropanol	79.0	21.0

From these data, calculate the conformational energy of the hydroxyl group in each solvent. Do you notice any correlation between the observed conformational preference and the properties of the solvent? Explain.

4. The preferred conformations of both 1-methyl-1-phenylcyclohexane and 2-methyl-2-phenyl-1,3-dioxane have the phenyl group in the axial orientation even though the conformational free energy of the phenyl group (2.9 kcal) is greater than that for a methyl group (1.8 kcal). Explain.

5. Draw clear conformational representations of the β-pyranose forms of each of the following carbohydrates:

(a) Fischer projection: CHO; HO–C–H; HO–C–H; H–C–OH; H–C–OH; CH_2OH

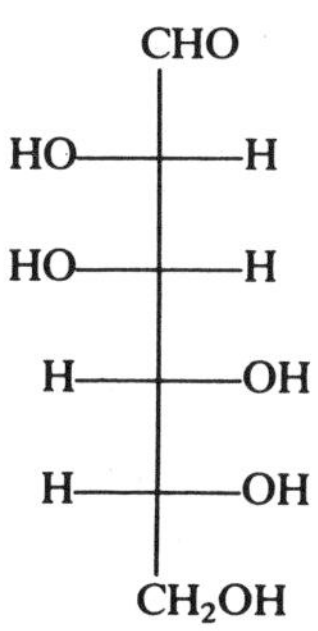

D-mannose

(b) Fischer projection: CHO; H–C–H; HO–C–H; CH_3O–C–H; H–C–OH; CH_3

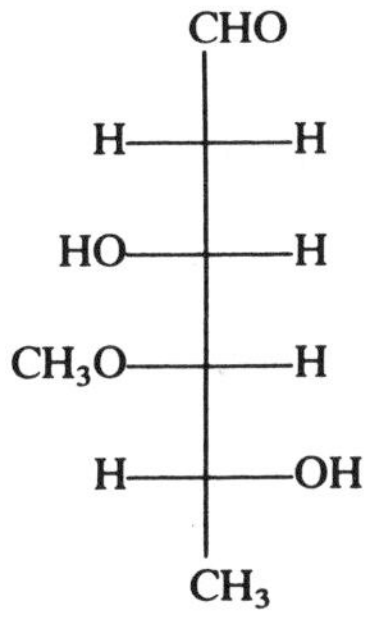

D-chromose A

(c) Fischer projection: CHO; H–C–H; CH_3–C–NH_2; H–C–OH; HO–C–H; CH_3

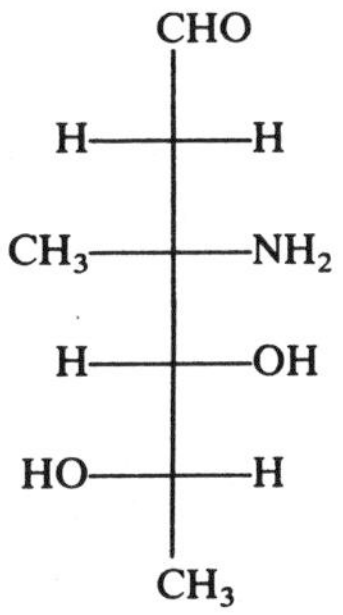

L-vancosamine

6. Explain the basis for the selective formation of the product shown over the alternative product.

(a) $CH_3O_2CC(=CH_2)CH_2CH_2CH_2OH \xrightarrow{^-OH}$ [lactone with exocyclic CH_2] not [tetrahydropyran with CO_2CH_3]

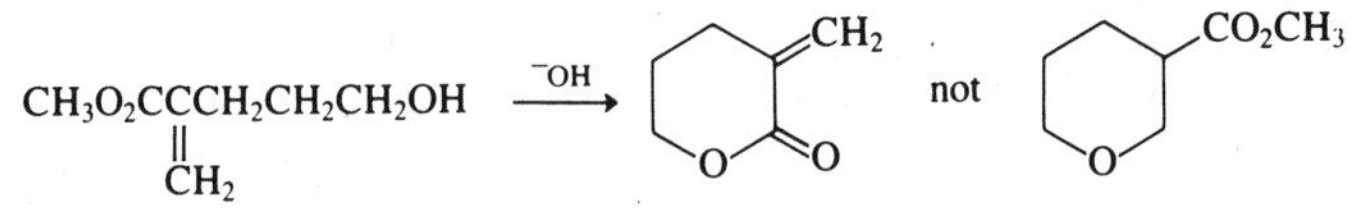

(b) [7,7-dimethylnorbornene, H_3C, CH_3] $\xrightarrow{RCO_3H}$ [exo epoxide] (88%) versus [endo epoxide] (12%)

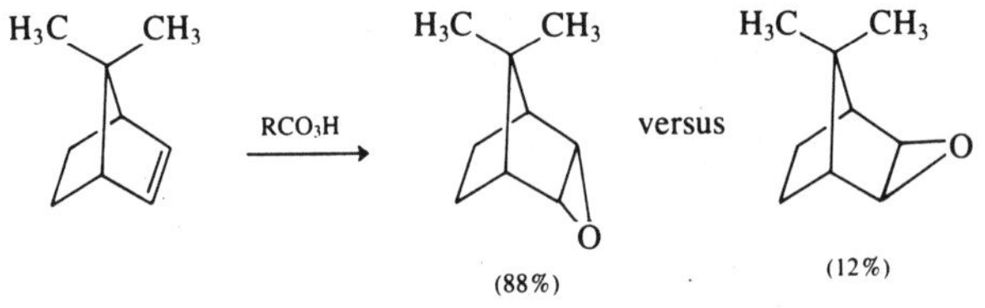

(c) CH_3—N [2-methylpiperidine, H, CH_3] $+BH_3 \longrightarrow$ [H_3B^-, CH_3, N^+, H, CH_3] preferred to [CH_3, N^+, H_3B^-, H, CH_3]

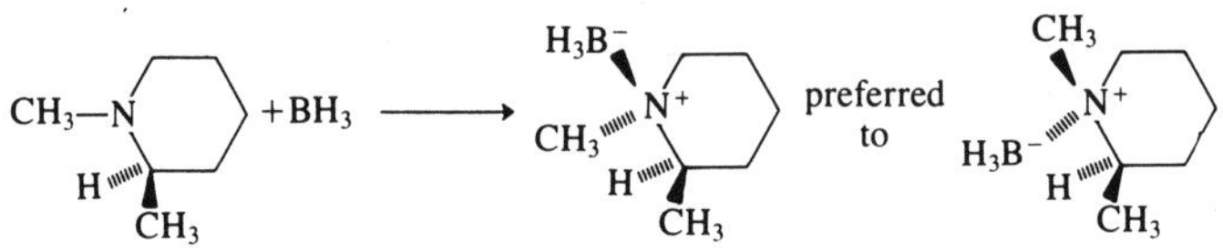

(d) [7-norbornanone with CH_3] $\xrightarrow{LiAlH_4}$ [H, OH, CH_3] rather than [HO, H, CH_3]

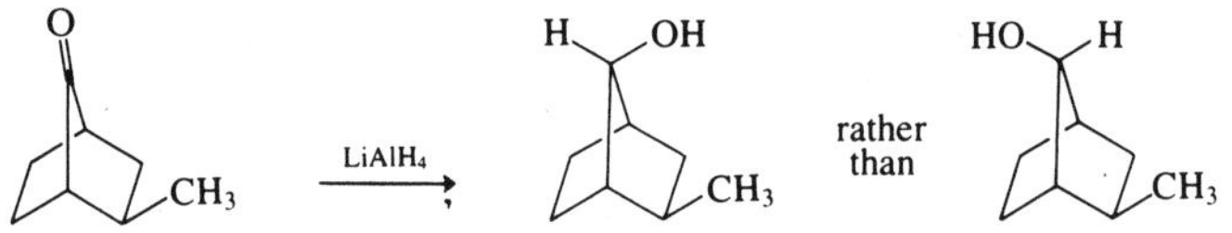

(e) [phthalimido penam with S, CH_3, CH_3, H, H, N, CO_2CH_3] $\xrightarrow{RCO_3H}$ [sulfoxide, O^-, S^+, CH_3, CH_3, CO_2CH_3] rather than the stereoisomeric sulfoxide

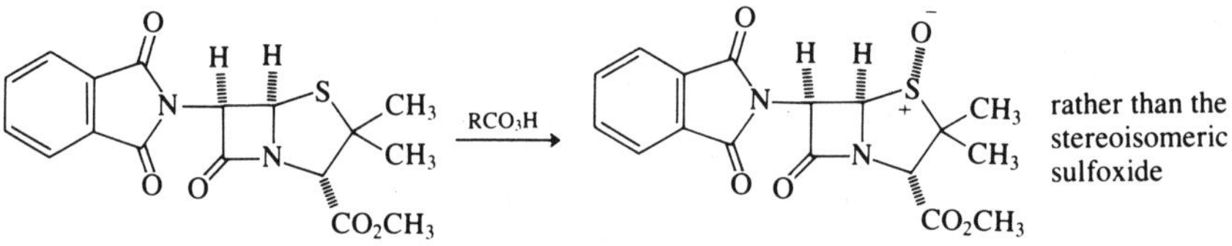

(f) $CH_3CCH_2CHCH_2CCH_3$ (with C=O groups, and $C(=O)CH_3$ substituent) $\xrightarrow{^-OH}$ [cyclohexenone with $CH_3C(=O)$ and CH_3] rather than [cyclopentenone with CH_3CCH_2 (C=O) and CH_3] or [cyclopentenone with CH_3CCH_2 (C=O) and CH_3]

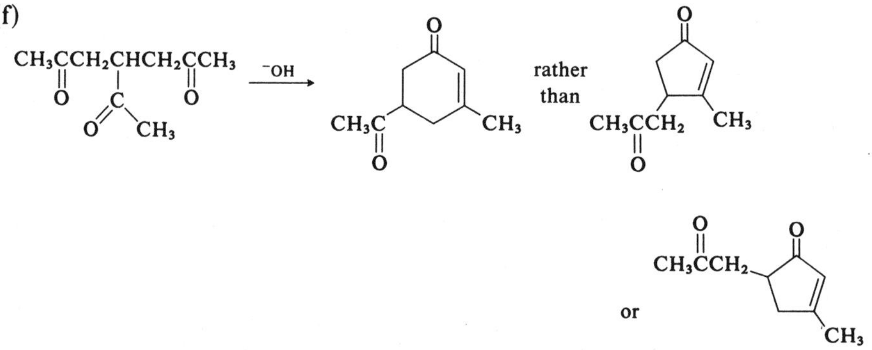

(g) [phenyl–CH_2CH_2–epoxide] $\xrightarrow{SnCl_4}$ [tetrahydronaphthalene with CH_2OH] rather than [benzocycloheptene with OH]

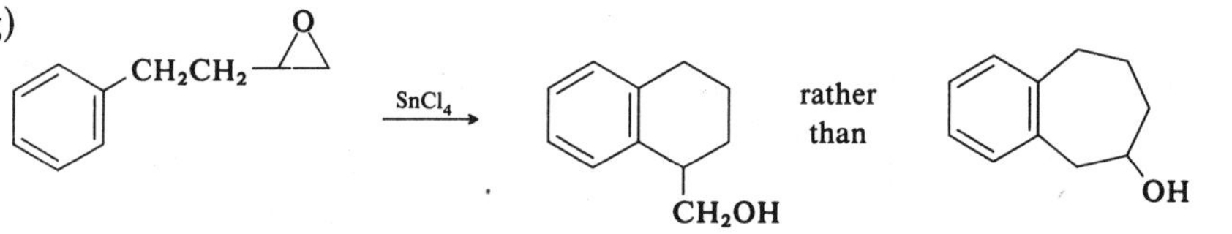

7. For the following pairs of reactions, indicate which you would expect to be more favorable and explain the basis of your prediction.

(a) Which isomer will solvolyze more rapidly in acetic acid?

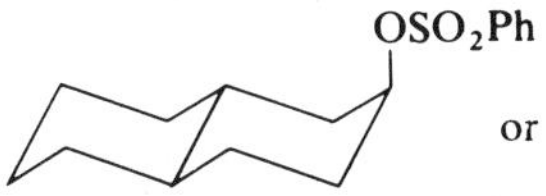
OSO2Ph

A

or

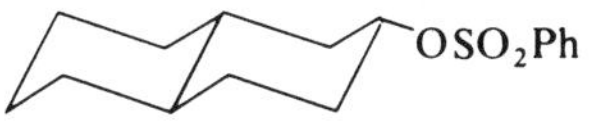
OSO2Ph

B

(b) Which will be the major reaction product?

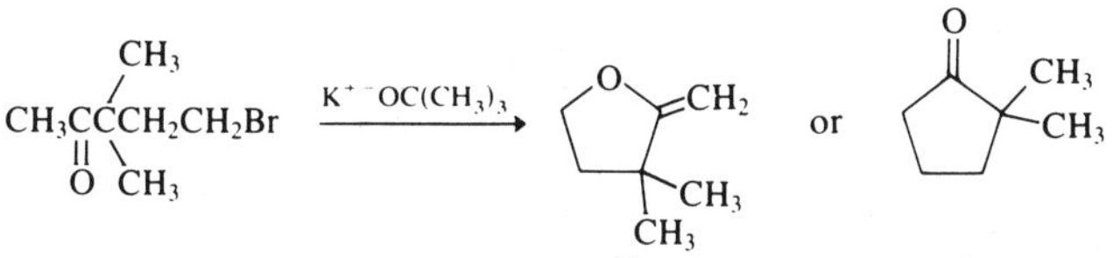
CH3
CH3CCCH2CH2Br
O CH3
K+ −OC(CH3)3
O
CH2
CH3
CH3
or
O
CH3
CH3

C **D**

(c) Which isomer will be converted to a quaternary salt more rapidly?

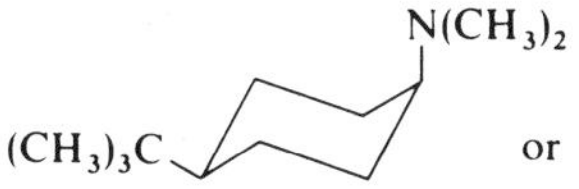
N(CH3)2
(CH3)3C

E

or

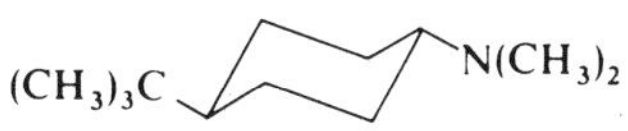
(CH3)3C
N(CH3)2

F

(d) Which lactone will be formed more rapidly?

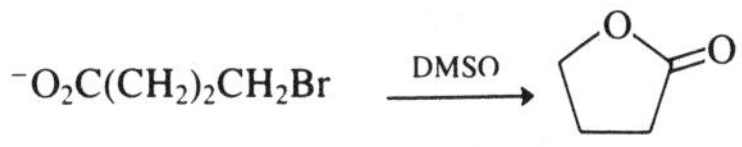
−O2C(CH2)2CH2Br
DMSO
O
O

G

or

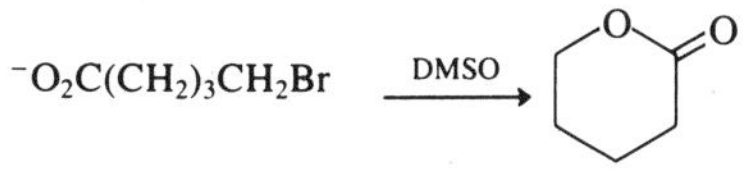
−O2C(CH2)3CH2Br
DMSO
O
O

H

(e) Which compound will undergo hydrolysis more rapidly?

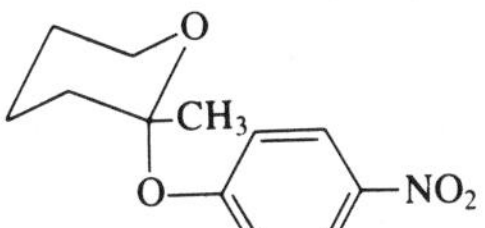
O
CH3
O
NO2

I

or

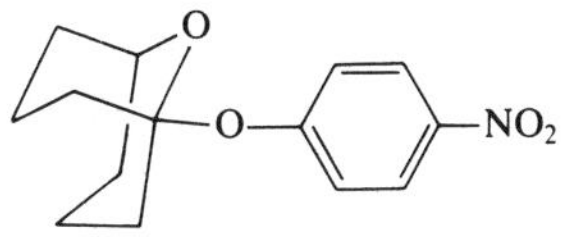
O
O
NO2

J

(f) Which compound will aromatize more rapidly by loss of ethoxide ion?

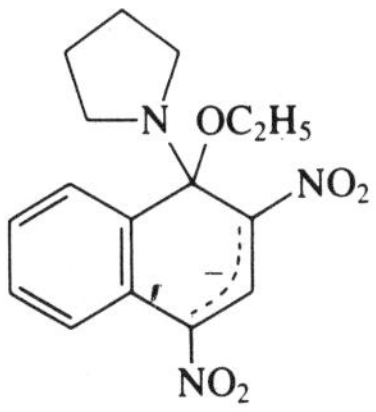
N OC2H5
NO2
NO2

K

or

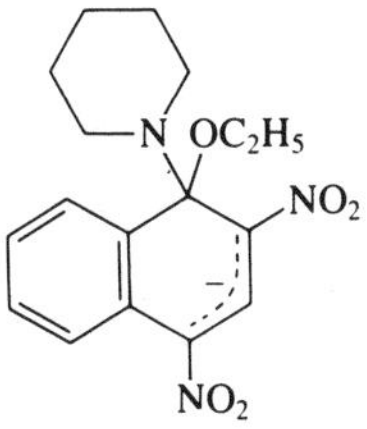
N OC2H5
NO2
NO2

L

(g) Which compound will be more rapidly oxidized by chromic acid?

HO — or — OH

M **N**

8. Predict the most stable conformation for each of the following molecules and explain the basis of your prediction.

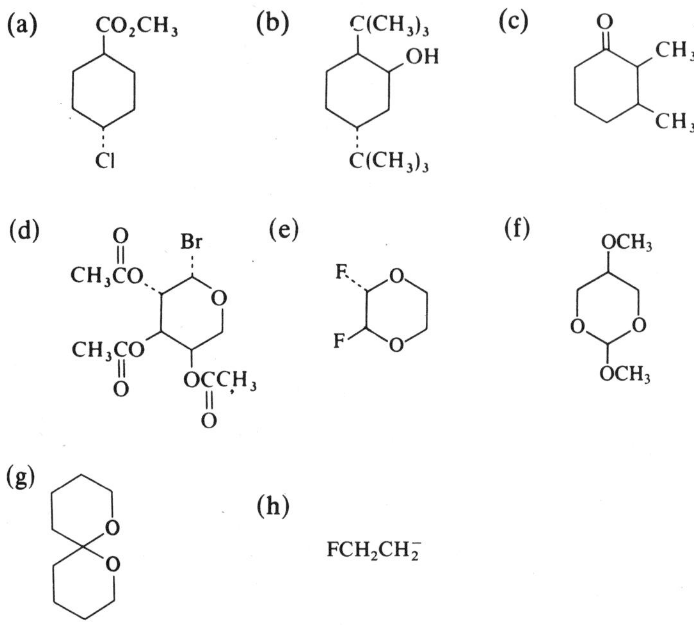

9. Given that the rotational barrier for a C—C bond in acetone is about 0.75 kcal, (a) sketch the relationship between energy and conformation using a Newman projection formula to define the angles of rotation, and (b) show how the energy-conformation curve will be perturbed by addition of first one (methyl ethyl ketone) and then two (methyl isopropyl ketone) methyl substituents on the methyl group undergoing rotation.

10. Using the data incorporated in Fig. 3.3 and assuming the additivity of *gauche* and eclipsing interactions of similar type, sketch the rotational energy profile you would expect for 2,3-dimethylbutane.

11. The following molecules present possibilities for stereoisomerism and/or the existence of different conformations. For each molecule predict which stereoisomer will be the most stable and predict its preferred conformation.

(a) OH (b) $CO_2C_2H_5$ (c) (d) $CH_3\overset{O}{\overset{\|}{C}}\underset{O}{\underset{\|}{C}}CH_3$

12. Consider the conformations possible for 3-substituted methylenecyclohexanes. Do you expect typical substituents to exhibit larger or smaller preferences for the equatorial orientation, as compared to the same substituent on a cyclohexane ring?

13. Discuss the aspects of conformation and stereochemistry that would have to be considered for complete description of the structure of molecules having the general structure **A**. How would the size of the $(CH_2)_n$ bridge affect conformational equilibria in these molecules?

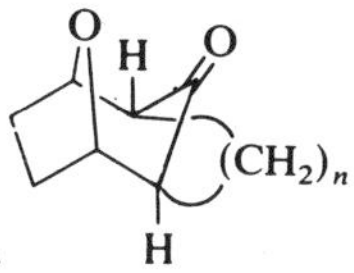

O H O
$(CH_2)_n$
A H

14. Predict the preferred conformation of the isomeric (*cis*- and *trans*-) 3-penten-2-ones, **A**, $R{=}CH_3$. How would you expect the conformational picture to change as R becomes progressively larger?

A $\overset{O}{\overset{\|}{RC}}CH{=}CHCH_3$

15. Two stereoisomers (**A** and **B**) of the structure shown below have been obtained and separated. The NMR spectrum of one isomer (**A**) shows two methyl peaks (doublets at 1.03 and 1.22 ppm) and two quartets (2.68 and 3.47 ppm) for the CH groups. The spectrum of the other isomer (**B**) shows single signals for the methyl (doublet at 1.25 ppm) and methine protons (broad quartet at 2.94 ppm). The spectra of both compounds show a change with temperature. For isomer **A**, at 95°C, the pairs of methyl doublets and the methine quartet both become single signals (still a doublet and a quartet, respectively). The low-temperature spectrum (−40°C) is unchanged from the room temperature spectrum. For isomer **B** at −40°C, the methyl signals split into two doublets of *unequal intensity* (1.38 and 1.22 ppm in the ratio 9:5). The methine signal also splits into two broad signals at 3.07 and 2.89 ppm, also in the ratio 9:5. From this information, assign the stereochemistry of isomers **A** and **B** and explain the cause of the temperature dependence of the NMR spectra of each isomer.

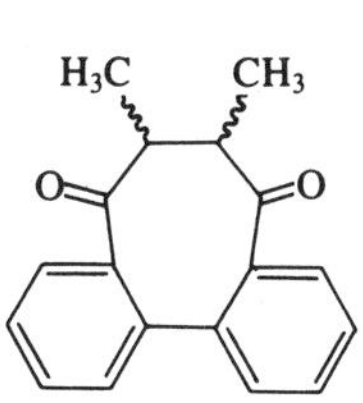

H_3C CH_3
O O

16. Estimate the energy difference between the stable and unstable chair conformations of each of the following trimethylcyclohexanes:

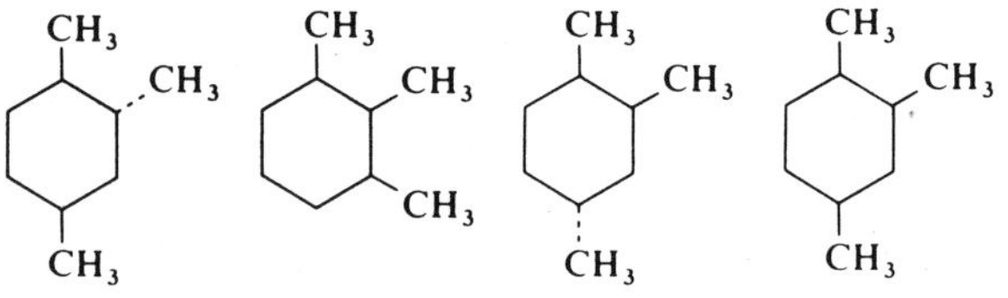

CH_3 CH_3 CH_3 CH_3 CH_3 CH_3 CH_3 CH_3 CH_3 CH_3 CH_3 CH_3

17. Predict the stereochemistry of each of the following reactions:

(a) 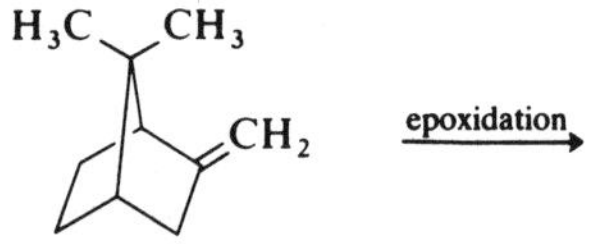 H_3C CH_3 CH_2 epoxidation

(b) 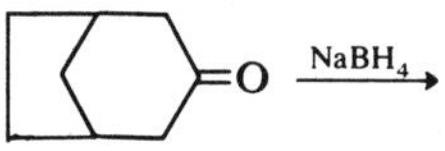 =O $NaBH_4$

(c) 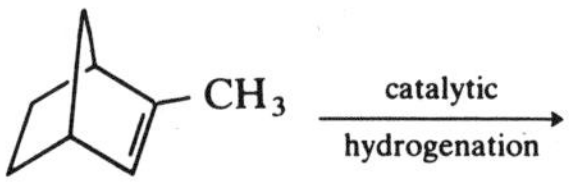 CH_3 catalytic hydrogenation

(d)  O= $LiAlH_4$

(e) 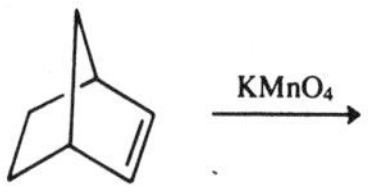 $KMnO_4$

4

Study and Description of Organic Reaction Mechanisms

Introduction

The chapters that follow this one will be devoted largely to the description of specific reactions. The development of a working understanding of organic chemistry requires the mastery of certain fundamental reaction types that occur in a wide variety of individual reactions. Most organic reactions occur in several steps; these steps constitute the *reaction mechanism.* Knowledge of the detailed mechanism of reactions often reveals close relationships between reactions that otherwise might appear to be unrelated. Consideration of reaction mechanism is also usually the basis for development of new transformations and improvement of existing procedures. In this chapter, the ways in which organic reactions can be studied in order to determine aspects of the reaction mechanism will be discussed. The types of experimental studies that provide data and the methods by which it is possible to develop information about reaction mechanism from the data will be considered.[1]

1. Extensive discussions of techniques for studying reaction mechanisms are presented in E. S. Lewis (ed.), *Investigation of Rates and Mechanisms of Reactions, Techniques of Organic Chemistry,* Third Edition, Vol. VI, Part I, Wiley-Interscience , New York, 1974; C. F. Bernasconi (ed.), *Investigation of Rates and Mechanisms of Reactions, Techniques of Organic Chemistry,* Fourth Edition, Vol. VI, Part I, Wiley-Interscience, New York, 1986.

4.1. Thermodynamic Data

Any reaction will have associated with it a change in enthalpy (ΔH), entropy (ΔS), and free energy (ΔG). The principles of thermodynamics assure us that ΔH, ΔS, and ΔG are independent of the reaction path. They are interrelated by the fundamental equation:

$$\Delta G = \Delta H - T\,\Delta S \tag{4.1}$$

Furthermore, the value of ΔG is related to the equilibrium constant K for the reaction:

$$\Delta G = -RT \ln K \tag{4.2}$$

Since these various quantities are characteristics of the reactants and products but are *independent* of the reaction path, they cannot provide insight into mechanisms. Information about ΔG, ΔH, and ΔS does, however, indicate the feasibility of any specific reaction. The enthalpy of a given reaction can be estimated from tabulated thermochemical data or from bond energy data such as those in Table 1.3 (p. 12). The example below indicates the use of bond energy data for estimating the enthalpy of a reaction.

Example 4.1. Calculate the enthalpy change associated with hydrogenation of butene.

$$CH_3CH{=}CHCH_3 + H_2 \longrightarrow CH_3CH_2CH_2CH_3$$

$$-\Delta H = \Sigma\,\text{bond energies}_{(\text{formed})} - \Sigma\,\text{bond energies}_{(\text{broken})}$$

Bonds formed:		Bonds broken:	
2 C—H	196.4	H—H	103.2
C—C	80.5	C=C	145
	276.9		248.2

$$\Delta H = -276.9 - (-248.2) = -28.7\ \text{kcal/mol}$$

The hydrogenation is therefore calculated to be exothermic by about 29 kcal/mol.

Calculations of this type can provide only an approximate indication of the enthalpy change that is associated with a given reaction. The generalized bond energies given in Table 1.3 assume that a bond energy is independent of the structure of the rest of the molecule. This is only approximately correct, as can be judged by observing the variations in C—H and C—C bond energies as a function of structure in Part B of Table 1.3.

More accurate calculation of the thermodynamic parameters of a reaction can be done on the basis of tabulated thermochemical data. There are extensive compilations of ΔH_f° and ΔG_f° for many compounds. The subscript f designates these as, respectively, the enthalpies and free energies of formation of the compound from its constituent elements. The superscript ° is used to designate data that refer to the substance in its standard state, i.e., the pure substance at 25°C and 1 atm. The compiled data can be used to calculate the enthalpy or free energy of a given reaction if the data are available for each reactant and product:

$$\Delta H^\circ = \Sigma\Delta H^\circ_{f_{\text{products}}} - \Sigma\Delta H^\circ_{f_{\text{reactants}}}$$

or

$$\Delta G^\circ = \Sigma\Delta G^\circ_{f_{\text{products}}} - \Sigma\Delta G^\circ_{f_{\text{reactants}}}$$

In the case of hydrogenation of 2-butane, ΔH°_f for butane (gas) is −30.15 kcal/mol, ΔH°_f for *trans*-2-butene (gas) is −2.67 kcal/mol, and ΔH°_f for H_2 is 0. Thus, ΔH° of the hydrogenation reaction at standard conditions is −27.5 kcal/mol.

If the thermodynamic data for a compound of interest have not been determined and tabulated, it may be possible to estimate ΔH_f or ΔG_f from tabulated data pertaining to individual structural fragments. Procedures have been developed for estimating thermodynamic characteristics of hydrocarbons and some derivatives by summing the contributions of the constituent groups.[2]

Estimation of the free-energy change associated with a reaction permits the calculation of the equilibrium position for a reaction and indicates the feasibility of a given chemical process. A positive ΔG° imposes a limit on the extent to which a reaction can occur. For example, as can be calculated using Eq. (4.2), a ΔG° of 1.0 kcal/mol limits conversion to product at equilibrium to 18%. An appreciably negative ΔG° indicates that the reaction is thermodynamically favorable.

Molecular orbital calculations provide another approach to obtaining estimates of thermodynamic data. The accuracy with which the various computational methods reproduce molecular energies differs. The total stabilization energies calculated even for small hydrocarbons, relative to the separate nuclei and electrons, are very large numbers (typically 50,000 and 100,000 kcal/mol for C_2 and C_4 species, respectively). This is the total energy that comes directly out of an MO calculation. The energy differences that are of principal chemical interest, such as ΔH for a reaction, are likely to be in the range of 0–50 kcal/mol. A very small error, relative to the total energy, in a MO calculation becomes a very large error in a calculated ΔH value. Fortunately, the absolute errors for compounds of similar type are likely to be comparable so that errors will cancel in calculation of the *energy differences* between two related molecules. Table 4.1 gives some calculated ΔH values for some simple reactions and the corresponding experimental values. It is clear from the variation in results among the calculations and the deviation from experimental data that the ability to produce reliable estimates depends upon the computational method that is chosen.

Calculations are frequently done on the basis of *isodesmic reactions* in order to provide for maximum cancellation of errors in the total energies. An isodesmic reaction is defined as a process in which the number of formal bonds of each type is kept constant; that is, the number of C—H, C=C, C=O, etc., bonds on each

2. G. J. Janz, *Thermodynamic Properties of Organic Compounds*, Academic Press, New York, 1967; D. R. Stull, E. F. Westrum, Jr., and G. C. Sinke, *The Chemical Thermodynamics of Organic Compounds*, John Wiley, New York, 1969.

Table 4.1. ΔH for Some Reactions Calculated by MO Methods[a]

Reaction	Calculation method					
	CNDO/2	MNDO	STO-3G	4-31G	6-31G	Exp
$CH_3{-}CH_3 + H_2 \rightarrow 2CH_4$					−22.0	−15
$CH_3CH{=}CH_2 \rightarrow$ △					+7.8	+7
$C_2H_2 + H_2 \rightarrow CH_2{=}CH_2$	−157	−42.7	−64.4	−57.3		−41
$C_2H_2 + C_2H_4 \rightarrow CH_2{=}CHCH{=}CH_2$	−203	−3.7	−57.9	−41.3		−41

a. Enthalpy given in kcal/mol. Data for CNDO/2, STO-3G, and 4-31G from T. A. Halgren, D. A. Kleier, J. H. Jr., L. D. Brown, and W. N. Lipscomb, *J. Am. Chem. Soc.* **100**, 6595 (1978); data for MNDO from M. J. S. D and G. P. Ford, *J. Am. Chem. Soc.* **101**, 5558 (1979); data for 6-31G from J. A. Pople, *J. Am. Chem. Soc* **97**, 5306 (1975).

Table 4.2. Calculated and Experimental ΔH Values for Some Isodesmic Reactions

Reaction	Calculated ΔH (4-31G)[a] (kcal/mol)	Calculated ΔH (6-31G)[b] (kcal/mol)	Experimental ΔH (kcal/mol)
$CH_3CH_2CH_3 + CH_4 \rightarrow 2CH_3CH_3$	1.0	0.8	2.6
△ $+ 3CH_4 \rightarrow 2CH_3CH_3 + CH_2{=}CH_2$	−58	−50.4	−43.9
$H_2C{=}C{=}O + CH_4 \rightarrow CH_2{=}CH_2 + H_2C{=}O$	12.8	13.3	15.0
$CH_3CN + CH_4 \rightarrow CH_3CH_3 + HCN$	12.0	11.7	14.4

a. Data from W. J. Hehre, R. Ditchfield, L. Radom, and J. A. Pople, *J. Am. Chem. Soc.* **92**, 4796
b. From W. J. Hehre, L. Radom, P. v. R. Schleyer, and J. Pople. *Ab Initio Molecular Orbital Theory*, Wiley, New York, 1986, pp 299–305.

side of the equation is identical.[3] Although the reaction may not correspond to any real chemical process, the calculation can provide a test of the reliability of the computational method because of the additivity of enthalpies of formation. The "experimental" ΔH of the process can be obtained by summation of the tabulated ΔH_f° values of the participating molecules. Table 4.2 illustrates some isodesmic reactions and shows ΔH values calculated at the 4-31G and 6-31G levels.

Of the semiempirical methods, only MINDO,[4] MNDO,[5] and AM-1[6] are claimed to reliably estimate energies, and the range of reliability is open to some discussion.[7] With the *ab initio* method, the 4-31G and 6-31G basis sets achieve a level of accuracy that permits comparison of energy data. Users of MO data, however, must critically assess the reliability of the method being applied in the *particular case* under study.

Whether ΔH for a projected reaction is based on bond energy data, tabulated thermochemical data, or MO computations, there remain some fundamental prob-

3. W. J. Hehre, R. Ditchfield, L. Radom, and J. A. Pople, *J. Am. Chem. Soc.* **92**, 4796 (1970).
4. R. C. Bingham, M. J. S. Dewar, and D. H. Lo, *J. Am. Chem. Soc.* **97**, 1294 (1975).
5. M. J. S. Dewar and G. P. Ford, *J. Am. Chem. Soc.* **101**, 5558 (1979).
6. M. J. S. Dewar, E. G. Zoebisch, E. F. Healy, and J. J. P. Stewart, *J. Am. Chem. Soc.* **107**, 3902 (1985).
7. J. A. Pople, *J. Am. Chem. Soc.* **97**, 5307 (1975); T. A. Halgren, D. A. Kleier, J. H. Hall, Jr., L. D. Brown, and W. L. Lipscomb, *J. Am. Chem. Soc.* **100**, 6595 (1978); M. J. S. Dewar and D. M. Storch, *J. Am. Chem. Soc.* **107**, 3898 (1985).

lems which prevent a final conclusion about a reaction's feasibility. In the first place, most reactions of interest occur in solution, and the enthalpy, entropy, and free energy associated with any reaction depend strongly on the solvent medium. There is only a limited amount of tabulated thermochemical data that is directly suitable for treatment of reactions in organic solvents. Molecular orbital calculations usually refer to the isolated (gas phase) molecule. Estimates of solvation effects must be made in order to apply either experimental or theoretical data to reactions occurring in solution.

There is an even more basic limitation to the usefulness of thermodynamic data for making predictions about reactions: Thermodynamics provides no information about the energy requirements of the pathways that a potential reaction can follow; that is, thermodynamics provides no information about the *rates of chemical reactions.* In the absence of a relatively low-energy pathway, two molecules that can potentially undergo a highly favorable reaction will coexist without reacting. Thus, even if a reaction is thermodynamically feasible, it may not occur at a significant rate. It is therefore extremely important to develop an understanding of reaction mechanisms and the energy requirements and rates of the various steps by which organic reactions proceed.

4.2. Kinetic Data

Kinetic data are capable of providing much detailed insight into reaction mechanisms. The rate of a given reaction can be determined by following the disappearance of a reactant or the appearance of product. The extent of reaction is often measured spectroscopically, since spectroscopic techniques provide a rapid, continuous means of monitoring changes in concentration. Numerous other methods are available, however, and may be preferable in certain cases. For example, continuous pH measurement or acid–base titration can be used to follow the course of reactions that consume or generate acid or base. Conductance measurements provide a means for determining the rate or reactions that generate ionic species; polarimetry is a convenient way of following reactions involving optically active materials. In general, any property that can be measured and related to the concentration of a reactant or product can be used to determine a reaction rate.

The goal of a kinetic study is to establish the quantitative relationship between the concentration of reactants and catalysts and the rate of the reaction. Typically, such a study involves rate measurements at enough different concentrations of each reactant so that the *kinetic order* with respect to each reactant can be assessed. A complete investigation allows the reaction to be described by a rate law, which is an algebraic expression containing one or more *rate constants* as well as the concentrations of all reactant species that are involved in the rate-determining step and steps prior to the rate-determining step. In the rate law, each concentration has an exponent that is the order of the reaction with respect to that component. The overall kinetic order of the reaction is equal to the sum of all the exponents in the rate expression. Several examples of rate laws which illustrate the variety observed are presented in Scheme 4.1. Some are simple; others are more complex.

The relationship between a kinetic expression and a reaction mechanism can be appreciated by considering the several individual steps that constitute the overall reaction mechanism. The expression for the rate of any *single step* in a reaction mechanism will contain a term for the concentration of each reacting species. Thus, for the reaction sequence

$$A + B \underset{k_{-1}}{\overset{k_1}{\rightleftharpoons}} C \xrightarrow{k_2} D \xrightarrow{k_3} E + F$$

Scheme 4.1. Some Representative Rate Laws

1[a] A (1,3-dichlorobicyclo[1.1.0]butane-type structure, Cl at top and bottom) $\xrightarrow{\Delta}$ $CH_2C(Cl){=}CH_2$ / $CH_2C(Cl){=}CH_2$

rate = k[**A**]

2[b] $CH_3CH(Cl)CH_2CH_2CH_2CH_3$ (**B**) + $NaOCH_3$ $\longrightarrow$ $CH_3CH{=}CHCH_2CH_2CH_3$ (mixture of isomers)

rate = k[**B**][$NaOCH_3$]

3[c] $CH_3Cl + GaCl_3^* \rightleftharpoons CH_3Cl^* + GaCl_2^*Cl$

rate = $k[CH_3Cl][GaCl_3^*]^2$

4[d] $(CH_3)_2C{=}CHCH_3$ (**C**) + HCl $\xrightarrow{CH_3NO_2}$ $(CH_3)_2C(Cl)CH_2CH_3$

rate = k[**C**][HCl]^2

5[e] $\frac{1}{4}(CH_3Li)_4$ + H_3C–$C_6H_3(CH_3)$–C(=O)–C_6H_4–SCH_3 (**D**) $\longrightarrow$ H_3C–$C_6H_3(CH_3)$–C(O^-Li^+)(CH_3)–C_6H_4–SCH_3

rate = $k[(CH_3Li)_4]^{1/4}$[**D**]

6[f] $F_2CHC(=O)OC_6H_5$ (**E**) + $CH_3(CH_2)_3NH_2$ (**F**) $\xrightarrow{\text{dioxane}}$ $F_2CHC(=O)NH(CH_2)_3CH_3$

rate = k_1[**E**][**F**] + k_2[**E**][**F**]^2

a. E. N. Cain and R. K. Solly, *J. Am. Chem. Soc.* **95**, 7884 (1973).
b. R. A. Bartsch and J. F. Bunnett, *J. Am. Chem. Soc.* **90**, 408 (1968).
c. F. P. DeHaan, H. C. Brown, D. C. Conway, and M. G. Gibby, *J. Am. Chem. Soc.* **91**, 4854 (1969).
d. Y. Pocker, K. D. Stevens, and J. J. Champoux, *J. Am. Chem. Soc.* **91**, 4199 (1969).
e. S. G. Smith, L. F. Charbonneau, D. P. Novak, and T. L. Brown, *J. Am. Chem. Soc.* **94**, 7059 (1972).
f. A. S. A. S. Shawali and S. S. Biechler, *J. Am. Chem. Soc.* **89**, 3020 (1967).

the rates for the successive steps are

$$\text{step 1:} \quad \frac{d[\mathrm{C}]}{dt} = k_1[\mathrm{A}][\mathrm{B}] - k_{-1}[\mathrm{C}]$$

$$\text{step 2:} \quad \frac{d[\mathrm{D}]}{dt} = k_2[\mathrm{C}]$$

$$\text{step 3:} \quad \frac{d[\mathrm{E}]}{dt} = \frac{d[\mathrm{F}]}{dt} = k_3[\mathrm{D}]$$

Let us further specify that the first step is a very rapid but unfavorable equilibrium, and that $k_2 \ll k_3$, i.e., that the second step is slow relative to the third step. Under these circumstances, the overall rate of the reaction will depend on the rate of the second step, and the second step is called the *rate-determining step.*

Kinetic data provide information only about the rate-determining step and steps preceding it. In the hypothetical reaction under consideration, the final step follows the rate-determining step, and since its rate will not affect the rate of the overall reaction, k_3 will not appear in the overall rate expression. The rate of the overall reaction is governed by the second step, which is the bottleneck in the process. The rate of this step is equal to k_2 multiplied by the molar concentration of intermediate C, which may not be directly measurable. It is therefore necessary to express the rate in terms of the concentration of reactants. In the case under consideration, this can be done by recognizing that [C] is related to [A] and [B] by an equilibrium constant

$$K = \frac{[\mathrm{C}]}{[\mathrm{A}][\mathrm{B}]}$$

Furthermore, K is related to k_1 and k_{-1} by the requirement that no net change in composition occur at equilibrium.

$$k_{-1}[\mathrm{C}] = k_1[\mathrm{A}][\mathrm{B}]$$

$$[\mathrm{C}] = \frac{k_1}{k_{-1}}[\mathrm{A}][\mathrm{B}]$$

The rate of step 2 can therefore be written in terms of [A] and [B]:

$$\frac{d[\mathrm{D}]}{dt} = k_2[\mathrm{C}] = k_2\frac{k_1}{k_{-1}}[\mathrm{A}][\mathrm{B}] = k_{\mathrm{obs}}[\mathrm{A}][\mathrm{B}]$$

Experimentally, it would be observed that the reaction rate would be proportional to both [A] and [B]. Kinetic data are normally handled using the integrated forms of the differential equations. The integrated rate equations for simple first-order and second-order reactions are

$$\text{first-order reaction:} \quad k = \frac{1}{t}\ln\left(\frac{c_0}{c}\right)$$

$$\text{second-order reaction:} \quad k = \frac{1}{t(a_0 - b_0)}\ln\frac{b_0(a)}{a_0(b)}$$

where a, b, and c refer to concentrations of reactants at time t, and a_0, b_0, and c_0 to initial concentrations. As reaction mechanisms become more complex, the mathematical form of their rate expressions and their solutions become increasingly complex. For the simple rate expressions, graphical or numerical analysis of the data directly provides a value for the appropriate rate constant. More complex rate expressions may be analyzed by a variety of graphical or analytical techniques, in conjunction with sufficient changes in the reactant concentrations and other variables to determine the kinetic expression and rate constants.

Most organic reactions involve more than one step. It is therefore necessary to consider the kinetic expressions that arise from some of the more important cases of multistep reactions. There may be a rapid equilibrium preceding the rate-determining step. Such a mechanism may operate, for example, in the reaction of an alcohol with hydrobromic acid to give an alkyl bromide:

$$\text{ROH} + \text{H}^+ \underset{k_{-1}}{\overset{\text{fast}\atop k_1}{\rightleftharpoons}} \text{R}\overset{+}{\text{O}}\text{H}_2$$

$$\text{R}\overset{+}{\text{O}}\text{H}_2 + \text{Br}^- \xrightarrow[k_2]{\text{slow}} \text{RBr} + \text{H}_2\text{O}$$

The overall rate being measured is that of step 2, but there may be no means of directly measuring $[\text{ROH}_2{}^+]$. The concentration of the protonated intermediate $\text{ROH}_2{}^+$ can be expressed in terms of the concentration of the starting material by taking into consideration the equilibrium constant, which relates [ROH], $[\text{Br}^-]$, and $[\text{H}^+]$:

$$K = \frac{[\text{ROH}_2{}^+]}{[\text{ROH}][\text{H}^+]}$$

$$[\text{ROH}_2{}^+] = K[\text{ROH}][\text{H}^+]$$

$$\text{rate} = k_2K[\text{ROH}][\text{H}^+][\text{Br}^-] = k_{\text{obs}}[\text{ROH}][\text{H}^+][\text{Br}^-]$$

A useful approach that is often used in analysis and simplification of kinetic expressions is the *steady-state approximation.* It can be illustrated with a hypothetical reaction scheme:

$$\text{A} + \text{B} \underset{k_{-1}}{\overset{k_1}{\rightleftharpoons}} \text{C}$$

$$\text{C} + \text{D} \xrightarrow{k_2} \text{E} + \text{F}$$

$$\overline{\text{A} + \text{B} + \text{D} \longrightarrow \text{E} + \text{F}}$$

If C is a reactive, unstable species, its concentration will never be very large. It must then be consumed at a rate that closely approximates the rate at which it is formed. Under these conditions, it is a valid approximation to set the rate of formation of C equal to its rate of destruction:

$$k_1[\text{A}][\text{B}] = k_2[\text{C}][\text{D}] + k_{-1}[\text{C}]$$

Rearrangement of this equation provides an expression for [C]:

$$\frac{k_1[A][B]}{k_2[D] + k_{-1}} = [C]$$

By substituting into the rate for the second step, the following expression is obtained:

$$\text{rate} = k_2[C][D] = k_2 \frac{k_1[A][B]}{k_2[D] + k_{-1}}[D]$$

If $k_2[D]$ is much greater than k_{-1}, the rate expression simplifies to

$$\text{rate} = \frac{k_2 k_1[A][B][D]}{k_2[D]} = k_1[A][B]$$

On the other hand, if $k_2[D]$ is much less than k_{-1}, the observed rate expression becomes

$$\text{rate} = \frac{k_1 k_2[A][B][D]}{k_{-1}}$$

The first situation corresponds to the first step being rate-determining. In the second case, it is the second step that is rate-determining, with the first step being a preequilibrium.

The normal course of a kinetic investigation involves the postulation of likely mechanisms and comparison of the observed rate law with those expected for the various mechanisms. Those mechanisms that are incompatible with the observed kinetics can be eliminated as possibilities. Let us consider aromatic nitration by nitric acid in an inert solvent as a typical example and restrict the mechanisms being considered to the three shown below. In an actual case, such arbitrary restriction would not be imposed, but instead all mechanisms compatible with existing information would be considered.

(A) $$2\,HONO_2 \underset{k_{-1}}{\overset{k_1}{\underset{\text{fast}}{\rightleftharpoons}}} H_2\overset{+}{O}NO_2 + NO_3^-$$

$$\text{benzene} + H_2\overset{+}{O}NO_2 \xrightarrow[k_2]{\text{slow}} \text{(H, NO}_2\text{ cyclohexadienyl cation)} + H_2O$$

$$\text{(H, NO}_2\text{ cyclohexadienyl cation)} \xrightarrow{\text{fast}} \text{nitrobenzene} + H^+$$

$$\text{rate} = k_2[H_2\overset{+}{O}NO_2][\text{benzene}] = \frac{k_2 k_1}{k_{-1}} \frac{[HONO_2]^2}{[NO_3^-]}[\text{benzene}]$$

$$= k_{obs} \frac{[HONO_2]^2}{[NO_3^-]}[\text{benzene}]$$

(B)

$$2\,HONO_2 \underset{k_{-1}}{\overset{\text{fast}\atop k_1}{\rightleftharpoons}} H_2\overset{+}{O}NO_2 + NO_3^-$$

$$H_2\overset{+}{O}NO_2 \xrightarrow[k_2]{\text{slow}} H_2O + NO_2^+$$

$$NO_2^+ + \text{benzene} \overset{\text{fast}}{\rightleftharpoons} \text{(H, NO}_2\text{ cyclohexadienyl cation)}$$

$$\text{(H, NO}_2\text{ cyclohexadienyl cation)} \xrightarrow{\text{fast}} \text{nitrobenzene} + H^+$$

$$\text{rate} = \frac{k_1 k_2}{k_{-1}} \frac{[HONO_2]^2}{[NO_3^-]} = k_{obs} \frac{[HONO_2]^2}{[NO_3^-]}$$

(C)

$$2\,HONO_2 \underset{k_{-1}}{\overset{\text{fast}\atop k_1}{\rightleftharpoons}} NO_2^+ + NO_3^- + H_2O$$

$$NO_2^+ + \text{benzene} \underset{k_{-2}}{\overset{\text{fast}\atop k_2}{\rightleftharpoons}} \text{(H, NO}_2\text{ cyclohexadienyl cation)} \quad \textbf{I}$$

$$\text{(H, NO}_2\text{ cyclohexadienyl cation)} \xrightarrow{\text{slow}\atop k_3} \text{nitrobenzene} + H^+$$

The third step is rate-controlling, so

$$\text{rate} = k_3[\mathbf{I}]$$

[**I**] can be expressed in terms of the rapid equilibria involved in its formation:

$$k_{-2}[\mathbf{I}] = k_2[NO_2^+][\text{benzene}]$$

$$[NO_2^+] = \frac{k_1[HNO_3]^2}{k_{-1}[NO_3^-][H_2O]}$$

$$\text{rate} = k_3 \frac{k_2[\text{benzene}]k_1[HNO_3]^2}{k_{-2}k_{-1}[NO_3^-][H_2O]}$$

$$\text{rate} = \frac{k_{obs}[\text{HNO}_3]^2[\text{benzene}]}{[\text{NO}_3^-][\text{H}_2\text{O}]}$$

$$= k_{obs}\frac{[\text{HNO}_3]^2[\text{benzene}]}{[\text{NO}_3^-]} \qquad \text{if } [\text{H}_2\text{O}] \gg [\text{benzene}]$$

Mechanism B has the distinctive feature that it is zero-order in the reactant benzene, since the rate-determining step occurs prior to the involvement of benzene. Mechanism B has, in fact, been established for nitration of benzene in several organic solvents, and the absence of a benzene concentration term in the rate law is an important part of the evidence for this mechanism.[8]

Mechanisms A and C, on the other hand, provide kinetic expressions that are rather similar in form, differing only in the inclusion of water in the expression for mechanism C. This might not be a detectable difference. If the concentration of water is several times larger than that of benzene, its overall concentration will change little during the course of the reaction. In this circumstance, the term for the concentration of water would disappear (by becoming a component of the observed rate constant k) so that the form of the kinetic expression alone would not distinguish between mechanisms A and C.

To illustrate the development of a kinetic expression from a postulated reaction mechanism, let us consider the base-catalyzed reaction of benzaldehyde and acetophenone.

$$\text{PhCH{=}O} + \text{Ph}\overset{\text{O}}{\overset{\|}{\text{C}}}\text{CH}_3 \rightarrow \text{Ph}\underset{\text{H}}{\overset{\text{OH}}{\text{C}}}\text{CH}_2\overset{\text{O}}{\overset{\|}{\text{C}}}\text{Ph} \rightarrow \text{PhCH{=}CH}\overset{\text{O}}{\overset{\|}{\text{C}}}\text{Ph} + \text{H}_2\text{O}$$

Based on general knowledge of base-catalyzed reactions of carbonyl compounds, a reasonable sequence of steps can be written, but the relative rates of the steps are an open question. Furthermore, it is known that reactions of this type are generally reversible so that the potential reversibility of each step must be taken into account. A completely reversible mechanism is as follows:

(1) $$\text{Ph}\overset{\text{O}}{\overset{\|}{\text{C}}}\text{CH}_3 + {}^-\text{OEt} \underset{k_{-1}}{\overset{k_1}{\rightleftharpoons}} \text{Ph}\overset{\text{O}}{\overset{\|}{\text{C}}}\text{CH}_2^- + \text{EtOH}$$

(2) $$\text{Ph}\overset{\text{O}}{\overset{\|}{\text{C}}}\text{CH}_2^- + \text{Ph}\overset{\text{O}}{\overset{\|}{\text{C}}}\text{H} \underset{k_{-2}}{\overset{k_2}{\rightleftharpoons}} \text{Ph}\overset{\text{O}^-}{\overset{|}{\text{C}}}\text{HCH}_2\overset{\text{O}}{\overset{\|}{\text{C}}}\text{Ph}$$

(3) $$\text{EtOH} + \text{Ph}\overset{\text{O}^-}{\overset{|}{\text{C}}}\text{HCH}_2\overset{\text{O}}{\overset{\|}{\text{C}}}\text{Ph} \underset{k_{-3}}{\overset{\text{fast}\; k_3}{\rightleftharpoons}} \text{Ph}\overset{\text{OH}}{\overset{|}{\text{C}}}\text{HCH}_2\overset{\text{O}}{\overset{\|}{\text{C}}}\text{Ph} + \text{EtO}^-$$

(4) $$\text{Ph}\overset{\text{OH}}{\overset{|}{\text{C}}}\text{HCH}_2\overset{\text{O}}{\overset{\|}{\text{C}}}\text{Ph} + \text{EtO}^- \xrightarrow{k_4} \text{PhCH{=}CH}\overset{\text{O}}{\overset{\|}{\text{C}}}\text{Ph} + {}^-\text{OH} + \text{EtOH}$$

8. J. H. Ridd, *Acc. Chem. Res.* **4**, 248 (1971); J. H. Ridd, in *Studies on Chemical Structure and Reactivity*, J. H. Ridd (ed.), John Wiley and Sons, New York, 1966, Chapter 7; J. G. Hoggett, R. B. Moodie, J. R. Penton, and K. Schofield, *Nitrogen and Aromatic Reactivity*, Cambridge University Press, Cambridge, 1971; K. Schofield, *Aromatic Nitration*, Cambridge University Press, Cambridge, 1980.

Since proton transfer reactions between oxygen atoms are usually very fast, step 3 can be assumed to be a rapid equilibrium. With the above mechanism assumed, let us examine the rate expressions that would result, depending upon which of the steps is rate-determining.

If step 1 is rate-controlling, the rate expression would be

$$\text{rate} = k_1[\text{PhCOCH}_3][^-\text{OEt}]$$

Under these conditions, the concentration of the second reactant, benzaldehyde, would not enter into the rate expression.

If step 1 is an equilibrium and step 2 is rate-controlling, we obtain the rate expression

$$\text{rate} = k_2[\text{PhCOCH}_2^-][\text{PhCHO}]$$

which on substituting in terms of the rapid prior equilibrium gives

$$\text{rate} = k_2K_1[\text{PhCOCH}_3][^-\text{OEt}][\text{PhCHO}]$$

since

$$[\text{PhCOCH}_2^-] = K_1[\text{PhCOCH}_3][^-\text{OEt}]$$

where K_1 is the equilibrium constant for the deprotonation in the first step.

If the final step is rate-controlling, the rate is

$$\text{rate} = k_4[^-\text{OEt}][\text{PhCH(OH)CH}_2\text{COPh}]$$

The concentration of the intermediate

$$\text{PhCH(OH)CH}_2\text{COPh}$$

can be expressed in terms of the three prior equilibria. Using **I** for the intermediate and **I**$^-$ for its conjugate base and neglecting [EtOH], since EtOH is the solvent and its concentration will remain constant, gives the relationships:

$$K_3 = \frac{[\mathbf{I}][^-\text{OEt}]}{[\mathbf{I}^-]} \qquad \text{and} \qquad [\mathbf{I}] = K_3\frac{[\mathbf{I}^-]}{[^-\text{OEt}]}$$

and, since $[\mathbf{I}^-] = K_2[\text{PhCOCH}_2^-][\text{PhCHO}]$, substituting for $[\mathbf{I}^-]$ gives

$$[\mathbf{I}] = K_3\frac{K_2[\text{PhCOCH}_2^-][\text{PhCHO}]}{[^-\text{OEt}]}$$

Substituting for $[\text{PhCOCH}_2^-]$ from the equilibrium expression for step 1 gives

$$[\mathbf{I}] = \frac{K_3K_2[\text{PhCHO}]}{[^-\text{OEt}]}K_1[\text{PhCOCH}_3][^-\text{OEt}] = K'[\text{PhCHO}][\text{PhCOCH}_3]$$

and this provides the final rate expression

$$\text{rate} = k_{\text{obs}}[^-\text{OEt}][\text{PhCHO}][\text{PhCOCH}_3]$$

The form of this third-order kinetic expression is identical to that in the case where the second step is rate-determining.

Experimental studies of this base-catalyzed condensation have revealed that it is third-order, indicating that either the second or fourth step must be rate-determining. Studies on the intermediate **I** obtained by an alternative synthesis have shown that k_4 is about four times as large as k_{-3} so that about 80% of the intermediate goes on to product. These reactions are faster than the overall reaction under the same conditions, so the second step must be rate-controlling.[9]

These examples illustrate the relationship between kinetic results and the determination of reaction mechanism. Kinetic results can exclude from consideration all mechanisms that require a rate law different from the observed one. It is often true, however, that related mechanisms give rise to identical predicted rate expressions. In this case, the mechanisms are "kinetically equivalent," and a choice between them is not possible on the basis of kinetic data. A further limitation on the information that kinetic studies provide should also be recognized. Although the data can give the *composition* of the activated complex for the rate-determining step and preceding steps, they provide no information about the *structure* of the intermediate. Sometimes the structure can be inferred from related chemical experience, but it is never established by kinetic data alone.

The nature of the rate constants k_r can be discussed in terms of *transition state theory.* This is a general theory for analyzing the energetic and entropic components of a reaction process. In transition state theory, a reaction is assumed to involve the attainment of an activated complex that goes on to product at an extremely rapid rate. The rate of decomposition of the activated complex has been calculated from the assumptions of the theory to be $6 \times 10^{12}\ \mathrm{s}^{-1}$ at room temperature and is given by the expression[10]

$$\text{rate of activated complex decomposition} = \frac{\kappa k T}{h} \tag{4.3}$$

in which κ is the transmission coefficient, which is usually taken to be 1, k is Boltzmann's constant, h is Planck's constant, and T is absolute temperature.

$$\text{rate of reaction} = \frac{\kappa k T}{h}[\text{activated complex}]$$

If the activated complex is considered to be in equilibrium with its component molecules, the attainment of the transition state (T.S.) can be treated as being

9. E. Coombs and D. P. Evans, *J. Chem. Soc.*, 1295 (1940); D. S. Noyce, W. A. Pryor, and A. H. Bottini, *J. Am. Chem. Soc.* 77, 1402 (1955).

10. For a complete development of these relationships, see M. Boudart, *Kinetics of Chemical Processes*, Prentice-Hall, Englewood Cliffs, New Jersey, 1968, pp. 35–46; I. Amdur and G. G. Hammes, *Chemical Kinetics, Principles and Selected Topics*, McGraw-Hill, New York, 1966, pp. 43–58; J. W. Moore and R. G. Pearson, *Kinetics and Mechanism*, Wiley, New York, 1981, pp. 159–169; M. M. Kreevoy and D. G. Truhlar, in *Investigation of Rates and Mechanisms of Reactions*, C. F. Bernasconi (ed.), *Techniques of Chemistry*, Fourth Edition, Vol. VI, Part 1, Wiley-Interscience, New York, 1986.

analogous to a bimolecular reaction:

$$\mathrm{A} + \mathrm{B} \rightarrow \mathrm{C}$$

$$\mathrm{A} + \mathrm{B} \rightleftharpoons \mathrm{T.S.} \rightarrow \mathrm{C}$$

$$K^{\ddagger} = \frac{[\mathrm{T.S.}]}{[\mathrm{A}][\mathrm{B}]}$$

The position of this equilibrium is related to the free energy required for attainment of the transition state. The superscript sign (‡) is used to specify that it is a process involving a transition state or "activated complex" that is under discussion:

$$\Delta G^{\ddagger} = -RT \ln K^{\ddagger}$$

This free energy is referred to as the *free energy of activation.* The rate of the reaction is then given by

$$\text{rate} = \frac{\kappa kT}{h}[\mathrm{T.S.}]$$

$$[\mathrm{T.S.}] = K^{\ddagger}[\mathrm{A}][\mathrm{B}]$$

Since

$$K^{\ddagger} = e^{-\Delta G^{\ddagger}/RT}$$

$$\text{rate} = \frac{\kappa kT}{h} e^{-\Delta G^{\ddagger}/RT}[\mathrm{A}][\mathrm{B}] \tag{4.4}$$

Comparison with the form of the expression for the rate of any single reaction step:

$$\text{rate} = k_r[\mathrm{A}][\mathrm{B}]$$

reveals that the magnitude of $\Delta G^{\ddagger}$ will be the factor that determines the magnitude of k_r at any given temperature.

Qualitative features of reaction mechanisms are often described in the context of transition state theory and illustrated with potential energy diagrams. The potential energy diagrams for a hypothetical one-step bimolecular reaction and for a two-step reaction are shown in Fig. 4.1. The lower diagram depicts a two-step reaction in which an intermediate having a finite lifetime is involved. Two transition states are then involved. The higher activation energy of the first transition state implies that the first step would be slower and therefore rate-determining. These two-dimensional diagrams are useful devices for qualitative discussion of reaction mechanisms. The line plots the free energy of the reaction complex as it progresses along the reaction coordinate from reactants to products.

Such diagrams make clear the difference between an intermediate and a transition state. An intermediate lies in a depression on the potential energy curve. Thus, it will have a finite lifetime. The actual lifetime will depend on the depth of the depression. A shallow depression implies a low activation energy for the subsequent step, and therefore a short lifetime. The deeper the depression, the longer will be the lifetime of the intermediate. The situation at a transition state is quite different.

It has only fleeting existence and represents an energy maximum on the reaction path.

There is one path between reactants and products that has a lower energy maximum than any other; this is the pathway that the reaction will follow. The line in a two-dimensional potential energy plot represents this lowest-energy pathway. It represents a path across an energy surface describing energy as a function of the spatial arrangement of the atoms involved in the reaction. The *principle of microscopic reversibility* arises directly from transition state theory. *The same pathway that is traveled in the forward direction of a reaction will be traveled in the reverse direction,*

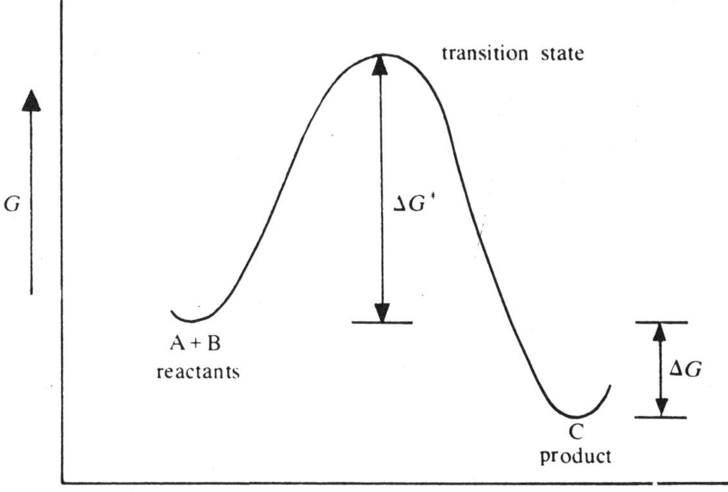

Reaction coordinate for a single-step reaction

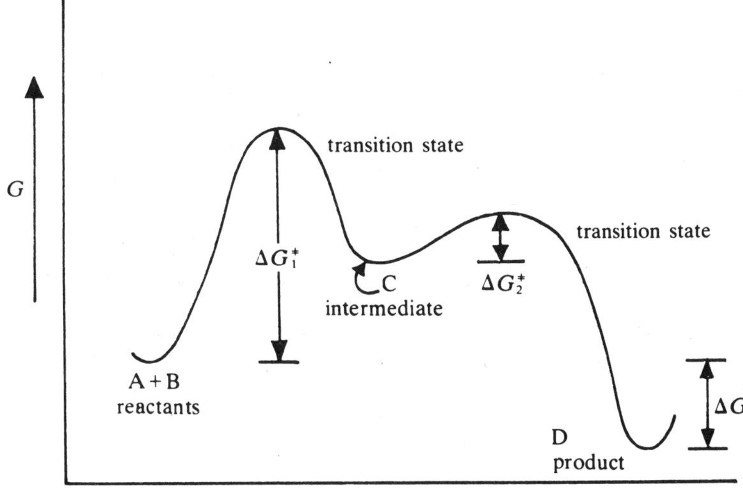

Reaction coordinate for a two-step reaction

Fig. 4.1. Potential energy diagrams for single-step and two-step reactions.

since it affords the lowest energy barrier for either process. Thus, information about the nature of a transition state or intermediate deduced by a study of a forward reaction is applicable to the discussion of the reverse process occurring under the same conditions.

Since transition states cannot be observed, there is no experimental method for establishing their structures. In recent years, theoretical descriptions of molecules have been applied to this problem. By applying one of the MO methods, structures can be calculated for successive geometries which gradually transform the reactants into products. Exploration of a range of potential geometries and calculation of the energy of the resulting ensembles can, in principle, locate and describe the minimum-energy pathway. To the extent that the calculations accurately reflect the molecular reality, this provides a structural description of the reaction path and transition state.

The temperature dependence of reaction rates permits evaluation of the enthalpy and entropy components of the free energy of activation. The terms in Eq. (4.4) corresponding to k_r can be expressed as

$$k_r = \frac{\kappa kT}{h}(e^{-\Delta H^{\ddagger}/RT})(e^{\Delta S^{\ddagger}/R}) \tag{4.5}$$

The term $(\kappa kT/h)\, e^{\Delta S^{\ddagger}/R}$ varies only slightly with T compared to $e^{-\Delta H^{\ddagger}/RT}$ because of the exponential nature of the latter. To a good approximation, then

$$\frac{k_r}{T} = C\,e^{-\Delta H^{\ddagger}/RT} \tag{4.6}$$

$$\ln\frac{k_r}{T} = \frac{-\Delta H^{\ddagger}}{RT} + C' \tag{4.7}$$

A plot of $\ln(k_r/T)$ versus $(1/T)$ is then a straight line, and its slope is $-\Delta H^{\ddagger}/R$. Once $\Delta H^{\ddagger}$ is determined in this manner, $\Delta S^{\ddagger}$ is available from the relationship

$$\Delta S^{\ddagger} = \frac{\Delta H^{\ddagger}}{T} + R\ln\frac{hk_r}{\kappa kT} \tag{4.8}$$

which can be obtained by rearranging Eq. (4.5).

The temperature dependence of reactions can also be expressed in terms of the Arrhenius equation

$$k_r = A\,e^{-E_a/RT} \tag{4.9}$$

$$\ln k_r = -E_a/RT + \ln A \tag{4.10}$$

Comparison of the form of Eq. (4.9) with Eq. (4.5) indicates that A in the Arrhenius equation corresponds to $(\kappa kT/h)e^{\Delta S^{\ddagger}/R}$. The Arrhenius equation shows that a plot of $\ln k_r$ versus $1/T$ will have the slope $-E_a/R$.[11] For reactions in solution at constant pressure, $\Delta H^{\ddagger}$ and E_a are related by

$$E_a = \Delta H^{\ddagger} + RT \tag{4.11}$$

11. For full consideration of the relationship between Eqs. (4.5) and (4.9), see I. Amdur and G. G. Hammes, *Chemical Kinetics, Principles and Selected Topics*, McGraw-Hill, New York, 1966, pp. 53–58.

The magnitudes of ΔH^{+} and ΔS^{+} reflect transition state structure. Atomic positions in the transition state do not correspond to those in the ground state. In particular, the reacting bonds will be partially formed and partially broken. This is reflected in the higher energy content of the activated complex and corresponds to the enthalpy of activation, ΔH^{+}. The entropy of activation is a measure of the degree of order produced in the formation of the activated complex. If translational, vibrational, or rotational degrees of freedom are lost in going to the transition state, there will be a decrease in the total entropy of the system. Conversely, an increase of translational, vibrational, or rotational degrees of freedom will result in a positive entropy of activation.

Wide variation in enthalpy and entropy of activation for different reaction systems is possible, as illustrated by the following two reactions.

Dimerization of cyclopentadiene[12]:

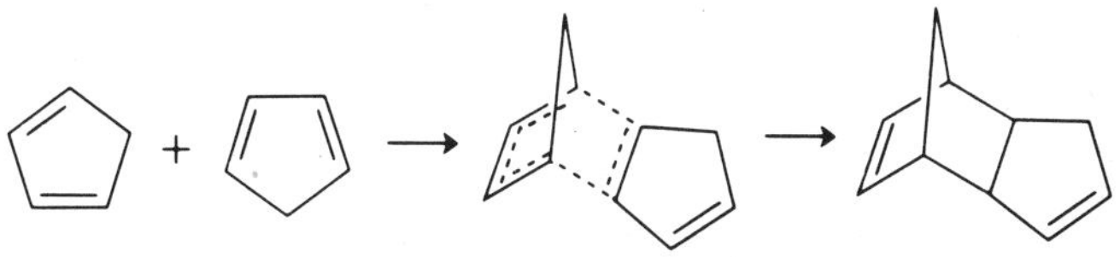

$$\text{gas phase } \Delta H^{+} = 15.5 \text{ kcal/mol}$$

$$\Delta S^{+} = -34 \text{ eu}$$

Decomposition of 1,1′-azobutane[13]:

$$\text{Bu—N=N—Bu} \rightarrow 2\text{Bu}\cdot + N_2$$

$$\text{gas phase } \Delta H^{+} = 52 \text{ kcal/mol}$$

$$\Delta S^{+} = +19 \text{ eu}$$

The relatively low ΔH^{+} term for the dimerization of cyclopentadiene is characteristic of concerted reactions (see Chapter 11), in which bond making accompanies bond breaking. It differs markedly from ΔH^{+} for the thermal decomposition of 1,1′-azobutane, in which the rate-determining step is a homolytic cleavage of a C—N bond, with little new bond making to compensate for the energy cost of the bond breaking. The entropy of activation, on the other hand, is more favorable in the 1,1′-azobutane decomposition, since a translational degree of freedom is being gained in the transition state as the molecular fragments separate. The dimerization of cyclopentadiene is accompanied by a very negative entropy of activation because of the loss of translational and rotational degrees of freedom. The two reacting molecules must attain a specific geometry to permit the bonding interactions that occur as the transition state is approached.

12. A. Wassermann, *Monatsh. Chem.* **83**, 543 (1952).
13. A. U. Blackham and N. L. Eatough, *J. Am. Chem. Soc.* **84**, 2922 (1962).

Unimolecular reactions that take place by way of cyclic transition states typically have negative entropies of activation because of the loss of rotational degrees of freedom associated with the highly ordered transition state. For example, thermal isomerization of allyl vinyl ether to 4-pentenal has $\Delta S^{\ddagger} = -8$ eu.[14]

$$\text{H}_2\text{C}{=}\text{CH}\text{–CH}_2\text{–O–CH}{=}\text{CH}_2 \ (\text{cyclic}) \longrightarrow [\text{cyclic transition state}] \longrightarrow \text{CH}_2{=}\text{CH–CH}_2\text{–CH}_2\text{–CH}{=}\text{O}$$

It is important to remember that the enthalpy and entropy of activation reflect the response of the reacting system *as a whole* to formation of the activated complex. As a result, the interpretation of these parameters is more complicated for reactions taking place in solution than for gas phase reactions. This complexity is particularly true for processes involving formation or destruction of charged species. The solvolysis of *t*-butyl chloride in 80% aqueous ethanol, for example, has as its rate-determining step unimolecular ionization of the carbon–chlorine bond to form chloride ion and the *t*-butyl cation. One might guess that this ionization should lead to a positive entropy of activation, since two particles are being generated. In fact, the entropy of activation is −6.6 eu. Because of its polar character, the transition state requires a greater ordering of solvent molecules than the nonpolar reactant.[15] It turns out to be generally true that reactions which generate charged species exhibit negative entropies of activation in solution. The reverse is true for reactions in which charged reactants lead to a neutral transition state.

4.3. Substituent Effects and Linear Free-Energy Relationships

In Chapter 1, Section 1.2 (p. 16), the effect of substituent groups on the acid strength of acetic acid derivatives was discussed qualitatively. It was noted in particular that the presence of groups more electronegative than hydrogen increased the acid strength relative to acetic acid. A number of important relationships between substituent groups and chemical properties have been developed. In many cases, such relationships can be expressed quantitatively and are useful both for interpretation of reaction mechanisms and for prediction of reaction rates and equilibria.

The most widely applied of these relationships is the *Hammett equation*, which relates rates and equilibria for many reactions of compounds containing substituted phenyl groups. It was noted in the 1930s that there is a relationship between the acid strengths of substituted benzoic acids and the rates of many other chemical reactions, for instance, the rates of hydrolysis of substituted ethyl benzoates. The correlation is illustrated graphically in Fig. 4.2, which shows $\log k/k_0$, where k_0 is the rate constant for hydrolysis of ethyl benzoate and k is the rate constant for the

14. F. W. Schuler and G. W. Murphy, *J. Am. Chem. Soc.* **72**, 3155 (1950).
15. E. Grunwald and S. Winstein, *J. Am. Chem. Soc.* **70**, 846 (1948).

substituted esters, plotted against $\log K/K_0$, where K and K_0 are the corresponding acid dissociation constants. Analogous plots for many other reactions of aromatic compounds show a similar linear correlation with the acid dissociation constants of the corresponding benzoic acids. Neither the principles of thermodynamics nor theories of reaction rates predict or require that there should be such linear relationships. There are, in fact, numerous reaction series that fail to show such correlations. Some insight into the origin of the correlation can be gained by considering the relationship between the correlation equation and the free-energy changes involved in the two processes. The line in Fig. 4.2 defines the following equation, in which m is the slope of the line:

$$m \log \frac{K}{K_0} = \log \frac{k}{k_0} \tag{4.12}$$

Substituting for K and k with the appropriate free energy or free energy of activation:

$$m(\log K - \log K_0) = \log k - \log k_0$$

$$m(-\Delta G/2.3RT + \Delta G_0/2.3RT) = -\Delta G^{\ddagger}/2.3RT + \Delta G_0^{\ddagger}/2.3RT$$

$$m(-\Delta G + \Delta G_0) = -\Delta G^{\ddagger} + \Delta G_0^{\ddagger}$$

$$m\Delta\Delta G = \Delta\Delta G^{\ddagger} \tag{4.13}$$

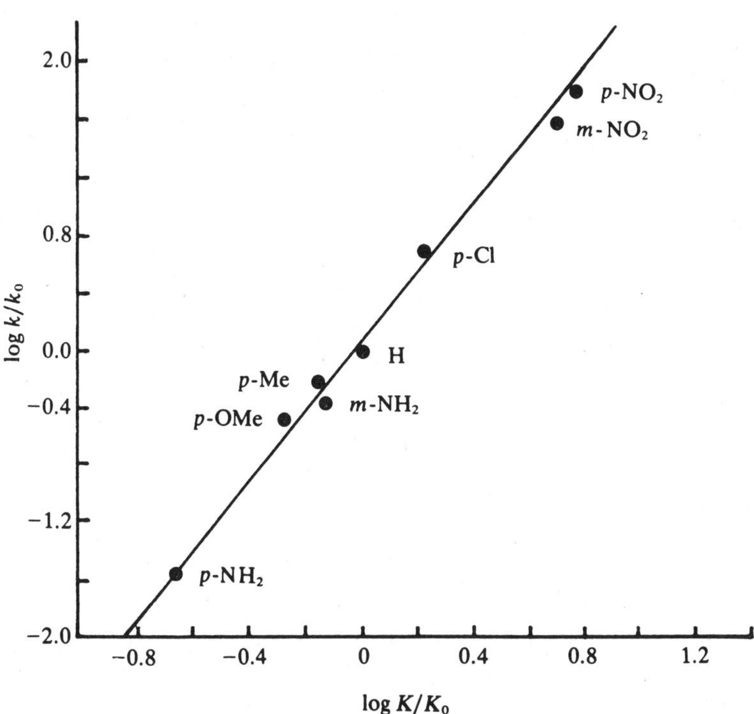

Fig. 4.2. Correlation of acid dissociation constants of benzoic acids with rates of alkaline hydrolysis of ethyl benzoates. [From L. P. Hammett, *J. Am. Chem. Soc.* **59**, 96 (1937).]

The linear correlation therefore indicates that the change in the free energy of activation for hydrolysis of ethyl benzoate on introduction of a series of substituent groups is *directly proportional* to the change in the free energy of ionization that is caused by the same series of substituents on benzoic acid. The various correlations arising from such directly proportional changes in free energies are called *linear free-energy relationships.*

Since ΔG and ΔG^{+} are combinations of enthalpy and entropy terms, a linear free-energy relationship between two reaction series can result from one of three circumstances: (1) ΔH is constant and the ΔS terms are proportional for the two series, (2) ΔS is constant and the ΔH terms are proportional, or (3) ΔH and ΔS are linearly related. Dissection of the free-energy changes into enthalpy and entropy components has often shown the third case to be true.[16]

The Hammett free-energy relationship is expressed in the following equations for equilibria and for rate data, respectively:

$$\log \frac{K}{K_0} = \sigma\rho \tag{4.14}$$

$$\log \frac{k}{k_0} = \sigma\rho \tag{4.15}$$

The numerical values of the terms σ and ρ are defined by selection of the reference reaction, the ionization of benzoic acids. This reaction is arbitrarily assigned a value of the *reaction constant,* ρ, of 1. The *substituent constant,* σ, can then be determined for a series of substituent groups by measurement of the acid dissociation constant of the substituted benzoic acids. The σ values so defined are used in the correlation of other reaction series, and the ρ values of the reactions are thus determined.

The relationship between Eqs. (4.12) and (4.14) is evident when the Hammett equation is expressed in terms of free energy. For the standard reaction, $\log[K/K_0] = \sigma\rho$:

$$-\Delta G/2.3RT + \Delta G_0/2.3RT = \sigma\rho = \sigma$$

since $\rho = 1$ for the standard reaction. Substituting into Eq. (4.12):

$$\begin{aligned} m\sigma &= -\Delta G^{+}/2.3RT + \Delta G_0{}^{+}/2.3RT \\ m\sigma &= \log k - \log k_0 \\ m\sigma &= \log \frac{k}{k_0} \\ m &= \rho \end{aligned} \tag{4.16}$$

The value of σ reflects the effect the substituent group has on the free energy of ionization of the substituted benzoic acid. The effect of the substituent is believed to represent a combination of factors. A substituent group can cause a polarization

16. P. D. Bolton, K. A. Fleming, and F. M. Hall, *J. Am. Chem. Soc.* **94**, 1033 (1972); J. E. Leffler, *J. Org. Chem.* **20**, 1202 (1955).

of charge density around the ring through the π system in both the reactant and the product. This will affect the position of the equilibrium. In the case of a reaction rate, the relative effects on the reactant and the transition state will determine the effect on the energy of activation. One mechanism for polarization and charge redistribution is the *resonance effect*, which is illustrated in Fig. 4.3a for several substituents. There is also an effect that originates with the bond dipoles between groups of differing electronegativity. Substituents with electronegativity greater than an aromatic carbon will place a net positive charge on the substituted carbon atom, whereas atoms less electronegative than an aromatic carbon will have the opposite effect. The resulting dipoles can perturb the electronic situation at the reaction site in two ways. The presence of the charge separation will influence the energy associated with development of charge elsewhere in the molecule. This is the result of through-space electrostatic interaction and is called a *field effect*. Depending on

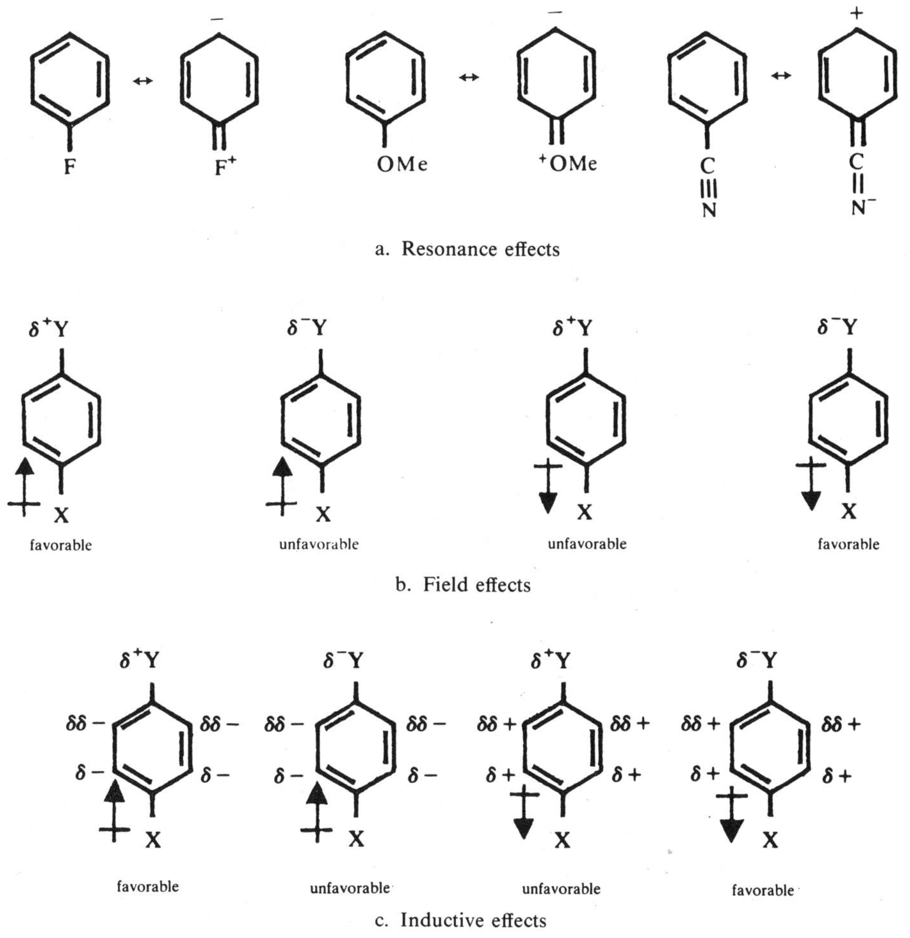

Fig. 4.3. Resonance, field, and inductive components of substituent effects in substituted benzenes.

the orientation of the dipole and of the charge developing at the reaction site, a substituent can either favor or disfavor the reaction, as illustrated in Fig. 4.3b. Another possible means of interaction of a substituent and the reaction site is called the *inductive effect.* This is transmission of bond dipoles through the intervening bonds by successive polarization of each bond. The experimental and theoretical results presently available indicate that the field effect outweighs the inductive effect as the primary means of transmission of the effect of bond dipoles.[17]

The Hammett equation in the form Eq. (4.14) or Eq. (4.15) is free of complications due to steric effects, since it is applied only to *meta* and *para* substituents. The geometry of the benzene ring ensures that groups in these positions cannot interact sterically with the site of reaction. Tables of σ values for many substituents have been collected; some values are shown in Table 4.3. The σ value for any substituent reflects the interaction of the substituent with the reacting site by a combination of resonance and field interactions. Table 4.4 shows a number of ρ values. The ρ value reflects the sensitivity of the particular reaction to substituent effects. The examples that follow illustrate some of the ways in which the Hammett equation can be used.

Example 4.2. The pK_a of *p*-chlorobenzoic acid is 3.98; that of benzoic acid is 4.19. Calculate σ for *p*-Cl.

$$\begin{aligned}\sigma = \log\frac{K_{p\text{-Cl}}}{K_{\text{H}}} &= \log K_{p\text{-Cl}} - \log K_{\text{H}} \\ &= -\log K_{\text{H}} - (-\log K_{p\text{-Cl}}) \\ &= \text{p}K_{a_{\text{H}}} - \text{p}K_{a_{p\text{-Cl}}} \\ &= 4.19 - 3.98 = 0.21\end{aligned}$$

Example 4.3. If the ρ value for alkaline saponification of methyl esters of substituted benzoic acids is 2.38, and the rate constant for saponification of methyl benzoate under the conditions of interest is $2 \times 10^{-4}\ M^{-1}\ \text{s}^{-1}$, calculate the rate constant for hydrolysis of methyl *m*-nitrobenzoate.

$$\log\frac{k_{m\text{-NO}_2}}{k_{\text{H}}} = \sigma_{m\text{-NO}_2}(\rho) = (0.70)(2.38) = 1.69$$

$$\frac{k_{m\text{-NO}_2}}{k_{\text{H}}} = 49$$

$$k_{m\text{-NO}_2} = 98 \times 10^{-4}\ M^{-1}\ \text{s}^{-1}$$

17. M. J. S. Dewar and P. J. Grisdale, *J. Am. Chem. Soc.* **84**, 3548 (1962); M. J. S. Dewar and A. P. Marchand, *J. Am. Chem. Soc.* **88**, 354 (1966); H. D. Holtz amd L. M. Stock, *J. Am. Chem. Soc.* **86**, 5188 (1964); C. L. Liotta, W. F. Fisher, G. H. Greene, Jr., and B. L. Joyner, *J. Am. Chem. Soc.* **94**, 4891 (1972); C. F. Wilcox and C. Leung, *J. Am. Chem. Soc.* **90**, 336 (1968); W. F. Reynolds, *Prog. Phys. Org. Chem.* **14**, 165 (1983).

Table 4.3. Substituent Constants[a]

Substituent group		σ_m	σ_p	σ^+	σ^-	$\mathscr{F}$	$\mathscr{R}$	σ_I	σ_R^0
Acetamido	CH_3CONH	**0.14**	**0.0**	−0.6	0.47	−0.27			
Acetoxy	CH_3CO_2	0.39	0.31	0.18		0.68	−0.07		
Acetyl	CH_3CO	**0.36**	**0.47**		0.82	0.53	0.20	0.20	0.16
Amino	NH_2	**−0.09**	**−0.30**	−1.3		0.04	−0.68	0.12	−0.50
Bromo	Br	0.37	**0.26**	0.15		0.72	−0.18	0.44	−0.16
t-Butyl	$(CH_3)_3C$	**−0.09**	**−0.15**	−0.26		−0.10	−0.14		
Carbomethoxy	CH_3O_2C	**0.35**	**0.44**		0.74			0.20	0.16
Carboxy	HO_2C	**0.35**	**0.44**		0.73	0.55	0.14		
Chloro	Cl	**0.37**	**0.24**	0.11		0.69	−0.16	0.46	−0.18
Cyano	CN	**0.62**	**0.70**		0.99	0.85	0.18	0.56	0.08
Ethoxy	C_2H_5O	0.1	−0.14	−0.82		0.36	−0.44		
Ethyl	C_2H_5	**−0.08**	**−0.13**	−0.30		−0.07	−0.11		
Fluoro	F	**0.34**	**0.15**	−0.07		0.71	−0.34	0.50	−0.31
Hydrogen	H	**0**	**0**	0	0	0	0	0	0
Hydroxy	OH	0.13	−0.38	−0.92		0.49	−0.64		
Methanesulfonyl	CH_3SO_2	**0.64**	**0.73**		1.05	0.90	0.21	0.60	0.12
Methoxy	CH_3O	**0.10**	−0.12	−0.78		0.41	−0.50	0.27	−0.42
Methyl	CH_3	**−0.06**	**−0.14**	−0.31		−0.05	−0.14	−0.04	−0.13
Nitro	NO_2	**0.71**	**0.81**		1.23	1.11	0.16	0.65	0.15
Phenyl	C_6H_5	**0.05**	**0.05**	−0.18	0.08	0.14	0.09		
Trifluoromethyl	CF_3	**0.46**	**0.53**		0.74	0.63	0.19	0.42	0.08
Trimethylammonio	$(CH_3)_3N^+$	0.99	0.96			1.46	0.0		
Trimethylsilyl	$(CH_3)_3Si$	−0.04	−0.07			−0.05	−0.04		

a. Values of σ_m, σ_p, σ^+, and σ^- from O. Exner in *Correlation Analysis in Chemistry*, N. B. Chapman and J. Shorter (eds.), Plenum Press, New York, 1978, Chap. 10. $\mathscr{F}$ and $\mathscr{R}$ from C. G. Swain and E. C. Lupton, *J. Am. Chem. Soc.* **90**, 4328 (1968). Values of σ_I and σ^0_R from J. Bromilow, R. T. C. Brownlee, V. O. Lopez, and R. W. Taft, *J. Org. Chem.* **44**, 4766 (1979). Values of σ_m and σ_p shown in boldface type are regarded as particularly reliable.

Example 4.4. Using data in Tables 4.3 and 4.4, calculate how much faster *p*-bromobenzyl chloride will solvolyze in water than *p*-nitrobenzyl chloride.

$$\log \frac{k_{p\text{-Br}}}{k_H} = (-1.31)(0.26), \qquad \log \frac{k_{p\text{-NO}_2}}{k_H} = (-1.31)(0.81)$$

$$\log k_{Br} - \log k_H = -0.34, \qquad \log k_{NO_2} - \log k_H = -1.06$$

$$\log k_{Br} + 0.34 = \log k_H, \qquad \log k_{NO_2} + 1.06 = \log k_H$$

$$\log k_{Br} + 0.34 = \log k_{NO_2} + 1.06$$

$$\log k_{Br} - \log k_{NO_2} = 0.72$$

$$\log \frac{k_{Br}}{k_{NO_2}} = 0.72$$

$$\frac{k_{Br}}{k_{NO_2}} = 5.25$$

Given in Table 4.3 in addition to the σ_m and σ_p values used with the classical Hammett equation are σ^+ and σ^-. These are substituent constant sets which reflect a recognition that the extent of resonance participation can vary for different reactions. The σ^+ values are used for reactions in which there is direct resonance interaction between an electron donor substituent and a cationic reaction center, whereas the σ^- set pertains to reactions in which there is a direct resonance interaction between the substituent and an electron-rich reaction site. These are cases where the resonance component of the substituent effect will be particularly important.

Table 4.4. Reaction Constants[a]

Reaction	ρ
$ArCO_2H \rightleftharpoons ArCO_2^- + H^+$, water	1.00
$ArCO_2H \rightleftharpoons ArCO_2^- + H^+$, EtOH	1.57
$ArCH_2CO_2H \rightleftharpoons ArCH_2CO_2^- + H^+$, water	0.56
$ArCH_2CH_2CO_2H \rightleftharpoons ArCH_2CH_2CO_2^- + H^+$, water	0.24
$ArOH \rightleftharpoons ArO^- + H^+$, water	2.26
$ArNH_3^+ \rightleftharpoons ArNH_2 + H^+$, water	3.19
$ArCH_2NH_3^+ \rightleftharpoons ArCH_2NH_2 + H^+$, water	1.05
$ArCO_2Et + {}^-OH \longrightarrow ArCO_2^- + EtOH$	2.61
$ArCH_2CO_2Et + {}^-OH \longrightarrow ArCH_2CO_2^- + EtOH$	1.00
$ArCH_2Cl + H_2O \longrightarrow ArCH_2OH + HCl$	−1.31
$ArC(Me)_2Cl + H_2O \longrightarrow ArC(Me)_2OH + HCl$	−4.48
$ArNH_2 + PhCOCl \longrightarrow ArNHCOPh + HCl$	−3.21

a. From P. R. Wells, *Linear Free Energy Relationships*, Academic Press, New York, 1968, pp. 12, 13.

Direct resonance interaction with a cationic center

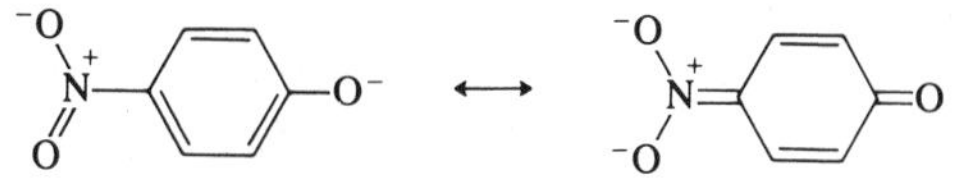

Direct resonance interaction with an anionic center

The underlying physical basis for the failure of Hammett σ_m and σ_p values to correlate certain reaction series is that all substituent interactions are some mixture of resonance and field effects. When direct resonance interaction is possible, the extent of the resonance increases and the substituent constants appropriate to a more "normal" mix of resonance and field effects then fail. There have been many attempts to develop sets of σ values that take into account extra resonance interactions.

One approach is to correct for the added resonance interaction. This is done in a modification of the Hammett equation known as the Yukawa–Tsuno equation[18]:

$$\log \frac{K}{K_0} = \rho\sigma + \rho(r)(\sigma^+ - \sigma) \tag{4.17}$$

The additional parameter r is adjusted from reaction to reaction; it reflects the extent of the additional resonance contribution. A large r corresponds to a reaction with a large resonance component, whereas when r goes to zero, the equation is identical to the original Hammett equation. When there is direct conjugation with an electron-rich reaction center, an equation analogous to Eq. (4.17) can be employed, but σ^- is used instead of σ^+.

The Yukawa–Tsuno relationship expanded to include both the σ^+ and σ^- constants is called the LArSR equation[19]:

$$\log \frac{k}{k_0} = \rho(\sigma^0 + r^+\Delta\sigma_R{}^+ + r^-\Delta\sigma_R{}^-)$$

In this equation, the substituent parameters $\Delta\sigma_R{}^+$ and $\Delta\sigma_R{}^-$ reflect the incremental resonance interaction with electron-demanding and electron-releasing reaction centers, respectively. The variables r^+ and r^- are established for a reaction series by regression analysis and are measures of the extent of the extra resonance contribution: the larger their value, the greater the extra resonance contribution. Since both donor and acceptor capacity will not contribute in a single reaction process, either r^+ or r^- would be expected to be zero.

18. Y. Yukawa and Y. Tsuno, *Bull. Chem. Soc. Jpn.* **32**, 971 (1959); J. Hine, *J. Am. Chem. Soc.* **82**, 4877 (1960); B. M. Wepster, *J. Am. Chem. Soc.* **95**, 102 (1973).
19. Y. Yukawa, Y. Tsuno, and M. Sawada, *Bull. Chem. Soc. Jpn.* **39**, 2274 (1966); Y. Yukawa, Y. Tsuno, and M. Sawada, *Bull. Chem. Soc. Jpn.* **45**, 1210 (1972).

Another approach to the treatment of the variability of resonance and field effects was devised by Swain and Lupton.[20] Their approach is to partition substituent effects into pure resonance and field contributions. The substituent constant would then be expressed as a sum of the field and resonance contributions.

$$\sigma = f\mathcal{F} + r\mathcal{R}$$

This treatment requires that *meta*- and *para*-substituted compounds be treated as separate reaction series, since they will have different relative resonance and field components, with resonance being stronger in the *para* series. Since the Swain–Lupton treatment uses four parameters, f, r, $\mathcal{F}$, and $\mathcal{R}$, in place of the single σ of the Hammett equation, the mathematical manipulations are somewhat more complex, but they can be easily handled with an appropriate calculator program. The result of the calculation is a "best-fit" correlation of the rate or equilibrium data with the terms f, r, $\mathcal{F}$, and $\mathcal{R}$. The computation also gives a "percent resonance" by comparing the magnitudes of f and r. The numerical reliability of both the substituent constants, $\mathcal{F}$ and $\mathcal{R}$, and the question of whether complete separation of resonance and field effects is achieved is a matter of some dispute.[21] The derived substituent constants are nevertheless qualitatively useful in comparing the properties of individual substituent groups and making at least a rough comparison of the interplay of resonance and field effects in different reactions. $\mathcal{F}$ and $\mathcal{R}$ values are included in Table 4.3 (p. 201). Certain substituents exhibit opposing resonance and field effects. Fluoro and methoxy groups are two examples (others can be located by examination of Table 4.3). This arises because these substituents act as electron donors by the resonance mechanism but the dipole associated with the substituent operates in the opposite direction.

The most elaborate and accurate treatment of substituent effects has been the development of a series of substituent constants, σ_R, chosen to reflect the resonance contribution of the substituent under various structural circumstances. This substituent constant is then used in conjunction with a second one σ_I, which reflects the inductive (and field) component of the overall substituent effect. The modified equation, called a *dual-substituent-parameter equation*, takes the form

$$\log\frac{K}{K_0} \quad \text{or} \quad \log\frac{k}{k_0} = \sigma_I\rho_I + \sigma_R\rho_R$$

where ρ_I and ρ_R are the reaction constants, which reflect the sensitivity of the system to inductive (and field) and resonance effects.[22] The σ_I values have been defined from studies in aliphatic systems where no resonance component should be present.

20. C. G. Swain and E. C. Lupton, *J. Am. Chem. Soc.* **90**, 4328 (1968); C. G. Swain, S. H. Unger, N. R. Rosenquist, and M. S. Swain, *J. Am. Chem. Soc.* **105**, 492 (1983); C. Hansch, A. Leo, S. H. Unger, K. H. Kim, D. Nikaitani, and E. J. Lien, *J. Med. Chem.* **16**, 1207 (1973).
21. W. F. Reynolds and R. D. Topsom, *J. Org. Chem.* **49**, 1989 (1984); A. J. Hoefnagel, W. Osterbeek, and B. M. Wepster, *J. Org. Chem.* **49**, 1993 (1984); M. Charton, *J. Org. Chem.* **49**, 1997 (1984); S. Marriott, W. F. Reynolds, and R. D. Topsom, *J. Org. Chem.* **50**, 741 (1985); C. G. Swain, *J. Org. Chem.* **49**, 2005 (1984).
22. S. Ehrenson, R. T. C. Brownlee, and R. W. Taft, *Prog. Phys. Org. Chem.* **10**, 1 (1973).

By properly scaling the σ_I values with σ values from aromatic systems, it is possible to assign values such that

$$\sigma = \sigma_I + \sigma_R$$

Statistical analysis of data from many reaction series has shown that no single σ_R is applicable to the entire range of reactions. This again reflects the fact that the resonance component is variable and responds to the nature of the particular reaction. Therefore, a series of four σ_R values was established, each of which applies to various reaction types, ranging from direct conjugation with electron-deficient reaction centers to the other extreme. We will discuss only one of these, $\sigma_R{}^0$, which applies in cases of minimal perturbation of the aromatic ring by charge development at the reaction site. The $\sigma_R{}^0$ values given in Table 4.3 are based on the use of ^{13}C chemical shifts as a measure of the sum of resonance and inductive effects. The chemical shift data of substituted benzenes were analyzed to provide the best correlation with the dual-substituent-parameter equation. In nonpolar solvents, which presumably best reflect the inherent molecular properties, $\rho_I = 3.74$ for cyclohexane and 3.38 for carbon tetrachloride. The corresponding values of ρ_R are 20.59 and 20.73. The relative magnitudes of ρ_I and ρ_R indicate that the ^{13}C chemical shift is more responsive to the resonance character of the substituent than to the inductive character.[23]

In general, the dissection of substituent effects need not be limited to resonance and field components, which are of special prominence in reactions of aromatic compounds. Any type of substituent interaction with a reaction center could be characterized by a substituent constant characteristic of the particular type of interaction and a reaction parameter indicating the sensitivity of the reaction series to that particular type of interaction. For example, it has been suggested that electronegativity and polarizability can be treated as substituent effects separate from field and resonance effects. This gives rise to the equation

$$\log \frac{k}{k_0} = \sigma_F \rho_F + \sigma_R \rho_R + \sigma_\chi \rho_\chi + \sigma_\alpha \rho_\alpha$$

where σ_F is the field, σ_R is the resonance, σ_χ is the electronegativity, and σ_α is the polarizability substituent constant.[24] We will, in general, emphasize the resonance and field components in our discussion of substituent effects.

The various multiparameter equations can usually improve the correlation of data that give poor correlations with the single-parameter Hammett equation. It must be recognized that this in part may be a direct consequence of the introduction of additional variable parameters. To derive solid mechanistic insight into the basis of improved correlation requires a critical appraisal of the results, and statistical analysis may be necessary. The details of critical statistical evaluation of free-energy

23. J. Bromilow, R. T. C. Brownlee, V. O. Lopez, and R. W. Taft, *J. Org. Chem.* **44**, 4766 (1979).
24. R. W. Taft and R. D. Topsom, *Prog. Phys. Org. Chem.* **16**, 1 (1987).

correlations have been extensively discussed, but this topic is beyond the scope of our coverage.[25]

It should be recognized that all of the approaches to linear free-energy relationships have been developed on an empirical basis. The reaction constants, substituent constants, and other parameters involved are specified by the definitions of the correlation equation. The data are then treated by the equation, and the variable parameters arise from the best fit to the correlation equation. In the case of the Hammett equation, each reaction series gives rise to its characteristic ρ value. For the Swain–Lupton equation, the *meta* series and the *para* series would each give rise to f and r.

We might ask if insight into substituent effects can be obtained by analyzing the structural perturbations produced by substituents by molecular orbital calculations. This is a fundamentally challenging task because of the uncertainty of the structure at the transition state and also a large task because of the large number of calculations that would be involved. In one approach to the problem, the stabilizing (or destabilizing) effect of substituents as a positive charge (representing an electrophile) or a negative charge (representing a nucleophile) approaches a benzene ring was calculated.[26] The effect of the substituents, as reflected in the calculated stabilization or destabilization, was parallel to that indicated by linear free-energy correlations. The amino, hydroxy, and fluoro groups, for example, were found to provide extra stabilization for the approach of an electrophile in comparison with other substituents where a strong resonance interaction would not be expected. For the approach of a negative charge, these substituents were destabilizing, but extra stabilization was found for groups such as nitro, cyano, and sulfonyl. While detailed calculations have been undertaken for only a limited number of substituents, it appears that the MO calculations give rise to the same patterns as found on the basis of empirical correlation. The MO calculations support the idea that substituent effects in aromatic compounds are a combination of field effects and resonance effects.[27]

Let us now consider how linear free-energy relationships can provide insight into reaction mechanisms. The choice of benzoic acid ionization as the reference reaction for the Hammett equation leads to $\sigma > 0$ for electron-withdrawing groups and $\sigma < 0$ for electron-releasing groups, since electron-withdrawing groups favor the ionization of the acid and electron-releasing groups have the opposite effect. Further inspection of the Hammett equation shows that ρ will be positive for all reactions that are favored by electron-withdrawing groups and negative for all

25. For detailed discussion of analysis of substituent effects, see J. Shorter, in *Correlation Analysis in Chemistry*, Plenum Press, New York, 1978; M. Charton, *Prog. Phys. Org. Chem.* **13**, 119 (1981); W. F. Reynolds, *Prog. Phys. Org. Chem.* **13**, 119 (1981); W. F. Reynolds, *Prog. Phys. Org. Chem.* **14**, 165 (1983).
26. E. R. Vorpagel, A. Streitwieser, Jr., and S. D. Alexandratos, *J. Am. Chem. Soc.* **103**, 3777 (1981).
27. R. D. Topsom, *Prog. Phys. Org. Chem.* **16**, 125 (1987); H. Agren and P. S. Bagus, *J. Am. Chem. Soc.* **107**, 134 (1985); R. D. Topsom, *Acc. Chem. Res.* **16**, 292 (1983); W. J. Reynolds, P. Dais, D. W. MacIntyre, R. D. Topsom, S. Marriott, E. v. Nagy-Felsobuki, and R. W. Taft, *J. Am. Chem. Soc.* **105**, 378 (1983); A. Pross and L. Radom, *Prog. Phys. Org. Chem.* **13**, 1 (1980).

reactions that are favored by electron-releasing groups. If the rates of a reaction series show a satisfactory correlation, both the sign and the magnitude of ρ provide information about the transition state for the reaction. In Example 4.3 (p. 200), the ρ value for saponification of substituted methyl benzoates is $+2.38$. This indicates that electron-withdrawing groups facilitate the reaction and that the reaction is somewhat more sensitive to substituent effects than the ionization of benzoic acids. The conclusion that the reaction is favored by electron-withdrawing substituents is in agreement with the accepted mechanism for ester saponification. The tetrahedral intermediate is negatively charged. Its formation should therefore be favored by substituents that can stabilize the developing charge.

$$\text{ArC(=O)OCH}_3 + {}^-\text{OH} \underset{}{\overset{\text{slow}}{\rightleftharpoons}} \text{ArC(O}^-\text{)(OH)OCH}_3 \xrightarrow{\text{several fast steps}} \text{ArCO}_2^- + \text{CH}_3\text{OH}$$

The solvolysis of diarylmethyl chlorides in ethanol shows a ρ value of -5.0, indicating that electron-releasing groups strongly facilitate the reaction. This ρ value provides support for a mechanism involving ionization of the halides in the rate-determining step. Electron-releasing groups can facilitate the ionization by a stabilizing interaction with the electron-deficient carbon that develops as ionization proceeds.

$$\text{ArC(Ph)(H)Cl} \xrightarrow{\text{slow}} \text{Ar}-\overset{+}{\text{C}}(\text{Ph})(\text{H}) + \text{Cl}^- \xrightarrow[\text{fast}]{\text{EtOH}} \text{ArC(Ph)(H)OC}_2\text{H}_5 + \text{HCl}$$

The relatively large ρ shows that the reaction is very sensitive to substituent effects and implies that there is a relatively large redistribution of charge in the transition state.

Not all reactions can be fitted by the Hammett equation or the multiparameter variants. There can be several reasons for this. The most common is that there is a change in mechanism as substituents vary. In a multistep reaction, for example, one step may be rate-determining in the region of electron-withdrawing substituents, but a different step may become rate-limiting as the substituents become electron-releasing. The rate of the semicarbazone formation reaction of benzaldehydes, for example, shows a nonlinear Hammett plot with ρ of about 3.5 for electron-releasing groups, but with ρ near -0.25 for electron-withdrawing groups.[28] The change in ρ is believed to be the result of a change in the rate-limiting step.

$$\text{ArCH=O} + \text{NH}_2\text{NHC(=O)NH}_2 \longrightarrow \text{ArCH(OH)NHNHC(=O)NH}_2$$

(rate-controlling for electron-releasing substituents)

$$\text{ArCH(OH)NHNHC(=O)NH}_2 \longrightarrow \text{ArCH=NNHC(=O)NH}_2 + \text{H}_2\text{O}$$

(rate-controlling for electron-attracting substituents)

28. D. S. Noyce, A. T. Bottini, and S. G. Smith, *J. Org. Chem.* **23**, 752 (1958).

Any reaction which shows a major shift in transition state structure over the substituent series would be expected to give a curved Hammett plot, since a variation in the extent of resonance participation would then be expected.

The substituent constants recorded in Table 4.3 (p. 201), provide valuable insight into the electronic nature of the various functional groups. By comparing σ with σ^+, one gains a qualitative impression of the ability of a given substituent to act as an electron-donating group by resonance. The $\mathscr{F}$ and σ_I values are indicators of the field and inductive characteristics of the various substituents. It should be noted that some groups, such as OMe and NH_2, that have strong electron-releasing capacity by resonance are electron-withdrawing when only field and inductive effects are considered. If $\mathscr{F}$ and $\mathscr{R}$ are taken as the best indicators of field versus resonance interactions, the substituents can be classified as in Table 4.5. By comparing σ_m values with σ_p values (Table 4.3), one can see that field effects usually dominate with *meta* substituents, while resonance effects are more important for *para* substituents. Thus σ_m constants of substituents such as hydroxy and methoxy have the positive signs while the σ_p constants are negative. This relationship is reasonable in terms of structure, sine the *para* substituents are more favorably situated for resonance interaction.

There is a system for designating the electron-releasing or -attracting properties of substituents. This system is illustrated in Table 4.5. The symbols $+M$ and $-M$ have been used to designate resonance interactions (the M comes from "mesomerism," a synonym for resonance) and the symbols $+I$ and $-I$ for combined field and inductive effects (I for inductive because the importance of field effects was recognized only after the system was established).

The development of linear free-energy relationships in purely aliphatic molecules is complicated because steric and conformation factors come into play along with electronic effects. A number of successful treatment of aliphatic systems have been developed by separating polar effects from steric effects. We will not

Table 4.5. Classification of Substituent Groups

Resonance:				
Electron-releasing ($-M$)	Electron-releasing ($-M$)			Electron-withdrawing ($+M$)
Field:				
Electron-releasing ($-I$)	Electron-withdrawing ($+I$)			Electron-withdrawing ($+I$)
Me	AcNH	Br	OH	Ac
Et	AcO	Cl	MeO	CN
$(Me)_3C$	NH_2	F	EtO	NO_2
			Ph	CF_3
				$(Me)_3N^+$

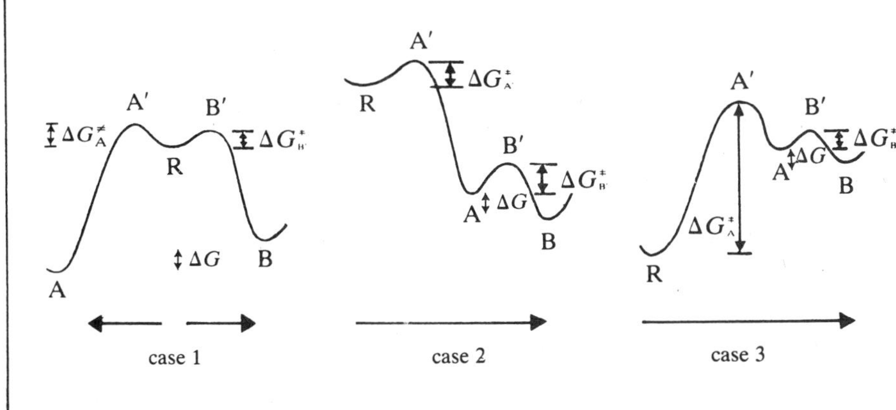

Fig. 4.4. Kinetic versus thermodynamic control.

discuss these methods but there are reviews available which can be consulted for information about this area.[29]

4.4. Basic Mechanistic Concepts: Kinetic versus Thermodynamic Control, Hammond's Postulate, and the Curtin–Hammett Principle

Use of two-dimensional potential energy diagrams can provide insight into the important general concepts listed in the heading of this section. There are many organic reactions in which the energy requirements for competing reaction paths are rather similar. It is important to be able to analyze the factors that may permit a particular reaction path to dominate.

4.4.1. Kinetic versus Thermodynamic Control

Product composition may be governed by the equilibrium thermodynamics of the system. When this is true, the product composition is governed by *thermodynamic control.* Alternatively, product composition may be governed by competing rates of formation of products. This is called *kinetic control.*

Let us consider cases 1–3 in Fig. 4.4. In case 1, $\Delta G^{\ddagger}$'s for formation of transition states A′ and B′ from the reactant R are much less than $\Delta G^{\ddagger}$'s for formation of A′ and B′ from A and B, respectively. If the latter two $\Delta G^{\ddagger}$'s are sufficiently large that the competitively formed products B and A do not return to R, the ratio of the products A and B at the end of the reaction will not depend on their relative

29. J. Hine, *Physical Organic Chemistry*, McGraw-Hill, New York, 1962, pp. 95–98; P. R. Wells, *Linear Free Energy Relationships*, Academic Press, New York, 1968, pp. 35–44; M. Charton, *Prog. Phys. Org. Chem.* **10**, 81 (1973).

stabilities, but only on their relative rates of formation. The formation of A and B is effectively irreversible in these circumstances. The energy plot in case 1 corresponds to this situation. This is a case of kinetic control. The relative amounts of products A and B will depend on the heights of the activation barriers $\Delta G_A{}^{\ddagger}$ and $\Delta G_b{}^{\ddagger}$.

In case 2, the lowest $\Delta G^{\ddagger}$ is that for formation of A′ from R. However, the $\Delta G^{\ddagger}$ for formation of B′ from A is not much larger. System 2 might be governed by either kinetic or thermodynamic factors. Conversion of R to A will be only slightly more favorable than conversion of A to B. If the reaction conditions are carefully adjusted, it will be possible for A to accumulate and not proceed to B. Under such conditions, A will be the dominant product and the reaction will be under kinetic control. Under somewhat more energetic conditions, for example, at a higher temperature, A will be transformed to B, and under these conditions the reaction will be under thermodynamic control. A and B will equilibrate, and the product ratio will depend on the equilibrium constant determined by ΔG.

In case 3, the barrier separating A and B is very small relative to that for formation of A′ from R. In this case, A and B will equilibrate more rapidly than R is converted to A. Adjustment of temperature or other reaction conditions would not change the A:B ratio at the end of the reaction very much.

The idea of kinetic versus thermodynamic control can be illustrated by discussing briefly the case of formation of enolate anions from unsymmetrical ketones. This is a very important case for synthesis and will be discussed more fully in Chapter 1 of Part B. Any ketone with more than one type of α-hydrogen can give rise to more than one enolate. Many studies have shown that the ratio of the possible enolates depends on the reaction conditions.[30] This can be illustrated for the case of methyl isopropyl ketone (3-methyl-2-butanone). If the base chosen is a strong, sterically hindered one and the solvent is aprotic, the major enolate formed is **3**. If a protic solvent is used or if a weaker base (one comparable in basicity to the ketone enolate) is used, the dominant enolate is **2**. Enolate **3** is the "kinetic enolate" while **2** is the thermodynamically favored enolate.

$$\underset{\mathbf{2}}{(^-O)(H_3C)C{=}C(CH_3)_2} \xleftarrow{B^-} \underset{\mathbf{1}}{CH_3\overset{O}{\overset{\|}{C}}CH(CH_3)_2} \xrightarrow{B^-} \underset{\mathbf{3}}{H_2C{=}C(O^-)CH(CH_3)_2}$$

The structural and mechanistic basis for the relationships between kinetic versus thermodynamic control and the reaction conditions is as follows. The α-hydrogens of the methyl group are less hindered sterically than the α-hydrogen of the isopropyl group. As a result, abstraction of one of these hydrogens as a proton is faster than

30. J. d'Angelo, *Tetrahedron* **32**, 2979 (1976); H. O. House, *Modern Synthetic Reactions*, Second Edition, W. A. Benjamin, Menlo Park, California, 1972.

for the isopropyl hydrogen. This effect is magnified when the base is sterically hindered so that it is particularly sensitive to the difference in the steric situation of the competing hydrogens. If the base is very strong, the enolate will not be reconverted to the ketone because the enolate will be too weak a base to regain the proton. These conditions correspond to case 1 in Fig. 4.4 and represent a case of kinetic control. If a weaker base is used or if the solvent is protic, protons can be transferred reversibly between the isomeric enolates and the base (because the base strengths of the enolate and the base are comparable). Under these conditions, the more stable enolate will be the predominant one because the enolates are equilibrated. The more substituted enolate **2** is the more stable of the pair, just as more substituted alkenes are more stable than terminal alkenes.

4.4.2. Hammond's Postulate

Because of the crucial role played by the energy of the transition state in determining the rates of chemical reactions, information about the structure of transition states is crucial to understanding reaction mechanisms. However, because transition states have only transitory existence, it has not been possible to make experimental measurements that would provide direct information about the structure of transition states. Hammond has discussed the circumstances under which it is valid to relate transition state structure to the structure of reactants, intermediates, and products.[31] His statements concerning transition state structure are known as *Hammond's postulates.* Discussing individual steps in a reaction mechanism, Hammond's postulate states "if two states, as for example, a transition state and an unstable intermediate, occur consecutively during a reaction process and have nearly the same energy content, their interconversion will involve only a small reorganization of molecular structure."

This statement can best be discussed with reference to potential energy diagrams. Case 1 in Fig. 4.5 represents a highly exothermic step with a low activation energy. It follows from Hammond's postulate that in this step the transition state will structurally resemble the reactant since they are close in energy and therefore interconverted by a small structural change. This is depicted in the potential energy diagram by a small displacement toward product along the reaction coordinate. Case 2 describes a step in which the transition state is a good deal higher in energy than either the reactant or the product. In this case, neither the reactant nor the product will be a good model of the transition state. Case 3 illustrates an endothermic step such as would occur in the formation of an unstable intermediate. In this case, the energy of the transition state is similar to that of the intermediate and the transition state should be similar in structure to the intermediate.

The significance of the concept incorporated in Hammond's postulate then is that, in appropriate cases, it permits discussion of transition state structure in terms

31. G. S. Hammond, *J. Am. Chem. Soc.* 77, 334 (1955).

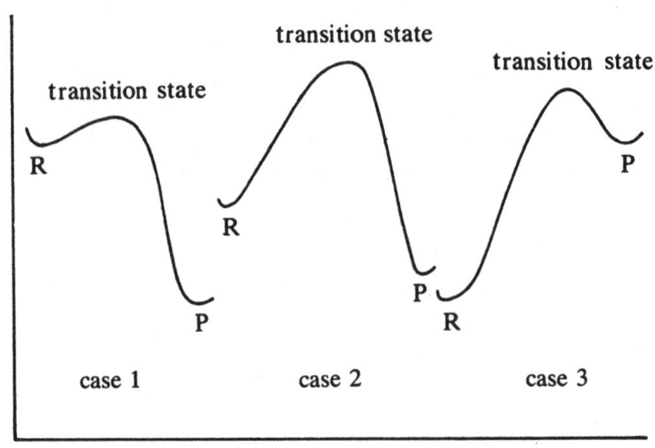

Fig. 4.5. Some typical potential energy diagrams that illustrate the application of Hammond's postulate.

of the reactants, intermediates, or products in a multistep reaction sequence. The postulate indicates that the cases where such comparison is appropriate are those in which the transition state is close in energy to the reactant, intermediate, or product.

The case of electrophilic aromatic substitution can illustrate a situation in which it is useful to discuss transition state structure in terms of a reaction intermediate. The *ortho–para-* and *meta*-directing effects of aromatic substituents were among the first structure–reactivity relationships to be developed in organic chemistry. Certain functional groups were found to activate aromatic rings toward substitution and to direct the entering electrophile to the *ortho* and *para* positions whereas others were deactivating and led to substitution in the *meta* position. The bromination of anisole (methoxybenzene), benzene, and nitrobenzene can serve as cases for discussion.

$C_6H_5OCH_3 \xrightarrow[\text{fast}]{Br_2}$ o-$BrC_6H_4OCH_3$ + p-$BrC_6H_4OCH_3$

$C_6H_6 \xrightarrow[\text{moderate}]{Br_2,\ FeBr_3}$ C_6H_5Br

$C_6H_5NO_2 \xrightarrow[\text{slow}]{Br_2,\ FeBr_3}$ m-$BrC_6H_4NO_2$

It can be demonstrated that the reactions are kinetically controlled. It is therefore the $\Delta G^{\ddagger}$ value that holds the key to the connection between the rate effects and the substituent's directing effects. But to discuss $\Delta G^{\ddagger}$ satisfactorily, we must know something about the reaction mechanism and the nature of the competing transition states. Electrophilic aromatic substitution will be discussed in detail in Chapter 10.

Evidence presented there will indicate that electrophilic aromatic substitution involves a distinct intermediate and two less well-defined states. The potential energy diagram in Fig. 4.6 is believed to be a good representation of the energy changes that occur during bromination. By application of the Hammon postulate, we can conclude that the rate-determining step involves formation of a transition state that should closely resemble the intermediate σ complex. It is therefore legitimate to discuss effects of substituents on the transition state in terms of the structure of this intermediate.

Since the product composition is kinetically controlled, the isomer ratio will be governed by the relative magnitudes of ΔG_o^{+}, ΔG_m^{+}, and ΔG_p^{+}, the energies of activation for the *ortho*, *meta*, and *para* transition states, respectively. In Fig. 4.7 a qualitative comparison of these ΔG^{+} values is made.

At the transition state, a considerable positive charge is present on the benzene ring, primarily at positions 2, 4, and 6 in relation to the entering bromine.

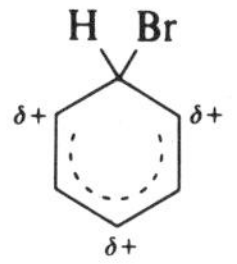

The electron-releasing methoxy group can interact directly to delocalize the charge and stabilize the intermediates leading to *o*- and *p*-bromoanisole. It cannot stabilize the intermediate leading to *m*-bromoanisole.

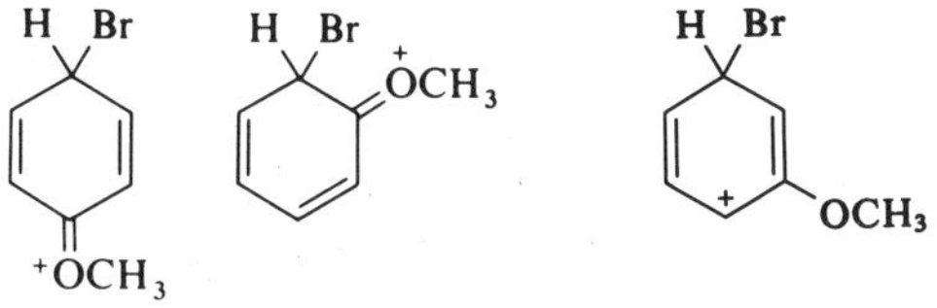

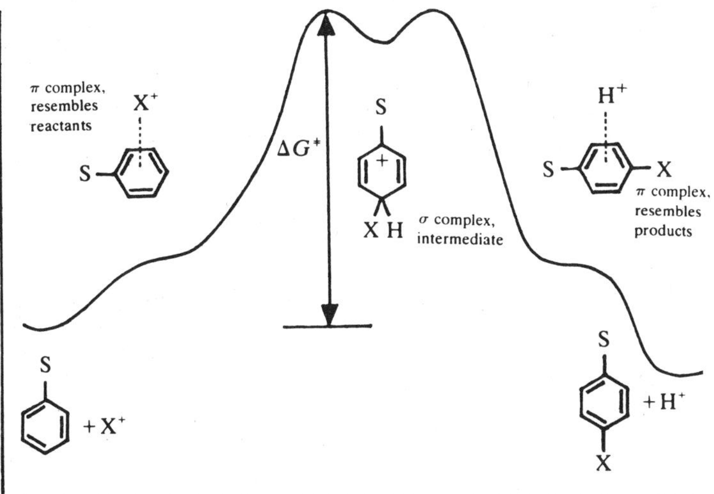

Fig. 4.6. Potential energy diagram for electrophilic aromatic substitution.

The *ortho* and *para* intermediates are therefore stabilized relative to benzene but the *meta* intermediate is not, as is illustrated in Fig. 4.7. As a result, anisole reacts faster than benzene, and the products are mainly the *ortho* and *para* isomers.

In the case of nitrobenzene, the electron-withdrawing nitro group is not able to stabilize the positive charge in the σ-complex intermediate. In fact, it strongly destabilizes the intermediate. This destabilization is greatest in the *ortho* and *para* intermediates, which place positive charge on the nitro-substituted carbon. The *meta* transition state is also destablized relative to benzene, but not as much as the *ortho* and *para* transition states. As a result, nitrobenzene is less reactive than benzene and the product is mainly the *meta* isomer

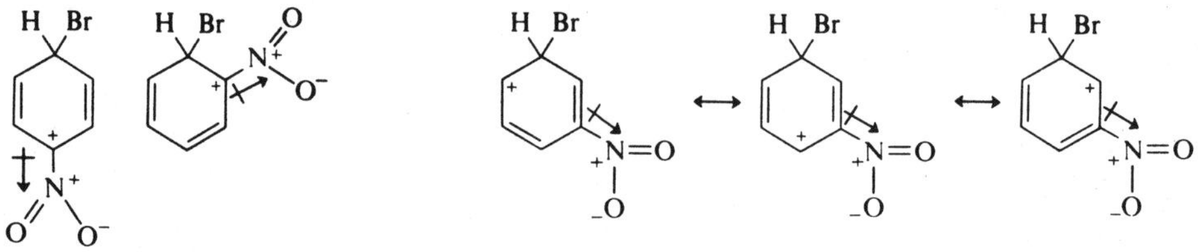

The substituent effects in aromatic electrophilic substitution can be analyzed in terms of resonance effects. In other systems, stereoelectronic effects or steric

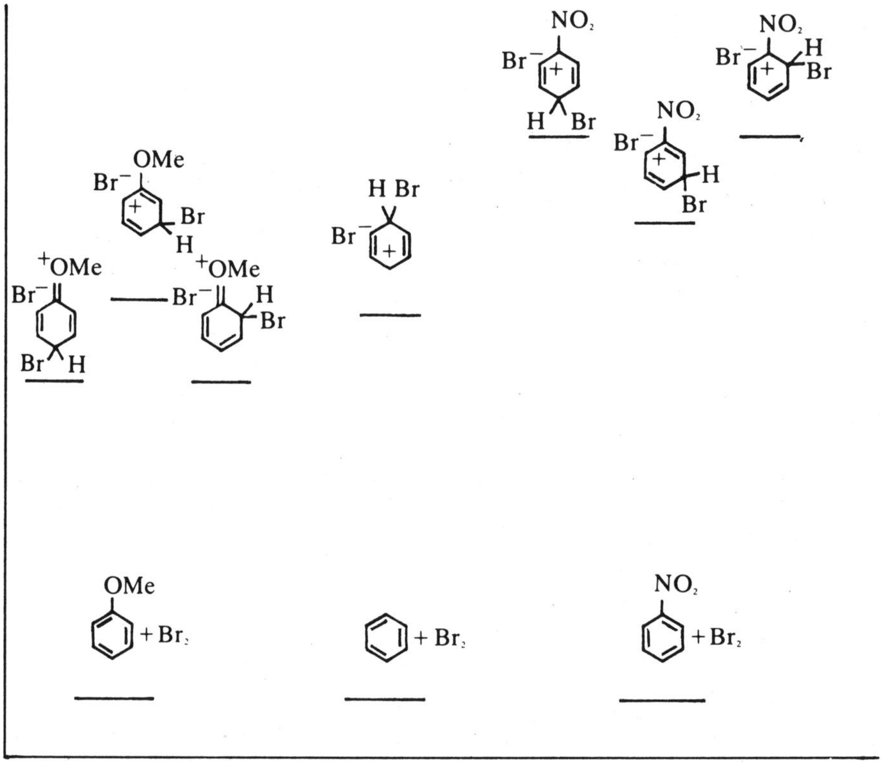

Fig. 4.7. Transition state energies in bromination.

effects might be more important. Whatever the nature of the substituent effects, the Hammond postulate insists that structural discussion of transition states in terms of reactants, intermediates, or products is valid only when their energies are similar.

4.4.3. The Curtin–Hammett Principle

In Chapter 3, equilibria among conformers of organic molecules were discussed. At this point, let us consider in a general way the effect that conformational equilibria can have on a chemical reaction. Under what circumstances can the position of the conformational equilibrium for a reactant determine which of two competing reaction paths will be followed? A potential energy diagram is shown in Fig. 4.8. In most cases, the energy of activation for a chemical reaction will be greater than that for a conformational equilibrium as is illustrated in the figure. If this is the case, $\Delta G_a^{\ddagger}$ and $\Delta G_b^{\ddagger} \gg \Delta G_c$. The conformers of the reactant are in equilibrium and are interconverted at a rate much faster than that at which the competing reactions occur.

$$\mathrm{B} \rightleftharpoons \mathrm{A} \qquad K_c = \frac{[\mathrm{A}]}{[\mathrm{B}]}$$

$$\text{rate of formation of product } P_{\mathrm{A}} = \frac{dP_{\mathrm{A}}}{dt} = k_a[\mathrm{A}] = k_a K_c[\mathrm{B}]$$

$$\text{rate of formation of product } P_{\mathrm{B}} = \frac{dP_{\mathrm{B}}}{dt} = k_b[\mathrm{B}]$$

$$\text{product ratio} = \frac{dP_{\mathrm{A}}/dt}{dP_{\mathrm{B}}/dt} = \frac{k_a K_c[\mathrm{B}]}{k_b[\mathrm{B}]} = \frac{k_a K_c}{k_b}$$

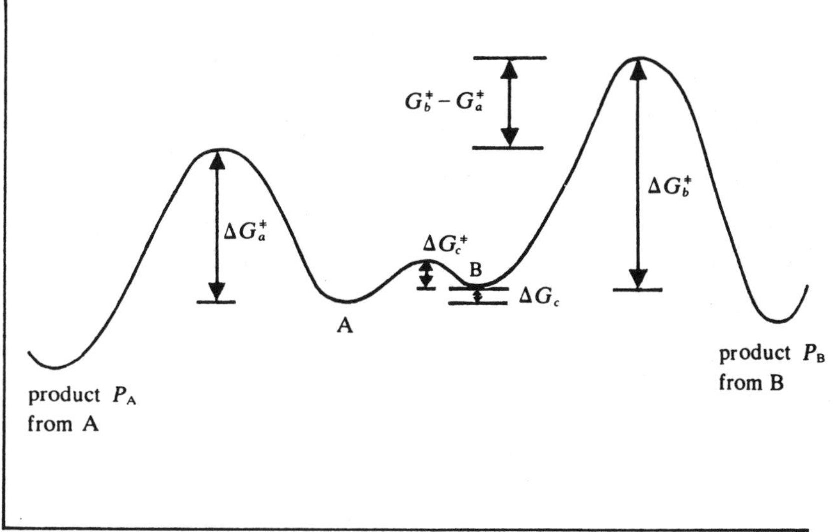

Fig. 4.8. Effect of conformation on product distribution.

According to transition state theory,

$$k_r = \frac{\kappa kT}{h} e^{-\Delta G^{\ddagger}/RT} \quad \text{and} \quad K_c = e^{-(-\Delta G_c)/RT}$$

$$\text{product ratio} = \frac{(\kappa kT/h)\, e^{-\Delta G_a^{\ddagger}/RT}\, e^{+\Delta G_c/RT}}{(\kappa kT/h)\, e^{-\Delta G_b^{\ddagger}/RT}}$$

$$= e^{(-\Delta G_a^{\ddagger} + \Delta G_b^{\ddagger} + \Delta G_c)/RT}$$

But from Fig. 4.8,

$$\Delta G_b^{\ddagger} - \Delta G_a^{\ddagger} + \Delta G_c = G_b^{\ddagger} - G_a^{\ddagger}$$

The product ratio is therefore not determined by ΔG_c but instead primarily on the relative energy of the two transition states leading to A and B.

The conclusion that the ratio of products formed from conformational isomers is not determined by the conformation population ratio is known as the *Curtin–Hammett principle*.[32] While the rate of the formation of the products is dependent upon the relative concentration of the two conformers, since $\Delta G_b^{\ddagger}$ is decreased relative to $\Delta G_a^{\ddagger}$ to the extent of the difference in the two conformational energies, the conformational preequilibrium is established rapidly, relative to the two competing product-forming steps.[33] The position of the conformational equilibrium cannot control the product ratio. The reaction may proceed through a minor conformation if it is the one which provides access to the lowest-energy transition state.

The same arguments can be applied to other energetically facile interconversions of two potential reactants. For example, many organic molecules undergo rapid proton shifts (tautomerism) and the chemical reactivity of the two isomers may be quite different. It is not valid, however, to deduce the ratio of two tautomers on the basis of subsequent reactions which have activation energies greater than that of the tautomerism. Just as in the case of conformational isomerism, the ratio of products formed in subsequent reactions will not primarily be controlled by the position of the facile equilibrium.

4.5. Isotope Effects

A special type of substituent effect that has proved very valuable in the study of reaction mechanisms is the replacement of an atom by one of its isotopes. Isotopic substitution has most often involved replacing protium by deuterium (or tritium),

32. D. Y. Curtin, *Rec. Chem. Prog.* **15**, 111 (1954); E. L. Eliel, *Stereochemistry of Carbon Compounds*, Mc-Graw-Hill, New York, 1962, pp. 151–152, 237–238.
33. For a more complete discussion of the relationship between conformational equilibria and reactivity, see J. I. Seeman, *Chem. Rev.* **83**, 83 (1983).

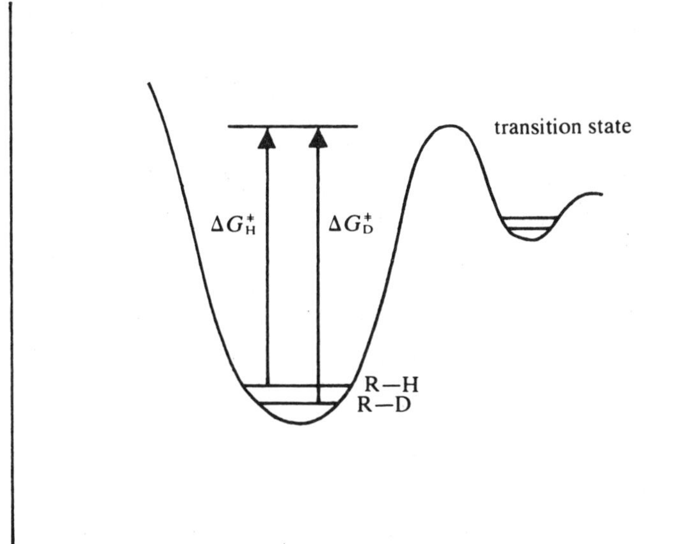

Fig. 4.9. Differing zero-point energies of protium- and deuterium-substituted molecules as the cause of primary kinetic isotope effects.

but the principle is applicable to nuclei other than hydrogen. The quantitative differences are largest, however, for hydrogen. Isotopic substitution has no effect on the qualitative chemical reactivity of the substrate, but it often has an easily measured effect on the rate at which reaction occurs. Let us consider how this modification of the rate arises. Initially, the discussion will concern *primary kinetic isotope effects*, those in which a bond to the isotopically substituted atom is broken in the rate-determining step. We will use C—H bonds as the specific topic of discussion, but the same concepts apply for other elements.

Any C—H bond has characteristic vibrations which impart some energy to the molecule in its normal state. This energy is called the *zero-point* energy. The energy associated with these vibrations is related to the mass of the vibrating atoms. Because of the greater mass of deuterium, the vibrations associated with a C—D bond contribute less to the zero-point energy than do those of the corresponding C—H bond. For this reason, substitution of protium by deuterium lowers the zero-point energy of a molecule. For a reaction involving cleavage of a bond to hydrogen (or deuterium), a vibrational degree of freedom in the normal molecule is converted to a translational degree of freedom on passing through the transition state. The energy difference due to this vibration disappears at the transition state. The transition state has the same energy for the protonated and deuterated species. Since the deuterated molecule had the lower zero-point energy, it necessarily has a higher activation energy to reach the transition state. This is illustrated in Fig. 4.9.

Just how large the rate difference is depends on the nature of the transition state. The maximum effect occurs when the hydrogen being transferred is bound about equally to two other atoms at the transition state. The calculated maximum for the isotope effect k_H/k_D involving C—H bonds is about 7 at room temperature.[34] When bond breaking is more or less than half complete at the transition state, the value is less and can be close to 1 if the transition state is very reactant-like or very product-like. Primary isotope effects can provide two very useful pieces of information about a reaction mechanism. First, the existence of a substantial isotope effect—that is, if k_H/k_D is 2 or more—is strong evidence that the bond to the isotopically substituted hydrogen atom is being broken in the rate-determining step. Second, the magnitude of the isotope effect provides a qualitative indication of where the transition state lies with respect to product and reactant. A relatively low primary isotope effect implies that the bond to hydrogen is either only slightly or nearly completely broken at the transition state. That is, the transition state must occur quite close to reactant or to product. An isotope effect near the theoretical maximum is good evidence that the transition state involves strong bonding of the hydrogen to both its new and old bonding partner.

Isotope effects may be observed even when the substituted hydrogen atom is not directly involved in the reaction. Such effects are called *secondary kinetic isotope effects.* Secondary isotope effects are smaller than primary ones and are usually in the range of $k_H/k_D = 0.7\text{–}1.5$. Secondary isotope effects may be normal ($k_H/k_D > 1$) or inverse ($k_H/k_D < 1$). They are also classified as α, β, etc., depending on whether the isotopic substitution is on the reacting carbon or farther away. Secondary isotope effects result from a tightening or loosening of the C—H bond at the transition state. The strength of the bond may change because of a hybridization change or a change in the extent of hyperconjugation, for example. If sp^3-hybridized carbon is converted to sp^2 as reaction occurs, a hydrogen bound to the carbon will experience decreased resistance to C—H bending. The freeing of the vibration for a C—H bond is greater than that for a C—D bond because the C—H bond is slightly longer and the vibration therefore has a larger amplitude. This will result in a normal isotope effect. Entry 5 in Scheme 4.2 is an example of such a reaction since it proceeds through a carbocation intermediate.

An inverse isotope effect will occur if coordination at the reaction center increases in the transition state. The bending will become more restricted. Entry 4 in Scheme 4.2 exemplifies such a case involving conversion of a tricoordinate carbonyl group to a tetravelent cyanohydrin. In this case the secondary isotope effect is 0.73.

Secondary isotope effects at the β position have been especially thoroughly studied in nucleophilic substitution reactions. When carbocations are involved as intermediates, substantial isotope effects are observed. This is because the hyperconjugative stabilization by the β-hydrogens weakens the C—H bond.[35] The observed

34. K. B. Wiberg, *Chem. Rev.* **55**, 713 (1955); F. H. Westheimer, *Chem. Rev.* **61**, 265 (1961).
35. V. J. Shiner, W. E. Buddenbaum, B. L. Murr, and G. Lamaty, *J. Am. Chem. Soc.* **90**, 809 (1968); A. J. Kresge and R. J. Preto, *J. Am. Chem. Soc.* **89**, 5510 (1967); G. J. Karabatsos, G. C. Sonnichsen, C. G. Papaioannou, S. E. Scheppele, and R. L. Shone, *J. Am. Chem. Soc.* **89**, 463 (1967); D. D. Sunko and W. J. Hehre, *Prog. Phys. Org. Chem.* **14**, 205 (1983).

Scheme 4.2. Some Representative Kinetic Isotope Effects

	Reaction	$k_H/k_D(°C)^a$
A. Primary kinetic isotope effects		
1[b]	$PhCH_2{-}H^* + Br\cdot \longrightarrow Ph{-}CH_2\cdot + H^*{-}Br$	4.6 (77)
2[c]	$(CH_3)_2C(H^*){-}C(=O){-}C(H^*)(CH_3)_2 + OH^- \longrightarrow (CH_3)_2C(H^*){-}C(O^-){=}C(CH_3)_2$	6.1 (25)
3[d]	2-H*-cyclopentyl-$\overset{+}{N}(CH_3)_3 + OH^- \longrightarrow$ cyclopentene $+ (CH_3)_3N$	4.0 (191)
B. Secondary kinetic isotope effects		
4[e]	$CH_3O{-}C_6H_4{-}\overset{*}{C}H{=}O + HCN \longrightarrow CH_3O{-}C_6H_4{-}C(H^*)(OH){-}C{\equiv}N$	0.73 (25)
5[f]	$H_3C{-}C_6H_4{-}C(H^*)_2{-}Cl \xrightarrow[CF_3CH_2OH]{H_2O} H_3C{-}C_6H_4{-}C(H^*)_2{-}OH$	1.30 (25)
6[g]	9,10-dihydro-9,10-ethanoanthracene ($CH_2^*CH_2^*$ bridge) $\longrightarrow$ anthracene $+ CH_2^*{=}CH_2^*$	1.37 (50)

a. Temperature of measurement is indicated in parentheses.
b. K. B. Wiberg and L. H. Slaugh, *J. Am. Chem. Soc.* **80**, 3033 (1958).
c. R. A. Lynch, S. P. Vincenti, Y. T. Lin, L. D. Smucker, and S. C. Subba Rao, *J. Am. Chem. Soc.* **94**, 8351 (1972).
d. W. H. Saunders, Jr., and T. A. Ashe, *J. Am. Chem. Soc.* **91**, 4473 (1969).
e. L. do Amaral, H. G. Bull, and E. H. Cordes, *J. Am. Chem. Soc.* **94**, 7579 (1972).
f. V. J. Shiner, Jr., M. W. Rapp, and H. R. Pinnick, Jr., *J. Am. Chem. Soc.* **92**, 232 (1970).
g. M. Taagepera and E. R. Thornton, *J. Am. Chem. Soc.* **94**, 1168 (1972).

secondary isotope effects are normal as would be predicted since the bond is weakened.

$$H{-}C{-}\overset{+}{C} \longleftrightarrow H^+\ C{=}C$$

Detailed analysis of isotope effects reveals that there are many other factors which can contribute to the overall effect in addition to the dominant change in bond vibrations. For that reason, it is not possible to quantitatively predict the magnitude of either primary or secondary isotope effects for a given reaction.

Furthermore, there is not a sharp numerical division between primary and secondary effects, especially in the range between 1 and 2. For these reasons, isotope effects are usually used in conjugation with other criteria in the description of reaction mechanisms.[36]

4.6. Isotopes in Labeling Experiments

A quite different use of isotopes in mechanistic studies involves their use as labels for ascertaining the location of a given atom involved in a reaction. As in kinetic experiments, the isotopic substitution will not qualitatively affect the course of the reaction. The nuclei most commonly used for isotopic tracer experiments in organic chemistry are deuterium, tritium, and the ^{13}C and ^{14}C isotopes of carbon. There are several means of locating isotopic labels. Deuterium can frequently be located by analysis of NMR spectra. In contrast to the normal ^{1}H isotope, deuterium does not show an NMR signal under the normal operating circumstances. The absence of a specific signal can therefore be used to locate deuterium. Both mass spectrometry and infrared spectroscopy also can be used to locate deuterium. Tritium and ^{14}C and other radioactive isotopes are usually located on the basis of the radioactivity. This is a very sensitive method. It should be borne in mind that, in most experiments in which radioactive labels are used, only a small fraction of the atoms at the site of substitution are the radioactive nuclide. The high sensitivity of radioactive atoms to detection makes the method practical. Usually, the location of ^{14}C requires a degradation process of some sort to separate the atoms that might conceivably be labeled. Carbon-13 has become an important isotope for tracer experiments relatively recently. Unlike normal ^{12}C, ^{13}C has a nuclear magnetic moment and can be detected in NMR spectrometers of the proper frequency. This avoids the necessity of developing a degradation scheme to separate specific carbon atoms.

There are many excellent examples of isotopic labeling experiments in both organic chemistry and biochemistry.[37] An interesting example is the case of hydroxylation of the amino acid phenylalanine which is carried out by the enzyme phenyl-

36. For more complete discussion of isotope effects, see W. H. Saunders, in *Investigation of Rates and Mechanisms of Reactions*, E. S. Lewis (ed.), *Techniques of Chemistry*, Third Edition, Vol. VI, Part 1, Wiley-Interscience, New York, 1974, pp. 211–255; L. Melander and W. H. Saunders, Jr., *Reaction Rates of Isotopic Molecules*, Wiley, New York, 1980; W. H. Saunders, in *Investigation of Rates and Mechanisms of Reactions*, C. F. Bernasconi (ed.), *Techniques of Chemistry*, Fourth Edition, Vol. VI, Part 1, Wiley-Interscience, New York, 1986, Chapter VIII.
37. For examples of use of isotopic labels in mechanistic studies, see V. F. Raaen, in *Investigation of Rates and Mechanisms of Reactions*, E. S. Lewis (ed.), *Techniques of Chemistry*, Vol VI, Part 1, Wiley-Interscience, New York, 1974, pp. 257–284, and *Isotopes in Organic Chemistry*, Vols. 1-4, E. Buncel and C. C. Lee, (eds.), Elsevier, New York, 1975–1978; C. Wentrup, in *Investigation of Rates and Mechanisms of Reactions*, C. F. Bernasconi (ed.), *Techniques of Chemistry*, Fourth Edition, Vol. VI, Part 1, Wiley-Interscience, New York, 1986, Chapter IX.

alanine hydroxylase.

$$\text{T}-\text{C}_6\text{H}_4-\text{CH}_2-\underset{\displaystyle \text{NH}_2}{\text{CH}}-\text{CO}_2\text{H} \xrightarrow[\text{hydroxylase}]{\text{phenylalanine}} \text{HO}-\text{C}_6\text{H}_3(\text{T})-\text{CH}_2-\underset{\displaystyle \text{NH}_2}{\text{CH}}-\text{CO}_2\text{H}$$

When this reaction was studied by use of tritium, the phenylaline was labeled with tritium at the 4-position of the phenyl ring. When the product, tyrosine, was isolated, it retained much of the original radioactivity, even though the 4-position was now substituted by a hydroxyl group. When this was studied in detail, it was found that the ^3H originally at the 4-position had rearranged to the 3-position in the course of oxidation. This hydrogen shift, called the *NIH shift*,[38] has subsequently been found to occur in many biological oxidations of aromatic compounds.

4.7. Characterization of Reaction Intermediates

Identification of the intermediates in a multistep reaction is a major objective of studies of reaction mechanisms. When the nature of each intermediate is fairly well understood, a great deal is known about the reaction mechanism. The amount of an intermediate present in a reacting system at any instant of time will depend on the rates of the steps by which it is formed and the rate of its subsequent reaction. A qualitative indication of the relationship between intermediate concentration and the kinetics of the reaction can be gained by considering a simple two-step reaction mechanism:

$$\text{reactants} \xrightarrow{k_1} \text{intermediate} \xrightarrow{k_2} \text{products}$$

In some reactions, the situation $k_1 > k_2$ exists. Under these conditions, the concentration of the intermediate will build up as it goes on more slowly to product. The possibility of isolating, or at least observing, the intermediate then exists. If both k_1 and k_2 are large, the reaction may proceed too rapidly to permit isolation of the intermediate, but spectroscopic studies, for example, should reveal the existence of two distinct stages for the overall reaction. It should be possible to analyze such a system and determine the two rate constants.

If the two steps are of about equal rates, only a small concentration of the intermediate will exist at any time. It is sometimes possible to interrupt such a reaction by lowering the temperature rapidly or adding a reagent that stops the reaction and isolate the intermediate. Intermediates can also be "trapped." In this approach, a compound that is expected to react specifically with the intermediate is added to the reaction system. If trapping occurs, the intermediate is diverted from

38. From its discovery at the National Institutes of Health (NIH); for an account of this discovery, see G. Guroff, J. W. Daly, D. M. Jerina, J. Renson, B. Witkop, and S. Udenfriend, *Science* **157**, 1524 (1967).

its normal course, and evidence for the existence of the intermediate is obtained if the structure of the trapped product is consistent with expectation.

Often it is more practical to study intermediates present in low concentration by spectroscopic methods. The theory and practice of instrumental methods of detection of intermediates will be discussed here only very briefly. Instrumental techniques have become very important in the study of reaction mechanisms. The most common methods in organic chemistry include ultraviolet-visible (UV-VIS), infrared (IR), nuclear magnetic resonance (NMR), and electron paramagnetic resonance (EPR) spectroscopy.

UV-VIS spectrometers can rapidly scan the electronic region of the spectrum and provide evidence of the development of characteristic *chromophores*. The major limitation imposed is that the compound to be detected must have a characteristic absorbance in the range 220–700 nm. Absorption of energy in this region is associated with promotion of electrons to higher energy states. In organic molecules, absorption of energy in this region usually requires the presence of multiple bonds, especially two or more such bonds in conjugation. Saturated molecules normally do not absorb significantly in this region. The amount of an intermediate that can be detected depends on how strongly it absorbs relative to other components of the reaction systems. In favorable cases, concentrations as low as 10^{-6} *M* can be detected.

Infrared (IR) spectrometers measure absorption of energy by excitation of molecular vibrations, including stretching, bending, and twisting of various parts of the molecule. Most organic molecules have a large number of bands in the IR spectrum. While it is usually not possible to assign every band to a specific vibration, individual bands can be highly characteristic of a specific molecule. Nearly all of the organic functional groups also have one or more regions of characteristic absorption. If it is suspected that a particular functional group is present in an intermediate, examination of the changes in the spectrum in the characteristic region of the functional group may permit detection. An example of IR detection of intermediates can be drawn from a study of the photochemical conversion of **1** to **3**. It was suspected that the ketene **2** might be an intermediate.[39]

Ketenes absorb near 2100–2130 cm^{-1}. When the photolysis was carried out and the IR spectrum of the solution monitored, it was found that a band appeared at 2118 cm^{-1}, grew, and then decreased as photolysis proceeded. The observation of this characteristic absorption constitutes good evidence for a ketone intermediate. As with UV-VIS spectroscopy, the amount of intermediate that can be detected depends both on the intensity of the absorption band and the presence of interfering bands. In general, IR requires somewhat higher concentration for detection than does UV-VIS.

71042 39. O. L. Chapman and J. D. Lassila, *J. Am. Chem. Soc.* **90**, 2449 (1968).

Either UV-VIS or IR spectroscopy can be combined with the technique of *matrix isolation* to detect and identify highly unstable intermediates. In this method, the intermediate is trapped in a solid inert matrix, usually one of the inert gases, at very low temperatures. Since each molecule is surrounded by inert gas atoms, there is no possibility for intermolecular reactions and the rates of intramolecular reactions are slowed by the low temperature. Matrix isolation is a very useful method for looking for intermediates in photochemical reactions. The method can also be used for gas phase reactions that can be conducted in such a way that the intermediates can be rapidly condensed into the matrix.

Nuclear magnetic resonance (NMR) is very widely used for detection of intermediates in organic reactions. Proton magnetic resonance is most useful because ^1H provides the greatest sensitivity of detection among the nuclei of interest in organic chemistry. Fluorine-19 and phosphorus-31 are other elements that provide high sensitivity. Carbon-13, oxygen-17, and nitrogen-15 provide relatively lower sensitivity, but isotopic enrichment is possible. Initially, NMR suffered from very low sensitivity relative to other spectroscopic methods. However, the development of high-field instruments and the use of Fourier transform methods have greatly increased sensitivity so that NMR can now be used to detect many reaction intermediates.

Free radicals and other intermediates with unpaired electrons can be detected in extremely low concentration by electron paramagnetic resonance (EPR). This technique measures the energy absorbed to reorient an electron's spin in a magnetic field. It provides structural information on the basis of splitting of the signal by adjacent nuclei, much as in NMR interpretation. EPR is not only extremely sensitive, but also very specific. Diamagnetic molecules present in solution give no signals, and the possibility for interference is therefore greatly decreased. The method can only be applied to reactions involving paramagnetic intermediates.

All other spectroscopic methods are equally applicable, in principle, to the detection of reaction intermediates so long as the method provides sufficient structural information to assist in the identification of the transient species. In the use of all methods, including those discussed above, it must be remembered that simple detection does not prove the species is an intermediate. It also must be shown that the species is converted to product. In favorable cases, this may be done by isolation or trapping experiments. More often, it may be necessary to determine the kinetic behavior of the appearance and disappearance of the intermediate and demonstrate that it is consistent with the species being an intermediate.

4.8. Catalysis by Acids and Bases

A detailed understanding of a reaction mechanism requires knowledge of the role catalysts play in the reaction. Catalysts cannot affect the position of equilibrium of a reaction. They function by increasing the rate of one or more steps in a reaction mechanism by providing a reaction path having a lower activation energy. The most general family of catalytic processes are those that involve transfer of a proton. Many reactions involving neutral reactants are strongly catalyzed by proton donors (Brønsted acids) or proton acceptors (Brønsted bases). Catalysis occurs when the conjugate base or conjugate acid of the substrate is a more reactive molecule than the neutral species. For example, reactions involving nucleophilic attack at carbonyl groups are often accelerated by acids. This type of catalysis occurs because the conjugate acid of the carbonyl compound is much more electrophilic than the neutral molecule.

$$\mathrm{R\overset{O}{\overset{\|}{C}}R} + \mathrm{H^+} \rightleftharpoons \mathrm{R\overset{^+OH}{\overset{\|}{C}}R}$$

$$\mathrm{R\overset{^+OH}{\overset{\|}{C}}R} + \mathbf{Nu}{:} \longrightarrow \mathrm{R-\underset{^+\mathbf{Nu}}{\underset{|}{\overset{OH}{\overset{|}{C}}}}-R} \longrightarrow \text{product}$$

Many important organic reactions involve nucleophilic carbon species (carbanions). The properties of carbanions will be discussed in detail in Chapter 7 and in Part B, Chapters 1 and 2. Most C—H bonds are very weakly acidic and have no tendency to ionize spontaneously to form carbanions. Reactions that involve carbanion intermediates are therefore usually carried out by reaction of the neutral organic molecule and the electrophile in the presence of a base that can generate the more reactive carbanion intermediate. Base-catalyzed condensation reactions of carbonyl compounds provide many examples of this type of reaction. The reaction between acetophenone and benzaldehyde which was considered in Section 4.2, for example, requires a basic catalyst to proceed, and the kinetics of the reaction show that the rate is proportional to the catalyst concentration. This is because the neutral acetophenone molecule is not nucleophilic and does not react with benzaldehyde. The enolate (carbanion) formed by deprotonation is much more nucleophilic.

$$\mathrm{Ph\overset{O}{\overset{\|}{C}}CH_3} + \mathrm{B^-} \rightleftharpoons \mathrm{Ph\overset{^-O}{\overset{|}{C}}{=}CH_2} + \mathrm{BH}$$

$$\mathrm{Ph\overset{^-O}{\overset{|}{C}}{=}CH_2} + \mathrm{PhCH{=}O} \rightarrow \mathrm{Ph\overset{O}{\overset{\|}{C}}CH_2\overset{O^-}{\overset{|}{C}}HPh} \rightarrow \text{product}$$

The role that acid and base catalysts play can be quantitatively studied by kinetic techniques. It is possible to recognize several distinct types of catalysis by acids and bases. The term *specific acid catalysis* is used when the reaction rate is dependent on the equilibrium for protonation of the reactant. This type of catalysis is independent of the concentration and specific structure of the various proton donors present in solution. Specific acid catalysis is governed by the *hydrogen ion concentration* (pH) of the solution. For example, for a series of reactions in an aqueous buffer system, the rate of the reaction would be a function of the pH, but not of the concentration or identity of the acidic and basic components of the buffer. The kinetic expression for any such reaction will include a term for hydrogen ion concentration, $[H^+]$. When the nature and concentration of proton donors present in solution affect the reaction rate, the term used is *general acid catalysis.* The kinetic expression for such a reaction will include terms for each of the potential proton donors that act as catalysts. The terms *specific base catalysis* and *general base catalysis* apply in the same way to base-catalyzed reactions.

Specific acid catalysis:

$$\text{rate} = k[\text{H}^+][\text{X}][\text{Y}], \qquad \text{where } [\text{X}][\text{Y}] = \text{concentration of the reactants}$$

General acid catalysis:

$$\text{rate} = k_1[\text{H}^+][\text{X}][\text{Y}] + k_2[\text{HA}^1][\text{X}][\text{Y}] + k_3[\text{HA}^2][\text{X}][\text{Y}],$$

where HA^1, HA^2 . . . are all kinetically significant proton donors

The experimental detection of general acid catalysis is done by rate measurements at constant pH but differing buffer concentration. Since under these circumstances $[H^+]$ is constant but the weak acid component(s) of the buffer (HA^1, HA^2, etc.) changes, the observation of a change in rate is evidence of general acid catalysis. If the rate remains constant, the reaction exhibits specific acid catalysis. Similarly, general base-catalyzed reactions will show a dependence of the rate on the concentration and identity of the basic constituents of the buffer system.

Specific acid catalysis is observed when a reaction proceeds only through a protonated intermediate which is in equilibrium with its conjugate base. Since the position of this equilibrium is a function of the concentration of solvated protons, only a single acid-dependent term appears in the kinetic expression. For example, in a two-step reaction involving rate-determining reaction of one reagent with the conjugate acid of a second, the kinetic expression will be as follows:

$$\text{B} + \text{H}^+ \xrightleftharpoons{\text{fast}} \text{BH}^+ \qquad \text{BH}^+ + \text{C} \xrightarrow{\text{slow}} \text{B—C} + \text{H}^+$$

$$\text{rate} = k_2[\text{BH}^+][\text{C}] = k_2[\text{C}]\text{K}[\text{B}][\text{H}^+]$$

$$= k_{\text{obs}}[\text{H}^+][\text{B}][\text{C}]$$

Several situations can lead to the observation of general acid catalysis. General acid catalysis can occur as a result of hydrogen bonding between the reactant R and a proton donor D—H to form a reactive complex {D-H . . . R} which then reacts with a substance Z:

$$\text{D—H} + \text{R} \rightleftharpoons \text{D—H}\cdots\text{R}$$

$$\text{D—H}\cdots\text{R} + \text{Z} \xrightarrow{\text{slow}} \text{D—H} + \text{R—Z}$$

Under these circumstances, a distinct contribution to the overall rate will be seen for each potential hydrogen bond donor D—H. General acid catalysis is also observed when a rate-determining proton transfer occurs from acids other than the solvated proton:

$$\text{R} + \text{HA} \xrightarrow{\text{slow}} {}^+\text{RH} + \text{A}^-$$

$${}^+\text{RH} \rightarrow \text{product}$$

Each acid HA^1, HA^2, etc., will then make a contribution to the overall rate of the reaction.

A kinetic expression which is equivalent to that for general acid catalysis also occurs if a prior equilibrium between reactant and the acids is followed by rate-controlling proton transfer. Each individual conjugate base will appear in the overall rate expression:

$$R + HA \rightleftharpoons {}^{+}RH + A^-$$

$$^{+}RH + A^- \xrightarrow{\text{slow}} \text{product} + HA$$

Notice that specific acid catalysis describes a situation where the reactant is in equilibrium with regard to proton transfer, and proton transfer is not rate-determining. On the other hand, each case that leads to general acid catalysis involves proton transfer in the rate-determining step. Because of these differences, the study of rates as a function of pH and buffer concentrations can permit conclusions about the nature of proton transfer processes and their relationship to the rate-determining step in a reaction.

As might be expected intuitively, there is a relationship between the effectiveness of general acid catalysts and the acidity of a proton donor as measured by its acid dissociation constant. This relationship is expressed by the following equation, which is known as the *Brønsted catalysis law*:

$$\log k_{\text{cat}} = \alpha \log K_a + b \tag{4.19}$$

An analogous equation holds for catalysis by bases. This equation requires that the free energies of activation for the catalytic step for a series of acids be directly proportional to the free energies of dissociation for the same series of acids. The proportionality constant σ is an indication of the sensitivity of the catalytic step to structural changes relative to the effect of the same structural changes on acid dissociation. It is often found that a single proportionality constant α is restricted to only structurally related types of acids and that α values of different magnitude are revealed by each type.

Figure 4.10 is plot of the Brønsted relationship for hydrolysis of an enol ether. The plot shows that the effectiveness of the various carboxylic acids as catalysts is related to their dissociation constants. In this particular case, the constant α is 0.79[40]:

OCH_3 (1-methoxycyclohexene) $+ HA \xrightarrow{\text{slow}}$ $^{+}OCH_3$, H (protonated cyclohexanone methyl ether) $+ A^-$

$^{+}OCH_3$ (cyclohexylidene) $+ H_2O \xrightarrow{\text{fast}}$ HO OCH$_3$ (cyclohexane) $\xrightarrow{\text{fast}}$ O (cyclohexanone) $+ CH_3OH$

40. A. J. Kresge, H. L. Chen, Y. Chiang, E. Murrill, M. A. Payne, and D. S. Sagatys, *J. Am. Chem. Soc.* **93**, 413 (1971).

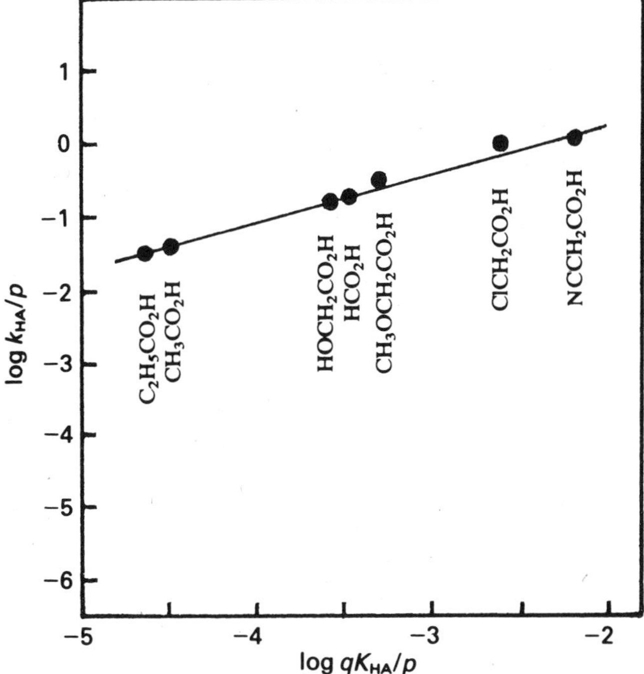

Fig. 4.10. Brønsted relation for the hydrolysis of methyl cyclohexenyl ether. (Adapted from Ref. 40 by permission of the American Chemical Society.)

Since α relates the sensitivity of the proton transfer to that of dissociation of the acid, it is frequently assumed that the value of α can be used as an indicator of transition state structure. The closer α approaches unity, the greater would be the degree of proton transfer in the transition state. There are limits to the generality of this interpretation, however.[41]

The details of proton transfer processes can also be probed by examination of *solvent isotope effects*, for example, by comparing the rates of a reaction in H_2O versus D_2O. The solvent isotope effect can be either normal or inverse, depending on the nature of the proton transfer process in the reaction mechanism. D_3O^+ is a stronger acid than H_3O^+. As a result, reactants in D_2O solution are somewhat more extensively protonated than in H_2O at identical acid concentration. A reaction that involves a rapid equilibrium protonation will proceed faster in D_2O than in H_2O, because of the higher concentration of the protonated reactant. On the other hand, if proton transfer is part of the rate-determining step, the reaction will be faster in H_2O than in D_2O because of the normal primary kinetic isotope effect of the type considered in Section 4.5.

41. A. J. Kresge, *J. Am. Chem. Soc.* **92**, 3210 (1970); R. A. Marcus, *J. Am. Chem. Soc.* **91**, 7224 (1969); F. G. Bordwell and W. J. Boyle, Jr., *J. Am. Chem. Soc.* **94**, 3907 (1972); D. A. Jencks and W. P. Jencks, *J. Am. Chem. Soc.* **99**, 7948 (1977); A. Pross, *J. Org. Chem.* **49**, 1811 (1984).

The interpretation of solvent isotope effects can be complicated by the large number of secondary isotope effects that can conceivably operate when it is the solvent molecule that is the site of isotopic substitution. The quantitative evaluation of solvent isotope effects is a very difficult problem. The relationship between the magnitude of the solvent isotope effect and the occurrence of equilibrium protonation as opposed to rate-limiting proton transfer is sufficiently general to be of value in mechanistic studies. As with nearly all mechanistic criteria, however, there are circumstances that permit exceptions, so corroborating evidence from other studies is always desirable.

Many organic reactions involve acid concentrations considerably higher than can be accurately measured on the pH scale, which applies to relatively dilute aqueous solutions. It is not difficult to prepare solutions in which the formal proton concentration is 10 *M* or more, but these formal concentrations are not a suitable measure of the *activity* of protons in such solutions. For this reason, it has been necessary to develop *acidity functions* to measure the proton-donating strength of concentrated acidic solutions. The activity of the hydrogen ion (solvated proton) can be related to the extent of protonation of a series of bases by the equilibrium expression for the protonation reaction:

$$\mathrm{B} + \mathrm{H}^+ \rightleftharpoons {}^+\mathrm{BH}$$

$$K = \frac{(a^+{}_{\mathrm{BH}})}{(a_{\mathrm{H}^+})(a_{\mathrm{B}})} = \frac{[{}^+\mathrm{BH}]\gamma_{\mathrm{BH}}^+}{a_{\mathrm{H}^+} + [\mathrm{B}]\gamma_{\mathrm{B}}}$$

where γ is the activity coefficient for the base and its conjugate acid. A common measure of acidity is referred to as h_0 and is defined by measuring the extent of protonation of a series of bases for which K has been measured. The relative concentration of the base and its conjugate acid then defines h_0 for any particular acidic solution.

$$h_0 = \frac{[{}^+\mathrm{BH}]\gamma_{\mathrm{BH}}^+}{K[\mathrm{B}]\gamma_{\mathrm{B}}}$$

The quantity H_0, defined as $-\log h_0$, is commonly tabulated, and it corresponds to the "pH" of very concentrated acidic solutions.

The problem of determining K independently of measurement of H_0 is the principal issue to be faced in establishing the H_0 scale for a series of acidic solutions. What is done is to measure K for some base in aqueous solution where $H_0 \approx \mathrm{pH}$. This base can then be used to find the H_0 of a somewhat more acidic solution. The K of a second, somewhat weaker base is then determined in the more acidic solution. This second base can then be used to extend H_0 into a still more acidic solution. The process is continued by using a series of bases to establish H_0 for successively more acidic solution. The H_0 is thereby referenced to the original aqueous measurement.[42] The assumption involved in this procedure is that the ratio of the activity

42. For reviews and discussion of acidity functions, see E. M. Arnett, *Prog. Phys. Org. Chem.* **1**, 223 (1963); C. H. Rochester, *Acidity Functions*, Academic Press, New York, 1970; R. A. Cox and K. Yates, *Can. J. Chem.* **61**, 225 (1983); C. D. Johnson and B. Stratton, *J. Org. Chem.* **51**, 4100 (1986).

coefficients for the series of bases and the series of cations does not change from solvent to solvent, that is,

$$\frac{\gamma_{\overset{+}{B_1H}}}{\gamma_{B_1}} = \frac{\gamma_{\overset{+}{B_2H}}}{\gamma_{B_2}} = \frac{\gamma_{\overset{+}{B_3H}}}{\gamma_{B_3}} \ldots \text{etc.}$$

Not unexpectedly, this procedure reveals some dependence on the particular type of base used, so no absolute H_0 scale can be established. Nevertheless, this technique provides a very useful measure of the relative hydrogen ion activity of concentrated acid solutions that can be used in the study of reactions that proceed only at high acid concentration. Table 4.6 gives H_0 values for some water–sulfuric acid mixtures.

4.9. Lewis Acid Catalysis

Lewis acids are defined as molecules which act as electron pair acceptors. The proton is an important special case, but many other species can play an important role in the catalysis of organic reactions. The most important in organic reactions are metal cations and covalent compounds of metals. Metal cations which play prominent roles as catalysts include the alkali metal monocations Li^+, Na^+, K^+, Cs^+, and Rb^+, the divalent ions Mg^{2+}, Ca^{2+}, and many of the transition metal cations. The most commonly employed of the covalent compounds include boron trifluoride, aluminum trichloride, titanium tetrachloride, and stannic tetrachloride. Various other derivatives of boron, aluminum, and titanium also are employed as Lewis acid catalysts.

The catalytic activity of metal ions originates in the formation of a donor–acceptor complex between the cation and the reactant, which must act as a Lewis base. The result of the complexation is that the donor atom becomes effectively more electronegative. All functional groups that have unshared electron pairs are

Table 4.6. H_0 as a Function of Composition of Aqueous Sulfuric Acid[a]

$\%H_2SO_4$	H_0	$\%H_2SO_4$	H_0
5	0.24	55	−3.91
10	−0.31	60	−4.46
15	−0.66	65	−5.04
20	−1.01	70	−5.80
25	−1.37	75	−6.56
30	−1.72	80	−7.34
35	−2.06	85	−8.14
40	−2.41	90	−8.92
45	−2.85	95	−9.85
50	−3.38	98	−10.41

a. From M. J. Jorgenson and D. R. Hartter, *J. Am. Chem. Soc.* **85**, 878 (1963).

potential electron donors, but especially prominent in reaction chemistry are carbonyl (sp^2) oxygens, hydroxyl and ether (sp^3) oxygens, and nitrogen- and sulfur-containing functional groups. Halogen substituents can act as donors to very strong Lewis acids. The presence of two potential donor atoms in a favorable geometric relationship permits formation of bidentate 'chelate" structures and may lead to particularly strong complexes.

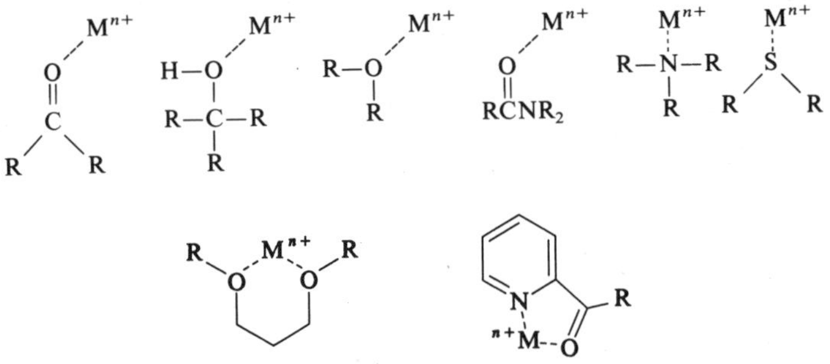

If the complexation process is regarded as resulting in a full covalent bond between the donor and the Lewis base, there is a net transfer of one unit of formal charge to the metal ion from the donor atom. This enhances the electronegativity of the donor atom. The complexes of carbonyl groups, for example, are more reactive to nucleophilic attack. Hydroxyl groups complexed to metal cations are stronger acids than the uncomplexed hydroxyl. Ether or sulfide groups complexed with metal ions are better leaving groups.

The strength of the complexation is a function both of the donor atom and the metal ion. The solvent medium is also an important factor because solvent molecules that have potential electron donor atoms can compete for the Lewis acid. Qualitative predictions about the strength of donor-acceptor complexation can be made on the basis of the hard-soft-acid-base concept. The better matched the donor and acceptor, the stronger is the complexation. Scheme 4.3 gives an ordering or hardness and softness for some neutral and ionic Lewis acids and bases.

Scheme 4.3. Relative Ordering of Hardness and Softness

	Lewis acids		Lewis bases	
	Cationic	Neutral	Neutral	Anionic
Hard	H^+	BF_3, $AlCl_3$	H_2O	F^-, SO_4^-
	Li^+, Na^+, Ca^{2+}	R_3B	Alcohols, ketones, ethers	Cl^-
				Br^-
	Zn^{2+}, Cu^{2+}		Amines (aliphatic)	N_3^-
			Amines (aromatic)	
	Pd^{2+}, Hg^{2+}, Ag^+			CN^-
	RS^+, RSe^+			I^-
Soft	I^+		Sulfides	S^{2-}

Compounds such as boron trifluoride and aluminum chloride form complexes by accepting an electron from the donor molecule. The same functional groups that act as lone pair donors to metal cations form complexes with boron trifluoride, aluminum trichloride, and related compounds.

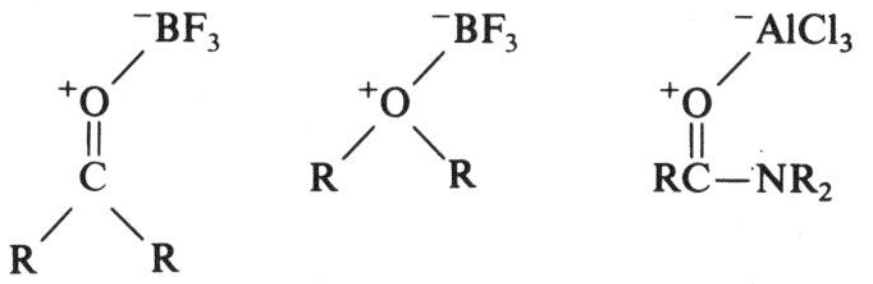

$$R_2C{=}\overset{+}{O}{-}\overset{-}{B}F_3 \qquad R_2\overset{+}{O}{-}\overset{-}{B}F_3 \qquad RC(NR_2){=}\overset{+}{O}{-}\overset{-}{Al}Cl_3$$

Since in this case the complex is formed between two neutral species, it, too, is neutral, but a formal positive charge develops on the donor atom, and a formal negative charge on the acceptor atom. The result is to increase the effective electronegativity of the donor atom and increase the electrophilicity of the complexed functional group.

Titanium tetrachloride and stannic chloride can form complexes that are related in character to both those formed by metal ions and those formed by neutral Lewis acids. Complexation can occur with displacement of a chloride from the metal coordination sphere or by an increase in the coordination number at the Lewis acid.

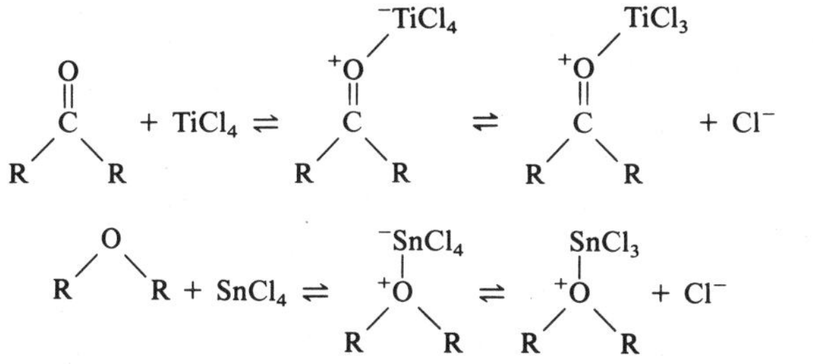

$$R_2C{=}O + TiCl_4 \rightleftharpoons R_2C{=}\overset{+}{O}{-}\overset{-}{Ti}Cl_4 \rightleftharpoons R_2C{=}\overset{+}{O}{-}TiCl_3 + Cl^-$$

$$R_2O + SnCl_4 \rightleftharpoons R_2\overset{+}{O}{-}\overset{-}{Sn}Cl_4 \rightleftharpoons R_2\overset{+}{O}{-}SnCl_3 + Cl^-$$

The crystal structure of the adduct of titanium tetrachloride and the ester formed from ethyl 2-hydroxypropanoate (ethyl lactate) and acrylic acid has been solved.[43] It is a chelated structure with the oxygen donor atoms being incorporated into the titanium coordination sphere along with the four chloride anions.

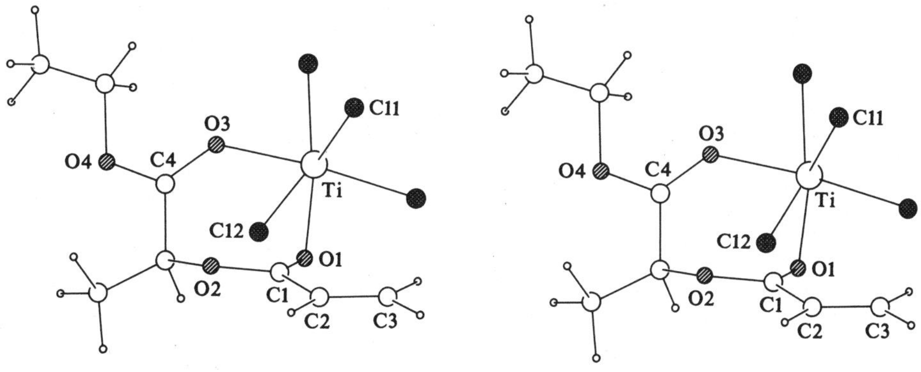

43. T. Poll, J. O. Melter, and G. Helmchen, *Angew. Chem. Int. Ed. Engl.* **24**, 112 (1985).

Diels–Alder reactions represent a type of reaction in which a Lewis acid is often used in catalytic quantities. The complexed ester (ethyl acrylate in the example given below) is substantially more reactive than the uncomplexed molecule, and the reaction proceeds through the complex. The reactive complex is regenerated by exchange of the Lewis acid from the adduct.

$$CH_2{=}CH{-}C({=}O)OC_2H_5 + AlCl_3 \rightleftharpoons CH_2{=}CH{-}C({=}\overset{+}{O}{-}\bar{A}lCl_3)OC_2H_5$$

$$CH_2{=}CH{-}C({=}\overset{+}{O}{-}\bar{A}lCl_3)OC_2H_5 + CH_2{=}CH{-}CH{=}CH_2 \rightarrow \text{(cyclohexenyl)}{-}C({=}\overset{+}{O}{-}\bar{A}lCl_3)OC_2H_5$$

$$\text{(cyclohexenyl)}{-}C({=}\overset{+}{O}{-}\bar{A}lCl_3)OC_2H_5 + CH_2{=}CH{-}C({=}O)OC_2H_5 \rightleftharpoons \text{(cyclohexenyl)}{-}C({=}O)OC_2H_5 + CH_2{=}CH{-}C({=}\overset{+}{O}{-}\bar{A}lCl_3)OC_2H_5$$

4.10. Solvent Effects

Most organic reactions are done in solution, and it is therefore important to recognize some of the general ways in which solvent can affect the course and rates of reactions. Some of the more common solvents can be roughly classified as in Table 4.7 on the basis of their structure and dielectric constants. There are important differences between *protic* solvents—solvents that contain relatively mobile protons such as those bonded to oxygen, nitrogen, or sulfur—and *aprotic* solvents, in which all hydrogen is bonded to carbon. Similarly, *polar* solvents, those that have high dielectric constants, have effects on reaction rates that are different from those of *nonpolar* solvent media.

When discussing solvent effects, it is important to distinguish between the macroscopic effects of the solvent and effects that depend upon details of structure. Macroscopic properties refer to properties of the bulk solvent. An important example is the dielectric constant, which is a measure of the ability of the bulk material to increase the capacitance of a condenser. In terms of structure, the dielectric constant is a function of both the permanent dipole of the molecule and its *polarizability.* Polarizability refers to the ease of distortion of the molecule's electron density. Dielectric constants increase with dipole moment and with polarizability. An important property of solvent molecules with regard to reactions is the response of the solvent to changes in charge distribution as the reaction occurs. The dielectric

Table 4.7. Dielectric Constants of Some Common Solvents[a]

Aprotic solvents				Protic solvents	
Nonpolar		Polar			
Hexane	1.9	Pyridine	12	Acetic acid	6.1
Carbon tetrachloride	2.2	Acetone	21	Trifluoroacetic acid	8.6
Dioxane	2.2	Hexamethyl phosphoramide	30	*tert*-Butyl alcohol	12.5
Benzene	2.3	Nitromethane	36	Ammonia	(22)
Diethyl ether	4.3	Dimethylformamide	37	Ethanol	24.5
Chloroform	4.8	Acetonitrile	38	Methanol	32.7
Tetrahydrofuran	7.6	Dimethyl sulfoxide	47	Water	78

a. Dielectric constant data are abstracted from the compilation of solvent properties in J. A. Riddick and W. B. Bunger (eds.), *Organic Solvents*, Vol. II of *Techniques of Organic Chemistry*, Third Edition, Wiley–Interscience, New York, 1970.

constant of a solvent is a good indicator of the ability of the solvent to accommodate separation of charge. It is not the only factor, however, since, being a macroscopic property, it conveys little information about the ability of the solvent molecules to interact with the solute molecules at close range. These direct solute–solvent interactions will depend on the specific structures of the molecules.

Let us consider how solvent might affect the solvolysis of *t*-butyl chloride. Much evidence, which will be discussed in detail in Chapter 5, indicates that the rate-determining step of the reaction is ionization of the carbon–chlorine bond to give a carbocation and the chloride ion. The transition state must reflect some of the charge separation that occurs in the ionization. Figure 4.11 gives a schematic interpretation of the solvation changes which would take place during the ionization of *t*-butyl chloride, with S representing surrounding solvent molecules.

The bulk dielectric constant may be a poor indicator of the ability of solvent molecules to facilitate the charge separation in the transition state. The fact that the carbon and chlorine remain partially bonded at the transition state prevents the solvent molecules from actually intervening between the developing centers of charge. Instead, the solvent molecules must stabilize the charge development by acting around the periphery of the activated complex. This interaction will depend

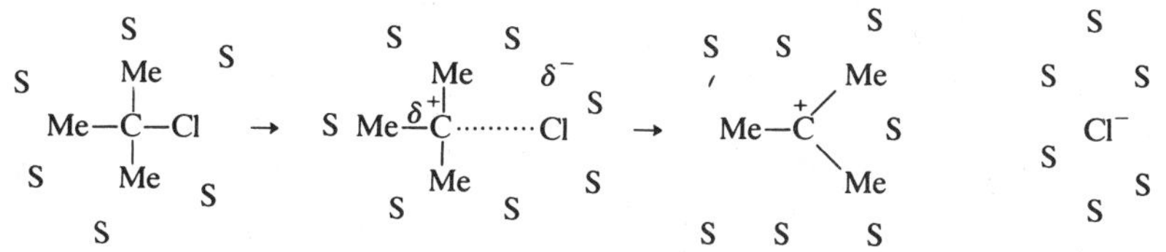

Fig. 4.11. Solvation changes during ionization of *t*-butyl chloride.

Table 4.8. Y Values for Some Solvent Systems[a]

Ethanol–Water		Methanol–Water			
(% ethanol)	Y	(% methanol)	Y	Other solvents	Y
100	−2.03	100	−1.09	Acetic acid	−1.64
80	0.0	80	0.38	Formic acid	2.05
50	1.65	50	1.97	*t*-Butyl alcohol	−3.2
20	3.05	10	3.28	90% Acetone–Water	−1.85
0	3.49			90% Dioxane–Water	−2.03

a. From A. H. Fainberg and S. Winstein, *J. Am. Chem. Soc.* **78**, 2770 (1956).

upon the detailed structure of the activated complex and the solvent. The ability of a number of solvents to stabilize the transition state of *t*-butyl chloride ionization has been measured by comparing the rate of the reaction in the various solvents. The reference solvent was taken as 80:20 ethanol-water. The Y value of other solvents is defined by the equation

$$\log \frac{k_{\text{solvent}}}{k_{80\%\ \text{ethanol}}} = Y$$

Table 4.8 lists the Y values for some alcohol–water mixtures and for some other solvents. Notice that among the solvents listed there is a spread of more than 10^6 in the measured rate of reaction between *t*-butyl alcohol and water. This large range of reaction rates demonstrates how important solvent effects can be. The Y value reflects primarily the ionization power of the solvent. It is largest for polar solvents such as water and formic acid and becomes progressively smaller and eventually negative as the solvent becomes less polar and contains more (or larger) nonpolar alkyl groups.

Solvents that fall in the nonpolar aprotic class are not effective at stabilizing the development of charge separation. These molecules have small dipole moments and do not have hydrogens capable of forming hydrogen bonds. Reactions that involve charge separation in the transition state therefore usually proceed much more slowly in this class of solvents than in protic or polar aprotic solvents. The reverse is true for reactions in which species having opposite charges come together in the transition state. Since in this case the transition state is less highly charged than the individual reactants, it is favored by the poor solvation, which leaves the oppositely charge reactants in a very reactive state. Arguing along these lines, the broad relationships between reactivity and solvent type shown in Scheme 4.4 can be deduced.

Many other measures of solvent polarity have been developed.[44] One of the most useful is based on shifts in the absorption spectrum of a reference dye. The positions of absorption bands in general are sensitive to solvent polarity because the electronic distribution, and therefore the polarity, of the excited state is different

44. C. Reichardt, *Angew. Chem. Int. Ed. Engl.* **18**, 98 (1979); C. Reichardt, *Solvent Effects in Organic Chemistry*, Verlag Chemie, Weinheim, 1979.

Table 4.9. $E_T(30)$, an Empirical Measure of Polarity, Compared with Dielectric Constant

	$E_T(30)$	ε		$E_T(30)$	ε
Water	63.1	78	Dimethylformamide	43.8	37
CF_3CH_2OH	59.5		Acetone	42.2	21
Methanol	55.5	32.7	CH_2Cl_2	41.1	8.9
80:20 ethanol–water	53.7		$CHCl_3$	39.1	4.8
Ethanol	51.9	24.5	$CH_3CO_2C_2H_5$	38.1	6.0
Acetic acid	51.2	6.1	Tetrahydrofuran	37.4	7.6
Isopropanol	48.6	19.9	Diethyl ether	34.6	4.3
Acetonitrile	46.7	38	Benzene	34.5	2.3
Nitromethane	46.3	36	Hexane	30.9	1.0
Dimethyl sulfoxide	45.0	47			

a. Data from C. Reichardt, *Angew. Chem. Int. Ed. Engl.* **18**, 98 (1979).

Scheme 4.4. Effect of Solvent Polarity on Reactions of Various Charge Types

$A^- + B^+ \longrightarrow \overset{\delta-}{A}\text{---}\overset{\delta+}{B} \longrightarrow A\text{–}B$	Favored by nonpolar solvent
$A\text{–}B \longrightarrow \overset{\delta-}{A}\text{---}\overset{\delta+}{B} \longrightarrow A^- + B^+$	Favored by polar solvent
$A + B \longrightarrow A\text{---}B \longrightarrow A\text{–}B$	Relatively insensitive to solvent polarity
$A\text{–}B^+ \longrightarrow \overset{\delta+}{A}\text{---}\overset{\delta+}{B} \longrightarrow A + B^+$	Slightly favored by polar solvent
$A^+ + B \longrightarrow \overset{\delta+}{A}\text{---}\overset{\delta+}{B} \longrightarrow A\text{–}B^+$	Slightly favored by nonpolar solvent

from that of the ground state. The shifts in the absorption maxima reflect the effect of solvent on the energy gap between the ground state and excited state molecules. An empirical solvent polarity measure called $E_T(30)$ is based on this concept.[45] Some values for common solvents are given in Table 4.9 along with the dielectric constants for the solvents. It can be seen that there is a rather different order of polarity given by these two quantities.

The electrostatic solvent effects discussed in the preceding paragraphs are not the only possible modes of interaction of solvent with reactants and transition states. Specific structural effects may cause either the reactants or transition state to be particularly strongly solvated. Figure 4.12 shows how such solvation can affect the relative energies of the ground state and transition state and cause rate variations from solvent to solvent. Unfortunately, no general theory for quantitatively predicting such specific effects has been developed to date.

Since a solvent may affect the rates of two competing reactions to different extents, a change in solvent may strongly modify the composition of a product

45. C. Reichardt and K. Dimroth, *Fortshr. Chem. Forsch.* **11**, 1 (1968); C. Reichardt, *Justus Liebigs Ann. Chem.* **752**, 64 (1971).

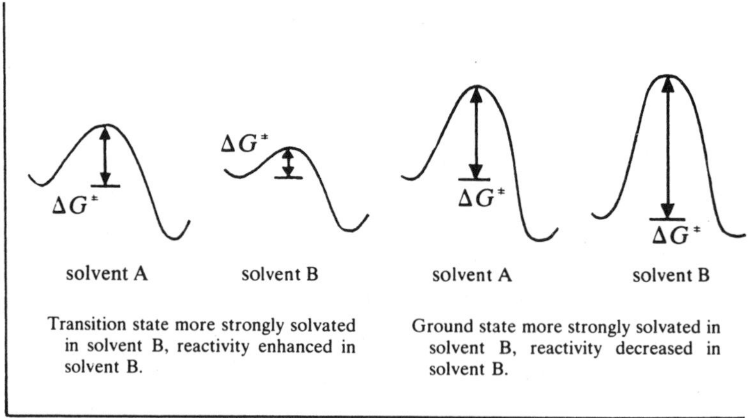

Fig. 4.12. Potential energy diagrams showing effect of preferential solvation of (a) transition state and (b) ground state on the activation energy.

mixture arising from competing reaction paths. Many such instances have been encountered in synthetic chemistry. An important example of solvent effects is the enhanced nucleophilicity of many anions in polar aprotic solvents as compared with protic solvents.[46] In protic solvents, anions are strongly solvated by hydrogen bonding. This is particularly true for anions that have a high concentration of charge on oxygen or nitrogen. This hydrogen bonding decreases the availability of the electrons of the anion to participate in reactions as a nucleophile.

HOR
ROH···A⁻···HOR
ROH

In aprotic solvents no hydrogens suitable for hydrogen bonding are present. As a result, the electrons of the anion are more easily available for reaction. Stated another way, the anion is at a higher energy level because of the absence of solvent stabilization. The polarity of the aprotic solvents is important because ionic compounds have very low solubility in nonpolar solvents. Dissolved ionic compounds are likely to be present as ion pairs or larger aggregates in which the reactivity of the anion is diminished by the electrostatic interaction with the cation. Energy must be expended against this electrostatic attraction to permit the anion to react as a nucleophile. Metal cations such as K^+ and Na^+ are strongly solvated by polar aprotic solvents such as dimethyl sulfoxide and dimethylformamide. The oxygen atoms in these molecules act as electron donors toward the cations. The dissolved salts are dissociated and as a result the anions are highly reactive because they are poorly solvated and not associated with cations.

$CH_3\overset{+}{S}(O^-)CH_3$ $\qquad$ $(CH_3)_2NCH(=O) \leftrightarrow (CH_3)_2\overset{+}{N}=CH(O^-)$

46. A. J. Parker, *Q. Rev. Chem. Soc.* **16**, 163 (1962); C. D. Ritchie, in *Solute-Solvent Interactions*, J. F. Coetzee and C. D. Ritchie (eds.), Marcel Dekker, New York, 1969, Chapter 4; E. Buncel and H. Wilson, *Adv. Phys. Org. Chem.* **14**, 133 (1977).

The realization that the nucleophilicity of anions is strongly enhanced in polar aprotic solvents has led to important improvements of several types of synthetic processes that involve nucleophilic substitutions or additions.

Particularly striking examples of the effect of specific solvation can be cited from among the *crown ethers.* These are macrocyclic polyethers that have the property of specifically solvating cations such as Na^+ and K^+.

18-crown-6

dibenzo-18-crown-6

When added to nonpolar solvents, the crown ethers increase the solubility of ionic materials. For example, in the presence of 18-crown-6, potassium fluoride is soluble in benzene and acts as a reactive nucleophile:

$$CH_3(CH_2)_7Br + KF \xrightarrow[\text{benzene}]{\text{18-crown-6}} CH_3(CH_2)_7F$$ Ref. 47

In the absence of the polyether, potassium fluoride is insoluble in benzene and unreactive toward alkyl halides. Similar enhancement of solubility and reactivity of other salts is observed in the presence of crown ethers. The solubility and reactivity enhancement results because the ionic compound is dissociated to a tightly complexed cation and a "naked" anion. Figure 4.13 shows the tight coordination that can be achieved with a typical crown ether. The complexed cation, since it is surrounded by the nonpolar crown ether, has high solubility in the nonpolar media. To maintain electroneutrality, the anion is also transported into the solvent. The cation is shielded from interaction with the anion by the surrounding crown ether molecule. As a result, the anion is unsolvated and at a relatively high energy and therefore highly reactive.

A closely related solvation phenomenon is the basis for *phase transfer catalysis.* Phase transfer catalysts are salts in which one of the ions (usually the cation) has large nonpolar substituent groups which confer good solubility in organic solvents. The most common examples are tetraalkylammonium and tetraalkylphosphonium ions. In two-phase systems consisting of water and a nonpolar organic solvent, these cations are extracted into the organic phase and, as a result, anions are also present in the organic phase. The anions are weakly solvated and again display high reactivity. Reactions are carried out between a salt containing the desired nucleophilic anion and an organic reactant, typically, an alkyl halide. The addition of the phase transfer catalysts causes migration of the anion into the organic phase, and, because of the high nucleophilicity of the anion, reaction occurs under excep-

47. C. Liotta and H. P. Harris, *J. Am. Chem. Soc.* **96**, 2250 (1974).

tionally mild conditions. Section 3.2 of Part B gives some specific examples of the use of phase transfer catalysis in nucleophilic displacement reactions.

It should always be borne in mind that solvent effects can modify the energy of both the reactants and the transition state. It is the *difference* in the two solvation effects that is the basis for changes in activation energies and reaction rates. Thus, although it is common to see solvent effects discussed solely in terms of reactant solvation or transition state solvation, this is usually an oversimplification. One case that illustrates this point is the hydrolysis of esters by hydroxide ion.

$$^{-}OH + CH_3\overset{O}{\overset{\|}{C}}OC_2H_5 \rightleftharpoons H_3C-\overset{O^{\delta-}}{\overset{\vdots}{\underset{HO^{\delta-}}{\underset{\vdots}{C}}}}-OC_2H_5 \longrightarrow \text{product}$$

The reaction is found to be much more rapid in dimethyl sulfoxide–water than in ethanol–water. Reactant solvation can be separated from transition state solvation by calorimetric measurement of the heat of solution of the reactants in each solvent system. The data in Fig. 4.14 compare the energies of the reactants and the transition state for ethyl acetate and hydroxide ion reacting in aqueous ethanol versus aqueous dimethyl sulfoxide. It can be seen that both the reactants and the transition state

Fig. 4.13. Space-filling molecular model depicting a metal cation complexed by 18-crown-6.

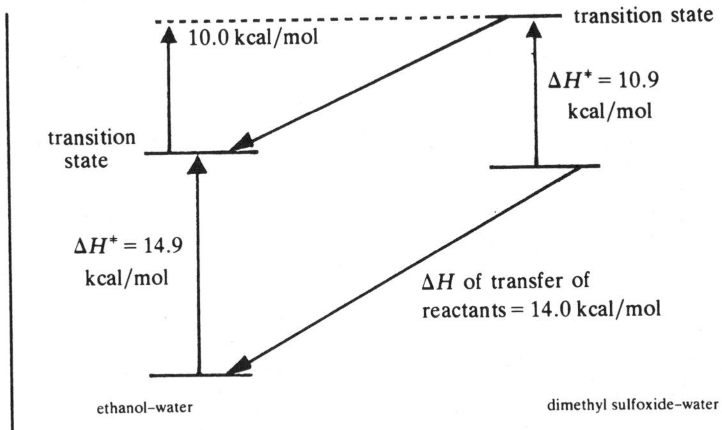

Fig. 4.14. Reactant and transition state solvation in the reaction of ethyl acetate with hydroxide ion. [From P. Haberfield, J. Friedman, and M. F. Pinkston, *J. Am. Chem. Soc.* **94**, 71 (1972).]

are more strongly solvated in the ethanol-water medium. The enhancement in reaction rate comes from the fact that the difference is greater for the small hydroxide ion than for the larger anionic species present at the transition state. It is generally true that solvation forces are strongest for the small, hard anions and decrease with size and softness.

4.11. Structural Effects in the Gas Phase

Having considered how solvents can strongly affect the properties of molecules in solution, let us consider some of the special features that arise in the gas phase, where solvation effects are totally eliminated. Although the majority of organic preparative reactions and mechanistic studies have been conducted in solution, some important reactions are carried out in the gas phase. Also, since current theoretical calculations generally do not treat solvent effects, experimental data from the gas phase are the most appropriate basis for comparison with theoretical results. Frequently, quite different trends in substituent effects are seen when systems in the gas phase are compared to similar systems in solution.

It is possible to measure equilibrium constants and heats of reaction in the gas phase by using mass spectrometers of special configuration.[48] With proton transfer reactions, for example, the equilibrium constant can be determined by measuring the ratio of two reactant species competing for protons. Table 4.10 compares ΔH_{gas}

48. Discussion of the techniques for gas phase equilibrium measurements can be found in T. A. Lehman and M. M. Bursey, *Ion Cyclotron Resonance Spectrometry*, Wiley-Interscience, New York, 1976, and in M. T. Bowers (ed.), *Gas Phase Ion Chemistry*, Vols. 1 and 2, Academic Press, New York, 1979.

Table 4.10. Comparison of Substituent Contributions to Phenol Ionization in the Gas Phase and Solution[a]

X	Substituent increment in kcal/mol[b]			
	ΔG_{gas}	ΔH_{gas}	ΔG_{H_2O}	ΔH_{H_2O}
m-CH_3	+0.4	+0.4	+0.18	+0.02
p-CH_3	+1.3	+1.3	+0.42	+0.02
m-Cl	−7.9	−7.9	−1.2	−0.3
p-Cl	−6.6	−6.6	−0.7	−0.2
p-NO_2	−25.8	−25.8	−3.8	−0.8

a. Data are from T. B. McMahon and P. Kebarle, *J. Am. Chem. Soc.* **99**, 2222 (1977).
b. The tabulated increments give the change in ΔG and ΔH resulting from replacement of hydrogen by the substituent specified.

with ΔH_{H_2O} for a series of phenol ionizations.

$$X\text{-}C_6H_4\text{-}OH + C_6H_5\text{-}O^- \rightleftharpoons X\text{-}C_6H_4\text{-}O^- + C_6H_5\text{-}OH$$

A key point to recognize is that the relative magnitude of the substituent effects is much larger in the gas phase. In general terms, this can be explained on the basis that all sovents effects have been removed. Whereas a phenolate anion in aqueous solution is stabilized by hydrogen bonding, there is no such stabilization in the gas phase. Since the solvent stabilization will be rather similar, on an absolute scale, for all the phenolate anions, the magnitudes of the substituent effects are "leveled" to some extent. In contrast, in the gas phase the importance of the internal substituent effect on stability of the anion is undiminished. The importance of the solvation can also be judged by noting that entropy makes a larger contribution to ΔG than enthalpy for the reaction series in solution. This reflects the extensive solvent organization that accompanies solvation.[49]

A comparison of phenol acidity in dimethyl sulfoxide versus the gas phase also shows an attenuation of substituent effects, but not nearly as much as in water. While the effect of substituents on ΔG for deprotonation in aqueous solution is about one-sixth that in the gas phase, the ratio for dimethyl sulfoxide is about one-third. This result points to hydrogen bonding of the phenolate anion by water as the major difference in the solvating properties of water and dimethyl sulfoxide.[50]

Another example of enhanced sensitivity to substituent effects in the gas phase can be seen in a comparison of the gas phase basicity for a series of substituted acetophenones and methyl benzoates. It was found that sensitivity of the free energy

49. L. P. Fernandez and L. G. Hepler, *J. Am. Chem. Soc.* **81**, 1783 (1959); C. L. Liotta, H. P. Hopkins, Jr., and P. T. Kasudia, *J. Am. Chem. Soc.* **96**, 7153 (1974).
50. M. Mashima, R. T. McIver, Jr., R. W. Taft, F. G. Bordwell and W. N. Olmstead, *J. Am. Chem. Soc.* **106**, 2717 (1984); M. Fujio, R. T. McIver, and R. W. Taft, *J. Am. Chem. Soc.* **103**, 4017 (1981).

Table 4.11. Acidities of Simple Alcohols in Solution[a]

	pK_a	
	H_2O	DMSO
H_2O	15.7	31.4
CH_3OH	15.5	29.0
C_2H_5OH	15.9	29.8
$(CH_3)_2CHOH$		30.2
$(CH_3)_3COH$		32.2

a. Data are from W. N. Olmstead, Z. Margolin, and F. G. Bordwell, *J. Org. Chem.* **45**, 3295 (1980).

to substituent changes was about four times that in solution, as measured by the comparison of ΔG for each substituent.[51] The gas phase data for both series were correlated by the Yukawa–Tsuno equation. For both series, the ρ value was about 12. However, the parameter r^+, which reflects the contribution of extra resonance effects, was greater in the acetophenone series than in the methyl benzoate series. This can be attributed to the substantial resonance stabilization provided by the methoxy groups in the esters. This diminishes the extent of conjugation with the substituents.

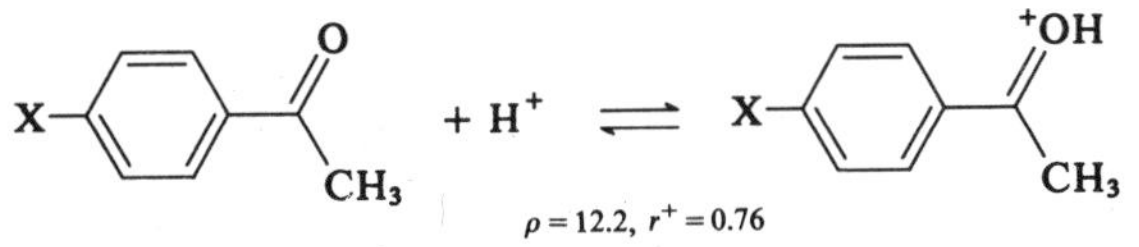

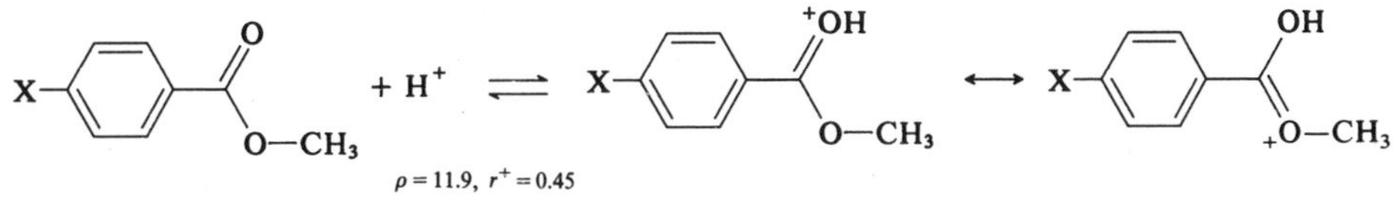

$\rho = 12.2,\ r^+ = 0.76$

$\rho = 11.9,\ r^+ = 0.45$

Another area of gas phase substituent effects which has attracted interest is the acidity of simple alcohols. In the gas phase, the order is *t*-BuOH > EtOH > MeOH ≫ H_2O.[52] This is different from the order in solution as revealed by the pK data in water and DMSO shown in Table 4.11.

These changes in relative acidity can again be traced to solvation effects. In the gas phase, any substituent effect can be analyzed directly in terms of its stabilizing or destabilizing effect on the anion. Two factors are believed to be of primary importance. The inductive effect arising from the H—C bond dipoles should be destabilizing since they tend to increase electron density at the carbon bonded to oxygen. This dipolar interaction is removed by replacing hydrogens with methyl

51. M. Mishima, M. Fujo, and Y. Tsuno, *Tetrahedron Lett.* **27**, 939, 951 (1986).
52. J. I. Brauman and L. K. Blair, *J. Am. Chem. Soc.* **92**, 5986 (1970); J. E. Bartmess and R. T. McIver, Jr., *Gas Phase Ion Chemistry*, Vol. 2, M. T. Bowers (ed.), Academic Press, New York, 1979.

substituents seen in many other contexts. This effect cannot be the dominant one, however, since it would lead to a gas phase acidity opposite to that observed. The dominant effect is believed to be polarizability. The methyl substituents, each consisting of four atoms, are better able to undergo local electronic distortion (induced polarizability) to accommodate the negative charge than is a single hydrogen atom. Thus, each substitution of a methyl for a hydrogen increases gas phase acidity.[53] An additional factor which determines the greater acidity of alcohols compared with water is the fact that the water O—H bond is somewhat stronger. In solution, these factors are outweighed by solvation effects, and the observed order of acidity is MeOH > H_2O > EtOH > *i*-PrOH > *t*-BuOH. This order results from the fact that the small anion can be better solvated than the more substituted ones.[54] The fact that water's position in the relative acidity order is most drastically changed in going from solution to the gas phase reflects the high sensitivity of the small hydroxide ion to solvation effects.

Within an isomeric series of alcohols, the gas phase acidity order is *sec* > *tert* > *pri* for C_4 and C_5 alcohols, although the differences between the isomeric *sec* and *tert* alcohols are small.[55] Analysis of a large number of alcohols showed the effect of adding an additional methyl substituent at the α, β, and γ positions. The effect at the α carbon was the greatest, but methyl groups at the β and γ positions were also stabilizing. This is again indicative of the importance of the entire molecule in stabilizing the charge in the gas phase.[56]

4.12. Stereochemistry

The study of the stereochemical course of organic reactions often leads to detailed insight into reaction mechanisms. Instrumental techniques including IR and NMR spectroscopy, optical rotatory dispersion, and circular dichroism make it possible to determine the stereochemical course of many reactions. Mechanistic postulates frequently make distinctive predictions about the stereochemical outcome of the reaction. Throughout the chapters dealing with specific types of reactions, consideration will be given to the stereochemistry of the reaction and its relationship to the reaction mechanism. As an example, the bromination of alkenes can be cited. A very simple mechanism for bromination is given below:

Br—Br
>C=C< → >C(Br)—C(Br)<

53. R. W. Taft, M. Taagepera, J. L. M. Abboud, J. F. Wolf, D. J. DeFrees, W. J. Hehre, J. E. Bartmess, and R. T. McIver, *J. Am. Chem. Soc.* **100**, 7765 (1978).
54. W. N. Olmstead, Z. Margolin, and F. G. Bordwell, *J. Org. Chem.* **45**, 3295 (1980).
55. G. Boaud, R. Houriet, and T. Gäuman, *J. Am. Chem. Soc.* **105**, 2203 (1983).
56. For a broad review of substituent and solvent effects on acidity and basicity, see R. W. Taft, *Prog. Phys. Org. Chem.* **14**, 247 (1983).

According to this mechanism, a molecule of bromine becomes complexed to the double bond of the alkene, and reorganization of the bonding electrons gives the product. This mechanism can be shown to be incorrect for most alkenes on the basis of stereochemistry. Most alkenes give bromination products in which the two added bromines are on opposite sides of the former carbon–carbon double bond. The above mechanism does not predict this.

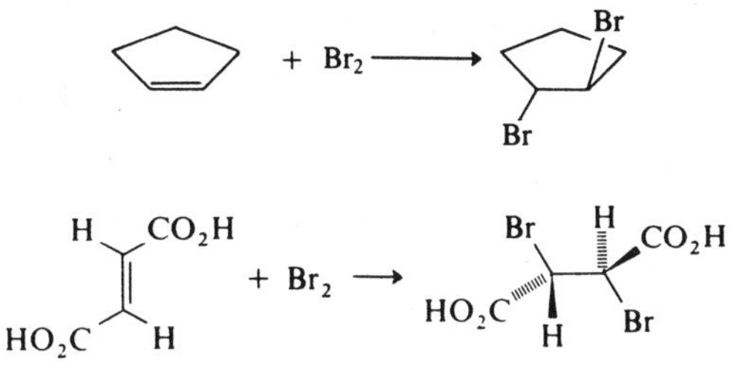

+ Br_2 ⟶ Br, Br

H, CO_2H, HO_2C, H + Br_2 ⟶ Br, H, CO_2H, HO_2C, H, Br

Another example of a reaction in which the stereochemistry of the process provides some valuable information about the mechanism is the thermal rearrangement of 1,5-dienes and substituted analogs.

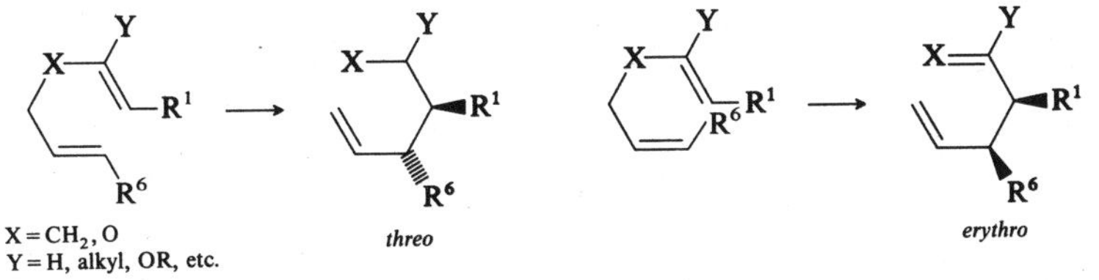

X = CH_2, O
Y = H, alkyl, OR, etc.

threo *erythro*

These reactions will be discussed in more detail under the topic of 3,3-sigmatropic rearrangements in Chapter 11. For the present, we simply want to focus on the fact that the reaction is *stereospecific*; the *E*-isomer will give one diastereomeric product while the related *Z*-isomer will give another one. The stereochemical relationship between reactants and products can be explained if the reaction occurs through a cyclic transition state in which the configurational relationships in the original double bonds are maintained.

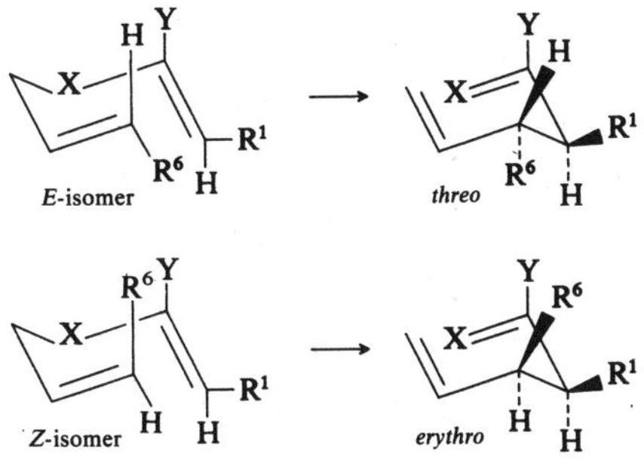

E-isomer ⟶ *threo*

Z-isomer ⟶ *erythro*

A large number of stereochemical results on the 3,3-sigmatropic family of reactions have been correlated by this type of analysis.[57] The basic assumption of the analysis is that the transition state is similar to a cyclohexane ring in its response to steric effects. The success of the model in interpretation of the stereochemical results supports its correctness. It can be concluded that the transition state must be a rather tightly organized arrangement of the atoms with strong bonding between both atoms 3 and 4 and atoms 1 and 6.

4.13. Conclusion

To conclude this chapter, it is important to emphasize a logical point about the determination of reaction mechanisms. *A proposed mechanism can never really be proven*; *rather, it is a case of alternative mechanisms being eliminated.* Having in mind a mechanism that explains all the facts does not constitute proof that the mechanism is correct. That conclusion is possible only when all alternatives have been excluded. A key stage in a mechanistic investigation then is the enumeration of the various possible mechanisms and the design of experiments that distinguish between them. The principal basis for enumerating mechanistic possibilities is accumulated chemical experience with related systems and the inherent structural features of the system. A chemist approaching a mechanistic study must cast as broad as possible vision on the problem so as not to exclude possibilities.

General References

General

C. F. Bernasconi (ed.), *Investigation of Rates and Mechanisms of Reactions, Techniques of Chemistry*, Fourth Edition, Vol. VI, Part 1, Wiley-Interscience, New York, 1986.
B. K. Carpenter, *Determination of Organic Reaction Mechanisms*, Wiley-Interscience, New York, 1984.
J. Hine, *Structural Effects on Equilibria in Organic Chemistry*, Wiley, New York, 1975.
J. A. Hirsch, *Concepts in Theoretical Organic Chemistry*, Allyn and Bacon, Boston, 1974.
E. S. Lewis (ed.), *Investigation of Rates and Mechanisms of Reactions, Techniques of Chemistry*, Third Edition,Vol. VI, Part 1, Wiley-Interscience, New York, 1974.
C. D. Ritchie, *Physical Organic Chemistry*, Marcel Dekker, New York, 1975.
K. B. Wiberg, *Physical Organic Chemistry*, Wiley, New York, 1964.

Thermodynamics

J. D. Cox and G. Pilcher, *Thermochemistry of Organic and Organometallic Compounds*, Academic Press, New York, 1970.

57. R. K. Hill, *Asymmetric Syntheses*, Vol. 3, J. D. Morrison (ed.), Academic Press, New York, 1984, Chapter 8.

G. J. Janz, *Thermodynamic Properties of Organic Compounds*, Academic Press, New York, 1967.
D. R. Stull, E. F. Westrum, Jr., and G. C. Sinke, *The Chemical Thermodynamics of Organic Compounds*, Wiley, New York, 1969.

Linear Free-Energy Relationships and Substituent Effects

C. D. Johnson, *The Hammett Equation*, Cambridge University Press, Cambridge, 1973.
P. R. Well, *Linear Free Energy Relationships*, Academic Press, New York, 1968.

Isotope Effects

C. J. Collins and N. S. Borman (eds.), *Isotope Effects on Chemical Reactions*, Van Nostrand Reinhold, New York, 1970.
L. Melander, *Isotope Effects on Reaction Rates*, Ronald Press, New York, 1960.
L. Melander and W. H. Saunders, Jr., *Reaction Rates of Isotopic Molecules*, Wiley, New York, 1980.

Solvent Effects

J. F. Coetzee and C. D. Ritchie, *Solute-Solvent Interactions*, Marcel Dekker, New York, 1969.

Catalysis

R. P. Bell, *The Proton in Chemistry*, Chapman and Hall, London, 1973.
W. P. Jencks, *Catalysis in Chemistry and Enzymology*, McGraw-Hill, New York, 1969.
C. H. Rochester, *Acidity Functions*, Academic Press, New York, 1970.

Problems

(*References for these problems will be found on page* 776.)

1. Measurement of the equilibrium constant for the interconversion of the dithiete **A** and the dithione **B** yielded the data given below. Calculate $\Delta G°$, $\Delta H°$, and $\Delta S°$.

A $\rightleftharpoons$ ArC(=S)C(=S)Ar **B** Ar = p-$Me_2NC_6H_4$

Temp. (°C)	K
−2.9	16.9
11.8	11.0
18.1	8.4
21.9	7.9
29.3	6.5
32.0	6.1
34.9	5.7
37.8	5.3
42.5	4.6

2. (a) Calculate the enthalpy and entropy of activation ($\Delta H^{\ddagger}$ and $\Delta S^{\ddagger}$) at 40°C for the acetolysis of *m*-chlorobenzyl *p*-toluenesulfonate from the data given:

Temp. (°C)	$k \times 10^5$ (s^{-1})
25.0	0.0136
40.0	0.085
50.1	0.272
58.8	0.726

(b) Calculate the activation parameters ΔE_a, $\Delta H^{\ddagger}$, and $\Delta S^{\ddagger}$ at 100°C from the data given for the reaction shown below:

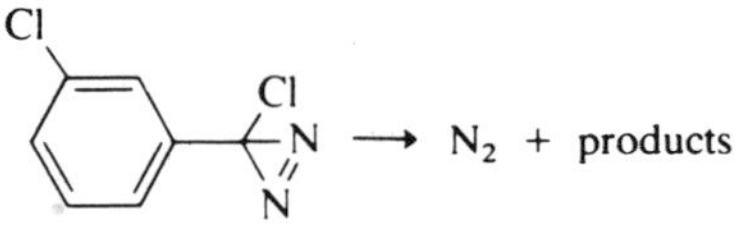

$\longrightarrow N_2$ + products

Temp. (°C)	$k \times 10^4$ (s^{-1})
60.0	0.30
70.0	0.97
75.0	1.79
80.0	3.09
90.0	8.92
95.0	15.90

3. 2-Vinylmethylenecyclopropane rearranges in the gas phase to 3-methylenecyclopentene. Two possible reaction mechanisms are:

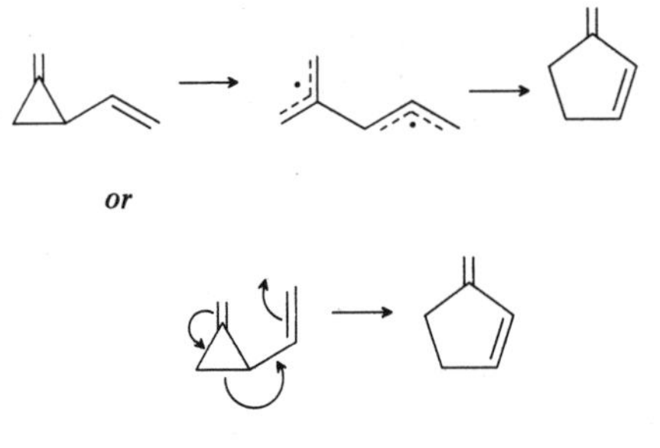

or

(a) Sketch a reaction energy diagram for each process.
(b) How might an isotopic labeling experiment distinguish between these mechanisms?

4. In Table 4.3 the phenyl group is assigned both a σ^+ and a σ^- value. Furthermore, the signs of σ^+ and σ^- are different. Discuss the reasons that the phenyl group has both a σ^+ and a σ^- value and explain why they are of a different sign.

5. Match the ρ values with the appropriate reactions. Explain your reasoning.
Reaction constants: +2.45, +0.75, −2.39, −7.29
Reactions:

(a) nitration of substituted benzenes

(b) ionization of substituted benzenthiols
(c) ionization of substituted benzenephosphonic acids
(d) reaction of substituted *N,N*-dimethylanilines with methyl iodide.

6. (a) Determine the value of ρ for the reaction shown from the data given:

$$\text{Y-C}_6\text{H}_3\text{(CH}_2\text{SO}_2\text{O)} + \text{OH}^- \longrightarrow \text{Y-C}_6\text{H}_3\text{(CH}_2\text{SO}_3^-\text{)(OH)}$$

Y	$k(M^{-1}\,s^{-1})$
H	37.4
CH_3O	21.3
CH_3	24.0
Br	95.1
NO_2	1430

(b) The pseudo-first-order rate constants for the acid-catalyzed hydration of substituted styrenes in 3.5 *M* $HClO_4$ at 25° are given. Plot the data against σ and σ^+ and determine ρ and ρ^+. Interpret the significance of the results.

Substituent	$k \times 10^8$ (sec^{-1})
p-CH_3O	488,00
p-CH_3	16,400
H	811
p-Cl	318
p-NO_2	1.44

(c) The basicity of a series of substituted benzyldimethylamines has been measured. Determine whether these basicity data are correlated by the Hammett equation. What is the value of ρ? What interpretation do you put on its sign?

$$\text{X-C}_6\text{H}_4\text{-CH}_2\overset{+}{\text{N}}\text{H(CH}_3)_2 \overset{K_a}{\rightleftharpoons} \text{X-C}_6\text{H}_4\text{-CH}_2\text{N(CH}_3)_2 + \text{H}^+$$

X	pK_a
p-CH_3O	9.32
p-CH_3	9.22
p-F	8.94
H	9.03
m-NO_2	8.19
p-NO_2	8.14
p-Cl	8.83
m-Cl	8.67

7. Write the rate law that would describe the rate of product formation for each of the following systems:

(a)

$$\text{(H)(Br)C=C(H)(Br)} + R_3N \underset{k_{-1}}{\overset{k_1}{\rightleftharpoons}} \text{(H)(Br)C=C}^-\text{(Br)} + R_3NH^+$$

$$\text{(H)(Br)C=C}^-\text{(Br)} \xrightarrow{k_2} H{-}C{\equiv}C{-}Br + Br^-$$

$$H{-}C{\equiv}C{-}Br + R_3N \xrightarrow{k_3} H{-}C{\equiv}C{-}\overset{+}{N}R_3 + Br^-$$

if the second step is rate-controlling and the first step is a preequilibrium.

(b)

$$(CH_3)_3CO_2C{-}C_6H_4{-}NO_2 \xrightarrow[\text{acetone}]{k_1,\ H_2O} (CH_3)_3C^+ + {}^-O_2C{-}C_6H_4{-}NO_2$$

$$(CH_3)_3C^+ + H_2O \xrightarrow{k_2} (CH_3)_3COH + H^+$$

and

$$(CH_3)_3C^+ \xrightarrow{k_3} (CH_3)_2C{=}CH_2 + H^+$$

if the competing product-forming steps are faster than the first step.

(c)

$$C_6H_5X + Br_2 \underset{k_{-1}}{\overset{k_1}{\rightleftharpoons}} \text{[X-C}_6\text{H}_5\text{(H)(Br)]}^+ + Br^-$$

$$\text{[X-C}_6\text{H}_5\text{(H)(Br)]}^+ \xrightarrow{k_2} Br{-}C_6H_4{-}X + H^+$$

$$Br^- + Br_2 \underset{k_{-3}}{\overset{k_3}{\rightleftharpoons}} Br_3^-$$

assuming that the σ complex is a steady-state intermediate. The final step is a rapid equilibrium that converts some of the initial Br_2 to unreactive Br_3^-. What is the rate expression if the intermediate goes to product much faster than it reverts to starting material and if the equilibrium constant for tribromide ion formation is large?

(d) $\mathrm{PhC(=O)CH(CH_3)_2 \underset{k_{-1}}{\overset{k_1}{\rightleftharpoons}} PhC(OH){=}C(CH_3)_2}$

$\mathrm{PhC(OH){=}C(CH_3)_2 + Cr(VI) \xrightarrow{k_2} products}$

where no assumption is made as to the relative magnitude of k_1, k_{-1}, and k_2.

8. The rates of brominolysis of a series of 1,2-diarylcyclopropanes under conditions where the rate is determined by Br· attack and leads to 1,3-dibromo-1,3-diarylpropanes are given below.

$$\text{Ar}^1\text{-cyclopropane-Ar}^2 \xrightarrow{\text{Br}\cdot} \text{Ar}^1\dot{\text{C}}\text{H}\text{CH}_2\text{CH(Br)Ar}^2 \text{ or } \text{Ar}^1\text{CH(Br)CH}_2\dot{\text{C}}\text{HAr}^2 \longrightarrow \text{Ar}^1\text{CH(Br)CH}_2\text{CH(Br)Ar}^2$$

Set up an equation which you would expect to correlate the observed rate of reaction with both the Ar^1 and the Ar^2 substituent. Check the performance of your equation by comparing the correlation with the data given below. Discuss the results of the correlation.

Ar^1	Ar^2	Relative rate	Ar^1	Ar^2	Relative rate
p-MeO	*p*-MeO	1.6×10^4	*p*-Cl	*p*-Ph	48
p-Ph	*p*-MeO	4.4×10^3	*p*-CN	*p*-Ph	45
H	*p*-MeO	2.5×10^3	*p*-Br	*p*-Cl	18
p-Cl	*p*-MeO	2.3×10^3	*p*-Br	*p*-Br	15
m-Br	*p*-MeO	2.1×10^3	*m*-Br	H	10
p-CN	*p*-MeO	1.6×10^3	*m*-Br	*p*-Cl	4.8
p-NO_2	*p*-MeO	1.3×10^3	*p*-CN	H	1.8
H	*p*-Ph	6.8×10^2	*p*-NO_2	H	0.88
p-Cl	*p*-Ph	4.3×10^2	*p*-Cl	*p*-CN	0.55
m-Br	*p*-Ph	1.8×10^2	*p*-Cl	*p*-NO_2	0.37
H	H	1.4×10^2	*p*-CN	*p*-CN	0.046

9. Predict whether normal or inverse isotope effects will be observed for each reaction below. Explain. Indicate any reactions in which you would expect $k_H/k_D > 2$. The isotopically substituted hydrogens are marked with asterisks.

(a) $\mathrm{CH_3CH_2^*CH(OSO_2Ar)CH_2^*CH_3 \xrightarrow[H_2O]{EtOH} CH_3CH_2^*\overset{+}{C}HCH_2^*CH_3}$

$\longrightarrow \mathrm{CH_3CH_2^*CH(OH)CH_2^*CH_3 + CH_3CH_2^*CH(OCH_2CH_3)CH_2^*CH_3}$

(b) $\mathrm{CH_3^*CH(NO_2)CH_3^* + {}^-OH \longrightarrow CH_3^*C(=NO_2^-)CH_3^* + H_2O}$

(c) $Ph_2C{=}C{=}O + PhCH{=}CH_2^* \longrightarrow$ [four-membered transition state: $Ph_2C{\cdots}C{=}O$ / $H{-}C{\cdots}CH_2^*$, Ph] $\longrightarrow$ [cyclobutanone: Ph_2, $=O$, Ph, H, H^*, H^*]

(d) $CH_3CH_2CH^*(OC_2H_5)_2 \underset{\text{fast}}{\overset{H^+}{\rightleftharpoons}} CH_3CH_2C(H^*)(\overset{+}{O}HC_2H_5){-}OC_2H_5 \xrightarrow{\text{slow}} CH_3CH_2C(H^*){=}\overset{+}{O}C_2H_5$

$CH_3CH_2C(H^*){=}\overset{+}{O}C_2H_5 + H_2O \xrightarrow{\text{fast}} CH_3CH_2CH^*{=}O + C_2H_5OH$

(e) $C_6H_5{-}CH_3^* + Br_2 \xrightarrow{150°C} C_6H_5{-}CH_2^*Br + H^*Br$

(f) PhLi + [fluorene with 9-position $^*H$ H^*] $\longrightarrow$ [fluorene with 9-position Li H^*] + PhH^*

(g) $H_2^*C{=}CHCH_2OC({=}S)Ph \xrightarrow{100°C} PhC({=}O)SCH_2^*CH{=}CH_2$

(h) $H_2^*C{=}CH_2^* + Ag^+ \rightleftharpoons \left[H_2^*C{=}CH_2^* \cdot Ag\right]^+$

10. Reactions of dialkylaluminum hydrides with acetylenes give addition products:

$$\underset{\mathbf{A}}{R_2AlH} + R'C{\equiv}CR'' \longrightarrow (R_2Al)(R')C{=}C(H)(R'')$$

The rate expression for the reaction is

$$-\frac{d[\mathbf{A}]}{dt} = k[\mathbf{A}][(R_2AlH)_3]^{1/3}$$

Propose a mechanism that could account for the overall four-thirds order kinetics and the appearance of the dialkylaluminum hydride concentration to the one-third power.

11. The Cannizzaro reaction is a disproportionation which takes place in strongly basic solution and converts benzaldehyde to benzyl alcohol and sodium benzoate.

$$2PhCH{=}O + NaOH \xrightarrow{H_2O} PhCO_2^-Na^+ + PhCH_2OH$$

Several mechanisms have been postulated, all of which propose a *hydride ion* transfer as a key step. On the basis of the following results, postulate one or more mechanisms which are consistent with all the data provided. Indicate the significance of each observation with respect to the mechanism(s) you postulate.

(1) When the reaction is carried out in D_2O, the benzyl alcohol contains no deuterium in the methylene group.
(2) When the reaction is carried out in $H_2{}^{18}O$, both the benzyl alcohol and sodium benzoate contain ^{18}O.
(3) The reaction rate is given by the expression

$$\text{rate} = k_{obs}[\text{PhCH}{=}\text{O}]^2[^-\text{OH}]$$

(4) The rates of substituted benzaldehydes are correlated by the Hammett equation with $\rho = +3.76$.
(5) The solvent isotope effect k_{D_2O}/k_{H_2O} is 1.90.

12. A mechanism for olefin arylation by palladium(II) is given below. The isotope effect k_H/k_D was found to be 5 when benzene-d_6 was used. When styrene-β,β-d_2 was used, no isotope effect was observed. Which step is rate-determining?

$$\text{C}_6\text{H}_6 + \text{Pd(O}_2\text{CCH}_3)_2 \xrightarrow{k_1} \text{Ph—PdO}_2\text{CCH}_3 + \text{CH}_3\text{CO}_2\text{H}$$

$$\text{Ph—PdO}_2\text{CCH}_3 + \text{PhCH}{=}\text{CH}_2 \xrightarrow{k_2} \text{Ph—Pd—CH}{=}\text{CHPh} + \text{CH}_3\text{CO}_2\text{H}$$

$$\text{Ph—Pd—CH}{=}\text{CHPh} \xrightarrow{k_3} \text{PhCH}{=}\text{CHPh} + \text{Pd}$$

13. A scale for solvent ionizing power, Y^+, applicable in solvolysis reactions of cationic substrates, has been developed. For example,

$$\text{S} + (\text{C}_2\text{H}_5)_3\text{O}^+\text{PF}_6^- \rightarrow (\text{C}_2\text{H}_5)_2\text{O} + \text{C}_2\text{H}_5\text{S}^+ + \text{PF}_6^-$$

S = solvent

The numerical values of Y^+ are found to be related to Y, the measure of solvent ionizing power for neutral substrates, by the equation

$$Y^+ = -0.09\ Y$$

Explain, in qualitative terms, (a) why Y^+ is negative with respect to Y and (b) why Y^+ is smaller in magnitude than Y (as is indicated by the coefficient of 0.09).

14. Two mechanisms are among those that have been postulated for decomposition of aryl diazonium salts in aqueous solution containing nucleophilic anions, A^-:

Mechanism A

$$\text{C}_6\text{H}_5\text{—}\overset{+}{\text{N}}{\equiv}\text{N} + \text{H}_2\text{O} \longrightarrow \left[\text{C}_6\text{H}_5\text{(}\cdots\overset{\delta^+}{\text{N}}{\equiv}\text{N)(}\cdots\overset{\delta^+}{\text{OH}_2}\text{)}\right] \longrightarrow \text{C}_6\text{H}_5\text{—OH} + \text{N}_2 + \text{H}^+$$

and

$$\text{C}_6\text{H}_5\text{—}\overset{+}{\text{N}}{\equiv}\text{N} + \text{A}^- \longrightarrow \left[\text{C}_6\text{H}_5\text{(}\cdots\overset{\delta^+}{\text{N}}{\equiv}\text{N)(}\cdots\overset{\delta^-}{\text{A}}\text{)}\right] \longrightarrow \text{C}_6\text{H}_5\text{—A} + \text{N}_2$$

Mechanism B

$$C_6H_5-\overset{+}{N}\equiv N \xrightarrow{\text{slow}} C_6H_5^+ + N_2$$

$$C_6H_5^+ + H_2O \xrightarrow{\text{fast}} C_6H_5-OH + H^+$$

and

$$C_6H_5^+ + A^- \xrightarrow{\text{fast}} C_6H_5-A$$

Indicate how each of the following techniques might be applied to distinguishing between these mechanisms:

(a) kinetic studies
(b) rate and product composition as a function of $[A^-]$
(c) solvent isotope effect studies
(d) isotope effect resulting from substitution of D for H at *ortho* positions
(e) substituent effect studies

15. Cycloheptatrienes are in many cases in rapid equilibrium with an isomeric bicyclo[4.1.0] heptadiene. The thermodynamics of the valence isomerism has been studied in a number of instances and some of the data are given below. Calculate the equilibrium constant for each case at 25°C. Calculate the temperature at which $K = 1$ for each system. Are the signs of the enthalpy and entropy as you would expect them to be? Can you discern any pattern of substituent effects from the data?

$CO_2C_2H_5$, Ar ⇌ $CO_2C_2H_5$, Ar

Ar	$\Delta H°$ (kcal/mol)	$\Delta S°$ (eu)
Phenyl	−5.4	−16.8
p-Nitrophenyl	−3.5	−11.0
p-Methoxyphenyl	−2.3	−7.4

16. Bicyclo[2.2.1]heptadiene rearranges at elevated temperatures to cycloheptatriene and toluene. The reaction is facilitated by substituents at C-7 such as phenyl and alkoxy, in which case cycloheptatrienes are the dominant products.

R → R

For R = *t*-butoxy, the rate data are given for several temperatures in decane. The reaction is about 50% faster in ethoxyethanol than in decane. Calculate the

Temp. (°C)	k (s^{-1})
139.8	7.28×10^{-6}
154.8	3.37×10^{-5}
170.3	1.43×10^{-4}

activation parameters at 150°C. Although precisely comparable data are not available, E_a for the gas phase isomerization of norbornadiene is ~50 kcal/mol. Draw a sketch showing the degree of transition state stabilization or destabilization caused by the alkoxy substituent. Is there any basis for regarding the bond cleavage in the rate-determining step to be heterolytic or homolytic? How do you propose that the effect of the substuent group R operates? Can you propose an experiment that might support your proposal?

17. A study of the aromatic nitration reaction in aqueous nitric acid revealed that when no aromatic substrate was present, an incorporation of ^{18}O from labeled water into nitric acid occurred. The rate of this exchange process was identical with the rate of nitration of several reactive aromatic hydrocarbons. Discuss how this result is consistent with mechanism B on page 188, but not with mechanisms A or C.

$$HNO_3 + H_2{}^{18}O \longrightarrow HN^{18}O_3 + H_2O$$

18. Comparison of several series of solvolysis reactions which proceed via carbocation intermediates reveals that an α-cyano group retards the reactions by a factor of about 10^3. A β-cyano group is even more strongly rate-retarding, with the factor being as high as 10^7. Why are both α- and β-cyano groups rate-retarding? What might cause the β-cyano group to be more rate-retarding than the α-cyano group?

19. Comparison of the gas phase acidity of benzoic acids with pK_a values of the same compounds in aqueous solution provides some interesting relationships.

    (1) The trend in acidity as a function of substituent is the same but the magnitude of the substituent effects is much larger in the gas phase. (The $\Delta\Delta G°$ for any given substituent is about ten times larger in the gas phase.)
    (2) Whereas acetic acid and benzoic acid are of comparable acidity in water, benzoic acid is much more acidic in the gas phase.
    (3) While the substituent effect in the gas phase is assumed to be nearly entirely an enthalpy effect, it can be shown that in solution the substituent effect is largely the result of changes in ΔS.

    Discuss how the change from gas phase to water solution can cause each of these effects.

20. It has been suggested that the chemical shift of aromatic ring carbons might provide a good indication of the intrinsic electron-releasing or electron-attracting capacity of substituents in circumstances where there is no perturbation by an approaching reagent. Such a perturbation is always present in substituent effects determined on the basis of reactivity. The measured chemical shifts from benzene for the carbon *para* to the substituent are given below. Plot these against σ, σ^+, and what conclusions do you draw from these plots? If you have access to an appropriate program and computer, determine the "percent resonance" associated with the chemical shift, as determined by the Swain-Lupton equation.

R	$\Delta\delta^a$	R	$\Delta\delta^a$
NH_2	−9.86	CN	3.80
OCH_3	−7.75	CCH_3	4.18
F	−4.49	CO_2CH_3	4.12
Cl	−2.05	SO_2CH_3	4.64
Br	−1.62	NO_2	5.53
CH_3	−2.89	CH=O	5.51
CF_3	3.19		

a. $\Delta\delta$ is the change in chemical shift in CCl_4 from benzene in ppm; a negative sign indicates increased shielding.

21. The ionization of a series of 4-substituted pyridines has been studied, and both equilibrium acidities (pK_a) and enthalpies of ionization have been recorded at 25°C:

$$R-C_5H_4\overset{+}{N}-H \rightleftharpoons R-C_5H_4N + H^+$$

R	pK_a	ΔH° (kcal/mol)
H	5.21	4.8
NH_2	9.12	11.3
OCH_3	6.58	6.8
CH_3	6.03	6.1
Cl	3.83	3.6
Br	3.75	3.5
CN	1.86	1.3

Calculate ΔS° for ionization of each compound. Comment on the contribution of ΔH° and ΔS° terms to the free energy of ionization. Test the data for linear free-energy correlations. Are the linear free-energy correlations dominated by entropy or enthalpy terms?

22. Allyl esters undergo rearrangement reactions at 300°C and above. Two examples are shown, one of which is "degenerate," since the product and reactant are identical:

$$CH_3CH{=}CHCH_2O\overset{\overset{\displaystyle O}{\|}}{C}CH_3 \longrightarrow CH_3\underset{\underset{\underset{\displaystyle O}{\|}}{OCCH_3}}{\underset{|}{C}}HCH{=}CH_2$$

$$H_2C{=}CHCH_2O\overset{\overset{\displaystyle O}{\|}}{C}CH_3 \longrightarrow CH_3\overset{\overset{\displaystyle O}{\|}}{C}OCH_2CH{=}CH_2$$

At least three distinct mechanisms can be written for these reactions. Write down some possible mechanisms, and suggest isotopic labeling studies that could distinguish among the possibilities you have proposed.

23. Estimates of the heat of solvation of various species in DMSO as compared to water have been made, and can be expressed as enthalpies of transfer. Some data are given below. Discuss their significance.

$$X^{\pm}{}_{H_2O} \longrightarrow X^{\pm}{}_{DMSO}$$

X	$\Delta H_{transfer}$ (kcal/mol)
K^+	−8.8
Na^+	−7.1
Cl^-	+4.9

5

Nucleophilic Substitution

Introduction

Nucleophilic substitution at carbon has received exceptionally detailed mechanistic study by organic chemists. The reaction is of broad synthetic utility, and many individual observations had accumulated before systematic efforts to characterize the reaction mechanism began. The task of developing a coherent mechanistic interpretation was undertaken by C. K. Ingold and E. D. Hughes in England in the 1930s. Their studies laid the basis for current understanding.[1] Since those initial investigations, organic chemists have continued to study substitution reactions, and the level of detailed information about this area is greater than for any of the other broad classes of reactions we will consider. From these accumulated data, a very satisfactory conceptual interpretation has developed. We can provide only a small selection of these details to illustrate the general concepts. The area of nucleophilic substitution also illustrates the fact that while a broad conceptual framework can outline the general features to be expected for a given system, the precise details will reveal aspects that are characteristic of specific systems. As the chapter unfolds, the reader should come to appreciate both the depth and breadth of the general conceptual understanding and the characteristics of some of the individual systems.

Nucleophilic substitution reactions may involve several different combinations of charged and uncharged species as reactants. The equations in Scheme 5.1 illustrate the four most common charge types.

The equations in Scheme 5.1 illustrate the relationship of reactants and products in nucleophilic substitution reactions, but say nothing about mechanism. In order to approach an understanding of the mechanisms of such reactions, let us begin with a review of the limiting cases as defined by Hughes and Ingold. These limiting

1. C. K. Ingold, *Structure and Mechanism in Organic Chemistry*, Second Edition, Cornell University Press, Ithaca, New York, 1969.

Scheme 5.1. Representative

A. Neutral substrate + neutral nucleophile

$RX + Y: \longrightarrow RY^+ + X:^-$

1[a] $CH_3CH_2I + (CH_3CH_2CH_2CH_2)_3P: \xrightarrow{\text{acetone}} (CH_3CH_2CH_2CH_2)_3\overset{+}{P}CH_2CH_3\ I^-$ quantitative

2[b] $PhC(CH_3)_2Cl + CH_3CH_2OH \longrightarrow PhC(CH_3)_2\overset{+}{O}(H)CH_2CH_3 \xrightarrow{-H^+} PhC(CH_3)_2OCH_2CH_3$ 87%

3[c] $CH_3CH(OSO_2C_6H_4CH_3)CH_2CH_3 + H_2O \xrightarrow{\text{acetone}} CH_3CH(\overset{+}{O}H_2)CH_2CH_3 \xrightarrow{-H^+} CH_3CH(OH)CH_2CH_3$ 77%

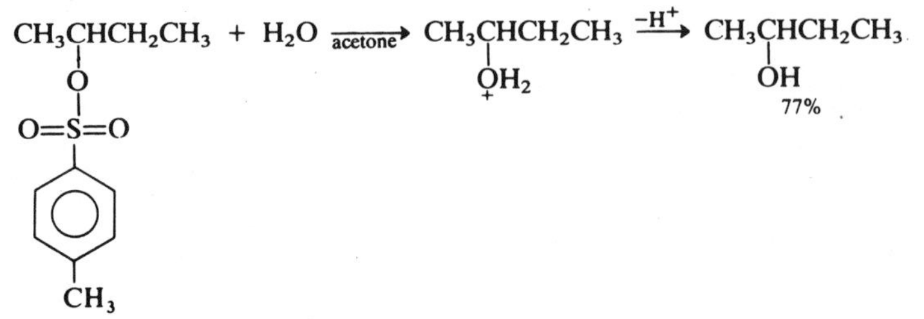

(The *p*-toluenesulfonate group is commonly referred to as *tosylate* and abbreviated –OTs.)

B. Neutral substrate + anionic nucleophile

$RX + Y:^- \longrightarrow RY + X:^-$

4[d] $CH_3CH(Br)CH_2C{\equiv}N \xrightarrow[\text{acetone}]{\text{NaI}} CH_3CH(I)CH_2C{\equiv}N$ 96%

5[e] (bicyclic)–$CH_2OTs \xrightarrow[\text{acetone}]{\text{LiBr}}$ (bicyclic)–CH_2Br 94%

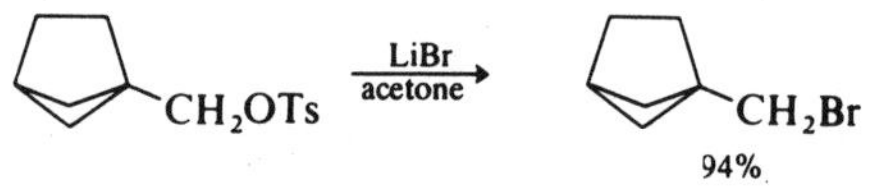

6[f] $CH_3CH(OTs)(CH_2)_5CH_3 + PhS^- \xrightarrow{\text{ethanol}} CH_3CH(SPh)(CH_2)_5CH_3$

cases are the *ionization mechanism* (S_N1, substitution-nucleophilic-unimolecular) and the *direct displacement mechanism* (S_N2, substitution-nucleophilic-bimolecular).

5.1. The Limiting Cases—Substitution by the Ionization (S_N1) Mechanism

The ionization mechanism for nucleophilic substitution proceeds by rate-determining heterolytic dissociation of the reactant to a tricoordinate carbocation (also

C. Cationic substrate + neutral nucleophile

$RX^+ + Y: \longrightarrow RY^+ + X:$

7[g] $PhCH(CH_3)\overset{+}{S}(CH_3)_2 + (H_2N)_2C{=}S \xrightarrow[\text{acetonitrile}]{60°C} PhCH(CH_3){-}S{-}C(\overset{+}{=}NH_2)NH_2$

8[h] $Ph_2C{=}\overset{+}{N}{=}\overset{-}{N}: \xrightarrow{TsOH} Ph_2CH{-}\overset{+}{N}{\equiv}N \xrightarrow{CH_3CH_2OH} Ph_2CH\overset{+}{O}(H)CH_2CH_3 \xrightarrow{-H^+} Ph_2CHOCH_2CH_3$

D. Cationic substrate + anionic nucleophile

$RX^+ + Y:^- \longrightarrow RY + X:$

9[i] $(CH_3CH_2)_3O^+ + (CH_3)_3CCO_2^- \longrightarrow (CH_3)_3CCO_2CH_2CH_3$ 90%

10[j] $CH_2{=}CHCH_2CH_2\overset{+}{S}(CH_3)C_6H_5 \xrightarrow[\text{dimethylformamide}]{NaI} CH_2{=}CHCH_2CH_2I$ 52%

11[k] N-(trimethylammoniomethyl)phthalimide, $NCH_2\overset{+}{N}(CH_3)_3 \xrightarrow[\text{dimethylformamide}]{NaCN}$ N-(cyanomethyl)phthalimide, NCH_2CN

a. S. A. Buckler and W. A. Henderson, *J. Am. Chem. Soc.* **82**, 5795 (1960).
b. R. L. Buckson and S. G. Smith, *J. Org. Chem.* **32**, 634 (1967).
c. J. D. Roberts, W. Bennett, R. E. McMahon, and E. W. Holroyd, Jr., *J. Am. Chem. Soc.* **74**, 4283 (1952).
d. M. S. Newman and R. D. Closson, *J. Am. Chem. Soc.* **66**, 1553 (1944).
e. K. B. Wiberg and B. R. Lowry, *J. Am. Chem. Soc.* **85**, 3188 (1963).
f. H. L. Goering, D. L. Towns, and B. Dittmar, *J. Org. Chem.* **27**, 736 (1962).
g. H. M. R. Hoffmann and E. D. Hughes, *J. Chem. Soc.*, 1259 (1964).
h. J. D. Roberts and W. Watanabe, *J. Am. Chem. Soc.* **72**, 4869 (1950).
i. D. J. Raber and P. Gariano, *Tetrahedron Lett.*, 4741 (1971).
j. E. J. Corey and M. Jautelat, *Tetrahedron Lett.*, 5787 (1968).
k. H. Hellman, I. Loschmann, and F. Lingens, *Chem. Ber.* **87**, 1690 (1954).

referred to as a *carbonium ion* or *carbenium ion*)[2] and the leaving group. This dissociation is followed by rapid combination of the highly electrophilic carbocation with a Lewis base present in the medium. A two-dimensional potential energy diagram representing this process for a neutral reactant and anionic nucleophile is shown in Fig. 5.1.

This mechanism has several characteristic consequences. Since the ionization step is rate-determining, the reaction will exhibit first-order kinetics, with the rate

2. Tricoordinate carbocations are frequently called *carbonium ions*; for a discussion of this terminology and a suggestion favoring the term *carbenium ion*, see G. A. Olah, *J. Am. Chem. Soc.* **94**, 808 (1972). Current practice seldom uses carbenium ion; instead, the terms methyl cation, butyl cation, etc., are used to describe the corresponding tricoordinate cations. *Chemical Abstracts* uses as specific names *methylium, ethylium, propylium.* We will use *carbocation* as a generic term for trivalent carbon cations.

of decomposition of the substrate being independent of the concentration or nature of the nucleophile.

$$RX \xrightarrow[\text{slow}]{k_1} R^+ + X:^-$$

$$R^+ + Y:^- \xrightarrow[\text{fast}]{k_2} RY$$

$$\text{rate} = -\frac{d[RX]}{dt} = -\frac{d[Y:^-]}{dt} = k_1[RX]$$

Application of Hammond's postulate indicates that the transition state should resemble the product of the first step, the carbocation intermediate. Ionization will be facilitated by factors that either lower the energy of the carbocation or raise the energy of the reactant. The rate of ionization will depend primarily on how reactant structure and solvent ionizing power affect these energies.

Ionization reaction rates are subject to both electronic and steric effects. The most important electronic effects are stabilization of the carbocation by electron release and the ability of the leaving group to accept the electron pair from the covalent bond of the reactant. Steric effects are pronounced because of the change in coordination that occurs on ionization. The three remaining substituents are spread apart as ionization occurs so that steric compression by bulky groups favors the ionization. Geometrical constraints that preclude planarity of the carbocation are unfavorable and will increase the energy of activation.

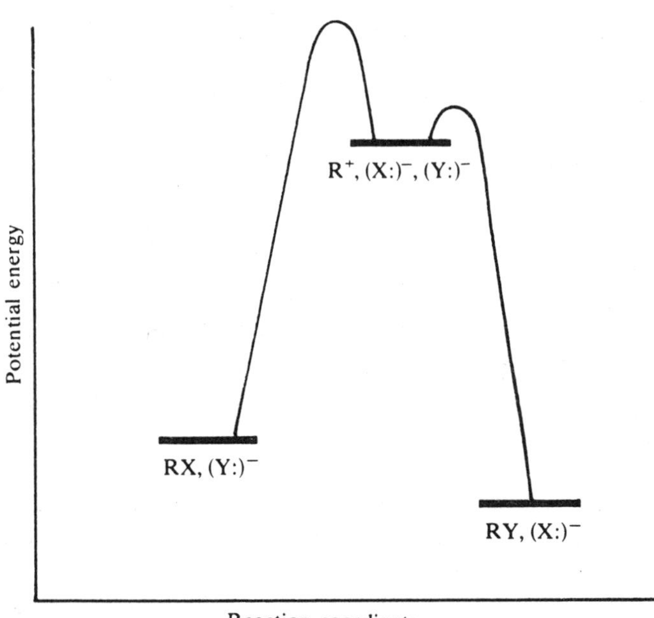

Fig. 5.1. Potential energy diagram for nucleophilic substitution by the ionization (S_N1) mechanism.

The ionization process is very sensitive to medium effects and depends on the charge type of the reactants. These relationships follow the general pattern for solvent effects discussed in Section 4.10. Ionization of a neutral substrate results in charge separation in the transition state, and the influence of solvent polarity will be greater at the transition state than for the reactants. Solvents of higher dielectric constant will lower the energy of the transition state more than will solvents of lower polarity. Ionization of cationic substrates such as alkyldiazonium ions or trialkylsulfonium ions leads to dispersal of charge in the transition state and should be enhanced by less polar solvents, because the reactants will be more strongly solvated than the transition state.

Stereochemical analysis has the potential to add detail to the mechanistic picture of the substitution reaction. The ionization mechanism results in formation of a carbocation intermediate which is achiral, because of the plane of symmetry associated with the sp^2 carbon. If the carbocation is sufficiently long-lived under the reaction conditions to diffuse away from the leaving group, it should become symmetrically solvated and produce racemic product. If this condition is not met, the solvation is dissymetric, and optically active product with net retention or inversion may be obtained even though a carbocation intermediate is formed. The extent of inversion or retention will depend upon the details of the system. Examples of this effect will be discussed in later sections of the chapter.

A further consequence of the ionization mechanism is that if the same carbocation can be generated from more than one precursor, its subsequent reactions should be independent of its origin. However, as in the case of stereochemistry, certainty about this must be tempered by the fact that the ionization initially produces an ion pair. If the subsequent reaction takes place from this ion pair, rather than from the completely dissociated ion, the leaving group may influence the course of the reaction.

5.2. The Limiting Cases—Substitution by the Direct Displacement (S_N2) Mechanism

The direct displacement mechanism is concerted, without an intermediate, and involves a single rate-determining transition state. According to this mechanism, the reactant is attacked by a nucleophile from the side opposite the leaving group, with bond making occurring simultaneously with bond breaking between the carbon atom and the leaving group. The transition state involves trigonal bipyramidal geometry at a pentacoordinate carbon. The nucleophile and the leaving group are both coordinated to the central carbon in the transition state. A potential energy diagram for direct displacement is given in Fig. 5.2.

The frontier molecular orbital approach provides a description of the bonding interactions that occur in the S_N2 process. The orbitals involved are depicted in Fig. 5.3. The frontier orbitals are a filled lone pair orbital on the approaching nucleophile Y: and the σ^* antibonding orbital associated with the carbon undergoing substitution

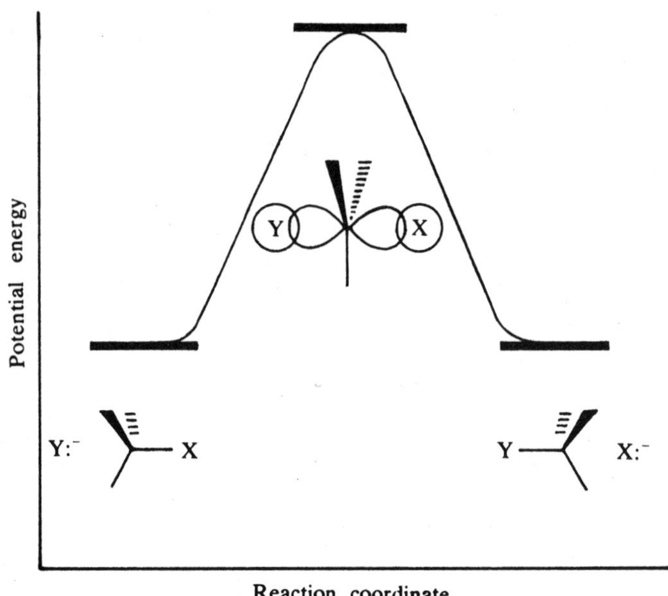

Fig. 5.2. Potential energy diagram for nucleophilic substitution by the direct displacement (S_N2) mechanism.

and the leaving group X. This antibonding orbital has a large lobe on carbon directed away from the C—X bond.[3] Back-side approach by the nucleophile is favored since the strongest initial interaction will be between the filled orbital on the nucleophile and the antibonding σ^* orbital.

Front-side approach is disfavored both because the density of the σ^* orbital is less in the region between the carbon and the leaving group and because front-side approach would involve both a bonding and an antibonding interaction with the σ^* orbital since it has a nodal surface between the atoms.

The molecular orbital picture also predicts that the reaction would proceed with inversion of configuration since the development of the transition state would be accompanied by rehybridization of the carbon to the trigonal bipyramidal geometry. As the reaction proceeded on to product, sp^3 hybridization would be reestablished, and the product would have the inverted configuration.

3. L. Salem, *Chem. Brit.* **5**, 449 (1969); L. Salem, *Electrons in Chemical Reactions: First Principles*, Wiley, New York, 1982, pp. 164–165.

The molecular orbital description of the S_N2 transition state presented in Fig. 5.3 shows a carbon *p* orbital interacting with two equivalent occupied orbitals, one from the leaving group and one from the nucleophile. These interacting orbitals give rise to three MOs describing the reacting bonds of the transition state. The nucleophile and leaving group orbitals are shown as *p* orbitals, but the qualitative picture would not change if they were sp^3 or some other hybrid orbitals. The HOMO at the transition state is π in character at the reacting carbon. The energy of this orbital should be lowered by conjugation with adjacent substituents. The S_N2 transition state should therefore be stabilized by substituents with a π orbital. The vinyl, phenyl, and carbonyl groups can all provide such stabilization, and, as we shall see later, each of these groups does enhance S_N2 reactivity.

The concerted displacement mechanism implies both kinetic and stereochemical consequences. The reaction will exhibit second-order kinetics, first-order in both reactant and nucleophile.

$$\mathrm{RX} + \mathrm{Y{:}^-} \xrightarrow{k} \mathrm{RY} + \mathrm{X{:}^-}$$

$$\text{rate} = -\frac{d[\mathrm{RX}]}{dt} = -\frac{d[\mathrm{Y{:}^-}]}{dt} = k[\mathrm{RX}][\mathrm{Y{:}^-}]$$

Since the nucleophile is intimately involved in the rate-determining step, not only will the rate depend on its concentration, but the nature of the nucleophile should be very important in determining the rate of the reaction. The identity and concentration of the nucleophile, by way of contrast, do not affect the rate in the ionization mechanism.

Since the degree of coordination increases at the reacting carbon atom, the rate of direct displacement would be expected to be sensitive to the steric bulk of the other substituents. The optimum substrate from a steric point of view would be $CH_3—X$ since it provides minimum steric resistance to approach of the nucleophile. Each replacement of hydrogen by a more bulky alkyl group should decrease the rate of reaction, and this is observed. As in the case of the ionization mechanism, the better the leaving group is able to accommodate an electron pair, the more facile should be the reaction. Since the nucleophile assists in the departure of the leaving group in the displacement mechanism, we would anticipate that the leaving group effect on rate would be less pronounced than in the ionization mechanism.

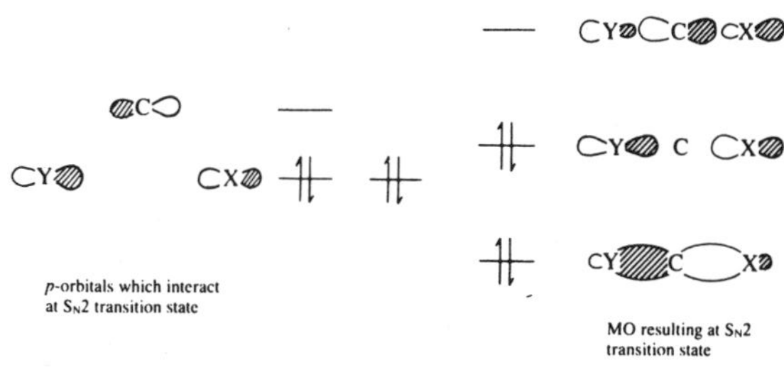

Fig. 5.3. MO description of the transition state for an S_N2 displacement at carbon.

The points we have emphasized in this brief overview of the S_N1 and S_N2 mechanisms are kinetics and stereochemistry. These have often been important pieces of evidence in ascertaining whether a particular nucleophilic substitution follows an ionization or direct displacement pathway. There are limitations to the generalization that reactions exhibiting first-order kinetics react by the S_N1 mechanism and those exhibiting second-order kinetics react by the S_N2 mechanism. Many nucleophilic substitutions are carried out under conditions in which the nucleophile is present in large excess. When this is the case, the concentration of the nucleophile is essentially constant during the reaction and the observed kinetics will become *pseudo-first-order*. This will be true, for example, when the solvent is the nucleophile. In this case, the kinetics of the reaction will provide no evidence as to whether the S_N1 or S_N2 mechanism operates.

Stereochemistry also frequently fails to provide a clear-cut distinction between the two limiting mechanisms. Many reactions proceed with partial inversion of configuration rather than complete racemization or inversion. Some reactions exhibit inversion of configuration but other features of the reaction suggest that an ionization mechanism must operate. Many systems exhibit "borderline" behavior in which it is difficult to distinguish between the ionization and the direct displacement mechanism. The types of reactants most likely to exhibit borderline behavior are secondary alkyl and primary and secondary benzylic systems. In the next section we will examine in more detail the characteristics of these borderline systems.

5.3. Detailed Mechanistic Description and Borderline Mechanisms

The ionization and direct displacement mechanisms are generally viewed as the extremes of a mechanistic continuum. At the S_N1 extreme, there is *no covalent interaction* between the reactant and the nucleophile in the transition state for cleavage of the bond to the leaving group. At the S_N2 extreme, the bond formation to the nucleophile is *synchronous* with the bond-breaking step. In between these two limiting cases lies the borderline area in which the degree of covalent interaction between the reactant and the nucleophile is intermediate between that in the two limiting cases. The concept of *ion pairs* is important in the detailed consideration of the borderline area. The concept of ion pairs was introduced by Saul Winstein, who proposed that there were two distinct types of ion pairs involved in substitution reactions.[4] The ion pair concept has been elaborated in the detailed interpretation of the substitution mechanism.[5]

4. S. Winstein, E. Clippinger, A. H. Fainberg, R. Heck, and G. C. Robinson, *J. Am. Chem. Soc.* **78**, 328 (1956); S. Winstein, B. Appel, R. Baker, and A. Diaz, *Chem. Soc. Spec. Publ.*, No. 19, 109 (1965).
5. J. M. Harris, *Prog. Phys. Org. Chem.* **11**, 89 (1974); D. J. Raber, J. M. Harris, and P. v. R. Schleyer, in *Ion Pairs*, M. Szwarc (ed.), John Wiley and Sons, New York, 1974, Chapter 3; T. W. Bentley and P. v. R. Schleyer, *Adv. Phys. Org. Chem.* **14**, 1 (1977).

Winstein suggested that two intermediates preceding the dissociated carbocation were required to reconcile data on kinetics, salt effects, and stereochemistry of solvolysis reactions. The process of ionization would initially generate a carbocation and counterion in proximity to each other. This species is called an *intimate ion pair* (or contact ion pair). This species can proceed to a *solvent-separated ion pair* in which one or more solvent molecules have inserted between the carbocation and leaving group but in which the ions have not diffused apart. The "free carbocation" is formed by diffusion away from the anion, which is called *dissociation.*

$$R{-}X \overset{\text{ionization}}{\rightleftharpoons} \underset{\substack{\text{intimate}\\\text{ion pair}}}{R^+X^-} \rightleftharpoons \underset{\substack{\text{solvent-}\\\text{separated}\\\text{ion pair}}}{R^+\|X^-} \overset{\text{dissociation}}{\rightleftharpoons} \underset{\substack{\text{dissociated}\\\text{ions}}}{R^+ + X^-}$$

Attack by a nucleophile or the solvent could occur at any of the ion pairs. Nucleophilic attack on the intimate ion pair would be expected to occur with inversion of configuration, since the leaving group should still shield the front side of the carbocation. At the solvent-separated ion pair stage, the nucleophile might approach from either face, particularly in the case where solvent is the nucleophile. Reactions through dissociated carbocations should occur with complete racemization. The identity and stereochemistry of the reaction products will be determined by the extent to which reaction occurs on the unionized reactant, the intimate ion pair, the solvent-separated ion pair, or the dissociated carbocation.

Various specific cases support this general scheme. For example, in 80% aqueous acetone, the rate constant for racemization of *p*-chlorobenzhydryl *p*-nitrobenzoate and the rate of equilibration of the ^{18}O in the carbonyl oxygen can both be measured.[6] At 100°C, $k_{eq}/k_{rac} = 2.3$.

$$Cl{-}C_6H_4{-}CH(C_6H_5)OC(={}^{18}O)C_6H_4NO_2 \xrightarrow{k_{eq}} Cl{-}C_6H_4{-}CH(C_6H_5){}^{18}OC(=O)C_6H_4NO_2$$

$$\underset{\text{optically active}}{Cl{-}C_6H_4{-}CH(C_6H_5)OC(=O)C_6H_4NO_2} \xrightarrow{k_{rac}} \underset{\text{racemic}}{Cl{-}C_6H_4{-}CH(C_6H_5)OC(=O)C_6H_4NO_2}$$

If it is assumed that ionization would result in complete randomization of the ^{18}O label in the carboxylate ion, k_{eq} is a measure of the rate of ionization with ion pair return and k_{rac} is a measure of the extent of racemization associated with ionization. The fact that the rate of equilibration exceeds that of racemization indicates that

6. H. L. Goering and J. F. Levy, *J. Am. Chem. Soc.* **86**, 120 (1964).

ion pair collapse occurs with predominant retention of configuration. When a nucleophile is added to the system (0.14 M NaN_3), k_{eq} is found to be unchanged but no racemization of reactant is observed. Instead, the intermediate which would return with racemization is captured by azide ion and converted to substitution product. This must mean that the intimate ion pair returns to reactant more rapidly than it is captured by azide ion, while the solvent-separated ion pair is captured by azide ion faster than it returns to racemic reactant.

Several other cases have been studied in which isotopic labeling reveals that the bond between the leaving group and carbon is able to break without net substitution having occurred. A particularly significant case, since it applies to secondary tosylates, which frequently exhibit borderline behavior, is isopropyl benzenesulfonate. During solvolysis of this compound in trifluoroacetic acid, it is found that exchange among the sulfonate oxygens occurs at about one-fifth the rate of solvolysis.[7]

$$(CH_3)_2CH{-}O{-}\overset{O^*}{\underset{O^*}{\overset{\|}{\underset{\|}{S}}}}{-}C_6H_5 \xrightarrow[CF_3CO_2Na,\ k = 36\times10^{-4}]{CF_3CO_2H} (CH_3)_2CHO_2CCF_3$$

$$k = 8\times10^{-4} \ \updownarrow$$

$$(CH_3)_2CH{-}O^*{-}\overset{O}{\underset{O^*}{\overset{\|}{\underset{\|}{S}}}}{-}C_6H_5 \qquad O^* = {}^{18}O \text{ label}$$

This implies that ion pair formation and recombination is occurring competitively with ion pair formation and substitution.

The ion pair return phenomenon can also be demonstrated by comparing the rate of loss of enantiomeric purity of reactant with the rate of product formation. For a number of systems, including l-arylethyl tosylates,[8] the rate of decrease of optical rotation is greater than the rate of product formation. This indicates the existence of an intermediate which can reform racemic reactant. The solvent-separated ion pair would be the most likely intermediate in the Winstein scheme to play this role.

$$\underset{ArSO_2O}{ArCHCH_3} \rightleftharpoons Ar\overset{+}{C}HCH_3\|{}^{-}O_3SAr \xrightarrow{Nu^-} \underset{Nu}{ArCHCH_3}$$

The "special salt effect" is another experimental observation that requires at least two ion pair intermediates to be adequately explained.[9] Addition of salts typically causes an increase in the rate of solvolysis of secondary alkyl arenesulfonates that is linear with salt concentration. This is due to the increase in dielectric constant of the medium. The effect that added lithium perchlorate has on certain

7. C. Paradisi and J. F. Bunnett, *J. Am. Chem. Soc.* **107**, 8223 (1985).
8. A. D. Allen, V. M. Kanagasabapathy, and T. T. Tidwell, *J. Am. Chem. Soc.* **107**, 4513 (1985).
9. S. Winstein and G. C. Robinson, *J. Am. Chem. Soc.* **80**, 169 (1958).

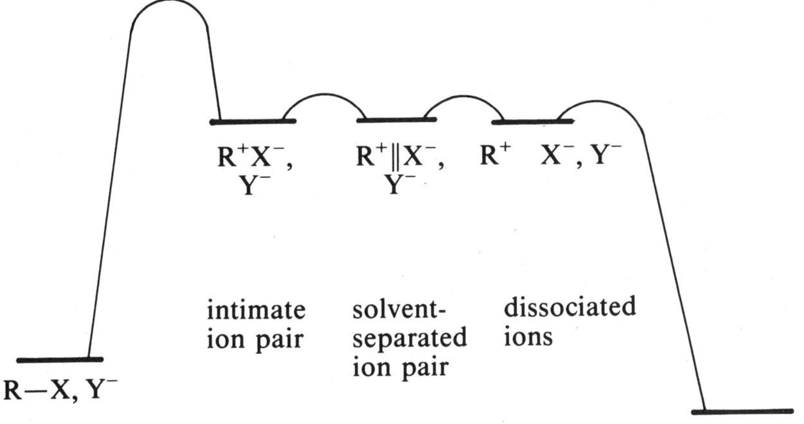

Fig. 5.4. Schematic relationship between reactants, intermediate species, and products in substitution proceeding through ion pairs.

substrates is anomalous in producing an initial sharp increase in the rate of solvolysis, followed by the normal linear increase at higher concentration. Winstein ascribed this to exchange between lithium perchlorate and the solvent-separated ion pair to form a new carbocation–perchlorate ion pair which was less likely to return to reactant than the carbocation–arenesulfonate ion pair. If the perchlorate ion pair can only go on to product, its formation would lead to an increase in the overall solvolysis rate.

The concept of ion pairs in nucleophilic substitution is now generally accepted. Presumably, the barriers separating the intimate, solvent-separated, and dissociated ions are quite small. The potential energy diagram in Fig. 5.4 depicts the three ion pair species as being roughly equivalent in energy and separated by small barriers.

Since ion pairs are undoubtedly important species, the question has arisen as to whether they might be intermediates in all nucleophilic substitution processes. R. A. Sneen and H. M. Robbins suggested that ion pairs might not only be involved in S_N1 and borderline processes but also in displacements exhibiting the stereochemical and kinetic characteristics of the S_N2 process.[10] They suggested the scheme shown below, in which SOH is a hydroxylic solvent and **Nu**⁻ is a nucleophilic anion. In this mechanism, reactions with S_N2 characteristics are postulated to occur by nucleophilic attack on the intimate ion pair.

$$\mathrm{RX} \rightleftharpoons \mathrm{R^+X^-} \rightleftharpoons \mathrm{R^+\|X^-} \rightleftharpoons \mathrm{R^+ + X^-}$$

$$\mathrm{R^+X^-} \rightarrow \mathrm{SOR}\quad \mathbf{NuR}$$

$$\mathrm{R^+\|X^-} \rightarrow \mathrm{ROS + SOR}\quad \mathbf{NuR}$$

$$\mathrm{R^+ + X^-} \rightarrow \mathrm{ROS + SOR}\quad \mathbf{RNu} + \mathbf{NuR}$$

10. R. A. Sneen and H. M. Robbins, *J. Am. Chem. Soc.* **94**, 7868 (1972); R. A. Sneen, *Acc. Chem. Res.* **6**, 46 (1973).

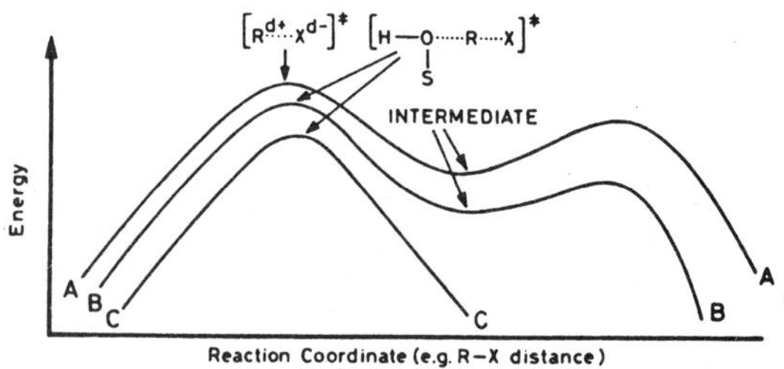

Fig. 5.5. Potential energy diagrams for substitution mechanisms. A is the S_N1 mechanism. B is the S_N2 mechanism with intermediate ion pair or pentacoordinate species. C is the classical S_N2 mechanism. Reproduced from T. W. Bentley and P. v. R. Schleyer, *Adv. Phys. Org. Chem.* **14**, 1 (1977).

According to this scheme, attack of either solvent or nucleophile on the intimate ion pair would occur with inversion of configuration. At the solvent-separated ion pair, reaction with solvent could occur with either retention or inversion, whereas attack by an external nucleophile would occur with inversion. At the dissociated ion pair stage, reaction with either solvent or nucleophile should occur with racemization, and this would correspond to the limiting ionization mechanism. The borderline behavior typical of many secondary alkyl systems would correspond to attack on the solvent-separated ion pair. The reservation most commonly expressed concerning the universality of the ion pair mechanism has to do with the intervention of ion pairs in methyl and other primary reactants. Since primary carbocations are highly unstable, it seems unlikely that the required ionization would occur at a rate sufficient to accommodate the reactivity of primary halides and sulfonates. Such substrates are *more reactive* than secondary systems in the presence of strong nucleophiles. The generally held view is that the direct displacement mechanism is a process which is distinct from ion pair substitution and is characterized by bonding to the nucleophile *before* ionization of the reactant.

The transition from S_N1 to S_N2 mechanisms can be described in terms of the shape of the potential energy diagrams for the reactions as illustrated in Fig. 5.5. Curves A and C represent the S_N1 and S_N2 limiting mechanisms, as described in the earlier sections. The transition from the S_N1 mechanism to the S_N2 mechanism involves greater and greater nucleophilic participation of the solvent or nucleophile in the transition state.[11] An ion pair with strong nucleophilic participation represents an intermediate stage between the S_N1 and S_N2 processes. This mechanism is designated the S_N2(intermediate) mechanism and pictures a carbocation-like transition state requiring back-side nucleophilic participation and therefore exhibiting second-order kinetics.

Jencks[12] has discussed how the transition from the S_N1 to the S_N2 mechanism is related to the stability and lifetime of the carbocation intermediate, as illustrated

11. T. W. Bentley and P. v. R. Schleyer, *Adv. Phys. Org. Chem.* **14**, 1 (1977).
12. W. P. Jencks, *Acc. Chem. Res.* **13**, 161 (1980).

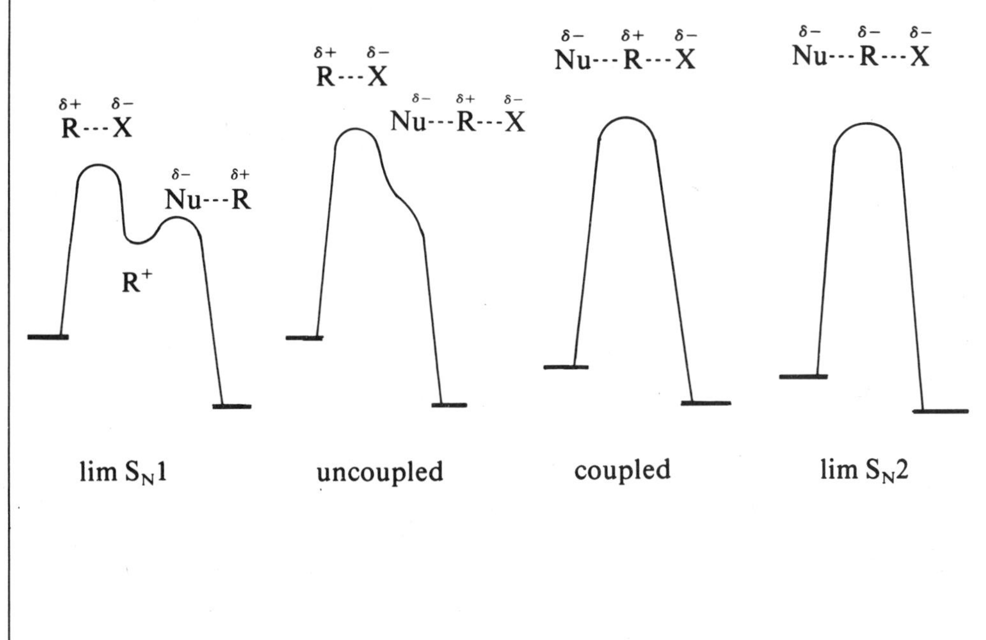

Fig. 5.6. Relationship between stability and potential lifetime of carbocation intermediate and mechanism for substitution.

in Fig. 5.6. In the S_N1 mechanism, the carbocation intermediate has a relatively long lifetime and is equilibrated with solvent prior to capture by a nucleophile. The reaction is clearly a stepwise one, and the energy minimum in which the carbocation intermediate resides is significant. As the stability of the carbocation decreases, its lifetime will become less. The barrier to capture by a nucleophile will become less and eventually disappear. This is described as the "uncoupled" mechanism. Ionization proceeds without nucleophilic participation but the carbocation does not exist as a free intermediate. Such a reaction would still exhibit S_N1 kinetics, since there is no nucleophilic participation in the ionization. At still lesser carbocation stability, the lifetime of the ion pair will be so short that the ion pair will always return to reactant unless a nucleophile is present to capture it as it is formed. At this stage, the reaction would exhibit second-order kinetics, since the nucleophile must be present for reaction to occur. Jencks describes this as the "coupled" substitution process. Finally, as the stability of the (potential) carbocation becomes so low that it cannot form without direct participation of the nucleophile, the limiting S_N2 mechanism is reached.

An example with the characteristics of the coupled displacement is the reaction of azide ion with substituted phenylethyl chlorides. Although the reaction exhibits second-order kinetics, it has a substantially positive ρ value, indicative of an electron deficiency at the transition state.[13] The physical description of this type of activated complex as an "exploded" S_N2 transition state is similar to that of the postulated intermediate in the S_N2(intermediate) mechanism. However, no discrete intermediate occupying an energy minimum is postulated.

All these descriptions of the species involved in borderline substitution picture the carbon undergoing substitution as having partial positive charge, either in an ion pair or as a carbon weakly bound to both departing leaving group and incoming nucleophile.

The importance of solvent participation in the borderline mechanisms should be noted. The types of solvents in which nucleophilic solvent participation is minimized are those in which high electronegativity of the constituent atoms reduces the Lewis basicity and polarizability of the solvent molecules. Trifluoroacetic acid and perfluoro alcohols are regarded as the least nucleophilic of the solvents commonly used in solvolysis studies.[14] These solvents are used to define the characteristics of reactions proceeding without nucleophilic solvent participation. Solvent nucleophilicity increases with the electron-releasing capacity of the molecule. The order trifluoroacetic acid $<$ trifluoroethanol $<$ acetic acid $<$ water $<$ ethanol gives a qualitative indication of the trend in solvent nucleophilicity. More will be said about solvent nucleophilicity in Section 5.5.

Substrate structure will also influence the degree of nucleophilic solvent participation. Solvation is minimized by steric hindrance, and the 2-adamantyl system is regarded as being a secondary substrate which cannot accommodate significant back-side nucleophilic participation.

X H H

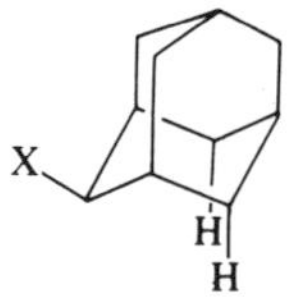

The 2-adamantyl system has been studied as a model reactant for defining the characteristics of ionization without nucleophilic participation. The degree of nucleophilic participation in other reactions can then be estimated by comparison with the 2-adamantyl system.[15]

5.4. Carbocations

It is clear that since carbocations are key intermediates in many nucleophilic substitution reactions, we will need to develop a grasp of their structural properties and the effect substituents have on their stability. The critical step in the ionization

13. J. P. Richard and W. P. Jencks, *J. Am. Chem. Soc.* **106**, 1383 (1984).
14. T. W. Bentley, C. T. Bowen, D. H. Morten, and P. v. R. Schleyer, *J. Am. Chem. Soc.* **103**, 5466 (1981).
15. F. L. Schadt, T. W. Bentley, and P. v. R. Schleyer, *J. Am. Chem. Soc.* **98**, 7667 (1976).

mechanism of nucleophilic substitution is the generation of the tricoordinate carbocation intermediate. For this mechanism to operate, it is essential that this species not be prohibitively high in energy. Carbocations are inherently high-energy species. The ionization of *t*-butyl chloride is endothermic by 153 kcal/mol in the gas phase.[16]

$$(CH_3)_3CCl \rightarrow (CH_3)_3C^+ + Cl^-$$

An activation energy of this magnitude would lead to an unobservably slow reaction at normal temperature. There is an abundance of evidence that carbocations can be intermediates in nucleophilic substitution reactions. Carbocation formation in solution is feasible because of the solvation of the ions that are produced. One of the earliest pieces of evidence for the existence of carbocation intermediates was the observation that triphenylmethyl chloride (trityl chloride) gave conducting solutions when dissolved in liquid sulfur dioxide, a polar non-nucleophilic solvent. Trityl chloride also reacted with Lewis acids, such as aluminum chloride, to give colored salt-like solids.[17]

In contrast to triphenylmethyl chloride, which has the properties of a covalent compound, triphenylmethyl perchlorate behaves as an ionic compound. The presence of triphenylmethyl cations in this solid has been confirmed by an X-ray crystal structure determination.[18] The central carbon is planar, but the three phenyl rings are at an angle of 54° to the plane of the trigonal carbon so that the overall cation has a propeller-like shape. The temperature-dependent NMR spectrum of the carbocation also indicates that it has this structure in solution.[19] The twisting of the aromatic rings with respect to each other is evidently the result of van der Waals repulsions between the *ortho* hydrogens.

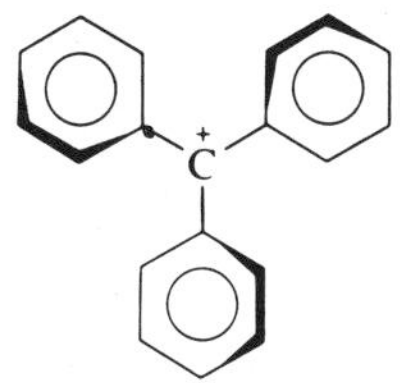

The triarylmethyl cations are particularly stable because of the conjugation with the aryl groups, which delocalizes the positive charge. Because of their stability and ease of generation, the triarylmethyl cations have been the subject of studies aimed at determining the effect of substituents on carbocation stability. Many of these studies used the characteristic ultraviolet absorption spectra of the cations to determine their concentration. In acidic solution, equilibrium is established between triarylcarbinols and the corresponding carbocation.

$$R^+ + H_2O \rightleftharpoons ROH + H^+$$

16. D. W. Berman, V. Anicich, and J. L. Beauchamp, *J. Am. Chem. Soc.* **101**, 1239 (1979).
17. Reviews of the arylmethyl cations are given by C. D. Nenitzescu, Chapter 1, Vol. I, 1968, and by H. H. Freeman, Chapter 28, Vol. IV, 1973, in *Carbonium Ions*, G. A. Olah and P. v. R. Schleyer (eds.), Wiley-Interscience, New York.
18. A. H. Gomes de Mesquita, C. H. MacGillavry, and K. Eriks, *Acta Crystallogr.* **18**, 437 (1965).
19. I. I. Schuster, A. K. Colter, and R. J. Kurland, *J. Am. Chem. Soc.* **90**, 4679 (1968).

Table 5.1. Values of pK_{R^+} for Some Carbocations[a]

Carbocation	pK_{R^+}
Triarylmethyl Cations	
Triphenylmethyl	−6.63
4,4′,4″-Trimethyltriphenylmethyl	−3.56
4-Methoxytriphenylmethyl	−3.40
4,4′-Dimethoxytriphenylmethyl	−1.24
4,4′,4″-Trimethoxytriphenylmethyl	+0.82
4,4′,4″-Trichlorotriphenylmethyl	−7.74
4-Nitrotriphenylmethyl	−9.15
4,4′,4″-Trinitrotriphenylmethyl	−16.27
4,4′,4″-Tri(dimethylamino)triphenylmethyl	+9.36
Sesquixanthydryl[b]	+9.05
Diarylmethyl Cations	
Diphenylmethyl	−13.3
4,4′-Dimethyldiphenylmethyl	−10.4
4,4′-Dimethoxydiphenylmethyl	−5.71
2,2′,4,4′,6,6′-Hexamethyldiphenylmethyl	−6.6
4,4′-Dichlorodiphenylmethyl	−13.96
Miscellaneous Carbonium Ions	
Tricyclopropylmethyl cation[c]	−2.3
Tropylium cation (cycloheptatrienyl cation)[a]	+4.7
Triphenylcyclopropenyl cation[e]	+3.1
Trimethylcyclopropenyl cation[f]	+7.8
Tricyclopropylcyclopropenyl cation[g]	+9.7

a. Unless otherwise indicated, the pK_{R^+} values are taken from N. C. Deno, J. J. Jaruzelski, and A. Schriesheim, *J. Am. Chem. Soc.* 77, 3044 (1955); for an extensive compilation of similar data, see H. H. Freedman, in *Carbonium Ions*, Vol. IV, G. A. Olah and P. v. R. Schleyer (eds.), Wiley-Interscience, New York, 1973, Chap. 28.
b. J. C. Martin and R. G. Smith, *J. Am. Chem. Soc.* **86**, 2252 (1964).
c. N. C. Deno, H. G. Richey, Jr., J. S. Liu, D. N. Lincoln, and J. O. Turner, *J. Am. Chem. Soc.* **87**, 4533 (1965).
d. W. E. Doering and L. H. Knox, *J. Am. Chem. Soc.* **76**, 3203 (1954).
e. R. Breslow, H. Höver, and H. W. Chang, *J. Am. Chem. Soc.* **83**, 2375 (1961).
f. J. Ciabattoni and E. C. Nathan III, *Tetrahedron Lett*, 4997 (1969).
g. K. Komatsu, I. Tomioka, and K. Okamoto, *Tetrahedron Lett.*, 947 (1980); R. A. Moss and R. C. Munjal, *Tetrahedron Lett.*, 1221 (1980).

The relative stability of the carbocation can be expressed in terms of its pK_{R^+}, which is defined as

$$pK_{R^+} = \log \frac{[R^+]}{[ROH]} + H_R$$

where H_R is an acidity function defined for the medium.[20] (See Section 4.8 to review the general principles of acidity functions.) In dilute aqueous solution, H_R is equivalent to pH, and pK_{R^+} is equal to the pH at which the carbocation and alcohol are present in equal concentrations.

By measuring the extent of carbocation formation at several acidities and applying the definition of pK_{R^+}, the values shown in Table 5.1 were determined.

20. N. C. Deno, J. J. Jaruzelski, and A. Schriesheim, *J. Am. Chem. Soc.* **77**, 3044 (1955).

The carbocations that can be studied in this way are all relatively stable carbocations. The data in Table 5.1 reveal that electron-releasing substituents on the aryl rings stabilize the carbocation (more positive pK_{R^+}) while electron-withdrawing groups such as nitro are destabilizing. This is what would be expected in view of the electron-deficient nature of the carbocation.

The diarylmethyl cations listed in Table 5.1 are 6–7 pK_{R^+} units less stable than the corresponding triarylmethyl cations. This indicates that the additional aryl groups have a cumulative, although not necessarily additive, effect on stability of the carbocation. Primary benzylic cations (monoarylmethyl cations) are generally not sufficiently stable for determination of pK_{R^+} values. A particularly stable benzylic ion, the 2,4,6-trimethylphenylmethyl cation, has a pK_{R^+} of −17.4.

One of the most important and general trends in organic chemistry is the increase in carbocation stability with additional alkyl substitution. This stability relationship is fundamental to understanding many aspects of reactivity, especially of nucleophilic substitution. In recent years, it has become possible to put the stabilization effect on a quantitative basis. One approach has been gas phase measurements which determine the proton affinity of alkenes leading to carbocation formation. From these data, the hydride affinity of the carbocation can be obtained.

$$R^+ + H^- \rightarrow R{-}H, \qquad -\Delta H^\circ = \text{hydride affinity}$$

These data provide a thermodynamic basis for comparison of the relative stability of nonisomeric carbocations. Some representative results are shown in Table 5.2.

Table 5.2. Hydride Affinity of Some Carbocations

Hydride affinity (kcal/mol)[a]			
CH_3^+	314[b]	$CH_2{=}CHCH_2^+$	256
$CH_3CH_2^+$	274[b]	$CH_2{=}CH\overset{+}{C}HCH_3$	237
$(CH_3)_2CH^+$	247[b]	$CH_2{=}CH\overset{+}{C}(CH_3)_2$	225
$(CH_3)_3C^+$	230[b]	$CH_3CH{=}CH\overset{+}{C}HCH_3$	225
$CH_2{=}CH^+$	287[c]	phenyl cation ($C_6H_5^+$)	298[c]
cyclopropenyl cation	223	$C_6H_5CH_2^+$	233
cyclopentadienyl cation	258	$C_6H_5\overset{+}{C}HCH_3$	226
cycloheptatrienyl cation	200	$C_6H_5\overset{+}{C}(CH_3)_2$	220

a. Except where noted, data are from D. H. Aue and M. T. Bowers, in *Gas Phase Ion Chemistry*, M. T. Bowers (ed.), Academic Press, New York, 1979.
b. F. A. Houle and J. L. Beauchamp, *J. Am. Chem. Soc.* **101**, 4067 (1979).
c. D. W. Berman, V. Anichich, and J. L. Beauchamp *J. Am. Chem. Soc.* **101**, 1239 (1979).

The stability order tertiary > secondary > primary > methyl is the same order as established by solvolysis rate measurements in solution.

There is also a less dramatic but consistent trend which reveals that within each structural class (primary, secondary, tertiary), larger ions are more stable than smaller ones, e.g., $t\text{-}C_4H_9^+ < t\text{-}C_5H_{11}^+ < t\text{-}C_6H_{13}^+$.[21] The same trend is observed for C_2 through C_5 primary cations.[22] The greater stability of the larger ions in the gas phase reflects their ability to disperse the positive charge over a larger number of atoms.

Since these stability measurements pertain to the gas phase, it is important to consider the effects solvation might have on the structure–stability relationships. It has been possible to obtain thermodynamic data for the ionization of alkyl chlorides by reaction with SbF_5, a Lewis acid, in the non-nucleophilic solvent SO_2ClF.[23] As long as subsequent reactions of the carbocation can be avoided, the thermodynamic characteristics of this reaction provide a measure of the relative ease of carbocation formation in solution.

$$RCl + SbF_5 \xrightarrow{SO_2ClF} R^+ + [SbF_5Cl]^-$$

It has been found that the solvation energies of the carbocations in this medium are small and do not differ much from one another, making comparison of the nonisomeric systems possible. There is an excellent correlation between these data and the gas phase data, in terms of both the stability order and the energy differences between different carbocations. A plot of the gas phase hydride affinity versus the ionization enthalpy gives a line of slope 1.63 with a correlation coefficient of 0.973. This result is in agreement with the expectation that the gas phase stability would be somewhat more sensitive to structure than the solution phase stability. The energy gap between tertiary and secondary ions is about 17 kcal/mol in the gas phase and about 9.5 kcal/mol in the SO_2ClF solution.

An independent measurement of the energy difference between secondary and tertiary cations in solution is available from calorimetric measurement of the enthalpy of isomerization of the *s*-butyl cation to the *t*-butyl cation. This value has been found to be 14.5 kcal/mol in SO_2ClF solution.[24]

$$CH_3CH_2\underset{+}{C}HCH_3 \xrightarrow{SO_2ClF} CH_3\overset{\overset{\displaystyle CH_3}{|}}{\underset{+}{C}}CH_3, \qquad \Delta H = -14.5\ \text{kcal/mol}$$

A wide range of carbocation stability data have been obtained by measuring the heat of ionization of a series of chlorides and carbinols in non-nucleophilic solvents in the presence of Lewis acids.[25] Some representative data are given in

21. F. P. Lossing and J. J. Holmes, *J. Am. Chem. Soc.* **106**, 6917 (1984).
22. J. C. Schultz, F. A. Houle, and J. L. Beauchamp, *J. Am. Chem. Soc.* **106**, 3917 (1984).
23. E. M. Arnett and N. J. Pienta, *J. Am. Chem. Soc.* **102**, 3329 (1980).
24. E. W. Bittner, E. M. Arnett, and M. Saunders, *J. Am. Chem. Soc.* **98**, 3734 (1976).
25. E. M. Arnett and T. C. Hofelich, *J. Am. Chem. Soc.* **105**, 2889 (1983).

Table 5.3. ΔH for Ionization of Chlorides and Alcohols in SO_2ClF over a Wide Structural Range[a]

Reactant	ΔH (kcal/mol) X = Cl	ΔH (kcal/mol) X = OH
$(CH_3)_2CH{-}X$	−15	
$CH_3CH_2CH(CH_3){-}X$	−16	
$(CH_3)_3C{-}X$	−25	−35
$(CH_3)_2C(Ph){-}X$	−30	−40
$(Ph)_2C(CH_3){-}X$		−37.5
$(Ph)_3C{-}X$		−49
(cyclopropyl)$_3C{-}X$		−59

a. Data from E. M. Arnett and T. C. Hofelich, *J. Am. Chem. Soc.* **105**, 2889 (1983).

Table 5.3. Included are data for the diarylmethyl and triarylmethyl systems, for which pK_{R^+} data are available (Table 5.1), and this gives some basis for comparison of the stability of secondary and tertiary alkyl carbocations with that of the more stable aryl-substituted ions.

Any structural effect which reduces the electron deficiency at the tricoordinate carbon will have the effect of stabilizing the carbocation. Allyl cations are stabilized by delocalization involving the adjacent double bond.

$$H_2C{=}CH{-}\overset{+}{C}H_2 \longleftrightarrow H_2\overset{+}{C}{-}CH{=}CH_2 \qquad H_2C\overset{+}{\cdots}CH\cdots CH_2$$

The π-electron delocalization requires proper orbital alignment. As a result, there is a significant barrier to rotation about the carbon–carbon bonds in the allyl cation. The exact height of the barrier depends upon the substituent groups a, b, and R. Some measured values are shown in Scheme 5.2.

$$b{-}C(a)\overset{+}{\cdots}C(R)\cdots C(a){-}b \rightleftharpoons b{-}C(a)\overset{+}{\cdots}C(R)\cdots C(b){-}a$$

Benzyl cation stability is strongly affected by the substituents on the benzene ring. A molecular orbital calculation estimating the stabilization has been done using STO-3G-level basis functions. The electron-donating *p*-amino and *p*-methoxy groups are found to stabilize a benzyl cation by 26 and 14 kcal/mol, respectively.

Scheme 5.2. Rotational Energy Barriers for Allyl Cations (kcal/mol)[a]

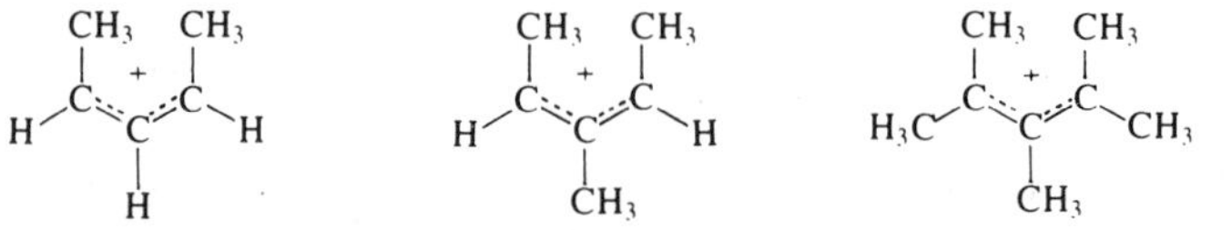

1,3-dimethylallyl cation (2-H)	1,2,3-trimethylallyl cation	1,1,2,3,3-pentamethylallyl cation
18.7	18.1	13.8

a. From J. M. Bollinger, J. M. Brinich, and G. A. Olah, *J. Am. Chem. Soc.* **92**, 4025 (1970).

On the other hand, electron-attracting groups such as *p*-cyano and *p*-nitro are destabilizing by 12 and 20 kcal/mol, respectively.[26]

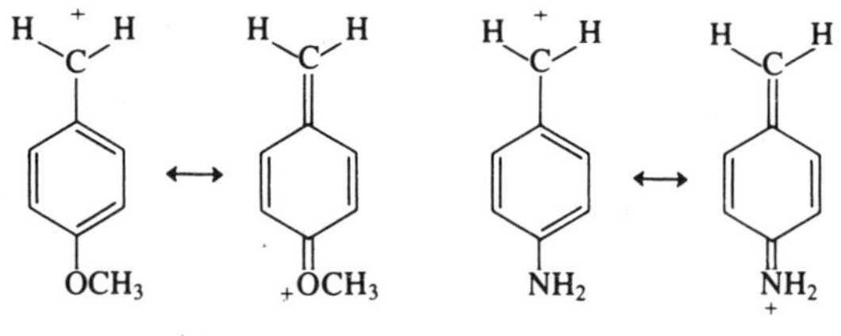

Stabilized

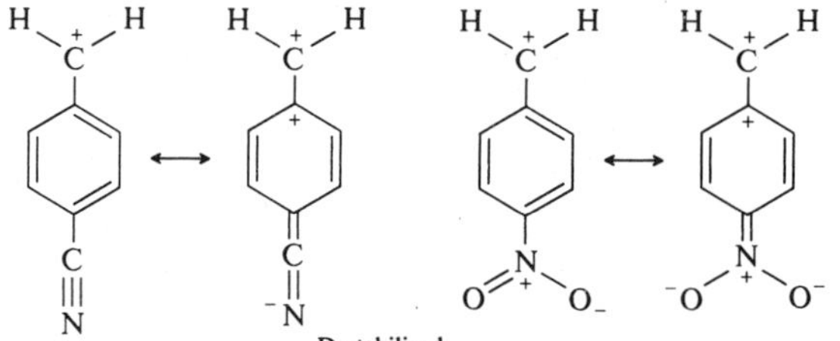

Destabilized

Adjacent atoms with one or more lone pairs of electrons strongly stabilize a carbocation. Table 1.13 (p. 26) shows the calculated stabilization of the methyl cation by such substituents. Alkoxy and dialkylamino groups are important examples of this effect.

$$CH_3\ddot{O}-\overset{+}{C}H_2 \leftrightarrow CH_3\overset{+}{O}=CH_2, \qquad (CH_3)_2\ddot{N}-\overset{+}{C}H_2 \leftrightarrow (CH_3)_2\overset{+}{N}=CH_2$$

Although these structures have a positive charge on a more electronegative atom, they benefit from an additional bond which satisfies the octet requirement of the tricoordinate carbon. These "carbocations" are well represented by the doubly bonded structures. One indication of the participation of adjacent oxygen sub-

26. W. J. Hehre, M. Taagepera, R. W. Taft, and R. D. Topsom, *J. Am. Chem. Soc.* **103**, 1344 (1981).

Table 5.4. Destabilization of 2-Substituted-2-Propyl Cation by Electron-Withdrawing Substituents

Z	Solvolysis rate relative to Z = H	Destabilization (kcal/mol) energy relative to Z = H
CN	$\sim 10^{-3}$ [a]	9.9[b]
CF_3	$\sim 10^{-6}$ [c]	37.3[b]
CH=O	—	6.1[b]

a. O. G. Gassman and J. J. Talley, *J. Am. Chem. Soc.* **102**, 1214 (1980).
b. M. N. Paddon-Row, C. Santiago, and K. N. Houk, *J. Am. Chem. Soc.* **102**, 6561 (1980).
c. K. M. Koshy and T. T. Tidwell, *J. Am. Chem. Soc.* **102**, 1216 (1980).

stituents is the existence of a barrier to rotation about the C—O bonds in this type of carbocation.

CH3 O+ C H H ⇌ CH3 +O C H H
A

CH3 O+ ⇌ +O CH3
B

The barrier in **A** is about 14 kcal/mol (ΔG^{+}) as measured by coalescence of the peaks due to the nonidentical vinyl protons in the NMR spectrum. The value of the barrier is sensitive to the nature of the solvent.[27] The gas phase barrier is calculated by MO methods to be 26 kcal/mol. The observed barrier for **B** is 19 kcal/mol.[28,29] The stabilizing effect of adjacent lone pairs persists even for fluorine. The gas phase stability order $F_2CH^+ > FCH_2^+ > F_3C^+ > CH_3^+$ reflects the stabilizing influence of the fluorine lone pairs counterbalanced by the inductive effect of fluorine.[30]

Electron-withdrawing groups that are substituted directly on the cationic site are destabilizing. Table 5.4 gives an indication of the relative retardation of the rate of ionization and the calculated destabilization for several substituents. The trifluoromethyl group, which exerts a powerful inductive effect, is strongly destabilizing both on the basis of the kinetic data and the MO calculations. The cyano and formyl groups are less so. In fact, the destabilizing effect of these groups is considerably less than would be predicted on the basis of their inductive substituent constants. Both the cyano and formyl groups can act as π donors, even though the effect is

27. D. Cremer, J. Gauss, R. F. Childs, and C. Blackburn, *J. Am. Chem. Soc.* **107**, 2435 (1985).
28. R. F. Childs and M. E. Hagar, *Can. J. Chem.* **58**, 1788 (1980).
29. There is another mechanism for equilibration of the cation pairs $\mathbf{A}_1 \rightleftharpoons \mathbf{A}_2$ and $\mathbf{B}_1 \rightleftharpoons \mathbf{B}_2$, namely, inversion at oxygen. However, the observed barrier represents at least the *minimum* for the C=O rotational barrier and therefore demonstrates that the C—O bond has double-bond character.
30. R. J. Blint, T. B. McMahon, and J. L. Beauchamp, *J. Am. Chem. Soc.* **96**, 1269 (1974).

to place partial positive charge on nitrogen and oxygen atoms, respectively. The relevant resonance structures are depicted below:

$$\geq\overset{+}{C}-C\equiv N: \leftrightarrow\ \geq C=C=\overset{+}{N}: \qquad \geq\overset{+}{C}-\underset{H}{C}=\ddot{O}: \leftrightarrow\ \geq C=\underset{H}{C}-\ddot{O}:^{+}$$

These interactions are reflected in MO energies, bond lengths, and charge distributions calculated for such cations.[31] The resonance structures are the nitrogen and oxygen analogs of the allyl cation. The effect of this π delocalization is to attenuate the inductive destabilization by these substituents.[32]

Several very stable carbocations are included in the "Miscellaneous" part of Table 5.1. These ions are remarkably stable, considering that they do not bear electron-releasing heteroatom substituents such as oxygen or nitrogen. The tricyclopropylmethyl cation, for example, is more stable than the triphenylmethyl cation.[33] The stabilization of carbocations by cyclopropyl substituents results from the interaction of the electrons in the cyclopropyl C—C bonds with the positive carbon. The electrons in these orbitals are at higher energy than normal σ electrons and are therefore particularly effective in interacting with the vacant p orbital of the carbocation. This stabilization involves interaction of the cyclopropyl bonding orbitals with the carbon p orbital. This interaction imposes a preference for the bisected conformation of the cyclopropylmethyl cation in comparison with the perpendicular conformation.

bisected conformation　　perpendicular conformation　　bisected conformation　　perpendicular conformation

Only the bisected conformation aligns the cyclopropyl C—C orbitals for effective overlap. Crystal structure determinations on two cyclopropylmethyl cations with additional stabilizing substituents, **C** and **D**, have confirmed the preference for the bisected geometry (Fig. 5.7).

C　　**D**

31. D. A. Dixon, P. A. Charlier, and P. G. Gassman, *J. Am. Chem. Soc.* **102**, 3957 (1980); M. N. Paddon-Row, C. Santiago, and K. N. Houk, *J. Am. Chem. Soc.* **102**, 6561 (1980); D. A. Dixon, R. A. Eades, R. Frey, P. G. Gassman, M. L. Hendewerk, M. N. Paddon-Row, and K. N. Houk, *J. Am. Chem. Soc.* **106**, 3885 (1984).
32. T. T. Tidwell, *Angew. Chem. Int. Ed. Engl.* **23**, 20 (1984); P. G. Gassman and T. T. Tidwell, *Acc. Chem. Res.* **16**, 279 (1983).
33. For a review of cyclopropylmethyl cations, see H. G. Richey, Jr., in *Carbonium Ions*, Vol. III, G. A. Olah and P. v. R. Schleyer (eds.), Wiley-Interscience, New York, 1972, Chapter 25.

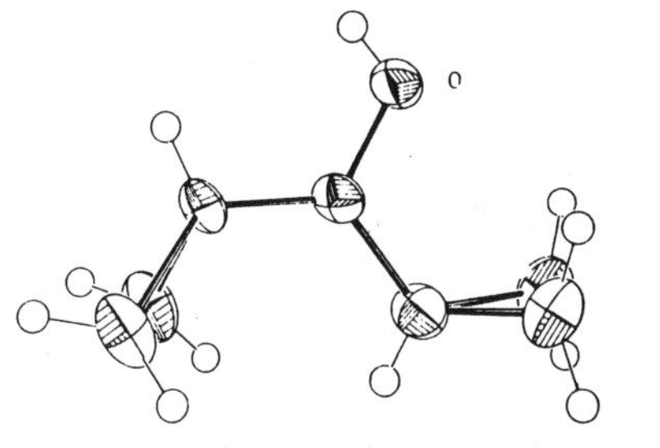

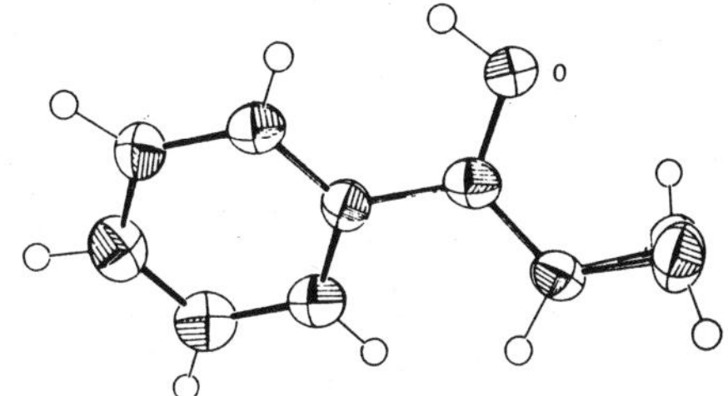

Fig. 5.7. Crystal structures of bis(cyclopropyl)hydroxymethyl cation and 1-cyclopropyl-1-phenylhydroxymethyl cation. Structural diagrams are reproduced from Ref. 34 with permission.

In ion **D**, where the phenyl group would be expected to be coplanar with the cationic center to maximize delocalization, the observed angle is 25–30°. This should permit partial benzylic stabilization. The planes of the cyclopropyl groups in both structures are at ~85° to the plane of the trigonal carbon, in agreement with expectation for the bisected ion.[34]

Solvolysis rate studies also indicate that there is greater stabilization by a cyclopropyl group in a bisected geometry. Tosylate **1**, in which the cyclopropane ring is locked into an orientation that affords a perpendicular arrangement, reacts 300 times more slowly than the model compound **2**. Tosylate **3**, which corresponds to the bisected geometry, undergoes acetolysis at least 10^5 times faster than the model 2-adamantyl tosylate, **4**.[35]

OTs — **1**; CH_3, CH_3, OTs — **2**; H, OTs — **3**; H, OTs — **4**

The tropylium ion and the cyclopropenyl ions included in Table 5.1 are examples of cations stabilized by being part of a delocalized aromatic system. These ions are aromatic according to Hückel's rule, with the cyclopropenium ion having two π electrons and the tropylium ion six. Both ring systems are planar and possess cyclic conjugation, as is required for aromaticity.

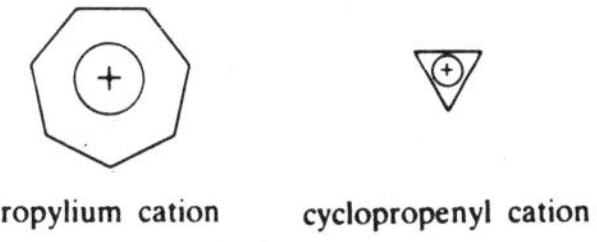

A major advance in the direct study of carbocations was made during the 1960s when methods for observation of the NMR spectra of the cations in "superacid"

34. R. F. Childs, R. Faggiani, C. J. Lock, M. Mahendran, and S. D. Zweep, *J. Am. Chem. Soc.* **108**, 1692 (1986).
35. J. E. Baldwin and W. D. Fogelsong, *J. Am. Chem. Soc.* **90**, 4303 (1968).

Scheme 5.3. Protonation and Ionization

Aliphatic alcohols in $FSO_3H-SbF_5-SO_2$

1[a] $ROH \xrightarrow{-60^\circ C} R\overset{+}{O}H_2$

R = methyl, ethyl, *n*-propyl, isopropyl, *n*-butyl, *sec*-butyl, *n*-amyl, isoamyl, neopentyl, *n*-hexyl, neohexyl

2[a] $(CH_3)_2CHCH_2OH \xrightarrow{-60^\circ C} (CH_3)_2CHCH_2\overset{+}{O}H_2 \xrightarrow{-30^\circ C} (CH_3)_3C^+$

3[a] $(CH_3)_3COH \xrightarrow{-60^\circ C} (CH_3)_3C^+$

Alkyl halides in antimony pentafluoride

4[b] $CH_3\underset{F}{CH}CH_2CH_3 \xrightarrow{-110^\circ C} CH_3\overset{+}{C}HCH_2CH_3 \xrightarrow{-40^\circ C} (CH_3)_3C^+$

5[c] $CH_3OCH_2Cl \xrightarrow{-60^\circ C} CH_3OCH_2^+$

Cyclopentylmethyl and cyclohexyl systems

6[d] cyclopentyl(H)CH₂Cl $\xrightarrow{SbF_5-SO_2,\ -60^\circ C}$

7[d] cyclopentyl(CH₃)Cl $\xrightarrow{SbF_5-SO_2,\ -60^\circ C}$

8[d] cyclohexyl(H)Cl $\xrightarrow{SbF_5-SO_2,\ -60^\circ C}$

9[d] cyclohexyl(H)OH $\xrightarrow{FSO_3H-SbF_5,\ -60^\circ C}$

→ 1-methylcyclopentyl cation ($^+$—CH_3)

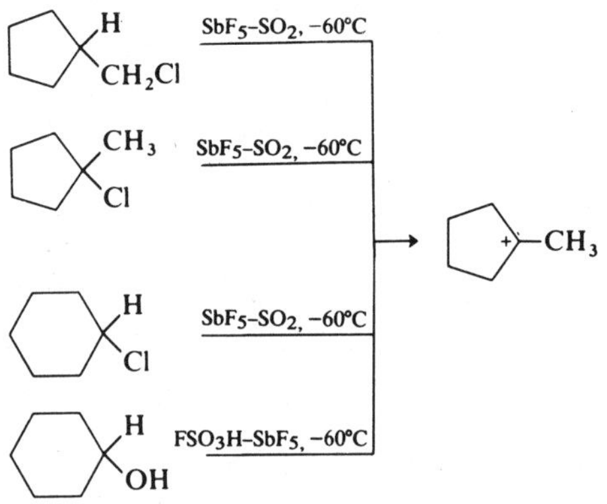

media were developed. The term superacid refers to media of very high proton-donating capacity, for example, more acidic than 100% sulfuric acid. The solution is essentially non-nucleophilic, so carbocations of only moderate stability can be generated and observed.[36] A convenient medium for these NMR measurements is $FSO_3H-SbF_5-SO_2$. The fluorosulfonic acid acts as a proton, and antimony pentafluoride is a powerful Lewis acid. This medium remains nonviscous at low temperature. This particular combination has been dubbed "magic acid" because of its powerful protonating ability.

Alkyl halides and alcohols, depending on the structure of the alkyl group, react with magic acid to give rise to carbocations. Primary and secondary alcohols are

36. A review of the extensive studies of carbocations in superacid media is available in G. A. Olah, G. K. SuryaPrakash, and J. Sommer, *Super Acids*, Wiley, New York, 1985.

Bicyclooctyl systems in SbF_5–SO_2ClF, −78°C

10[e]

Cl

11[e]

12[e]

Benzylic and cyclopropylcarbinyl systems

13[f] OCH_3 … CH_2OH $\xrightarrow{SbF_5-SO_2ClF,\ -78°C}$ OCH_3 … $\overset{+}{C}H_2$

14[g] CH_3–C(OH)–CH_3 $\xrightarrow{FSO_3H-SbF_5-SO_2,\ -75°C}$ C+ (CH_3)(CH_3)

a. G. A. Olah, J. Sommer, and E. Namanworth, *J. Am. Chem. Soc.* **89**, 3576 (1967).
b. M. Saunders, E. L. Hagen, and J. Rosenfeld, *J. Am. Chem. Soc.* **90**, 6882 (1968).
c. G. A. Olah and J. M. Bollinger, *J. Am. Chem. Soc.* **89**, 2993 (1967).
d. G. A. Olah, J. M. Bollinger, C. A. Cupas, and J. Lukas, *J. Am. Chem. Soc.* **89**, 2692 (1967).
e. G. A. Olah and G. Liang, *J. Am. Chem. Soc.* **93**, 6873 (1971).
f. G. A. Olah, R. D. Porter, C. L. Juell, and A. M. White, *J. Am. Chem. Soc.* **94**, 2044 (1972).
g. C. U. Pittman, Jr., and G. A. Olah, *J. Am. Chem. Soc.*, **87**, 2998 (1965).

protonated at −60°C but do not ionize. Tertiary alcohols ionize, giving rise to the cation. As the temperature is increased, carbocation formation also occurs from secondary alcohols. *s*-Butyl alcohol ionizes with rearrangement to the *t*-butyl cation. At −30°, the protonated primary alcohol isobutanol ionizes, also forming the *t*-butyl cation. Protonated *n*-butanol is stable to 0°C, at which point it too gives rise to the *t*-butyl cation. It is typically observed that ionization in superacids gives rise to the most stable of the isomeric carbocations which could be derived from the alkyl group. The *t*-butyl cation is generated from C_4 systems while C_5 and C_6 alcohols give rise to the *t*-pentyl and *t*-hexyl ions, respectively. These and related observations are illustrated in Scheme 5.3.

Entries 6–9 and 10–12 in Scheme 5.3 further illustrate the tendency for rearrangement to the most stable cation to occur. The tertiary 1-methylcyclopentyl cation is the only ion observed upon ionization of a variety of five- and six-membered ring derivatives. The tertiary bicyclo[3.3.0]octyl cation is formed from all bicyclooctyl precursors. The tendency to rearrange to the thermodynamically stable ions by multiple migrations is a consequence of the very low nucleophilicity of the solvent system. In the absence of nucleophilic capture by solvent, the ions have a long lifetime and undergo extensive skeletal rearrangement and accumulate as the most stable isomer.

Studies in superacid media provided early information concerning the preferred conformation of the cyclopropylmethyl cation.[37] The NMR spectrum of the 2-cyclopropyl-2-propyl cation revealed that the two methyl groups were nonequivalent. This is consistent with the bisected, but not the perpendicular, conformation. Furthermore, the existence of the two peaks requires a significant barrier to rotation since, otherwise, an averaged peak would be observed in the NMR spectrum. Later work established the rotational barrier to be about 14 kcal/mol.[38]

bisected conformation

perpendicular conformation

Up to this point in our discussion, we have considered only carbocations in which the cationic carbon can be sp^2-hybridized so that it is planar. When this hybridization cannot be achieved, the carbocation will have higher energy. In a classic experiment, Bartlett and Knox demonstrated that the tertiary chloride 1-chloroapocamphane was inert to nucleophilic substitution.[39] Starting material was recovered unchanged even after refluxing for 48 h in ethanolic silver nitrate. The unreactivity of this compound is attributed to the structure of the bicyclic system, which prevents rehybridization to a planar sp^2 carbon. Direct displacement by back-side attack is also precluded, because of the bridgehead location of the C—Cl bond.

The apocamphyl structure is particularly rigid, and bridgehead carbocations become accessible in more flexible structures. The relative solvolysis rates of the bridgehead bromides 1-bromoadamantane, 1-bromobicyclo[2.2.2]octane, and 1-

37. C. U. Pittman, Jr., and G. A. Olah, *J. Am. Chem. Soc.* **87**, 5123 (1965).
38. D. S. Kabakoff and E. Namanworth, *J. Am. Chem. Soc.* **92**, 3234 (1970).
39. P. D. Bartlett and L. H. Knox, *J. Am. Chem. Soc.* **61**, 3184 (1939).

bromobicyclo[2.2.1]heptane illustrate this trend. The relative rates for solvolysis in 80% ethanol at 25°C are shown.[40]

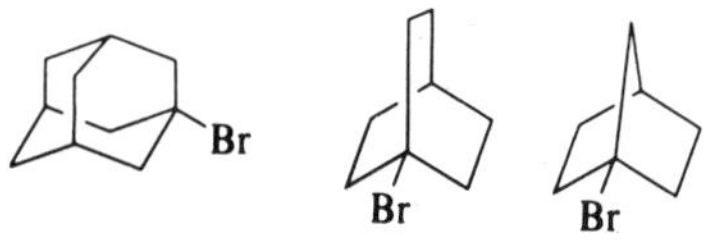

Br Br Br

1 10^{-3} 10^{-10}

The relative reactivity of tertiary bridgehead systems toward solvolysis is well correlated with the increased strain, calculated by molecular mechanics, resulting from conversion of the ring structure to a carbocation.[41]

Carbocations in which the cationic carbon is *sp*-hybridized are of higher energy than those in which the cationic center is sp^2. This is because of the higher electronegativity with increasing *s* character. It has been estimated that the vinyl cation, $CH_2{=}CH^+$, lies between the ethyl cation and the methyl cation in stability. The intermediacy of substituted vinyl cations in solvolysis reactions has been demonstrated, but direct observation has not been possible for simple vinyl cations.[42] Most examples of solvolytic generation of vinyl cations involve very reactive leaving groups, especially trifluoromethylsulfonates (triflates). Typical products include allenes, acetylenes, and vinyl esters.[43]

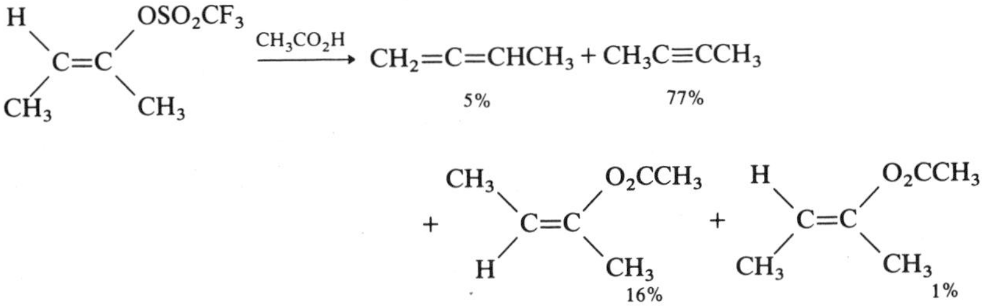

$$(CH_3)(H)C{=}C(OSO_2CF_3)(CH_3) \xrightarrow{CH_3CO_2H} \underset{5\%}{CH_2{=}C{=}CHCH_3} + \underset{77\%}{CH_3C{\equiv}CCH_3}$$

$$+ \underset{16\%}{(CH_3)(H)C{=}C(O_2CCH_3)(CH_3)} + \underset{1\%}{(H)(CH_3)C{=}C(O_2CCH_3)(CH_3)}$$

Ref. 44

The phenyl cation is an extremely unstable cation, as is reflected by the high hydride affinity shown in Table 5.2. In this case, the ring geometry opposes rehybridization so the vacant orbital retains sp^2 character. Since the empty orbital is in the nodal plane of the ring, it receives no stabilization from the π electrons.

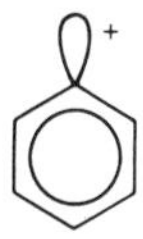

+

40. For a review of bridgehead carbocations, see R. C. Fort, Jr., in *Carbonium Ions*, Vol. IV, G. A. Olah and P. v. R. Schleyer (eds.), Wiley-Interscience, New York, 1973, Chapter 32.
41. P. Müller and J. Mareda, *Helv. Chim. Acta* **70**, 1017 (1987).
42. H.-U. Siehl and M. Hanack, *J. Am. Chem. Soc.* **102**, 2686 (1980).
43. For reviews of vinyl cations, see Z. Rappoport, in *Reactive Intermediates*, R. A. Abramovitch (ed.), Vol. 3, Plenum Press, New York, 1983; P. J. Stang, *Prog. Phys. Org. Chem.* **10**, 205 (1973); G. Modena and U. Tonellato, *Adv. Phys. Org. Chem.* **9**, 185 (1971).
44. R. H. Summerville, C. A. Senkler, P. v. R. Schleyer, T. E. Dueber, and P. J. Stang, *J. Am. Chem. Soc.* **96**, 1100 (1974).

Phenyl cations are formed by thermal decomposition of aryldiazonium ions.[45] The cation is so extremely reactive that under some circumstances it can recapture the nitrogen generated in the decomposition.[46] Attempts to observe formation of phenyl cations by ionization of aryl triflates have only succeeded when especially stabilizing groups, such as trimethylsilyl groups, are present at the 2- and 6- positions of the aromatic ring.[47]

5.5. Nucleophilicity and Solvent Effects

The term *nucleophilicity* is generally accepted to refer to the effect of a Lewis base on the *rate* of a nucleophilic substitution reaction and may be contrasted with *basicity*, which is defined in terms of the position of an equilibrium reaction with a proton or some other acid. Nucleophilicity then is used to describe trends in the *kinetic* aspects of reactions. The relative nucleophilicity of a given species may differ from substrate to substrate. It has not been possible to devise an absolute scale of nucleophilicity. The situation is analogous to that for basicity, in which the basicity is defined with respect to some specific acid. We need to gain some impression of the structural features that govern nucleophilicity and to understand the relationship between nucleophilicity and basicity.[48]

The factors that influence nucleophilicity have usually been assessed in the context of the limiting S_N2 case, since it is here that the properties of the nucleophile will be most apparent. The rate of an S_N2 reaction is directly related to the effectiveness of the nucleophile in displacing the leaving group. In contrast, the effect of nucleophilicity will not be evident in the rate of an S_N1 reaction. The nature of the nucleophile will only affect the product distribution resulting from partitioning of the carbocation intermediate among the available pathways.

Many properties have an influence on nucleophilicity. Those considered to be most generally significant are (1) the solvation energy of the nucleophile; (2) the strength of the bond being formed to carbon; (3) the size of the nucleophile; (4) the electronegativity of the attacking atom; and (5) the polarizability of the attacking atom.[49] Let us consider how each of these factors affects nucleophilicity.

45. C. G. Swain, J. E. Sheats, and K. G. Harbison, *J. Am. Chem. Soc.* **97**, 783 (1975).
46. R. G. Bergstrom, R. G. M. Landells, G. W. Wahl, Jr., and H. Zollinger, *J. Am. Chem. Soc.* **98**, 3301 (1976).
47. Y. Apeloig and D. Arad, *J. Am. Chem. Soc.* **107**, 5285 (1985); Y. Himeshima, H. Kobayashi, and T. Sonoda, *J. Am. Chem. Soc.* **107**, 5286 (1985).
48. For general reviews of nucleophilicity, see R. F. Hudson, in *Chemical Reactivity and Reaction Paths*, G. Klopman (ed.), John Wiley and Sons, New York, 1974, Chapter 5; J. M. Harris and S. P. McManus (eds.), *Nucleophilicity*, Advances in Chemistry Series, Vol. 215, American Chemical Society, Washington, D.C. 1987.
49. A. Streitwieser, Jr., *Solvolytic Displacement Reactions*, McGraw-Hill, New York, 1962; J. F. Bunnett, *Annu. Rev. Phys. Chem.* **14**, 271 (1963).

1. A high solvation energy lowers the ground state energy relative to the transition state, in which the charge is more diffuse. This results in an increased activation energy. Viewed from another perspective, the solvation shell must be disrupted to arrive at the transition state, and this desolvation energy contributes to the activation energy.
2. A stronger bond between the nucleophilic atom and carbon will be reflected in a more stable transition state and therefore a reduced activation energy. Since the S_N2 process is concerted, the strength of the partially formed new bond will be reflected in the energy of the transition state.
3. A bulky nucleophile will be less reactive than a smaller one because of the nonbonded repulsions that develop in the transition state. The trigonal bipyramidal geometry of the S_N2 transition state is sterically more demanding than the tetrahedral reactant.
4. A more electronegative atom binds its electrons more tightly than a less electronegative one. Since the S_N2 process requires donation of electron density to an antibonding orbital of the reactant, high electronegativity is unfavorable.
5. Polarizability describes the ease of distortion of the electron cloud of the attacking atom of the nucleophile. Again, since the S_N2 process requires bond formation using an electron pair from the nucleophile, the more easily distorted the atom, the better its nucleophilicity. Polarizability increases with atomic number going down in the periodic table.

Empirical measures of nucleophilicity may be obtained by comparing relative rates of reaction of a standard reactant with various nucleophiles. One measure of nucleophilicity is the *nucleophilic constant* (n), defined originally by Swain and Scott.[50] Taking methanolysis of methyl iodide as the standard reaction, n is defined as

$$n_{CH_3I} = \log (k_{nucleophile}/k_{CH_3OH}) \qquad \text{in } CH_3OH, 25°C$$

Table 5.5 lists the nucleophilic constants for a number of species according to this definition.

It is apparent from Table 5.5 that nucleophilicity toward methyl iodide does not correlate directly with basicity. Azide ion, phenoxide ion, and bromide are all equivalent in nucleophilicity but differ greatly in basicity. Conversely, azide ion and acetate ion and nearly identical in basicity, but azide ion is 30 times (1.5 log units) more nucleophilic. Among neutral nucleophiles, while triethylamine is *more basic* than triethylphosphine (pK_a of the conjugate acid is 10.70 versus 8.69), the phosphine is more nucleophilic ($n = 8.7$ versus 6.7, a factor of 100). Correlation with basicity is better if the attacking atom is the same. Thus, for the series of oxygen nucleophiles $CH_3O^- > C_6H_5O^- > CH_3CO_2^- > NO_3^-$, nucleophilicity parallels basicity.

Nucleophilicity usually decreases going across a row in the periodic table. For example, $HO^- > F^-$ or $C_6H_5S^- > Cl^-$. This order is primarily determined by elec-

50. C. G. Swain and C. B. Scott, *J. Am. Chem. Soc.* **75**, 141 (1953).

Table 5.5. Nucleophilic Constants of Various Nucleophiles[a]

Nucleophile	n_{CH_3I}	pK_a of conjugate acid
CH_3OH	0.0	−1.7
NO_3^-	1.5	−1.3
F^-	2.7	3.45
$CH_3CO_2^-$	4.3	4.8
Cl^-	4.4	−5.7
$(CH_3)_2S$	5.3	
NH_3	5.5	9.25
N_3^-	5.8	4.74
$C_6H_5O^-$	5.8	9.89
Br^-	5.8	−7.7
CH_3O^-	6.3	15.7
HO^-	6.5	15.7
NH_2OH	6.6	5.8
NH_2NH_2	6.6	7.9
$(CH_3CH_2)_3N$	6.7	10.70
CN^-	6.7	9.3
$(CH_3CH_2)_3As$	7.1	
I^-	7.4	−10.7
HO_2^-	7.8	
$(CH_3CH_2)_3P$	8.7	8.69
$C_6H_5S^-$	9.9	6.5
$C_6H_5Se^-$	10.7	
$(C_6H_5)_3Sn^-$	11.5	

a. Data from R. G. Pearson and J. Songstad, *J. Am. Chem. Soc.* **89**, 1827 (1967); R. G. Pearson, H. Sobel, and J. Songstad, *J. Am. Chem. Soc.* **90**, 319 (1968); P. L. Bock and G. M. Whitesides, *J. Am. Chem. Soc.* **96**, 2826 (1974).

tronegativity. Nucleophilicity also usually increases going down the periodic table, as, for example, $I^- > Br^- > Cl^- > F^-$ and $C_6H_5Se^- > C_6H_5S^- > C_6H_5O^-$. Three factors work together to determine this order. Electronegativity decreases going down the periodic table. Probably more important are the greater polarizability and weaker solvation of the heavier atoms, which have more diffuse electron distribution.

There is clearly a conceptual relationship between the properties called nucleophilicity and basicity. Both describe a process involving formation of a new bond to an electrophile by donation of an electron pair. The pK_a values in Table 5.5 refer to basicity toward a proton. There are many reactions in which a given chemical species might act either as a nucleophile or as a base. Scheme 5.4 lists some examples. It is therefore of great interest to be able to predict which chemical species $Y:^-$ will act as nucleophiles and which will act as bases under a given set of circumstances. The definition of basicity is based on the ability of a substance to remove protons and refers to an *equilibrium*.

$$B: + H_2O \rightleftharpoons \overset{+}{B}H + {}^-OH, \qquad K_b = \frac{[\overset{+}{B}H][{}^-OH]}{[B:]}$$

Scheme 5.4. Competition between Nucleophilicity and Basicity

S_N1 Substitution	Y: acts as a nucleophile	$Y:^- + R_2\overset{+}{C}CHR'_2 \rightarrow R_2C(Y)CHR'_2$
versus		
E_1 Elimination	Y: acts as a base	$Y:^- + R_2\overset{+}{C}CHR'_2 \rightarrow R_2C{=}CR'_2 + HY$
S_N2 Substitution	Y: acts as a nucleophile	$Y:^- + R_2CHCH_2Br \rightarrow R_2CHCH_2Y + Br^-$
versus		
E_2 Elimination	Y: acts as a base	$Y:^- + R_2CHCH_2Br \rightarrow RCH{=}CH_2 + HY + Br^-$
Nucleophilic addition at a carbonyl carbon	Y: acts as a nucleophile	$Y:^- + R_2CHC({=}O)R' \rightarrow R_2CHC(O^-)(Y)R'$
versus		
Enolate formation	Y: acts as a base	$Y:^- + R_2CHC({=}O)R' \rightarrow R_2C{=}C(O^-)R' + HY$

Scales for bases that are too weak to study in aqueous solution employ other solvents but are related to the equilibrium in aqueous solution. These equilibrium constants provide a measure of *thermodynamic basicity*, but we also need to have some concept of *kinetic basicity*. For the reactions in Scheme 5.4, for example, it is important to be able to make generalizations about the rates of competing reactions.

The most useful qualitative approach for making predictions of this type is the hard-soft-acid-base (HSAB) concept.[51] This concept proposes that reactions will occur most readily between species that are matched in hardness and softness. Hard nucleophiles prefer hard electrophiles, while soft nucleophiles prefer soft electrophiles. This concept can be applied to the problem of competition between nucleophilic substitution and deprotonation in the reaction of anions with alkyl halides, for example. The sp^3 carbon is a soft electrophile whereas the proton is a hard electrophile. Thus, according to the HSAB theory, a soft anion should act primarily as a nucleophile, giving the substitution product, while a hard anion is more likely to abstract a proton, giving the elimination product. The property of softness correlates with high polarizability and low electronegativity. Species in Table 5.5 which exhibit high nucleophilicity toward methyl iodide include CN^-, I^-, and $C_6H_5S^-$. Hardness reflects a high charge density and is associated with small highly electronegative species. Examples from Table 5.5 include F^- and CH_3O^-. Table 5.6 classifies some representative chemical species with respect to softness and hardness.

The soft nucleophile-soft electrophile combination is normally associated with a late transition state where the strength of the newly forming bond contributes significantly to the stability of the transition state. The hard nucleophile-hard electrophile combination implies an early transition state with electrostatic attraction

51. R. G. Pearson and J. Songstand, *J. Am. Chem. Soc.* **89**, 1827 (1967); R. G. Pearson, *J. Chem. Educ.* **45**, 581, 643 (1968); T. L. Ho, *Chem. Rev.* **75**, 1 (1975).

Table 5.6. Hardness and Softness of Some Common Ions and Molecules

	Bases (Nucleophiles)	Acids (Electrophiles)
Soft:	RSH, RS^-, I^-, R_3P $^-C\equiv N$, $:\bar{C}\equiv O^+$, RCH=CHR benzene	I_2, Br_2, RS—X, RSe—X, RCH_2—X Cu(I), Ag(I), Pd(II), Pt(II), Hg(II) zerovalent metal complexes
Borderline:	Br^-, N_3^-, $ArNH_2$ pyridine	Cu(II), Zn(II), Sn(II) R_3C^+, R_3B
Hard:	H_2O, HO^-, ROH, RO^-, RCO_2^- F^-, Cl^-, NO_3^-, NH_3, RNH_2	H—X, H^+, Li^+, Na^+, K^+ Mg^{2+}, Ca^{2+}, Al(III), Sn(IV), Ti(IV) R_3Si—X,

contributing more than bond formation. The reaction pathway is chosen early on the reaction coordinate and primarily on the basis of charge distribution.

Another significant structural effect which imparts high nucleophilicity is the *alpha effect.* It is observed that atoms which are directly bonded to an atom with one or more unshared pairs of electrons tend to be stronger nucleophiles than would otherwise be expected. Examples in Table 5.5 include HOO^-, which is more nucleophilic than HO^-, and NH_2NH_2 (hydrazine) and NH_2OH (hydroxylamine), both of which are more nucleophilic than ammonia. Various explanations have been put forward for the alpha effect.[52] One view is that the ground state of the nucleophile is destabilized by lone pair–lone pair repulsions which are decreased as bond formation occurs in the transition state. In MO terms, this would imply a relatively high energy of the nucleophile HOMO that participates in bond formation.[53] Another view is that the adjacent electron pair can act to stabilize charge deficiency at the transition state. As discussed in Section 5.3, there are many S_N2 reactions in which the transition state is electron poor. The alpha effect seems to be quite solvent sensitive,[54] and gas phase reactions show no alpha effect.[55] This suggests that solvation may play a major role in the origin of the alpha effect.

The nucleophilicity of anions, in general, depends very much on the degree of solvation. Much of the data that form the basis for quantitative measurement of nucleophilicity is for reactions in hydroxylic solvents. In protic, hydrogen-bonding solvents, anions are subject to strong interactions with solvent. Hard nucleophiles are more strongly solvated by protic solvents than soft nucleophiles, and this difference contributes to the greater nucleophilicity of soft anions in such solvents. Nucleophilic substitution reactions often occur more readily in polar aprotic solvents than they do in protic solvents. This is because anions are weakly solvated in such

52. G. Klopman, K. Tsuda, J. B. Louis, and R. E. Davis, *Tetrahedron* **26**, 4549 (1970); W. B. England, P. Kovacic, S. M. Hanrah, and M. B. Jones, *J. Org. Chem.* **45**, 2057 (1980).
53. M. M. Heaton, *J. Am. Chem. Soc.* **100**, 2004 (1978).
54. E. Buncel and I.-H. Um, *J. Chem. Soc., Chem. Commun.*, 595 (1986).
55. C. H. DePuy, E. W. Della, J. Filley, J. J. Grabowski, and V. M. Bierbaum, *J. Am. Chem. Soc.* **105**, 2481 (1983).

solvents (see Section 4.10). The cations associated with nucleophilic anions are strongly solvated in solvents such as *N,N*-dimethylformamide (DMF), dimethyl sulfoxide (DMSO), hexamethylphosphoramide (HMPA), *N*-methylpyrrolidinone, and sulfolane.[56] As a result, the anions are dissociated from the cations, which further enhances their nucleophilicity.

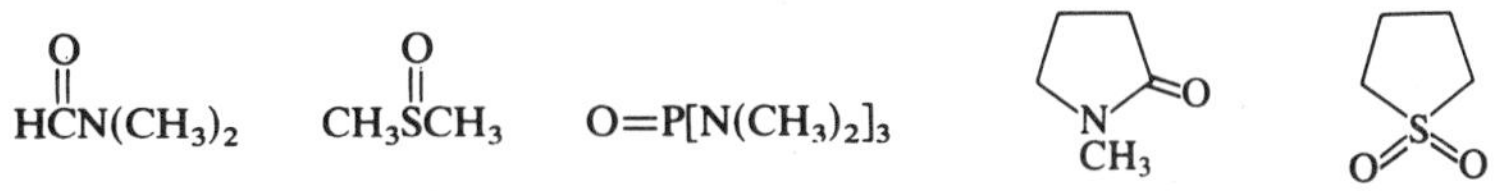

$HC(=O)N(CH_3)_2$ $CH_3S(=O)CH_3$ $O{=}P[N(CH_3)_2]_3$ N-CH_3 (pyrrolidinone, =O) (sulfolane, SO_2)

In the absence of the solvation typical of protic solvents, the relative nucleophilicity of anions changes. Hard nucleophiles increase in reactivity more than do soft nucleophiles. As a result, the relative reactivity order changes. In methanol, for example, the relative reactivity order is $N_3^- > I^- > CN^- > Br^- > Cl^-$. In DMSO the order becomes $CN^- > N_3^- > Cl^- > Br^- > I^-$.[57] In methanol the reactivity order is dominated by solvent effects, and the more weakly solvated N_3^- and I^- ions are the most reactive. The iodide ion is large and very polarizable. The anionic charge on the azide ion is dispersed by delocalization. When the effect of solvation is diminished in DMSO, other factors become more important. These presumably include the strength of the bond being formed, which would account for the reversed order of the halides in the two series. There is also evidence that S_N2 transition states are better solvated in aprotic dipolar solvents than in protic solvents.

In interpreting many aspects of displacement reactions, particularly solvolysis, it is important to be able to characterize the nucleophilicity of the solvent. Assessment of solvent nucleophilicity can be done by comparing rates of a standard substitution process in various solvents. One such procedure is based on the Winstein–Grunwald equation:

$$\log(k/k_0) = lN + mY$$

where N and Y are measures of the solvent nucleophilicity and ionizing power, respectively. The variables l and m are characteristic of specific reactions.[58] The value of N, the indicator of solvent nucleophilicity, can be determined by specifying a standard substrate for which l is assigned the value 1.00 and a standard solvent for which N is assigned the value 0.00. 2-Adamantyl tosylate has been taken as a standard substrate for which nucleophilic participation of the solvent is considered to be negligible, and 80:20 ethanol–water is taken as the standard solvent. The resulting solvent characteristics are called N_{Tos} and Y_{Tos}. Some representative values for solvents frequently used in solvolysis studies are given in Table 5.7.

56. T. F. Magnera, G. Caldwell, J. Sunner, S. Ikuta, and P. Kebarle, *J. Am. Chem. Soc.* **106**, 6140 (1984).
57. R. L. Fuchs and L. L. Cole, *J. Am. Chem. Soc.* **95**, 3194 (1973); R. Alexander, E. C. F. Ko, A. J. Parker, and T. J. Broxton, *J. Am. Chem. Soc.* **90**, 5049 (1968); D. Landini, A. Maia, and F. Montanari, *J. Am. Chem. Soc.* **100**, 2796 (1978).
58. S. Winstein, E. Grunwald, and H. W. Jones, *J. Am. Chem. Soc.* **73**, 2700 (1951); F. L. Schadt, T. W. Bentley, and P. v. R. Schleyer, *J. Am. Chem. Soc.* **98**, 7667 (1976).

Table 5.7. Solvent Nucleophilicity (N_{Tos}) and Ionization (Y_{Tos}) Parameters[a]

Solvent	N_{Tos}	Y_{Tos}
Ethanol	+0.09	−1.75
Methanol	+0.01	−0.92
50% Aqueous ethanol	−0.20	1.29
Water	−0.26	
Acetic acid	−2.05	−0.61
Formic acid	−2.05	3.04
Trifluoroethanol	−2.78	1.80
97% $(CF_3)_2CHOH-H_2O$	−3.93	1.83
Trifluoroacetic acid	−4.74	4.57

a. From F. L. Schadt, T. W. Bentley, and P. v. R. Schleyer, *J. Am. Chem. Soc.* **98**, 7667 (1976).

5.6. Leaving Group Effects

The nature of the leaving group will influence the rate of nucleophilic substitution proceeding by either the direct displacement or the ionization mechanism. Since the leaving group departs with the pair of electrons from its covalent bond to the reacting carbon atom, a correlation with electronegativity is expected. Provided the reaction series consists of structurally similar leaving groups, such relationships are observed. For example, a linear relationship has been demonstrated between the ionization of substituted benzoic acids and the rate of reaction of substituted arenesulfonates with ethoxide ion in ethanol (Hammett-type equation).[59] While the qualitative trend in reactivity of leaving group also holds for less similar systems, no generally applicable quantitative system for specifying leaving group ability has been established.

Table 5.8 lists estimated relative rates of solvolysis of 1-phenylethyl esters and halides in 80% aqueous ethanol at 75°C.[60] The reactivity of the leaving groups generally parallels their electron-attracting capacity. Trifluoroacetate, for example, is about 10^6 times as reactive as acetate, and *p*-nitrobenzenenesulfonate is about 10 times more reactive than *p*-toluenesulfonate. The order of reactivity of the halide leaving groups is $I^- > Br^- > Cl^- \gg F^-$. This order is opposite to that of the electronegativity and is dominated by the strength of the bond to carbon, which ranges from ~50 kcal for the C—I bond to ~100 kcal for the C—F bond.

Sulfonate esters are especially useful substrates in nucleophilic substitution reactions used in synthesis. They have a high level of reactivity and, unlike alkyl halides, they may be prepared from alcohols by reactions that do not directly involve the carbon atom at which substitution is to be effected. These properties are particularly important in cases where the stereochemical and structural integrity of

59. M. S. Morgan and L. H. Cretcher, *J. Am. Chem. Soc.* **70**, 375 (1948).
60. D. S. Noyce and J. A. Virgilio *J. Org. Chem.* **37**, 2643 (1972).

Table 5.8. Relative Solvolysis Rates of 1-Phenylethyl Esters and Halides[a,b]

Leaving group	k_{rel}
$CF_3SO_3^-$ (triflate)	1.4×10^8
p-Nitrobenzenesulfonate	4.4×10^5
p-Toluenesulfonate	3.7×10^4
$CH_3SO_3^-$ (mesylate)	3.0×10^4
I^-	91
Br^-	14
$CF_3CO_2^-$	2.1
Cl^-	1.0
F^-	9×10^{-6}
p-Nitrobenzoate	5.5×10^{-6}
$CH_3CO_2^-$	1.4×10^{-6}

a. Data from D. S. Noyce and J. A. Virgilio, *J. Org. Chem.* **37**, 2643 (1972).
b. In 80% aqueous ethanol at 75°C.

Table 5.9. Relative Solvolysis Rates of Ethyl Sulfonates and Halides[a]

Derivatives compared	k_{rel}	Solvent, 25°C
Triflate/tosylate	3×10^4	Acetic acid
Triflate/brosylate	5×10^3	Acetic acid
Triflate/iodide	4.5×10^5	Ethanol
Triflate/bromide	1.5×10^5	80% Ethanol

a. From A. Streitwieser, Jr., C. L. Wilkins, and E. Kiehlmann, *J. Am. Chem. Soc.* **90**, 1598 (1968).

the reactant must be maintained. Sulfonate esters are usually prepared by reaction of an alcohol with a sulfonyl halide in the presence of pyridine:

$$ROH + R'SO_2Cl \xrightarrow{\text{pyr}} ROSO_2R'$$

Tertiary alcohols are converted to sulfonate esters with difficulty, and because of their high reactivity, they are often difficult to isolate.[61] Because of these problems tertiary alcohols are frequently converted to *p*-nitrobenzoate esters for solvolytic studies.

Trifluoromethanesulfonate (triflate) ion is an exceptionally good leaving group. It can be used for nucleophilic substitution reaction on unreactive substrates. Acetolysis of cyclopropyl triflate, for example, occurs 10^5 times faster than acetolysis of cyclopropyl tosylate.[62] Table 5.9 gives a comparison of the triflate group with some other common leaving groups.

It would be anticipated that the limiting S_N1 and S_N2 mechanisms would differ in their sensitivity to the nature of the leaving group. The ionization mechanism

61. H. M. R. Hoffmann, *J. Chem. Soc.*, 6748 (1965).
62. T. M. Su, W. F. Sliwinski, and P. v. R. Schleyer, *J. Am. Chem. Soc.* **91**, 5386 (1969).

Table 5.10. Tosylate/Bromide Rate Ratios for Solvolysis of RX in 80% Ethanol[a]

R	k_{OTs}/k_{Br}
Methyl	11
Ethyl	10
Isopropyl	40
t-Butyl	4000
1-Adamantyl	9750

a. From J. L. Fry, C. J. Lancelot, L. K. M. Lam, J. M. Harris, R. C. Bingham, D. J. Raber, R. E. Hall, and P. v. R. Schleyer, *J. Am. Chem. Soc.* **92**, 2539 (1970).

should exhibit a greater dependence on leaving group ability because it requires cleavage of the bond to the leaving group without assistance by the nucleophile. Table 5.10 presents data on the variation of the relative leaving group abilities of tosylate and bromide as a function of substrate structure. The dependence is as expected, with smaller differences in reactivity between tosylate and bromide being observed for systems that react by the S_N2 mechanism.

A poor leaving group can be made more reactive by coordination to an electrophilic species. Hydroxide is a very poor leaving group. Normally, alcohols therefore do not undergo direct nucleophilic substitution. It has been estimated that the reaction

$$CH_3OH + Br^- \rightarrow CH_3Br + HO^-$$

is endothermic by 16 kcal/mol.[63] Since the activation energy for the reverse process is about 21 kcal/mol, the reaction would have an activation energy of 37 kcal/mol. As predicted by this activation energy, the reaction is too slow to detect at normal temperature. The reaction, however, is greatly accelerated in acidic solution. Protonation of the hydroxyl group provides the much better leaving group water, which is about as good a leaving group as bromide ion. The practical result is that primary alcohols can be converted to alkyl bromides by heating with sodium bromide and sulfuric acid or with concentrated hydrobromic acid. The leaving group ability of the halogens is enhanced by coordination to metal ions. Silver salts are frequently used to accelerate substitution of unreactive halides.

One of the best leaving groups is molecular nitrogen attached to alkyl groups in alkyl diazonium ions. Diazonium ions are generated by nitrosation of primary amines. The diazonium ions generated from alkyl amines are very unstable and immediately decompose with loss of nitrogen.

$$RNH_2 + HONO \rightarrow R{-}\underset{H}{N}{-}N{=}O + H_2O$$

$$R{-}\underset{H}{N}{-}N{=}O \rightarrow RN{=}NOH \xrightarrow{H^+} R{-}\overset{+}{N}{\equiv}N + H_2O$$

$$R{-}\overset{+}{N}{\equiv}N + \mathbf{Nu^-{:}} \rightarrow R{-}\mathbf{Nu} + N_2$$

63. R. A. Ogg, Jr., *Trans. Faraday Soc.* **31**, 1385 (1935).

Table 5.11. Rate Constants for Nucleophilic Substitution in Primary Alkyl Substrates[a]

	10^5k for RCH_2-				
Reaction	R = H—	CH_3—	CH_3CH_2—	$(CH_3)_2CH$—	$(CH_3)_3C$—
$RCH_2Br + LiCl$, acetone	600	9.9	6.4	1.5	0.00026
$RCH_2Br + Bu_3P$, acetone	26,000	154	64	4.9	
$RCH_2Br + NaOCH_3$, methanol	8140	906	335	67	
RCH_2OTs, acetic acid	0.052	0.044		0.018	0.0042

a. From M. Charton, *J. Am. Chem. Soc.* **97**, 3694 (1975).

Because a neutral molecule is eliminated, rather than an anion, there is no electrostatic attraction (ion pairing) between the products of the dissociation step. As a result, the carbocations generated by diazonium ion decomposition frequently exhibit rather different behavior from those generated from halides or sulfonates under solvolytic conditions.[64]

5.7. Steric and Strain Effects on Substitution and Ionization Rates

Examples of effects of substrate structure on the rate of nucleophilic substitution reactions have appeared in the preceding sections of this chapter. The general trends of reactivity of primary, secondary, and tertiary systems and the special reactivity of allylic and benzylic systems have been discussed in several contexts. This section will emphasize the role that steric effects can play in nucleophilic substitution reactions.

Reactions with good nucleophiles in solvents of low ionizing power are sensitive to the degree of substitution at the carbon atom undergoing reaction. Reactions that proceed by the direct displacement mechanism are retarded by increased steric repulsions at the transition state. This is the principal cause for the relative reactivities of methyl, ethyl, and 2-propyl chloride, which are, for example, in the ratio 93 : 1 : 0.0076 toward iodide ion in acetone.[65] A statistical analysis of rate data for 18 sets of nucleophilic substitution reactions of substrates of the type RCH_2Y, where Y is a leaving group and R is H or alkyl, indicated that steric effects of R were the dominant factor in determining rates.[66] Table 5.11 records some of the data. Notice that the fourth entry, involving solvolysis in acetic acid, shows a diminished sensitivity to steric effects. This reflects a looser transition state with less nucleophilic participation than the other examples. This reaction, which would proceed through a transition state with a large degree of carbocation character, involves a weaker

64. C. J. Collins, *Acc. Chem. Res.* **4**, 315 (1971); A. Streitwieser, Jr., *J. Org. Chem.* **22**, 861 (1957).
65. J. B. Conant and R. E. Hussey, *J. Am. Chem. Soc.* **47**, 476 (1925).
66. M. Charton, *J. Am. Chem. Soc.* **97**, 3694 (1975).

interaction with the nucleophile. The relative rates of formolysis of alkyl bromides at 100°C are methyl, 0.58; ethyl, 1.000; 2-propyl, 26.1; and *t*-butyl, 10^8.[67] This order is clearly dominated by carbocation stability. The effect of substituting a methyl group for hydrogen can be seen on the basis of this type of data to depend on the extent of nucleophilic participation in the transition state. A large CH_3/H rate ratio is expected if nucleophilic participation is weak and stabilization of the cationic nature of the transition state is important. A low ratio is expected when nucleophilic participation is strong.

The relative rate of acetolysis of *t*-butyl bromide as compared to 2-propyl bromide at 25°C is $10^{3.7}$, while that of 2-methyl-2-adamantyl bromide as compared to 2-adamantyl bromide is $10^{8.1}$:

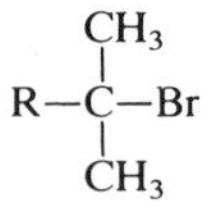

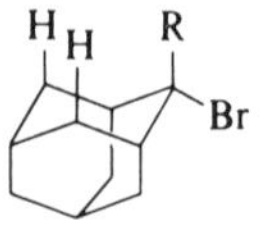

$$R-C(CH_3)_2-Br \qquad k_{rel}\ \frac{R=CH_3}{R=H},\ 10^{3.7}$$

$$\text{2-R-2-adamantyl bromide} \qquad k_{rel}\ \frac{R=CH_3}{R=H},\ 10^{8.1}$$

Ref. 68

The reason that the adamantyl system is much more sensitive to the substitution of CH_3 for H is that it has no nucleophilic solvent participation while the 2-propyl system has much stronger solvent participation. As was discussed earlier, the adamantyl structure effectively shields the back side of the reacting carbon.

Steric effects of another kind become important in highly branched substrates, in which ionization is facilitated by relief of steric crowding in going from the tetrahedral ground state to the transition state for ionization.[69] The relative hydrolysis rates in 80% aqueous acetone of *t*-butyl *p*-nitrobenzoate and 2,3,3-trimethyl-2-butyl *p*-nitrobenzoate are 1:4.4:

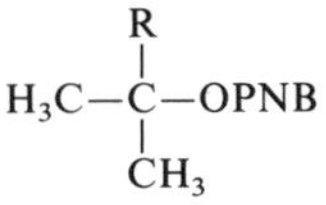

$$H_3C-C(R)(CH_3)-OPNB \qquad k_{rel}\ \frac{R=t\text{-butyl}}{R=CH_3},\ 4.4$$

The cause of this effect has been called *B-strain* (back strain), and in this example only a modest rate enhancement is observed. As the size of the groups is increased, the effect on rate becomes larger. When all three of the groups in the above example are *t*-butyl, the solvolysis occurs 13,500 times faster than in *t*-butyl *p*-nitrobenzoate.[70]

Large B-strain effects are observed in rigid systems such as the 2-alkyl-2-adamantyl *p*-nitrobenzoates. Table 5.12 shows the pertinent data. The repulsive van der Waals interaction between the substituent and the *syn*-axial hydrogens is relieved as the hybridization at C-2 goes from sp^3 to sp^2. As the alkyl group becomes more

67. L. C. Bateman and E. D. Hughes, *J. Chem. Soc.*, 1187 (1937); 945 (1940).
68. J. L. Fry, J. M. Harris, R. C. Bingham, and P. v. R. Schleyer, *J. Am. Chem. Soc.* **92**, 2540 (1970).
69. H. C. Brown, *Science* **103**, 385 (1946); E. N. Peters and H. C. Brown, *J. Am. Chem. Soc.* **97**, 2892 (1975).
70. P. D. Bartlett and T. T. Tidwell, *J. Am. Chem. Soc.* **90**, 4421 (1968).

Table 5.12. Relative Hydrolysis Rates of 2-Alkyl-2-adamantyl *p*-Nitrobenzoates[a]

R	k_{rel}, 25°C[b]
CH_3–	2.0
CH_3CH_2–	15.4
$(CH_3)_3CCH_2$–	20.0
$(CH_3)_2CH$–	67.0
$(CH_3)_3C$–	4.5×10^5

a. From J. L. Fry, E. M. Engler, and P. v. R. Schleyer, *J. Am. Chem. Soc.* **94**, 4628 (1972).
b. Relative to *t*-butyl *p*-nitrobenzoate = 1.

sterically demanding, the ground state energy is increased more than the transition state energy by the steric repulsion and as a result reactivity is enhanced.

Another feature of systems that are subject to B-strain is their reluctance to form strained products. The cationic intermediates usually escape to elimination products in preference to capture of a nucleophile. Rearrangements are also common. 2-Methyl-2-adamantyl *p*-nitrobenzoate gives 82% methylene adamantane by elimination and 18% 2-methyl-2-adamantol by substitution in aqueous acetone. Elimination accounts for 95% of the product from 2-neopentyl-2-adamantyl *p*-nitrobenzoate. The major product (83%) from 2-*t*-butyl-2-adamantyl *p*-nitrobenzoate is the rearranged alkene **5**:

$C(CH_3)_3$, OPNB → CH_3, $C{=}CH_2$, CH_3 **5** Ref. 71

The role of flexibility and strain in determining the reactivity of tertiary bridgehead systems was mentioned briefly on p. 282. This relationship has been tested extensively, and it has been shown that reactivity of bridgehead substrates is correlated with ring strain.[72] This result implies that the increased energy associated with a nonplanar carbocation is proportional to the strain energy present in the ground state reactant.

71. J. L. Fry, E. M. Engler, and P. v. R. Schleyer, *J. Am. Chem. Soc.* **94**, 4628 (1972).
72. T. W. Bentley and K. Roberts, *J. Org. Chem.* **50**, 5852 (1985); R. C. Bingham and P. v. R. Schleyer, *J. Am. Chem. Soc.* **93**, 3189 (1971).

Table 5.13. α-Substituent Effects[a]

$X\text{-}CH_2Cl + I^- \rightarrow X\text{-}CH_2I + Cl^-$

X	Relative rate	X	Relative rate
$CH_3CH_2CH_2$–	1	$PhC(=O)$–	3.2×10^4
$PhSO_2$–	0.25	N≡C–	3×10^3
$CH_3C(=O)$–	3.5×10^4	$C_2H_5OC(=O)$–	1.7×10^3

a. Data from F. G. Bordwell and W. T. Branner, Jr., *J. Am. Chem. Soc.* **86**, 4545 (1964).

5.8. Substituent Effects on Reactivity

In addition to steric effects, there are other important substituent effects which determine both the rate and mechanism of nucleophilic substitution reactions. We mentioned on p. 275 that arylmethyl (benzylic) and allylic cations are stabilized by electron delocalization. It is therefore easy to understand why substitution reactions of the ionization type proceed more rapidly in such systems than in simple alkyl systems. It has been observed as well that direct displacement reactions also take place particularly rapidly in benzylic and allylic systems. Allyl chloride is 33 times more reactive than ethyl chloride toward iodide ion in acetone, for example.[73] These enhanced rates reflect stabilization of the S_N2 transition state through overlap between the adjacent π orbitals of the p-type orbital which develops at the α-carbon in the transition state.[74] The π systems of the allylic and benzylic groups provide extended conjugation.

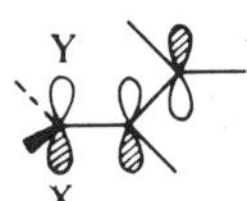

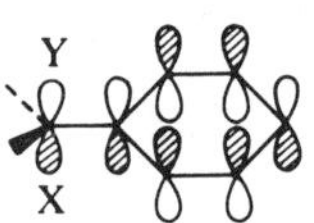

Substitution reactions by the ionization mechanism proceed very slowly on α-halo derivatives of ketones, aldehydes, acids, esters, nitriles, and related compounds. As discussed on p. 277, such substituents strongly destabilize a carbocation intermediate. Substitution by the direct displacement mechanism, however, proceeds especially readily in these systems. Table 5.13 gives some representative relative rate accelerations. Steric effects may be responsible for part of the observed acceleration, since an sp^2 carbon, such as in a carbonyl group, will provide less steric resistance to the incoming nucleophile than an alkyl group. The major effect is believed to be of an electronic nature. The adjacent π-LUMO of the carbonyl group can interact

73. J. B. Conant and R. E. Hussey, *J. Am. Chem. Soc.* **47**, 476 (1925).
74. A. Streitwieser, Jr., *Solvolytic Displacement Reactions*, McGraw-Hill, New York, 1962, p. 13; F. Carrion and M. J. S. Dewar, *J. Am. Chem. Soc.* **106**, 3531 (1984).

with the electron density that is built up at the pentacoordinate carbon. This can be described in resonance terminology as a contribution from an enolate-like structure to the transition state. In molecular orbital terminology, the low-lying LUMO has a stabilizing interaction with the developing *p* orbital of the transition state.[75]

resonance representation of electronic interaction with carbonyl group at the transition state for substitution which delocalizes negative charge

MO representation of stabilization by interaction with π^* orbital

It should be noted that not all electron-attracting groups enhance reactivity. The sulfonyl and trifluoro groups, which cannot participate in this type of conjugation, retard the rate of S_N2 substitution at an adjacent carbon.[76]

The extent of the rate enhancement due to adjacent substituents is dependent on the nature of the transition state. The most important factor is the nature of the π-type orbital which develops at the trigonal bipyramidal carbon in the transition state. If this carbon is cationic in character, electron donation from adjacent substituents becomes stabilizing. If bond formation at the transition state is far advanced, electron withdrawal should be more stabilizing. Substituents such as carbonyl therefore have their greatest effect on reactions with strong nucleophiles. Adjacent alkoxy substituents can stabilize S_N2 transition states that are cationic in character. Since the vinyl and phenyl groups can stabilize either type of transition state, the allyl and benzyl systems show enhanced reactivity toward both strong and weak nucleophiles.[77]

5.9. Stereochemistry of Nucleophilic Substitution

Studies of the stereochemical course of nucleophilic substitution reactions have proven to be a powerful tool for investigation of nucleophilic substitution reactions. Bimolecular direct displacement reactions are expected to result in 100% inversion of configuration. The stereochemical outcome of the ionization mechanism is less

75. R. D. Bach, B. A. Coddens, and G. J. Wolber, *J. Org. Chem.* **51**, 1030 (1986); F. Carrion and M. J. S. Dewar, *J. Am. Chem. Soc.* **106**, 3531 (1984); S. S. Shaik, *J. Am. Chem. Soc.* **105**, 4359 (1983); D. McLennon and A. Pross, *J. Chem. Soc., Perkin Trans. 2*, 981 (1984); T. I. Yousaf and E. S. Lewis, *J. Am. Chem. Soc.* **109**, 6137 (1987).
76. F. G. Bordwell and W. T. Brannen, *J. Am. Chem. Soc.* **86**, 4645 (1964).
77. D. N. Kost and K. Aviram, *J. Am. Chem. Soc.* **108**, 2006 (1986); S. S. Shaik, *J. Am. Chem. Soc.* **105**, 4359 (1983).

predictable since it depends on whether reaction occurs via one of the ion pair intermediates or through a dissociated ion. Borderline mechanisms may also show variable stereochemistry, depending upon the lifetime of the intermediates and the extent of internal return. It is important to dissect the overall stereochemical outcome into the various steps of such reactions.

Table 5.14 presents data on some representative nucleophilic substitution processes. The first entry illustrates the use of 1-butyl-1-*d* *p*-bromobenzenesulfonate to demonstrate that primary systems react with inversion even under solvolysis conditions in formic acid. The observation of inversion indicates a high degree of solvent participation, even with this weakly nucleophilic solvent.

Neopentyl (2,2-dimethylpropyl) systems are resistant to nucleophilic substitution reactions. They are primary, and so do not readily form carbocation intermediates, but the *t*-butyl substituent effectively hinders back-side attack. The rate of reaction of neopentyl bromide with iodide ion is 470 times less than that of *n*-butyl bromide.[78] Usually the neopentyl system reacts with rearrangement to the *t*-pentyl system. Use of good nucleophiles in polar aprotic solvents permits direct displacement to occur. Entry 2 shows that such a reaction with azide ion as the nucleophile proceeds with complete inversion of configuration.

On the other hand, the primary benzyl system in entry 3 exhibits high, but not complete, inversion. This is attributed to racemization of the reactant by ionization and internal return. Entry 4 shows that reaction of a secondary 2-octyl system with the moderately good nucleophile acetate ion occurs with complete inversion.

The results cited in entry 5 serve to illustrate the importance of solvation of ion pair intermediates in reactions of secondary substrates. The results show that partial racemization occurs in aqueous dioxane but that an added nucleophile (azide ion) results in complete inversion, both in the product resulting from reaction with azide ion and the alcohol resulting from reaction with water. The alcohol of retained configuration is attributed to an intermediate oxonium ion resulting from reaction of the ion pair with the dioxane solvent. This would react with water to give a product of retained configuration. When azide ion is present, dioxane does not effectively compete for the ion pair intermediate and all of the alcohol arises from the inversion mechanism.[79]

$$\mathrm{ROBs} \longrightarrow \mathrm{R^{+}\,{}^{-}OBs}$$

$$\mathrm{R^{+}\,{}^{-}OBs} \xrightarrow{H_2O} \mathrm{H_2\overset{+}{O}R} \xrightarrow{-H^{+}} \mathrm{HOR} \quad \text{inverted alcohol}$$

$$\mathrm{R^{+}\,{}^{-}OBs} \xrightarrow{\text{dioxane}} \mathrm{O(CH_2CH_2)_2\overset{+}{O}{-}R} \xrightarrow{H_2O} \mathrm{ROH} \quad \text{retained and/or racemized alcohol}$$

$$\mathrm{R^{+}\,{}^{-}OBs} \xrightarrow{N_3^{-}} \mathrm{N_3R} \quad \text{inverted azide}$$

Nucleophilic substitution in cyclohexyl systems is quite slow and is often accompanied by extensive elimination. The stereochemistry of substitution has been determined using a deuterium-labeled substrate (entry 6). In the example shown,

78. P. D. Bartlett and L. J. Rosen, *J. Am. Chem. Soc.* **64**, 543 (1942).
79. H. Weiner and R. A. Sneen, *J. Am. Chem. Soc.* **87**, 292 (1965).

the substitution process occurs with complete inversion of configuration. By NMR analysis it can be determined that there is about 15% of rearrangement by hydride shift accompanying solvolysis in acetic acid. This increases to 35% in formic acid and 75% in trifluoroacetic acid. The extent of rearrangement increases with decreasing solvent nucleophilicity, as would be expected.

$$\text{(cyclohexyl-OTs, H, D, H)} \xrightarrow{-\text{TsO}^-} \text{(cyclohexyl cation}^+\text{, H, D, H)}$$

Stabilization of a carbocation intermediate by benzylic conjugation, as in the 1-phenylethyl system shown in entry 8, leads to substitution with diminished stereospecificity. A thorough analysis of stereochemical, kinetic, and isotope effect data on solvolysis reactions of 1-phenylethyl chloride has been carried out.[80] The system has been analyzed in terms of the fate of the intimate ion pair and solvent-separated ion pair intermediates.

$$\begin{array}{ccccccc} RX & \underset{80}{\rightleftharpoons} & R^+X^- & \overset{13}{\rightleftharpoons} & R^+ \| X^- & \overset{0}{\rightleftharpoons} & R^+ + X^- \\ & & \uparrow\downarrow & & & & \\ XR & \overset{6}{\rightleftharpoons} & X^-R^+ & \overset{1}{\rightleftharpoons} & X^- \| R^+ & & \end{array}$$

From this analysis, it has been estimated that for every 100 molecules of 1-phenylethyl chloride that undergo ionization to an intimate ion pair (in trifluoroethanol), 80 return to starting material of retained configuration, 6 return to inverted starting material, and 13 go on to the solvent-separated ion pair.

As is evident from the result shown for the tertiary benzylic substrate 2-phenyl-2-butyl *p*-nitrobenzoate in entry 9, the simple expectation of complete racemization is not rigorously realized. In weakly nucleophilic media such as potassium acetate in acetic acid, this ideal is almost achieved, with just a slight excess of inversion. Use of a better nucleophile such as azide ion, however, leads to product with a significant (56%) degree of inversion. This can be attributed to nucleophilic attack on an ion pair intermediate prior to symmetrical solvation or dissociation. More surprising is the observation of net retention of configuration in the hydrolysis of optically active 2-phenyl-2-butyl *p*-nitrobenzoate in aqueous acetone. It is possible that this is the result of preferential solvent collapse from the front side at the solvent-separated ion pair stage. The bulky tertiary system may hinder solvation from the rear side. It is also possible that hydrogen bonding between a water molecule and the anion of the ion pair facilitates capture of a water molecule from the front side of the ion pair.

Nucleophilic substitution reactions that occur under conditions of amine diazotization often differ significantly in stereochemistry, as compared with that seen in halide or sulfonate solvolysis. Diazotization generates an alkyl diazonium

80. V. J. Shiner, Jr., S. R. Hartshorn, and P. C. Vogel, *J. Org. Chem.* **38**, 3604 (1973).

Table 5.14. Stereochemical Course of Nucleophilic Substitution Reactions

	Substrate[a]	Reaction conditions	Product[a]	Stereochemistry	Ref.
Primary					
1	$CH_3CH_2CH_2CHDOBs$	Acetic acid, 99°C Formic acid, 99°C	$CH_3CH_2CH_2CHDOAc$ $CH_3CH_2CH_2CHDOCHO$	96 ± 8% inversion 99 ± 6% inversion	b b
2	$(CH_3)_3CCHDOTs$	Sodium azide in hexa-methylphosphoramide, 90°C	$(CH_3)_3CCHDN_3$	98 ± 2% inversion	c
3	$C_6H_5CHDOTs$	Acetic acid, 25°C	$C_6H_5CHDOAc$	82 ± 1% inversion	d
Secondary					
4	$CH_3CH(OTs)(CH_2)_5CH_3$	Tetraethylammonium acetate in acetone, reflux	$CH_3CH(OAc)(CH_2)_5CH_3$	100% inversion	d
5	$CH_3CH(OBs)(CH_2)_5CH_3$	75% aqueous dioxane, 65°C	$CH_3CH(OH)(CH_2)_5CH_3$	77% inversion	e
		75% aqueous dioxane containing 0.06 *M* sodium azide, 65°C	$CH_3CH(OH)(CH_2)_5CH_3$	100% inversion	e
			$CH_3CH(N_3)(CH_2)_5CH_3$	100% inversion	e
6	H, OTs; D, D; H, D	Acetic acid	H, OAc; D, D; H, D	100% inversion	f

	Reactant	Conditions	Product	Stereochemistry	Ref.
7	D-cyclopentyl OBs (cis)	80% ethanol– water	D-cyclopentanol (OH, trans) + D-cyclopentyl OC_2H_5 (trans)	>97% inverson	g
8	$C_6H_5CH(Cl)CH_3$	Potassium acetate in acetic acid, 50°C	$C_6H_5CH(OAc)CH_3$	15% inversion	h
		Tetraethylammonium acetate in acetone, 50°C	$C_6H_5CH(OAc)CH_3$	65% inversion	h
		60% aqueous ethanol	$C_6H_5CH(OH)CH_3$	33% inversion	i
Tertiary					
9	$CH_3CH_2C(C_6H_5)(OPNB)CH_3$	Potassium acetate in acetic acid, 23°C	$CH_3CH_2C(C_6H_5)(OAc)CH_3$	5±2% inversion	j
		Sodium azide in methanol, 65°C	$CH_3CH_2C(C_6H_5)(N_3)CH_3$	56±1% inversion	j
			$CH_3CH_2C(C_6H_5)(OCH_3)CH_3$	14% inversion	j

continued

Table 5.14. continued.

Substrate	Reaction conditions	Product[a]	Stereochemistry	Ref.
	90% aqueous acetone	C_6H_5 on $CH_3CH_2C(OH)CH_3$	38% retention	k

a. Abbreviations used: OBs = *p*-bromobenzenesulfonate; OTs = *p*-toluenesulfonate; OAc = acetate; OPNB = *p*-nitrobenzoate.
b. A. Streitwieser, Jr., *J. Am. Chem. Soc.* **77**, 1117 (1955).
c. B. Stephenson, G. Solladie, and H. S. Mosher, *J. Am. Chem. Soc.* **94**, 4184 (1972).
d. A. Streitwieser, Jr., T. D. Walsh, and J. R. Wolfe, Jr., *J. Am. Chem. Soc.*, **87**, 3682 (1965).
e. H. Weiner and R. A. Sneen, *J. A. Chem. Soc.* **87**, 287 (1965).
f. J. B. Lambert, G. J. Putz, and C. E. Mixan, *J. Am. Chem. Soc.* **94**, 5132 (1972); see also J. E. Nordlander, and T. J. McCrary, *J. Am. Chem. Soc.* **94**, 5133 (1972).
g. K. Humski, V. Sendijarevic, and V. J. Shiner, *J. Am. Chem. Soc.* **98**, 2865 (1976); K. Humski, V. Sendijarevic, and V. J. Shiner, *J. Am. Chem. Soc.* **95**, 7722 (1973).
h. J. Steigman and L. P. Hammett, *J. Am. Chem. Soc.* **59**, 2536 (1937).
i. V. J. Shiner, Jr., S. R. Hartshorn, and P. C. Vogel, *J. Org. Chem.* **38**, 3604 (1973).
j. L. H. Sommer and F. A. Carey, *J. Org. Chem.* **32**, 800 (1967).
k. H. L. Goering and S. Chang, *Tetrahedron Lett.*, 3607 (1965).

ion, which rapidly decomposes to a carbocation, molecular nitrogen, and water:

$$R{-}NH_2 \rightarrow R{-}\underset{H}{N}{-}N{=}O \rightarrow R{-}N{=}N{-}OH \xrightarrow{H^+} R{-}\overset{+}{N}{\equiv}N + H_2O \rightarrow R^+ + N_2$$

Thus, in contrast to an ionization process from a neutral substrate, which initially generates an intimate ion pair, deamination reactions generate a cation which does not have an anion closely associated with it. The stereochemistry of substitution is shown for four representative systems in Table 5.15. Displacement on the primary 1-butyl system is much less stereospecific than the 100% inversion observed on acetolysis of the corresponding brosylate (entry 1, Table 5.14). Similarly, the 2-butyl diazonium ion affords 2-butyl acetate with only 28% net inversion of configuration. Small net retention is seen in the deamination of 1-phenylethylamine. The tertiary benzylic amine 2-phenyl-2-butylamine reacts with 24% net retention. These results indicate that the lifetime of the carbocation is so short that a symmetrically solvated state is not reached. Instead, the composition of the product is determined by a nonselective collapse of the solvent shell.

An analysis of the stereochemistry of deamination has also been done using the conformationally rigid 2-decalylamines:

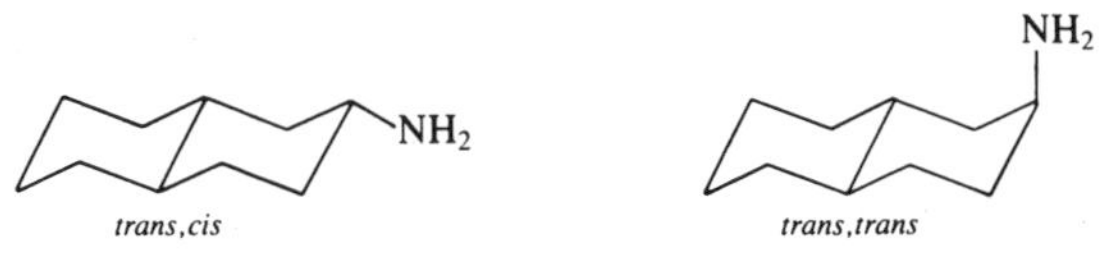

Table 5.15. Stereochemical Course of Deamination Reactions in Acetic Acid

$$RNH_2 \xrightarrow{NaNO_2,\, CH_3CO_2H} ROAc$$

	Amine	Stereochemistry of acetate ester formation
1[a]	$CH_3CH_2CH_2CHDNH_2$	69% inversion
2[b]	$CH_3CH(NH_2)CH_2CH_3$	28% inversion
3[c]	$C_6H_5CH(NH_2)CH_3$	10% retention
4[d]	$C_6H_5C(CH_3)(NH_2)CH_2CH_3$	24% retention

a. A. Streitwieser, Jr., and W. D. Schaeffer, *J. Am. Chem. Soc.* **79**, 2888 (1957).
b. K. B. Wiberg, Dissertation, Columbia University, 1950.
c. R. Huisgen and C. Ruchardt, *Justus Liebigs Ann. Chem.* **601**, 21 (1956).
d. E. H. White and J. E. Stuber, *J. Am. Chem. Soc.* **85**, 2168 (1963).

Table 5.16. Product Composition from Deamination of Stereoisomeric Amines

	Product composition[a]			
	Alcohol		Ester	
	Ret	Inv	Ret	Inv
cis-4-*t*-Butylcyclohexylamine (ax)[b]	33	8	25	33
trans-4-*t*-Butylcyclohexylamine (eq)[b]	43	2	43	12
trans,trans-2-Decalylamine (ax)[c]	26	2	32	40
trans,cis-2-Decalylamine (eq)[c]	18	1	55	26

a. Composition of total of alcohol and acetate ester. Considerable, and variable, amounts of alkene are also formed.
b. H. Maskill and M. C. Whiting, *J. Chem. Soc., Perkin Trans. 2,* 1462 (1976).
c. T. Cohen, A. D. Botelhjo, and E. Jankowski, *J. Org. Chem.* **45**, 2839 (1980).

In solvent systems containing low concentrations of water in acetic acid, dioxane, or sulfolane, most of the alcohol is formed by capture of water with retention of configuration. This result has been rationalized as involving a solvent-separated ion pair which would arise as a result of concerted protonation and nitrogen elimination.[81]

$$\begin{matrix} CH_3CO_2H \quad HO_2CCH_3 \\ R{-}N{=}N{-}OH \\ CH_3CO_2H \quad H{-}O_2CCH_3 \end{matrix} \rightarrow \begin{matrix} CH_3CO_2H \quad HO_2CCH_3 \\ R^+ \quad N{\equiv}N \quad OH_2 \\ CH_3CO_2H \quad {}^-O_2CCH_3 \end{matrix} \rightarrow \begin{matrix} CH_3CO_2H \quad HO_2CCH_3 \\ R{-}OH \\ CH_3CO_2H \quad HO_2CCH_3 \end{matrix}$$

In this process, the water molecule formed in the elimination step is captured primarily from the front side, leading to net retention of configuration for the alcohol. For the ester, the extent of retention and inversion is more similar, although it varies among the four systems. It is clear that the two stereoisomeric amines *do not form the same intermediate*, even though a simple mechanistic interpretation would suggest both would form the 2-decalyl cation. The collapse of the ions to product is evidently so rapid that there is not time for relaxation of the initially formed intermediates to reach a common structure. Similar results have been found for *cis*- and *trans*-4-*t*-butylcyclohexylamine. These data are included in Table 5.16.

A few nucleophilic substitution reactions have been observed to proceed with a high degree of retention of configuration. One example is reaction of alcohols with thionyl chloride, which under some conditions gives predominantly product of retained configuration. This reaction is believed to involve formation of a chlorosulfite ester. This can then react with chloride to give inverted product.

$$R{-}OH + Cl{-}\overset{\overset{O}{\|}}{S}{-}Cl \rightarrow R{-}O{-}\overset{\overset{O}{\|}}{S}{-}Cl + Cl^- \qquad Cl^- \; R{-}O{-}\overset{\overset{O}{\|}}{S}{-}Cl \longrightarrow Cl{-}R + SO_2 + HCl$$

81. (a) H. Maskill and M. C. Whiting, *J. Chem. Soc., Perkin Trans. 2,* 1462 (1976); (b) T. Cohen, A. D. Botelhjo, and E. Jankowski, *J. Org. Chem.* **45**, 2839 (1980).

When the reaction is done in dioxane solution, an oxonium ion is formed from the solvent and the chlorosulfite ester. The oxonium ion then undergoes substitution by chloride. Two inversions are involved so that the result is overall retention.[82]

5.10. Neighboring-Group Participation

When a molecule that is a potential substrate for nucleophilic substitution also carries a group that can act as a nucleophile, it is often observed that the kinetics and stereochemistry of nucleophilic substitution are strongly affected. The involvement of nearby nucleophilic substituents in a substitution process is called *neighboring-group participation.*[83]

A classic example of neighboring-group participation involves the solvolysis of compounds in which an acetoxy substituent is present near a carbon that is undergoing nucleophilic substitution. For example, the rates of solvolysis of the *cis* and *trans* isomers of 2-acetoxycyclohexyl *p*-toluenesulfonate differ by a factor of about 670, the *trans* compound being the more reactive[84]:

Besides the pronounced difference in rate, the isomeric compounds exhibit a marked difference in the stereochemistry of solvolysis. The diacetate obtained from the *cis* isomer is the *trans* compound (inverted stereochemistry), whereas retention of

82. E. S. Lewis and C. E. Boozer, *J. Am. Chem. Soc.* **74**, 308 (1952).
83. B. Capon, *Q. Rev. Chem. Soc.* **18**, 45 (1964); B. Capon and S. P. McManus, *Neighboring Group Participation*, Plenum Press, New York, 1976.
84. S. Winstein, E. Grunwald, R. E. Buckles, and C. Hanson, *J. Am. Chem. Soc.* **70**, 816 (1948).

configuration is observed for the *trans* isomer.

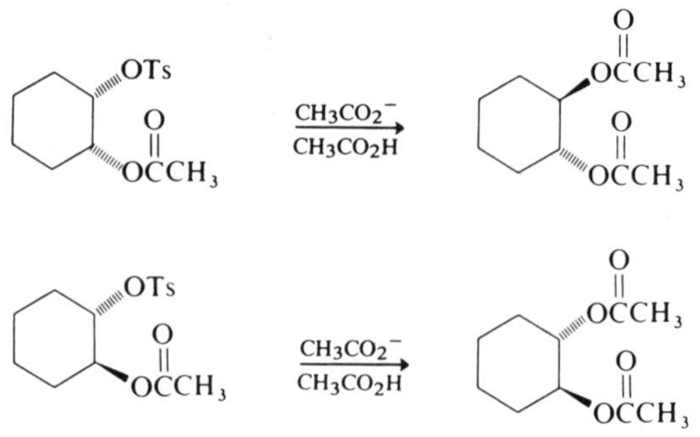

$CH_3CO_2^-$ / CH_3CO_2H

$CH_3CO_2^-$ / CH_3CO_2H

These results can be explained by the *participation* of the *trans* acetoxy group in the ionization process. The assistance provided by the acetoxy carbonyl group facilitates the ionization of the tosylate group, accounting for the rate enhancement. The acetoxonium ion intermediate is subsequently opened by nucleophilic attack with inversion at one of the two equivalent carbons, leading to the observed *trans* product.[85]

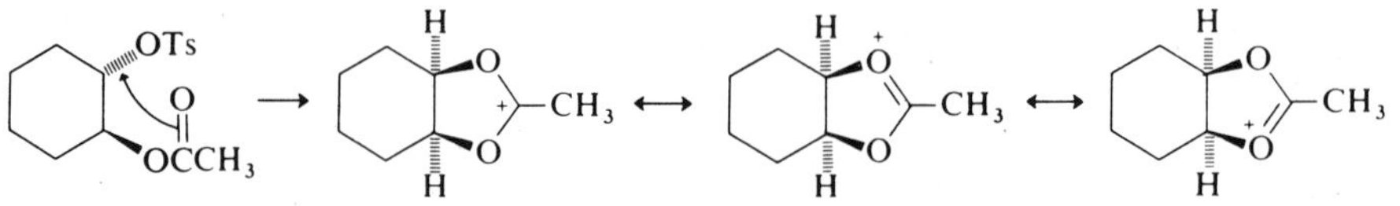

When optically active *trans*-2-acetoxycyclohexyl tosylate is solvolyzed, the product is racemic *trans*-diacetate. This is consistent with the proposed mechanism, since the acetoxonium intermediate is achiral and can only give rise to racemic material.[86] Additional evidence for this interpretation comes from the isolation of a cyclic orthoester when the solvolysis is carried out in ethanol. In this solvent the acetoxonium ion is captured by the solvent.

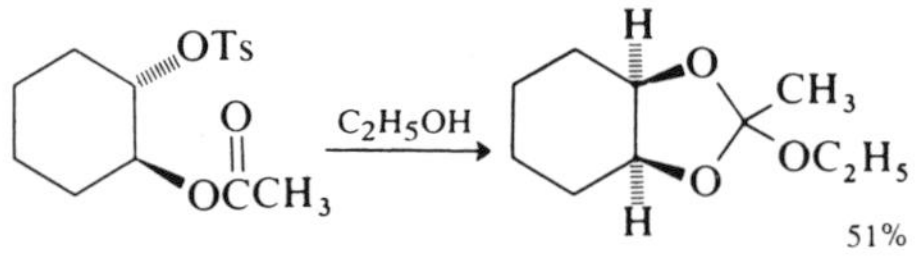

C_2H_5OH — 51% — Ref. 87

The hydroxy group can act as an intramolecular nucleophile. Solvolysis of 4-chlorobutanol in water gives as the product the cyclic ether tetrahydrofuran.[88]

85. S. Winstein, C. Hanson, and E. Grunwald, *J. Am. Chem. Soc.* **70**, 812 (1948).
86. S. Winstein, H. V. Hess, and R. E. Buckles, *J. Am. Chem. Soc.* **64**, 2796 (1942).
87. S. Winstein and R. E. Buckles, *J. Am. Chem. Soc.* **65**, 613 (1943).
88. H. W. Heine, A. D. Miller, W. H. Barton, and R. W. Greiner, *J. Am. Chem. Soc.* **75**, 4778 (1953).

The reaction is much faster than solvolysis of 3-chloropropanol under similar conditions.

$$Cl(CH_2)_4OH \xrightarrow{H_2O} \text{(tetrahydrofuran)} + HCl$$

In basic solution the alkoxide ions formed by deprotonation are still more effective nucleophiles. In ethanol containing sodium ethoxide, 2-chloroethanol reacts about 5000 times faster than ethyl chloride. The product is ethylene oxide, confirming the involvement of the oxygen atom.

$$HOCH_2CH_2Cl \rightleftharpoons {}^{-}OCH_2CH_2Cl \longrightarrow H_2C\overset{O}{—}CH_2 + Cl^-$$

As would be expected, the effectiveness of neighboring-group participation depends on the ease with which the molecular geometry required for participation can be achieved. The rate of cyclization of ω-hydroxyalkyl halides, for example, shows a strong dependence on the length of the chain separating the two substituents. Some data are given in Table 5.17. The maximum rate occurs for the 4-hydroxybutyl system, involving a five-membered ring. As discussed in Section 3.9, intramolecular processes involving five-membered ring formation are often the most rapid.

Like the unionized hydroxyl group, an alkoxy group is a weak nucleophile. Nevertheless, it can operate as a neighboring nucleophile. For example, solvolysis of the isomeric *p*-bromobenzenesulfonate esters **6** and **7** leads to identical product mixtures, suggesting the involvement of a common intermediate. This could be explained by involvement of the cyclic oxonium ion that would result from

Table 5.17. Solvolysis Rates of ω-Chloroalcohols[a]

ω-Chloroalcohols	Approximate relative rate
$Cl(CH_2)_2OH$	2000
$Cl(CH_2)_3OH$	1
$Cl(CH_2)_4OH$	5700
$Cl(CH_2)_5OH$	20

a. B. Capon, *Q. Rev. Chem. Soc.* **18**, 45 (1964); W. H. Richardson, C. M. Golino, R. H. Wachs, and M. B. Yelvington, *J. Org. Chem.* **36**, 943 (1971).

Table 5.18. Relative Solvolysis Rates of Some ω-Methoxyalkyl *p*-Bromobenzenesulfonates in Acetic Acid[a]

$CH_3(CH_2)_2OSO_2Ar$	1.00
$CH_3O(CH_2)_2OSO_2Ar$	0.28
$CH_3O(CH_2)_3OSO_2Ar$	0.67
$CH_3O(CH_2)_4OSO_2Ar$	657
$CH_3O(CH_2)_5OSO_2Ar$	123
$CH_3O(CH_2)_6OSO_2Ar$	1.16

a. From S. Winstein, E. Allred, R. Heck, and R. Glick, *Tetrahedron* **3**, 1 (1958).

intramolecular participation.[89]

$CH_3OCH_2CH_2CH_2CH(OSO_2Ar)CH_3$ (**6**) *or* $ArSO_3CH_2CH_2CH_2CH(OCH_3)CH_3$ (**7**) → cyclic oxonium ion (O^+–CH_3, 2-CH_3) $\xrightarrow{ROH}$ $ROCH_2CH_2CH_2CH(OCH_3)CH_3$ (**8**) + $CH_3OCH_2CH_2CH_2CH(OR)CH_3$ (**9**)

The occurrence of nucleophilic participation is also indicated by a rate enhancement (~4000-fold) relative to solvolysis of *n*-butyl *p*-bromobenzenesulfonate. The solvolysis rates of a series of ω-methoxyalkyl *p*-bromobenzenesulfonates have been determined. A maximum rate is again observed where participation of a methoxy group via a five-membered ring is possible (see Table 5.18).

Transannular participation of ether oxygen has also been identified by kinetic studies of a series of cyclic ethers. The relative rates for compounds **10–13** show that there is a large acceleration in the case of replacement of the 5-CH_2 group by an ether oxygen.[90]

	10	**11**	**12**	**13**
relative rate:	1.0	0.014	0.14	4.85×10^4

The huge difference in rate that results from the alternative placements of oxygen in the eight-membered rings reflects the relative stability of the various oxonium

89. E. L. Allred and S. Winstein, *J. Am. Chem. Soc.* **89**, 3991 (1967).
90. L. A. Paquette and M. K. Scott, *J. Am. Chem. Soc.* **94**, 6760 (1972).

ions that result from participation. The ion **16** is much more favorable than **14** or **15**.

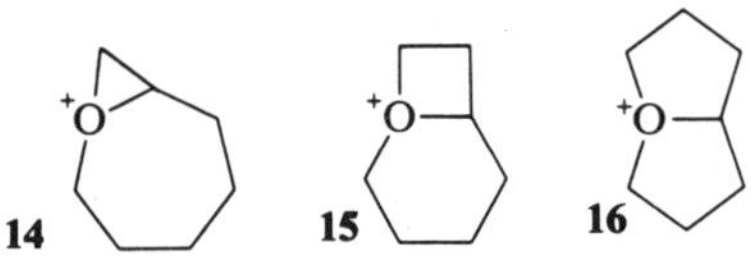

The rate retardation evident for **11** and **12** can be attributed to an unfavorable inductive effect of the C—O bond.

In general, any system that has a potentially nucleophilic substituent group situated properly for back-side displacement of a leaving group at another carbon atom of the molecule can be expected to display neighboring-group participation. The extent of the rate enhancement will depend on how effectively the group acts as an internal nucleophile. The existence of participation may be immediately obvious from the structure of the product if some derivative of the cyclic intermediate is stable. In other cases, demonstration of kinetic acceleration or stereochemical consequences may provide the basis for identifying neighboring-group participation.

The π electrons of carbon–carbon double bonds can also become involved in nucleophilic substitution processes. This can result in facilitation of the ionization step and may lead to a carbocation having special stability. Solvolysis reactions of the *syn* and *anti* isomers of 7-substituted norbornenes provide some dramatic examples of the influence of participating double bonds on reaction rates and stereochemistry. The *anti*-tosylate is more reactive by a factor of about 10^{11} toward acetolysis than the saturated analog. The acetolysis product, *anti*-7-acetoxynorbornene, is the product of retention of configuration. These results can be explained by participation of the π electrons of the double bond to give the ion **17**, which would be stabilized by delocalization of the positive charge.[91]

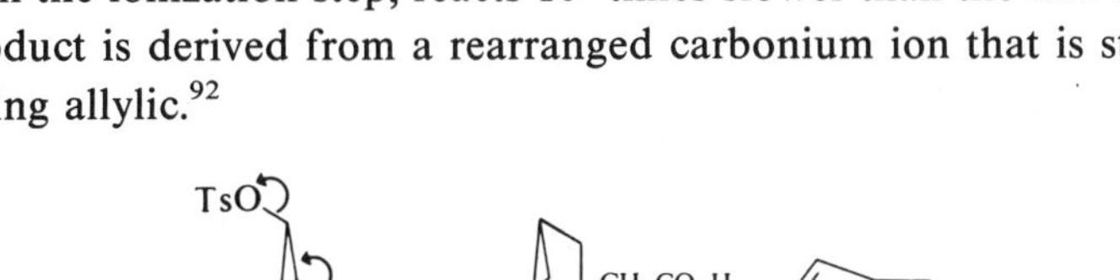

In contrast, the *syn* isomer, in which the double bond is not in a position to participate in the ionization step, reacts 10^7 times slower than the *anti* isomer. The reaction product is derived from a rearranged carbonium ion that is stabilized by virtue of being allylic.[92]

91. S. Winstein, M. Shavatsky, C. Norton, and R. B. Woodward, *J. Am. Chem. Soc.* **77**, 4183 (1955); S. Winstein and M. Shavatsky, *J. Am. Chem. Soc.* **78**, 592 (1956); S. Winstein, A. H. Lewin, and K. C. Pande, *J. Am. Chem. Soc.* **85**, 2324 (1963).
92. S. Winstein and E. T. Stafford, *J. Am. Chem. Soc.* **79**, 505 (1957).

The extent of participation of the carbon–carbon double bond in the ionization of *anti*-7-norbornenyl systems is a function of the substitution at C-7. The placement of an aryl substituent at C-7 diminishes the relative rate acceleration due to participation by the double bond. Evidently, the extent of participation is a function of the stability of the potential carbocation. When an aryl group is present at C-7, the resulting benzyl-type stabilization decreases the importance of participation by the double bond. The degree of stabilization is sensitive to substituents on the phenyl ring. For *p*-methoxyphenyl, phenyl, and *p*-trifluoromethylphenyl, the rate factor for the unsaturated relative to the saturated system is 3, 40, and 3.5×10^4, respectively.[93] The double bond clearly has a much larger effect on the poorly stabilized *p*-trifluoromethyl-substituted system. This dependence of the extent of participation on other stabilizing features is a general trend and has been observed with other types of carbocations.[94]

X
O_2CAr
dioxane
H_2O
X
OH

Participation of π electrons from an adjacent double bond controls the stereochemistry of substitution in the case of cyclopent-3-enyl tosylates, even though no strong rate enhancement is observed. The stereochemistry has been demonstrated by solvolysis of a stereospecifically labeled analog.[95] The product of formolysis is formed with complete retention of configuration, in contrast to the saturated system, which reacts with complete inversion under similar conditions.[96] The retention of configuration is explained by a structure similar to that shown in the case of the *anti*-7-norbornenyl cation.

D D
OTs
D
H
D
HCO_2H
D D
O_2CH
D
H
D

δ^+ δ^-
OTs
HCO_2H
O_2CH

Evidently, since there is no appreciable rate acceleration, this participation is not very strong at the transition state. Instead, the bridging must arise after the ionization is essentially complete. In fact, when more nucleophilic solvents are used, e.g., acetic acid, participation is not observed and the product is 100% of inverted configuration.

Participation of carbon–carbon double bonds in solvolysis reactions is revealed in some cases by isolation of products with new carbon–carbon σ bonds. A par-

93. P. G. Gassman and A. F. Fentiman, Jr., *J. Am. Chem. Soc.* **91**, 1545 (1969); **92**, 2549 (1970).
94. H. C. Brown, *The Nonclassical Ion Problem*, Plenum Press, New York, 1977, pp. 163–175.
95. J. B. Lambert and R. B. Finzel, *J. Am. Chem. Soc.* **105**, 1954 (1983).
96. K. Humski, V. Sendijarevič, and V. J. Shiner, Jr., *J. Am. Chem. Soc.* **95**, 7722 (1973).

ticularly significant case is the formation of the bicyclo[2.2.1]heptane system during solvolysis of 2-cyclopent-3-enylethyl tosylate[97]:

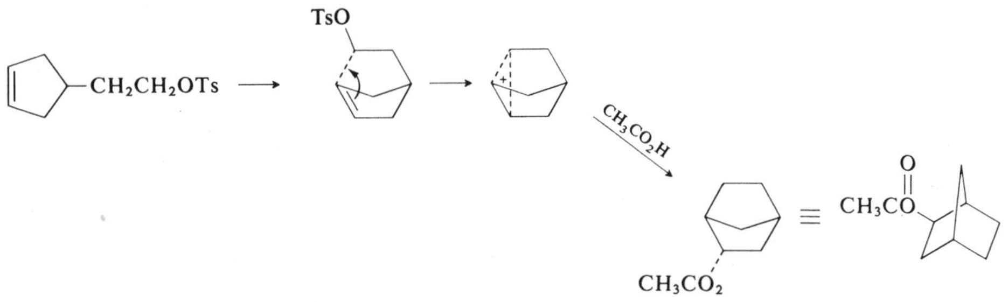

CH_2CH_2OTs TsO CH_3CO_2H O CH_3CO CH_3CO_2

In this case, the participation leads to the formation of the norbornyl cation, which is captured as the acetate. More will be said later about this important cation in Section 5.12.

A system in which the details of aromatic π-electron participation has been thoroughly probed is the case of the "phenonium" ions, the species resulting from participation by a β-phenyl group.

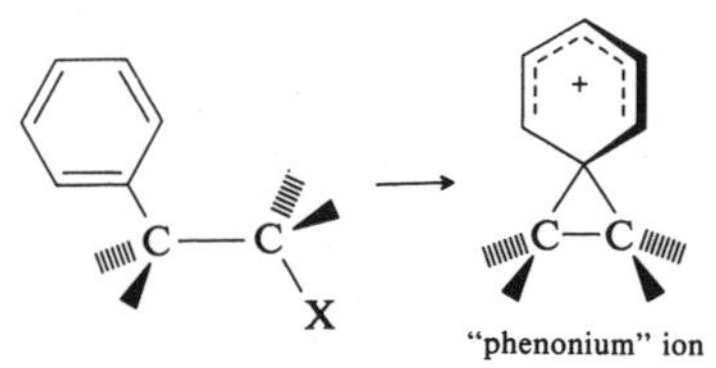

X

"phenonium" ion

Such participation leads to a bridged ion with the positive charge delocalized into the aromatic ring. Evidence for this type of participation was first obtained in a study in which the stereochemistry of solvolysis of 3-phenyl-2-butyl tosylates was studied. The *erythro* isomer gave largely retention of configuration, a result that can be explained via the bridged ion intermediate. The *threo* isomer, where participation leads to an achiral intermediate, gave racemic *threo* product.[98]

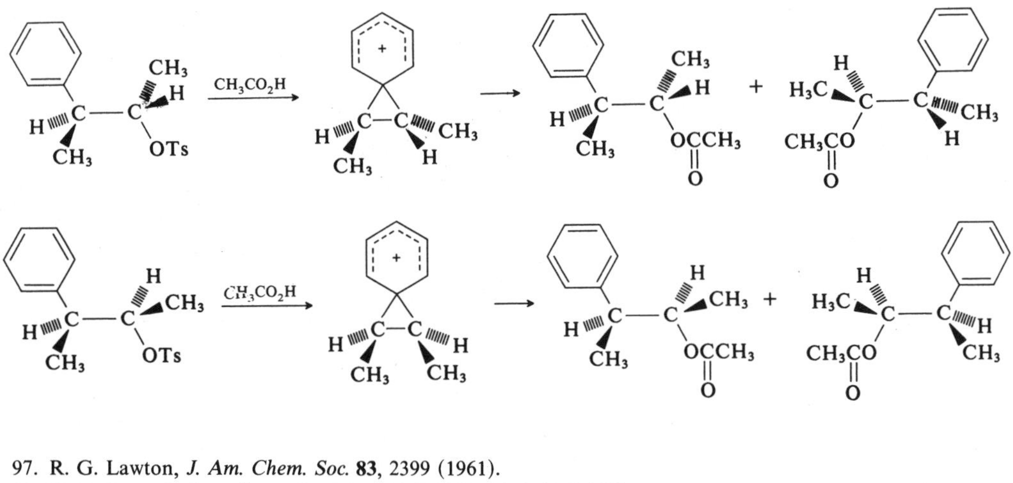

CH_3 H OTs CH_3CO_2H CH_3 H $OCCH_3$ O + H_3C CH_3CO O CH_3 H

H CH_3 OTs CH_3CO_2H H CH_3 $OCCH_3$ O + H_3C CH_3CO O H CH_3

97. R. G. Lawton, *J. Am. Chem. Soc.* **83**, 2399 (1961).
98. D. J. Cram, *J. Am. Chem. Soc.* **71**, 3863 (1949); **74**, 2129 (1952).

Table 5.19. Extent of Aryl Rearrangement in 2-Phenylethyl Tosylate Solvolysis[a]

Solvent	Rearrangement (%)
C_2H_5OH	0.3
CH_3CO_2H	5.5
$H_2O:HCO_2H$ (10:90)	40
HCO_2H	45
CF_3CO_2H	100

a. C. C. Lee, G. P. Slater, and J. W. T. Spinks, *Can. J. Chem.* **35**, 1417 (1957); J. E. Nordlander and W. G. Deadman, *Tetrahedron Lett.*, 4409 (1967).

Both primary and secondary carbocations with β-phenyl substituents usually give evidence of aryl participation. For example, isotopically labeled carbons are scrambled to some extent during solvolysis of β-phenylethyl tosylates. A bridged-ion intermediate or rapidly reversible rearrangement of a primary carbocation could account for the randomization of the label. The extent of label scrambling increases as solvent nucleophilicity decreases. The data are shown in Table 5.19. This trend can be attributed to competition between S_N2 displacement by solvent and ionization with participation of the aryl group. While substitution in more nucleophilic solvents such as ethanol proceeds almost exclusively by direct displacement, the non-nucleophilic solvent trifluoroacetic acid leads to complete randomization of the label.

$$\mathrm{Ph^*CH_2CH_2OTs} \xrightarrow[k_s]{\mathrm{SOH}} \mathrm{Ph^*CH_2CH_2OS + TsOH}$$

$$\mathrm{Ph^*CH_2CH_2OTs} \xrightarrow{k_\Delta} [\text{bridged ion}, \mathrm{H_2C{-}CH_2}] \xrightarrow{\mathrm{SOH}} \mathrm{Ph^*CH_2CH_2OS + PhCH_2{}^*CH_2OS + TsOH}$$

The relative importance of aryl participation is a function of the substituents on the aryl ring. The extent of participation can be quantitatively measured by comparing the rate of direct displacement, k_s, with the rate of aryl-assisted solvolysis, designated k_Δ.[99] The relative contributions to individual solvolyses can be dissected by taking advantage of the higher sensitivity to aryl substituent effects of the assisted mechanism. In systems with electron-withdrawing substituents, the aryl ring does not participate effectively and only the process described by k_s contributes to the rate. Such compounds give a Hammett correlation with ρ values (-0.7 to -0.8) characteristic of a weak substituent effect. Compounds with electron-releasing substituents deviate from the correlation line because of the aryl participation. The extent of reaction proceeding through the k_s process can be estimated from the correlation line for electron-withdrawing substituents. Table 5.20 gives data indicating the extent of aryl participation under a variety of conditions. This method of analysis also confirms that the relative extent of participation of the β-phenyl groups

99. A. Diaz, I. Lazdins, and S. Winstein, *J. Am. Chem. Soc.* **90**, 6546 (1968).

Table 5.20. Extent of Solvolysis with Aryl Participation as a Function of Substituent and Solvent for 1-Aryl-2-propyl Tosylates

	Solvent		
X	80% EtOH[a]	CH_3CO_2H[b]	HCO_2H[b]
NO_2	0	—	—
CF_3	0	—	—
Cl	7	—	—
H	21	38	78
CH_3	63	71	94
OCH_3	93	94	99

a. D. J. Raber, J. M. Harris, and P. von R. Schleyer, *J. Am. Chem. Soc.* **93**, 4829 (1971).
b. C. C. Lancelot and P. von R. Schleyer, *J. Am. Chem. Soc.* **91**, 4296 (1969).

is highly dependent on the solvent.[100] In solvents of good nucleophilicity (e.g., ethanol), the normal solvent displacement mechanism makes a larger contribution. As solvent nucleophilicity decreases, the relative extent of aryl participation increases.

The bridged form of the β-phenylethyl cation can be observed in superacid media and can be characterized by carbon-13 and proton NMR spectroscopy.[101] The bridged ion subsequently rearranges to the more stable α-methylbenzyl cation, with E_a for rearrangement being 13 kcal/mol. With more substituted cations, the rearrangement to benzyl cations occurs too rapidly for the bridged ion to be observed.[102]

5.11. Rearrangements of Carbocations

The discussion of the behavior of carbocation intermediates in superacid media and the discussion of neighboring-group participation have already provided examples of skeletal rearrangements. This is a characteristic feature of the chemistry of carbocations. Rearrangements can occur by shift of an alkyl group, aryl group, or hydrogen. Rearrangement creates a new cation with the positive charge located on the atom from which the migration occurred. 1,2-Shifts are the most common type of rearrangement.[103]

$$R_3C\overset{+}{C}HR' \longrightarrow R_2\overset{+}{C}CHR'(R) \quad \text{alkyl shift}$$

$$R_2\overset{+}{C}C(H)HR' \longrightarrow R_2\overset{+}{C}CH_2R' \quad \text{hydride shift}$$

100. F. L. Schadt III, C. J. Lancelot, and P. v. R. Schleyer, *J. Am. Chem. Soc.* **100**, 228 (1978).
101. G. A. Olah, R. J. Spear, and D. A. Forsyth, *J. Am. Chem. Soc.* **98**, 6284 (1976).
102. G. A. Olah, R. J. Spear, and D. A. Forsyth, *J. Am. Chem. Soc.* **99**, 2615 (1977).
103. V. G. Shubin, *Top. Curr. Chem.* **116–117**, 267 (1984).

A thermodynamic driving force exists for rearrangement of the carbon skeleton in the direction of forming a more stable structure. Activation energies for skeletal migrations are not large, and it is not uncommon to observe overall rearrangements that must have involved individual steps that proceeded from a more stable to a less stable species. Thus, while rearrangement of a tertiary cation to a secondary cation is endothermic by 10–15 kcal/mol, this barrier is not prohibitive if this rearrangement can lead to an ion of still greater stability. Formation of primary cations by rearrangement is less likely to occur, since the primary ions are ~20 and 30–35 kcal/mol higher in energy than secondary and tertiary cations, respectively.[104] The barriers to conversion of ions to more stable ions are apparently very low, and such rearrangements occur very rapidly. For example, in superacid media at −160°C, the equilibration of the five methyl groups of the 2,3,3-trimethylbutyl cation is so rapid that the barrier must be less than 5 kcal/mol[105]:

$$(CH_3)_3C-\overset{+}{C}(CH_3)_2 \rightleftharpoons (CH_3)_2\overset{+}{C}-C(CH_3)_3$$

The barrier to the hydride and methyl shifts that interconvert the methyl groups in the *t*-pentyl cation is 10–15 kcal/mol. This shift involves the formation of secondary ions as transient intermediates.[106]

The question of relative preference for migration of different groups, "migratory aptitude," is a complex one, and there is no absolute order. In general, aryl groups and branched alkyl groups migrate in preference to unbranched alkyl groups but, since the barriers involved are low, selectivity is not high. Very often, the preferred migration involves that group which is best positioned from a stereoelectronic point of view. The course of migration if also influenced by strain. In general, a shift that will reduce strain is favored.

The preferred alignment of orbitals for a 1,2-hydride or 1,2-alkyl shift involves coplanarity of the *p* orbital at the carbonium ion center and the σ orbital of the migrating group.

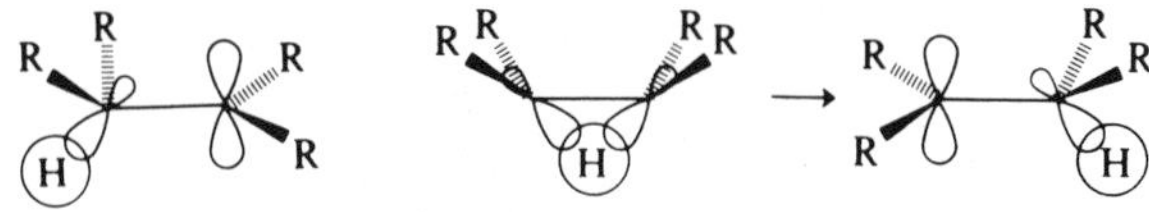

The transition state involves a three-center, two-electron bond. This corresponds to a symmetrically bridged structure, which in some cases may actually be an intermediate. The migration process can be concerted with the formation of the carbocation; that is, the migration can begin before the bond to the leaving group at the adjacent carbon atom is completely broken. The phenonium ion case discussed in the previous section is one example. When migration is concerted, the group that is aligned *anti* to the leaving group will migrate preferentially. In some cases, this alignment will be controlled by conformational equilibria. Conformation can also be a factor in migrations under nonconcerted conditions when the barrier to migration is of the

104. J. C. Schultz, F. A. Houle, and J. L. Beauchamp, *J. Am. Chem. Soc.* **106**, 3917 (1984).
105. G. A. Olah and A. M. White, *J. Am. Chem. Soc.* **91**, 5801 (1969).
106. M. Saunders and E. L. Hagen, *J. Am. Chem. Soc.* **90**, 2436 (1968).

same magnitude as the conformational barrier. Rearrangements following deamination reactions seem to be particularly likely to be governed by the conformation of the reactant, and this may reflect the high energy of the cations generated by deamination.[107]

The extent to which rearrangement occurs depends on the structure of the cation and the nature of the reaction medium. Capture of carbocations by nucleophiles is a process with a very low activation energy, so that only very rapid rearrangements can occur in the presence of nucleophiles. In contrast, in non-nucleophilic media, in which the carbocations have a longer lifetime, many rearrangement steps may occur. This accounts for the fact that the most stable possible ion is usually the one observed in superacid systems.

The occurrence and extent of rearrangement of carbocations can be studied effectively by isotopic labeling. One case which has been studied in this way is the 2-butyl cation. When 2-butyl tosylate is solvolyzed in acetic acid, only 9% rearrangement occurs in the 2-butyl acetate that is isolated.[108] Thus, under these conditions most of the reaction proceeds by direct participation of the solvent.

$$CH_3CH_2CH(OTs)^{14}CH_3 \xrightarrow{CH_3CO_2H} CH_3CH_2CH(O_2CCH_3)^{14}CH_3\ (91\%) + CH_3CH(O_2CCH_3)CH_2{}^{14}CH_3\ (9\%)$$

When 2-butyl tosylate is solvolyzed in the less nucleophilic trifluoroacetic acid, a different result emerges. The extent of migration approaches the 50% that would result from equilibration of the two possible secondary cations.[109]

$$CH_3CH_2CD(OTs)CD_3 \xrightarrow{CF_3CO_2H} CH_3CH_2CD(O_2CCF_3)CD_3\ (49\%) + CH_3CH(O_2CCF_3)CDHCD_3\ (45\%)$$
$$+ CH_3CHDCH(O_2CCF_3)CD_3\ (4\%) + CH_3CD(O_2CCF_3)CH_2CD_3\ (2\%)$$

Two hydride shifts resulting in interchange of the C-2 and C-3 hydrogens account for the two minor products.

$$CH_3CH_2CD(OTs)CD_3 \longrightarrow CH_3C(H)\overset{H}{\overset{+}{-}}C(D)CD_3 \rightleftarrows CH_3C(H)\overset{D}{\overset{+}{-}}C(H)CD_3 \rightleftarrows CH_3C(D)\overset{H}{\overset{+}{-}}C(H)CD_3$$
$$\searrow\ CH_3CHDCH(O_2CCF_3)CD_3 + CH_3CD(O_2CCF_3)CH_2CD_3$$

107. J. A. Berson, J. W. Foley, J. M. McKenna, H. Junge, D. S. Donald, R. T. Luibrand, N. G. Kundu, W. J. Libbey, M. S. Poonian, J. J. Gajewski, and J. B. E. Allen, *J. Am. Chem. Soc.* **93**, 1299 (1971).
108. J. D. Roberts, W. Bennett, R. E. McMahon, and E. W. Holroyd, Jr., *J. Am. Chem. Soc.* **74**, 4283 (1952).
109. J. J. Dannenberg, B. J. Goldberg, J. K. Barton, K. Dill, D. H. Weinwurzel, and M. O. Longas, *J. Am. Chem. Soc.* **103**, 7764 (1981); J. J. Dannenberg, J. K. Barton, B. Bunch, B. J. Goldberg, and T. Kowalski, *J. Org. Chem.* **48**, 4524 (1983); A. D. Allen, I. C. Ambridge, and T. T. Tidwell, *J. Org. Chem.* **48**, 4527 (1983).

In many systems the occurrence of rearrangement processes is evident from the structure of the products. Many neopentyl systems, for example, react to give *t*-pentyl products. This is very likely to occur under solvolytic conditions but can be avoided by adjusting reaction conditions to favor direct substitution, for example, by use of an aprotic dipolar solvent to enhance the reactivity of the nucleophile.[110]

The 4-cyclohexenyl ion provides another example of the dependence of the extent of rearrangement on reaction conditions. Acetolysis of the tosylate gives some product resulting from hydride shift to the more stable allylic ion as well as a trace of the bicyclo[3.1.0]hexane product arising from participation of the double bond.[111]

OTs $\xrightarrow{CH_3CO_2H}$ O_2CCH_3 (64%) + O_2CCH_3 (30%) + CH_3CO_2 (trace)

Deamination of the corresponding amine gives the allylic alcohol resulting from hydride shift as the main product and an increased amount of the cyclization product.

NH_2 $\xrightarrow{HONO}$ OH (21%) + OH (68%) + HO (11%)

Formation of the 3-cyclohexenyl cation from the alcohol in superacid media is followed by extensive rearrangement, giving the methylcyclopentenyl ion, which is tertiary and allylic.[112]

OH $\xrightarrow[SbF_5-SO_2ClF,\ -78°C]{FSO_3H}$ CH_3 (+)

This trend of increasing amount and extent of rearrangement can be readily interpreted. In the acetolysis, a large part of the reaction must be occurring via direct nucleophilic participation by the solvent or rapid ion pair capture so that only a relatively small amount of hydride shift occurs. As is characteristic of deamination, the carbocation formed from the amine is somewhat less tightly solvated at formation, and, as a result, the extent of rearrangement is greater. Finally, in the non-nucleophilic superacid media, the cations are relatively long-lived and undergo several rearrangements, eventually leading to the most stable accessible ion.

Not all carbocation rearrangements can be adequately accounted for by 1,2 shifts. For example, the ring contraction of a cyclohexyl cation to a methylcyclopentyl

110. B. Stephenson, G. Solladie, and H. S. Mosher, *J. Am. Chem. Soc.* **94**, 4184 (1972).
111. M. Hanack and W. Keverie, *Chem. Ber.* **96**, 2937 (1963).
112. G. A. Olah, G. Liang, and Y. K. Mo, *J. Am. Chem. Soc.* **94**, 3544 (1972).

cation is thermodynamically favorable, but would require a substantial energy of activation if the rearrangement proceeded through a primary cyclopentylmethyl cation.

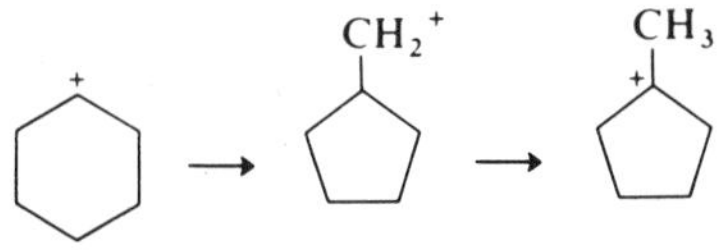

It is believed that a more correct description of the process involves migration through a pentacoordinate carbon species referred to as a *protonated cyclopropane.* A "corner-protonated cyclopropane" is a structure in which an alkyl group acts as a bridge in an electron-deficient carbocation structure. The cyclohexyl → methylcyclopentyl rearrangement is postulated to occur by rearrangement between two such structures.

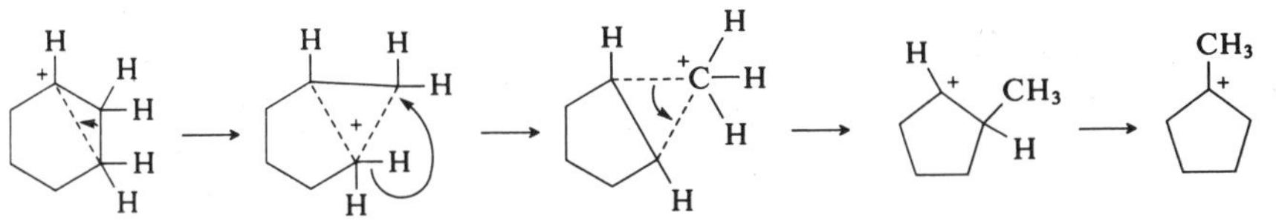

Shifts of hydride between carbon atoms separated by several atoms are possible, and particularly clear-cut examples have been found in medium-sized rings. For example, solvolysis of cyclononyl-1-^{14}C tosylate can be shown by degradation of the product cyclononene to occur with about 20% of the ^{14}C becoming located at C-5, C-6, and C-7 positions.

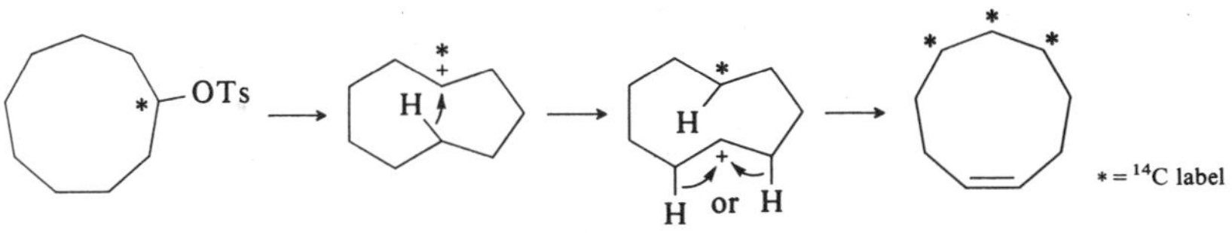

∗ = ^{14}C label

This result can be explained by a "transannular" 1,5-hydride shift. Many similar processes have been documented.[113]

Reaction of cyclooctene with trifluoroacetic acid occurs by a hydride shift process.

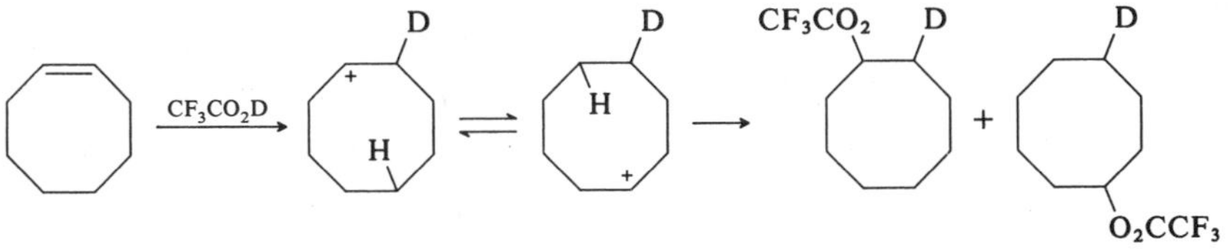

113. V. Prelog and J. G. Trayham, in *Molecular Rearrangements*, P. de Mayo (ed.), Wiley-Interscience, New York, 1963, p. 593.

The main reaction path is *stereospecific*, with the trifluoroacetate being added *syn* to the proton. This implies that the reaction proceeds through a discrete hydride-bridged intermediate.[114]

Hydride-bridged ions of this type are evidently quite stable in favorable cases and can be observed under stable ion conditions. The hydride-bridged cyclooctyl and cyclononyl cations can be observed at −150°C but rearrange, even at that temperature, to methylcycloheptyl and methylcyclooctyl ions, respectively.[115] However, a hydride-bridged ion in which the bridging hydride is located in a bicyclic cage is stable in trifluoroacetic acid at room temperature.[116]

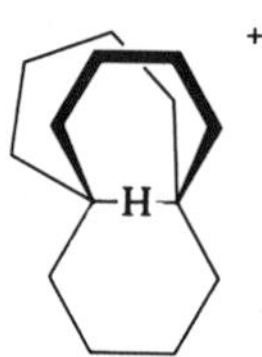

In some cases, NMR studies in superacid media have permitted the observation of successive intermediates in a series of rearrangements. An example is the series of cations originating with the bridgehead ion **E**, generated by ionization of the corresponding chloride. Rearrangement eventually proceeds to the tertiary ion **J**. The bridgehead ion is stable below −75°C. The unrearranged methyl ether is obtained if the solution is quenched with sodium methoxide in methanol at −90°C. At about −65°C, ion **E** rearranges to ion **I**. This is believed to involve the methyl-bridged ion **G** as an intermediate. Ion **I** is stable below −30°C, but above −30°C **J** is formed. This latter rearrangement involves a sequence of steps again including a methyl-bridged species.[117] This multistep sequence terminating in the most stable $C_9H_{15}^+$ ion is quite typical of carbocations in superacid media. In the presence of

114. J. E. Nordlander, K. D. Kotian, D. E. Raff II, F. G. Njoroge, and J. J. Winemiller, *J. Am. Chem. Soc.* **106**, 1427 (1984).
115. R. P. Kirchen and T. S. Sorensen, *J. Am. Chem. Soc.* **101**, 3240 (1979).
116. J. E. McMurry and C. N. Hodge, *J. Am. Chem. Soc.* **106**, 6450 (1984).
117. G. A. Olah, G. Liang, J. R. Wiseman, and J. A. Chong, *J. Am. Chem. Soc.* **94**, 4927 (1972).

nucleophilic anions or solvent, rearrangement usually does not proceed all the way to the most stable ion, because nucleophilic trapping captures one or more of the rearranged species.

SbF_5, FSO_3H, SO_2ClF; CH_3O^-; E; F; G; H; I; J

Carbocation rearrangement is particularly facile if a functional group can stabilize the rearrangement product. A good example of this is the rearrangement of carbocations having a hydroxyl group on an adjacent carbon atom. Migration in this case generates a protonated carbonyl compound and is very favorable. These rearrangements are referred to as "pinacol rearrangements."[118]

$$R_2C(OH)-\overset{+}{C}R_2 \longrightarrow RC(=\overset{+}{O}H)-CR_3 \xrightarrow{-H^+} RC(=O)CR_3$$

The rearrangement is frequently observed when geminal diols react with acid. The structure of the products from unsymmetrical diols can be predicted on the basis of ease of carbocation formation. For example, 1,1-diphenyl-2-methyl-1,1-propanediol rearranges to 3,3-diphenyl-2-butanone because the diarylcarbinol is most readily ionized.[119] Synthetically useful examples of this type of rearrangement are discussed in Section 10.1.2 of Part B.

$$Ph_2C(OH)C(CH_3)_2(OH) \xrightarrow{H^+} Ph_2C(OH)C(CH_3)_2(\overset{+}{O}H_2) \longrightarrow Ph_2\overset{+}{C}C(CH_3)_2(OH) \longrightarrow Ph_2C(CH_3)C(=\overset{+}{O}H)CH_3 \longrightarrow Ph_2C(CH_3)C(=O)CH_3 + H^+$$

5.12. The Norbornyl Cation and Other Nonclassical Carbocations

Throughout the discussion of carbocation structure and reactivity, we have encountered examples of bridged species that require expansion of bonding concepts

118. Y. Pocker, in *Molecular Rearrangements*, P. de Mayo (ed.), Wiley-Interscience, New York, 1963, pp. 15–25.
119. W. M. Schubert and P. H. LeFevre, *J. Am. Chem. Soc.* **94**, 1639 (1972).

beyond the two-center, two-electron bonds which suffice for most closed-shell organic molecules. These bridged structures, which involve delocalization of σ electrons and formation of three-center two-electron bonds, are called *nonclassical ions.* The case for the importance of such bridged structures largely originated with a specific structure, the norbornyl cation, and the issue of whether it has a classical or nonclassical (bridged) structure.[120] The special properties of this intermediate were first recognized on the basis of the thorough studies of Saul Winstein and his collaborators. The behavior of norbornyl systems in solvolytic displacement reactions was suggestive of neighboring-group participation. Evidence for both enhanced rate and abnormal stereochemistry was developed by study of the acetolysis of *exo*-2-norbornyl sulfonates.

The acetolyses of both *exo*-2-norbornyl brosylate and *endo*-2-norbonyl brosylate produce exclusively *exo*-2-norbornyl brosylate. The *exo*-brosylate is more reactive than the *endo* isomer by a factor of 350.[121] Furthermore, optically active *exo*-brosylate gave completely racemic *exo*-acetate, and the *endo*-brosylate gave acetate that was at least 93% racemic.

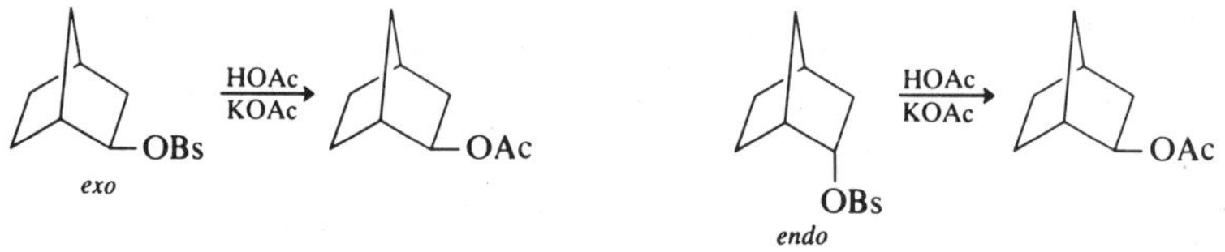

OBs (*exo*) —HOAc, KOAc→ OAc   OBs (*endo*) —HOAc, KOAc→ OAc

Both acetolyses were considered to proceed by way of a rate-determining formation of a carbocation. The rate of ionization of the *endo*-brosylate was considered normal, since its reactivity was comparable to that of cyclohexyl brosylate. Elaborating on a suggestion made earlier concerning rearrangement of camphene hydrochloride,[122] Winstein proposed that ionization of the *exo*-brosylate was assisted by the C-1—C-6 bonding electrons and led directly to the formation of a nonclassical ion as an *intermediate.*

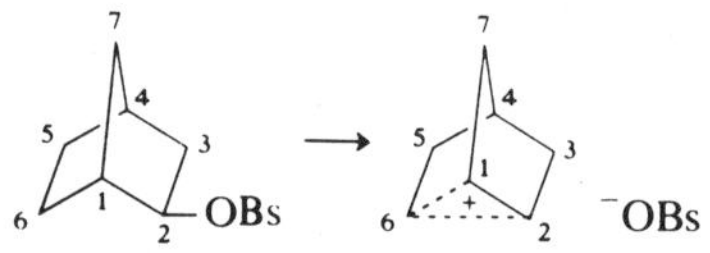

(7, 4, 5, 3, 6, 1, 2) OBs → (7, 4, 5, 3, 1, 6, +, 2) $^{-}$OBs

This intermediate serves to explain the formation of racemic product, since it is achiral. The cation has a plane of symmetry passing through C-4, C-5, C-6, and the

120. H. C. Brown, *The Nonclassical Ion Problem,* Plenum Press, New York, 1977; H. C. Brown, *Tetrahedron* **32**, 179 (1976); P. D. Bartlett, *Nonclassical Ions,* W. A. Benjamin, New York, 1965; S. Winstein, in *Carbonium Ions,* Vol. III, G. A. Olah and P. v. R. Schleyer (eds.), Wiley-Interscience, New York, 1972, Chapter 22; G. D. Sargent *ibid.,* Chapter 24; C. A. Grob, *Angew. Chem. Int. Ed. Eng.* **21**, 87 (1982).
121. S. Winstein and D. S. Trifan, *J. Am. Chem. Soc.* **71**, 2953 (1949); **74**, 1147, 1154 (1952); S. Winstein, E. Clippinger, R. Howe, and E. Vogelfanger, *J. Am. Chem. Soc.* **87**, 376 (1965).
122. T. P. Nevell, E. deSalas, and C. L. Wilson, *J. Chem. Soc.* 1188 (1939).

midpoint of the C-1—C-2 bond. The plane of symmetry is seen more easily in an alternative, but equivalent, representation.

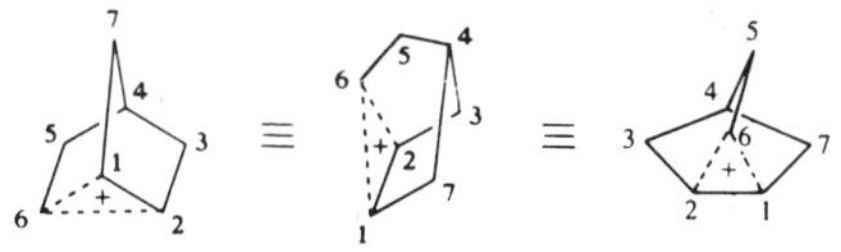

Carbon-6, which bears two hydrogens, is pentacoordinate and serves as the bridging atom in the cation.

Attack by acetate at C-1 or C-2 would be equally likely and would result in equal amounts of the enantiomeric acetates. The acetate ester would be *exo* since reaction must occur from the direction opposite that of the bridging interaction. The nonclassical ion can be formed directly only from the *exo*-brosylate because it has the proper *anti* relationship between the C-1—C-6 bond and the leaving group. The bridged ion can be formed from the *endo*-brosylate only after an unassisted ionization. This would explain the rate difference between the *exo* and *endo* isomers.

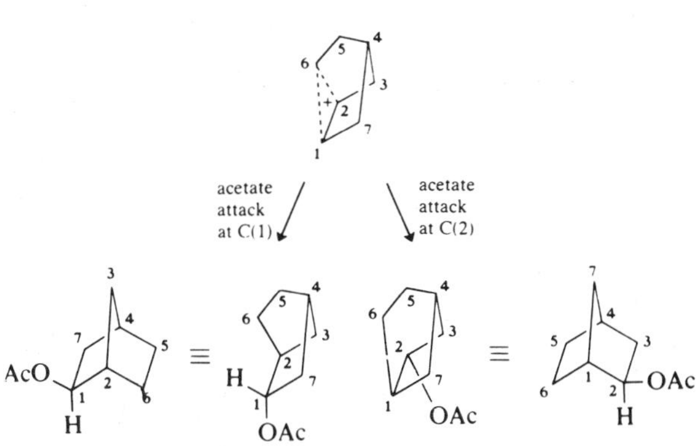

acetate attack at C(1)

acetate attack at C(2)

The nonclassical ion concept proved to be an intriguing one, and many tests for the intermediacy of the nonclassical norbornyl ion were employed.[123] Other bicyclic systems were also investigated to explore the generality of the concept of nonclassical ions. While the classical ion in the norbornyl system is chiral and the nonclassical ion is achiral, the situation is reversed in the bicyclo[2.2.2]octane system.

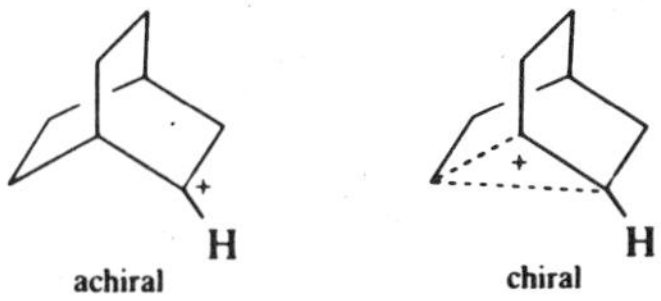

achiral

chiral

When bicyclo[2.2.2]octyl brosylate was solvolyzed in acetic acid containing sodium acetate, the products were a mixture of bicyclo[2.2.2]octyl acetate and

123. Much of the extensive work on the norbornyl cation is discussed in a series of reviews in *Accounts of Chemical Research*: C. A. Grob, *Acc. Chem. Res.* **16**, 426 (1983); H. C. Brown, *Acc. Chem. Res.* **16**, 432 (1983); G. A. Olah, G. K. SuryaPrakash, and M. Saunders, *Acc. Chem. Res.* **16**, 440 (1983); C. Walling, *Acc. Chem. Res.* **16**, 448 (1983).

bicyclo[3.2.1]octyl acetate, each of which was optically active. The stereochemistry of formation of bicyclo[2.2.2]octyl acetate was found to be 82 ± 15% *retention* of configuration, a result which is in accord with involvement of a bridged-ion intermediate.[124] The achiral classical cation could not have been the major intermediate.

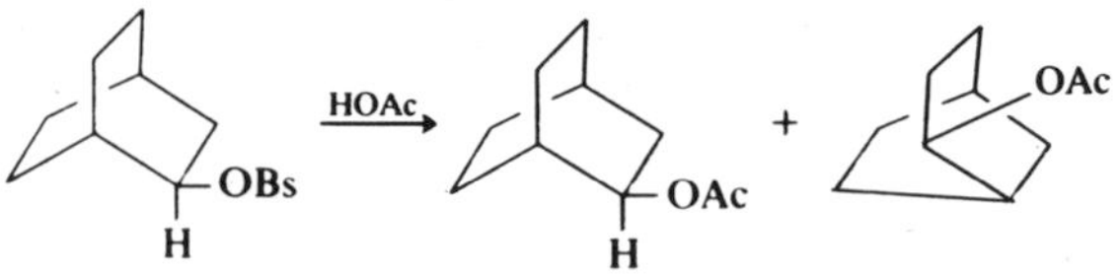

OBs — HOAc → OAc + OAc

The description of the nonclassical norbornyl cation developed by Winstein implies that the nonclassical ion is stabilized, relative to a secondary ion, by C—C σ bond delocalization. H. C. Brown of Purdue University put forward an alternative interpretation.[125] He concluded that all the available data were consistent with describing the intermediate as a rapidly equilibrating classical ion. The 1,2 shift which interconverts the two ions was presumed to be rapid relative to capture of the nucleophile. Such a rapid rearrangement would account for the isolation of racemic product, and Brown proposed that the rapid migration would lead to preferential approach of the nucleophile from the *exo* direction.

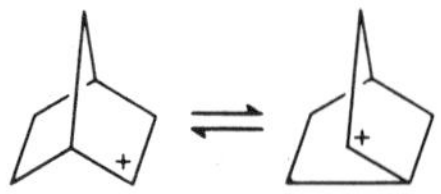

The two competing descriptions of the norbornyl cation have been very extensively tested. In essence, the question that is raised has to do with the relative energy of the bridged structure. Is it lower in energy than the classical ion and therefore an *intermediate* to which the classical ion would collapse, or is it a transition state (or higher-energy intermediate) in the rapid isomerization of two classical structures? Figure 5.8 illustrates the potential energy diagrams corresponding to the various possibilities.

Brown's approach was to show that the properties ascribed to the nonclassical ion could be duplicated in systems involving classical carbocations. Since the rate enhancement of solvolysis of the *exo* isomer was an important part of the argument for σ-bond participation, he also analyzed the relative reactivities of various systems. He argued that the *exo*-norbornyl system should be compared to the cyclopentyl system rather than to cyclohexyl brosylate. The torsional relationship of the leaving group and adjacent substituents is eclipsed in norbornyl sulfonates, and strain is relieved on ionization. Cyclohexyl brosylate, in contrast, is completely staggered in the ground state, and so no strain relief accompanies ionization. In the cyclopentyl

124. H. M. Walborsky, M. E. Baum, and A. A. Youssef, *J. Am. Chem. Soc.* **83**, 988 (1961).
125. H. C. Brown, *The Transition State, Chemical Society Special Publication*, No. 16, 140 (1962); *Chem. Brit.*, 199 (1966); *Tetrahedron* **32**, 179 (1976).

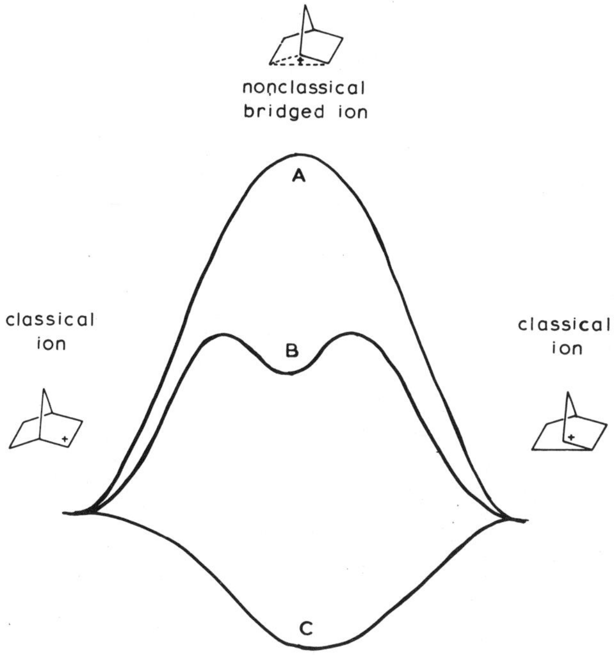

Fig. 5.8. Contrasting potential energy diagrams for stable and unstable bridged norbornyl cation. (A) Bridged ion is a transition state for rearrangement between classical structures. (B) Bridged ion is an intermediate in rearrangement of one classical structure to the other. (C) Bridged nonclassical ion is the only stable structure.

system there is similar relief of eclipsing interactions. Since norbornyl brosylate is only 14 times more reactive to acetolysis than cyclopentyl brosylate, Brown and co-workers concluded that there was little evidence of rate acceleration.[126]

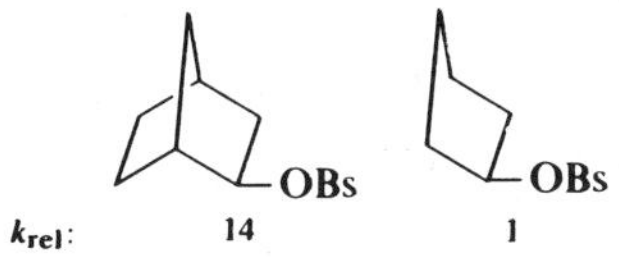

Another characteristic property associated with the nonclassical norbornyl cation, the high *exo*/*endo* rate ratio, was also scrutinized. This was done by examining the rate ratio for related tertiary cations that have classical structures, such as the 2-phenylnorbornyl cation.[127] It was found that the *exo* and *endo* 2-phenylnorbornyl *p*-nitrobenzoates exhibited an *exo*/*endo* rate ratio of 140:1 in aqueous

126. H. C. Brown, F. J. Chloupek, and M.-H. Rei, *J. Am. Chem. Soc.* **86**, 1247 (1964).
127. D. G. Farnum and G. Mehta, *J. Am. Chem. Soc.* **91**, 3256 (1969).

dioxane. Brown concluded that the high *exo*/*endo* rate ratio could not be a distinctive property of the nonclassical ion.[128]

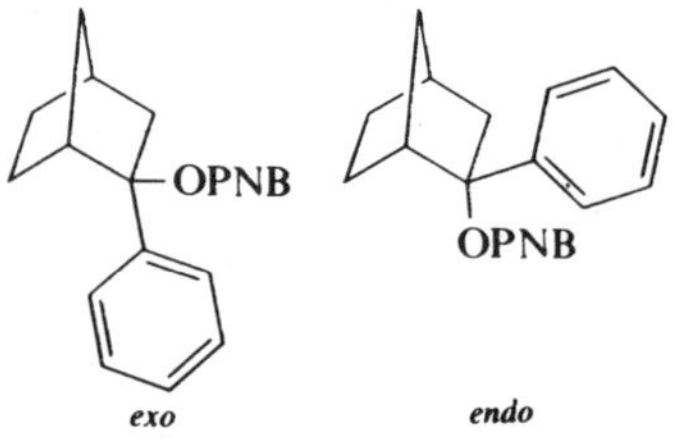

OPNB OPNB $\frac{k_{exo}}{k_{endo}} = 140$

exo *endo*

Stereoselective capture of the 2-phenyl-2-norbornyl cation from the *exo* direction was also demonstrated. Hydrolysis of 2-phenyl-*exo*-2-chloronorbornane in aqueous acetone gave 2-phenyl-*exo*-norborneol as the only product.[128]

Since the kinetic and stereochemical properties that were the primary justification for the nonclassical σ-bridged structure could be observed with a classical ion, Brown concluded that the involvement of the bridged nonclassical ion was not required to explain the results.

The arguments discussed to this point, both for and against the nonclassical structure, rest on indirect evidence derived from interpretation of the kinetic and stereochemical characteristics of the substitution reactions. When techniques for direct observation of carbocations became available, the norbornyl cation was subjected to intense study from this perspective.

The norbornyl cation was generated in SbF_5–SO_2–SOF_2, and the temperature dependence of the proton magnetic resonance spectrum was examined.[129] Subsequently, the ^{13}C NMR spectrum was studied, and the proton spectrum was determined at higher field strength. These studies excluded rapidly equilibrating classical ions as a description of the norbornyl cation under stable ion conditions.[130] It was determined also that 3,2- and 6,2-hydride shifts were occurring under stable ion conditions. Activation energies of 10.8 and 5.9 kcal/mol were measured for these processes, respectively.

The resonances observed in the ^{13}C spectrum have been assigned. None of the signals appears at a position near that where the C-2 carbon of the classical secondary 2-propyl cation is found. Instead, the resonances for the norbornyl cation appear at relatively high field and are supportive of the pentacoordinate nature of the bridged-ion structure.[131]

128. H. C. Brown, F. J. Chloupek, and M.-H. Rei, *J. Am. Chem. Soc.* **86**, 1246, 1248 (1964).
129. P. v. R. Schleyer, W. E. Watts, R. C. Fort, Jr., M. B. Comisarow, and G. A. Olah, *J. Am. Chem. Soc.* **86**, 5679 (1964); M. Saunders, P. v. R. Schleyer, and G. A. Olah, *J. Am. Chem. Soc.* **86**, 5680 (1964).
130. G. Olah, G. K. SuryaPrakash, M. Arvanaghi, and F. A. L. Anet, *J. Am. Chem. Soc.* **104**, 7105 (1982).
131. G. A. Olah, G. Liang, G. D. Mateescu, and J. L. Riemenschneider, *J. Am. Chem. Soc.* **95**, 8698 (1973); G. A. Olah, *Acc. Chem. Res.* **9**, 41 (1976).

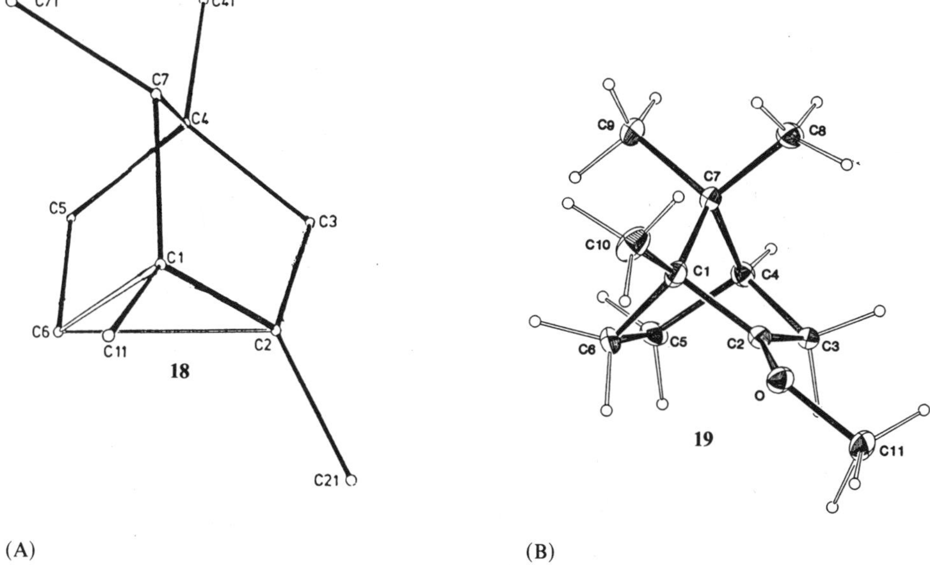

Fig. 5.9. Crystal structures of substituted norbornyl cations. (A) 1,2,4,7-Tetramethylnorbornyl cation; reproduced from Ref. 137. (B) 2-Methoxy-1,7,7-trimethylnorbornyl cation; reproduced from Ref. 138.

Other specialized NMR techniques have also been applied to the problem and confirm the conclusion that the spectra observed under stable ion conditions cannot be the result of averaged spectra of two rapidly equilibrating ions.[132]

These results, which pertain to *stable ion conditions*, provide a strong case that the most stable structure for the norbornyl cation is the symmetrically bridged nonclassical ion. How much stabilization does the σ bridging provide? An estimate based on molecular mechanics calculations and a thermodynamic cycle suggest a stabilization of about 6 ± 1 kcal/mol.[133] Molecular orbital methods suggest a range of 8–15 kcal/mol for the stabilization of the nonclassical structure relative to the classical secondary ion.[134] An experimental value based on mass spectrometric measurements is 11 kcal/mol.[135] Gas phase hydride affinity and chloride affinity data also show the norbornyl cation to be especially stable.[136]

X-ray crystal structure determinations have been completed on two salts containing bicyclo[2.2.1]heptyl cations (Fig. 5.9). Both are more stable than would be the 2-norbornyl cation itself; **18** is tertiary whereas **19** contains a stabilizing methoxy

132. C. S. Yannoni, V. Macho, and P. C. Myhre, *J. Am. Chem. Soc.* **104**, 7105 (1982); M. Saunders and M. R. Kates, *J. Am. Chem. Soc.* **102**, 6867 (1980); M. Saunders and M. R. Kates, *J. Am. Chem. Soc.* **105**, 3571 (1983).
133. P. v. R. Schleyer and J. Chandrasekhar, *J. Org. Chem.* **46**, 225 (1981).
134. H. J. Kohler and H. Lischka, *J. Am. Chem. Soc.* **101**, 3479 (1979); D. W. Goetz, H. B. Schlegel, and L. C. Allen, *J. Am. Chem. Soc.* **99**, 8118 (1977); K. Raghavachari, R. C. Haddon, P. v. R. Schleyer, and H. Schaefer, *J. Am. Chem. Soc.* **105**, 5915 (1983); M. Yoshimine, A. D. McLean, B. Liu, D. J. DeFrees, and J. S. Binkley, *J. Am. Chem. Soc.* **105**, 6185 (1983).
135. M. C. Blanchette, J. L. Holmes, and F. P. Lossing, *J. Am. Chem. Soc.* **109**, 1392 (1987).
136. R. B. Sharma, D. K. S. Sharma, K. Hiraoka, and P. Kebarle, *J. Am. Chem. Soc.* **107**, 3747 (1985).

group. The crystal structure of **18** shows an extremely long (1.74 Å) C—C bond between C-1 and C-6. The C-1—C-2 bond is shortened to 1.44 Å. The distance between C-2 and C-6 is shortened from 2.5 Å in norbornane to 2.09 Å.[137] These structural changes can be depicted as a partially bridged structure.

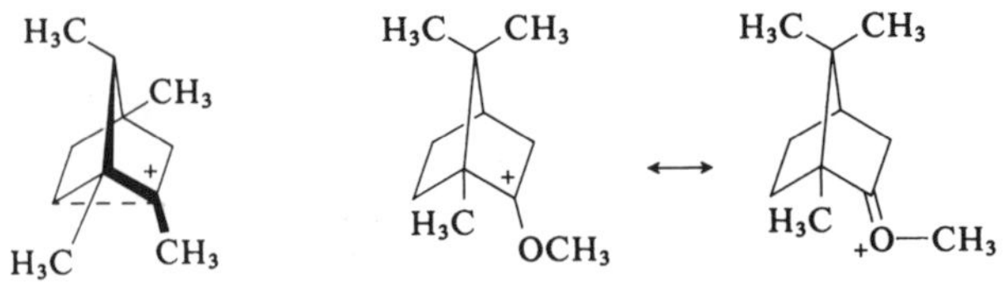

18 **19**

The evidence for this tendency toward bridging in a tertiary cation is supportive of the bridged structure for the less stable secondary cation.

The ion **19** shows some expected features, such as a shortened C-2—O bond, corresponding to resonance interaction with the methoxy group. The C-1—C-6 bond is abnormally long (1.60 Å), although not as long as in **18**.[138] This and other features of the structure are consistent with some σ delocalization even in this cation, which should be strongly stabilized by the methoxy group.

Let us now return to the question of solvolysis and how it relates to the structure under stable ion conditions. To relate the structural data to solvolysis conditions, the primary issues that must be considered are the extent of solvent participation in the transition state and the nature of solvation of the cationic intermediate. The extent of solvent participation has been probed by comparison of solvolysis characteristics in trifluoroacetic acid with the solvolysis in acetic acid. The *exo-endo* reactivity ratio in trifluoroacetic acid is 1120, compared to 280 in acetic acid. While the *endo* isomer shows solvent sensitivity typical of normal secondary tosylates, the *exo* isomer reveals a reduced sensitivity. This indicates that the transition state for solvolysis of the *exo* isomer possesses a greater degree of charge dispersal, which would be consistent with a bridged structure. This fact, along with the rate enhancement of the *exo* isomer, indicates that the σ participation commences prior to the transition state being attained, so that it can be concluded that bridging is a characteristic of the solvolysis intermediate, as well as of the stable ion structure.[139]

Another line of evidence that bridging is important in the transition state for solvolysis has to do with substituent effects for groups placed at C-4, C-5, C-6, and C-7 on the norbornyl system. The solvolysis rate is most strongly affected by C-6 substituents, and the *exo* isomer is more sensitive to these substituents than is the *endo* isomer. This implies that the transition state for solvolysis is especially sensitive to C-6 substituents, as would be expected if the C-1—C-6 bond participates in solvolysis.[140]

137. T. Laube, *Angew. Chem. Int. Ed. Engl.* **26**, 560 (1987).
138. L. K. Montgomery, M. P. Grendez, and J. C. Huffman, *J. Am. Chem. Soc.* **109**, 4749 (1987).
139. J. E. Nordlander, R. R. Gruetzmacher, W. J. Kelly, and S. P. Jindal, *J. Am. Chem. Soc.* **96**, 181 (1984).
140. F. Fuso, C. A. Grob, P. Sawlewicz, and G. W. Yao, *Helv. Chim. Acta* **69**, 2098 (1986); P. Flury and C. A. Grob, *Helv. Chim. Acta* **66**, 1971 (1983).

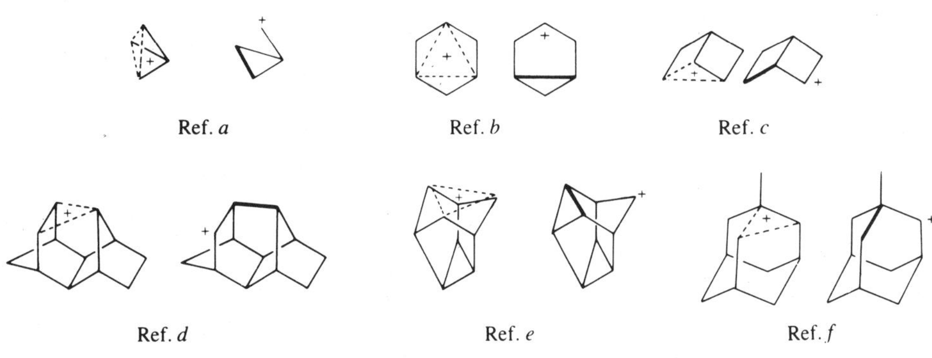

a. M. Saunders and H.-U. Siehl, *J. Am. Chem. Soc.* **102**, 6868 (1980); J. S. Starat, J. D. Roberts, G. K. Surya Prakash, D. J. Donovan, and G. A. Olah, *J. Am. Chem. Soc.* **100**, 8016, 8018 (1978).
b. G. A. Olah, G. K. Surya Prakash, T. N. Rawdah, D. Whittaker, and J. C. Rees, *J. Am. Chem. Soc.* **101**, 3935 (1979).
c. R. N. McDonald and C. A. Curi, *J. Am. Chem. Soc.* **101**, 7116, 7118 (1979).
d. S. Winstein and R. L. Hansen, *Tetrahedron Lett.*, **No. 25**, 4 (1960).
e. R. M. Coates and E. R. Fretz, *J. Am. Chem. Soc.* **99**, 297 (1977); H. C. Brown and M. Ravindranathan, *J. Am. Chem. Soc.* **99**, 299 (1977).
f. J. E. Nordlander and J. E. Haky, *J. Am. Chem. Soc.* **103**, 1518 (1981).

Many other cations besides the norbornyl cation have nonclassical structures.[141] Scheme 5.5 shows some examples which have been characterized by structural studies or by evidence derived from solvolysis reactions. To assist in interpretation of the nonclassical structures, the bond representing the bridging electron pair is darkened in a corresponding classical structure. Not surprisingly, the borderline between classical structures and nonclassical structures is blurred. There are two fundamental factors which prevent an absolute division. (1) The energies of the two (or more) possible structures may be so close as to prevent a clear distinction as to stability. (2) The molecule may adopt a geometry which is intermediate between a classical geometry and a symmetrical bridged structure.

To summarize, it now appears that nonclassical or bridged structures are either readily attainable intermediates or transition states for many cations and are intimately involved in rearrangement processes. For others, such as the norbornyl cation, the bridged structure is the most stable structure. As a broad generalization, tertiary cations are nearly always more stable than related bridged ions and therefore have classical structures. Primary carbocations can be expected to undergo rearrangement to more stable secondary or tertiary ions, with bridged ions being likely transition states (or intermediates) on the rearrangement path. The energy balance between classical secondary structure and bridged structures will depend on the individual system. Bridged structures are most likely to be stable where a strained bond can participate in bridging or where solvation of the positive charge is difficult. Because of poor solvation, bridged structures are particularly likely to be favored in superacid media and in the gas phase.

141. V. A. Barkhash, *Top. Curr. Chem.* **116–117**, 1 (1984); G. A. Olah and G. K. SuryaPrakash, *Chem. Brit.* **19**, 916 (1983).

General References

Nucleophilic Substitution Mechanisms

C. A. Bunton, *Nucleophilic Substitution at a Saturated Carbon Atom*, Elsevier, New York, 1963.
T. L. Ho, *Hard and Soft Acids and Bases Principle in Organic Chemistry*, Academic Press, New York, 1977.
A. Streitwieser, Jr., *Solvolytic Displacement Reactions*, McGraw-Hill, New York, 1962.
E. R. Thornton, *Solvolysis Mechanisms*, Ronald Press, New York, 1964.

Carbocations

D. Bethell and V. Gold, *Carbonium Ions, an Introduction*, Academic Press, London, 1967.
S. P. McManus and C. U. Pittman, Jr., in *Organic Reactive Intermediates*, S. P. McManus (ed.), Academic Press, New York, 1973, Chapter 4.
G. A. Olah, *Carbocations and Electrophilic Reactions*, Wiley, New York, 1974.
G. A. Olah and P. v. R. Schleyer (eds.), *Carbonium Ions*, Vols. I-IV, Wiley-Interscience, New York, 1968-1973.
M. Saunders, J. Chandrasekhar, and P. v. R. Schleyer, in *Rearrangements in Ground and Excited States*, P. de Mayo (ed.), Academic Press, 1980, Chapter 1.

Problems

(*References for these problems will be found on page* 777.)

1. Provide an explanation for the relative reactivity or stability relationships revealed by the following sets of data.

   (a) For solvolysis of $R-C(CH_3)_2-O-C(=O)-C_6H_4-NO_2$

   in aqueous acetone, the relative rates are: $R=CH_3$, 1; *i*-Pr, 2.9; *t*-Bu, 4.4; Ph, 10^3; cyclopropyl, 5×10^5.

   (b) For solvolysis of

   $$Ar-C(CH_3)(X)-OSO_2R,$$

   the Hammett ρ^+ reaction constant varies with the substituent X as shown below:

X	ρ^+
CH_3	−4.5
CF_3	−6.9
CH_3SO_2	−8.0

(c) The hydride affinities measured for the methyl cation with increasing fluorine substitution are in the order CH_3^+ (312) > CH_2F^+ (290) > CHF_2^+ (284) < CF_3^+ (299).

(d) The rates of solvolysis of a series of 2-alkyl-2-adamantyl *p*-nitrobenzoates are: R = CH_3, $1.4 \times 10^{-10}\ s^{-1}$; C_2H_5, $1.1 \times 10^{-9}\ s^{-1}$; *i*-$C_3H_7$, $5.0 \times 10^{-9}\ s^{-1}$; *t*-$C_4H_9$, $3.4 \times 10^{-5}\ s^{-1}$; $(CH_3)_3CCH_2$, $1.5 \times 10^{-9}\ s^{-1}$.

(e) The relative rates of methanolysis of a series of alkyl chlorides having ω-phenylthio substituents as a function of chain length are: $n = 1$, 3.3×10^4; $n = 2$, 1.5×10^2; $n = 3$, 1.0; $n = 4$, 1.3×10^2; $n = 5$, 4.3.

2. Which reaction in each pair would be expected to be faster? Explain.

(a) 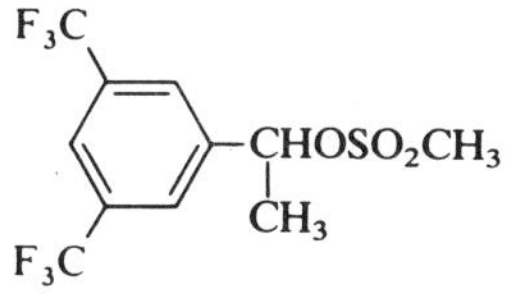 3,5-$(F_3C)_2C_6H_3$–$CH(CH_3)OSO_2CH_3$ *or* 3,5-$(F_3C)_2C_6H_3$–$CH(CH_3)OSO_2CH_2CF_3$ — solvolysis in 80% ethanol

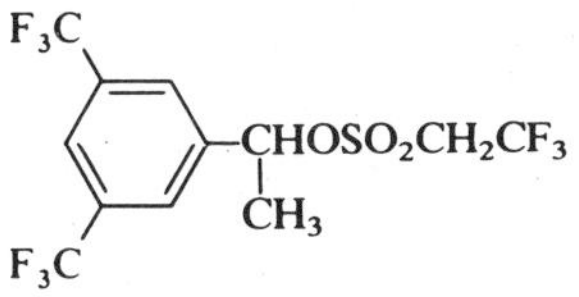

(b) 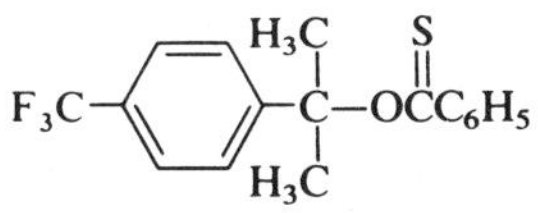 F_3C–C_6H_4–$C(CH_3)_2$–$OC(=S)C_6H_5$ *or* H_3C–C_6H_4–$C(CH_3)_2$–$OC(=S)C_6H_5$ — solvolysis in 100% ethanol

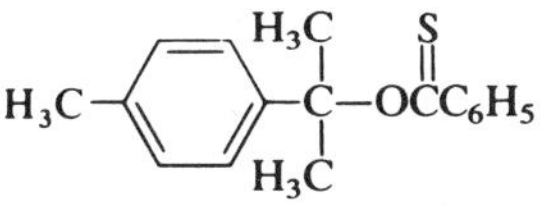

(c)  OTs (bridgehead) *or* H, OTs (7-position) — solvolysis in acetic acid

(d) 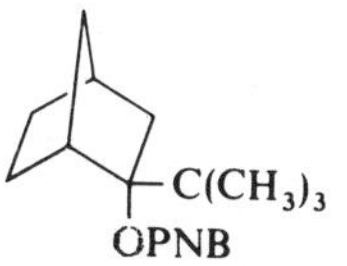 $C(CH_3)_3$, OPNB *or* CH_3, OPNB — solvolysis in aqueous acetone

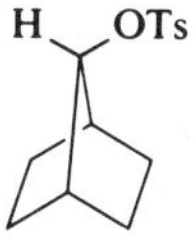

PNB = *p*-nitrobenzoate

(e) 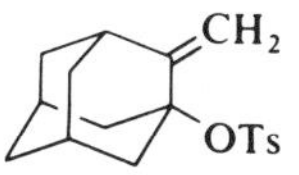 CH_2, OTs *or* OTs — solvolysis in acetic acid

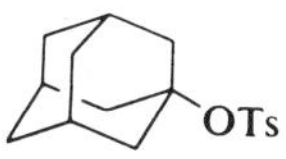

(f) $C_6H_5SO_2CH_2CH_2Cl$ *or* $C_6H_5SO_2CH_2Cl$ — reaction with KI in acetone

(g) $(CH_3)_3CCH_2OTs$ *or* CH_3, CH_2OTs (cyclopropane) — solvolysis in aqueous dioxane

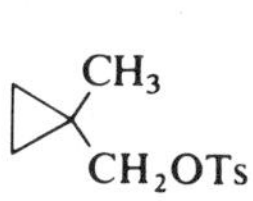

(h) 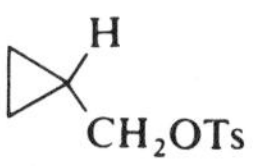 H, CH_2OTs (cyclopropane) *or* CH_3, CH_2OTs (cyclopropane) — solvolysis in aqueous dioxane

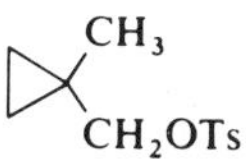

(i) $H_2C{=}CHCH_2CH_2OTs$ *or* $H_3CCH{=}CHCH_2OTs$ solvolysis in 98% formic acid

(j) 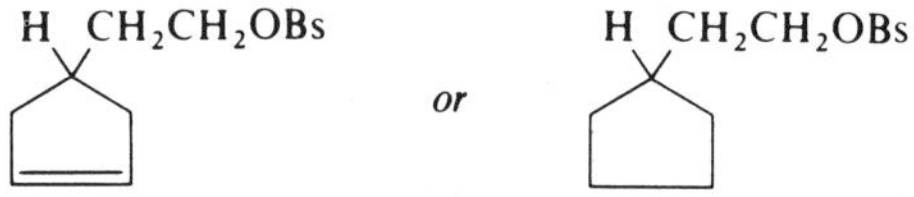 [cyclopentenyl-CH_2CH_2OBs] *or* [cyclopentyl-CH_2CH_2OBs] solvolysis in acetic acid

(k) $PhS(CH_2)_3Cl$ *or* $PhS(CH_2)_4Cl$ solvolysis in aqueous dioxane

(l) 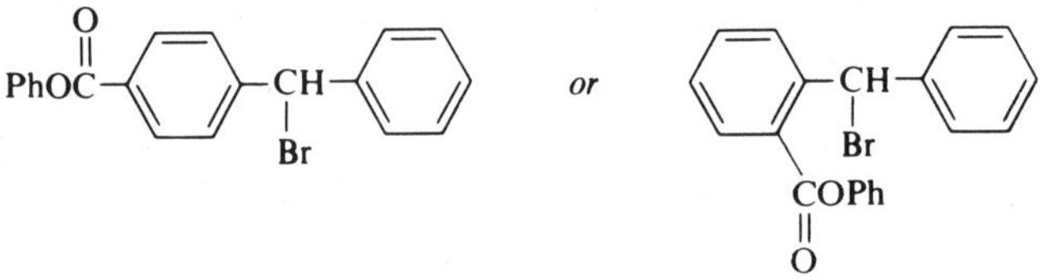 [p-PhOC(O)-C_6H_4-CH(Br)-Ph] *or* [o-PhOC(O)-C_6H_4-CH(Br)-Ph] solvolysis in acetic acid

3. Suggest reasonable mechanisms for each of the following reactions. The starting materials were the racemic substance, except where noted otherwise.

(a) $PhCH_2Cl + P(OCH_3)_3 \longrightarrow PhCH_2P(O)(OCH_3)_2 + CH_3Cl$

(b) 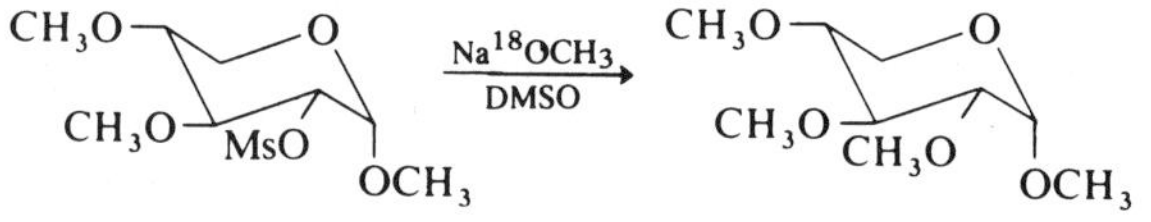 [MsO-substituted dimethoxy methyl pyranoside] $\xrightarrow[\text{DMSO}]{Na^{18}OCH_3}$ [trimethoxy methyl pyranoside]

no ^{18}O label in product

(c) 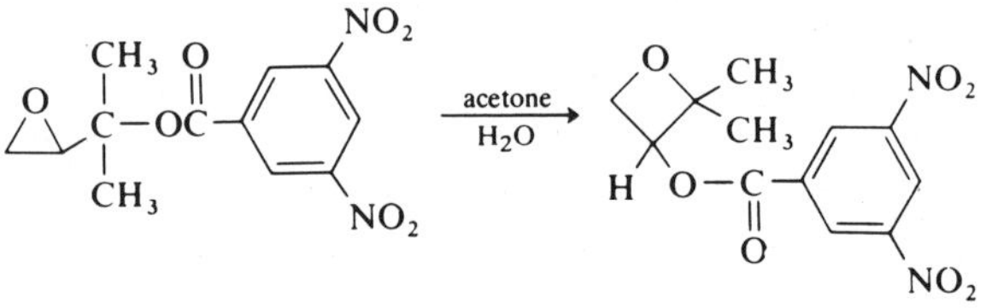 [epoxide-$C(CH_3)_2$-OC(O)-3,5-dinitrophenyl] $\xrightarrow[H_2O]{\text{acetone}}$ [oxetane with CH_3, CH_3, H, O-C(O)-3,5-dinitrophenyl]

(d) 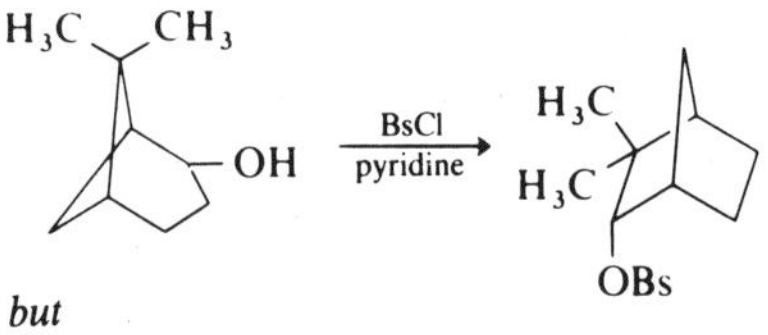 [7,7-dimethylnorbornan-2-ol (exo OH)] $\xrightarrow[\text{pyridine}]{\text{BsCl}}$ [rearranged brosylate with H_3C, H_3C, OBs]

but

[7,7-dimethylnorbornan-2-ol (endo OH)] $\xrightarrow[\text{pyridine}]{\text{BsCl}}$ [7,7-dimethylnorbornyl OBs]

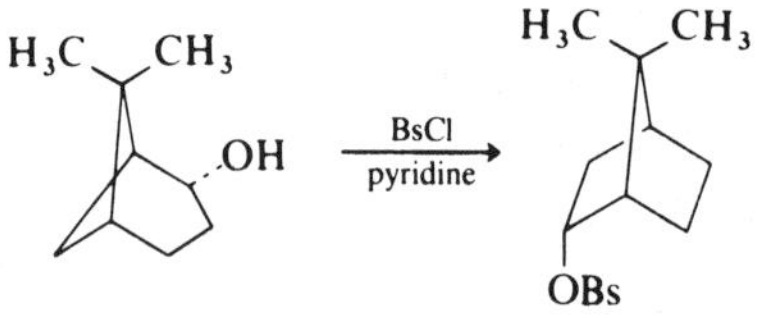

(e) $HC{\equiv}CCH_2CH_2CH_2Cl \xrightarrow{CF_3CO_2H} H_2C{=}C(Cl)CH_2CH_2CH_2OC(O)CF_3$

(f) H_3C CH_3 OBs $\xrightarrow{CH_3CO_2H}$ H_3C CH_3 $OCCH_3$ (O) + H_3C H_3C $OCCH_3$ (O) + H_3C H_3C CH_3CO (O)

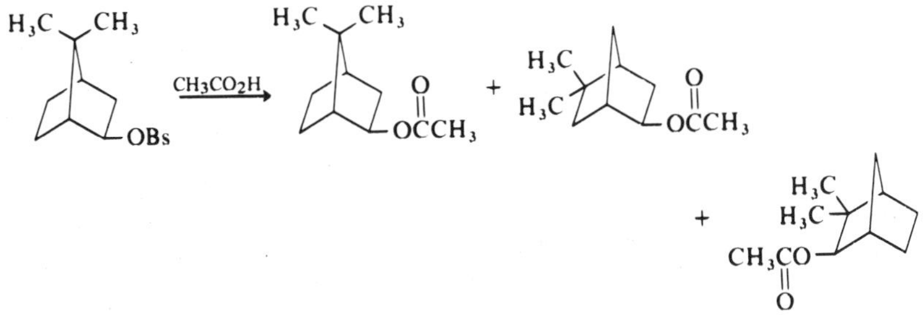

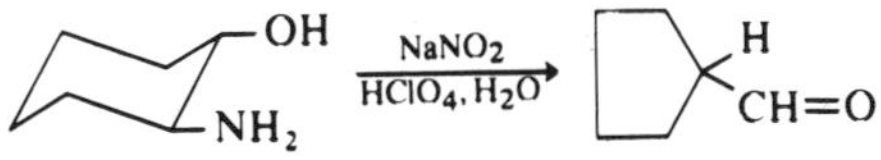

(g) OH NH_2 $\xrightarrow[HClO_4, H_2O]{NaNO_2}$ H CH=O

(h) $(CH_3)_3C$ OH NH_2 $\xrightarrow[HClO_4, H_2O]{NaNO_2}$ $(CH_3)_3C$ =O

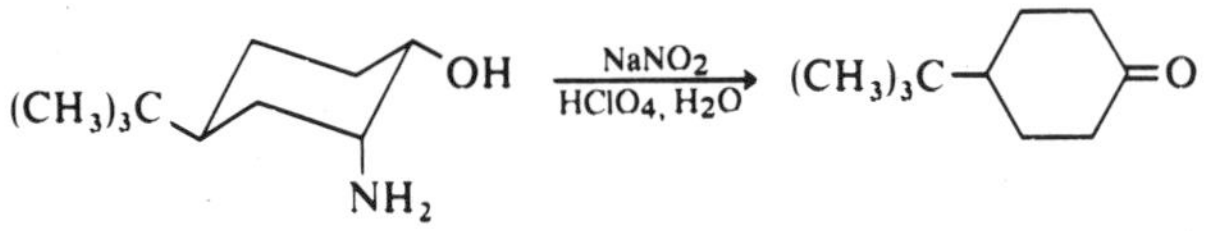

(i) O Br $\longrightarrow$ HO CH=O

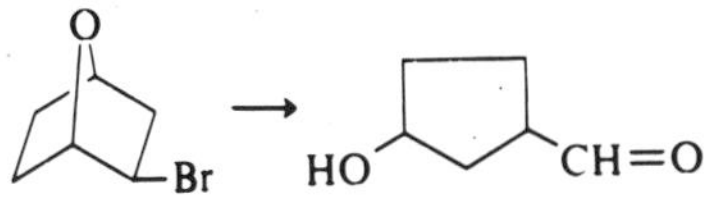

(j) N CH_3 OH $\xrightarrow[heat]{(CH_3CO)_2O, H^+}$ N CH_3 $OCCH_3$ (O)

optically active — racemic

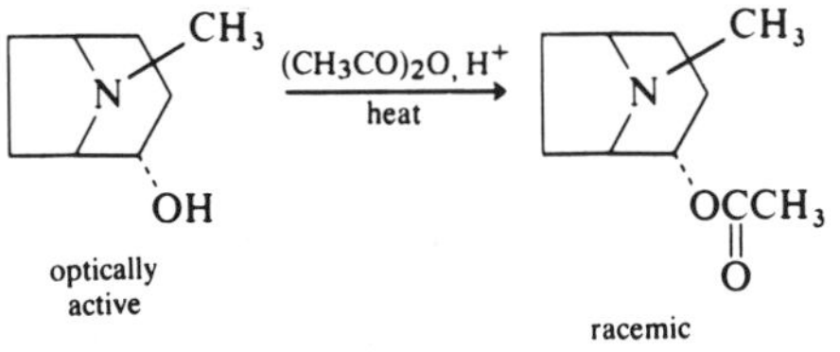

(k) $(PhC)_2NCH_2CH_2Cl$ (O) $\xrightarrow[CH_3C\equiv N, heat]{H_2O}$ $PhCNHCH_2CH_2OCPh$ (O) (O)

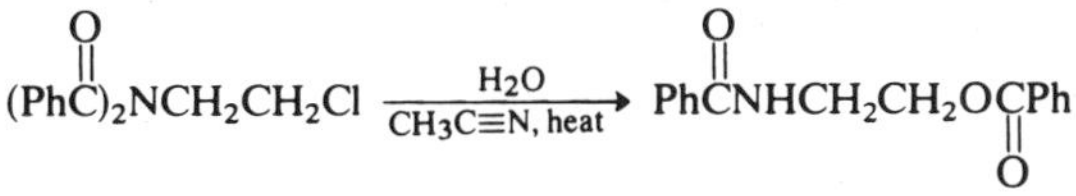

(l) H_3C CH_3 H_3C OH Ph O $\xrightarrow[0°C]{H_2SO_4}$ O O H CH_3 CH_3 Ph CH_3

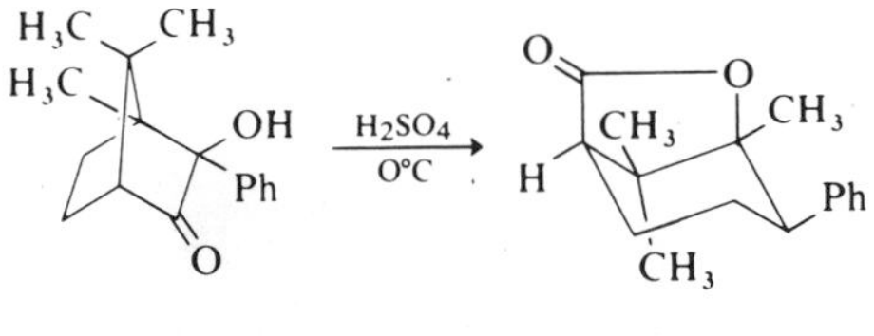

(m) CH_3SO_2O $\xrightarrow{HOAc}$ O_2CCH_3

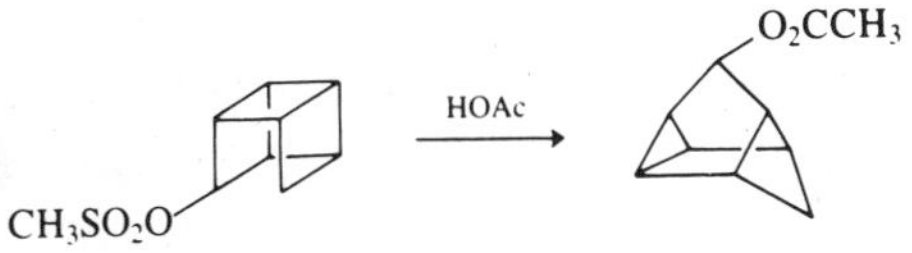

4. The solvolysis of $2R,3S$-3-(4-methoxyphenyl)but-2-yl *p*-toluensulfonate in acetic acid can be followed by several kinetic measurements: (a) rate of decrease of observed rotation (k_α); rate of release of the leaving group (k_t); and, when ^{18}O-labeled sulfonate is used, the rate of equilibration of the sulfonate oxygens

(k_{ex}). At 25°C, the rate constants are

$$k_\alpha = 25.5 \times 10^{-6}\ \mathrm{s}^{-1}; \qquad k_t = 5.5 \times 10^{-6}\ \mathrm{s}^{-1}; \qquad k_{ex} = 17.2 \times 10^{-6}\ \mathrm{s}^{-1}.$$

Describe the nature of the process measured by each of these rate constants, and devise a mechanism which includes each of these processes. Rationalize the order of the rates $k_\alpha > k_{ex} > k_t$.

5. Both *endo*- and *exo*-norbornyl *p*-bromobenzenesulfonates react with $R_4P^{+-}N_3$ (R = long-chain alkyl) in toluene to give azides of inverted configuration. The yield from the *endo* reactant is 95%, and from the *exo* reactant, the yield is 80%. In the case of the *exo* reactant, the remaining material is converted to nortricyclane (tricyclo[2.2.1.0^{2,6}]heptane). The measured rates of azide formation are first-order in both reactant and azide ion. No rearrangement of deuterium is observed in the azides when deuterium-labeled reactants are used. What conclusion about the mechanism of the substitution process do you draw from these results? How do the reaction conditions relate to the mechanism you have suggested? How is the nortricyclane formed?

6. The following observations have been made concerning the reaction of *Z*-1-phenyl-1,3-butadiene (**A**) and *Z*-4-phenyl-3-buten-1-ol (**B**) in 3–7 *M* H_2SO_4 and 0.5–3 *M* $HClO_4$:

   (a) Both compounds are converted to the corresponding *E*-isomer by acid-catalyzed processes with rates given by

   $$\text{rate} = k[\text{reactant}][\text{H}^+]$$

   when $[H^+]$ is measured by the H_0 acidity function.

   (b) The rate of isomerization of **A** is slower in deuterated (D_2SO_4–D_2O) media by a factor of 2–3. For **B**, the rate of isomerization is faster in D_2SO_4–D_2O by a factor of 2.5.

   (c) When ^{18}O-labeled **B** is used, the rate of loss of ^{18}O to the solvent is equal to the rate of isomerization.

   (d) The activation energies are 19.5 ± 1 kcal/mol for **A** and 22.9 ± 0.7 kcal/mol for **B**.

   Write a mechanism for each isomerization which is consistent with the information given.

7. Treatment of 2-(*p*-hydroxyphenyl)ethyl bromide with basic alumina produces a white solid: mp, 40–43°C; IR, 1640 cm^{-1}; UV, 282 nm in H_2O, 261 nm in ether; NMR, two singlets of equal intensity at 1.69 and 6.44 ppm from TMS. *Anal.*: C, 79.97; H, 6.71. Suggest a reasonable structure for this product and a rationalization for its formation.

8. The solvolysis of the tosylate of 3-cyclohexenol has been studied in several solvents. The rate of solvolysis is not very solvent sensitive, being within a factor of 5 for all solvents. The product distribution is solvent sensitive, however, as shown below.

Solvent (ROH)	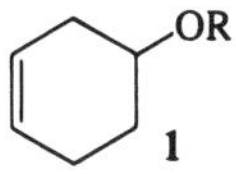 1	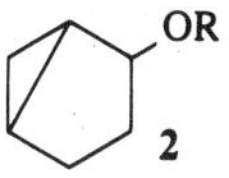 2	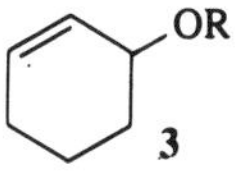 3	Cyclohexadienes
acetic acid	20%	*a*	10%	70%
formic acid	58%	*a*	*a*	42%
$CF_3CH(OH)CF_3$	10%	65%	*a*	25%

a. Minor product, less than 3% yield.

Furthermore, the stereochemistry of the product of structure **1** changes as the solvent is changed. In aqueous dioxane the reaction proceeds with complete inversion but in hexafluoropropanol with 100% retention. In acetic acid the reaction occurs mainly with inversion (83%) but in formic acid the amount of retention (40%) is comparable to inversion (60%). Discuss these results, particularly with respect to the change of product composition and stereochemistry as a function of solvent.

9. Each of the following carbocation ions can rearrange to a cation with special stabilization. Indicate likely routes for the rearrangement to a more stable species for each ion.

(a) 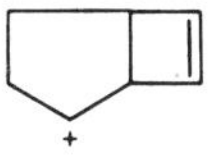 (b) 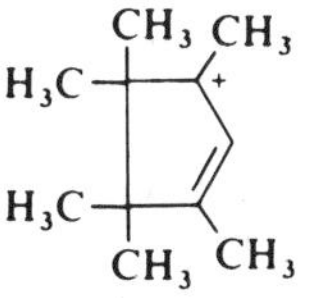 H_3C, CH_3, CH_3, H_3C, CH_3, CH_3 (c) 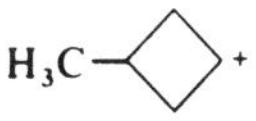 H_3C (d)

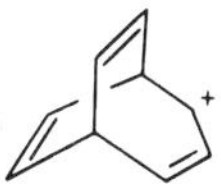

10. In the discussion of the *syn*- and *anti*-7-norbornenyl tosylates, it was pointed out that, relative to 7-norbornyl tosylate, the reactivities of the *syn* and *anti* isomers were 10^4 and 10^{11}, respectively. The high reactivity of the *anti* isomer was attributed to participation of the carbon–carbon double bond. What is the source of the 10^4 factor of acceleration in the *syn* isomer relative to the saturated model?

11. Indicate the structure of the ion you expect to be formed as the stable species when each of the following compounds is dissolved in superacid media at −30°C:

(a) 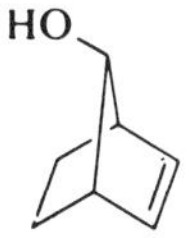 HO (b) 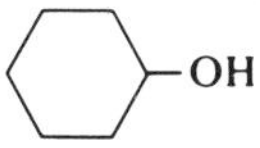 OH (c) 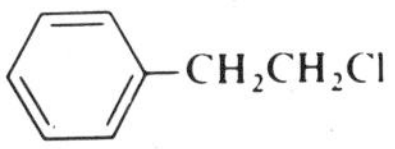 CH_2CH_2Cl

(d) 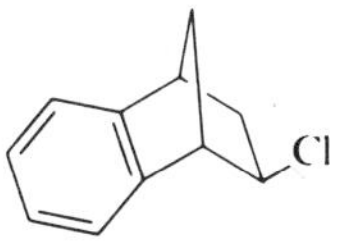 Cl (e) 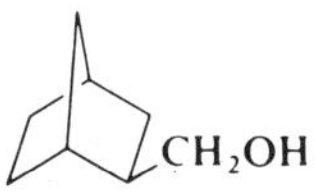 CH_2OH (f)

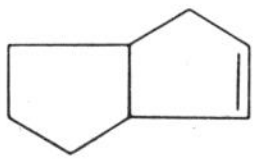

12. The behavior of compounds **A** and **B** on solvolysis in acetic acid containing acetate ion has been studied. The solvolysis of **A** is about 13 times faster than that of **B**. Kinetic studies in the case of **A** show that **A** is racemized competitively with solvolysis. A single product is formed from **A**, but **B** gives a mixture. Explain these results.

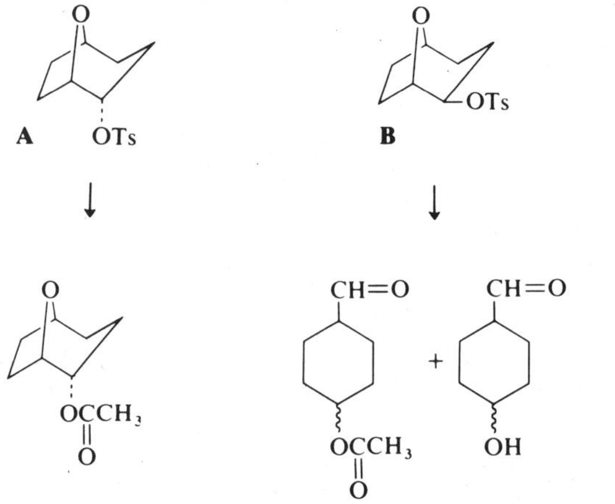

A → (acetate product, $OCCH_3$ with C=O)

B → 4-(CH=O)cyclohexyl $OCCH_3$ + 4-(CH=O)cyclohexanol (OH)

13. A series of ^{18}O-labeled sulfonate esters was prepared, and the extent of ^{18}O scrambling which accompanies solvolysis was measured. The rate of ^{18}O exchange was compared with the rate of solvolysis.

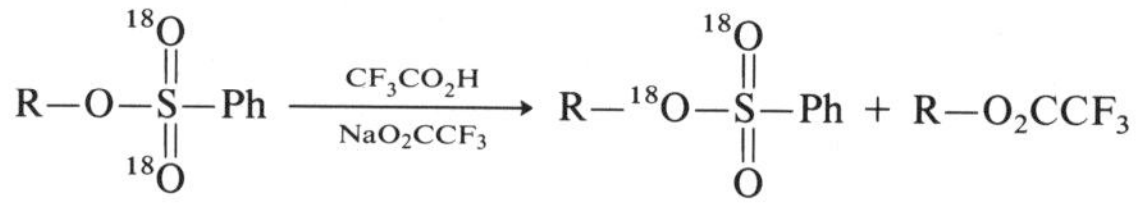

$$R{-}O{-}S(={}^{18}O)_2{-}Ph \xrightarrow[NaO_2CCF_3]{CF_3CO_2H} R{-}^{18}O{-}S(={}^{18}O)(=O){-}Ph + R{-}O_2CCF_3$$

R	k_{sol}	k_{ex}
$(CH_3)_2CH—$	3.6×10^{-5}	7.9×10^{-6}
cyclopentyl—	3.8×10^{-3}	8.5×10^{-4}
2-adamantyl	1.5×10^{-3}	1.8×10^{-3}
$(CH_3)_3CCHCH_3$[a]	7.3×10^{-3}	negligible

a. Solvolysis product is $(CH_3)_2C(O_2CCF_3)CH(CH_3)_2$.

Discuss the variation in the ratio $k_{sol} : k_{ex}$. Offer an explanation for the absence of exchange in the 3,3-dimethyl-2-butyl case.

14. Studies of the solvolysis of 1-phenylethyl chloride and its *p*-substituted derivatives in aqueous trifluorethanol containing azide anion as a potential nucleophile provide details relative to the mechanism of nucleophilic substitution in this system.

(a) The reaction is independent of the azide ion concentration for *para* substituents that have σ^+ values more negative than −0.3 but is first-order in $[N_3^-]$ for substituents with σ^+ more positive than −0.08.

(b) When other good nucleophiles are present that can compete with azide ion, e.g., $CH_3CH_2CH_2SH$, substrates undergoing solvolysis at rates that are zero-order in $[N_3^-]$ show little selectivity between the nucleophiles.

(c) For substrates that solvolyze at rates independent of $[N_3^-]$, the ratio of 1-arylethyl azide to 1-arylethanol in the product increases as the σ^+ of the substituent becomes more negative.

(d) The major product in reactions in which the solvolysis is first-order in $[N_3^-]$ is the 1-arylethyl azide.

Consider these results with respect to the mechanisms outlined in Fig. 5.6 (p. 269). Delineate the types of substituted 1-arylethyl halides which react with azide ion according to each of these mechanisms on the basis of the data given above.

15. Offer a mechanistic interpretation of each of the following phenomena:

(a) Although there is a substantial difference in the rate at which **A** and **B** solvolyze (**A** reacts 4.4×10^4 times faster in acetic acid), both compounds give products of completely retained configuration.

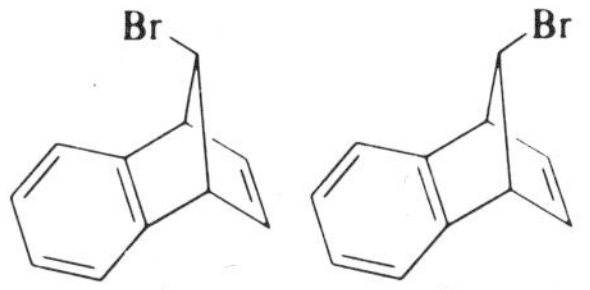

Br Br

A **B**

(b) The solvolysis of **C** is much more sensitive to substituent effects than that of **D**.

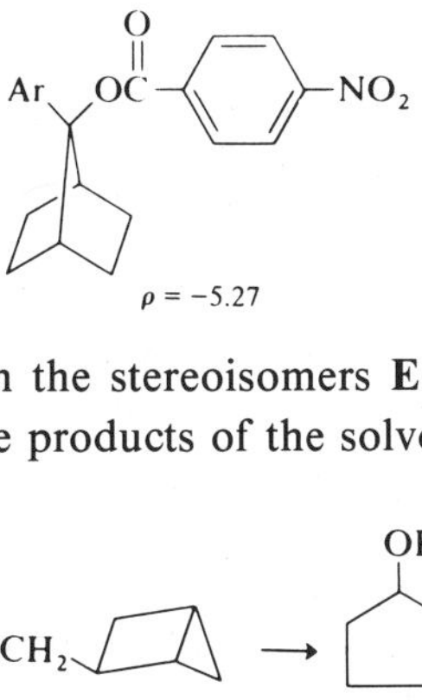

Ar OC(=O)–C₆H₄–NO₂

C $\rho = -5.27$

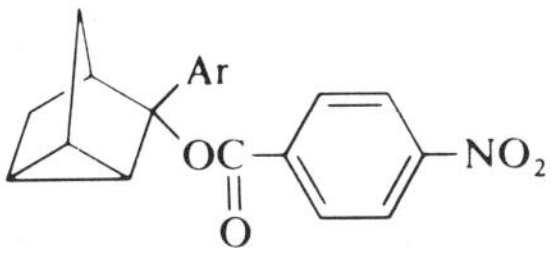

D $\rho = -3.27$

(c) Although the stereoisomers **E** and **F** solvolyze in acetone at comparable rates, the products of the solvolysis reactions are very different.

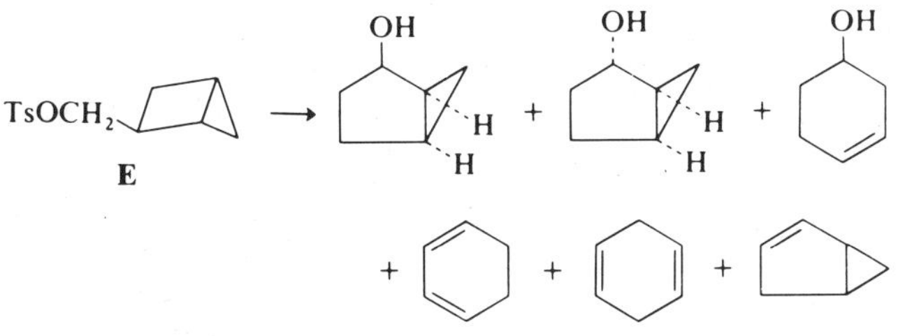

$TsOCH_2$ **E** → OH, H, H + OH, H, H + OH + + +

$TsOCH_2$ **F** → H, HO, H, H

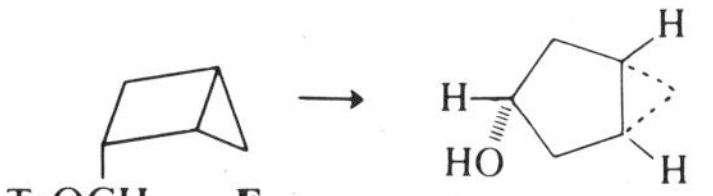

(d) Solvolysis of *endo*-2-chloro-7-thiabicyclo[2.2.1]heptane occurs 4.7×10^9 times faster than that of the *exo* isomer. The product from either isomer in the presence of sodium acetate is the *endo* acetate.

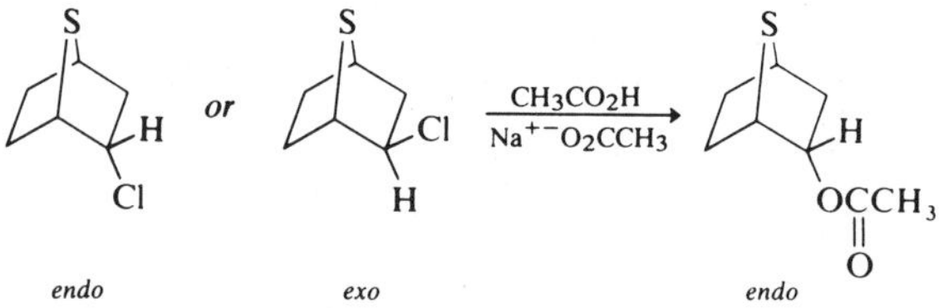

(e) Solvolysis of 2-octyl *p*-bromobenzenesulfonate in 80% methanol: 20% acetone gives, in addition to the expected methyl 2-octyl ether, a 15% yield of 2-octanol. The 2-octanol could be shown not to result from the presence of adventitious water in the medium.

(f) Addition of CF_3CHN_2 to fluorosulfonic acid at −78°C gives a solution the ^1H NMR spectrum of which shows a quartet ($J_{HF} = 6.1$ Hz) at δ 6.3 ppm from external TMS. On warming to −20°C, this quartet disappears and is replaced by another one ($J_{HF} = 7.5$ Hz) at δ 5.50 ppm.

(g) 2-*t*-Butyl-*exo*-norbornyl *p*-nitrobenzoate is an extremely reactive compound, undergoing solvolysis 2.8×10^6 times faster than *t*-butyl *p*-nitrobenzoate. The *endo* isomer is about 500 times less reactive. In contrast to the unsubstituted norbornyl system, which gives almost exclusively *exo* product, both *t*-butyl isomers give about 5% of the *endo* product.

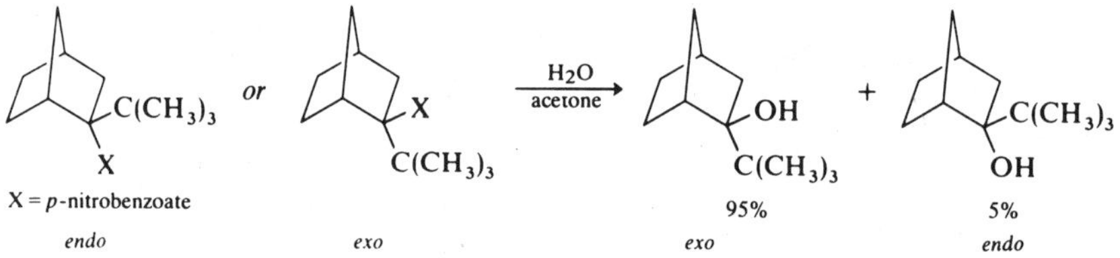

(h) Solvolysis of 2,4,6-trimethylbenzyl chloride in 80% aqueous ethanol is characterized by $\Delta S^{\ddagger} = -11.0$ eu. Solvolysis of 2,4,6-tri-*t*-butylbenzyl chloride, however, has $\Delta S^{\ddagger} = +0.3$ eu. Suggest an explanation for the difference in the entropy of activation for solvolysis of these two systems.

(i) Solvolysis of the *p*-nitrobenzoates of both the *syn* and *anti* isomers of 2-hydroxybicyclo[6.1.0]nonane gives as the major solvolysis product the corresponding alcohol of retained stereochemistry when carried out in buffered aqueous acetone.

(j) Solvolysis of compounds **G** and **H** gives a product mixture which is quite similar for both compounds.

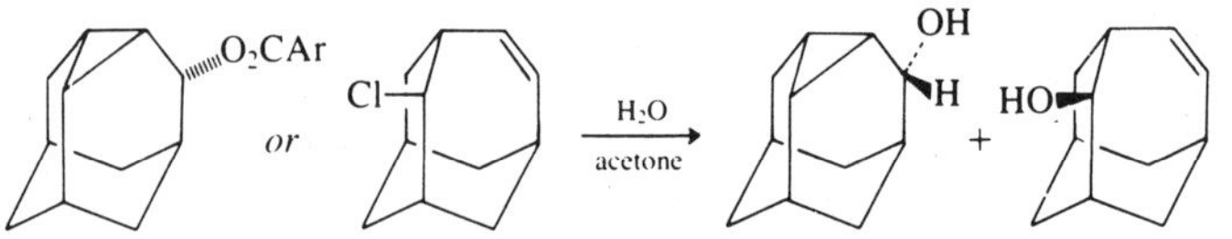

G, Ar = 3,5-dinitrophenyl **H** 83%−85% 8%−12%

On the other hand, compound **I** gives a completely different mixture.

$\xrightarrow[\text{acetone}]{H_2O}$

I 39% 51%

(k) The isomeric tosylates **J** and **K** give an identical product mixture consisting of the alcohol **L** and ether **M** when solvolyzed in aqueous ethanol.

J L, R=H M, $R{=}C_2H_5$ K

16. Both experimental studies on gas phase ion stability and MO calculations indicate that the two vinyl cations shown below benefit from special stabilization. Indicate what structural features present in these cations can provide this stabilization.

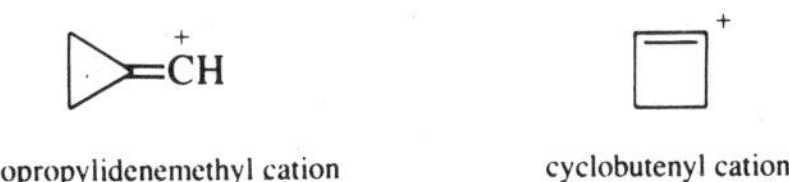

cyclopropylidenemethyl cation cyclobutenyl cation

17. The rates of solvolysis of four isomeric tricyclooctane derivatives have been determined. After correction for leaving group and temperature effects, the relative reactivities are as shown.

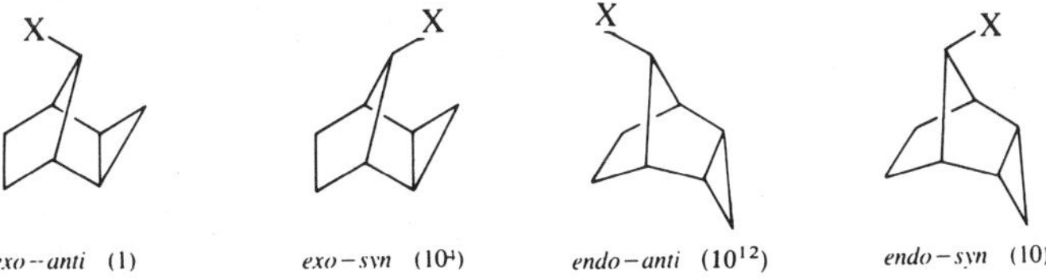

exo–anti (1) *exo–syn* (10^4) *endo–anti* (10^{12}) *endo–syn* (10)

In aqueous dioxane the *endo-anti* isomer gave a product mixture consisting of alcohol **A** and the corresponding ester (derived from capture of the leaving group *p*-nitrobenzoate). The other isomers gave much more complex product mixtures which were not completely characterized. Explain the trend in rates and discuss the structural reason for the stereochemical course of the reaction in the case of the *endo-anti* isomer.

A

18. The ^{13}C NMR chemical shift of the trivalent carbon is a sensitive indicator of carbocation structure. Given below are the data for three carbocations with varying aryl substituents. Generally, the greater the chemical shift, the lower is the electron density at the carbon atom.

Aryl substituent(s)	A	B (chemical shifts)	C
3,5-di-CF_3	287	283	73
4–CF_3	284	278	81
H	272	264	109
4–CH_3	262	252	165
4–OCH_3	235	230	220

How do you explain the close similarity in the substituent group trends for ions **A** and **B** as contrasted to the opposing trend in **C**?

19. A variety of kinetic data permit the assignment of relative reactivities toward solvolysis of a series of systems related to the norbornane skeleton. Offer a general discussion of the structural effects that are responsible for the observed relative rates.

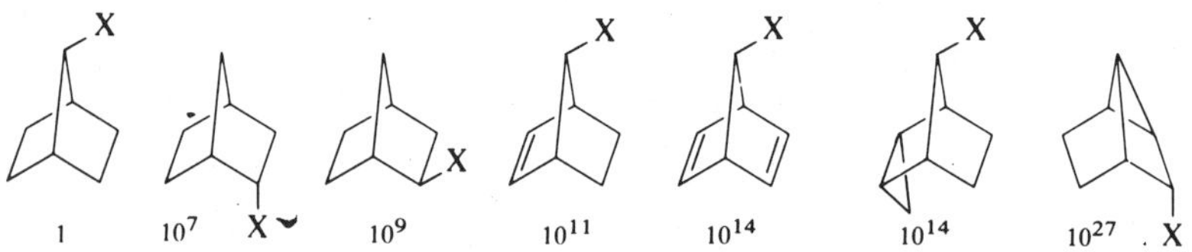

20. The comparison of activation parameters for reactions in two different solvents requires consideration of differences in solvation of both the reactants and the transition states. This can be done with potential energy diagrams such as that illustrated below, where A and B refer to two different solvents. By thermodynamic methods, it is possible to establish $\Delta H_{\text{transfer}}$ values which correspond to the enthalpy change associated with transfer of a solute from one solvent to another.

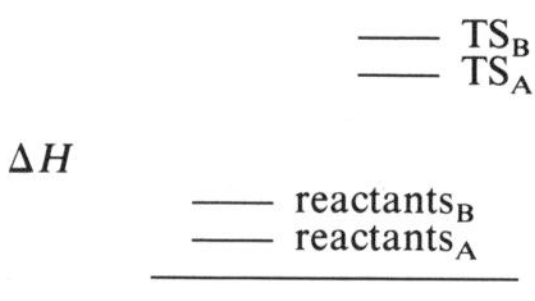

$\Delta H_{\text{transfer}}$ data for *n*-hexyl tosylate and several ions are given in Table 5.p20A. In Table 5.p20B, activation parameters are given for the reaction of these ions with *n*-hexyl tosylate by the S_N2 mechanism. Use these data to construct a chart such as that above for each of the nucleophiles. (Show the enthalpy in kcal/mol on each chart.) Use these charts to interpret the relative reactivity data given in Table 5.p20C. Discuss each of the following aspects of the data.

(a) Why is Cl^- more reactive in DMSO than Br^- while the reverse is true in MeOH?

Table 5.p20A. Enthalpies of Transfer of Ions and of *n*-Hexyl Tosylate from Methanol to Dimethyl Sulfoxide at 25°C

Species	$\Delta\Delta H_s$ (kcal/mol)
n-$C_6H_{13}OTs$	−0.4
Cl^-	6.6
N_3^-	3.6
Br^-	2.3
SCN^-	1.0
I^-	−1.1

(b) Why does the rate of thiocyanate (SCN^-) ion change the least of the five nucleophiles in going from MeOH to DMSO?

Table 5.p20B. Activation Parameters for Nucleophile–Hexyl Tosylate Reactions

Nu^-	Solvent	$\Delta H^{\ddagger}$ (kcal/mol)	$\Delta S^{\ddagger}$ (eu)	$\Delta G^{\ddagger}$ (kcal/mol)
Cl^-	MeOH	24.3 ± 0.2	−4.2	25.5
	DMSO	20.2 ± 0.1	−4.4	21.6
N_3^-	MeOH	21.2 ± 0.1	−8.2	23.5
	DMSO	18.6 ± 0.1	−7.8	21.0
Br^-	MeOH	22.9 ± 0.2	−6.4	24.9
	DMSO	20.5 ± 0.1	−5.6	22.0
SCN^-	MeOH	19.8 ± 0.2	−15.8	24.2
	DMSO	20.0 ± 0.2	−12.4	23.7
I^-	MeOH	22.4 ± 0.1	−6.0	23.9
	DMSO	20.9 ± 0.2	−5.8	22.8

(c) Why does thiocyanate ion have the most negative entropy of activation?

Table 5.p20C. Rates[a] of Displacement on Hexyl Tosylate by Nucleophiles

Nu^-	Solvent	k (40°C)	k (30°C)	k (20°C)
Cl^-	DMSO	50.4 ± 0.4	16.7 ± 0.1	5.06 ± 0.04
	MeOH	0.0852 ± 0.0004	0.0226 ± 0.0003	0.00550 ± 0.00007
N_3^-	DMSO	135 ± 0.7	48.3 ± 0.2	16.1 ± 0.1
	MeOH	1.66 ± 0.01	0.514 ± 0.002	0.152 ± 0.001
Br^-	DMSO	17.8 ± 0.1	5.69 ± 0.04	1.75 ± 0.01
	MeOH	0.250 ± 0.002	0.0721 ± 0.0007	0.0191 ± 0.0002
SCN^-	DMSO	1.11 ± 0.01	0.365 ± 0.003	0.115 ± 0.002
	MeOH	0.481 ± 0.005	0.165 ± 0.002	0.0512 ± 0.0005
I^-	DMSO	5.50 ± 0.07	1.75 ± 0.02	16.0 ± 0.1 (50°)
	MeOH	0.956 ± 0.007	0.275 ± 0.001	0.0767 ± 0.0006

a. $k_2 \times 10^4$ liter/(mol·s). Error limits are standard deviations.

(d) Do you see any correlation between softness (which is in the order I^- > ^-SCN > N_3^- > Br^- > Cl^-) and the effect of solvent on rate?

6

Polar Addition and Elimination Reactions

Addition and elimination processes are the reverse of one another. There is a close relationship between the two reactions, and in many systems the reaction can occur in either sense. For example, hydration of alkenes and dehydration of alcohols are both familiar reactions that are related as an addition–elimination pair.

$$\mathrm{RCH{=}CHR' + H_2O \xrightarrow{H^+} RCH(OH)CH_2R'}$$

$$\mathrm{RCH(OH)CH_2R' \xrightarrow{H^+} RCH{=}CHR' + H_2O}$$

The addition and elimination reactions can proceed by similar mechanistic paths but in opposite directions. In these circumstances, mechanistic conclusions about the addition reaction are applicable to the elimination reaction and vice versa. The principle of microscopic reversibility states that the mechanism (pathway) traversed in a reversible reaction is the same in the reverse as in the forward direction. Thus, if an addition–elimination system proceeds by a reversible mechanism, the intermediates and transition states involved in the addition process are the same as in the elimination reaction. The reversible acid-catalyzed reaction of alkenes with water represents such a system.

The initial discussion will focus on addition reactions. The discussion here is restricted to reactions that involve polar or ionic mechanisms. There are other important classes of addition reactions which are discussed elsewhere; these include concerted addition reactions proceeding through nonpolar transition states (Chapter 11), radical additions (Chapter 12), photochemical additions (Chapter 13), and nucleophilic addition to electrophilic alkenes (Part B, Chapter 1, Section 1.10).

Several limiting generalized mechanisms can be described for polar additions.

(A)

$$E{-}Y \longrightarrow E^+ + Y^-$$

$$E^+ + \gt C{=}C\lt \longrightarrow \gt\underset{E}{C}{-}\overset{+}{C}\lt$$

$$\gt\underset{E}{C}{-}\overset{+}{C}\lt + Y^- \longrightarrow \gt\underset{E}{C}{-}\underset{Y}{C}\lt$$

(B)

$$E{-}Y + \gt C{=}C\lt \longrightarrow \gt\underset{E}{C}{-}\overset{+}{C}\lt + Y^- \longrightarrow \gt\underset{E}{C}{-}\underset{Y}{C}\lt$$

(C)

$$2E{-}Y + \gt C{=}C\lt \longrightarrow \underset{Y-E}{\gt C}{=}\overset{Y-E}{C\lt} \longrightarrow \gt\underset{E}{C}{-}\overset{Y}{C}\lt + E{-}Y$$

Mechanism A implies that a carbocation is generated which is free of the counterion Y^- at its formation. This mechanism involves prior dissociation of the electrophilic reagent. Mechanism B also involves a carbocation intermediate but one that is generated in the presence of an anion and exists initially as an ion pair. Depending on the mutual reactivity of the two ions, they might or might not become free of one another before combining to give product. Both mechanisms A and B would be referred to as Ad_E2 reactions; that is, they are *bimolecular electrophilic additions.* Mechanism C is a process that has been observed for several electrophilic additions. It implies transfer of the electrophilic and nucleophilic components of the reagent from two separate molecules. It would be described as a *termolecular electrophilic addition,* Ad_E3. Examples of each of these types of processes will be encountered as specific reactions are discussed in the sections that follow. The discussion here will focus on a few reactions that have received the most detailed mechanistic study. Synthetically important polar additions are described in Chapter 4 of Part 4.

6.1. Addition of Hydrogen Halides to Alkenes

The addition of hydrogen halides to alkenes has been studied from a mechanistic point of view over a period of many years. One of the first aspects of the mechanism to be established was its regioselectivity, that is, the direction of addition. A reaction is described as *regioselective* if an unsymmetrical alkene gives a predominance of one of the two possible addition products; the term *regiospecific* is used if one product is formed exclusively.[1]

1. A. Hassner, *J. Org. Chem.* **33**, 2684 (1968).

$$\mathrm{RCH{=}CH_2 + X{-}Y \longrightarrow \underset{\substack{|\\ X}}{RCHCH_2Y}\ (major) + \underset{\substack{|\\ Y}}{RCHCH_2X}\ (minor)}$$

regioselective reaction

$$\mathrm{RCH{=}CH_2 + X{-}Y \longrightarrow \underset{\substack{|\\ X}}{RCHCH_2Y}}$$

regiospecific reaction

In the addition of hydrogen halides, it is generally found that the halogen atom becomes attached to the most substituted carbon atom of the alkene. The statement of this general observation is called *Markownikoff's rule.* The basis for this regioselectivity lies in the relative ability of the carbon atoms to accept positive charge. The addition of hydrogen halide is initiated by an electrophilic attack involving transfer of a proton to the alkene. The new C—H bond is formed from the π electrons of the carbon–carbon double bond. It is easy to see that if a carbocation is formed, the halide would be added to the more substituted carbon, since addition of the proton at the less substituted carbon atom provides the more stable carbocation intermediate.

$$\mathrm{R_2C{=}CHR \xrightarrow{HX} R_2\overset{+}{C}CH_2R \xrightarrow{X^-} \underset{\substack{|\\ X}}{R_2CCH_2R}}$$

more favorable

$$\mathrm{R_2C{=}CHR \xrightarrow{HX} R_2CH\overset{+}{C}HR \xrightarrow{X^-} \underset{\substack{|\\ X}}{R_2CHCHR}}$$

less favorable

As will be indicated when the mechanism is discussed in more detail, discrete carbocations are not formed in all cases. An unsymmetrical alkene will nevertheless follow the Markownikoff rule, because the partial positive charge that develops will be located primarily at the carbon most able to accommodate the electron deficiency, that is, the more substituted one.

The regioselectivity of addition of hydrogen bromide to alkenes can be complicated if a free-radical chain addition occurs in competition with the ionic addition. The free-radical reaction is readily initiated by peroxide impurities or by light and leads to the anti-Markownikoff addition product. The mechanism of this reaction will be considered fully in Chapter 12. Conditions which minimize the competing radical addition include use of high-purity reagent and solvent, exclusion of light, and addition of free-radical inhibitors.[2]

The studies that have been applied to determining mechanistic details of hydrogen halide addition to alkenes have focused on the kinetics and stereochemistry of the reaction and on the effect of added nucleophiles. The kinetic studies often reveal complex rate expressions that demonstrate that more than one process contributes to the overall reaction rate. For addition of hydrogen bromide or

2. D. J. Pasto, G. R. Meyer, and B. Lepeska, *J. Am. Chem. Soc.* **96**, 1858 (1974).

hydrogen chloride to alkenes, an important contribution to the overall rate is often made by a third-order process:

$$\text{rate} = k[\text{alkene}][\text{HX}]^2$$

Among the cases in which this type of kinetics has been observed are the addition of hydrogen chloride to 2-methyl-1-butene, 2-methyl-2-butene, 1-methylcyclopentene,[3] and cyclohexene.[4] The addition of hydrogen bromide to cyclopentene also follows a third-order rate expression.[2] The transition state associated with the third-order rate expression involves proton transfer to the alkene from one hydrogen halide molecule and capture of the halide ion from the second:

$$\text{H–X} \cdots \text{C=C} \cdots \text{H–X} \longrightarrow H^+ + \text{H–C–C–X} + X^-$$

The reaction presumably involves interaction of a complex formed between the alkene and hydrogen halide with the second hydrogen halide molecule, since there is little likelihood of productive termolecular collisions.

$$\text{C=C} + \text{H–X} \underset{}{\overset{\text{fast}}{\rightleftharpoons}} \text{C=C}\cdots\text{H–X}$$

$$\text{C=C}\cdots\text{H–X} + \text{HX} \xrightarrow{\text{slow}} \text{H–C–C–X} + \text{HX}$$

The stereochemistry of addition of hydrogen halides to unconjugated alkenes is predominantly *anti.* This is true for addition of hydrogen bromide to 1,2-dimethylcyclohexene,[5] cyclohexene,[6] 1,2-dimethylcyclopentene,[7] cyclopentene,[2] *cis-* and *trans*-2-butene,[2] and 3-hexene,[2] among others. *Anti* stereochemistry is also dominant for addition of hydrogen chloride to 1,2-dimethylcyclohexene[8] and 1-methylcyclopentene.[3] Temperature and solvent can modify the stereochemistry, however. For example, although the addition of hydrogen chloride to 1,2-dimethylcyclohexene is *anti* near room temperature, *syn* addition dominates at −78°C.[9]

Anti stereochemistry can be explained by a mechanism in which the alkene interacts simultaneously with the proton-donating hydrogen halide and with a source of halide ion, either a second molecule of hydrogen halide or a free halide ion. The

3. Y. Pocker, K. D. Stevens, and J. J. Champoux, *J. Am. Chem. Soc.* **91**, 4199 (1969); Y. Pocker and K. D. Stevens, *J. Am. Chem. Soc.* **91**, 4205 (1969).
4. R. C. Fahey, M. W. Monahan, and C. A. McPherson, *J. Am. Chem. Soc.* **92**, 2810 (1970).
5. G. S. Hammond and T. D. Nevitt, *J. Am. Chem. Soc.* **76**, 4121 (1954).
6. R. C. Fahey and R. A. Smith, *J. Am. Chem. Soc.* **86**, 5035 (1964); R. C. Fahey, C. A. McPherson, and R. A. Smith, *J. Am. Chem. Soc.* **96**, 4534 (1974).
7. G. S. Hammond and C. H. Collins, *J. Am. Chem. Soc.* **82**, 4323 (1960).
8. R. C. Fahey and C. A. McPherson, *J. Am. Chem. Soc.* **93**, 1445 (1971).
9. K. B. Becker and C. A. Grob, *Synthesis*, 789 (1973).

anti stereochemistry is consistent with the expectation that the attack of halide ion would be from the side opposite that from which proton delivery occurs.

HX + [alkene with H–X above] ⟶ [alkene with H---X above, H–X below] ⟶ [anti adduct with H and X] + H^+ + X

A significant variation in the stereochemistry is observed when the double bond is conjugated with a group that can stabilize a carbocation intermediate. Most of the examples involve an aryl substituent. Examples of alkenes that give primarily *syn* addition are *cis*- and *trans*-1-phenylpropene,[10] *cis*- and *trans*-*β*-*t*-butylstyrene,[11] 1-phenyl-4-*t*-butylcyclohexene,[12] and indene.[13] The mechanism proposed for these additions features an ion pair as the key intermediate. Because of the greater stability of the carbocations in these systems, concerted attack by halide ion is not required for carbon–hydrogen bond formation. If the ion pair formed by alkene protonation collapses to product faster than rotation takes place, the result will be *syn* addition, since the proton and halide ion are initially on the same side of the molecule.

$$\mathrm{ArCH{=}CHR + HX \longrightarrow Ar\overset{+}{C}H{-}CH(H)R\ \ \bar{X} \longrightarrow ArCH(X)CH_2R}$$

Kinetic studies of the addition of hydrogen chloride to styrene support the conclusion that an ion pair mechanism operates when aromatic conjugation is a factor. The reaction is first-order in hydrogen chloride, indicating that only one molecule of hydrogen chloride participates in the rate-determining step.[14]

There is usually a competing reaction with solvent when hydrogen halide additions to alkenes are carried out in nucleophilic solvents:

$$\mathrm{PhCH{=}CH_2 \xrightarrow[HOAc]{HCl} PhCH(Cl)CH_3 + PhCH(OCOCH_3)CH_3} \qquad \text{Ref. 14}$$

$$\text{cyclohexene} \xrightarrow[HOAc]{HCl} \text{chlorocyclohexane (Cl)} + \text{cyclohexyl acetate } (\mathrm{OCOCH_3}) \qquad \text{Ref. 4}$$

It is not difficult to incorporate this result into the general mechanism for hydrogen halide additions. These products are formed as the result of solvent competing with halide ion as the nucleophilic component in the addition. Solvent addition can occur via the concerted mechanism or by capture of a carbocation intermediate.

10. M. J. S. Dewar and R. C. Fahey, *J. Am. Chem. Soc.* **85**, 3645 (1963).
11. R. J. Abraham and J. R. Monasterios, *J. Chem. Soc., Perkin Trans. 2*, 574 (1975).
12. K. D. Berlin, R. D. Lyerla, D. E. Gibbs, and J. P. Devlin, *J. Chem. Soc., Chem. Commun.*, 1246 (1970).
13. M. J. S. Dewar and R. C. Fahey, *J. Am. Chem. Soc.* **85**, 2248 (1963).
14. R. C. Fahey and C. A. McPherson, *J. Am. Chem. Soc.* **91**, 3865 (1969).

Addition of a halide salt increases the likelihood of capture of a carbocation intermediate by halide ion. The effect of added halide salt can be detected kinetically. For example, the presence of tetramethylammonium chloride increases the rate of addition of hydrogen chloride to cyclohexene.[8] Similarly, lithium bromide increases the rate of addition of hydrogen bromide to cyclopentene.[5]

Skeletal rearrangements have been observed in hydrogen halide additions when hydrogen or carbon migration leading to a more stable carbocation can occur.

$$(CH_3)_2CHCH{=}CH_2 \xrightarrow[CH_3NO_2]{HCl} (CH_3)_2CHCH(Cl)CH_3 \ (40\%) + (CH_3)_2CCl CH_2CH_3 \ (60\%)$$ Ref. 3

$$(CH_3)_3CCH{=}CH_2 \xrightarrow[CH_3NO_2]{HCl} (CH_3)_2CClCH(CH_3)_2 \ (83\%) + (CH_3)_3CCH(Cl)CH_3 \ (17\%)$$ Ref. 3

Even though the rearrangements suggest that discrete carbocation intermediates are involved, these reactions frequently show kinetics consistent with the presence of at least two hydrogen chloride molecules in the rate-determining transition state. A termolecular mechanism in which the second hydrogen chloride molecule assists in the ionization of the electrophile has been suggested.[3]

$$H{-}Cl\cdots H{-}Cl$$
$$(CH_3)_2CHCH{=}CH_2 \rightarrow (CH_3)_2CH\overset{+}{C}HCH_3 + [Cl{-}H{-}Cl]^-$$
$$(CH_3)_2CH\overset{+}{C}HCH_3 \rightarrow (CH_3)_2\overset{+}{C}CH_2CH_3$$
$$(CH_3)_2CH\overset{+}{C}HCH_3 + [Cl{-}H{-}Cl]^- \rightarrow (CH_3)_2CHCH(Cl)CH_3 + HCl$$

and

$$(CH_3)_2\overset{+}{C}CH_2CH_3 + [Cl{-}H{-}Cl]^- \rightarrow (CH_3)_2CClCH_2CH_3 + HCl$$

The addition of hydrogen halides to dienes can result in either 1,2- or 1,4 addition. The extra stability of the allyl cation that can be formed by proton transfer to a diene makes the ion pair mechanism favorable. 1,3-Pentadiene, for example, gives a mixture of products favoring the 1,2-addition product by a ratio of from 1.5:1 to 3.4:1, depending on the temperature and solvent.[15]

$$CH_3CH{=}CHCH{=}CH_2 \xrightarrow{DCl} CH_3CH(Cl)CH{=}CHCH_2D \ (22\text{–}38\%) + CH_3CH{=}CHCH(Cl)CH_2D \ (62\text{–}78\%)$$

With 1-phenyl-1,3-butadiene, the addition is exclusively at the 3,4 double bond. This reflects the greater stability of this product, which retains a styrene-type conjugation. Initial protonation at C-4 is favored by the fact that the resulting

15. J. E. Nordlander, P. O. Owuor, and J. E. Haky, *J. Am. Chem. Soc.* **101**, 1288 (1979).

carbocation benefits from both allylic and benzylic stabilization.

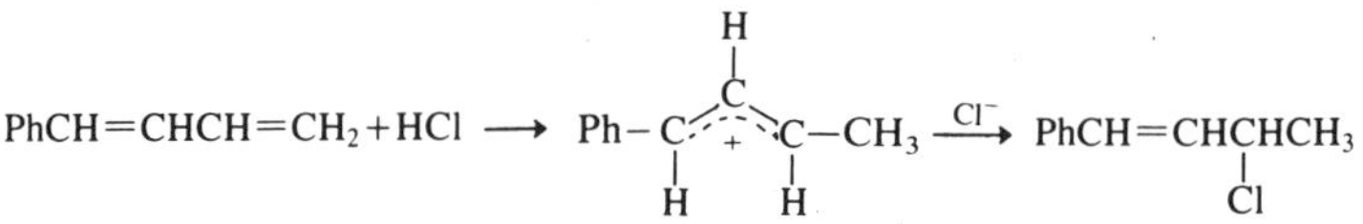

$$PhCH{=}CHCH{=}CH_2 + HCl \longrightarrow Ph{-}\overset{H}{C}{\cdots}\overset{H}{C}{\cdots}\overset{H}{C}{-}CH_3 \xrightarrow{Cl^-} PhCH{=}CH\underset{Cl}{C}HCH_3$$

The kinetics of this reaction are second-order, as could be expected for the formation of a stable carbocation by an Ad_E2 mechanism.[16]

The additions of hydrogen chloride and hydrogen bromide to norbornene are interesting cases because such factors as the stability and facile rearrangement of the norbornyl cation come into consideration. Addition of deuterium bromide to norbornene gives *exo*-norbornyl bromide. Degradation to locate the deuterium atom shows that about half of the product has formed via the bridged norbornyl cation, which leads to deuterium at both the 3- and 7-positions. The *exo* orientation of the bromine atom and the redistribution of the deuterium indicate the involvement of the bridged ion.

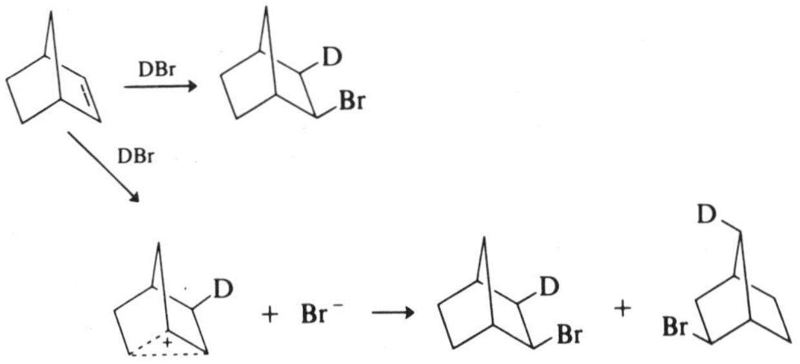

DBr, DBr, D, Br, D, + Br⁻, D, Br, +, D, Br

Similar studies have been carried out on the addition of hydrogen chloride to norbornene.[17]

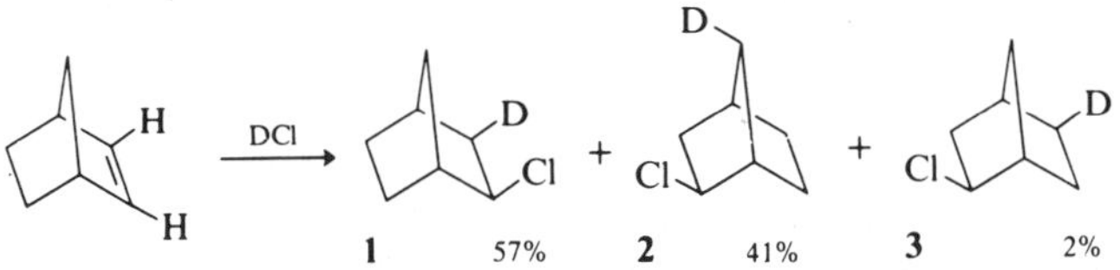

H, H, DCl, D, Cl, +, D, Cl, +, D, Cl

1 57% **2** 41% **3** 2%

Again, the chloride is almost exclusively the *exo* isomer. The distribution of deuterium in the product was determined by NMR. The fact that **1** and **2** are formed in unequal amounts excludes the bridged ion as the only intermediate.[18] The excess of **1** over **2** indicates that some *syn* addition occurs by ion pair collapse before the bridged ion achieves symmetry with respect to the chloride ion. If the amount of **2** is taken as an indication of the extent of bridged-ion involvement, one would

16. K. Izawa, T. Okuyama, T. Sakagami, and T. Fueno, *J. Am. Chem. Soc.* **95**, 6752 (1973).
17. H. Kwart and J. L. Nyce, *J. Am. Chem. Soc.* **86**, 2601 (1964); J. K. Stille, F. M. Sonnenbeg, and T. H. Kinstle, *J. Am. Chem. Soc.* **88**, 4922 (1966).
18. H. C. Brown and K.-T. Liu, *J. Am. Chem. Soc.* **97**, 600 (1975).

conclude that 82% of the reaction proceeds through this intermediate, which must give equal amounts of **1** and **2**.

6.2. Acid-Catalyzed Hydration and Related Addition Reactions

The formation of alcohols by acid-catalyzed addition of water to alkenes is a fundamental organic reaction. At the most rudimentary mechanistic level, it can be viewed as involving a carbocation intermediate. The alkene is protonated and the carbocation is then captured by water.

$$RCH{=}CH_2 \xrightarrow{H^+} R\overset{+}{C}HCH_3 \xrightarrow{H_2O} \underset{\displaystyle OH}{\underset{|}{RCHCH_3}} + H^+$$

This mechanism explains the observed formation of the more highly substituted alcohol from unsymmetrical alkenes (Markownikoff's rule). A number of other points must be considered in order to provide a more complete picture of the mechanism. Is the protonation step reversible? Is there a discrete carbocation intermediate, or does the nucleophile become involved before proton transfer is complete? Can other reactions of the carbocation, such as rearrangement, compete with capture by water?

Much of the mechanistic work on hydration reactions has been done with conjugated alkenes, particularly styrenes. With styrenes, the rate of hydration is increased by electron-releasing substituents, and there is an excellent correlation with σ^+.[19] A substantial solvent isotope effect, $k_{H_2O}/k_{D_2O} = 2\text{–}4$, is observed. Both of these observations are in accord with a rate-determining protonation to give a carbocation intermediate. Capture of the resulting cation by water is evidently fast relative to deprotonation. This has been demonstrated by showing that in the early stages of hydration of styrene deuterated at C-2, there is no loss of deuterium from the unreacted alkene that is recovered by quenching the reaction.

$$PhCH{=}CD_2 \overset{H^+}{\rightleftharpoons} Ph\overset{+}{C}HCD_2H \underset{}{\overset{-H^+}{\underset{slow}{\rightleftharpoons}}} PhCH{=}CHD$$

$$Ph\overset{+}{C}HCD_2H \xrightleftharpoons[fast]{H_2O} \underset{\displaystyle OH}{\underset{|}{PhCHCD_2H}}$$

The overall process is reversible, however, and some styrene remains in equilibrium with the alcohol, so exchange eventually occurs.

19. W. M. Schubert and J. R. Keefe, *J. Am. Chem. Soc.* **94**, 559 (1972); W. M. Schubert and B. Lamm, *J. Am. Chem. Soc.* **88**, 120 (1966); W. K. Chwang, P. Knittel, K. M. Koshy, and T. T. Tidwell, *J. Am. Chem. Soc.* **99**, 3395 (1977).

Table 6.1. Rates of Hydration of Some Alkenes in Aqueous Sulfuric Acid[a]

Alkene	k_2 (M^{-1} s^{-1})	k_{rel}
$H_2C{=}CH_2$	1.46×10^{-15}	1
$CH_3CH{=}CH_2$	2.38×10^{-8}	1.6×10^{7}
$CH_3(CH_2)_3CH{=}CH_2$	4.32×10^{-8}	3.0×10^{7}
$(CH_3)_2C{=}CHCH_3$	2.14×10^{-3}	1.5×10^{12}
$(CH_3)_2C{=}CH_2$	3.71×10^{-3}	2.5×10^{12}
$PhCH{=}CH_2$	2.4×10^{-6}	1.6×10^{9}

a. W. K. Chwang, V. J. Nowlan, and T. T. Tidwell, *J. Am. Chem. Soc.* **99**, 7233 (1977).

Alkenes lacking phenyl substituents appear to react by a similar mechanism. Both the observation of general acid catalysis[20] and a solvent isotope effect[21] are consistent with rate-limiting protonation with simple alkenes such as 2-methylpropene and 2,3-dimethyl-2-butene. The observation of general acid catalysis rules out an alternative mechanism for alkene hydration, namely, water attack on an alkene–proton complex. The preequilibrium for formation of such a complex would be governed by the acidity of the solution, and so this mechanism would exhibit specific acid catalysis.

$$RCH{=}CH_2 + H^+ \rightleftharpoons RCH\overset{H^+}{\overset{\vdots}{=}}CH_2$$

$$RCH\overset{H^+}{\overset{\vdots}{=}}CH_2 + H_2O \xrightarrow{\text{slow}} \underset{\displaystyle OH}{RCHCH_3}$$

Relative rate data in aqueous sulfuric acid for a series of simple alkenes reveal that the reaction is accelerated by alkyl substituents. This is expected since alkyl groups both increase the electron density of the double bond and stabilize the carbocation intermediate. Table 6.1 gives some representative data. These same reactions show solvent isotope effects consistent with the reaction proceeding through a rate-determining protonation.[22]

Other nucleophilic solvents can add to alkenes under the influence of strong acid catalysts. The mechanism is presumably analogous to that for hydration, with the solvent replacing water as the nucleophile. In the presence of the strongest acid catalysts, reaction probably occurs via discrete carbocation intermediates, whereas in the case of weaker acids, reaction of the solvent with an alkene–acid complex may be involved. In the addition of acetic acid ito *cis*- or *trans*-2-butene, the use of DBr as the catalyst results in stereospecific *anti* addition, whereas use of a stronger

20. A. J. Kresge, Y. Chiang, P. H. Fitzgerald, R. S. McDonald, and G. H. Schmid, *J. Am. Chem. Soc.* **93**, 4907 (1971).
21. V. Gold and M. A. Kessick, *J. Chem. Soc.* 6718 (1965).
22. V. J. Nowlan and T. T. Tidwell, *Acc. Chem. Res.* **10**, 252 (1977).

acid, trifluoromethanesulfonic acid, leads to loss of stereospecificity. This difference in stereochemistry can be explained by a stereospecific Ad_E3 mechanism in the case of hydrogen bromide and an Ad_E2 mechanism in the case of trifluoromethanesulfonic acid.[23] The dependence on acid strength would reflect the degree to which nucleophilic solvent participation was required to complete proton transfer.

$$CH_3CH{=}CHCH_3 \overset{DBr}{\rightleftharpoons} CH_3CH{=}CHCH_3 \cdots D{-}Br$$

$$CH_3CH{=}CHCH_3 \cdots D{-}Br + CH_3CO_2H \rightarrow CH_3CH(O_2CCH_3){-}CHDCH_3$$

anti-addition

$$CH_3CH{=}CHCH_3 + CF_3SO_3D \rightarrow CH_3\overset{+}{C}HCHDCH_3$$

$$CH_3\overset{+}{C}HCHDCH_3 + CH_3CO_2H \rightarrow CH_3CH(O_2CCH_3){-}CHDCH_3$$

nonstereospecific addition

Strong acids also catalyze the addition of alcohols to alkenes to give ethers, and the mechanistic studies which have been done indicate that the reaction closely parallels the hydration process.[24]

Trifluoroacetic acid adds to alkenes without the necessity of a stronger acid catalyst.[25] The mechanistic features of this reaction are similar to those of addition of water catalyzed by strong acids. For example, there is a substatial isotope effect when CF_3CO_2D is used,[26] and the reaction rates of substituted styrenes are correlated with σ^+.[27]

The reactivity of carbon–carbon double bonds toward acid-catalyzed addition of water is greatly increased by electron-releasing substituents, and the reaction of vinyl ethers with water in acid solution is an example that has been extensively studied. With these substrates, the initial addition products are unstable hemiacetals, which decompose to a ketone and alcohol. Nevertheless, the hydration step is rate-determining, and so the kinetic results pertain to this step. The mechanistic features that have been examined are similar to those for hydration of simple alkenes.

23. D. J. Pasto and J. F. Gadberry, *J. Am. Chem. Soc.* **100**, 1469 (1978).
24. N. C. Deno, F. A. Kish, and H. J. Peterson, *J. Am. Chem. Soc.* **87**, 2157 (1965).
25. P. E. Peterson and G. Allen, *J. Am. Chem. Soc.* **85**, 3608 (1963); A. D. Allen and T. T. Tidwell, *J. Am. Chem. Soc.* **104**, 3145 (1982).
26. J. J. Dannenberg, B. J. Goldberg, J. K. Barton, K. Dill, D. M. Weinwurzel, and M. O. Longas, *J. Am. Chem. Soc.* **103**, 7764 (1981).
27. A. D. Allen, M. Rosenbaum, N. O. L. Seto, and T. T. Tidwell, *J. Org. Chem.* **47**, 4234 (1982).

Proton transfer is the rate-determining step, as is demonstrated by general acid catalysis and solvent isotope effect data.[28]

$$\underset{\displaystyle R'}{R''CH{=}\underset{|}{C}OR} \xrightarrow[\text{slow}]{H^+} \underset{\displaystyle R'}{R''CH_2\underset{|}{C}{=}\overset{+}{O}R} \xrightarrow{H_2O} \underset{\displaystyle R'}{R''CH_2\overset{\displaystyle OH}{\overset{|}{\underset{|}{C}}}OR} + H^+ \longrightarrow R''CH_2\underset{\displaystyle O}{\underset{\|}{C}}R' + ROH$$

Strained alkenes show enhanced reactivity toward acid-catalyzed hydration. *trans*-Cyclooctene is about 2500 times as reactive as the *cis* isomer.[29] This reflects the higher ground state energy of the strained alkene.

6.3. Addition of Halogens

Alkene chlorinations and brominations are among the most general of organic reactions, and mechanistic study of these reactions has provided much detailed insight into the addition reactions of alkenes.[30] Two of the principal points at issue in the description of the mechanism for a given reaction are: (1) Is there a discrete positively charged intermediate, or is the addition concerted? (2) If there is a positively charged intermediate, is it a carbocation or a cyclic halonium ion? Stereochemical studies have provided much of the data pertaining to these points. The results of numerous stereochemical studies can be generalized as follows. For brominations, *anti* addition is preferred for alkenes that do not contain substituent groups that would strongly stabilize a carbocation intermediate. When the alkene is conjugated with an aryl group, the extent of *syn* addition becomes much larger, and *syn* addition can become the dominant pathway. Chlorination is not as stereospecific as bromination but tends to follow the same pattern. Some specific cases are given in Table 6.2.

Interpretaton of reaction stereochemistry has focused attention on the role played by cyclic halonium ions.

$$RCH{=}CH_2 \xrightarrow{Br_2} \underset{\text{bridged bromonium ion}}{R{-}\underset{\displaystyle H}{\underset{|}{C}}\overset{\overset{+}{Br}}{\diagup\!\diagdown}CH_2} \rightleftharpoons \underset{\text{open carbocation}}{R\overset{+}{C}H{-}CH_2Br}$$

28. A. J. Kresge and H. J. Chen, *J. Am. Chem. Soc.* **94**, 2818 (1972); A. J. Kresge, D. S. Sagatys, and H. L. Chen, *J. Am. Chem. Soc.* **99**, 7228 (1977).
29. Y. Chiang and A. J. Kresge, *J. Am. Chem. Soc.* **107**, 6363 (1985).
30. Reviews: D. P. de la Mare and R. Bolton, in *Electrophilic Additions to Unsaturated Systems*, second edition, Elsevier, New York, 1982, pp. 136–197; G. H. Schmidt and D. G. Garratt, in *The Chemistry of Double Bonded Functional Groups*, Supplement A, Part 2, S. Patai (ed.), Wiley-Interscience, New York, 1977, Chapter 9.

Table 6.2. Stereochemistry of Halogenation

Alkene	Solvent	Ratio *anti*:*syn*	Ref.
Bromination			
cis-2-Butene	CH_3CO_2H	>100:1	a
trans-2-Butene	CH_3CO_2H	>100:1	a
Cyclohexene	CCl_4	very large	b
Z-1-Phenylpropene	CCl_4	83:17	c
E-1-Phenylpropene	CCl_4	88:12	c
E-2-Phenylbutene	CH_3CO_2H	68:32	a
Z-2-Phenylbutene	CH_3CO_2H	63:37	a
cis-Stilbene	CCl_4	>10:1	d
	CH_3NO_2	1:9	d
Chlorination			
cis-2-Butene	none	>100:1	e
	CH_3CO_2H	>100:1	f
trans-2-Butene	none	>100:1	e
	CH_3CO_2H	>100:1	f
Cyclohexene	none	>100:1	g
E-1-Phenylpropene	CCl_4	45:55	f
	CH_3CO_2H	41:59	f
Z-1-Phenylpropene	CCl_4	32:68	f
	CH_3CO_2H	22:78	f
cis-Stilbene	$ClCH_2CH_2Cl$	92:8	h
trans-Stilbene	$ClCH_2CH_2Cl$	65:35	h

a. J. H. Rolston and K. Yates, *J. Am. Chem. Soc.* **91**, 1469, 1477 (1969).
b. S. Winstein, *J. Am. Chem. Soc.* **64**, 2792 (1942).
c. R. C. Fahey and H.-J. Schneider, *J. Am. Chem. Soc.* **90**, 4429 (1968).
d. R. E. Buckles, J. M. Bader, and R. L. Thurmaier, *J. Org. Chem.* **27**, 4523 (1962).
e. M. L. Poutsma, *J. Am. Chem. Soc.* **87**, 2172 (1965).
f. R. C. Fahey and C. Schubert, *J. Am. Chem. Soc.* **87**, 5172 (1965).
g. M. L. Poutsma, *J. Am. Chem. Soc.* **87**, 2161 (1965).
h. R. E. Buckles and D. F. Knaack, *J. Org. Chem.* **25**, 20 (1960).

If the addition of Br^+ to the alkene results in a bromnium ion, the *anti* stereochemistry can be readily explained. Nucleophilic ring opening by bromide ion would occur by back-side attack at carbon, with rupture of one of the C—Br bonds, giving overall *anti* additions.

$$\text{bromonium ion} + Br^- \longrightarrow \text{anti-dibromide}$$

On the other hand, a freely rotating open carbocation would be expected to give both the *syn* and *anti* addition products. If the principal intermediate were an ion pair that collapsed faster than rotation about the C—C bond, *syn* addition could predominate.

Br—Br, C=C → Br^- Br, $\overset{+}{C}$—C → Br Br, C—C

Whether or not a bridged intermediate or a carbocation is involved in bromination depends primarily on the stability of the potential cation. Aliphatic systems normally go through the bridged intermediate, but styrenes are borderline cases. When the phenyl ring has electron-releasing substituents, there is sufficient stabilization to permit carbocation formation, while electron-attracting groups favor the bridged intermediate.[31] As a result, styrenes with electron-attracting substituents give a higher proportion of the *anti* addition product.

The stereochemistry of chlorination can be explained in similar terms. Chlorine would be expected to be a somewhat poorer bridging group than bromine because it is less polarizable and more reluctant to become positively charged. Comparison of the data for *E*- and *Z*-1-phenylpropene in bromination and chlorination confirms this expectation (see Table 6.2). Although *anti* addition is dominant for bromination, *syn* addition is slightly preferred for chlorination. Styrenes generally appear to react with chlorine via ion pair intermediates.[32] For nonconjugated alkenes, stereospecific *anti* addition is usually observed for both halogens.

There is direct evidence for the existence of bromonium ions. The bromonium ion related to propene can be observed by NMR spectroscopy when 1-bromo-2-fluoropropane is subjected to superacid conditions. The terminal bromine adopts a bridging position in the resulting cation.

$$CH_3\underset{F}{\underset{|}{C}}HCH_2Br \xrightarrow[SO_2,\ -60°C]{SbF_5} CH_3CH\underset{\overset{+}{Br}}{—}CH_2 + SbF_6^- \qquad \text{Ref. 33}$$

Bromonium ions can also be produced by an electrophilic attack on alkenes by a species that should generate a positive bromine.

$$(CH_3)_2C{=}C(CH_3)_2 + Br{-}C{\equiv}\overset{+}{N}{-}\overset{-}{S}bF_5 \longrightarrow (CH_3)_2C\overset{\overset{+}{Br}}{—}C(CH_3)_2 + CNSbF_5 \qquad \text{Ref. 34}$$

The highly hindered alkene adamantylideneadamantane, forms a bromonium ion which crystallizes as a tribromide salt. An X-ray crystal structure (Fig. 6.1) has

31. M. F. Ruasse, A. Argile, and J. E. Dubois, *J. Am. Chem. Soc.* **100**, 7645 (1978).
32. K. Yates and H. W. Leung, *J. Org. Chem.* **45**, 1401 (1980).
33. G. A. Olah, J. M. Bollinger, and J. Brinich, *J. Am. Chem. Soc.* **90**, 2587 (1968).
34. G. A. Olah, P. Schilling, P. W. Westerman, and H. C. Lin, *J. Am. Chem. Soc.* **96**, 3581 (1974).

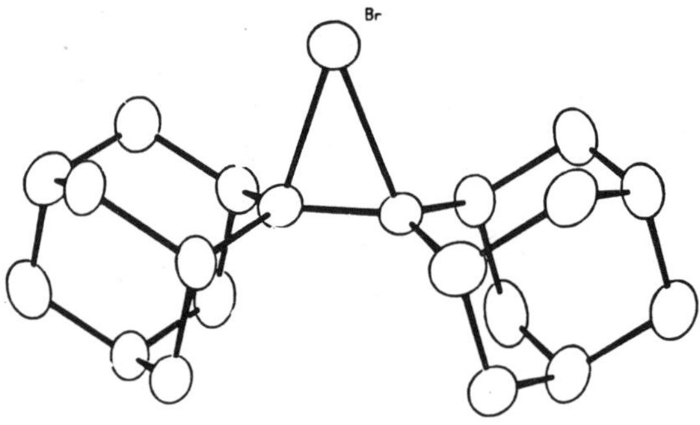

Fig. 6.1. Crystal structure of bromonium ion from adamantylideneadamantane. [Reproduced from Ref. 35.]

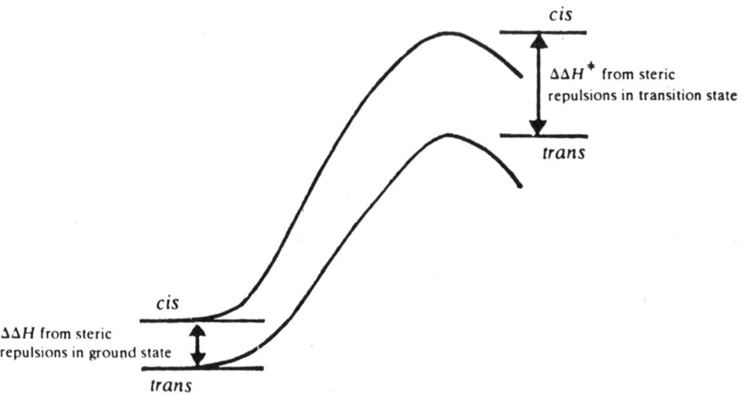

Fig. 6.2. Enthalpy differences of starting alkenes and transition states in bromination.

confirmed the cyclic nature of the bromonium ion species.[35] This particular bromonium ion does not react further because of extreme steric hindrance to back-side approach by bromide ion.

An interpretation of activation parameters has led to the conclusion that the bromination transition state resembles a three-membered ring, even in the case of alkenes that eventually react via open carbocation intermediates. It was found that for *cis–trans* pairs of alkenes the difference in enthalpy at the transition state for bromination was *greater than* the enthalpy difference for the isomeric alkenes, as shown in Fig. 6.2. This finding indicates that the steric repulsions between *cis* groups increase on going from reactant to transition state. This would be consistent with a cyclic transition state in which the *cis* substituents were still eclipsed and somewhat closer together than in the alkene.[36]

35. H. Slebocka-Tilk, R. G. Ball, and R. S. Brown, *J. Am. Chem. Soc.* **107,** 4504 (1985).
36. K. Yates and R. S. McDonald, *J. Org. Chem.* **38,** 2465 (1973).

The kinetics of brominations are often complex, with at least three terms making contributions under given conditions:

$$\text{rate} = k_1[\text{alkene}][\text{Br}_2] + k_2[\text{alkene}][\text{Br}_2]^2 + k_3[\text{alkene}][\text{Br}_2][\text{Br}^-]$$

In methanol, pseudo-second-order kinetics are observed when a high concentration of Br^- is present.[37] Under these conditions, the dominant contribution to the overall rate comes from the third term of the general expression. The occurrence of third-order terms suggests the possibility of a mechanism similar to the Ad_E3 mechanism for addition of hydrogen halides to alkenes, namely, attack of halide ion on an alkene–halogen complex.

$$\text{>C=C<} + Br_2 \rightleftharpoons \text{>C}\overset{Br_2}{\overset{\vdots}{=}}\text{C<}$$

$$\text{>C}\overset{Br_2}{\overset{\vdots}{=}}\text{C<} + Br^- \longrightarrow \text{>}\underset{Br}{C}\text{—}\overset{Br}{C}\text{<} + Br^-$$

As in the case of hydrogen halide additions, this mode of attack should lead to *anti* addition.

In nonpolar solvents, the observed rate is frequently found to be described as a sum of the first two terms in the general expression. The second-order term is interpreted as the collapse of an alkene–halogen complex to an ion pair, which then goes on to product. The cationic intermediate can have the bromonium ion structure.

$$\text{>C=C<} \xrightarrow{Br_2} \text{>C}\overset{Br_2}{\overset{\vdots}{=}}\text{C<} \longrightarrow \text{>C}\overset{Br^+\ Br^-}{\text{—}}\text{C<} \longrightarrow \underset{Br}{C}\text{—}\overset{Br}{C}$$

There is good evidence that the initial complex is a charge-transfer complex. The formation of the charge-transfer complex can be observed spectroscopically under appropriate conditions, as can its subsequent disappearance with kinetics corresponding to the formation of bromination product.[38,39]

Several mechanisms have been considered for the term that is overall third-order and second-order in bromine.[40]

(1) $$\text{>C=C<} \xrightarrow{Br_2} \text{>C}\overset{Br\text{—}Br\ \ Br\text{—}Br}{\overset{\vdots}{=}}\text{C<} \xrightarrow{\text{slow}} \text{>C}\overset{Br^+\ Br_3^-}{\text{—}}\text{C<} \longrightarrow \underset{Br}{C}\text{—}\overset{Br}{C} + Br_2$$

37. J.-E. Dubois and G. Mouvier, *Tetrahedron Lett.*, 1325 (1963); *Bull. Soc. Chim. Fr.* 1426 (1968).
38. S. Fukuzumi and J. K. Kochi, *J. Am. Chem. Soc.* **104**, 7599 (1982).
39. G. Belluci, R. Bianchi, and R. Ambrosetti, *J. Am. Chem. Soc.* **107**, 2464 (1985).
40. (a) G. Belluci, R. Bianchi, R. A. Ambrosetti, and G. Ingrosso, *J. Org. Chem.* **50**, 3313 (1985); G. Belluci, G. Berti, R. Bianchi, G. Ingrosso, and R. Ambrosetti, *J. Am. Chem. Soc.* **102**, 7480 (1980). (b) K. Yates and R. S. McDonald, *J. Org. Chem.* **38**, 2465 (1973). (c) S. Fukuzumi and J. K. Kochi, *Int. J. Chem. Kinetics* **15**, 249 (1983).

(2) $$\mathrm{C{=}C} \underset{}{\overset{Br_2}{\rightleftharpoons}} \mathrm{C{-}C}\ (\text{bridged } Br^+)\ Br^- \xrightarrow[\text{slow}]{Br_2} \mathrm{BrC{-}CBr} + Br_2$$

(3) $$\mathrm{C{=}C} \xrightarrow{2Br_2} \mathrm{C{=}C}\cdot Br_4 \xrightarrow{\text{slow}} \mathrm{C{-}C}\ (\text{bridged } Br^+)\ Br_3^- \longrightarrow \mathrm{BrC{-}CBr} + Br_2$$

The first possibility envisages essentially the same mechanism as for the second-order process but with Br_2 replacing solvent in the rate-determining conversion to an ion pair. The second mechanism pictures Br_2 attack on a reversibly formed ion pair intermediate. The third mechanism postulates collapse of a ternary complex that is structurally similar to the initial charge-transfer complex but has 2:1 bromine:alkene stoichiometry. There are very striking similarities between the second-order and third-order reactions in terms of the magnitude of ρ values and product distribution.[40b] In fact, there is a quantitative correlation between the rates of the two reactions over a broad series of alkenes, which can be expressed as

$$\Delta G_3^{\ddagger} = \Delta G_2^{\ddagger} + \text{constant}$$

where $\Delta G_3^{\ddagger}$ and $\Delta G_2^{\ddagger}$ are the free energies of activation for the third-order and second-order processes, respectively.[40c] These correlations suggest that the two mechanisms must be very similar, and it has been argued on this basis that the third of the mechanisms shown above is the most likely one.[40c]

In summary, it appears that bromination usually involves a charge-transfer complex which collapses to an ion pair intermediate. The cation can be a carbocation, as in the case of styrenes, or a bromonium ion. The complex also can evidently be captured by bromine when it is present in sufficiently high concentration.

$$\mathrm{C{=}C} \overset{Br_2}{\rightleftharpoons} \mathrm{C{=}C}\cdot Br_2 \xrightarrow{Br^-} \mathrm{BrC{-}CBr} + Br^-$$

$\mathrm{C{=}C}\cdot Br_2 \xrightarrow{Br_2}$ bridged Br^+ Br_3^- $\longrightarrow$ $\mathrm{BrC{-}CBr} + Br_2$

$\mathrm{C{=}C}\cdot Br_2 \longrightarrow$ bridged Br^+ Br^- $\longrightarrow$ $\mathrm{BrC{-}CBr}$

$\mathrm{C{=}C}\cdot Br_2 \longrightarrow \mathrm{{}^{+}C{-}CBr}\ Br^- \longrightarrow \mathrm{BrC{-}CBr}$

Table 6.3. Relative Reactivity of Alkenes toward Halogenation

Alkenes	Relative reactivity		
	Chlorination[a]	Bromination[b]	Bromination[c]
Ethylene		0.01	0.0045
1-Butene	1.00	1.00	1.00
3,3-Dimethyl-1-butene	1.15	0.27	1.81
cis-2-Butene	63	27	173
trans-2-Butene	50	17.5	159
2-Methylpropene	58	57	109
2-Methyl-2-butene	11,000	1,380	
2,3-Dimethyl-2-butene	430,000	19,000	

a. M. L. Poutsma, *J. Am. Chem. Soc.* **87**, 4285 (1965); solvent is excess alkene.
b. J. E. Dubois and G. Mouvier, *Bull. Soc. Chim. Fr.* 1426 (1968); solvent is methanol.
c. A. Modro, G. H. Schmid, and K. Yates, *J. Org. Chem.* **42**, 3637 (1977); solvent is carbon tetrachloride.

Chlorination generally exhibits second-order kinetics, first-order in both alkene and chlorine.[41] The reaction rate also increases with alkyl substitution, as would be expected for an electrophilic process. The magnitude of the rate increase is quite large, as shown in Table 6.3.

In chlorination, loss of a proton can be a competitive reaction of the cationic intermediate. This process leads to formation of products resulting from net substitution with double-bond migration:

$$\mathrm{R(R_2CH)\overset{+}{C}{-}CR_2Cl} \quad \text{or} \quad \mathrm{R(R_2CH)C{-}CR_2\ (bridged\ Cl^+)} \longrightarrow \mathrm{R_2C{=}C(R){-}CR_2Cl + H^+}$$

Isobutylene and tetramethylethylene give products of this type.

$$\mathrm{(CH_3)_2C{=}CH_2 + Cl_2 \longrightarrow H_2C{=}C(CH_3)CH_2Cl} \quad 87\%$$

Ref. 42

$$\mathrm{(CH_3)_2C{=}C(CH_3)_2 + Cl_2 \longrightarrow H_2C{=}C(CH_3)C(Cl)(CH_3)_2} \quad 100\%$$

Alkyl migrations can also occur.

$$\mathrm{(CH_3)_3CCH{=}CH_2 + Cl_2 \longrightarrow H_2C{=}C(CH_3)CH(CH_3)CH_2Cl} \quad {\sim}10\%$$

Ref. 42

$$\mathrm{(CH_3)_3CCH{=}CHC(CH_3)_3 + Cl_2 \longrightarrow CH_2{=}C(CH_3)CH(CH_3)CH(Cl)C(CH_3)_3} \quad 46\%$$

Ref. 43

41. G. H. Schmid, A. Modro, and K. Yates, *J. Org. Chem.* **42**, 871 (1977).
42. M. L. Poutsma, *J. Am. Chem. Soc.* **87**, 4285 (1965).
43. R. C. Fahey, *J. Am. Chem. Soc.* **88**, 4681 (1966).

The relative reactivities of some alkenes toward chlorination and bromination are given in Table 6.3. The relative reactivities are solvent dependent.[44] The reaction is faster in the more polar solvents, and, in all media, reactivity increases with additional substitution of electron-releasing alkyl groups at the double bond.[45] Quantitative estimation of the solvent effect using the Winstein–Grunwald Y values indicates that the transition state has a high degree of ionic character. The Hammett correlation for bromination of styrenes is best with σ^+ substituent constants and gives $\rho = -4.8$.[46] All these features are in accord with an electrophilic mechanism.

Much less detail is available about the mechanism of fluorination and iodination of alkenes. Elemental fluorine reacts violently with alkenes, giving mixtures including products resulting from degradation of the carbon chain. Electrophilic additions of fluorine to akenes can be achieved with xenon diffuoride[47] or other electrophilic fluorine derivatives[48] or by use of highly dilute elemental fluorine at low temperature.[49] Under the latter conditions, *syn* stereochemistry is observed. The reaction is believed to proceed by rapid formation and then collapse of a β-fluorocarbocation–fluoride ion pair. Both from the stereochemical results and theoretical calculations,[50] it appears unlikely that a bridged fluoronium ion is involved. Acetyl hypofluorite, which is prepared by reaction of fluorine with sodium acetate at $-75°C$ in halogenated solvents,[51] reacts with alkenes to give β-acetoxyalkyl fluorides.[52] The reaction gives predominantly *syn* addition, which would be consistent with rapid collapse of a β-fluorocarbocation–acetate ion pair.

There have also been relatively few mechanistic studies of the addition of iodine. One significant feature of iodination is that it is easily reversible, even in the presence of excess alkene.[53] The addition is stereospecifically *anti* , but it is not entirely clear whether a polar or a radical mechanism is involved.[54]

As with other electrophiles, halogens can react with conjugated dienes to give 1,2- or 1,4-addition products. When molecular bromine is used as the brominating

44. F. Garnier and J.-E. Dubois, *Bull. Soc. Chim. Fr.* 3797 (1968); A. Modro, G. H. Schmid, and K. Yates, *J. Org. Chem.* **42**, 3673 (1977).
45. F. Garnier, R. H. Donnay, and J.-E. Dubois, *J. Chem. Soc., Chem. Commun.*, 829 (1971); M.-F. Ruasse and J.-E. Dubois, *J. Am. Chem. Soc.* **97**, 1977 (1975).
46. K. Yates, R. S. McDonald, and S. A. Shapiro, *J. Org. Chem.* **38**, 2460 (1973).
47. M. Zupan and A. Pollak, *J. Chem. Soc., Chem. Commun.*, 845 (1973); M. Zupan and A. Pollak, *Tetrahedron Lett.*, 1015 (1974).
48. For reviews of fluorinating agents, see A. Haas and M. Lieb, *Chimia* **39**, 134 (1985); W. Dmowski, *J. Fluorine Chem.* **32**, 255 (1986); H. Vyplel, *Chimia* **39**, 305 (1985).
49. S. Rosen and M. Brand, *J. Org. Chem.* **51**, 3607 (1986).
50. W. J. Hehre and P. C. Hiberty, *J. Am. Chem. Soc.* **96,** 2665 (1974).
51. O. Lerman, Y. Tov, D. Hebel, and S. Rozen, *J. Org. Chem.* **49**, 806 (1984).
52. S. Rozen, O. Lerman, M. Kol. and D. Hebel, *J. Org. Chem.* **50**, 4753 (1985).
53. P. W. Robertson, J. B. Butchers, R. A. Durham, W. B. Healy, J. K. Heyes, J. K. Johannesson, and D. A. Tait, *J. Chem. Soc.*, 2191 (1950).
54. M. Zanger, and J. L. Rabinowitz, *J. Org. Chem.* **40**, 248 (1975); R. L. Ayres, C. J. Michejda, and E. P. Rack, *J. Am. Chem. Soc.* **93**, 1389 (1971); P. S. Skell and R. R. Pavlis, *J. Am. Chem. Soc.* **86**, 2956 (1964).

agent in chlorinated hydrocarbon solvent, the 1,4-addition product dominates by ~7:1 in the case of butadiene.[55]

$$\mathrm{CH_2{=}CHCH{=}CH_2} \xrightarrow[25^\circ C]{Br_2} \underset{\substack{|\\ \mathrm{Br}\\ 12\%}}{\mathrm{BrCH_2CHCH{=}CH_2}} + \underset{88\%}{\mathrm{BrCH_2CH{=}CHCH_2Br}}$$

The product distribution can be shifted to favor the 1,2 product by use of such milder brominating agents as the pyridine-bromine complex or the tribromide ion, Br_3^-. It is believed that molecular bromine reacts through a cationic intermediate, whereas in the case of the less reactive brominating agents a process more like the Ad_E3 *anti*-addition mechanism is involved.

$$\mathrm{CH_2{=}CHCH{=}CH_2} \xrightarrow{Br_2} \mathrm{BrCH_2{-}\overset{\delta+}{CH}{\cdots}\overset{\delta+}{CH}{\cdots}CH_2}\ \mathrm{Br^-} \rightarrow \text{products}$$

$$\mathrm{CH_2{=}CHCH{=}CH_2} \overset{Br_3^-}{\rightleftharpoons} \mathrm{\overset{Br_3^-}{\vdots}{CH_2{=}CHCH{=}CH_2}}\ (\mathrm{Br^-}) \rightarrow \underset{\substack{|\\ \mathrm{Br}}}{\mathrm{BrCH_2CHCH{=}CH_2}}$$

The stereochemistry of both chlorination and bromination of several cyclic and acyclic dienes has been determined. The results show that bromination is often stereospecifically *anti* for the 1,2-addition process, whereas *syn* addition is preferred for 1,4 addition. Comparable results for chlorination show much less stereospecificity.[56] It appears that chlorination proceeds primarily through ion pair intermediates, whereas in bromination a stereospecific *anti*-1,2-addition process may compete with a process involving a carbocation intermediate. The latter can presumably give *syn* or *anti* product.

6.4. Electrophilic Additions Involving Metal Ions

Certain metal cations are capable of electrophilic attack on alkenes. Addition is completed when a nucleophile, from either the solvent or the metal ion's coordination sphere, acts as a nucleophile toward the alkene-cation complex.

$$\mathrm{M}^{n+} + \;\rangle\mathrm{C{=}C}\langle\; \rightleftharpoons \;\rangle\overset{\mathrm{M}^{n+}}{\overset{\vdots}{\mathrm{C{=}C}}}\langle$$

$$\rangle\overset{\mathrm{M}^{n+}}{\overset{\vdots}{\mathrm{C{=}C}}}\langle + \mathrm{Nu^-} \longrightarrow \left[-\underset{\substack{|\\ \mathrm{Nu}}}{\overset{\diagdown}{\mathrm{C}}}-\overset{\substack{\mathrm{M}\\ |}}{\mathrm{C}}\diagdown \right]^{(n-1)+}$$

55. G. Bellucci, G. Berti, R. Bioanchini, G. Ingrosso, and K. Yates, *J. Org. Chem.* **41**, 334 (1976).
56. G. E. Heasley, D. C. Hayes, G. R. McClung, D. K. Strickland, V. L. Heasley, P. D. Davis, D. M. Ingle, K. D. Rold, and T. L. Ungermann, *J. Org. Chem.* **41**, 334 (1976).

The best studied of these reactions involve the mercuric ion, Hg^{2+}, as the cation.[57] While the same process occurs for other transition metal cations, the products often go on to react further. The mercuration products are stable, and this allows a relatively uncomplicated study of the addition reaction itself. The usual nucleophile is the solvent, either water or an alcohol, but in less nucleophilic solvents, other nucleophiles can compete for the complex. The term *oxymercuration* is used to refer to reactions in which water or an alcohol acts as the nucleophile.

$$\mathrm{ROH + RCH{=}CHR \xrightarrow{Hg^{2+}} RCH(OR){-}CHR(Hg^{+}) + H^{+}}$$

In interesting contrast to protonation and halogenation reactions, the mercuration reaction is not accelerated by alkyl substituents on the alkene. For example, 1-pentene is about 10 times more reactive than *Z*-2-pentene and 40 times more reactive than *E*-2-pentene.[58] This reversal of reactivity has been attributed to steric effects that evidently outweigh the normal electron-releasing effect of alkyl substituents. When steric factors are taken into account, the reactivity trends are similar to those for other electrophilic additions.[59] As expected for an electrophilic reaction, the ρ value is negative.[60] A bridged mercurinium ion is considered to be formed in the rate-determining step. The addition of the nucleophile follows Markownikoff's rule, and the regioselectivity of oxymercuration is ordinarily very high.

$$\mathrm{RCH{=}CH_2 \xrightarrow{Hg^{2+}} RCH{-}CH_2\ (bridged\ by\ Hg^{2+}) \xrightarrow{NuH} R(H)(Nu)C{-}CH_2Hg^{+} + H^{+}}$$

A mercurinium ion has both similarities and differences as compared with the intermediates that have been described for other electrophilic additions. The proton which initiates acid-catalyzed addition processes is hard acid and has no unshared electrons. It can form either a carbocation or a hydrogen-bridged cation. Either species is *electron deficient* and highly reactive.

$$\mathrm{H^{+} + {>}C{=}C{<} \longrightarrow {>}\overset{+}{C}{-}C(H){<}\quad or\quad {>}C\overset{+}{\cdots}C{<}\ (H\text{-bridged})}$$

The positive bromine which leads to bromonium ion intermediates is softer and also has unshared electron pairs which can permit a total of *four* electrons to participate in the bridged bromonium ion intermediate. This would be expected to

57. W. Kitching, *Organomet. Chem. Rev.* **3**, 61 (1968); R. C. Larock, *Solvomercuration/Demercuration Reactions in Organic Synthesis*, Springer-Verlag, New York, 1986.
58. H. C. Brown and P. J. Geoghegan, Jr., *J. Org. Chem.* **37**, 1937 (1972).
59. S. Fukuzumi and J. K. Kochi, *J. Am. Chem. Soc.* **103**, 2783 (1981).
60. A. Lewis and J. Arozo, *J. Org. Chem.* **46**, 1764 (1981); A. Lewis, *J. Org. Chem.* **49**, 4682 (1984).

lead to a more strongly bridged and more stable species than is possible for the proton. Thus, the bromonium ion can be represented as having two covalent bonds to bromine and is electrophilic but not electron deficient.

$$:\ddot{Br}^+ + \;\rangle C{=}C\langle \;\longrightarrow\; -\overset{\overset{+}{Br}}{C{-}C}-$$

The electrophile in oxymercuration reactions, ^+HgX or Hg^{2+}, is a soft acid and strongly polarizing. It polarizes the π electrons of an alkene to the extent that a three-center, two-electron bond is formed between mercury and the two carbons of the double bond. A three-center, two-electron bond implies weaker bridging in the mercurinium ion than in the three-center, four-electron bonding of the bromonium ion. Oxymercuration of simple alkenes is usually a stereospecific *anti* addition. This result is consistent with the involvement of a mercurinium intermediate which is opened by nucleophilic attack.

$$RCH{=}CHR + Hg^{2+} \longrightarrow RH\overset{Hg^{2+}}{C{-}C}HR \xrightarrow{Nu^-} RH\overset{Hg^+}{C}{-}\underset{Nu}{C}HR$$

The reactivity of mercury salts is a function of both the solvent and the counterion in the mercury salt.[61] Mercuric chloride, for example, is unreactive, and usually mercuric acetate is used. When higher reactivity is required, salts of electronegatively substituted carboxylic acids such as mercuric trifluoroacetate can be used. Mercuric nitrate and mercuric perchlorate are also highly reactive. Soft anions reduce the reactivity of the Hg^{2+} ion by coordination, which reduces the electrophilicity of the cation. The harder oxygen anions leave the mercuric ion in a more reactive state. Organomercury compounds have a number of valuable synthetic applications which will be discussed in Section 7.3.3. of Part B.

6.5. Additions to Alkynes and Allenes

Since the HOMO of alkynes are also π-orbitals, it is not surprising that there is a good deal of similarity between the reactivity of alkenes and alkynes toward electrophilic reagents.[62] The fundamental questions about additions to alkynes include the following: how reactive are alkynes in comparison with alkenes; what is the stereochemistry of additions to alkynes; and what is the regiochemistry of additions to alkynes? The important role of halonium ions and mercurinium ions

61. H. C. Brown, J. T. Kurek, M.-H. Rei, and K. L. Thompson, *J. Org. Chem.* **49**, 2551 (1984).
62. G. H. Schmid, *The Chemistry of the Carbon-Carbon Triple Bond*, Part, 1, S. Patai (ed.), Wiley, New York, 1978, Chapter 3.

also raises the question of whether similar entities can be involved with acetylenes, where the ring would have to include a double bond:

$$\text{bridged } Br^{+} \text{ and } Hg^{2+} \text{ ions: } R\text{–C=C–R (ring including double bond)}$$

The three basic mechanisms that have been considered to be involved in electrophilic additions to alkynes are shown below. The first involves a discrete vinyl cation. In general, it could lead to either of the two stereoisomeric addition products. The second mechanism is a termolecular process which would be expected to lead to stereospecific *anti* addition. The third mechanism postulates a bridged-ion intermediate. Mechanisms A and C are of the Ad_E2 type while mechanism B would be classified as Ad_E3.

A. $R\text{–C}\equiv\text{C–R} + X\text{–}Y \rightarrow R\text{–}\overset{+}{C}\text{=C(X)R} \quad Y^- \rightarrow (R)(Y)C\text{=C}(X)(R) + (Y)(R)C\text{=C}(X)(R)$

B. $R\text{–C}\equiv\text{C–R} + 2\,X\text{–}Y \rightarrow (R)(Y)C\text{=C}(X)(R)\ [X^+,\ Y^-] \rightarrow (R)(Y)C\text{=C}(X)(R) + X\text{–}Y$

C. $R\text{–C}\equiv\text{C–R} + X\text{–}Y \rightleftharpoons R\text{–C(}\overset{+}{X}\text{)=C–R}\ \ Y^- \longrightarrow (R)(Y)C\text{=C}(X)(R)$

Further details must be added for a complete description, but these outlines encompass most reactions of alkynes with simple electrophiles.

Hydrogen chloride adds to aryl acetylenes in acetic acid to give mixtures of α-chlorostyrenes and the corresponding vinyl acetate.[63] A vinyl cation, which would be stabilized by the aryl substituent, is believed to be an intermediate. The ion pair formed by protonation can either collapse to give the vinyl halide or capture solvent to give the acetate. Aryl-substituted acetylenes give mainly the *syn* addition product.

$$Ar\text{–C}\equiv\text{C–R} + HCl \rightarrow Ar\text{–}\overset{+}{C}\text{=C(R)H} \quad Cl^- \rightarrow (Ar)(Cl)C\text{=C}(R)(H);\ \xrightarrow{HOAc} (Ar)(CH_3CO_2)C\text{=C}(R)(H)$$

63. R. C. Fahey and D.-J. Lee, *J. Am. Chem. Soc.* **90**, 2124 (1968).

Alkyl-substituted acetylenes can react by either the Ad_E3 or the Ad_E2 mechanisms. Reactions proceeding through a vinyl cation would not be expected to be stereospecific, since the cation is expected to adopt *sp* hybridization. The Ad_E3 mechanism leads to *anti* addition. The preference for one or the other mechanism depends on the individual structure and the reaction conditions.

Alkynes can be hydrated in concentrated aqueous acid solutions. The initial product is an enol, which isomerizes to the more stable ketone.

$$CH_3C{\equiv}CH \xrightarrow{H^+} CH_3\overset{+}{C}{=}CH_2 \xrightarrow{H_2O} CH_3\underset{OH}{\underset{|}{C}}{=}CH_2 \rightarrow CH_3\underset{O}{\underset{\|}{C}}CH_3$$

Solvent isotope effects are indicative of a rate-determining protonation.[64] Alkyne reactivity increases with addition of electron-donating substituents. The reactivity of alkynes is somewhat more sensitive to substituent effects than is the case for alkenes.[65] These reactions are believed to proceed by rate-determining proton transfer to give a vinyl cation. A hydrogen bridged structure is not regarded as energetically feasible. Various MO calculations place the bridged ion 30–45 kcal/mol above the vinyl cation in energy.[66]

$$R{-}\overset{\overset{\overset{+}{H}}{\cdot\,\cdot}}{C{=}C}{-}H \quad \text{less stable than} \quad R{-}\overset{+}{C}{=}CH_2$$

Alkynes react when heated with trifluoroacetic acid to give addition products. Mixtures of *syn* and *anti* addition products are obtained.[67] These reactions are similar to acid-catalyzed hydration and proceed through a vinyl cation intermediate.

$$R{-}C{\equiv}C{-}R + CF_3CO_2H \rightarrow (R)(CF_3CO_2)C{=}C(R)(H) + (R)(CF_3CO_2)C{=}C(H)(R)$$

Alkynes undergo addition reactions with halogens. The reaction has been thoroughly examined from a mechanistic point of view. In the presence of excess halogen, tetrahaloalkanes are formed, but mechanistic studies can be carried out with a limited amount of halogen so that the initial addition step can be characterized. In general, halogenation of acetylenes is slower than for the corresponding alkenes. We will discuss the reason for this shortly. The reaction shows typical characteristics of an electrophilic reaction. For example, the rates of chlorination of substituted phenylacetylenes are correlated by σ^+ with $\rho = -4.2$. In acetic acid the reaction is overall second-order, first-order in both reactants. The addition is not stereospecific,

64. P. Cramer and T. T. Tidwell, *J. Org. Chem.* **46**, 2683 (1981).
65. A. D. Allen, Y. Chiang, A. J. Kresge, and T. T. Tidwell, *J. Org. Chem.* **47**, 775 (1982).
66. H.-J. Kohler and H. Lischka, *J. Am. Chem. Soc.* **101**, 3479 (1979).
67. P. E. Peterson and J. E. Dudley, *J. Am. Chem. Soc.* **88**, 4990 (1966); R. H. Summerville and P. v. R. Schleyer, *J. Am. Chem. Soc.* **96**, 1110 (1974).

and a considerable amount of solvent capture product is formed. All of these features are consistent with reaction proceeding through a vinyl cation intermediate.[68]

$$\mathrm{ArC{\equiv}CH + Cl_2 \xrightarrow{CH_3CO_2H} Ar\overset{+}{C}{=}CHCl \;\; Cl^- \rightarrow (Ar)(Cl)C{=}C(H)(Cl) + (Ar)(Cl)C{=}C(Cl)(H) + (Ar)(CH_3CO_2)C{=}C(H)(Cl) + (Ar)(CH_3CO_2)C{=}C(Cl)(H)}$$

For alkyl-substituted acetylenes, there is a difference in stereochemistry between mono- and disubstituted derivatives. The former give *syn* addition while the latter react by *anti* addition. The disubstituted (internal) compounds are considerably (~100 times) more reactive than the monosubstituted (terminal) ones. This result suggests that the transition state of the rate-determining step is stabilized by *both* alkyl substituents and points to a structure with bridged character. This would be consistent with the overall stereochemistry of the reaction.

$$\mathrm{R{-}C{\equiv}C{-}R \xrightarrow{Cl_2} R{-}C{=}C{-}R\ (bridged\ \overset{+}{Cl}) \xrightarrow{Cl^-} (R)(Cl)C{=}C(Cl)(R)}$$

The monosubstituted intermediate does not seem to be effectively bridged since *syn* addition predominates. A very short-lived vinyl cation appears to be the best description of the intermediate in this case.[69]

The rates of bromination of a number of alkynes have been measured under conditions that permit comparison with the corresponding alkenes. The rate of bromination of styrene exceeds that of phenylacetylene by about 10^3.[70] For dialkylacetylene–disubstituted alkene comparisons, the ratios range from 10^3 to 10^7, being greatest in the least nucleophilic solvents.[71] Bromination of alkyl-substituted alkynes shows rate enhancement by both alkyl substituents, and this must be taken to indicate that the transition state has bridged character.[72]

The stereochemistry of addition is usualy *anti* for alkyl-substituted alkynes, whereas the addition to aryl-substituted compounds is not stereospecific. This suggests a termolecular mechanism in the alkyl case, as opposed to an aryl-stabilized vinyl cation intermediate in the aryl case.[73] Aryl-substituted alkynes can be shifted

68. K. Yates and T. A. Go, *J. Org. Chem.* **45**, 2377 (1980).
69. K. Yates and T. A. Go, *J. Org. Chem.* **45**, 2385 (1980).
70. M.-F. Ruasse and J.-E. Dubois, *J. Org. Chem.* **42,** 2689 (1977).
71. K. Yates, G. H. Schmid, T. W. Regulski, D. G. Garratt, H.-W. Leung, and R. McDonald, *J. Am. Chem. Soc.* **95,** 160 (1973); J. M. Kornprobst and J.-E. Dubois, *Tetrahedron Lett.*, 2203 (1974); G. Modena, F. Rivetti, and U. Tonellato, *J. Org. Chem.* **43**, 1521 (1978).
72. G. H. Schmid, A. Mondro, and K. Yates, *J. Org. Chem.* **45**, 665 (1980).
73. J. A. Pincock and K. Yates, *Can. J. Chem.* **48**, 3332 (1970).

toward *anti* addition by including bromide salts in the reaction medium. Under these conditions, a species preceding the vinyl cation must be intercepted by bromide ion. This species can be represented as a complex of molecular bromine with the alkyne. An overall mechanistic summary is shown in the following equations.

$$ArC{\equiv}CH \xrightarrow{Br_2} ArC{\equiv}CH\cdot Br_2 \longrightarrow \text{bridged bromonium } (Br^+\ Br^-) \longrightarrow (E)\text{-}ArC(Br){=}C(H)Br + (Z)\text{-}ArC(Br){=}C(Br)H$$

$$ArC{\equiv}CH\cdot Br_2 \xrightarrow{Br^-} (E)\text{-}ArC(Br){=}CHBr \qquad ArC{\equiv}CH\cdot Br_2 \longrightarrow Ar\overset{+}{C}{=}CHBr \xrightarrow{Br^-} \text{mixture}$$

This scheme represents an alkyne–bromine complex as an intermediate in all alkyne brominations. This is analogous to the case of alkenes. This complex may dissociate to a vinyl cation when the cation is sufficiently stable, as is the case when there is an aryl substituent. It may collapse to a bridged bromonium ion or undergo reaction with a nucleophile. These are the dominant reactions for alkyl-substituted alkynes and lead to stereospecific *anti* addition. Reactions proceeding through vinyl cations are exprected to be nonstereospecific.

Chlorination of 1-hexyne in acetic acid leads mainly to 1,1-dichlorohexan-2-one via chlorination and deacetylation of the initial product, 2-acetoxy-1-chlorohexene.

$$HC{\equiv}C(CH_2)_3CH_3 \rightarrow ClCH{=}C(OCCH_3)(CH_2)_3CH_3 \quad (OCCH_3 = O\text{-}C(=O)CH_3)$$

$$ClCH{=}C(CH_2)_3CH_3(O\text{-}C(=O)CH_3) + Cl{-}Cl \longrightarrow Cl_2CHC(=O)(CH_2)_3CH_3$$

The corresponding intermediate in the chlorination of 3-hexyne, *E*-3-acetoxy-4-chlorohexene, can be isolated. In dichloromethane, both 1-hexyne and 3-hexyne give mixtures of the expected dichlorohexenes, with the *E*-isomer predominating.[74]

Acetylenes react with mercuric acetate in acetic acid to give addition products. For 3-hexyne the product has *E*-stereochemistry, but the *Z*-isomer is isolated from diphenylacetylene.[75]

$$C_2H_5C{\equiv}CC_2H_5 \xrightarrow{Hg(OAc)_2} (C_2H_5)(CH_3CO_2)C{=}C(HgO_2CCH_3)(C_2H_5)$$

$$PhC{\equiv}CPh \xrightarrow{Hg(OAc)_2} (Ph)(CH_3CO_2)C{=}C(Ph)(HgO_2CCH_3)$$

74. G. E. Heasley, C. Codding, J. Sheehy, K. Gering, V. L. Heasley, D. F. Shellhamer, and T. Rempel, *J. Org. Chem.* **50**, 1773 (1985).
75. R. D. Bach, R. A. Woodard, T. J. Anderson, and M. D. Glick, *J. Org. Chem.* **47**, 3707 (1982).

The kinetics of the addition reaction are first-order in both alkyne and mercuric acetate.[76]

We can understand many of the general characteristics of electrophilic additions to alkynes by recognizing the possibility for both bridged ions and vinyl cations as intermediates. Reactions proceeding through vinyl cations can be expected to be nonstereospecific, with the precise stereochemistry depending upon the lifetime of the vinyl cation and the identity and concentration of the potential nucleophiles. Stereospecific *anti* addition can be expected from processes involving nucleophilic attack on either a bridged-ion intermediate or an alkyne–electrophile complex. These general mechanisms can also help to explain the relative reactivity of alkenes and alkynes in comparable addition processes. In general, reactions which proceed through vinyl cations, such as those involving rate-determining protonation, are only moderately slower for alkynes as compared to similar alkenes. This rate difference can be attributed to the relatively higher energy of vinyl cations compared to cations with sp^2 hybridization. It has been estimated that this difference is around 10–15 kcal/mol, a significant but not enormous difference.[77] This difference is also partially compensated by the higher ground state energy of acetylenes. Reactions which proceed through transition states leading to bridged intermediates typically show much larger rate retardation for the alkyne addition. Bromination is the best-studied example of this type. This presumably reflects the greater destabilization of bridged species by strain in the case of alkynes. Bridged intermediates derived from alkynes must incorporate a double bond in the three-membered ring.[78] The bridged bromonium ion can also be considered an antiaromatic 4π electron species.

$:\overset{+}{Br}:$ bridging C—C — favorable $\qquad$ $:\overset{+}{Br}:$ bridging C=C — less favorable

The activation energies for additions to alkynes through bridged intermediates are thus substantially greater than for alkenes.

Electrophilic additions to allenes represent an interesting reaction type which is related to additions to both alkenes and alkynes.[79] An allene could, for example, conceivably be protonated at either a terminal sp^2 carbon or the central sp carbon.

$$RCH{=}C{=}CHR \xrightarrow{H^+} R\overset{+}{C}H{-}CH{=}CHR \quad \text{versus} \quad RCH{=}C{=}CHR \xrightarrow{H^+} RCH_2{-}\overset{+}{C}{=}CHR$$

The allylic carbocation resulting from protonation of the center carbon might seem the obvious choice, but, in fact, the kinetically favored protonation leads to the vinyl cation intermediate. The reason for this is stereoelectronic. The allene structure

76. M. Bassetti and B. Floris, *J. Org. Chem.* **51**, 4140 (1986).
77. K. Yates, G. H. Schmid, T. W. Regulski, D. G. Garratt, H.-W. Leung, and R. McDonald, *J. Am. Chem. Soc.* **95**, 160 (1973); Z. Rappoport, in *Reactive Intermediates*, Vol. 3, R. A. Abramovitch (ed.), Plenum Press, New York, 1985, Chapter 7.
78. G. Melloni, G. Modena, and U. Tonellato, *Acc. Chem. Res.* **8**, 227 (1981).
79. For a review of electrophilic additions to allenes, see W. Smadja, *Chem. Rev.* **83**, 263 (1983).

is nonplanar, so that an initial protonation of the center carbon leads to a twisted structure which is devoid of allylic conjugation.

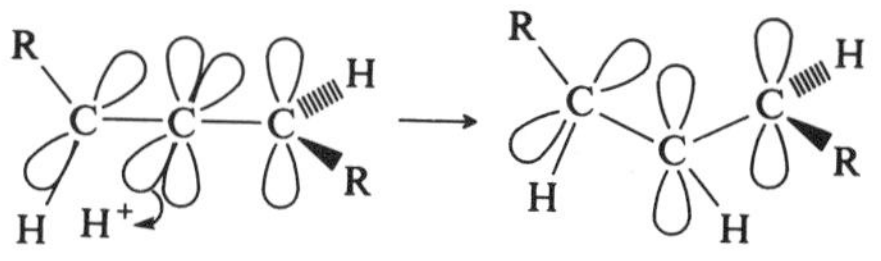

This twisted cation is about 17 kcal/mol higher in energy than the ion formed by protonation at a terminal carbon.[80]

Addition of hydrogen halides to simple allenes initially gives the vinyl halide, and if the second double bond reacts, a geminal dihalide is formed.[81]

$$RCH{=}C{=}CH_2 \xrightarrow{HX} RCH_2{-}C(X){=}CH_2 \xrightarrow{HX} RCH_2{-}CX_2{-}CH_3$$

Strong acids in aqueous solution convert allenes to ketones via an enol intermediate. This process also involves protonation at a terminal carbon.

$$CH_2{=}C{=}CH_2 \xrightarrow[H_2O]{H^+} CH_3{-}C(OH){=}CH_2 \rightarrow CH_3CCH_3 \ (C{=}O)$$

The kinetic features of this reaction, including the solvent isotope effect, are consistent with a rate-determining protonation to form a vinyl cation.[82]

Allenes react with other typical electrophiles such as the halogens and mercuric ion. In systems where bridged-ion intermediates would be expected, nucleophilic capture generally occurs at the allylic position. This pattern is revealed, for example, in the products of solvent capture in halogen additions[83] and by the structures of mercuration products.[84]

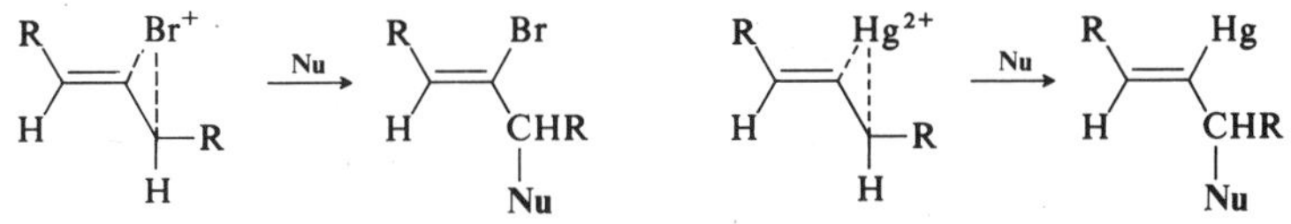

80. L. Radom, P. C. Hariharan, J. A. Pople, and P. v. R. Schleyer, *J. Am. Chem. Soc.* **95**, 6531 (1973).
81. T. L. Jacobs and R. N. Johnson, *J. Am. Chem. Soc.* **82**, 6397 (1960); R. S. Charleston, C. K. Dalton, and S. R. Schraeder, *Tetrahedron Lett.*, 5147 (1969); K. Griesbaum, W. Naegle, and G. G. Wanless, *J. Am. Chem. Soc.* **87**, 3151 (1965).
82. P. Cramer and T. T. Tidwell, *J. Org. Chem.* **46**, 2683 (1981).
83. H. G. Peer, *Recl. Trav. Chim. Pays-Bas* **81**, 113 (1962); W. R. Dolbier, Jr. and B. H. Al-Sader, *Tetrahedron Lett.*, 2159 (1975).
84. W. Waters and E. F. Kieter, *J. Am. Chem. Soc.* **89**, 6261 (1967).

6.6. The E2, E1, and E1cb Mechanisms

An elimination reaction—the expulsion of another molecule from a substrate—can be classified according to the relative placement of the carbon atoms from which elimination occurs.

$$\text{R—C(H)(R)—X} \longrightarrow \text{R—}\ddot{\text{C}}\text{—R} + \text{HX} \quad \alpha \text{ elimination}$$

$$\text{R—CH(H)—C(H)(X)—R} \longrightarrow \text{RCH=CHR} + \text{HX} \quad \beta \text{ elimination}$$

$$\text{R—CH}_2\text{—CH}_2\text{—CH(X)—R} \longrightarrow \text{R-cyclopropane-R} + \text{HX} \quad \gamma \text{ elimination}$$

The products of α eliminations are unstable divalent carbon species. They will be discussed in Part B, Chapter 10. In the present chapter, attention will be focused on β-elimination reactions.[85] Some representative examples of β-elimination reactions are given in Scheme 6.1.

The β eliminations can be further subdivided by closer examination of the mechanisms involved. Three distinct mechanisms are outlined below:

E2 Mechanism

$$\text{RCH}_2\text{CH(X)R}' + \text{B}^- \longrightarrow [\text{B}^{\delta-}\cdots\text{H}\cdots\text{C(R)(H)}\cdots\text{C(H)(R}')\cdots\text{X}^{\delta-}] \longrightarrow \text{RCH=CHR}' + \text{BH} + \text{X}^-$$

E1 Mechanism

$$\text{RCH}_2\text{CH(X)R}' \longrightarrow \text{RCH}_2\overset{+}{\text{C}}\text{HR}' \; \text{X}^- \xrightarrow{\text{B}^-} \text{RCH=CHR}' + \text{BH}$$

E1cb Mechanism

$$\text{RCH}_2\text{CH(X)R}' + \text{B}^- \longrightarrow \text{R}\overset{-}{\text{C}}\text{HCH(X)R}' + \text{BH} \longrightarrow \text{RCH=CHR}' + \text{X}^-$$

85. Reviews: E. Baciocchi, in *Chemistry of Halides, Pseudo-Halides and Azides*, Part 2, S. Patai and Z. Rappoport (eds.), Wiley-Interscience, New York, 1983, Chapter 23; W. H. Saunders, Jr., and A. F. Cockerill, *Mechanisms of Elimination Reactions*, Wiley, New York, 1973; D. J. McLennan, *Tetrahedron* **31**, 2999 (1975).

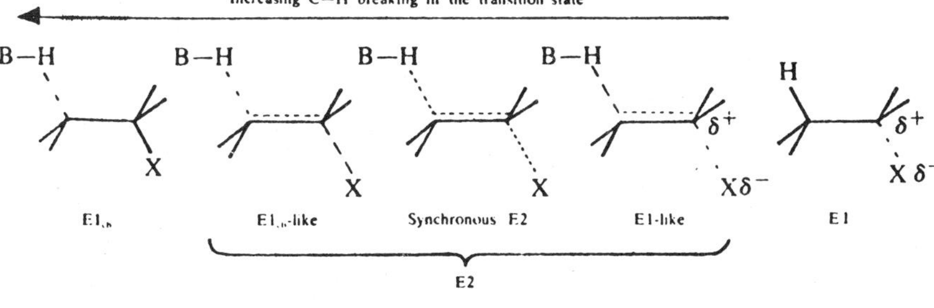

Fig. 6.3. Variable transition state theory of elimination reactions. J. F. Bunnett, *Angew. Chem. Int. Ed. Engl.* **1**, 225 (1962); J. F. Bunnett, *Surv. Prog. Chem.* **5**, 53 (1969); W. H. Saunders, Jr., and A. F. Cockerill, *Mechanisms of Elimination Reactions*, Wiley, New York, 1973, pp. 48–55; D. J. McLennan, *Tetrahedron* **31**, 2999 (1975); W. H. Saunders, Jr., *Acc. Chem. Res.* **9**, 19 (1976).

As depicted, the E2 mechanism involves a bimolecular transition state in which abstraction of a proton β to the leaving group is concerted with departure of the leaving group. In contrast, the rate-determining step in the E1 mechanism is the unimolecular ionization of the substrate. It will be recognized that this is the same process as the rate-determining step in the S_N1 mechanism. Elimination is completed by rapid removal of a β proton. The E1cb mechanism, like the E1, involves two steps, but the order is reversed. Proton abstraction precedes expulsion of the leaving group. The correlation of many features of β-elimination reactions is greatly aided by recognition that these three mechanisms represent *variants of a continuum of mechanistic possibilities.* Many β-elimination reactions occur via mechanisms that are intermediate between the limiting mechanistic types. This idea, called the *variable E2 transition state theory*, is outlined in Fig. 6.3.

We will shortly discuss the most important structure–reactivity features of the E2, E1, and E1cb mechanisms. The variable transition state theory allows discussion of reactions proceeding through transition states of intermediate character in terms of the limiting mechanistic types. The most important structural features to be considered in such a discussion are (1) the nature of the leaving group, (2) the nature of the base, (3) electronic and steric effects of substituents in the reactant molecule, and (4) solvent effects.

There is another useful way of depicting the ideas embodied in the variable transition state theory of elimination reactions. This is to construct a *three-dimensional* potential energy diagram.[86] Suppose that we consider the case of an ethyl halide. The two stepwise reaction paths both require the formation of high-energy intermediates. The E1 mechanism requires formation of a carbocation

86. R. A. More O'Ferral, *J. Chem. Soc., B*, 274 (1970).

Scheme 6.1. Some Examples

Dehydrohalogenations

1[a] $CH_3(CH_2)_5CH_2CH_2Br \xrightarrow{K^+ t\text{-}BuO^-} CH_3(CH_2)_5CH{=}CH_2$ 85%

2[b] 1,2-dibromocyclohexane $\xrightarrow{K^+ i\text{-}PrO^-}$ 1,3-cyclohexadiene 34%–40%

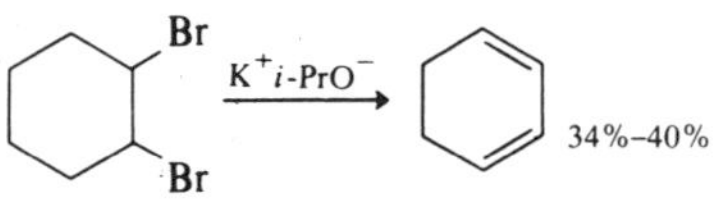

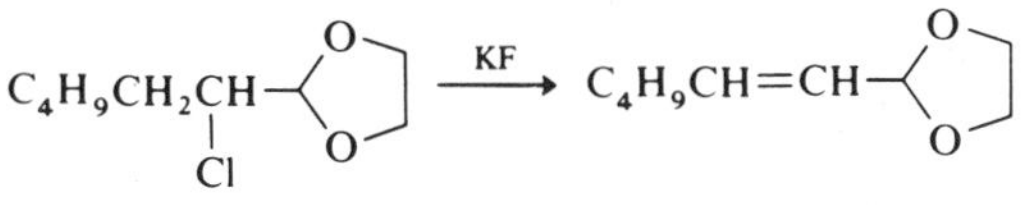

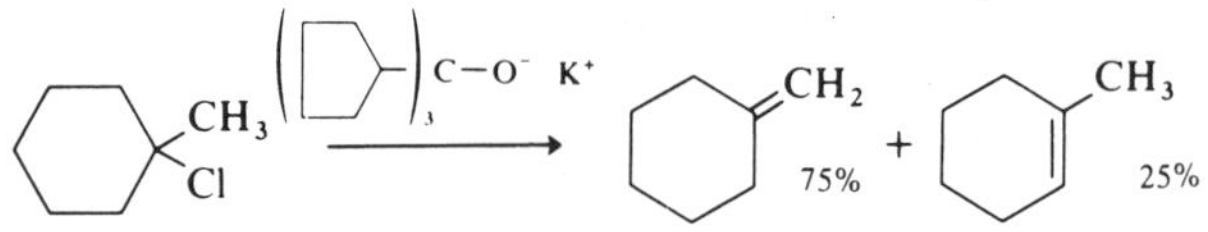

3[c] $C_4H_9CH_2CH(Cl)$–(1,3-dioxolan-2-yl) $\xrightarrow{KF}$ $C_4H_9CH{=}CH$–(1,3-dioxolan-2-yl)

4[d] 1-chloro-1-methylcyclohexane $\xrightarrow{((C_5H_9)C)_3C{-}O^-\ K^+}$ methylenecyclohexane 75% + 1-methylcyclohexene 25%

5[e] 2-chloro-2-methylcyclohexanone $\xrightarrow[DMF]{LiCl}$ 2-methyl-2-cyclohexenone

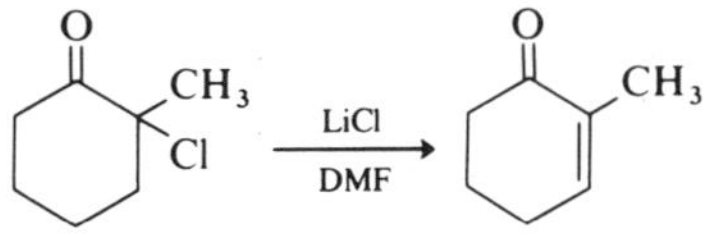

6[f] $PhCH(Br)CH(Br)CCH_3$ (C=O) $\xrightarrow{NaOAc}$ $PhCH{=}C(Br)CCH_3$ (C=O) 64%–73%

Dehydrohalogenations to acetylenes

7[g] $PhCH(Br)CH_2Br \xrightarrow[NH_3]{NaNH_2} PhC{\equiv}CH$ 45%–52%

8[h] $CH_3C(Cl){=}CHCH_2OH \xrightarrow[NH_3]{NaNH_2} \xrightarrow{NH_4Cl} CH_3C{\equiv}CCH_2OH$ 75%–85%

whereas the E1cb proceeds via a carbanion intermediate.

$$CH_3CH_2X \rightarrow CH_3CH_2^+ + X^- \qquad \text{E1 mechanism}$$

$$CH_3CH_2X \xrightarrow{B^-} {}^-CH_2CH_2X + BH \qquad \text{E1cb mechanism}$$

In the absence of other stabilizing substituent groups, both a primary carbocation and a primary carbanion are highly unstable intermediates. If we construct a three-dimensional diagram in which progress of C—H bond breaking is one dimension, progress of C—X bond breaking is the second, and the energy of the reacting system is the third, we obtain a diagram such as in Fig. 6.4. In Fig. 6.4(A) only the two horizontal (bond-breaking) dimensions are shown. We see that the E1 mechanism corresponds to complete C—X cleavage before C—H cleavage starts.

Eliminations using sulfonates

9[i] cyclopentyl–$OSO_2C_7H_7$ $\xrightarrow{K^+ t\text{-}BuO^-}$ cyclopentene 76%

10[j] $Ph_2CHCH_2OSO_2C_7H_7 \xrightarrow{NaOMe} Ph_2C{=}CH_2$ 92%

11[k] $HC{\equiv}CCH_2CH(OSO_2C_7H_7)CH_3 \xrightarrow[H_2O]{KOH} HC{\equiv}CCH{=}CHCH_3$ 91%

Eliminations involving quaternary ammonium hydroxides

12[l] $(CH_3)_3CCH_2CH_2\overset{+}{N}(CH_3)_3\ {}^-OH \xrightarrow{\Delta} (CH_3)_3CCH{=}CH_2$ 81%

13[m] 4-methylcyclohexyl–$(CH_3)_2C\overset{+}{N}(CH_3)_3$ $\xrightarrow[H_2O]{\Delta}$ 4-methylcyclohexyl–$C(CH_3){=}CH_2$ 98%

a. P. Veeravagu, R. T. Arnold, and E. W. Eigemann, *J. Am. Chem. Soc.* **86**, 3072 (1964).
b. J. P. Schaeffer and L. Endres, *Org. Synth.* **47**, 31 (1967).
c. E. Elkik, *Bull. Soc. Chim. Fr.*, 283 (1968).
d. S. A. Acharya and H. C. Brown, *J. Chem. Soc., Chem. Commun.*, 305 (1968).
e. E. W. Warnhoff, D. G. Martin, and W. S. Johnson, *Org. Synth.* **IV**, 162 (1963).
f. N. H. Cromwell, D. J. Cram, and C. E. Harris, *Org. Synth.* **III**, 125 (1953).
g. K. N. Campbell and B. K. Campbell, *Org. Synth.* **IV**, 763 (1963).
h. P. J. Ashworth, G. H. Mansfield, and M. C. Whiting, *Org. Synth.* **IV**, 128 (1963).
i. C. H. Snyder and A. R. Soto, *J. Org. Chem.* **29**, 742 (1964).
j. P. J. Hamrick, Jr., and C. R. Hauser, *J. Org. Chem.* **26**, 4199 (1961).
k. G. Eglinton and M. C. Whiting, *J. Chem. Soc.*, 3650 (1950).
l. A. C. Cope and D. L. Ross, *J. Am. Chem. Soc.* **83**, 3854 (1961).
m. L. C. King, L. A. Subluskey, and E. W. Stern, *J. Org. Chem.* **21**, 1232 (1956).

The E1cb mechanism corresponds to complete C—H cleavage before C—X cleavage begins. In Fig. 6.4(B), the energy dimension is added. The front right and back left corners correspond to the E1 and E1cb intermediates, respectively.

Because of the high energy of both the E1 and E1cb intermediates, the lowest-energy path will be the concerted E2 path, more or less diagonally across the energy surface. This pathway is of lower energy because the partially formed double bond provides some compensation for the energy required to break the C—H and C—X bonds.

If substituents are added to the ethyl group that would stabilize the carbocation intermediate, this would cause a lowering of the front right corner of the diagram, which indicates the energy of the carbocation intermediate. Similarly, if a substituent is added that would stabilize a carbanion intermediate, the back left corner of the

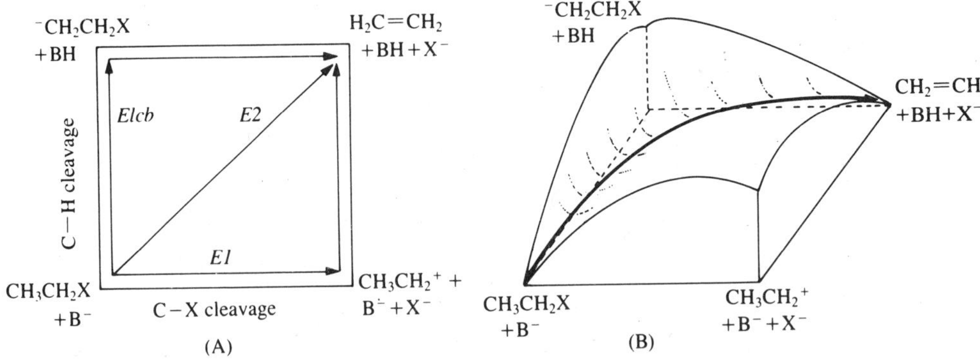

Fig. 6.4. Three-dimensional (More O'Ferrall) diagrams depicting transition state locations for E1, E1cb, and E2 mechanisms.

diagram would be of lower energy. For this reason, substituents that would stabilize carbocation character will move the E2 transition state to a point where it more closely resembles the E1 transition state. A structural change that effects stabilization of carbanion character would cause the E2 transition state to become more similar to the E1cb transition state. In the E1-like transition state, C—X bond cleavage will be more advanced than C—H cleavage, while in the E1cb-like transition state, the C—H bond breaking will be more advanced. Figure 6.5 illustrates how these changes can be depicted with this type of energy diagram.

We will now use these general ideas to discuss specific structural effects that favor the various possible mechanisms for elimination reactions. We have a back-

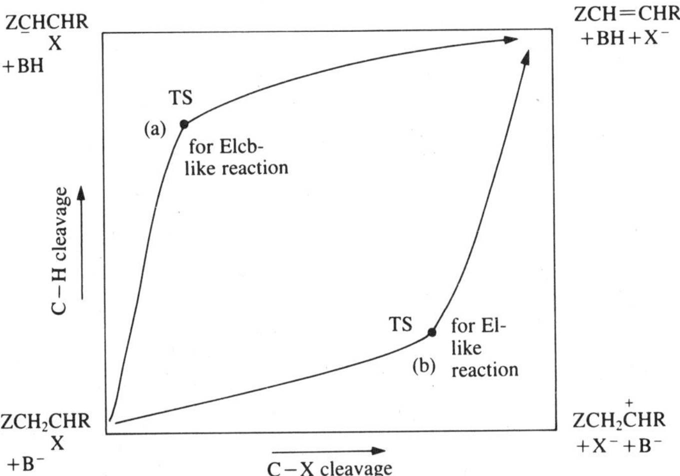

Fig. 6.5. Representation of changes in transition state character in the variable transition state E2 elimination reaction, showing displacement of transition state location as a result of substituent effects: (a) substituent Z stabilizes carbanion character of E1cb-like transition state; (b) substituent R stabilizes carbocation character of E1-like transition state.

ground that is pertinent to the structure–reactivity effects in E1 reactions from the discussion of S_N1 reactions in Chapter 5. Ionization is favored by (1) electron-releasing groups that stabilize the positive charge in the carbocation intermediate; (2) readily ionized, i.e., "good," leaving groups; and (3) solvents of high ionizing strength. The base plays no role in the rate-determining step in the E1 mechanism, but its identity cannot be ignored. Once ionization has occurred, the cationic intermediate is subject to two competing reactions: nucleophilic capture (S_N1) or proton removal (E1). Stronger bases favor the E1 path over the S_N1 path.

E2 reactions are distinguished from E1 reactions in that the base is present in the transition state for the rate-determining step. The reactions therefore exhibit overall second-order kinetics. The precise nature of the transition state is a function of variables such as the strength of the base, the identity of the leaving group, and the solvent. For example, an elimination reaction proceeding by an E2 transition state will be moved in the E1cb direction by an increase in base strength or by a change to a poorer leaving group. On the other hand, a good leaving group in a highly ionizing solvent will result in an E2 transition state that resembles the transition state for an E1 process, with extensive weakening of the bond to the leaving group.

Reactions that proceed by the E1cb mechanism are limited to reactants having substituent groups that can effectively stabilize the intermediate carbanion. This mechanism is not observed with simple alkyl halides or sulfonates. It is more likely to be involved when the leaving group is β to a carbonyl, nitro, cyano, sulfonyl, or other carbanion-stabilizing group.

The nature of the transition state is of great importance, since it controls the direction of β elimination in compounds in which the double bond can be introduced in one of several positions. These orientation effects are discussed in the next section.

6.7. Orientation Effects in Elimination Reactions

The most useful generalizations and predictions regarding regioselectivity in elimination reactions are drawn from the variable transition state theory. As shown in Fig. 6.3, this theory proposes that the transition states in E2 reactions may vary over a mechanistic range spanning the gap between the E1 and E1cb extremes. As long as the base is present in the transition state, the reaction will exhibit second-order kinetics. In all such cases, the cleavage of the C—H bond and the C—X bond must be concerted. The relative extent of breaking of the two bonds at the transition state may differ a great deal, however, depending on the nature of the leaving group X and the ease of removal of the hydrogen as a proton. If one examines the orientation effects in E1 and E1cb eliminations, it is seen that quite different structural features govern the direction of elimination for these two mechanisms. The variable transition state theory of E2 reactions suggests that E2 elimination proceeding through "E1-like" transition states will follow the orientation preferences of E1 eliminations whereas E2 eliminations proceeding through "E1cb-like" transition states will show

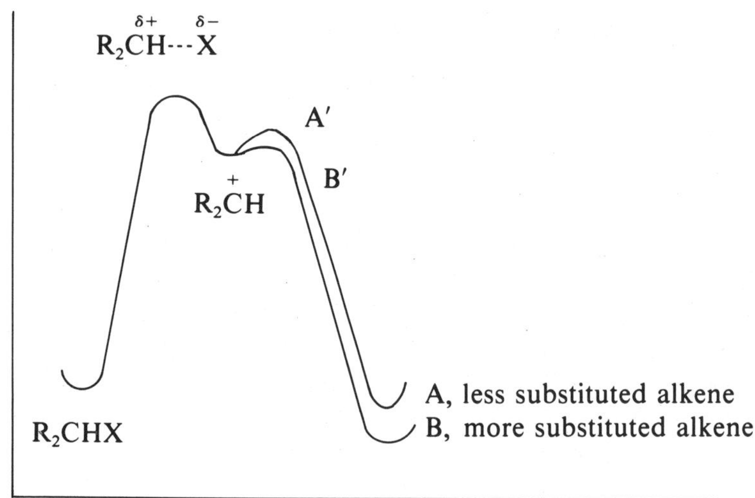

Fig. 6.6. Product-determining step for E1 elimination.

selectivity similar to that found for E1cb reactions. It is therefore instructive to consider these mechanisms before discussing the E2 case.

In the E1 mechanism, the leaving group has completely ionized before C—H bond breaking occurs. The direction of the elimination therefore depends on the structure of the carbocation and the identity of the base involved in the proton transfer that follows C—X heterolysis. Because of the relatively high energy of the carbocation intermediate, quite weak bases can effect proton removal. The solvent may often serve this function. The counterion formed in the ionization step may also act as the proton acceptor:

$$\text{B:}\ \text{H}{-}\text{CH}_2{-}\overset{+}{\text{C}}\text{HCH}_2\text{CH}_2\text{CH}_3 \longleftarrow \text{CH}_3\overset{+}{\text{C}}\text{HCH}_2\text{CH}_2\text{CH}_3 \longrightarrow \text{CH}_3\overset{+}{\text{C}}\text{H}{-}\underset{\text{H}\ \ \text{:B}}{\text{C}}\text{HCH}_2\text{CH}_3$$

$$\downarrow \qquad\qquad\qquad\qquad \downarrow$$

$$\text{H}_2\text{C}{=}\text{CHCH}_2\text{CH}_2\text{CH}_3 \qquad\qquad \text{CH}_3\text{CH}{=}\text{CHCH}_2\text{CH}_3$$

The product composition of the alkenes formed in an E1 elimination reaction usually favors the more substituted, and therefore more stable, alkene. One possible explanation for this observation is that the energies of the transition states parallel those of the isomeric alkenes. However, since the activation energy for proton removal from a carbocation is low, the transition state should resemble the carbocation intermediate more than the alkene product as shown in Fig. 6.6. In the carbocation, there will be hyperconjucation involving each β-hydrogen.[87] Since the hyperconjugation structures possess some double-bond character, it may be that the interaction with hydrogen will be greatest at more highly substituted carbon. This structural

87. P. B. D. de la Mare, *Pure Appl. Chem.* **56**, 1755 (1984).

effect in the carbocation intermediate could then govern the direction of elimination.

$$\xrightarrow{-H^+}$$

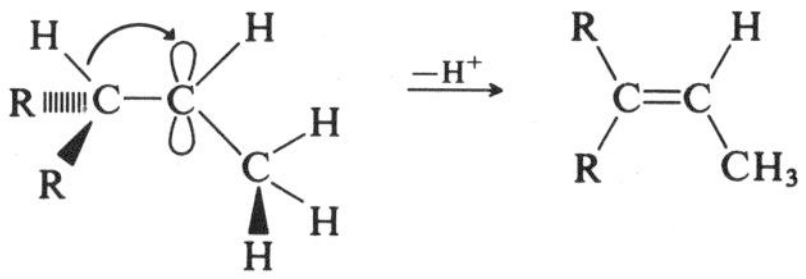

In the E1cb mechanism, the direction of elimination is governed by the ease of removal of the individual β protons, which in turn is determined by the polar and resonance effects of nearby substitutents and by the degree of steric hindrance to approach of base to the proton. Alkyl substituents will tend to retard proton abstraction both electronically and sterically. Preferential proton abstraction from less hindered positions leads to the formation of the less substituted alkene. This orientational preference is opposite to that of the E1 reaction.

The preferred direction of elimination via the E2 mechanism depends on the precise nature of the transition state. The two extreme transition states for the E2 elimination will resemble the E1 and E1cb mechanisms in their orientational effects. At the "E1cb-like" end of the E2 range, a highly developed bond is present between the proton and the base. The leaving group remains tightly bound to carbon, and there is relatively little development of the carbon–carbon double bond. When the transition state of an E2 reaction has extensive E1cb character, the direction of the elimination is governed by the ease of proton removal. In this case, the less substituted of the possible alkene products usually dominates. At the "E1-like" end of the spectrum, the transition state is characterized by well-advanced cleavage of the C—X bond and a largely intact C—H bond. "E1-like" transition states for E2 reactions lead to formation of the more highly substituted of the possible alkenes. In a more synchronous E2 reaction, the new double bond is substantially formed at the transition state at the expense of partial rupture of both the C—H and C—X bonds. E2 elimination gives mainly the more substituted alkene. This is because the transition states leading to the possible alkenes will reflect the partial double-bond character, and the greater stability of the more substituted double bond will favor the corresponding transition state.

Prior to development of the mechanistic ideas outlined above, it was recognized by experience that some types of elimination reactions gave the more substituted alkene as the major product. Such eliminations were said to follow the "Saytzeff rule." This behavior is characteristic of E1 reactions and E2 reactions involving relatively good leaving groups, such as halides and sulfonates. These are now recognized as reactions which proceed with C—X cleavage being well advanced in the transition state. E2 reactions involving poor leaving groups, particularly those involving quaternary ammonium salts, were said to follow the "Hofmann rule" and gave mainly the less substituted alkene. We would now recognize that such reactions would proceed through transition states with E1cb character.

The data recorded in Table 6.4 for the 2-hexyl system illustrate two general trends that have been recognized in other systems as well. First, poorer leaving groups favor elimination according to the "Hofmann rule," as shown, for example,

Table 6.4. Product Ratios for Some E2 Eliminations

Substrate: $CH_3CH_2CH_2CH_2CH(X)CH_3$[a]	Base, solvent	Percent composition of alkene: 1-Hexene	2-Hexene *trans*	2-Hexene *cis*
X = I	MeO^-, MeOH	19	63	18
Cl	MeO^-, MeOH	33	50	17
F	MeO^-, MeOH	69	21	9
$OSO_2C_7H_7$	MeO^-, MeOH	33	44	23
I	*t*-BuO^-, *t*-BuOH	78	15	7
Cl	*t*-BuO^-, *t*-BuOH	91	5	4
F	*t*-BuO^-, *t*-BuOH	97	1	1
$OSO_2C_7H_7$	*t*-BuO^-, *t*-BuOH	83	4	14

a. From R. A. Bartsch and J. F. Bunnett, *J. Am. Chem. Soc.* **91**, 1376 (1967).

by the increasing amount of terminal olefin in the halogen series as the leaving group is changed from iodide to fluoride. Poorer leaving groups move the transition state in the E1cb direction. A higher negative charge must build up on the β-carbon to induce loss of the leaving group. This charge buildup is accomplished by more complete proton abstraction.

Comparison of the data for methoxide with those for *t*-butoxide in Table 6.4 illustrates the second general trend: Stronger bases favor formation of the less substituted alkene.[88] A stronger base leads to an increase in the carbanion character at the transition state and thus shifts it in the E1cb direction. A linear correlation between the strength of the base and the difference in $\Delta G^{\ddagger}$ for the formation of 1-butene versus 2-butene has been established.[88b] Some of the data are given in Table 6.5.

The direction of elimination is also affected by steric effects, and if both the base and the reactant are highly branched, steric factors will lead to preferential removal of the less hindered hydrogen.[89] Thus, when 4-methyl-2-pentyl iodide reacts with very hindered bases such as potassium tricyclohexylmethoxide, there is preferential formation of the terminal alkene. In this case, potassium *t*-butoxide favors the internal alkene, although by a smaller ratio than for less branched alkoxides.

$$(CH_3)_2CHCH_2CH(I)CH_3 \rightarrow (CH_3)_2CHCH{=}CHCH_3 + (CH_3)_2CHCH_2CH{=}CH_2$$

base		
$K^{+-}OC(C_6H_{11})_3$	42%	58%
$K^{+-}OC(CH_3)_3$	61%	39%
$K^{+-}OCH_2CH_2CH_3$	75%	25%

88. (a) D. H. Froemsdorf and M. D. Robbins, *J. Am. Chem. Soc.* **89**, 1737 (1967); I. N. Feit and W. H. Saunders, Jr., *J. Am. Chem. Soc.* **92**, 5615 (1970); (b) R. A. Bartsch, G. M. Pruss, B. A. Bushaw, and K. E. Wiegers, *J. Am. Chem. Soc.* **95**, 3405 (1973); (c) R. A. Bartsch, K. E. Wiegers, and D. M. Guritz, *J. Am. Chem. Soc.* **96**, 430 (1974).
89. R. A. Bartsch, R. A. Read, D. T. Larsen, D. K. Roberts, K. J. Scott, and B. R. Cho, *J. Am. Chem. Soc.* **101**, 1176 (1979).

Table 6.5. Orientation in E2 Elimination as a Function of Base Strength

Base (potassium salt)	pK	% 1-Butene from 2-iodobutane[a]	% 1-Butene from 2-butyl tosylate[b]
p-Nitrobenzoate	8.9	5.8	*c*
Benzoate	11.0	7.2	*c*
Acetate	11.6	7.4	*c*
Phenolate	16.4	11.4	30.6
2,2,2-Trifluoroethoxide	21.6	14.3	46.0
Methoxide	29.0	17.0	*c*
Ethoxide	29.8	17.1	56.0
t-Butoxide	32.2	20.7	58.5

a. From R. A. Bartsch, G. M. Pruss, B. A. Bushaw, and K. E. Wiegers, *J. Am. Chem. Soc.* **95**, 3405 (1973). The p*K* values refer to DMSO solution.
b. R. A. Bartsch, R. A. Read, D. T. Larsen, D. K. Roberts, K. J. Scott, and B. R. Cho, *J. Am. Chem. Soc.* **101**, 1176 (1979).
c. Not reported.

The leaving group also affects the amount of internal versus terminal alkene that is formed. The poorer the leaving group, the more E1cb-like is the transition state. This trend is illustrated for the case of the 2-butyl system by the data in Table 6.6. Positively charged leaving groups, such as in dimethylsulfonium and trimethylammonium salts, may favor a more E1cb-like transition state because their inductive and field effects increase the acidicity of the β protons.

6.8. Stereochemistry of E2 Elimination Reactions

Two elements of stereochemistry enter into determining the ratio of isomeric alkenes formed in E2 reactions. First, elimination may proceed in *syn* or *anti* fashion:

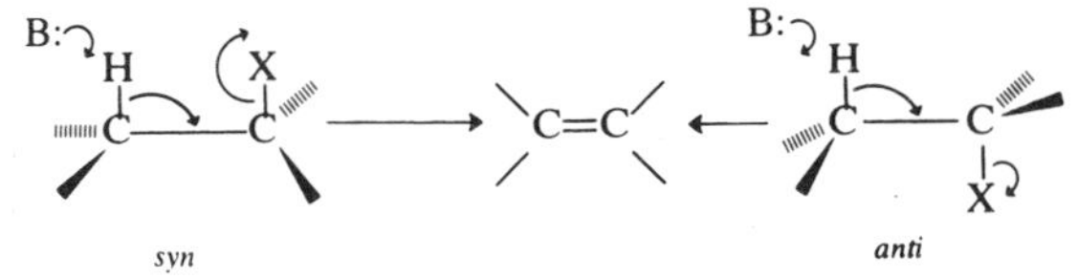

syn *anti*

Second, in many case, the product alkene may be a mixture of the *cis and trans* isomers. The product ratio therefore depends on these stereochemical details of the elimination. The stereochemical aspects of elimination reactions have been of interest because of the insight the data provide into the reaction mechanism.

In most cases, E2 elimination proceeds via a transition state involving the *anti* arrangement. Nevertheless, *syn* elimination is possible, and, when special structural features retard *anti* elimination, *syn* elimination becomes the dominant mode. In acyclic systems, the extent of *anti* versus *syn* elimination can be determined by use of stereospecifically deuterated substrates or by use of diastereomeric reactants which will give a different product by *syn* and *anti* elimination. The latter approach

Table 6.6. Orientation of Elimination in the 2-Butyl System under Various E2 Conditions

	1-Butene (%)	2-Butene (%)	Ref.
$CH_3CH(I)CH_2CH_3 \xrightarrow[DMSO]{PhCO_2^-}$	7	93	a
$CH_3CH(I)CH_2CH_3 \xrightarrow[DMSO]{C_2H_5O^-}$	17	83	a
$CH_3CH(I)CH_2CH_3 \xrightarrow[DMSO]{(CH_3)_3CO^-}$	21	79	b
$CH_3CH(Br)CH_2CH_3 \xrightarrow[DMSO]{(CH_3)_3CO^-}$	33	67	b
$CH_3CH(Cl)CH_2CH_3 \xrightarrow[DMSO]{(CH_3)_3CO^-}$	43	57	b
$CH_3CH(Br)CH_2CH_3 \xrightarrow[C_2H_5OH]{C_2H_5O^-}$	19	81	c
$CH_3CH(OSO_2C_7H_7)CH_2CH_3 \xrightarrow[C_2H_5OH]{C_2H_5O^-}$	35	65	d
$CH_3CH(OSO_2C_7H_7)CH_2CH_3 \xrightarrow[DMSO]{(CH_3)_3CO^-}$	61	39	d
$CH_3CH(\overset{+}{S}(CH_3)_2)CH_2CH_3 \xrightarrow[C_2H_5OH]{C_2H_5O^-}$	74	26	e
$CH_3CH(\overset{+}{N}(CH_3)_3)CH_2CH_3 \xrightarrow{^-OH}$	95	5	f

a. R. A. Bartsch, B. M. Pruss, B. A. Bushaw, and K. E. Wiegers, *J. Am. Chem. Soc.* **95**, 3405 (1973).
b. D. L. Griffith, D. L. Meges, and H. C. Brown, *J. Chem. Soc., Chem. Commun.*, 90 (1968).
c. M. L. Dhar, E. D. Hughes, and C. K. Ingold, *J. Chem. Soc.*, 2058 (1948).
d. D. H. Froemsdorf and M. D. Robbins, *J. Am. Chem. Soc.* **89**, 1737 (1967).
e. E. D. Hughes, C. K. Ingold, G. A. Maw and L. I. Woolf, *J. Chem. Soc.*, 2077 (1948).
d. A. C. Cope, N. A. LeBel, H.-H. Lee, and W. R. Moore, *J. Am. Chem. Soc.* **79**, 4720 (1957).

showed that elimination from 3-phenyl-2-butyl tosylate is a stereospecific *anti* process.[90]

$$\text{CH}_3(\text{Ph})(\text{H})\text{C—C}(\text{OTs})(\text{H})\text{CH}_3 \xrightarrow[\text{EtOH}]{\text{NaOEt}} (\text{Ph})(\text{CH}_3)\text{C=C}(\text{H})(\text{CH}_3) \qquad \text{Ph}(\text{CH}_3)(\text{H})\text{C—C}(\text{OTs})(\text{H})\text{CH}_3 \xrightarrow[\text{EtOH}]{\text{NaOEt}} (\text{CH}_3)(\text{Ph})\text{C=C}(\text{H})(\text{CH}_3)$$

The occurrence of *syn* elimination in 3-decyl systems has been demonstrated using diastereomeric deuterium-labeled substrates. Stereospecifically labeled 5-substituted decane derivatives were prepared and subjected to appropriate elimination conditions. By comparison of the amount of deuterium in the *E*-and *Z*-isomers

90. W.-B. Chiao and W. H. Saunders, *J. Org. Chem.* **45**, 1319 (1980).

Table 6.7. Extent of *Syn* Elimination as a Function of the Leaving Group in the 5-Decyl System[a]

	Percent *syn* elimination			
	E product		*Z* product	
Leaving group	DMSO	Benzene	DMSO	Benzene
Cl	6	62	7	39
OTs	4	27	4	16
$^{+}N(CH_3)_3$	93	92	76	84

a. Data from M. Pankova, M. Svoboda, and J. Zavada, *Tetrahedron Lett.* 2465 (1972). The base used was potassium *t*-butoxide.

of the product, it is possible to determine the extent of *syn* and *anti* elimination.[91]

Data obtained for several different leaving groups are shown in Table 6.7. The results show that *syn* elimination is extensive for quaternary ammonium salts. With better leaving groups, the extent of *syn* elimination is small in the polar solvent DMSO but quite significant in benzene. The factors which promote *syn* elimination will be discussed shortly.

Cyclohexyl systems have a very strong preference for *anti* elimination via conformations in which both the departing proton and the leaving group occupy

91. M. Pankova, M. Svoboda, and J. Zavada, *Tetrahedron Lett.*, 2465 (1972). The analysis of the data also requires that account be taken of (a) isotope effects and (b) formation of 4-decene. The method of analysis is described in detail by J. Sicher, J. Zavada, and M. Pankova, *Collect. Czech. Chem. Commun.* **36**, 3140 (1971).

axial positions. This orientation permits the alignment of the involved orbitals so that *anti* elimination can occur.

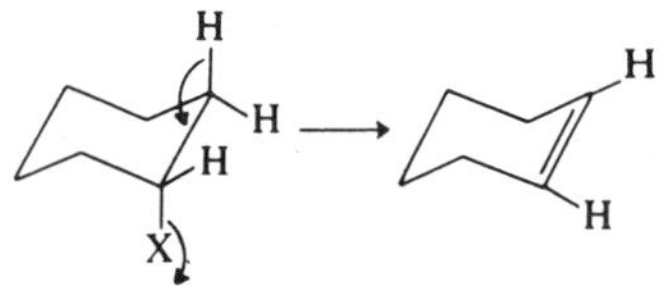

For example, *cis*-4-*t*-butylcyclohexyl bromide undergoes E2 elimination at a rate about 500 times greater than the *trans* isomer.[92]

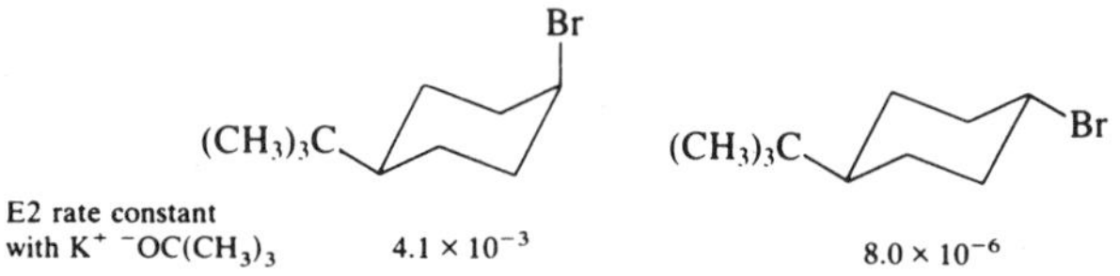

E2 rate constant with $K^+ \ ^-OC(CH_3)_3$ 4.1×10^{-3} 8.0×10^{-6}

Other cyclic systems are not so selective. In the decomposition of *N,N,N*-trimethylcyclobutylammonium hydroxide, elimination is 90% *syn*.[93] The cyclobutyl ring resists the conformation required for *anti* elimination. The more flexible five-membered ring analog undergoes about 50% *syn* elimination. Elimination from the *N,N,N*-trimethylnorbornylammonium ion is exclusively *syn*.[94] This is another case where the rigid ring prohibits attainment of an *anti*-elimination process. There is also a steric effect operating against removal of an *endo* proton, which is required for *anti* elimination.

Although there is usually a preference for *anti* elimination in acyclic systems, *syn* elimination is competitive in some cases. Table 6.8 summarizes some data *syn* versus *anti* elimination in acyclic systems.

The general trend revealed by these and other data is that *anti* stereochemistry is normally preferred for reactions involving good leaving groups such as bromide and tosylate. With poorer leaving groups (e.g., fluoride, trimethylamine), *syn* elimination becomes important. The amount of *syn* elimination is small in the 2-butyl system, but it becomes a major pathway with 3-hexyl compounds and longer chains. *Syn* elimination is especially prevalent in the medium-sized alicyclic systems.[95]

92. J. Zavada, J. Krupicka, and J. Sicher, *Collect. Czech. Chem. Commun.* **33**, 1393 (1968).
93. M. P. Cooke, Jr., and J. L. Coke, *J. Am. Chem. Soc.* **90**, 5556 (1968).
94. J. P. Coke and M. P. Cooke, *J. Am. Chem. Soc.* **89**, 6701 (1967).
95. J. Sicher, *Angew. Chem. Int. Ed. Engl.* **11**, 200 (1972).

Table 6.8. Stereochemistry of E2 Eliminations for Some Acyclic Substrates

Substrate	Base, solvent	% *anti*	% *syn*	Ref.
$CH_3CHCHCH_3$ D Br	$K^{+-}OC(CH_3)_3$, $(CH_3)_3COH$	100	0	a
$CH_3CHCHCH_3$ D $OSO_2C_7H_7$	$K^{+-}OC(CH_3)_3$, $(CH_3)_3COH$	>98	<2	b
$CH_3CHCHCH_3$ D $^+N(CH_3)_3$	$K^{+-}OC(CH_3)_3$, DMSO	100	0	c
$CH_3CH_2CHCHCH_2CH_3$ D $^+N(CH_3)_3$	$K^{+-}OC(CH_3)_3$, $(CH_3)_3COH$			d
$CH_3CH_2CHCHCH_2CH_3$ D F	$K^{+-}OC(CH_3)_3$, $(CH_3)_3COH$	32	68	e
$CH_3(CH_2)_3CHCH(CH_2)_3CH_3$ D $^+N(CH_3)_3$	$K^{+-}OC(CH_3)_3$, DMSO	24	76	f
$CH_3(CH_2)_3CHCH(CH_2)_3CH_3$ D $OSO_2C_7H_7$	$K^{+-}OC(CH_3)_3$, $(CH_3)_3COH$	93	7	g
$CH_3(CH_2)_3CHCH(CH_2)_3CH_3$ D Cl	$K^{+-}OC(CH_3)_3$, benzene	62	38	h
$CH_3(CH_2)_3CHCH(CH_2)_3CH_3$ D F	$K^{+-}OC(CH_3)_3$, benzene	<20	>80	h
$CH_3(CH_2)_3CHCH(CH_2)_3CH_3$ D CL	$K^{+-}OC(CH_3)_3$, DMSO	93	7	h
$CH_3(CH_2)_3CHCH(CH_2)_3CH_3$ D F	$K^{+-}OC(CH_3)_3$, DMSO	80	20	h

a. R. A. Bartsch, *J. Am. Chem. Soc.* **93**, 3683 (1971).
b. D. H. Froemsdorf, W. Dowd, W. A. Gifford, and S. Meyerson, *J. Chem. Soc., Chem. Commun.*, 449 (1968).
c. D. H. Froemsdorf, H. R. Pinnick, Jr., and S. Meyerson, *J. Chem. Soc., Chem. Commun.*, 1600 (1968).
d. D. S. Bailey and W. H. Saunders, Jr., *J. Am. Chem. Soc.* **92**, 6904 (1970).
e. J. K. Borchardt, J. C. Swanson, and W. H. Saunders, Jr., *J. Am. Chem. Soc.* **96**, 3918 (1974).
f. J. Sicher, J. Závada, and M. Pánková, *Collect. Czech. Chem. Commun.* **36**, 3140 (1971).
f. J. Závada, M. Pánková, and J. Sicher, *J. Chem. Soc., Chem. Commun.*, 1145 (1968).
h. M. Pánková, M. Svoboda, and J. Závada, *Tetrahedron Lett.*, 2465 (1972).

The factors that determine whether *syn* or *anti* elimination predominates are still subject to investigation. One factor that is believed to be important is whether the base is free or present as an ion pair.[96] The evidence is that an ion pair promotes *syn* elimination of anionic leaving groups. This effect can be explained by proposing

96. R. A. Bartsch, G. M. Pruss, R. L. Buswell, and B. A. Bushaw, *Tetrahedron Lett.*, 2621 (1972); J. K. Borchardt, J. C. Swanson, and W. H. Saunders, J., *J. Am. Chem. Soc.* **96**, 3918 (1974).

a transition state in which the anion functions as a base and the cation assists in the departure of the leaving group.

$$\text{H}\cdots\overset{-}{\text{O}}(\text{R})\cdots\overset{+}{\text{M}}\cdots\text{X (C—C)} \longrightarrow \text{C=C} + \text{ROH} + \text{M}^+$$

This interpretation is in agreement with the solvent effect that is evident in the 5-decyl system, as revealed in Table 6.7. The extent of *syn* elimination is much higher in the nondissociating solvent benzene than in DMSO. The ion pair interpretation is also favored by the fact that addition of specific metal-ion-complexing agents (crown ethers) that would favor dissociation of the ion pair leads to diminished amounts of *syn* elimination.[97] A theory based on steric effects was also considered, but it suggested that only relatively large leaving groups would show *syn* elimination. Later studies revealed that extensive *syn* elimination occurred even with the small fluoride ion as the leaving group.[85] Another factor that affects the *syn* : *anti* ratio is the strength of the base. Strong bases are more likely to exhibit a high proportion of *syn* elimination.[98]

Steric effects also play a significant role. With *N*-(β,β-disubstituted-ethyl)-*N*,*N*,*N*-trimethylammonium ions, *syn* elimination is important when the β substituents are aryl or branched. As the β groups become less bulky, the amount of *syn* elimination diminishes. This effect is demonstrated by the data below.[99]

$$R^1-\underset{H}{\overset{R^2}{C}}-CHD\overset{+}{N}(CH_3)_3 \longrightarrow (R^1)(R^2)C=C(H)(D)$$

R^1	R^2	% *syn*	$\in$ *anti*
Ph	MeO—Ph	62	38
Ph	$(CH_3)_2CH$	68	32
Ph	Me	26	79
n-Bu	H	<5	>95

The dependence on steric bulk is attributed to the steric requirements imposed by the bulky trimethylamine leaving group. In the transition state for *anti* elimination, steric repulsion is increased as R^1 and R^2 increase in size. When the repulsion is

97. R. A. Bartsch, E. A. Mintz, and R. M. Parlman, *J. Am. Chem. Soc.* **96**, 3918 (1974).
98. K. C. Brown and W. H. Saunders, Jr., *J. Am. Chem. Soc.* **92**, 4292 (1970); D. S. Bailey and W. H. Saunders, Jr., *J. Am. Chem. Soc.* **92**, 6904 (1970).
99. Y.-T. Tao and W. H. Saunders, Jr., *J. Am. Chem. Soc.* **105**, 3183 (1983).

sufficiently large, the transition state for *syn* elimination is preferred.

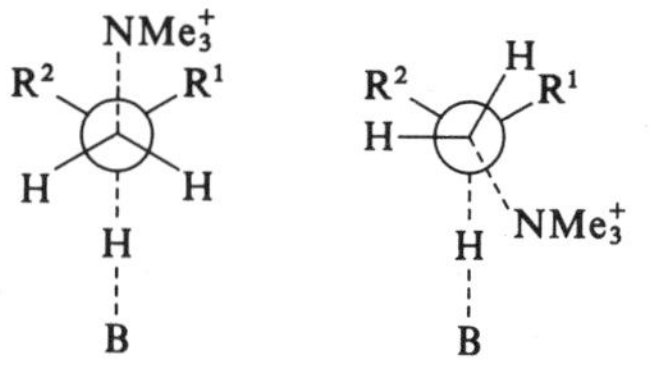

The proportion of *cis* and *trans* isomers of internal alkenes formed during elimination reactions depends on the identity of the leaving group. Halides usually give predominantly the *trans* alkenes.[100] Bulkier groups, particularly arenesulfonates, give higher proportions of the *cis* alkene. Sometimes, more *cis* isomer is formed than *trans*. The normal preference for *trans* alkene probably reflects the greater stability of the *trans* alkene; that is, the unfavorable steric repulsions present in the *cis* alkene are also present in the E2 transition state leading to the *cis* alkene. High *cis* : *trans* ratios are attributed to a second steric effect that becomes important only when the leaving group is large. The conformations leading to the *cis* and the *trans* alkene by *anti* elimination are depicted below.

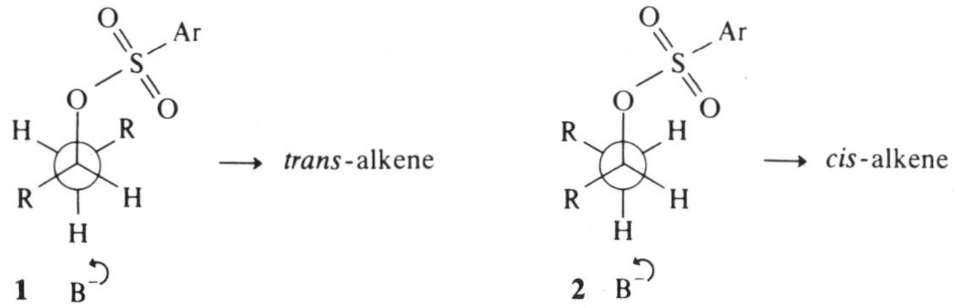

1 → *trans*-alkene   **2** → *cis*-alkene

When the leaving group and base are both large, conformation **2** is favored because it permits the leaving group to occupy a position removed from both alkyl substituents. *Anti* elimination through a transition state arising from conformation **2** gives the *cis* alkene.

6.9. Dehydration of Alcohols

The dehydration of alcohols is an important elimination reaction that takes place under acidic rather than basic conditions. It involves an E1 mechanism.[101] The function of the acidic reagent is to convert the hydroxyl group to a better leaving group by protonation.

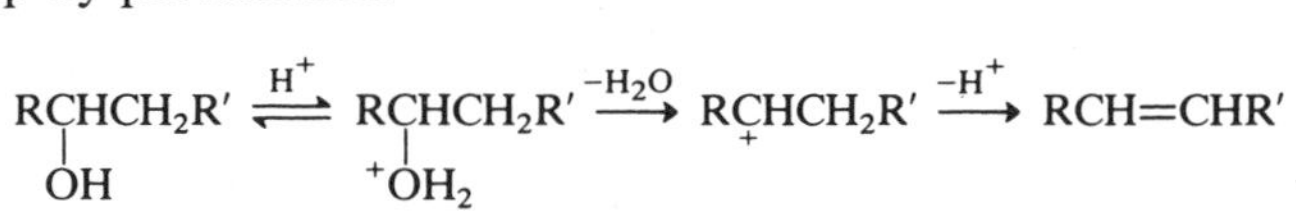

$$\underset{\displaystyle\text{OH}}{\text{RCHCH}_2\text{R}'} \xrightleftharpoons{\text{H}^+} \underset{\displaystyle{}^+\text{OH}_2}{\text{RCHCH}_2\text{R}'} \xrightarrow{-\text{H}_2\text{O}} \underset{+}{\text{RCHCH}_2\text{R}'} \xrightarrow{-\text{H}^+} \text{RCH=CHR}'$$

100. H. C. Brown and R. L. Kliminsch, *J. Am. Chem. Soc.* **87**, 5517 (1965); I. N. Feit and W. H. Saunders, Jr., *J. Am. Chem. Soc.* **92**, 1630 (1970).
101. D. V. Banthorpe, *Elimination Reactions*, Elsevier, New York, 1963, pp 145–156.

Since a carbocation or closely related species is the intermediate, the elimination step would be expected to favor the more substituted alkene as discussed on p. 374. The E1 mechanism also explains the general trends in relative reactivity. Tertiary alcohols are the most reactive, and reactivity decreases going to secondary and primary alcohols. Also in accord with the E1 mechanism is the fact that rearranged products are found in cases where a carbocation intermediate could be expected to rearrange.

$$R_3C\underset{\displaystyle OH}{\underset{|}{C}}HR' + H^+ \rightleftharpoons R_3C\underset{\displaystyle ^+OH_2}{\underset{|}{C}}HR' \rightleftharpoons R_3C\overset{+}{C}HR' + H_2O$$

$$R_3\overset{+}{C}CHR' \longrightarrow R_2\overset{+}{C}\overset{\displaystyle R}{\overset{|}{C}}HR' \longrightarrow R_2C{=}\overset{\displaystyle R}{\overset{|}{C}}R'$$

For many alcohols, exhange of the hydroxyl group with solvent competes with dehydration.[102] This exchange indicates that the carbocation can undergo S_N1 capture in competition with elimination. Under conditions where proton removal is rate-determining, it would be expected that a significant isotope effect would be seen. This is in fact observed.

$$Ph\underset{\displaystyle HO}{\underset{|}{C}}H\underset{\displaystyle H^*}{\underset{|}{C}}HPh \xrightarrow[H_2O]{H_2SO_4} PhCH{=}CHPh \qquad k_H/k_D = 1.8$$ Ref. 103

6.10. Eliminations Not Involving C—H Bonds

The discussion of β-elimination processes thus far has focused on those that involve abstraction of a proton bound to carbon. It is the electrons in the C—H σ bond, however, that are primarily involved in the elimination process. Compounds bearing substituents other than protons that are attached to the carbon framework by a σ bond that can release electrons should undergo similar eliminations. Many such processes are known, and frequently the reactions are stereospecific.

Vicinal dibromides may be debrominated by treating them with certain reducing agents, including iodide ion and zinc. The stereochemical course of such reactions in 1,1,2-tribromocyclohexane was determined using a ^{82}Br-labeled sample prepared by *anti* addition of ^{82}Br to bromocyclohexene. Exclusive *anti* elimination would give unlabeled bromocyclohexene, while ^{82}Br-labeled product would result from *syn* elimination. Debromination with sodium iodide was found to be cleanly an *anti* elimination, while debromination with zinc gives mainly, but not entirely, *anti*

102. C. A. Bunton and D. R. Llewellyn, *J. Chem. Soc.*, 3402 (1957); J. Manassen and F. S. Klein, *J. Chem. Soc.*, 4203 (1960).
103. D. S. Noyce, D. R. Hartter, and R. M. Pollack, *J. Am. Chem. Soc.* **90**, 3791 (1968).

elimination.[104]

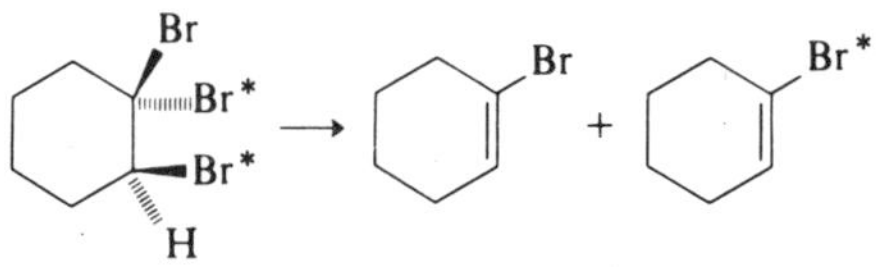

Reagent	*anti*	*syn*
Sodium iodide/MeOH	100%	0%
Zinc/EtOH	89%	11%

The iodide-induced reduction is essentially the reverse of a halogenation. Application of the principle of microscopic reversibility would suggest that the reaction would proceed through a bridged intermediate as shown below.[105]

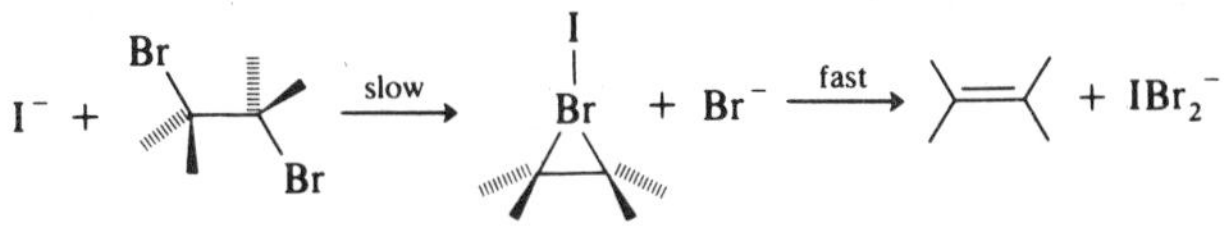

$$I^- + \text{BrC–CBr} \xrightarrow{\text{slow}} \text{I–Br}^+\text{(bridged)} + Br^- \xrightarrow{\text{fast}} \text{C=C} + IBr_2^-$$

The rate-determining expulsion of bromide ion through a bridged intermediate requires an *anti* orientation of the two bromides. The nucleophilic attack of iodide at one bromide enhances its nucleophilicity and permits formation of the bridged ion. The stereochemical preference in noncyclic systems is also *anti*, as indicated by the fact that *meso*-stilbene dibromide yields *trans*-stilbene, while *d,l*-stilbene dibromide gives mainly *cis*-stilbene under these conditions.[105]

The zinc-induced debromination could proceed by formation of an organozinc intermediate, with the loss of stereospecificity occurring during the formation of this intermediate. Similar nonstereospecific debrominations occur with one-electron donors, such as chromium(II) salts, and have been interpreted as resulting from a free-radical intermediate.[106] The organozinc and organochromium reagents which are postulated as intermediates in these reductive eliminations are representatives of a general structural type M—C—C—X, in which M is a metal and X is a leaving group. These structures are in general very prone to elimination with formation of a double bond.

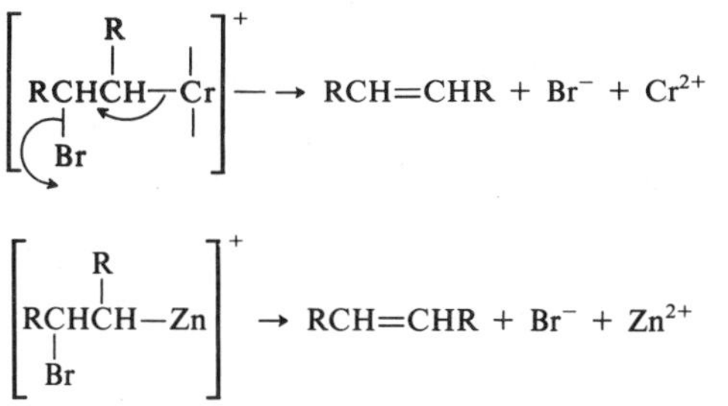

$$\left[\text{RCH(Br)CH(R)–Cr}\right]^+ \longrightarrow \text{RCH=CHR} + Br^- + Cr^{2+}$$

$$\left[\text{RCH(Br)CH(R)–Zn}\right]^+ \rightarrow \text{RCH=CHR} + Br^- + Zn^{2+}$$

104. C. L. Stevens and J. A. Valicenti, *J. Am. Chem. Soc.* **87**, 838 (1965).
105. C. S. T. Lee, I. M. Mathai, and S. I. Miller, *J. Am. Chem. Soc.* **92**, 4602 (1970).
106. J. K. Kochi and D. M. Singleton, *J. Am. Chem. Soc.* **90**, 1582 (1968).

One example of elimination reactions of this type is acid-catalyzed deoxymercuration.[107] The β-oxyorganomercurials are much more stable than similar reagents derived from more electropositive metals but are much more reactive than simple alcohols. For example, $CH_3CH(OH)CH_2HgI$ is converted to propene under acid-catalyzed conditions at a rate 10^{11} times greater than that for dehydration of 2-propanol under the same condition. These reactions are pictured as proceeding through a bridged mercurinium ion by a mechanism which is the reverse of oxymercuration.

One of the pieces of evidence in favor of this mechanism is the fact that the $\Delta H^{\ddagger}$ for deoxymercuration of *trans*-2-methoxycyclohexylmercuric iodide is about 8 kcal/mol less than for the *cis* isomer. Only the *trans* isomer can undergo elimination by an *anti* process through a chair conformation.

Comparing the rates of acid-catalyzed β elimination of compounds of the type MCH_2CH_2OH yields the reactivity order for β substituents $IHg \approx Ph_3Pb \approx Ph_3Sn > Ph_3Si > H$. The relative rates are within a factor of 10 for the first three, but these are 10^6 greater than for Ph_3Si and 10^{11} greater than for a proton. There are two factors involved in these very large rate accelerations. One is bond energies. The relevant values are $Hg—C = 27 < Pb—C = 31 < Sn—C = 54 < Si—C = 60 < H—C = 96$.[108] The metal substituents also have a very strong stabilizing effect for carbocation character at the β-carbon. This stabilization can be pictured as an orbital-orbital interaction in which the electron-rich carbon-metal bond donates electron density to the adjacent p orbital or as formation of a bridged species.

107. M. M. Kreevoy and F. R. Kowitt, *J. Am. Chem. Soc.* **82**, 739 (1960).
108. D. D. Davis and H. M. Jacocks III, *J. Organomet. Chem.* **206**, 33 (1981).

There are a number of synthetically important β-elimination processes involving organosilicon[109] and organotin[110] compounds. Treatment of β-hydroxyalkylsilanes or β-hydroxyalkylstannanes with acid results in stereospecific *anti* eliminations which are much more rapid than for compounds lacking the group IV substituent.

$$CH_3CH_2CH_2(H)((CH_3)_3Si)C{-}C(OH)(H)CH_2CH_2CH_3 \xrightarrow{H_2SO_4} (H)(CH_3CH_2CH_2)C{=}C(H)(CH_2CH_2CH_3)$$ Ref. 111

$$Ph_3Sn(H)(CH_3)C{-}C(H)(CH_3)(OH) \xrightarrow{H^+} (H)(H_3C)C{=}C(H)(CH_3)$$ Ref. 112

β-Halosilanes also undergo facile elimination when treated with methoxide ion.

$$C_4H_9CH(Br)CH(Br)Si(CH_3)_3 \xrightarrow{NaOCH_3} C_4H_9CH{=}CHBr$$ Ref. 113

Fluoride-induced β eliminations of silanes having leaving groups in the β position are important processes in synthetic chemistry, as for example in the removal of β-t:imethylsilylethoxy groups.

$$RCO_2CH_2CH_2Si(CH_3)_3 + R_4N^+F^- \rightarrow RCO_2^-\,{}^+NR_4 + CH_2{=}CH_2 + FSi(CH_3)_3$$ Ref. 114

These reactions proceed by alkoxide or fluoride attack at silicon, which results in C—Si bond cleavage and elimination of the leaving group from the β-carbon. These reactions are stereospecific *anti* eliminations.

$$RCH(X){-}CH(Si(CH_3)_3)R + Nu^- \rightarrow RCH{=}CHR + X^- + (CH_3)_3SiNu$$

β-Elimination reactions of this type can also be effected by converting a β-hydroxy group to a good leaving group. For example, conversion of β-

109. A. W. P. Jarvie, *Organomet. Chem. Rev., Sect. A* **6**, 153 (1970); W. P. Weber, *Silicon Reagents for Organic Synthesis*, Springer-Verlag, Berlin, 1983; E. W. Colvin, *Silicon in Organic Synthesis*, Butterworths, London, 1981.
110. M. Pereyre, J.-P. Quintard, and A. Rahm, *Tin in Organic Synthesis*, Butterworths, London, 1987.
111. P. F. Hudrlick and D. Peterson, *J. Am. Chem. Soc.* **97**, 1464 (1975).
112. D. D. Davis and C. E. Gray, *J. Org. Chem.* **35**, 1303 (1970).
113. A. W. P. Jarvie, A. Holt, and J. Thompson, *J. Chem. Soc., B*, 852 (1969); B. Miller and G. J. McGarvey, *J. Org. Chem.* **43**, 4424 (1978).
114. P. Sieber, *Helv. Chim. Acta* **60**, 2711 (1977).

hydroxyalkylsilanes to the corresponding methanesulfones leads to rapid elimination.[115]

$$(CH_3)_3SiCH_2\underset{\displaystyle OH}{\underset{|}{C}}R_2 \xrightarrow{CH_3SO_2Cl} H_2C{=}CR_2$$

The ability to promote β elimination and the electron donor capacity of the β-metalloid substituents can be exploited in a very useful way in synthetic chemistry.[116] Vinylstannanes and vinylsilanes react readily with electrophiles. The resulting intermediates then undergo elimination of the stannyl or silyl substituent, so that the net effect is replacement of the stannyl or silyl group by the electrophile. The silyl and stannyl substituents are crucial to these reactions in two ways. In the electrophilic addition step, they promote addition and strongly control the regiochemistry. A silyl or stannyl substituent strongly stabilizes carbocation character at the β-carbon atom and thus directs the electrophile to the α-carbon. Molecular orbital calculations indicate a stabilization of 38 kcal/mol, which is about the same as the value calculated for an α-methyl group.[117] The reaction is then completed by the elimination step, in which the carbon–silicon or carbon–tin bond is broken.

$$E^+ + RCH{=}CHMR_3 \rightarrow R\overset{+}{C}H{-}\underset{\displaystyle E}{\underset{|}{C}}HMR_3 \rightarrow RCH{=}CHE$$

An example is the replacement of a trimethylsilyl substituent by acetyl by reaction with acetyl chloride.

4,4-dimethyl-1-(trimethylsilyl)cyclohexene (ring bearing $Si(CH_3)_3$ on C=C, CH_3, CH_3) $+ CH_3COCl \longrightarrow$ 1-acetyl-4,4-dimethylcyclohexene ($\overset{O}{\overset{\|}{C}}CH_3$ on C=C; CH_3, CH_3) 77% Ref. 118

Allylsilanes and allylstannanes are also reactive toward electrophiles and usually undergo a concerted elimination of the silyl substituent.

$$(CH_3)_3SiCH_2CH{=}CH_2 + I_2 \rightarrow CH_2{=}CHCH_2I \qquad \text{Ref. 119}$$

$$(CH_3)_3SiCH_2CH{=}CHC_6H_{13} + (CH_3)_3CCl \xrightarrow{TiCl_4} CH_2{=}CH\overset{\displaystyle C(CH_3)_3}{\overset{|}{C}}HC_6H_{13} \qquad \text{Ref. 120}$$

$$(CH_3)_3SnCH_2\underset{\displaystyle CH_2}{\underset{\|}{C}}CH{=}CH_2 + BrCH_2CH{=}C(CH_3)_2 \rightarrow CH_2{=}\underset{\displaystyle CH_2{=}CH}{\underset{|}{C}}CH_2CH_2CH{=}C(CH_3)_2 \qquad \text{Ref. 121}$$

$$(CH_3)_3SnCH_2CH{=}CH_2 + (MeO)_2CHCH_2CH_2Ph \xrightarrow{(Et_2Al)SO_4} CH_2{=}CHCH_2\underset{\displaystyle OMe}{\underset{|}{C}}HCH_2CH_2Ph \qquad \text{Ref. 122}$$

115. F. A. Carey and J. R. Toler, *J. Org. Chem.* **41**, 1966 (1976).
116. T. H. Chan and I. Fleming, *Synthesis*, 761 (1979); I. Fleming, *Chem. Soc. Rev.* **10**, 83 (1981).
117. S. E. Wierschke, J. Sandrasekhar, and W. L. Jorgensen, *J. Am. Chem. Soc.* **107**, 1496 (1985).
118. I. Fleming and A. Pearce, *J. Chem. Soc., Chem. Commun.*, 633 (1975).
119. D. Grafstein, *J. Am. Chem. Soc.* **77**, 6650 (1955).
120. I. Fleming and I. Paterson, *Synthesis*, 445 (1979).
121. J. P. Godschalx and J. K. Stille, *Tetrahedron Lett.* **24**, 1905 (1983).
122. A. Hosomi, H. Iguchi, M. Endo, and H. Sakurai, *Chem. Lett.*, 977 (1969).

Theoretical investigations indicate that there is a ground state interaction between the alkene π orbital and the carbon–silicon bond which raises the energy of the π HOMO and enhances reactivity.[123] Furthermore, this stereoelectronic interaction promotes attack of the electrophile *anti* to the silyl substituent.

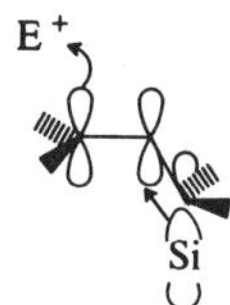

Further examples of these synthetically useful reactions can be found in Sections 9.2 and 9.3 of Part B.

General References

Polar Addition Reactions

P. B. D. de la Mare and R. Bolton, *Electrophilic Additions to Unsaturated Systems*, Second Edition, Elsevier, New York, 1982.
R. C. Fahey, in *Topics in Stereochemistry*, Vol. 3., E. L. Eliel and N. L. Allinger (eds.), Wiley-Interscience, New York, 1968, pp. 237–342.
G. H. Schmid, in *The Chemistry of the Carbon–Carbon Triple Bond*, Part 1, S. Patai (ed.), Wiley, New York, 1978, Chapter 8.
G. H. Schmid and D. G. Garratt, in *The Chemistry of Double-Bonded Functional Groups*, Part 2, S. Patai (ed.), Wiley, New York, 1977, Chapter 9.

Elimination Reactions

A. F. Cockerill and R. G. Harrison, *The Chemistry of Double-Bonded Functional Groups*, Part 1, S. Patai (ed.), Wiley, New York, 1977, Chapter 4.
W. H. Saunders, J., and A. F. Cockerill, *Mechanisms of Elimination Reactions*, Wiley, New York, 1973.

Problems

(*References for these problems will be found on page* 778.)

1. Which compound in each of the following pairs will react faster with the indicated reagent?

   (a) 1-hexene *or trans*-3-hexene with bromine in acetic acid

123. S. D. Kahn, C. F. Pau, A. R. Chamberlin, and W. J. Hehre, *J. Am. Chem. Soc.* **109**, 650 (1987).

(b) 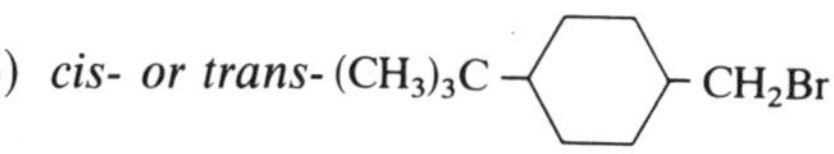 *cis-* or *trans-* $(CH_3)_3C$–(cyclohexane-1,4-diyl)–CH_2Br with potassium *t*-butoxide in *t*-butyl alcohol

(c) 2-phenylpropene *or* 4-isopropenylbenzoic acid with sulfuric acid in water

(d) 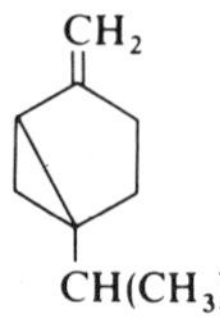 [bicyclic structure with exocyclic CH_2 and $CH(CH_3)_2$] *or* [bicyclic structure with ring CH_3 and $CH(CH_3)_2$] toward acid-catalyzed hydration

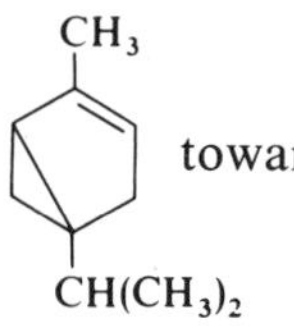

(e) $CH_3CH(CH_2)_3CH_3$ with $SO_2C_6H_5$ *or* $CH_3CH(CH_2)_3CH_3$ with $OSO_2C_6H_5$ with potassium *t*-butoxide in *t*-butyl alcohol

(f) 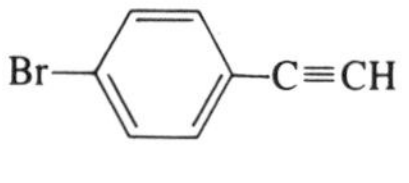 Br–C_6H_4–C≡CH *or* CH_3–C_6H_4–C≡CH with chlorine in acetic acid

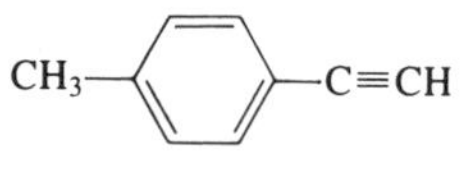

(g) 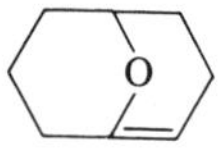 [bicyclic enol ether with ring O] *or* [cyclohexene]–OC_2H_5 toward acid-catalyzed hydration

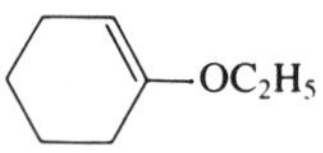

2. Predict the structure, including stereochemistry, of the product(s) of the following reactions. If more than one product is expected, indicate which will be the major product and which the minor product.

(a) $PhC(CH_2OH)_2CH_2CH{=}CH_2 \xrightarrow[CCl_4]{Br_2} C_{12}H_{15}O_2Br$

(b) 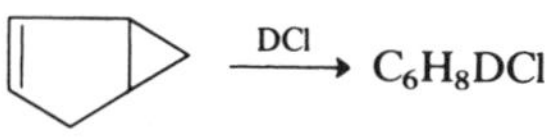 [bicyclo[3.1.0]hex-2-ene] $\xrightarrow{DCl} C_6H_8DCl$

(c) *erythro* Cl–C_6H_4–$\overset{O}{\overset{\|}{C}}$CH(Cl)CH(Cl)–$C_6H_4$–Cl $\xrightarrow[\text{ethanol}]{C_2H_5O^-Na^+} C_{15}H_9Cl_3O$

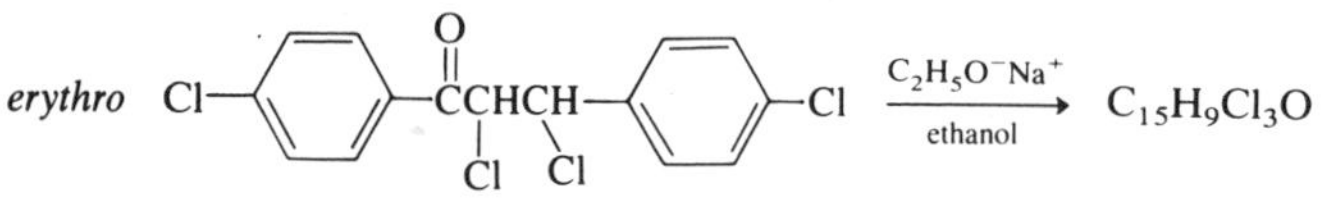

(d) 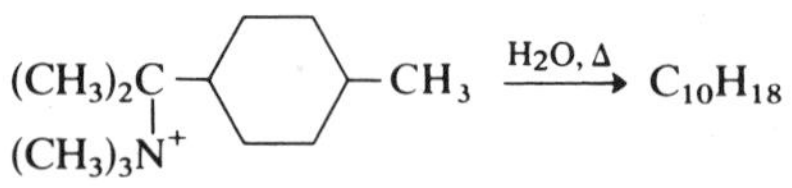 $(CH_3)_2C(N^+(CH_3)_3)$–(cyclohexane-1,4-diyl)–$CH_3 \xrightarrow{H_2O,\ \Delta} C_{10}H_{18}$

(e) 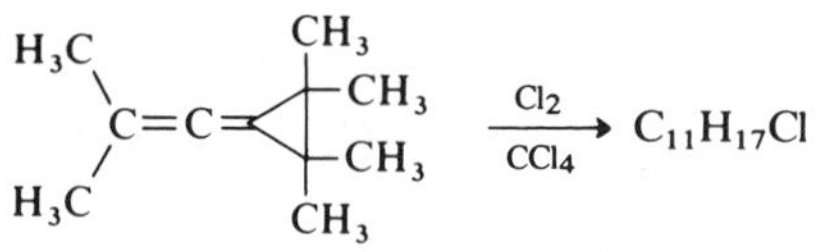 $(H_3C)_2C{=}C{=}$[cyclopropane with four CH_3] $\xrightarrow[CCl_4]{Cl_2} C_{11}H_{17}Cl$

(f) 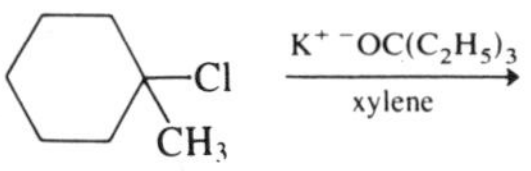 [1-chloro-1-methylcyclohexane] $\xrightarrow[\text{xylene}]{K^+\ ^-OC(C_2H_5)_3}$

(g) $CH_3-C_6H_4-CH{=}CH_2 \xrightarrow[\text{(2) NaCl}]{\text{(1) } Hg(O_2CCH_3)_2,\ CH_3OH}$

(h) $PhC{\equiv}CCH_2CH_3 \xrightarrow[CH_3CO_2H,\ 25°C]{Cl_2}$

3. The reactions of the *cis* and *trans* isomers of 4-*t*-butylcyclohexyltrimethylammonium chloride with potassium *t*-butoxide in *t*-butanol have been compared. The *cis* isomer gives 90% 4-*t*-butylcyclohexene and 10% *N,N*-dimethyl-4-*t*-butylcyclohexylamine, while the *trans* isomer gives only the latter product in quantitative yield. Explain the different behavior of the two isomers.

4. For E2 eliminations in 2-phenylethyl systems with several different leaving groups, both the primary isotope effect and Hammett ρ values for the reactions are known. Deduce from these data the relationship between the location on the E2 transition state spectrum and the nature of the leaving group; i.e., deduce which system has the most E1-like transition state and which has the most E1cb-like. Explain your reasoning.

$$PhCH_2CH_2X \xrightarrow{C_2H_5O^-} PhCH{=}CH_2$$

X	k_H/k_D	ρ
Br	7.11	2.1
$OSO_2C_7H_7$	5.66	2.3
$^+S(CH_3)_2$	5.07	2.7
$^+N(CH_3)_3$	2.98	3.7

5. When 2-bromo-2-methylpentane is dissolved in DMF, the formation of 2-methyl-1-pentene (A) and 2-methyl-2-pentene (B) occurs. The ratio of alkenes formed is not constant throughout the course of the reaction, however. Initially, the A:B ratio is about 1:1, but this drops to about 1:4 by the time the reaction is 25% complete, and then remains fairly constant. In a similar reaction, but with NaBr present in excess, the A:B ratio is constant at about 1:5 throughout the reaction. Suggest an explanation for this phenomenon.

6. For the reactions given below, predict the effect on the rate of the isotopic substitution which is described. Explain the basis of your prediction.

   (a) The effect on the rate of dehydration of 1,2-diphenylethanol of introduction of deuterium at C—2.
   (b) The effect on the rate of dehydration of 1,2-diphenylethanol of using $D_2O-D_2SO_4$ in place of $H_2O-H_2SO_4$ as the reaction medium.
   (c) The effect on the rate of bromination of styrene when deuterium is introduced on the α carbon.

7. Predict the effect on the 1-butene: *cis*-2-butene: *trans*-2-butene product ratio when the E2 elimination of *erythro*-3-*deuterio*-2-bromobutane is compared with that of 2-bromobutane. Which alkene(s) will increase in relative amount and which will decrease in relative amount? Explain the basis of your prediction.

8. Arrange the following compounds in order of increasing rate of acid-catalyzed hydration: ethylene, 2-cyclopropylpropene, 2-methylpropene, propene, 1-cyclopropyl-1-methoxyethene. Explain the basis of your prediction.

9. Discuss the factors which are responsible for the stereochemistry observed for the following reactions.

(a) $(CH_3)_3C$–cyclohexene (D, Ph, D, D) $\xrightarrow{HCl}$ $(CH_3)_3C$–cyclohexane (Ph, D, Cl, D, D, H)

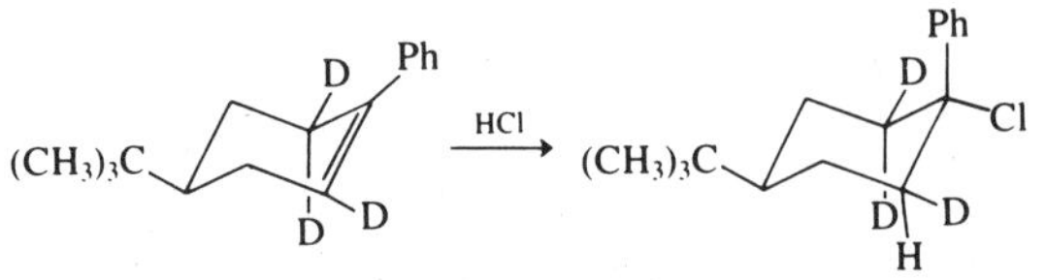

(b) Ph(H)C=C(H)C(CH₃)₃ $\xrightarrow{DBr}$ Ph, H, Br / H, D, $C(CH_3)_3$ (81%) + Ph, H, Br / H, $C(CH_3)_3$, D (19%)

(c) 2-methylenenorbornane (CH_2) $\xrightarrow{CF_3CO_2H}$ O_2CCF_3, CH_3

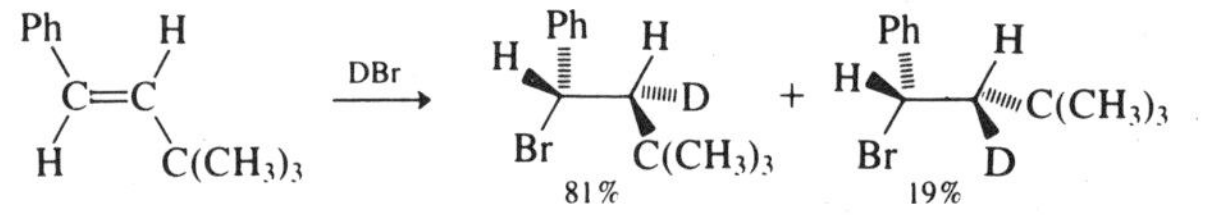

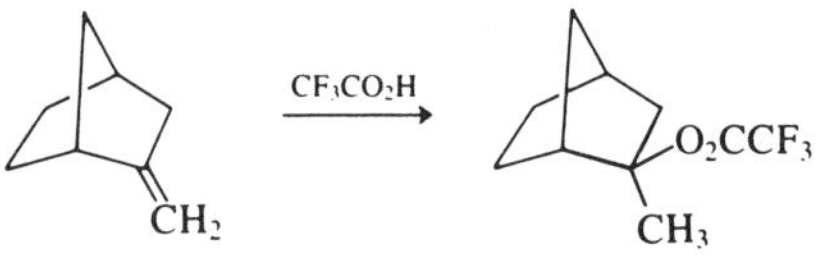

(d) $(CH_3)_3C$–cyclohexene (CH_3) $\xrightarrow{Br_2}$ $(CH_3)_3C$–cyclohexane (Br, CH_3, Br)

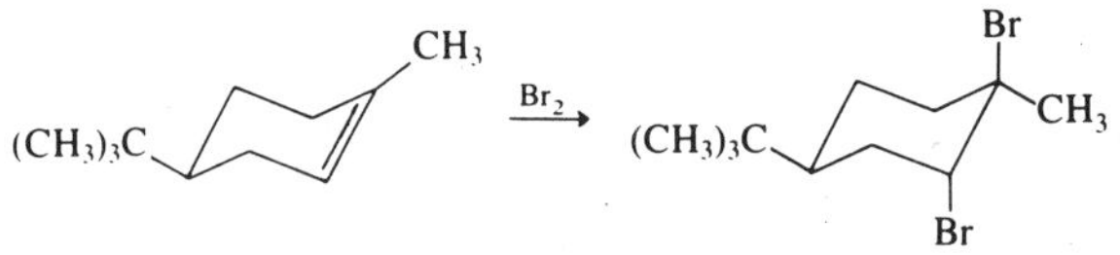

10. Explain the mechanistic basis of the following observations and discuss how the observation provides information about the reaction mechanism.
   (a) When substituted 1-aryl-2-methyl-2-propyl chlorides react with sodium methoxide, a mixture of terminal and internal alkene is formed:

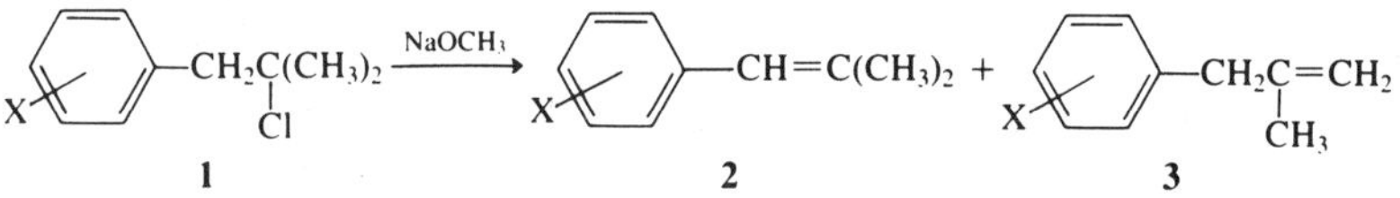

$$\text{X-}C_6H_4\text{-}CH_2C(CH_3)_2Cl \ (\mathbf{1}) \xrightarrow{NaOCH_3} \text{X-}C_6H_4\text{-}CH{=}C(CH_3)_2 \ (\mathbf{2}) + \text{X-}C_6H_4\text{-}CH_2C(CH_3){=}CH_2 \ (\mathbf{3})$$

   By using the product ratio, the overall rate can be dissected into the individual rates for formation of **2** and **3**. These rates are found to be substituent dependent for formation of **2** ($\rho = +1.4$) but substituent independent for formation of **3** ($\rho = -0.1 \pm 0.1$). The reactions are both second-order, first-order in base and first-order in substrate.
   (b) When 1,3-pentadiene reacts with DCl in forms more *E*-4-chloro-5-*deuterio*-2-pentene than *E*-4-chloro-1-*deuterio*-2-pentene.

(c) When indene (**4**) is brominated in carbon tetrachloride, it gives some *syn* addition (~15%), but indenone (**5**) gives only *anti* addition under the same conditions.

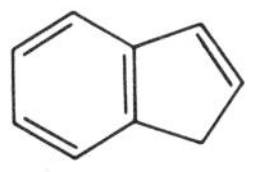

4 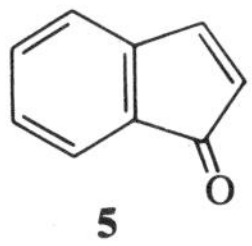 **5**

(d) The acid-catalyzed hydration of allene gives acetone, not allyl alcohol or propionaldehyde.

(e) In the addition of hydrogen chloride to cyclohexene in acetic acid, the ratio of cyclohexyl acetate to cyclohexyl chloride drops significantly when tetramethylammonium chloride is added in increasing concentration. This effect is not observed with styrene.

(f) The ρ values for base-catalyzed elimination of HF from a series of 1-aryl-2-fluoroethanes increases from the mono- to the di- and trifluoro compounds as shown by the data below:

$ArCH_2CF_3$	$ArCH_2CHF_2$	$ArCH_2CH_2F$
$\rho = +4.04$	$\rho = +3.56$	$\rho = +3.24$

11. Suggest reasonable mechanisms for each of the following reactions:

(a) LiCl, DMF, reflux 15 hr.

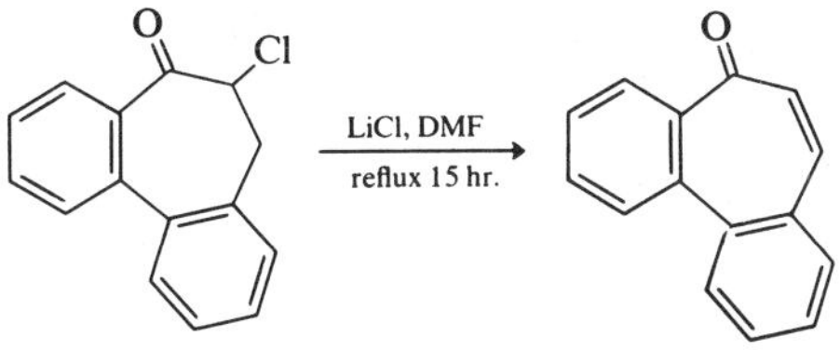

(b) $C_2H_5(H)C{=}C(Si(CH_3)_3)(C_2H_5)$ $\xrightarrow[(2)\ CH_3O^-Na^+,\ CH_3OH]{(1)\ Br_2,\ CH_2Cl_2}$ $C_2H_5(H)C{=}C(C_2H_5)(Br)$

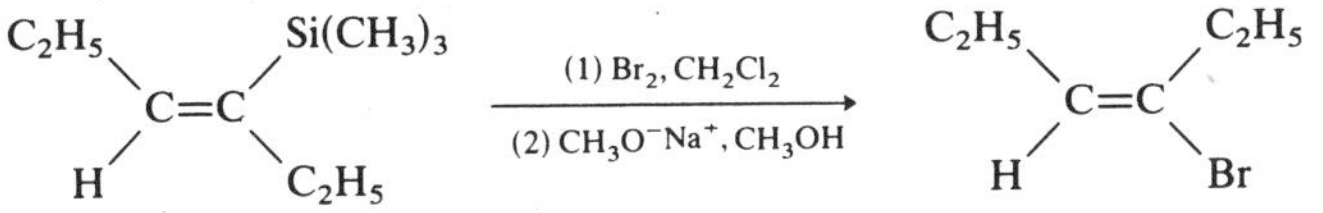

(c) *threo*-5-trimethylsilyl-4-octanol $\xrightarrow[CH_3CO_2H]{NaO_2CCH_3}$ *cis*-4-octene

(d) optically active $\xrightarrow{\Delta}$ racemic

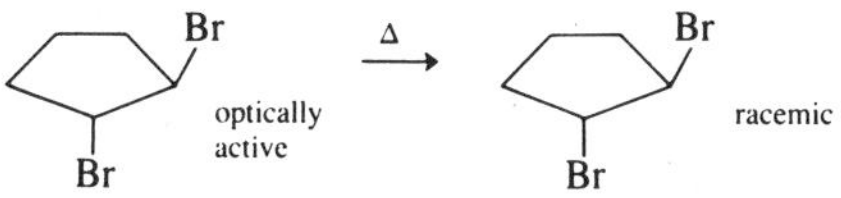

(e) $\xrightarrow{Br_2}$

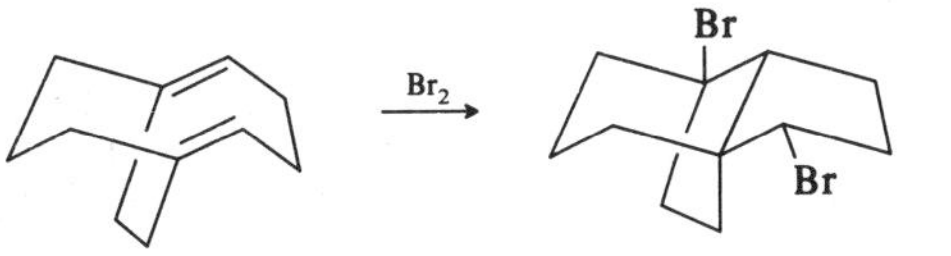

12. The rates of bromination of dialkylacetylenes are roughly 100 times greater than for the corresponding monosubstituted alkenes. For hydration, however, the rates of reaction are less than 10 times greater for disubstituted derivatives. Account for this observation by comparison of the mechanisms for bromination and hydration.

13. The bromination of 3-aroyloxycyclohexenes gives rise to a mixture of stereoisomeric and positionally isomeric addition products. The product composition for Ar = phenyl is shown. Account for the formation of each of the products and describe the factors which will affect the product ratio.

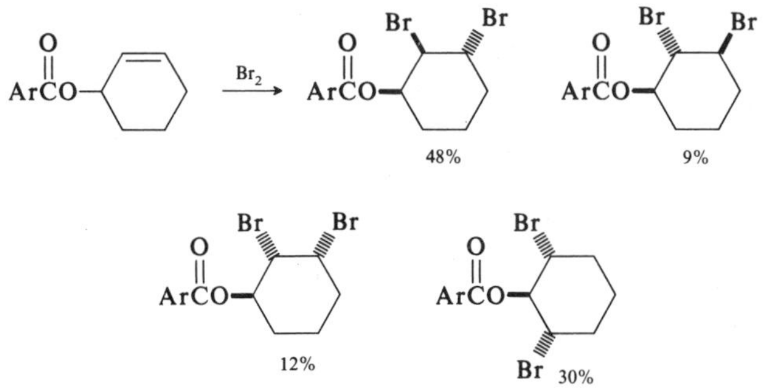

48%   9%   12%   30%

14. The reaction of substituted 1-arylethyl chlorides with $K^{+-}OC(CH_3)_3$ in DMSO does not follow a Hammett correlation. Instead, the reactivity order is p-NO_2 > p-MeO > p-CF_3 > p-CH_3 > H > p-Cl. What explanation can you offer for the failure to observe a Hammett relationship?

15. The ratio of terminal to internal alkene from decomposition of some sulfonium salts under alkaline conditions is as indicated:

$$(CH_3)_2S^+-C(CH_3)_2-(CH_2)_nX$$

n	X	term : inter
6	H	93 : 7
2	OH	100 : 0
3	OH	89 : 11
2	OPh	50 : 50
3	OPh	25 : 75

What explanation can you offer for the change in product ratio?

16. The Hammett correlation of the acid-catalyzed dehydration of 1,2-diarylethanols has been studied.

$$ArCH(OH)CH_2Ar' \xrightarrow{H^+} ArCH{=}CHAr'$$

The equation that correlates the data resulting from substitution in the Ar and Ar′ rings is

$$\log k = -3.78(\sigma^+_{Ar} + 0.23\sigma_{Ar'}) - 3.18$$

Give a rationalization for the form of this correlation equation. What information does it give regarding involvement of the Ar′ ring in the rate-determining step?

17. The addition of hydrogen chloride to olefins in nitromethane follows the rate expression

$$\text{Rate} = k[\text{HCl}]^2[\text{alkene}]$$

Two other features of the reaction that have been established are the following: (1) When DC1 is used instead of HC1, unreacted olefin recovered by stopping the reaction at 50% completion contains no deuterium; (2) added chloride salts ($R_4N^+Cl^-$) decrease the reaction rate, but other salts ($R_4N^+ClO_4^-$) do not. Write a mechanism for this reaction that is in accord with the data given.

18. In the bromination of styrene, a $\rho\sigma^+$ plot is noticeably curved. If the extremes of the curves are taken to represent straight lines, the curve can be resolved into two Hammett relationships with $\rho = -2.8$ for electron-attracting substituents and $\rho = -4.4$ for electron-releasing substituents. When the corresponding β-methylstryrenes are examined, a similarly curved $\sigma\rho$ plot is obtained. Furthermore, the stereospecificity of the reaction in the case of the β-methylstyrenes varies with the aryl substituents. The reaction is a stereospecific *anti* addition for strongly electron-attracting substituents but becomes only weakly stereoselective for electron-releasing substituents, e.g., 63% *anti*, 37% *syn*, for *p*-methoxy. Discuss the possible mechanistic basis for the Hammett plot curvature and its relationship to the stereochemical results.

19. The second-order rate constants for hydration and the kinetic solvent isotope effect for hydration of several 2-substituted 1,3-butadienes are given below. Discuss the information these data provide about the hydration mechanism.

$$CH_2{=}C(R){-}CH{=}CH_2 \xrightarrow[H_2O]{H^+} CH_3C(R)(OH)CH{=}CH_2 + CH_3C(R){=}CHCH_2OH$$

R	k_2 ($M^{-1}\,s^{-1}$) (25°C)	k_{H^+/D^+}
cyclopropyl	1.22×10^{-2}	1.2
CH_3	3.19×10^{-5}	1.8
Cl	2.01×10^{-8}	1.4
H	3.96×10^{-8}	1.8
C_2H_5O	6×10^1	—

20. The reaction of both *E*- and *Z*-2-butene with acetic acid to give 2-butyl acetate can be catalyzed by various strong acids. Using DBr, DCl, and CH_3SO_3D, in CH_3CO_2D, it was possible to demonstrate that the reaction proceeded largely with *anti* addition (84% ± 2%). If the reaction was stopped short of completion, there was not interconversion of *Z*-2-butene with either *E*-2-butene or 1-butene. When CF_3SO_3D was used as the catalyst, several features of the reaction changed.

(1) the recovered butene showed small amounts of conversion to 1-butene and partial isomerization to the stereoisomeric 2-butene.
(2) the recovered 2-butene contained small amounts of deuterium.
(3) the stereoselectivity was somewhat reduced (60%–70% *anti* addition).

How do you account for the changes which occur when CF_3SO_3D is used as a catalyst, as compared with the other acids?

7

Carbanions and Other Nucleophilic Carbon Species

Introduction

This chapter is concerned with carbanions, which are the conjugate bases (in the Brønsted sense) of organic molecules that are formed by deprotonation of a carbon atom. Carbanions may vary widely in stability, depending on the ability of substituent groups to stabilize negative charge. In the absence of substituents that are effective at delocalizing the charge, proton abstraction from a C—H bond is difficult. Carbanions are very useful in synthesis, since formation of new carbon-carbon bonds often requires a nucleophilic carbon species. Extensive study has been devoted to improving methods of generating carbanions and developing an understanding of substituent effects on stability and reactivity.

7.1. Acidity of Hydrocarbons

In the discussion of the relative acidity of carboxylic acids in Chapter 1, the thermodynamic acidity, expressed as the acid dissociation constant, was taken as the measure of acidity. It is straightforward to measure dissociation constants of such acids in aqueous media by measurement of the titration curve with a pH-sensitive electrode (pH meter). Determination of the relative acidity of most carbon acids is more difficult. Because most are very weak acids, very strong bases are required to cause deprotonation. Water and alcohols are far more acidic than most hydrocarbons and are unsuitable solvents for generation of hydrocarbon anions. Any strong base will deprotonate the solvent rather than the hydrocarbon. For this

Table 7.1. Values of H_- for Some Representative Solvent–Base Systems[a]

Solution	H_-[b]
5 *M* KOH	15.5
10 *M* KOH	17.0
15 *M* KOH	18.5
0.01 *M* NaOMe in 1 : 1 DMSO–MeOH	15.0
0.01 *M* NaOMe in 10 : 1 DMSO–MeOH	18.0
0.01 *M* NaOEt in 20 : 1 DMSO–EtOH	21.0

a. Values are rounded to the nearer 0.5 pH unit; this is typical of the range of disagreement using different indicator series.
b. Selected values from J. R. Jones, *The Ionization of Carbon Acids*, Academic Press, New York, 1973, Chap. 6.

reason, very weakly acidic solvents such as dimethyl sulfoxide and cyclohexylamine are used in the preparation of strongly basic carbanions. A further feature of dimethyl sulfoxide is its high polarity and cation-solvating ability, which facilities dissociation of ion pairs so that the equilibrium data obtained refer to the free ions, rather than to ion aggregates.

The basicity of a base–solvent system may be specified by a basicity constant H_-, analogous to the Hammett acidity function H_0. The value of H_- corresponds essentially to the pH of strongly basic nonaqueous solutions. The larger the value of H_-, the greater is the proton-abstracting ability of the medium. Use of a series of overlapping indicators permits assignment of H_- values to base–solvent systems and allows pH values to be determined over a range of 0–30 pK units. The indicators employed include substituted anilines and arylmethanes, which have significantly different electronic (UV-VIS) spectra in their neutral and anionic forms. The assumptions and procedures used to assign H_- values are similar to those involved in establishing H_0 scales, as discussed in Section 4.8. Table 7.1 presents H_- values for some representative solvent–base systems.

The acidity of a hydrocarbon can be determined in an analogous way.[1] If the electronic spectra of the neutral and anionic forms are sufficiently different, the concentrations of each can be determined directly, and the equilibrium constant for

$$\mathrm{RH} + \mathrm{B}^- \rightleftharpoons \mathrm{R}^- + \mathrm{BH}$$

is related to pK by the equation

$$\mathrm{p}K_{\mathrm{RH}} = H_- + \log\frac{[\mathrm{RH}]}{[\mathrm{R}^-]}$$

A measurement of the ratio $[\mathrm{RH}]:[\mathrm{R}^-]$ at a known H_- yields the pK. If, as is frequently the case, the electronic spectra of the hydrocarbon and its anion are not sufficiently different, one of the indicators is used and its spectrum is monitored.

1. D. Dolman and R. Stewart, *Can. J. Chem.* **45**, 911 (1967); E. C. Steiner and J. M. Gilbert, *J. Am. Chem. Soc.* **87**, 382 (1965); K. Bowden and R. Stewart, *Tetrahedron* **21**, 261 (1965).

The equilibrium established between the indicator and the hydrocarbon in the basic medium

$$RH + In^- \rightleftharpoons R^- + HIn$$

then provides a way to relate the concentrations that are not directly measured, [RH] and [R^-], to quantities that are, [HIn] and [In^-].

When the acidities of hydrocarbons are discussed in terms of the relative stabilities of neutral and anionic forms, particularly with respect to the extent of electron delocalization in the anion, the appropriate data are equilibrium acidity measurements. We have just seen how such data may be obtained, but, in many instances, it is not possible to obtain equilibrium data. In such cases, it may be possible to compare the rates of deprotonation, that is, the *kinetic acidity.* Such comparison can be made between different protons in the same compound or in two different compounds. This is usually done by following an isotopic exchange. In the presence of a source of deuterons, the rate of incorporation of deuterium into an organic molecule is a measure of the rate of carbanion formation.[2]

$$R{-}H + B^- \rightleftharpoons R^- + BH$$

$$R^- + S{-}D \rightleftharpoons R - D + S^-$$

$$S^- + BH \rightleftharpoons SH + B^-$$

It has been found that there is often a correlation between the rate of proton abstraction (kinetic acidity) and the thermodynamic stability of the carbanion (thermodynamic acidity). Because of this relationship, kinetic measurements can be used to construct orders of hydrocarbon acidities. These kinetic measurements have the advantage of not requiring the presence of a measurable concentration of the carbanion at any time; instead, the relative ease of carbanion formation is judged from the rate at which exchange occurs. This method is therefore applicable to very weak acids, for which no suitable base will generate a measurable carbanion concentration.

The kinetic method of determining relative acidity suffers from one serious complication, however. This complication has to do with the fate of the ion pair that is formed immediately on abstraction of the proton.[3] If the ion pair separates and diffuses into the solution rapidly, so that each deprotonation results in exchange, the exchange rate is an accurate measure of the rate of deprotonation. Under many conditions of solvent and base, however, an ion pair may return to reactants at a rate exceeding protonation of the carbanion by the solvent. This phenomenon is called *internal return*:

$$R_3C{-}H + M^+B^- \underset{\text{internal return}}{\overset{\text{ionization}}{\rightleftharpoons}} [R_3C^-M^+ + BH] \xrightarrow{\text{separation}} R_3C^- + M^+ + BH$$

$$R_3C^- \xrightarrow[\text{exchange}]{S\text{-}D} R_3CD + S^-$$

2. A. I. Shatenshtein, *Adv. Phys. Org. Chem.* **1**, 155 (1963).
3. W. T. Ford, E. W. Graham, and D. J. Cram, *J. Am. Chem. Soc.* **89**, 4661 (1967); D. J. Cram, C. A. Kingsbury, and B. Rickborn, *J. Am. Chem. Soc.* **83**, 3688 (1961).

When internal return occurs, a deprotonation has escaped detection because exchange has not resulted. One experimental test for the occurrence of internal return is racemization at chiral carbanionic sites that occurs without exchange. Even racemization cannot be regarded as an absolute measure of the deprotonation rate since, under some conditions, hydrogen–deuterium exchange has been shown to occur with retention of configuration. Because of these uncertainties about the fate of ion pairs, it is important that a linear relationship between exchange rates and equilibrium acidity be established for representative examples of the compounds under study. This provides a basis for using kinetic acidity data for compounds of that structural type.

In general, the extent of ion pairing is primarily a function of the ability of the solvent to solvate the ionic species present in solution. Ion pairing is greatest in nonpolar solvents such as ethers. In dipolar aprotic solvents, especially dimethyl sulfoxide, ion pairing is much less likely to be significant.[4] The structure of the solvent in which the extent or rate of deprotonation is determined will have a significant effect on the apparent acidity of the hydrocarbon. The identity of the cation present can also have a significant effect if ion pairs are present. The pK values determined depend on the solvent and other conditions of the measurement. The numerical values are not absolute but nevertheless provide a useful measure of relative acidity. The two solvents which have been used for most quantitative measurements on weak carbon acids are cyclohexylamine and dimethyl sulfoxide.

An extensive series of hydrocarbons has been studied in cyclohexylamine, using cesium cyclohexylamide as base. For many of the compounds studied, spectroscopic measurements were used to determine the relative extent of deprotonation of two hydrocarbons and thus establish relative acidity.[5] For other hydrocarbons, the acidity was derived by kinetic measurements. It was shown that the rate of tritium exchange for a series of related hydrocarbons is linearly related to the equilibrium acidities of these hydrocarbons in the solvent system. This method was used to extend the scale to hydrocarbons such as toluene for which the exchange rate, but not equilibrium data, can be obtained.[6] Representative values of some hydrocarbons with pK values ranging from 16 to above 40 are given in Table 7.2.

Table 7.2 also lists some hydrocarbon acidity values determined in DMSO. It is not expected that the values in different solvent media will be numerically identical. The same relative order of acidity is normally observed for hydrocarbons of similar structural type.

For synthetic purposes, carbanions are most frequently generated in ether solvents, frequently tetrahydrofuran or dimethoxyethane. There is relatively little quantitative data available on hydrocarbon acidity in such solvents. Table 7.2

4. E. M. Arnett, T. C. Moriarity, L. E. Small, J. P. Rudolph, and R. P. Quirk, *J. Am. Chem. Soc.* **95**, 1492 (1973); T. E. Hogen-Esch and J. Smid, *J. Am. Chem. Soc.* **88**, 307 (1966).
5. A. Streitwieser, Jr., J. R. Murdoch, G. Hafelinger, and C. J. Chang, *J. Am. Chem. Soc.* **95**, 4248 (1973); A. Streitwieser, Jr., E. Ciuffarin, and J. H. Hammons, *J. Am. Chem. Soc.* **89**, 63 (1967); A. Steitwieser, Jr., E. Juaristi, and L. L. Nebenzahl, in *Comprehensive Carbanion Chemistry*, Part A, E. Buncel and T. Durst (eds.), Elsevier, New York, 1980, Chapter 7.
6. A. Streitwieser, Jr., M. R. Granger, F. Mares, and R. A. Wolf, *J. Am. Chem. Soc.* **95**, 4257 (1973).

Table 7.2. Acidities of Some Hydrocarbons

	Hydrocarbon	pK^a		
		Cs^+ (cyclohexylamine)[b]	Cs^+ (THF)[c]	K^+ (DMSO)[d]
1	$PhCH_2-\underline{H}$	41.2		
2	$(H_3C-C_6H_4)_2CH-\underline{H}$	35.1	32.9	
3	$(Ph)_2CH-\underline{H}$	33.4	33.0	32.3
4	$(Ph)_3C-\underline{H}$	31.4	31.0	30.6
5	dibenzocycloheptatriene CH–$\underline{H}$	31.2		
6	fluorene CH–$\underline{H}$	22.7	22.4	22.6
7	indene CH–$\underline{H}$	19.9		20.1
8	9-phenylfluorene C(Ph)–$\underline{H}$	18.5	18.5	17.9
9	cyclopentadiene (H, $\underline{H}$)	16.6[e]		18.1

a. Values refer to indicated solvent medium containing salt of cation specified.
b. From A. Streitweiser, Jr., J. R. Murdoch, G. Häfelinger, C. J. Chang, *J. Am. Chem. Soc.* **95**, 4248 (1973); A. Streitwieser, Jr., E. Ciuffarin, and J. H. Hammons, *J. Am. Chem. Soc.* **89**, 63 (1967); A. Streitwieser, Jr., and F. Guibe, *J. Am. Chem. Soc.* **100**, 4532 (1978).
c. From D. A. Bors, M. J. Kaufman, and A. Streitwieser, Jr., *J. Am. Chem. Soc.* **107**, 6975 (1985).
d. From C. D. Ritchie and R. E. Uschold, *J. Am. Chem. Soc.* **90**, 2821 (1968); F. G. Bordwell, J. E. Bartmess, G. E. Drucker, Z. Margolin, and W. S. Matthews, *J. Am. Chem. Soc.* **97**, 3226 (1975); W. S. Matthews, J. E. Bares, J. E. Bartmess, F. G. Bordwell, F. J. Cornforth, G. E. Drucker, Z. Margolin, R. J. McCallum, G. J. McCollum, and N. R. Vanier, *J. Am. Chem. Soc.* **97**, 7006 (1975); F. G. Bordwell, G. E. Drucker, and H. E. Fried, *J. Org. Chem.* **46**, 632 (1981).
e. From A. Streitwieser, Jr., and L. L. Nebenzahl, *J. Am. Chem. Soc.* **98**, 2188 (1976); in water, the pK_a of cyclopentadiene is 16.0.

contains a few entries for Cs^+ salts. The numerical values are scaled with reference to the pK of 9-phenylfluorene.[7]

Some of the relative acidities in Table 7.2 can be easily rationalized. The order of decreasing acidity $Ph_3CH > Ph_2CH_2 > PhCH_3$, for example, reflects the ability of each successive phenyl group to delocalize the negative charge on carbon and thereby stabilize the carbanion. The much greater acidity of fluorene relative to dibenzocycloheptatriene (entries 5 and 6) reflects the aromatic stabilization of the cyclopentadienide ring in the anion of fluorene. Cyclopentadiene is an exceptionally acidic hydrocarbon, comparable in acidity to simple alcohols, and this reflects the aromatic stabilization of the anion.

Allylic conjugation provides carbanion stabilization, and values of 43 (in cyclohexylamine)[8] and 47–48 (in THF–HMPA)[9] have been determined for propene. On the basis of exchange rates with cesium cyclohexylamide, cyclohexene and cycloheptene have been found to have pK values of about 45 in cyclohexylamine.[10] The hydrogens on the sp^2 carbons in benzene and ethylene would be expected to be more acidic than the hydrogens in saturated hydrocarbons. A pK of 43 has been estimated for benzene on the basis of extrapolation from a series of fluorobenzenes.[11] Electrochemical measurements have been used to establish a lower limit of about 46 for the pK of ethylene.[9]

For saturated hydrocarbons, exchange is too slow and reference points are so uncertain that direct determination of pK values by exchange measurements is not feasible. The most useful approach to obtaining pK data for such hydrocarbons involves making a measurement of the electrochemical potential for the reaction

$$R^{\cdot} + e^- \rightleftharpoons R^-$$

From this value and known C—H bond dissociation energies, pK values can be calculated. The electrochemical measurements can be made on halides or on alkyllithium compounds. This type of approach has some significant uncertainties but nevertheless can provide a least a semiquantitative estimate of acidities of very weakly acidic hydrocarbons. The pK for isobutane obtained in this way is 71.[12] The necessary electrochemical measurements cannot be made directly for methane, but an extrapolation from toluene and diphenylmethane leads to a range of 52–62 for the pK of methane.[9]

The acetylenes as a group are among the most acidic of the hydrocarbons. For example, in DMSO, phenylacetylene is found to have a pK near 26.5.[13] In cyclohexylamine, the value is given as 23.2.[14] An estimate of the pK in aqueous solution of 20 is based on a Brønsted relationship.[15] The relatively high acidity of acetylenes

7. D. A. Bors, M. J. Kaufman, and A. Streitwieser, Jr., *J. Am. Chem. Soc.* **107**, 6975 (1985).
8. D. W. Boerth and A. Streitwieser, Jr., *J. Am. Chem. Soc.* **103**, 6443 (1981).
9. B. Jaun, J. Schwarz, and R. Breslow, *J. Am. Chem. Soc.* **102**, 5741 (1980).
10. A. Streitwieser, Jr., and D. W. Boerth, *J. Am. Chem. Soc.* **100**, 755 (1978).
11. A. Streitwieser, Jr., P. J. Scannon, and H. M. Niemeyer, *J. Am. Chem. Soc.* **94**, 7936 (1972).
12. R. Breslow and R. Goodin, *J. Am. Chem. Soc.* **98**, 6076 (1976).
13. F. G. Bordwell and W. S. Matthews, *J. Am. Chem. Soc.* **96**, 1214 (1974).
14. A. Streitwieser, Jr., and D. M. E. Rueben, *J. Am. Chem. Soc.* **93**, 1794 (1971).
15. D. B. Dahlberg, M. A. Kuzemko, Y. Chiang, A. J. Kresge, and M. F. Powell, *J. Am. Chem. Soc.* **105**, 5387 (1983).

is associated with the large degree of *s* character of the C—H bond. The *s* character is 50%, as opposed to 25% in sp^3 bonds. The electrons in orbitals with high *s* character experience decreased shielding from the nuclear charge. The carbon is therefore effectively more electronegative, as viewed from the proton sharing an *sp* hybrid orbital, and hydrogens on *sp* carbons exhibit exceptional acidity. This same effect accounts for the relatively high acidity of the hydrogens on cyclopropane rings,[16] which have increased *s* character in the C—H bonds.

Knowledge of the structure of carbanions is important to understanding the stereochemistry of their reactions. Theoretical calculations at the *ab initio* level indicate a pyramidal geometry at the carbanionic carbon in methyl anion and ethyl anion. The optimum H—C—H angle in these two carbanions is calculated to be 97–100°. An interesting effect is observed in that the proton affinity (basicity) of methyl anion decreases in a regular manner as the H—C—H angle is decreased.[17] This increase in acidity with decreasing internuclear angle has a parallel in small-ring compounds, in which the acidity of hydrogens is substantially greater than in compounds having tetrahedral geometry at carbon. Pyramidal geometry at carbanions can also be predicted on the basis of qualitative considerations of the orbital occupied by the unshared electron pair. In a planar carbanion, the lone pair would occupy a *p* orbital. In a pyramidal geometry, the orbital would have substantial *s* character. Since the electron pair would be of lower energy in an orbital with some *s* character, it would be predicted that a pyramidal geometry would be favored.

The stereochemistry observed in hydrogen-exchange reactions of carbanions is very dependent on the conditions under which the anion is formed and trapped by proton transfer. The dependence on solvent, counterion, and base is the result of the importance of ion-pairing effects. The base-catalyzed cleavage of **1** is noteworthy. The anion of **1** cleaves at elevated temperature to 2-butanone and 2-phenyl-2-butyl anion, which, under the conditions of the reaction, abstracts a proton from the solvent. Use of optically active **1** allows the stereochemical features of the anion to be probed by measuring the enantiomeric purity of the 2-phenylbutane product.

$$\underset{\textbf{1}}{CH_3CH_2\overset{CH_3}{\underset{Ph}{C}}-\overset{OH}{\underset{CH_3}{C}}-CH_2CH_3} \xrightarrow{\text{base}} CH_3CH_2\underset{Ph}{\bar{C}}CH_3 + CH_3\underset{\|}{C}CH_2CH_3 \ (C{=}O) \xrightarrow[\text{B-H}]{\text{S-H}} CH_3CH_2\overset{H}{\underset{Ph}{C}}CH_3$$

Retention of configuration was observed in solvents of low dielectric constant, while increasing amounts of inversion occurred as the proton-donating ability and dielectric constant of the solvent increased. Cleavage of **1** with potassium *t*-butoxide in benzene gave 2-phenylbutane with 93% net retention of configuration. The stereochemical course changed to 48% net inversion of configuration when potassium

16. A. Streitwieser, Jr., R. A. Caldwell, and W. R. Young, *J. Am. Chem. Soc.* **91**, 529 (1969).
17. A. Streitwieser, Jr., and P. H. Owens, *Tetrahedron Lett.*, 5221 (1973); A. Streitwieser, Jr., P. H. Owens, R. A. Wolf, and J. E. Williams, Jr., *Am. Chem. Soc.* **96**, 5448 (1974); E. D. Jemmis, V. Buss, P. v. R. Schleyer, and L. C. Allen, *J. Am. Chem. Soc.* **98**, 6483 (1976).

hydroxide in ethylene glycol was used. In dimethyl sulfoxide with potassium *t*-butoxide as base, completely racemic 2-phenylbutane was formed.[18] The retention in benzene presumably reflects a short lifetime for the carbanion in a tight ion pair. Under these conditions, the carbanion does not become symmetrically solvated before proton transfer from either the protonated base or the ketone occurs. The solvent benzene would not be an effective proton donor. In ethylene glycol, the solvent provides a good proton source, and, since net inversion is observed, the protonation must be occurring on an unsymmetrically solvated species that favors back-side protonation. The racemization that is observed in dimethyl sulfoxide indicates that the carbanion has a sufficient lifetime to become symmetrically solvated.

The stereochemistry of hydrogen–deuterium exchange at the chiral carbon in 2-phenylbutane has also been studied. When potassium *t*-butoxide is used as the base, the exchange occurs with retention of configuration in *t*-butanol, but racemization occurs in DMSO.[19] The retention of configuration is visualized as occurring through an ion pair in which a solvent molecule coordinated to the metal ion acts as the proton donor.

$$\mathrm{H_3C(CH_2CH_3)(Ph)C{-}H + (DOR)_2K^+\ {}^-OR \longrightarrow H_3C(CH_3CH_2)(Ph)C^-\cdots K^+(DOR)_2(HOR) \longrightarrow H_3C(CH_3CH_2)(Ph)C{-}D + {}^-OR\ K^+(DOR)(HOR)}$$

In DMSO, symmetrical solvation is achieved prior to deuteration, and complete racemization occurs.

The organometallic derivatives of lithium, magnesium, and other strongly electropositive metals have some of the properties expected for salts of carbanions but are also significantly covalent in character. Because of the very weak acidity of most hydrocarbons, the simple organolithium compounds are usually not prepared by proton transfer reactions. Instead, the most general preparative methods start with the corresponding halogen compound.

$$\mathrm{CH_3I + 2\,Li \longrightarrow CH_3Li + LiI}$$

$$\mathrm{CH_3(CH_2)_3Br + 2\,Li \longrightarrow CH_3(CH_2)_3Li + LiBr}$$

$$\mathrm{C_6H_5{-}Br + 2\,Li \longrightarrow C_6H_5{-}Li + LiBr}$$

There are other preparative methods which will be considered in Part B, Chapter 7.

18. D. J. Cram, A. Langemann, J. Allinger, and K. R. Kopecky, *J. Am. Chem. Soc.* **81**, 5740 (1959).
19. D. J. Cram, C. A. Kingsbury, and B. Rickborn, *J. Am. Chem. Soc.* **83**, 3688 (1961).

Although these compounds have some covalent character, organolithium compounds react as would be expected of the carbanions derived from simple hydrocarbons. Organolithium compounds are rapidly protonated by any molecule having an acidic —OH, —NH, or —SH group to form the hydrocarbon. All the organolithium compounds derived from saturated hydrocarbons are extremely strong bases. Accurate p*K* values are not known but would range upward from the estimate of ≈52–62 for methane. The order of basicity $CH_3Li < CH_3(CH_2)_3Li < (CH_3)_3CLi$ would be predicted on the basis of the electron-releasing effect of alkyl substituents and is consistent with increasing reactivity in proton abstraction reactions in the order $CH_3Li < CH_3(CH_2)_3Li < (CH_3)_3CLi$. Phenyl-, methyl-, *n*-butyl-, and *t*-butyllithium are certainly all stronger bases than the anions of the hydrocarbons listed in Table 7.2. Unlike proton transfers involving oxygen, nitrogen, or sulfur atoms, proton transfer between carbon atoms is usually not a fast reaction. Thus, even though *t*-butyllithium is thermodynamically capable of deprotonating toluene, for example, the reaction is quite slow in a hydrocarbon solvent medium. In part, the reason is that the organolithium compounds exist as tetramers, hexamers, and higher aggregates in hydrocarbon and ether solvents.[20] The nature of the species present in solution can be studied by low-temperature NMR spectroscopy. *n*-Butyllithium, for example, in THF is present as a tetramer–dimer mixture.[21] The tetrameric species is dominant.

$$[(BuLi)_4{\cdot}(THF)_4] + 4THF \rightleftharpoons 2[(BuLi)_2{\cdot}(THF)_4]$$

In solutions of *n*-propyllithium in cyclopropane at 0°C, the hexamer is the main species, but higher aggregates are present at lower temperatures.[20] The reactivity of the organolithium compounds is increased by adding molecules capable of solvating the organometallic species. Tetramethylethylenediamine (TMEDA) has been commonly used for organolithium systems. This tertiary amine can chelate lithium. These complexes generally are able to effect deprotonation at accelerated rates.[22]

$$(RLi)_4 + 4\,(CH_3)_2NCH_2CH_2N(CH_3)_2 \rightleftharpoons 2\,[(CH_3)_2NCH_2CH_2N(CH_3)_2Li(\mu\text{-}R)_2LiN(CH_3)_2CH_2CH_2N(CH_3)_2]$$

In the case of phenyllithium, it has been possible to demonstrate by NMR studies that the compound is tetrameric in 1:2 ether–cyclohexane but dimeric in

20. G. Fraenkel, M. Henrichs, J. M. Hewitt, B. M. Su, and M. J. Geckle, *J. Am. Chem. Soc.* **102**, 3345 (1980); G. Frankel, M. Henrichs, M. Hewitt, and B. M. Su, *J. Am. Chem. Soc.* **106**, 255 (1984).
21. D. Seebach, R. Hassig, and J. Gabriel, *Helv. Chem. Acta* **66**, 308 (1983); J. F. McGarrity and C. A. Ogle, *J. Am. Chem. Soc.* **107**, 1805 (1985).
22. G. G. Eberhardt and W. A. Butte, *J. Org. Chem.* **29**, 2928 (1964); R. West and P. C. Jones, *J. Am. Chem. Soc.* **90**, 2656 (1968).

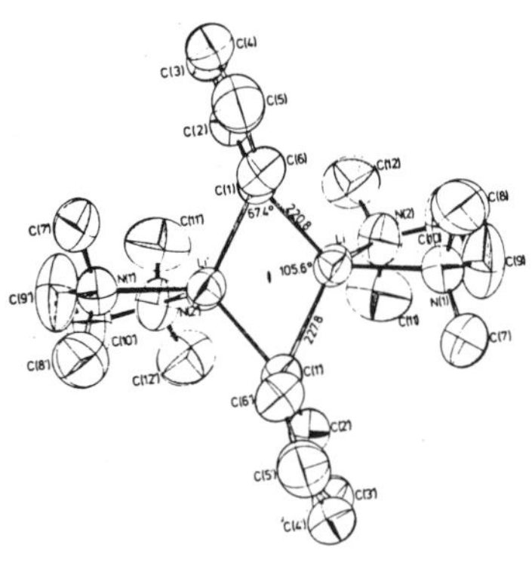

(A) $[(PhLi)_2{\cdot}(TMEDA)_2]$

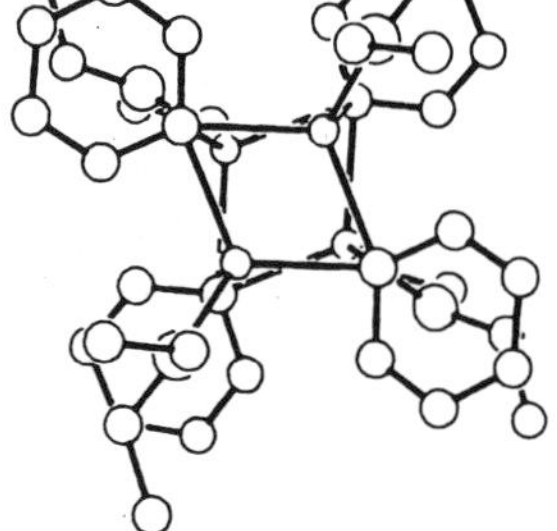

(B) $[(PhLi)_4{\cdot}(Et_2O)_4]$

Fig. 7.1. Crystal structures of phenyllithium: (A) dimeric structure incorporating tetramethylethylenediamine; (B) tetrameric structure incorporating diethyl ether. [Reproduced from Refs. 24 and 25 with permission of the publishers.]

1:9 TMEDA–cyclohexane.[23] X-ray crystal structure determinations have been done on both a dimeric structure and a tetramer. A dimeric structure crystallizes from hexane containing TMEDA.[24] This structure is shown in Fig. 7.1(A). A tetrameric structure incorporating four ether molecules forms from ether–hexane solution.[25] This structure is shown in Fig. 7.1(B). There is a good correspondence between the structures which crystallize and those indicated by the NMR studies.

Tetrameric structures based on distorted cubic structures are also found for $(CH_3Li)_4$ and $(C_2H_5Li)_4$.[26] These tetrameric structures can also be represented as based on tetrahedra of lithium ions with each face occupied by a carbanion.

R Li R
Li Li
R Li R

The THF solvate of lithium *t*-butylacetylide is another example of a tetrameric structure.[27]

Both 2,2′-dilithiobiphyl[28] and lithium phenylacetylide[29] possess dimeric structures in complexes with diamines.

23. L. M. Jackman and L.M. Scarmoutzos, *J. Am. Chem. Soc.* **106**, 4627 (1984).
24. D. Thoennes and E. Weiss, *Chem. Ber.* **111**, 3157 (1978).
25. H. Hope and P. P. Power, *J. Am. Chem. Soc.* **105**, 5320 (1983).
26. E. Weiss and E. A. C. Lucken, *J. Organomet. Chem.* **2**, 197 (1964); E. Weiss and G. Hencken, *J. Organomet. Chem.* **21**, 265 (1970); H. Köster, D. Thoennes, and E. Weiss, *J. Organomet. Chem.* **160**, 1 (1978); H. Dietrich, *Acta Crystallogr.* **16**, 681 (1963); H. Dietrich, *J. Organomet. Chem.* **205**, 291 (1981).
27. W. Neuberger, E. Weiss, and P. v. R. Schleyer, quoted in Ref. 30.
28. U. Schubert, W. Neugebauer, and P. v. R. Schleyer, *J. Chem. Soc., Chem. Commun.*, 1184 (1982).
29. B. Schubert and E. Weiss, *Chem. Ber.* **116**, 3212 (1983).

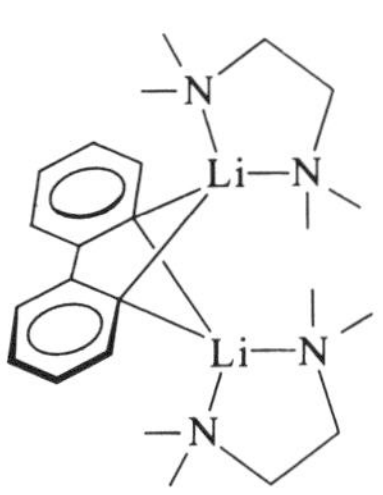

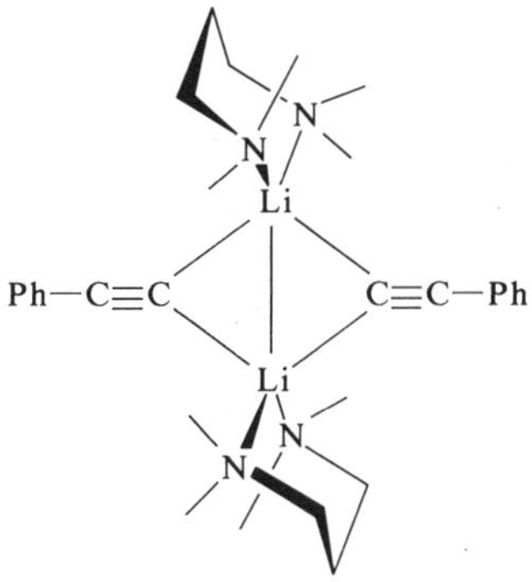

These and many other organolithium structures have been compared in a review of this topic.[30]

The relative slowness of abstraction of protons from carbon acids by organolithium reagents is probably due to the compact character of the carbon–lithium clusters. Since the electrons associated with the carbanion are tightly associated with the cluster of lithium cations, some activation energy is required to break the bond before the carbanion can act as a base. This relative sluggishness of organometallic compounds as bases permits important reactions in which the organometallic species acts as a nucleophile in preference to functioning as a strong base. The addition of organolithium and organomagnesium compounds to carbonyl groups in aldehydes, ketones, and esters is an important example. As will be seen in the next section, most carbonyl compounds are much more acidic than hydrocarbons. Nevertheless, in most cases, the proton transfer reaction is slower than nucleophilic attack at the carbonyl group. It is this feature of the reactivity of organometallics that permits the very extensive use of organometallic compounds in organic synthesis. The reactions of organolithium and organomagnesium compounds with carbonyl compounds in a synthetic context will be discussed in Part B, Chapter 7.

7.2. Carbanions Stabilized by Functional Groups

Functional groups that permit the negative charge of a carbanion to be delocalized to a more electronegative atom such as oxygen cause very large increases in the acidity of C—H bonds. Among the functional groups that exert a strong stabilizing effect on carbanions are carbonyl, nitro, sulfonyl, and cyano groups. Perhaps the best basis for comparing these groups is the data on the various substituted methanes.[31] Bordwell and co-workers determined the relative acidities

30. W. N. Setzer and P. v. R. Schleyer, *Adv. Organomet. Chem.* **24**, 353 (1985).
31. F. G. Bordwell and W. S. Matthews, *J. Am. Chem. Soc.* **96**, 2116 (1974); W. S. Matthews, J. E. Bares, J. E. Bartmess, F. G. Bordwell, F. J. Cornforth, G. E. Drucker, Z. Margolin, R. J. McCallum, G. J. McCollum, and N. R. Vanier, *J. Am. Chem. Soc.* **97**, 7006 (1975).

Table 7.3. Equilibrium Acidities of Substituted Methanes in Dimethyl Sulfoxide[a]

Compound	pK
CH_3NO_2	17.2
CH_3COPh	24.7
CH_3COCH_3	26.5
CH_3SO_2Ph	29.0
$CH_3CO_2C_2H_5$	30.5[b]
$CH_3SO_2CH_3$	31.1
CH_3CN	31.3
$CH_3CON(C_2H_5)_2$	34.5[b]

a. Except where noted otherwise, data are from W. S. Matthews, J. E. Bares, J. E. Bartmess, F. G. Bordwell, F. J. Cornforth, G. E. Drucker, Z. Margolin, R. J. McCallum, G. J. McCollum, and N. R. Vanier, *J. Am. Chem. Soc.* **97**, 7006 (1975).
b. Accurate to ±0.5, F. G. Bordwell and H. E. Fried, *J. Org. Chem.* **46**, 4327 (1981).

of the substituted methanes with reference to aromatic hydrocarbon indicators in DMSO. The data are given in Table 7.3. The ordering $NO_2 > C{=}O > CO_2R \approx SO_2 \approx CN > CONR_2$ for anion stabilization is established by these data. Both dipolar and resonance effects are involved in the ability of these functional groups to stabilize the negative charge.

$$H_2\bar{C}{-}\overset{+}{N}(=O)O^- \longleftrightarrow H_2C{=}\overset{+}{N}(O^-)O^- \qquad H_2\bar{C}{-}S(=O)_2{-}R \longleftrightarrow H_2C{=}S(O^-)(=O){-}R \longleftrightarrow H_2C{=}S(=O)(O^-){-}R$$

$$H_2\bar{C}{-}C(=O)R \longleftrightarrow H_2C{=}C(O^-)R \qquad H_2\bar{C}{-}C{\equiv}N \longleftrightarrow H_2C{=}C{=}N^-$$

The presence of two such groups further stabilizes the negative charge. 2,4-Pentanedione, for example, has a pK around 9. Most β-diketones are sufficiently acidic that the derived carbanions can be generated using the conjugate bases of hydroxylic solvents such as water or alcohols, which have pK values of 15–20. This ability to generate carbanions stabilized by electron-attracting groups is very important from a synthetic point of view, and the synthetic aspects of the chemistry of these carbanions will be discussed in Part B, Chapters 1 and 2. Stronger bases are required for compounds that have a single stabilizing functional group. Alkali metal salts of ammonia or amines or sodium hydride are sufficiently strong bases to form cabanions from most ketones, aldehydes, and esters. The anion of diisopropylamine is also a popular strong base for use in synthetic procedures. It is prepared from the amine by reaction of an alkyllithium reagent.

Carbanions derived from carbonyl compounds are often referred to as *enolate* anions. The reason for this name is that the resonance-stabilized enolate anion is the conjugate base of both the keto and enol tautomers of carbonyl compounds:

$$\underset{\text{keto}}{\mathrm{R\overset{O}{\overset{\|}{C}}CH_2R'}} \rightleftharpoons \underset{\text{enol}}{\mathrm{R\overset{OH}{\overset{|}{C}}{=}CHR'}}$$

$$\mathrm{R\overset{O^-}{\overset{|}{C}}{=}CHR'} \leftrightarrow \underset{\text{enolate}}{\mathrm{R\overset{O}{\overset{\|}{C}}{-}\underline{C}HR'}}$$

There have been numerous studies of the rates of deprotonation of carbonyl compounds. These data are of interest not only because they define the relationship between thermodynamic and kinetic acidity in these compounds, but also because they are necessary for understanding mechanisms of reactions in which enolates are involved as intermediates. Rates of enolate formation can be measured conveniently by following isotopic exchange using either deuterium or tritium.

$$\mathrm{R_2CH\overset{O}{\overset{\|}{C}}R'} \overset{B^-}{\rightleftharpoons} \mathrm{R_2C{=}\overset{O^-}{\overset{|}{C}}R'} + \mathrm{BH}$$

$$\mathrm{R_2C{=}\overset{O^-}{\overset{|}{C}}R'} + \mathrm{S{-}D} \rightleftharpoons \mathrm{R_2\underset{D}{\underset{|}{C}}\overset{O}{\overset{\|}{C}}R} + \mathrm{S^-}$$

Another technique is to measure the rate of halogenation of the carbonyl compound. Ketones and aldehydes in their carbonyl forms do not react rapidly with the halogens, but the enolate is rapidly attacked. The rate of halogenation is therefore a measure of the rate of deprotonation.

$$\mathrm{R_2CH\overset{O}{\overset{\|}{C}}R'} + \mathrm{B^-} \xrightarrow{\text{slow}} \mathrm{R_2C{=}\overset{O^-}{\overset{|}{C}}R'} + \mathrm{BH}$$

$$\mathrm{R_2C{=}\overset{O^-}{\overset{|}{C}}R'} + \mathrm{X_2} \xrightarrow{\text{fast}} \mathrm{R_2\underset{X}{\underset{|}{C}}\overset{O}{\overset{\|}{C}}R'} + \mathrm{X^-}$$

Table 7.4 gives data on the rates of deuteration of some simple alkyl ketones. From these data, the order of reactivity toward deprotonation is $CH_3 > RCH_2 > R_2CH$. Steric hindrance to the approach of the base is probably the major factor in establishing this order. The importance of steric effects can be seen by comparing the CH_2 group in 2-butanone with the more hindered CH_2 group in 4,4-dimethyl-2-pentanone. The two added methyl groups on the adjacent carbon decrease the rate of proton removal by a factor of about 100. The rather slow rate of exchange at the CH_3 group of 4,4-dimethyl-2-pentanone must also reflect a steric factor arising from the bulky nature of the neopentyl group. If bulky groups interfere with effective solvation of the developing negative charge on oxygen, the rate of proton abstraction will be reduced.

$$CH_3C(=O)CH_2C(CH_3)_3$$

Structural effects on the rates of deprotonation of ketones have also been studied using very strong bases under conditions where complete conversion to the enolate occurs. In solvents such as THF or dimethoxyethane (DME) bases such as lithium diisopropylamide (LDA) and potassium hexamethyldisilylamide (HMDS) give solutions of the enolates whose compositions reflect the relative rates of removal of the different protons in the carbonyl compound (kinetic control). The least hindered

Table 7.4. Relative Rates of Base-Catalyzed Deuteration of Some Ketones[a]

Ketone	Relative rate
$CH_3C(=O)CH_2\text{–}\underline{H}$	100
$CH_3C(=O)CH(\underline{H})CH_3$	41.5
$\underline{H}\text{–}CH_2C(=O)CH_2CH_3$	45
$CH_3C(=O)C(\underline{H})(CH_3)_2$	<0.1
$\underline{H}\text{–}CH_2C(=O)CH(CH_3)_2$	45
$CH_3C(=O)CH(\underline{H})C(CH_3)_3$	0.45
$\underline{H}\text{–}CH_2C(=O)CH_2C(CH_3)_3$	5.1

a. In aqueous dioxane with sodium carbonate as base. The data of C. Rappe and W. H. Sachs, *J. Org. Chem.* **32**, 4127 (1967), given on a per-group basis, have been converted to a per-hydrogen basis.

proton is removed most rapidly under these conditions so that for unsymmetrical ketones the major enolate is the less substituted one. Scheme 7.1 shows some representative data.

Equilibrium between the various enolates of a ketone can be established by the presence of an excess of the ketone, which permits proton transfer. Equilibration is also favored by the presence of coordinating solvents such as HMPA. The composition of the equilibrium mixture is usually more closely balanced than for kinetically controlled conditions. In general, the more highly substituted enolate is the preferred isomer, but if the alkyl groups are sufficiently branched as to interfere with solvation of the enolate, there can be exceptions. The equilbrium ratios of enolates for several ketone–enolate systems are shown in Scheme 7.1.

Nitroalkanes show a related relationship between kinetic acidity and thermodynamic acidity. Additional alkyl substituents on nitromethane retard the rate of proton removal although the equilibrium is more favorable for the more highly substituted derivatives.[32] The alkyl groups have a strong stabilizing effect on the nitronate ion, but unfavorable steric effects are dominant at the transition state for proton removal. As a result, kinetic and thermodynamic acidity show opposite responses to alkyl substitution.

	Kinetic acidity k (M^{-1}min^{-1})	Thermodynamic acidity pK
CH_3NO_2	238	10.2
$CH_3CH_2NO_2$	39.1	8.5
$(CH_3)_2CHNO_2$	2.08	7.7

The cyano group is also effective at stabilizing negative charge on carbon. It has been possible to synthesize a number of hydrocarbon derivatives that are very highly substituted with cyano groups. Table 7.5 gives pK values for some of these compounds. As can be seen, the highly substituted derivatives are very strong acids.

Second-row elements, particularly phosphorus and sulfur, stabilize adjacent carbanions. The pK values of some pertinent compounds are given in Table 7.6.

The conjugate base of 1,3-dithiane has proven valuable in synthetic applications as a nucleophile (Part B, Chapter 13). The anion is generated by deprotonation (lithiation) using n-butyllithium.

$$\text{1,3-dithiane (C-2: H, H)} + \text{BuLi} \xrightarrow{\text{THF}} \text{2-lithio-1,3-dithiane (C-2: H, Li)} + \text{BuH}$$

The pK of 1,3-dithiane is 31 (in cyclohexylamine). There are several factors that can contribute to the anion-stabilizing effect of sulfur substituents. Inductive effects can contribute but cannot be the dominant factor since oxygen substituents

32. F. G. Bordwell, W. J. Boyle, Jr., and K. C. Yee, *J. Am. Chem. Soc.* **92**, 5926 (1970).

Scheme 7.1. Composition of Ketone–Enolate Mixtures Formed under Kinetic and Thermodynamic Conditions[a]

Ketone	Conditions	Enolate mixture		
$CH_3CH_2\overset{O}{\overset{\|\|}{C}}CH_3$		CH_3H_2C–C(O^-)=CH_2	H_3C–CH=C(O^-)–CH_3	CH_3–CH=C(O^-)–CH_3
	Kinetic (LDA, 0°)	71%	13%	16%
$(CH_3)_2CH\overset{O}{\overset{\|\|}{C}}CH_3$		H_3C–CH(CH_3)–C(O^-)=CH_2	H_3C–C(CH_3)=C(O^-)–CH_3	
	Kinetic (K HMDS, −78°)	99%	1%	
	Thermodynamic (KH)	88%	12%	
$CH_3(CH_2)_3\overset{O}{\overset{\|\|}{C}}CH_3$		$CH_3(CH_2)_3$–C(O^-)=CH_2	$CH_3(CH_2)_2$–CH=C(O^-)–CH_3	$CH_3(CH_2)_2$–CH=C(O^-)–CH_3
	Kinetic (LDA, −78°)	100%	0%	0%
	Thermodynamic (KH, 20°)	42%	46%	12%
$PhCH_2\overset{O}{\overset{\|\|}{C}}CH_3$		$PhCH_2$–C(O^-)=CH_2	Ph–CH=C(O^-)–CH_3	Ph–CH=C(O^-)–CH_3
	Kinetic (LDA, 0°)	14%	86% (combined)	
	Thermodynamic (NaH)	2%	98% (combined)	

Scheme 7.1.—continued.

Ketone	Conditions	Enolate mixture	
		O^-, CH_3	O^-, CH_3
O, CH_3	Kinetic (LDA, 0°)	99%	1%
	Thermodynamic (NaH)	26%	74%
		O^-, $CH(CH_3)_2$	O^-, $CH(CH_3)_2$
O, $CH(CH_3)_2$	Kinetic ($LiCPh_3$)	100%	0%
	thermodynamic ($KCPh_3$)	35%	65%
		O^-, CH_3	O^-, CH_3
O, CH_3	Kinetic ($LiCPh_3$)	82%	18%
	Thermodynamic ($KCPh_3$)	52%	48%
		O^-	O^-
O	Kinetic (LDA)	98%	2%
	Thermodynamic (NaH)	50%	50%

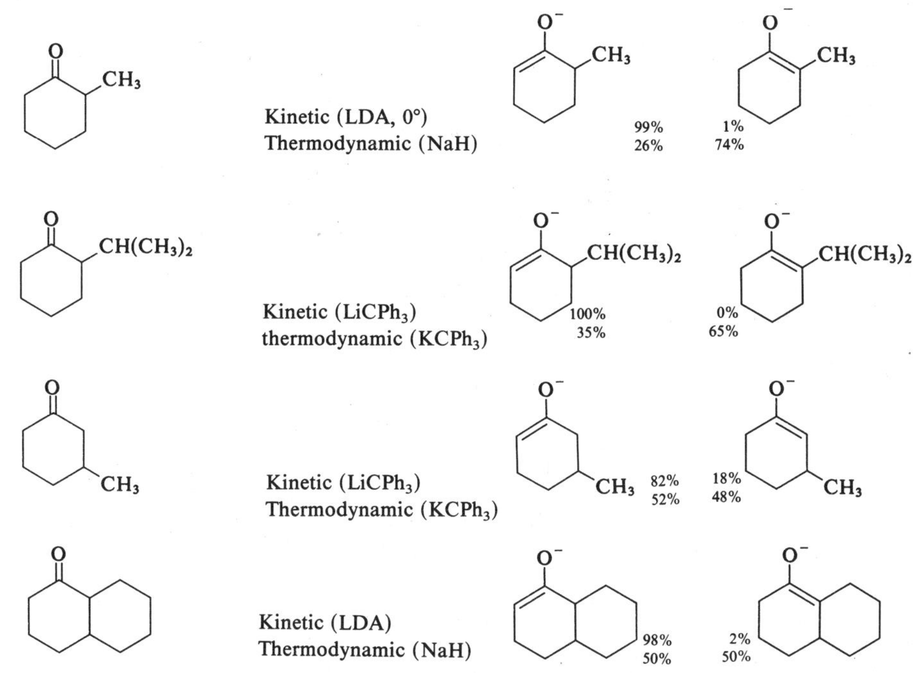

a. Selected from a more complete compilation by D. Caine, *Carbon-Carbon Bond Formation*, R. L. Augustine (ed.), Marcel Dekker, New York, 1979.

Table 7.5. Acidities of Some Cyano Compounds[a]

Compound	pK
CH_3CN	>25.0
$NCCH_2CN$	11.2
$(NC)_3CH$	−5.0
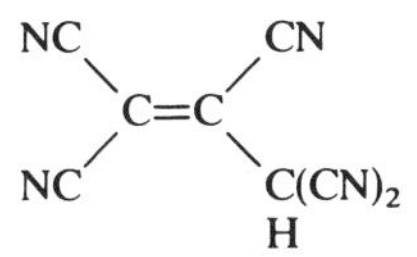 NC, CN, C=C, NC, $C(CN)_2$ H	<−8.5
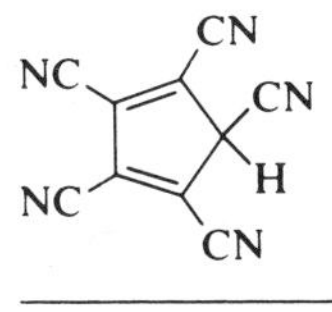 CN, NC, CN, H, NC, CN	<−11.0

a. Selectred from Tables 5.1 and 5.2 in J. R. Jones, *The Ionization of Carbon Acids*, Academic Press, New York, 1973, pp. 64, 65.

Table 7.6. Acidities of Some Compounds with Sulfur and Phosphorus Substituents

Compound	pK (DMSO)	Ref.
$PhCH_2SPh$	30.8	a
$PhSO_2CH_3$	29.0	b
$PhSO_2CH_2Ph$	23.4	b
$PhCH(SPh)_2$	23.0	a
$(PhS)_2CH_2$	38.0	a
$PhSO_2CH_2SPh$	20.3	b
$PhSO_2CH_2PPh_2$	20.2	b
$H_5C_2O_2CCH_2PPh_3$	9.2[c]	d
$PhCOCH_2PPh_3$	6.0[c]	d

a. F. G. Bordwell, J. E. Bares, J. E. Bartmess, G. E. Drucker, J. Gerhold, G. J. McCollum, M. Van Der Puy, N. R. Vanier, and W. S. Matthews, *J. Org. Chem.* **42**, 326 (1977).
b. F. G. Bordwell, W. S. Matthews, and N. R. Vanier, *J. Am. Chem. Soc.* **97**, 442 (1975).
c. In methanol.
d. A. J. Speziale and K. W. Ratts, *J. Am. Chem. Soc.* **85**, 2790 (1963).

do not have a comparable stabilizing effect. Delocalization can be described as involving $3d$ orbitals on sulfur or the σ^* orbital of the C—S bond.[33] Molecular orbital calculations favor the latter interpretation. Whatever the structural basis is, there is no question that thio substituents enhance the acidity of hydrogens on the adjacent carbons. In hydrocarbons the phenylthio group increases the K by at least 15 pK units. The effect is from 5–10 pK units in carbanions stabilized by another electron-accepting group.[34]

Carbanions derived from sulfoxides have an interesting stereochemical feature. Because of the chiral nature of the adjacent sulfur atom, the two faces of the carbanion are nonequivalent. Conversely, there is a preference for removal from the sulfoxide of one of the two diastereotopic protons on an adjacent methylene group.[35]

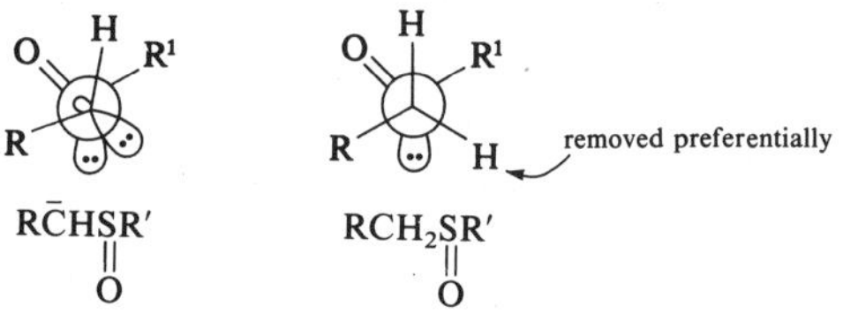

$R\bar{C}HS(=O)R'$ $\qquad$ $RCH_2S(=O)R'$ (H removed preferentially)

33. W. T. Borden, E. R. Davidson, N. H. Andersen, A. D. Deniston, and N. D. Epiotis, *J. Am. Chem. Soc.* **100**, 1604 (1978); A. Streitwieser, Jr., and S. P. Ewing, *J. Am. Chem. Soc.* **97**, 190 (1975); A. Streitwieser, Jr., and J. E. Williams, Jr., *J. Am. Chem. Soc.* **97**, 191 (1975); N. D. Epiotis, R. L. Yates, F. Bernardi, and S. Wolfe, *J. Am. Chem. Soc.* **98**, 5435 (1976); J.-M. Lehn and G. Wipff, *J. Am. Chem. Soc.* **98**, 7498 (1976); D. A. Bors and A. Streitwieser, Jr., *J. Am. Chem. Soc.* **108**, 1397 (1986).
34. F. G. Bordwell, J. E. Bares, J. E. Bartmess, G. E. Drucker, J. Gerhold, G. J. McCollum, V. Van Der Puy, N. R. Vanier, and W. S. Matthews, *J. Org. Chem.* **42**, 326 (1977); F. G. Bordwell, M. Van Der Puy, and N. R. Vanier, *J. Org. Chem.* **41**, 1885 (1976).
35. T. Durst, R. R. Fraser, M. R. McClory, R. B. Swingle, R. Viau, and Y. Y. Wigfield, *Can. J. Chem.* **48**, 2148 (1970); T. Durst, R. Viau, and M. R. McClory, *J. Am. Chem. Soc.* **93**, 3077 (1971).

This result can be explained in terms of a preference for the carbanion structure in which the carbanion is pyramidal with the electron pair *anti* to the sulfoxide oxygen. Protonation of such carbanions is highly stereoselective and occurs with retention of configuration.

Another important group of nucleophilic carbon species is the phosphorus and sulfur ylides. These species have achieved great synthetic importance, and their reactivity will be considered in some detail in Part B, Chapters 1 and 2. Here, we will discuss the structures of some of the best-known ylides. *Ylide* is the name given to molecules for which one of the contributing structures has opposite charges on adjacent atoms when the atoms have octets of electrons. Since we are dealing with nucleophilic carbon species, our interest is in ylides with negative charge on carbon. The three groups of primary importance are phosphonium ylides, sulfoxonium ylides, and sulfonium ylides.

$$R_2\overset{-}{C}-\overset{+}{P}R'_3 \longleftrightarrow R_2C{=}PR'_3$$

phosphonium ylide

$$R_2\overset{-}{C}-\overset{+}{S}(=O)R'_2 \longleftrightarrow R_2C{=}S(=O)R'_2$$

sulfoxonium ylide

$$R_2\overset{-}{C}-\overset{+}{S}R'_2 \longleftrightarrow R_2C{=}SR'_2$$

sulfonium ylide

The question of which resonance structure is the principal contributor has been a point of considerable discussion. Since the nonpolar *ylene* resonance structures have 10 electrons at the phosphorus of sulfur atom, these structures imply participation of *d* orbitals on the heteroatoms. Structural studies indicate that the dipolar ylide structure is probably the main contributor.[36] Molecular orbital calculations confirm the stabilizing effect that the second-row elements phosphorus and sulfur have in ylides, relative to the corresponding first-row elements nitrogen and oxygen.[37]

The ylides are formed by deprotonation of the corresponding "onium salts."

$$R_2CH-\overset{+}{P}R'_3 + B^- \rightleftharpoons R_2\overset{-}{C}-\overset{+}{P}R'_3$$

$$R_2CH-\overset{+}{S}R'_2 + B^- \rightleftharpoons R_2\overset{-}{C}-\overset{+}{S}R'_2$$

The stability of the resulting neutral species is increased by substituent groups that can help to stabilize the electron-rich carbon. Phosphonium ions with acylmethyl substituents, for example, are quire acidic. A series of aroylmethyl phosphonium ions have pK values of 4–7, with the precise value depending on the aryl substituents.[38]

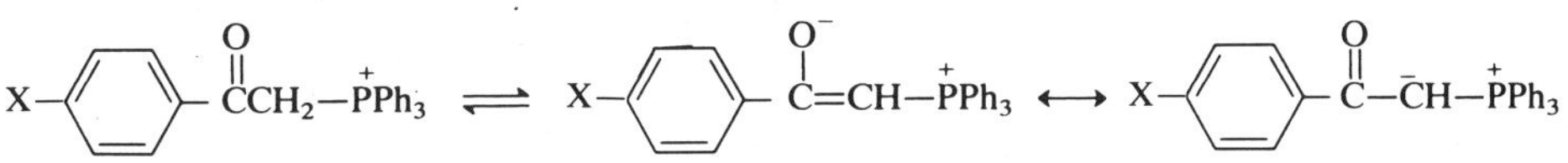

$$X-C_6H_4-\overset{O}{\overset{\|}{C}}CH_2-\overset{+}{P}Ph_3 \rightleftharpoons X-C_6H_4-\overset{O^-}{\overset{|}{C}}{=}CH-\overset{+}{P}Ph_3 \longleftrightarrow X-C_6H_4-\overset{O}{\overset{\|}{C}}-\overset{-}{C}H-\overset{+}{P}Ph_3$$

36. H. Schmidbaur, W. Buchner, and D. Scheutzow, *Chem. Ber.* **106**, 1251 (1973).
37. F. Bernardi, H. B. Schlegel, H.-H. Whangbo, and S. Wolfe, *J. Am. Chem. Soc.* **99**, 5633 (1977).
38. S. Fliszar, R. F. Hudson and G. Salvadori, *Helv. Chim. Acta* **46**, 1580 (1963).

In the absence of the carbonyl or similar stabilizing group, the onium salts are much less acidic. Strong bases such as amide ion or the anion of DMSO are required to deprotonate alkylphosphonium salts.

$$RCH_2\overset{+}{P}R'_3 \xrightarrow[\text{base}]{\text{strong}} R\overset{-}{C}H{-}\overset{+}{P}R_3$$

Similar considerations apply to the sulfoxonium and sulfonium ylides. These ylides are formed by deprotonation of the corresponding positively charged sulfur-containing cations.

$$\underset{\text{sulfonium salt}}{R'_2\overset{+}{S}CH_2R} \xrightarrow{\text{base}} \underset{\text{sulfonium ylide}}{R_2\overset{+}{S}{-}\overset{-}{C}HR} \qquad \underset{\text{sulfoxonium salt}}{R'_2\overset{O}{\overset{\|}{S^+}}{-}CH_2R} \xrightarrow{\text{base}} \underset{\text{sulfoxonium ylide}}{R'_2\overset{O}{\overset{\|}{S^+}}{-}\overset{-}{C}HR}$$

The additional electronegative oxygen atom in the sulfoxonium salts stabilizes these ylides considerably, relative to the sulfonium ylides.[39]

7.3. Enols and Enamines

The study of the chemistry of carbonyl compounds has shown that they can act as carbon nucleophiles in the presence of acid catalysts as well as bases. The nucleophilic reactivity of carbonyl compounds in acidic solution is due to the presence of the enol tautomer. Enolization in acidic solution is catalyzed by O-protonation. Subsequent deprotonation at carbon gives the enol.

$$R\overset{O}{\overset{\|}{C}}CH_2R + H^+ \rightleftharpoons R\overset{+OH}{\overset{\|}{C}}CH_2R \rightleftharpoons R\overset{OH}{\overset{|}{C}}{=}CHR + H^+$$

Like simple alkenes, enols are nucleophilic by virtue of their π electrons. Enols are much more reactive than simple alkenes, however, because the hydroxyl group can participate as an electron donor during the reaction process.

$$R\overset{HO}{\overset{|}{C}}{=}CHR + E^+ \longrightarrow R\overset{HO^{\delta+}}{\overset{\vdots}{C}^{\delta+}}{\cdots}\underset{E^{\delta+}}{CHR} \longrightarrow R\overset{H\overset{+}{O}}{\overset{\|}{C}}{-}\underset{\underset{E}{|}}{CHR} \longrightarrow R\overset{O}{\overset{\|}{C}}\underset{\underset{E}{|}}{CHR} + H^+$$

Enols are not as reactive as enolate anions. This lower reactivity simply reflects the presence of the additional proton in the enol, which decreases the nucleophilicity of the enol relative to the enolate.

A number of studies of the acid-catalyzed mechanism of enolization have been done. The case of cyclohexanone is illustrative.[40] The reaction is catalyzed by various

39. E. J. Corey and M. Chaykovsky, *J. Am. Chem. Soc.* **87**, 1353 (1965).
40. G. E. Lienhard and T.-C. Wang, *J. Am. Chem. Soc.* **91**, 1146 (1969).

carboxylic acids and substituted ammonium ions. The effectiveness of these proton donors as catalysts correlates with their pK_a values. When plotted according to the Brønsted catalysis law, the value of the slope α is 0.74. When deuterium or tritium is introduced in the α position, there is a marked decrease in the rate of acid-catalyzed enolization: $k_H/k_D \approx 5$. This kinetic isotope effect indicates that the C—H bond cleavage is part of the rate-determining step. The generally accepted mechanism for acid-catalyzed enolization pictures the rate-determining step as deprotonation of the protonated ketone.

$$\text{cyclohexanone} + HA \underset{}{\overset{\text{fast}}{\rightleftharpoons}} \text{protonated cyclohexanone } (HO^+) \cdots H \; A^- \underset{}{\overset{\text{slow}}{\rightleftharpoons}} \text{cyclohexenol } (HO) + HA$$

Rates of enolization have usually been measured in one of two ways. One method involves measuring the rate of halogenation of the ketone. In the presence of a sufficient concentration of bromine or iodine, halogenation is much faster than enolization or its reverse and can therefore serve to measure the rate of enolization.

$$RCCH_2R\ (\text{C=O}) \underset{k_{-1}}{\overset{k_1,\ HA}{\rightleftharpoons}} RC(OH){=}CHR \xrightarrow[\text{fast}]{k_2,\ X_2} RC(=O)CHXR + HX$$

$$k_2 \gg k_1, k_{-1}$$

It is also possible to measure the rate of enolization by isotopic exchange. Much of the early work was done using the halogenation technique, but because proton NMR spectroscopy provides a very convenient method for following hydrogen-deuterium exchange, this is now the preferred method. Data for several ketones are given in Table 7.7.

A point of contrast with the data for base-catalyzed removal of a proton is the tendency for acid-catalyzed enolization to result in preferential formation of the more substituted enol. For 2-butanone, the ratio of exchange at CH_2 to that at CH_3 is 4.2:1, after making the statistical correction for the number of hydrogens. The preference for acid-catalyzed enolization to give the more substituted enol is usually rationalized in terms of the stabilizing effect that alkyl groups have on carbon–carbon double bonds. To the extent that the transition state resembles product,[41] alkyl groups would be expected to stabilize the more branched transition state. There is an opposing steric effect that appears to be significant for 4,4-dimethyl-2-pentanone, in which the methylene group that is flanked by a *t*-butyl group is less reactive than the methyl group. The overall range of reactivity differences in acid-catalyzed exchange is much less than for base-catalyzed exchange, however (compare Tables 7.4 and 7.7).

The amount of enol present in equilibrium with a carbonyl group is influenced by other substituent groups. In the case of compounds containing a single ketone,

41. C. G. Swain, E. C. Stivers, J. F. Reuwer, Jr., and L. J. Schaad, *J. Am. Chem. Soc.* **80**, 5885 (1958).

Table 7.7. Relative Rates of Acid-Catalyzed Enolization for Some Ketones[a]

Ketone	Relative rate
$CH_3C(=O)CH_2{-}\underline{H}$	100
$CH_3C(=O)CH(\underline{H})CH_3$	220
$\underline{H}{-}CH_2C(=O)CH_2CH_3$	76
$CH_3C(=O)CH(\underline{H})CH_2CH_3$	171
$CH_3C(=O)C(\underline{H})(CH_3)_2$	195
$\underline{H}{-}CH_2C(=O)CH(CH_3)_2$	80
$CH_3C(=O)CH(\underline{H})C(CH_3)_3$	46
$\underline{H}{-}CH_2C(=O)CH_2C(CH_3)_3$	105

a. In D_2O–dioxane with DCl catalyst. The data of C. Rappe and W. H. Sachs, *J. Org. Chem.* **32**, 3700 (1967), given on a per-group basis, have been converted to a per-hydrogen basis.

aldehyde, or ester function, there is very little of the enol present at equilibrium. When two such groups are close to one another in a molecule, however, particularly if they are separated by a single carbon atom, a major amount of the enol may be present. The enol forms of β-diketones and β-ketoesters are stabilized by intramolecular hydrogen bonds and by conjugation of the carbon–carbon double bond with the carbonyl group.

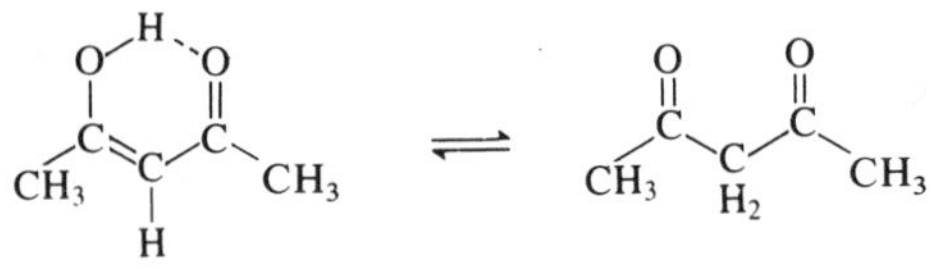

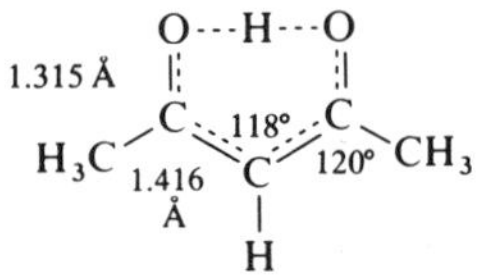

$$CH_3C(OH)=CHC(=O)CH_3 \rightleftharpoons CH_3C(=O)CH_2C(=O)CH_3$$

(Malonaldehyde enol, with O···H···O hydrogen bond, C–O 1.315 Å, C–C 1.416 Å, angles 118° and 120°; H_3C and CH_3 substituents, central C–H.)

The simplest compound with this type of enolic structure is malonaldehyde. The bond lengths determined by microwave spectroscopy on a deuterated analog have provided the bond length data shown below.[42] The barrier for shift of the

42. S. L. Baughcum, R. W. Duerst, W. F. Rowe, Z. Smith, and E. B. Wilson, *J. Am. Chem. Soc.* **103**, 6296 (1981).

enolic hydrogen (or deuterium) between the two oxygen atoms is about 4–5 kcal.[43] The structural data given above for the enol form of 2,4-pentanedione were obtained by an electron diffraction study.[44] In this case, the data pertain to the time-averaged structure resulting from proton transfer between the two hydrogen-bonded oxygens.

0.969 D 1.708
O 147° O
1.310 | || 1.234
D C 1.348 C 1.454 C H
D

Table 7.8 gives data on the amount of enol present at equilibrium for some representative compounds.

The precise percent of enol present at equilibrium is solvent dependent. For example, for ethyl acetoacetate, the amount of enol is higher in nonpolar solvents (15–30%) such as carbon tetrachloride and benzene than in more polar solvents such as water and acetone (5% enol in acetone, 1% enol in water).[45] The strong intramolecular hydrogen bond in the enol form minimizes the molecular dipole by reducing the negative charge on the oxygen of the carbonyl group. In more polar solvents, this stabilization is less important, and in protic solvents such as water, hydrogen bonding by the solvent is dominant.

O O H_3C O C_2H_5 ⟷ O O^- H_3C O^+ C_2H_5 ⇌ O H O H_3C OC_2H_5

Enols of simple ketones can be generated in high concentration as metastable species by special techniques.[46] Vinyl alcohol, the enol of acetaldehyde, can be generated by hydrolysis of any of several ortho ester derivatives.[47]

$$RCO_2\underset{OCH_3}{\overset{H}{C}}OCH{=}CH_2 \xrightarrow[-20^\circ]{H_2O{-}CH_3CN} RCO_2H + HCO_2CH_3 + HOCH{=}CH_2$$

The enol can be observed by NMR and at −20°C has a half-life of several hours. At +20°C the half-life is only 10 minutes. The presence of bases causes very rapid isomerization to acetaldehyde via the enolate. Solvents have a significant effect on the lifetime of such unstable enols. Solvents such as DMF and DMSO, which are known to slow rates of proton exchange by hydrogen bonding, increase the lifetime of unstable enols.[48]

43. S. L. Baughcum, Z. Smith, E. B. Wilson, and and R. W. Duerst, *J. Am. Chem. Soc.* **106**, 2260 (1984).
44. A. H. Lowrey, C. George, P. D'Antonio, and J. Karle, *J. Am. Chem. Soc.* **93**, 6399 (1971).
45. K. D. Grande and S. M. Rosenfeld, *J. Org. Chem.* **45**, 1626 (1980); S. G. Mills and P. Beak, *J. Org. Chem.* **50**, 1216 (1985).
46. B. Capon, B. Z. Guo, F. C. Kwok, A. F. Siddhanta, and C. Zucco, *Acc. Chem. Res.* **21**, 135 (1988).
47. B. Capon, D. S. Rycroft, T. W. Watson, and C. Zucco, *J. Am. Chem. Soc.* **103**, 1761 (1981).
48. E. A. Schmidt and H. M. R. Hoffmann, *J. Am. Chem. Soc.* **94**, 7832 (1972).

Table 7.8. Equilibrium Constants for Enolization of Some Carbonyl Compounds.

Compound	K = enol/keto	Ref.
$CH_3CH{=}O \rightleftharpoons CH_2{=}CHOH$	10^{-5}	a
$(CH_3)_2CHCH{=}O \rightleftharpoons (CH_3)_2C{=}CHOH$	1.4×10^{-4}	b
$CH_3C(=O)CH_3 \rightleftharpoons H_2C{=}C(OH)CH_3$	8×10^{-8} 6×10^{-9}	a c,d
$CH_3CH_2C(=O)CH_2CH_3 \rightleftharpoons CH_3CH{=}C(OH)CH_2CH_3$	2×10^{-8} 3.6×10^{-8}	a c
O (cyclopentanone) $\rightleftharpoons$ OH (cyclopentenol)	1×10^{-7} 1.7×10^{-8}	a c
O (cyclohexanone) $\rightleftharpoons$ OH (cyclohexenol)	5×10^{-6} 3.9×10^{-7}	a c
$PhC(=O)CH_3 \rightleftharpoons PhC(OH){=}CH_2$	2×10^{-7} 1.2×10^{-8}	a c
$CH_3C(=O)CH_2C(=O)OC_2H_5 \rightleftharpoons CH_3C(OH){=}CHC(=O)OC_2H_5$ (OH····O)	7×10^{-2} (H_2O) 3×10^{-1} (CCl_4)	d d
$CH_3C(=O)CH_2C(=O)CH_3 \rightleftharpoons CH_3C(OH){=}CHC(=O)CH_3$ (OH····O)	2.3×10^{-1} (H_2O) 29 (CCl_4)	d d
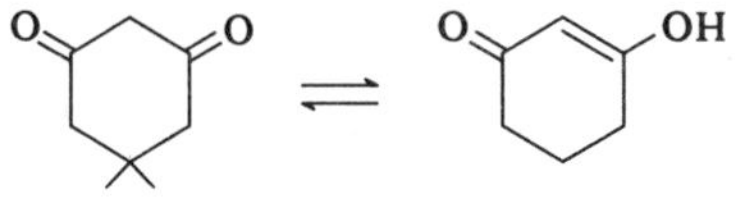 O, O (dione) $\rightleftharpoons$ O, OH (enol)	20 (H_2O)	d
O, O=CH (keto) $\rightleftharpoons$ OH····O=CH (76%) $\rightleftharpoons$ O····HO–CH (24%)	>50	e

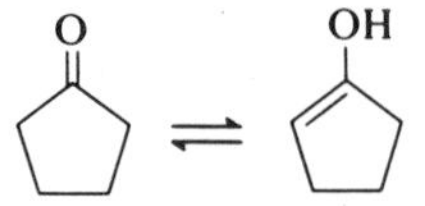

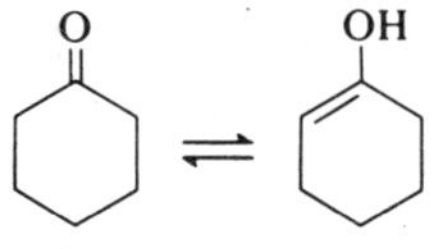

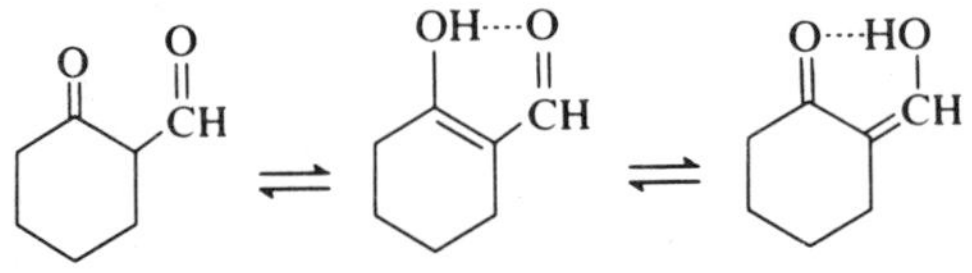

a. In water; J. P. Guthrie and P. A. Cullimore, *Can. J. Chem.* **57**, 240 (1979).
b. In water; Y. Chiang, A. J. Kresge, and P. A. Walsh, *J. Am. Chem. Soc.* **108**, 6314 (1986).
c. In water; J. E. Dubois, M. El-Alauoi, and J. Toullec, *J. Am. Chem. Soc.* **103**, 5393 (1981); J. Toullec, *Tetrahedron Lett.*, **25**, 4401 (1984).
d. S. G. Mills and P. Beak, *J. Org. Chem.* **50**, 1216 (1985).
e. E. W. Garbisch, *J. Am. Chem. Soc.* **85**, 1696 (1963).

Solutions of unstable enols of simple ketones and aldehydes can be generated in water by addition of a solution of the enolate to water.[49] The initial protonation takes place on oxygen, generating the enol, which is then ketonized at a rate that depends on the solution pH. The ketonization exhibits both acid and base catalysis.[50] Acid catalysis involves C-protonation with concerted O-deprotonation.

$$H_2O \curvearrowright H{-}O(H)C{=}CH_2 + H{-}A \rightarrow HC(=O)CH_3 + H_3O^+ + A^-$$

In agreement with expectation for a rate-determining proton transfer, the reaction shows general acid catalysis. Base-catalyzed ketonization occurs by C-protonation of the enolate.

$$HO(H)C{=}CH_2 + B^- \rightleftharpoons {}^-O(H)C{=}CH_2 \xrightarrow{H_2O} HC(=O){-}CH_3 + {}^-OH$$

As would be expected on the basis of electronegativity arguments, enols are much more acidic than the corresponding keto forms. It has been possible to determine the pK of the enol form of acetophenone as being 4.6×10^{-11}. The pK of the keto form is 18.2.[51] Since the enolate is the same for both equilibria, the pK values are related to the enol–keto equilibrium.

$$\begin{array}{ccc} & PhC(=O)CH_3 & \\ K = 10^{-7.9} \;\swarrow\!\nearrow & & \searrow\!\nwarrow\; K = 10^{-18.4} \\ CH_2{=}C(OH)Ph & \underset{K = 10^{-10.5}}{\rightleftharpoons} & CH_2{=}C(O^-)Ph \end{array}$$

Similar measurements have been made for the equilibria involving acetone and its enol, 2-hydroxypropene.[52]

$$\begin{array}{ccc} & CH_3C(=O)CH_3 & \\ K = 10^{-8.2} \;\swarrow\!\nearrow & & \searrow\!\nwarrow\; K = 10^{-19.2} \\ CH_2{=}C(OH)CH_3 & \underset{K = 10^{-11}}{\rightleftharpoons} & CH_2{=}C(O^-)CH_3 + H^+ \end{array}$$

49. Y. Chiang, A. J. Kresge, and P. A. Walsh, *J. Am. Chem. Soc.* **104**, 6122 (1982); Y. Chiang, A. J. Kresge, and P. A. Walsh, *J. Am. Chem. Soc.* **108**, 6314 (1986).
50. B. Capon and C. Zucco, *J. Am. Chem. Soc.* **104**, 7567 (1982).
51. Y. Chiang, A. J. Kresge, and J. Wirz, *J. Am. Chem. Soc.* **106**, 6392 (1984).
52. Y. Chiang, A. J. Kresge, Y. S. Tang, and J. Wirz, *J. Am. Chem. Soc.* **106**, 460 (1984).

The accessibility of enols and enolates, respectively, in acidic and basic solutions of carbonyl compounds makes possible a wide range of reactions that depend on the nucleophilicity of such species. The reactions will be discussed in Chapter 8 and in Chapters 1 and 2 of Part B.

Amino substituents on a carbon–carbon double bond enhance the nucleophilicity of the β carbon to an even greater extent than the hydroxyl group in enols. This is because of the greater electron-donating power of nitrogen. Compounds containing such substituents are called *enamines.*

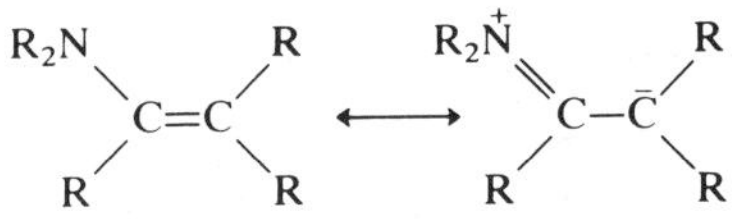

$$R_2N(R)C{=}CR_2 \longleftrightarrow R_2\overset{+}{N}{=}C(R){-}\overset{-}{C}R_2$$

An interesting and useful property of enamines of 2-alkylcyclohexanones is the fact that there is a substantial preference for the less substituted isomer to be formed. This tendency is especially pronounced for enamines derived from cyclic secondary amines such as pyrrolidine. This preference can be traced to a strain effect called $A^{1,3}$ or allylic strain. In order to accommodate conjugation between the nitrogen lone pair and the carbon–carbon double bond, the nitrogen substituent must be complanar with the double bond. This creates a steric repulsion when the enamine bears a β substituent and leads to a preference for the unsubstituted enamine.

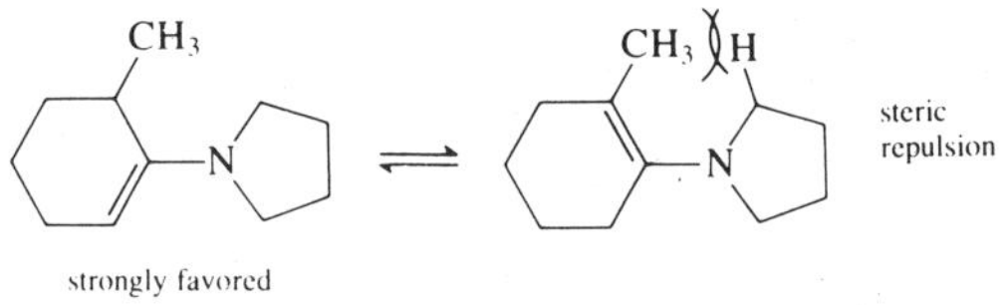

CH₃ N ⇌ CH₃ H N steric repulsion

strongly favored

Because of the same preference for coplanarity in the enamine system, α-alkyl substituents adopt an axial conformation to minimize steric interaction with the amino group.[53]

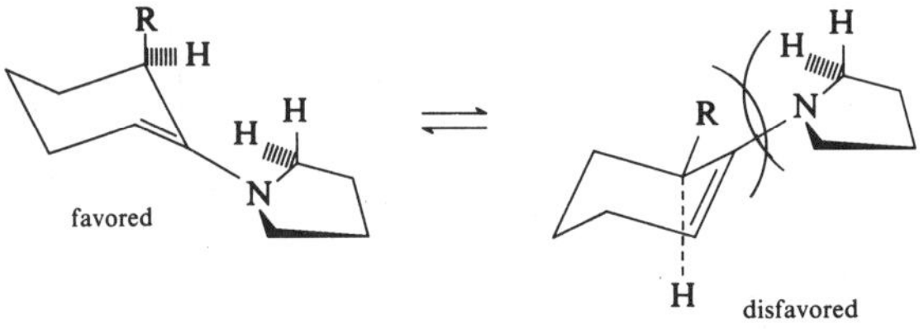

R H H H N ⇌ H H R N H

favored disfavored

The preparation of enamines with be discussed in Chapter 8, and their application as carbon nucleophiles in synthesis is discussed in Chapter 1 of Part B.

53. A. G. Cook, *Enamines*, Second Edition, Marcel Dekker, New York, 1988, Chapter 1.

7.4. Carbanions as Nucleophiles in S_N2 Reactions

Carbanions are very useful intermediates in the formation of carbon–carbon bonds. This is true both for unstabilized structures found in organometallic reagents and stabilized structures such as enolates. Carbanions can participate as nucleophiles in both addition and substitution reactions. At this point we will discuss aspects of the reactions of carbanions as nucleophiles in reactions that proceed by the S_N2 mechanism. Other synthetic applications of carbanions will be discussed more completely in Part B.

Carbanions are classified as soft nucleophiles. It would be expected that they would be good nucleophiles in S_N2 reactions, and this is generally true. The reactions of aryl-, alkenyl-, and alkyllithium reagents with primary alkyl halides and tosylates appear to proceed by S_N2 mechanisms. Similar reactions occur between arylmagnesium halides (Grignard reagents) and alkyl sulfates and sulfonates. Some examples of these reactions are given in Scheme 7.2.

Evidence for an S_N2-type mechanism in the reaction of allyl- and benzyllithium reagents has been obtained from stereochemical studies. With 2-bromobutane, both of these reagents react with complete inversion of configuration.[54] *n*-Butyllithium, however, gives largely racemic product, indicating some complicating process must also occur.[55] A general description of the mechanism for reaction of organolithium compounds with alkyl halides must take account of the structural nature of the organometallic compound. It is known that halide anions are accommodated into typical organolithium cluster structures and can replace solvent molecules as ligands. A similar process in which the alkyl halide becomes complexed at lithium would provide an intermediate structure that could account for the subsequent alkylation. This process is represented below for a tetrameric strcture, with the organic group simply represented by C.

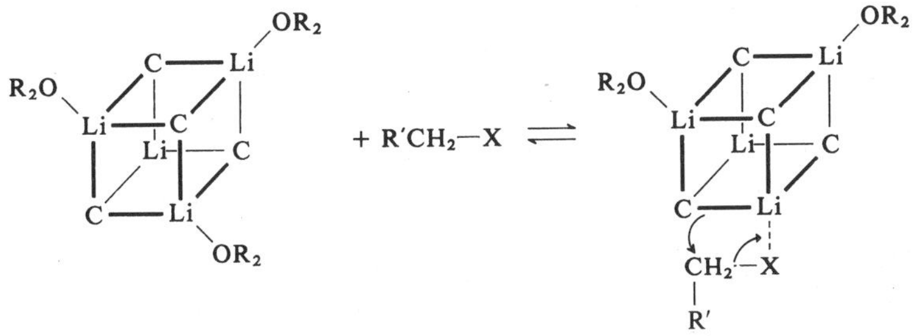

$$+ \; R'CH_2{-}X \rightleftharpoons$$

In general terms, the reactions of organolithium reagents with alkylating agents could conceivably occur at any of the aggregation stages present in solution, and there could be reactivity differences among these stages. There has been little detailed

54. L. H. Sommer and W. D. Korte, *J. Org. Chem.* **35**, 22 (1970).
55. H. D. Zook and R. N. Goldey, *J. Am. Chem. Soc.* **75**, 3975 (1953).

Scheme 7.2. Alkylation of Some Organometallic Reagents

A. Organolithium Reagents

1[a] (1-lithiocyclooctene) + CH_3I → 1-methylcyclooctene 72%

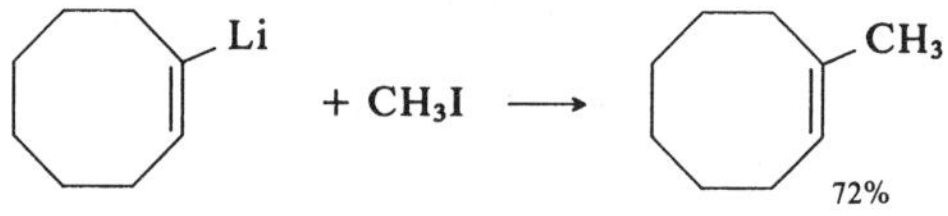

2[b] $CH_3(H)C{=}C(CH_3)Li$ + $CH_3(CH_2)_6CH_2I$ → $CH_3(H)C{=}C(CH_3)CH_2(CH_2)_6CH_3$ 77%

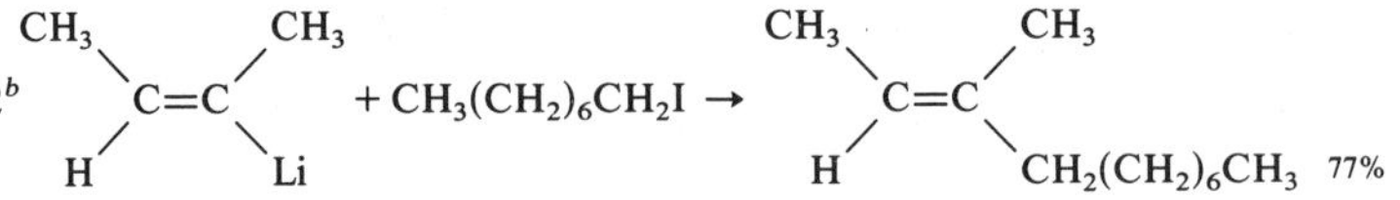

3[c] $(CH_3)_3CSi(CH_3)_2O$–$C_6H_2(OCH_3)_2$–Li + $BrCH_2CH{=}C(CH_3)_2$ → $(CH_3)_3CSi(CH_3)_2O$–$C_6H_2(OCH_3)_2$–$CH_2CH{=}C(CH_3)_2$ 60%

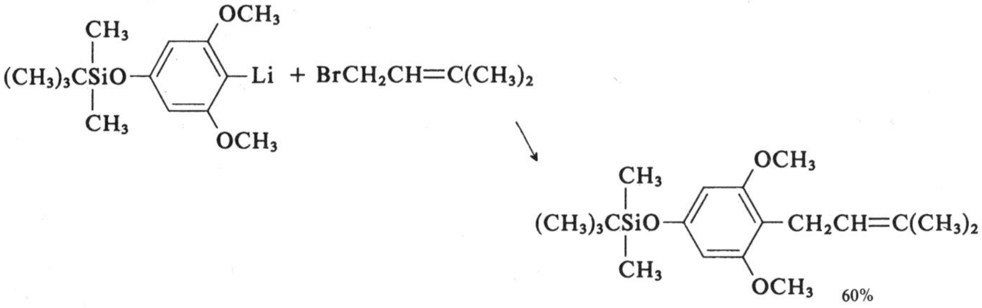

4[d] $CH_3(CH_2)_2C{\equiv}CLi + C_2H_5Br \rightarrow CH_3(CH_2)_2C{\equiv}CCH_2CH_3$ 65%

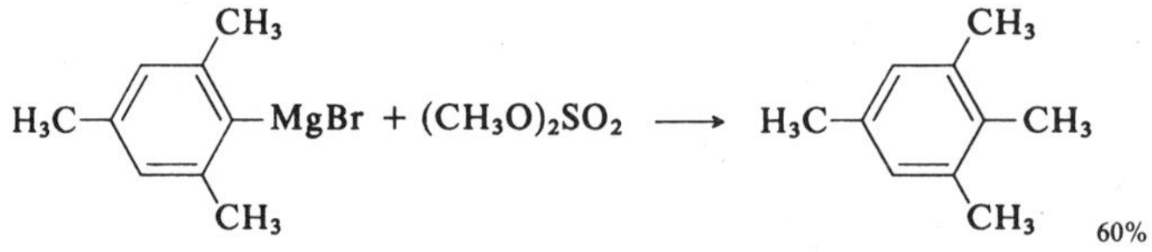

B. Organomagnesium Compounds

5[e] $C_6H_5CH_2MgCl$ + $(C_2H_5O)_2SO_2$ → $C_6H_5CH_2CH_2CH_3$ 70%

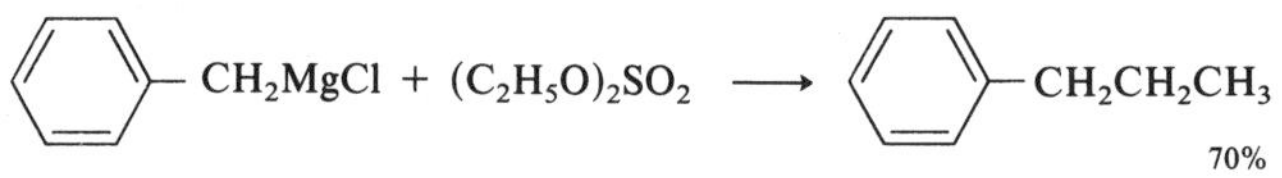

6[f] H_3C–$C_6H_2(CH_3)_2$–MgBr + $(CH_3O)_2SO_2$ → H_3C–$C_6H_2(CH_3)_2$–CH_3 60%

7[g] (1,1,4,4-tetramethyltetralin-6-yl)MgBr + $BrCH_2CH{=}CH_2$ → (1,1,4,4-tetramethyltetralin-6-yl)$CH_2CH{=}CH_2$ 79%

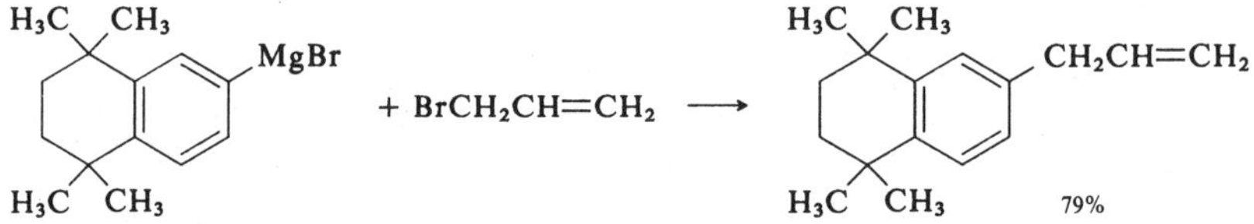

a. H. Neumann and D. Seebach, *Chem. Ber.* **111**, 2785 (1978).
b. J. Millon, R. Lorne, and G. Linstrumelle, *Synthesis*, 434 (1975).
c. T. L. Shih, M. J. Wyvratt, and H. Mrozik, *J. Org. Chem.* **52**, 2029 (1987).
d. A. J. Quillinan and F. Scheinman, *Org. Synth.* **58**, 1 (1978).
e. H. Gilman and W. E. Catlin, *Org. Synth., Coll. Vol. I*, 471 (1941).
f. L. I. Smith, *Org. Synth., Coll. Vol. II*, 360 (1943).
g. J. Eustache, J. M. Bernardon, and B. Shroot, *Tetrahedron Lett* **28**, 4681 (1987).

mechanistic study which would distinguish among these possibilities.

$$
\begin{array}{ccccc}
(RLi)_4 & \rightleftharpoons & 2\,(RLi)_2 & \rightleftharpoons & 4\,RLi \\
\downarrow{\scriptstyle R'X} & & \downarrow{\scriptstyle R'X} & & \downarrow{\scriptstyle R'X} \\
R{-}R' & & R{-}R' & & R{-}R'
\end{array}
$$

The reaction of phenyllithium and allyl chloride using a ^{14}C label reveals that allylic rearrangement occurs. About three-fourths of the product results from bond formation at C—3 rather than C—1. This can be accounted for by a cyclic transition state.[56]

$$\text{Ph—Li} + \text{Cl—}\overset{*}{\text{C}}\text{H}_2\text{CH=CH}_2 \longrightarrow [\text{cyclic TS}] \longrightarrow \text{PhCH}_2\text{CH=}\overset{*}{\text{C}}\text{H}_2$$

The portion of the product formed by reaction at C—1 in allylic systems may form by direct substitution, but it has also been suggested that a cyclic transition state involving an aryllithium dimer might be involved.

$$\text{Ph—Li} + \text{Cl—}\overset{*}{\text{C}}\text{H}_2\text{CH=CH}_2 \longrightarrow [\text{cyclic TS}] \longrightarrow \text{Ph}\overset{*}{\text{C}}\text{H}_2\text{CH=CH}_2$$

These mechanisms ascribe importance to the Lewis acid-Lewis base interaction between the allyl halide and the organolithium reagent. When substitution is complete, the halide ion is incorporated into the lithium cluster in place of one of the carbon ligands. Substituted allylic halides give product mixtures resulting from bond formation both at C—1 and C—3 of the allylic system, with the ratio of products favoring reaction at the less substituted site.

From a synthetic point of view, direct alkylation of lithium and magnesium organometallic compounds has been supplanted by transition metal-catalyzed processes. We will discuss these reactions in Part B, Chapter 8.

The alkylation reactions of enolate anions of both ketones and esters have been extensively utilized in synthesis. Both very stable enolates, such as those derived from β-ketoesters, β-diketones, and malonate esters, as well as less stable enolates, such as those derived from monofunctional ketones, esters, and nitriles, are reactive. Many aspects of the relationships between reactivity, stereochemistry, and mechanism have been clarified. A starting point for the discussion of these reactions is the

56. R. M. Magid and J. G. Welch, *J. Am. Chem. Soc.* **90**, 5211 (1968); R. M. Magid, E. C. Nieh, and R. D. Gandour, *J. Org. Chem.* **36**, 2099 (1971).

structure of the enolates. Because of the delocalized nature of enolates, an electrophile can conceivably attack at oxygen or at carbon.

Soft electrophiles will prefer carbon, and it is found experimentally that most alkyl halides react to give C-alkylation. Because of the π character of the HOMO of the enolate, there will be a stereoelectronic preference for attack of the electrophile approximately perpendicular to the plane of the enolate. Thus, the transition state for an S_N2 alkylation of an enolate can be represented as below.

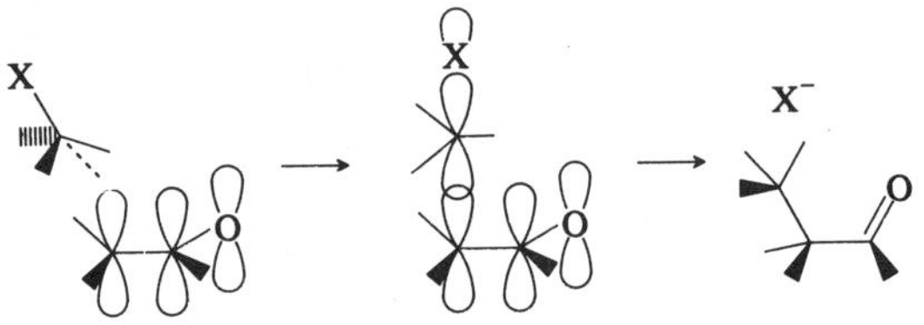

A more detailed representation of the reaction requires more intimate knowledge of the enolate structure. The structures of several lithium enolates of ketones have been determined by X-ray crystallography. They reveal aggregated structure in which oxygen and lithium occupy alternating corners of distorted cubes. Figure 7.2 illustrates some of the observed structures. Figure 7.2(A) shows an unsolvated enolate of methyl *t*-butyl ketone (pinacolone).[57] The structures in Fig. 7.2(B) and 7.2(C) are the THF solvates of the enolates of methyl *t*-butyl ketone and cyclopentanone, respectively.[58] Each of these structures consists of clusters of four enolate anions and four lithium cations arranged with lithium and oxygen at alternating corners of a distorted cube. The structure in Fig. 7.2(D) includes only two enolate anions. Four lithium ions are present along with two diisopropylamide ions. An interesting feature of this structure is the coordination of the remote siloxy oxygen atom to one of the lithium cations.[59] This is an example of the type of Lewis acid–Lewis base interaction that is frequently invoked as a force in organizing transition state structure in the reactions of lithium enolates. A common feature of all four of the structures is the involvement of the enolate oxygen in multiple contacts with lithium cations in the cluster. An approaching electrophile will clearly be somewhat hindered from direct approach to oxygen in such structures.

Several ester enolates have also been examined by X-ray crystallography.[60] The enolates of *t*-butyl propionate and *t*-butyl 2-methylpropionate were obtained as TMEDA solvates of enolate dimers. Methyl 3,3-dimethylbutanoate was obtained as a THF-solvated tetramer.

57. P. G. Williard and G. B. Carpenter, *J. Am. Chem. Soc.* **107**, 3345 (1985).
58. R. Amstutz, W. B. Schweizer, D. Seebach, and J. D. Dunitz, *Helv. Chem. Acta* **64**, 2617 (1981).
59. P. G. Williard and M. J. Hintze, *J. Am. Chem. Soc.* **109**, 5539 (1987).
60. D. Seebach, R. Amstutz, T. Laube, W. B. Schweizer, and J. D. Dunitz, *J. Am. Chem. Soc.* **107**, 5403 (1985).

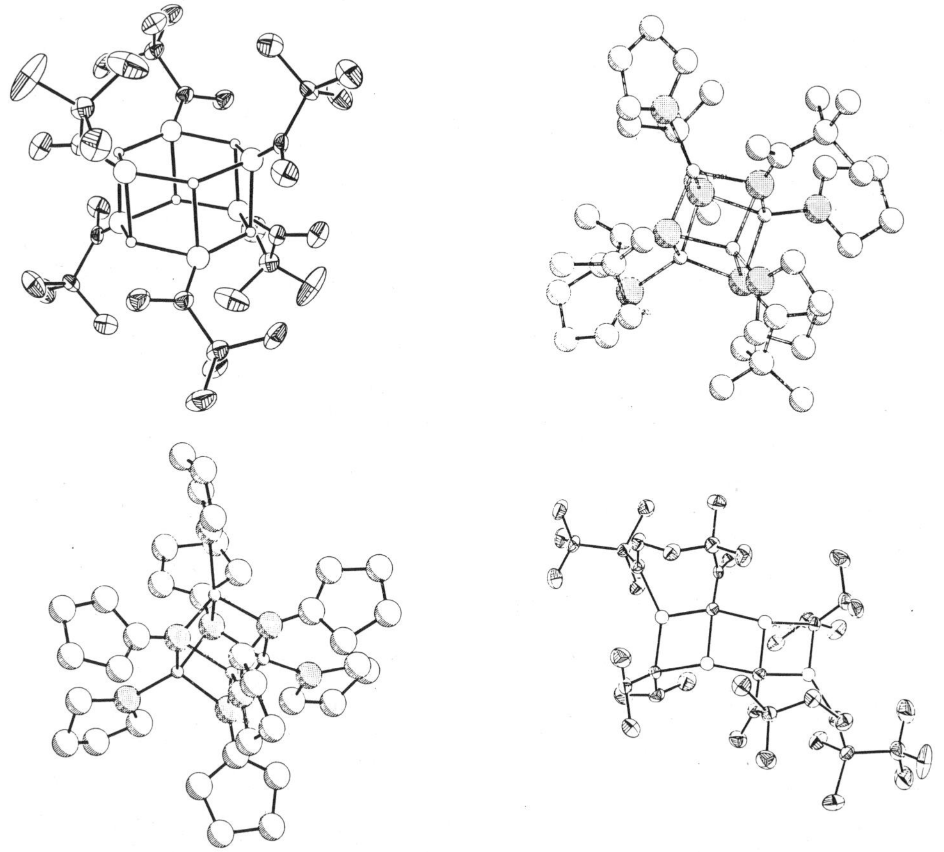

Fig. 7.2. Crystal structures of some lithium enolates to ketones of ketones. (A) Unsolvated hexameric enolate of methyl *t*-butyl ketone. (B) Tetrahydrofuran solvate of tetramer of enolate of methyl *t*-butyl ketone. (C) Tetrahydrofuran solvate of tetramer of enolate of cyclopentanone. (D) Dimeric enolate of 3,3-dimethyl-4-(*t*-butyldimethylsiloxy)-2-pentanone. Structural diagrams are reproduced from Refs. 57–59.

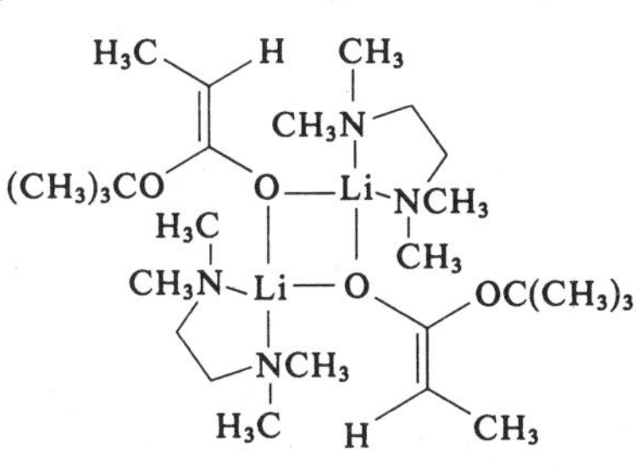

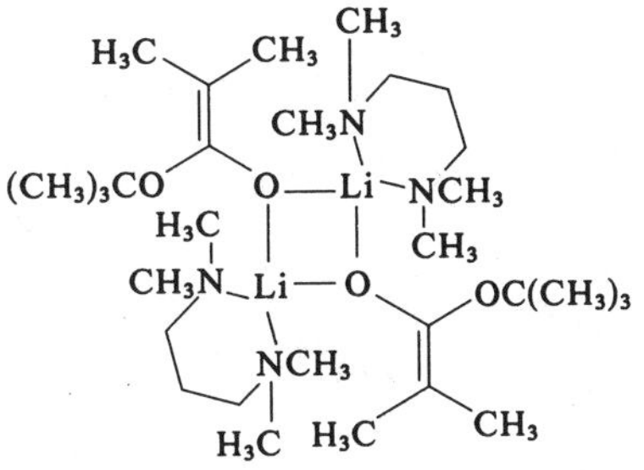

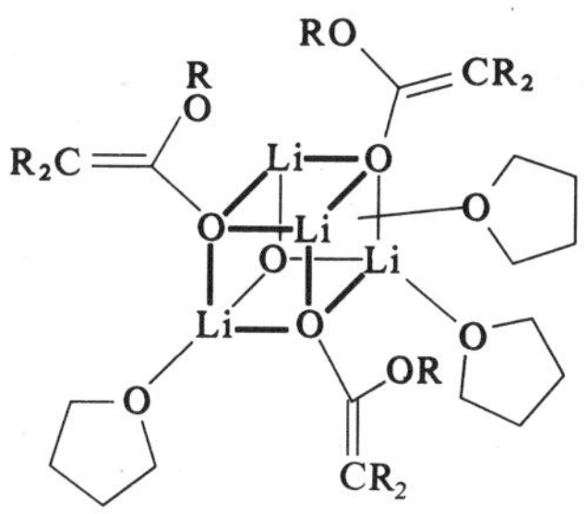

Studies of ketone enolates in solution indicate both tetrameric and dimeric clusters can exist. Tetrahydrofuran, a solvent in which many synthetic reactions are performed, favors tetrameric structures for the lithium enolate of isobutyrophenone, for example.[61]

One of the general features of the reactivity of enolate anions is the sensitivity of both the reaction rate and the ratio of C- versus O-alkylation to the degree of aggregation of the enolate. For example, addition of HMPA frequently increases the rate of enolate alkylation reactions.[62] Use of dipolar aprotic solvents such as DMF and DMSO in place of the THF also leads to rate acceleration.[63] These effects can be attributed, at least in part, to dissociation of the lithium enolate aggregates. Similar effects are observed when crown ethers or similar cation-complexing agents are added to reaction mixtures.[64]

The order of enolate reactivity also depends on the metal cation that is present. The general order is $BrMg^+ < Li^+ < Na^+ < K^+$. This order, too, is the order of greater dissociation of the enolate–cation ion pairs and ion aggregates. Carbon-13 chemical shift data can provide an indication of electron density at the nucleophilic carbon in enolates. These shifts have been found to be both cation dependent and solvent dependent. Apparent electron density increases in the order $K^+ > Na^+ > Li^+$ and THF/HMPA > DME > THF > ether.[65] There is a good correlation with observed reactivity under the corresponding conditions.

The leaving group in the alkylating reagent has a major influence on whether C- or O-alkylation occurs. With the lithium enolate of acetophenone, for example, C-alkylation is the major product with methyl iodide but C- and O-alkylation occur to approximately equal extents with dimethyl sulfate. The C:O ratio is shifted more to O-alkylation by addition of HMPA or other cation-complexing agents. Thus, with four equivalents of HMPA the C:O ratio for methyl iodide drops from >200:1 to 10:1 while with dimethyl sulfate the C:O ratio changes from 1.2:1 to 0.2:1 when HMPA is added.[66] It has been shown that the C:O ratio *changes* during the course of the reaction. This is attributed to the formation of the anion from the alkylation reagent. The anion can displace an enolate anion at one of the anionic sites in the lithium cluster, giving a cluster of somewhat different reactivity.[67]

The C- versus O-alkylation ratio has also been studied for the potassium salt of ethyl acetoacetate as a function of both solvent and leaving group.[68]

61. L. M. Jackman and N. Szeverenyi, *J. Am. Chem. Soc.* **99**, 4954 (1977); L. M. Jackman and B. C. Lange, *Tetrahedron* **33**, 2737 (1977).
62. L. M. Jackman and B. C. Lange, *J. Am. Chem. Soc.* **103**, 4494 (1981); C. L. Liotta and T. C. Caruso, *Tetrahedron Lett.* **26**, 1599 (1985).
63. H. D. Zook and J. A. Miller, *J. Org. Chem.* **36**, 1112 (1971); H. E. Zaugg, J. F. Ratajczyk, J. E. Leonard, and A. D. Schaeffer, *J. Org. Chem.* **37**, 2249 (1972); H. E. Zaugg, *J. Am. Chem. Soc.* **83**, 837 (1961).
64. A. L. Kurts, S. M. Sakembaeva, J. P. Beletskaya, and O. A. Reutov, *Zh. Org. Khim.* (Engl. Transl.) **10**, 1588 (1974).
65. H. O. House, A. V. Prabhu, and W. V. Phillips. *J. Org. Chem.* **41**, 1209 (1976).
66. L. M. Jackman and B. C. Lange, *J. Am. Chem. Soc.* **103**, 4494 (1981).
67. L. M. Jackman and T. S. Dunne, *J. Am. Chem. Soc.* **107**, 2805 (1985).
68. A. L. Kurts, A. Masias, N. K. Gerkina, I. P. Beletskaya, and O. A. Reutov, *Dokl. Akad. Nauk SSSR* (Engl. Transl.) **187**, 595 (1969); A. L. Kurts, N. K. Gerkina, A. Masias, I. P. Beletskaya, and O. A. Reutov, *Tetrahedron* **27**, 4777 (1971).

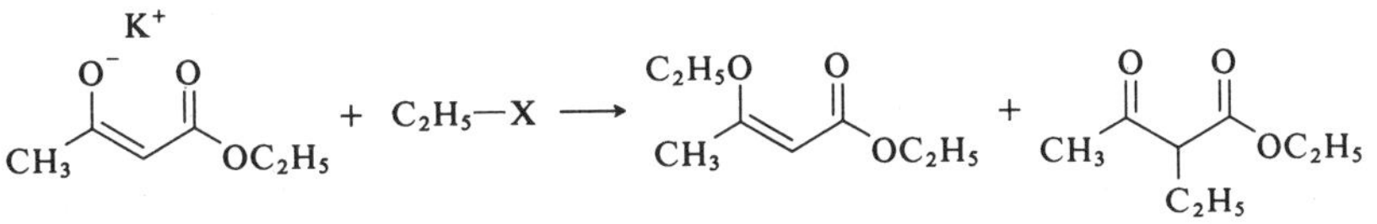

Leaving group, X	Solvent	C:O ratio
$OSO_2C_2H_5$	t-BuOH	100:0
$OSO_2C_2H_5$	THF	100:0
$OSO_2C_2H_5$	HMPA	17:83
$OSO_2C_7H_7$	HMPA	12:88
Cl	HMPA	40:60
Br	HMPA	61:39
I	HMPA	87:13

These data show that a change from a hard leaving group (sulfonate, sulfate) to a softer leaving group (bromide, iodide) favors carbon alkylation.

One major influence on the C:O ratios is presumably the degree of aggregation. The reactivity at oxygen should be enhanced by dissociation since the electron density will be less tightly associated with the cation. It also appears that the nature of the aggregate, that is, the anions incorporated into it, may also be a major influence on reactivity.

Steric and stereoelectronic effects control the direction of approach of an electrophile to the enolate. Electrophiles approach from the side of the enolate that is least hindered. Numerous examples of such effects have been observed.[69] In ketone and ester enolates that are exocyclic to a conformationally biased cyclohexane ring, there is a slight preference for the electrophile to approach from the equatorial direction.[70]

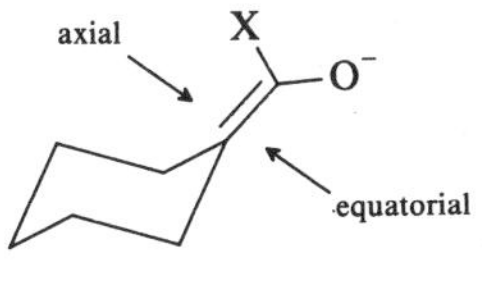

If the axial face is further hindered by addition of a substituent, the selectivity is increased.

Endocyclic cyclohexanone enolates with 2-alkyl groups show a small preference (1:1–5:1) for approach of the electrophile from the direction which permits mainten-

69. Reviews: D. A. Evans, in *Asymmetric Synthesis*, Vol. 3, J. D. Morrison (ed.), Academic Press, New York, 1984, Chapter 1; D. Caine, in *Carbon-Carbon Bond Formation*, R. L. Augustine (ed.), Marcel Dekker, New York, 1979.
70. A. P. Krapcho and E. A. Dundulis, *J. Org. Chem.* **45**, 3236 (1980); H. O. House and T. M. Bare, *J. Org. Chem.* **33**, 943 (1968).

ance of the chair conformation.[71]

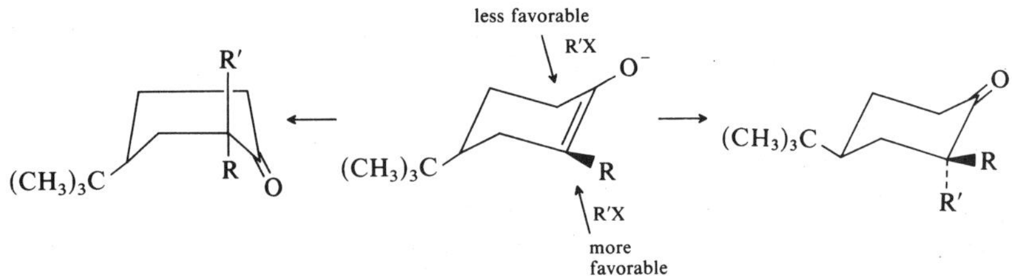

The 1(9)-enolate of 1-decalone exhibits a preference for alkylation to form a *cis* ring juncture.[72]

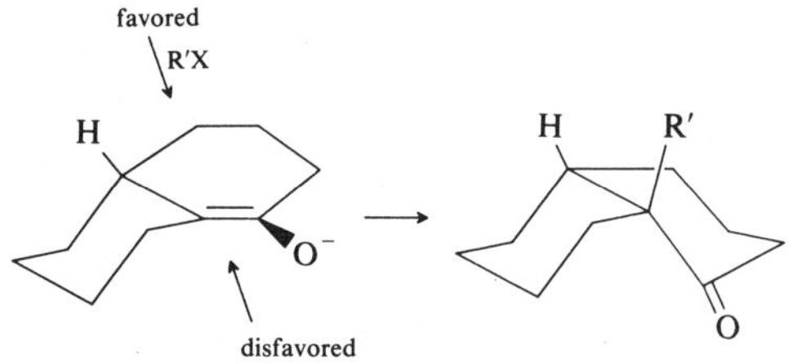

This is the result of a steric differentiation on the basis of the electrophile approaching from the side of the enolate occupied by the smaller hydrogen, rather than the alkyl group, at the 10-position.

In general, the stereoselectivity of enolate alkylation can be predicted and interpreted on the basis of the stereoelectronic requirement for approximately perpendicular approach to the enolate in combination with selection between the two faces on the basis of steric factors.

General References

E. Buncel, *Carbanions: Mechanistic and Isotopic Aspects*, Elsevier, Amsterdam, 1975.

E. Buncel and T. Durst (eds.), *Comprehensive Carbanion Chemistry*, Elsevier, New York, 1981.

D. J. Cram, *Fundamentals of Carbanion Chemistry*, Academic Press, New York, 1965.

H. F. Ebel, *Die Acidität der CH-Säuren*, George Thieme Verlag, Stuttgart, 1969.

J. R. Jones, *The Ionization of Carbon Acids*, Academic Press, New York, 1973.

E. M. Kaiser and D. W. Slocum, in *Organic Reactive Intermediates*, S. P. McManus (ed.), Academic Press, New York, 1973, Chapter 5.

M. Szwarc, *Ions and Ion Pairs in Organic Reactions*, Wiley, New York, 1972.

J. Toullec, *Adv. Phys. Org. Chem.* **18**, 1 (1982).

71. H. O. House, B. A. Tefertiller, and H. D. Olmstead, *J. Org. Chem.* **33**, 935 (1968); H. O. House and M. J. Umen, *J. Org. Chem.* **38**, 1000 (1973); J. M. Conia and P. Briet, *Bull. Soc. Chim. Fr.* 3881, 3886 (1966); C. Djerassi, J. Burkevich, J. W. Chamberlin, D. Elad, Y. Toda, and G. Stork, *J. Am. Chem. Soc.* **86**, 465 (1964); C. Agami, J. Levisalles, and B. Lo Cicero, *Tetrahedron* **35**, 961 (1979).
72. H. O. House and B. M. Trost, *J. Org. Chem.* **41**, 2502 (1965).

Problems

(References for these problems will be found on page 779.)

1. Predict the order of increasing thermodynamic acidity in each series of compounds:

   (a) benzene, 1,4-cyclohexadiene, cyclopentadiene, cyclohexane

   (b) CH_3CN, CH_3NO_2, $CH_3C(=O)CH_3$, $CH_3S(=O)_2CH_3$, $CH_3S(=O)CH_3$

   (c) $PhCH_3$, $PhSiH_3$, $PhS(=O)_2CH_3$, $PhCH_2S(=O)_2CH_3$

   (d) $CH_3C(=O)CH_2C(=O)CH_3$, $CH_3CH_2C(=O)CH_2C(=O)CH_2CH_3$, $PhC(=O)CH_2C(=O)CH_3$, $PhC(=O)CH_2C(=O)CF_3$

   (e) 9-(*m*-chlorophenyl)fluorene, 9-(*p*-methoxyphenyl)fluorene, 9-phenylfluorene, 9-(*m*-methoxyphenyl)fluorene, 9-(*p*-methylphenyl)fluorene.

2. Indicate which proton is the most acidic in each of the following molecules. Explain your reasoning.

   (a) $H_3CC{\equiv}CH$

   (b)

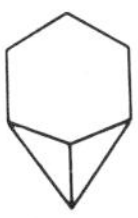

   (c) 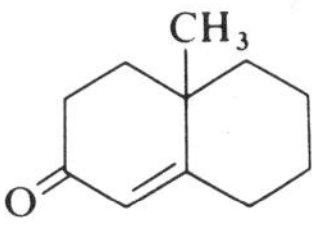 CH_3, O

   (d) O=, $-CO_2C_2H_5$, CH_3

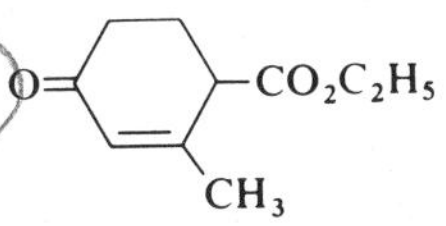

   (e) 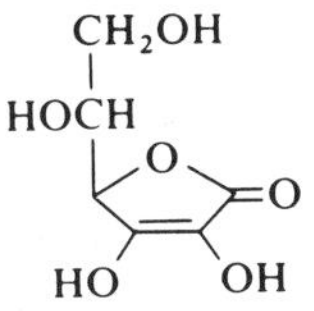 CH_2OH, HOCH, O, =O, HO, OH

   (f) 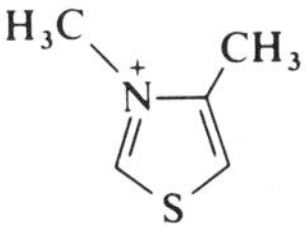 H_3C, CH_3, $\overset{+}{N}$, S

   (g) $CH_3S(=O)CH_2SCH_3$

   (h)  H_3C CH_3

   (i) $(CH_3)_2CH\overset{+}{N}(O^-){=}NOCH_3$

   (j) 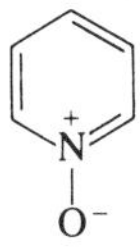 $\overset{+}{N}$, O^-

3. Offer an explanation for the following observations.

   (a) Exchange rates indicate the hydrocarbon cubane to be much more acidic than cyclobutane, and even more acidic than cyclopropane.

(b) Phenyl cyclopropyl ketone ($pK = 28.2$) is less acidic than acetophenone ($pK = 24.7$) and undergoes C—H exchange more slowly than phenyl isopropyl ketone.

(c) The order of acidity for cyclopentadiene, indene, and fluorene is as indicated by the pK data given:

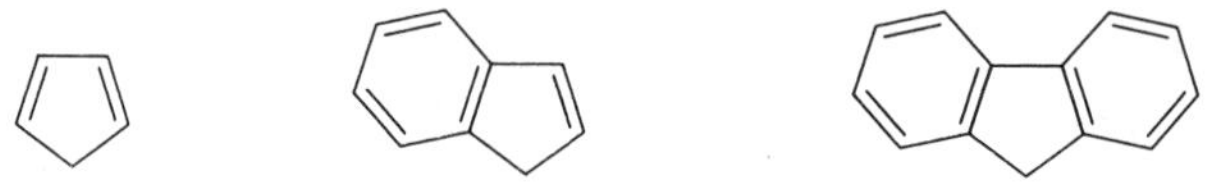

	cyclopentadiene	indene	fluorene
pK	18.0	20.1	22.6

(d) The rotational barrier for the allyl anion in THF, as measured by NMR, is a function of the cation that is present:

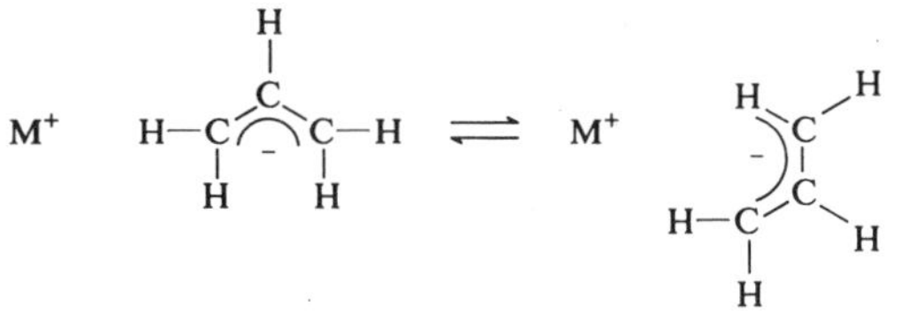

M^+	$\Delta G^{\ddagger}$
Li	10.7
K	16.7
Cs	18.0

4. (a) The relative rates of hydroxide ion-catalyzed deuterium exchange at C-3 (the CH_2 α to the C=O) have been measured for the bicyclic ketones shown below. Analyze the factors that would be involved in the relative ease of exchange in these compounds.

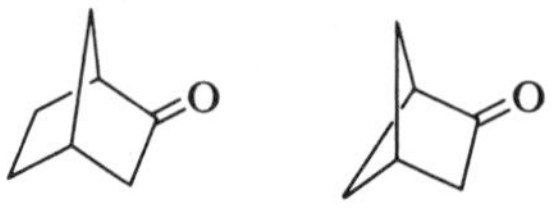

Rate constant for exchange $k = 5.6 \times 10^{-2}$ $k = 6.04 \times 10^{-6}$

(b) Treatment of (+)-camphenilone with potassium *t*-butoxide in *t*-butyl alcohol-O-*d* at 185°C results in H–D exchange accompanied by racemization at an equal rate. Prolonged reaction periods result in the introduction of three deuterium atoms. Suggest a mechanism to account for these observations.

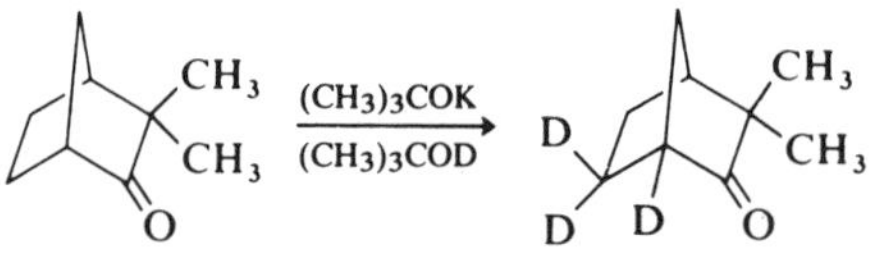

$(CH_3)_3COK$, $(CH_3)_3COD$

5. Using data from Tables 7.1 (p. 398) and 7.2 (p. 401), estimate the extent of deprotonation for each hydrocarbon-solvent-base combination. Discuss the uncertainties involved in your calculations.
(a) indene by 0.01 *M* $NaOCH_3$ in 1:1 DMSO-C_2H_5OH
(b) fluorene by 0.01 *M* $NaOC_2H_5$ in 20:1 DMSO—C_2H_5OH
(c) triphenylmethane by 5 *M* $KOCH_3$ in CH_3OH

6. The rates of abstraction of axial and equatorial protons from 4-*t*-butylcyclohexanone have been measured by an NMR technique. The rate of removal of an axial proton is 5.5 times faster than for an equatorial proton. What explanation can you offer for this difference?

7. The following tables gives exchange rates in methanolic sodium methoxide for a number of hydrocarbons and equilibrium acidities for some. Determine whether there is a correlation between kinetic and thermodynamic acidity in this series of compounds. If so, predict the thermodynamic acidity of the hydrocarbons for which no values are listed.

Compound	k (exchange) ($M^{-1}\,s^{-1}$)	pK
9-Phenylfluorene	173×10^{-4}	18.5
Indene	50×10^{-4}	19.9
3,4-Benzfluorene	90.3×10^{-4}	
1,2-Benzfluorene	31.9×10^{-4}	20.3
2,3-Benzfluorene	2.15×10^{-4}	
Fluorene	3.95×10^{-4}	22.7

8. The acidity of various substituted acetophenones has been measured in DMSO. Would you expect the ρ value for a Hammett correlation to be positive or negative? Would you expect the best correlation with σ, σ^+, or σ^-? Justify your prediction, considering each of the σ values explicitly. The data are given below. Check your prediction by plotting the pK versus σ, σ^+, and σ^-.

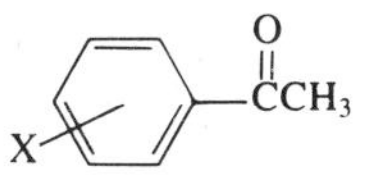

$$X{-}C_6H_4{-}\overset{O}{\overset{\|}{C}}CH_3$$

X	pK_{DMSO}	X	pK_{DMSO}	X	pK_{DMSO}
p-$(CH_3)_2N$	27.48	H	24.70	*m*-Cl	23.18
p-CH_3	25.70	*p*-F	24.45	*m*-Br	23.19
m-$(CH_3)_2N$	25.19	*m*-CH_3O	24.52	*m*-CF_3	22.76
p-CH_3	25.19	*p*-Br	23.81	*p*-CF_3	22.69
m-CH_3	24.95	*p*-Cl	23.78	*p*-CN	22.04
p-Ph	24.51	*m*-F	23.45		

9. Suggest mechanisms for each of the following reactions:

(a) $$Ph_3C\overset{O}{\overset{\|}{C}}Ph \xrightarrow[2)\ H_3O^+]{1)\ LiAlH_4,\ pyridine} Ph_3CH + PhCH_2OH$$

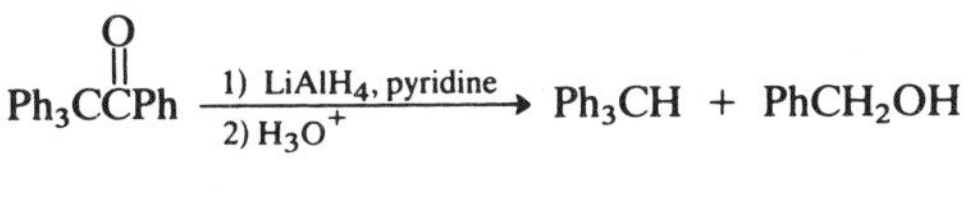

(b) 1,4-dioxane ring bearing H, HO; H, Ph; H, Ph; H, OH (2,5-dihydroxy-3,6-diphenyl-1,4-dioxane)

$$\xrightarrow[heat]{pyridine} Ph\overset{O}{\overset{\|}{C}}CH_2OH + Ph\underset{OH}{\underset{|}{C}}HCH{=}O$$

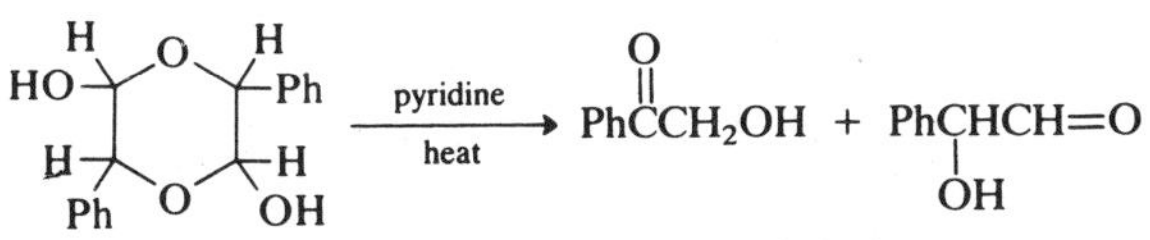

(c) Treatment of either diastereomer shown below with 0.025 M Na^+ $^-CH_2SOCH_3$ in DMSO produces the same equilibrium mixture of 72% *trans* and 28% *cis*.

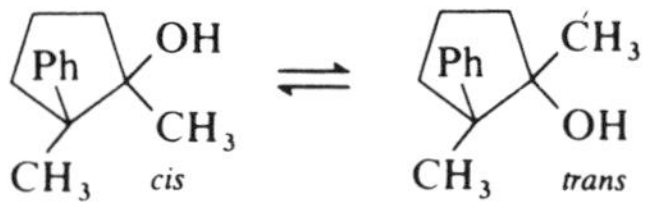

(d) The hemiacetal of aflatoxin B_1 racemizes readily in basic solution.

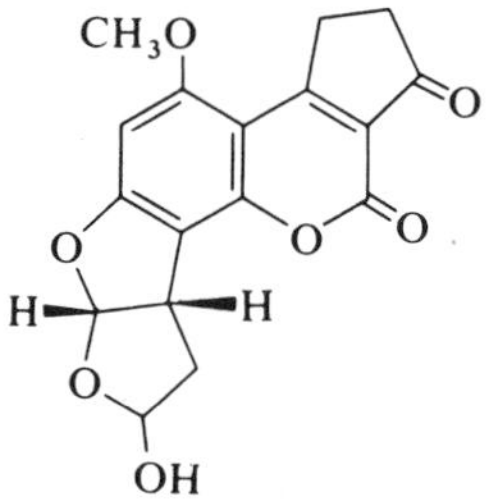

(e)

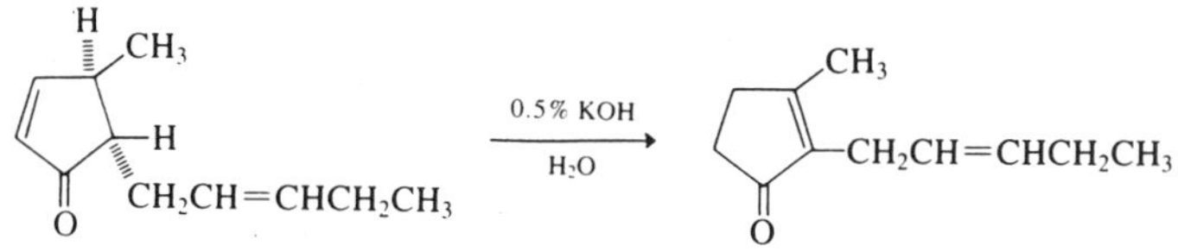

10. Meldrum's acid, pK 7.3, is exceptionally acidic in comparison to an acyclic analog such as dimethyl malonate, pK 15.9. For comparison, 5,5-dimethyl-1,3-cyclohexanedione is only moderately more acidic than 2,4-pentanedione (11.2 versus 13.3). The pK values are those for DMSO solution. It is also found that the enhanced acidity of Meldrum's acid derivatives decreases as the ring size is increased. Analyze factors that could contribute to the enhanced acidity of Meldrum's acid.

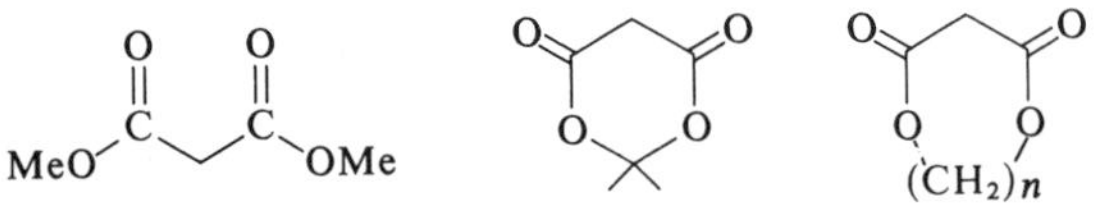

$n = 5$, 13.0
$n = 8$, 15.2
$n = 12$, 15.1

11. In some solvents, it can be shown that the equilibrium k_1/k_{-1} is fast relative to the process governed by k_2:

$$R_3C{-}H + B^- \underset{k_{-1}}{\overset{k_1}{\rightleftharpoons}} [R_3C^- + H{-}B]$$

$$[R_3C^- + H{-}B] + D{-}B \xrightarrow{k_2} [R_3C^- + D{-}B] + H{-}B$$

$$[R_3C^- + D{-}B] \xrightarrow{k_3} R_3CD + B^-$$

This is referred to as *internal return*; i.e., the base returns the proton to the carbanion faster than exchange of the protonated base with other solvent molecules occurs. If internal return is important under a given set of conditions, how would the correlation between kinetics of exchange and equilibrium acidity be affected? How could the occurrence of internal return be detected experimentally?

12. The pK_a values of the conjugate acids of several enamines derived from isobutyraldehyde have been reported. Rationalize the observed variation with the structure of the amino constituent.

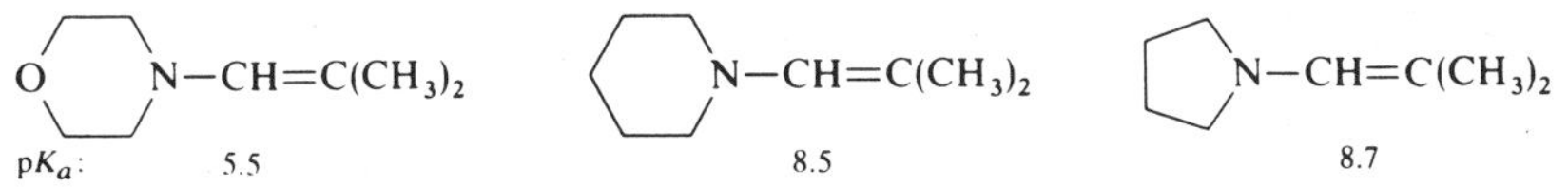

	Morpholine: O N—CH=C(CH$_3$)$_2$	Piperidine: N—CH=C(CH$_3$)$_2$	Pyrrolidine: N—CH=C(CH$_3$)$_2$
pK_a:	5.5	8.5	8.7

13. Metal ions, in particular Zn^{2+}, Ni^{2+}, and Cu^{2+}, enhance the rate of general base-catalyzed enolization of 2-acetylpyridine by several orders of magnitude. Account for this effect.

14. The C-2 equatorial proton is selectively removed when 1,3-dithianes are deprotonated. Furthermore, if the resulting carbanion is protonated, there is a strong preference for equatorial protonation, even if this leads to a less stable axial orientation for the 2-substituent.

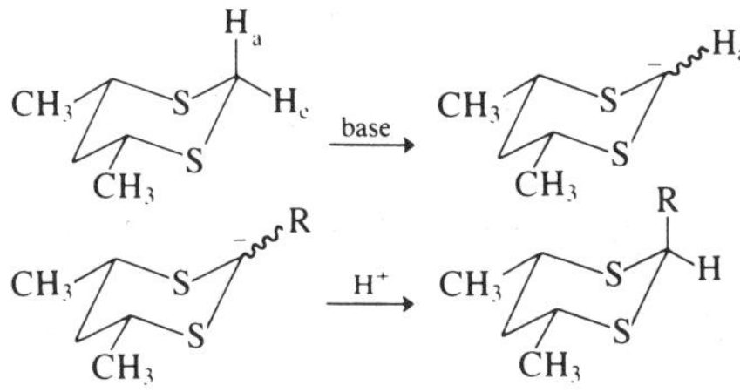

Discuss the relevance of these observations to the structure of sulfur-stabilized carbanions and rationalize your conclusion about the structure of the carbanions in MO terms.

15. (a) It is found that when 2-methyl-2-butene is converted to a dianion it first gives the 2-methylbutadiene dianion **A** but this is converted to the more stable anion **B**, which can be referred to as "methyltrimethylene-methane dianion."

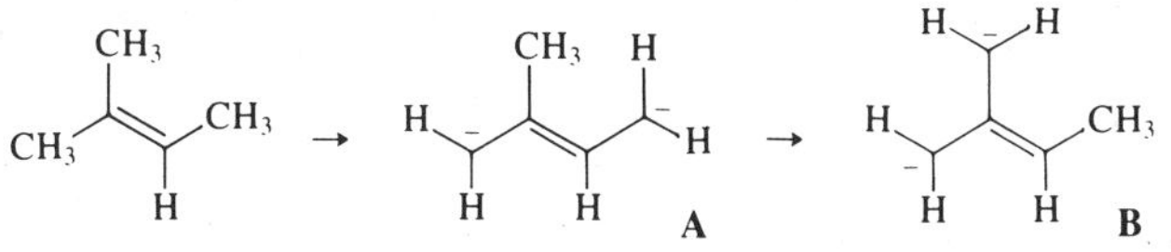

Does simple Hückel MO theory offer an explanation for this result?

(b) The Hückel MO diagrams for several conceivable dianions which might be formed by double deprotonation of 2-methyl-1,5-hexadiene are given. On the basis of these diagrams, which of the dianions would be expected to be the most stable species?

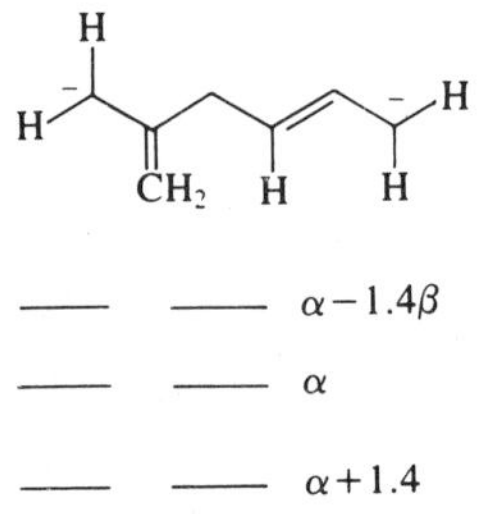

α − 1.4β
α
α + 1.4

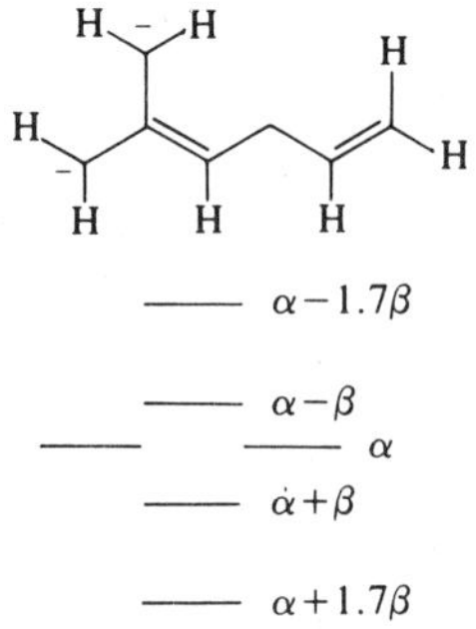

α − 1.7β
α − β
α
α + β
α + 1.7β

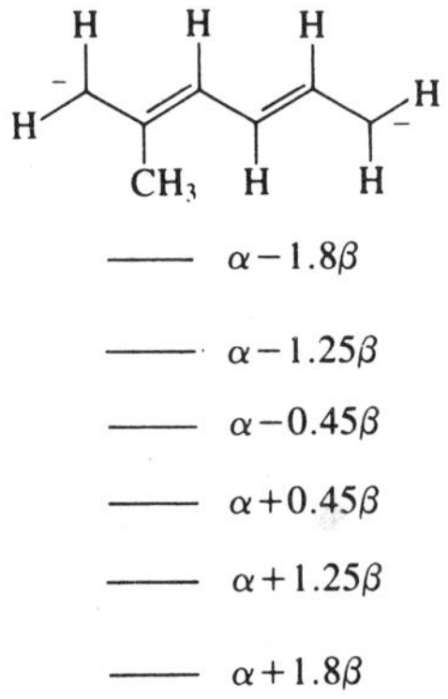

α − 1.8β
α − 1.25β
α − 0.45β
α + 0.45β
α + 1.25β
α + 1.8β

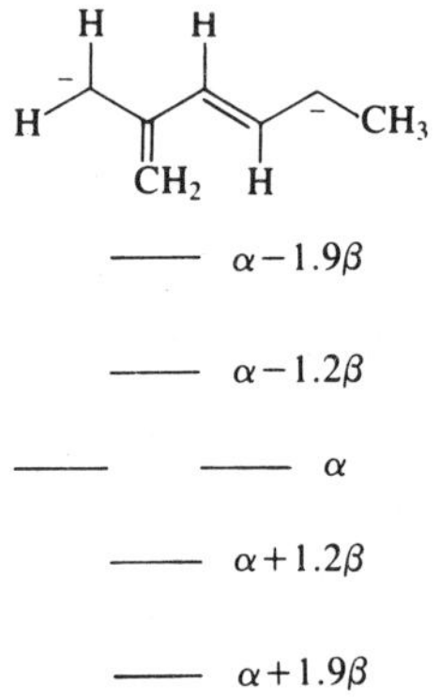

α − 1.9β
α − 1.2β
α
α + 1.2β
α + 1.9β

16. Which of the two plausible structures given for methylketene dimer is more consistent with its observed pK_a of 2.8? Why?

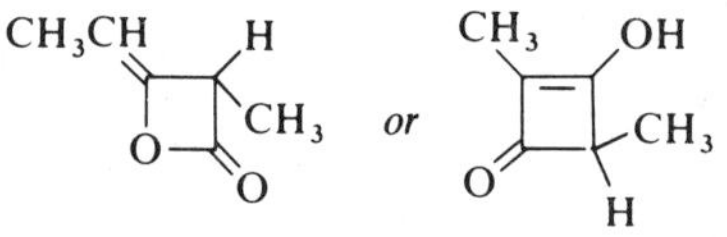

17. Predict the products of each of the following reactions:

(a) + Br_2 $\xrightarrow{\text{ether}}$

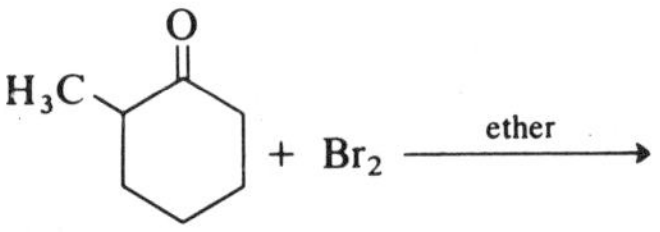

(b) + Br_2 $\xrightarrow[H_2O]{NaO_2CCH_3}$

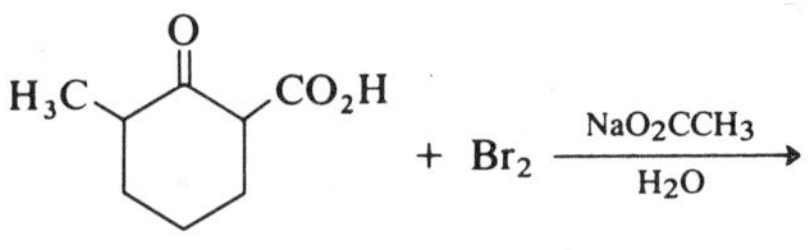

(c) + Cl_2 $\xrightarrow[CH_3CO_2H]{NaO_2CCH_3}$

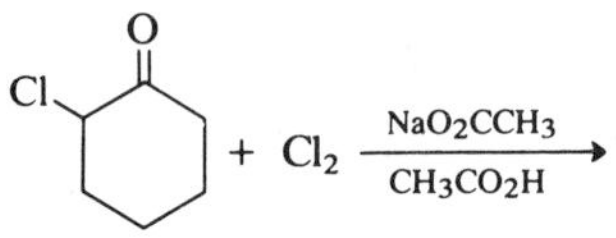

(d) 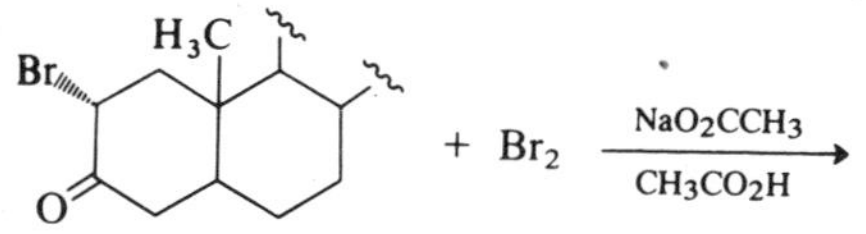 [structure] + Br_2 $\xrightarrow[CH_3CO_2H]{NaO_2CCH_3}$

(e) 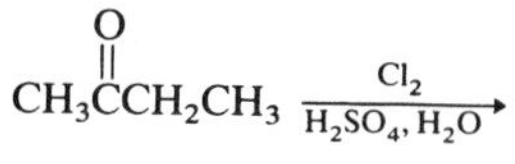 $CH_3CCH_2CH_3$ (C=O) $\xrightarrow[H_2SO_4,\ H_2O]{Cl_2}$

18. Predict the structure and stereochemistry of the products that would be obtained under the specified reaction conditions. Explain the basis of your prediction.

(a) 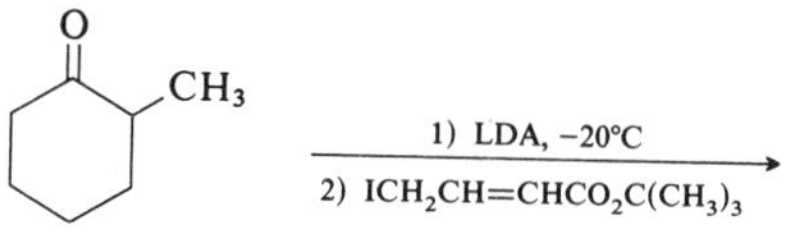 2-methylcyclohexanone $\xrightarrow[2)\ ICH_2CH{=}CHCO_2C(CH_3)_3]{1)\ LDA,\ -20°C}$

(b) [structure with CH_3, $C(CH_3)_3$] $\xrightarrow[2)\ CH_3I]{1)\ LDA,\ -40°C}$

(c) [structure: O^-Li^+, CH_3, $(CH_3)_3C$] $\xrightarrow{CH_3I}$

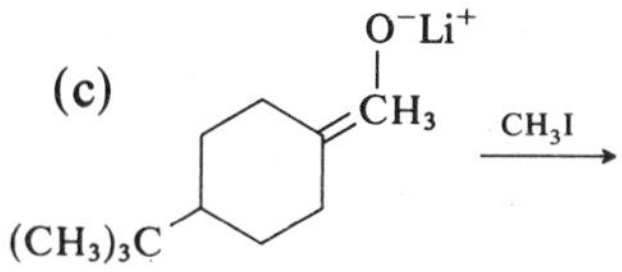

(d) 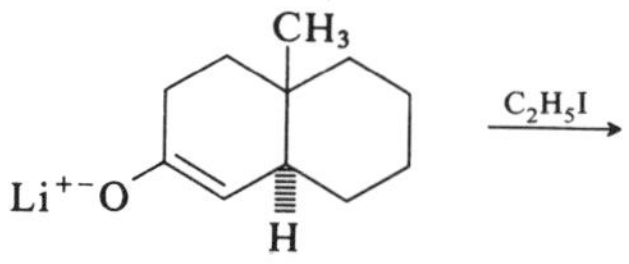 [structure: CH_3, $Li^{+-}O$, H] $\xrightarrow{C_2H_5I}$

(e) 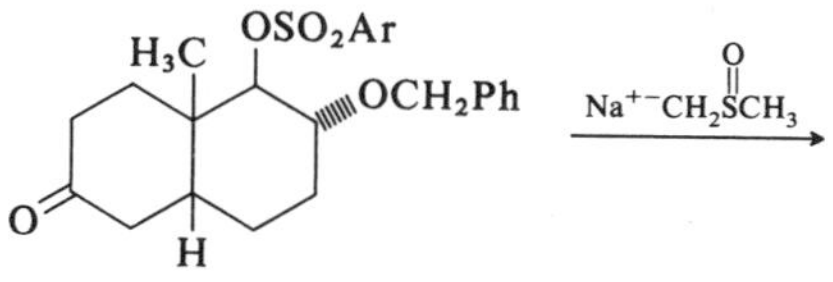 [structure: H_3C, OSO_2Ar, OCH_2Ph, O, H] $\xrightarrow{Na^{+-}CH_2SCH_3}$ (S=O)

(f) 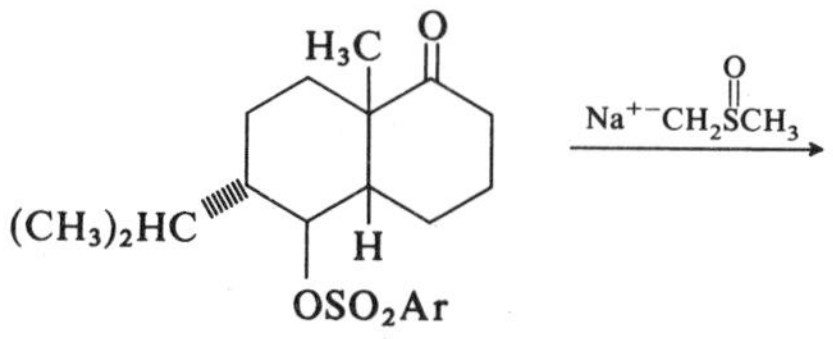 [structure: H_3C, O, $(CH_3)_2HC$, H, OSO_2Ar] $\xrightarrow{Na^{+-}CH_2SCH_3}$ (S=O)

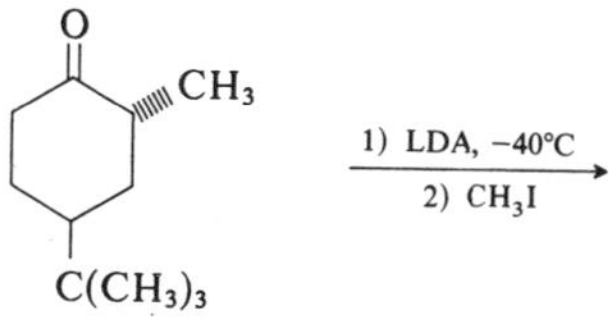

19. The alkylation of 3-methyl-2-cyclohexenone with several dibromides led to the products shown below. Discuss the course of each reaction and suggest an explantion for the dependence of the product structure on the identity of the dihalide.

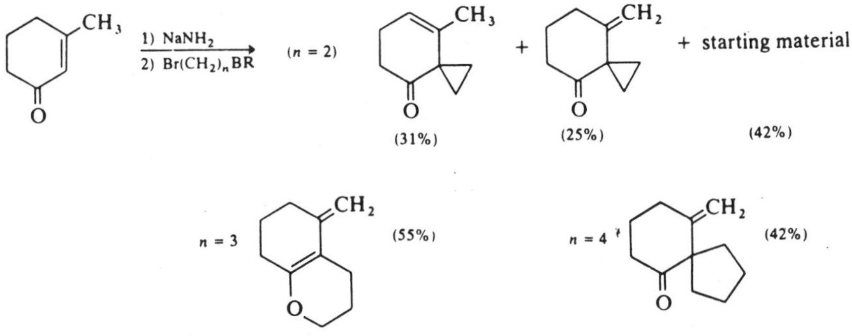

1) $NaNH_2$ 2) $Br(CH_2)_nBR$

(n = 2): (31%) + (25%) + starting material (42%)

n = 3 (55%)

n = 4 (42%)

20. The stereochemistry of base-catalyzed deuterium exchange has been examined for **A** where X = CN and X = $\overset{\underset{\|}{O}}{C}$Ph:

A → **B**

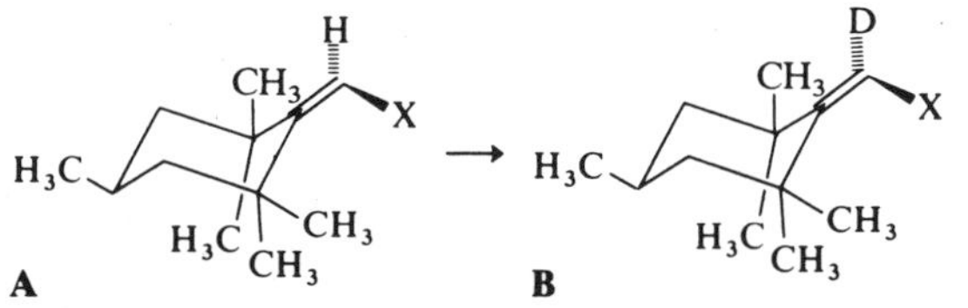

When X = CN, the isotopic exchange occurs with 99% retention of configuration, but when X = $\overset{\underset{\|}{O}}{C}$Ph, only about 30% net retention is observed. Explain.

21. The distribution of α-bromoketones formed in the reaction of acetylcyclopentane with bromine was studied as a function of deuterium substitution. On the basis of the data given below, calculate the primary kinetic isotope effect (k_H/k_D) for enolization of acetylcyclopentane.

Br_2, CCl_4

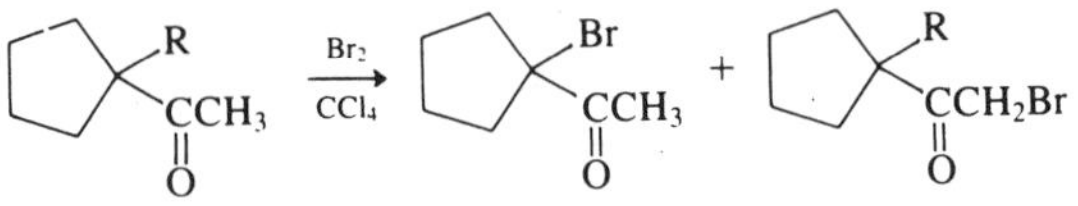

R = H	94%	6%
R = D	80%	20%

8

Reactions of Carbonyl Compounds

The carbonyl group is one of the most prevalent of the functional groups. The carbonyl group is involved in many synthetically important reactions. Reactions involving carbonyl groups are also exceptionally important in biological processes. Most of the reactions of aldehydes, ketones, esters, amides, and the other carboxylic acid derivatives directly involve the carbonyl group. In Chapter 7, the role of the carbonyl group in stabilizing carbanions was discussed. In this chapter, the primary topic for discussion will be the characteristic mechanistic patterns of reactions at carbonyl centers. The first two chapters of Part B deal mainly with the use of carbonyl compounds in synthesis to form carbon–carbon bonds.

In many reactions at carbonyl groups, a key step is addition of a nucleophile, generating a tetracoordinate carbon atom. The overall course of the reaction is then determined by the fate of this *tetrahedral intermediate.*

$$\mathrm{RC(=O)R'} + \mathbf{Nu}: \longrightarrow \mathrm{RC(O^-)(^+\mathbf{Nu})R'}$$

The reactions of the specific classes of carbonyl compounds are related by the decisive importance of tetrahedral intermediates, and differences in reactivity can often be traced to structural features present in the intermediates.

8.1. Hydration and Addition of Alcohols to Aldehydes and Ketones

For most simple carbonyl compounds, the equilibrium constant for addition of water to the carbonyl group is unfavorable.

Table 8.1. Hydration of Carbonyl Compounds

$$K = \frac{[\text{hydrate}]}{[\text{carbonyl}]}^{a}$$

Carbonyl compound	K (in water, 25°C)
CH_2O	2.28×10^{3} [b]
CH_3CHO	1.06[b]
CH_3CH_2CHO	0.85[b]
$(CH_3)_2CHCHO$	0.61[b]
$(CH_3)_3CCHO$	0.23[b]
CF_3CHO	2.9×10^{4} [b]
C_6H_5CHO	8×10^{-3} [c]
CH_3COCH_3	1.4×10^{-3} [b]
FCH_2COCH_3	0.11[c]
$ClCH_2COCH_3$	0.11[b]
CF_3COCH_3	35[b]
CF_3COCF_3	1.2×10^{6} [b]
$C_6H_5COCH_3$	9.3×10^{-6} [c]
$C_6H_5COCF_3$	78[b]
$CH_3COCOCH_3$	0.6[d]
$CH_3COCO_2CH_3$	0.8[d]

a. $K = K_{eq}[H_2O] = 55.5\ K_{eq}$
b. J. P. Guthrie, *Can. J. Chem.* **53**, 898 (1975).
c. J. P. Guthrie, *Can. J. Chem.* **56**, 962 (1978).
d. T. J. Burkey and R. C. Fahey, *J. Am. Chem. Soc.* **105**, 868 (1983).

$$RC(=O)R' + H_2O \rightleftharpoons RC(OH)_2R' \qquad K < 1$$

Exceptions are formaldehyde, which is nearly completely hydrated in aqueous solution, and aldehydes and ketones with highly electronegative groups such as trichloroacetaldehyde and hexafluoroacetone. The data given in Table 8.1 illustrate that the equilibrium constant for hydration decreases with increasing alkyl substitution.

Although the equilibrium constant for hydration is unfavorable, the equilibrium between an aldehyde or ketone and its hydrate is established rapidly and can be detected by isotopic exchange, using water labeled with ^{18}O, for example.

$$R_2C{=}O + H_2O^* \rightleftharpoons R_2C(^*OH)OH \rightleftharpoons R_2C{=}O^* + H_2O$$

The hydration reaction has been extensively studied because it is the mechanistic prototype for many reactions at carbonyl centers that involve more complex molecules.[1] For acetaldehyde, the half-life of the exchange reaction is on the order of one minute under neutral conditions, but the reaction if considerably faster in

1. R. P. Bell, *Adv. Phys. Org. Chem.* **4**, 1 (1966); W. P. Jencks, *Chem. Rev.* **72**, 705 (1972).

acidic or basic media. The second-order rate constant for acid-catalyzed hydration of acetaldehyde is on the order of 500 $M^{-1}\,s^{-1}$.[2] Acid catalysis involves either protonation or hydrogen bonding at the carbonyl oxygen. Both specific acid catalysis and general acid catalysis can be observed.[3] (Review Section 4.8 for the discussion of specific and general acid catalysis.)

Specific acid-catalyzed hydration

$$RCH{=}O + H^+ \rightleftharpoons RCH{=}\overset{+}{O}H$$

$$RCH{=}\overset{+}{O}H + H_2O \overset{\text{slow}}{\rightleftharpoons} \underset{{}^{+}OH_2}{RCHOH} \rightleftharpoons RCH(OH)_2 + H^+$$

General acid-catalyzed hydration

$$RCH{=}O + HA \rightleftharpoons RCH{=}O\cdots HA$$

$$RCH{=}O\cdots HA + H_2O \overset{\text{slow}}{\rightleftharpoons} \underset{{}^{+}OH_2}{RCHOH} + A^- \rightleftharpoons RCH(OH)_2 + HA$$

Basic catalysts function by deprotonating water to give the more nucleophilic hydroxide ion.

Base-catalyzed hydration

$$B^- + H_2O + RCH{=}O \rightleftharpoons \underset{OH}{RCHO^-} + BH$$

$$\underset{OH}{RCHO^-} + H_2O \rightleftharpoons RCH(OH)_2 + {}^{-}OH$$

$$BH + {}^{-}OH \rightleftharpoons B^- + H_2O$$

Molecular orbital (STO-3G) calculations on the gas phase hydrogen reaction of formaldehyde identify a concerted process involving two water molecules as a low-energy mechanism for hydration.

$$H_2C{=}O \cdots (H_2O)_2 \longrightarrow H_2C(OH)_2 + H{-}O{-}H$$

The calculated activation energy for this process in water is 16 kcal/mol, which is in good agreement with the experimentally observed value.[4]

Aldehydes and ketones undergo reversible addition reactions with alcohols. The product of addition of one mole of alcohol to an aldehyde or ketone is referred

2. P. Greenzaid, Z. Luz, and D. Samuel, *J. Am. Chem. Soc.* **89**, 756 (1967).
3. L. H. Funderburk, L. Aldwin, and W. P. Jencks, *J. Am. Chem. Soc.* **100**, 5444 (1978); R. A. McClelland and M. Coe, *J. Am. Chem. Soc.* **105**, 2718 (1983).
4. I. H. Williams, D. Spangler, D. A. Femec, G. M. Maggiora, and R. L. Schowen, *J. Am. Chem. Soc.* **105**, 31 (1983).

to as a *hemiacetal* or *hemiketal*, respectively. Dehydration followed by addition of a second molecule of alcohol gives an *acetal* or *ketal*. This second phase of the process can be catalyzed only by acids, since a necessary step is elimination of hydroxide (as water) from the tetrahedral intermediate. There is no low-energy mechanism for base assistance of this elimination step. Since there is no mechanism for basic catalysis, acetals and ketals are stable toward hydrolysis in alkaline aqueous solution but are hydrolyzed rapidly in acidic solution.

$$R_2C{=}O + R'OH \rightleftharpoons R_2\overset{OH}{\overset{|}{C}}OR' \quad \textit{hemiketal}$$

$$R_2\overset{OH}{\overset{|}{C}}OR' + H^+ \rightleftharpoons R_2\overset{^+OH_2}{\overset{|}{C}}OR'$$

$$R_2\overset{^+OH_2}{\overset{|}{C}}OR' \rightleftharpoons R_2C{=}\overset{+}{O}R' + H_2O$$

$$R_2C{=}\overset{+}{O}R' + R'OH \rightleftharpoons R_2\underset{H\overset{+}{O}R'}{\underset{|}{C}}OR' \rightleftharpoons R_2C(OR')_2 + H^+ \quad \textit{ketal}$$

The equilibrium constants for addition of alcohols to carbonyl compounds to give hemiacetals or hemiketals show the same response to structural features as for the hydration reaction. Equilibrium constants for addition of methanol to acetaldehyde in both water and chloroform solution are near 0.8 M^{-1}. The comparable value for addition of water is about 0.02 M^{-1}.[5] The overall equilibrium constant for formation of the dimethyl acetal of acetaldehyde is 1.58 M^{-1}. Because the position of the equilibrium does not strongly favor product, the formation of acetals and ketals must be carried out in such a way as to drive the reaction to completion. One approach is to use a dehydrating reagent or azeotropic distillation so that the water that is released is irreversibly removed from the system.

Because of the unfavorable equilibrium constant in aqueous solution and the relative facility of the hydrolysis, acetals and ketals are rapidly converted to aldehydes and ketones in acidic aqueous solution.

$$R_2C(OR')_2 + H_2O \xrightarrow{H^+} R_2C{=}O + 2R'OH$$

The mechanism of this hydrolysis reaction has been studied in great detail.[6] The mechanism is the reverse of that for acetal or ketal formation.

$$R_2C(OR')_2 + H^+ \rightleftharpoons R_2C\begin{matrix} \diagup OR' \\ \diagdown \underset{^+H}{O}R' \end{matrix} \rightleftharpoons R_2C{=}\overset{+}{O}R' + R'OH$$

5. R. Bone, P. Cullis, and R. Wolfenden, *J. Am. Chem. Soc.* **105**, 1339 (1983).
6. E. H. Cordes and H. G. Bull, *Chem. Rev.* **74**, 581 (1974).

$$R_2C{=}\overset{+}{O}R' + H_2O \rightleftharpoons R_2C\begin{matrix}OR' \\ \overset{+}{O}H_2\end{matrix} \rightleftharpoons R_2C\begin{matrix}OR \\ OH\end{matrix} + H^+$$

$$R_2C\begin{matrix}OR' \\ OH\end{matrix} + H^+ \rightleftharpoons R_2C{=}\overset{+}{O}H + R'OH \qquad R_2C{=}\overset{+}{O}H \rightleftharpoons R_2C{=}O + H^+$$

Some of the evidence that has helped to establish the general mechanism is as follows:

1. Isotopic labeling experiments have established that C—O bond rupture occurs between the carbonyl carbon and oxygen; therefore, substitution at the alcohol C—O bond is not involved.
2. For most acetals and ketals, the reaction is *specific-acid-catalyzed.* This is consistent with the existence of a preequilibrium in which the ketal is protonated. The role of the proton is to assist the departure of the alkoxy group by converting it to a better leaving group. In essence, this cleavage step is an S_N1 reaction with the remaining alkoxy group stabilizing the carbocation formed by ionization.
3. Hammett treatments show good correlations with large negative ρ values for the hydrolysis of acetals of aromatic aldehydes. This is consistent with the development of a positive charge at the carbonyl center in the rate-determining step.
4. Solvent isotope effects are usually in the range $k_{D_3O^+}/k_{H_3O^+} = 2\text{–}3$. These values reflect the greater equilibrium acidity of deuterated acids (Section 4.8) and indicate that the initial protonation is a fast preequilibrium.

Acetal and ketal hydrolyses usually exhibit specific acid catalysis, in agreement with a mechanism involving rate-determining cleavage of the conjugate acid of the reactant. General acid catalysis is observed, however, in certain acetals and ketals in which special structural features reduce the energy required for C—O bond cleavage.[7] Thus, hydrolysis of each of the acetals shown in Scheme 8.1 exhibits general acid catalysis, and each acetal has a special structural feature that facilitates C—O bond heterolysis. Easing the energy requirement for C—O bond cleavage permits the proton transfer step to become partially rate-determining, which results in the observation of general acid catalysis.

Three-dimensional potential energy diagrams of the type introduced in connection with the variable E2 transition state theory for elimination reactions can be used to consider structural effects on the reactivity of carbonyl compounds and the tetrahedral intermediates involved in carbonyl group reactions. Many of these reactions involve the formation or breaking of two separate bonds. This is the case in the first stage of acetal hydrolysis, which involves both a proton transfer and breaking of a C—O bond. The overall reaction might take place in several ways.

7. T. H. Fife, *Acc. Chem. Res.* **5**, 264 (1972).

Scheme 8.1. Acetals and Ketals That Exhibit General Acid Catalysis in Hydrolysis

	Acetal or ketal	Special structural feature
1[a]	7,7-Diethoxycycloheptatriene ($C_7H_6(OC_2H_5)_2$)	Very stable carbocation (stabilized by both alkoxy function and aromaticity)
2[b]	2-(4-Nitrophenoxy)tetrahydropyran ($-NO_2$)	Good leaving group
3[c]	2-Ethoxyoxetane ($-OC_2H_5$)	Ring strain relieved in cleavage step
4[d]	$(Ar)_2C(OC_2H_5)_2$	Aryl substituents stabilize carbocation ion
5[e]	$PhCH[OC(CH_3)_3]_2$	Aryl stabilization and relief of steric strain

a. E. Anderson and T. H. Fife, *J. Am. Chem. Soc.* **91**, 7163 (1969).
b. T. H. Fife and L. H. Brod, *J. Am. Chem. Soc.* **92**, 1681 (1970).
c. R. F. Atkinson and T. C. Bruice, *J. Am. Chem. Soc.* **96**, 819 (1974).
d. R. H. DeWolfe, K. M. Ivanetich, and N. F. Perry, *J. Org. Chem.* **34**, 848 (1969).
e. E. Anderson and T. H. Fife, *J. Am. Chem. Soc.* **93**, 1701 (1971).

There are two mechanistic extremes:

1. The proton could be completely transferred and then the departing alcohol molecule would leave to form a carbocation in a distinct second step. This is the usual specific-acid-catalyzed mechanism.
2. The acetal might undergo ionization with formation of an alkoxide ion and a carbocation. In a second step the alkoxide would be protonated. This mechanism is extremely rare if not impossible, because an alkoxide is a poor leaving group.

There could be an intermediate mechanism between these extremes. This is a general acid-catalyzed mechanism in which the proton transfer and the C—O bond rupture occur as a *concerted* process. The concerted process need not be perfectly synchronous; that is, proton transfer might be more complete at the transition state than C—O rupture, or vice versa. These ideas are represented in the three-dimensional energy diagram in Fig. 8.1.

The two paths around the edge of the diagram represent the stepwise processes described as the mechanistic extremes 1 and 2. We know that process 2 represented by path (a) is a high-energy process so that the upper-left corner of the diagram would have a high energy. The lines designated (b) and (c) indicate concerted but nonsynchronous mechanisms in which there is both partial proton transfer and partial C—O bond rupture at the transition state. In path (b), C—O cleavage is more complete than proton transfer at the transition state, while the reverse is true for path (c). Both these paths represent concerted, general-acid-catalyzed processes.

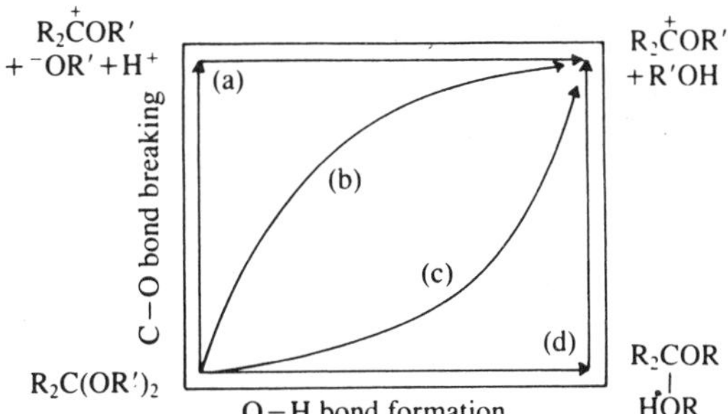

Fig. 8.1. Representation of transition states for the first stage of acetal hydrolysis. (a) Initial C—O bond breaking; (b) concerted mechanism with C—O bond breaking leading O—H bond formation; (c) concerted mechanism with proton transfer leading C—O bond breaking; (d) initial proton transfer.

Path (d) represents the specific-acid-catalyzed process in which proton transfer precedes C—O cleavage.

If it is possible to estimate or calculate the energy of the reacting system at various stages, the energy dimension can be added as in Fig. 8.2. The actual mechanism will then be the process which proceeds over the lowest-energy barrier. The energy dimension can be shown as contours. The diagram in Fig. 8.2 shows the initial ionization to an alkoxide and carbocation as very high in energy. The stepwise path of protonation followed by ionization is shown with small barriers with the protonated ketal as an intermediate. The lowest-energy path is shown as a concerted process represented by the dashed line. The transition state, which lies at the highest-energy point on this line, would exhibit more complete proton transfer than C—O cleavage.

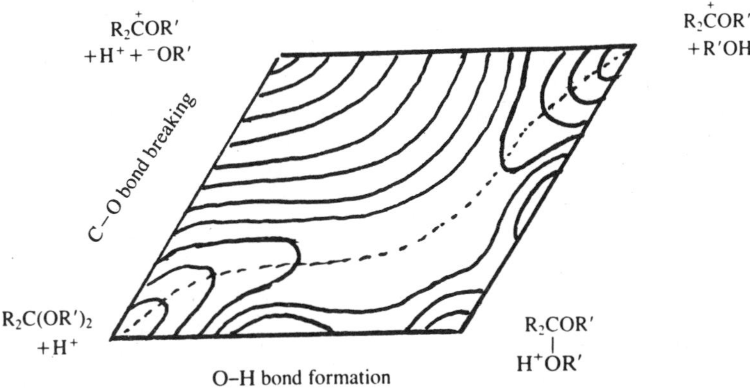

Fig. 8.2. Contour plot showing a favored concerted mechanism for the first step in acetal hydrolysis, in which proton transfer is more complete in the transition state than C—O bond breaking.

Structural effects can now be discussed by asking how they will affect the position of the transition state on the potential energy surface. The stepwise path via the protonated acetal should be followed in the case of alcohols that are poor leaving groups. As the alcohol becomes more acidic and its conjugate base becomes a better leaving group, the transition state will be shifted to a point where C—O bond breaking has begun before proton transfer is complete. This would mean that the mechanism has become concerted, although the transition state would still have much of the character of a carbocation. Three-dimensional reaction energy diagrams can be used to describe how structural changes can affect the nature of the transition state. Just as the two-dimensional diagrams can give meaning to such phrases as an "early" or a "late" transition state, the three-dimensional diagrams are illustrative of statements such as "C—O cleavage is more advanced than proton transfer."

Consideration of the types of acetals shown in Scheme 8.1, which exhibit general acid catalysis, indicates why the concerted mechanism operates in these molecules. The developing aromatic character of the cation formed in the case of entry 1 will lower the energy requirement for C—O bond rupture. The bond can begin to break before protonation is complete. Entry 2 represents a case where a good leaving group (a stable phenolate anion) reduces the energy requirement for C—O bond cleavage. In entry 3 the four-membered ring is broken in the reaction. Cleavage in this case is facilitated by release of strain energy. Entries 4 and 5 are similar to entry 1 because the aryl groups provide stabilization for developing cationic character.

The second step in acetal and ketal hydrolysis is conversion of the hemiacetal or hemiketal to the carbonyl compound. The mechanism of this step is similar to that of the first step. Usually, the second step is faster than the initial one.[8] Hammett σ–ρ plots and solvent isotope effects both indicate that the transition state has less cationic character than is the case for the first step. These features of the mechanism suggest that a concerted removal of the proton at the hydroxyl group occurs as the alcohol is eliminated.

$$\text{Ph—}\overset{\delta+}{\text{C}}\text{(H)(O---H---}\overset{\delta+}{\text{O}}\text{H}_2\text{)(OC}_2\text{H}_5\text{---H---}\overset{\delta+}{\text{O}}\text{H}_2\text{)} \longrightarrow \text{PhCH=O} \quad \text{H}_3\text{O}^+ \quad \text{HOC}_2\text{H}_5 \quad \text{H}_2\text{O}$$

This would disperse the positive charge over several atoms and diminish the sensitivity of the reaction to substituent effects. The ρ values that are observed are consistent with this interpretation. While ρ is -3.25 for acetal hydrolysis, it is only -1.9 for hemiacetal hydrolysis.[9]

In contrast to acetals, which are base stable, hemiacetals undergo base-catalyzed hydrolysis. In the alkaline pH range, the mechanism shifts toward a base-catalyzed

8. Y. Chiang and A. J. Kresge, *J. Org. Chem.* **50**, 5038 (1985).
9. T. J. Przystas and T. H. Fife, *J. Am. Chem. Soc.* **103**, 4884 (1981).

elimination.

$$\mathrm{Ph{-}\underset{OC_2H_5}{\overset{OH}{C}}{-}H + {}^{-}OH \rightleftharpoons Ph{-}\underset{OC_2H_5}{\overset{O^-}{C}}{-}H \rightarrow PhCH{=}O + C_2H_5O^-}$$

There are two competing substituent effects on this reaction. Electron-attracting groups favor the deprotonation but disfavor the elimination step. The observed substituent effects are small, and under some conditions the Hammett plot is nonlinear.[10]

8.2. Addition–Elimination Reactions of Aldehydes and Ketones

The mechanistic pattern established by study of hydration and alcohol addition reactions of ketones and aldehydes is followed in a number of other reactions of carbonyl compounds. Reactions at carbonyl centers usually involve a series of addition and elimination steps proceeding through tetrahedral intermediates. These steps can be either acid-catalyzed or base-catalyzed. The overall result of the reaction is determined by the reactivity of these tetrahedral intermediates.

In general terms, there are three possible mechanisms for addition of a nucleophile and a proton to give a tetrahedral intermediate in a carbonyl addition reaction.

(a) Protonation followed by nucleophilic attack on the protonated carbonyl group:

$$\mathrm{H^+ + \;>C{=}O \rightarrow \;>C{=}\overset{+}{O}{-}H}$$

$$\mathrm{\mathbf{Nu}{:}^- + \;>C{=}\overset{+}{O}{-}H \rightarrow \mathbf{Nu}{-}\overset{|}{\underset{|}{C}}{-}O{-}H}$$

(b) Nucleophilic addition at the carbonyl group followed by protonation:

$$\mathrm{\mathbf{Nu}{:}^- + \;>C{=}O \rightarrow \mathbf{Nu}{-}\overset{|}{\underset{|}{C}}{-}O^-}$$

$$\mathrm{\mathbf{Nu}{-}\overset{|}{\underset{|}{C}}{-}O^- + H^+ \rightarrow \mathbf{Nu}{-}\overset{|}{\underset{|}{C}}{-}O{-}H}$$

(c) Concerted proton transfer and nucleophilic attack:

$$\mathrm{\mathbf{Nu}{:}^- \;>C{=}O \;\; H^+ \rightarrow \mathbf{Nu}{-}\overset{|}{\underset{|}{C}}{-}O{-}H}$$

10. R. L. Finley, D. G. Kubler, and R. A. McClelland, *J. Org. Chem.* **45**, 644 (1980).

There are examples of each of these mechanisms, and a three-dimensional potential energy diagram can provide a useful general framework within which to consider the specific addition reactions that we will encounter. The breakdown of a tetrahedral intermediate involves the same processes but operates in the opposite direction, so the principles that are developed will apply equally well to the reactions of the tetrahedral intermediates. Let us examine the three general mechanistic cases in relation to the energy diagram in Fig. 8.3.

Case (a) should be favored for weak nucleophiles. The protonated carbonyl compound will be much more highly reactive toward such nucleophiles. Case (b) should be favored for strong nucleophiles. Such nucleophiles will generally be more basic than carbonyl groups and would be protonated in preference to the carbonyl group. In such systems, proton donors would diminish the overall reaction rate by decreasing the amount of nucleophile that is available for reaction. The reactions of primary amines are generally not catalyzed by acids because the protonated amines are not nucleophilic toward the carbonyl group and the carbonyl group does not compete with the amine as a site for protonation. The concerted mechanism (c) is observed for less basic nucleophiles. The simultaneous transfer of the proton at the carbonyl oxygen assists in achieving addition by species that are not sufficiently nucleophilic to react by mechanism (b). The general trend then is that the weaker and less basic the nucleophile, the more important is partial or complete protonation of the carbonyl group. If we consider the reverse process, the same general relationships will hold. The good leaving groups can be expected to follow path (a), poor leaving groups will follow path (b), and intermediate cases are likely to react by the concerted mechanism.

Certain nucleophilic species add to carbonyl groups to give tetrahedral intermediates that are unstable and break down to form a new double bond. An important group of such reactions are those with compounds containing primary amino groups. Scheme 8.2 lists some of the more familiar classes of such reactions. In general,

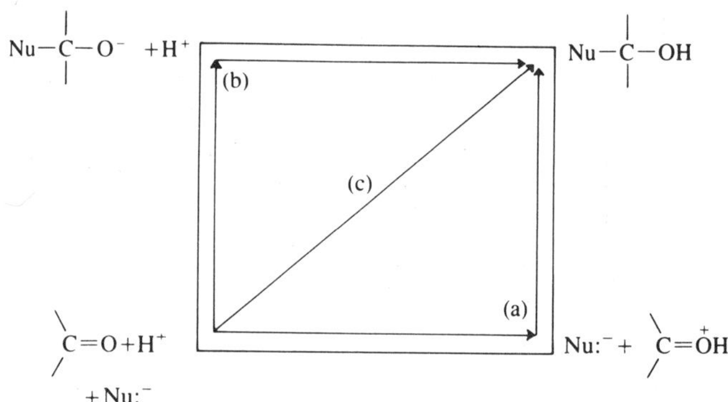

Fig. 8.3. Three-dimensional potential energy diagram for addition of a proton and nucleophile to a carbonyl group. (a) Proton transfer complete before nucleophilic addition begins; (b) nucleophilic addition complete before proton transfer begins; (c) concerted proton transfer and nucleophilic addition.

Scheme 8.2. Some Addition–Elimination Reactions of Aldehydes and Ketones

$$R_2C{=}O + R'NH_2 \rightleftharpoons R_2C{=}NR'$$

imine (also often called Schiff base, especially when the amine is an aniline derivative)

$$R_2C{=}O + HONH_2 \rightleftharpoons R_2C{=}NOH$$

oxime

$$R_2C{=}O + NH_2NHR' \rightleftharpoons R_2C{=}NNHR'$$

hydrazone

$$R_2C{=}O + NH_2NH\overset{\overset{O}{\|}}{C}NH_2 \rightleftharpoons R_2C{=}NNH\overset{\overset{O}{\|}}{C}NH_2$$

semicarbazone

these reactions are reversible, and mechanistic information can be obtained by study of either the forward or the reverse process.

The hydrolysis of simple imines occurs readily in aqueous acid and has been studied in great detail by kinetic methods. The precise mechanism is a function of the reactant structure and the pH of the solution. The overall mechanism consists of an addition of water to the C=N bond, followed by expulsion of the amine from a tetrahedral intermediate.[11]

$$RCH{=}NR' + H^+ \rightleftharpoons RCH{=}\overset{+}{N}HR'$$

$$RCH{=}\overset{+}{N}HR' + H_2O \rightleftharpoons \underset{^{+}OH_2}{RCHNHR'} \rightleftharpoons \underset{OH}{RCH\overset{+}{N}H_2R'}$$

or

$$RCH{=}\overset{+}{N}HR' + {}^-OH \rightleftharpoons \underset{OH}{RCHNHR'}$$

$$\underset{OH}{RCHNHR'} + H^+ \rightleftharpoons \underset{OH}{RCH\overset{+}{N}H_2R'}$$

$${}^-OH + \underset{OH}{RCH\overset{+}{N}H_2R'} \rightleftharpoons H_2O + \underset{{}^-O}{RCH\overset{+}{N}H_2R'} \rightleftharpoons \underset{\|\atop O}{RCH} + H_2NR'$$

The relative rates of the various steps are a function of the pH of the solution and the basicity of the imine. In the alkaline range, the rate-determining step is usually nucleophilic attack by hydroxide ion on the protonated C=N bond. At intermediate pH values, water replaces hydroxide as the dominant nucleophile. In acidic solution, the rate-determining step becomes the breakdown of the tetrahedral

11. J. Hine, J. C. Craig, Jr., J. G. Underwood II, and F. A. Via, *J. Am. Chem. Soc.* **92**, 5194 (1970); E. H. Cordes and W. P. Jencks, *J. Am. Chem. Soc.* **85**, 2843 (1963).

intermediate. A mechanism of this sort, in which the overall rate is sensitive to pH, can be usefully studied by constructing a pH–rate profile, which is a plot of the observed rate constants versus pH. Figure 8.4 shows the pH–rate profiles for hydrolysis of a series of imines derived from substituted aromatic aldehydes and *t*-butylamine. The form of pH–rate profiles can be predicted on the basis of the detailed mechanism of the reaction. The value of the observed rate can be calculated quantitatively as a function of pH if a sufficient number of the individual rate constants and the acid dissociation constants of the participating species are known. Agreement between the calculated and observed pH–rate profiles can serve as a sensitive test of the adequacy of the postulated mechanism. Alternatively, one may begin with the experimental pH–rate profile and deduce details of the mechanism from it.

Complete understanding of the shape of the curves in Fig. 8.4 requires a kinetic expression somewhat more complicated than we wish to deal with here. The nature of the extremeties of the curves can be understood, however, on the basis of qualitative arguments. The rate decreases with decreasing pH in the acidic region because formation of the zwitterionic tetrahedral intermediate is required for

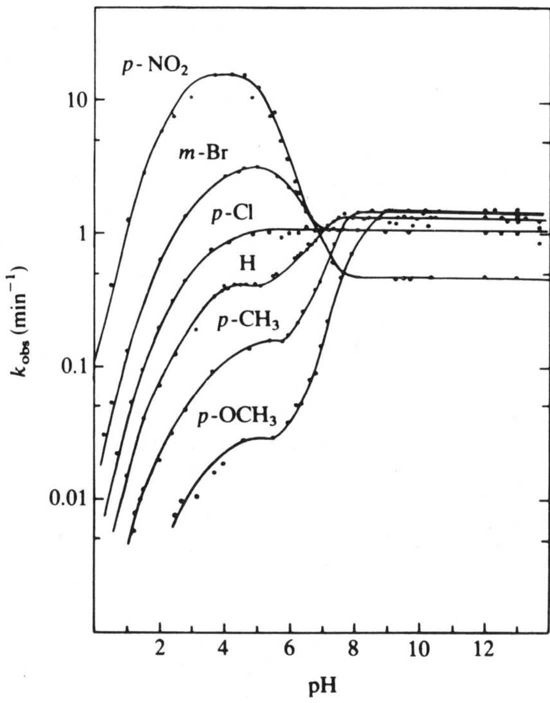

Fig. 8.4. Logarithm of the first-order rate constants for the hydrolysis of substituted benzylidene-1,1-dimethyl-ethylamines as a function of pH. [Reproduced from *J. Am. Chem. Soc.* **85**, 2843 (1963) by permission of the American Chemical Society.]

explusion of the amine. The concentration of the zwitterionic species decreases with increasing acidity, since its concentration is governed by an acid–base equilibrium:

$$K = \frac{[H^+][RCH(O^-)\overset{+}{N}H_2R]}{[RCH(OH)\overset{+}{N}H_2R]}$$

As the hydrogen ion concentration increases, the concentration of the reactive form of the intermediate decreases.

In the alkaline region, the rate is pH independent. In this region, the rate-controlling step is attack of the hydroxide ion on the protonated imine. The concentration of both of these species is pH dependent, but in opposite, compensating, ways. The overall rate is therefore pH independent in the alkaline range. (Work Problem 6 to establish that this is so.)

The formation of imines takes place by a mechanism that is the reverse of the hydrolysis. Preparative procedures often ensure completion of the reaction by removing water as it is formed by azeotropic distillation or by the use of a dehydrating agent.

The other C=N systems included in Scheme 8.2 are more stable to aqueous hydrolysis than are the imines. For many cases, the equilibrium constants for formation are high, even in aqueous solution. The additional stability can be attributed to the participation of the atom adjacent to the nitrogen in delocalized bonding.

$$R_2C{=}N{-}\ddot{O}H \longleftrightarrow R_2\bar{C}{-}N{=}\overset{+}{O}H$$

$$R_2C{=}N{-}\ddot{N}H_2 \longleftrightarrow R_2\bar{C}{-}N{=}\overset{+}{N}H_2$$

The formation of oximes, hydrazones, and related imine derivatives is usually catalyzed by both general acids and general bases. General base catalysis of dehydration of the tetrahedral intermediate involves nitrogen deprotonation concerted with elimination of hydroxide ion.[12]

$$R_2C{=}O + NH_2R' \underset{}{\overset{fast}{\rightleftharpoons}} R_2C(OH)NHR'$$

$$\ddot{B}\; H{-}N(R'){-}C(R)_2{-}OH \xrightarrow{slow} \overset{+}{B}H + R'N{=}CR_2 + \bar{O}H$$

12. W. P. Jencks, *Prog. Phys. Org. Chem.* **2**, 63 (1964); J. M. Sayer, M. Peskin, and W. P. Jencks, *J. Am. Chem. Soc.* **95**, 4277 (1973).

General acid catalysis of the breakdown of the carbinolamine intermediate occurs by assistance of the expulsion of water.

$$\underset{\underset{R'}{|}}{\overset{}{R_2C(\text{–}OH)(\text{–}NH)}} \cdots H\text{–}A \longrightarrow R_2C{=}\overset{+}{N}R(H) + H_2O + A^-$$

As with simple imines, the identity of the rate-limiting step changes with solution pH. As the pH decreases, the rate of the addition decreases because protonation of the amino compound reduces the concentration of the nucleophilic unprotonated form. Thus, while the dehydration step is normally rate-determining in neutral and basic solution, addition becomes rate-determining in acidic solutions.

Secondary amines cannot form imines, and dehydration proceeds to give carbon–carbon double bonds bearing amino substituents (enamines). Enamines were mentioned in Chapter 7 as examples of nucleophilic carbon species, and their synthetic utility is illustrated in Part B, Chapter 1. The equilibrium for the reaction between secondary amines and carbonyl compounds ordinarily lies far to the left in aqueous solution, but the reaction can be driven forward by dehydration methods.

$$R''_2NH + RCH_2\overset{O}{\overset{\|}{C}}R' \rightleftharpoons RCH_2\overset{OH}{\underset{NR''_2}{C}}R' \rightleftharpoons RCH{=}\underset{NR''_2}{C}R' + H_2O$$

The mechanism of hydrolysis of enamines has been studied kinetically over a range of pH. In alkaline solution, rate-determining C-protonation is followed by attack of hydroxide ion on the resulting iminium ion. The carbinolamine intermediate then breaks down as in imine hydrolysis. In the neutral and weakly acidic pH range, water attack on the C-protonated enamine becomes rate-limiting. As in imine hydrolysis, decomposition of the tetrahedral intermediate becomes rate-limiting in strongly acidic solution.[13]

$$RCH{=}\underset{R'}{C}NR''_2 + H^+ \longrightarrow RCH_2\underset{R'}{C}{=}\overset{+}{N}R''_2$$

$$RCH_2\underset{R'}{C}{=}\overset{+}{N}R''_2 + {}^-OH \longrightarrow RCH_2\overset{OH}{\underset{R'}{C}}NR''_2 \longrightarrow RCH_2\overset{O}{\overset{\|}{C}}R' + HNR''_2$$

Certain reactions between carbonyl compounds and nucleophiles are catalyzed by amines. Some of these reactions are of importance for forming carbon–carbon bonds, and these are discussed in Section 2.2 of Part B. The mechanistic principle can be illustrated by considering the catalysis of the reaction between ketones and

13. P. Y. Sollenberger and R. B. Martin, *J. Am. Chem. Soc.* **92**, 4261 (1970); W. Maas, M. J. Janssen, E. J. Stamhuis, and H. Wynberg, *J. Org. Chem.* **32**, 1111 (1967); E. J. Stamhuis and W. Maas, *J. Org. Chem.* **30**, 2156 (1965).

hydroxylamine by aniline derivatives.

$$\mathrm{ArCH{=}O + NH_2OH \xrightarrow{Ar'NH_2} ArCH{=}NOH + H_2O}$$

Analysis of the kinetics of this catalysis points to the protonated imine as a key intermediate.

$$\mathrm{ArCH{=}O + Ar'NH_2 + H^+ \rightarrow ArCH{=}\overset{+}{N}HAr' + H_2O}$$

$$\mathrm{ArCH{=}\overset{+}{N}HAr' + NH_2OH \xrightarrow{fast} ArCH{=}NOH + Ar\overset{+}{N}H_3}$$

Because the imine is much more basic than the original carbonyl compound, it is more extensively protonated at any given pH than is the aldehyde. The protonated imine is also more reactive as an electrophile than the neutral aldehyde. The four distinct electrophiles present in the system are

$$\mathrm{ArCH{=}\overset{+}{O}H > ArCH{=}\overset{+}{N}HAr' > ArCH{=}O > ArCH{=}NAr'}$$

$$\xrightarrow[\text{decreasing reactivity}]{}$$

The protonated imine is the dominant reactive form. Although the protonated aldehyde is more reactive, its concentration is very low since it is much less basic than the imine or the reactant hydroxylamine. On the other hand, even though the aldehyde may be present in a greater concentration than the protonated imine, its reactivity is sufficiently less so that the iminium ion is the major reactant.[14]

8.3. Addition of Carbon Nucleophiles to Carbonyl Groups

The addition of carbon nucleophiles, including organometallic compounds, enolates, and enols, to carbonyl groups is one of the main methods of formation of carbon–carbon bonds. Such reactions are extremely important in synthesis and will be discussed extensively in Part B. Here, we will examine some of the fundamental mechanistic aspects of addition of carbon nucleophiles to carbonyl groups.

Organolithium and organomagnesium reagents are highly reactive toward most carbonyl compounds. With aldehydes and ketones, the tetrahedral adduct is stable, and alcohols are isolated after hydrolysis of the product alkoxide salt.

$$\mathrm{R'M + R_2C{=}O \longrightarrow R'{-}CR_2{-}O^-M^+ \xrightarrow{H^+} R'{-}CR_2{-}OH}$$

In the case of esters, carboxylate anions, amides, and acid chlorides, the tetrahedral adduct may undergo elimination. The elimination forms a ketone, permitting a subsequent addition to occur. The rate at which breakdown of the tetrahedral adduct

14. E. H. Cordes and W. P. Jencks, *J. Am. Chem. Soc.* **84**, 826 (1962); J. Hine, R. C. Dempsey, R. A. Evangelista, E. T. Jarvi, and J. M. Wilson, *J. Org. Chem.* **42**, 1593 (1977).

occurs is a function of the reactivity of the heteroatom substituent as a leaving group. The order of stability of the tetrahedral intermediates is shown below.

$$\mathrm{R{-}\overset{O}{\overset{\|}{C}}{-}O^-} \xrightarrow{R'M} \mathrm{R{-}\underset{R'}{\underset{|}{\overset{O^-}{\overset{|}{C}}}}{-}O^-} \rightarrow \mathrm{R{-}\overset{O}{\overset{\|}{C}}{-}R'}$$

$$\mathrm{R{-}\overset{O}{\overset{\|}{C}}{-}NR_2} \xrightarrow{R'M} \mathrm{R{-}\underset{R'}{\underset{|}{\overset{O^-}{\overset{|}{C}}}}{-}NR_2} \rightarrow \mathrm{R{-}\overset{O}{\overset{\|}{C}}{-}R'}$$

$$\mathrm{R{-}\overset{O}{\overset{\|}{C}}{-}OR} \xrightarrow{R'M} \mathrm{R{-}\underset{R'}{\underset{|}{\overset{O^-}{\overset{|}{C}}}}{-}OR} \rightarrow \mathrm{R{-}\overset{O}{\overset{\|}{C}}{-}R'}$$

$$\mathrm{R{-}\overset{O}{\overset{\|}{C}}{-}Cl} \xrightarrow{R'M} \mathrm{R{-}\underset{R'}{\underset{|}{\overset{O^-}{\overset{|}{C}}}}{-}Cl} \rightarrow \mathrm{R{-}\overset{O}{\overset{\|}{C}}{-}R'}$$

decreasing stability of intermediate →

Im most cases, the product ratio can be controlled by choice of reaction conditions. Ketones are isolated under conditions where the tetrahedral intermediate is stable until hydrolyzed, whereas alcohols are formed when the tetrahedral intermediate decomposes in the presence of unreacted organometallic reagent. Examples of synthetic application of this class of reactions will be discussed in Section 7.2 of Part B.

The reaction of organolithium reagents with simple carbonyl compounds is very fast, and there is relatively little direct kinetic evidence concerning the details of the reaction. It would be expected that one important factor in determining reactivity would be the degree of aggregation of the organolithium reagent. It has been possible to follow the reaction of benzaldehyde with *n*-butyllithium at −85°C, using NMR techniques that are capable of monitoring rapid reactions. The reaction occurs over a period of 50–300 ms. It has been concluded that the dimer of *n*-butyllithium is more reactive than the tetramer by a factor of about 10. As the reaction proceeds, the product alkoxide ion is incorporated into butyllithium aggregates. This gives rise to an additional species with a different reactivity.[15]

$$[\mathrm{BuLi}]_4 \rightleftharpoons 2[\mathrm{BuLi}]_2$$

$$[\mathrm{BuLi}]_4 \xrightarrow{\mathrm{PhCH{=}O}} \mathrm{Ph{-}\underset{O^-Li^+\cdot(BuLi)_3}{\underset{|}{\overset{H}{\overset{|}{C}}}}{-}Bu} \qquad [\mathrm{BuLi}]_2 \xrightarrow{\mathrm{PhCH{=}O}} \mathrm{Ph{-}\underset{O^-Li^+\cdot BuLi}{\underset{|}{\overset{H}{\overset{|}{C}}}}{-}Bu}$$

The rate laws for the reaction of several aromatic ketones with alkyllithium reagents has been examined. The reaction of 2,4-dimethyl-4′-methylthiobenzophenone with methyllithium in ether exhibits the rate law

$$\text{rate} = k[\mathrm{MeLi}]^{1/4}[\text{ketone}]$$

15. J. F. McGarrity, C. A. Ogle, Z. Brich, and H.-R. Loosli, *J. Am. Chem. Soc.* **107**, 1810 (1985).

This is consistent with a mechanism in which monomeric methyllithium in equilibrium with the tetramer is the reactive nucleophile.[16]

$$[\text{MeLi}]_4 \rightleftharpoons 4\,\text{MeLi}$$

$$\text{MeLi} + \text{ketone} \xrightarrow{\text{slow}} \text{product}$$

Most other studies have led to considerably more complex behavior. The rate data for reaction of 3-methyl-1-phenylbutanone with *s*-butyllithium and *n*-butyllithium in cyclohexane can be fit to a mechanism involving product formation both through a complex of the ketone with alkyllithium aggregate and through reaction with dissociated alkyllithium.[17] Evidence for the initial formation of a complex can be observed in the form of a shift in the carbonyl absorption band in the infrared spectrum. Complex formation presumably involves a Lewis acid–base interaction between the carbonyl oxygen and lithium ions in the alkyllithium cluster.

In general terms, it appears likely that alkyllithium reagents have the possibility of reacting through any of several aggregated and dissociated forms. The precise distribution will depend on the specific conditions of reaction.

$$[\text{R'Li}]_n \rightleftharpoons \frac{n}{2}[\text{R'Li}]_2 \rightleftharpoons n\text{R'Li}$$

Each form reacts with $R_2C{=}O$ to give $R'-C(R)_2-O^-Li^+$, accompanied by $(R'Li)_{n-1}$, $(R'Li)$, and nothing, respectively.

A cyclic mechanism can be written for reaction with a dimeric species.

$$R_2C{=}O\cdots Li(R')_2Li \longrightarrow R_2C(R')O^-\ Li^+ \quad R'Li$$

Molecular orbital modeling of the reaction of organolithium compounds with carbonyl groups has examined the interaction of formaldehyde with the dimer of methyllithium. The reaction is predicted to proceed by initial complexation of the carbonyl group at lithium, followed by a rate-determining step involving formation of the new carbon–carbon bond. The cluster then reorganizes to incorporate the newly formed alkoxide ion.[18]

16. S. G. Smith, L. F. Charbonneau, D. P. Novak, and T. L. Brown, *J. Am. Chem. Soc.* **94**, 7059 (1972).
17. M. A. Al-Aseer and S. G. Smith, *J. Org. Chem.* **49**, 2608 (1984).
18. E. Kaufman, P. v. R. Schleyer, K. N. Houk, and Y.-D. Wu, *J. Am. Chem. Soc.* **107**, 5560 (1985).

The transition state is reached very early in the second step with only slight formation of the C—C bond.

The kinetics of addition of alkyllithium reagents has been studied using a series of ethyl benzoates. The rates show a rather complex dependence on both alkyllithium concentration and the nature of aryl substituents in the ester. The rapid formation of an initial ester–alkyllithium complex can be demonstrated. It is believed that product can be formed by reaction of both aggregated and monomeric alkyllithium reagent. *N,N,N,N*-Tetramethylethylenediamine greatly accelerates the reaction, presumably by dissociating the organometallic aggregate (see Section 7.1).

$$(\mathrm{RLi})_n + n\mathrm{TMEDA} \rightleftharpoons \frac{n}{2}[(\mathrm{RLi})_2\cdot 2\mathrm{TMEDA}]$$

$$[(\mathrm{RLi})_2\cdot 2\mathrm{TMEDA}] + \mathrm{R'O_2CPh} \rightarrow \left[\mathrm{R}-\underset{\mathrm{Ph}}{\overset{\mathrm{O-R'}}{\mathrm{C}}}-\mathrm{O^-Li^+\cdot RLi\cdot 2TMEDA}\right]$$

The kinetics of reaction of Grignard reagents with ketones are also subject to a number of complications. The purity of the magnesium metal used in the preparation of the Grignard reagent is crucial since trace transition metal impurities have a major effect on the observed reaction rates. One of the most thorough studies available involves the reaction of methylmagnesium bromide with 2-methylbenzophenone in diethyl ether.[19] The results suggest the reaction mechanism is similar to that discussed for alkyllithium reactions. There is initial complexation between the

19. E. C. Ashby, J. Laemmle, and H. M. Neumann, *J. Am. Chem. Soc.* **94**, 5421 (1972).

ketone and Grignard reagent. The main Grignard species CH_3MgBr is in equilibrium with $(CH_3)_2Mg$, and this species can contribute to the overall rate. Finally, the product alkoxide complexes with the Grignard reagent to give another reactive species. The general mechanistic scheme is outlined below.

$$2RMgX \rightleftharpoons MgX_2 + R_2Mg$$

$$RMgX + R'_2C{=}O \rightarrow [R'_2C{=}O{\cdot}\underset{X}{Mg}{-}R] \rightarrow R{-}\overset{R'}{\underset{R'}{C}}{-}O^{-+}MgX$$

$$R{-}\overset{R'}{\underset{R'}{C}}{-}O^{-+}MgX + RMgX \rightarrow \left[R{-}\overset{R'}{\underset{R'}{C}}{-}O^{-}\overset{+}{Mg}\begin{smallmatrix}X\\ \\ X\end{smallmatrix}Mg{-}R\right]$$

$$\left[R{-}\overset{R'}{\underset{R'}{C}}{-}O^{-}\overset{+}{Mg}\begin{smallmatrix}X\\ \\ X\end{smallmatrix}Mg{-}R\right] + R_2C{=}O \longrightarrow R{-}\overset{R'}{\underset{R'}{C}}{-}O^{-}\begin{smallmatrix}MgX^{+}\\ \\ MgX^{+}\end{smallmatrix}{}^{-}O{-}\overset{R'}{\underset{R'}{C}}{-}R$$

There is another mechanism for addition of organometallic reagents to carbonyl compounds. This involves a discrete electron transfer step.[20]

$$(R{-}M)_n + O{=}CR'_2 \rightarrow [(R{-}M)_n\,O{=}CR'_2] \rightarrow [(R{-}\overset{+}{M})_n\,{}^{-}O{\dot{-}}CR'_2] \rightarrow (R{-}M)_{n-1} + M^{+-}O{-}\overset{R'}{\underset{R'}{C}}{-}R$$

The distinguishing feature of this mechanism is the second step, in which an electron is transferred from the organometallic reagent to the carbonyl compound to give the radical anion of the carbonyl compound. Subsequent collapse of the ion pair gives the same product as is formed in the normal mechanism. The electron transfer mechanism would be expected to be favored by structural features that stabilize the radical anion. Aryl ketones and diones fulfill this requirement, and much evidence for the electron transfer mechanism has been accumulated for such ketones. In several cases, it is possible to observe the intermediate radical anion by EPR spectroscopy.[21]

The stereochemistry of organometallic additions with carbonyl compounds fits into the general pattern for nucleophilic attack discussed in Chapter 3. With 4-*t*-butylcyclohexanone, there is a preference for equatorial attack but the selectivity

20. E. C. Ashby, *Pure Appl. Chem.* **52**, 545 (1980); E. C. Ashby, J. Laemmle, and H. M. Neumann, *Acc. Chem. Res.* **7**, 272 (1974).
21. K. Maruyama and T. Katagiri, *J. Am. Chem. Soc.* **108**, 6263 (1986); E. C. Ashby and A. B. Goel, *J. Am. Chem. Soc.* **103**, 4983 (1981).

is low. Enhanced steric factors promote exclusive equatorial addition.

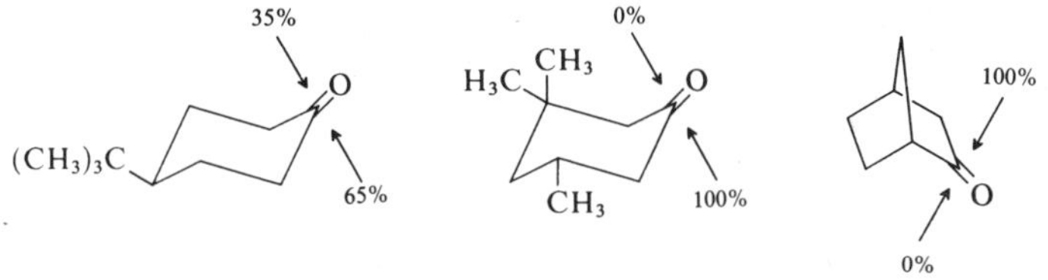

35% 0% 100%
$(CH_3)_3C$ O 65% H_3C CH_3 O CH_3 100% O 0% Ref. 22

Stereoselectivity in approach of methyllithium to cyclic ketones.

Addition of Grignard reagents to ketones and aldehydes was one of the reactions that led to the formulation of Cram's rule.[23] (See Section 3.10.) Many ketones and aldehydes have subsequently been subjected to studies to determine the degree of stereoselectivity. Cram's rule is obeyed when no special complexing functional groups are present near the reaction site. One such series of studies is summarized in Table 8.2.

The addition reaction of enolates with carbonyl compounds is of very broad scope and is of great synthetic importance. Essentially all of the enolates introduced in Chapter 7 are capable of adding to carbonyl groups. The reaction is known as the *generalized aldol condensation.*

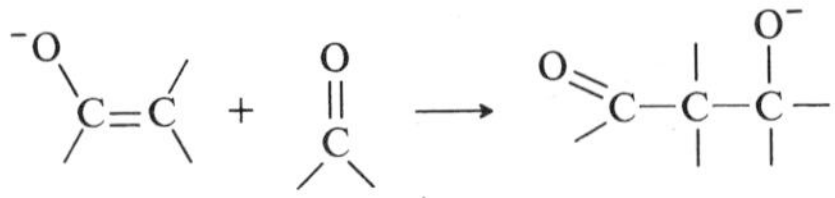

$$^{-}O\text{-}C{=}C< \; + \; >C{=}O \longrightarrow O{=}C\text{-}C\text{-}C\text{-}O^{-}$$

Enolates of aldehydes, ketones, and esters and the carbanions of nitriles and nitro compounds as well as phosphorus- and sulfur-stabilized carbanions undergo the reaction. The synthetic application of this group of reactions will be discussed in detail in Chapter 2 of Part B. In this section, we will discuss the fundamental mechanistic issues using the reaction of ketone enolates with aldehydes and ketones.

The aldol condensation can be carried out under either of two broad sets of conditions which lead to the product being determined by either kinetic or thermodynamic factors. To achieve *kinetic control,* the enolate which is to serve as the nucleophile is generated stoichiometrically, usually with lithium as the counterion in an aprotic solvent. Under these conditions, enolates are both structurally and stereochemically stable. The electrophilic carbonyl compound is then added. Under these conditions, the structure of the reaction product is determined primarily by two factors: (1) the structure of the initial enolate and (2) the stereoselectivity of the addition to the electrophilic carbonyl group.

The other broad category of reaction conditions results in *thermodynamic control.* This can result from several factors. Aldol condensations can be effected for many compounds using less than a stoichiometric amount of base. Under these conditions,

22. E. C. Ashby and S. A. Noding, *J. Org. Chem.* **44**, 4371 (1979).
23. D. J. Cram and F. A. Abd Elhafez, *J. Am. Chem. Soc.* **74**, 5828 (1952); D. J. Cram and J. D. Knight, *J. Am. Chem. Soc.* **74**, 5835 (1952); F. A. Abd Elhafez and D. J. Cram, *J. Am. Chem. Soc.* **74**, 5846 (1952).

Table 8.2. Stereoselectivity in Addition of Organometallic Reagents to Some Chiral Aldehydes and Ketones[a]

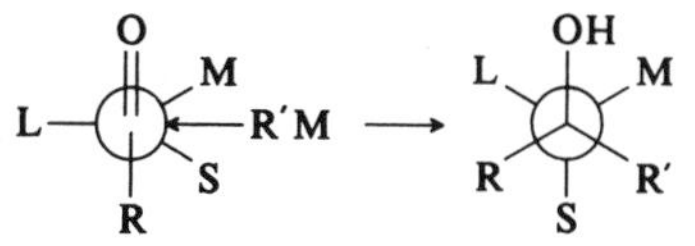

R	L	M	S	R'M	Percent of favored product
H	Ph	CH_3	H	CH_3MgBr	71
H	Ph	CH_3	H	PhMgBr	78
H	t-C_4H_9	CH_3	H	PhMgBr	98
CH_3	Ph	CH_3	H	C_2H_5Li	93
CH_3	Ph	CH_3	H	C_2H_5MgBr	88
CH_3	Ph	CH_3	H	t-C_4H_9MgBr	96
C_2H_5	Ph	CH_3	H	CH_3MgBr	86
C_2H_5	Ph	CH_3	H	CH_3Li	94
C_2H_5	Ph	CH_3	H	PhLi	85
i-C_3H_7	Ph	CH_3	H	CH_3MgBr	90
i-C_3H_7	Ph	CH_3	H	CH_3Li	96
i-C_3H_7	Ph	CH_3	H	PhLi	96
t-C_4H_9	Ph	CH_3	H	CH_3MgBr	96
t-C_4H_9	Ph	CH_3	H	CH_3Li	97
t-C_4H_9	Ph	CH_3	H	PhLi	98
Ph	Ph	CH_3	H	CH_3MgBr	87
Ph	Ph	CH_3	H	CH_3Li	97
Ph	Ph	CH_3	H	t-C_4H_9MgBr	96

a. Data from O. Arjona, R. Perez-Ossorio, A. Perez-Rubalcaba, and M. L. Quiroga, *J. Chem. Soc., Perkin Trans. 2,* 597 (1981); C. Alvarez-Ibarra, P. Perez-Ossorio, A. Perez-Rubalcaba, M. L. Quiroga, and M. J. Santesmases, *J. Chem. Soc., Perkin Trans. 2,* 1645 (1983).

the aldol reaction is reversible and the product ratio will be determined by the relative stability of the various possible products. Thermodynamic conditions also permit equilibration between all the enolates of the nucleophile. The conditions which permit equilibration include higher reaction temperatures, the presence of protic solvents, and the use of less tightly coordinating cations.

The fundamental mechanistic concept around which the course of aldol reactions under conditions of kinetic control has been interpreted involves the postulation of a cyclic transition state in which both the carbonyl and the enolate oxygen are coordinated to a Lewis acid. We will use the Li^+ cation in our discussion, but another metal cation or electron-deficient atom may play the same role.

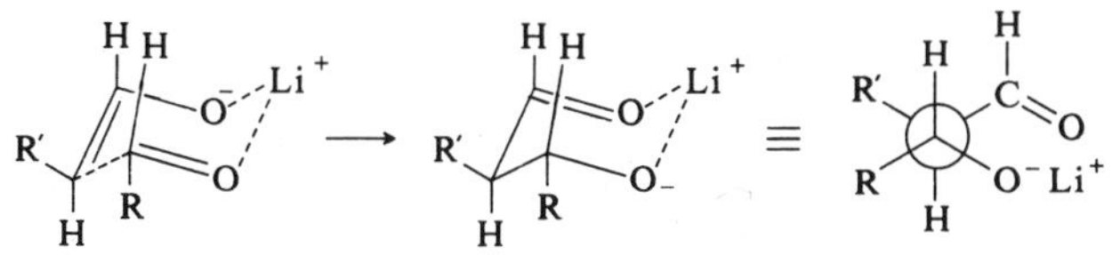

According to this concept, the aldol condensation normally occurs through a chairlike transition state. It is further assumed that the structure of this transition state is sufficiently similar to that of chair cyclohexane that the conformational concepts developed for cyclohexane derivatives can be applied. Thus, in the example above, the reacting aldehyde is shown with R rather than H in the equatorial-like position. The differences in stability of the various transition states, and therefore the product ratios, are governed by the steric interactions between substituents.

A consequence of this mechanism is that the reaction if *stereospecific* with respect to the *E*- or *Z*-configuration of the enolate. The *E*-enolate will give the *anti* aldol product while the *Z*-enolate will give the *syn* aldol.

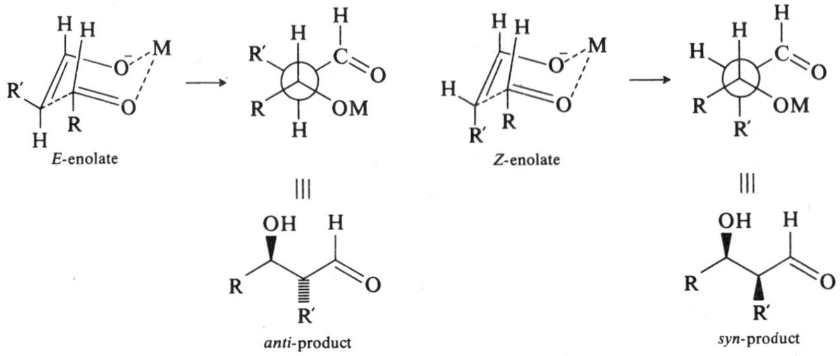

E-enolate

Z-enolate

anti-product

syn-product

Numerous observations on the reactions of enolates of both ketones and esters are consistent with this general concept.[24] However, the prediction or interpretation of the *specific ratio* of *syn* to *anti* product from any given reaction process requires assessment of several variables. These include: (1) What is the stereochemistry of the enolate that is reacting? (2) How strong is the selectivity between the two faces of the carbonyl group? This is a function of the steric interactions in the diastereomeric transition states. (3) Does the Lewis acid promote a tight coordination with both the carbonyl and enolate oxygen atoms? (4) Are there any special structural features, such as additional Lewis base coordination sites, which could override steric effects? (5) Are the reaction conditions such as to promote kinetic control? Chapter 2 of Part B will give a more complete discussion of the ways in which these factors can be controlled to provide specific reaction products.

When the aldol reaction is carried out under thermodynamic conditions, the product selectivity is often not as high as under kinetic conditions. All the regioisomeric and stereoisomeric enolates may participate as nucleophiles. The adducts can return to reactants, and so only the difference in stability of the stereoisomeric *anti* and *syn* products will influence the product composition.

24. C. H. Heathcock, in *Asymmetric Syntheses*, Vol. 3, J. D. Morrison (ed.), Academic Press, New York, 1984, Chapter 2; C. H. Heathcock, in *Comprehensive Carbanion Chemistry*, Part B, E. Buncel and T. Durst (eds.), Elsevier, Amsterdam, 1984, Chapter 4; T. Mukaiyama, *Org. React.* **28**, 203 (1982); D. A. Evand, J. V. Nelson, and T. R. Taber, *Top Stereochem.* **13**, 1 (1982); A. T. Nielsen and W. J. Houlihan, *Org. React.* **16**, 1 (1968).

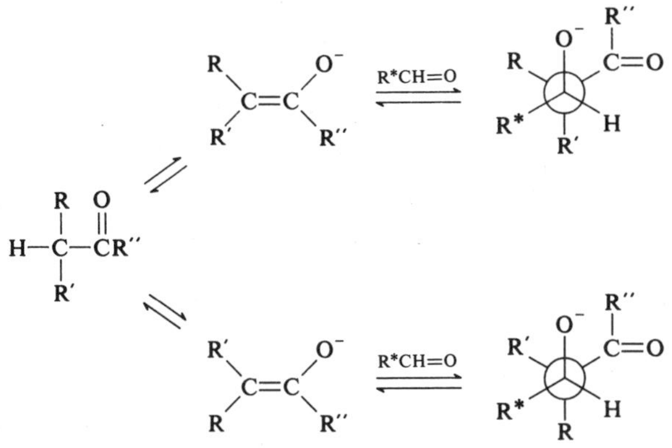

It is also possible to carry out the aldol condensation under acidic conditions. The reactive nucleophile is then the enol. The mechanism, as established in detail for acetaldehyde,[25] involves nucleophilic attack of the enol on the protonated carbonyl tautomer.

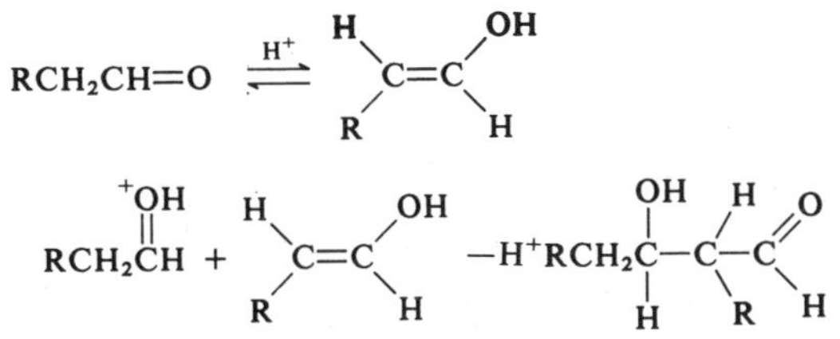

There has been little study of the stereochemistry of the reaction under acidic conditions. When regioisomeric enols are possible, acid-catalyzed reactions tend to go through the more substituted of the enols.

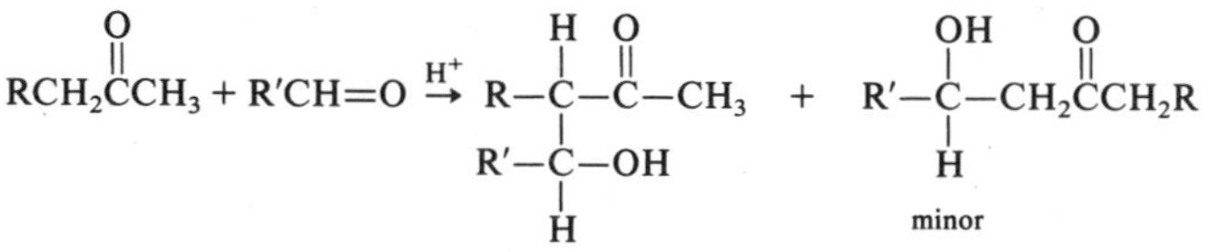

major

minor

Under both basic and acidic conditions, the aldol condensation can proceed to dehydrated product.

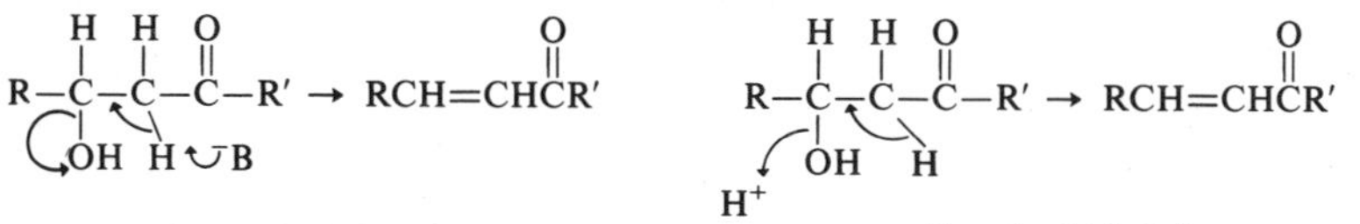

base-catalyzed dehydration

acid-catalyzed dehydration

25. L. M. Baigrie, R. A. Cox, H. Slebocka-Tilk, M. Tencer, and T. T. Tidwell, *J. Am. Chem. Soc.* **107**, 3640 (1985).

The dehydration reactions require somewhat higher activation energies than the addition step and are not usually observed under strictly kinetic conditions. Detailed rate studies have provided rate and equilibrium constants for the individual steps in some cases. The results for the acetone-benzaldehyde system in the presence of hydroxide ion are given below.

$$CH_3\overset{O}{\overset{\|}{C}}CH_3 + PhCH{=}O \underset{k_{-1}}{\overset{K_1,\ k_1}{\rightleftharpoons}} CH_3\overset{O}{\overset{\|}{C}}CH_2\overset{HO}{\overset{|}{C}}HPh \underset{k_{-2}}{\overset{K_2,\ k_2}{\rightleftharpoons}} CH_3\overset{O}{\overset{\|}{C}}CH{=}CHPh$$

Ref. 26

$K_1 = 0.83M^{-1}$ $\qquad$ $K_2 = 24$

$k_1 = 0.17M^{-2}\,s^{-1}$ $\qquad$ $k_2 = 7.5 \times 10^{-3}M^{-1}\,s^{-1}$

$k_{-1} = 1.4 \times 10^{-2}M^{-1}\,s^{-1}$ $\qquad$ $k_{-2} = 3.2 \times 10^{-4}M^{-1}\,s^{-1}$

8.4. Reactivity of Carbonyl Compounds toward Addition

We would like at this point to consider some general relationships concerning the reactivity of carbonyl compounds. The discussion of the nucleophilic addition reactions in Sections 8.2 and 8.3 indicates that many factors influence the overall rate of a reaction under typical conditions. Among the crucial factors are (1) the role of protons or other Lewis acids in activating the carbonyl group toward nucleophilic attack, (2) the reactivity of the nucleophilic species and its influence on the reaction mechanism and (3) the stability of the tetrahedral intermediate and the extent to which it proceeds to product rather than reverting to starting material. Since consideration of all of these factors complicates the interpretation of the inherent reactivity of the carbonyl compound itself, an irreversible process where the addition product is stable affords the most direct means of comparing the reactivity of different carbonyl compounds. Under these conditions, the relative rates of reaction of different carbonyl compounds can be directly compared. One such reaction is hydride reduction. In particular, reduction by sodium borohydride in alcohol solvents is a fast, irreversible reaction which has provided a convenient basis for comparing the reactivity of different carbonyl compounds.[27]

$$R\overset{O}{\overset{\|}{C}}R' + BH_4^- \longrightarrow R\underset{H}{\underset{|}{\overset{OBH_3^-}{\overset{|}{C}}}}R' \longrightarrow R\overset{OH}{\overset{|}{C}}HR'$$

The reaction is second-order overall, with the rate given by $k[R_2C{=}O][NaBH_4]$. The interpretation of the rates is complicated slightly by the fact that the alkoxy-borohydrides produced by the first addition can also function as reducing agents, but this has little apparent effect on the relative reactivity of the carbonyl compounds. Table 8.3 presents some of the rate data obtained from these studies.

26. J. P. Guthrie, J. Cossar, and K. F. Taylor, *Can. J. Chem.* **62**, 1958 (1984).
27. H. C. Brown, O. H. Wheeler, and K. Ichikawa, *Tetrahedron* **1**, 214 (1957); H. C. Brown and K. Ichikawa, *Tetrahedron* **1**, 221 (1957).

Table 8.3. Rates of Reduction of Aldehydes and Ketones by Sodium Borohydride

Carbonyl compound	$k_2 \times 10^4$ $(M^{-1}\,s^{-1})^a$
Benzaldehyde	12,400[b]
Benzophenone	1.9
Acetophenone	2.0
Acetone	15.1
Cyclobutanone	264
Cyclopentanone	7
Cyclohexanone	161

a. In isopropyl alcohol at 0°C.
b. Extrapolated from data at lower temperatures.

Reductions by $NaBH_4$ are characterized by low enthalpies of activation (8–13 kcal/mol) and large negative entropies of activation (−28 to −40 eu). Aldehydes are substantially more reactive than ketones as can be seen by comparison of the rates for benzaldehyde and acetophenone. This relative reactivity is characteristic for nearly all carbonyl addition reactions. The lesser reactivity of ketones can be attributed primarily to steric effects. Not only does the additional substituent increase the steric restrictions to approach of the nucleophile, but also it gives rise to larger steric interaction in the tetrahedral product as the hybridization changes from trigonal to tetrahedral.

Among the cyclic ketones, the reactivity of cyclobutanone is enhanced because of the strain of the four-membered ring, which is decreased in going from sp^2 to sp^3 hybridization. The greater reactivity of cyclohexanone as compared to cyclopentanone is also quite general for carbonyl addition reactions. The major factor responsible for the large difference in this case is the change in torsional strain as addition occurs. As the hybridization goes from sp^2 to sp^3, the torsional strain is increased in cyclopentanone. The opposite is true for cyclohexanone. The equatorial hydrogens are nearly eclipsed with the carbonyl oxygen in cyclohexanone, but the chair structure of cyclohexanol allows all bonds to be in staggered arrangements.

eclipsed

staggered

The borohydride reduction rate data are paralleled by many other carbonyl addition reactions. In fact, for a series of ketones, most of which are cyclic, a linear free-energy correlation of the form

$$\log k = A \log k_0 + B$$

Table 8.4. Relative Reactivity of Some Ketones toward Addition of Nucleophiles

Ketone	B = log relative reactivity[a]
Cyclobutanone	0.09
Cyclohexanone	0.00[a]
4-*t*-butylcyclohexanone	−0.008
Adamantanone	−0.46
Cycloheptanone	−0.95
Cyclopentanone	−1.18
Acetone	−1.19
Bicyclo[2.2.1]heptan-2-one	−1.48
3,3,5,5-Tetramethylcyclohexanone	−1.92

a. A. Finiels and P. Geneste, *J. Org. Chem.* **44**, 1577 (1979); reactivity relative to cyclohexanone as a standard.

exists for nucleophiles such as NH_2OH, CN^-, $HOCH_2CH_2S^-$, and SO_3^-.[28] These nucleophiles span a wide range of reactivity and represent nitrogen, carbon, and sulfur atoms acting as the nucleophile. The free-energy relationship implies that in this series of ketones the same structural features govern reactivity toward each of the nucleophiles. To a good approximation, the parameter $A = 1$, which reduces the correlation to

$$\log(k/k_0) = B$$

This equation implies that the relative reactivity is independent of the specific nucleophile and that relative reactivity is insensitivie to changes in position of the transition state. Table 8.4 lists some of the B values for some representative ketones. The parameter B indicates relative reactivity on a log scale. Cyclohexanone is seen to be a particularly reactive ketone, being almost as reactive as cyclobutanone and more than 10 times as reactive as acetone.

The same structural factors come into play in determining the position of equilibria in reversible additions to carbonyl compounds. The best studied of such equilibrium processes is probably addition of cyanide to give cyanohydrins.

$$R_2C{=}O + HCN \rightleftharpoons N{\equiv}C{-}\overset{\displaystyle R}{\underset{\displaystyle R}{\overset{|}{\underset{|}{C}}}}{-}OH$$

For cyclopentanone, cyclohexanone, and cycloheptanone the K values for addition are 48, 1000, and 8 M^{-1}, respectively.[29] For aromatic aldehydes, the equilibria are affected by the electronic nature of the aryl substituent. Electron donors disfavor addition by stabilizing the aldehyde whereas electron-accepting substituents have

28. A. Finiels and P. Geneste, *J. Org. Chem.* **44**, 1577 (1979).
29. V. Prelog and M. Kobelt, *Helv. Chim. Acta* **32**, 1187 (1949).

the opposite effect.

$$\text{X}-\text{C}_6\text{H}_4-\text{CH}{=}\text{O} + \text{HCN} \underset{k_r}{\overset{k_a}{\rightleftharpoons}} \text{X}-\text{C}_6\text{H}_4-\text{CH(OH)CN}$$ Ref. 30

K is correlated by Hammett equation with σ^+, $\rho = 1.01$

k_a is correlated by Hammett equation with σ^+, $\rho = 1.18$

There are large differences in the reactivity of the various carboxylic acid derivatives, such as amides, esters, and acyl chlorides. One important factor is the resonance stabilization provided by the heteroatom. This is in the order N > O > Cl. Electron donation reduces the electrophilicity of the carbonyl group, and the corresponding stabilization is lost in the tetrahedral intermediate.

$$\text{R(H}_2\text{N)C}{=}\text{O} \leftrightarrow \text{R(H}_2\text{N)}\overset{+}{\text{C}}-\text{O}^- \leftrightarrow \text{R(H}_2\overset{+}{\text{N}}{=})\text{C}-\text{O}^- \qquad \text{R(RO)C}{=}\text{O} \leftrightarrow \text{R(RO)}\overset{+}{\text{C}}-\text{O}^- \leftrightarrow \text{R(R}\overset{+}{\text{O}}{=})\text{C}-\text{O}^-$$

The high reactivity of the acyl chlorides also reflects the electron-withdrawing inductive effect of the chlorine, which more than outweighs the small π-donor effect.

8.5. Ester Hydrolysis

Esters can be hydrolyzed in either basic or acidic solution. In acidic solution, the reaction is reversible.

$$\text{RCOR}' + \text{H}_2\text{O} \overset{\text{H}^+}{\rightleftharpoons} \text{RCOH} + \text{R}'\text{OH}$$

The position of the equilibrium depends on the relative concentration of water and the alcohol. In aqueous solution, hydrolysis occurs. In alcohlic solution, the equilibrium is shifted in favor of the ester. In alkaline aqueous solution, ester hydrolysis is essentially irreversible:

$$\text{RCOR}' + {}^-\text{OH} \rightarrow \text{RCO}_2^- + \text{R}'\text{OH}$$

The carboxylic acid is converted to its anion under these conditions, and the position of the equilibrium lies far to the right. The mechanistic designations $\text{A}_{\text{AC}}2$ and $\text{B}_{\text{AC}}2$ are given to the acid- and base-catalyzed hydrolysis mechanisms, respectively. The A denotes acid catalysis, while B indicates base catalysis. The AC designation indicates acyl–oxygen bond cleavage occurs. The digit 2 has its usual significance,

30. W.-M. Ching, and R. G. Kallen, *J. Am. Chem. Soc.* **100**, 6119 (1978).

indicating the bimolecular nature of the rate-determining step.

$A_{AC}2$ mechanism

$$\mathrm{RC(=O)OR'} + \mathrm{H^+} \rightleftharpoons \mathrm{RC(={}^{+}OH)OR'}$$

$$\mathrm{RC(={}^{+}OH)OR'} + \mathrm{H_2O} \rightleftharpoons \mathrm{RC(OH)(H_2O^+)OR'}$$

$$\mathrm{RC(OH)(H_2O^+)OR'} \rightleftharpoons \mathrm{RC(OH)(HO)\overset{+}{O}(H)R'} \rightleftharpoons \mathrm{RC(=O)OH} + \mathrm{R'OH} + \mathrm{H^+}$$

$B_{AC}2$ mechanism

$$\mathrm{RC(=O)OR'} + {}^{-}\mathrm{OH} \rightleftharpoons \mathrm{RC(O^-)(OH)OR'}$$

$$\mathrm{RC(O^-)(OH)OR'} \rightleftharpoons \mathrm{RC(=O)OH} + \mathrm{R'O^-} \longrightarrow \mathrm{RC(=O)O^-} + \mathrm{R'OH}$$

Esters without special structural features hydrolyze by these mechanisms. Among the evidence supporting these mechanisms are kinetic studies that show the expected dependence on hydrogen ion or hydroxide ion concentration and isotopic labeling studies that prove that it is the acyl-oxygen, not the alkyl-oxygen, bond that is cleaved during hydrolysis.[31] Acid-catalyzed hydrolysis of esters is accompanied by some exchange of oxygen from water into the ester. This exchange occurs by way of the tetrahedral intermediate when expulsion of water is competitive with expulsion of the alcohol.

$$\mathrm{RC(=O)OR'} + \mathrm{H_2\overset{*}{O}} \rightleftharpoons \mathrm{RC(OH)(\overset{*}{O}H)OR'} \rightleftharpoons \mathrm{RC(=\overset{*}{O})OR'} + \mathrm{H_2O}$$

Substituent effects come into play at several points in the ester hydrolysis mechanism. In the base-catalyzed reaction, electron-withdrawing substituents in either the acyl or the alkoxy group facilitate hydrolysis. Since the tetrahedral intermediate formed in the rate-determining step is negatively charged, it and the transition state leading to it are stabilized by electron withdrawal. If the carbonyl group is conjugated with an electron-releasing group, reactivity is decreased by

31. M. L. Bender, *Chem. Rev.* **60**, 53 (1960).

ground state stabilization. The partitioning of the tetrahedral intermediate between reversion to starting material by loss of hydroxide ion and formation of product by expulsion of the alkoxide is strongly affected by substituents in the alkoxy group. Electron-withdrawing groups shift the partitioning in favor of loss of the alkoxide and favor hydrolysis. For this reason, exchange of carbonyl oxygen with solvent does not occur in basic hydrolyses when the alkoxy group is a good leaving group. This has been demonstrated, for example, for esters of phenols. Because phenols are much stronger acids than alcohols, their conjugate bases are better leaving groups than simple alkoxide ions. Aryl esters are hydrolyzed faster than alkyl esters and without observable exchange of carbonyl oxygen with solvent.

$$\mathrm{RC(=O)OAr} + {}^{-}\mathrm{OH} \underset{\leftarrow\cdots}{\xrightarrow{\text{slow}}} \mathrm{RC(O^-)(OH)OAr} \xrightarrow{\text{fast}} \mathrm{RC(=O)OH} + \mathrm{ArO^-} \xrightarrow{^{-}\mathrm{OH}} \mathrm{RCO_2^-} + \mathrm{ArO^-}$$

Even simple alkyl benzoate esters give only a small amount of exchange under basic hydrolysis conditions. This means that reversal of the hydroxide addition must be slow relative to the forward breakdown of the tetrahedral intermediate.[32]

These substituent effects can be summarized in a general way for the B_{AC} mechanism by noting the effect of substituents on each step of the mechanism:

$$\mathrm{RC(=O)OR'} + {}^{-}\mathrm{OH} \rightleftarrows \mathrm{R{-}C(O^-)(OH){-}OR'}$$

favored by electron-attracting substituents in both R and R′

$$\mathrm{R{-}C(O^-)(OH){-}OR'} \rightarrow \mathrm{R{-}C(=O){-}OH} + {}^{-}\mathrm{OR'}$$

strongly favored by electron-attracting substituents in R′

It is possible to shift ester hydrolysis away from the normal $A_{AC}2$ or $B_{AC}2$ mechanisms by structural changes in the reactant. When the ester is derived from a tertiary alcohol, acid-catalyzed hydrolysis often occurs by a mechanism involving alkyl–oxygen fission. The change in mechanism is due to the stability of the tertiary carbocation that can be formed by alkyl–oxygen cleavage.[33] When this mechanism occurs, alkenes as well as alcohols may be produced, since the carbocation can react by substitution or elimination. This mechanism is referred to as $A_{AL}1$, reflecting the fact that the alkyl–oxygen bond is broken.

$$\mathrm{RC(=O)OCR'_3} + \mathrm{H^+} \rightleftharpoons \mathrm{RC(={}^{+}OH)OCR'_3} \xrightarrow{\text{slow}} \mathrm{RCO_2H} + \mathrm{R'_3C^+}$$

$$\mathrm{R'_3C^+} + \mathrm{H_2O} \longrightarrow \mathrm{R'_3COH} \qquad \textit{and} \qquad \mathrm{R'_3C^+} \longrightarrow \text{alkene} + \mathrm{H^+}$$

The change of mechanism of tertiary alkyl esters is valuable in synthetic methodology since it permits certain esters to be hydrolyzed very selectively. The usual situation

32. R. A. McClelland, *J. Am. Chem. Soc.* **106**, 7579 (1984).
33. A. G. Davies and J. Kenyon, *Q. Rev. Chem. Soc.* **9**, 203 (1955).

involves the use of *t*-butyl esters, which can be cleaved to carboxylic acids by action of acids such as *p*-toluenesulfonic acid or trifluoroacetic acid under anhydrous conditions where other esters are stable.

In the preceding paragraphs, the ester hydrolysis mechanisms discussed pertained to aqueous solutions of strong acids and strong bases. These are conditions under which specific acid catalysis or specific base catalysis is expected to be dominant. In media in which other acids or bases are present, the possible occurrence of general-acid-catalyzed and general-base-catalyzed hydrolysis must be considered. General base catalysis has been observed in the case of esters in which the acyl group carries electron-attracting substituents.[34] The transition state for esters undergoing hydrolysis by a general-base-catalyzed mechanism involves partial proton transfer from the attacking water molecule to the general base during formation of the tetrahedral intermediate.

$$B^{-}\ H{-}\underset{H}{O}\ \ R\overset{}{C}(=O)OR \xrightarrow{\text{slow}} B{-}H + HO{-}\underset{OR}{\overset{R}{C}}{-}O^{-} \xrightarrow{\text{fast}} RCOH(=O) + {}^{-}OR \longrightarrow RCO_2^{-} + ROH$$

Ester hydrolysis can also be promoted by *nucleophilic catalysis*. If a component of the reaction system is a more effective nucleophile toward the carbonyl group than hydroxide ion or water under a given set of conditions, an acyl transfer reaction can take place to form an intermediate.

$$\mathbf{HNu} + R\overset{O}{\overset{\|}{C}}OR' \longrightarrow R\overset{O}{\overset{\|}{C}}\mathbf{Nu} + R'OH$$

$$R\overset{O}{\overset{\|}{C}}\mathbf{Nu} + H_2O \longrightarrow RCO_2H + \mathbf{HNu}$$

If this intermediate, in turn, is more rapidly attacked by water or hydroxide ion than the original ester, the overall reaction will be faster in the presence of the nucleophile than in its absence. These are the requisite conditions for nucleophilic catalysis. Esters of relatively acidic alcohols (in particular, phenols) are hydrolyzed by the nucleophilic catalysis mechanism in the presence of imidazole.[35]

$$\text{imidazole (HN–N)} + R\overset{O}{\overset{\|}{C}}OAr \longrightarrow \text{N-acylimidazole } (N{-}\overset{O}{\overset{\|}{C}}R) + ArOH$$

$$\text{N-acylimidazole } (N{-}\overset{O}{\overset{\|}{C}}R) + H_2O \longrightarrow \text{imidazole (N–NH)} + R\overset{O}{\overset{\|}{C}}OH$$

34. W. P. Jencks and J. Carriuolo, *J. Am. Chem. Soc.* **83**, 1743 (1961).
35. T. C. Bruice and G. L. Schmir, *J. Am. Chem. Chem. Soc.* **79**, 1663 (1957); M. L. Bender and B. W. Turnquest, *J. Am. Chem. Soc.* **79**, 1652, 1656 (1957).

Carboxylate anions can also serve as nucleophilic catalysts.[36] In this case, an anhydride is the reactive intermediate.

$$\mathrm{RC(=O)OR' + R''CO_2^- \longrightarrow RC(=O)OC(=O)R'' + R'O^-}$$

$$\mathrm{RC(=O)OC(=O)R'' + H_2O \longrightarrow RCO_2H + R''CO_2H}$$

The nucleophilic catalysis mechanism only operates when the alkoxy group being hydrolyzed is not much more basic than the nucleophilic catalyst. This relationship can be understood by considering the tetrahedral intermediate generated by attack of the potential catalyst on the ester:

$$\mathrm{R''CO_2^- + RC(=O)OR' \longrightarrow RC(O^-)(OR')\!-\!OC(=O)R''}$$

The relative leaving group abilities of $R'O^-$ and $R''CO_2^-$ are strongly correlated with the basicity of the two anions. If $R''CO_2^-$ is a much better leaving group than $R'O^-$, it will be eliminated preferentially form the tetrahedral intermediate and no hydrolysis will occur.

The preceding discussion has touched on the most fundamental aspects of ester hydrolysis mechanisms. Much effort has been devoted to establishing some of the finer details, particularly concerning proton transfers during the formation and breakdown of the tetrahedral intermediates. These studies have been undertaken in part because of the fundamental importance of hydrolytic reactions in biological systems. These biological hydrolyses are catalyzed by enzymes. The detailed mechanistic studies of ester hydrolysis lay the groundwork for understanding the catalytic mechanisms of the hydrolytic enzymes. Discussion of the biological mechanisms and their relationship to the fundamental mechanistic studies is available in several books that discuss enzyme catalysis in terms of molecular mechanisms.[37]

Esters react with alcohols in either acidic or basic solution to exchange alkoxy groups (ester interchange) by mechanisms which parallel those described for hydrolysis. The alcohol or alkoxide takes the role of the nucleophilic species:

$$\mathrm{RC(=O)OR' + R''OH \overset{R''O^-}{\rightleftharpoons} RC(=O)OR'' + R'OH}$$

$$\mathrm{RC(=O)OR' + R''OH \overset{H^+}{\rightleftharpoons} RC(=O)OR'' + R'OH}$$

36. V. Gold, D. G. Oakenfull, and T. Riley, *J. Chem. Soc., B*, 515 (1968).
37. T. C. Bruice and S. J. Benkovic, *Bioorganic Mechanisms*, Vol. 1, W. A. Benjamin, New York, 1966, pp. 1–258; W. P. Jencks, *Catalysis in Chemistry and Enzymology*, McGraw-Hill, New York, 1969; M. L. Bender, *Mechanisms of Homogeneous Catalysis from Protons to Proteins*, Wiley-Interscience, New York, 1971; C. Walsh, *Enzymatic Reaction Mechanisms*, W. H. Freeman, San Francisco, 1979; A. Fersht, *Enzyme Structure and Mechanism*, Second Edition, W. H. Freeman, New York, 1985.

As in the case of hydrolysis, there has been a good deal of study of substituent effects, solvent effects, isotopic exchange, kinetics, and the catalysis of these processes.[38] The alcoholysis reaction is reversible in both acidic and basic solution, in contrast to hydrolysis. The key intermediate is the tetrahedral addition product. Its fate is determined largely by the relative basicity of the two alkoxy groups. A tetrahedral intermediate generated by addition of methoxide ion to a *p*-nitrophenyl ester, for example, breaks down exclusively by elimination of much less basic *p*-nitrophenoxide ion.

$$CH_3OC(O^-)(R)(O\text{-}C_6H_4\text{-}p\text{-}NO_2) \longrightarrow CH_3OC(=O)R + O_2N\text{-}C_6H_4\text{-}O^-$$

In general, the equilibrium in a base-catalyzed alcohol exchange reaction lies in the direction of incorporation of the less acidic alcohol into the ester.

8.6. Aminolysis of Esters

Esters react with ammonia and amines to give amides. The mechanism is similar to that of hydroxide-ion-catalyzed ester hydrolysis and involves nucleophilic attack of the amine at the carbonyl group, followed by expulsion of alkoxide ion from the tetrahedral intermediate. The identity of the rate-determining step depends primarily on the leaving group ability of the alkoxy group.[39] With relatively good leaving groups such as phenolate or acidic alcohols such as trifluoroethanol, the slow step is expulsion of the oxygen leaving group from a zwitterionic tetrahedral intermediate **A**. With poorer leaving groups, breakdown of the tetrahedral intermediate occurs only after formation of the anionic species **B**. For such systems, the deprotonation of **A** is rate-determining.

$$RC(=O)OR' + R''NH_2 \rightleftharpoons \underset{\mathbf{A}}{R''\overset{+}{N}H_2C(O^-)(R)OR'} \rightleftharpoons R''NHC(OH)(R)OR'$$

$$\mathbf{A} \underset{H^+}{\rightleftharpoons} \underset{\mathbf{B}}{R''NHC(O^-)(R)OR'} \underset{H^+}{\rightleftharpoons} R''NHC(OH)(R)OR'$$

$$R''NHC(O^-)(R)OR' \longrightarrow R''NHC(=O)R + R'O^-$$

Aminolysis of esters often reveals general base catalysis and in particular a contribution to the reaction rate from terms that are second-order in the amine. The

38. C. G. Mitton, R. L. Schowen, M. Gresser, and J. Shapely, *J. Am. Chem. Soc.* **91**, 2036 (1969); C. G. Mitton, M. Gresser, and R. L. Schowen, *J. Am. Chem. Soc.* **91**, 2045 (1969).
39. F. M. Menger and J. H. Smith, *J. Am. Chem. Soc.* **94**, 3824 (1972); A. C. Satterthwait and W. P. Jencks, *J. Am. Chem. Soc.* **96**, 7018 (1974).

general base is believed to function by deprotonating the zwitterionic tetrahedral intermediate.[40] Deprotonation of the nitrogen facilitates breakdown of the tetrahedral intermediate, since the increased electron density at nitrogen favors expulsion of an anion.

$$\mathrm{B{:}\longrightarrow H{-}\overset{+}{N}(H)(R'')\!-\!C(O^-)(R)OR' \longrightarrow \overset{+}{B}H + R''NH\!-\!C(O^-)(R)OR' \longrightarrow R''NHCR(=O) + {}^-OR'}$$

Detailed mechanistic studies have been carried out on aminolysis of substituted aryl acetates and aryl carbonates.[41] Aryl esters are considerably more reactive than alkyl esters because the less basic phenoxide ions are better leaving groups than alkoxide ions. The tetrahedral intermediate formed in aminolysis can exist in several different forms differing in extent and site of protonation.

$$\mathrm{RNH_2 + R'\overset{O}{\overset{\|}{C}}OR'' \rightarrow \underset{\mathbf{A}}{R\overset{+}{N}H_2{-}C(O^-)(R'){-}OR''} \rightleftarrows \underset{\mathbf{C}}{RNH{-}C(OH)(R'){-}OR''}}$$

$$\mathrm{A \underset{-H^+}{\rightleftarrows} B;\quad C \underset{-H^+}{\rightleftarrows} D;\quad A \overset{H^+}{\rightleftarrows} D \text{ and } B \overset{H^+}{\rightleftarrows} C}$$

$$\mathrm{\underset{\mathbf{B}}{RNH{-}C(O^-)(R'){-}OR''} \overset{H^+}{\rightleftarrows} \underset{\mathbf{D}}{R\overset{+}{N}H_2{-}C(OH)(R'){-}OR''}}$$

In **A** and **D** the best leaving group is the neutral amine, whereas in **B** and **C** the group $R''O^-$ would be expected to be a better leaving group than RNH^-. Furthermore, in **B** and **C** the lone pair on nitrogen can assist in elimination. In **A** the negatively charged oxygen also has the capacity to assist by "pushing" with re-formation of the carbonyl group. Precisely how the intermediate proceeds to product depends upon pH and the identity of the groups RNH_2 and $R''O^-$. When $R''O^-$ is a relatively poor leaving group, as would be the case for alkyl esters, reaction usually occurs through **B** or **C**.

$$\mathrm{\underset{\mathbf{B}}{RNH{-}C(O^-)(R'){-}OR''} \rightarrow RNH{-}\overset{O}{\overset{\|}{C}}{-}R' + R''O^-}$$

or

$$\mathrm{\underset{\mathbf{C}}{R\ddot{N}H{-}C(OH)(R'){-}OR''} \rightarrow R\overset{+}{N}H{=}C(OH){-}R' + {}^-OR'' \rightarrow RNH{-}\overset{O}{\overset{\|}{C}}{-}R' + R''OH}$$

40. W. P. Jencks and M. Gilchrist, *J. Am. Chem. Soc.* **88**, 104 (1966); J. F. Kirsch and A. Kline, *J. Am. Chem. Soc.* **91**, 1841 (1969).
41. W. P. Jencks and M. Gilchrist, *J. Am. Chem.. Soc.* **90**, 2622 (1968); A. Satterthwait and W. P. Jencks, *J. Am. Chem. Soc.* **96**, 7018 (1974); A. Satterthwait and W. P. Jencks, *J. Am. Chem. Soc.* **96**, 7031 (1974); M. J. Gresser and W. P. Jencks, *J. Am. Chem. Soc.* **99**, 6970 (1977).

When the leaving group is better, breakdown can occur directly from **A**. This is the case when $R''O^-$ is a phenolate anion. The mechanism will also depend upon the pH and the presence of general acids and bases since the position of the equilibria among the tetrahedral intermediates and rates of breakdown will be determined by these factors.

Insight into the factors that govern breakdown of tetrahedral intermediates has also been gained by studying the hydrolysis of amide acetals. If the amine is expelled, an ester is formed, whereas elimination of an alcohol gives an amide.

$$R-C(OR')_2-NR''_2 \xrightarrow{H_2O} RC(=O)NR''_2 + 2R'OH \qquad \text{alcohol elimination}$$

$$R-C(OR')_2-NR''_2 \xrightarrow{H_2O} RC(=O)OR' + R'OH + R''_2NH \qquad \text{amine elimination}$$

The pH of the solution is of overwhelming importance in determining the course of these hydrolysis.[42] In basic solution, oxygen elimination is dominant. This is because the unprotonated nitrogen substituent is a poor leaving group and is also more effective at stabilizing the intermediate resulting from alkoxide elimination.

$$R''_2\ddot{N}-C(R)(OR')-OR' \rightarrow R''_2\overset{+}{N}=C(R)(OR') + {}^-OR' \xrightarrow{H_2O} RC(=O)NR''_2 + 2R'OH$$

In acidic solution, nitrogen is protonated and becomes a better leaving group and also loses its ability to assist in the elimination of the alkoxide. Under these circumstances, nitrogen elimination is favored.

$$R-C(\ddot{O}R')(OR')-\overset{+}{N}HR''_2 \rightarrow R-C(=\overset{+}{O}R')(OR') + R''_2NH \xrightarrow{H_2O} RC(=O)OR' + R'OH + R''_2NH$$

In analyzing the behavior of these types of tetrahedral intermediates, it must always be kept in mind that proton transfer reactions are usually fast relative to other steps. This permits the possibility that a minor species in equilibrium with the major species may be the primary intermediate. Detailed studies of kinetics, solvent isotope effects, and the nature of catalysis are the best tools for investigating the various possibilities.

It is useful to recognize that the dissociation of tetrahedral intermediates in carbonyl chemistry is related to the generation of carbocations by ionization pro-

42. R. A. McClelland, *J. Am. Chem. Soc.* **100**, 1844 (1978).

cesses. Thus, the question of which substituent on a tetrahedral intermediate is the best leaving group is similar to the issues raised in comparing the reactivity of S_N1 reactants on the basis of leaving group ability. Poorer leaving groups can react in the case of tetrahedral intermediates because of the assistance provided by the oxygen or nitrogen substituents that are present. For example, the iminium ions and protonated carbonyl compounds that are generated by breakdown of tetrahedral intermediates can be recognized as being examples of stabilized carbocations.

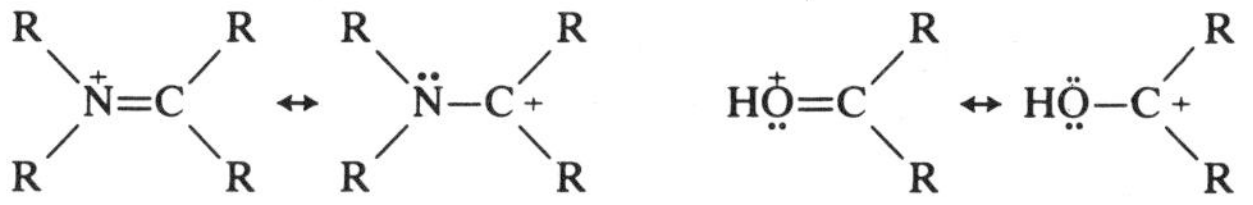

$$R_2\overset{+}{N}{=}CR_2 \leftrightarrow R_2\ddot{N}{-}\overset{+}{C}R_2 \qquad H\overset{+}{\ddot{O}}{=}CR_2 \leftrightarrow H\ddot{\underset{..}{O}}{-}\overset{+}{C}R_2$$

Keeping these relationships in mind should be helpful in understanding the reactivity of tetrahedral intermediates.

8.7. Amide Hydrolysis

The hydrolysis of amides to carboxylic acids and amines requires considerably more vigorous conditions than ester hydrolysis.[43] The reason is that the electron-releasing nitrogen substituent imparts a very significant ground state stabilization to the carbonyl group, and this stabilization is lost in the transition state leading to the tetrahedral intermediate.

$$R\overset{O}{\overset{\|}{C}}{-}NR_2 \longleftrightarrow R\overset{O^-}{\overset{|}{C}}{=}\overset{+}{N}R_2$$

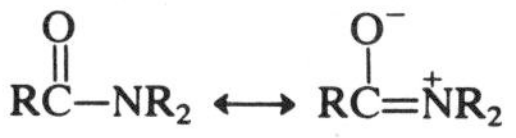

In basic solution, a mechanism similar to the $B_{AC}2$ mechanism for ester hydrolysis is believed to operate[44]:

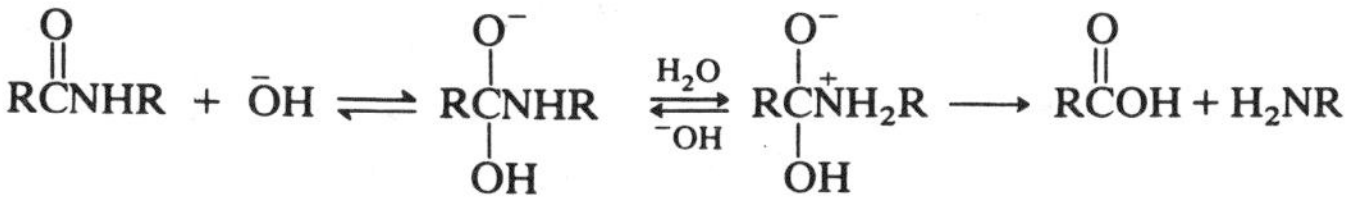

$$R\overset{O}{\overset{\|}{C}}NHR + {}^-OH \rightleftharpoons R\underset{OH}{\underset{|}{\overset{O^-}{\overset{|}{C}}}}NHR \underset{^-OH}{\overset{H_2O}{\rightleftharpoons}} R\underset{OH}{\underset{|}{\overset{O^-}{\overset{|}{C}}}}\overset{+}{N}H_2R \longrightarrow R\overset{O}{\overset{\|}{C}}OH + H_2NR$$

$$RCO_2H + {}^-OH \rightleftharpoons RCO_2^- + H_2O$$

The principal difference lies in the poorer ability of amide ions to act as leaving groups, compared to alkoxides. As a result, protonation at nitrogen is required for breakdown of the tetrahedral intermediate. Also, exchange between the carbonyl oxygen and water is extensive because reversion of the tetrahedral intermediate to reactants is faster than decomposition to products.

43. C. O'Connor, *Q. Rev. Chem. Soc.* **24**, 553 (1970).
44. M. L. Bender and R. J. Thomas, *J. Am. Chem. Soc.* **83**, 4183 (1961).

In some amide hydrolyses, the breakdown of the tetrahedral intermediate in the forward direction may proceed through formation of a dianion.[45]

$$\mathrm{RC(=O)NHR'} + {}^{-}\mathrm{OH} \rightleftharpoons \mathrm{RC(O^-)(OH)NHR'}$$

$$\mathrm{RC(O^-)(OH)NHR'} + {}^{-}\mathrm{OH} \xrightarrow{\text{slow}} \mathrm{RC(O^-)(O^-)NHR'} \longrightarrow \mathrm{RCO_2^-} + {}^{-}\mathrm{NHR}$$

$$\mathrm{R\bar{N}H} + \mathrm{H_2O} \longrightarrow \mathrm{RNH_2} + {}^{-}\mathrm{OH}$$

This variation from the ester hydrolysis mechanism also reflects the poorer leaving ability of amide ions as compared to alkoxide ions. The evidence for the involvement of the dianion comes from kinetic studies and from solvent isotope effects, which suggest that a rate-limiting proton transfer is involved.[46] The reaction is also higher than first-order in hydroxide ion under these circumstances, which is consistent with the dianion mechanism.

The mechanism for acid-catalyzed hydrolysis of amides involves attack by water on the protonated amide. An important feature of the chemistry of amides is that the most basic site in an amide is the carbonyl oxygen. Very little of the N-protonated form is present.[47] The major factor that contributes to the stability of the O-protonated form is the π-electron delocalization over the O—C—N system. No such delocalization is possible in the N-protonated form.

$$\mathrm{R{-}C({=}\overset{+}{O}H)NH_2} \leftrightarrow \mathrm{R{-}C(OH){=}\overset{+}{N}H_2} \qquad \mathrm{R{-}C({=}O)\overset{+}{N}H_3}$$

The usual hydrolysis mechanism in strongly acidic solution involves addition of water to the O-protonated amide followed by breakdown of the tetrahedral intermediate.

$$\mathrm{RC({=}\overset{+}{O}H)NH_2} + \mathrm{H_2O} \rightleftharpoons \mathrm{RC(OH)(\overset{+}{O}H_2)NH_2} \rightleftharpoons \mathrm{RC(OH)(OH)\overset{+}{N}H_3} \rightleftharpoons \mathrm{RC({=}\overset{+}{O}H)OH} + \mathrm{NH_3} \longrightarrow \mathrm{RCO_2H} + {}^{+}\mathrm{NH_4}$$

By using *N*-acyltrialkylammonium ions as models of the N-protonated amide, it has been possible to show that it would be kinetically impossible for acid-catalyzed

45. R. M. Pollack and M. L. Bender, *J. Am. Chem. Soc.* **92**, 7190 (1970).
46. R. L. Schowen, H. Jayaraman, L. Kershner, and G. W. Zuorick, *J. Am. Chem. Soc.* **88**, 4008 (1966).
47. R. J. Gillespie and T. Birchall, *Can. J. Chem.* **41**, 148, 2642 (1963); A. R. Fersht, *J. Am. Chem. Soc.* **93**, 3504 (1971); R. B. Martin, *J. Chem. Soc., Chem. Commun.*, 793 (1972); A. J. Kresge, P. H. Fitzgerald, and Y. Chiang, *J. Am. Chem. Soc.* **96**, 4698 (1974).

hydrolysis to proceed via the N-protonated form.[48] There is almost no exchange of oxygen with water during acid-catalyzed hydrolysis of amides.[49] Since a tetrahedral intermediate is involved, the lack of exchange requires that the intermediate must dissociate exclusively by elimination of the nitrogen substituent. This requirement is reasonable, since the amino group is the most basic site and is the preferred site of protonation in the tetrahedral intermediate. The protonated amine will be a better leaving group than hydroxide ion.

8.8. Acylation of Nucleophilic Oxygen and Nitrogen Groups

The conversion of alcohols to esters by O-acylation and of amines to amides by N-acylation are fundamental organic reactions. These reactions are the reverse of the hydrolytic procedures discussed in the preceding sections. In Section 3.4 of Part B, these reactions are discussed from the point of view of synthetic applications and methods.

Although the previous two sections of this chapter emphasized hydrolytic processes, two mechanisms that led to O- or N-acylation were considered. In the discussion of acid-catalyzed ester hydrolysis, it was pointed out that this reaction is reversible. Thus, it is possible to acylate alcohols by reaction with a carboxylic acid. To drive the reaction forward, the alcohol is usually used in large excess, and it may also be necessary to remove water as it is formed. This can be done by azeotropic distillation in some cases.

$$\mathrm{R\overset{O}{\overset{\|}{C}}OH + R'OH \overset{H^+}{\rightleftharpoons} R\overset{O}{\overset{\|}{C}}OR' + H_2O}$$

The second reaction that should be recalled is the aminolysis of esters. This reaction leads to the formation of amides by N-acylation.

$$\mathrm{R\overset{O}{\overset{\|}{C}}OR' + R''NH_2 \longrightarrow R\overset{O}{\overset{\|}{C}}NHR'' + R'OH}$$

The equilibrium constant for this reaction is ordinarily favorable, but the reaction is rather slow.

The most common O- and N-acylation procedures use acylating agents that are more reactive than carboxylic acids or their esters. Carboxylic acid chlorides and anhydrides react rapidly with most unhindered hydroxy and amino groups to

48. A. Williams, *J. Am. Chem. Soc.* **97**, 5281 (1975).
49. R. A. McClelland, *J. Am. Chem. Soc.* **97**, 5281 (1975); for cases in which some exchange does occur, see H. Slebocka-Tilk, R. S. Brown, and J. Olekszyk, *J. Am. Chem. Soc.* **109**, 4620 (1987).

give esters and amides, respectively.

$$R'OH + RCCl(=O) \longrightarrow RC(=O)OR' + HCl$$

$$R'OH + (RCO)_2O \longrightarrow RC(=O)OR' + RCO_2H$$

$$R'NH_2 + RCCl(=O) \longrightarrow RC(=O)NHR' + HCl$$

$$R'NH_2 + (RCO)_2O \longrightarrow RC(=O)NHR' + RCO_2H$$

The general mechanisms are well known.[50] The nucleophilic species undergoes addition at the carbonyl group, followed by elimination of the halide or carboxylate group. Acyl halides and anhydrides are reactive acylating reagents because of a combination of the inductive effect of the halogen or oxygen substituent on the reactivity of the carbonyl group and the ease with which the tetrahedral intermediate can expel such relatively good leaving groups.

$$R'OH + RC(=O)X \rightarrow RC(O^-)(X)(\overset{+}{O}HR') \rightarrow RC(=O)OR' + HX$$

$$R'_2NH + RC(=O)X \longrightarrow RC(O^-)(X)(\overset{+}{N}HR'_2) \longrightarrow RC(=O)NR'_2 + HX$$

X = halogen or carboxylate

Acylation of alcohols is usually performed in the presence of an organic base such as pyridine. The base serves two purposes. It neutralizes the protons generated in the reaction and prevents the development of high acid concentrations. Pyridine also becomes directly involved in the reaction as a *nucleophilic catalyst.*

$$RC(=O)Cl + :NC_5H_5 \longrightarrow RC(=O){-}\overset{+}{N}C_5H_5 + Cl^-$$

$$R'OH + RC(=O){-}\overset{+}{N}C_5H_5 \longrightarrow RC(O^-)(\overset{+}{N}C_5H_5)(\overset{+}{O}HR') \longrightarrow RC(=O)OR' + H\overset{+}{N}C_5H_5$$

Pyridine is more nucleophilic than an alcohol toward the carbonyl center of an acid chloride. The product that results, an acylpyridinium ion, is, in turn, more reactive toward an alcohol than the original acid chloride. The conditions required for

50. D. P. N. Satchell, *Q. Rev. Chem. Soc.* **17**, 160 (1963).

nucleophilic catalysis therefore exist, and acylation of the alcohol by acid chloride is faster in the presence of pyridine than in its absence. Among the evidence that supports this mechanism is spectroscopic observation of the acetylpyridinium ion.[51]

With more strongly basic tertiary amines such as triethylamine, another mechanism can come into play. It has been found that when methanol deuterated on oxygen reacts with acid chlorides in the presence of triethylamine, some deuterium is found α to the carbonyl group in the ester.

$$CH_3CH_2CH_2\overset{O}{\overset{\|}{C}}Cl + CH_3OD \xrightarrow[\text{octane}]{Et_3N} \underset{67\%}{CH_3CH_2CH_2\overset{O}{\overset{\|}{C}}OCH_3} + \underset{33\%}{CH_3CH_2\underset{D}{\underset{|}{C}}H\overset{O}{\overset{\|}{C}}OCH_3}$$

This finding suggests that some of the ester is formed via a ketene intermediate.[52]

$$RCH_2\overset{O}{\overset{\|}{C}}Cl + Et_3N \longrightarrow RCH{=}C{=}O + Et_3\overset{+}{N}H + Cl^-$$

$$RCH{=}C{=}O + CH_3OD \longrightarrow R\underset{D}{\underset{|}{C}}H\overset{O}{\overset{\|}{C}}OCH_3$$

Ketenes undergo rapid addition by nucleophilic attack at the *sp*-carbon atom. The reaction of tertiary amines and acid halides, in the absence of nucleophiles, is a general preparative route to ketenes.[53]

Kinetic studies of the reaction of alcohols with acid chlorides in polar solvents in the absence of basic catalysts generally reveal terms both first-order and second-order in alcohol.[54] Transition states in which the second alcohol molecule acts as a proton acceptor have been proposed.

$$R{-}\overset{O}{\overset{\|}{C}}{-}Cl \;\; (R'\underset{H}{\underset{|}{O}}\cdots HOR') \longrightarrow R'\overset{+}{O}H_2 + R\overset{O^-}{\overset{|}{C}}\underset{OR'}{\underset{|}{Cl}} \longrightarrow R\overset{O}{\overset{\|}{C}}OR' + Cl^-$$

In addition to acid chlorides and acid anhydrides, there are a number of other types of compounds that are reactive acylating agents. Many have been developed to facilitate the synthesis of polypeptides, in which mild conditions and high selectivity are required. An important group of reagents for converting carboxylic acids to active acylating agents are the carbodiimides, such as dicyclohexyl-

51. A. R. Fersht and W. P. Jencks, *J. Am. Chem. Soc.* **92**, 5432, 5442 (1970).
52. W. E. Truce and P. S. Bailey, *J. Org. Chem.* **34**, 1341 (1969).
53. R. N. Lacey, in *The Chemistry of Alkenes*, S. Patai (ed.), Wiley-Interscience Publishers, New York, 1964, pp. 1168–1170; W. E. Hanford and J. C. Sauer, *Org. React.* **3**, 108 (1947).
54. D. N. Kevill and F. D. Foss, *J. Am. Chem. Soc.* **91**, 5054 (1969); S. D. Ross, *J. Am. Chem. Soc.* **92**, 5998 (1970).

carbodiimide. The mechanism for carbodiimide-promoted amide bond formation is shown below.

$$RCOH(=O) + R'N{=}C{=}NR' \longrightarrow RC(=O)O{-}C(NHR'){=}NR'$$

$$RC(=O)OC(NHR'){=}NR' + R''NH_2 \longrightarrow RC(O^-)(\overset{+}{N}H_2R''){-}O{-}C(NHR'){=}NR' \longrightarrow RC(=O)NHR'' + R'NHC(=O)NHR'$$

The first step is addition of the carboxylic acid to the C=N bond of the carbodiimide, which generates an O-acylated urea derivative. This is a reactive acylating agent because there is a strong driving force for elimination of the urea unit, with formation of the very stable amide carbonyl group.[55] The amine reacts with the active acylating agent. In the absence of an amine, the acid would be converted to the anhydride with a second molecule of the carboxylic acid serving as the nucleophile.

Carboxylic acids react with trifluoroacetic anhydride to give mixed anhydrides that are especially useful for the acylation of hindered alcohols and phenols.

$$2,4,6\text{-}(CH_3)_3C_6H_2CO_2H + 2,4,6\text{-}(CH_3)_3C_6H_2OH \xrightarrow{(CF_3C(=O))_2O} 2,4,6\text{-}(CH_3)_3C_6H_2C(=O)O\text{-}C_6H_2(CH_3)_3\text{-}2,4,6$$

The active acylating agent may be the protonated mixed anhydride,[56] or, alternatively, the anhydride may dissociate to the acylium and trifluoroacetate ions[57]:

$$RC(=O)OC(=O)CF_3 \overset{H^+}{\rightleftharpoons} RC(=\overset{+}{O}H)OC(=O)CF_3 \quad \text{or} \quad RC(=O)OC(=O)CF_3 \overset{H^+}{\rightleftharpoons} RC{\equiv}\overset{+}{O} + CF_3CO_2H$$

Either mechanism explains why trifluoroacetylation of the nucleophile does not occur. Protonation of the anhydride would occur selectively at the more electron-rich carbonyl oxygen, rather than at the carbonyl flanked by the very electron-withdrawing trifluoromethyl group. Similarly, cleavage of the unsymmetrical anhydride would occur to give the more stable acylium ion. The trifluoroacetylium ion would be highly unstable.

Another useful family of acylating agents are the enol esters. The acetate of the enol form of acetone, isopropenyl acetate, is the most commonly used member of this group of compounds. These compounds act as acylating agents in the presence of a trace amount of an acid catalyst and are reactive toward weak nucleophiles

55. D. F. DeTar and R. Silverstein, *J. Am. Chem. Soc.* **88**, 1013, 1020 (1966).
56. R. C. Parish and L. M. Stock, *J. Org. Chem.* **30**, 927 (1965).
57. J. M. Tedder, *Chem. Rev.* **55**, 787 (1955).

such as hindered hydroxyl groups.

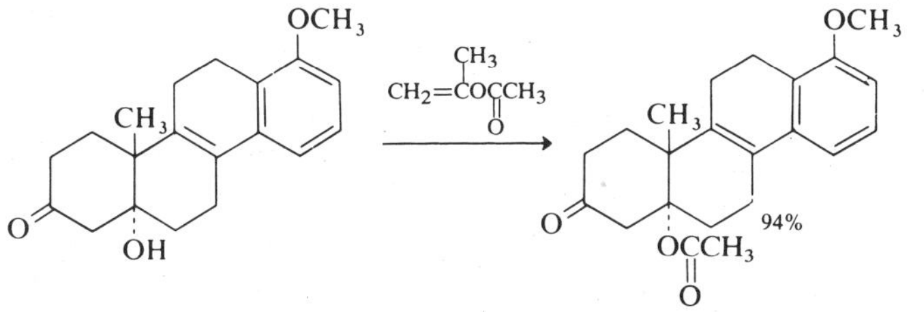

$$\text{steroid-OH} \xrightarrow{CH_2{=}C(CH_3)OC(O)CH_3} \text{steroid-OC(O)CH}_3 \quad 94\%$$

Ref. 58

The active acylating agent is presumably the C-protonated enol ester. This species would be highly reactive owing to the presence of a positively charged oxygen. An alternative possibility is that the protonated enol ester decomposes to acetone and an acylium ion, which then acts as the acylating agent.

$$(CH_3)_2C{=}\overset{+}{O}{-}\overset{O}{\overset{\|}{C}}CH_3 \rightleftharpoons CH_3\overset{O}{\overset{\|}{C}}CH_3 + \overset{+}{O}{\equiv}CCH_3$$

Section 3.4 of Part B gives additional examples of synthetically useful acylating reagents.

8.9. Intramolecular Catalysis

The reactions of carbonyl compounds have provided the testing ground for developing the facts and interpretation of intramolecular catalysis. Studies in intramolecular catalysis have been designed to determine how much more efficiently a given functional group acts as a catalyst when it is part of the reacting molecule and located in such a position that encounter between the catalytic group and the reaction center is facilitated. These studies are important to understanding biological mechanisms because enzymes are believed to act as exceedingly efficient catalysts by bringing together at the "active site" various basic, acidic, or nucleophilic groups in such a geometry that the particular reaction is highly favorable. This section will illustrate some of the facts that have emerged from these studies and the mechanistic conclusions that have been drawn.

It was pointed out in the mechanistic discussion concerning acetal and ketal hydrolysis that general acid catalysis occurs only for acetals and ketal having special structural features. Usually, specific acid catalysis operates. The question of whether general acid catalysis could be observed in intramolecular reactions has been of interest because intramolecular general acid catalysis is postulated to play a part in the mechanism of action of the enzyme lysozyme, which hydrolyzes the acetal linkage present in certain polysacharides. One group of molecules that has been examined as a model system are acetals derived from *o*-hydroxybenzoic acid (sali-

58. W. S. Johnson, J. Ackerman, J. F. Eastham, and H. A. DeWalt, Jr., *J. Am. Chem. Soc.* **78**, 6302 (1956).

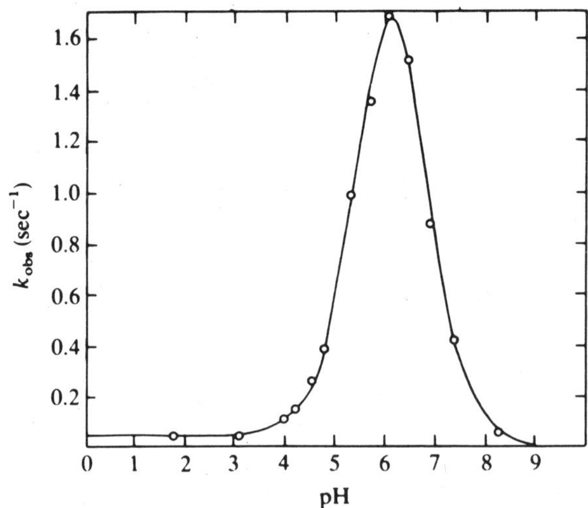

Fig. 8.5. pH-Rate profile for release of salicyclic acid from benzaldehyde disalicyl acetal. [From E. Anderson and T. H. Fife, *J. Am. Chem. Soc.* **95**, 6437 (1973). Reproduced by permission of the American Chemical Society.]

cyclic acid):

The pH-rate profile (see Fig. 8.5) indicates that of the species that are available, the monoanion of the acetal is the most reactive. The concentration of this species is at a maximum in the intermediate pH range. The neutral molecule decreases in concentration with increasing pH; the converse is true of the dianion.

The transition state for the rapid hydrolysis of the monoanion has been depicted as involving an intramolecular general acid catalysis by the carboxylic acid group, with participation by the anionic carboxylate group which becomes bound at the developing electrophilic center.

A mixed acetal of benzaldehyde, methanol, and salicyclic acid has also been studied.[59] It, too, shows a marked rate enhancement attributable to intramolecular general acid catalysis.

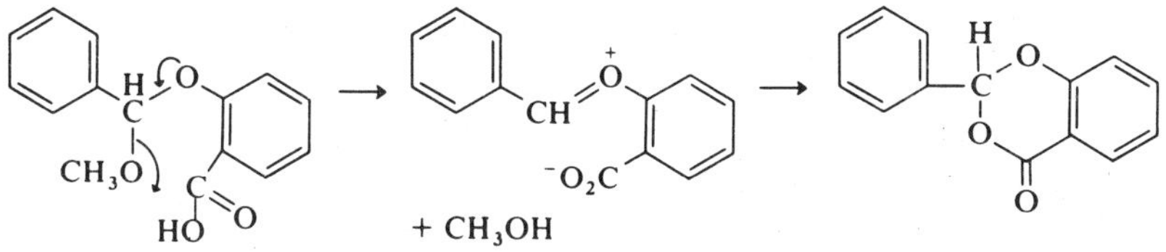

The case of intramolecular participation in ester hydrolysis has been extensively studied using acetylsalicyclic acid (aspirin) and its derivatives. The kinetic data show that the anion is hydrolyzed more rapidly than the neutral species, indicating that the carboxylate group becomes involved in the reaction in some way. Three mechanisms can be considered:

I. Nucleophilic catalysis

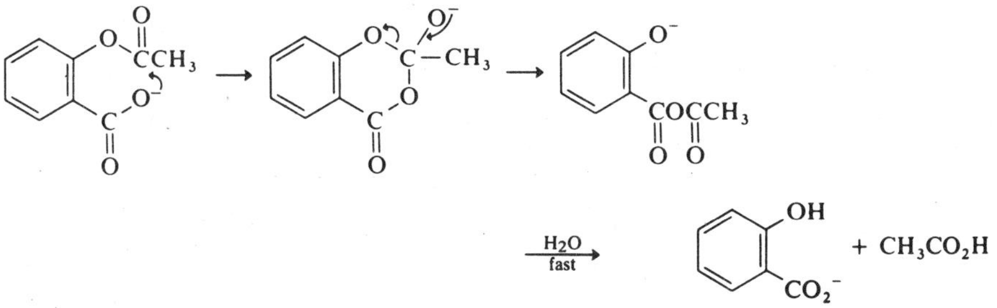

$\xrightarrow[\text{fast}]{H_2O}$ + CH_3CO_2H

II. General base catalysis

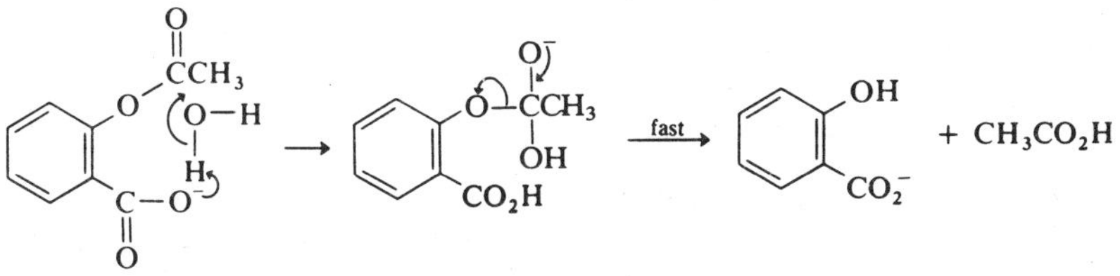

$\xrightarrow{\text{fast}}$ + CH_3CO_2H

III. General acid catalysis of hydroxide ion attack

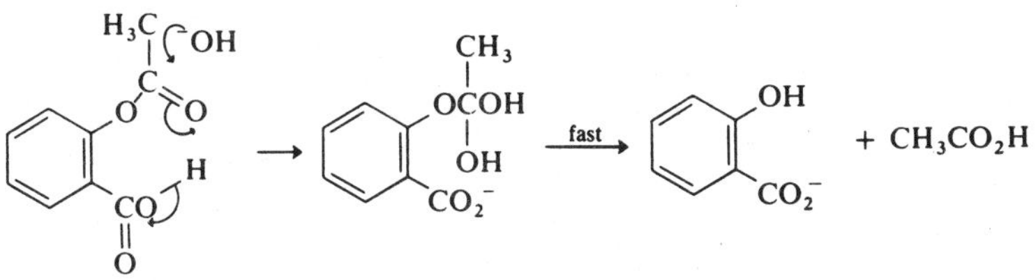

$\xrightarrow{\text{fast}}$ + CH_3CO_2H

59. T. H. Fife and E. Anderson, *J. Am. Chem. Soc.* **93**, 6610 (1971).

Mechanism III cannot be distinguished from the first two on the basis of kinetics alone, because the reactive species shown is in rapid equilibrium with the anion and therefore equivalent to it in terms of reaction kinetics.

$$\text{C}_6\text{H}_4(\text{OCOCH}_3)(\text{CO}_2^-) + \text{H}_2\text{O} \rightleftharpoons \text{C}_6\text{H}_4(\text{OCOCH}_3)(\text{CO}_2\text{H}) + {}^-\text{OH}$$

Mechanism I has been ruled out by an isotopic labeling experiment. The mixed anhydride of salicyclic acid and acetic acid is an intermediate if nucleophilic catalysis occurs by mechanism I. This molecule is known to hydrolyze in water with about 25% incorporation of solvent water into the salicyclic acid.

$$\text{C}_6\text{H}_4(\text{OH})(\text{COOCOCH}_3) + \text{H}_2\overset{*}{\text{O}} \longrightarrow \underset{25\%}{\text{C}_6\text{H}_4(\text{OH})(\text{CO}\overset{*}{\text{O}}\text{H})} + \text{CH}_3\text{CO}_2\text{H} + \underset{75\%}{\text{C}_6\text{H}_4(\text{OH})(\text{CO}_2\text{H})} + \text{CH}_3\text{C}\overset{*}{\text{O}}_2\text{H}$$

Hydrolysis of aspirin in $H_2{}^{18}O$ leads to no incorporation of ^{18}O into the product salicyclic acid, ruling out the anhydride as an intermediate and thereby excluding mechanism I.[60] The general acid catalysis of mechanism III can be ruled out on the basis of failure of other nucleophiles to show evidence for general acid catalysis by the neighboring carboxylic acid group. Since there is no reason to believe hydroxide should be special in this way, mechanism III is eliminated. Thus, mechanism II, general base catalysis of hydroxide ion attack, is believed to be the correct description of the hydrolysis of aspirin.

The extent to which intramolecular nucleophilic catalysis of the type depicted in mechanism I is important is a function of the leaving ability of the alkoxy group. This has been demonstrated by the study of the hydrolysis of a series of monoesters of phthalic acid.

$$\text{C}_6\text{H}_4(\text{CO}_2^-)(\text{CO}_2\text{R}) + \text{HO}^- \longrightarrow \text{C}_6\text{H}_4(\text{CO}_2^-)_2 + \text{ROH}$$

Nucleophilic participation is important only for esters of alcohols that have $pK_a \leq 13$. Specifically, phenyl and trifluoroethyl esters show nucleophilic catalysis, but methyl and 2-chloroethyl esters do not.[61] This result reflects the fate of the tetrahedral intermediate that results from nucleophilic participation. For relatively acidic

60. A. R. Fersht and A. J. Kirby, *J. Am. Chem. Soc.* **89**, 4857 (1967).
61. J. W. Thanassi and T. C. Bruice, *J. Am. Chem. Soc.* **88**, 747 (1966).

alcohols, the alkoxide group can be eliminated, leading to hydrolysis via nucleophilic catalysis.

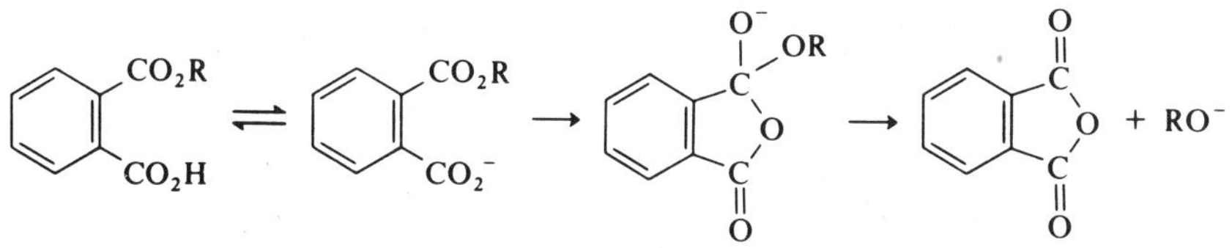

For less acidic alcohols, nucleophilic participation is ineffective because of the low tendency for such alcohols to function as leaving groups. The tetrahedral intermediate formed by intramolecular addition simply returns to starting material because the carboxylate is a much better leaving group than the alkoxide. In contrast to aspirin itself, acetyl salicylates with electron-withdrawing groups (*o*- and *p*-nitro analogs) hydrolyze via the nucleophilic catalysis mechanism.[62]

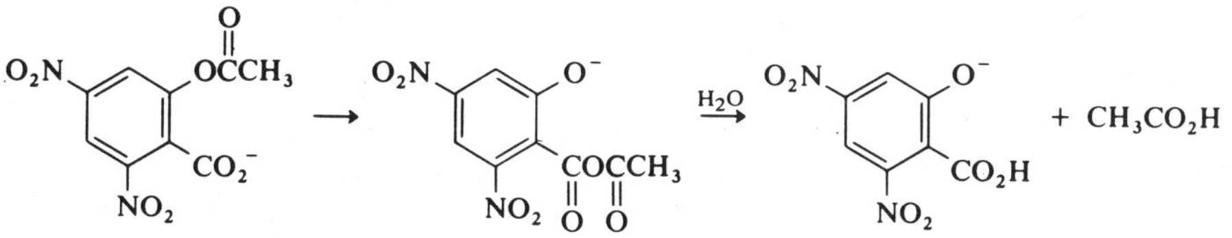

Intramolecular catalysis of ester hydrolysis by nitrogen nucleophiles is also important. The role of imidazole rings in intramolecular catalysis has received particularly close scrutiny. There are two major reasons for this. One is that the imidazole ring of the histidine residue in proteins is believed to frequently be involved in enzyme-catalyzed hydrolyses. Secondly, the imidazole ring has several possible catalytic functions including action as a general acid in the protonated form, action as a general base in the neutral form, and action as a nucleophile in the neutral form. A study of a number of derivatives of structure **3** was undertaken to distinguish between the importance of these various possible mechanisms as a function of pH.[63]

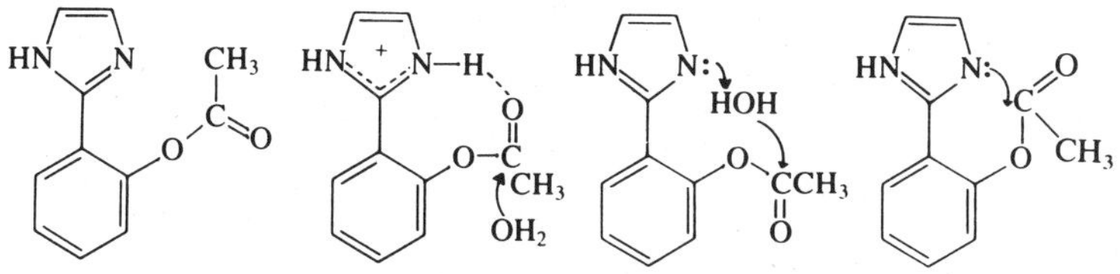

3

general acid catalysis

general base catalysis

nucleophilic catalysis

The relative importance of the potential catalytic mechanisms depends on pH, which also determines the concentration of the other participating species such as water,

62. A. R. Fersht and A. J. Kirby, *J. Am. Chem. Soc.* **89**, 5960 (1967); **90**, 5818 (1968).
63. G. A. Rogers and T. C. Bruice, *J. Am. Chem. Soc.* **96**, 2463 (1974).

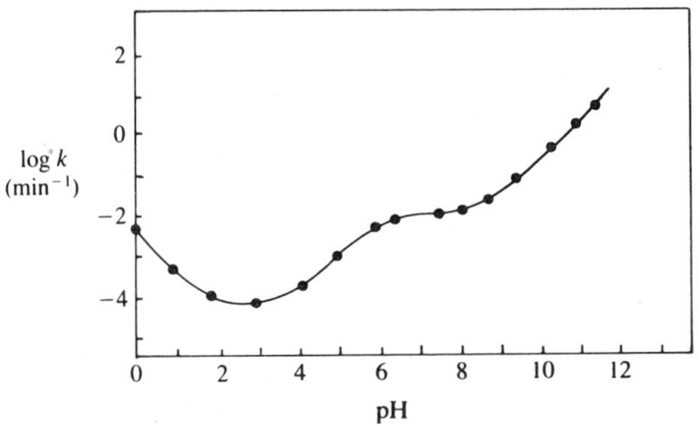

Fig. 8.6. pH-Rate profile for compound **3**. Reproduced with permission from Ref. 63.

hydronium ion, and hydroxide ion. At low pH, the general acid catalysis mechanism dominates, and comparison with analogous systems where the intramolecular proton transfer is not available suggests that the intramolecular catalysis results in a 25–100-fold rate enhancement. At neutral pH, the intramolecular general base mechanism begins to operate. It is estimated that the catalytic effect for this mechanism is a factor of about 10^4. Although the nucleophilic catalysis mechanism was not observed in the parent compound, it occurred in certain substituted derivatives. The change in mechanism with pH for compound **3** gives rise to the pH–rate profile shown in Fig. 8.6.

The rates at the extremeties $pH < 2$ and $pH > 9$ are proportional to $[H^+]$ and $[^-OH]$, respectively, and represent the specific proton-catalyzed and hydroxide-catalyzed mechanisms. The region at pH 2–4 is the area where intramolecular general acid catalysis operates. At pH 6–8, the intramolecular general base catalysis mechanism is dominant. In the absence of the intramolecular catalytic mechanisms, the rates of the H^+- and ^-OH-catalyzed reactions would decrease in proportion to the concentration of the catalytic species to a minimum value representing the "uncatalyzed water hydrolysis." An estimate of the effectiveness of the intramolecular mechanisms can be made by extrapolating the lines which are proportional to $[H^+]$ and $[^-OH]$. The extent to which the actual rate lies above these extrapolated lines in the pH range 2–7 represents the contribution from the intramolecular catalysis.

Intramolecular participation of the *o*-hydroxy group in aminolysis of phenyl salicylate has been established by showing that such compounds are more reactive than analogs lacking the hydroxyl substituent. This reaction exhibits overall third-order kinetics, second-order in the reacting amine. Similar kinetics are observed in the aminolysis of simple esters. Both intermolecular general base catalysis (by the second amine molecule) and intramolecular general acid catalysis (by the hydroxyl group) apparently occur.[64]

64. F. M. Menger and J. H. Smith, *J. Am. Chem. Soc.* **91**, 5346 (1969).

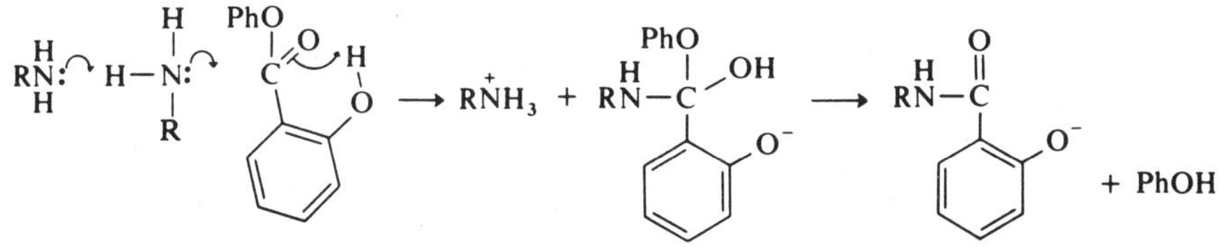

This mechanism can reduce the activation energy of the reaction in at least two ways. The partial transfer of a proton to the carbonyl oxygen increases the electrophilicity of the carbonyl. Likewise, partial deprotonation of the amino group increases its nucleophilicity.

Certain molecules that can permit concerted proton transfers are efficient catalysts for reaction at carbonyl centers. An example that can be cited is the catalytic effect that 2-pyridone has on the aminolysis of esters. Although neither a strong base ($pK_{a_H}{}^+ = 0.75$) nor a strong acid ($pK_a = 11.6$), 2-pyridone is an effective catalyst of the reaction of *n*-butylamine with 4-nitrophenyl acetate.[65] The overall rate is more than 500 times greater when 2-pyridone acts as the catalyst than when a second molecule of butylamine (acting as a general base) is present in the transition state. 2-Pyridone has been called a *tautomeric catalyst* to emphasize its role in proton transfer. Such molecules are also called *bifunctional catalysts*, since two atoms in the molecule are involved in the proton transfer process.

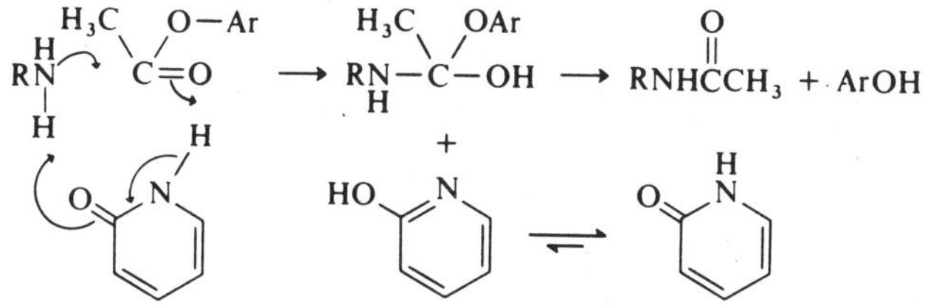

2-Pyridone also catalyzes epimerization of the anomeric position of the tetramethyl ether of glucose. The mechanism involves two double proton transfers. The first leads to a ring-opened intermediate, and the second results in ring closure to the isomerized product.

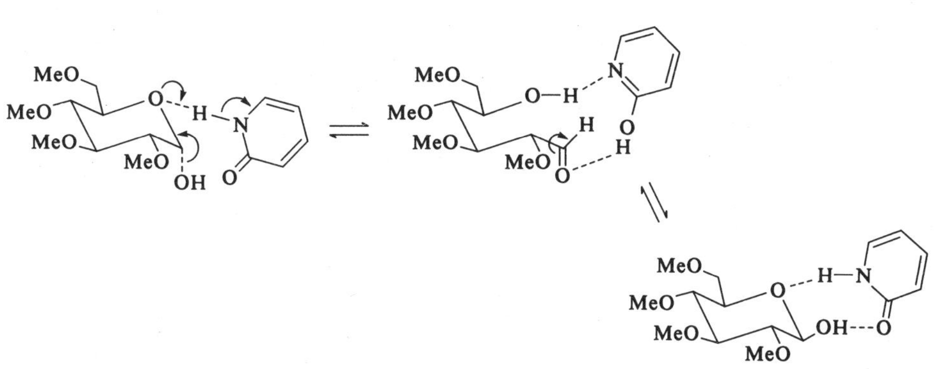

65. P. R. Rony, *J. Am. Chem. Soc.* **91**, 6090 (1969).

Another type of bifunctional catalysis has been noted with α, ω-diamines in which one of the amino groups is primary and the other tertiary. These substituted diamines are from several times to as much as 100 times more reactive toward imine formation that similar monofunctional amines.[66] This is attributed to a catalytic intramolecular proton transfer.

$$R_2C{=}O + H_2N(CH_2)_n\overset{+}{N}H(CH_3)_2 \rightleftharpoons R_2C(O^-)\text{-}\overset{+}{N}H_2\text{-}(CH_2)_n\text{-}\overset{+}{N}H(CH_3)_2 \rightleftharpoons R_2C(OH)\text{-}NH\text{-}(CH_2)_n\text{-}\overset{+}{N}H(CH_3)_2$$

$$\rightleftharpoons R_2C(\overset{+}{O}H_2)\text{-}NH\text{-}(CH_2)_n\text{-}N(CH_3)_2 \xrightarrow{-H_2O} R_2C{=}N(CH_2)_n\overset{+}{N}H(CH_3)_2$$

The rate enhancement is greatest for $n = 2$ (1000) but still significant for $n = 3$ (a factor of 10). As the chain is lengthened to $n = 4$ and $n = 5$, the rate enhancement, if any, is minor. This relationship reflects the fact that when $n = 4$ or 5, the transition state for the intramolecular proton transfer would have to involve rings of nine and ten atoms, respectively, and would not be geometrically advantageous. The particularly rapid reaction when $n = 2$ corresponds to the possibility for a proton transfer via a seven-membered cyclic transition state. Assuming that the proton is transferred in a colinear fashion through a hydrogen bond, this represents a favorable transition state geometry.

$$R_2C(\overset{\delta+}{O}H\cdots H\cdots \overset{\delta+}{N}(CH_3)_2)\text{-}NH\text{-}CH_2\text{-}CH_2$$

These examples serve to introduce the idea of intramolecular catalysis and the fact that favorable juxtaposition of acidic, nucleophilic, or basic sites can markedly accelerate some of the common reactions of carbonyl compounds. It is believed that nature has evolved a similar strategy of optimal placement of functional groups to achieve the catalytic activity of enzymes. The functional groups employed to accomplish this are those present on the amino acid residues found in proteins. The acidic sites available include phenolic or carboxylic acid groups from tyrosine, glutamic acid, and aspartic acid. Basic sites include the imidazole ring in histidine and the ε-amino group of lysine. This latter group and the amidine group in arginine are normally protonated at physiological pH and can serve as cationic centers or general acids, as well. Thiol (cysteine) and hydroxyl (threonine and serine) groups and the deprotonated carboxyl groups of glutamic acid and aspartic acid are potential

66. J. Hine, R. C. Dempsey, R. A. Evangelista, E. T. Jarvi, and J. M. Wilson, *J. Org. Chem.* **42**, 1593 (1977); J. Hine and Y. Chou, *J. Org. Chem.* **46**, 649 (1981).

nucleophilic sites. The student interested in examining further the mechanism of enzymatic reactions will find these topics covered in detail in several of the texts listed as general references.

General References

M. L. Bender, *Mechanisms of Homogeneous Catalysis from Protons to Proteins*, Wiley-Interscience, New York, 1971.
T. C. Bruice and S. J. Benkovic, *Bioorganic Mechanisms*, W. A. Benjamin, New York, 1966.
W. P. Jencks, *Catalysis in Chemistry and Enzymology*, McGraw-Hill, New York, 1969.
A. J. Kirby and A. R. Fersht, in *Progress in Bioorganic Chemistry*, Vol. 1, E. T. Kaiser and R. J. Kezdy (eds.), Wiley-Interscience, New York, 1971, pp. 1–82.
S. Patai (ed.), *The Chemistry of the Carbonyl Group*, Wiley-Interscience, New York, 1969.
S. Patai (ed.), *The Chemistry of Carboxylic Acids and Esters*, Wiley-Interscience, New York, 1969.
J. E. Zabricky (ed.), *The Chemistry of Amides*, Wiley-Interscience, New York, 1970.

Problems

(*References for these problems will be found on page* 780.)

1. The hydrates of aldehydes and ketones are considerably more acidic than normal alcohols (p$K \approx$ 16–19). How would you account for this fact? Some reported values are shown below. Explain the order of relative acidity.

Hydrate	pK
$H_2C(OH)_2$	13.4
$Cl_3CCH(OH)_2$	10.0
$PhC(OH)_2(CF_3)$	10.0
$O_2N-C_6H_4-C(OH)_2(CF_3)$	9.2

2. Suggest explanations for each of the following observations.

   (a) The equilibrium constant for cyanohydrin formation from 3,3-dimethyl-2-butanone is 40 times larger than that for acetophenone.

   (b) The rate of release of *p*-nitrophenoxide from compound **A** is independent of pH in aqueous solution of pH > 10.

$$\text{2-HS-C}_6\text{H}_4\text{-N(CH}_3\text{)-C(=O)-O-C}_6\text{H}_4\text{-NO}_2$$

A

(c) Ester **B** undergoes alkaline hydrolysis 8300 times faster than ester **C** in aqueous dioxane.

O=CH CO$_2$CH$_3$ CO$_2$CH$_3$

B **C**

(d) Under comparable conditions, the general-base-catalyzed elimination of bisulfite ion from **D** is about ten times greater than from **E**.

OH OH
CH$_3$O—C$_6$H$_4$—CCH$_3$ CH$_3$O—C$_6$H$_4$—CH
SO$_3^-$ SO$_3^-$

D **E**

(e) The rate of isotopic exchange of the carbonyl group in tropone (**F**) and 2,3-diphenylcyclopropenone (**G**) is much less than for acetophenone.

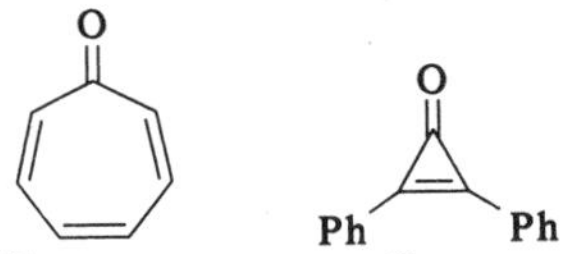

O O

Ph Ph
F **G**

3. Arrange the carbonyl compounds in each group in order of decreasing rate of hydrolysis of their respective diethyl acetals or ketals. Explain your reasoning.
   (a) acetaldehyde, chloroacetaldehyde, crotonaldehyde
   (b) acetaldehyde, formaldehyde, acetone
   (c) cyclopentanone, cyclohexanone, camphor
   (d) acetone, methyl *t*-butyl ketone, methyl neopentyl ketone
   (e) benzaldehyde, *p*-methoxybenzaldehyde, butyraldehyde.

4. The acid-catalyzed hydrolysis of 2-alkoxy-2-phenyl-1,3-dioxolanes has been studied. The initial step is rate-determining under certain conditions and is described by the rate law given below, which reveals general acid catalysis.

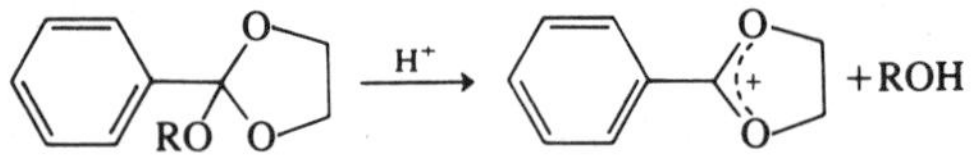

RO O $\xrightarrow{H^+}$ O + +ROH

$$k_{obs} = k_{H^+}[H^+] + k_{H_2O}[H_2O] + k_{HA}[HA]$$

By determining k_{HA} for several different buffer catalysts and each of several alkoxy leaving groups, it was determined that there was a relationship between the Brønsted coefficient α and the structure of the alkoxy leaving group. The data are given and show that α decreases as the alkoxy group becomes less basic.

RO—	α
$Cl_2CHCH_2O—$	0.69
$ClCH_2CH_2O—$	0.80
$CH_3OCH_2CH_2O—$	0.85
$CH_3O—$	0.90

What information is provided by the fact that the Brønsted α decreases as the acidity of the alcohol increases? Discuss these results in terms of a three-dimensional potential energy diagram with the extent of O—H bond formation and the extent of C—O bond breaking taken as the reaction progress coordinates.

5. Each of the following molecules has been considered to be capable of some form of intramolecular catalysis of ester hydrolysis. For each substrate indicate one or more mechanisms by which intramolecular catalysis might occur. Depict a transition state arrangement that shows this catalysis.

(a) 8-quinolyl ester: $OCCH_3$ (C=O) on quinoline ring with N

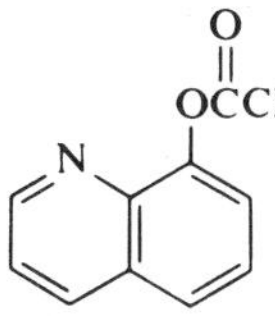

(d) phenyl ester $OCCH_3$ (C=O) with ortho imidazolyl (HN, N)

(b) $OCCH_2CH_2CO_2H$ (C=O) on benzene with ortho OH

(e) CO_2CH_3 on benzene with ortho CPh (C=O)

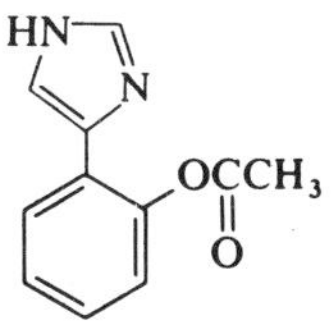

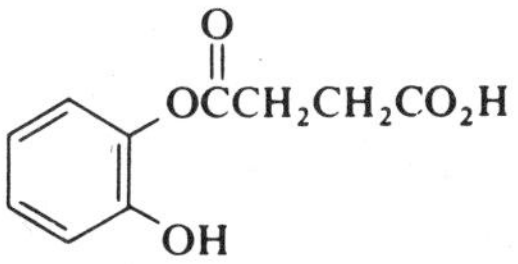

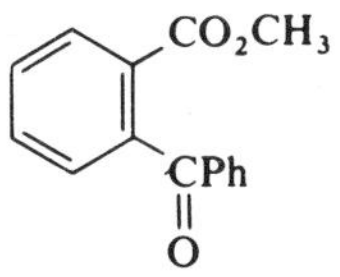

(c) $NHCCH_3$ (C=O) on benzene with ortho CO_2CH_3

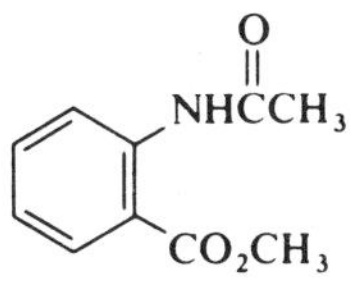

(f) $OC(CH_2)_3N(CH_3)_2$ (C=O) on benzene

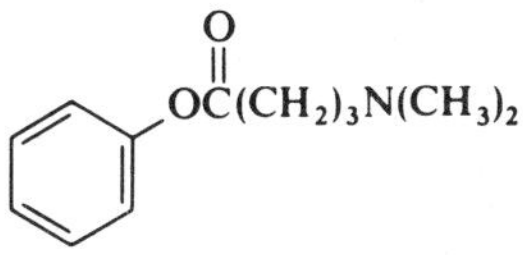

6. Consider the alkaline pH region of the pH-rate profile in Fig. 8.4 (p. 450), which indicates a rate independent of pH. The rate-controlling reaction in this region is

$$HO^- + PhCH{=}\overset{+}{N}HC(CH_3)_3 \longrightarrow PhCH(OH)NHC(CH_3)_3$$

Show that the rate of this reaction is pH independent, despite the involvement of two species, the concentrations of which are pH dependent.

7. Derive the general expression for the observed rate constant for hydrolysis of **A** as a function of pH. Assume, as is the case experimentally, that intramolecular

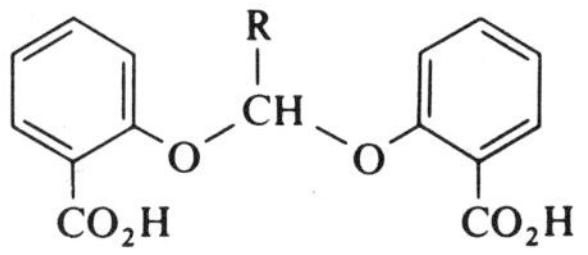

A

general acid catalysis completely outweighs intermolecular catalysis by hydronium ion in the pH range of interest. Does the form of your expression agree with the pH-rate profile given for this reaction in Fig. 8.5 (p. 480)?

8. Optically pure dipeptide is obtained when the *p*-nitrophenyl ester of *N*-benzolyl-L-leucine is coupled with glycine ethyl ester in ethyl acetate:

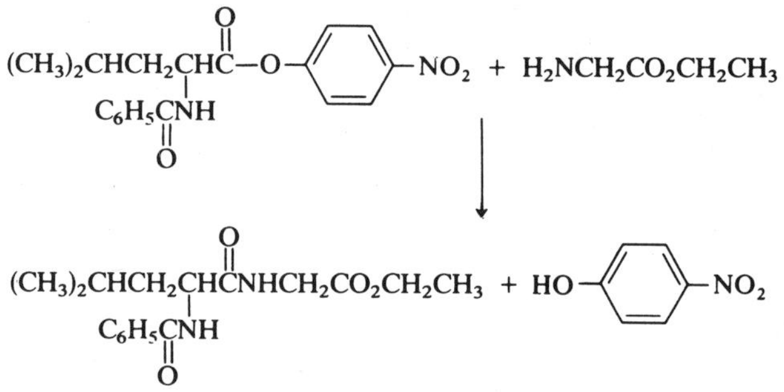

$(CH_3)_2CHCH_2CH(NHCOC_6H_5)C(=O)\text{—O—}C_6H_4\text{—}NO_2 + H_2NCH_2CO_2CH_2CH_3$

$\downarrow$

$(CH_3)_2CHCH_2CH(NHCOC_6H_5)C(=O)NHCH_2CO_2CH_2CH_3 + HO\text{—}C_6H_4\text{—}NO_2$

If, however, the *p*-nitrophenyl ester of *N*-benzoyl-L-leucine is treated with l-methylpiperidine in chloroform for 30 min, then coupled with glycine ethyl ester, the dipeptide isolated is almost completely racemic. Furthermore, treatment of the *p*-nitrophenyl ester of *N*-benzoyl-L-leucine with 1-methylpiperidine alone leads to the formation of a crystalline material, $C_{13}H_{15}NO_2$, having strong IR bands at 1832 and 1664 cm^{-1}. Explain these observations, and suggest a reasonable structure for the crystalline product.

9. Offer as complete as possible explanations, based on structural and mechanistic concepts, of the following observations.

   (a) The bicyclic lactam **A** hydrolyzes 10^7 times faster than the related monocyclic compound **B**.

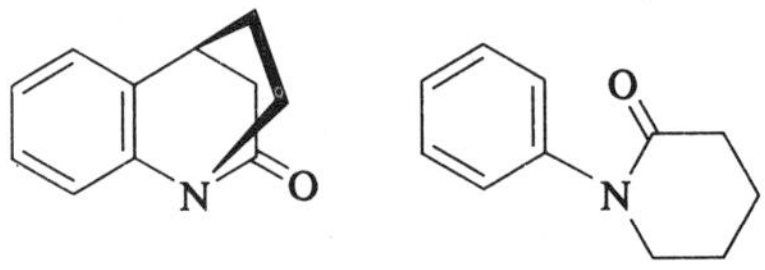

A **B**

(b) Leaving groups, X, solvolyze from structure **C** at a rate which is 10^{-13} less than that for monocyclic model **D**.

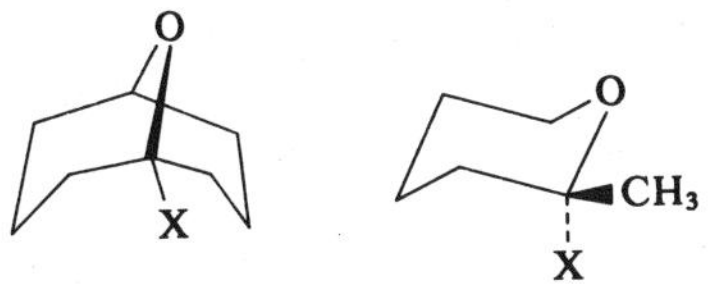

C D

10. Analyze the factors which would determine stereoselectivity in the addition of organometallic compounds to the following carbonyl compounds. Predict the major product.

(a) CH_3, $(CH_3)_2CH$, =O $\xrightarrow{n\text{-BuLi}}$

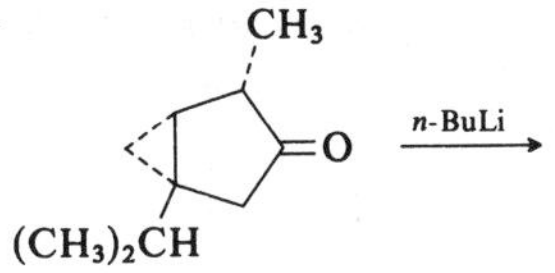

(b) =O $\xrightarrow{\text{PhLi}}$

(c) $PhCH(CH_3)C(=O)CH(CH_3)_2 \xrightarrow{\text{PhMgBr}}$

11. Indicate which compound in each of the following pairs will have the more negative standard free-energy change for hydrolysis at pH 7:

(a) $CH_3CO_2CH_3$ *or* $CH_3C(=O)SCH_3$

(b) $CH_3CO_2CH_3$ *or* $CH_3CO_2PO_3H_2$

(c) $CH_3CO_2CH_3$ *or* $H_2NCH_2CO_2CH_3$

(d) $CH_3C(=O)CH_2C(=O)SCH_3$ *or* $CH_3C(=O)SCH_3$

(e) $CH_3CO_2CH_3$ *or* $CH_3C(=O)$–N (imidazole, N)

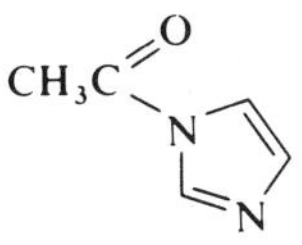

(f) $CH_3CO_2CH_3$ *or* $CH_3C(=O)N(CH_3)_2$

12. Sodium acetate reacts with *p*-nitrophenyl benzoates to give mixed anhydrides if the reaction is conducted in a polar aprotic solvent in the presence of a crown ether. The reaction is strongly accelerated by quaternary nitrogen groups substituted at the *ortho* position. Explain the basis for the enhanced reactivity of these compounds.

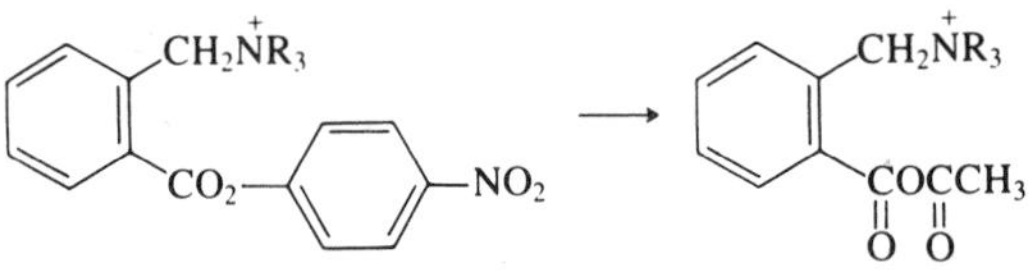

$CH_2\overset{+}{N}R_3$ ... CO_2–C_6H_4–NO_2 → $CH_2\overset{+}{N}R_3$ ... $\underset{O}{\overset{\|}{C}}O\underset{O}{\overset{\|}{C}}CH_3$

Introduction of $^+NR_3$ for H accelerates the reaction by a factor of $> 10^3$

13. The kinetics of the hydrolysis of some imines derived from benzophenone and primary amines revealed the normal dependence of mechanism on pH with rate-determining nucleophilic attack at high pH and rate-determining decomposition of the tetrahedral intermediate at low pH. The simple primary amines show a linear correlation between the rate of nucleophilic addition and the basicity of the amine. Several diamines which were included in the study, in particular **A**, **B**, and **C**, all showed a positive (more reactive) deviation from the correlation line for the simple amines. Why might these amines be more reactive than predicted on the basis of their basicity?

$H_2NCH_2CH_2N(C_2H_5)_2$ (**A**) $H_2NCH_2CH_2N$(morpholine, O) (**B**) $H_2NCH_2CH_2$–(2-pyridyl, N) (**C**)

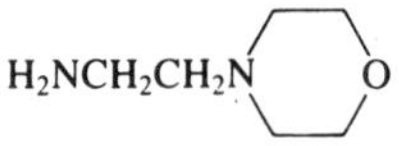

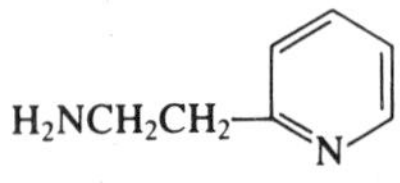

14. The following data give the dissociation constants for several acids that catalyze hydration of acetaldehyde. Also given are the rate constants for the hydration reaction catalyzed by each acid. Treat the data according to the Brønsted equation, and comment on the mechanistic significance of the result.

Acid	K	k_{hydr}
Formic	1.77×10^{-4}	1.74
Phenylacetic	4.9×10^{-5}	0.91
Acetic	1.75×10^{-5}	0.47
Pivalic	9.4×10^{-6}	0.33

15. 1,1-Diphenylthioalkanes react with mercuric fluoride to give 1-fluoro-1-phenylthioalkanes. Provide a detailed description of a likely mechanism for this reaction. Consider such questions as: (1) Is an S_N1 or an S_N2 process most likely to be involved? (2) Would NaF cause the same reaction to occur? (3) Why is only one of the phenylthio groups replaced?

$(PhS)_2C(R)H \xrightarrow[CH_3CN]{HgF_2} (PhS)(F)C(R)H$

16. The acid-catalyzed hydrolysis of thioacetanilide can follow two courses.

$$C_6H_5NHC(=S)CH_3 \xrightarrow[H_2O]{H^+} \begin{cases} \text{A: } CH_3C(=S)OH \rightleftharpoons CH_3C(SH)=O + H_2N\text{–}C_6H_5 \\ \text{or} \\ \text{B: } CH_3C(=O)NH\text{–}C_6H_5 + H_2S \end{cases}$$

The product analysis permits determination of the amount of product formed by each path, as a function of the acidity of the solution. The results are as shown:

H_2SO_4 (% by weight)	3.2	6.1	12	18	36	48
% following path A	50	55	65	75	96	100

Provide a mechanism in sufficient detail to account for the change in product ratio with acid strength.

17. The rates of hydrolysis of the ester group in compounds **A** and **B** have been compared. The effect of an added metal ion (Ni^{2+}) on the rate of hydrolysis has been studied, and the observed rate constants for attack by ^{-}OH are tabulated. Suggest the most favorable transition state structure for the addition step of the hydrolysis reaction with each substrate under each set of conditions. Discuss the relationship between the structures of these transition states and the relative rates of attack by hydroxide ion.

A: $^{-}O_2C$–(pyridine-2,6-diyl)–C(=O)–O–C_6H_5

k_{HO^-}: $3.0 \times 10^2\ M^{-1}\ s^{-1}$
k_{HO^-} in presence of excess Ni^{2+}: $2.8 \times 10^6\ M^{-1}\ s^{-1}$

B: $^{-}O_2C$–(pyridine-2,6-diyl)–C(=O)–O–C_6H_4–2-CO_2H

k_{HO^-}: $7.1 \times 10^1\ M^{-1}\ s^{-1}$
k_{HO^-} at pH values where salicylic acid group is not ionized: $2.7 \times 10^5\ M^{-1}\ s^{-1}$
k_{HO^-} when salicylic acid group is not ionized and excess Ni^{2+} is present: $2.7 \times 10^7\ M^{-1}\ s^{-1}$

18. Data pertaining to substituent effects on the acid-catalyzed hydrolysis of mixed methyl aryl acetals of benzaldehyde are given below. The reactions exhibited general acid catalysis, and the Brønsted α values are tabulated. Discuss the information provided by these data about the transition state for the first hydrolysis step, making reference to a diagram showing the location of the transition state as a function of O—H bond formation and C—O bond breaking.

$$\text{Ar}-\underset{\text{OCH}_3}{\overset{\text{OAr}'}{\text{CH}}} + H_2O \xrightarrow{H^+} \text{ArCH=O} + \text{Ar'OH} + CH_3OH$$

Series I, substituent in Ar			Series II, substituent in Ar′		
X	k_{cat}[a]	α	X	k_{cat}[a]	α
m-NO_2	2.7×10^{-4}	1.05	m-NO_2	8.85×10^{-2}	0.49
m-F	2.2×10^{-3}	0.92	m-Br	4.7×10^{-2}	0.65
m-CH_3O	9.6×10^{-3}	0.78	m-F	2.45×10^{-2}	0.67
H	1.3×10^{-2}	0.77	m-CH_3O	2.55×10^{-2}	0.71
p-CH_3	1.1×10^{-1}	0.72	H	1.3×10^{-2}	0.77
p-CH_3O	2.8×10^{-1}	0.68	p-CH_3	1.3×10^{-2}	0.88
			p-CH_3O	1.65×10^{-2}	0.96

a. Rate constant for catalysis by acetic acid.

19. The hydrolysis of the lactone **A** shows a significant catalysis by acetate ion in acetate buffer, with the rate expression being

$$k_{obs} = 1.6 \times 10^{-6} + 6.4 \times 10^{-4}[H^+] + 2.08 \times 10^{-5}[OAc^-] + 49[^-OH]$$

This results in a pH-rate profile as shown in Fig. 8.P19, with the acetate catalysis being significant in the pH range 3–6. Discuss how this catalysis by acetate ion might occur. What are the most likely mechanisms for hydrolysis at pH < 2 and pH > 7, where the rates are linear in $[H^+]$ and $[^-OH]$, respectively?

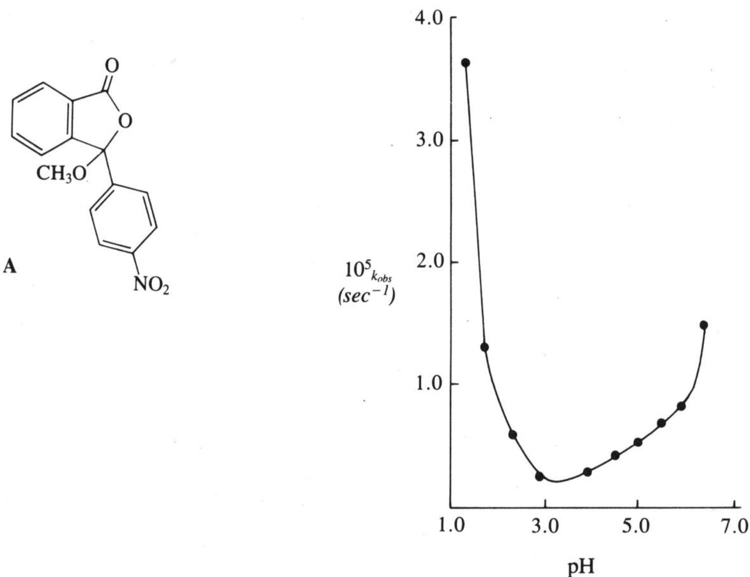

Fig. 8.P19. pH-Rate profile for hydrolysis of **A** in buffered aqueous solution at 70°C.

20. Some data on substituent effects for the reaction of trifluoroacetanildes with methoxide ion in methanol and methanol-O-*d* are given below. Calculate the isotope effect for each system. Plot the rate data against appropriate Hammett substituent constants. What facets of the data are in agreement with a normal addition-elimination sequence passing through a tetrahedral intermediate? What facets of the data indicate additional complications? Can you propose a mechanism that is consistent with all the data given?

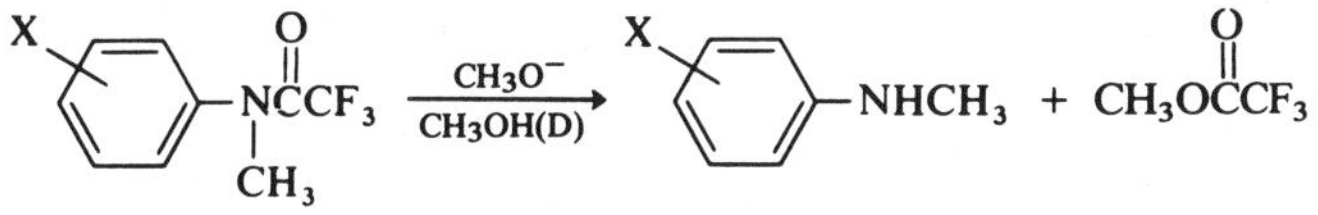

$$X\text{-}C_6H_4\text{-}N(CH_3)C(=O)CF_3 \xrightarrow[CH_3OH(D)]{CH_3O^-} X\text{-}C_6H_4\text{-}NHCH_3 + CH_3OC(=O)CF_3$$

X	k_{CH_3OH} [a]	k_{CH_3OD} [a]
m-NO_2	5.75	8.13
m-Br	0.524	0.464
p-Cl	0.265	0.274
p-Br	0.349	0.346
m-Cl	0.513	0.430
m-OCH_3	0.110	0.101
H	0.104	0.0899
m-CH_3	0.0833	0.0595
p-CH_3	0.0729	0.0451
p-OCH_3	0.0564	0.0321

a. Second-order rate constants in $M^{-1}\ sec^{-1}$

21. The halogenation of simple ketones such as acetone can proceed through the enol or enolate. By applying the steady-state condition to the enolate, derive a kinetic expression for reaction of acetone with any halogenating agent X-Y in a buffered solution where both C-protonation and O-protonation of the enolate can compete with halogenation. Show that this rate expression predicts that halogenation will be zero-order in halogenating agent under some conditions but first-order in halogenating agent under other conditions.

22. The order of reactivity toward hydrolysis of the cyclic acetals shown below is **A** ≪ **B** ≪ **C**. Offer an explanation for this difference in reactivity.

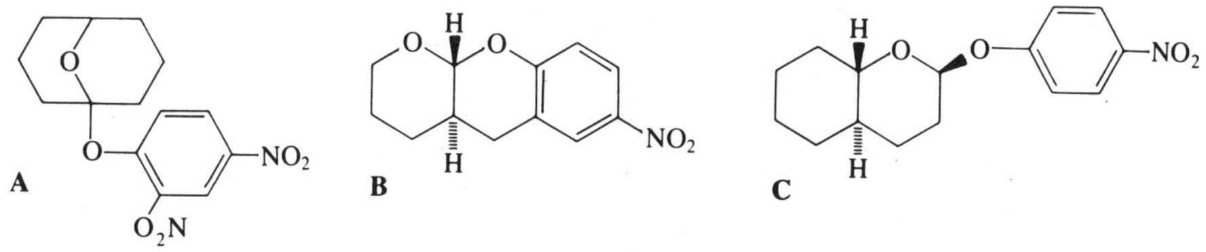

A B C

23. Examine the structure of the reactants given and the pH-rate profiles (Figs. 8.P23a-d) of the reactions in question. Offer explanations for the response of the observed reaction rate to the pH for each case.

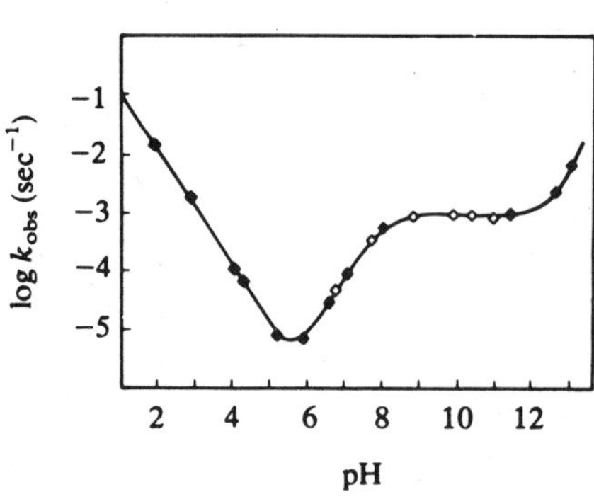

Fig. 8.P23a. (Reproduced from problem reference 23a by permission of the American Chemical Society.)

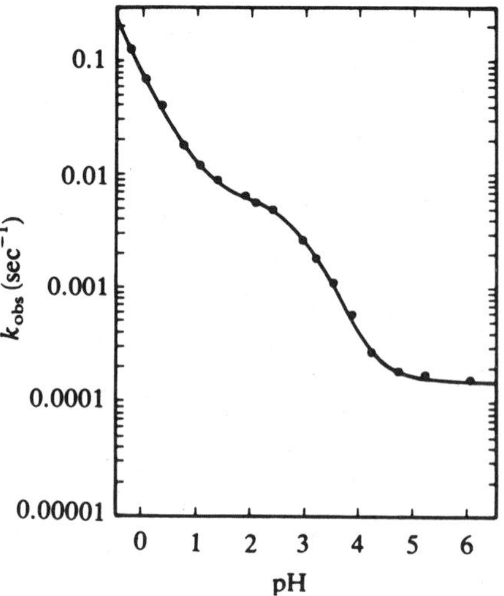

Fig. 8.P23b. (Reproduced from problem reference 23b by permission of the American Chemical Society.)

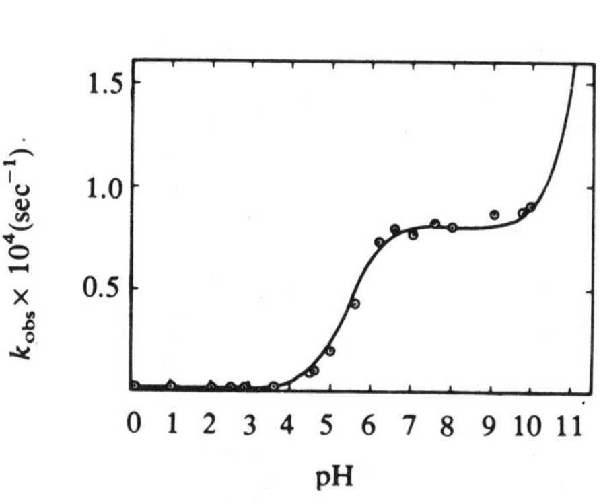

Fig. 8.P23c. (Reproduced from problem reference 23c by permission of the American Chemical Society.)

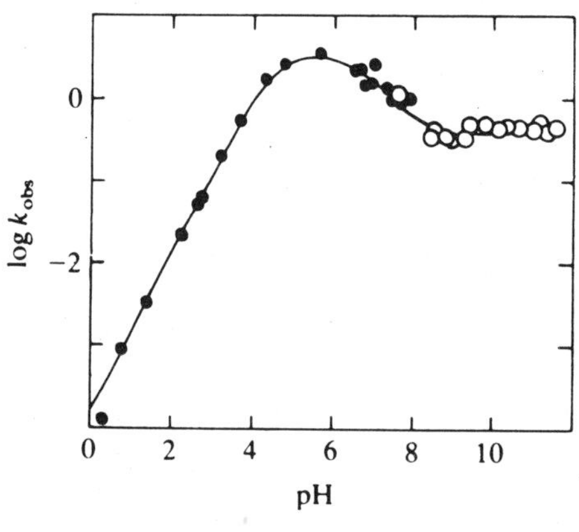

Fig. 8.P23d. (Reproduced from problem reference 23d by permission of the American Chemical Society.)

(a) $CH_3C(=O)O$, Cl, CH_2, OH, Cl, Cl, Cl, Cl, Cl — ester hydrolysis

(b) HO_2C, CH=CH, HO, H_3C, CH_3 — lactonization

(c) O_2N, $OCO_2C_2H_5$, OH — ester hydrolysis

(d) $(CH_3)_2CHCH{=}NCH_3$ — hydrolysis

24. The rates of both formation and hydrolysis of dimethyl acetals of *p*-substituted benzaldehydes are substituent dependent. Do you expect k_{form} to increase or decrease with increasing electron-attracting capacity of the *para* substituent? Do you expect the k_{hydrol} to increase or decrease with the electron-attracting power of the substituent? How do you expect K, the equilibrium constant for acetal formation, to vary with the nature of the substituent?

25. Consider the kinetic effect that would be observed in the reaction of semicarbazide with benzaldehyde:

$$\mathrm{Ph\overset{*}{C}H{=}O + H_2NNH\overset{O}{\overset{\|}{C}}NH_2 \longrightarrow Ph\overset{*}{C}H{=}NNH\overset{O}{\overset{\|}{C}}NH_2}$$

Would you expect to find k_H/k_D to be normal or inverse? Would you expect k_H/k_D to be constant, or would it vary with pH?

26. Figure 8.P26 gives the pH-rate profile for conversion of the acid **A** to the anhydride **B** in aqueous solution. The reaction shows no sensitivity to buffer concentration. Notice that the reaction rate increases with the size of the alkyl substituent, and in fact the derivative with $R^1 = R^2 = CH_3$ is still more reactive. Propose a mechanism which is consistent with the pH-rate profile and the structure of the initially formed product (which is subsequently hydrolyzed to the diacid). How do you account for the effect of the alkyl substituents on the rate?

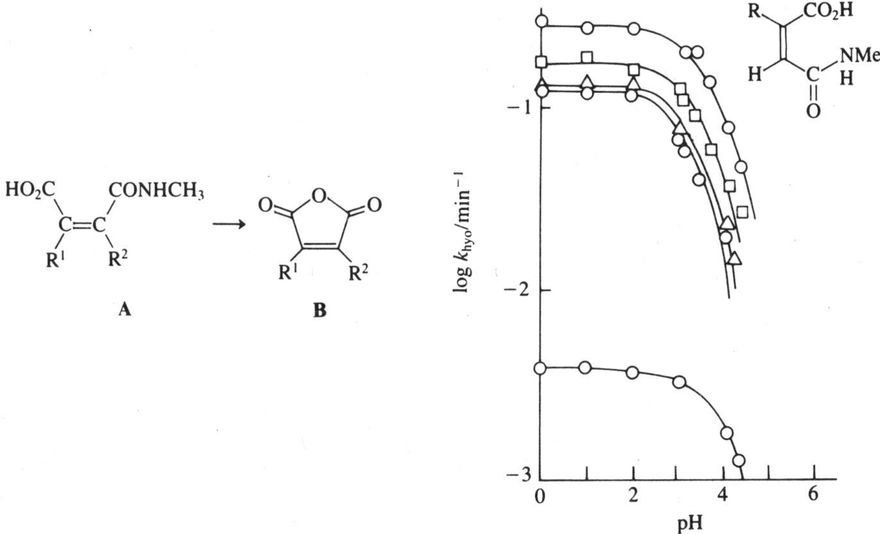

Fig. 8.P26. pH-Rate profiles for the hydrolysis of alkyl-*N*-methylmaleamic acids at 39°C and ionic strength 1.0. In increasing order of reactivity, R = H, Me, Et, *i*-Pr, and *t*-Bu.

27. Assume that the usual mechanism for hydrolysis of an imine, Im, is operative, i.e., that the hydrolysis occurs through a tetrahedral intermediate, TI:

    Assume that the steady-state approximation can be applied to the intermediate TI. Derive the kinetic expression for hydrolysis of the imine. How many variables must be determined to construct the pH–rate profile? What simplifying assumptions are justified at very high and very low pH values? What are the kinetic expressions that result from these assumptions?

28. Give a specific structure, including all stereochemical features, for the product expected for each of the following reactions.

(a) $CH_3CH_2CH_2CH{=}O \xrightarrow[80°C]{1M\ NaOH} C_8H_{17}O$

(b) cyclodecane-1,6-dione $\xrightarrow[100°C,\ 1\ h]{5\%\ Na_2CO_3} C_{10}H_{14}O$

(c) $CH_3CH_2\overset{O}{\overset{\|}{C}}Ph \xrightarrow[\text{2) } PhCH{=}O,\ -78°D \text{, quench after 10 seconds}]{\text{1) } LiN(i\text{-}Pr)_2,\ -78°C} C_{16}H_{16}O_2$

(d) $CH_3\overset{O}{\overset{\|}{C}}CH_2CH_3 + PhCH{=}O$ $\xrightarrow{NaOH} C_{11}H_{14}O_2$; $\xrightarrow{HCl} C_{11}H_{12}O$

9

Aromaticity

9.1. The Concept of Aromaticity

The meaning of the word *aromaticity* has evolved as understanding of the reason for the special properties of benzene and other aromatic molecules has deepened. Originally, aromaticity was associated with a special chemical reactivity.[1] The aromatic hydrocarbons were considered to be those unsaturated systems that underwent substitution reactions in preference to addition. Later, the idea of special stability came to play a larger role. Benzene can be shown to be much lower in enthalpy than predicted by summation of the normal bond energies for the C=C, C—C, and C—H bonds in the Kekulé representation of benzene. Aromaticity is now generally associated with this property of lowered molecular energy. A major contribution to the stability of the aromatic systems is considered to be due to the delocalization of electrons in these molecules.

Currently, aromaticity is usually described in MO terminology. Cyclic structures that have a particularly stable arrangement of occupied π-molecular orbitals are called aromatic. A simple expression of the relationship between an MO description of structure and aromaticity is known as the *Hückel rule.* It is derived from Hückel molecular orbital (HMO) theory and states that *planar, monocyclic, completely conjugated hydrocarbons will be aromatic when the ring contains* $(4n + 2)\pi$ *electrons.* HMO calculations assign the π-orbital energies of the cyclic unsaturated systems of ring size 3–9 as shown in Fig. 9.1. (See Section 1.5, p. 37, to review the basis of HMO theory.)

Orbitals below the dashed reference line are bonding orbitals; when they are filled, the molecule is stabilized. The orbitals that fall on the reference line are nonbonding; placing electrons in these orbitals has no effect on the total bonding

1. For a historical account of early consideration of aromaticity, see J. P. Snyder, *Nonbenzenoid Aromatics*, Vol. 1, Academic Press, New York, 1969, Chapter 1.

energy of the molecule. The orbitals above the reference line are antibonding; the presence of electrons in these orbitals destabilizes the molecule. The dramatic difference between the properties of cyclobutadiene (extremely unstable) and of benzene (very stable) are explicable in terms of these energy level diagrams:

$\alpha - 2\beta$

α

$\alpha + 2\beta$

cyclobutadiene
$E = 4\alpha + 4\beta$

$\alpha - 2\beta$

$\alpha - \beta$

$\alpha + \beta$

$\alpha + 2\beta$

benzene
$E = 6\alpha + 8\beta$

Cyclobutadiene has two bonding electrons, but the other two electrons are unpaired because of the degeneracy of the two nonbonding orbitals. The two electrons in the nonbonding levels do not contribute to the stabilization of the molecule. Furthermore, these electrons, since they occupy a high-energy orbital, are particularly available for chemical reactions. As we shall see in a moment, experimental evidence indicates that cyclobutadiene is rectangular rather than square. This modifies the orbital picture from the simple Hückel pattern, which assumes a square geometry. The two nonbonding levels are no longer degenerate, so cyclobutadiene is not

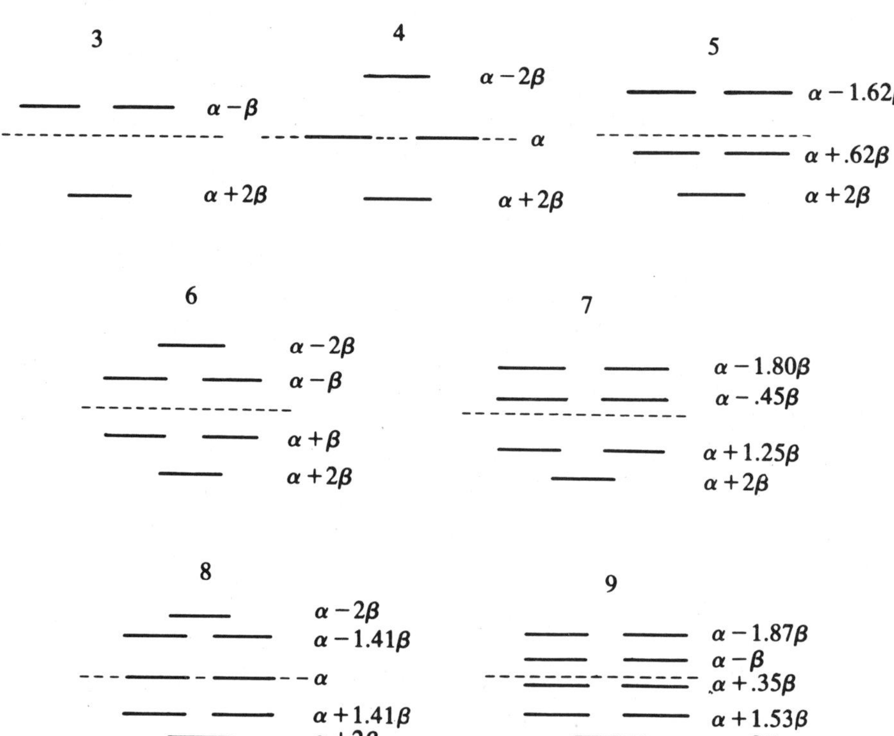

Fig. 9.1. HMO energies for conjugated ring systems of three to nine carbon atoms.

predicted to have unpaired electrons. Nevertheless, higher-level calculations agree with the Hückel concept in predicting cyclobutadiene to be an extremely unstable molecule with a high-lying occupied orbital.

Simple Hückel calculations on benzene, in contrast, place all the π electrons in bonding MOs. The π-electron energy of the benzene molecule is calculated by summing the energies of the six π electrons, which is $6\alpha + 8\beta$, lower by 2β than the value of $6\alpha + 6\beta$ for three isolated double bonds. Thus, the HMO method predicts a special stabilization for benzene.

The pattern of two half-filled degenerate levels persists for larger rings containing $4n$ π electrons. In contrast, all $4n + 2$ systems are predicted to have all electrons paired in bonding MOs. This provides the theoretical basis of the Hückel rule.

As indicated in Chapter 1, the simple HMO theory is based on rather drastic assumptions. More elaborate MO treatments indicate that the most stable geometry for cyclobutadiene is rectangular.[2] Although several derivatives of cyclobutadiene are known and will be discussed shortly, cyclobutadiene itself has been observed only as a "matrix-isolated" species. Several compounds when photolyzed at very low temperature (~10 K) in solid argon give rise to cyclobutadiene. Analysis of the infrared spectrum of the product and deuterated analogs generated from appropriately labeled precursors has confirmed the theoretical conclusion that cyclobutadiene is a rectangular molecule.[3]

Attempts to describe just how stable a given aromatic molecule is in terms of simple HMO calculations have centered on the *delocalization energy*. The total π-electron energy of a molecule is expressed in terms of the energy parameters α and β, which arise in simple HMO calculations. This energy value can be compared to that for a hypothetical localized version of the same molecule. The HMO energy for the π electrons of benzene is $6\alpha + 8\beta$. The same quantity for the hypothetical localized model, cyclohexatriene, is $6\alpha + 6\beta$, the sum of three isolated C—C bonds. The difference of 2β is called the *delocalization energy* or *resonance energy*. Although this quantity is often useful for comparing related systems, it should be remembered that it is not a measurable physical quantity; rather, it is a comparison between a real molecule and a hypothetical one. Most estimates of the stabilization of benzene are in the range 20–40 kcal/mol and depend on the choice of properties assigned to the hypothetical cyclohexatriene point of reference.

There have been two general approaches to determining the amount of stabilization that results from aromatic delocalization. One is to use experimental thermodynamic measurements. Bond energies, as was mentioned in Chapter 1, are nearly additive when there are no special interactions between the various bond types. Thus, it is possible to calculate such quantities as the heat of combustion or heat of hydrogenation of "cyclohexatriene" by assuming that it is a compound with no interaction between the conjugated double bonds. For example, a very simple

2. J. A. Jafri and M. D. Newton, *J. Am. Chem. Soc.* **100**, 5012 (1978); W. T. Borden, E. R. Davidson, and P. Hart, *J. Am. Chem. Soc.* **100**, 388 (1978); H. Kollmar and V. Staemmler, *J. Am. Chem. Soc.* **99**, 3583 (1977); M. J. S. Dewar and A. Kormornicki, *J. Am. Chem. Soc.* **99**, 6174 (1977).
3. S. Masamune, F. A. Souto-Bachiller, T. Machiguchi, and J. E. Bertie, *J. Am. Chem. Soc.* **100**, 4889 (1978); B. A. Hess, Jr., P. Carsky, and L. J. Schaad, *J. Am. Chem. Soc.* **105**, 695 (1983).

calculation of the heat of hydrogenation for cyclohexatriene would be to multiply the heat of hydrogenation of cyclohexane by 3, i.e., $3 \times 28.6 = 85.8$ kcal/mol. The actual heat of hydrogenation of benzene is 49.8 kcal/mol, suggesting a total stabilization or delocalization energy of 36.0 kcal/mol. There are other, more elaborate, ways of approximating the thermodynamic properties of the hypothetical cyclohexatriene. The difference between the calculated and corresponding measured thermodynamic property of benzene is taken to be the aromatic stabilization. For benzene, these values are usually around 30 kcal/mol, but they cannot be determined in an absolute sense since the value is established by the properties assigned to the cyclohexatriene model.

The second general approach to estimating aromatic stabilization is to use molecular orbital methods. This has already been illustrated by the discussion of treatment of benzene according to simple HMO theory, which assigns the stabilization energy a value of 2β units. More advanced MO methods can assign the stabilization energy in a more quantitative way. The most successful method is to perform calculations on the aromatic compound and on a linear, conjugated polyene containing the same number of double bonds.[4] This method assigns a resonance stabilization of zero to the polyene, even though it is known by thermodynamic criteria that conjugated polyenes do have some stabilization relative to isomeric compounds with isolated double bonds. Using this definition, semiempirical MO calculations assign the stabilization energy of benzene a value of about 20 kcal/mol, relative to 1,3,5-hexatriene. The use of polyenes as reference compounds has proven to give better agreement with experimental trends in stability than comparison with the sums of isolated double bonds.

Another molecular orbital approach is to calculate the energy of a model in which the π bonds are constrained to be localized double bonds by the definition of the wave function. The calculated energy of this model can then be compared with the computed energy of the molecule in which delocalization is permitted.[5] By this definition, butadiene has a resonance stabilization of about 9.3 kcal/mol, while benzene has a resonance energy of about 56 kcal/mol. To compare this value with that obtained with the polyene reference, one must subtract a correction for the butadiene resonance energy (3×9.3), which gives a value of about 28 kcal/mol as the resonance stabilization of benzene.

The isodesmic reaction approach has also been applied to calculation of the resonance stabilization of benzene.

$$\text{(benzene)} + 3\ CH_2{=}CH_2 \rightarrow 3\ CH_2{=}CH{-}CH{=}CH_2$$

This approach can be taken using either experimental thermochemical data or energies obtained by MO calculations.[6] If the resonance energy of butadiene is assigned as zero, the above reaction gives the resonance energy of benzene as

4. M. J. S. Dewar and C. de Llano, *J. Am. Chem. Soc.* **91**, 789 (1969).
5. H. Kollmar, *J. Am. Chem. Soc.* **101**, 4832 (1979).
6. P. George, M. Trachtman, C. W. Bock, and A. M. Brett, *J. Chem. Soc., Perkin Trans. 2,* 1222 (1976); P. George, M. Trachtman, C. W. Bock, and A. M. Brett, *Tetrahedron* **32**, 1357 (1976).

21.2 kcal/mol. If butadiene is considered to have a resonance energy, the computation must be modified to reflect that fact. Using 7.2 kcal/mol as the butadiene resonance energy gives a value of 42.8 kcal/mol as the benzene resonance energy.

Both thermochemical and molecular orbital approaches agree that benzene is an especially stable molecule and are reasonably consistent with one another in the stabilization energy that is assigned. It is very significant that MO calculations also show a destabilization of certain conjugated cyclic polyenes, cyclobutadiene in particular. The instability of cyclobutadiene has precluded any thermochemical evaluation of the extent of destabilization. Compounds that are destabilized relative to conjugated but noncyclic polyene models are called antiaromatic.[7]

There are also physical measurements that can give evidence of aromaticity. The determination of the bond lengths in benzene by electron diffraction is a classic example of use of the bond-length criterion of aromaticity. Spectroscopic methods or X-ray diffraction can also provide bond-length data. Aromatic molecules consistently show bond lengths in the range 1.38–1.40 Å, and the bond lengths are quite uniform around the ring. In contrast, localized polyenes show alternation between typical sp^2–sp^2 single-bond and sp^2–sp^2 double-bond lengths along the conjugated chain.

NMR spectroscopy also provides an experimental tool capable of assessing aromaticity. Aromatic compounds are characterized by the capacity to exhibit a *diamagnetic ring current.* Qualitatively, this ring current can be viewed as the migration of the delocalized electrons in the π system under the influence of the magnetic field in an NMR spectrometer. The ring current effect is responsible for a large magnetic anisotropy in aromatic compounds. The induced ring current gives rise to a local magnetic field perpendicular to the ring that is opposed to the direction of the applied magnetic field. Nuclei in a cone above and below the plane of an aromatic ring are shielded by the induced field and appear at relatively high field in the NMR spectrum, while nuclei in the plane of the ring—that is, the atoms bound directly to the ring—occur at downfield positions. The occurence of these chemical shift phenomena can be taken as evidence for aromaticity. This criterion must be applied with some care, and a recent discussion of the relationship between ring current and aromaticity can be consulted to pursue this topic in more depth.[8]

9.2. The Annulenes

The term *annulene* has been coined to refer to the completely conjugated monocyclic polyenes.[9] The synthesis of annulenes has now been extended well beyond the first two members of the series, [4]annulene (cyclobutadiene) and [6]annulene (benzene). The generality of the Hückel rule can be tested by considering the properties of the members of the annulene series.

7. R. Breslow, *Acc. Chem. Res.* **6**, 393 (1973).
8. R. C. Haddon, *J. Am. Chem. Soc.* **101**, 1722 (1979); J. Aihara, *J. Am. Chem. Soc.* **103**, 5704 (1981).
9. F. Sondheimer, *Pure Appl. Chem.* **28**, 331 (1971); *Acc. Chem. Res.* **5**, 81 (1972).

The smallest member, cyclobutadiene, was the objective of synthesis for many years. The first success was achieved when cyclobutadiene released from stable iron complexes was trapped with various reagents.[10]

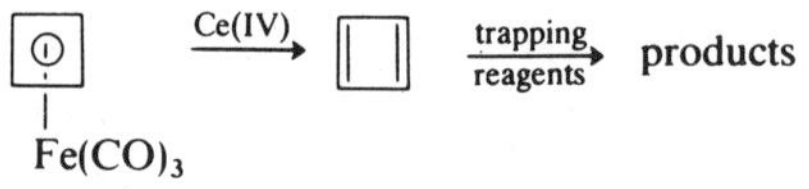

Dehalogenation of *trans*-3,4-dibromocyclobutene was shown to generate a species with the same reactivity.[11]

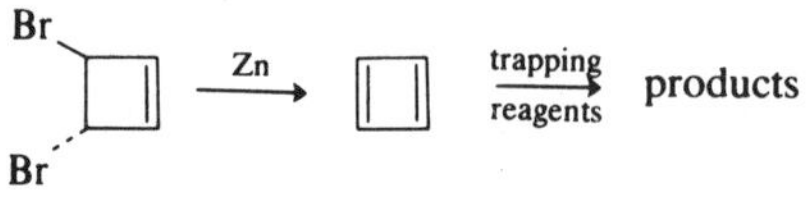

Various trapping agents react with cyclobutadiene to give Diels–Alder adducts.[12]:

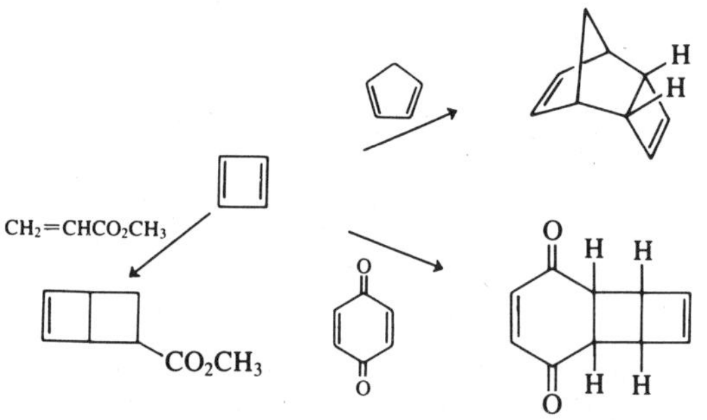

In the absence of trapping agents, a characteristic dimer is produced.

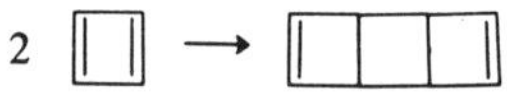

This dimerization is an extremely fast reaction and limits the lifetime of cyclobutadiene, except at very low temperatures.

Cyclobutadiene can also be prepared by photolysis of several different precursors at very low temperature in solid inert gases.[13] These methods provide cyclo-

10. L. Watts, J. D. Fitzpatrick, and R. Pettit, *J. Am. Chem. Soc.* **87**, 3253 (1965).
11. E. K. G. Schmidt, L. Brener, and R. Pettit, *J. Am. Chem. Soc.* **92**, 3240 (1970).
12. L. Watts, J. D. Fitzpatrick, and R. Pettit, *J. Am. Chem. Soc.* **88**, 623 (1966); J. C. Barborak, L. Watts, and R. Pettit, *J. Am. Chem. Soc.* **88**, 1328 (1966); D. W. Whitman and B. K. Carpenter, *J. Am. Chem. Soc.* **102**, 4272 (1980).
13. G. Maier and M. Schneider, *Angew. Chem. Int. Ed. Engl.* **10**, 809 (1971); O. L. Chapman, C. L. McIntosh, and J. Pacansky, *J. Am. Chem. Soc.* **95**, 614 (1973); O. L. Chapman, D. De La Cruz, R. Roth, and J. Pacansky, *J. Am. Chem. Soc.* **95**, 1337 (1973); C. Y. Lin and A. Krantz, *J. Chem. Soc., Chem. Commun.*, 1111 (1972); G. Maier, H. G. Hartan, and T. Sayrac, *Angew. Chem. Int. Ed. Engl.* **15**, 226 (1976); H. W. Lage, H. P. Reisenauer, and G. Maier, *Tetrahedron Lett.* **23**, 3893 (1982).

butadiene in a form that is convenient for spectroscopic study. Under these conditions, cyclobutadiene begins to react to give the dimer at around 35°K.

While simple HMO theory assumes a square geometry for cyclobutadiene, most MO methods predict a rectangular structures as the minimum-energy geometry. With very high level calculations, good agreement is obtained between the calculated minimum-energy structure, which is rectangular, and observed spectroscopic properties.[14]

1.567
1.346

The rectangular structure is still calculated to be strongly destabilized (antiaromatic) with respect to a polyene model. With 6-31G* calculations, for example, cyclobutadiene is found to have a negative resonance energy of −54.7 kcal/mol, relative to 1,3-butadiene. In addition, 30.7 kcal/mol of strain is found, giving a total destabilization of 85.4 kcal/mol.[15]

A number of alkyl-substituted cyclobutadienes have been prepared by related methods.[16] Increasing alkyl substitution enhances the stability of the compounds. The tetra-*t*-butyl derivative is stable up to at least 105°C but is still very reactive toward oxygen.[17] This reactivity reflects the high energy of the HOMO. The chemical behavior of the cyclobutadienes as a group is in excellent accord with what would be expected on the basis of the theoretical picture of the structure of these compounds.

[6]Annulene is benzene. Its properties are so familiar to students of organic chemistry that not much need be said here. It is the parent compound of a vast series of derivatives. The benzene ring shows exceptional stability, both with regard to thermodynamic stability and in terms of its diminished reactivity in comparison with conjugated polyenes. As was discussed earlier, a stabilization on the order of 30 kcal/mol is found by thermodynamic methods. Benzene is much less reactive toward electrophiles than are conjugated polyenes.

The next higher annulene, cyclooctatetraene, is nonaromatic.[18] The bond lengths around the ring alternate as expected for a polyene. The C=C bonds are 1.334 Å while the C—C bonds are 1.462 Å in length.[19] Thermodynamic data provide no evidence of any special stability.[20] Cyclooctatetraene is readily isolable, and the chemical reactivity is normal for a polyene. The structure determination shows that the molecule is tub-shaped[19] and therefore is not a system to which the Hückel rule applies.

14. B. A. Hess, Jr., P. Carsky, and L. J. Schaad, *J. Am. Chem. Soc.* **105**, 695 (1983); H. Kollmar and V. Staemmler, *J. Am. Chem. Soc.* **100**, 4304 (1978).
15. B. A. Hess, Jr., and L. J. Schaad, *J. Am. Chem. Soc.* **105**, 7500 (1983).
16. G. Maier, *Angew. Chem. Int. Ed. Engl.* **13**, 425 (1974); S. Masamune, *Tetrahedron* **36**, 343 (1980).
17. G. Maier, S. Pfriem, U. Schafer, and R. Matusch, *Angew. Chem. Int. Ed. Engl.* **17**, 520 (1978).
18. G. Schröder, *Cyclooctatetraene*, Verlag Chemie, Weinheim, 1965; G. I. Fray and R. G. Saxton, *The Chemistry of Cyclooctatetraene and Its Derivatives*, Cambridge University Press, Cambridge, 1978.
19. M. Traetteberg, *Acta Chem. Scand.* **20**, 1724 (1966).
20. R. B. Turner, B. J. Mallon, M. Tichy, W. v. E. Doering, W. Roth, and G. Schröder, *J. Am. Chem. Soc.* **95**, 8605 (1973).

NMR studies have revealed that two dynamic processes can occur in cyclooctatetraenes. One is a conformational flip:

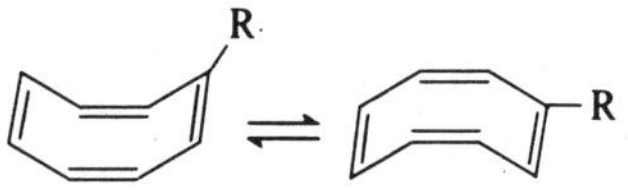

The other is π-bond migration:

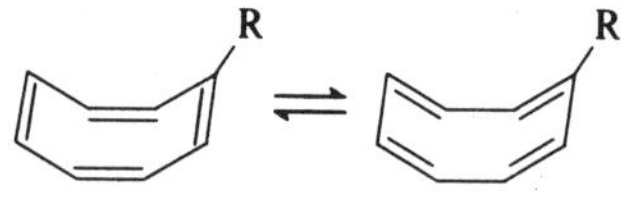

The transition states for these two processes should correspond to the localized and delocalized planar forms, respectively, of cyclooctatetraene. Therefore, comparison of the energy of the two transition states might indicate the extent of stabilization or destabilization of the planar [8]annulene system.[21]

planar localized transition state for conformational flip

planar delocalized transition state for bond reorganization

Using alkoxy-substituted cyclooctatetraenes for the measurements, ΔH^{+} for the ring flip and bond switch have been placed in the range of 10.9–12.1 and 14.9–15.8 kcal/mol, respectively. These data imply an unfavorable delocalization energy of about 4 kcal/mol.

Because of the localized nature of the double bonds in cyclooctatetraene, it is possible to isolate two *different* tetramethyl derivatives with four adjacent methyl groups. These are the 1,2,3,4- and 2,3,4,5-isomers.

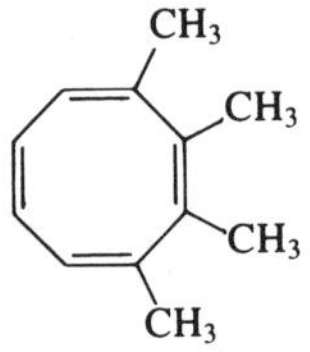

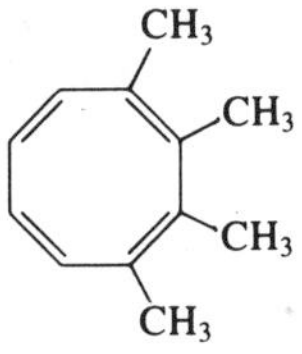

The compounds are not interconverted at room temperature, but on heating, equilibration can occur. After 6 h at 160°C, a 7:3 mixture favoring **A** is obtained from either isomer. The activation energy for the process is 33 kcal/mol.[22] The reason the process is much slower than in the case of the alkoxy-substituted

21. F. A. L. Anet, A. J. R. Bourn, and Y. S. Lin, *J. Am. Chem. Soc.* **86**, 3576 (1964); L. A. Paquette, *Pure Appl. Chem.* **54**, 987 (1982).
22. L. A. Paquette and J. M. Photis, *J. Am. Chem. Soc.* **98**, 4936 (1976); L. A. Paquette and J. M. Gardlik, *J. Am. Chem. Soc.* **102**, 5033 (1980).

cyclooctatetraene is the additional steric repulsions between the methyl groups in the planar transition state.

Larger annulenes permit the incorporation of *trans* double bonds into the rings. Beginning with [10]annulene, isomeric structures are possible. According to the Hückel rule, [10]annulene should possess aromatic stabilization if it were planar. However, all the isomeric 1,3,5,7,9-cyclodecapentaenes suffer serious steric strain that prevents the planar geometry from being adopted. The *Z,E,Z,Z,E*-isomer, which has minimal bond-angle strain, suffers a severe nonbonded repulsion between the two internal hydrogens.

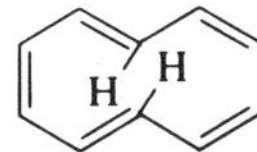

The *Z,Z,Z,Z,Z*-isomer is required by geometry to have bond angles of 144° to maintain planarity and is therefore enormously destabilized by distortion of the normal trigonal bond angle. The most stable structure is a twisted form of the *E,Z,Z,Z,Z*-isomer. Two other structures, the *Z,Z,Z,Z,Z*- and *Z,E,Z,Z,E*-isomers, are found by MO calculations to be about 2.4 kcal/mol higher in energy.[23]

Two of the isomeric [10]annulenes, as well as other products, are formed by photolysis of *cis*-9,10-dihydronaphthalene.[24]

Neither compound exhibits properties that would suggest aromaticity. The NMR spectra are consistent with polyene structures. Both compounds are thermally unstable and revert back to dihydronaphthalenes.

6°C

−15°C

As with cyclooctatetraene, then, [10]annulene does not exist in a planar form and the Hückel rule cannot be applied.

A number of structures have been prepared that avoid the steric problems associated with the 1,3,5,7,9-cyclodecapentaenes. In compound **1**, the steric problem

23. L. Farnell, J. Kao, L. Radom, and H. F. Schaefer III, *J. Am. Chem. Soc.* **103**, 2147 (1981).
24. S. Masamune, K. Hojo, G. Gigam, and D. L. Rabenstein, *J. Am. Chem. Soc.* **93**, 4966 (1971); S. Masamune and N. Darby, *Acc. Chem. Res.* **5**, 272 (1972).

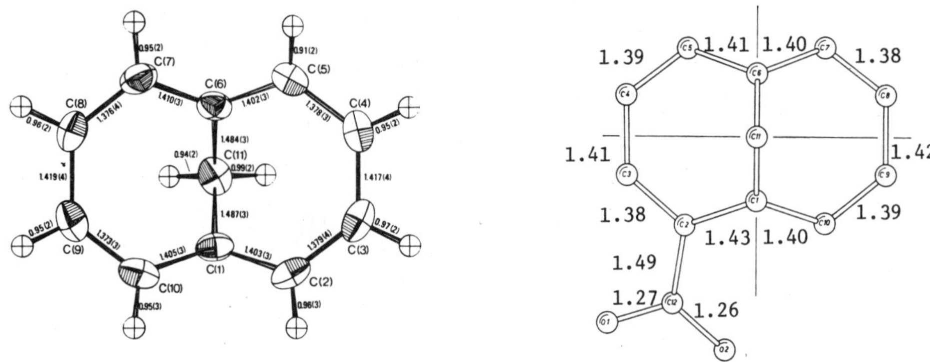

Fig. 9.2. X-ray crystal structures of (A) 1,6-methanodeca-1,3,5,7,9-pentaene and (B) 1,6-methanodeca-1,3,5,7,9-pentaene-2-carboxylic acid. [Structures are reproduced from Refs. 27 and 28.]

is avoided with only a slight loss of planarity in the π system.[25]

H H 1 H H 2 CO_2H

The NMR spectrum of this compound shows a diamagnetic ring current of the type expected in an aromatic system.[26] X-ray crystal structure determinations on **1**[27] and its carboxylic acid derivative **2**[28] (Fig. 9.2) both reveal a pattern of bond lengths which is very similar to that in naphthalene (see p. 524).[29]

Most molecular orbital methods find a bond alternation pattern in the minimum-energy structure, but calculations that include electron correlation lead to a delocalized minimum-energy structure.[30] Thus, while the π system in **1** is not completely planar, it appears to be sufficiently close to provide a delocalized 10-electron π system. A resonance energy of 17.2 kcal/mol has been calculated on the basis of an experimental heat of hydrogenation.[31]

The deviation from planarity that is present in a structure such as **1** raises the question of how severely a conjugated system can be distorted from the ideal coplanar alignment of *p* orbitals and still retain aromaticity. This problem has been analyzed by determining the degree of rehybridization necessary to maximize *p*-orbital overlap in **1**. It is found that rehybridization to incorporate fractional amounts of *s* character can improve orbital alignment substantially. Orbitals with about 6%

25. E. Vogel and H. D. Roth, *Angew Chem. Int. Ed. Engl.* **3**, 228 (1964).
26. E. Vogel, *Pure Appl. Chem.* **20**, 237 (1969).
27. R. Bianchi, T. Pilati, and M. Simonetta, *Acta Crystallogr. Sect B* **B36**, 3146 (1980).
28. M. Dobler and J. D. Dunitz, *Helv. Chim. Acta* **48**, 1429 (1965).
29. O. Bastainsen and P. N. Skancke, *Adv. Chem. Phys.* **3**, 323 (1961).
30. R. C. Haddon and K. Raghavarchari, *J. Am. Chem. Soc.* **107**, 289 (1985); L. Farnell and L. Radom, *J. Am. Chem. Soc.* **104**, 2650 (1982).
31. W. R. Roth, M. Böhm, H. W. Lennartz, and E. Vogel, *Angew. Chem. Int. Ed. Engl.* **22**, 1007 (1983).

s character are suggested to be involved for **1**.[32] Thus, a relatively small amount of rehybridization greatly improves orbital overlap in such twisted systems.

[12]Annulene is a very unstable compound that undergoes cyclization to bicyclic isomers and can be kept only at very low temperature.[33] The NMR spectrum has been studied at low temperature.[34] Besides indicating the double-bond geometry shown in the structure below, the spectrum reveals a *paramagnetic* ring current, the opposite to what is observed for aromatic systems. This feature is quite characteristic of the [4*n*]annulenes and has been useful in characterizing the aromaticity or lack of it in annules.[35]

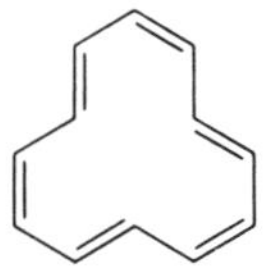

[14]Annulene was first prepared in 1960.[36] Its NMR spectrum has been investigated and shows that two geometric isomers are in equilibrium.[37]

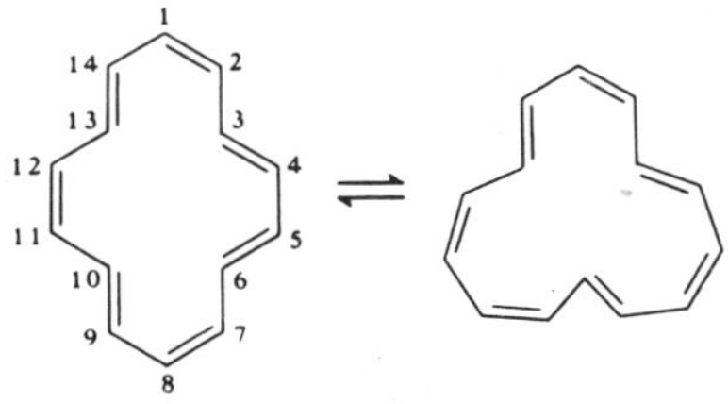

The spectrum also reveals a significant diamagnetic (aromatic) ring current. The internal hydrogens (C-3, C-6, C-10, C-13) are very far upfield ($\delta = -0.61$ ppm).[36] The interconversion of the two forms involves a configurational change from *E* to *Z* at one double bond at least. The activation energy for this process is only about 10 kcal/mol. The crystal structure for [14]annulene shows the *Z,E,E,Z,E,Z,E*-form to be present in the solid.[38] The bond lengths around the ring range from 1.35 to 1.41 Å but do not show the alternating pattern of short and long bonds expected for a localized polyene. There is some distortion from planarity, especially at carbon atoms 3, 6, 10, and 13, which is caused by nonbonded repulsions between the internal hydrogens.

A 14-electron π can be generated in circumstances in which the steric problem associated with the internal hydrogens of [14]annulene can be avoided. This can

32. R. C. Haddon, *Acc. Chem. Res.* **21**, 243 (1988).
33. J. F. M. Oth, H. Rottele, and G. Schröder, *Tetrahedron Lett.*, 61 (1970).
34. J. F. M. Oth, J.-M. Gilles, and G. Schröder, *Tetrahedron Lett.*, 67 (1970).
35. R. C. Haddon, *Tetrahedron* **28**, 3613, 3635 (1972).
36. F. Sondheimer and Y. Gaoni, *J. Am. Chem. Soc.* **82**, 5765 (1960).
37. J. F. M. Oth, *Pure Appl. Chem.* **25**, 573 (1971).
38. C. C. Chiang and I. C. Paul, *J. Am. Chem. Soc.* **94**, 4741 (1972).

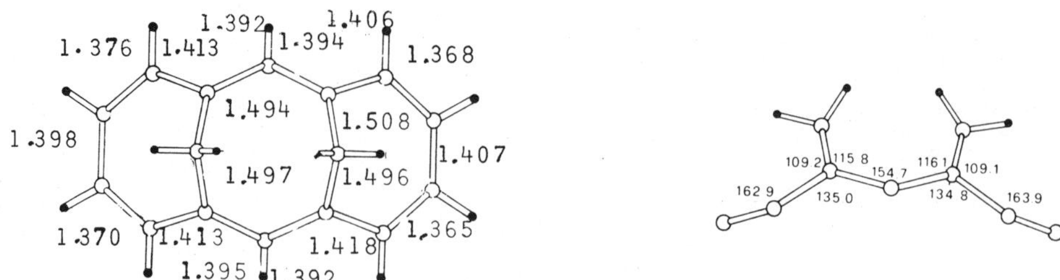

Fig. 9.3. (A) Carbon framework from X-ray crystal structure of *syn*-tricyclo[8.4.1.1^{3,8}]hexadeca-1,3,5,7,9,11,13-heptaene. (B) Side view showing deviation from planarity of annulene ring. [Reproduced from Ref. 44.]

be achieved in systems in which the annulene ring is built around a saturated core:

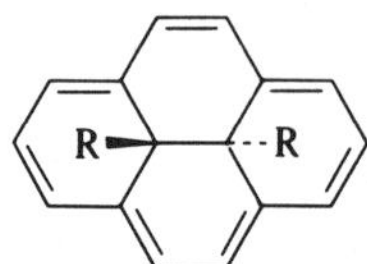

Several derivatives of this ring system have been synthesized.[39] These compounds exhibit properties indicating that the conjugated system is aromatic. They exhibit NMR shifts characteristic of a diamagnetic ring current. Typical aromatic substitution reactions can be carried out.[40] An X-ray crystal structure ($R = C_2H_5$) shows that the bond lengths are in the aromatic range (1.39–1.40 Å), and there is no strong alternation around the ring.[41] The peripheral atoms are not precisely planar, but the maximum deviation from the average plane is only 0.23 Å.

Another family of 14-π-electron systems is derived from structure **3**.[42]

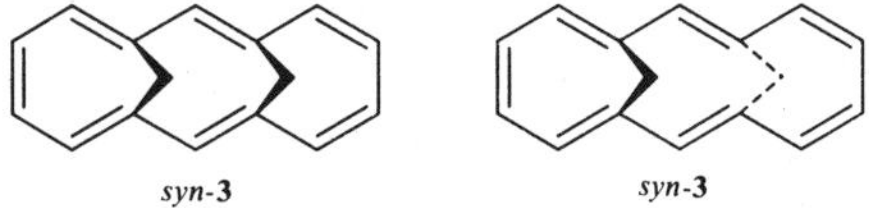

syn-**3** *syn*-**3**

The *syn* isomer can achieve a conjugated system with angles of up to 35° between adjacent *p* orbitals. The *anti* isomer is much more twisted.[43] An X-ray crystal structure determination has been done on the *syn* isomer (Fig. 9.3). It shows C—C bond lengths between 1.368 and 1.418 Å for the conjugated system.[44] The spectro-

39. R. H. Mitchell and V. Boekelheide, *J. Am. Chem. Soc.* **96**, 1547 (1974); V. Boekelheide and T. A. Hylton, *J. Am. Chem. Soc.* **92**, 3669 (1970); H. Blaschke, C. E. Ramey, I. Calder, and V. Boekelheide, *J. Am. Chem. Soc.* **92**, 3675 (1970); V. Boekelheide and J. B. Phillips, *J. Am. Chem. Soc.* **89**, 1695 (1967).
40. J. B. Phillips, R. J. Molyneux, E. Sturm, and V. Boekelheide, *J. Am. Chem. Soc.* **89**, 1704 (1967).
41. A. W. Hanson, *Acta Crystallogr.* **23**, 476 (1967).
42. E. Vogel, *Pure Appl. Chem.* **28**, 355 (1971).
43. E. Vogel, J. Sombroek, and W. Wagemann, *Angew. Chem. Int. Ed. Engl.* **14**, 564 (1975); E. Vogel, U. Haberland, and H. Günther, *Angew. Chem. Int. Ed. Engl.* **9**, 513 (1970).
44. R. Destro, T. Pilati, and M. Simonetta, *Acta Crystallogr., Sect. B* **B33**, 940 (1977).

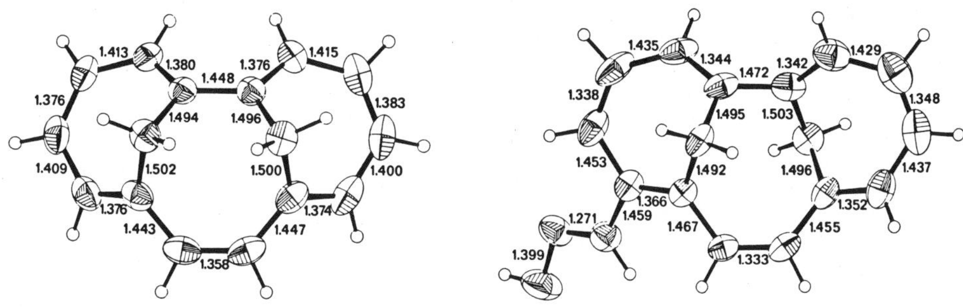

Fig. 9.4. (A) X-ray crystal structure of *syn*-tricyclo[8.4.1.$1^{4,9}$]hexadeca-2,4,6,8,10,12,14-heptaene. (B) X-ray crystal structure of *anti* stereoisomer of tricyclo[8.4.1.$1^{4,9}$]hexadeca-2,4,6,8,10,12,14-heptaene-5-carboxylic acid. [Reproduced from Ref. 46.]

scopic properties of the *syn* isomer are consistent with considering it to be a delocalized annulene.[45]

An isomeric system is related to the benzenoid hydrocarbon phenanthrene. Both the *syn* and *anti* stereoisomers have been synthesized.[46]

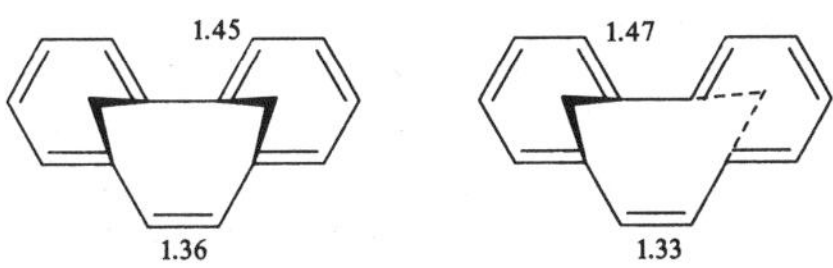

The *syn* isomer shows evidence of a diagmagnetic ring current, both from the relatively low field position of the vinylic hydrogens and the upfield shift of the methylene hydrogens. The *anti* isomer shows much less pronounced shifts. X-ray crystal structures are available for both the *syn* and *anti* rings as shown in Fig. 9.4. The X-ray crystal structure of the *syn* isomer shows a moderate level of bond alternation, ranging from 1.36 to 1.45 Å. In the *anti* structure, bond alternation is more pronounced, with the bonds in the center ring being essentially localized bond lengths.

The Hückel rule predicts nonaromaticity for [16]annulene. The compound has been synthesized and thoroughly characterized.[47] The bond lengths show significant alternation (C=C, 1.34 Å; C—C, 1.46 Å), and the molecule is less planar than [14]annulene.[48] These structural data are consistent with regarding [16]annulene as being nonaromatic.

45. J. Dewey, H. M. Deger, W. Fröhlich, B. Dick, K. A. Klingensmith, G. Hohlneicher, E. Vogel, and J. Michl, *J. Am. Chem. Soc.* **102**, 6412 (1980).
46. E. Vogel, W. Püttmann, W. Duchatsch, T. Schieb, H. Schmickler, and J. Lex, *Angew. Chem. Int. Ed. Engl.* **25**, 720 (1986); E. Vogel, T. Schieb, W. H. Schulz, K. Schmidt, H. Schmickler, and J. Lex, *Angew. Chem. Int. Ed. Engl.* **25**, 723 (1986).
47. I. Calder, Y. Gaoni, and F. Sondheimer, *J. Am. Chem. Soc.* **90**, 4946 (1968); G. Schröder and J. F. M. Oth, *Tetrahedron Lett.*, 4083 (1966).
48. S. M. Johnson and I. C. Paul, *J. Am. Chem. Soc.* **90**, 6555 (1968).

[18]Annulene offers a particularly significant test of the Hückel rule. The internal cavity in [18]annulene is large enough to minimize steric interaction between the internal hydrogens in a geometry that is free of angle strain.

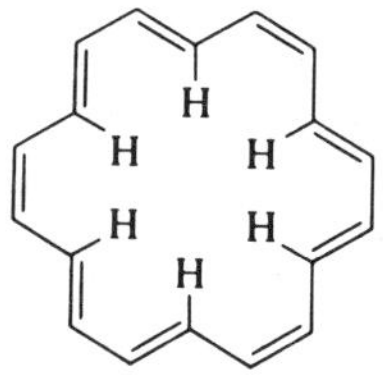

The properties of [18]annulene are consistent with its being aromatic. The X-ray crystal structure shows the molecule to be close to planarity, with the maximum deviation from the plane being 0.085 Å.[49] The bond lengths are in the range 1.38–1.42 Å, and the pattern is short, short, long, rather than alternating. The NMR spectrum is indicative of an aromatic ring current.[50] The chemical reactivity of the molecule would also justify its classification as aromatic.[51]

There are also examples of [18]annulene systems constructed around a saturated central core, such as in compound **4**.[52] In this compound, the internal protons are at very high field (−6 to −8 ppm), whereas the external protons are far downfield (~9.5 ppm).

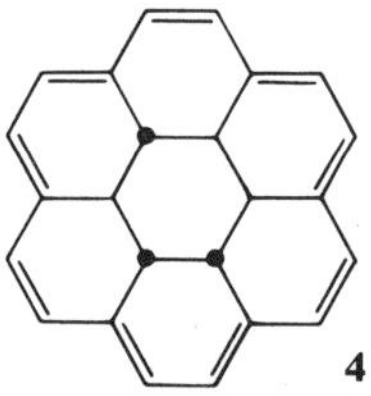

4

The chemical shift data can be used as the basis for calculating the diamagnetic ring current and then comparing that with the maximum ring current expected for a completely delocalized π system. By this criterion, the flexible [18]annulene maintains only about half (0.56) of the maximum ring current, whereas the rigid ring gives a value of 0.88, indicating more effective conjugation in this system.

The synthesis of annulenes has been carried forward to larger rings as well. [20]Annulene,[53] [22]annulene,[54] and [24]annulene[55] have all been reported. The NMR spectra of these compounds are consistent with regarding [22]annulene as

49. J. Bregman, F. L. Hirshfeld, D. Rabinovich, and G. M. J. Schmidt, *Acta Crystallogr.* **19**, 227 (1965); F. L. Hirshfeld and D. Rabinovich, *Acta Crystallogr.* **19**, 235 (1965).
50. Y. Gaoni, A. Melera, F. Sondheimer, and R. Wolovsky, *Proc. Chem. Soc.*, 397 (1965).
51. I. C. Calder, P. J. Garratt, H. C. Longuet-Higgins, F. Sondheimer, and R. Wolovsky, *J. Chem. Soc., C*, 1041 (1967).
52. T. Otsubo, R. Gray, and V. Boekelheide, *J. Am. Chem. Soc.* **100**, 2449 (1978).
53. B. W. Metcalf and F. Sondheimer, *J. Am. Chem. Soc.* **93**, 6675 (1971).
54. R. M. McQuilkin, B. W. Metcalf, and F. Sondheimer, *J. Chem. Soc., Chem. Commun.*, 338 (1971).
55. I. C. Calder and F. Sondheimer, *J. Chem. Soc., Chem. Commun.*, 904 (1966).

aromatic while the [20] and [24] analogs are not. In each case, there is some ambiguity as to the preferred conformation in solution, and the NMR spectra are temperature dependent. While the properties of these molecules have not been studied as completely as those of the smaller systems, they are consistent with the predictions of the Hückel rule.

It has been pointed out that a different array of atomic orbitals might be conceived of in large conjugated rings. The array, called a *Möbius twist*, results in there being one point in the ring at which the atomic orbitals would show a phase reversal.[56]

If the ring were sufficiently large that the twist between individual orbitals were small, such a system would not necessarily be less stable than the normal array of atomic orbitals. This same analysis points out that in such an array the Hückel rule is reversed and aromaticity is predicted for the $4n$ π-electron systems. So far, no ground state molecule in which the twisted conjugation would exist has been made, so the prediction remains to be tested. Its correctness is strongly suggested, however, by the fact that transition states with twisted orbital arrays appear to be perfectly acceptable in many organic reactions.[57] We will return to this topic in Chapter 11. The rules for aromaticity can be generalized to include Möbius orbital arrays:

Hückel orbital array	Möbius orbital array
$4n+2$ aromatic	$4n$ aromatic
$4n$ antiaromatic	$4n+2$ antiaromatic

9.3. Aromaticity in Charged Rings

There are also striking examples of stability relationships due to aromaticity in charged ring systems. The HMO energy levels that apply to fully conjugated planar three- to nine-membered rings were given in Fig. 9.1 on p. 500. These energy levels are applicable to charged species as well as to the neutral annulenes. A number of cations and anions that are completely conjugated planar structures are shown in Scheme 9.1. Among these species, the Hückel rule would predict aromatic stability

56. E. Helibronner, *Tetrahedron Lett.*, 1923 (1964).
57. H. E. Zimmerman, *Acc. Chem. Res.* **4**, 272 (1971).

Scheme 9.1. Completely Conjugated Cyclic Cations and Anions

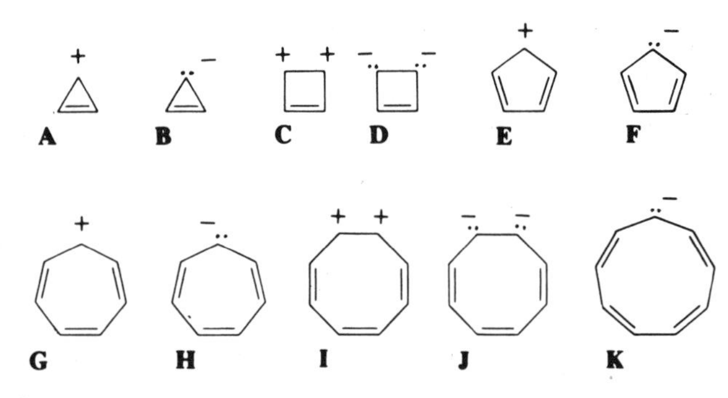

for cyclopropenium ion (**A**), cyclobutadiene dication (**C**), cyclobutadiene dianion (**D**), cyclopentadienide anion (**F**), cycloheptatrienyl cation (tropylium ion, **G**), the dications and dianions derived from cyclooctatetraene (**I, J**), and the cyclononatetraenide anion (**K**). The other species shown, having $4n$ π electrons, would be expected to be very unstable. Let us examine what is known about the chemistry of some of these species.

The cyclopropenium ion and a number of derivatives have been generated by ionization procedures:

$$C_3Cl_4 + SbCl_5 \longrightarrow C_3Cl_3^+ \; SbCl_6^- \qquad \text{Ref. 58}$$

$$C_3H_3Cl + SbCl_5 \longrightarrow C_3H_3^+ \; SbCl_6^- \qquad \text{Ref. 59}$$

The 1,2,3-tri-*t*-butylcyclopropenium cation is so stable that the perchlorate salt can be recrystallized from water.[60] An X-ray study of triphenylcyclopropenium perchlorate has verified the existence of the carbocation as a discrete species.[61] Quantitative estimation of the stability of the unsubstituted ion can be made in terms of its pK_{R^+} value of -7.4, which is intermediate between those of such highly stabilized ions as triphenylmethyl cation and that of the *bis*(4-methoxyphenyl)methyl cation.[62] An MO calculation on the isodesmic reaction

$$C_3H_3^+ + CH_3CH{=}CH_2 \longrightarrow C_3H_4 + {}^+CH_2CH{=}CH_2$$

58. S. W. Tobey and R. West, *J. Am. Chem. Soc.* **86**, 1459 (1964); R. West, A. Sado, and S. W. Tobey, *J. Am. Chem. Soc.* **88**, 2488 (1966).
59. R. Breslow, J. T. Groves, and G. Ryan, *J. Am. Chem. Soc.* **89**, 5048 (1967).
60. J. Ciabattoni and E. C. Nathan III, *J. Am. Chem. Soc.* **91**, 4766 (1969).
61. M. Sundaralingam and L. H. Jensen, *J. Am. Chem. Soc.* **88**, 198 (1966).
62. R. Breslow and J. T. Groves, *J. Am. Chem. Soc.* **92**, 984 (1970).

yields a ΔH of +38.2 kcal/mol, while experimental data on the heats of formation of the various species indicate $\Delta H = +31$ kcal/mol.[63] The heterolytic gas phase bond dissociation energy to form cyclopropenium ion from cyclopropene is 225 kcal/mol. This compares with 256 kcal/mol for formation of the allyl cation from propene or 252 kcal/mol for formation of the secondary propyl cation from propane.[64]

In contrast, the less strained four-π-electron cyclopentadienyl cation is very unstable. Its pK_{R^+} has been estimated as −40, using an electrochemical cycle.[65] The heterolytic bond dissociation energy to form the cation from cyclopentadiene is 258 kcal/mol, which is substantially more than for formation of an allylic cation from cyclopentene but only slightly more than the 252 kcal/mol for formation of an unstabilized secondary carbocation.[64] Solvolysis of cyclopentadienyl halides assisted by silver ion is extremely slow, even though the halide is doubly allylic.[66] When the bromide and antimony pentafluoride react at −78°C, the EPR spectrum observed indicates that the cyclopentadienyl cation is a triplet.[67] Similar studies indicate that pentachlorocyclopentadienyl cation is also a triplet, but the ground state of the pentaphenyl derivative is a singlet.

The relative stability of the anions derived from cyclopropene and cyclopentadiene by deprotonation is just the reverse of the situation for the cations. Cyclopentadiene is one of the most acidic hydrocarbons known, with a pK_a of 16.0.[68] The pK's of triphenylcyclopropene and trimethylcyclopropene have been estimated as 50 and 62, respectively, using electrochemical cycles.[69] The unsubstituted compound would be expected to fall somewhere between and thus must be roughly 40 powers of 10 less acidic than cyclopentadiene. Thus the six-electron cyclopentadienide ion is enormously stabilized relative to the four-π-electron cyclopropenide ion, in agreement with the Hückel rule.

The Hückel rule predicts aromaticity for the six-π-electron cation derived from cycloheptatriene by hydride abstraction and antiaromaticity for the planar eight-π-electron anion that would be formed by deprotonation. The cation is indeed very stable, with a pK_{R^+} of +4.7.[70] Salts containing the cation can be isolated as a result of a variety of preparative procedures.[71] On the other hand, the pK_a of cycloheptatriene has been estimated at 36.[72] This value does not indicate strong destabilization. Thus, the seven-membered eight-π-electron anion is probably nonplanar. This would

63. L. Radom, P. C. Hariharan, J. A. Pople, and P. v. R. Schleyer, *J. Am. Chem. Soc.* **98**, 10 (1976).
64. F. P. Lossing and J. L. Holmes, *J. Am. Chem. Soc.* **106**, 6917 (1984).
65. R. Breslow and S. Mazur, *J. Am. Chem. Soc.* **95**, 584 (1973).
66. R. Breslow and J. M. Hoffman, Jr., *J. Am. Chem. Soc.* **94**, 2110 (1972).
67. M. Saunders, R. Berger, A. Jaffe, J. M. McBride, J. O'Neill, R. Breslow, J. M. Hoffman, Jr., C. Perchonock, E. Wasserman, R. S. Hutton, and V. J. Kuck, *J. Am. Chem. Soc.* **95**, 3017 (1973).
68. A. Streitwieser, Jr., and L. L. Nebenzahl, *J. Am. Chem. Soc.* **98**, 2188 (1976).
69. R. Breslow and W. Chu, *J. Am. Chem. Soc.* **95**, 411 (1973).
70. W. v. E. Doering and L. H. Knox, *J. Am. Chem. Soc.* **76**, 3203 (1954).
71. T. Nozoe, *Prog. Org. Chem.* **5**, 132 (1961); K. M. Harmon, in *Carbonium Ions*, Vol. IV, G. A. Olah and P. v. R. Schleyer, (eds.), Wiley-Interscience, New York, 1973, Chapter 2.
72. R. Breslow and W. Chu, *J. Am. Chem. Soc.* **95**, 411 (1973).

be similar to the situation in the nonplanar eight-π-electron hydrocarbon cyclooctatetraene.

The cyclononatetraenide anion is generated by treatment of the halide **5** with lithium metal[73]:

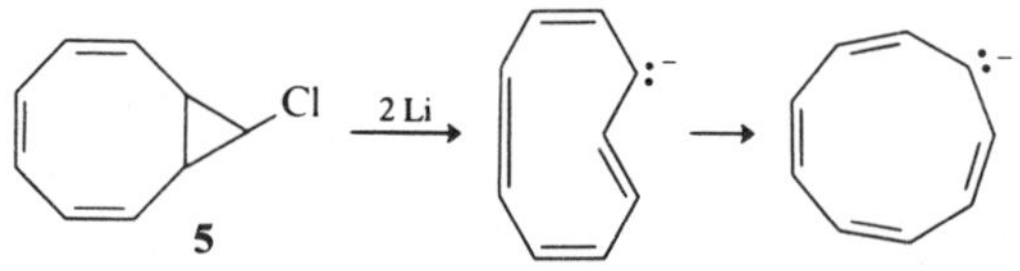

An isomeric form of the anion that is initially formed is converted to the all-*cis* system rapidly at room temperature.[74a] Data on the equilibrium acidity of the parent hydrocarbon are not available, so the stability of the anion cannot be judged quantitatively. The NMR spectrum of the anion, however, is indicative of aromatic character.[74b]

Several doubly charged ions are included in Scheme 9.1; some have been observed experimentally. The NMR spectrum recorded following ionization of 3,4-dichloro-1,2,3,4-tetramethylcyclobutene in SbF_5–SO_2 at −75°C has been attributed to the tetramethyl derivative of the cyclobutadienyl dication.[75]

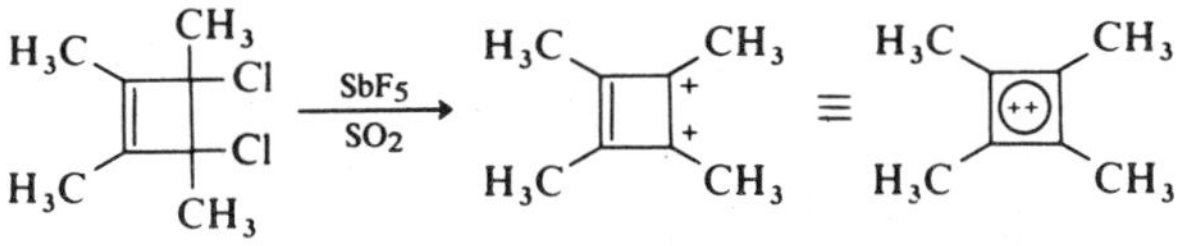

It is difficult to choose a reference compound against which to judge the stability of the dication. That it can be formed at all, however, is suggestive of special stabilization associated with the two-π-electron system. Aromaticity would also be predicted by Hückel's rule for the dianion formed by adding two electrons to the π system of cyclobutadiene. In this case, however, four of the six electrons would occupy nonbonding orbitals. There is some evidence that this species may have a finite existence.[76] Reaction of 3,4-dichlorocyclobutene with sodium naphthalenide, followed a few minutes later by methanol-O-*d*, gives a low yield of 3,4-di-*deuterio*-cyclobutene. The inference is that the dianion $[C_4H_2^{2-}]$ is present. As yet, however, no direct experimental observation of this species has been accomplished.

Cyclooctatetraene is reduced by alkali metals to a dianion.

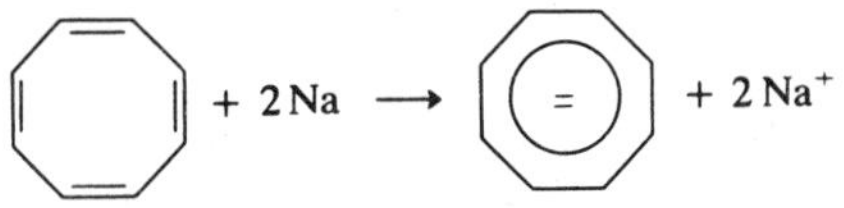

73. T. J. Katz and P. J. Garratt, *J. Am. Chem. Soc.* **86**, 5194 (1964); E. A. LaLancette and R. E. Benson, *J. Am. Chem. Soc.* **87**, 1941 (1965).
74. (a) G. Boche, D. Martens, and W. Danzer, *Angew. Chem. Int. Ed. Engl.* **8**, 984 (1969); (b) S. Fliszar, G. Cardinal, and M. Bernaldin, *J. Am. Chem. Soc.* **104**, 5287 (1982); S. Kuwajima and Z. G. Soos, *J. Am. Chem. Soc.* **108**, 1707 (1986).
75. G. A. Olah, J. M. Bollinger, and A. M. White, *J. Am. Chem. Soc.* **91**, 3667 (1969); G. A. Olah and G. D. Mateescu, *J. Am. Chem. Soc.* **92**, 1430 (1970).
76. J. S. McKennis, L. Brener, J. R. Schweiger, and R. Pettit, *J. Chem. Soc., Chem. Commun.*, 365 (1972).

The NMR spectrum is indicative of a planar aromatic structure.[77] It has been demonstrated that the dianion is more stable than the radical anion formed by one-electron reduction, since the radical anion disproportionates to cyclooctatetraene and the dianion.

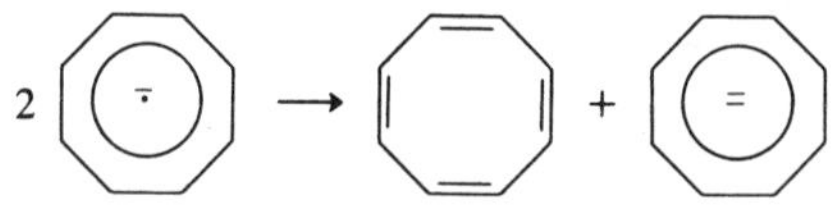

The crystal structure of the potassium salt of 1,3,5,7-tetramethylcyclooctatetraene dianion has been determined by X-ray diffraction.[78] The eight-membered ring is planar, with "aromatic" C—C bond lengths of about 1.41 Å without significant alternation. The spectroscopic and structural studies lead to the conclusion that the cyclooctatetraene dianion is a stabilized delocalized structure.

A dication derived from 1,3,5,7-tetramethylcyclooctatetraene is formed at −78°C in SO_2ClF by reaction with SbF_5. Both the proton and the carbon NMR spectrum indicate that the ion is a symmetrical, diamagnetic species, and the chemical shifts are consistent with an aromatic ring current. At about −20°C, this dication undergoes a chemical transformation to a more stable dication[79]:

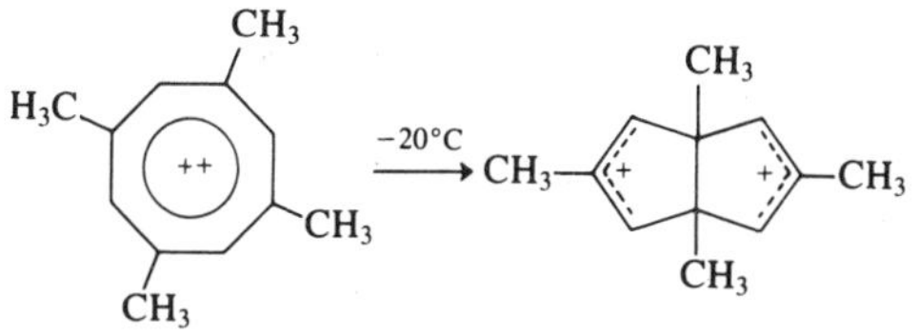

Reduction of the nonaromatic polyene [12]annulene, either electrochemically or with lithium metal, generates a 14-π-electron dianion[80]:

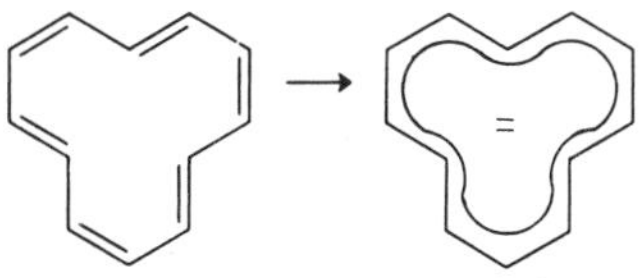

The NMR spectrum of the resulting dianion shows a diamagnetic ring current indicative of aromatic character, even though steric interactions among the internal hydrogens must prevent complete coplanarity. In contrast to the neutral [12]annulene, which is thermally unstable above −50°C, the dianion remains stable at 30°C. The dianion of [16]annulene has also been prepared and shows properties consistent with its being regarded as aromatic.[81]

77. T. J. Katz, *J. Am. Chem. Soc.* **82**, 3784 (1960).
78. S. Z. Goldberg, K. N. Raymond, C. A. Harmon, and D. H. Templeton, *J. Am. Chem. Soc.* **96**, 1348 (1974).
79. G. A. Olah, J. S. Staral, G. Liang, L. A. Paquette, W. P. Melega, and M. J. Carmody, *J. Am. Chem. Soc.* **99**, 3349 (1977).
80. J. F. M. Oth and G. Schröeder, *J. Chem. Soc., B*, 904 (1971).
81. J. F. M. Oth, G. Anthoine, and J.-M. Gilles, *Tetrahedron Lett.*, 6265 (1968).

Table 9.1. Hückel's Rule Relationships for Charged Species

Compound	π-Electrons
Aromatic species	
Cyclopropenium cation	2
Cyclopentadienide anion	6
Cycloheptatrienyl cation	6
Cyclooctatetraene dianion	10
Cyclononatetraenide anion	10
[12]Annulene dianion	14
Antiaromatic species	
Cyclopropenide anion	4
Cyclopentadienyl cation	4
Nonaromatic species	
Cycloheptatrienyl anion	8

The pattern of experimental results on charged species with cyclic conjugated systems is summarized in Table 9.1. It is consistent with the applicability of Hückel's rule to charged, as well as neutral, conjugated planar cyclic structures.

9.4. Homoaromaticity

Homoaromaticity is a term used to describe systems in which a stabilized cyclic conjugated system is formed by bypassing one saturated atom.[82] The resulting stabilization would, in general, be expected to be reduced because of poorer overlap of the orbitals. The properties of several such cationic species, however, suggest that substantial stabilization does result. The cyclooctatrienyl cation is an example.

6

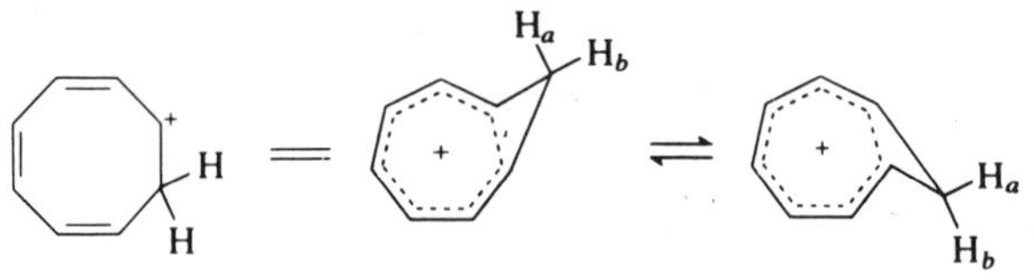

A significant feature of the NMR spectrum of this cation is the fact that the protons *a* and *b* exhibit sharply different chemical shifts. Proton *a* is 5.8 ppm upfield from *b*, indicating the existence of an aromatic ring current.[83a] The fact that the two protons exhibit separate signals also establishes that there is a substantial barrier for the conformational process that interchanges H_a and H_b. Molecular orbital calculations that include the effects of electron correlation find a minimum energy corresponding to the homoconjugated structure.[83b]

82. S. Winstein, *Q. Rev. Chem. Soc.* **23**, 141 (1969); L. A. Paquette, *Angew. Chem. Int. Ed. Engl.* **17**, 106 (1978).
83. (a) P. Warner, D. L. Harris, C. H. Bradley, and S. Winstein, *Tetrahedron Lett.*, 4013 (1970); C. E. Keller and R. Pettit, *J. Am. Chem. Soc.* **88**, 604, 606 (1966); R. F. Childs, *Acc. Chem. Res.* **17**, 347 (1984); (b) R. C. Haddon, *J. Am. Chem. Soc.* **110**, 1108 (1988).

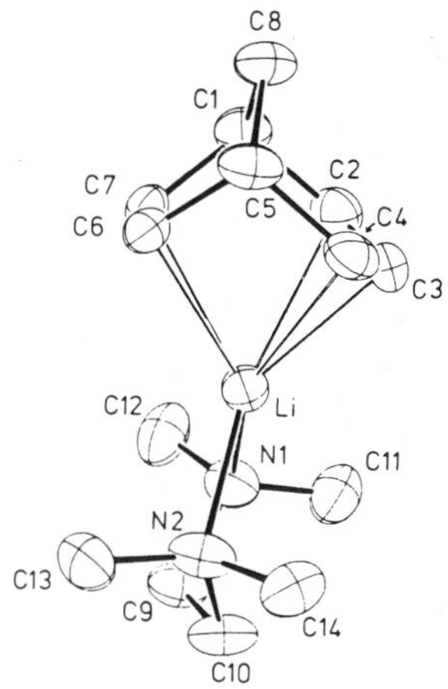

Fig. 9.5. Structure of TMEDA complex of lithium bicyclo-[3.2.1]octa-2,6-dienide. [Reproduced from Ref. 89.]

The cyclobutenyl cation is the homoaromatic analog of the very stable cyclopropenium cation. This ion can be prepared from 3-acetoxycyclobutene using "super-acid" conditions.[84]

O_2CCH_3 —($HOSO_2F$; SbF_5-SO_2ClF, $-78°C$)→ **7**

The temperature-dependent NMR spectrum of the ion can be analyzed to show that there is a barrier (8.4 kcal/mol) for the ring flip which interchanges the two hydrogens of the methylene group. The ^{13}C NMR chemical shift is also compatible with the homoaromatic structure. Molecular orbital calculations are successful in reproducing the structural characteristics of the cation only when the orbitals used are very polarizable and the effects of electron correlation are taken into account.[85]

The existence of stabilizing homoconjugation in anions has been more difficult to establish. Much of the discussion has revolved about anion **8**. The species was proposed to have aromatic character on the basis of the large upfield shift of the CH_2 group that would lie in the shielding region generated by a diamagnetic ring current.[86] The ^{13}C NMR spectrum can also be interpreted in terms of homoaromaticity.[87] Both gas phase and solution measurements suggest that the parent hydrocarbon is more acidic than would be anticipated if there were no special stabilization of the anion.[88] An X-ray crystal structure of a monomeric TMEDA complex of the lithium salt has been done (Fig. 9.5).[89] The lithium is not symmetrically disposed

84. G. A. Olah, J. S. Staral, R. J. Spear, and G. Liang, *J. Am. Chem. Soc.* **97**, 5489 (1975).
85. R. C. Haddon and K. Raghavachari, *J. Am. Chem. Soc.* **105**, 1188 (1983).
86. S. Winstein, M. Ogliaruso, M. Sakai, and J. M. Nicholson, *J. Am. Chem. Soc.* **89**, 3656 (1967).
87. M. Cristl, H. Leininger, and D. Bruecker, *J. Am. Chem. Soc.* **105**, 4843 (1983).
88. R. E. Lee and R. R. Squires, *J. Am. Chem. Soc.* **108**, 5078 (1986); W. N. Washburn, *J. Org. Chem.* **48**, 4287 (1983).
89. N. Hertkorn, F. H. Kohler, G. Müller, and G. Reber, *Angew. Chem. Int. Ed. Engl.* **25**, 468 (1986).

Scheme 9.2. Stabilization Energies of

	Benzene	Naphthalene	Anthracene	Naphthacene
HMO	2.00β	3.68β	5.31β	6.93β
HMO′	0.39β	0.55β	0.66β	0.76β
RE	0.38β	0.59β	0.71β	0.83β
SCF-MO	0.869 eV	1.323 eV	1.600 eV	1.822 eV

	Phenanthrene	Triphenylene	Pyrene	Perylene
HMO	5.44β	7.27β	6.50β	8.24β
HMO′	0.77β	1.01β	0.82β	0.96β
RE	0.85β	1.13β	0.95β	1.15β
SCF-MO	1.933 eV	2.654 eV	2.10 eV	2.619 eV

	Butalene	Pentalene	Azulene	Heptalene
HMO	1.66β	2.45β	3.26β	3.61β
HMO′	−0.48β	−0.14β	0.23β	−0.048β
RE	−0.34β	−0.09β	0.27β	−0.01β
SCF-MO	−0.28 eV	−0.006 eV	0.169 eV	−0.004 eV

toward the anion but is closer to one carbon of the allyl system. There is no indication of flattening of the homoconjugated atoms, and the C-6—C-7 bond distance is in the normal double-bond range (1.354 Å). In contrast to results for the homoaromatic cations **6** and **7**, MO calculations fail to reveal substantial stabilization of the anion **8**.[90] The final reconciliation of the divergent indications of the degree of delocalization and stabilization of this anion will have to await further work.

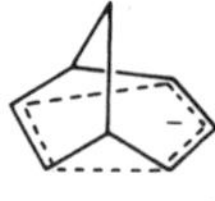

8

90. J. B. Grutzner and W. L. Jorgenson, *J. Am. Chem. Soc.* **103**, 1372 (1981); E. Kaufman, H. Mayr, J. Chandrasekhar, and P. v. R. Schleyer, *J. Am. Chem. Soc.* **103**, 1375 (1981); R. Lindh, B. O. Roos, G. Jonsäll, and P. Ahlberg, *J. Am. Chem. Soc.* **108**, 6554 (1986).

	Methylene-cyclopropene	Fulvene	Calicene	Fulvalene
HMO	0.96β	1.46β	——	2.80β
HMO′	0.02β	-0.012β	0.34β	-0.33β
RE		-0.01β	0.39β	-0.29β
SCF-MO	——	——	——	——

	Benzocyclo-butadiene	Biphenylene	Acenaphthylene	
HMO	2.38β	4.50β	4.61β	——
HMO′	-0.22β	0.32β	0.47β	-0.22β
RE	-0.16β	0.42β	0.57β	
SCF-MO	——	1.346 eV	1.335 eV	——

a. Stabilization energies given are from the following sources:

HMO: C.A. Coulson, A. Streitwieser, Jr., M. D. Poole, and J. I. Brauman, *Dictionary of π-Electron Calculations*, W. H. Freeman, San Franscisco, 1965.
HMO′: B. A. Hess, Jr., and L. J. Schaad, Jr., *J. Am. Chem. Soc.* **93**, 305, 2413 (1971); *J. Org. Chem.* **36**, 3418 (1971); **37**, 4179 (1972).
RE: A. Moyano and J. C. Paniagua, *J. Org. Chem.* **51**, 2250 (1986).
SCF-MO: M. J. S. Dewar and C. de Llano, *J. Am. Chem. Soc.* **91**, 789 (1969). 1 eV = 23 kcal/mol.

9.5. Fused-Ring Systems

Many completely conjugated hydrocarbons can be built up from the annulenes and related structural fragments. Scheme 9.2 gives the structures, names, and stabilization energies of a variety of such hydrocarbons. Derivatives of these hydrocarbons having substituent groups or with heteroatoms in place of one or more carbon atoms constitute another important class of organic compounds.

It is of interest to be able to predict the stability of such fused-ring compounds. Because Hückel's rule applies only to monocyclic systems, it cannot be applied to the fused-ring compounds, and there have been many efforts to develop relationships which would predict the stability of these compounds. The underlying concepts should be the same as for monocyclic systems; stabilization would result from particularly stable arrangements of molecular orbitals whereas instability would be associated with unpaired electrons or electrons in high-energy orbitals.

Table 9.2. Energy Values for Reference Bond Types

Bond type	Hess-Schaad value (β)	Bond type	Moyano-Paniagua value (β)
$H_2C{=}CH$	2.000	$H_2C{=}CH{-}CH{=}$	2.2234
HC=CH	2.070	$H_2C{=}CH{-}C{=}$	2.2336
$H_2C{=}C$	2.000	=CH—CH=CH—CH=	2.5394
HC=C	2.108	=CH—CH=CH—C=	2.5244
C=C	2.172	=C—CH=CH—C=	2.4998
HC—CH	0.466	$H_2C{=}C{-}$	2.4320
HC—C	0.436	—CH=CH—	2.7524
C—C	0.436	—C=C—	2.9970

The same approximations discussed in Section 1.5 permit HMO calculations foı conjugated systems of the type shown in Scheme 9.2, and many of the results have been tabulated.[91] However, attempts to correlate stability with the Hückel delocalization energy relative to isolated double bonds give poor correlation with the observed chemical properties of the compounds. With the choice of a polyene as the reference state, much better agreement between calculated stabilization energy and experimental chemical properties is achieved. A series of energy terms corresponding to the structural units in the reference polyene have been established empirically.[92] The difference between the energy of the conjugated hydrocarbon by HMO calculation and the sum of the energy terms for the appropriate structural units gives a stabilization energy. For azulene, for example, the HMO calculation gives an energy of $10\alpha + 13.36\beta$. The energy for the localized model is obtained by summing contributions for the component bond types: 3(HC=CH) + 2(HC=C) + 3(HC—CH) + 2(HC—C) + 1(C—C) = 13.13β. The difference, 0.23β, is the stabilization or resonance energy assigned to azulene by this method. For comparison of nonisomeric molecules, the Hess-Schaad treatment uses resonance energy per electron, which is obtained simply by dividing the calculated stabilization energy by the number of π electrons. Although the resulting stabilization energies are based on a rudimentary HMO calculation, they are in good qualitative agreement with observed chemical stability. The stabilizations have been calculated for most of the molecules in Scheme 9.2 and are listed as HMO′.

The energy parameters for the reference polyene used by Hess and Schaad were developed on a strictly empirical basis. Subsequently, Moyano and Paniagua developed an alternative set of reference bond energies on a theoretical basis.[93] These values are shown along with the Hess-Schaad values in Table 9.2. The stabilizations calculated for the various hydrocarbons in Scheme 9.2 with this point of reference are those listed as RE (for resonance energy). The Hess-Schaad HMO′

91. E. Heilbronner and P. A. Straub, *Hückel Molecular Orbitals*, Springer-Verlag, Berlin, 1966; C. A. Coulson and A. Streitwieser, *Dictionary of π-Electron Calculations*, W. H. Freeman, San Francisco, 1965.
92. B. A. Hess, Jr., and L. J. Schaad, *J. Am. Chem. Soc.* **93**, 305, 2413 (1971); *J. Org. Chem.* **36**, 3418 (1971); **37**, 4179 (1972).
93. A. Moyano and J. C. Paniagua, *J. Org. Chem.* **51**, 2250 (1986).

Table 9.3. Rates of Diels–Alder Addition of Linear Polycyclic Aromatic Hydrocarbons[a]

	k (M^{-1} s^{-1}) (80°C, in toluene)		
Dienophile	Anthracene	Naphthacene	Pentacene
Benzoquinone		44	181
Maleic anhydride	5	294	4,710
N-Phenylmaleimide	10	673	19,280

a. V. D. Samuilov, V. G. Uryadov, L. F. Uryadova, and A. J. Konolova, *Zh. Org. Khim.* (Engl. Transl.), **21**, 1137 (1985).

and the RE values are in generally good agreement with observed stability. Both calculations give negative stabilization for benzocyclobutadiene, for example.[94]

The values listed in Scheme 9.2 as SCF-MO are from an early semiempirical SCF calculation. This was the first instance in which a polyene was chosen as the reference state.[95] The energy assigned the polyene was calculated by the same computational method.

All these approaches agree that benzene and the structures that can be built up by fusing together benzenoid rings are strongly stabilized relative to the reference polyenes. The larger rings tend to have lower resonance energies per π electron than does benzene. This feature is in agreement with trends in stability observed experimentally. The smaller ring structures are the most stable, and when many rings are fused together, there is a tendency for them to react readily to give the smaller and more stable structures.

+ X—Y ⟶ (Y H / X H)

SCF stabilization energy = 41.9 kcal

SCF stabilization energy = 50.4 kcal

This trend is revealed, for example, by the rates of Diels–Alder addition reactions of anthracene, naphthacene, and pentacene, in which three, four, and five rings, respectively, are linearly fused. The rate data are shown in Table 9.3.

The same trend can be seen in the E_a and resonance energy gained when cycloreversion of the adducts **9–12** yields aromatic compounds, as shown in Scheme 9.3.

There is evidence that aromatic segments can exist as part of larger conjugated units, resulting in an aromatic segment in conjugation with a "localized" double

94. There are a number of other systems for comparing the stability of conjugated cyclic compounds with reference polyenes. For example, see L. J. Schaad and B. A. Hess, Jr., *Pure Appl. Chem.* **54**, 1097 (1982); J. Aihara, *Pure Appl. Chem.* **54**, 1115 (1982); K. Jug, *J. Org. Chem.* **48**, 1344 (1983); W. Gründler, *Monatsh. Chem.* **114**, 155 (1983).
95. M. J. S. Dewar and C. de Llano, *J. Am. Chem. Soc.* **91**, 789 (1969).

Scheme 9.3. Correlation between E_a for Retro-Diels–Alder Reaction and Resonance Stabilization of Aromatic Products

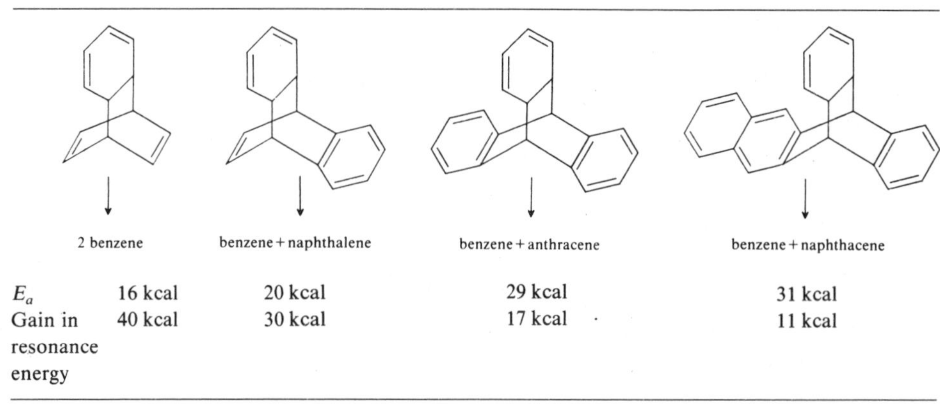

E_a	16 kcal	20 kcal	29 kcal	31 kcal
Gain in resonance energy	40 kcal	30 kcal	17 kcal	11 kcal

bond. For example, in acenaphthylene, the double bond in the five-membered ring is both structurally and chemically similar to a normal localized double bond. The resonance energy given in Scheme 9.2, 0.57β, is slightly less than that for napththalene (0.59β). The additional double bond of acenaphthylene has only a small effect on the stability of the conjugated system. The molecular structure determined at 80 K by neutron diffraction shows bond lengths for the aromatic portion that are very similar to those of naphthalene.[96] The double bond is somewhat longer than a normal double bond, and this may reflect the strain imposed by the naphthalene framework on the double bond.

1.42 1.37 1.42 1.41

naphthalene

1.395 1.466 1.381 1.386 1.424 1.433 1.382

acenaphthylene

The various approaches to prediction of relative stability diverge more widely when nonbenzenoid systems are considered. The simple Hückel method using total π delocalization energies relative to an isolated double-bond reference energy ($\alpha + \beta$) fails. This approach predicts stabilization of the same order of magnitude for such unstable systems as pentalene and fulvalene as it does for much more stable aromatics. The HMO′, RE, and SCF–MO methods, which use polyene reference energies, do much better. All show drastically reduced stabilization for such systems and, in fact, indicate destabilization of systems such as butalene and pentalene (Scheme 9.2).

It is of interest to consider at this point some of the specific molecules in Scheme 9.2 and compare their chemical properties with the calculated stabilization energies.

96. R. A. Wood, T. R. Welberry, and A. D. Rae, *J. Chem. Soc., Perkin Trans. 2*, 451 (1985).

Benzocyclobutadiene has been generated in a number of ways, including dehalogenation of dibromobenzocyclobutene.[97]

Br Zn Br

The compound is highly reactive, dimerizing or polymerizing readily.[98]

2 → Ref. 98c

Benzocyclobutadiene is very reactive as a dienophile in the Diels–Alder reaction.

+ → Ref. 98d

The high reactivity of benzocyclobutadiene has precluded detailed structural studies, but the reactivity confirms the prediction that the presence of the cyclobutadiene moiety is strongly destabilizing, since the compound is much more reactive than the noncyclic analog styrene.

Azulene is one of the few nonbenzenoid hydrocarbons that appears to have appreciable aromatic stabilization. There is some divergence on this point between the SCF–MO and HMO′ results in Scheme 9.2. The latter estimates a resonance energy about half that for the isomeric naphthalene, whereas the SCF–MO method assings a resonance energy which is only about one-seventh that of naphthalene. Naphthalene is more stable than azulene by about 38.5 kcal/mol. Molecular mechanics calculations attribute about 12.5 kcal/mol of the difference to strain and about 26 kcal/mol to greater resonance stabilization of naphthalene.[99] Based on heats of hydrogenation, the experimental stabilization energy of azulene is about 16 kcal/mol.[100] The parent hydrocarbon and many of its derivatives are well-characterized compounds with considerable stability. The structure of azulene has been determined by both X-ray crystallography and electron diffraction measurements.[101] The peripheral bond lengths are in the aromatic range and show no regular alterna-

97. M. P. Cava and D. R. Napier, *J. Am. Chem. Soc.* **78**, 500 (1956); **79**, 1701 (1957).
98. (a) M. P. Cava and M. J. Mitchell, *Cyclobutadiene and Related Compounds*, Academic Press, New York, 1967, pp. 192–216; (b) P. Gandhi, *J. Sci. Ind. Res.* **41**, 495 (1982); (c) M. P. Cava and D. R. Napier, *J. Am. Chem. Soc.* **80**, 2255 (1958); (d) M. P. Cava and M. J. Mitchell, *J. Am. Chem. Soc.* **81**, 5409 (1959).
99. N. L. Allinger and Y. H. Yu, *Pure Appl. Chem.* **55**, 191 (1983).
100. W. R. Roth, M. Boehm, H. W. Lennartz, and E. Vogel, *Angew. Chem. Int. Ed. Engl.* **22**, 1007 (1983).
101. A. W. Hanson, *Acta Crystallogr.* **19**, 19 (1965); O. Bastiansen and J. L. Derissen, *Acta Chem. Scand.* **20**, 1319 (1966).

tion. The bond shared by the two rings is significantly longer, indicating that it has predominantly single-bond character.

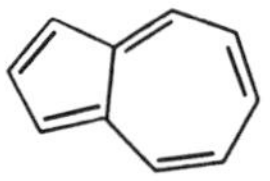

azulene

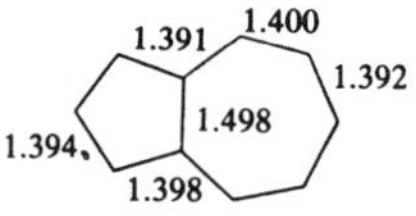

X-ray bond lengths

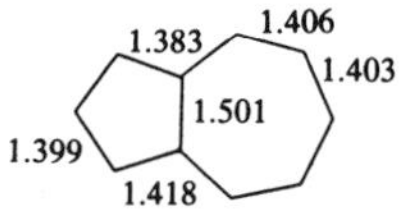

1.391 1.400 1.392 1.394 1.498 1.398

electron-diffraction bond lengths

1.383 1.406 1.403 1.399 1.501 1.418

An interesting structural question revolves around the contribution of a dipolar structure that pictures the molecule as the fusion of a cyclopentadienide anion and a cycloheptatrienyl cation:

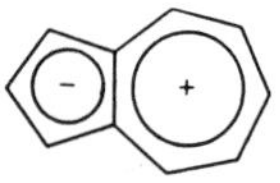

The molecule does have an appreciable dipole moment (0.8 D).[102] The essentially single-bond nature of the shared bond indicates, however, that the conjugation is principally around the periphery of the molecule.

Several MO calculations have been applied to azulene. At the MNDO and STO-3G levels, structures with considerable bond alternation are found as the minimum-energy structures. Calculations that include electron correlation effects give a delocalized π system as the minimum-energy structure.[103]

The significant resonance stabilization of azulene can be contrasted with the situation for pentalene and heptalene, both of which are indicated to be destabilized relative to a reference polyene.

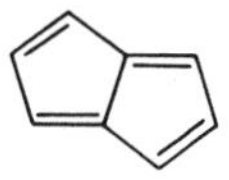

pentalene

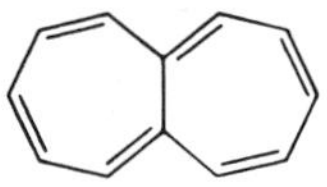

heptalene

Preparation of pentalene by a Hofmann elimination at 20°C is followed by immediate dimerization. Low-temperature photolysis produces a new species believed to be pentalene, but the compound reverts to dimer at −100°C.[104]

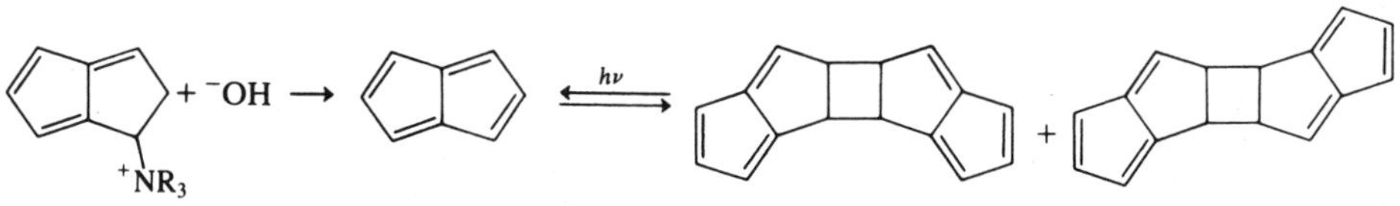

$^{+}NR_3$ + ^{-}OH → ⇌ ($h\nu$) +

102. H. J. Tobler, A. Bauder, and H. H. Günthard, *J. Mol. Spectrosc.* **18**, 239 (1965); G. W. Wheland and D. E. Mann, *J. Chem. Phys.* **17**, 264 (1949).
103. C. Glidewell and D. Lloyd, *Tetrahedron* **40**, 4455 (1984); R. C. Haddon and K. Raghavachari, *J. Am. Chem. Soc.* **104**, 3516 (1982).
104. K. Hafner, R. Dönges, E. Goedecke, and R. Kaiser, *Angew. Chem. Int. Ed. Engl.* **12**, 337 (1973).

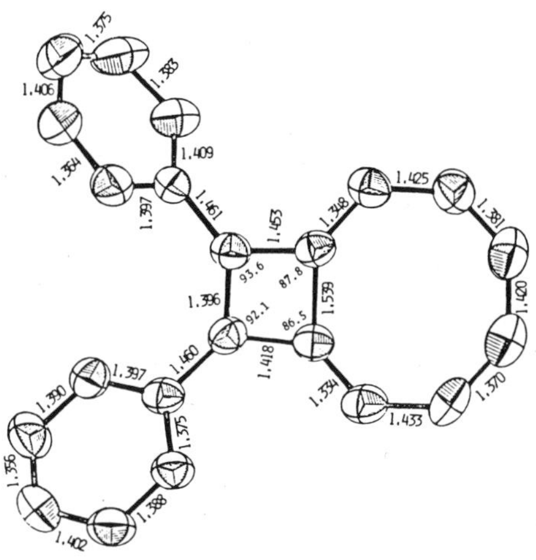

Fig. 9.6. Crystal structure of 9,10-diphenyl-bicyclo[6.2.0]deca-1,3,5,7,9-pentaene. [Reproduced from Ref. 106.]

Heptalene readily polymerizes and is sensitive to oxygen. The NMR spectrum does not indicate the presence of an aromatic ring current. The conjugate acid of heptalene, however, is very stable (even at pH 7 in aqueous solution), reflecting the stability of the cation, which is a tropylium ion.[105]

H^+

H H

Another structure with a 10-electron conjugated system is bicyclo[6.2.0]deca-1,3,5,7,9-pentaene. The crystal structure of the 9,10-diphenyl derivative shows the conjugated system to be nearly planar (Fig. 9.6).[106]

There is significant bond alternation, however. The bond at the ring fusion is quite long. A molecular mechanics calculation on this molecule which includes an SCF–MO treatment of the planar conjugated system concludes that the molecule is slightly destabilized (4 kcal/mol) relative to a polyene reference.[107]

The possibility of extra stabilization in conjugated systems that have conjugated components exocyclic to the ring has also been examined. Some representative structures are shown in Scheme 9.4.

105. H. J. Dauben, Jr., and D. J. Bertelli, *J. Am. Chem. Soc.* **83**, 4657, 4659 (1961).
106. C. Kabuto and M. Oda, *Tetrahedron Lett.* 103 (1980).
107. N. L. Allinger and Y. H. Yuh, *Pure Appl. Chem.* **55**, 191 (1983).

Scheme 9.4. Completely Conjugated Hydrocarbons Incorporating Exocyclic Double Bonds

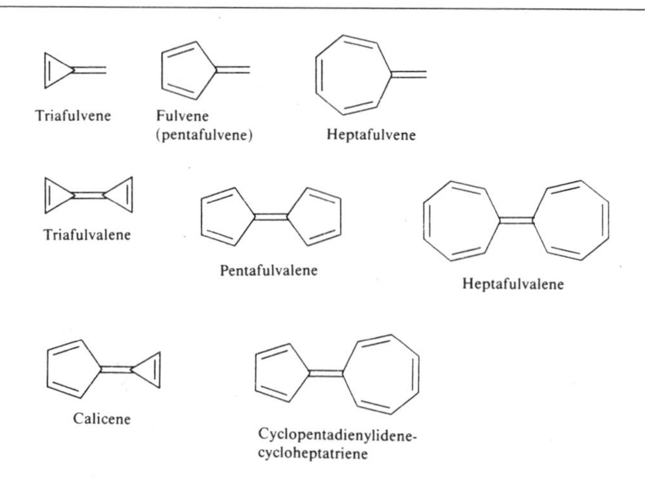

Cyclopropenes and cyclopentadienes with exocyclic double bonds have the possibility of dipolar resonance structures which provide aromatic character in the cyclic structure:

fulvene

For methylenecyclopropene, a microwave structure determination has established bond lengths that show the strong alternation anticipated for a localized structure.[108] The molecule does have a significant (1.90 D) dipole moment, implying a contribution from the dipolar resonance structure. The net stabilization calculated at the 6-31G* level is small and comparable to the stabilization of 1,3-butadiene. The molecular geometry of dimethylfulvene has been examined by electron diffraction methods. Strong bond-length alternation indicative of a localized structure is found.[109]

The fulvalene systems are not predicted to be aromatic by any of the quantitative estimates of stability. Even simple resonance considerations would suggest polyene behavior, since only dipolar resonance structures can be drawn in addition to the

108. T. D. Norden, S. W. Staley, W. H. Taylor, and M. D. Harmony, *J. Am. Chem. Soc.* **108**, 7912 (1986).
109. J. F. Chiang and S. H. Bauer, *J. Am. Chem. Soc.* **92**, 261 (1970).

single nonpolar structure.

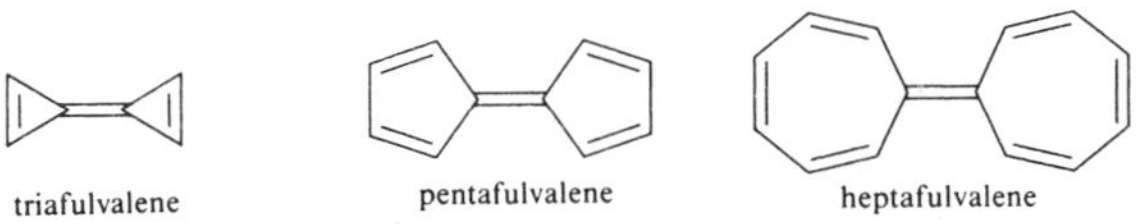

triafulvalene pentafulvalene heptafulvalene

Triafulvalene (cyclopropenylidenecyclopropene) has not been isolated. A substantial number of pentafulvalene derivatives have been prepared.[110] The chemical properties of these molecules are those of reactive polyenes. The NMR spectrum of pentafulvalene is characteristic of a localized system.[111] Heptafulvalene (cycloheptatrienylidenecycloheptatriene) is a well-characterized compound with the properties expected for a polyene.[112]

Because the five-membered ring is a substituted cyclopentadienide anion in some dipolar resonance structures, it might be expected that exocyclic groups that could strongly stabilize a positive charge might lead to a large contribution from dipolar structures and enhanced stability. The structures **13** and **14** would appear to be cases in which a large dipolar contribution would be feasible.

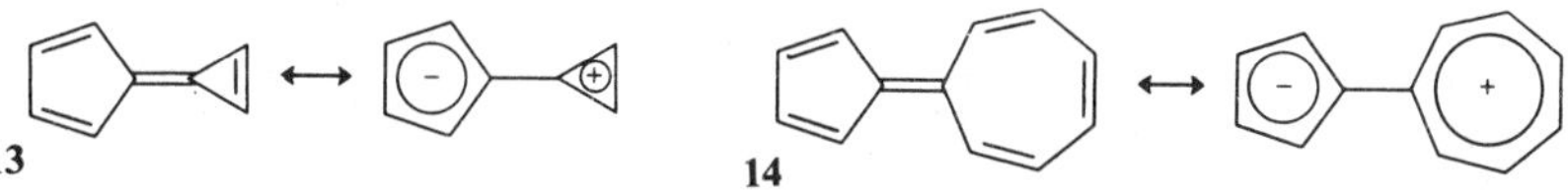

13 **14**

The stability of such dipolar systems depends on the balance between the energy required to separate unlike charges and the aromaticity associated with Hückel $4n+2$ systems. Phenyl-substituted analogs are known, and the large measured dipole moments suggest considerable charge separation.

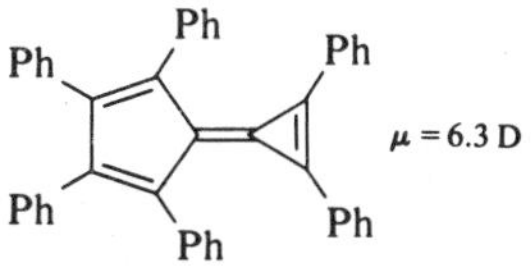

Ph Ph Ph Ph Ph Ph $\mu = 6.3$ D Ref. 113

Some alkyl derivatives have been prepared. Their chemical behavior is that of highly reactive polyenes. One interesting property does appear in the NMR spectra, which reveal a reduced barrier to rotation about the double bond between the two rings.[114] This property suggests that rotation about this bond takes place easily through a

110. E. D. Bergmann, *Chem. Res.* **68**, 41 (1968).
111. E. Escher, P. Bönzil, A. Otter, and M. Neuenschwander, *Magn. Reson. Chem.* **24**, 350 (1986).
112. T. Nozoe and I. Murata, *Int. Rev. Sci., Org. Chem. Ser. Two* **3**, 197 (1976).
113. E. D. Bergmann and I. Agranat, *J. Chem. Soc., Chem. Commun.* 512 (1965).
114. A. S. Kende, P. T. Izzo, and W. Fulmor, *Tetrahedron Lett.*, 3697 (1966); H. Prinzbach, *Pure Appl. Chem.* **28**, 281 (1971).

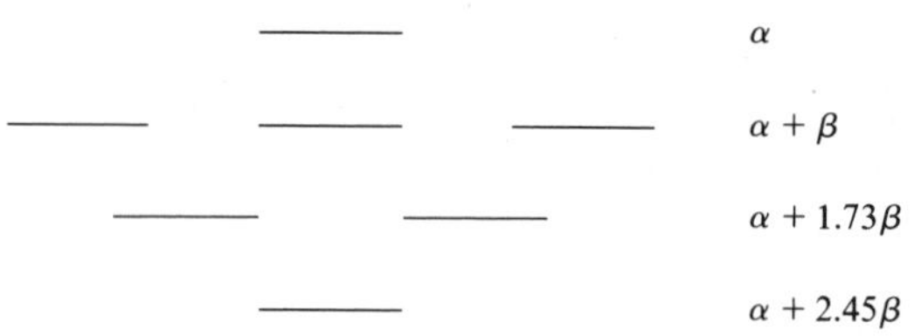

α

$\alpha + \beta$

$\alpha + 1.73\beta$

$\alpha + 2.45\beta$

Fig. 9.7. Hückel molecular orbitals for phenalene.

transition state in which the two charged aromatic rings are twisted out of conjugation.

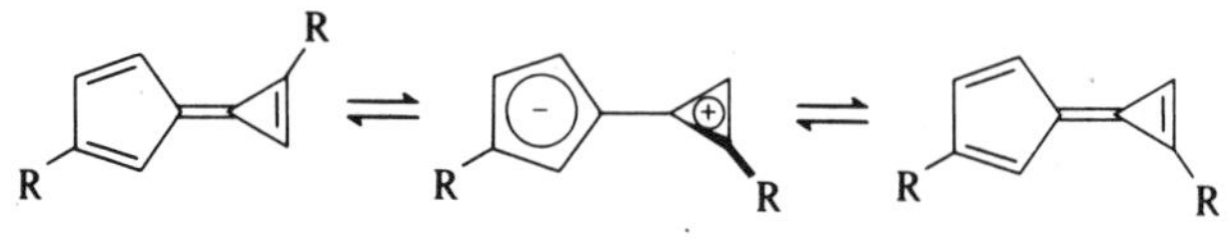

STO-3G and 3-21G molecular orbital calculations indicate a rotational barrier that is substantially reduced relative to the corresponding barrier in ethylene. The transition state for the rotation is calculated to have a charge separation of the type suggested by the dipolar resonance structure.[115]

The hydrocarbon phenalene is the precursor of both a highly stabilized anion and a highly stabilized cation. The Hückel MO diagram is shown in Fig. 9.7. The single orbital at the nonbonding level is the LUMO in the cation and the HOMO in the anion. The stabilization energy calculated for both would be the same and is 0.41β by the HMO′ comparison.[116]

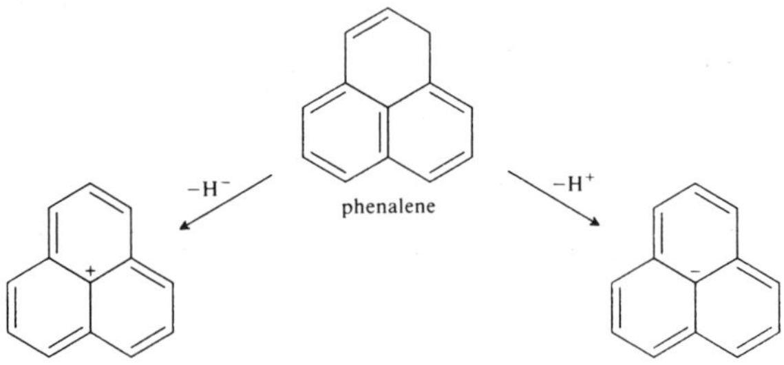

$-H^-$ phenalene $-H^+$

The pK for conversion of phenalene to its anion is 19.[117] The cation is estimated to have a pK_{R^+} of about 0–2.[118] Several methods for generating the phenalenyl cation have been developed.[119]

In general conclusion, the HMO′ and SCF methods both appear to make reasonably accurate predictions about the stabilization in conjugated molecules.

115. B. A. Hess, Jr., L. J. Schaad, C. S. Ewig, and P. Carsky, *J. Comput. Chem.* **4**, 53 (1982).
116. J. Aihara, *Bull. Chem. Soc. Jpn.* **51**, 3540 (1978); P. Ilic and N. Trinjastic, *J. Org. Chem.* **45**, 1738 (1980).
117. A. Streitwieser, Jr., J. M. Word, F. Guibe, and J. S. Wright, *J. Org. Chem.* **46**, 2588 (1981); R. A. Cox and R. Stewart, *J. Am. Chem. Soc.* **98**, 488 (1976).
118. D. Menche, H. Strauss, and E. Heilbronner, *Helv. Chim. Acta* **41**, 57 (1958).
119. I. Murata, in *Topics in Nonbenzenoid Aromatic Chemistry*, T. Nozoe, R. Breslow, K. Hafner, S. Ito, and I. Murata (eds.), Hirokawa, Tokyo, 1976.

The stabilization is general for benzenoid compounds, but quite restricted in nonbenzenoid systems. Since the HMO′ method of estimating stability is based on the ideas of HMO theory, its general success vindicates the ability of this very simplified MO approach to provide some insight into the structural nature of the annulenes and other conjugated polyenes. More sophisticated MO methods, of course, are now accessible and should be applied for more detailed analysis of the structures of these molecules.

9.6. Heterocyclic Rings

Certain structural units containing heteroatoms can be substituted into conjugated systems in such a way that the system remains conjugated and isoelectronic with the original hydrocarbon. The most common examples are —CH=N— and —N=N— double bonds and divalent sp^2 —O—, —S—, and —NR— units. Each of these structural fragments can replace a —CH=CH— unit in a conjugated system and contribute two π electrons. Scheme 9.5 gives some of the common structures that are isoelectronic with benzene and naphthalene.

Molecular orbital calculations on compounds in which a —CH=N— unit replaces —CH=CH— indicate that the resonance stabilization is very similar to that of the original compound. For the —O—, —S—, and —NR— fragments, the resonance stabilization is somewhat reduced but nevertheless high enough for the resulting compounds to be considered aromatic in character. These compounds are called *heteroaromatic* to recognize both the heterocyclic structure and the relationship to benzene and other aromatic structures.

Additional heteroaromatic structures can be built up by fusing benzene rings to the aromatic heterocyclic rings or by fusing together heterocyclic rings. Examples of this type are included in Scheme 9.5. When benzene rings are fused to the heterocyclic five-membered rings, the structures from fusion at the 2,3-positions are much more stable than those from fusion at the 3,4-positions. The π-electron system in the 3,4-fused compounds is more similar to a peripheral 10-π-electron system than to the 10-electron system for naphthalene. As a result, these compounds have a strong tendency to undergo reactions that restore benzene conjugation in the carbocyclic ring. The isobenzofuran structure **15** is known to be an exceptionally reactive diene, for example. Isoindole, **16**, readily tautomerizes to the benzenoid imine **17**.

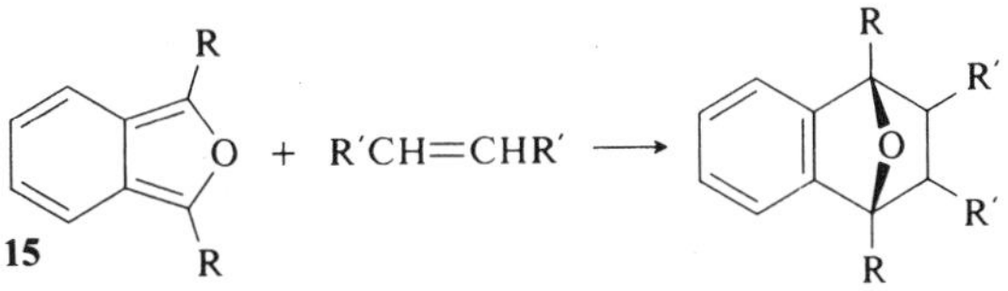

15 + R′CH=CHR′ ⟶

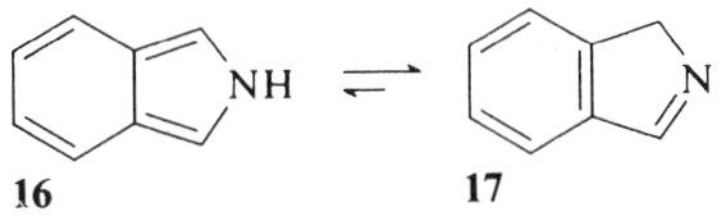

16 ⇌ **17**

Scheme 9.5. Heteroaromatic Structures Isoelectronic with Benzene or Naphthalene

A. Structures Isoelectronic with Benzene

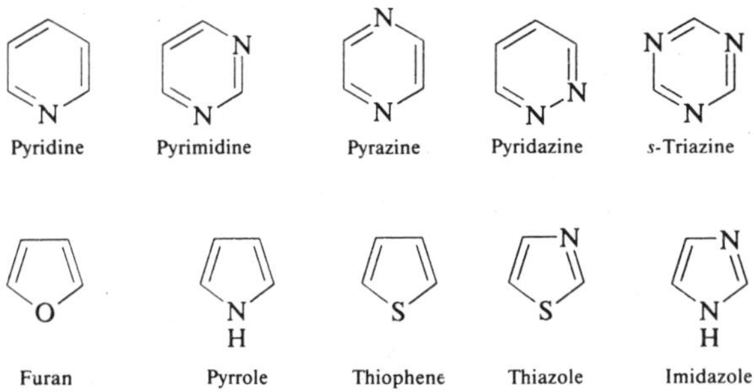

B. Structures Isoelectronic with Naphthalene

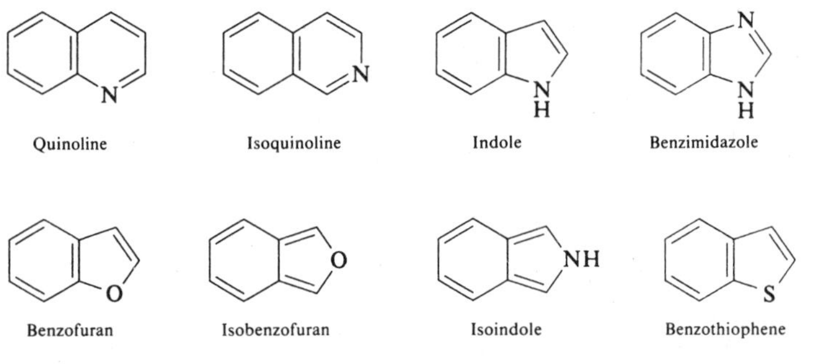

General References

E. Clar, *Polycyclic Hydrocarbons*, Academic Press, New York, 1964.
P. J. Garratt, *Aromaticity*, Wiley, New York, 1986.
D. Lloyd, *Nonbenzenoid Conjugated Carbocyclic Compounds*, Elsevier, Amsterdam, 1984.
J. P. Snyder, *Nonbenzenoid Aromatics*, Vols. 1 and 2, Academic Press, New York, 1969.

Problems

(*References for these problems will be found on page* 781.)

1. The reaction of *o*-diphenylcyclobutadiene (generated *in situ* by oxidation of its iron tricarbonyl complex) with *p*-benzoquinone yields **A** as the exclusive product. With tetracyanoethylene, however, **B** and **C** are formed in a 1:7 ratio. Discuss these results, and explain how they relate to the question of the square versus rectangular shape of cyclobutadiene.

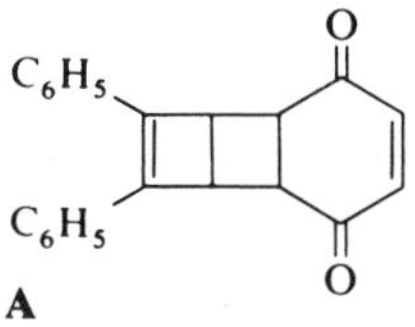

A 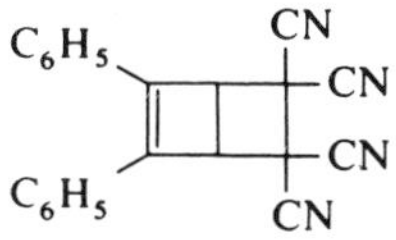 B 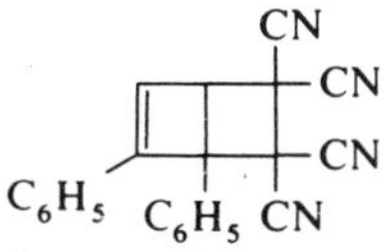 C

2. A single resonance structure is shown below for each of several molecules. Consider other resonance structures. Comment on those that would be expected to make a major stabilizing contribution to the molecule in question.

(a) 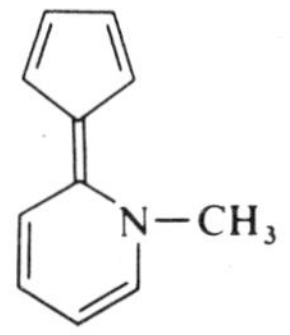 (c)

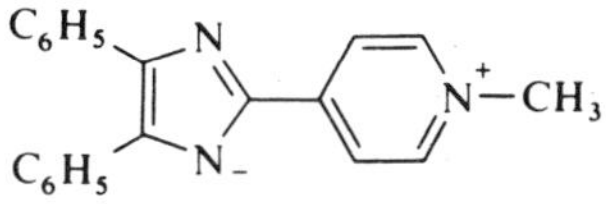

(b) 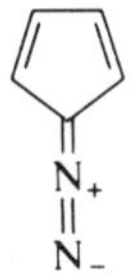 (d)

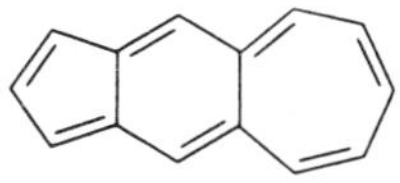

3. (a) A synthesis of tropone (cycloheptatrienone) entails treating 1-methoxycycloheptatriene with bromine. A salt is produced that yields tropone on treatment with aqueous sodium bicarbonate. What is the salt? Write a mechanism for its formation.
   (b) The optically active dichlorophenylcyclobutenone **A** undergoes racemization in acetic acid at 100°C. Suggest an experiment to determine if the enol (a hydroxycyclobutadiene) is an intermediate.

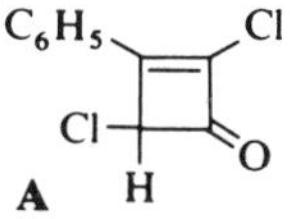

A

4. Predict whether the following systems would be expected to show strong (aromatic or homoaromatic) stabilization, weak stabilization by conjugation (nonaromatic), or destabilization (antiaromatic) relative to localized model structures. Explain the basis for your prediction.

(a) 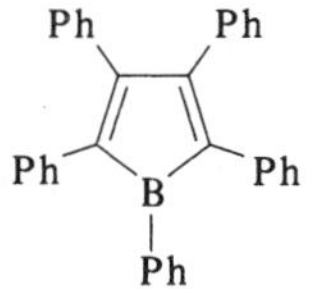 (b) 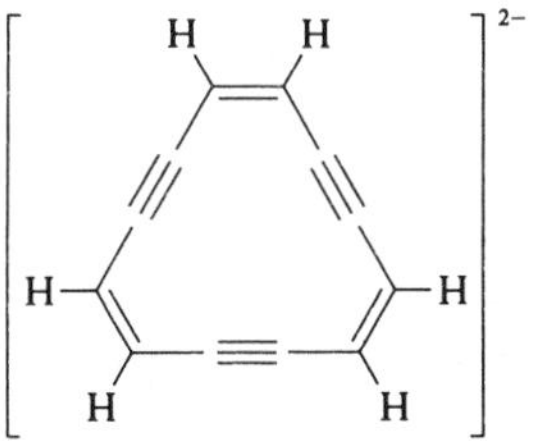 (c)

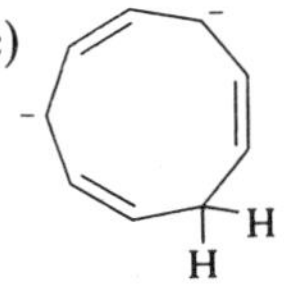

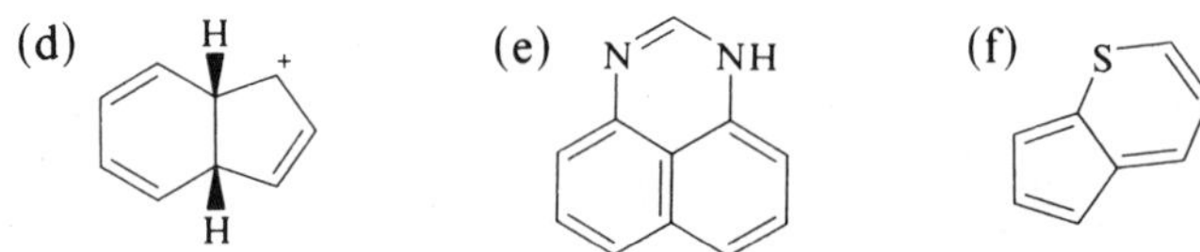

5. Bicyclo[6.2.0]deca-2,4,6,8,10-pentaene has been synthesized, and a number of molecular orbital and molecular mechanics calculations have been performed to determine whether it is aromatic or antiaromatic. Consider the structure and discuss the following points.

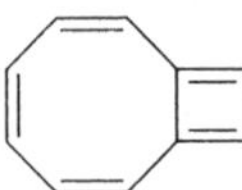

(a) What aspects of the structure suggest antiaromaticity might be observed?
(b) What aspects of the structure suggest aromaticity might be observed?
(c) What are some of the experimental and theoetical criteria which could be applied to assess aromaticity or antiaromaticity? Cite at least three such probes, and indicate the nature of the observation and how it would be interpreted.

6. Using the empirically chosen energy equivalents for contributing bond types given on p. 520 and a standard compilation of simple HMO calculations, calculate resonance energies for the following molecules according to the modified procedure of Hess and Schaad. Do you find any discrepancies between predicted and observed stability?

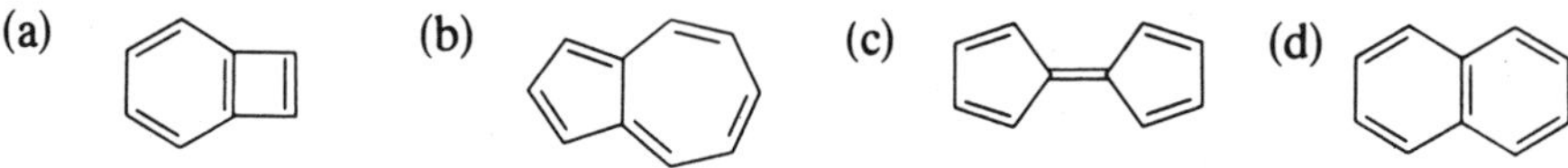

7. The relative basicity of carbonyl oxygen atoms can be measured by studying the strength of hydrogen bonding with the carbonyl compound to a hydrogen donor such as phenol. In carbon tetrachloride, the K_{eq} for 1:1 complex formation for the compounds shown have been measured. Rationalize the observed order of basicity.

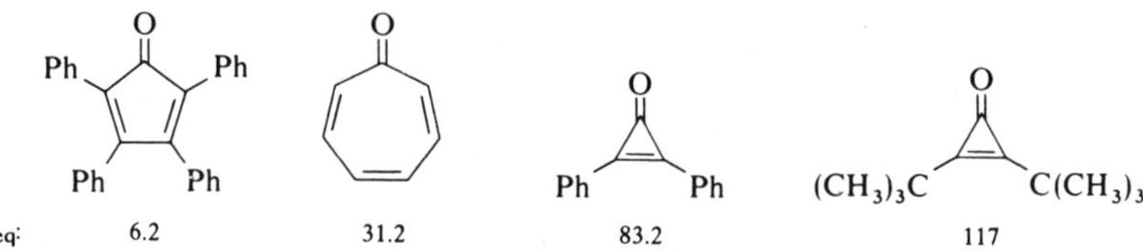

8. One criterion of aromaticity is the "ring current" which is revealed by a chemical shift difference between protons in the plane of the conjugated system and those above or below the plane. The chemical shifts of two isomeric hydrocarbons

are given below. In qualitative terms, which appears to be more aromatic? (Because the chemical shift is a sensitive function of the geometric relationship to the ring current, a quantitative calculation would be necessary to confirm the correctness of this qualitative impression.) Does Hückel MO theory predict a difference in the aromaticity of these two compounds?

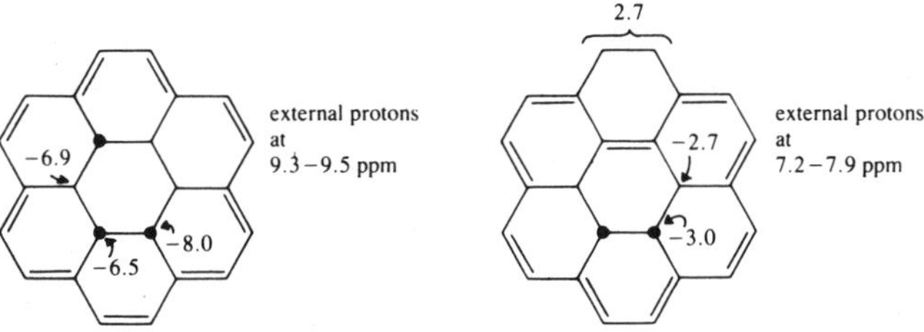

chemical shift in ppm from tetramethylsilane

9. Offer an explanation for the following observations.

(a) Hydrocarbon **A** (p$K \approx 14$) is much more acidic than **B** (p$K \approx 22$).

A **B**

(b) The hydrocarbon **C** has an unusually small separation of its oxidation and reduction potentials, as established by electrochemical measurements. It is both easily reduced and easily oxidized. Both mono- and dications and mono- and dianions can be readily formed.

C

(c) The barrier for rotation about the marked bond in **D** is only about 14 kcal/mol.

D

(d) The hydrocarbon **E** is easily reduced to a dianion. The proton NMR spectrum of the dianion shows an average *downfield* shift, relative to the hydrocarbon. The center carbon shows a very large upfield shift in the ^{13}C NMR spectrum.

E

10. The Hückel molecular orbitals for acenaphthylene are shown below. The atomic coefficients for the orbital which is the LUMO in the neutral compound and the HOMO in the dianion are given to the right.

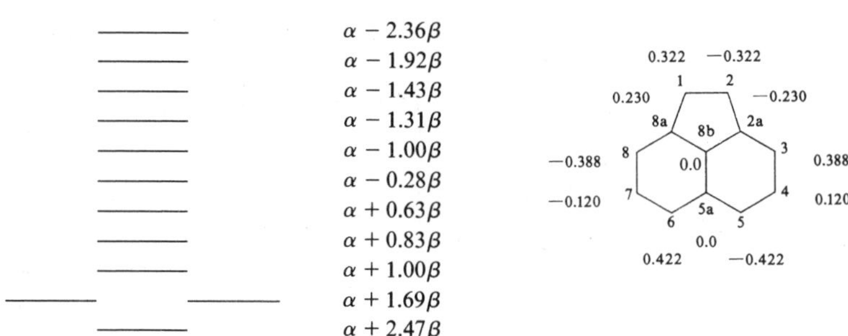

Comment on the aromaticity, antiaromaticity, or nonaromaticity of acenaphthylene and its dianion on the basis of the following physical measurements.

(a) The bond lengths of acenaphthylene are shown below. Compare them with the bond lengths for naphthalene given on p. 524. What conclusions do you draw about the aromaticity of acenaphthylene?

(b) Both X-ray and NMR data indicate that the C-1—C-2 bond lengthens significantly in the dianion as shown below. There is also a different pattern of bond length alternation. What conclusions do you draw about the aromaticity of the acenaphthylene dianion?

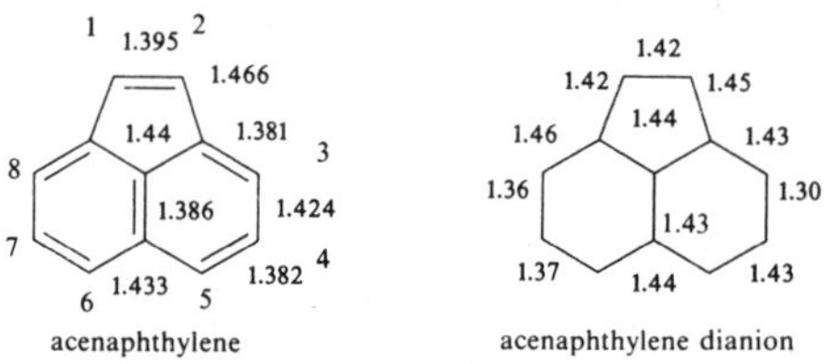

(c) The ^{1}H and ^{13}C NMR shifts for acenaphthylene and its dianion (Na^{+} counterion) are given below. What conclusions about charge density and aromaticity can be drawn from these data?

		2,3	3,8	4,7	5,6	2a, 8a	5a	8b
^{1}H	neutral	7.04	7.65	7.50	7.78			
	dianion	4.49	4.46	5.04	3.34			
^{13}C	neutral	129.9	124.7	128.3	127.8	140.7	129.1	129.3
	dianion	86.1	97.0	126.8	82.6	123.4	149.3	137.7

11. There have been extensive physical and chemical studies of cyclopropenone, cyclopentadienone, and cycloheptatrienone (tropone). The results of these studies can be briefly summarized as follows:
    (a) Cyclopropenone appears to be stabilized by 20 ± 5 kcal/mol relative to a localized model structure.
    (b) Cyclopentadienone is a kinetically unstable molecule.
    (c) Tropane is estimated to be stabilized by less than 10 kcal/mol relative to localized models. It is a nonplanar molecule.

    Derive the π-MO pattern for these three molecules by treating them as derivatives of the three-, five, and seven-membered cyclic conjugated systems. Explain the relationship between the derived MO pattern and the observed properties and stabilities of the molecules.

12. The orbital coefficients for the MO of energy α for the phenalenyl system described in Fig. 9.7 are as shown below. Predict the general appearance of the NMR spectra of the anion and cation derived from phenalene.

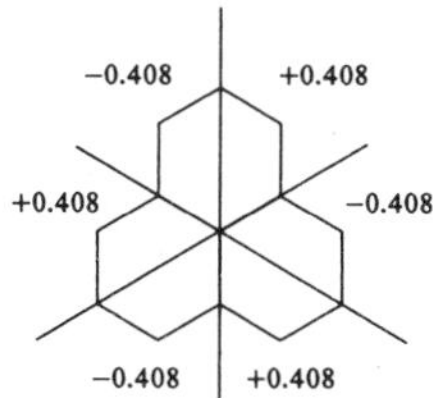

13. The ^{13}C NMR spectrum of octalene is temperature dependent. At −150°C there are signals for 14 different carbons. At −100°C, these collapse to seven different signals. Above 80°C, all but one of the remaining signals becomes broad. Although not attained experimentally, because of decomposition, it would be expected that only four different signals would be observed at still higher temperature. (1) Show that these data rule out structures **A** and **B** for the room temperature structure of octalene and favor structure **C**; (2) indicate the nature of the dynamic process that converts the 14-line spectrum to the 7-line spectrum; (3) indicate the nature of the process which would be expected to convert the 7-line spectrum to a 4-line spectrum.

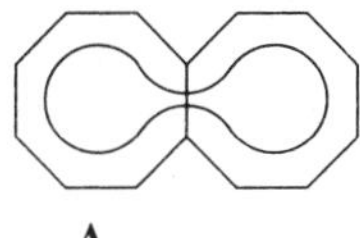
A

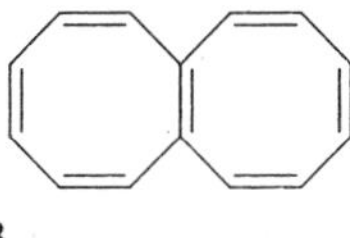
B

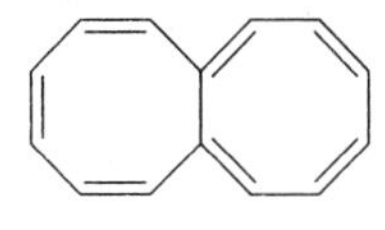
C

14. When the alcohol **1** is dissolved in fluorosulfonic acid at −136°C and then allowed to warm to −110°C, it gives rise to a cation having a ^{13}C NMR spectrum consisting of five lines in the intensity ratio 2:1:2:2:2. Suggest possible structures for this cation, and discuss any stabilizing features which might favor a particular structure.

OH

1

15. (a) The heats of combustion, ΔH_c, the heats of hydrogenation for addition of one mole of H_2, ΔH_{H_2}, and the estimated stabilization energies (S.E.) for benzene and cyclooctatetraene are given below. The heat of combustion of [16]annulene is also given. Estimate the stabilization energy of [16]annulene. Does this value agree with the prediction of simple Hückel MO theory?

	Benzene	Cyclooctatetraene	[16]Annulene
ΔH_c (kcal/mol)	781	1086	2182
ΔH_{H_2} (kcal/mol)	−5.16	25.6	28
S.E.[a] (kcal/mol)	36	4	?

a. Estimated stabilization resulting from conjugation in kcal/mol.

(b) The enthalpies of the reaction of the cyclooctatetraene and [16]annulene dianions (HC^{2-}) with water have been measured.

$$2Na^+ + (C_nH_n)^{2-} + 2H_2O_{(1)} \rightarrow C_nH_{n+2} + 2NaOH$$

$$\Delta H = -33.3 \text{ kcal/mol for cyclooctatetraene}$$

$$\Delta H = -81.1 \text{ kcal/mol for [16]annulene}$$

Using these data and the enthalpy value for the reaction

$$2Na_{(s)} + H_2O_{(1)} \rightarrow 2NaOH_{(aq)} + H_2, \qquad \Delta H = -88.2 \text{ kcal/mol}$$

calculate ΔH for the reaction

$$2Na_{(s)} + C_nH_n \rightarrow 2Na^+ + (C_nH_n)^{2-}$$

How do you interpret the difference in the heat of reaction for the two hydrocarbons in the reaction to form the respective dianions?

10

Aromatic Substitution

10.1. Electrophilic Aromatic Substitution Reactions

Electrophilic aromatic substitution reactions are important for synthetic purposes, and they also represent one of the most thoroughly studied classes of organic reactions from a mechanistic point of view. The synthetic aspects of these reactions are discussed in Part B. The discussion here will emphasize the mechanisms of several of the most completely studied reactions. These mechanistic ideas lay the groundwork for the extensive study that has been done on structure–reactivity relationships in aromatic electrophilic substitution. This topic will be discussed in Section 10.2.

A wide variety of electrophilic species can attack aromatic rings and effect substitution. Usually, it is a substitution of some other group for hydrogen that is observed, but this is not always the case. Scheme 10.1 lists some of the specific electrophilic species that are capable of carrying out substitution for hydrogen. Some indication of the relative reactivity of the electrophiles is given as well. Most of these electrophiles will not be treated in detail until Part B. Nevertheless, it is important to recognize that the scope of electrophilic aromatic substitution is very broad.

The reactivity of a particular electrophile determines which aromatic compounds can be successfully substituted. Those electrophiles grouped in the first category are sufficiently reactive to attack almost all aromatic compounds, even those having strongly electron-attracting substituents. Those in the second group react readily with benzene and derivatives having electron-releasing substituents but are not generally reactive toward aromatic rings with electron-withdrawing substituents. Those classified in the third group are reactive only toward aromatic compounds that are much more reactive than benzene. These broad groupings can provide a general guide to the feasibility of a given electrophilic aromatic substitution.

Scheme 10.1 Electrophilic Species

Electrophile	Typical mode of generation	Ref.
A. Electrophiles capable of substituting both activated and deactivated aromatic rings		
$O{=}\overset{+}{N}{=}O$	$2H_2SO_4 + HNO_3 \rightleftharpoons NO_2^+ + 2HSO_4^- + H_3O^+$	a
Br_2 or $Br_2\text{–}MX_n$	$Br_2 + MX_n \rightleftharpoons Br_2\text{–}MX_n$	b
$Br\overset{+}{O}H_2$	$BrOH + H_3O^+ \rightleftharpoons Br\overset{+}{O}H_2 + H_2O$	b
Cl_2 or $Cl_2\text{–}MX_n$	$Cl_2 + MX_n \rightleftharpoons Cl_2\text{–}MX_n$	b
$Cl\overset{+}{O}H_2$	$ClOH + H_3O^+ \rightleftharpoons Cl\overset{+}{O}H_2 + H_2O$	b
SO_3	$H_2S_2O_7 \rightleftharpoons H_2SO_4 + SO_3$	c
RSO_2^+	$RSO_2Cl + AlCl_3 \rightleftharpoons RSO_2^+ + AlCl_4^-$	d
B. Electrophiles capable of substituting activated but not deactivated aromatic rings		
R_3C^+	$R_3CX + MX_n \rightleftharpoons R_3C^+ + [MX_{n+1}]^-$	e
	$R_3COH + H^+ \rightleftharpoons R_3C^+ + H_2O$	f
	$R_2C{=}CR_2' + H^+ \rightleftharpoons R_2\overset{+}{C}CHR_2'$	g

a. G. A. Olah and S. J. Kuhn, in *Friedel–Crafts and Related Reactions*, Vol. III, G. A. Olah (ed.), Interscience, New York, 1964, Chap. XLIII.
b. H. P. Braendlin and E. T. McBee, in *Friedel–Crafts and Related Reactions*, Vol. III, G. A. Olah (ed.), Interscience, New York, 1964, Chap. XLVI.
c. K. L. Nelson, in *Friedel–Crafts and Related Reactions*, Vol. III, G. A. Olah (ed.), Interscience, New York, 1964, Chap. XLVII.
d. F. R. Jensen and G. Goldman, in *Friedel–Crafts and Related Reactions*, Vol. III, G. A. Olah (ed.), Interscience, New York, 1964, Chap. XL.
e. F. A. Drahowzal, in *Friedel–Crafts and Related Reactions*, Vol. II, G. A. Olah (ed.), Interscience, New York, 1964, Chap. XVII.
f. A. Schreisheim, in *Friedel–Crafts and Related Reactions*, Vol. II, G. A. Olah (ed.), Interscience, New York, 1964, Chap. XVIII.

Despite the wide range of electrophilic species and the variety of aromatic ring systems that can undergo substitution, a single broad mechanistic picture encompasses the large majority of electrophilic aromatic substitution reactions. As would be expected, the identity of the rate-determining step and the shape of the potential energy surface are specific to the reagents that are involved, but the series of steps and nature of the intermediates are very similar across a wide range of reactivity. This permits discussion of electrophilic aromatic substitution in terms of a general mechanism, which is outlined in Scheme 10.2.

A nonspecific complexation of the electrophile with the π-electron system of the aromatic ring is the first step. The species formed, called the π complex, may or may not be involved directly in the substitution mechanism. π-Complex formation is, in general, rapidly reversible, and in many cases the equilibrium constant is small. The π complex is a donor–acceptor-type complex, with the π electrons of the aromatic ring donating electron density to the electrophile. No specific positional selectivity is usually associated with the π complex.

Electrophile	Typical mode of generation	Ref.
$RCH_2X\text{-}MX_n$	$RCH_2X + MX_n \rightleftharpoons RCH_2X\text{-}MX_n$	e
$RC{\equiv}O^+$	$RC(=O)X + MX_n \rightleftharpoons RC{\equiv}O^+ + [MX_{n+1}]^-$	h
$RC(=O)X\text{-}MX_n$	$RC(=O)X + MX_n \rightleftharpoons RC(=O)X\text{-}MX_n$	h
H^+	$HX \rightleftharpoons H^+ + X^-$	i
$R_2C{=}\overset{+}{O}H$	$R_2C{=}O + H^+ \rightleftharpoons R_2C{=}\overset{+}{O}H$	j
$R_2C{=}\overset{+}{O}\text{-}\overset{-}{M}X_n$	$R_2C{=}O + MX_n \rightleftharpoons R_2C{=}\overset{+}{O}\text{-}\overset{-}{M}X_n$	j
C. Electrophiles capable of substituting only strongly activated aromatic rings		
$HC{\equiv}\overset{+}{N}H$	$HC{\equiv}N + HX \rightleftharpoons HC{\equiv}\overset{+}{N}H + X^-$	k
$N{\equiv}O^+$	$HNO_2 + H^+ \longrightarrow N{\equiv}O^+ + H_2O$	l
$Ar\overset{+}{N}{\equiv}N$	$ArNH_2 + HNO_2 + H^+ \longrightarrow Ar\overset{+}{N}{\equiv}N + 2H_2O$	m

g. S. H. Patinkin and B. S. Friedman, in *Friedel-Crafts and Related Reactions*, Vol. II, G. A. Olah (ed.), Interscience, New York, 1964, Chap. XIV.
h. P. H. Gore, in *Friedel-Crafts and Related Reactions*, Vol. III, G. A. Olah (ed.), Interscience, New York, 1964, Chap. XXXI.
i. R. O. C. Norman and R. Taylor, *Electrophilic Substitution in Benzenoid Compounds*, Elsevier, New York, 1965, Chap. 8.
j. J. E. Hofmann and A. Schriesheim, in *Friedel-Crafts and Related Reactions*, G. A. Olah (ed.), Interscience, New York, 1964, Vol. II, Chap. XIX.
k. W. Ruske, in *Friedel-Crafts and Related Reactions*, Vol. III, G. A. Olah (ed.), Interscience, New York, 1964, Chap. XXXII.
l. B. C. Challis, R. J. Higgins, and A. J. Lawson, *J. Chem. Soc., Perkin Trans. 2*, 1831 (1972).
m. H. Zollinger, *Azo and Diazo Chemistry*, translated by H. E. Nursten, Interscience, New York, 1961, Chap. 10.

In order for a substitution to occur, a "σ complex" must be formed. The term *σ complex* is used to describe an *intermediate* in which the carbon at the site of substitution is bonded to both the electrophile and the group to be displaced. The term implies that a σ bond has been formed at the site of substitution. The intermediate is a cyclohexadienyl cation. Its fundamental structural characteristics can be described in simple MO terms. The σ complex is a four-π-electron delocalized system which is electronically equivalent to a pentadienyl cation. The π-molecular orbitals and energy levels for the pentadienyl cation are shown in Fig. 10.1. The LUMO has nodes at C-2 and C-4 of the pentadienyl structure, and these positions correspond to the positions *meta* to the site of substitution on the aromatic ring. As a result, the positive charge of the cation is located at the positions *ortho* and *para* to the site of substitution.

Formation of the σ complex can be reversible. The partitioning of the σ complex forward to product or back to reactants depends on the ease with which the electrophile can be eliminated relative to a proton. For most electrophiles, it is

Scheme 10.2. Generalized Mechanism for Electrophilic Aromatic Substitution

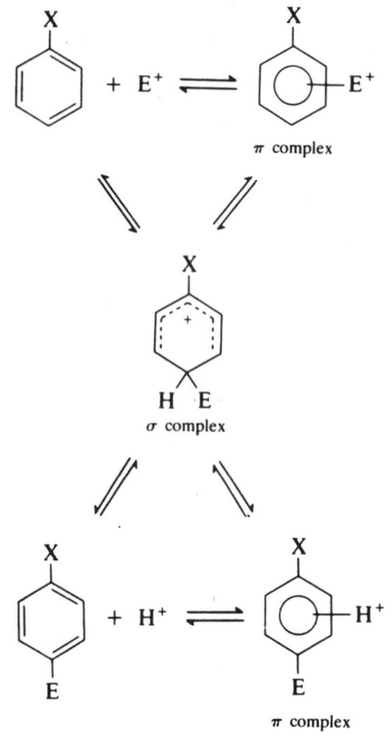

easier to eliminate the proton, in which case the formation of the σ complex is essentially irreversible. Formation of the σ complex is usually, but not always, the rate-determining step in electrophilic aromatic substitution. There may also be a π complex involving the aromatic ring and the departing electrophile. This would be logical on the basis of the principle of overall reversibility of the process. There is little direct evidence on this point however.[1]

Let us now consider some of the evidence for this general mechanism. Such evidence has, of course, been gathered by study of specific reaction mechanisms. Only some of the most clear-cut examples are cited here. Additional evidence will be mentioned when individual mechanisms are discussed in Section 10.4. A good example of a reaction that has been studied with a focus on the identity and mode of generation of the electrophile is aromatic nitration. Primarily on the basis of kinetic studies, it has been possible to show that the active electrophile in nitration is often the nitronium ion, NO_2^+. In some cases, the generation of the electrophile is the rate-determining step. Several lines of evidence have been used to establish the role of the nitronium ion. Reaction of nitric acid with concentrated sulfuric acid leads to formation of the nitronium ion. It can be detected spectroscopically, and

1. For additional discussion of the role of σ and π complexes in aromatic substitution, see G. A. Olah, *Acc. Chem. Res.* **4**, 240 (1971); J. H. Ridd, *Acc. Chem. Res.* **4**, 248 (1971).

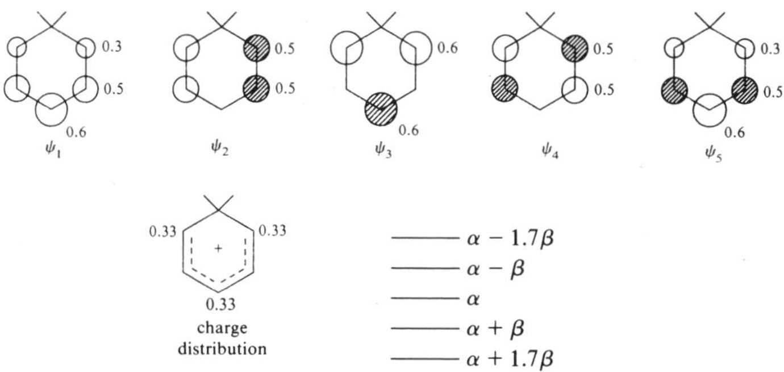

Fig. 10.1. π-Molecular orbitals and energy levels for the pentadienyl cation.

the freezing point depression of the solution is consistent with the following equation:

$$2H_2SO_4 + HNO_3 \rightarrow NO_2^+ + H_3O^+ + 2HSO_4^-$$

Solid salts in which the nitronium ion is the cation can be prepared with unreactive anions; examples are $NO_2^+BF_4^-$ and $NO_2^+PF_6^-$.

Two major types of rate laws have been found to describe the kinetics of most aromatic nitration reactions. With relatively unreactive substrates, second-order kinetics, first-order in the nitrating reagent and first-order in the aromatic, are observed. This rate law corresponds to rate-limiting attack of the electrophile on the aromatic reactant. With more reactive aromatics, this step can become faster than formation of the active electrophile. When formation of the active electrophile becomes the rate-determining step, the concentration of the aromatic reactant no longer appears in the observed rate expression. Under these conditions, different aromatic substrates undergo nitration at the same rate, corresponding to the rate of formation of the active electrophile.

An important general point to be drawn from the specific case of nitration is that the active electrophile is usually some species that is more reactive than the added reagents. The formation of the active electrophile may or may not be the rate-determining step. Scheme 10.1 shows the structures of some of the electrophilic species that are involved in typical electrophilic aromatic substitution processes and the reactions involved in their formation.

There are several lines of evidence for formation of σ complexes as intermediates in electrophilic aromatic substitution. One particularly informative approach involves measurement of isotope effects on the rate of substitution. If removal of the proton at the site of substitution were concerted with introduction of the electrophile, a primary isotope effect would be observed in reactions in which electrophilic attack on the ring is rate-determining. This is not the case for nitration nor for several other types of aromatic substitution reactions. Nitration of aromatic substrates partially labeled by tritium shows no selectivity between protium- and tritium-substituted sites.[2] Similarly, the rate of nitration of nitrobenzene is identical to that

2. L. Melander, *Acta Chem. Scand.* **3**, 95 (1949); *Ark. Kemi* **2**, 211 (1950).

for penta-*deuterio*-nitrobenzene.[3] The lack of an isotope effect indicates that the proton is lost in a fast step subsequent to the rate-determining step. This means that proton loss must occur from some intermediate formed before the cleavage of the C—H bond begins. The σ-complex intermediate fits this requirement. There are some electrophilic aromatic substitution reactions that show k_H/k_D values between 1 and 2, and there are a few others with k_H/k_D values in the range indicating a primary isotope effect.[4] The existence of these isotope effects is compatible with the general mechanism if the proton removal is rate-limiting (or partially rate-limiting). Many of the modest kinetic isotope effects ($k_H/k_D \approx 1.2$–2.0) have been interpreted in terms of comparable rates for formation and destruction of the σ-complex intermediate.

The observation of an isotope effect will depend on the relative magnitudes of the rate constants k_1, k_{-1}, and k_2. When formation of the σ complex is rate-determining, no isotope effect will be observed.

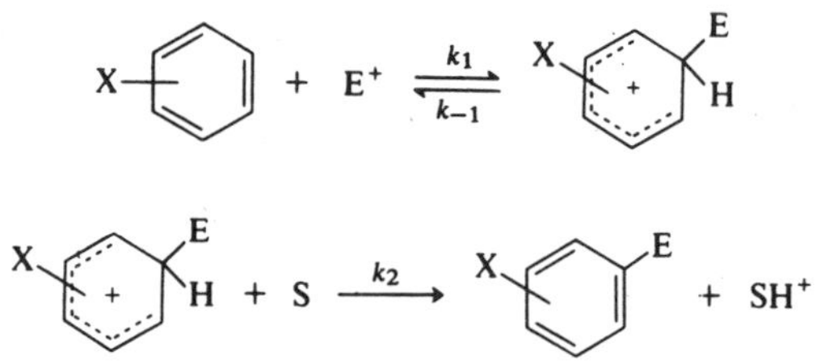

If the second step is rate-determining, an isotope effect in the range characteristic of primary isotope effects would be expected. In substitutions involving the weakly electrophilic aryl diazonium ions, such primary isotope effects have been observed. This is consistent with rate-determining deprotonation of the σ complex. The same types of reaction show general base catalysis, which would be consistent with a rate-determining deprotonation step.

The case for the generality of the σ-complex mechanism is further strengthened by numerous studies showing that such compounds can exist as stable entities under suitable conditions. Substituted benzenium ions (an alternative name for the σ complex) can be observed by NMR techniques under stable ion conditions. They are formed by protonation of the aromatic substrate.[5]

3. T. G. Bonner, F. Bower, and G. Williams, *J. Chem. Soc.* 2650 (1953).
4. H. Zollinger, *Adv. Phys. Org. Chem.* **2**, 163 (1964).
5. G. A. Olah, R. H. Schlosberg, R. D. Porter, Y. K. Mo, D. P. Kelly, and G. Mateescu, *J. Am. Chem. Soc.* **94**, 2034 (1972).

F + HF–SbF$_5$ $\xrightarrow{SO_2}$ [F, +, H H] SbF_6^- Ref. 6

CH_3 + HF–SbF$_5$ $\xrightarrow{SO_2}$ [CH_3, +, H H] SbF_6^- Ref. 7

Salts formed by alkylation of benzene derivatives have also been characterized.

CH_3 + C_2H_5F + BF_3 ⟶ [CH_3, +, H CH_2CH_3] BF_4^- Ref. 8

Under normal reaction conditions these types of benzenium ions are short-lived intermediates. The fact that the structures are stable in non-nucleophilic media clearly demonstrates the feasibility of such intermediates.

The existence of σ-complex intermediates can be inferred from experiments in which they are trapped by nucleophiles under special circumstances. For example treatment of the acid **1** with bromine gives the cyclohexadienyl lactone **2**. This product results from capture of the σ complex by intramolecular nucleophilic attack by the carboxylate group.

H_3C–C$_6H_4$–$OC(CH_3)_2CO_2^-$ (**1**) $\xrightarrow{Br_2}$ [Br, H_3C, +, $OC(CH_3)_2$, $C(=O)O^-$] ⟶ [Br, H_3C, O, CH_3, CH_3, O, O] (**2**)

Ref. 9

A number of examples of nucleophilic capture of σ complexes have also been uncovered in the study of nitration of alkylated benzenes in acetic acid. For example, nitration of **3** at 0°C leads to formation of **4** with acetate acting as the nucleophile.[10]

CH_3–C$_6H_4$–$CH(CH_3)_2$ (**3**) + NO_2^+ ⟶ [H_3C NO_2, +, $CH(CH_3)_2$] $\xrightarrow{CH_3CO_2H}$ [H_3C NO_2, $(CH_3)_2CH$ $OCCH_3$ (=O)] (**4**)

6. G. A. Olah and T. E. Kiovsky, *J. Am. Chem. Soc.* **89**, 5692 (1967).
7. G. A. Olah, *J. Am. Chem. Soc.* **87**, 1103 (1965).
8. G. A. Olah and S. J. Kuhn, *J. Am. Chem. Soc.* **80**, 6541 (1958).
9. E. J. Corey, S. Barcza, and G. Klotmann, *J. Am. Chem. Soc.* **91**, 4782 (1969).
10. R. C. Hahn and D. L. Strack, *J. Am. Chem. Soc.* **96**, 4335 (1974).

This type of addition process is particularly likely to be observed when the electrophile attacks a position that is already substituted, since facile rearomatization by deprotonation is then blocked. Attack at a substituted position is called *ipso* attack. Addition products have also been isolated, however, when initial electrophilic attack has occurred at an unsubstituted position. The extent of addition in competition with substitution tends to increase on going to naphthalene and the larger polycyclic aromatic ring systems.[11]

The general mechanistic framework outlined in the preceding paragraphs must be elaborated by other details to fully describe the mechanisms of the individual electrophilic substitutions. The question of the identity of the active electrophile in each reaction is important. We have discussed the case of nitration, in which, under many circumstances, the electrophile is the nitronium ion. Similar questions arise in most of the other substitution processes. Other matters that are important include the ability of the electrophile to select among the alternative positions on a substituted aromatic ring. The relative reactivities of different substituted benzenes toward various electrophiles have also been important in developing a firm understanding of electrophilic aromatic substitution. The next section considers some of the structure–reactivity relationships that have proven to be informative.

10.2. Structure–Reactivity Relationships

The effect that substituents already on the ring have on electrophilic aromatic substitution reactions is an area of structure–reactivity relationships that has been studied since about 1870. The classification of substituents as activating and *ortho, para*-directing or deactivating and *meta*-directing became clear from the early studies. An understanding of the origin of these substituent effects became possible when ideas about electronic interactions and resonance theory were developed. Activating, *ortho, para*-directing substituents are those that can serve as electron donors and stabilize the transition state leading to σ-complex formation. Both saturated and unsaturated hydrocarbon groups and substituents with an unshared electron pair on the atom adjacent to the ring fall in this group. The stabilizing effects of these types of substituents can be expressed in terms of resonance structures. Direct resonance stabilization is only possible when the substituent is *ortho* or *para* to the incoming electrophile. As a result, the transition states for *ortho* and *para* substitution are favored over that for *meta* substitution.

11. P. B. de la Mare, *Acc. Chem. Res.* 7, 361 (1974).

stabilization by alkyl groups

stabilization by heteroatoms with unshared pairs

stabilization by vinyl groups

Because the substituent groups have a direct resonance interaction with the charge that develops in the σ complex, quantitative substituent effects exhibit a high resonance component. Hammett equations usually correlate best with the σ^+ substituent constants (see Section 4.3).[12]

Electron-attracting groups retard electrophilic substitution. Substituents falling in this group include the various substituents in which a carbonyl group is directly attached to the ring and substituents containing electronegative elements that do not have a lone pair on an atom adjacent to the ring. The classification of specific substituents given in Table 4.5 on p. 208 indicates which are electron-attracting. Because of the direct conjugation with the *ortho* and *para* positions, electrophilic attack occurs primarily at the *meta* position, which is *less* deactivated than the *ortho* and *para* positions.

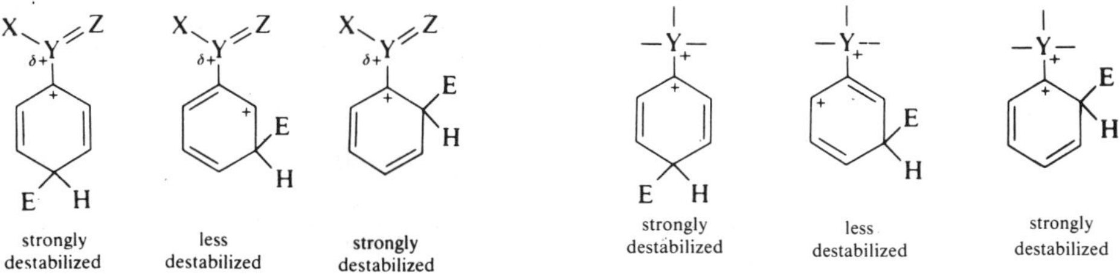

destabilization by electron-attracting multiple bonds

destabilization by electronegative heteroatoms with no unshared pairs

A few substituents, most notably chlorine and bromine, decrease the rate of reaction but nevertheless direct incoming electrophiles to the *ortho* and *para* positions. This is the result of the opposing influence of field and resonance effects for these substituents. The halogens are more electronegative than carbon, and, as a result, the electron density in the ring is diminished. The carbon–halogen bond dipole opposes the development of positive charge in the ring. Overall reactivity toward electrophiles is therefore reduced. The unshared electron pairs on the

12. H. C. Brown and Y. Okamoto, *J. Am. Chem. Soc.* **80**, 4979 (1958).

halogen, however, can preferentially stabilize the *ortho* and *para* transition states by resonance. As a result, the substituents are deactivating but *ortho, para*-directing.

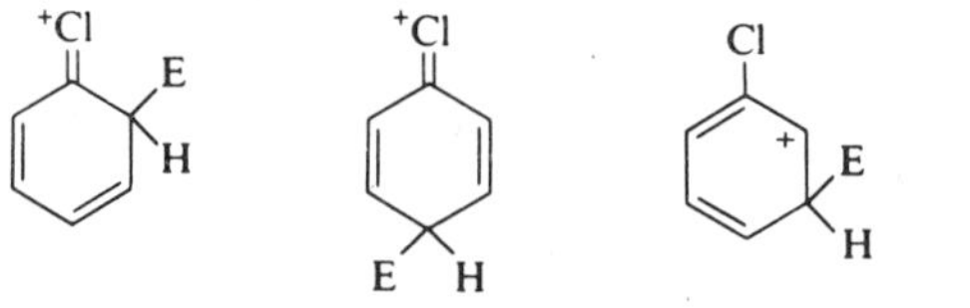

stabilizing resonance participation by chlorine

no resonance interaction with chlorine

The *ortho, para*- versus *meta*-directing and activating versus deactivating effect of substituents can also be described in terms of PMO theory. The discussion can focus either on the σ complex or on the aromatic reactant. According to the Hammond postulate, it would be most appropriate to focus on the intermediate in the case of steps that have relatively high energies of activation and a late transition state. In such cases, the transition state should closely resemble the σ complex. For highly reactive electrophiles, where the activation energy is very low, it may be more appropriate to regard the transition state as closely resembling the reactant aromatic. Let us examine the PMO description of substituent effects from both these perspectives.

If the transition state resembles the intermediate σ complex, the structure involved is a substituted pentadienyl cation. The electrophile has localized one pair of electrons to form the new σ bond. The Hückel orbitals for the pentadienyl system are shown in Fig. 10.1. A substituent can stabilize the pentadienyl cation by electron donation. The LUMO is ψ_3. This orbital has its highest coefficients at carbons 1, 3, and 5 of the pentadienyl system. These are the positions *ortho* and *para* to the position occupied by the electrophile. Electron donor substituents at the 2- or 4-position will stabilize the system much less because of the nodes at these carbons in the LUMO.

If we consider a π-acceptor substituent, we see that such a substituent will strongly destabilize the system when it occupies the 1-, 3-, or 5-position on the pentadienyl cation. The destabilizing effect would be less at the 2- or 4-position. The conclusions drawn from this PMO interpretation are the same as for resonance arguments. Donor substituents will be most *stabilizing* in the transition state leading to *ortho, para* substitution. Acceptor substituents will be least *destabilizing* in the transition state leading to *meta* substitution.

The effect of the bond dipole associated with electron- withdrawing groups can also be expressed in terms of its interaction with the cationic σ complex. The atoms with the highest coefficients in the LUMO ψ_3 are the most positive. The unfavorable interaction of the bond dipole will therefore be greatest at these positions. This effect operates with substituents such as carbonyl, cyano, and nitro groups. With ether and amino substituents, the unfavorable dipole interaction is overwhelmed by the stabilizing effect of the lone pair electrons on ψ_3.

Table 10.1. Energy Changes for Isodesmic Proton Transfer Reactions of Substituted Benzenes[a]

	ΔE (kcal/mol)	
Substituent	*meta*	*para*
NO_2	−17.9	−22.1
CN	−14.0	−13.8
CF_3	−7.5	−8.4
F	−7.5	3.7
CH_3	2.0	8.5
OCH_3		15.7
OH	−5.3	16.0
NH_2	0.6	27.2

a. From STO-3G calculations reported by J. M. McKelvey, S. Alexandratos, A. Streitwieser, Jr., J.-L. M. Abboud, and W. J. Hehre, *J. Am. Chem. Soc.* **98**, 244 (1976).

The effect of substituents has been probed by MO calculations at the STO-3G level.[13] An isodesmic reaction corresponding to transfer of a proton from a substituted σ complex to an unsubstituted one will indicate the stabilizing or destabilizing effect of the substituent. The results are given in Table 10.1.

The calculated energy differences give a good correlation with σ^+. The ρ parameter ($\rho = -17$) is larger than that observed experimentally for proton exchange ($\rho = \sim -8$). A physical interpretation of this trend would be that the theoretical results pertain to the gas phase where the effect of substituents would be at a maximum because of the absence of any leveling effect due to solvation.

Both HMO calculations and more elaborate MO methods can be applied to the issue of the position of electrophilic substitution in aromatic molecules. The most direct approach is to calculate the localization energy. This is the energy difference between the aromatic molecule and the σ-complex intermediate. In simple Hückel calculations, the localization energy is just the difference between the energy calculated for the initial π system and that remaining after two electrons and the carbon atom at the site of substitution have been removed from the conjugated system.

$6\alpha+8\beta \xrightarrow{E^+} 4\alpha+5.46\beta$ localization energy $= 2\alpha+2.54\beta$

13. J. M. McKelvey, S. Alexandratos, A. Streitwieser, Jr., J.-L. M. Abboud, and W. J. Hehre, *J. Am. Chem. Soc.* **98**, 244 (1976).

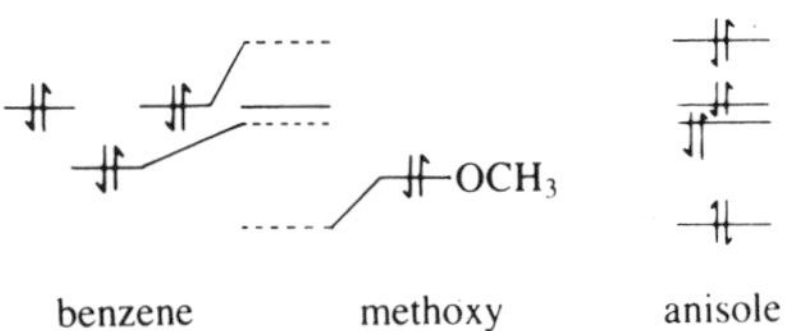

Fig. 10.2. MO diagram for anisole by application of perturbation for a methoxy substituent.

Comparison of localization energies has frequently been applied to prediction of the relative positional reactivity in polycyclic aromatic hydrocarbons. Simple HMO calculations have only marginal success. CNDO/2 and all-electron SCF calculations give results which show good correlation with experimental data on the rate of proton exchange.[14]

Now let us turn to the case of a highly reactive electrophile, where we expect an early transition state. In this case, the charge density and coefficients of the HOMO characteristic of the aromatic reactant would be expected to be major features governing the orientation of electrophilic attack. The transition state should resemble the reactants, and, according to frontier orbital theory, the electrophile should attack the position with the largest coefficient of the HOMO. The case of methoxybenzene (anisole) can be taken as an example of a reactive molecule. MO calculations place the lone pair oxygen orbital lower in energy than the aromatic π orbitals, leading to the MO diagram in Fig. 10.2.

The degeneracy of the two highest-lying occupied π orbitals is broken because the methoxy group interacts preferentially with one. The other has a node at the site of methoxy substitution. Figure 10.3 gives the coefficients for the two highest occupied π orbitals, as calculated by the CNDO method. We see that the HOMO has its highest coefficients at the *ipso*, *ortho*, and *para* positions. As indicated in Fig. 10.2, the energy of this orbital is raised by its interaction with the electron donor substituent. Figure 10.4 shows the distribution of π electrons from all the orbitals based on STO-3G calculations. The electron-donating substituents show increased electron density at the *ortho* and *para* positions. Both the HOMO coefficients and the total charge distribution predict preferential attack by the electrophile *ortho* and *para* to donor substituents.

Some examples of electron-attracting substituents are also shown in Fig. 10.3 and 10.4. As expected, the acceptor substituents lower the energies of the π orbitals. The HOMO distribution remains high at the *para* position, however. The total charge distribution shows greater depletion at the *ortho* and *para* positions than at the *meta* position. The lower energy of the HOMO is consistent with decreased reactivity for rings with electron-accepting substituents. The distribution of the HOMO would however erroneously predict *para* substitution if frontier orbital theory were used. Aromatic rings with acceptor substituents are relatively unreactive and therefore less likely to have very early transition states. For such compounds,

14. A. Streitwieser, Jr., P. C. Mowery, R. G. Jesaitis, and A. Lewis, *J. Am. Chem. Soc.* **92**, 6529 (1970).

	CH_3	OCH_3	CHO	NO_2
HOMO: substituent	H 0.20 (H—C—H)	0.58 O, 0.56 CH_3	H 0.27 C, 0.53 O	0.28 O, 0.28 O, 0.26 N
HOMO: ipso	0.52	0.38	0.46	0.51
HOMO: ortho	0.31	0.31	0.28	0.23
HOMO: meta	0.24	0.18	0.20	0.26
HOMO: para	0.53	0.47	0.45	0.50
HOMO energy	−0.48	−0.46	−0.50	−0.57
next highest: ipso	0.0	0.0	0.0	0.0
next highest: ortho	0.50	0.50	0.50	0.50
next highest: meta	0.50	0.50	0.50	0.50
next highest: para	0.0	0.0	0.0	0.0
next highest energy	−0.51	−0.52	−0.54	−0.60

Fig. 10.3. Orbital coefficients for HOMO and next highest π orbital for some substituted benzenes. (From CNDO/2 calculations. *Ortho* and *meta* coefficients have been averaged in the case of the unsymmetrical methoxy and formyl substituents. Orbital energies are given in atomic units.)

considerations of the stability of the σ-complex intermediate, which predict *meta* substitution, are likely to be most appropriate.

Substituents that are not directly bound to the aromatic ring can also influence the course of electrophilic aromatic substitution. Several alkyl groups bearing electron-attracting substituents are *meta*-directing and deactivating. Some examples are given in Table 10.2. In these molecules, stabilization of the *ortho* and *para* σ complex by electron release from the alkyl group is opposed by the electron-withdrawing effect of the electronegative substituent. Both the reduced electron density at the alkyl substituent and the bond dipoles in the substituent would reduce electron donation by the methylene group.

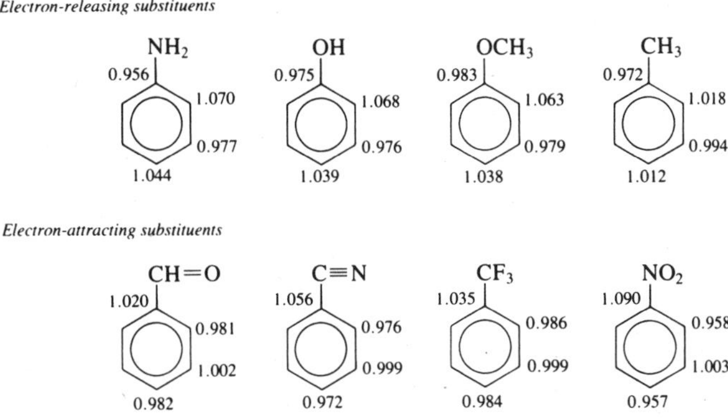

Fig. 10.4. Total π-electron density for some substituted benzenes. [From STO-3G calculations as reported by W. J. Hehre, L. Radom, and J. A. Pople, *J. Am. Chem. Soc.* **94**, 1496 (1972).]

Table 10.2. Percent *meta* Nitration for Some Alkyl Groups with Electron-Withdrawing Substituents[a]

$CH_2CO_2C_2H_5$	$CHCl_2$	CH_2CCl_3	CH_2NO_2	CCl_3	$CH_2\overset{+}{N}(CH_3)_3$
11%	34%	37%	55%	64%	85%

a. From C. K. Ingold, *Structure and Mechanism in Organic Chemistry*, Second Edition, Cornell University Press, Ithaca, New York, 1969, pp. 275, 281; F. DeSarlo, G. Grynkiewicz, A. Rici, and J. H. Ridd, *J. Chem. Soc. B*, 719 (1971).

The relationships between substituents and the typical electrophilic substitution reactions, such as those listed in Scheme 10.1, can be summarized as follows:

1. The hydroxyl and amino groups are highly activating *ortho*, *para*-directing groups. Such compounds are attacked by all the electrophilic reagents tabulated in Scheme 10.1 (p. 540). With some electrophilic reagents, all available *ortho* and *para* positions are rapidly substituted.
2. The alkyl, amido, and alkoxy groups are activating and *ortho*, *para*-directing, but not as strongly so as hydroxyl or amino groups. Synthetically useful conditions for selective substitution are available for essentially all the electrophile in Scheme 10.1 except for very weak electrophiles such as NO^+ or PhN_2^+.
3. The halogens as mentioned earlier, are somewhat unique substituents, being deactivating but *ortho*, *para*-directing. In general, halogenated aromatics will react succesfully with electrophiles listed in categories A and B.
4. The carbonyl group in aldehydes, ketones, acids, esters, and amides is deactivating and *meta*-directing. There are distinct limitations on the type of substitution reactions that are satisfactory for these deactivating substituents. In general, only those electrophiles in category A in Scheme 10.1 react readily.
5. The cyano, nitro, and quaternary ammonium groups are strongly deactivating and *meta*-directing. Electrophilic substitution involving reactants having these substituents usually requires especially vigorous conditions and fails completely with all but strong electrophiles.

Since nitration has been studied for a wide variety of aromatic compounds, it is a useful reaction with which to illustrate the directing effect of substituent groups. Table 10.3 presents some of the data. A variety of reaction conditions are represented, so direct comparison is not always valid, but the trends are nevertheless clear. It is important to remember that other electrophiles, while following the same qualitative trends, will show large quantitative differences in position selectivity.

The effect of substituents on electrophilic substitution can be placed on a quantitative basis by use of *partial rate factors.* The reactivity of each position in a substituted aromatic compound can be compared with benzene by measuring the overall rate, relative to benzene, and dissecting the total rate by dividing it among the *ortho*, *meta*, and *para* products. Correction for the statistical factor arising from the relative number of available positions permits the partial rate factors to provide meaningful comparisons of the reactivity at each position on a substituted ring with that at a single position on benzene:

$$\text{partial rate factor} = f = \frac{(6)(k_{\text{subs}})(\text{fraction } z \text{ product})}{(y)(k_{\text{benz}})}$$

where y is the number of equivalent positions. A partial rate factor calculation for nitration of toluene is given in Example 10.1.

Example 10.1. The nitration of toluene is 23 times as fast as for benzene in nitric acid-acetic anhydride. The product ratio is 63% *ortho*, 34% *para*, and 3% *meta*. Calculate the partial rate factor at each position.

$$f_{ortho} = \frac{(6)}{(2)} \times \frac{(23)}{(1)} \times (0.63) = 43.5$$

$$f_{meta} = \frac{(6)}{(2)} \times \frac{(23)}{(1)} \times (0.03) = 2.1$$

$$f_{para} = \frac{(6)}{(1)} \times \frac{(23)}{(1)} \times (0.34) = 46.9$$

Table 10.3. Isomer Proportions in the Nitration of Some Substituted Benzenes[a]

	Product (%)		
Substituent	*o*	*m*	*p*
$\overset{+}{N}H_3$	3–5	35–50	50–60
$\overset{+}{N}(CH_3)_3$	0	89	11
$CH_2\overset{+}{N}(CH_3)_3$	0	85	15
$\overset{+}{S}(CH_3)_2$	4	90	6
NO_2	5–8	91–93	0–2
CO_2H	15–20	75–85	~1
$C{\equiv}N$	15–17	81–83	~2
$CO_2C_2H_5$	24–28	66–73	1–6
$COCH_3$	26	72	0–2
F	9–13	0–1	86–91
Cl	30–35	~1	64–70
Br	36–43	1	56–62
I	38–45	1–2	54–60
CCl_3	7	64	29
CF_3	6	91	3
$CH_2C{\equiv}N$	24	20	56
CH_2NO_2	22	55	23
CH_2OCH_3	51	7	42
CH_3	56–63	2–4	34–41
CH_2CH_3	46–50	2–4	46–51
OCH_3	30–40	0–2	60–70

a. Data are from Tables 9.1, 9.2, 9.3, 9.4, 9.5, and 9.6 in J. G. Hoggett, R. B. Moodie, J. R. Penton, and K. Schofield, *Nitration and Aromatic Reactivity*, Cambridge University Press, Cambridge, 1971.

Partial rate factors give insight into two related aspects of reactivity. They reveal the selectivity of a given electrophile for different *substrates.* Some reactions exhibit high selectivity; that is, there are large differences in rate of reaction depending on the identity of the ring substituent. In general, low substrate selectivity is correlated with high electrophile reactivity, and vice versa. Clearly, when substrate selectivity is high, the partial rate factors for the substituted aromatic compound will be very different form unity. The partial rate factors will also reveal *positional* selectivity within the substituted aromatic. This selectivity also varies for different electrophiles and provides some insight into the details of the mechanism. In general, there is a correlation between position and substrate selectivity. Electrophiles that show high substrate selectivity generally exhibit low *ortho* : *para* ratios and negligible amounts of *meta* substitution. High substrate selectivity is accompanied by high position selectivity. Very reactive electrophiles tend to show low position and substrate selectivity. Table 10.4 gives some data on the selectivity of some representative aromatic substitution reactions. The most informative partial rate factor in terms of substrate is f_p, since the partial rate factors for *ortho* substitution contain variable steric components. Based on f_p as the criterion, halogenation and Friedel–Crafts acylation exhibit high selectivity, protonation and nitration are intermediate, and Friedel–Crafts alkylation shows low selectivity.

Reactivity and selectivity are largely determined by the position of the transition state on the reaction coordinate. With highly reactive electrophiles, the transition state will come early on the reaction coordinate as in Fig. 10.5(A). The transition state then resembles the reactants more closely than the σ complex. The positive charge on the ring is small, and, as a result, the interaction with the substituent

Table 10.4. Selectivity in Some Electrophilic Aromatic Substitution Reactions[a]

	Partial rate factors for toluene		
Reaction	f_o	f_m	f_p
Nitration			
$HNO_3(CH_3NO_2)$	38.9	1.3	45.7
Halogenation			
$Cl_2(CH_3CO_2H)$	617	5	820
$Br_2(CH_3CO_2H–H_2O)$	600	5.5	2420
Protonation			
$H_2O–H_2SO_4$	83	1.9	83
$H_2O–CF_3CO_2H–H_2SO_4$	330	7.2	313
Acylation			
$PhCOCl(AlCl_3, PhNO_2)$	32.6	5.0	831
$CH_3COCl(AlCl_3, ClCH_2CH_2Cl)$	4.5	4.8	749
Alkylation			
$CH_3Br(GaBr_3)$	9.5	1.7	11.8
$(CH_3)_2CHBr(GaBr_3)$	1.5	1.4	5.0
$PhCH_2Cl(AlCl_3)$	4.2	0.4	10.0

a. From L. M. Stock and H. C. Brown, *Adv. Phys. Org. Chem.* **1**, 35 (1963).

^{+}E H
Early transition state with little σ-bond formation
X
(A)
+ E^{+}
X
E H
+
X
σ-complex
E
+ H^{+}
X

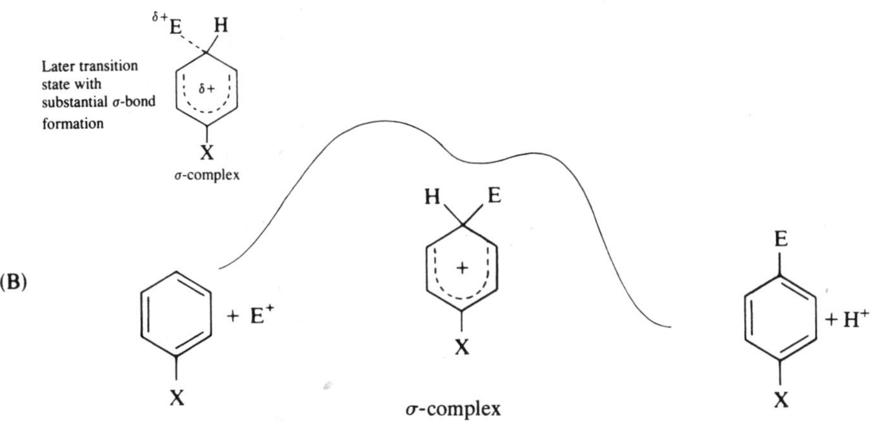

Fig. 10.5. Transition states for highly reactive (A) and less reactive (B) electrophiles.

group is relatively weak. With a less reactive electrophile, the transition state comes later, as in Fig. 10.5(B). The bond to the electrophile is more completely formed and a substantial positive charge is present on the ring. Thus situation results in stronger substituent effects. These arguments follow the general lines of Hammond's postulate (Section 4.4.2). Molecular orbital calculations at the STO-3G level reproduce these qualitative expectations by revealing greater stabilization of the *ortho* and *para* positions in toluene with a closer approach of an electrophile.[15]

Hammett correlations also permit some insight into the reactivity and selectivity of electrophiles in aromatic substitution reactions. In general, the standard Hammett σ substituent constant gives poor correlations with reactions involving electrophilic

15. C. Santiago, K. N. Houk, and C. L. Perrin, *J. Am. Chem. Soc.* **101**, 1337 (1979).

Table 10.5. Values of ρ for Some Electrophilic Aromatic Substitution Reactions[a]

Reaction	ρ
Bromination (CH_3CO_2H)	−13.1
Chlorination (CH_3NO_2)	−13.0
Chlorination (CH_3CO_2H–H_2O)	−8.8
Proton exchange (H_2SO_4–CF_3CO_2H–H_2O)	−8.6
Acetylation (CH_3COCl, $AlCl_3$, $C_2H_4Cl_2$)	−8.6
Nitration (H_2SO_4–HNO_3)	−6.4
Chlorination ($HOCl$, H^+)	−6.1
Alkylation (C_2H_5Br, $GaBr_3$)	−2.4

a. From P. Rys, P. Skrabal, and H. Zollinger, *Angew. Chem. Int. Ed. Engl.* **11**, 874 (1972).

aromatic substitution. The σ^+ values, which reflect an increased importance of direct resonance interaction, give better correlations and, indeed, were developed as a result of the poor correlations observed with σ in electrophilic aromatic substitution. It has been suggested that one could judge the position of a transition state on the reaction coordinate by examining the slope (ρ) of the correlation line between rate of substitution and σ^+.[16] The rationale is along the following lines. A numerically large value for the slope suggests a strong substituent effect, that is, a late transition state which resembles the σ complex. A small value indicates a weak substituent effect and implies an early transition state. Table 10.5 gives some ρ values for typical electrophilic substitution reactions. The data indicate that halogenation reactions show the characteristics of a highly selective electrophile, nitration and Friedel–Crafts acylation represent reactions of modest selectivity, and Friedel–Crafts alkylation is an example of a reaction of low selectivity. This is in good general agreement with the selectivity as measured by f_p indicated in Table 10.4.

Isotope effects are also useful in providing insight into other aspects of the mechanisms of individual electrophilic aromatic substitution reactions. In particular, since primary isotope effects are expected only when the breakdown of the σ complex to product is rate-determining, the observation of a substantial k_H/k_D points to a rate-determining deprotonation. Some typical isotope effects are summarized in Table 10.6. While isotope effects are rarely observed for nitration and halogenation, Friedel–Crafts acylation, sulfonation, nitrosation, and diazo coupling provide examples in which the rate of proton abstraction can affect the rate of substitution. Only in the case of the reactions involving weak electrophiles, namely, nitrosation and diazo coupling, are the isotope effects in the range expected for a fully rate-controlling deprotonation.

Figure 10.6 summarizes the general ideas presented in Section 10.2. At least four types of energy profiles can exist for individual electrophilic aromatic substitution reactions. Case A is the case of rate-determining generation of the electrophile. It is most readily identified by kinetics. A rate law independent of the concentration of the aromatic is diagnostic of this case. Case B represents rate-determining

16. P. Rys, P. Skrabal, and H. Zollinger, *Angew. Chem. Int. Ed. Engl.* **11**, 874 (1972).

Table 10.6. Kinetic Isotope Effects in Some Electrophilic Aromatic Substitution Reactions

Reaction and substrates	Electrophilic reagents	k_H/k_D or k_H/k_T	Ref.
Nitration			
Benzene-*t*	HNO_3–H_2SO_4	<1.2	a
Toluene-*t*	HNO_3–H_2SO_4	<1.2	a
Nitrobenzene-d_5	HNO_3–H_2SO_4	1	a
Halogenation			
Benzene-d_6	HOBr, $HClO_4$	1	a
Anisole-*d*	Br_2	1.05	a
Acylation			
Benzene-d_6	$CH_3C{\equiv}O^+SbF_6^-$, CH_3NO_2	2.25	b
Benzene-d_6	$PhC{\equiv}O^+SbF_6^-$, CH_3NO_2	1.58	b
Sulfonation			
Benzene-d_6	$ClSO_3H$, CH_3NO_2	1.7	c
Benzene-d_6	$ClSO_3H$, CH_2Cl_2	1.6	c
Nitrobenzene-d_5	H_2SO_4–SO_3	1.6-1.7	a
Nitrosation			
Benzene-d_6	HNO_2, D_2SO_4	8.5	d
Diazo coupling			
1-Naphthol-4-sulfonic acid-2-*d*	PhN_2^+	1.0	a
2-Naphthol-8-sulfonic acid-1-*d*	PhN_2^+	6.2	a

a. From a more extensive compilation by H. Zollinger, *Adv. Phys. Org. Chem.* **2**, 163 (1964).
b. From G. A. Olah, J. Lukas, and E. Lukas, *J. Am. Chem. Soc.* **91**, 5319 (1969).
c. From M. P. van Albada and H. Cerfontain, *Rev. Trav. Chim.* **91**, 499 (1972).
d. From B. C. Challis, R. J. Higgins, and A. J. Lawson, *J. Chem. Soc. Perkin Trans. 2,* 1831 (1972).

σ-complex formation with an electrophile of low selectivity. The rate law in such a case should have terms in both the electrophile and the aromatic. Furthermore, low selectivity, as indicated by low ρ values and low partial rate factors, is expected when this energy profile is applicable. Case C is rate-determining α-complex formation with a more selective electrophile having a later transition state. Finally, there is case D, in which the proton removal and rearomatization are rate-limiting. This case can be recognized by the observation of a primary kinetic isotope effect at the site of substitution.

10.3. Reactivity of Polycyclic and Heteroaromatic Compounds

The polycyclic aromatic hydrocarbons, particularly naphthalene, anthracene, and phenanthrene, as well as simple substituted derivatives, undergo the various types of aromatic substitution and are generally more reactive than benzene. The reason for this is that the activation energy for formation of the σ complex is lower than for benzene because more of the initial resonance stabilization is retained in

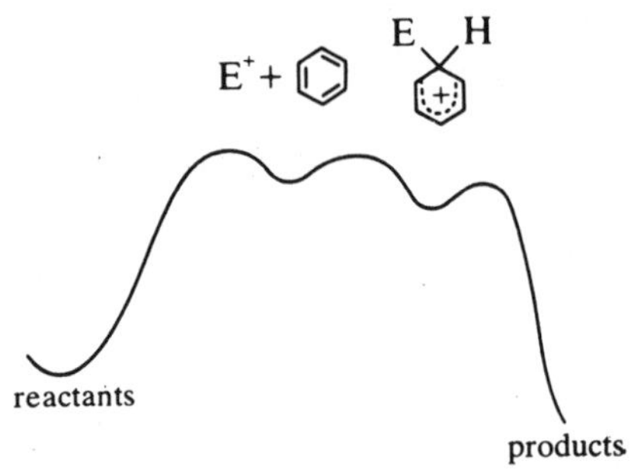

(A) rate-controlling formation of the electrophile

(B) rate-controlling σ-complex formation (nonselective electrophile)

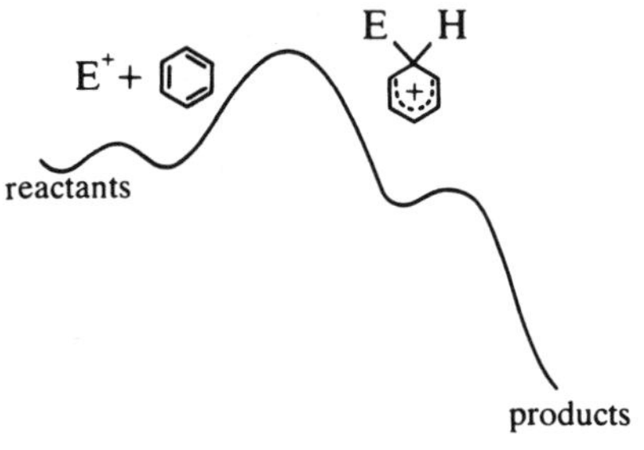

(C) rate-controlling σ-complex formation (selective electrophile)

(D) rate-controlling deprotonation

Fig. 10.6. Various potential energy profiles for electrophilic aromatic substitution.

intermediates that have a fused benzene ring.

Molecular orbital calculations provide estimates of the localization energies. For benzene, naphthalene, and anthracene these are, respectively, 36.3, 15.4, and 8.4 kcal/mol.[17]

The relative stability of the intermediates determines the position of substitution under kinetically controlled conditions. For naphthalene, the preferred site for electrophilic attack is the 1-position. Two factors can result in substitution at the

17. A. Streitwieser, Jr., P. C. Mowery, R. G. Jesaitis, and A. Lewis, *J. Am. Chem. Soc.* **92**, 6529 (1970).

2-position. If the electrophile is very bulky, the hydrogen on the adjacent ring may cause a steric preference for attack at C-2. Under conditions of reversible substitution, where relative thermodynamic stability is the controlling factor, 2-substitution is frequently preferred. An example of this behavior is in sulfonation, where low-temperature reaction gives the 1-isomer but at elevated temperatures the 2-isomer is formed.[18]

Phenanthrene and anthracene both react preferentially in the center ring. This behavior is expected from simple resonance considerations. The σ complexes that result from substitution in the center ring have two intact benzene rings. The total resonance stabilization of these intermediates is larger than that of the napththalene system that results if substitution occurs at one of the terminal rings.

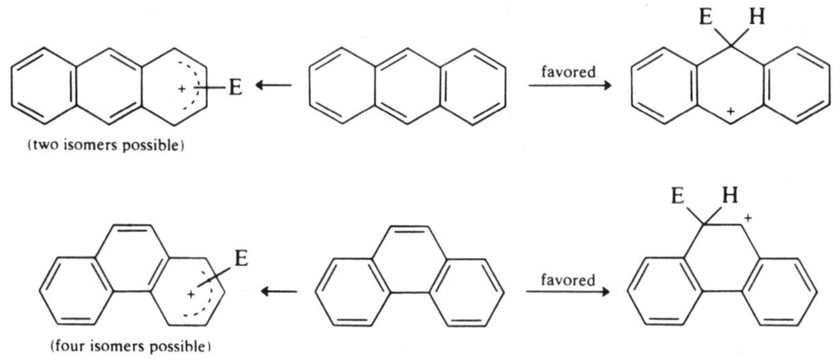

(two isomers possible)

(four isomers possible)

Both phenanthrene and anthracene have a tendency to undergo addition reactions under the conditions involved in certain electrophilic substitutions. Halogenation and nitration may proceed in part via addition intermediates.[19] For example, in the nitration of anthracene in the presence of hydrochloric acid, an intermediate addition product can be isolated.[20]

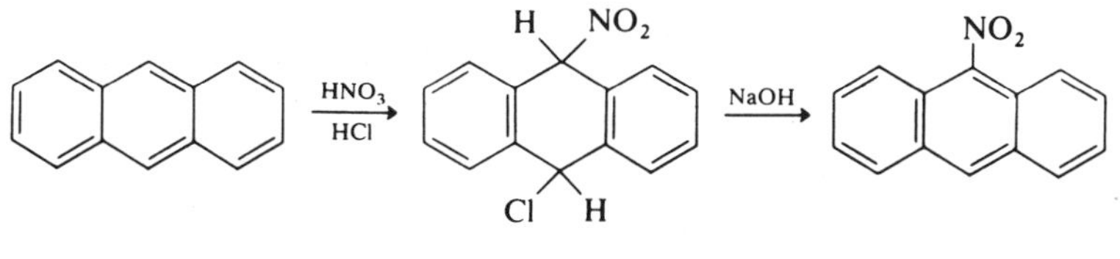

The heteroaromatic compounds can be divided into two broad groups, called *π-excessive* and *π-deficient*, depending on whether the heteroatom acts as an electron donor or an electron acceptor. Furan, pyrrole, thiophene, and other heterocyclics incorporating the oxygen, nitrogen, or sulfur atom in a structure in which it contributes two π electrons are in the π-excessive group. This classification is suggested

18. H. Cerfontain, *Mechanistic Aspects in Aromatic Sulfonation and Desulfonation*, Interscience, New York, 1968, pp. 68–69.
19. P. B. D. de la Mare and J. H. Ridd, *Aromatic Substitution*, Academic Press, New York, 1959, p. 174.
20. C. E. Braun, C. D. Cook, C. Merritt, Jr., and J. E. Rousseau, *Org. Synth.* **IV**, 711 (1965).

by resonance structures and confirmed by various molecular orbital methods.[21]

The reactivity order is pyrrole > furan > thiophene, which indicates the order N > O > S in electron-donating capacity.[22] The N > O order is as expected on the basis of electronegativity, and O > S probably reflects the better overlap of the oxygen $2p$ orbital, as compared to the sulfur $3p$ orbital, with the carbon $2p$ orbitals of the ring.

Structures, such as pyridine, which incorporate the —N=CH— unit are π-deficient and are deactivated to electrophile attack. Again a resonance interpretation is evident.

The nitrogen, being more electronegative than carbon, is a net acceptor of π electron density. There is another important factor in the low reactivity of pyridine derivatives toward electrophilic substitution. The —N=CH— unit is basic since the electron pair on nitrogen is not part of the aromatic π system. The nitrogen will be protonated or complexed with a Lewis acid under many of the conditions typical of electrophilic substitution reactions. The formal positive charge present at nitrogen in such species further reduces the reactivity toward electrophiles.

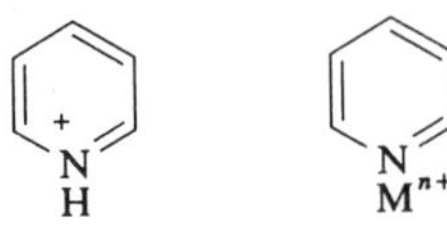

The position selectivity for electrophilic substitution in the simple five-membered heteroaromatic rings is usually 2 > 3. This reflects the more favorable

21. N. D. Epiotis, W. P. Cherry, F. Bernardi, and W. J. Hehre, *J. Am. Chem. Soc.* **98**, 4361 (1976); W. Adam and A. Grimison, *Theor. Chim. Acta* **7**, 342 (1967); D. W. Genson and R. E. Christoffersen, *J. Am. Chem. Soc.* **94**, 6904 (1972); N. Bodor, M. J. S. Dewar, and A. J. Harget, *J. Am. Chem. Soc.* **92**, 2929 (1970).
22. S. Clementi, F. Genel, and G. Marino, *J. Chem. Soc., Chem. Commun.*, 498 (1967).

conjugation in intermediate **A** than in intermediate **B**. In structure **A** the remaining C=C bond can delocalize the positive charge more effectively than in **B**.

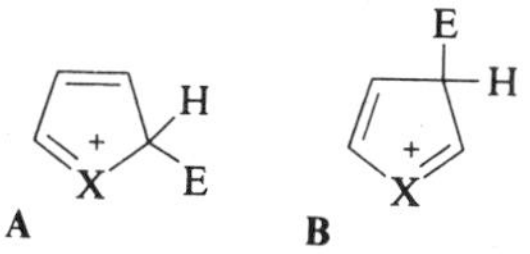

A B

Substituents on the ring can easily override this directive influence, however. For pyridine, the reactivity toward electrophilic substitution is 3 > 4, 2. The ring nitrogen acts as a strongly destabilizing "internal" electron-withdrawing substituent in the 2- and 4-intermediates. The nitrogen also deactivates the 3-position, but less so than the 2- and 4-positions.

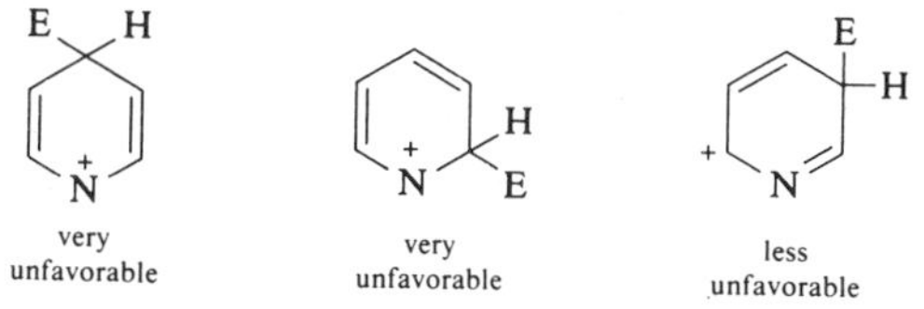

very unfavorable — very unfavorable — less unfavorable

10.4. Specific Substitution Mechanisms

At this point, attention can be given to specific electrophilic substitution reactions. The kinds of data that have been especially pertinent to elucidating mechanistic detail include linear free-energy relationships, kinetic studies, isotope effects, and selectivity patterns. In general, the basic questions that need to be asked about each mechanism are: (1) What is the active electrophile? (2) Which step in the general mechanism for electrophilic aromatic substitution is rate-determining? (3) What are the orientation and selectivity patterns?

10.4.1. Nitration

A substantial body of data including reaction kinetics, isotope effects, and structure–reactivity relationships has permitted a quite thorough understanding of the steps in aromatic nitration.[23] As anticipated from the general mechanism for electrophilic substitution, there are three distinct steps:

1. Generation of the Electrophile

$$2H_2SO_4 + HNO_3 \rightleftharpoons NO_2^+ + 2\,HSO_4^- + H_3O^+$$

or

$$2HNO_3 \rightleftharpoons NO_2^+ + NO_3^- + H_2O$$

23. L. M. Stock, *Prog. Phys. Org. Chem.* **12**, 21 (1976).

2. Attack on the Aromatic Ring

$$NO_2^+ + R\text{-}C_6H_5 \longrightarrow R\text{-}[C_6H_5(H)(NO_2)]^+$$

3. Deprotonation

$$R\text{-}[C_6H_5(H)(NO_2)]^+ \longrightarrow R\text{-}C_6H_4NO_2 + H^+$$

Conditions under which each of the first two steps is rate-determining have been recognized. The third step is usually very fast.

The existence of the nitronium ion in sulfuric–nitric acid mixtures can be demonstrated both by cryoscopic measurements and by spectroscopy. An increase in the strong acid concentration increases the rate of reaction by shifting the equilibrium of step 1 to the right. Addition of a nitrate salt has the opposite effect, by suppressing the preequilibrium dissociation of nitric acid. It is possible to prepare crystalline salts of nitronium ion such as nitronium tetrafluoroborate. Solutions of these salts in organic solvents rapidly nitrate aromatic compounds.[24] The general features of the σ-complex formation step can be successfully modeled by *ab initio* MO calculations at the STO-3G level.[25]

There are three general types of kinetic situations which have been observed for aromatic nitration. Aromatics of modest reactivity exhibit second-order kinetics in mixtures of nitric acid with the stronger acids sulfuric or perchloric acid.[26] Under these conditions, the formation of the nitronium ion is a rapid preequilibrium, and step 2 of the nitration mechanism is rate-controlling. If nitration is conducted in inert organic solvents, such as nitromethane or carbon tetrachloride in the absence of a strong acid, the rate of formation of nitronium ion is slowed and becomes rate-limiting.[27] Finally, some very reactive aromatics, including alkylbenzenes, can react so rapidly under conditions where nitronium ion concentration is high that the rate of nitration becomes governed by encounter rates. Under these circumstances, mixing and diffusion control the rate of reaction, and there are no differences in reactivity between different substrates.

With very few exceptions, the final step in the nitration mechanism, the deprotonation of the σ complex, is fast and therefore has no effect on the observed kinetics. The fast deprotonation can be confirmed by the absence of an isotope effect when deuterium or tritium is introduced at the substitution site. Several compounds such as benzene, toluene, bromobenzene, and fluorobenzene have been

24. S. J. Kuhn and G. A. Olah, *J. Am. Chem. Soc.* **83**, 4564 (1961); G. A. Olah and S. J. Kuhn, *J. Am. Chem. Soc.* **84**, 3684 (1962).
25. P. Politzer, K. Jayasuriya, P. Sjoberg, and P. R. Laurence, *J. Am. Chem. Soc.* **107**, 1174 (1985).
26. J. G. Hoggett, R. B. Moodie, J. R. Penton, and K. Schofield, *Nitration and Aromatic Reactivity*, Cambridge University Press, Cambridge, 1971, Chapter 2.
27. E. D. Hughes, C. K. Ingold, and R. I. Reed, *J. Chem. Soc.*, 2400 (1950); R. G. Coombes, *J. Chem. Soc., B*, 1256 (1969).

subjected to this test and found not to exhibit isotope effects during nitration.[28] The only case where a primary isotope effect indicating rate-controlling deprotonation has been seen is with 1,3,5-tri-*t*-butylbenzene, where steric hindrance evidently makes deprotonation the slow step.[29]

The question of what other species can be the active electrophile in nitration arises in the case of nitration in solutions of nitric acid in acetic anhydride. Acetyl nitrate is formed in such solutions, and so the question arises as to whether it is the actual nitrating species.

$$HNO_3 + (CH_3CO)_2O \rightarrow CH_3\overset{\overset{\displaystyle O}{\|}}{C}ONO_2 + CH_3CO_2H$$

The solutions are very potent nitrating mixtures and effect nitrations at higher rates than solutions of nitric acid in inert organic solvents.

The identification of the nitrating species can be approached by comparing selectivity with that of nitration under conditions known to involve the nitronium ion. Examination of part B of Table 10.7 shows that the position selectivity exhibited by acetyl nitrate toward toluene and ethylbenzene is not dramatically different from that observed with nitronium ion. The data for 2-propylbenzene suggest a lower $o:p$ ratio for acetyl nitrate nitrations. This could indicate a larger steric factor for nitration by acetyl nitrate.

The selectivity data for nitration must be treated with special caution because of the possibility of encounter control. An example of this can be seen in Table 10.7 where no difference in reactivity between mesitylene and xylene is found in H_2SO_4–HNO_3, whereas in HNO_3–CH_3NO_2, the rates of nitrations differ by more than a factor of 2. Encounter control prevails in the former case. In general, nitration is a relatively unselective reaction, with toluene partial rate factors being about 50–60, as shown in Table 10.7. When the aromatic reactant carries an electron-attracting group, the selectivity increases since the transition state occurs later. For example, while toluene is ~20 times more reactive than benzene, *p*-nitrotoluene is ~200 times more reactive than nitrobenzene. Because of the later transition state in the latter case, the effect of the methyl substituent is magnified.

An alternative mechanism for nitration has been considered which involves an electron transfer step between the encounter of the nitronium ion by the aromatic and formation of the σ complex.[30]

$$O{=}\overset{+}{N}{=}O + C_6H_5X \longrightarrow O{=}\dot{\underset{+}{N}}{-}\ddot{\underset{\cdot\cdot}{O}}{:}^{-} \;\; [C_6H_5X]^{+\cdot} \longrightarrow [O{=}\overset{+}{N}(O^-)(H)C_6H_5X]^{+} \longrightarrow O_2N{-}C_6H_4{-}X + H^+$$

28. G. A. Olah, S. J. Kuhn, and S. H. Flood, *J. Am. Chem. Soc.* **83**, 4571, 4581 (1961); H. Suhr and H. Zollinger, *Helv. Chim. Acta* **44**, 1011 (1961); L. Melander, *Acta Chem. Scand.* **3**, 95 (1949); *Ark. Kemi* **2**, 211 (1950).
29. P. C. Myhre, M. Beug, and L. L. James, *J. Am. Chem. Soc.* **90**, 2105 (1968).
30. C. L. Perrin, *J. Am. Chem. Soc.* **99**, 5516 (1977).

Table 10.7. Relative Reactivity and Position Selectivity for Nitration of Some Aromatic Compounds

A. Relative reactivity of some hydrocarbons

Substrate	H_2SO_4–HNO_3–H_2O[a]	HNO_3–CH_3NO_2[b]	HNO_3–$(CH_3CO)_2O$[c]
Benzene	1	1	1
Toluene	17	25	27
p-Xylene	38	139	92
m-Xylene	38	146	—
o-Xylene	38	139	—
Mesitylene	36	400	1750

B. Partial rate factors for some monoalkylbenzenes

Substrate	H_2SO_4–HNO_3 in sulfolane[d]			HNO_3–CH_3NO_2[e,f]			HNO_3–$(CH_3CO)_2O$[g]		
	f_o	f_m	f_p	f_o	f_m	f_p	f_o	f_m	f_p
Toluene	52.1	2.8	58.1	49	2.5	56	49.7	1.3	60.0
Ethylbenzene	36.2	2.6	66.4	32.7	1.6	67.1	31.4	2.3	69.5
2-Propylbenzene	17.9	1.9	43.3	—	—	—	14.8	2.4	71.6
t-Butylbenzene	—	—	—	5.5	3.7	71.4	4.5	3.0	75.5

C. Relative reactivity and isomer distribution for nitrobenzene and the nitrotoluenes[h]

Substrate	Relative reactivity	Product %		
		o	*m*	*p*
Nitrobenzene	1	7	92	1
o-Nitrotoluene	545	29	1	70
m-Nitrotoluene	138	38	1	60
p-Nitrotoluene	217	100	0	—

a. From R. G. Coombes, R. B. Moodie, and K. Schofield, *J. Chem. Soc. B*, 800 (1968).
b. From J. G. Hoggett, R. B. Moodie, and K. Schofield, *J. Chem. Soc. B*, 1 (1969).
c. From A. R. Cooksey, K. J. Morgan, and D. P. Morrey, *Tetrahedron* **26**, 5101 (1970).
d. From G. A. Olah, S. J. Kuhn, S. H. Flood, and J. C. Evans, *J. Am. Chem. Soc.* **84**, 3687 (1962).
e. From L. M. Stock, *J. Org. Chem.* **26**, 4120 (1961).
f. From G. A. Olah and H. C. Lin, *J. Am. Chem. Soc.* **96**, 549 (1974); *o, m, p* designations refer to the methyl substituent.
g. From J. R. Knowles, R. O. C. Norman, and G. K. Radda, *J. Chem. Soc.*, 4885 (1960).
h. From G. A. Olah and H. C. Lin, *J. Am. Chem. Soc.* **96**, 549 (1974); *o, m, p* designations for the nitrotoluenes are relative to the methyl groups.

This mechanism would attribute position selectivity to a different structural feature than does the general mechanism for electrophilic aromatic substitution. If the radical pair intermediate were involved, position selectivity would be determined by the collapse to the σ complex. The product distribution should then be governed by the distribution of unpaired spin in the aromatic radical cation. This would be expected to exhibit the normal *ortho, para* preference but might differ quantitatively from the usual mechanism. Detailed consideration of the electron transfer step has, however, led to the conclusion that this mechanism is unlikely for most benzene derivatives.[31]

10.4.2. Halogenation

Substitution for hydrogen by halogen is a synthetically important electrophilic aromatic substitution reaction. The reactivity of the halogens increases in the order $I_2 < Br_2 < Cl_2$. The molecular halogens are only reactive enough to halogenate quite reactive aromatics. Many reactions are run in the presence of Lewis acids, in which case a complex of the halogen with the Lewis acid is probably the active electrophile.

Bromine and iodine form complexes with the corresponding halide ions. These anionic complexes are less reactive than the free halogen but are capable of substituting highly reactive molecules. They present a complication in kinetic studies, since the concentration of halide ion increases during the course of halogenation and successively more of the halogen will be present as the complex ion. The hypohalous acids, ClOH, BrOH, and IOH, are weak halogenating agents but are much more reactive in acidic solution, where acid catalysis occurs. Halogenation is also effected by the hypohalites of carboxylic acids such as acetyl hypochlorite and trifluoroacetyl hypobromite.[32]

$$Cl_2 + Hg(OAc)_2 \rightleftharpoons HgClOAc + CH_3CO_2Cl$$

$$Br_2 + Hg(O_2CCF_3)_2 \rightleftharpoons HgBr(O_2CCF_3) + CF_3CO_2Br$$

The latter is an extremely reactive species. The inductive effect of the trifluoroacetyl group makes it a good leaving group and facilitates cleavage of the O—Br bond. The acyl hypohalites are also the active halogenating species in solutions of the hypohalous acids in carboxylic acids, where they exist in equilibrium.

$$HOCl + RCO_2H \rightleftharpoons R\overset{\overset{O}{\|}}{C}OCl + H_2O$$

$$HOBr + RCO_2H \rightleftharpoons R\overset{\overset{O}{\|}}{C}OBr + H_2O$$

Molecular chlorine is believed to be the active electrophile in uncatalyzed chlorination of aromatic compounds. Simple second-order kinetics are observed in acetic acid.[33] The reaction is much slower in nonpolar solvents such as dichloromethane and carbon tetrachloride. Chlorination in nonpolar solvents is catalyzed by added acid. The catalysis by acids is probably the result of assistance

31. L. Eberson and F. Radner, *Acta Chem. Scand.* **B38**, 861 (1984); **B39**, 357 (1985).
32. P. B. D. de la Mare, I. C. Hilton, and S. Varma, *J. Chem. Soc.*, 4044 (1960); J. R. Barnett, L. J. Andrews, and R. M. Keefer, *J. Am. Chem. Soc.* **94**, 6129 (1972).
33. L. M. Stock and F. W. Baker, *J. Am. Chem. Soc.* **84**, 1661 (1962).

by protonation in the cleavage of the Cl—Cl bond.[34]

$$R\text{-}C_6H_5 + Cl\text{—}Cl\cdots HX \longrightarrow R\text{-}C_6H_5(Cl)^{+} + Cl^- + HX$$

Chlorination in acetic acid is characterized by a large ρ value (~ -9 to -10), and for toluene the partial rate factor f_p is 820. Both values indicate a late transition state which should resemble the σ-complex intermediate.

For preparative purposes, a Lewis acid such as $AlCl_3$ or $FeCl_3$ is often used to catalyze chlorination. Chlorination of benzene by $AlCl_3$ is overall third-order[35]:

$$\text{rate} = k[\text{ArH}][Cl_2][AlCl_3]$$

This rate law could correspond to formation of a Cl_2–$AlCl_3$ complex that acts as the active halogenating agent but is also consistent with a rapid equilibrium involving formation of Cl^+:

$$Cl_2 + AlCl_3 \rightleftharpoons Cl_2\text{–}AlCl_3 \rightleftharpoons Cl^+ + AlCl_4^-$$

There is, however, no direct evidence for the formation of Cl^+, and it is much more likely that the complex is the active electrophile. The substrate selectivity under catalyzed conditions ($k_{tol}/k_{benz} = 160$) is lower than in uncatalyzed chlorinations, as would be expected for a more reactive electrophile. The effect of the Lewis acid is to weaken the Cl—Cl bond, which will lower the activation energy for σ-complex formation.

Hypochlorous acid is a weak chlorinating agent. In acidic solution, it is converted to a much more active chlorinating agent. Although early mechanistic studies suggested that Cl^+ might be formed under these conditions, it has since been shown that this is not the case. Detailed kinetic analysis of the chlorination of methoxybenzene has revealed a rather complex rate law[36]:

$$\text{rate} = k_1[\text{HOCl}]^2 + k_2[H_3O^+][\text{HOCl}]^2 + k_3[\text{ArH}][H_3O^+][\text{HOCl}]$$

Some of the terms are independent of the concentration of the aromatic reactant. This rate law can best be explained in terms of the formation of Cl_2O, the anhydride of hypochlorous acid:

$$2\text{HOCl} \xrightarrow{H^+} Cl_2O + H_2O$$

Both Cl_2O and $[H_2OCl]^+$ apparently are active electrophiles under these conditions. The terms involving Cl_2O are zero-order in the aromatic reactant because the rate of formation of Cl_2O is lower than the rate of the subsequent reaction with the aromatic. Thermodynamic considerations argue strongly against rate-determining

34. L. J. Andrews and R. M. Keefer, *J. Am. Chem. Soc.* **81**, 1063 (1959); R. M. Keefer and L. J. Andrews, *J. Am. Chem. Soc.* **82**, 4547 (1960).
35. S. Y. Caille and R. J. P. Corriu, *Tetrahedron* **25**, 2005 (1969).
36. C. G. Swain and D. R. Crist, *J. Am. Chem. Soc.* **94**, 3195 (1972).

cleavage of $[H_2OCl]^+$ to H_2O and Cl^+. The estimated equilibrium constant for this dissociation is so small that the concentration of Cl^+ would be around 10^{-40}—, which is far too low to account for the observed reaction rate.[37]

Molecular bromine is believed to be the reactive brominating agent in uncatalyzed brominations. The bromination of benzene and toluene is first-order in both bromine and the aromatic substrate in trifluoroacetic acid solution,[38] but the rate law becomes more complicated in the presence of water.[39] The bromination of benzene in aqueous acetic acid exhibits a first-order dependence on bromine concentration when bromide ion is present. The observed rate is dependent on bromide ion concentration. The detailed kinetics are consistent with a rate-determining formation of the σ-complex when bromide ion concentration is low, but with a shift to reversible formation of the σ complex with rate-determining deprotonation at high bromide ion concentration.[40]

$$\text{R}-C_6H_5 + Br_2 \rightleftharpoons \text{R}-[C_6H_5(H)(Br)]^+ + Br^- \xrightarrow{-H^+} \text{R}-C_6H_4-Br$$

Bromination is characterized by high substrate selectivity.[41] The data in Table 10.4 (p. 554) show that, for toluene, f_p is around 2,500, as compared to about 50 for nitration. The very large stabilizing effect of electron donor substituents is also evident in the large negative ρ value (-12).[42] The fact that substituents can strongly influence both the rate and orientation implies that the transition state comes late in the reaction and resembles the σ complex.

Bromination has been shown not to exhibit a primary kinetic isotope effect in the case of benzene,[43] bromobenzene,[44] toluene,[45] or methoxybenzene.[46] There are several examples of substrates which do show significant isotope effects, including substituted anisoles,[46] *N,N*-dimethylanilines,[47] and 1,3,5-trialkylbenzenes.[48] The observation of isotope effects in highly substituted systems seems to be the result of steric factors that can operate in two ways. There may be resistance to the bromine taking up a position coplanar with adjacent substituents in the aromatization step. This would favor return of the σ complex to reactants. In addition, the steric bulk of several substituents may hinder solvent or other base from assisting in the proton removal. Either factor would allow deprotonation to become rate-controlling.

37. E. Berliner, *J. Chem. Educ.* **43**, 124 (1966).
38. H. C. Brown and R. A. Wirkkala, *J. Am. Chem. Soc.* **88**, 1447 (1966).
39. W. M. Schubert and D. F. Gurka, *J. Am. Chem. Soc.* **91**, 1443 (1969).
40. E. Berliner and J. C. Powers, *J. Am. Chem. Soc.* **83**, 905 (1961); W. M. Schubert and J. L. Dial, *J. Am. Chem. Soc.* **97**, 3877 (1975).
41. L. M. Stock and H. C. Brown, *Adv. Phys. Org. Chem.* **1**, 35 (1963).
42. H. C. Brown and L. M. Stock, *J. Am. Chem. Soc.* **79**, 1421 (1957).
43. P. B. D. de la Mare, T. M. Dunn, and J. T. Harvey, *J. Chem. Soc.*, 923 (1957).
44. L. Melander, *Acta Chem. Scand.* **3**, 95 (1949); *Ark. Kemi* **2**, 211 (1950).
45. R. Josephson, R. M. Keefer, and L. J. Andrews, *J. Am. Chem. Soc.* **83**, 3562 (1961).
46. J.-J. Aaron and J.-E. Dubois, *Bull. Soc. Chim. Fr.*, 603 (1971).
47. J.-E. Dubois and R. Uzan, *Bull. Soc. Chim. Fr.*, 3534 (1968); A. Nilsson, *Acta Chem. Scand.* **21**, 2423 (1967); A. Nilsson and K. Olsson, *Acta Chem. Scand.* **14**, 219 (1960).
48. P. C. Myhre, *Acta Chem. Scand.* **14**, 219 (1960).

Bromination is catalyzed by Lewis acids, and a study of the kinetics of bromination of benzene and toluene in the presence of aluminium chloride has been reported.[49] Toluene is found to be about 35 times more reactive than benzene under these conditions. The catalyzed reaction thus shows a good deal less substrate selectivity than the uncatalyzed reaction, as would be expected on the basis of the greater reactivity of the aluminum chloride–bormine complex.

Bromination can also be carried out using solutions of acetyl hypobromite or trifluoroacetyl hypobromite.[50] Acetyl hypobromite is considered to be the active halogenating species in solutions of hypobromous acid in acetic acid.

$$CH_3CO_2H + HOBr \rightleftharpoons CH_3\overset{\overset{\displaystyle O}{\|}}{C}OBr + H_2O$$

This reagent can also be formed by reaction of bromine with mercuric acetate.

$$Hg(OAc)_2 + Br_2 \rightleftharpoons Hg(OAc)Br + CH_3CO_2Br$$

Both of the above equilibria lie to the left, but acetyl hypobromite is sufficiently reactive that it is the principal halogenating species in both solutions. The reactivity of the acyl hypohalites as halogenating agents increases with the ability of the carboxylate to function as a leaving group. This is, of course, correlated with the acidity of the carboxylic acid. The estimated order of reactivity of Br_2, CH_3CO_2Br, and CF_3CO_2Br is $1:10^6:1^{10}$.[51] It is this exceptionally high reactivity of the hypobromites that permits them to be the reactive halogenating species in solutions where they are present in relatively low equilibrium concentration.

Molecular iodine is not a very powerful halogenating agent. Only very reactive aromatics such as anilines or phenolate anions are reactive toward iodine. Iodine monochloride can be used as an iodinating agent. The greater electronegativity of the chlorine ensures that the iodine will be the electrophilic entity in the substitution reaction. Iodination by iodine monochloride can be catalyzed by Lewis acids, such as $ZnCl_2$.[52] Iodination can also be carried out with acetyl hypoiodite and trifluoroacetyl hypoiodite. The methods of formation of these reagents are similar to those for the hypobromites.[53]

Direct fluorination of aromatics is not a preparatively important reaction because it can occur with explosive violence. Mechanistic studies have been done at very low temperatures and with low fluorine concentrations. For toluene, the f_p and f_m values are 8.2 and 1.55, respectively, indicating that fluorine is a very unselective electrophile. The ρ value in a Hammett correlation with σ^+ is -2.45. Thus, fluorination exhibits the characteristics that would be expected for a very

49. S. Y. Caille and R. J. P. Corriu, *Tetrahedron* **25**, 2005 (1969).
50. P. B. D. de la Mare and J. L. Maxwell, *J. Chem. Soc.*, 4829 (1962); Y. Hatanaka, R. M. Keefer, and L. J. Andrews, *J. Am. Chem. Soc.* **87**, 4280 (1965).
51. P. B. D. de la Mare, I. C. Hilton, and S. Varma, *J. Chem. Soc.*, 4044 (1960); J. R. Bennett, L. J. Andrews, and R. M. Keefer, *J. Am. Chem. Soc.* **94**, 6129 (1972).
52. R. M. Keefer and L. J. Andrews, *J. Am. Chem. Soc.* **78**, 5623 (1956).
53. E. M. Chen, R. M. Keefer, and L. J. Andrews, *J. Am. Chem. Soc.* **89**, 428 (1967).

Table 10.8. Partial Rate Factors for Hydrogen Exchange in Some Substituted Aromatic Compounds

X	f_o	f_m	f_p	Ref.
CH_3	330	7.2	313	a
F	0.136	—	1.79	b
Cl	0.035	—	0.161	b
OPh	6900	~0.1	31,000	c
Ph	133	<1	143	d

a. From C. Eaborn and R. Taylor, *J. Chem. Soc.*, 247 (1961).
b. From C. Eaborn and R. Taylor, *J. Chem. Soc.*, 2388 (1961).
c. From R. Baker and C. Eaborn, *J. Chem. Soc.*, 5077 (1961).
d. From C. Eaborn and R. Taylor, *J. Chem. Soc.*, 1012 (1961).

reactive electrophile.[54] A number of reagents in which fluorine is bound to a very electronegative group also serve as fluorinating agents. These include CF_3OF, CF_3CO_2F, CH_3CO_2F, and $HOSO_2OF$.[55] These reagents will be discussed in Section 11.1.2 of Part B.

10.4.3. Protonation and Hydrogen Exchange

Hydrogen exchange resulting from reversible protonation of an aromatic ring can be followed by the use of isotopic labels. Both deuterium and tritium have been employed, and the experiment can be designed to follow either the incorporation or the release of the labeled hydrogen. The study of the mechanism of electrophilic hydrogen exchange is somewhat simplified by the fact that the proton must be the active electrophile. The principle of microscopic reversibility implies that the transition state must occur on a symmetrical potential energy surface, since the attacking electrophile is chemically identical to the displaced proton. The transition states involve partial transfer of a proton to (or from) a solvent molecule(s) to the aromatic ring. The intermediate σ complex is a cyclohexadienyl cation. As mentioned earlier, these cations are stable in strongly acidic non-nucleophilic media and can be subjected to spectroscopic characterization.

Partial rate factors for a number of substituted aromatic compounds have been measured. They reveal activation of *ortho* and *para* positions by electron-releasing groups. Some typical data are given in Table 10.8. The k_{tol}/k_{benz} ratio of around 300 indicates considerable substrate selectivity. The f_p value for toluene varies somewhat, depending on the reaction medium, but generally is about 10^2.[56] The ρ value for hydrogen exchange in H_2SO_4–CF_3CO_2H–H_2O is -8.6.[57] A similarly large

54. F. Cacace, P. Giacomello, and A. P. Wolff, *J. Am. Chem. Soc.* **102**, 3511 (1980).
55. A. Haas and M. Lieb, *Chimia* **39**, 134 (1985).
56. L. M. Stock and H. C. Brown, *Adv. Phys. Org. Chem.* **1**, 35 (1963).
57. P. Rys, P. Skrabal, and H. Zollinger, *Angew. Chem. Int. Ed. Engl.* **11**, 874 (1972).

ρ value of -7.5 has been observed in aqueous sulfurinc acid.[58] As seen for other electrophilic aromatic substitution reactions, the best correlation is with σ^+.

Among the many experimental results pertaining to hydrogen exchange, a most important one is that general acid catalysis has been demonstrated.[59] This finding is in accord with a rate-limiting step involving proton transfer. Since proton removal is partially rate-determining, hydrogen exchange exhibits an isotope effect. A series of experiments using both deuterium and tritium labels arrived at $k_H/k_D = 9.0$ for the proton loss step for 1,3,5-trimethoxybenzene.[60] A substantial isotope effect has also been observed for the exchange process with azulene.[61]

Data on hydrogen exchange have been particularly interesting for comparison with theoretical predictions of aromatic reactivity. Because the electrophile is well-defined and small, calculations on the stability of the competing intermediates by MO methods is more secure than in cases where the nature of the electrophile is less certain.

10.4.4. Friedel–Crafts Alkylation and Related Reactions

The Friedel–Crafts reaction is a very important method for introducing alkyl substituents on an aromatic ring. It involves generation of a carbocation or related electrophilic species. The most general method of generating these electrophiles involves reaction between an alkyl halide and a Lewis acid. The most common Friedel–Crafts catalyst for preparative work is $AlCl_3$, but other Lewis acids such as SbF_5, $TiCl_4$, $SnCl_4$, and BF_3 can also promote reaction. Alternative routes to alkylating species include protonation and dehydration of alcohols and protonation of alkenes.

There are relatively few kinetic data on the Friedel–Crafts reaction. Alkylation of benzene or toluene with methyl bromide or ethyl bromide with gallium bromide as catalyst is first-order in each reactant and in catalysts.[62] With aluminum bromide as catalyst, the rate of reaction changes with time, apparently because of heterogeneity of the reaction mixture.[63] The initial rate data fit the following kinetic expression:

$$\text{rate} = k[\text{EtBr}][\text{benzene}][\text{AlBr}_3]^2$$

58. S. Clementi and A. R. Katritzky, *J. Chem. Soc., Perkin Trans 2,* 1077 (1973).
59. A. J. Kresge and Y. Chiang, *J. Am. Chem. Soc.* **83**, 2877 (1961); A. J. Kresge, S. Slae, and D. W. Taylor, *J. Am. Chem. Soc.* **92**, 6309 (1970).
60. A. J. Kresge and Y. Chiang, *J. Am. Chem. Soc.* **89**, 4411 (1967).
61. L. C. Gruen and F. A. Long, *J. Am. Chem. Soc.* **89**, 1287 (1967).
62. S. U. Choi and H. C. Brown, *J. Am. Chem. Soc.* **85**, 2596 (1963).
63. B. J. Carter, W. D. Covey, and F. P. DeHaan, *J. Am. Chem. Soc.* **97**, 4783 (1975); cf. S. U. Choi and H. C. Brown, *J. Am. Chem. Soc.* **81**, 3315 (1959); F. P. DeHaan and H. C. Brown, *J. Am. Chem. Soc.* **91**, 4844 (1969); H. Jungk, C. R. Smoot, and H. C. Brown, *J. Am. Chem. Soc.* **78**, 2185 (1956).

The reaction rates of toluene and benzene with 2-propyl chloride in nitromethane can be fit to a third-order rate law[64]:

$$\text{rate} = k[\text{AlCl}_3][i\text{-PrCl}][\text{ArH}]$$

The same rate law pertains to *t*-butyl chloride.[65] The reaction of benzyl chloride and toluene shows a second-order dependence on titanium chloride concentration under conditions where there is a large excess of hydrocarbon.[66]

$$\text{rate} = k[\text{PhCH}_2\text{Cl}][\text{TiCl}_4]^2$$

Rates which are *independent* of aromatic substrate concentration have been found for reaction of benzyl chloride catalyzed by $TiCl_4$ or SbF_5 in nitromethane.[67] This can be interpreted as resulting from rate-determining formation of the electrophile, presumably an ion pair of the benzyl cation.

All these kinetic results can be accommodated by a general mechanistic scheme which incorporates the following fundamental components: (1) complexation of the alkylating agent and the Lewis acid; (2) electrophilic attack on the aromatic substrate to form the σ complex; and (3) deprotonation. In many systems, there may be an ionization of the complex between the alkylating agent and the Lewis acid to yield a carbocation. This step accounts for the fact that frequently rearrangement of the alkyl group is observed during Friedel–Crafts alkylation.

$$\text{(1)}\quad \text{R—X} + \text{MY}_n \rightleftharpoons \text{R—}\overset{+}{\text{X}}\text{—}\overset{-}{\text{M}}\text{Y}_n$$
$$\text{(1a)}\quad \text{R—}\overset{+}{\text{X}}\text{—}\overset{-}{\text{M}}\text{Y}_n \rightleftharpoons \text{R}^+ + [\text{MY}_n\text{X}]^-$$

(2) Z–C₆H₅ + R–X⁺–M⁻Yₙ or R⁺ ⟶ Z–(C₆H₅)⁺(R)(H)

(3) Z–(C₆H₅)⁺(R)(H) ⟶ Z–C₆H₄–R + H⁺

Absolute rate data for the Friedel–Craft reactions are difficult to obtain. The reaction is complicated by sensitivity to moisture and heterogeneity. For this reason, most of the structure-reactivity trends have been developed using competitive methods, rather than by direct measurements. Relative rates are established by allowing the electrophile to compete for an excess of the two reagents. The product ratio establishes the relative reactivity.

64. F. P. DeHaan, G. L. Delker, W. D. Covey, J. Ahn, R. L. Cowan, C. H. Fong, G. Y. Kim, A. Kumar, M. P. Roberts, D. M. Schubert, E. M. Stoler, Y. J. Suh, and M. Tang, *J. Org. Chem.* **51**, 1587 (1986).
65. F. P. DeHaan, W. H. Chan, J. Chang, D. M. Ferrara, and L. A. Wamschel, *J. Org. Chem.* **51**, 1591 (1986).
66. F. P. DeHaan, W. D. Covey, R. L. Ezeele, J. E. Margetan, S. A. Pace, M. J. Sollenberger, and D. S. Wolfe, *J. Org. Chem.* **49**, 3954 (1984).
67. F. P. DeHaan, G. L. Delker, W. D. Covey, J. Ahn, M. S. Anisman, E. C. Brehm, J. Chang, R. M. Chicz, R. L. Cowan, D. M. Ferrara, C. H. Fong, J. D. Harper, C. D. Irani, J. Y. Kim, R. W. Meinhold, K. D. Miller, M. P. Roberts, E. M. Stoler, Y. J. Suh, M. Tang, and E. L. Williams, *J. Am. Chem. Soc.* **106**, 7038 (1984).

Table 10.9. Substrate and Position Selectivity in Friedel–Crafts Alkylation Reactions

	Electrophilic reagents	k_{tol}/k_{benz}	Toluene $o:p$ ratio
1[a]	CH_3Br–$AlBr_3$	2.5–4.1	1.9
2[b]	C_2H_5Br–$GaBr_3$	6.5	—
3[c]	$(CH_3)_2CHBr$–$AlCl_3$	1.9	1.2
4[d]	$(CH_3)_2CHCl$–$AlCl_3$	2.0	1.5
5[e]	$(CH_3)_3CCl$–$AlCl_3$	25	0
6[f]	$(CH_3)_3CBr$–$SnCl_4$	16.6	0
7[f]	$(CH_3)_3CBr$–$AlCl_3$	1.9	0
8[g]	$PhCH_2Cl$–$AlCl_3$	3.2	0.82
9[h]	$PhCH_2Cl$–$AlCl_3$	2–3	0.9
10[i]	$PhCH_2Cl$–$TiCl_4$	6.3	0.74
10[i]	p-$MeOPhCH_2Cl$–$TiCl_4$	97	0.40

a. From H. C. Brown and H. Jungk, *J. Am. Chem. Soc.* **77**, 5584 (1955).
b. From S. U. Choi and H. C. Brown, *J. Am. Chem. Soc.* **85**, 2596 (1963).
c. From G. A. Olah, S. H. Flood, S. J. Kuhn, M. E. Moffatt, and N. A. Overchuck, *J. Am. Chem. Soc.* **86**, 1046 (1964).
d. From F. P. DeHaan, G. L. Delker, W. D. Covey, J. Ahn, R. L. Cowan, C. H. Fong, G. Y. Kim, A. Kumar, M. P. Roberts, D. M. Schubert, E. M. Stoler, Y. J. Suh, and M. Tang, *J. Org. Chem.* **51**, 1587 (1986).
e. From F. P. DeHaan, W. H. Chan, J. Chang, D. M. Ferrara, and L. A. Wainschel, *J. Org. Chem.* **51**, 1591 (1986).
f. From G. A. Olah, S. H. Flood, and M. E. Moffatt, *J. Am. Chem. Soc.* **86**, 1060 (1964).
g. From G. A. Olah, S. J. Kuhn, and S. H. Flood, *J. Am. Chem. Soc.* **84**, 1688 (1962).
h. From F. P. DeHaan, W. D. Covey, R. L. Ezelle, J. E. Margetan, S. A. Pace, M. J. Sollenberger, and D. S. Wolf, *J. Org. Chem.* **49**, 3954 (1984).
i. From G. A. Olah, S. Kobayashi, and M. Tashiro, *J. Am. Chem. Soc.* **94**, 7448 (1972).

A study of alkylations with a group of substituted benzyl halides and a range of Friedel–Crafts catalysts has provided insight into the trends in selectivity and orientation that accompany changes in both the alkyl group and the catalysts.[68] There is a marked increase in substrate selectivity on going from *p*-nitrobenzyl chloride to *p*-methoxybenzyl chloride. For example, with titanium tetrachloride as the catalyst, k_{tol}/k_{benz} increases from 2.5 to 97. This increase in substrate selectivity is accompanied by an increasing preference for *para* substitution. With *p*-nitrobenzyl chloride, the $o:p$ ratio is 2:1 (the statistically expected ratio), whereas with the *p*-methoxy compound, the *para* product dominates by 2.3:1. There is a clear trend within the family of substituted benzyl chlorides of increasing selectivity with the increasing electron donor capacity of the substituent. All of the reactions, however, remain in a region that constitutes rather low selectivity. Thus, the position of the transition state for substitution by a benzylic cation must come quite early. The substituents on the ring undergoing substitution have a relatively weak orienting effect on the attacking electrophile. With benzylic cations stabilized by donor substituents, the transition state comes later and the selectivity is somewhat higher. Toluene–benzene reactivity ratios under a number of Friedel–Crafts conditions are recorded in Table 10.9. As would be expected on the basis of the low substrate selectivity, position selectivity is also modest. As shown by the isomer ratios in Table 10.9, the amount of *ortho* product is often comparable to that of *para* product.

68. G. A. Olah, S. Kobayashi, and M. Tashiro, *J. Am. Chem. Soc.* **94**, 7448 (1972).

Steric effects play a major role in determining the $o:p$ ratio in Friedel–Crafts alkylations. The amount of *ortho* substitution of toluene decreases as the size of the entering alkyl group increases along the series methyl, ethyl, 2-propyl.[69] No *ortho* product is found when the entering group is *t*-butyl.[70]

A good deal of experimental care is often required to ensure that the product mixture at the end of a Friedel–Crafts reaction is determined by *kinetic control.* The strong Lewis acid catalysts can catalyze the isomerization of alkylbenzenes, and if isomerization takes place, the product composition is not informative about the position selectivity of electrophilic attack. Isomerization increases the amount of the *meta* isomer in the case of dialkylbenzenes, because this isomer is thermodynamically the most stable.[71]

Alcohols and alkenes can also serve as sources of electrophiles in Friedel–Crafts reactions in the presence of strong acids:

$$R_3COH + H^+ \rightarrow R_3\overset{+}{C}OH_2 \rightarrow R_3C^+$$

$$R_2C{=}CHR' + H^+ \rightarrow R_2\overset{+}{C}CH_2R'$$

The generation of carbocations from these sources is well documented. The reaction of aromatics with alkenes in the presence of Lewis acid catalysts is the basis for the industrial production of many alkylated aromatic compounds. Styrene, for example, is prepared by dehydrogenation of ethylbenzene made from benzene and ethylene.

10.4.5. Friedel–Crafts Acylation and Related Reactions

Friedel–Crafts acylation usually involves the reaction of an acyl halide, a Lewis acid catalyst, and the aromatic substrate. Two possible electrophiles can be envisaged. A discrete positively charged acylium ion can be formed and act as the electrophile, or the active electrophile could be a complex formed between the acyl halide and the Lewis acid catalyst.

$$R\overset{O}{\overset{\|}{C}}X + MX_n \rightleftharpoons RC{\equiv}O^+ + [MX_{n+1}]^-$$

$$RC{\equiv}O^+ + C_6H_6 \longrightarrow [C_6H_6(H)(C(=O)R)]^+$$

or

$$R\overset{O}{\overset{\|}{C}}X + MX_n \rightleftharpoons R\overset{O}{\overset{\|}{C}}{-}\overset{+}{X}{-}\overset{-}{M}X_n$$

$$C_6H_6 + R\overset{O}{\overset{\|}{C}}{-}\overset{+}{X}{-}\overset{-}{M}X_n \longrightarrow [C_6H_6(H)(C(=O)R)]^+ + [MX_{n+1}]^-$$

69. R. H. Allen and L. D. Yats, *J. Am. Chem. Soc.* **83**, 2799 (1961).
70. G. A. Olah, S. H. Flood, and M. E. Moffatt, *J. Am. Chem. Soc.* **86**, 1060 (1964).
71. D. A. McCaulay and A. P. Lien, *J. Am. Chem. Soc.* **74**, 6246 (1952).

Table 10.10. Substrate and Position Selectivity in Friedel–Crafts Acylation Reactions

	Electrophilic reagents	k_{tol}/k_{benz}	Toluene $o:p$ ratio
1[a]	Acetyl chloride–$AlCl_3$	134	0.012
2[b]	Propionyl chloride–$AlCl_3$	106	0.033
3[c]	$CH_3C{\equiv}O^+SbF_6^-$	125	0.014
4[d]	Formyl fluoride–BF_3	35	0.82
5[d]	2,4-Dinitrobenzoyl chloride–$AlCl_3$	29	0.78
6[d]	Pentafluorobenzoyl chloride–$AlCl_3$	16	0.61
7[d]	Benzoyl chloride–$AlCl_3$	153	0.09
8[d]	*p*-Methylbenzoyl chloride–$AlCl_3$	164	0.08
9[d]	*p*-Methoxybenzoyl chloride–$AlCl_3$	233	0.2

a. From G. A. Olah, M. E. Moffatt, S. J. Kuhn, and B. A. Hardie, *J. Am. Chem. Soc.* **86**, 2198 (1964).
b. From G. A. Olah, J. Lukas, and E. Lukas, *J. Am. Chem. Soc.* **91**, 5319 (1969).
c. From G. A. Olah, S. J. Kuhn, S. H. Flood, and B. A. Hardie, *J. Am. Chem. Soc.* **86**, 2203 (1964).
d. From G. A. Olah and S. Kobayashi, *J. Am. Chem. Soc.* **93**, 6964 (1971).

The formation of acyl halide–Lewis acid complexes can be demonstrated readily. Acetyl chloride, for example, forms both 1:1 and 1:2 complexes with $AlCl_3$ which can be observed by NMR.[72] The existence of acylium ions has been demonstrated by X-ray diffraction studies on crystalline salts. For example, crystal structure determinations have been reported for *p*-methylphenylacylium[73] and methylacylium[74] (acetylium) ions as SbF_6 salts. There is also a good deal of evidence from NMR measurements which demonstrates that acylium ions can exist in non-nucleophilic solvents.[75] The positive charge on acylium ions is delocalized onto the oxygen atom. This delocalization is demonstrated in particular by the short O—C bond lengths in acylium ions, which imply a major contribution from the structure having a triple bond:

$$R\overset{+}{C}{=}O \leftrightarrow RC{\equiv}O^+$$

As is the case with Friedel-Crafts alkylations, direct kinetic measurements are difficult, and not many data are available. Rate equations of the form

$$\text{rate} = k_1[\text{RCOCl—AlCl}_3][\text{ArH}] + k_2[\text{RCOCl—AlCl}_3]^2[\text{ArH}]$$

have been reported for reaction of benzene and toluene with both acetyl and benzoyl chloride.[76] The available kinetic data do not permit unambiguous conclusions about the identity of the active electrophile. Most mechanistic discussions have depended on competitive rate data and on structure-reactivity relationships.

72. B. Glavincevski and S. Brownstein, *J. Org. Chem.* **47**, 1005 (1982).
73. B. Chevrier, J.-M. LeCarpentier, and R. Weiss, *J. Am. Chem. Soc.* **94**, 5718 (1972).
74. F. P. Boer, *J. Am. Chem. Soc.* **90**, 6706 (1968).
75. N. C. Deno, C. U. Pittman, Jr., and M. J. Wisotsky, *J. Am. Chem. Soc.* **86**, 4370 (1964); G. A. Olah and M. B. Comisarow, *J. Am. Chem. Soc.* **88**, 4442 (1966).
76. R. Corriu, M. Dore, and R. Thomassin, *Tetrahedron* **27**, 5601, 5819 (1971).

Selectivity in Friedel–Crafts acylation, with regard to both substrate and position, is moderate. Some representative data are collected in Table 10.10. It can be seen that the toluene:benzene reactivity ratio is generally between 100 and 200. A progression from low substrate selectivity (entries 5 and 6) to higher substrate selectivity (entries 8 and 9) has been demonstrated for a series of aroyl halides.[77] Electron-attracting groups on the aroyl chloride lead to low selectivity, presumably because of the increased reactivity of such electrophiles. Electron-releasing groups diminish reactivity and increase selectivity. The *p*-methoxy compound is somewhat anomalous. Although substrate selectivity increases as expected, the $o:p$ ratio is higher than for the unsubstituted system. In general, Friedel–Crafts acylation shows a strong preference for *para* over *ortho* substitution, although highly reactive acylating reagents such as perfluorobenzoyl chloride are an exception. Friedel–Crasts acylation is, in general, a more selective reaction than Friedel–Crafts alkylation. The implication is that the acylium ions are less reactive electrophiles than the cationic intermediates involved in the alkylation process.

One other feature of the data in Table 10.10 is worthy of further comment. Notice that alkyl (acetyl, propionyl) substituted acylium ions exhibit a smaller $o:p$ ratio than the various aroyl systems. If steric factors were dominating the position selectivity, one would expect the opposite result. A possible explanation for this feature of the data could be that the aryl compounds are reacting via free acylium ions, whereas the alkyl systems may involve more bulky acid chloride–catalyst complexes. Steric factors clearly enter into determining the $o:p$ ratio. The hindered 2,4,6-trimethylbenzoyl group is introduced with a 50:1 preference for the *para* position.[77] Similarly, in the benzoylation of alkylbenzenes by benzoyl chloride–aluminum chloride, the amount of *ortho* product decreases (10.3%, 6.0%, 3.1%, 0.6%, respectively) as the branching of the alkyl group is increased along the series methyl, ethyl, 2-propyl, *t*-butyl.[78]

Friedel–Crafts acylation sometimes shows a modest kinetic isotope effect.[79] This observation suggests that the proton removal is not much faster than the formation of the σ complex and that the formation of the σ complex may be reversible under some conditions.

A number of variations of the Friedel–Crafts reaction conditions are possible. Acid anhydrides can serve as the acylating agent in place of acid chlorides. Also, the carboxylic acid can be used directly, particularly in combination with strong acids. For example, mixtures of carboxylic acids with polyphosphoric acid, in which a mixed anhydride is presumably formed *in situ*, are reactive acylating agents.[80] Similarly, carboxylic acids dissolved in trifluoromethanesulfonic acid can carry out Friedel–Crafts acylation. The reactive electrophile under these conditions is believed to be the protonated mixed anhydride.[81] In these procedures, the leaving group

77. G. A. Olah and S. Kobayashi, *J. Am. Chem. Soc.* **93**, 6964 (1971).
78. G. A. Olah, J. Lukas, and E. Lukas, *J. Am. Chem. Soc.* **91**, 5319 (1969).
79. G. A. Olah, S. J. Kuhn, S. H. Flood, and B. A. Hardie, *J. Am. Chem. Soc.* **86**, 2203 (1964); D. B. Denney and P. P. Klemchuk, *J. Am. Chem. Soc.* **80**, 3285, 6014 (1958).
80. T. Katuri and K. M. Damodaran, *Can. J. Chem.* **47**, 1529 (1969).
81. R. M. G. Roberts and A. R. Sardi, *Tetrahedron* **39**, 137 (1983).

from the acylating agent is different, but other aspects of the reaction are similar to those under the usual conditions. Synthetic applications of Friedel–Crafts acylation are discussed in Chapter 11 of Part B.

10.4.6. Coupling with Diazonium Compounds

Among the reagents that would be classified as weak electrophiles, the best studied are the aromatic diazonium ions. These reagents react only with aromatic substrates having strong electron-donor substituents. The products are azo compounds. The aryl diazonium ions are usually generated by diazotization of aromatic amines. The mechanism of diazonium ion formation is discussed more completely in Part B, Section 11.2.

$Ar\overset{+}{N}\equiv N$ + O⁻ ⇌ ArN=N H O⁻ + → N=NAr OH

The aryl diazonium are stable in solution only near room temperature or below, and this also limits the range of compounds that can be successfully substituted by diazonium ions.

Kinetic investigations have revealed second-order kinetic behavior for substitution by diazonium ions in a number of instances. In the case of phenols, it is the conjugate base that undergoes substitution.[82] This finding is entirely reasonable, since the deprotonated oxy group is a better electron donor than the neutral hydroxyl substituent. The reactivity of the diazonium ion depends on the substituent groups which are present. Reactivity is increased by electron-attracting groups and decreased by electron donors.[83]

The most unique feature of the mechanism for diazonium coupling is that proton loss can be clearly demonstrated to be the rate-determining step in some cases. This feature is revealed in two ways. First, diazonium couplings of several naphthosulfonate ions exhibit primary isotope effects in the range 4–6 when deuterium is present at the site of substitution, clearly indicating that cleavage of the C—H bond is rate-determining. Second, these reactions can also be shown to be general-base-catalyzed. This, too, implies that proton removal is rate-deter-

82. R. Wistar and P. D. Bartlett, *J. Am. Chem. Soc.* **63**, 413 (1941).
83. A. F. Hegarty, in *The Chemistry of the Diazonium and Diazo Groups*, S. Patai (ed.), John Wiley, New York, 1978, Chapter 12.

mining.[84]

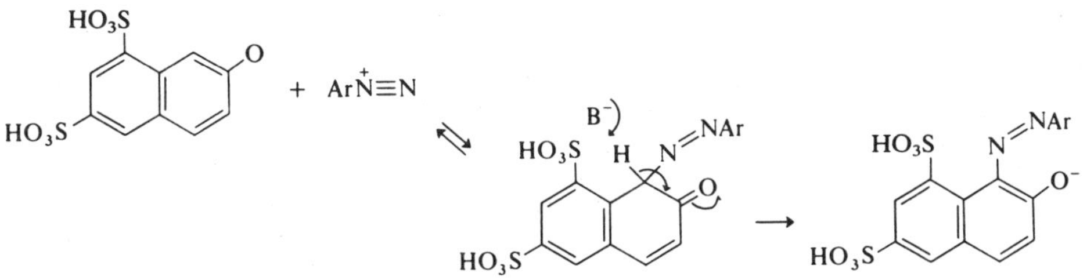

Because of the limited range of aromatic compounds that react with diazonium ions, selectivity data comparable to those discussed for other electrophilic substitutions are not available. Diazotization, since it involves a weak electrophile, would be expected to reveal high substrate and position selectivity.

10.4.7. Substitution of Groups Other Than Hydrogen

The general mechanism for electrophilic substitution suggests that groups other than hydrogen could be displaced, provided the electrophile attacked at the substituted carbon. Substitution at a site already having a substituent is called *ipso* substitution and has been observed in a number of circumstances. The ease of removal of a substituent depends on its ability to accommodate a positive charge. This factor will determine whether the newly attached electrophile or the substituent will be eliminated from the σ complex on rearomatization.

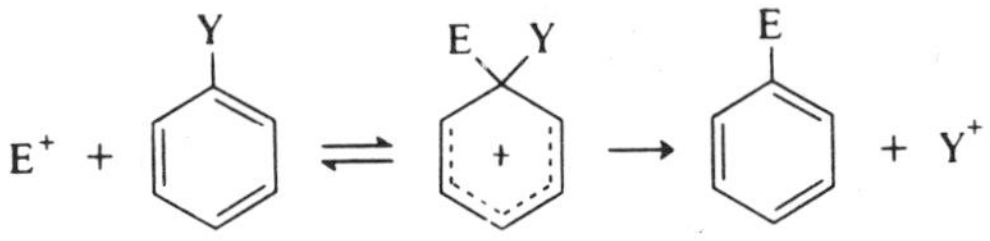

One of the more frequently encountered examples of substituent replacement involves cleavage of a highly branched alkyl substituent. The alkyl group is expelled as a carbocation, and, for this reason, substitution is most common for branched alkyl groups. The nitration of 1,4-bis-(2-propyl)benzene provides an example:

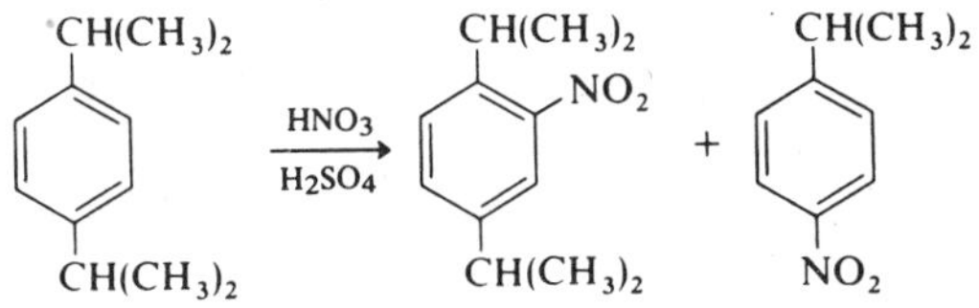

Ref. 85

84. H. Zollinger, *Azo and Diazo Chemistry*, translated by H. E. Nursten, Interscience, New York, 1961, Chapter 10; H. Zollinger, *Adv. Phys. Org. Chem.* **2**, 163 (1964); H. Zollinger, *Helv. Chim. Acta* **38**, 1597 (1955).
85. G. A. Olah and S. J. Kuhn, *J. Am. Chem. Soc.* **86**, 1067 (1964).

The replacement of bromine and iodine during aromatic nitration has also been observed. *p*-Bromoanisole and *p*-iodoanisole, for example, both give 30–40% of *p*-nitroanisole, a product resulting from displacement of halogen on nitration.

$$p\text{-}XC_6H_4OCH_3 + HNO_3 \longrightarrow p\text{-}O_2NC_6H_4OCH_3 + 2\text{-}NO_2\text{-}4\text{-}XC_6H_3OCH_3$$

Ref. 86

Because of the greater resistance to elimination of chlorine as a positively charged species, *p*-chloroanisole does not undergo dechlorination under similar conditions.

Cleavage of *t*-butyl groups has also been observed in halogenation reactions. Minor amounts of dealkylated products are formed during chlorination and bromination of *t*-butylbenzene.[87] The amount of dealkylation increases greatly in the case of 1,3,5-tri-*t*-butylbenzene, and the principal product of bromination is 3,5-dibromo-*t*-butylbenzene.[88]

The most thoroughly studied group of aromatic substitutions involving replacement of a substituent group in preference to a hydrogen are electrophilic substitutions of arylsilanes:

$$Ar{-}SiR_3 + E^+ + X^- \rightarrow Ar{-}E + R_3SiX$$

The silyl group directs electrophiles to the substituted position. That is, it is an *ipso*-directing group. Because of the polarity of the carbon–silicon bond, the substituted position is relatively electron-rich. The ability of silicon substituents to stabilize carbocation character at β-carbon atoms (see Section 6.10, p. 384) also promotes *ipso* substitution. The silicon substituent is easily removed from the σ complex by reaction with a nucleophile.

$$C_6H_5{-}SiR_3 \longrightarrow [C_6H_5(SiR_3)(E)]^+ \longrightarrow C_6H_5{-}E$$

The reaction exhibits other characteristics typical of an electrophilic aromatic substitution.[89] Examples of electrophiles which can effect substitution for silicon include protons, the halogens, as well as acyl, nitro, and sulfonyl groups.[90] The fact that these reactions occur very rapidly has made them attractive for situations where substitution must be done under very mild conditions.[91]

86. C. L. Perrin and G. A. Skinner, *J. Am. Chem. Soc.* **93**, 3389 (1971).
87. P. B. D. de la Mare and J. T. Harvey, *J. Chem. Soc.*, 131 (1957); P. B. D. de la Mare, J. T. Harvey, M. Hassan, and S. Varma, *J. Chem. Soc.*, 2756 (1958).
88. P. D. Bartlett, M. Roha, and R. M. Stiles, *J. Am. Chem. Soc.* **76**, 2349 (1954).
89. F. B. Deans and C. Eaborn, *J. Chem. Soc.*, 2299 (1959).
90. F. B. Deans, C. Eaborn, and D. E. Webster, *J. Chem. Soc.*, 3031 (1959); C. Eaborn, Z. Lasocki, and D. E. Webster, *J. Chem. Soc.*, 3034 (1959); C. Eaborn, *J. Organomet. Chem.* **100**, 43 (1975); J. D. Austin, C. Eaborn, and J. D. Smith, *J. Chem. Soc.*, 4744 (1963); F. B. Deans and C. Eaborn, *J. Chem. Soc.*, 498 (1957); R. W. Bott, C. Eaborn, and T. Hashimoto, *J. Chem. Soc.*, 3906 (1963).
91. S. R. Wilson and L. A. Jacob, *J. Org. Chem.* **51**, 4833 (1986).

Trialkyltin substituents are also powerful *ipso*-directing groups. As the substituent atom becomes more metallic and less electronegative, electron density at carbon increases, as does the stabilization of β-carbocation character. Acidic cleavage of arylstannanes is formulated as an electrophilic aromatic substitution proceeding through an *ipso*-oriented σ complex.[92]

$$\text{C}_6\text{H}_5\text{-SnR}_3 \xrightarrow{\text{H-X}} [\text{C}_6\text{H}_5(\text{H})(\text{SnR}_3)]^+\ \text{X}^- \longrightarrow \text{C}_6\text{H}_5\text{-H} + \text{R}_3\text{SnX}$$

10.5. Nucleophilic Aromatic Substitution by Addition–Elimination

Neither of the major mechanisms for nucleophilic substitution in saturated compounds is accessible for substitution on aromatic rings. A back-side S_N2-type reaction is precluded by the geometry of the benzene ring. The back lobe of the sp^2 orbital is directed toward the center of the ring. Any inversion mechanism is also precluded by the geometry of the ring. An S_N1 mechanism is very costly in terms of energy because a cation directly on a benzene ring is very unstable. From the data in Table 5.2 (p. 273), it is clear that a phenyl cation is less stable than even a primary carbocation. This is again a consequence of the geometry and hybridization of the aromatic carbon atoms. A carbocation must be localized in an sp^2 orbital. This orbital is orthogonal to the π system so there is no stabilization available from the π electrons.

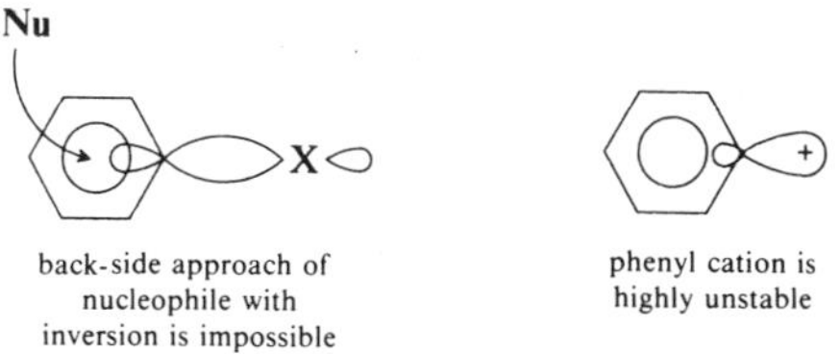

back-side approach of nucleophile with inversion is impossible

phenyl cation is highly unstable

There are several mechanisms by which net nucleophilic aromatic substitution can occur. In this section, we will discuss the addition–elimination mechanism and the elimination–addition mechanism. Substitutions via organometallic intermediates and via aryl diazonium ions will be considered in Part B.

The addition–elimination mechanism[93] uses one of the vacant π^* orbitals as the initial point of attack by the nucleophile. This permits bonding of the nucleophile to the aromatic ring without displacement of any of the existing substituents. If attack occurs at a position occupied by a potential leaving group, net substitution

92. C. Eaborn, I. D. Jenkins, and D. R. M. Walton, *J. Chem. Soc., Perkin Trans 2,* 596 (1974).
93. Reviews: C. F. Bernasconi, in *MTP Int. Rev. Sci., Organic Series One*, Vol. 3, H. Zollinger (ed.), Butterworths, London, 1973; J. A. Zoltewicz, *Top. Curr. Chem.* **59**, 33 (1975); J. Miller, *Aromatic Nucleophilic Substitution*, Elsevier, Amsterdam, 1968.

can occur by a second step in which the leaving group is expelled.

Nu^- Y— → Nu, Y → Nu— + Y^-

The addition intermediate is isoelectronic with a pentadienyl anion.

The HOMO of the pentadienyl anion is ψ_3, which has its electron density primarily at the carbons *ortho* and *para* to the position of substitution. The intermediate is therefore strongly stabilized by an electron-accepting group *ortho* or *para* to the site of substitution. Such substituents therefore activate the ring to nucleophilic substitution. The most powerful effect is exerted by a nitro group, but cyano and carbonyl groups are also favorable. Generally speaking, nucleophilic aromatic substitution is an energetically demanding reaction, even when electron-attracting substituents are present. The process disrupts the aromatic π system. Without electron-attracting groups present, nucleophilic aromatic substitution occurs only under extreme reaction conditions.

The role of the leaving group in determining the reaction rate is somewhat different than in S_N2 and S_N1 substitution at alkyl groups. In those cases, the bond strength is frequently the dominant factor so that the order of reactivity of the halogens is I > Br > Cl > F. In nucleophilic aromatic substitution, the formation of the addition intermediate is usually the rate-determining step so the ease of C—X bond breaking does not affect the rate. When this is the case, the order of reactivity is often F > Cl > Br > I.[94] This order is the result of the polar effect of the halogen. The stronger bond dipole associated with the more electronegative halogens favors the addition step and thus increases the overall rate of reaction.

Groups other than halogen can act as leaving groups. Alkoxy groups are very poor leaving groups in S_N2 reactions but can act as leaving groups in aromatic substitution. The reason is the same as for the inverted order of reactivity for the halogens. The rate-determining step is the addition, and the alkoxide can be eliminated in the energetically favorable rearomatization. Nitro[95] and sulfonyl[96] groups

94. G. P. Briner, J. Miller, M. Liveris, and P. G. Lutz, *J. Chem. Soc.*, 1265 (1954); G. Bartoli and P. E. Todesco, *Acc. Chem. Res.* **10**, 125 (1977).
95. J. R. Beck, *Tetrahedron* **34**, 2057 (1978).
96. A. Chisari, E. Maccarone, G. Parisi, and G. Perrini, *J. Chem. Soc., Perkin Trans. 2*, 957 (1982).

can also be displaced.

$$O_2N-C_6H_3(NO_2)-SO_2Ph + HN(CH_2)_5 \longrightarrow O_2N-C_6H_3(NO_2)-N(CH_2)_5$$ Ref. 97

$$C_6H_3(NO_2)_2-CN + CH_3O^- \longrightarrow C_6H_3(NO_2)(OCH_3)-CN$$ Ref. 98

The addition intermediates can frequently be detected spectroscopically and sometimes can be isolated.[99] They are called *Meisenheimer complexes.* Especially in the case of adducts stabilized by nitro groups, the intermediates are often strongly colored.

$$Nu^- \; CH_3O-C_6H_4-NO_2 \longrightarrow (CH_3O)(Nu)C_6H_4{=}\overset{+}{N}(O^-)_2 \longrightarrow Nu-C_6H_4-NO_2 + CH_3O^-$$

The range of nucleophiles that have been observed to participate in nucleophilic aromatic substitution is similar to those which participate in S_N2 reactions and includes alkoxides,[100] phenoxides,[101] sulfides,[102] fluoride ion,[103] and amines.[104] Substitution by carbanions is somewhat less common. This may be because there are frequently complications resulting from electron transfer processes, especially with nitroaromatics.

Solvent effects on nucleophilic aromatic substitutions are similar to those discussed for S_N2 reactions. Dipolar aprotic solvents,[105] crown ethers,[106] and phase

97. J. F. Bunnett, E. W. Garbisch, Jr., and K. M. Pruitt, *J. Am. Chem. Soc.* **79**, 385 (1957).
98. J. R. Beck, R. L. Sobczak, R. G. Suhr, and J. A. Vahner, *J. Org. Chem.* **39**, 1839 (1974).
99. E. Buncel, A. R. Norris, and K. E. Russel, *Q. Rev. Chem. Soc.* **22**, 123 (1968); M. J. Strauss, *Chem. Rev.* **70**, 667 (1970); C. F. Bernasconi, *Acc. Chem. Res.* **11**, 147 (1978).
100. J. P. P. Idoux, M. L. Madenwald, B. S. Garcia, D. L. Chu, and J. T. Gupton, *J. Org. Chem.* **50**, 1876 (1985).
101. R. O. Brewster and T. Groening, *Org. Synth.* **II**, 445 (1943).
102. M. T. Bogert and A. Shull, *Org. Synth* **I**, 220 (1941); N. Kharasch and R. B. Langford, *Org. Synth.* **V**, 474 (1973); W. P. Reeves, T. C. Bothwell, J. A. Rudis, and J. V. McClusky, *Synth. Commun.* **12**, 1071 (1982).
103. W. M. S. Berridge, C. Crouzel, and D. Comar, *J. Labelled Compd. Radiopharm.* **22**, 687 (1985).
104. H. Bader, A. R. Hansen, and F. J. McCarty, *J. Org. Chem.* **31**, 2319 (1966); F. Pietra and F. Del Cima, *J. Org. Chem.* **33**, 1411 (1968); J. F. Pilichowski and J. C. Gramain, *Synth. Commun.* **14**, 1247 (1984).
105. F. Del Cima, G. Biggi, and F. Pietra, *J. Chem. Soc., Perkin Trans. 2,* 55 (1973); M. Makosza, M. Jagusztyn-Grochowska, M. Ludwikow, and M. Jawdosiuk, *Tetrahedron* **30**, 3723 (1974); M. Prato, U. Quintily, S. Salvagno, and G. Scorrano, *Gazz. Chim. Ital.* **114**, 413 (1984).
106. J. S. Bradshaw, E. Y. Chen, R. H. Holes, and J. A. South, *J. Org. Chem.* **37**, 2051 (1972); R. A. Abramovitch and A. Newman, Jr., *J. Org. Chem.* **39**, 2690 (1974).

transfer catalysts[107] can all enhance the rate of substitution by providing the nucleophile in a reactive state with weak solvation.

One of the most significant examples of aromatic nucleophilic substitution is the reaction of amines with 2,4-dinitrofluorobenzene. This reaction was used by Sanger[108] to develop a method for identification of the N-terminal amino acid in a protein.

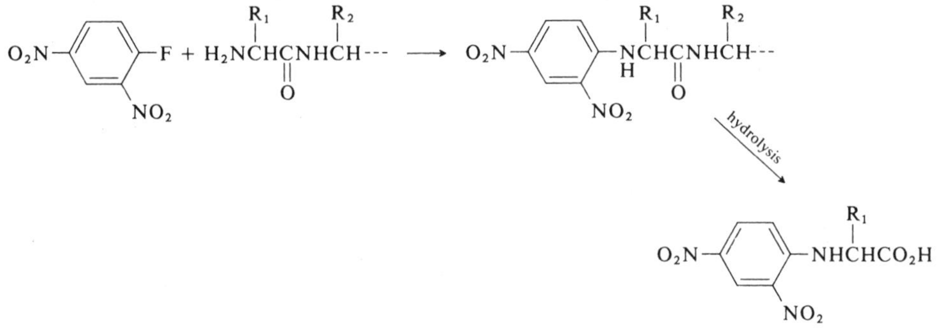

This process opened the way for structural characterization of proteins and other biopolymers.

Particularly detailed mechanistic studies have been carried out on the reaction of amines with nitroaryl halides and nitroaryl ethers. Such reactions can be formulated in terms of the general addition–elimination mechanism.

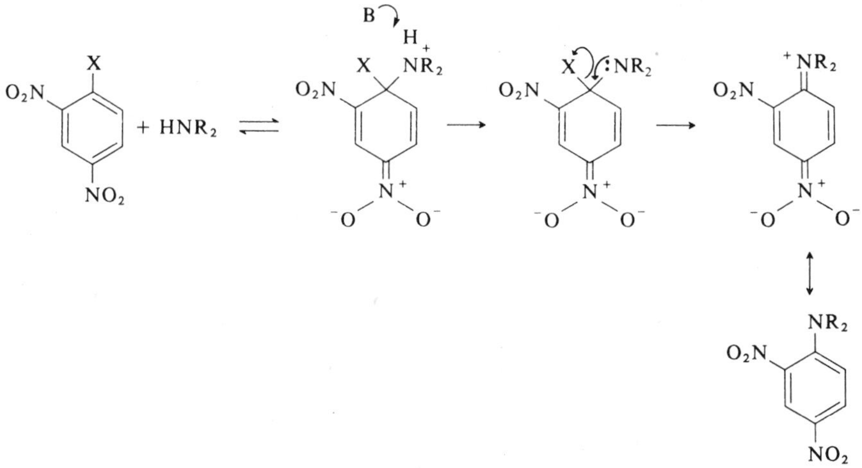

Detailed kinetic results can clarify the importance of the deprotonation step in the mechanism. It is sometimes the rate-determining step.[109]

107. M. Makosza, M. Jagusztyn-Grochowska, M. Ludwikow, and M. Jawdosiuk, *Tetrahedron* **30**, 3723 (1974).
108. F. Sanger, *Biochem. J.* **45**, 563 (1949).
109. C. F. Bernasconi, R. H. de Rossi, and P. Schmid, *J. Am. Chem. Soc.* **99**, 4090 (1977); E. Buncel, C. Innis, and I. Onyido, *J. Org. Chem.* **51**, 3680 (1986).

The pyridine family of heteroaromatic nitrogen compounds are reactive toward nucleophilic substitution at the C-2 and C-4 positions. The nitrogen atom serves to activate the ring toward nucleophilic attack by stabilizing the addition intermediate. This kind of substitution reaction is especially important in the chemistry of pyrimidines.

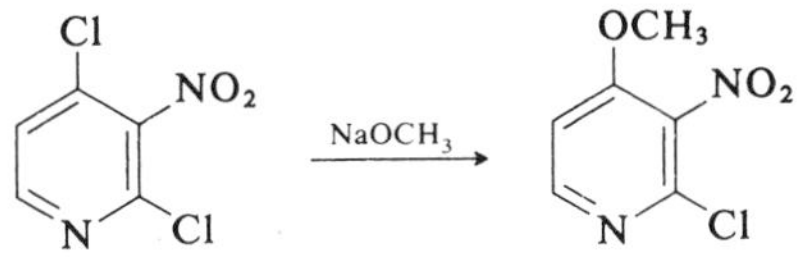

Ref. 110

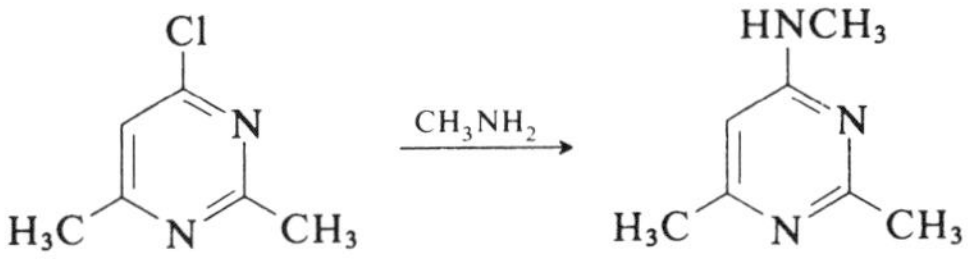

Ref. 111

A variation of the aromatic nucleophilic substitution process in which the leaving group is part of the entering nucleophile has been developed and is called *vicarious nucleophilic aromatic substitution.*

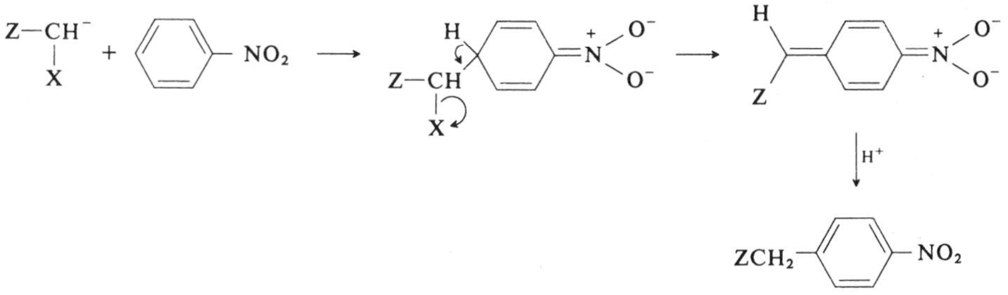

The combinations Z = CN, RSO_2, CO_2R, and SR and X = F, Cl, Br, I, ArO, ArS, and $(CH_3)_2NCS_2$ are among those that have been demonstrated.[112]

10.6. Nucleophilic Aromatic Substitution by the Elimination–Addition Mechanism

The elimination-addition mechanism involves a highly unstable intermediate, which is referred to as *dehydrobenzene* or *benzyne.*[113]

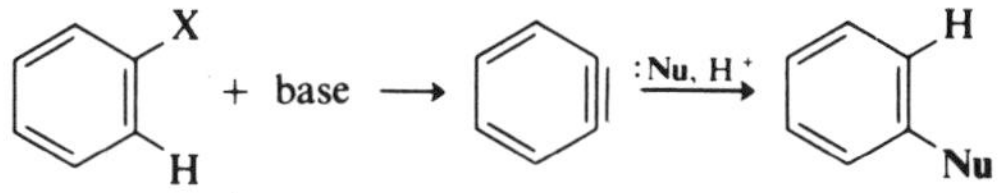

110. J. A. Montgomery and K. Hewson, *J. Med. Chem.* **9**, 354 (1966).
111. D. J. Brown, B. T. England, and J. M. Lyall, *J. Chem. Soc., C*, 226 (1966).
112. M. Makosza and J. Winiarski, *J. Org. Chem.* **45**, 1574 (1980); M. Makosza, J. Golinski, and J. Baron, *J. Org. Chem.* **49**, 1488 (1984); M. Makosza and J. Winiarski, *J. Org. Chem.* **49**, 1494 (1984); M. Makosza and J. Winiarski, *J. Org. Chem.* **49**, 5272 (1984).
113. R. W. Hoffmann, *Dehydrobenzene and Cycloalkynes*, Academic Press, New York, 1967.

A characteristic feature of this mechanism is the substitution pattern in the product. The entering nucleophile need not always enter at the carbon to which the leaving group was bound.

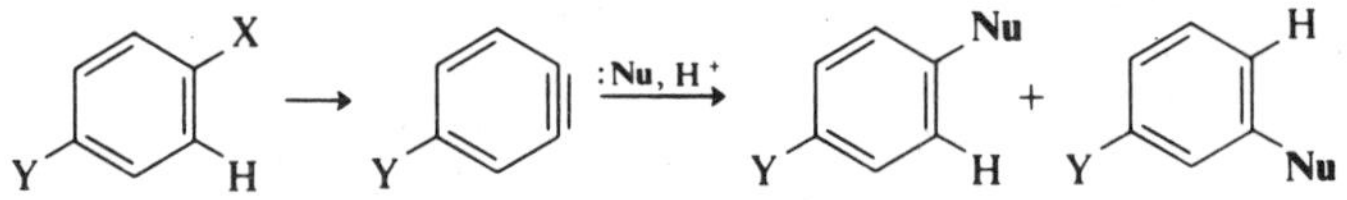

Benzyne has been observed spectroscopically in an inert matrix at very low temperatures.[114] For these studies the molecule was generated photolytically.

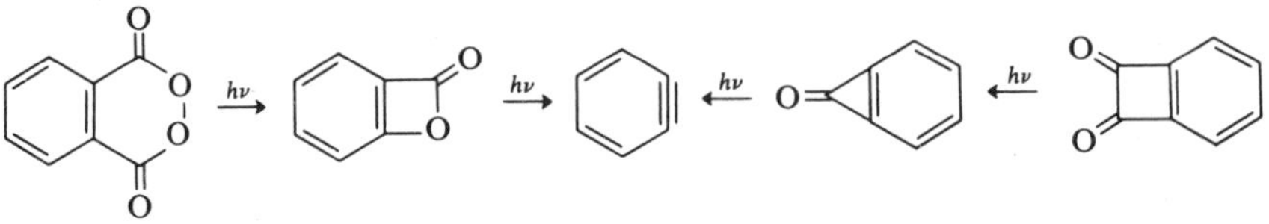

There have been several representations of the bonding in benzyne. The one most generally used pictures benzyne as being similar to benzene, but with an additional weak bond in the plane of the ring formed by overlap of the two sp^2 orbitals.[115] Molecular orbital calculations indicate that there is additional bonding between the "dehydro" carbons, though the strength of the bond is much less than that of a normal triple bond.[116]

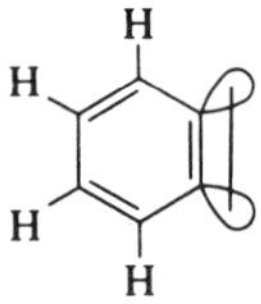

Analysis of the infrared spectrum of the matrix-isolated species gives a bond length of 1.35 Å, about 0.05 Å shorter than a normal benzene C—C bond length.[117]

An early case in which the existence of benzyne as a reaction intermediate was established was in the reaction of chlorobenzene with potassium amide. Carbon-14 label in the starting material was found to be distributed in the aniline as expected for a benzyne intermediate.[118]

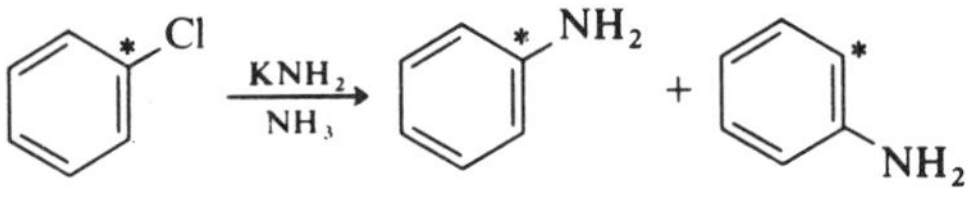

114. O. L. Chapman, K. Mattes, C. L. McIntosh, J. Pacansky, G. V. Calder, and G. Orr, *J. Am. Chem. Soc.* **95**, 6134 (1973).
115. H. E. Simmons, *J. Am. Chem. Soc.* **83**, 1657 (1961).
116. R. Hoffmann, A. Inamura, and W. J. Hehre, *J. Am. Chem. Soc.* **90**, 1499 (1968); D. L. Wilhite and J. L. Whitten, *J. Am. Chem. Soc.* **93**, 2858 (1971); J. O. Noell and M. D. Newton, *J. Am. Chem. Soc.* **101**, 51 (1979); C. W. Bock, P. George, and M. Trachtman, *J. Phys. Chem.* **88**, 1467 (1984).
117. J. W. Laing and R. S. Berry, *J. Am. Chem. Soc.* **98**, 660 (1976).
118. J. D. Roberts, D. A. Semenow, H. E. Simmons, Jr., and L. A. Carlsmith, *J. Am. Chem. Soc.* **78**, 601 (1956).

The elimination–addition mechanism is facilitated by electronic effects that favor removal of a hydrogen from the ring by strong base. Relative reactivity also depends on the halide. The order Br > I > Cl > F has been established in the reaction of aryl halides with KNH_2 in liquid ammonia.[119] This order has been interpreted as representing a balance between two effects. The inductive order favoring proton removal would be F > Cl > Br > I, but this is largely overwhelmed by the order of leaving group ability I > Br > Cl > F, which reflects bond strengths. With organometallic compounds as bases in aprotic solvents, the acidity of the hydrogen is the dominant factor and the reactivity order is F > Cl > Br > I.[120]

Addition of nucleophiles such as ammonia or alcohols or their conjugate bases to benzynes takes place very rapidly. These nucleophilic additions are believed to involve capture of the nucleophile, followed by protonation to give the substituted benzene.[121]

:Nu⁻ → Nu → (H−B) Nu, H + B⁻

The regiochemistry of the nucleophilic addition is influenced by an adjacent substituent. Electron-attracting groups tend to favor addition of the nucleophile at the more distant end of the "triple bond," since this permits maximum stabilization of the developing negative charge. Electron-donating groups have the opposite effect. Selectivity is usually not high, however, and formation of both possible products from monosubstituted benzynes is common.[122]

Nu EWG   ERG Nu

There are several methods for generation of benzyne in addition to base-catalyzed elimination of hydrogen halide from a halobenzene, and some of these are more generally applicable for preparative work. Probably the most convenient method is diazotization of *o*-aminobenzoic acids.[123] Concerted loss of nitrogen and carbon dioxide follows diazotization and generates benzyne. Benzyne can be formed in this manner in the presence of a variety of compounds with which it reacts rapidly.

CO_2H, NH_2 $\xrightarrow{HONO}$ $C(=O)-O^-$, $\overset{+}{N}\equiv N$ $\longrightarrow$ benzyne $+ CO_2 + N_2$

119. F. W. Bergstrom, R. E. Wright, C. Chandler, and W. A. Gilkey, *J. Org. Chem.* **1**, 170 (1936).
120. R. Huisgen and J. Sauer, *Angrw Chem.* **72**, 91 (1960).
121. J. F. Bunnett, D. A. R. Happer, M. Patsch, C. Pyun, and H. Takayama, *J. Am. Chem. Soc.* **88**, 5250 (1966); J. F. Bunnett and J. K. Kim. *J. Am. Chem. Soc.* **95**, 2254 (1973).
122. E. R. Biehl, E. Nieh, and K. C. Hsu, *J. Org. Chem.* **34**, 3595 (1969).
123. M. Stiles, R. G. Miller, and U. Burckhardt, *J. Am. Chem. Soc.* **85**, 1792 (1963); L. Friedman and F. M. Longullo, *J. Org. Chem.* **34**, 3595 (1969).

Oxidation of 1-aminobenzotriazole also serves as a source of benzyne under mild conditions. An oxidized intermediate decomposes with loss of two molecules of nitrogen.[124]

$$\longrightarrow \quad \longrightarrow \quad + 2N_2$$

Another heterocyclic molecule that can serve as a benzyne precursor is benzothiadiazole-1,1-dioxide, which decomposes with elimination of sulfur dioxide and nitrogen.[125]

$$\longrightarrow \quad + SO_2 + N_2$$

Benzyne can also be generated from *o*-dihaloaromatics. Reaction of lithium amalgam or magnesium results in formation of a transient organometallic compound that decomposes with elimination of lithium halide. 1-Bromo-2-fluorobenzene is the usual starting material in this procedure.[126]

$$\xrightarrow{\text{Li-Hg}} \quad \longrightarrow$$

Benzyne is capable of dimerizing, so that in the absence of either a nucleophile or a reactive unsaturated compound, biphenylene is formed.[127] The lifetime of benzyne is estimated to be on the order of a few seconds in solution near room temperature.[128]

When benzene is generated in the presence of unsaturated molecules, additions at the highly strained "triple bond" occur. Among the types of compounds which give Diels–Alder addition products are furans, cyclopentadienones, and anthracene.

Ref. 129

124. C. D. Campbell and C. W. Rees, *J. Chem. Soc., C*, 742, 752 (1969).
125. G. Wittig and R. W. Hoffmann, *Org. Synth.* **47**, 4 (1967); G. Wittig and R. W. Hoffmann, *Chem. Ber.* **95**, 2718, 2729 (1962).
126. G. Wittig and L. Pohmer, *Chem. Ber.* **89**, 1334 (1956); G. Wittig, *Org. Synth.* **IV**, 964 (1963).
127. F. M. Logullo, A. H. Seitz, and L. Friedman, *Org. Synth.* **48**, 12 (1968).
128. F. Gavina, S. V. Luis, and A. M. Costero, *Tetrahedron* **42**, 155 (1986).
129. G. Wittig and L. Pohmer, *Angew. Chem.* **67**, 348 (1955).

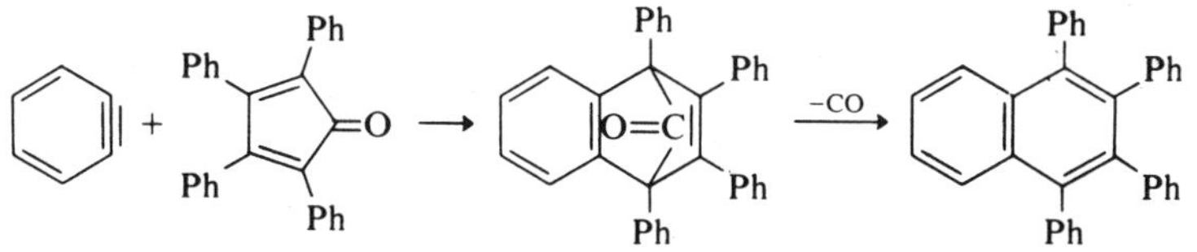

Ref. 130

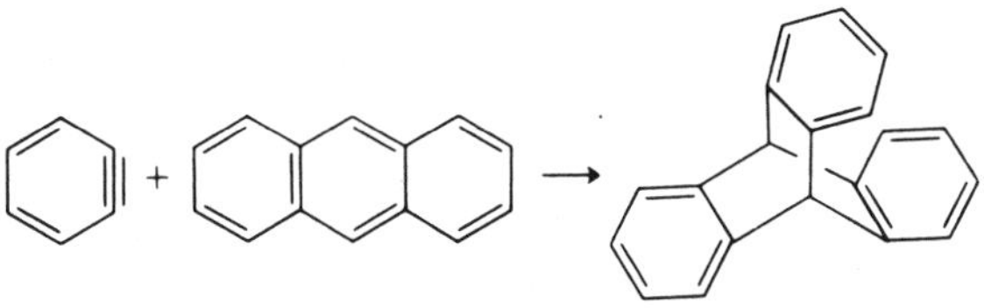

Ref. 131

General References

R. W. Hoffmann, *Dehydrobenzene and Cycloalkynes*, Academic Press, New York, 1967.
J. G. Hoggett, R. B. Moodie, J. R. Penton, and K. S. Schofield, *Nitration and Aromatic Reactivity*, Cambridge University Press, Cambridge, 1971.
C. K. Ingold, *Structure and Mechanism in Organic Chemistry*, Cornell University Press, Ithaca, New York, 1969, Chapter VI.
J. Miller, *Aromatic Nucleophilic Substitution*, Elsevier, Amsterdam, 1968.
R. O. C. Norman and R. Taylor, *Electrophilic Substitution in Benzenoid Compounds*, Elsevier, Amsterdam, 1965.
G. A. Olah, *Friedel-Crafts Chemistry*, Wiley, New York, 1973.
S. Patai (ed.), *The Chemistry of Diazonium and Diazo Groups*, Wiley, New York, 1978.
R. M. Roberts and A. A. Khalaf, *Friedel-Crafts Alkylation Chemistry*, Marcel Dekker, New York, 1984.
L. M. Stock, *Aromatic Substitution Reactions*, Prentice-Hall, Englewoods Cliffs, New Jersey, 1968.

Problems

(*References for these problems will be found on page* 782.)

1. Predict qualitatively the isomer ratio for the nitration of each of the following compounds.

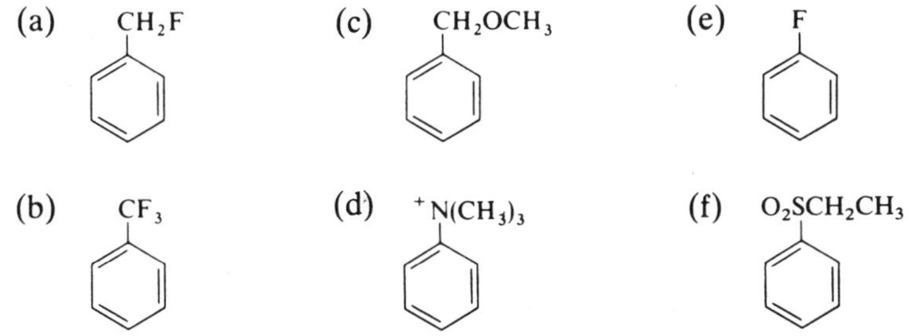

(a) CH_2F (c) CH_2OCH_3 (e) F

(b) CF_3 (d) $^+N(CH_3)_3$ (f) $O_2SCH_2CH_3$

130. L. F. Fieser and M. J. Haddadin, *Org. Synth.* **46**, 107 (1966).
131. L. Friedman and F. M. Logullo, *J. Org. Chem.* **34**, 3089 (1969).

2. While *N,N*,-dimethylaniline is an extremely reactive aromatic substrate and is readily attacked by such weak electrophiles as aryl diazonium ions and nitrosyl ion, this reactivity is greatly diminished by introduction of an alkyl substituent in the *ortho* position. Explain.

3. Toluene is 17 times more reactive than benzene and isopropylbenzene is 14 times more reactive than benzene when nitration is carried out in the organic solvent sulfolane. The $o:m:p$ ratio for toluene is 62:3:35, and for isopropylbenzene it is 43:5:52. Calculate the partial rate factors for each position in toluene and isopropylbenzene. Discuss the significance of the partial rate factors. Compare the reactivity at the various positions of each molecule, and explain any differences you consider to be significant.

4. Some bromination rate constants are summarized below. Compare the correlation of the rate data with σ and σ^+ substituent constants. What is the value of ρ? What is the mechanistic significance of these results?

X–C₆H₅ + Br₂ ⟶ X–C₆H₄–Br + HBr

X	k ($M^{-1}\,\mathrm{sec}^{-1}$)
H	2.7×10^{-6}
CH_3	1.5×10^{-2}
OCH_3	9.8×10^{3}
OH	4.0×10^{4}
$N(CH_3)_2$	2.2×10^{8}

5. Compare the results given below for the alkylation of *p*-xylene under a variety of conditions. Explain the reasons for the variation in product composition with temperature and with the use of 1- versus 2-propyl chloride.

A: H_3C, CH_3, $CH(CH_3)_2$ substituted benzene; **B**: H_3C, CH_3, $(CH_2)_2CH_3$ substituted benzene; **C**: H_3C, CH_3, $CH(CH_3)_2$ substituted benzene

		A	B	C
1-propyl chloride	0°C	27%	73%	0%
1-propyl chloride	50°C	31%	53%	16%
2-propyl chloride	0°C	100%	0%	0%
2-propyl chloride	50°C	62%	0%	38%

6. The table below gives first-order rate constants for reaction of substituted benzenes with *m*-nitrobenzenesulfonyl peroxide. From these data calculate the overall relative reactivity and partial rate factors. Does this reaction fit the pattern of an electrophilic aromatic substitution? If so, does the active electrophile exhibit low, moderate, or high substrate and position selectivity?

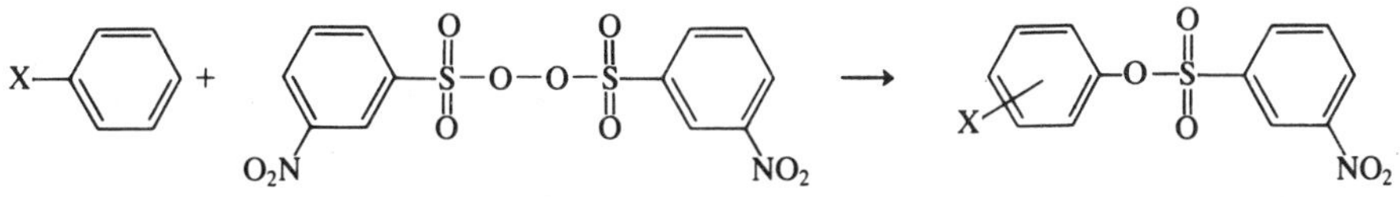

X	k (sec^{-1})	Product composition		
		o	*m*	*p*
H	8.6×10^{-5}	—	—	—
Br	4.8×10^{-5}	21	3	76
CH_3	1.7×10^{-3}	32	3	65
CH_3O	4.3×10^{-2}	14	0	86
CH_3O_2C	9.1×10^{-6}	24	67	9

7. Give the products to be expected from each of the following reactions:

(a) OH, Br_2, $CHCl_3$ (b) I, CH_3, $SO_3H + H_2O$, H_2SO_4, $HgSO_4$

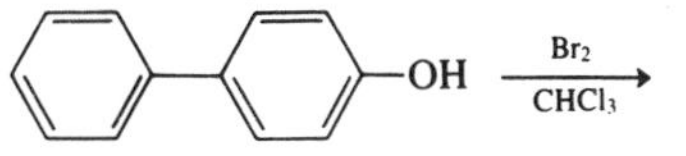

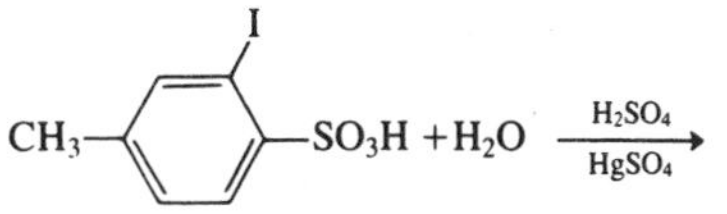

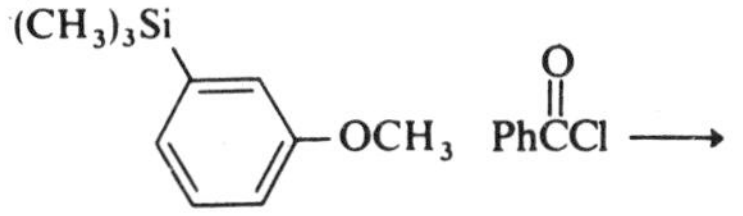

(c) $(CH_3)_3Si$, OCH_3, PhCCl (d) O, H^+

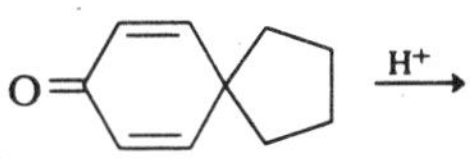

8. In 100% sulfuric acid, the cyclization shown below occurs:

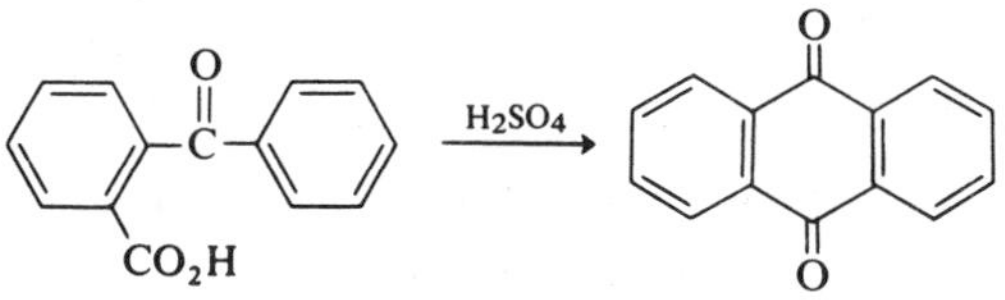

O, C, CO_2H, H_2SO_4, O, O

When one of the *ortho* hydrogens is replaced by deuterium, the rate drops from $1.53 \times 10^{-4}\ s^{-1}$ to $1.38 \times 10^{-4}\ s^{-1}$. What is the kinetic isotope effect? The product from such a reaction contains 60% of the original deuterium. Give a mechanism for this reaction that is consistent with both the kinetic isotope effect and the deuterium retention data.

9. Reaction of 3,5,5-trimethyl-2-cyclohexen-1-one with $NaNH_2$ (3 equiv) in THF generates its enolate. When bromobenzene is then added to this solution and stirred for 4 h, the product **A** is isolated in 30% yield. Formulate a mechanism for this transformation.

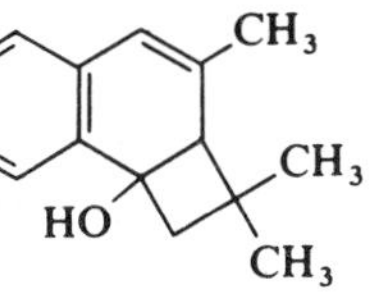

CH_3, CH_3, HO, CH_3, **A**

10. Various phenols can be selectively hydroxymethylated at the *ortho* position by heating with paraformaldehyde and phenylboronic acid.

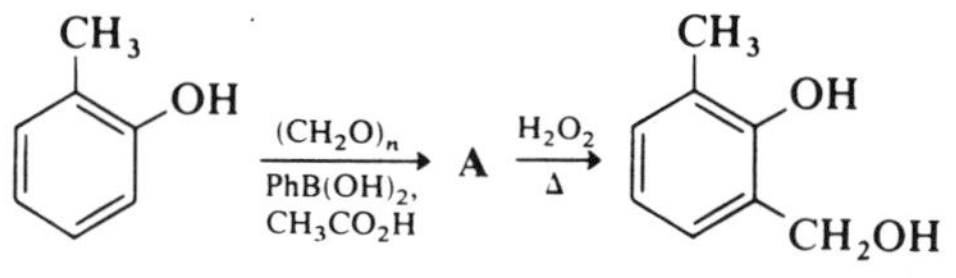

$$\text{2-methylphenol} \xrightarrow[\text{PhB(OH)}_2,\ CH_3CO_2H]{(CH_2O)_n} \mathbf{A} \xrightarrow[\Delta]{H_2O_2} \text{2-methyl-6-(hydroxymethyl)phenol}$$

An intermediate **A**, having the formula $C_{14}H_{13}O_2B$ for the case above, can be isolated after the first step. Postulate a structure for the intermediate and comment on its role in the reaction.

11. When compound **B** is dissolved in FSO_3H at −78°C, NMR shows that a carbocation is formed. If the solution is then allowed to warm to −10°C, a different ion forms. The first ion gives compound **C** when quenched with base, while the second gives **D**. What are the structures of the two carbocations, and why do they give different products on quenching?

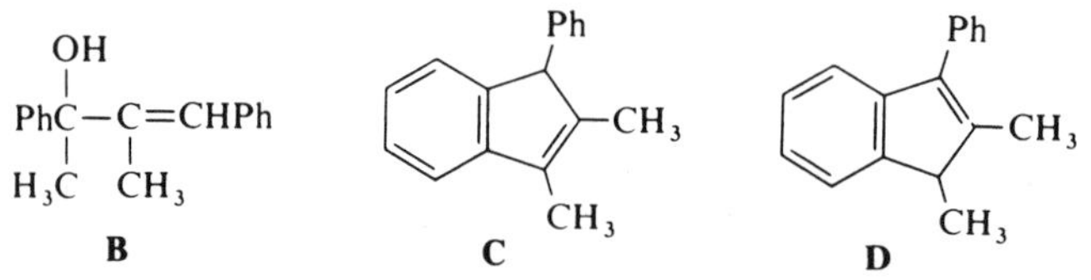

B: PhC(OH)(CH₃)–C(CH₃)=CHPh

C: 1-phenyl-2,3-dimethylindene (Ph, CH_3, CH_3)

D: 3-phenyl-1,2-dimethylindene (Ph, CH_3, CH_3)

12. Alkyl groups which are *para* to strong π-donor substituents such as hydroxy or methoxy can be removed from aromatic rings under acidic conditions, if the alkyl group is capable of forming a stable carbocation:

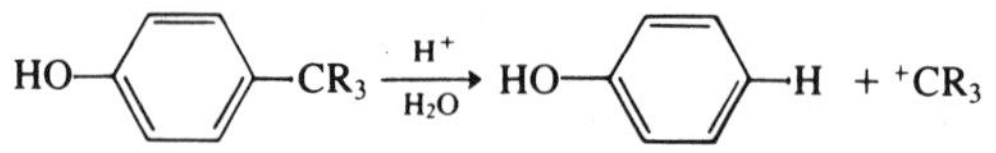

$$HO\text{–}C_6H_4\text{–}CR_3 \xrightarrow[H_2O]{H^+} HO\text{–}C_6H_4\text{–}H + {}^+CR_3$$

For the equation above, when R = CH_3, the solvent isotope effect is $k_H/k_D = 0.1$. When R = Ph, $k_H/k_D = 4.3$. How do you account for the difference in the isotope effect for the two systems, and, particularly, what is the probable cause of the inverse isotope effect in the case of R = CH_3?

13. Acylation of 1,4-dimethoxynaphthalene with acetic anhydride (1.2 equiv) and aluminum chloride (2.2 equiv) in ethylene dichloride (60°C, 3 h) gives two products, 6-acetyl-1,4-dimethoxynaphthalene (30%) and 1-hydroxy-2-acetyl-4-methoxynaphthalene (50%). Suggest a rationalization for the formation of these two products and in particular for the differing site of substitution in the two products.

14. The solvolysis of 4-arylbutyl arenesulfonates in non-nucleophilic media leds to the formation of tetralins:

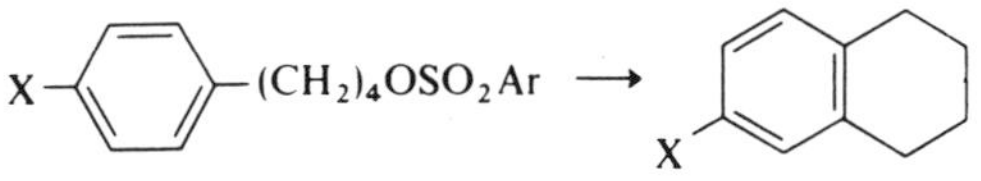

$$X-C_6H_4-(CH_2)_4OSO_2Ar \longrightarrow$$

Two σ intermediates are conceivable. **A** would lead directly to product on deprotonation, while **B** could give product by rearrangement to **A**, followed by deprotonation:

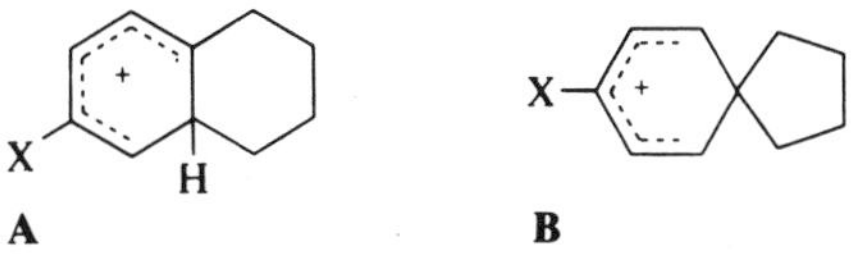

X H
A

X
B

Devise an experiment that would permit one to determine how much product is formed via **A** and how much via **B**. How would you expect the relative importance of the alternative routes to be related to the identity of the substituent group X?

15. The complex kinetic expression for chlorination of anisole by hypochlorous acid (p. 566) becomes simpler for both less reactive and more reactive substrates. For benzene, the expression is

$$\text{rate} = k[\text{benzene}][\text{ClOH}][\text{H}^+]$$

For p-dimethyoxybenzene, it is

$$\text{rate} = k[\text{ClOH}][\text{H}^+]$$

What is the reason for this dependence of the form of the rate expression on the reactivity of the aromatic compound?

16. The reactivities of chlorobenzene and bromobenzene relative to benzene are 0.033 and 0.030, respectively, using acetyl nitrate as the nitration reagent. The product ratios are: chlorobenzene, $o:m:p$, 30%, 1%, 69%; bromobenzene, $o:m:p$, 37%, 1%, 62%. Calculate the partial rate factors.

17. The chlorination of a series of compounds having electron-withdrawing substituents has been studied. The relative rates of chlorination and the isomer distribution are known. The data give a satisfactory correlation with the Hammett equation using σ^+, but no rate measurement for benzene under precisely comparable conditions is possible. How could you estimate f_o, f_m, and f_p for chlorination from the available data?

$$\rho = -6.6$$

$$o:m:p \text{ ratio for benzonitrile} = 34:55:11$$

18. *Ipso* substitution, in which the electrophile attacks a position already carrying a substituent, is relatively rare in electrophilic aromatic substitution and was not explicitly covered in Section 10.2 in the discussion of substituent effects on reactivity and selectivity. Using qualitative MO concepts, discuss the effect of the following types of substituents on the energy of the transition state for *ipso* substitution.

(a) A π-donor substituent which is more electronegative than carbon, e.g., F or CH_3O.
(b) A π-acceptor substituent which is more electronegative than carbon, e.g., NO_2 or CN.
(c) A group without a strong π-conjugating capacity which is more electronegative than carbon, e.g., $^{+}N(CH_3)_3$.
(d) A group without a strong π-conjugating capacity which is less electronegative than carbon, e.g., $(CH_3)_3Si$.

According to this analysis, which types of groups will most favor *ipso* substitution? Can you cite any experimental evidence to support this conclusion?

19. The nitration of 2,4,6-tri-*t*-butyltoluene gives rise to three products. The distribution is changed when the 3- and 5-positions are deuterated:

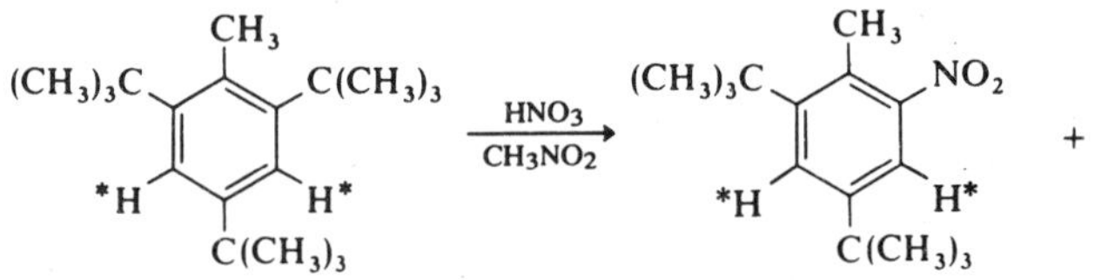

CH3 (CH3)3C C(CH3)3 *H H* C(CH3)3 —HNO3 / CH3NO2→ CH3 (CH3)3C NO2 *H H* C(CH3)3 +

*H=H 40.3%
*H=D 42.4%

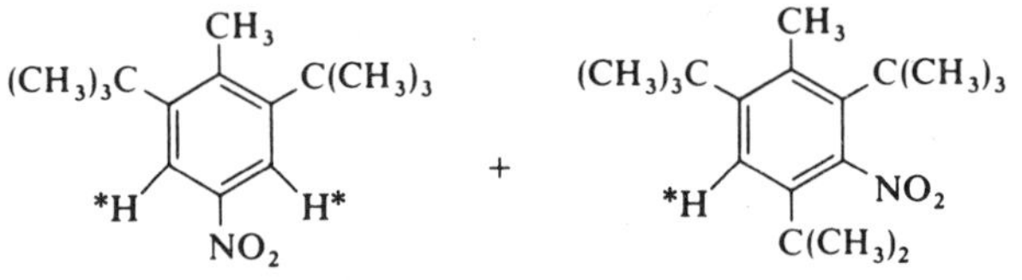

CH3 (CH3)3C C(CH3)3 *H H* NO2 + CH3 (CH3)3C C(CH3)3 *H NO2 C(CH3)2

*H=H	51.0%	*H=H	8.7%
*H=D	54.6%	*H=D	2.7%

Indicate mechanisms that would account for the formation of each product. Show how the isotopic substitution could cause a change in product composition. Does your mechanism predict that the isotopic substitution would give rise to a primary or secondary deuterium kinetic isotope effect? Calculate the magnitude of the kinetic isotope effect from the data given.

20. (a) Under several reaction conditions designed to determine the products of cyclization under Friedel–Crafts conditions, six-membered cyclic products were found to be favored over seven-membered ring products. Write a

detailed mechanism for each of the reactions shown below, and comment on the significance of the apparently general preference for formation of a six-membered ring over a seven-membered ring.

H_3C CH_3 Cl CH_3 $\xrightarrow[0°C,\ 4\ h]{FeCl_3,\ CH_3NO_2}$ H_3C CH_3 CH_2CH_3

H_3C CH_3 OH $-CH_3$ CH_3 $\xrightarrow[10°C,\ 1\ h]{HF,\ CCl_4}$ H_3C CH_3 $CH(CH_3)_2$

H_3C CH_3 Cl $-CH_3$ C_2H_5 $\xrightarrow[0°C,\ 1\ h]{AlCl_3,\ CS_2}$ H_3C CH_3 $CH_3CHCH_2CH_3$

(b) Examine the data below for cyclization of a variety of phenylalkanols in 85% H_3PO_4 at elevated temperatures. What general conclusions do you draw about the preferences for ring closure (as a function of ring size) under these conditions?

(1) $PhCH_2CH_2CH_2OH$ $\xrightarrow[>200°C]{H_3PO_4}$ mainly isomeric phenylpropenes (89%)

(2) $PhC(CH_2)(CH_3)CH_2CH_2OH$ $\xrightarrow[>200°C]{H_3PO_4}$ mainly 2-methyl-3-phenyl-2-butene (82%) plus some 1,1-dimethylindane (18%)

(3) $PhCH_2CH_2CH_2CH_2OH$ $\xrightarrow[>200°C]{H_3PO_4}$ mainly tetralin (80%)

(4) $PhCH_2CH_2CH(OH)CH_3$ $\xrightarrow[>200°C]{H_3PO_4}$ mainly phenylbutenes (100%)

(5) $m\text{-}CH_3PhCH_2CH_2C(CH_3)_2OH$ $\xrightarrow[>200°C]{H_3PO_4}$ mainly 1,1,5- and 1,1,7-trimethylindane

21. Explain each of the following reaction processes by presenting a detailed stepwise mechanism to show how the observed products are formed.

(a) The reaction of 2,6-di-*t*-butylphenoxide with *o*-nitroaryl halides gives 2,6-di-*t*-butyl-4-(*o*-nitrophenyl)phenols in 60%–90% yield. 1,4-Dinitrobenzene reacts under similar conditions to give 2,6-di-*t*-butyl-4-(*p*-nitrophenyl)phenol.

(b) 2-(3-Chlorophenyl)-4,4-dimethyloxazoline reacts with alkyllithium reagents to give 2-(2-alkylphenyl)-4,4-dimethyloxazolines.

(c) Nitrobenzene reacts with cyanomethyl phenyl sulfide in the presence of sodium hydroxide in dimethyl sulfoxide to give a mixture of 2- and 4-nitrophenylacetonitrile.

(d) N_2^+ CO_2^- + O $\xrightarrow{80°C}$ O

(e) Reaction of benzene with 3,3,3-trifluoropropene in the presence of aluminum chloride and a trace of moisture gives 3,3,3-trifluoropropylbenzene.

11

Concerted Reactions

There are many reactions in organic chemistry that give no evidence of involving intermediates when they are subjected to the usual probes for studying reaction mechanisms. Highly polar transition states do not seem to be involved either, since the rates of the reactions are insensitive to solvent polarity. Efforts to detect free-radical intermediates by physical or chemical means have not been successful, and the reaction rates are neither increased by initiators nor decreased by inhibitors of free-radical reactions. This lack of evidence for intermediates leads to the conclusion that the reactions are single-step processes in which bond making and bond breaking both contribute to the structure at the transition state, although not necessarily to the same degree. Such processes are called *concerted reactions.* There are numerous examples of both unimolecular and bimolecular concerted reactions.

An important class of concerted reactions is the *pericyclic reactions.*[1] A pericyclic reaction is characterized as a change in bonding relationship that takes place as a continuous concerted reorganization of electrons. The word "concerted" specifies that there is a single transition state and therefore no intermediates are involved in the process. To maintain continuous electron flow, pericyclic reactions occur through *cyclic transition states.* Furthermore, the cyclic transition state must correspond to an arrangement of the participating orbitals that can maintain a bonding interaction between the reaction components throughout the course of the reaction. We shall see shortly that these requirements make pericyclic reactions highly predictable, in terms of such features as relative reactivity, stereospecificity, and regioselectivity.

The key to understanding the mechanism of the pericyclic reactions was the recognition by Woodward and Hoffmann[2] that the pathways of such reactions were determined by the symmetry properties of the orbitals that are directly involved. Their recognition that the symmetry of each participating orbital must be conserved

1. R. B. Woodward and R. Hoffmann, *The Conservation of Orbital Symmetry*, Academic Press, New York, 1970.
2. R. B. Woodward and R. Hoffmann, *J. Am. Chem. Soc.* **87**, 395 (1965).

during the concerted process dramatically transformed the understanding of this family of reactions and stimulated much experimental work to test and extend their theories.[3] The success of the theory emphasized the potential that systematic analysis of orbital properties had for deepening the understanding of organic reaction mechanisms. Woodward and Hoffmann's approach led to other related interpretations of orbital properties that are also successful in predicting and explaining the course of concerted thermal reactions.[4]

11.1. Electrocyclic Reactions

There are several general classes of pericyclic reactions for which orbital symmetry factors determine both the stereochemistry and the relative reactivity. The first class that we will consider is the *electrocyclic reactions.* An electrocyclic reaction is defined as the formation of a single bond between the ends of a linear system of π electrons and the reverse process. An example is the thermal ring-opening of cyclobutenes to butadienes:

CH_3 H CH_3 H —175°C→ H H CH_3 H CH_3 H CH_3 H H CH_3 —175°C→ H H H CH_3 H CH_3 Ref. 5

It is not surprising that thermolysis of cyclobutenes leads to ring opening since the strain in the four-membered ring is relieved. The activation energy for simple alkyl-substituted cyclobutenes is in the range of 30–35 kcal/mol.[6] *What is particularly significant about these reactions is that they are stereospecific. cis*-3,4-Dimethylcyclobutene is converted to *E,Z*-2,5-hexadiene, while *trans*-3,4-dimethylcyclobutene yields the *E,E*-isomer. The stereospecificity of such processes is very high. In the ring opening of *cis*-3,4-dimethylcyclobutene, for example, only 0.005% of the minor product, *E,E*-2,4-hexadiene, is formed.[7]

The reason for the observed stereospecificity is that the groups bonded to the breaking bond all rotate in the same sense during the ring-opening process. Such motion, in which either all the substituents rotate clockwise or all rotate counterclockwise is called the *conrotatory* mode. When such motion is precluded by some structural feature, ring opening requires a higher temperature. In the bicycloheptene shown below, the five-membered ring prevents completion of a conrotatory ring

3. For reviews of several concerted reactions within the general theory of pericyclic reactions, see A. P. Marchand and R. E. Lehr (eds.), *Pericyclic Reactions*, Vols. I and II, Academic Press, New York, 1977.
4. H. C. Longuet-Higgins and E. W. Abrahamson, *J. Am. Chem. Soc.* **87**, 2045 (1965); M. J. S. Dewar, *Angew. Chem. Int. Ed. Engl.* **10**, 761 (1971); M. J. S. Dewar, *The Molecular Orbital Theory of Organic Chemistry*, McGraw-Hill, New York, 1969; H. E. Zimmerman, *Acc. Chem. Res.* **4**, 272 (1971).
5. R. E. K. Winter, *Tetrahedron Lett.*, 1207 (1965).
6. W. Kirmse, N. G. Rondan, and K. N. Houk, *J. Am. Chem. Soc.* **106**, 7989 (1984).
7. J. I. Brauman and W. C. Archie, Jr., *J. Am. Chem. Soc.* **94**, 4262 (1972).

opening because it would lead to a *cis,trans*-cycloheptadiene. The reaction takes place only at very high temperature, 400°C, and probably involves the diradical shown as an intermediate.

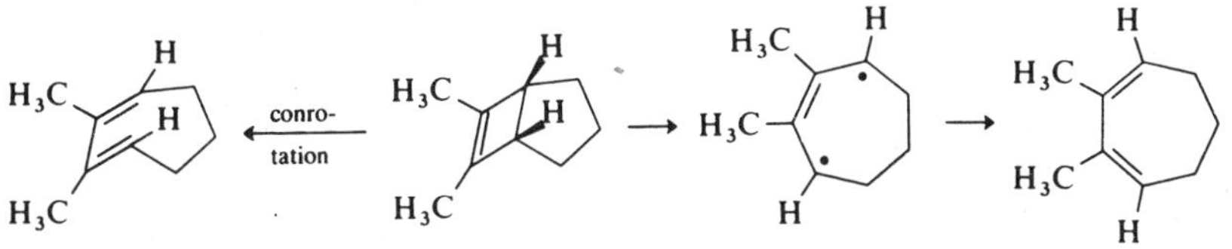

Ref. 8

The principle of microscopic reversibility requires that the reverse process, ring closure of a butadiene to a cyclobutene, must also be a conrotatory process. Usually, this is thermodynamically unfavorable, but a case in which the ring closure is energetically favorable is conversion of *trans,cis*-2,4-cyclooctadiene (**1**) to bicyclo[4.2.0]oct-7-ene (**2**). The ring closure is favorable in this case because of the strain associated with the *trans* double bond.

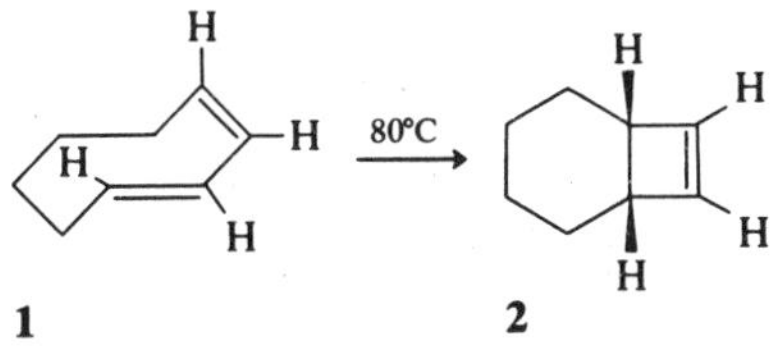

Ref. 9

Electrocyclic reactions of 1,3,5-trienes lead to 1,3-cyclohexadienes. These ring closures also exhibit a high degree of stereospecificity. The ring closure is normally the favored reaction in this case because of the greater thermodynamic stability of the cyclic compound, which has six σ bonds and two π bonds, compared with five σ and three π bonds for the triene. The stereospecificity is illustrated with octatrienes **3** and **4**. *E,Z,E*-2,4,6-Octatriene (**3**) cyclizes only to *cis*-5,6-dimethyl-1,3-cyclohexadiene, while the *E,Z,Z*-2,4,6-octatriene (**4**) leads exclusively to the *trans* cyclohexadiene isomer.[10]

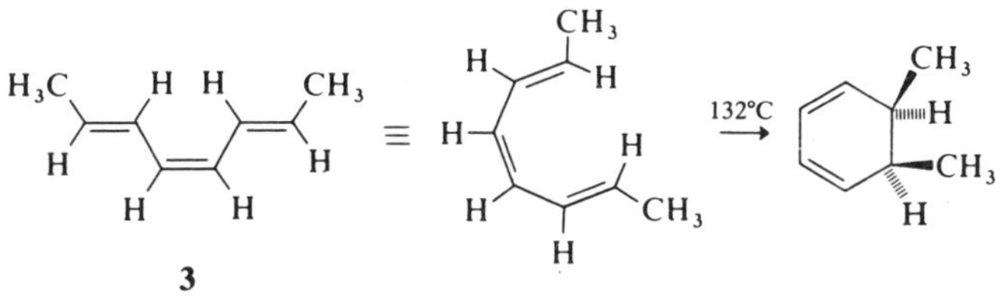

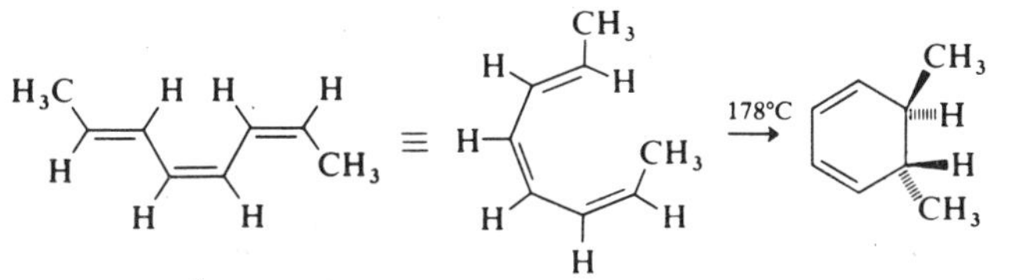

8. R. Criegee and H. Furrer, *Chem. Ber.* **97**, 2949 (1964).
9. K. M. Schumate, P. N. Neuman, and G. J. Fonken, *J. Am. Chem. Soc.* **87**, 3996 (1965); R. S. H. Liu, *J. Am. Chem. Soc.* **89**, 112 (1967).
10. E. N. Marvell, G. Caple, and B. Schatz, *Tetrahedron Lett.*, 385 (1965); E. Vogel, W. Grimme, and E. Dinne, *Tetrahedron Lett.*, 391 (1965); J. E. Baldwin and V. P. Reddy, *J. Org. Chem.* **53**, 1129 (1988).

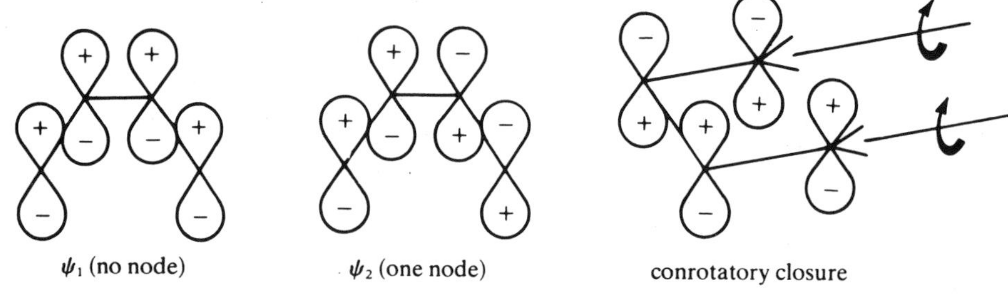

Fig. 11.1. Symmetry properties for the π system of a conjugated diene.

A point of particular importance regarding the stereochemistry of this reaction is that the groups at the termini of the triene system rotate in the opposite sense during the cyclization process. This mode of electrocyclic reaction is called *disrotatory.*

A complete mechanistic description of these reactions must explain not only their high degree of stereospecificity, but also why four-π-electron systems undergo conrotatory reactions whereas six-π-electron systems undergo disrotatory reactions. Woodward and Hoffmann proposed that the stereochemistry of the reactions is controlled by the symmetry properties of the highest occupied molecular orbital (HOMO) of the reacting system.[11] The idea that the HOMO should control the course of the reaction is an example of frontier orbital theory, which holds that it is the electrons of highest energy, that is, those in the HOMO, that are of prime importance.[12] The occupied orbitals of the π system of 1,3-butadiene are shown in Fig. 11.1. The HOMO is ψ_2.

Why do the symmetry properties of ψ_2 determine the stereochemistry of the electrocyclic reaction? For convenience, let us examine the microscopic reverse of the ring opening. The stereochemical feature of the reaction will be the same in either the forward or the reverse direction of the reaction. For bonding to occur between C-1 and C-4 of the conjugated system, the positive lobe on C-1 must overlap with the positive lobe on C-4 (or negative with negative, since the signs are arbitrary). This overlap of lobes of the same sign can be accomplished only by a conrotatory motion. Disrotatory motion causes overlap of orbitals of opposite sign. This is an antibonding overlap and would preclude bond formation. Other conjugated dienes will have identical orbital symmetries so that the conrotatory mode will be preferred for all thermal electrocyclic processes of 1,3-dienes.

The π orbitals for the 1,3,5-triene system are shown in Fig. 11.2. The analysis according to frontier orbital theory proceeds in the same way but leads to the conclusion that a bonding interaction between C-1 and C-6 of the triene will require a disrotatory motion. This is because the HOMO, ψ_3, has positive lobes on the same face of the π system, and these must come together to permit bond formation. The

11. R. B. Woodward and R. Hoffmann, *J. Am. Chem. Soc.* **87**, 395 (1965).
12. K. Fukui and H. Fujimoto, in *Mechanisms of Molecular Migrations*, Vol. 2, B. S. Thyagarajan (ed.), Interscience, New York, 1968, p. 117; K. Fukui, *Acc. Chem. Res.* **4**, 57 (1971); K. Fukui, *Angew. Chem. Int. Ed. Engl.* **21**, 801 (1982).

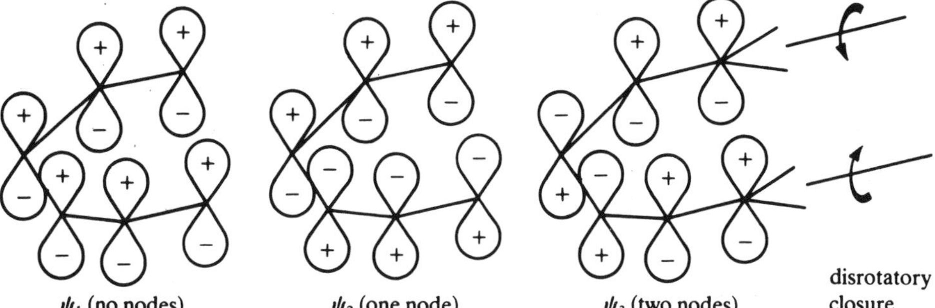

Fig. 11.2. Symmetry properties of hexatriene molecular orbitals.

symmetry properties of other six-π-electron triene systems will be the same, so we expect disrotatory ring closure (or opening) to be observed.

When we recall the symmetry patterns for linear polyenes which were discussed in Chapter 1, we can further generalize the predictions based on the symmetry of the polyene HOMO. Systems with $4n$ π electrons will undergo electrocyclic reactions by conrotatory motion, whereas systems with $4n + 2$ π electrons will react by the disrotatory mode.

An addition dimension was introduced into the analysis of concerted reactions with the use of orbital correlation diagrams.[13] This approach focuses attention on the orbital symmetries of both reactants and products and considers the symmetry properties of all the orbitals. In any concerted process, the orbitals of the starting material must be transformed into orbitals of product having the same symmetry. If this process of orbital conversion leads to the ground state electronic configuration of the product, the process should have a relatively low activation energy and is called an *allowed* process. If, on the other hand, the orbitals of the reactant are transformed into a set of orbitals that does not correspond to the ground state of the product, a high-energy transition state occurs and the reaction is called *forbidden*, since it would lead to an excited state of the product.

The cyclobutene–butadiene interconversion can serve as an example of the reasoning employed in construction of an orbital correlation diagram. For this reaction, the four π orbitals of butadiene are converted smoothly into the two π and two σ orbitals of the ground state of cyclobutene. The analysis is done as shown in Fig. 11.3. The π orbitals of butadiene are ψ_1, ψ_2, ψ_3, and ψ_4. For cyclobutene, the four orbitals are σ, π, σ^*, and π^*. Each of the orbitals is classified with respect to the symmetry elements that are maintained in the course of the transformation. The relevant symmetry features depend on the structure of the reacting system. The most common elements of symmetry to be considered are planes of symmetry and rotation axes. An orbital is classified as symmetric (S) if it is unchanged by reflection in a plane of symmetry or by rotation about an axis of symmetry. If the orbital changes sign (phase) at each lobe as a result of the symmetry operation, it is called

13. R. Hoffmann and R. B. Woodward, *J. Am. Chem. Soc.* **87**, 2046 (1965).

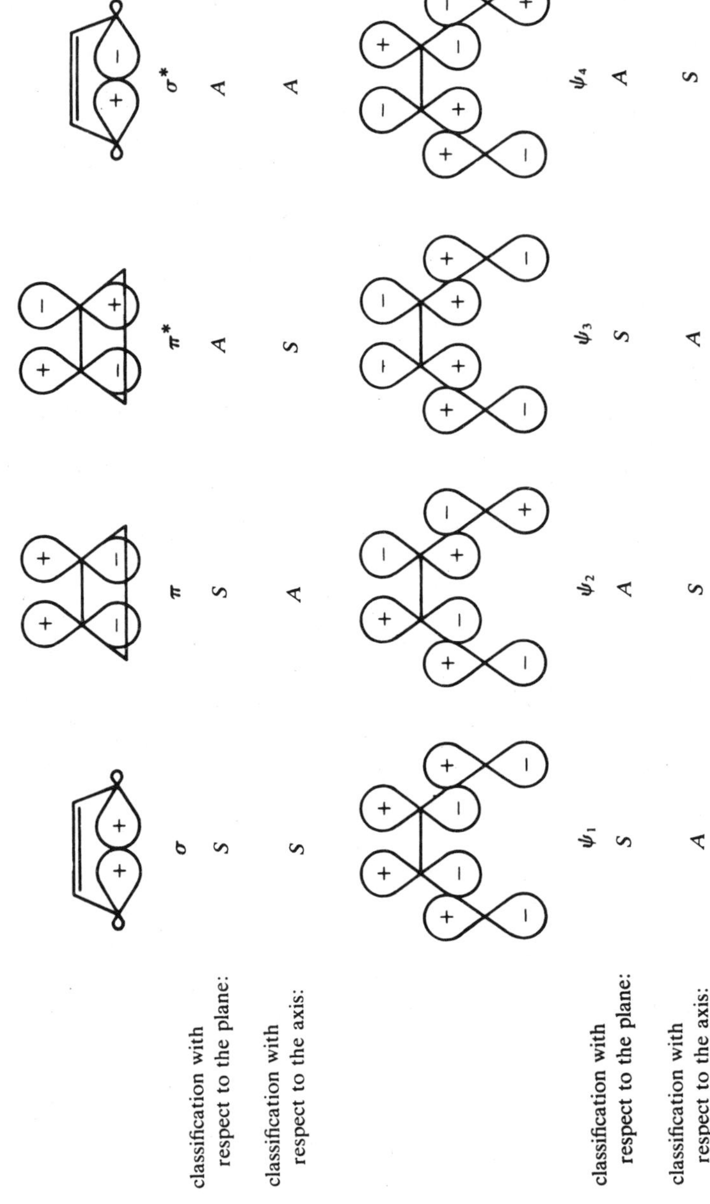

Fig. 11.3. Symmetry properties of cyclobutene and butadiene orbitals.

antisymmetric (A). Proper molecular orbitals must be either symmetric or antisymmetric. If an orbital is not sufficiently symmetric to be either S or A, it must be adapted by combination with other orbitals to meet this requirement.

There are two elements of symmetry that are common to both *s-cis*-butadiene and cyclobutene. These are a plane of symmetry and a twofold axis of rotation. The plane of symmetry is maintained during a disrotatory transformation of butadiene to cyclobutene. In the conrotatory transformation, the axis of rotation is maintained throughout the process. Therefore, to analyze the disrotatory process, the orbitals must be classified with respect to the plane of symmetry, and to analyze the conrotatory process, they must be classified with respect to the axis of rotation.

disrotatory
plane of symmetry

conrotatory
axis of symmetry

Both the disrotatory process and the conrotatory process can be analyzed by comparing the symmetry classification of reactant and product orbitals given in Fig. 11.3. The orbitals are arranged according to energy in Fig. 11.4, and the states of like symmetry for the disrotatory process are connected. It is evident that not all of the ground state orbitals of cyclobutene correlate with ground state orbitals of butadiene. The bonding π orbital of cyclobutene is transformed into an antiboding orbital (ψ_3) of butadiene. Considering the reverse process, ψ_2 of butadiene is transformed into the antiboding π^* orbital of cyclobutene. Because of the failure of the orbitals of the ground state molecules to correlate, the transformation would have to attain a very high energy transition state, and the reaction process is said to be *symmetry-forbidden.*

Analysis of the conrotatory process is carried out in exactly the same fashion. In this case, the element of symmetry that is maintained throughout the reaction process is the twofold rotation axis. The resulting correlation diagram is shown in

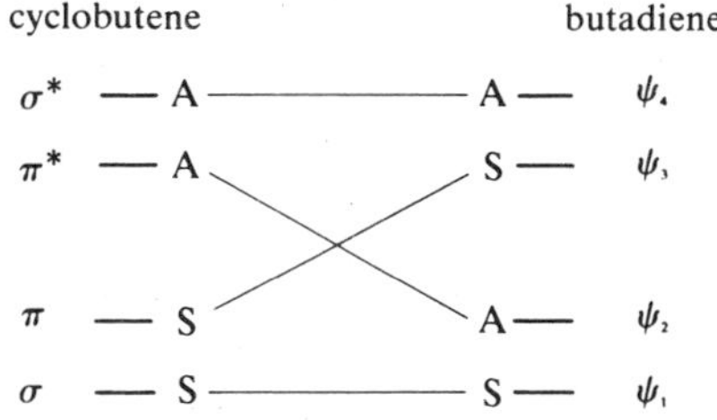

Fig. 11.4. Correlation diagram for cyclobutene and butadiene orbitals (symmetry-forbidden disrotatory reaction).

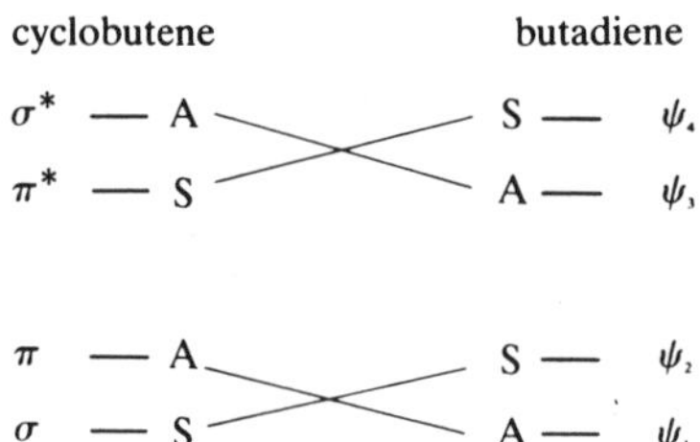

Fig. 11.5. Correlation diagram for cyclobutene and butadiene orbitals (symmetry-allowed conrotatory reaction).

Fig. 11.5. This reaction is *symmetry-allowed*, since the bonding orbitals of butadiene correlate with the bonding orbitals of cyclobutene, and vice versa. Detailed MO analysis of the transition state for conrotatory ring opening using high-level MO methods fully supports the conclusion that the reaction proceeds by a concerted process.[14]

Correlation diagrams can be constructured in an analogous fashion for the disrotatory and conrotatory modes for interconversion of hexatriene and cyclohexadiene. They lead to the prediction that the disrotatory mode is an allowed process while the conrotatory process is forbidden. This is in agreement with the experimental results on this reaction. Other electrocyclization reactions can be analyzed by the same process. Substituted derivatives of polyenes obey the orbital symmetry rules, even in cases where the substitution pattern does not correspond in symmetry to that of the orbital system. It is the symmetry of the participating orbitals, not of the molecule as a whole, that is crucial to the analysis.

There is another useful viewpoint of concerted reactions that is based on the idea that transition states can be classified as aromatic or antiaromatic, just as is the case for ground state molecules.[15] A stabilized aromatic transition state will lead to a low activation barrier, that is, an allowed reaction. An antiaromatic transition state will result in a high energy barrier and correspond to a forbidden process. The analysis of concerted reactions by this process consists in examining the array of orbitals that would be present in the transition state and classifying the system as aromatic or antiaromatic.

For the butadiene–cyclobutene interconversion, the transition states for conrotatory and disrotatory interconversion are shown below. The array of orbitals represents the *basis set orbitals*, that is, the total set of $2p$ orbitals involved in the reaction process, not the individual molecular orbitals. Each of the orbitals is π in character, and the phase difference is represented by shading to distinguish from the + and − designations used for molecular orbitals. The tilt at C-1 and C-4 as the butadiene system rotates toward the transition state is different for the disrotatory

14. N. G. Rondan and K. N. Houk, *J. Am. Chem. Soc.* **107**, 2099 (1985); J. Breulet and H. F. Schaefer III, *J. Am. Chem. Soc.* **106**, 221 (1984).
15. M. J. S. Dewar, *The Molecular Orbital Theory of Organic Chemistry*, McGraw-Hill, New York, 1969; *Angew. Chem. Int. Ed. Engl.* **10**, 761 (1971).

and conrotatory modes.

basis set orbitals for conrotatory closure

basis set orbitals for disrotatory closure

We will consistently assign phases to the basis set orbitals in such a way as to minimize the phase changes in the transition state. It has, however, been shown that this is not necessary, and the correct classification of the transition state will be achieved for any combination of basis set orbitals. When the array of orbitals corresponding to the transition state being analyzed has been drawn, two features must be determined to decide if it is aromatic or antiaromatic. These features are the topology of the orbital array and the number of electrons involved. The topology can be of the *Hückel type* or *Möbius type*. A Hückel system has zero (or any even number) of phase changes around the orbital array. For the butadiene–cyclobutene transition state, the conrotatory closure results in a Möbius system, while a disrotatory transition state gives a Hückel system. The second important feature of the transition state is the number of participating electrons. In the present case, it is four. The same rules of aromaticity apply as for ground state molecules. A Hückel system is aromatic when it has $4n + 2$ electrons. A Möbius system is aromatic when it has $4n$ electrons. In the case of the butadiene–cyclobutene interconversion, it is then the conrotatory transition state that is the favored aromatic transition state.

There are two stereochemically distinct possibilities for the conrotatory process. A substituent group might move toward or away from the breaking bond:

Steric factors would ordinarily be expected to induce a preference for the larger group to move outward, thus generating the *E*-isomer. It was observed, however, that in the case of 1,2,3,4-tetrafluoro-*trans*-3,4-bis(trifluoromethyl)cyclobutene, opening occurred with an inward rotation of the trifluoromethyl groups.[16]

Molecular orbital calculations for the case of Y = CH=O found that the formyl group preferred to rotate inward, and this was confirmed experimentally.[17] A general

16. W. R. Dolbier, Jr., H. Koroniak, D. J. Burton, and P. Heinze, *Tetrahedron Lett.* **27**, 4387 (1986).
17. K. Rudolf, D. C. Spellmeyer, and K. N. Houk, *J. Org. Chem.* **52**, 3708 (1987).

theoretical analysis indicates that the preference is for donor substituents to rotate outward whereas acceptor substituents should prefer to rotate inward.[18]

Orbital symmetry analysis of the hexatriene–cyclohexadiene system leads to the conclusion that the disrotatory process will be favored. This basis set orbitals for the conrotatory and disrotatory transition states are shown below:

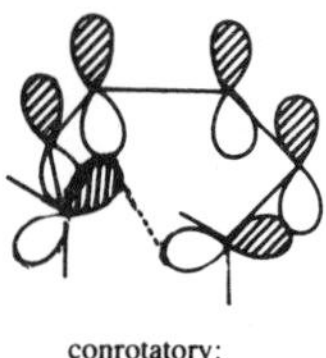

conrotatory:
Möbius system

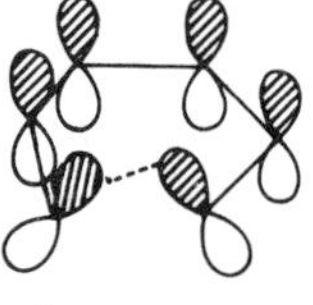

disrotatory:
Hückel system

Here, with six electrons involved, it is in the disrotatory mode (Hückel system) that the transition state is stabilized. The chart that follows gives a general summary of the relationship between transition state topology, the number of electrons, and the stability of the transition state.

Electrons	Hückel (disrotatory)	Möbius (conrotatory)
2	aromatic	antiaromatic
4	antiaromatic	aromatic
6	aromatic	antiaromatic
8	antiaromatic	aromatic

We have now considered three viewpoints from which thermal electrocyclic processes can be analyzed: symmetry characteristics of the frontier orbitals, orbital correlation diagrams, and transition state aromaticity. All arrive at the same conclusions about stereochemistry of such processes. *Processes involving 4n + 2 electrons will be disrotatory and involve a Hückel-type transition state, whereas those involving 4n electrons will be conrotatory and the orbital array will be of the Möbius type.* These general principles serve to explain and correlate many specific observations in organic chemistry made both before and after the orbital symmetry rules were formulated. We will discuss a few representative examples in the following paragraphs.

The bicyclo[2.2.0]hexa-2,5-diene ring system is a valence isomer of the benzene ring and is often referred to as *Dewar benzene.* After many attempts to prepare Dewar benzene derivatives failed, a pessimistic opinion existed that all such efforts would be fruitless because Dewar benzene would be so unstable as to immediately revert to benzene. Then, in 1962, van Tamelen and Pappas isolated a stable Dewar benzene derivative prepared by photolysis of 1,2,4-tri-*t*-butylbenzene.[19]

18. N. G. Rondan and K. N. Houk, *J. Am. Chem. Soc.* **107**, 2099 (1985); W. Kirmse, N. G. Rondan, and K. N. Houk, *J. Am. Chem. Soc.* **106**, 7989 (1984).
19. E. E. van Tamelen, S. P. Pappas, and K. L. Kirk, *J. Am. Chem. Soc.* **93**, 6092 (1971); this paper contains references to the initial work and describes subsequent studies.

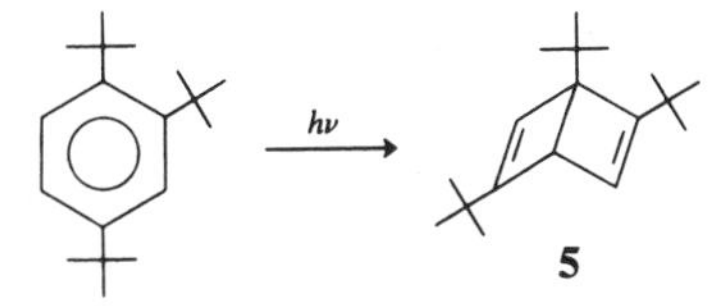

The compound was reasonably stable, reverting to the aromatic starting material only on heating. Part of the stability of this Dewar benzene derivative could be attributed to steric factors. The t-butyl groups are farther apart in the Dewar benzene structure than in the aromatic structure. The unsubstituted Dewar benzene was prepared in 1963.

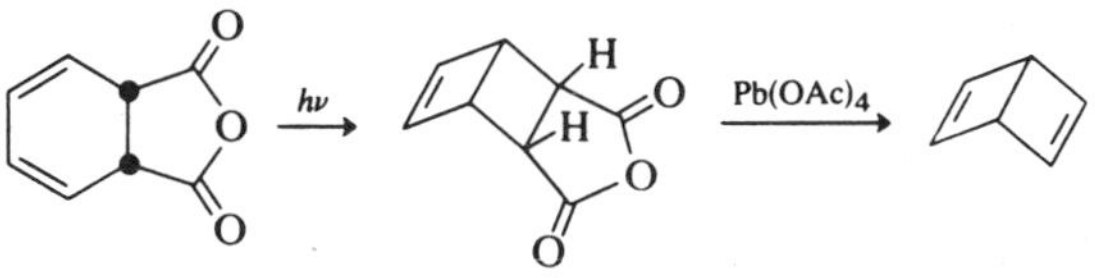

This compound is less stable than **5** and reverts to benzene with a half-life of about two days at 25°C, with $\Delta H^{+} = 23$ kcal/mol.[20] The observed kinetic stability of Dewar benzene is surprisingly high when one considers that its conversion to benzene is exothermic by 71 kcal/mol. The stability of Dewar benzene is intimately related to the orbital symmetry requirements for concerted electrocyclic transformations. The concerted, thermal pathway should be conrotatory, since the reaction is the ring opening of a cyclobutene, and therefore leads not to benzene, but to a highly strained Z,Z,E-cyclohexatriene. A disrotatory process, which would lead directly to benzene, is forbidden.

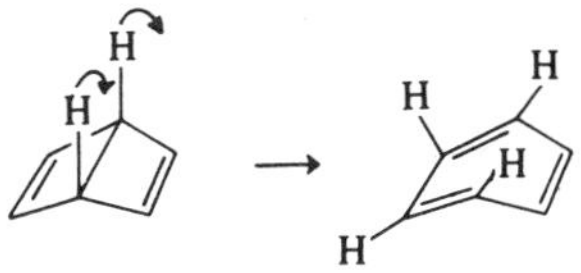

An especially interesting case of hexatriene–cyclohexadiene interconversion is the rapid equilibrium between cycloheptatrienes and bicyclo[4.1.0]hepta-2,4-dienes[21]:

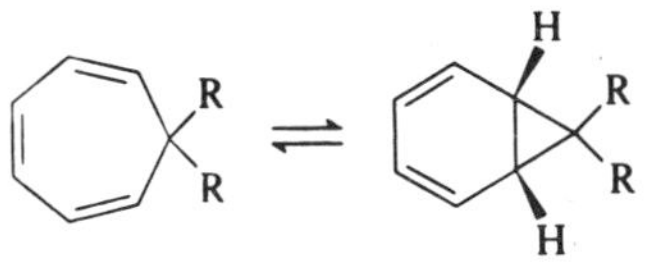

The energy requirement for this electrocyclic transformation is so low that the process is very rapid at room temperature. Low-temperature NMR measurements

20. M. J. Goldstein and R. S. Leight, *J. Am. Chem. Soc.* **99**, 8112 (1977).
21. G. Maier, *Angew. Chem. Int. Ed. Engl.* **6**, 402 (1967).

give a value of about 7 kcal/mol for the activation energy in the case where $R = CO_2CH_3$.[22] This transformation is an example of *valence tautomerism*, a rapid process involving only reorganization of bonding electrons. The reason the process is much more rapid than electrocyclization of acyclic trienes is that the ring holds the reacting termini together, reducing the negative entropy of activation. In contrast to the ring opening of Dewar benzene, disrotatory opening of the bicyclo[4.1.0]hepta-2,4-diene is allowed by orbital symmetry rules and is easily accommodated by the ring geometry. For unsubstituted bicyclo[4.1.0]hepta-2,4-diene, the equilibrium constant for ring closure is small, about 3×10^{-3} at 100°C. Alkyl groups do not have much of an effect on the position of equilibrium, but electron-withdrawing groups such as cyano and trifluoromethyl shift the equilibrium more in favor of the bicyclic ring.[23]

The prediction on the basis of orbital symmetry analysis that cyclization of eight-π-electron systems will be conrotatory has been confirmed by study of isomeric 2,4,6,8-decatetraenes. Electrocyclic reaction occurs near room temperature and establishes an equilibrium that favors the cyclooctatriene product. At slightly more elevated temperatures, the hexatriene system undergoes a subsequent disrotatory cyclization, establishing equilibrium with the corresponding bicyclo[4.2.0]octa-2,4-diene.[24]

a b c d ⇌ a b d c ⇌ a b d c

a = d = CH_3; b = c = H
b = c = CH_3; a = d = H
b = d = CH_3; a = c = H

The Woodward–Hoffmann orbital symmetry rules are not limited in application to the neutral systems that have been discussed up to this point. They also apply to charged systems, just as the Hückel rules can be applied to charged cyclic systems. The conversion of a cyclopropyl cation to an allyl cation is the simplest possible case of an electrocyclic process, since it involves only two π electrons.[25] Because of the strain imposed by the small internuclear angle in the cyclopropyl ring, cyclopropyl cations do not form easily, and cyclopropyl halides and sulfonates are quite unreactive under ordinary solvolytic conditions. For example, solvolysis of cyclopropyl tosylate in acetic acid requires a temperature of 180°C. The product is

22. M. Görlitz and H. Günther, *Tetrahedron* **25**, 4467 (1969).
23. P. Warner and S.-L. Lu, *J. Am. Chem. Soc.* **95**, 5099 (1973); P. M. Warner and S.-L. Lu, *J. Am. Chem. Soc.* **102**, 331 (1980); K. Takeuchi, H. Fujimoto, and K. Okamoto, *Tetrahedron Lett.* **22**, 4981 (1981).
24. R. Huisgen, A. Dahmen, and H. Huber, *Tetrahedron Lett.*, 1461 (1969); R. Huisgen, A. Dahmen, and H. Huber, *J. Am. Chem. Soc.* **89**, 7130 (1967); A. Dahmen and R. Huisgen, *Tetrahedron Lett.*, 1465 (1969).
25. P. v. R. Schleyer, W. F. Sliwinski, G. W. Van Dine, U. Schöllkopf, J. Paust, and K. Fellenberger, *J. Am. Chem. Soc.* **94**, 125 (1972); W. F. Sliwinski, T. M. Su, and P. v. R. Schleyer, *J. Am. Chem. Soc.* **94**, 133 (1972).

allyl acetate rather than cyclopropyl acetate.[26] This transformation might occur by formation of the cyclopropyl cation, followed by ring opening to the allyl cation:

Formation of allylic products is characteristic of solvolytic reactions of other cyclopropyl halides and sulfonates. Similarly, diazotization of cyclopropylamine in aqueous solution gives allyl alcohol.[27] The ring opening of a cyclopropyl cation is an electrocyclic process of the $4n + 2$ type, where n equals zero. It should therefore be a disrotatory process. There is another facet to the stereochemistry in substituted cyclopropyl systems. Note that for a *cis*-2,3-dimethylcyclopropyl cation, for example, two different disrotatory modes are possible, leading to conformationally distinct allyl cations.

The disrotatory mode in which the methyl groups move away from each other would be more favorable for steric reasons. If the ring opening occurs through a discrete cyclopropyl cation, the W-shaped allylic cation should be formed in preference to the sterically less favorable U-shaped cation. This point was investigated by comparing the rates of solvolysis of the cyclopropyl tosylates **6–8**:

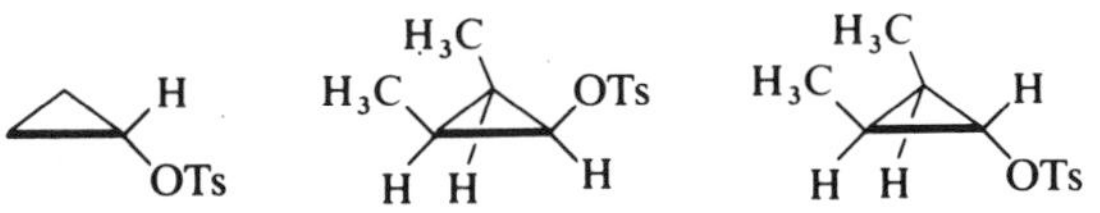

	6	**7**	**8**
relative rate:	1	4	41,000

Some very significant conclusions can be drawn from the data. If formation of the cyclopropyl cation were the rate-determining step, **7** should be more reactive than **8**, because the steric interaction between the tosylate leaving group and the methyl substituents in **7** will be relieved as ionization occurs. Since **7** is 10,000 times less reactive than **8**, some other factor must be determining the relative rates of reaction, and it is doubtful that rate-limiting ionization to a cyclopropyl cation is occurring. The results can be explained, as first proposed by DePuy,[28] if the ionization and ring opening are part of a single, concerted process. In such a process, the ionization

26. J. D. Roberts and V. C. Chambers, *J. Am. Chem. Soc.* **73**, 5034 (1951).
27. P. Lipp, J. Buckremer, and H. Seeles, *Justus Liebigs Ann. Chem.* **499**, 1 (1932); E. J. Corey and R. F. Atkinson, *J. Org. Chem.* **29**, 3703 (1964).
28. C. H. DePuy, *Acc. Chem. Res.* **1**, 33 (1968).

would be assisted by the electrons in the cleaving C-2—C-3 bond. These electrons would provide maximum assistance if positioned toward the back side of the leaving group. This, in turn, requires that the substituents *anti* to the leaving group rotate outward as the ionization proceeds. This concerted process would explain why **7** reacts more slowly than **8**. In **7**, such a rotation would move the methyl groups together, resulting in increased steric interaction and the formation of the U-shaped allylic anion. In **8**, the methyl groups would move away from one another and form the W-shaped allylic ion.

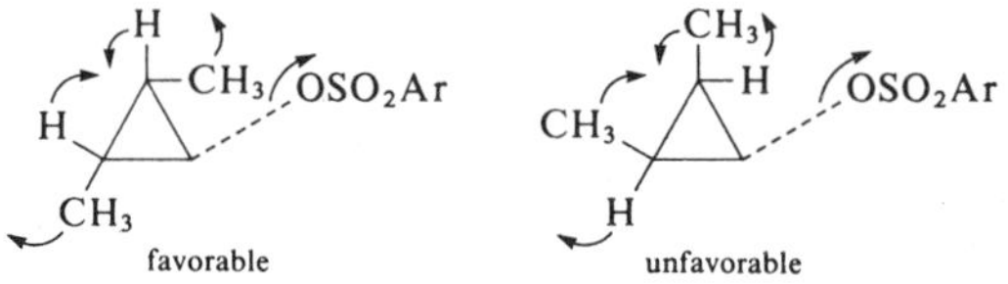

favorable

unfavorable

This interpretation is supported by results on the acetolysis of the bicyclic tosylates **9** and **10**. With **9**, after three months in acetic acid at 150°C, 90% of the starting material was recovered. This means that both ionization to a cyclopropyl cation and a concerted ring opening must be extremely slow. The preferred disrotatory ring-opening process would lead to an impossibly strained structure, the *trans*-cyclohexenyl cation. In contrast, the stereoisomer **10** reacts at least 2×10^6 times more rapily since it can proceed to a stable *cis*-cyclohexenyl cation.[29]

When the size of the fused ring permits ring opening to a *trans*-allylic cation, as in the case of compound **11**, solvolysis proceeds at a reasonable rate.[30]

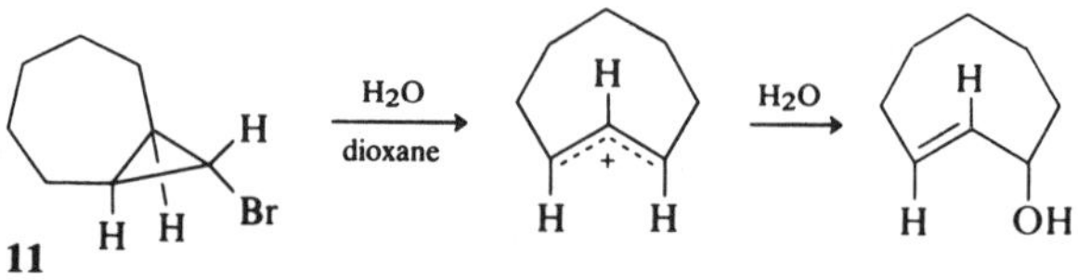

An example of preferred conrotatory cyclization of four-π-electron cation systems can be found in the acid-catalyzed cyclization of the dienone **12**, which proceeds through the 3-hydroxypentadienyl cation **13**. The stereochemistry is that expected for a conrotatory process.[31]

29. P. v. R. Schleyer, W. F. Sliwinski, G. W. Van Dine, U. Schöllkopf, J. Paust, and K. Fellenberger, *J. Am. Chem. Soc.* **94**, 125 (1972).
30. G. H. Whitham and M. Wright, *J. Chem. Soc., C*, 883 (1971).
31. R. B. Woodward, in *Aromaticity*, Chemical Society Special Publication No. 21, 217 (1969).

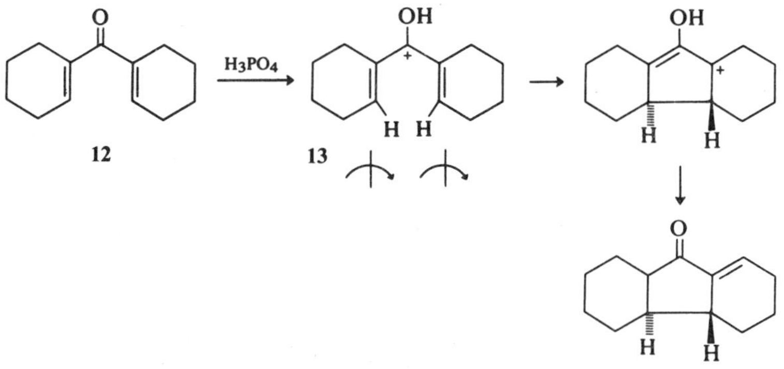

There are also examples of electrocyclic processes involving anionic species. Since the pentadienyl anion is a six-π-electron system, thermal cyclization to a cyclopentenyl anion should be disrotatory. Examples of this electrocyclic reaction are rare. NMR studies of pentadienyl anions indicate that they are stable and do not tend to cyclize.[32] Cyclooctadienyllithium provides an example where cyclization does occur, with the first-order rate constant being $8.7 \times 10^{-3}\ \text{min}^{-1}$. The stereochemistry of the ring-closure is consistent with the expected disrotatory nature of the reaction.

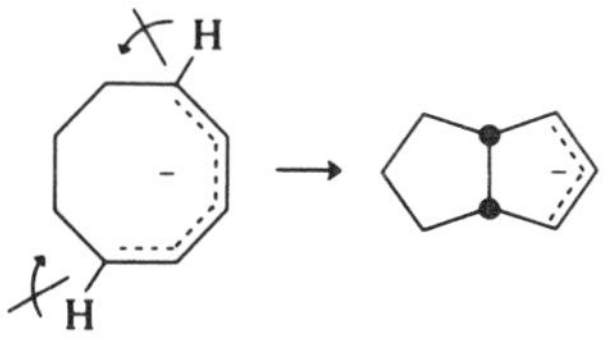

In contrast to pentadienyl anions, heptatrienyl anions have been found to cyclize readily to cycloheptadienyl anions.[33] The transformation of heptatrienyl anion to cycloheptadienyl anion proceeds with a half-life of 13 min at −13°C. The Woodward–Hoffmann rules predict this would be a conrotatory closure.[34]

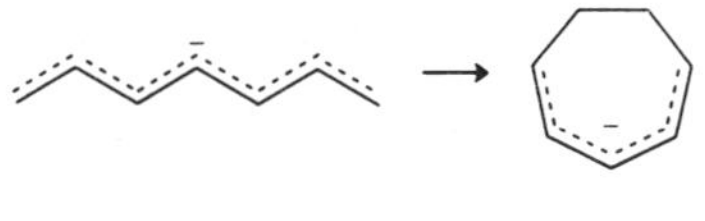

11.2. Sigmatropic Rearrangements

A group of reactions called *sigmatropic rearrangements* are closely related to electrocyclic transformations in being concerted processes governed by orbital

32. R. B. Bates, D. W. Gosselink, and J. A. Kaczynski, *Tetrahedron Lett.*, 199, 205 (1967); R. B. Bates and D. A. McCombs, *Tetrahedron Lett.*, 977 (1969).
33. E. A. Zeuch, D. L. Crain, and R. F. Kleinschmidt, *J. Org. Chem.* **33**, 771 (1968); R. B. Bates, W. H. Deines, D. A. McCombs, and D. E. Potter, *J. Am. Chem. Soc.* **91**, 4608 (1969).
34. S. W. Staley, in *Pericyclic Reactions*, Vol. 1, A. P. Marchand and R. E. Lehr (eds.), Academic Press, New York, 1977, Chapter 4.

Scheme 11.1. Examples of Sigmatropic Rearrangements

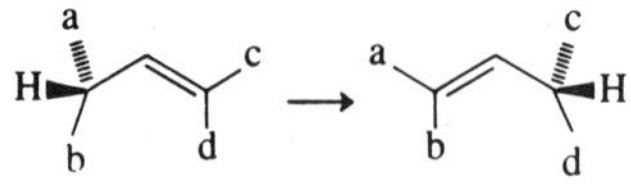

1,3-suprafacial shift of hydrogen

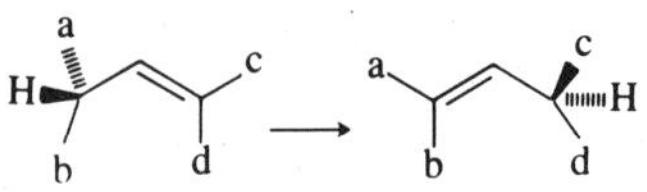

1,3-antarafacial shift of hydrogen

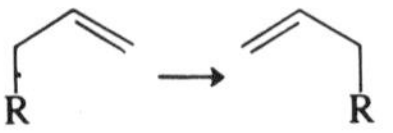

1,3-shift of alkyl group

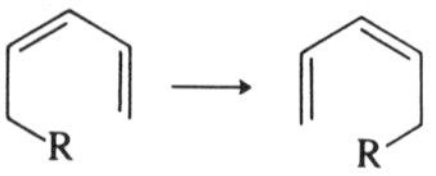

1,5-shift of alkyl group

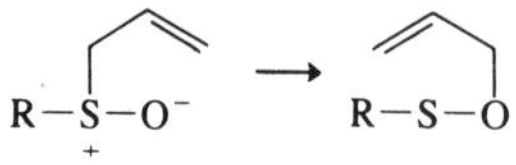

2,3-sigmatropic rearrangement of an allyl sulfoxide

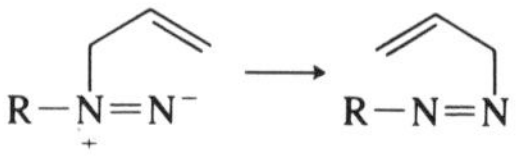

2,3-sigmatropic rearrangement of an allyl diazene

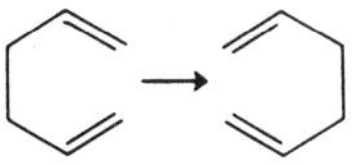

3,3-sigmatropic rearrangement of a 1,5-hexadiene

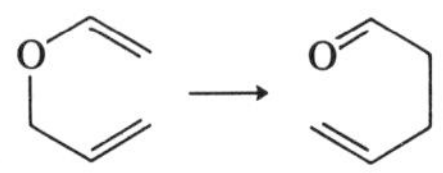

3,3-sigmatropic rearrangement of an allyl vinyl ether

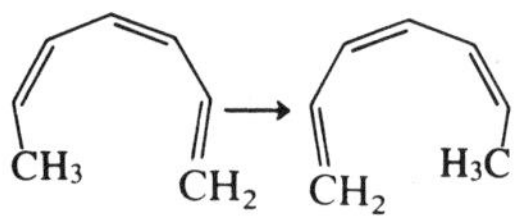

1,7-sigmatropic shift of hydrogen

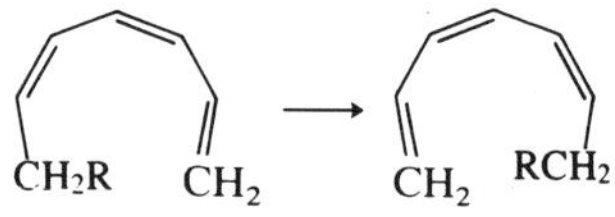

1,7-sigmatropic shift of an alkyl group

symmetry.[35] Sigmatropic processes involve a concerted reorganization of electrons during which a group attached by a σ bond migrates to a more distant terminus of an adjacent π-electron system. There is a simultaneous shift of the π electrons. Sigmatropic rearrangements are further described by stating the relationship between the reacting centers in the migrating fragment and the π system. The order $[i, j]$ specifies the number of atoms in the migrating fragment and the number of atoms in the π system that are directly involved in the bonding changes. This classification system is illustrated by the examples in Scheme 11.1. As with other concerted reactions, the topological properties of the interacting orbitals dictate the facility of the various sigmatropic rearrangements and the stereochemistry of the process. First, it must be recognized that there are two topologically distinct processes by which a sigmatropic migration can occur. If the migrating group remains associated with the same face of the conjugated π system throughout the process, the migration is termed *suprafacial.* The alternative mode involves a process in which the migrating group moves to the opposite face of the π system during the course of the migration and is called *antarafacial.*

35. R. B. Woodward and R. Hoffmann, *J. Am. Chem. Soc.* **87**, 2511 (1965).

The orbital symmetry requirements of sigmatropic reactions are analyzed by considering the interactions between the orbitals of the π system and those of the migrating fragment. The simplest case, the 1,3-sigmatropic shift of a hydrogen, is illustrated in the first entry in Scheme 11.1. A frontier orbital analysis of this process treats the system as a hydrogen atom interacting with an allyl radical. The frontier orbitals are the hydrogen 1*s* and the allyl ψ_2 orbitals. These interactions are depicted below for both the suprafacial and antarafacial modes.

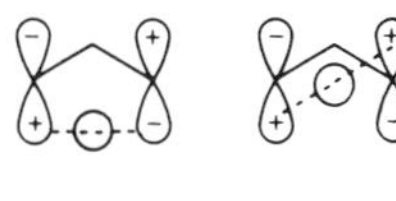

suprafacial antarafacial

A bonding interaction can be maintained only in the antarafacial mode. The 1,3-suprafacial shift of hydrogen is therefore forbidden by orbital symmetry considerations. The allowed antarafacial process is symmetry-allowed, but it involves such a contorted geometry that this shift, too, would be expected to be energetically difficult. As a result, orbital symmetry considerations reveal that 1,3-shifts of hydrogen are unlikely processes.

A similar analysis of the 1,5-sigmatropic shift of hydrogen leads to the opposite conclusion. The suprafacial mode is allowed while the antarafacial mode is forbidden. The relevant frontier orbitals in this case are the hydrogen 1*s* orbital and ψ_3 of the pentadienyl radical.

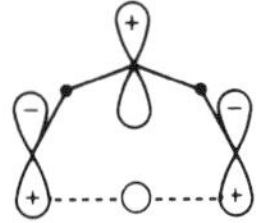

thermally allowed 1,5-suprafacial hydrogen shift in 1,3-pentadiene

An alternative analysis involves drawing the basis set atomic orbitals and classifying the resulting system as Hückel or Möbius in character. When this classification has been done, the electrons involved in the process are counted to determine if the transition state is aromatic or antiaromatic. This analysis is illustrated in Fig. 11.6.

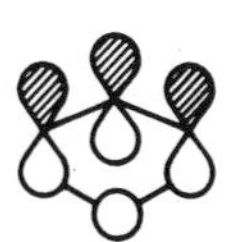

suprafacial 1,3
Hückel system
4 electrons
antiaromatic
forbidden

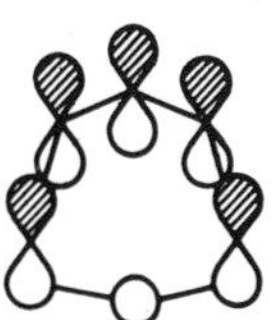

suprafacial 1,5
Hückel system
6 electrons
aromatic
allowed

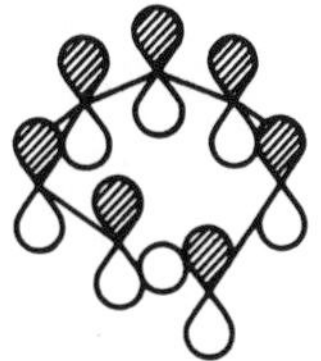

antarafacial 1,7
Möbius system
8 electrons
aromatic
allowed

Fig. 11.6. Classification of sigmatropic hydrogen shifts with respect to basis set orbitals.

1,3 suprafacial
retention
Hückel system
4 electrons
antiaromatic
forbidden

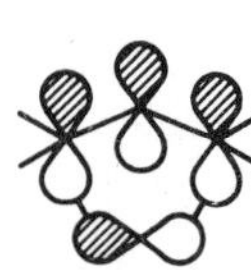

1,3 suprafacial
inversion
Möbius system
4 electrons
aromatic
allowed

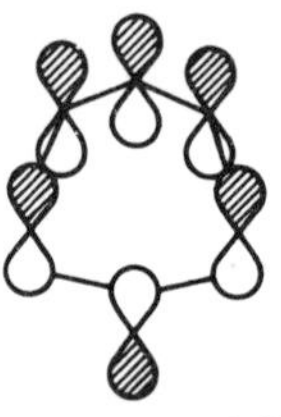

1,5 suprafacial
retention
Hückel system
6 electrons
aromatic
allowed

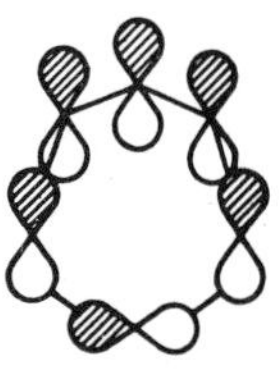

1,5 suprafacial
inversion
Möbius system
6 electrons
antiaromatic
forbidden

Fig. 11.7. Classification of sigmatropic shifts of alkyl groups with respect to basis set orbitals.

The conclusions reached are the same as for the frontier orbital approach. The suprafacial 1,3-shift of hydrogen is forbidden but the suprafacial 1,5-shift is allowed. Proceeding to a 1,7-shift of hydrogen, it is found that the antarafacial shift is allowed. These conclusions based on transition state aromaticity are supported by MO calculations at the 6-31* level, which conclude that 1,5-shifts should be suprafacial while 1,7-shifts should be antarafacial.[36] Theoretical calculations also find that the 1,3-shift of hydrogen should be antarafacial, but, in agreement with expectations based on molecular geometry, the transition state which is found is so energetic that it is close to a stepwise bond dissocation process.[37]

Sigmatropic migration involving alkyl group shifts can also occur:

R → R (1,3 alkyl shift) *or* R → R (1,5 alkyl shift)

When an alkyl group migrates, there is an additional stereochemical feature to consider. The shift can occur with retention or inversion at the migrating center. The analysis of sigmatropic shifts of alkyl groups is illustrated in Fig. 11.7. The allowed processes include the suprafacial 1,3-shift with inversion and the suprafacial 1,5-shift with retention.

Sigmatropic rearrangements of order [3,3] are very common:

→

The transition state for such processes can be considered to be two interacting allyl fragments. When the process is suprafacial in both groups, an aromatic transition

36. B. A. Hess, Jr., L. J. Schaad, and J. Pancir, *J. Am. Chem. Soc.* **107**, 149 (1985).
37. F. Bernardi, M. A. Robb, H. B. Schlegel, and G. Tonachini, *J. Am. Chem. Soc.* **106**, 1198 (1984).

Scheme 11.2. Generalized Selection Rules for Sigmatropic Processes

Selection rules for sigmatropic shifts of order [*i*, *j*]				
A. Order [1, j]				
1 + j	supra/retention	supra/inversion	antara/retention	antara/inversion
4*n*	forbidden	allowed	allowed	forbidden
4*n* + 2	allowed	forbidden	forbidden	allowed
B. Order [i,j]				
i + j	supra/supra	supra/antara	antara/antara	
4*n*	forbidden	allowed	forbidden	
4*n* + 2	allowed	forbidden	allowed	

state results and the process is thermally allowed. Usually a chairlike transition state is involved, but a boatlike conformation is also possible.

Generalization of the Woodward–Hoffmann rules for sigmatropic processes leads to the selection rules in Scheme 11.2.[38]

With these generalized rules as a unifying theoretical framework, we can consider specific examples of sigmatropic rearrangements. In accord with the theoretical concepts, there are many examples of sigmatropic 1,5-hydrogen migrations in molecules containg a pentadienyl fragment. The activation energies for such reactions are usually in the vicinity of 35 kcal/mol so the reactions usually require moderately elevated temperatures.[39] Two examples are given below.

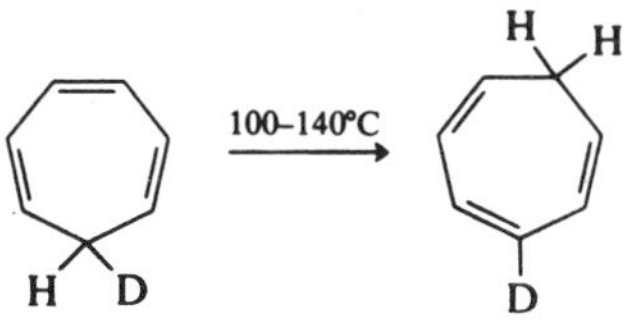

Ref. 40

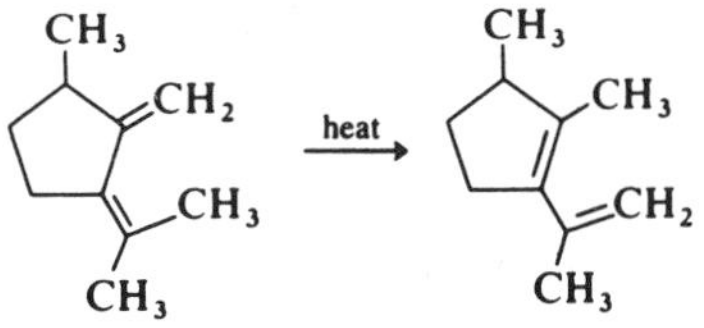

Ref. 41

38. R. B. Woodward and R. Hoffmann, *J. Am. Chem. Soc.* **87**, 2511 (1965).
39. W. R. Roth and J. König, *Justus Liebigs Ann. Chem.* **699**, 24 (1966).
40. A. P. ter Borg, H. Kloosterziel, and N. Van Meurs, *Proc. Chem. Soc.*, 359 (1962).
41. J. Wolinsky, B. Chollar, and M. D. Baird, *J. Am. Chem. Soc.* **84**, 2775 (1962).

The conversion of **14** to **15** provides confirmation of the prediction that 1,3-alkyl shifts should occur with inversion of configuration[42]:

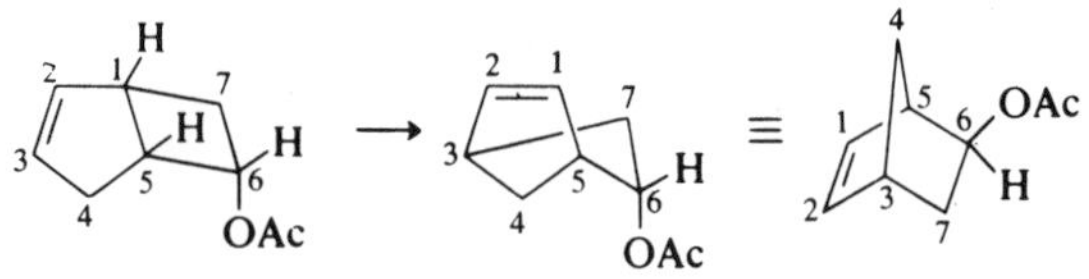

In the absence of the orbital symmetry considerations, one might assume that the C-3—C-7 bond would form as the C-1—C-7 bond breaks by simply "sliding over." This would be a violation of orbital symmetry rules since, as indicated in Scheme 11.2, a 1,3-suprafacial shift of an alkyl group must proceed with inversion at the migrating center. This point was investigated using **14** labeled with deuterium at C-7. In the starting material, the deuterium was *trans* to the acetoxy group, while in the product, it was found to be exclusively *cis*. This result establishes that inversion of configuration occurred at C-7 during the migration, in accord with the stereochemistry required by the Woodward–Hoffmann rules.

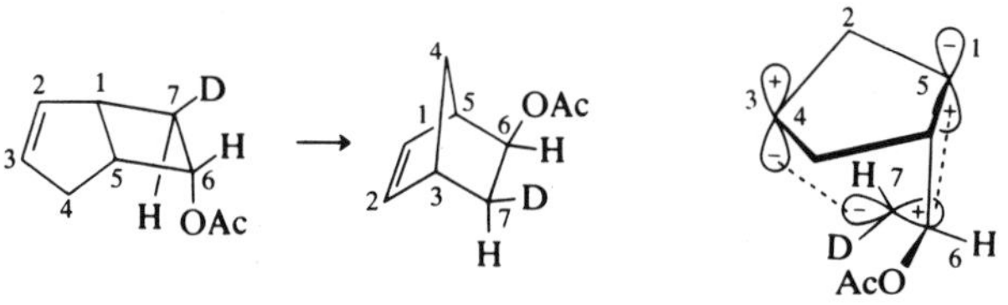

Suprafacial 1,3-shifts with inversion of configuration at the migrating carbon have also been observed in the thermal conversion of bicyclo[2.1.1]hexenes to bicyclo[3.1.0]hexenes.[43]

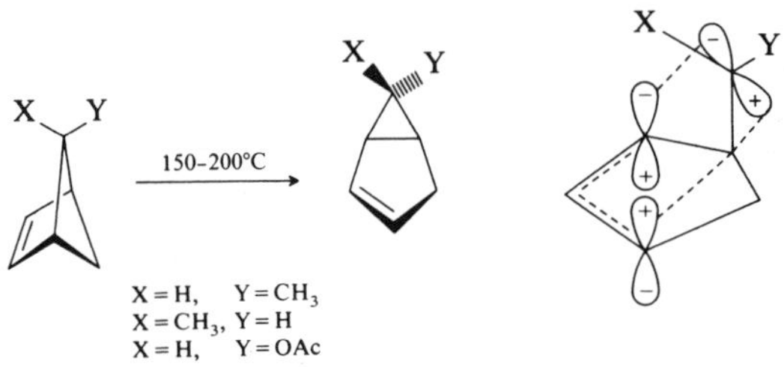

150–200°C

X = H, Y = CH_3
X = CH_3, Y = H
X = H, Y = OAc

The thermal rearrangements of methyl-substituted cycloheptatrienes have been proposed to proceed by sigmatropic migration of the norcaradiene valence tautomer.[44]

42. J. A. Berson, *Acc. Chem. Res.* **1**, 152 (1968); J. A. Berson and G. L. Nelson, *J. Am. Chem. Soc.* **89**, 5503 (1967).
43. W. R. Roth and A. Friedrich, *Tetrahedron Lett.*, 2607 (1969); S. Masamune, S. Takada, N. Nakatasuka, R. Vukov, and E. N. Cain, *J. Am. Chem. Soc.* **91**, 4322 (1969).
44. J. A. Berson and M. R. Willcott III, *Rec. Chem. Prog.* **27**, 139 (1966); *J. Am. Chem. Soc.* **88**, 2494 (1966).

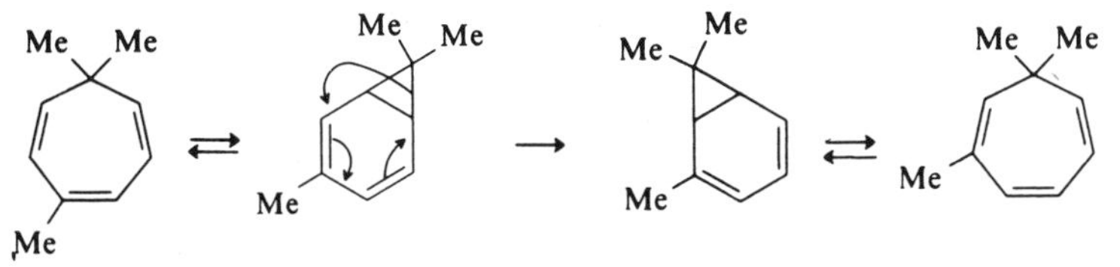

These are suprafacial sigmatropic shifts of order [1,5] and should occur with retention of configuration at the migrating carbon. This stereochemical course has been established for the 1,5-alkyl shift that converts **16** to **17**.[45] The product that is isolated, **18**, results from a subsequent 1,5-hydrogen shift, but this does not alter the stereochemistry at the migrating carbon. The configuration of the migrating carbon is retained, as expected.

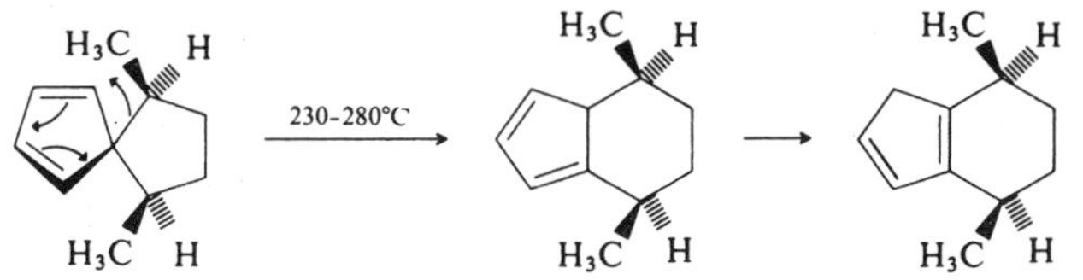

Like the thermal 1,3-hydrogen shift, a 1,7-hydrogen shift is allowed when antarafacial but forbidden when suprafacial. Because a π system involving seven carbon atoms is more flexible than one involving only three carbon atoms, the geometrical restrictions on the antarafacial transition state are not as restrictive as in the 1,3 case. For the conversion of *Z*,*Z*-1,3,5-octatriene to *Z*,*Z*,*E*-2,4,6-octatriene, the energy of activation is 20 kcal/mol. The *Z*,*Z*,*Z*-isomer is also formed, but with a slightly higher activation energy. The primary kinetic isotope for the transferred hydrogen is around 7.[46]

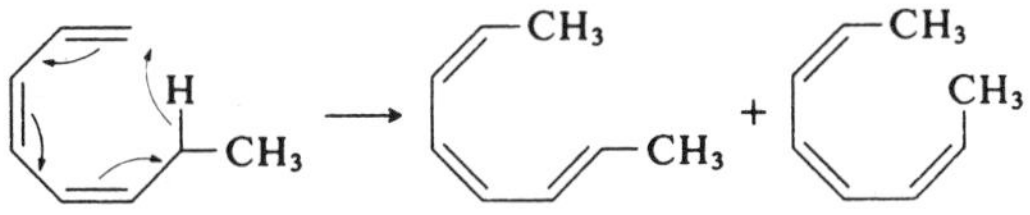

More complex structures such as **19** exhibit similar activation energies. This compound has also been used to demonstrate that the stereochemistry is *antarafacial*, as predicted.[47]

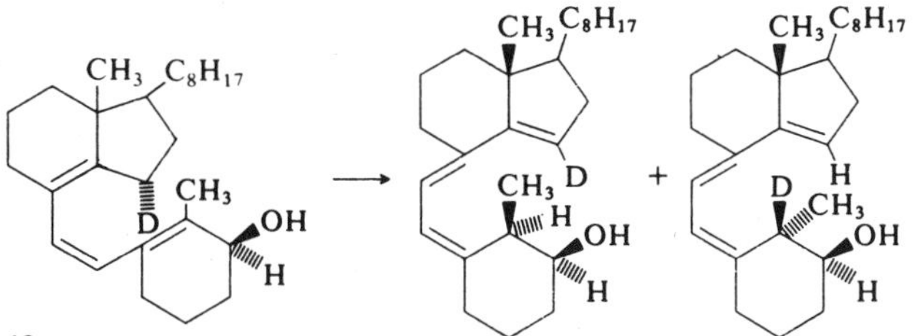

45. M. A. M. Boersma, J. W. de Haan, H. Kloosterziel, and L. J. M. van de Ven, *J. Chem. Soc., Chem. Commun.*, 1168 (1970).
46. J. E. Baldwin and V. P. Reddy, *J. Am. Chem. Soc.* **109**, 8051 (1987).
47. C. A. Hoeger, A. D. Johnston, and W. H. Okamura, *J. Am. Chem. Soc.* **109**, 4690 (1987); W. H. Okamura, C. A. Hoeger, K. J. Miller, and W. Reischl, *J. Am. Chem. Soc.* **110**, 973 (1988).

An especially important case is the thermal equilibrium between precalciferol (pre-vitamin D_2, **22**) and calciferol (vitamin D_2, **23**).[48,49]

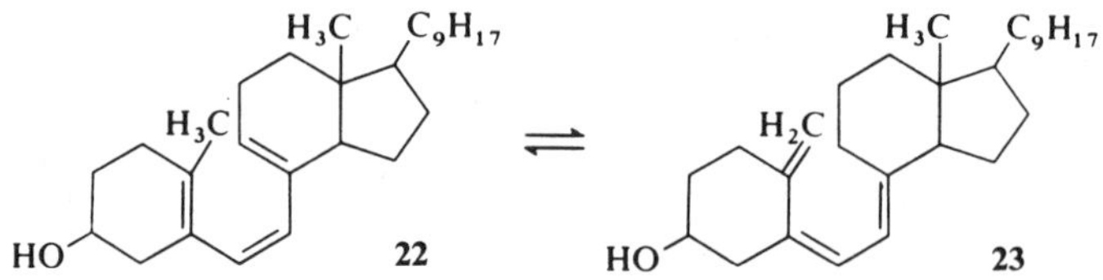

The most important sigmatropic rearrangements from the synthetic point of view are the [3,3] processes. The thermal rearrangement of 1,5-dienes by [3,3] sigmatropy is called the *Cope rearrangement.* The reaction establishes equilibrium between the two interrelated 1,5-dienes and proceeds in the thermodynamically favored direction. The conversion of **24** to **25** provides an example:

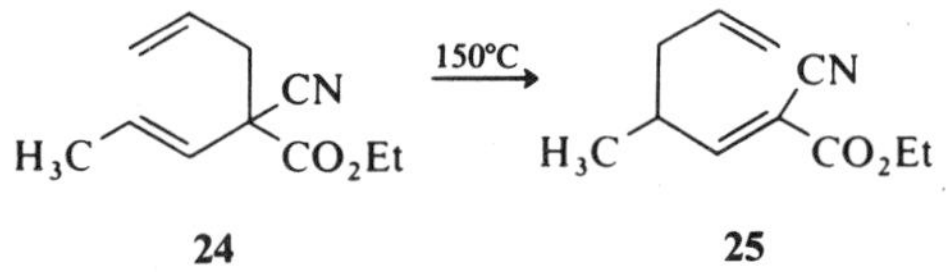

Ref. 50

The equilibrium in this case is controlled by the conjugation present in the product.

The rearrangement of the simplest possible case, 1,5-hexadiene, has been studied with the use of deuterium labeling. The activation enthalpy is 33.5 kcal/mol, and the entropy of activation is −13.8 eu.[51] The substantially negative entropy reflects the formation of the cyclic transition state.

Conjugated substituents at C-2, C-3, C-4, or C-5 accelerate the rearrangement.[52] Donor substituents at C-2 and C-3 also have an accelerating effect.[53] The effect of substituents can be rationalized in terms of the stabilization of the transition state by depicting their effect on two interacting allyl systems.

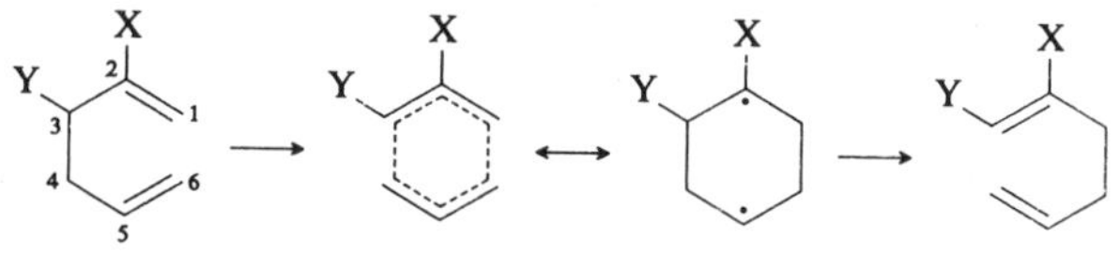

The transition state involves six partially delocalized electrons being transformed from one 1,5-diene system to another. The transition state could range in character

48. J. L. M. A. Schlatmann, J. Pot, and E. Havinga, *Rec. Trav. Chim.* **83**, 1173 (1964).
49. For a historical review of this reaction, see L. Fieser and M. Fieser, *Steroids*, Reinhold, New York, 1959, Chapter 4.
50. A. C. Cope and E. M. Hardy, *J. Am. Chem. Soc.* **62**, 441 (1940).
51. W. v. E. Doering, V. G. Toscano, and G. H. Beasley, *Tetrahedron* **27**, 5299 (1971).
52. M. J. S. Dewar and L. E. Wade, *J. Am. Chem. Soc.* **95**, 290 (1973); **99**, 4417 (1977); R. Wehrli, H. Schmid, D. E. Bellus, and H. J. Hansen, *Helv. Chim. Acta* **60**, 1325 (1977).
53. M. Dollinger, W. Henning, and W. Kirmse, *Chem. Ber.* **115**, 2309 (1982).

from a 1,4-diradical to two nearly independent allyl radicals, depending on whether bond making or bond breaking is more advanced.[54] The net effect of any substituent will be determined by whether it stabilizes the transition state or the ground state more effectively.[55]

The Cope rearrangement usually proceeds through a chairlike transition state. The stereochemical features of the reaction can usually be predicted and analyzed on the basis of a chair transition state that minimizes steric interactions between the substituents. Thus, compound **26** reacts primarily through transition state **27a** to give **28** as the major product. Minor product **29** is formed through the less favorable transition state **27b**.

Ph, CH_3, CH_3 **26** → Ph, CH_3, CH_3 **27a** → Ph, CH_3, H, CH_3 **28**

Ph, CH_3, CH_3 **26** → Ph, CH_3, CH_3 **27b** → Ph, CH_3, CH_3 **29**

When enantiomerically pure **26** is used, the product is >95% optically pure and has the chirality shown above.[56] This result establishes that chirality is maintained throughout the course of the reaction. This is a general feature of [3,3]-sigmatropic shifts and has made them valuable reactions in enantiospecific syntheses.[57]

There is a second possible transition state for the Cope rearrangement, in which the transition state adopts a boatlike geometry:

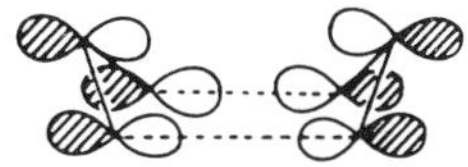

It is generally assumed that this transition state is higher in energy than the chair transition state. There have been several studies aimed at determining the energy difference between the two transition states. One study involved 1,1,1,8,8,8-*deuterio*-4,5-dimethyloctadienes. The chair and boat transition states predict different

54. J. J. Gajewski and N. D. Conrad, *J. Am. Chem. Soc.* **100**, 6268 (1978); J. J. Gajewski and K. E. Gilbert, *J. Org. Chem.* **49**, 11 (1984).
55. For analysis of substituent effects in molecular orbital terminology, see B. K. Carpenter, *Tetrahedron* **34**, 1877 (1978); F. Delbecq and N. T. Anh, *Nouv. J. Chim.* **7**, 505 (1983).
56. R. K. Hill and N. W. Gilman, *J. Chem. Soc., Chem. Commun.*, 619 (1967).
57. R. K. Hill, in *Asymmetric Synthesis*, Vol. 3, J. D. Morrison (ed.), Academic Press, New York, 1984, Chapter 8.

stereoisomeric products.

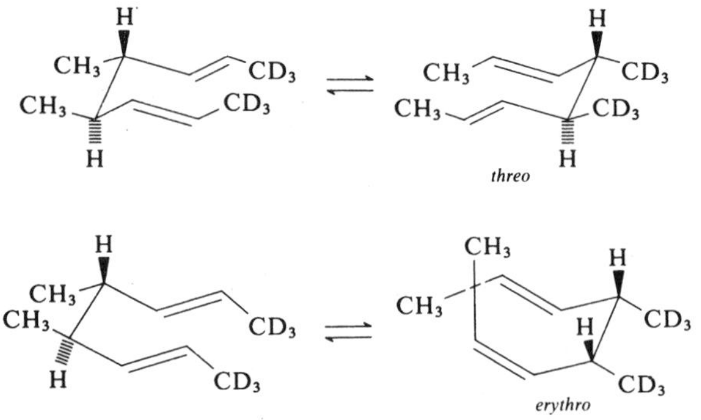

Although the process is further complicated by *cis-trans* isomerizations not considered in the above structures, it was possible by analysis of the product ratio to determine that the boat transition state is about 6 kcal/mol less stable than the chair.[58] This study also demonstrated that dissociation to two independent allyl radicals is only slightly higher in energy than the boat transition state. Related experiments on deuterated 1,5-hexadiene itself indicated a difference of 5.8 kcal in $\Delta G^{\neq}$ for the chair and boat transition states.[59]

Another approach to determining the energy difference between the chair and boat transition states is based on measurement of the activation parameters for the isomeric alkenes **30** and **31**.[60] These two compounds are diastereomeric. While **30** can attain a chairlike transition state, **31** can achieve bonding between the 1,6-carbons only in a boatlike transition state, **31bt**.

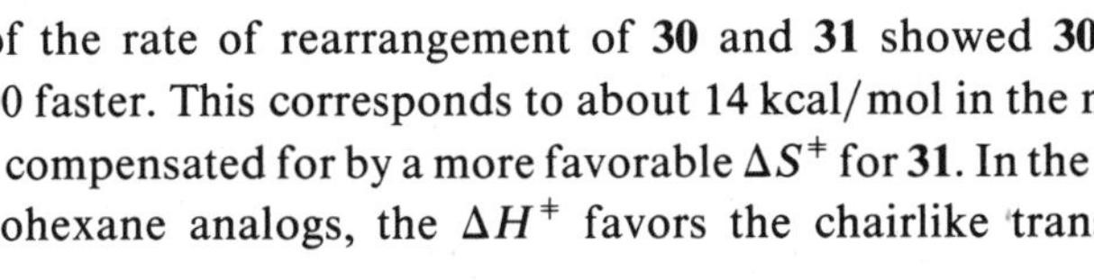

Comparison of the rate of rearrangement of **30** and **31** showed **30** to react by a factor of 18,000 faster. This corresponds to about 14 kcal/mol in the measured $\Delta H^{\neq}$ but is partially compensated for by a more favorable $\Delta S^{\neq}$ for **31**. In the corresponding methylenecyclohexane analogs, the $\Delta H^{\neq}$ favors the chairlike transition state by 16 kcal/mol.

Some particularly striking examples of Cope rearrangements can be found in the rearrangement of *cis*-divinylcyclopropanes. But before we go into these, let us

58. J. J. Gajewski, C. W. Benner, and C. M. Hawkins, *J. Org. Chem.* **52**, 5198 (1987).
59. M. J. Goldstein and M. S. Benzon, *J. Am. Chem. Soc.* **94**, 7147 (1972).
60. K. J. Shea and R. B. Phillips, *J. Am. Chem. Soc.* **102**, 3156 (1980).

examine vinylcyclopropane itself, which is known to rearrange thermally to cyclopentene.[61]

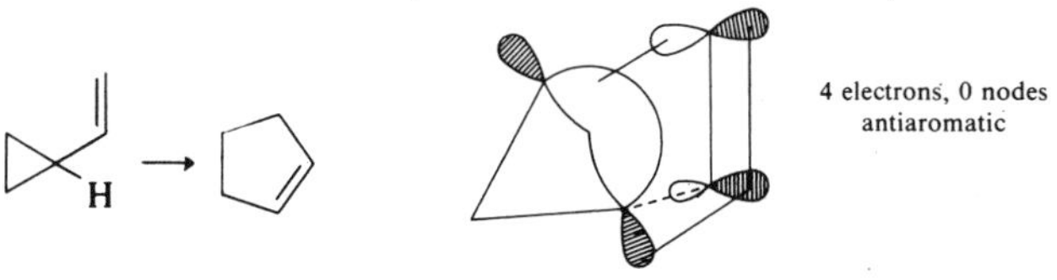

4 electrons, 0 nodes
antiaromatic

The most geometrically accessible transition state corresponds to a forbidden 1,3-suprafacial alkyl shift with retention of configuration. The rearrangement requires a temperature of at least 200–300°C.[62] The measured activation energy is 50 kcal/mol, which is consistent with a stepwise reaction beginning with rupture of a cyclopropane bond. Additional evidence in support of a nonconcerted sequence can be found in the observation that *cis–trans* isomerization of **32** occurs faster than the rearrangement. This isomerization presumably occurs by a reversible bond cleavage.

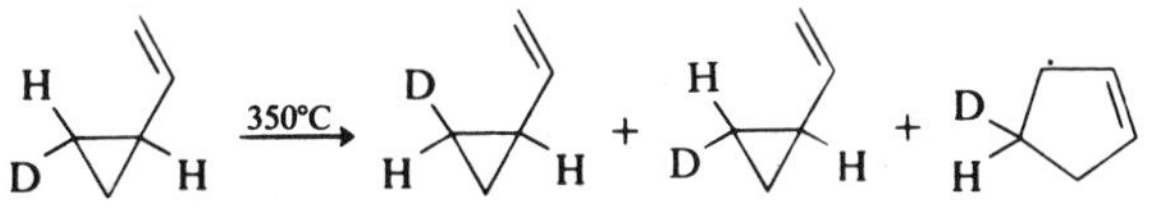

350°C

32 → 47% **33** + 47% **34** + 6% **35**

Ref. 63

A dramatic difference in reactivity is evident when *cis*-divinylcyclopropane is compared with vinylcyclopropane. *cis*-Divinylcyclopropane can only be isolated at low temperature because it very rapidly undergoes Cope rearrangement to 1,4-cycloheptatriene.[64] At 0°C, $\Delta H^{\ddagger}$ is 18.8 kcal/mol and $\Delta S^{\ddagger}$ is -9.4 eu.

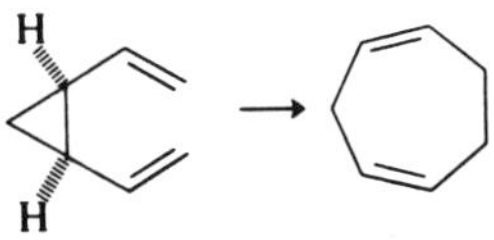

Because of unfavorable molecular geometry, the corresponding rearrangement of *trans*-divinylcyclopropane to cycloheptatriene cannot be concerted, and it requires temperatures on the order of 190°C to occur at a significant rate. The very low energy requirement for the Cope rearrangement of *cis*-divinylcyclopropane reflects several favorable circumstances. The *cis* orientation provides a favorable geometry for interaction of the diene termini so the loss in entropy in going to the transition state is smaller than for an acyclic diene. The bond being broken is strained and

61. C. G. Overberger and A. E. Borchert, *J. Am. Chem. Soc.* **82**, 1007 (1960).
62. T. Hudlicky, T. M. Kutchan, and S. M. Naqui, *Org. React.* **33**, 247 (1984).
63. M. R. Willcott and V. H. Cargle, *J. Am. Chem. Soc.* **89**, 1007 (1960).
64. J. M. Brown, B. T. Bolding, and J. F. Stofko, Jr., *J. Chem. Soc., Chem. Commun.*, 319 (1973); M. Schneider, *Angew. Chem. Int. Ed. Engl.* **14** 707 (1975); M. P. Schneider and A. Rau, *J. Am. Chem. Soc.* **101**, 4426 (1979).

this reduces the enthalpy of activation. The importance of the latter factor can be appreciated on the basis of comparison with *cis*-divinylcyclobutane and *cis*-divinylcyclopentane. The former compound has $\Delta H^{\ddagger} = 23$ kcal/mol for rearrangement to cyclooctadiene.[65] *cis*-Divinylcyclopentane does not rearrange to cyclononadient, even at 250°C.[66] In the latter case, the rearrangement is presumably thermodynamically disfavored.

Divinylcyclopropane rearrangements can proceed with even greater ease if the entropy of activation is made still less negative by incorporating both vinyl groups into a ring. An example of this is found in the degenerate homotropilidene rearrangement. A *degenerate rearrangement* is a reaction process in which no overall change in structure occurs. The product of rearrangement is structurally identical to the starting material. Depending on the rate at which the reaction occurs, the existence of a degenerate rearrangement can be detected by use of isotopic labels or by interpretation of the temperature dependence of the NMR spectrum. In the case of homotropilidene, **36**, the occurrence of a dynamic equilibrium is evident from the NMR spectrum. At low temperature, the rate of interconversion is slow enough that the spectrum is consistent with the presence of four vinyl protons, two allylic protons, and four cyclopropyl protons. As the temperature is raised and the rate of the rearrangement increases, it is observed that two of the vinyl protons remain essentially unchanged in terms of their chemical shift, while the signals of the other two vinyl protons coalesce with two of the cyclopropyl protons. Coalescence is also observed between the signals of the methylene protons and the two remaining cyclopropyl protons.[67] This means that the sets of protons for which the signals coalesce are undergoing sufficiently rapid interchange with one another to result in an averaged signal.

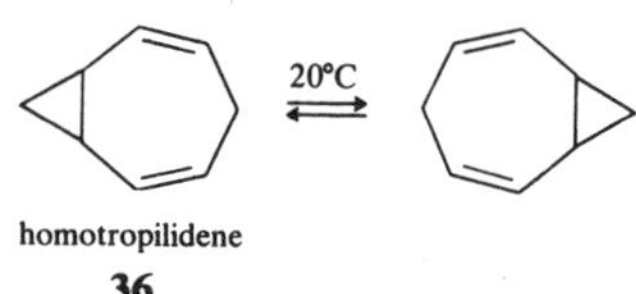

homotropilidene
36

Many other examples of this type of rearrangement are known, one of the most interesting being the case of bullvalene, which is converted into itself with a first-order rate constant of $3.4 \times 10^3\ s^{-1}$ at 25°C.[68] At 100°C, the NMR spectrum of bullvalene exhibits a single peak at 4.22 ppm. This result indicates the "fluxional" nature of the molecule. Because of the threefold axis of symmetry present in bullvalene, the degenerate rearrangement results in all of the carbons having an identical averaged

65. E. Vogel, *Justus Liebigs Ann. Chem.* **615**, 1 (1958); G. S. Hammond and C. D. DeBoer, *J. Am. Chem. Soc.* **86**, 899 (1964).
66. E. Vogel, W. Grimme, and E. Dinné, *Angew. Chem.* **75**, 1103 (1963).
67. G. Schröder, J. F. M. Oth, and R. Merényi, *Angew. Chem. Int. Ed. Engl.* **4**, 752 (1965); H. Günther, J. B. Pawliczek, J. Ulmen, and W. Grimme, *Angew. Chem. Int. Ed. Engl.* **11**, 517 (1972); W. v. E. Doering and W. R. Roth, *Tetrahedron* **19**, 715 (1963).
68. G. Schröder and J. F. M. Oth, *Angew. Chem. Int. Ed. Engl.* **6**, 414 (1967).

environment. The energy of activation for the rearrangement has been determined to be 13.9 kcal/mol.[69]

structures indicating changing environment of carbons in bullvalene

Other degenerate rearrangements have been discovered which are even faster than that of bullvalene. Barbaralane, **37**, rearranges to itself with a rate constant of $1.7 \times 10^7\ s^{-1}$ at 25°C.[70] The energy of activation of this rearrangement is only 7.7 kcal/mol. The lowered energy requirement can be attributed to an increase in ground state energy due to strain. Barbaralane is less symmetrical than bullvalene. There are four different kinds of carbons and protons in the averaged structure. Only the methylene group labeled d is not affected by the degenerate rearrangement.

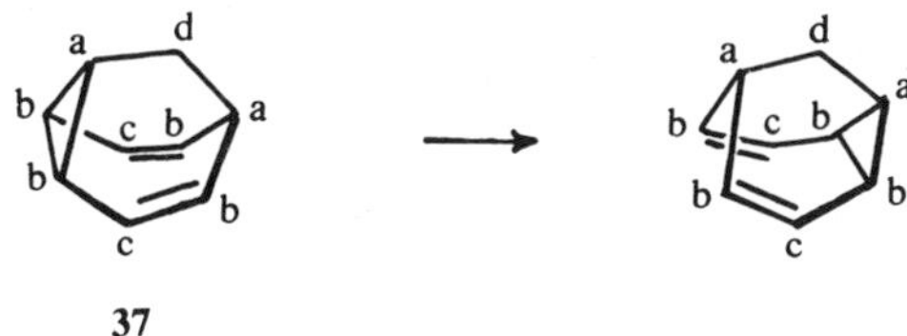

37

A further reduction in the barrier and increase in rate is seen with semibullvalene, **38**, in which strain is increased still more. The $\Delta G^{\neq}$ for this rearrangement is 5.5 kcal/mol at −143°C.[71]

38

The [3,3]-sigmatropic reaction pattern is quite general for other systems that incorporate one or more heteroatoms in place of carbon in the 1,5-hexadiene unit. The most useful and widely studied of these reactions is the Claisen rearrangement, in which an oxygen atom is present in the reacting system.[72] The simplest example of a Claisen rearrangement is the thermal conversion of allyl vinyl ether to 4-pentenal:

$$H_2C{=}CHCH_2OCH{=}CH_2 \longrightarrow [H_2C\cdots O\cdots CH,\ HC\cdots CH_2\cdots CH_2] \longrightarrow H_2C{=}CHCH_2CH_2CH{=}O$$

69. R. Pouke, H. Zimmerman, and Z. Luz, *J. Am. Chem. Soc.* **106**, 5391 (1984).
70. W. v. E. Doering, B. M. Ferrier, E. T. Fossel, J. H. Hartenstein, M. Jones, Jr., G. Klumpp, R. M. Rubin, and M. Saunders, *Tetrahedron* **23**, 3943 (1967); H. Günther, J. Runsink, H. Schmickler, and P. Schmitt, *J. Org. Chem.* **50**, 289 (1985).
71. A. K. Cheng, F. A. L. Anet, J. Mioduski, and J. Meinwald, *J. Am. Chem. Soc.* **96**, 2887 (1974).
72. G. B. Bennett, *Synthesis*, 589 (1977); S. J. Rhoads and N. R. Raulins, *Org. React.* **22**, 1 (1975).

This reaction occurs with an energy of activation of 30.6 kcal/mol and an entropy of activation of −7.7 eu at 180°C.[73]

Extensive studies on Claisen rearrangements of allyl ethers of phenols have provided further evidence bearing on [3,3]-sigmatropic rearrangements.[74] For example, an important clue as to the mechanism of the Claisen rearrangement was obtained by use of ^{14}C-labeled allyl phenyl ether. It was found that the rearrangement was specific with respect to which carbon atom of the allyl group became bonded to the ring, leading to the proposal of the following mechanism[75]:

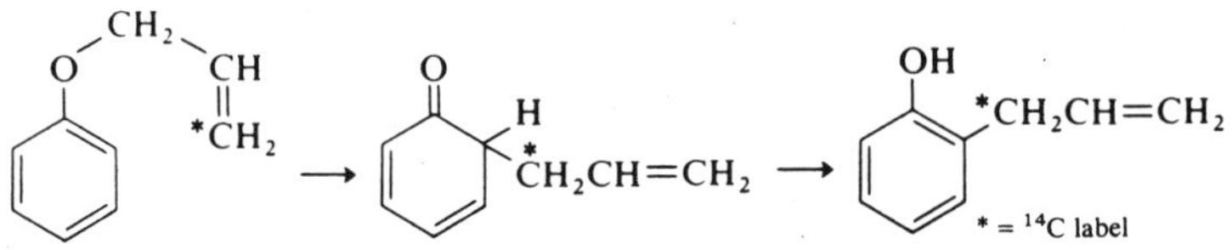

* = ^{14}C label

The intramolecular nature of the rearrangement was firmly established by a crossover experiment in which **39** and **40** were heated simultaneously and found to yield the same product as when they were heated separately. There was no evidence for the formation of the crossover products **43** and **44**.[76]

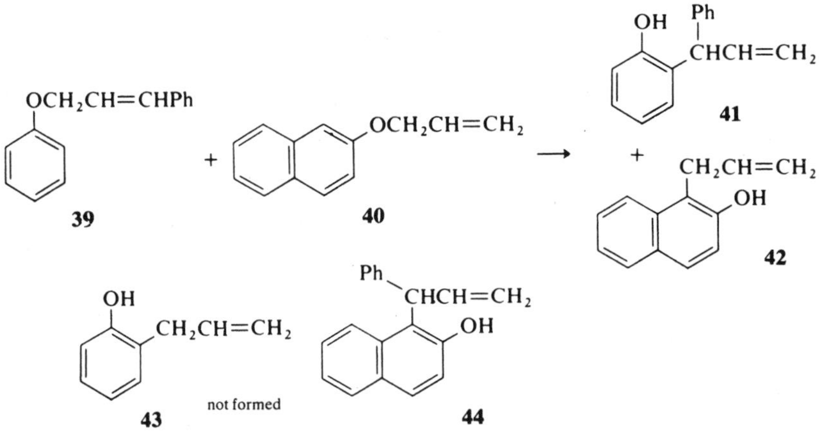

not formed

The stereochemical features of the Claisen rearrangement are very similar to those described for the Cope rearrangement, and reliable stereochemical predictions can be made on the basis of the preference for a chairlike transition state. The major product will have the *E*-configuration at the newly formed double bond because of the preference for placing the larger substituent in the pseudoequatorial position in the transition state.[77]

73. F. W. Schuler and G. W. Murphy, *J. Am. Chem. Soc.*, **72**, 3155 (1950).
74. D. S. Tarbell, *Org. React.* **2**, 1 (1944).
75. J. P. Ryan and P. R. O'Connor, *J. Am. Chem. Soc.* **74**, 5866 (1952).
76. C. D. Hurd and L. Schmerling, *J. Am. Chem. Soc.* **59**, 107 (1937).
77. R. Marbet and G. Saucy, *Helv. Chim. Acta* **50**, 2095 (1967); A. W. Burgstahler, *J. Am. Chem. Soc.* **82**, 4681 (1960).

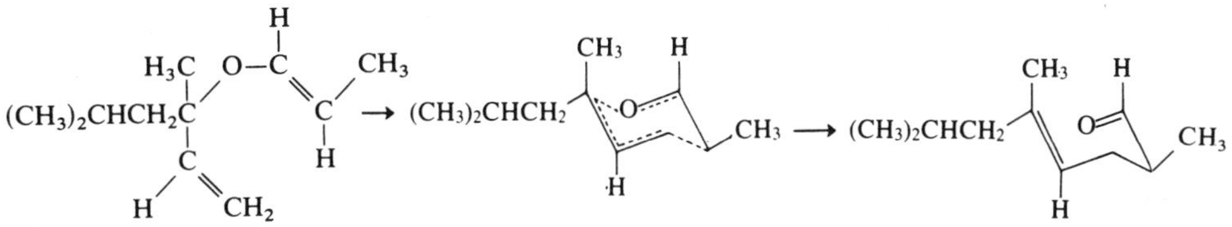

Studies of chiral substrates have also demonstrated that chirality is maintained in the reaction.[78] Further examples of the synthetic application of the Claisen rearrangement are discussed in Section 6.5 of Part B.

Like the Cope rearrangement, the Claisen rearrangement is sensitive to substituents on the reacting system. Cyano groups promote the rearrangement by a factor of 10^2 at positions 2 and 4 and have smaller effects at the other positions, as shown below.[79]

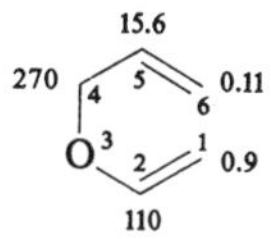

A donor substituent, for example, trimethylsilyloxy, at C-2 is strongly accelerating.[80] As in the case of the Cope rearrangement, the interpretation of these substituent effects is best approached by considering the effect on transition state stability. A model has been developed which treats the transition state as intermediate in character between a 2,5-diradical with strong 3–4 and 6–1 bonding and a pair of weakly interacting allyl-like radicals. By estimating the effect of the substituent on each structure and introducing a contribution for the overall free-energy change of the process, the following equation gives a good correlation with experimental activation energies[81]:

$$\Delta G^{\ddagger} = \frac{\Delta G_x^{\ddagger} \Delta G_t^{\ddagger}}{\Delta G_x^{\ddagger} + \Delta G_y^{\ddagger} + \Delta G_r}$$

where $\Delta G_x^{\ddagger}$ is the energy required for 3–4 bond scission, $\Delta G_y^{\ddagger}$ is the energy released by 1–6 bond formation, and ΔG_r is the overall free-energy change of the reaction. The correlation can be improved by including a weighting factor whose physical significance is presumably related to the extent of bond making versus bond breaking in the transition state.

There are also concerted rearrangements which exhibit the [2,3]-sigmatropic reactivity pattern. The most well developed of these reactions are rearrangements

78. H. L. Goering, and W. I. Kimoto, *J. Am. Chem. Soc.* **87**, 1748 (1965).
79. C. J. Burrows and B. K. Carpenter, *J. Am. Chem. Soc.* **103**, 6983 (1981).
80. J. J. Gajewski and J. Emrani, *J. Am. Chem. Soc.* **106**, 5733 (1984); S. E. Denmark and M. A. Marmata, *J. Am. Chem. Soc.* **104**, 4972 (1982).
81. J. J. Gajewski and K. E. Gilbert, *J. Org. Chem.* **49**, 11 (1984).

of nitrogen[82] and sulfur[83] ylides and rearrangements of allyl sulfoxides[84] and selenoxides.[85] One requirement for a [2,3]-sigmatropic process is that the atom at the allylic position be able to act as a leaving group when the adjacent atom begins bonding to the allyl system:

This reaction is most facile in systems where the atoms X and Y bear formal charges, as in the case of ylides and sulfoxides:

$$CH_2{=}CHCH_2{-}\overset{+}{S}(R){-}O^- \rightarrow RSO{-}CH_2CH{=}CH_2$$

$$CH_2{=}CHCH_2{-}\overset{+}{S}(R'){-}{}^-CHR \rightarrow R'SCH(R)CH_2CH{=}CH_2$$

$$CH_2{=}CHCH_2{-}\overset{+}{N}(R')_2{-}{}^-CHR \rightarrow (R')_2NCH(R)CH_2CH{=}CH_2$$

Other examples of [2,3]-sigmatropic rearrangements involve amine oxides and diazenes:

$$PhCH{=}CH{-}CH(CH_3){-}\overset{+}{N}(CH_3)_2{-}O^- \xrightarrow[24\text{ days}]{-20°C} PhCH(O{-}N(CH_3)_2){-}CH{=}CHCH_3$$ Ref. 86

$$CH_3N(NH_2){-}CH_2CH{=}CH_2 \xrightarrow[Et_3N]{(CH_3)_3COCl} CH_3{-}\overset{+}{N}(={}^-N{:}){-}CH_2{-}CH{=}CHCH_3 \rightarrow CH_3{-}N{=}N{-}CH(CH_3)CH{=}CH_2$$

Ref. 87

Some of the most useful synthetic applications of these reactions are for ring expansion:

$$\xrightarrow[-40°C]{K^+\ {}^-OC(CH_3)_3}$$ 85% Ref. 88

82. E. Vedejs, J. P. Hagen, B. L. Roach, and K. L. Spear, *J. Org. Chem.* **43**, 1185 (1978).
83. B. M. Trost and L. S. Melvin, Jr., *Sulfur Ylides*, Academic Press, New York, 1975.
84. D. A. Evans, and G. C. Andrews, *Acc. Chem. Res.* **7**, 147 (1974).
85. K. B. Sharpless and R. F. Lauer, *J. Am. Chem. Soc.* **95**, 2697 (1973); D. L. J. Clive, *Tetrahedron* **34**, 1049 (1978).
86. Y. Yamamoto, J. Oda, and Y. Inouye, *J. Org. Chem.* **41**, 303 (1976).
87. J. E. Baldwin, J. E. Brown, and G. Höfle, *J. Am. Chem. Soc.* **93**, 788 (1971).
88. V. Ceré, C. Paolucci, S. Pollicino, E. Sandri, and A. Fava, *J. Org. Chem.* **43**, 4826 (1978).

Further examples of synthesis using [2,3]-sigmatropic reactions are given in Section 6.6 of Part B.

11.3. Cycloaddition Reactions

The principles of conservation of orbital symmetry also apply to intermolecular cycloaddition and to the reverse, concerted fragmentation of one molecule into two or more smaller components (cycloreversion). The most important cycloaddition reaction from the point of view of synthesis is the *Diels–Alder reaction.* This reaction has been the object of extensive theoretical and mechanistic study, as well as synthetic application.[89] The Diels–Alder reaction is the addition of an alkene to a diene to form a cyclohexene. It is called a [4 + 2]-cycloaddition reaction because four π electrons from the diene and the two π electrons from the alkene (which is called the *dienophile*) are directly involved in the bonding change. All available data are consistent with describing the reaction as a concerted process. In particular, the reaction is stereospecific and is a *syn* (suprafacial) addition with respect to both the alkene and the diene. This stereospecificity has been demonstrated with many substituted dienes and alkenes and also holds for the simplest possible example of the reaction, that of ethylene with butadiene[90]:

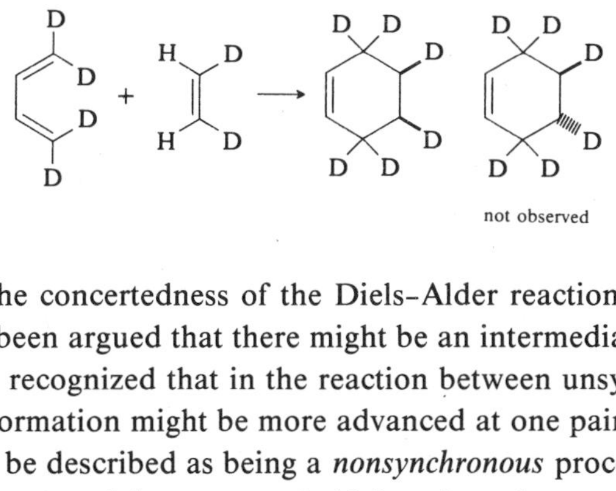

not observed

The issue of the concertedness of the Diels–Alder reaction has been debated extensively. It has been argued that there might be an intermediate that is diradical in character.[91] It is recognized that in the reaction between unsymmetrical alkenes and dienes, bond formation might be more advanced at one pair of termini than at the other. This can be described as being a *nonsynchronous* process. Loss of stereospecificity is to be expected, however, only if there is an intermediate in which one bond is formed and the other is not, so that there can be rotation at the unbound

89. M. C. Kloetzel, *Org. React.* **4**, 1 (1948); H. L. Holmes, *Org. React.* **4**, 60 (1958); S. Seltzer, *Adv. Alicyclic Chem.* **2**, 1 (1968); J. G. Martin and R. K. Hill, *Chem. Rev.* **61**, 537 (1961); A. Wasserman, *Diels–Alder Reactions*, Elsevier, London, 1965; J. Hamer (ed.), *1,4-Cycloaddition Reactions—The Diels–Alder Reaction in Heterocyclic Syntheses*, Academic Press, New York, 1967; J. Sauer and R. Sustmann, *Angew. Chem. Int. Ed. Engl.* **19**, 779 (1980); R. Gleiter and M. C. Böhm, *Pure Appl. Chem.* **55**, 237 (1983); R. Gleiter and M. C. Böhm, in *Stereochemistry and Reactivity of Systems Containing π Electrons*, W. H. Watson (ed.), Verlag Chemie, Deerfield Beach, Florida, 1983.
90. K. N. Houk, Y.-T. Lin, and F. K. Brown, *J. Am. Chem. Soc.* **108**, 554 (1986).
91. M. J. S. Dewar, S. Olivella, and J. P. Stewart, *J. Am. Chem. Soc.* **108**, 5771 (1986).

termini.

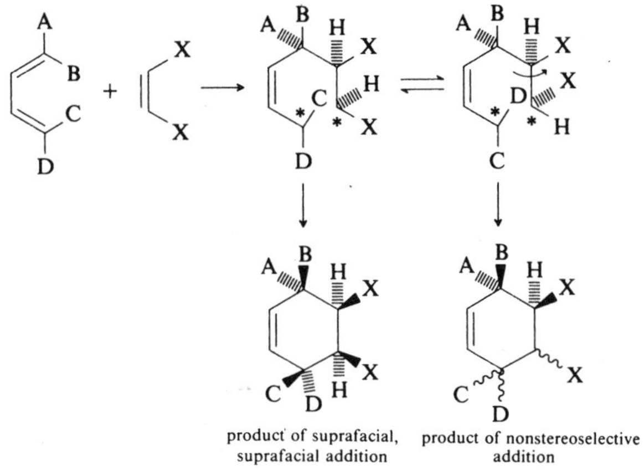

product of suprafacial, suprafacial addition   product of nonstereoselective addition

Diels–Alder reactions are almost always stereospecific, which implies that if an intermediate exists, it cannot have a lifetime sufficient to permit rotation. The prevailing view is that the majority of Diels–Alder reactions are concerted processes, and most current theoretical analyses agree with this view.[92]

Another stereochemical feature of the Diels–Alder reaction is addressed by the *Alder rule.* The empirical observation is that if two isomeric adducts are possible, the one that has an unsaturated substituent(s) on the alkene oriented toward the newly formed cyclohexene double bond is the preferred product. The two alternative transition states are referred to as the *endo* and *exo* transition states.

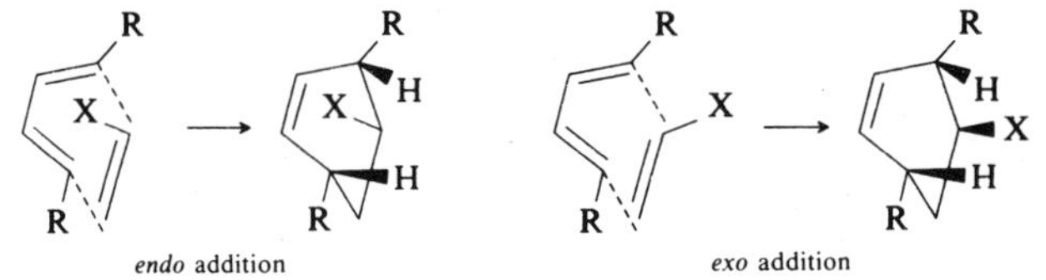

endo addition   *exo* addition

For example, the addition of dienophiles to cyclopentadiene usually favors the *endo* stereoisomer, even though this is the sterically more congested case.

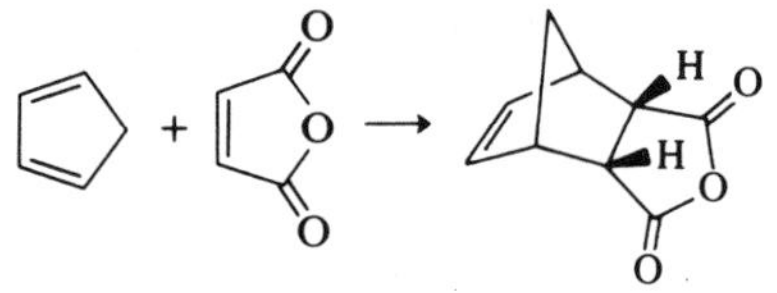

The preference for the *endo* mode of addition is not restricted to cyclic dienes such as cyclopentadiene. With the use of deuterium labels, it has been found that the addition of 1,3-butadiene and maleic anhydride leads to 85% yield of the product

92. J. J. Gajewski, K. B. Peterson, and J. R. Kagel, *J. Am. Chem. Soc.* **109**, 5545 (1987); K. N. Houk, Y.-T. Lin, and F. K. Brown, *J. Am. Chem. Soc.* **108**, 554 (1986).

arising from the *endo* transition state.[93]

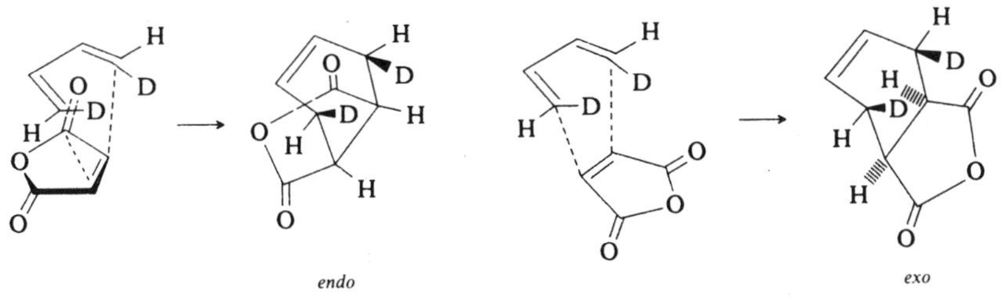

endo *exo*

In general, stereochemical predictions based on the Alder rule can be made by aligning the diene and dienophile in such a way that any unsaturated substituents on the dienophile overlap with the diene π system. The steroselectivity predicted by the Alder rule is independent of the requirement for suprafacial-suprafacial cycloaddition, since both the *endo* and *exo* transition states meet this requirement.

There are probably several factors which contribute to determining the *endo* : *exo* ratio in any specific case. These include steric effects, dipole-dipole interactions, and London dispersion forces. Molecular orbital interpretations emphasize *secondary orbital interactions* between the π orbitals on the dienophile substituent(s) and the developing π bond between C-2 and C-3 of the diene. There are quite a few exceptions to the Alder rule, and in most cases the preference for the *endo* isomer is relatively modest. For example, while cyclopentadiene reacts with methyl acrylate in decalin solution to give mainly the *endo* adduct (75%), the *endo* : *exo* ratio is solvent sensitive and ranges up to 90% *endo* in methanol. When a methyl substituent is added to the dienophile (methyl methacryalate), the *exo* product predominates.[94]

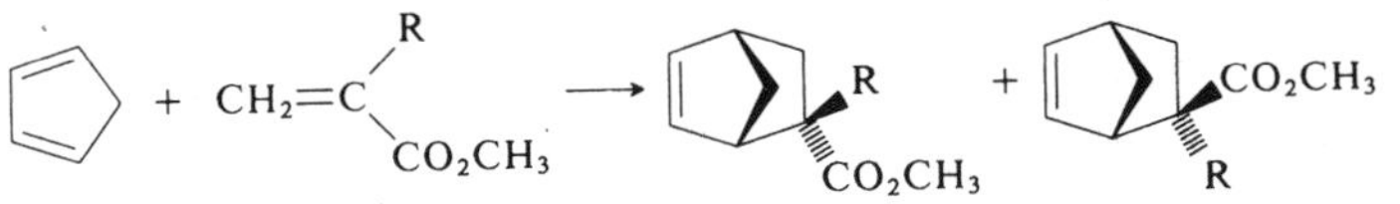

R = H	75–90%	25–10%
R = CH_3	22–40%	78–60%

How do orbital symmetry requirements relate to [4 + 2] and other cycloaddition reactions? Let us construct a correlation diagram for the addition of butadiene and ethylene to give cyclohexene. For concerted addition to occur, it may be assumed that the diene must adopt an *s-cis* conformation. Since the electrons that are involved are the π electrons in both the diene and the dienophile, it is expected that the reaction must occur via a face-to-face rather than an edge-to-edge orientation. When this orientation of the reacting complex and transition state is adopted, it can be seen that a plane of symmetry perpendicular to the planes of the reacting molecules

93. L. M. Stephenson, D. E. Smith, and S. P. Current, *J. Org. Chem.* **47**, 4170 (1982).
94. J. A. Berson, Z. Hamlet, and W. A. Mueller, *J. Am. Chem. Soc.* **84**, 297 (1962).

is maintained during the course of the cycloaddition:

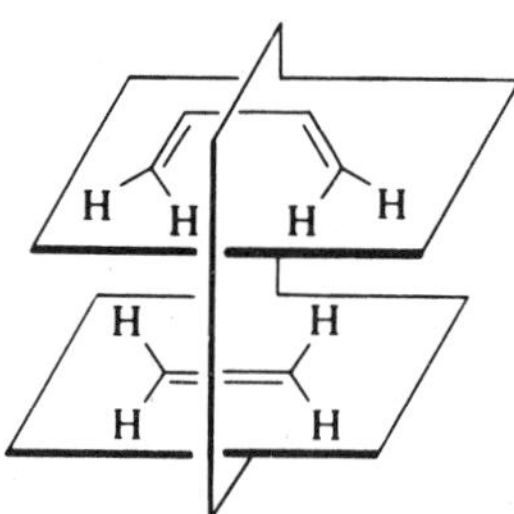

An orbital correlation diagram can be constructed by examining the symmetry of the reactant and product orbitals with respect to this plane. The orbitals are classified by symmetry with respect to this plane in Fig. 11.8. For the reactants ethylene and butadiene, the classifications are the same as for the consideration of electrocyclic reactions on p. 600. An additional feature must be taken into account in the case of cyclohexene. The cyclohexene orbitals σ_1, σ_2, σ_1^*, and σ_2^* are called *symmetry-adapted orbitals.* We might be inclined to think of the σ and σ^* orbitals as localized between specific pairs of carbon atoms. This is not the case for the MO treatment, and localized orbitals would fail the test of being either symmetric or antisymmetric with respect to the plane of symmetry. In the construction of orbital correlation diagrams, *all* orbitals involved must be either symmetric or antisymmetric with respect to the element of symmetry being considered.

When the orbitals have been classified with respect to symmetry, they can be arranged according to energy and the correlation lines can be drawn as in Fig. 11.9. From the orbital correlation diagram it can be concluded that the thermal concerted cycloaddition reaction between butadiene and ethylene is allowed. All bonding levels of the reactants correlate with product ground state orbitals. Extension of orbital correlation analysis to cycloaddition reactions with other numbers of π electrons leads to the conclusion that the suprafacial-suprafacial addition is allowed for systems with $4n + 2$ π electrons but forbidden for systems with $4n$ π electrons.

The same conclusions are drawn by analysis of the frontier orbitals involved in cycloadditions. For the most common case of the Diels–Alder reaction, which involves dienophiles with electron-attracting substituents, the frontier orbitals are ψ_2 of the diene (which is the HOMO) and π^* of the dienophile (which is the LUMO). Reaction occurs by interaction of the HOMO and LUMO, which can be seen from the illustration below to be allowed.

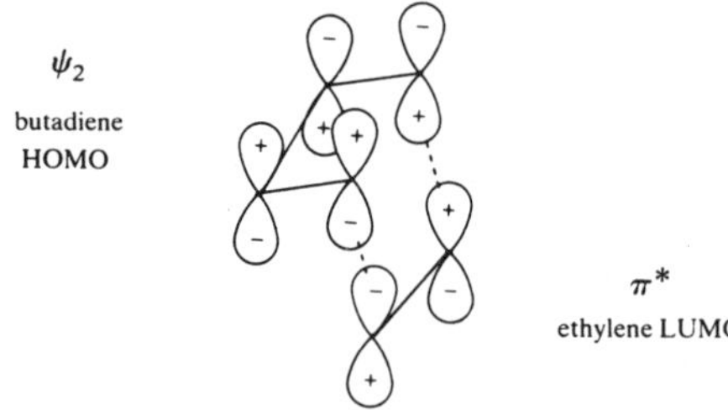

The selection rules for cycloaddition reactions can also be derived from consideration of the aromaticity of the transition state. The transition states for [2 + 2] and [4 + 2] cycloadditions are depicted in Fig. 11.10. For the [4 + 2] suprafacial-suprafacial cycloaddition, the transition state is aromatic. For [2 + 2] cycloaddition, the suprafacial-suprafacial mode is antiaromatic but the suprafacial-antarafacial mode is aromatic. In order to specify the topology of cycloaddition reactions,

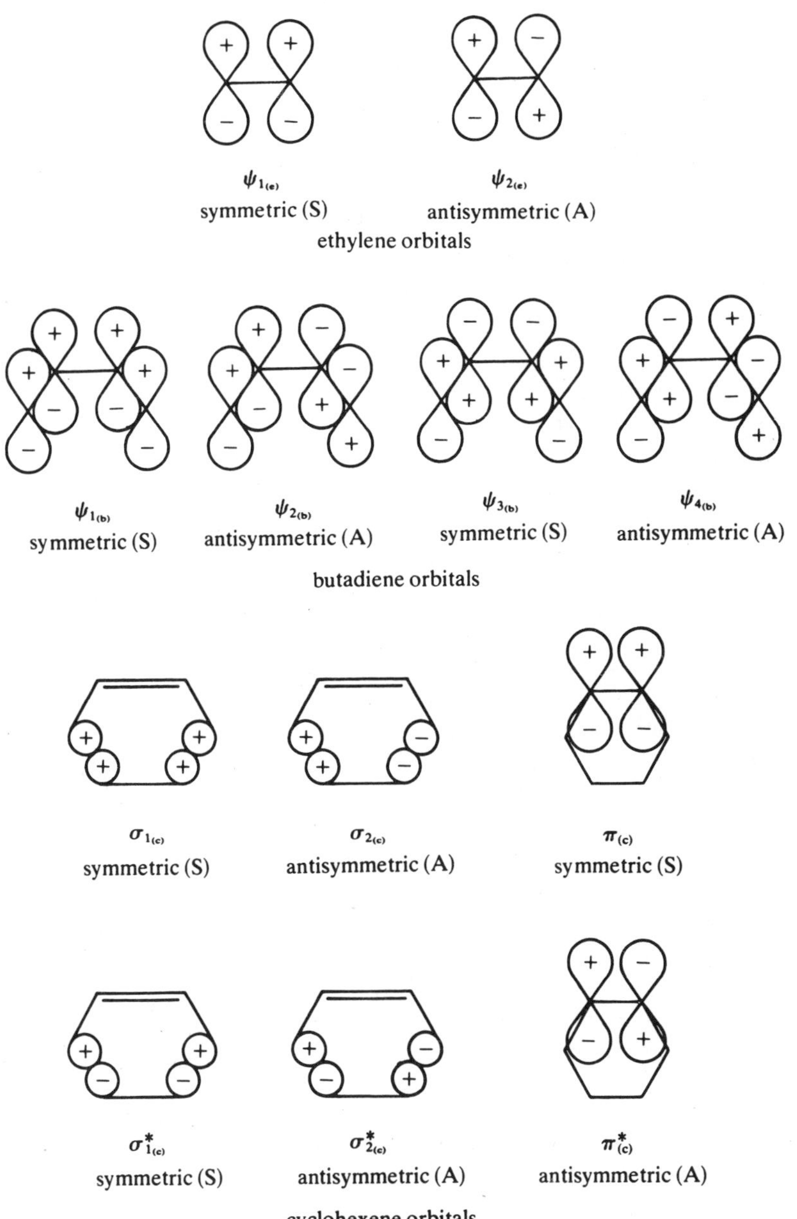

Fig. 11.8. Symmetry properties of ethylene, butadiene, and cyclohexene orbitals with respect to cycloaddition.

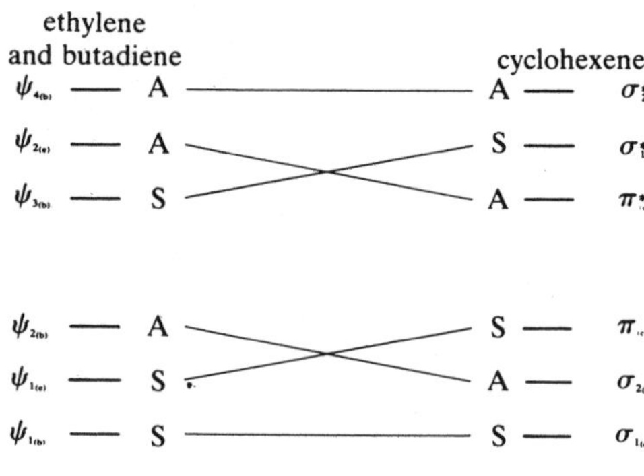

Fig. 11.9. Correlation diagram for ethylene, butadiene, and cyclohexene orbitals.

subscripts are added to the numerical classification. Thus, a Diels–Alder reaction is a $[4\pi_s + 2\pi_s]$ cycloaddition. The allowed [2 + 2] cycloaddition process would be more completely described by the designation $[2\pi_s + 2\pi_a]$, where the subscripts a and s indicate the antarafacial or suprafacial nature of the addition at each component. The generalized Woodward–Hoffmann rules for cycloaddition are summarized below. These can be arrived at by orbital correlation analysis, by frontier orbital theory, or by classification of the transition states.

Selection rules for $m + n$ cycloadditions

$m + n$	supra/supra	supra/antara	antara/antara
$4n$	forbidden	allowed	forbidden
$4n + 2$	allowed	forbidden	allowed

It has long been known that the Diels–Alder reaction is particularly efficient and rapid when the dienophile contains one or more electron-attracting groups and is favored still more if the diene also contains electron-releasing groups. These substituent effects are illustrated by the data in Tables 11.1 and 11.2.

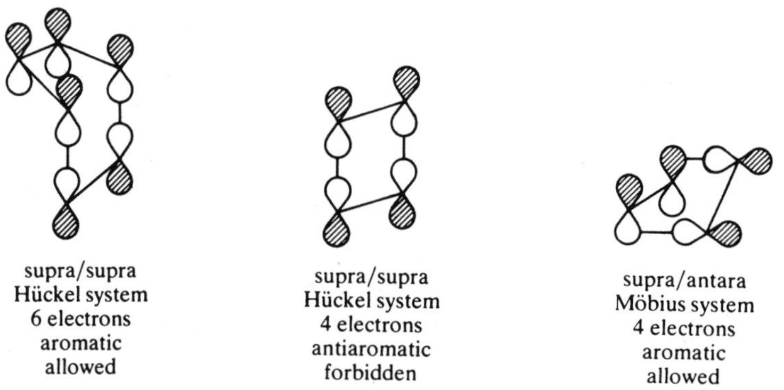

Fig. 11.10. Classification of cycloaddition reactions with respect to basis set orbitals.

Table 11.1. Relative Reactivity toward Cyclopentadiene in the Diels–Alder Reaction

Dienophile	Relative rate[a]
Tetracyanoethylene	4.3×10^7
1,1-Dicyanoethylene	4.5×10^5
Maleic anhydride	5.6×10^4
p-Benzoquinone	9.0×10^3
Maleonitrile	91
Fumaronitrile	81
Dimethyl fumarate	74
Dimethyl maleate	0.6
Methyl acrylate	1.2
Acrylonitrile	1.0

a. From second-order rate constants in dioxane at 20°C as reported by J. Sauer, H. Wuest, and A. Mielert, *Chem. Ber.* **97**, 3183 (1964).

The regiochemistry of the Diels–Alder reaction is also sensitive to the nature of substituents on the diene and dienophile. The combination of an electron donor in the diene and an electron acceptor in the dienophile gives rise to cases A and B. The preferred orientations are shown in Scheme 11.3. There are also examples where the substituents are reversed so that the electron donor substituent is on the dienophile and the electron-accepting substituent is on the diene. These are called *inverse electron demand Diels–Alder* reactions. This gives rise to combinations C and D in Scheme 11.3. Cases where both diene and dienophile have the same type

Table 11.2. Relative Reactivity of Substituted Butadienes in the Diels–Alder Reaction[a]

Diene	Dienophile: Tetracyano-ethylene	Dienophile: Maleic anhydride
$CH_2{=}CHCH{=}CH_2$	1	1
$CH_3CH{=}CHCH{=}CH_2$	103	3.3
$CH_2{=}C(CH_3)CH{=}CH_2$	45	2.3
$CH_3CH{=}CHCH{=}CHCH_3$	1.66×10^3	
$PhCH{=}CHCH{=}CH_2$	385	1.65
$CH_2{=}C(Ph)CH{=}CH_2$	191	8.8
$CH_3OCH{=}CHCH{=}CH_2$	5.09×10^4	12.4
$CH_2{=}C(OCH_3)CH{=}CH_2$	1.75×10^3	
$CH_3OCH{=}CHCH{=}CHOCH_3$	4.98×10^4	
Cyclopentadiene	2.1×10^6	1.35×10^3

a. C. Rücker, D. Lang, J. Sauer, H. Friege, and R. Sustmann, *Chem. Ber.* **113**, 1663 (1980).

Scheme 11.3. Regioselectivity of the Diels–Alder Reaction

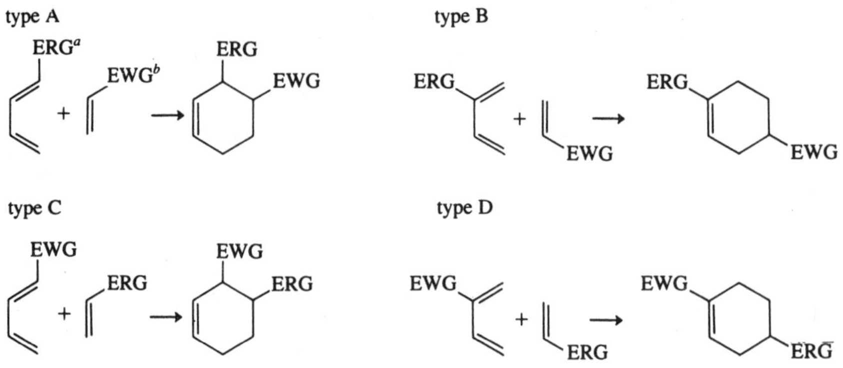

type A

ERG[a] + EWG[b] → ERG, EWG

type B

ERG + EWG → ERG, EWG

type C

EWG + ERG → EWG, ERG

type D

EWG + ERG → EWG, ERG

a. ERG, electron-releasing group.
b. EWG, electron-withdrawing group.

of substituent lead to poor reactivity and are rare. Both the reactivity data in Tables 11.1 and 11.2 and the regiochemical relationships in Scheme 11.3 can be understood on the basis of frontier orbital theory.

In reactions of types A and B illustrated in Scheme 11.3 the frontier orbitals will be the diene HOMO and the dienophile LUMO. This is illustrated in Fig. 11.11. This will be the strongest interaction because the donor substituent on the diene will raise the diene orbitals in energy whereas the acceptor substituent will lower the dienophile orbitals. The strongest interaction will be between ψ_2 and π^*. In reactions of types C and D, the pairing of diene LUMO and dienophile HOMO will be expected to be the strongest interaction because of the substituent effects, as illustrated in Fig. 11.11.

The regiochemical relationships summarized in Scheme 11.3 can be understood by considering the atomic coefficients of the frontier orbitals. Figure 11.12 gives the approximate energies and orbital coefficients for the various classes of dienes and dienophiles.

The regiochemistry can be predicted on the basis of the generalization that the strongest interaction will be between the centers on the frontier orbitals having the

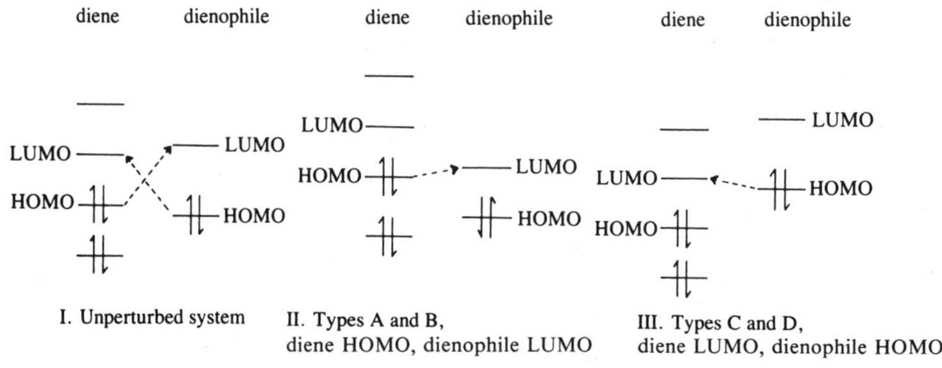

Fig. 11.11. Frontier orbital interactions in Diels–Alder reactions.

largest orbital coefficients. For dienophiles with electron-withdrawing substituents, π^* has its largest coefficient on the β-carbon atom. For dienes with electron donor substituents at C-1 of the diene, the HOMO has its largest coefficient at C-4. This is the case designated A in Scheme A. The prediction of regiochemistry that follows is that C-2 of the dienophile will interact most strongly with C-4 of the diene. This is the preferred regiochemistry for the type A Diels–Alder addition. A similar analysis of each of the other combinations in Scheme 11.3 using the orbitals in Fig. 11.12 leads to the prediction of the favored regiochemistry. Notice that in the type A and C reactions this leads to preferential formation of the more sterically congested 1,2-disubstituted cyclohexene. These frontier orbital relationships provide an excellent predictive basis for Diels–Alder reactions.[95]

Frontier orbital concepts can also serve to explain the very strong catalysis of certain Diels–Alder reactions by Lewis acids.[96] A variety of Lewis acids are effective catalysts including $SnCl_4$, $ZnCl_2$, $AlCl_3$, and derivatives of $AlCl_3$ such as $(C_2H_5)_2AlCl$. The types of dienophiles that are subject to catalysis are typically those with carbonyl activating groups. Lewis acids are known to form complexes at the carbonyl oxygen, and this has the effect of increasing the electron-withdrawing capacity of the substituent group:

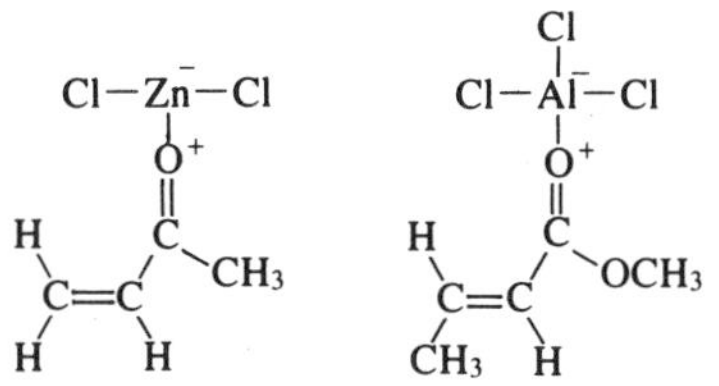

This complexation accentuates both the energy and orbital distortion effects of the substituent and enhances both the reactivity and the selectivity of the dienophile, relative to the uncomplexed compound.[97]

Cycloadditions of the [4 + 2] type are not restricted to the reactions of combinations of neutral dienes and dienophiles. There are examples of corresponding reactions involving ionic intermediates. The addition of 2-methylallyl cation to cyclopentadiene is an example.[98]

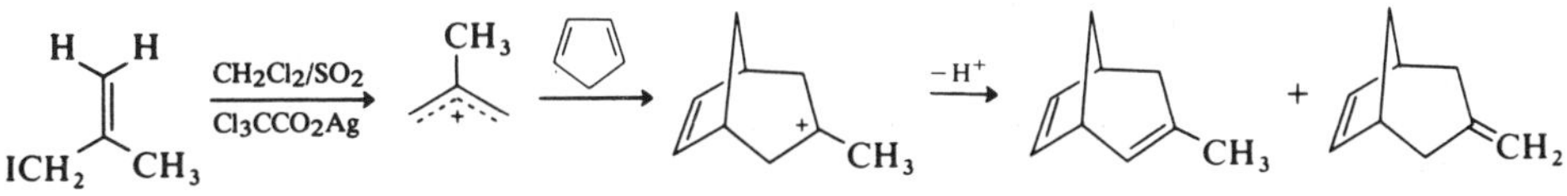

95. For general discussion on the development and application of frontier orbital concepts in cycloaddition rections, see K. N. Houk, *Acc. Chem Res.* **8**, 361 (1975); K. N. Houk, *J. Am. Chem. Soc.* **95**, 4092 (1973); K. N. Houk, *Top. Curr. Chem.* **79**, 1 (1979); R. Sustmann and R. Schubert, *Angew. Chem. Int. Ed. Engl.* **11**, 840 (1972); J. Sauer and R. Sustmann, *Angew. Chem. Int. Ed. Engl.* **19**, 779 (1980).
96. P. Laszlo and J. Lucche, *Actual. Chim.*, 42 (1984).
97. K. N. Houk and R. W. Strozier, *J. Am. Chem. Soc.* **95**, 4094 (1973).
98. H. M. R. Hoffmann, D. R. Joy, and A. K. Suter, *J. Chem. Soc., B*, 57 (1968); H. M. R. Hoffmann and D. R. Joy, *J. Chem. Soc., B*, 1182 (1968).

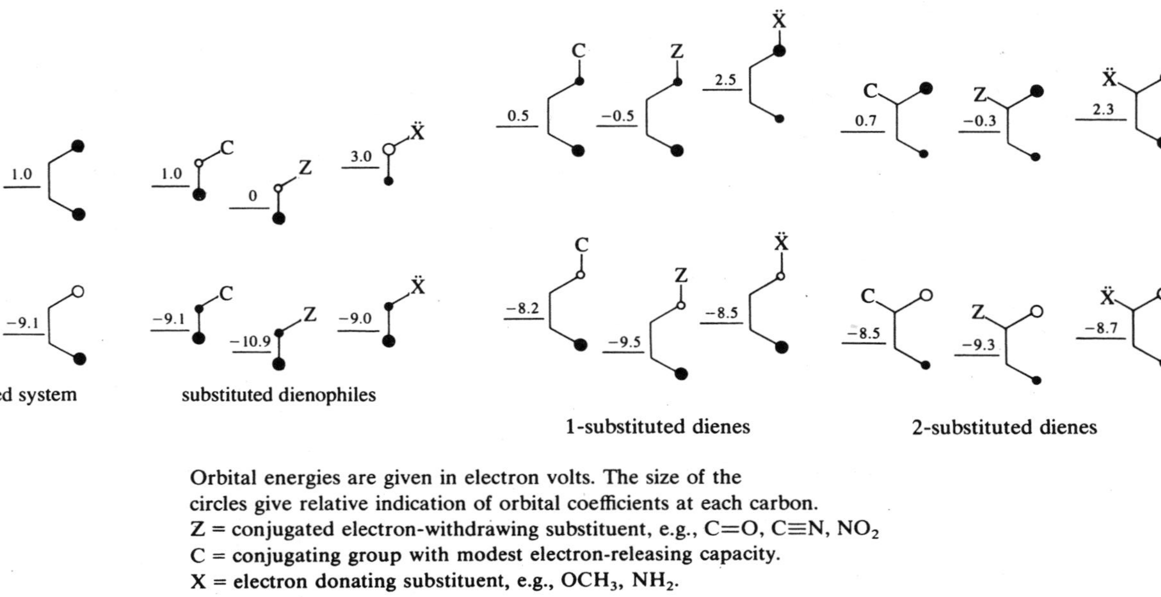

Fig. 11.12. Coefficients and relative energies of dienophile and diene molecular orbitals. Reproduced from K. N. Houk, *J. Am. Chem. Soc.* **95**, 4092 (1973).

Scheme 11.4. Some 1,3-Dipoles

Nitrile oxide	$R{-}C{\equiv}\overset{+}{N}{-}\ddot{O}{:}^{-} \longleftrightarrow R{-}\overset{+}{C}{=}N{-}\ddot{O}{:}^{-}$
Azides	$R{-}\overset{-}{\ddot{N}}{-}N{=}\overset{+}{N}{:} \longleftrightarrow R{-}\overset{-}{\ddot{N}}{-}\overset{+}{N}{\equiv}N{:}$
Diazomethane	$^{-}{:}CH_2{-}N{=}\overset{+}{N}{:} \longleftrightarrow {}^{-}{:}CH_2{-}\overset{+}{N}{\equiv}N{:}$
Nitrones	$R_2C{=}\overset{+}{N}(R){-}\overset{-}{\ddot{O}}{:} \longleftrightarrow R_2\overset{+}{C}{-}\ddot{N}(R){-}\overset{-}{\ddot{O}}{:}$
Nitrilimines	$R{-}\overset{+}{C}{=}N{-}\overset{-}{\ddot{N}}{-}R \longleftrightarrow R{-}C{\equiv}\overset{+}{N}{-}\overset{-}{\ddot{N}}{-}R$

A similar transformation results when trimethylsilyloxy-substituted allylic halides react with silver perchlorate in nitromethane. The resulting allylic cation gives cycloaddition reactions with dienes such as cyclopentadiene. The isolated products result from desilylation of the initial adducts.[99]

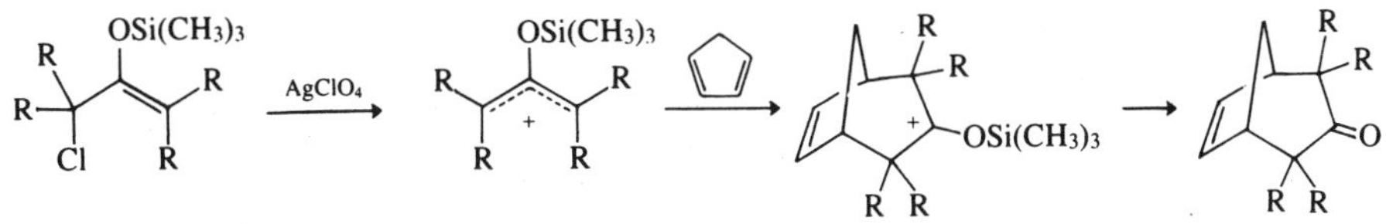

OSi(CH3)3, R, R, Cl, R —AgClO4→ allylic cation with OSi(CH3)3 —cyclopentadiene→ bicyclic cation with OSi(CH3)3 → bicyclic ketone (=O)

There also exists a large class of reactions known as 1,3-*dipolar cycloaddition reactions* that are analogous to the Diels–Alder reaction in that they are concerted $[4\pi + 2\pi]$ cycloadditions.[100] These reactions can be represented as in the following diagram. The entity $a{-}b{-}c$ is called the *1,3-dipolar molecule* and $d{-}e$ is the *dipolarophile.*

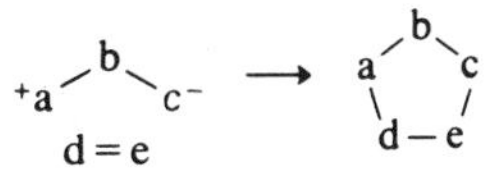

$$^{+}a{-}b{-}c^{-} \quad d{=}e \longrightarrow \text{five-membered ring } a{-}b{-}c{-}e{-}d$$

The 1,3-dipolar molecules are isoelectronic with the allyl anion and have four π electrons. Some typical 1,3-dipolar species are shown in Scheme 11.4. It should be noted that all have one or more resonance structures showing the characteristic 1,3-dipole. The dipolarophiles are typically alkenes or alkynes, but all that is essential is a π bond. The reactivity of dipolarophiles depends both on the substituents present on the π bond and on the nature of the 1,3-dipole involved in the reaction. Because of the wide range of structures that can serve either as a 1,3-dipole or as a dipolarophile, the 1,3-dipolar cycloaddition is a very useful reaction for the construction of five-membered heterocyclic rings.

99. N. Shimizu, M. Tanaka, and Y. Tsuno, *J. Am. Chem. Soc.* **104**, 1330 (1982).
100. R. Huiusgen, *Angew. Chem. Int. Ed. Engl.* **2**, 565 (1963); R. Huisgen, R. Grashey, and J. Sauer, in *The Chemistry of the Alkenes*, S. Patai (ed.), Interscience Publishers, London, 1965, pp 806–878; A. Padwa, *1,3-Dipolar Addition Chemistry*, John Wiley, New York, 1984.

The stereochemistry of the 1,3-dipolar cycloaddition reaction is analogous to that of the Diels–Alder reaction and is a stereospecific *syn* addition. Diazomethane, for example, adds stereospecifically to the diesters **45** and **46** to yield the pyrazolines **47** and **48**, respectively.

45 $\xrightarrow{CH_2N_2}$ **47**

46 $\xrightarrow{CH_2N_2}$ **48**

Ref. 101

When both the 1,3-dipole and the dipole are unsymmetrical, there are two possible orientations for addition. Both steric and electronic factors must play a role in determining the regioselectivity of the addition. The most generally satisfactory interpretation of the regiochemistry of dipolar cycloadditions is based on frontier orbital concepts. As with the Diels–Alder reaction, the most favorable orientation is that which involves complementary interaction between the frontier orbitals of the 1,3-dipole and the dipolarophile. Although most dipolar cycloadditions are of the type in which the LUMO of the dipolarophile interacts with the HOMO of the 1,3-dipole, there are a significant number of systems in which the relationship is reversed. There are also some in which the two possible HOMO–LUMO interactions are of comparable magnitude.

The analysis of the regioselectivity of a 1,3-dipolar cycloaddition therefore requires information about the energy and atomic coefficients of the frontier orbitals of the 1,3-dipole and the dipolarophile. A range of 1,3-dipoles have been examined by means of CNDO/2 calculations. Some of the results are shown in Fig. 11.13. By using these orbital coefficients and by calculation or estimation of the relative energies of the interacting orbitals, it is possible to make predictions of the regiochemistry of 1,3-dipolar cycloaddition reactions.[102] The most important dipolarophiles are the same types of compounds that are dienophiles. The relative magnitudes of orbital coefficients for dienophiles indicated in Fig. 11.12 can be used in analysis of 1,3-dipolar cycloaddition reactions. Figure 11.14 gives estimates of the energies of the HOMO and LUMO orbitals of some representative 1,3-dipoles. In conjunction with the orbital energies given in Fig. 11.12, this information allows conclusions to be drawn as to which HOMO–LUMO combination will interact most strongly for a given pair of reactants.

101. K. v. Auwers and E. Cauer, *Justus Liebigs Ann. Chem.* **470**, 284 (1929); K. v. Auwers and F. König, *Justus Liebigs Ann. Chem.* **496**, 252 (1932); T. L. van Auken and K. L. Rinehart, *J. Am. Chem. Soc.* **84**, 3736 (1962).
102. K. N. Houk, J. Sims, R. E. Duke, Jr., R. W. Strozier, and J. K. George, *J. Am. Chem. Soc.* **95**, 7287 (1973); R. Sustmann and H. Trill, *Angew. Chem. Int. Ed. Engl.* **11**, 838 (1972); K. N. Houk, *Top. Curr. Chem.* **79**, 1 (1979).

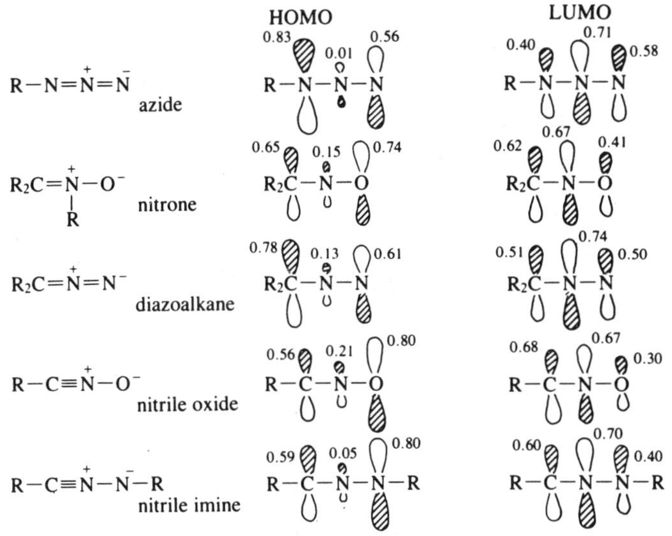

Fig. 11.13. Orbital coefficients for HOMO and LUMO π MOs of some common 1,3-dipoles. [From K. N. Houk, J. Sims, R. E. Duke, Jr., R. W. Strozier, and J. K. George, *J. Am. Chem. Soc.* **95**, 7287 (1973).]

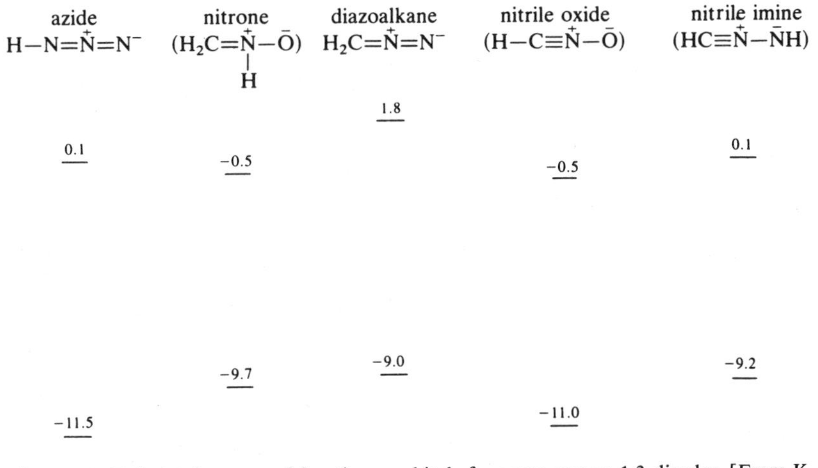

Fig. 11.14. Estimated energy of frontier π orbitals for some comon 1,3-dipoles. [From K. N. Houk, J. Sim, R. E. Duke, Jr., R. W. Strozier, and J. K. George, *J. Am. Chem. Soc.* **95**, 7287 (1973).]

$[2\pi + 2\pi]$ Additions are allowed only when the reaction is antarafacial for one of the components. There are relatively few systems which can meet the geometrical limits imposed by this restriction. Most examples of $[2\pi + 2\pi]$ additions are reactions involving ketenes.[103] Ketenes are ideal components in reactions of this type, since

103. W. T. Brady and R. Roe, *J. Am. Chem. Soc.* **93**, 1662 (1971); W. T. Brady, in *The Chemistry of Ketenes, Allenes and Related Compounds*, S. Patai (ed.), John Wiley, Chichester, UK, 1980, Chapter 8.

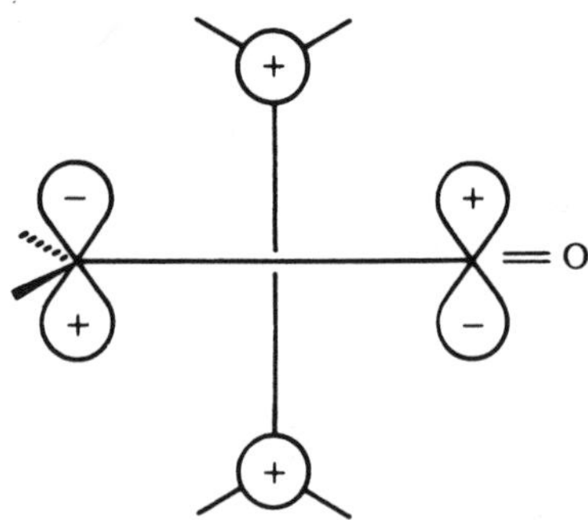

Fig. 11.15. Concerted cycloaddition of a ketene and an olefin. The orbitals represented are the HOMO of the olefin and the LUMO of the ethylenic portion of the ketene.

the linear geometry associated with the *sp*-hybridized carbon atom leads to a minimum of steric repulsion in the antarafacial transition state. Molecular orbital modeling of the reaction suggests a concerted, nonsynchronous process with bond formation at the *sp* carbon of the ketene leading bond formation at the terminal carbon.[104] The idealized geometry for the antarafacial addition of a ketene to an alkene is shown in Fig. 11.15.

The experimental results concerning the stereochemistry of [2 + 2] cycloadditions are in accord with the stereochemical predictions that can be made on the basis of this model. For example, *E*- and *Z*-2-butene give stereoisomeric products with ethoxyketene.[105] For monosubstituted alkenes, the substituent is vicinal and *cis* to the ethoxy group in the cyclobutanone product. This is exactly the stereochemistry predicted by the model in Fig. 11.15 since is maximizes the separation of the alkyl and ethoxy substituents in the transition state.

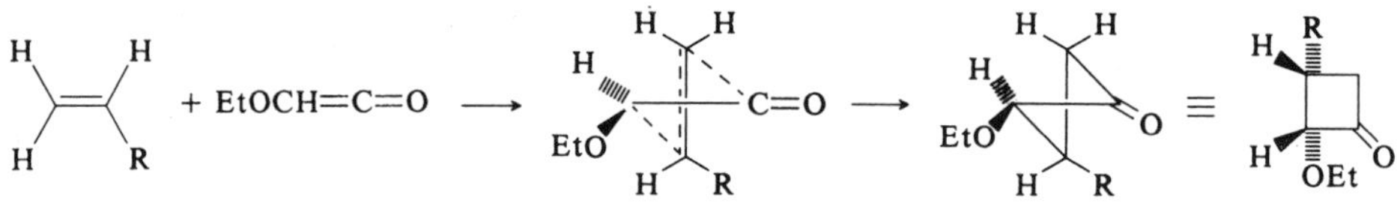

+ EtOCH=C=O ⟶ ⟶ ≡

There is an alternative description of the [2 + 2] cycloaddition reactions of ketenes. This formulation is $[2\pi_s + (2\pi_s + 2\pi_s)]$.[106] The basis set orbital array is shown below. This system has Hückel-type topology, and, with six π electrons involved, it would be an allowed process.

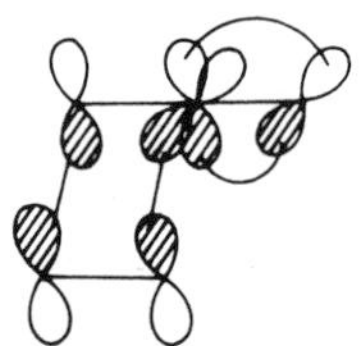

104. L. A. Burke, *J. Org. Chem.* **50**, 3149 (1985).
105. T. DoMinh and O. P. Strausz, *J. Am. Chem. Soc.* **92**, 1766 (1970).

This analysis is equally compatible with available data, and some predictions of stereoselectivity and reactivity based on this model are in better accord with experimental results than predictions derived from the $[2\pi_s + 2\pi_a]$ transition state.

Fluorinated alkenes, tetrafluoroethylene in particular, react with other alkenes to give cyclobutanes.[107]

$$CF_2{=}CF_2 + CH_2{=}CHSi(CH_3)_3 \rightarrow$$ [tetrafluorocyclobutane with $Si(CH_3)_3$ substituent] Ref. 108

With dienes, tetrafluoroethylene reacts by this route to the exclusion of the [4 + 2] process.

$$CF_2{=}CF_2 + CH_2{=}CH{-}CH{=}CH_2 \xrightarrow{125°C}$$ [tetrafluoro-vinylcyclobutane] (90%) Ref. 109

These reactions are interpreted as stepwise processes.

$$CF_2{=}CF_2 + CH_2{=}CH{-}CH{=}CH_2 \rightarrow \cdot CF_2CF_2CH_2CH{=}CH{-}\dot{C}H_2 \updownarrow \cdot CF_2CF_2CH_2\dot{C}H{-}CH{=}CH_2 \rightarrow$$ [tetrafluorocyclobutane with $CH{=}CH_2$ substituent]

The exceptional reactivity of the double bond in tetrafluoroethylene is the result of a strong stabilizing effect of the fluorine substituents on the radical intermediate.

The predictions of the Woodward–Hoffmann rules that thermal concerted cycloadditions are allowed for combinations in which $4n + 2$ π electrons are involved has stimulated the search for combinations with ten and larger numbers of participating electrons. An example of a [6 + 4] cycloaddition is the reaction of tropone with 2,5-dimethyl-3,4-diphenylcyclopentadienone[110]:

[2,5-dimethyl-3,4-diphenylcyclopentadienone (Ph, Ph, CH_3, CH_3, =O) + tropone (O) → bridged adduct (Ph, Ph, CH_3, CH_3, O, O)]

106. D. J. Pasto, *J. Am. Chem. Soc.* **101**, 37 (1979).
107. P. D. Bartlett, *Q. Rev. Chem. Soc.* **24**, 473 (1970); J. D. Roberts and C. M. Sharts, *Org. React.* **12**, 1 (1962).
108. D. C. England, F. J. Weigert, and J. C. Calabrese, *J. Org. Chem.* **49**, 4816 (1984).
109. D. D. Coffman, P. L. Barrick, R. D. Cramer, and M. S. Raasch, *J. Am. Chem. Soc.* **71**, 490 (1949).
110. K. N. Houk and R. B. Woodward, *J. Am. Chem. Soc.* **92**, 4145 (1970).

Heptafulvene derivatives undergo [8 + 2] cycloadditions.

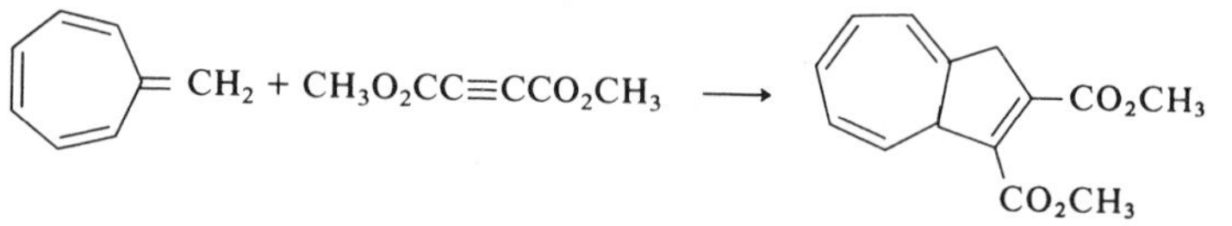

$CH_2 + CH_3O_2CC{\equiv}CCO_2CH_3 \longrightarrow$ CO_2CH_3 CO_2CH_3 Ref. 111

OCH_3 + O CH_3 H_3C O $\longrightarrow$ O CH_3 H_3C O Ref. 112

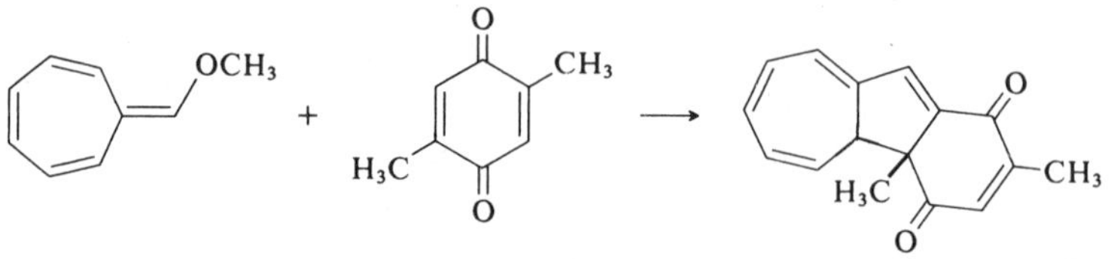

Flexible six-π-electron and eight-π-electron systems would impose an entropic barrier to concerted ten-π-electron concerted reactions. Most of the examples, as in the cases above, involve cyclic systems in which the two termini of the conjugated system are held close together.

General References

M. J. S. Dewar, *The Molecular Orbital Theory of Organic Chemistry*, McGraw-Hill, New York, 1969.
M. J. S. Dewar and R. C. Dougherty, *The PMO Theory of Organic Chemistry*, Plenum Press, New York, 1975.
I. Fleming, *Frontier Orbitals and Organic Chemical Reactions*, Wiley-Interscience, New York, 1976.
T. L. Gilchrist and R. C. Storr, *Organic Reactions and Orbital Symmetry*, Second Edition, Cambridge University Press, Cambridge, 1979.
J. B. Hendrickson, *Angew. Chem. Int. Ed. Engl.* **13**, 47 (1974).
W. C. Herndon, *Chem. Rev.* **72**, 157 (1972).
K. N. Houk, *Acc. Chem. Res.* **11**, 361 (1975).
R. E. Lehr and A. P. Marchand, *Orbital Symmetry, A Problem-Solving Approach*, Academic Press, New York, 1972.
A. P. Marchand and R. E. Lehr, *Pericyclic Reactions*, Vols. I and II, Academic Press, New York, 1977.
E. N. Marvell, *Thermal Electrocyclic Reactions*, Academic Press, New York, 1980.
S. J. Rhoads and N. R. Raulins, *Org. React.* **22**, 1 (1974).
L. Salem, *Electrons in Chemical Reactions*, Wiley, New York, 1982.
R. B. Woodward and R. Hoffman, *The Conservation of Orbital Symmetry*, Academic Press, New York, 1970.
H. E. Zimmerman, *Acc. Chem. Res.*, **4**, 272 (1971).

Problems

(*References for these problems will be found on page* 782.)

1. Show, by constructing a correlation diagram, whether each of the following disrotatory cyclizations is symmetry-allowed:

   (a) pentadienyl cation to cyclopentenyl cation
   (b) pentadienyl anion to cyclopentenyl anion

111. W. v. E. Doering and D. W. Wiley, *Tetrahedron* **11**, 183 (1960).
112. J. Bindl, Y. Burgemeister, and J. Daub, *Justus Liebigs Ann. Chem.*, 1346 (1985).

2. Which of the following reactions are allowed according to the orbital symmetry conservation rules? Explain.

(a)

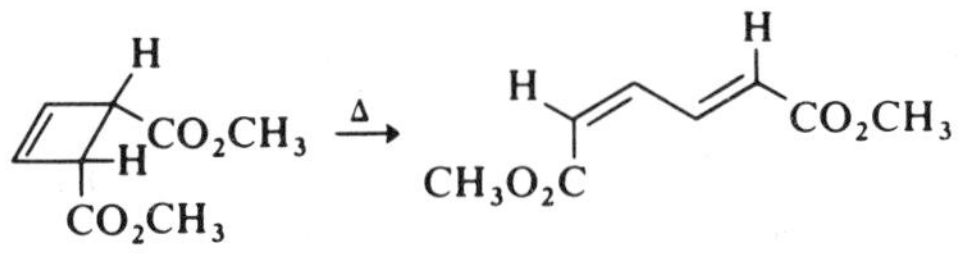

(b)

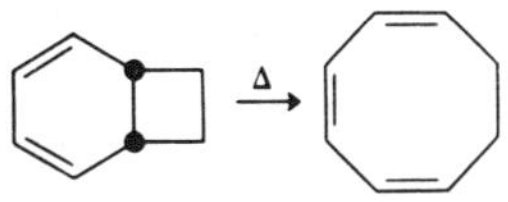

(c)

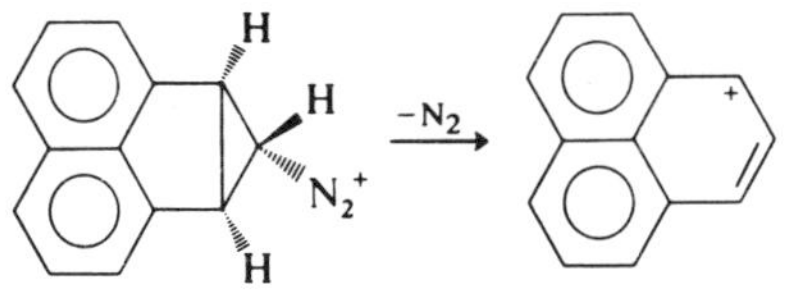

(d)

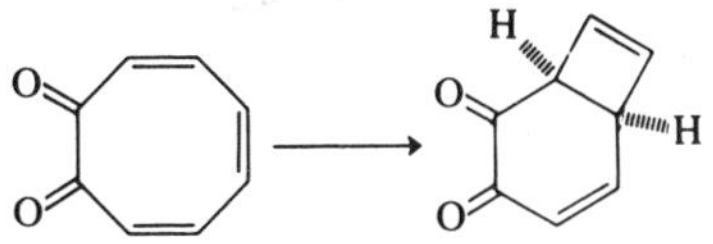

(e)

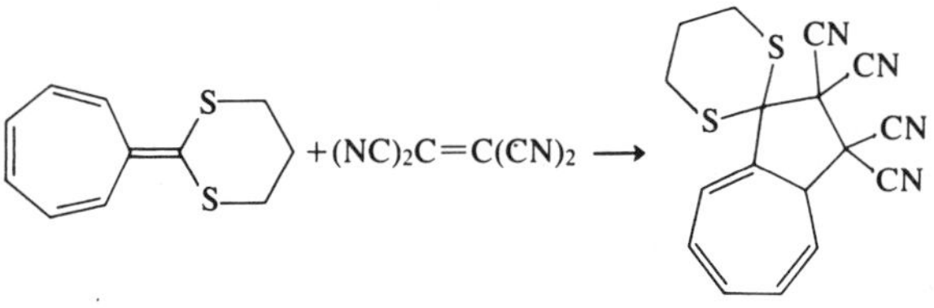

(f)

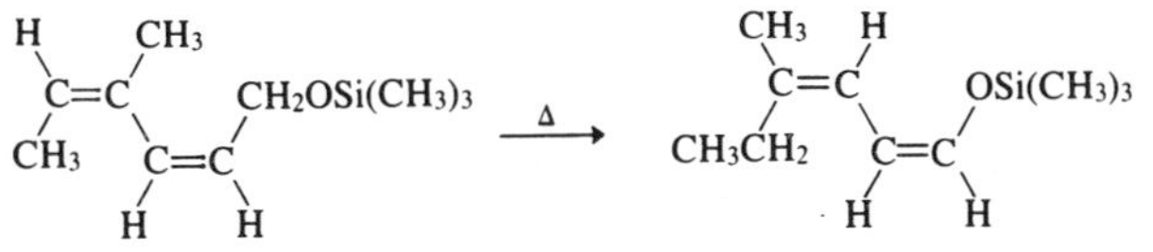

3. All *cis*-cyclononatetraene undergoes a spontaneous electrocyclic ring closure at 25°C to afford a single product. Suggest a structure for this product. Also, describe an alternative symmetry-allowed electrocyclic reaction that would lead to an isomeric bicyclononatriene. Explain why the product of this alternative reaction pathway is not formed.

4. Offer a mechanistic explanation of the following observations.

(a) The 3,5-dinitrobenzoate esters of the epimers shown below both yield 3-cyclopenten-1-ol on hydrolysis in dioxane-water. The relative rates, however, differ by a factor of 10 million! Which is more reactive and why?

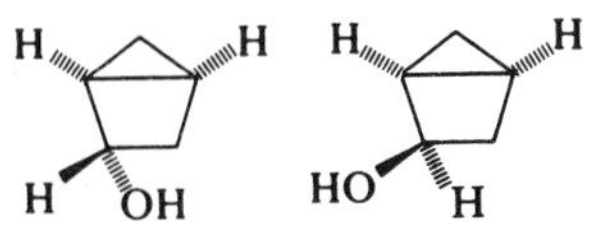

(b) Optically active **A** racemizes on heating at 50°C with a half-life of 24 h.

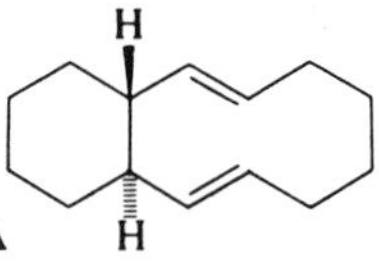

A

(c) The anions of various 2-vinylcyclopropanols undergo very facile vinylcyclopropane rearrangement to give 3-cyclopenten-1-ols. For example:

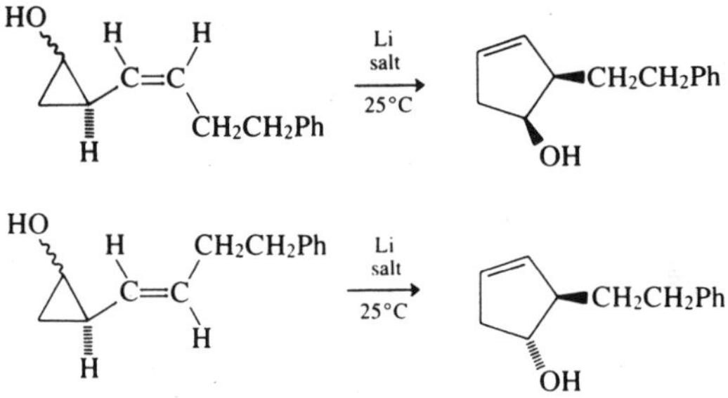

Offer an explanation for the facility of the reaction, as compared to the vinylcyclopropane rearrangement of hydrocarbons, which requires a temperature above 200°C. Consider concerted reaction pathways which would account for the observed stereospecificity of the reaction.

(d) On being heated at 320–340°C, compound **B** produces 1,4-dimethoxynaphthalene and 1-acetoxybutadiene.

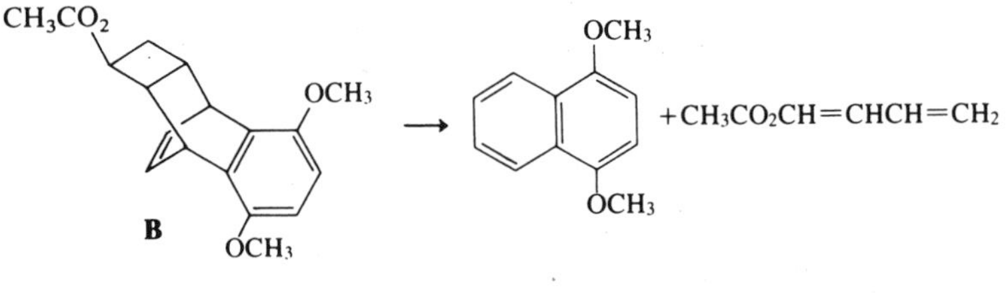

CH_3CO_2 … OCH_3 … **B** … OCH_3 → (1,4-dimethoxynaphthalene) $+ CH_3CO_2CH{=}CHCH{=}CH_2$

(e) It has been found that compounds **C** and **D** are opened at −25°C to allylic anions in the presence of strong bases such as lithium di-*t*-butylamide. In contrast, **E** opens only slowly at 25°C.

C 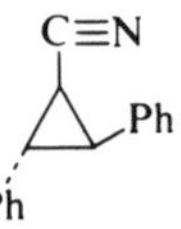 **D** 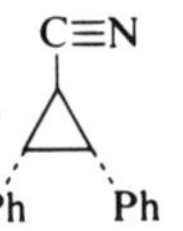 **E**

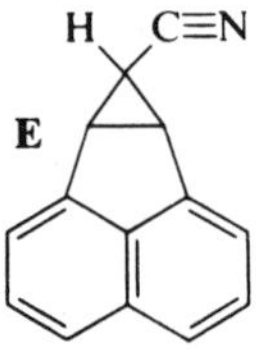

(f) When the 1,6-methanodeca-1,3,5,7,9-pentaene structure is modified by fusion of two benzene rings as in **Fa**, it exists as the valence tautomer **Fb**.

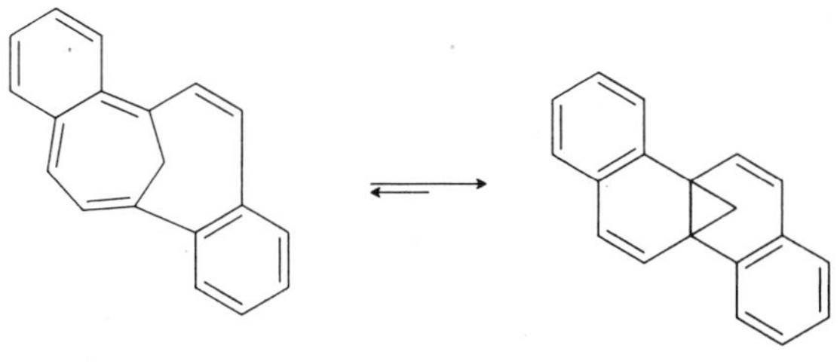

Fa ⇌ Fb

5. Suggest a mechanism by which each transformation could occur. More than one step is involved in each case.

(a) $\xrightarrow[\text{1–3 sec}]{420°\text{C}}$

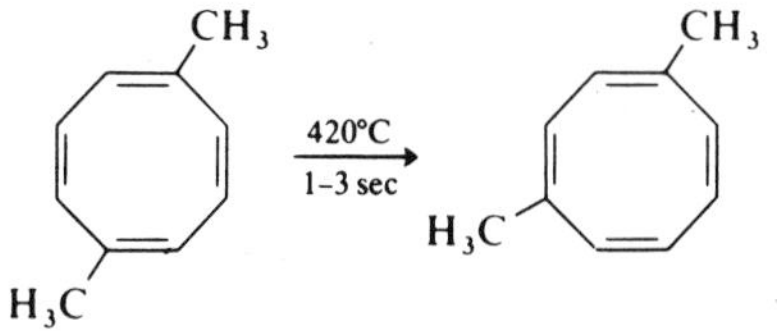

(b) + $H_2C{=}C(OCH_3)CH_2Br$ $\xrightarrow[Na_2CO_3]{Ag^{+}\,{}^{-}O_2CCF_3}$ $\xrightarrow[H_2O]{H^+,}$

(c) $\xrightarrow{150°\text{C}}$

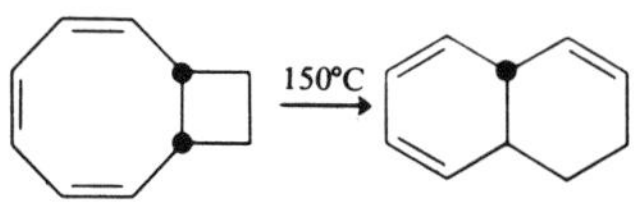

(d) $PhCH_2N{=}CHPh$ $\xrightarrow[\text{3) } H_2O]{\text{1) } LiN(i\text{-}C_3H_7)_2/THF \quad \text{2) } trans\text{-}PhCH{=}CHPh}$

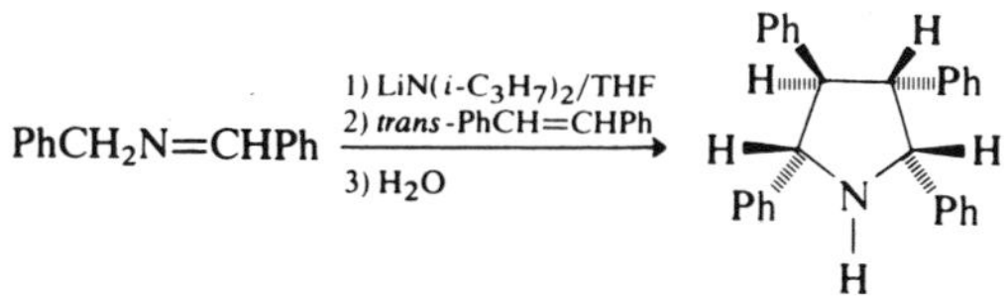

(e) $\xrightarrow{100°\text{C}}$ $CH_3CH{=}C(CH_3)CH{=}O$

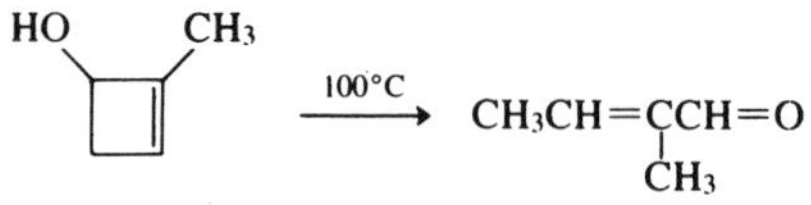

(f) $\xrightarrow{370°\text{C}}$

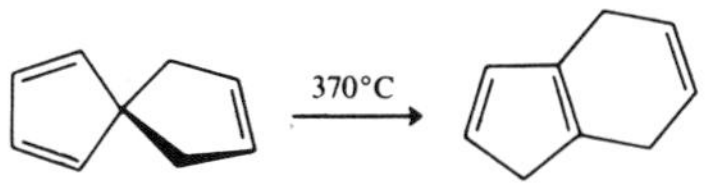

(g) + $CH_3O_2CC{\equiv}CCO_2CH_3$ $\xrightarrow{\Delta}$

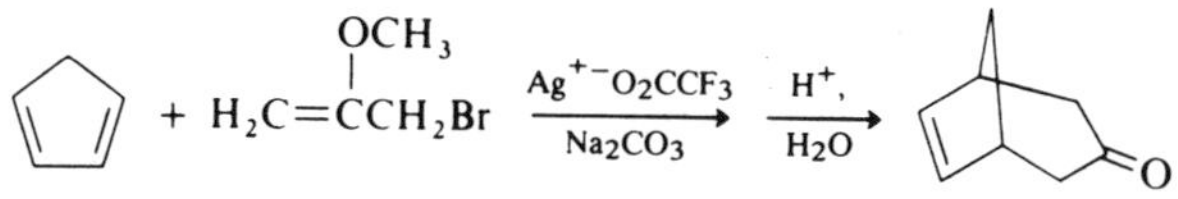

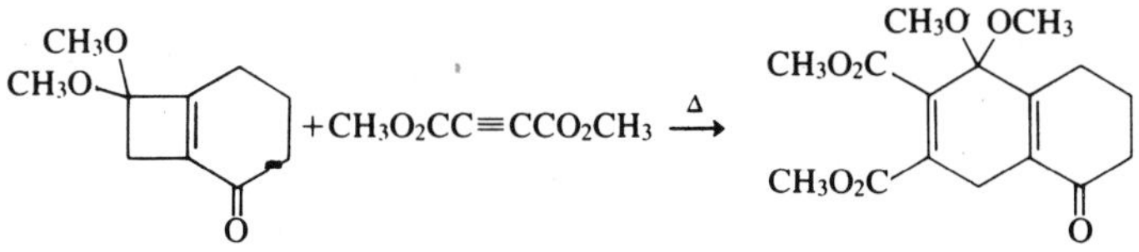

(h) $\xrightarrow{200°C}$

(i) $\xrightarrow[4\ hr]{180°C}$

6. Predict which is the more likely mode of ring closure for oxonin:
   (a) cyclization of the tetracene unit of a bicyclo[6.1.0] system, or
   (b) cyclization of a triene unit to a bicyclo[4.3.0] system.

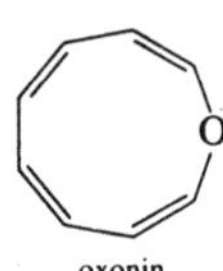

oxonin

7. Give the structure, including stereochemistry, of the products expected for the following reactions.

(a) $+ \ Ph_2C{=}C{=}O \longrightarrow$

(b) $(CH_3)_2C{=}C{=}O + C_2H_5OC{\equiv}CH \xrightarrow{0°C}$

(c) $\xrightarrow[20\ hr,\ 25°C]{KH, THF}$

(d) $PhCH_2CHSCH{=}CH_2$ (with $CH{=}CH_2$ on the CH) $\xrightarrow[DME,\ reflux,\ 12\ hr]{H_2O}$ $(C_{11}H_{14}O)$

(e) $(CH_3)_3CCHC{=}C(CH_3)_2$ (with Cl and $OSi(CH_3)_3$) $+$ $\xrightarrow[CaCO_3,\ CH_3NO_2]{AgClO_4}$

(f) $\xrightarrow{K^{+\,-}OC(CH_3)_3}$

(g) 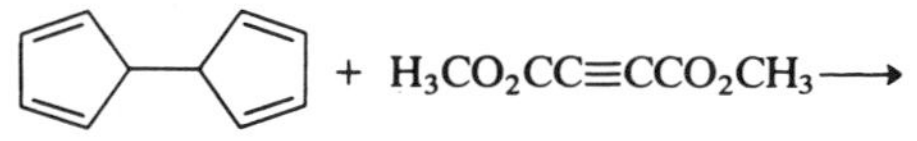 + $H_3CO_2CC{\equiv}CCO_2CH_3 \longrightarrow$

(h) N≡C, C≡N, N≡C, C≡N $\xrightarrow{\Delta}$

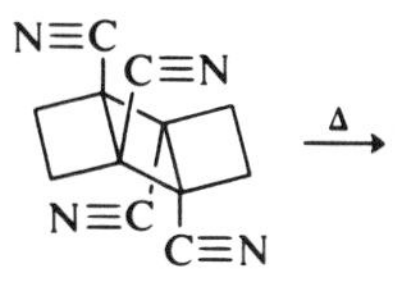

(i) 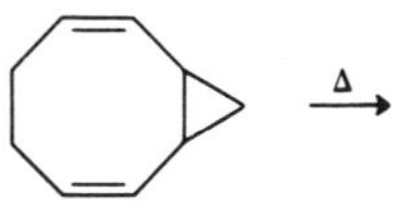 $\xrightarrow{\Delta}$

(j) 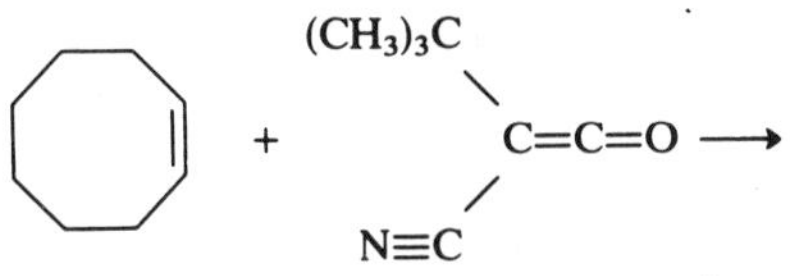 + $(CH_3)_3C$, N≡C C=C=O ⟶

(k) 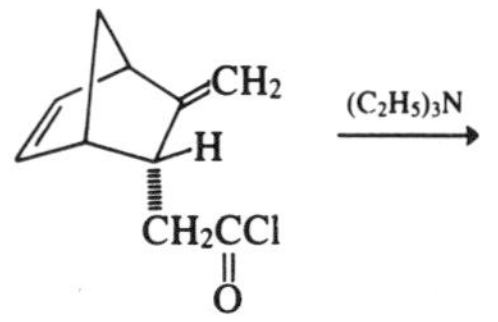 CH₂, H, CH₂CCl, O $\xrightarrow{(C_2H_5)_3N}$

8. In the series of oxepins **A**(n = 3, 4, and 5), only the compound with $n = 5$ undergoes rearrangement (at 60°C) to the isomeric oxepin **B**. The other two compounds (n = 3 or 4) are stable even at much higher temperature. When **B** (n = 3) was synthesized by an alternate route, it showed no tendency to revert to **A** (n = 3). Explain these observations.

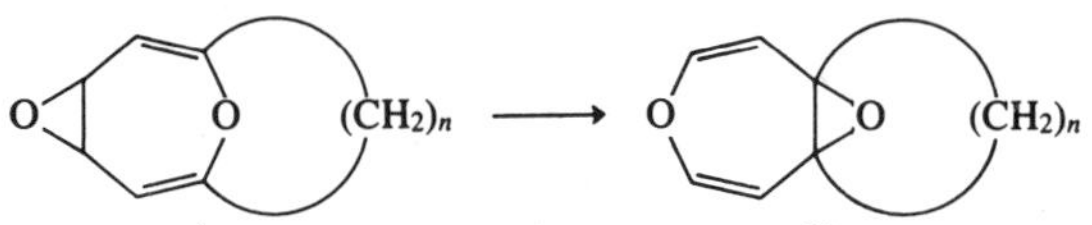

O O $(CH_2)_n$ ⟶ O O $(CH_2)_n$

A **B**

9. Bromocyclooctatetraene rearranges to *trans*-β-bromostyrene. The rate of the rearrangement is solvent dependent, the first-order rate constant increasing from $\sim 10^{-7}\ s^{-1}$ in cyclohexane to $\sim 10^{-3}\ s^{-1}$ in acetonitrile at 80°C. In the presence of lithium iodide, the product is *trans*-β-iodostyrene, although *trans*-β-bromostyrene is unaffected by lithium iodide under the conditions of the reaction. Suggest a mechanism for this rearrangement.

10. The BF_3-catalyzed rearrangement of phenyl pentadienyl ether has been shown to proceed strictly intramolecularly and with the isotopic pattern shown. Analyze the possibilities for concerted rearrangement in FMO terms, and specify those

mechanisms which are consistent with the observed result. Discuss the role of the catalyst in the reaction.

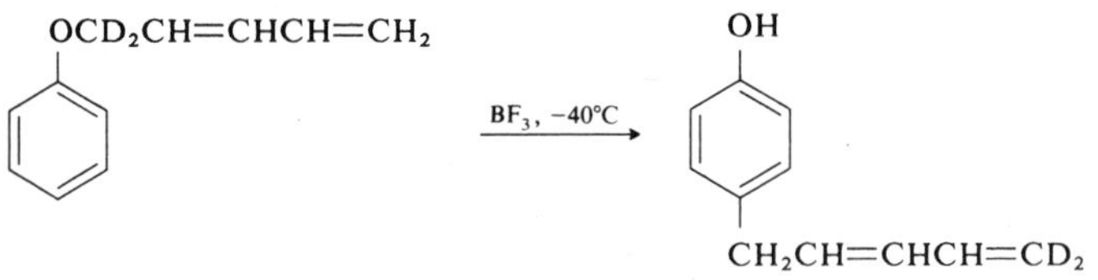

11. At room temperature the diene **1** and maleic anhydride react to give **A** and **B**. When the reaction mixture is heated to 160°C, **C** and **D** are formed, and as heating is continued the **C**:**D** ratio decreases. Explain the reason for the different products at the two reaction temperatures and the reason for the change of product ratio with time at the higher temperature.

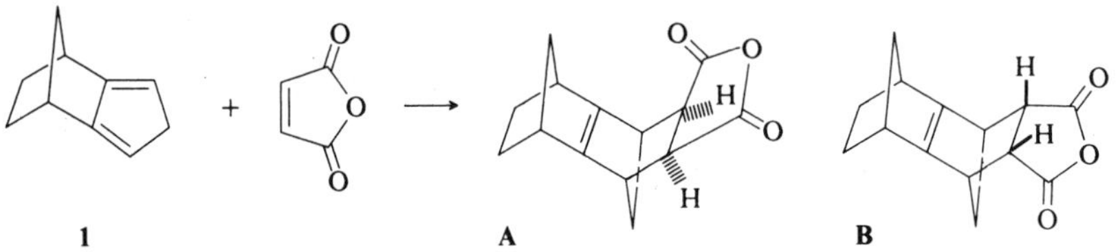

1 **A** **B**

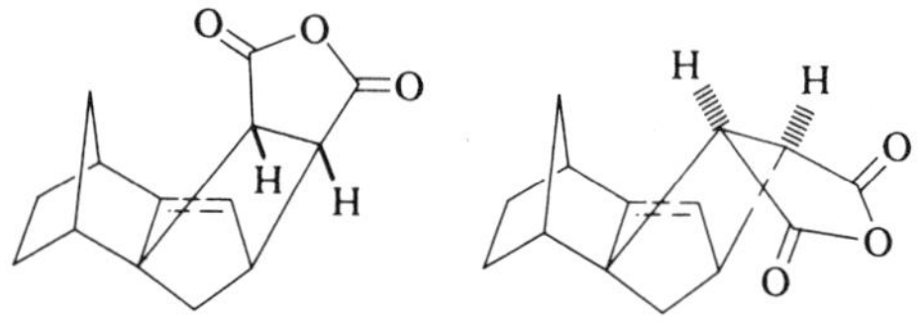

C **D**

12. Classify the following reactions as electrocyclizations, sigmatropic rearrangements, cycloadditions, etc., and give the correct symbolism for the electrons involved in each concerted process. Some of the reactions proceed by two sequential processes.

(a)

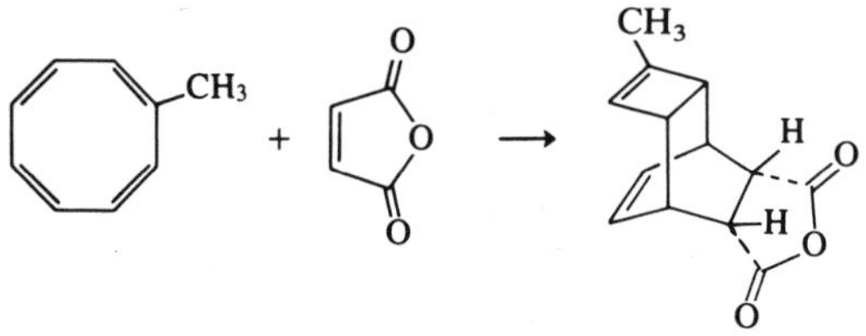

(b)

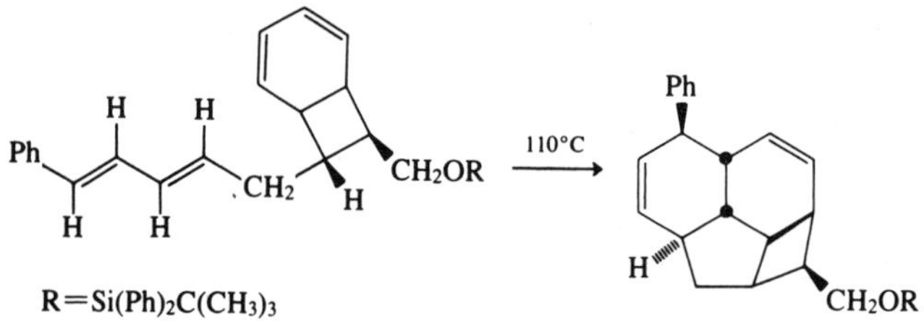

(c) [structure: 1-methoxy-2-(phenylthio)cyclobutene + maleic anhydride, $\xrightarrow{\Delta}$ bicyclic anhydride product with SPh and OCH_3]

SPh, OCH₃, O, Δ

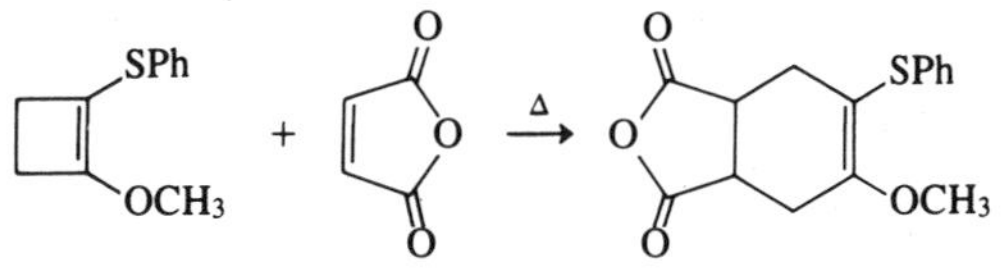

(d) [structure: $HOCH_2$–CH=CH–CH=CH–CH=CH–CH_2OH triene $\xrightarrow{<25°C}$ bicyclo[4.2.0]octadiene with two CH_2OH groups]

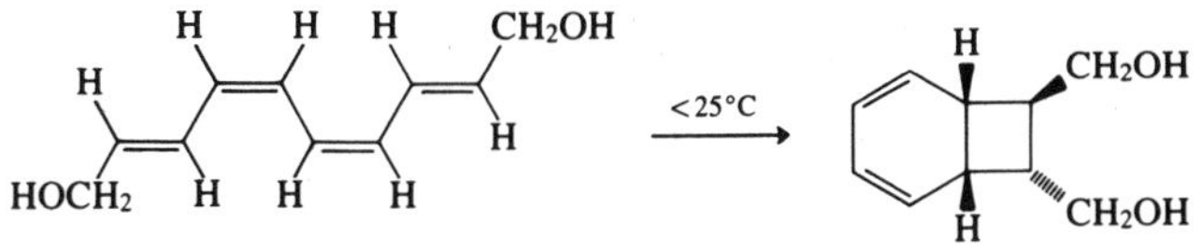

(e) $CH_2{=}C(CH_3)CH_2$–(2-methylenecyclopentyl) $\xrightarrow{\Delta}$ $CH_2{=}C(CH_3)CH_2CH_2$–(1-cyclopentenyl)

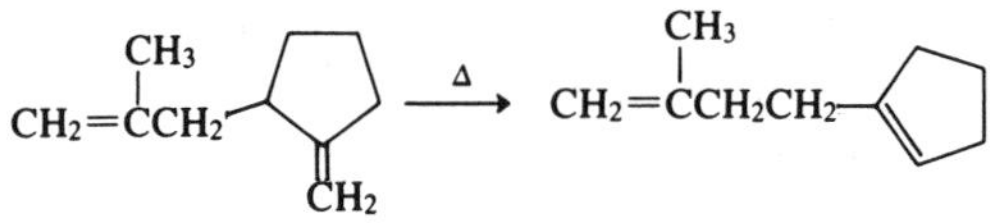

13. The "ene" reaction is a concerted reaction in which addition of an alkene to an electrophilic olefin occurs with migration of a hydrogen and the alkene double bond. For example:

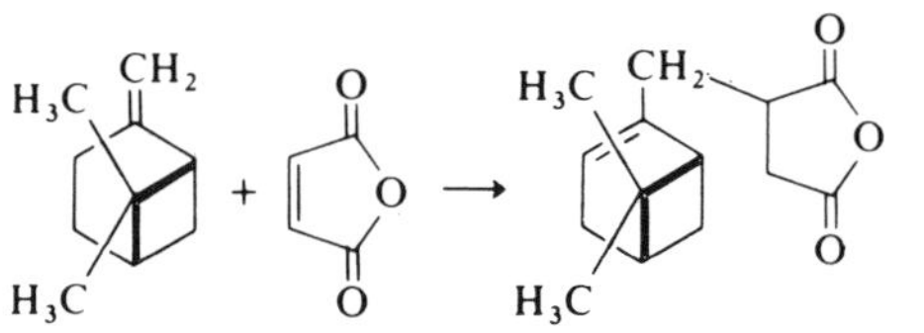

[structure: β-pinene + maleic anhydride → ene adduct]

Depict the orbital array through which this process could occur in a concerted manner.

14. Predict the regiochemistry and stereochemistry of the following cycloaddition reactions and indicate the basis for your prediction.

(a) [structure: $CH_3CH{=}CH{-}CH{=}CH_2$ (trans) + maleic anhydride $\longrightarrow$]

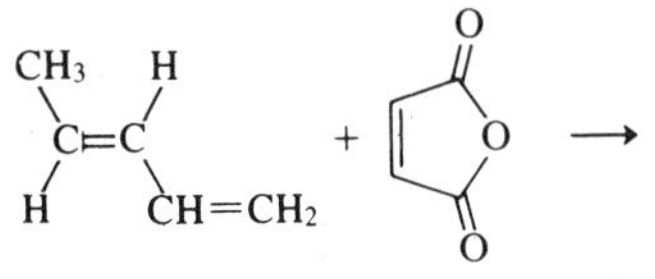

(b) [structure: $CH_3O_2CNH{-}CH{=}CH{-}CH{=}CH_2$] $+ CH_2{=}CHCO_2CH_3 \longrightarrow$

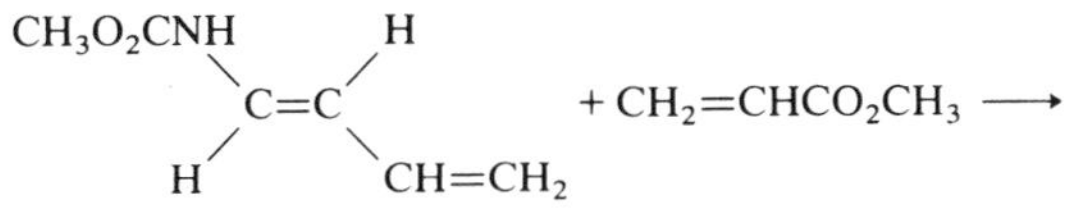

(c) [structure: $CH_3CH{=}CH{-}CH{=}CH{-}CN$ + 1-(1-cyclopentenyl)pyrrolidine $\longrightarrow$]

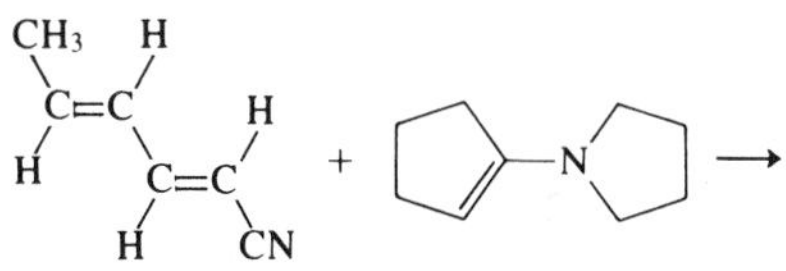

(d) $CH_2{=}CHCH{=}CH_2$ + [2-methoxy-6-methyl-1,4-benzoquinone: O, O, CH_3, CH_3O] $\longrightarrow$

(e) PhS, OCH_3, C=C, H [1-cyclohexenyl] $+ CH_2{=}CH\overset{O}{\overset{\|}{C}}CH_3$ $\xrightarrow{EtAlCl_2}$

15. On treatment with potassium metal, *cis*-bicyclo[6.1.0]nona-2,4,6-triene gives a monocyclic dianion. The *trans* isomer under similar conditions gives only a bicyclic monoanion (radical anion). Explain how the stereochemistry of the ring junction can control the course of these chemical reductions.

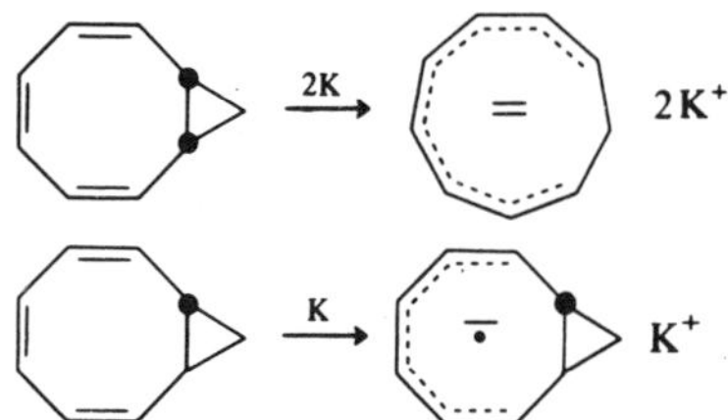

16. The following compounds are capable of degenerate rearrangement at the temperature given. Identify chemical processes which are consistent with the temperature and which would lead to degenerate rearrangement. Indicate by an appropriate labeling scheme the carbons and hydrogens which become equivalent as a result of the rearrangement process you have suggested.

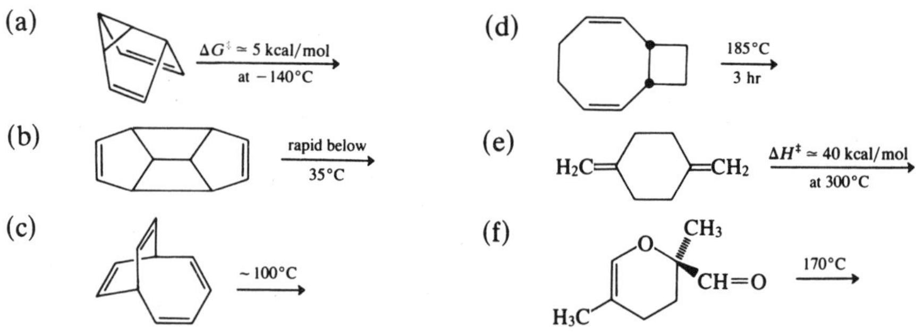

17. On heating at 225°C, 5-allylcyclohexa-1,3-diene, **A**, undergoes intramolecular cycloaddition to give the tricyclic nonene **B**. The mechanism of formation of **B** was probed using the deuterium-labeled sample of **A** which is shown. Indicate the position of deuterium labels in product **B** if the reaction proceeds by (a) a [2 + 2] cycloaddition or (b) a [4 + 2] cycloaddition.

$CH_2CH{=}CH_2$ **A** $\xrightarrow{\Delta}$ **B** $CD_2CH{=}CH_2$ (deuterium-lableled **A**)

18. The thermal behavior of **A** and **B** above 150°C has been studied. Both in the gas phase and in solution, each compound yields a 3:5 mixture of *trans,cis*-1,5-cyclooctadiene (**C**) and *cis-cis*-1,5-cyclooctadiene (**D**). When hexachlorocyclopentadiene is present, compound **E** is found in place of **C**, but the amount of **D** formed is about the same as in its absence. Formulate a description of the thermolysis mechanism that is consistent with these facts and the general theory of thermal electrolytic reactions.

A B E

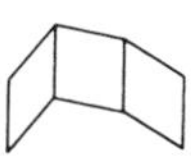

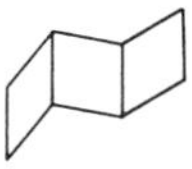

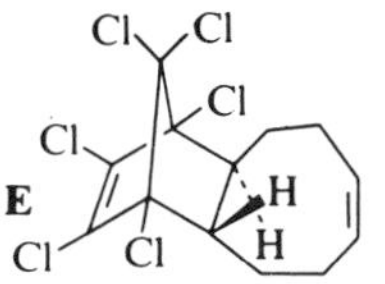

19. Suggest mechanisms for the following reactions. Classify the orbital symmetry-controlled process as clearly as you can with respect to type.

(a) 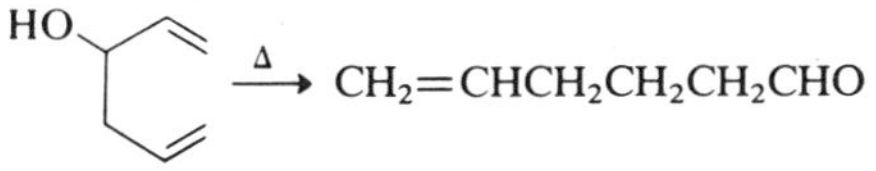 HO $\xrightarrow{\Delta}$ $CH_2{=}CHCH_2CH_2CH_2CHO$

(b) $(CH_2{=}CH)_2CHOH + CH_2{=}CHOEt \xrightarrow{Hg^{2+}} CH_2{=}CHCH{=}CHCH_2CH_2CHO$

(c) 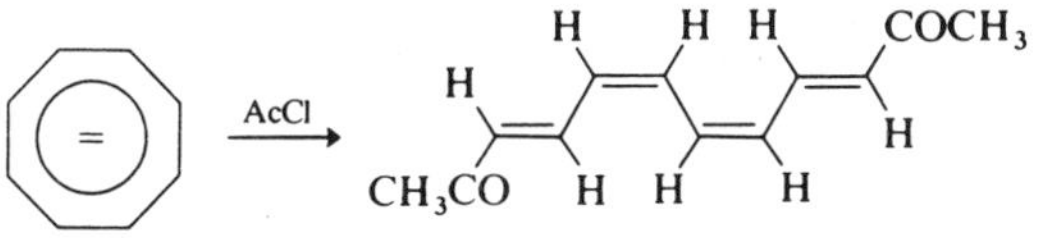 $\xrightarrow{AcCl}$ H H H COCH$_3$ H H CH$_3$CO H H H

(d) 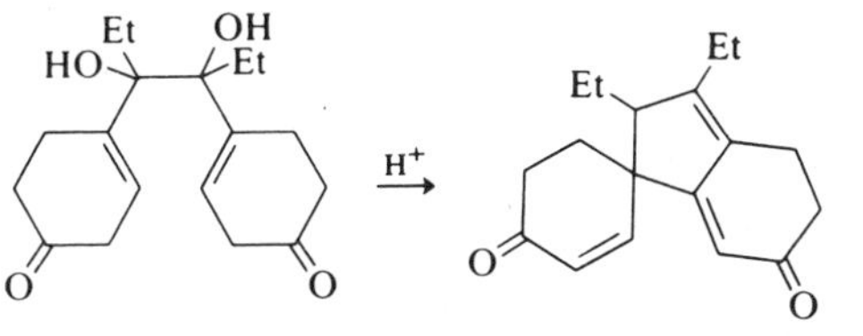 Et OH HO Et O O $\xrightarrow{H^+}$ Et Et O O

(e) 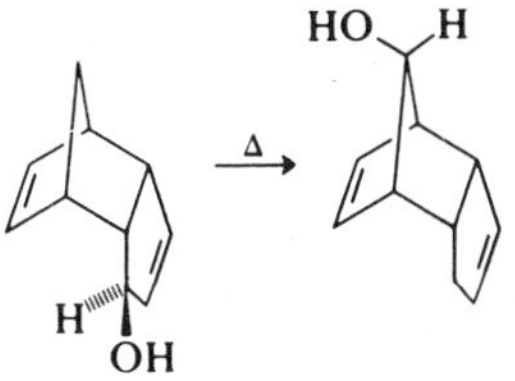 $\xrightarrow{\Delta}$ HO H H OH

(f) 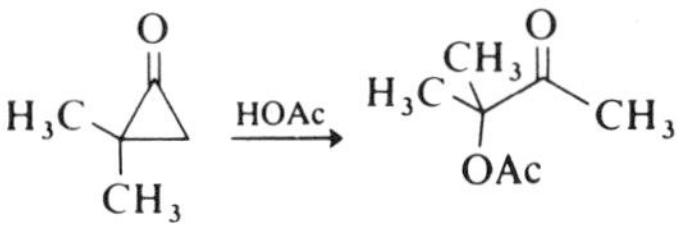 O H$_3$C CH$_3$ $\xrightarrow{HOAc}$ CH$_3$ O H$_3$C CH$_3$ OAc

(g) 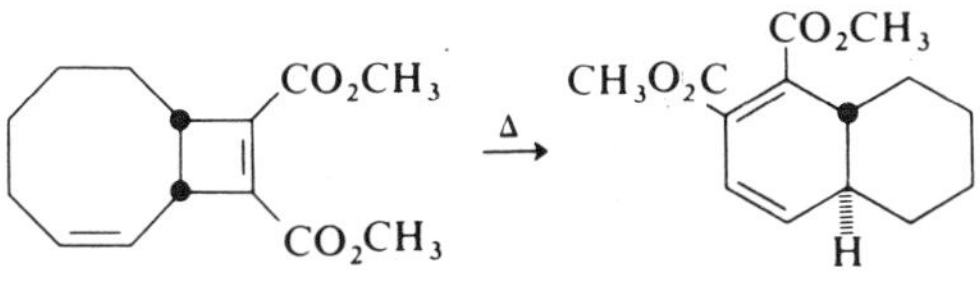 CO$_2$CH$_3$ CO$_2$CH$_3$ $\xrightarrow{\Delta}$ CO$_2$CH$_3$ CH$_3$O$_2$C H

(h) NC, CN, CN, CN, O; H, H, CH_3, CH_3 → NC, CN, NC, CN, O, H, H, H_3C, CH_3

(i) $(t\text{-Bu})_2C{=}C{=}CH_2$ + CH_3CO_3H → O, t-Bu, t-Bu

(j) $\xrightarrow{\Delta}$

(k) $\xrightarrow{\Delta}$

(l) N, $C_2H_5O_2CH_2-S^+$, CH_3 $\xrightarrow{Et_3N}$ SCH_3, O, N

(m) CN, CH_3 $\xrightarrow{180°C}$ NC, CH_3

(n) $=CH_2$ + N, N, O, O, N—Ph → N, N, O, O, N, Ph

(o) CH_3O_2C, O, O, ArSe, H → HO, CO_2CH_3, O, O, H

(p) N, S^+, CH_2Ph $\xrightarrow{Et_3N}$ H, N, S, Ph

(q) CH_2, CH_2COCl $\xrightarrow{Et_3N}$ CH_2, O

12

Free-Radical Reactions

12.1. Generation and Characterization of Free Radicals

12.1.1. Background

A free-radical reaction is a chemical process in which molecules having unpaired electrons are involved. The radical species could be a starting compound or a product, but in organic chemistry the most common cases are reactions that involve radicals as intermediates. Most of the reactions discussed to this point have been heterolytic processes involving polar intermediates and/or transition states in which all electrons remained paired throughout the course of the reaction. In radical reactions, *homolytic* bond cleavages occur. The generalized reactions shown below illustrate the formation of alkyl, vinyl, and aryl free radicals by hypothetical homolytic processes.

$$Y\cdot + R_3C\vdots X \longrightarrow R_3C\cdot + X{-}Y$$

$$R_2C{=}C\begin{matrix} R \\ \vdots X \end{matrix} \longrightarrow R_2C{=}\dot{C}{-}R + X\cdot$$

$$C_6H_5{-}X{-}Y\vdots Z \longrightarrow C_6H_5{-}X{-}Y\cdot + Z\cdot \longrightarrow C_6H_5\cdot + X{=}Y$$

The idea that substituted carbon atoms with seven valence electrons could be involved in organic reactions took firm hold in the 1930s. Two experimental studies have special historical significance in the development of the concept of free-radical reactions. The work of Gomberg around 1900 provided evidence that when triphenylmethyl chloride was treated with silver metal, the resulting solution contained $Ph_3C\cdot$ in equilibrium with a less reactive molecule. It was originally thought that the more stable molecule was hexaphenylethane, but eventually this was shown not to be so. The dimeric product is actually a cyclohexadiene derivative.[1]

$$(Ph)_3C(H)\text{-cyclohexadienylidene}{=}C(Ph)_2 \rightleftharpoons 2\,Ph_3C\cdot$$

The dissociation constant is small, only about $2 \times 10^{-4}\ M$ at room temperature. The presence of the small amount of the radical at equilibrium was deduced from observation of reactions that could not be reasonably attributed to a normal hydrocarbon.

The second set of experiments was carried out in 1929 by Paneth. The decomposition of tetramethyllead was carried out in such a way that the decomposition products were carried by a flow of inert gas over a film of lead metal. The lead was observed to disappear, with re-formation of tetramethyllead. The conclusion was reached that methyl radicals must exist long enough in the gas phase to be transported from the point of decomposition to the lead film.

$$Pb(CH_3)_{4(g)} \xrightarrow{450°C} Pb_{(s)} + 4\,CH_3\cdot_{(g)}$$

$$4\,CH_3\cdot_{(g)} + Pb_{(s)} \xrightarrow{100°C} Pb(CH_3)_{4(g)}$$

Since these early experiments, a great deal of additional information about the existence and properties of free-radical intermediates has been developed. In this chapter, we will discuss the structure of free radicals and some of the special properties associated with free radicals. We will also discuss some of the key chemical reactions in which free-radical intermediates are involved.

12.1.2. Stable and Persistent Free Radicals

Most organic free radicals have very short lifetimes, but various structural features enhance stability. Radicals without special stabilization rapidly dimerize or disproportionate. The usual disproportionation process involves transfer of a hydrogen from the carbon β to the radical site, leading to formation of an alkane and an alkene.

Dimerization $\quad 2\ {-}\overset{|}{\underset{|}{C}}\cdot \rightarrow {-}\overset{|}{\underset{|}{C}}{-}\overset{|}{\underset{|}{C}}{-}$

Disproportionation $\quad 2\ {-}\overset{|}{\underset{H}{C}}{-}\overset{|}{\underset{|}{C}}\cdot \rightarrow {-}\overset{|}{\underset{H}{C}}{-}\overset{|}{\underset{H}{C}}{-} + \;\rangle C{=}C\langle$

1. H. Lankamp, W. Th. Nauta, and C. MacLean, *Tetrahedron Lett.*, 249 (1968); J. M. McBride, *Tetrahedron* **30**, 2009 (1974); K. J. Skinner, H. S. Hochester, and J. M. McBride, *J. Am. Chem. Soc.* **96**, 4301 (1974).

Radicals also rapidly abstract hydrogen or other atoms from many types of solvents, and most radicals are highly reactive toward oxygen.

Hydrogen atom abstraction $\quad -\overset{|}{\underset{|}{C}}\cdot + H-Y \rightarrow -\overset{|}{\underset{|}{C}}-H + Y\cdot$

Addition to oxygen $\quad -\overset{|}{\underset{|}{C}}\cdot + O_2 \rightarrow -\overset{|}{\underset{|}{C}}-O-O\cdot$

A few free radicals are indefinitely stable. Entries 1, 4, and 6 in Scheme 12.1 are examples. These molecules are just as stable under ordinary conditions of temperature and atmosphere as typical closed-shell molecules. Entry 2 is somewhat less stable to oxygen, although it can exist indefinitely in the absence of oxygen. The structures shown in entries 1, 2, and 4 all permit extensive delocalization of the unpaired electron into aromatic rings. These highly delocalized radicals show no tendency toward dimerization or disproportionation. Radicals that have long lifetimes and are resistant to dimerization or other routes for bimolecular self-annihilation are called *stable free radicals.* The term *inert free radical* has been suggested for species such as entry 4, which is unreactive under ordinary conditions and is thermally stable even at 300°C.[2]

Entry 3 in Scheme 12.1 has only alkyl substituents and yet has a significant lifetimes in the absence of oxygen. The tris(*t*-butyl)methyl radical has an even longer lifetime, with a half-life of about 20 min at 25°C.[3] The steric hindrance provided by the *t*-butyl substituents greatly retards the rates of dimerization and disproportionation of these radicals. They remain highly reactive toward oxygen, however. The term *persistent radicals* is used to describe these species, since their extended lifetimes have more to do with kinetic factors than with inherent stability.[4] Entry 5 is a sterically hindered perfluorinated radical, which is even more stable than similar alkyl radicals.

There are only a few functional groups that contain an unpaired electron and yet are stable in a wide variety of structural environments. The best example is the nitroxide group, and there are numerous specific nitroxide radicals which have been prepared and characterized.

$$R_2\overset{\cdot}{N}^{+}-\ddot{\underset{\cdot\cdot}{O}}:^{-} \longleftrightarrow R_2\ddot{N}-\ddot{\underset{\cdot\cdot}{O}}\cdot$$

Many of these compounds are very stable under normal conditions, and heterolytic reactions can be carried out on other functional groups in the molecule without destroying the nitroxide group.[5]

2. M. Ballester, *Acc. Chem. Res.* **18**, 380 (1985).
3. G. D. Mendenhall, D. Griller, D. Lindsay, T. T. Tidwell, and K. U. Ingold, *J. Am. Chem. Soc.* **96**, 2441 (1974).
4. For a review of various types of persistent radicals, see D. Griller and K. U. Ingold, *Acc. Chem. Res.* **9**, 13 (1976).
5. For reviews of the preparation, reactions, and uses of nitroxide radicals, see J. F. W. Keana, *Chem. Rev.* **78**, 37 (1978); L. J. Berliner (ed.), *Spin-Labelling*, Vol. 2, Academic Press, New York, 1979.

Scheme 12.1. Stability of Some Free Radicals

Structure	Conditions for stability
1[a]	Indefinitely stable as a solid, even in the presence of air.
2[b] Ph, Ph, Ph, Ph, Ph	Crystalline substance is not rapidly attacked by oxygen, although solutions are air sensitive; the compound is stable to high temperature in the absence of oxygen.
3[c] $(CH_3)_3C$–$\dot{C}$(H)–$C(CH_3)_3$	Stable in dilute solution ($<10^{-5}$ M) below -30°C in the absence of oxygen, $t_{1/2}$ of 50 sec at 25°C.
4[d] Cl, Cl, Cl, Cl, Cl, Cl, Cl, Cl, Cl, Cl, Cl, Cl, Cl, Cl, Cl	Stable in solution for days, even in the presence of air. Indefinitely stable in solid state. Thermally stable up to 300°C.
5[e] $(CF_3)_2CF$, $\dot{C}$–CF_2CF_3, $(CF_3)_2CF$	Stable to oxygen; thermally stable to 70°C.
5[e] $(CH_3)_3C$, $\cdot$O–, –CH=, =O, $C(CH_3)_3$, $(CH_3)_3C$, galvinoxyl, $C(CH_3)_3$	Stable to oxygen; stable to extended storage as a solid. Slowly decomposes in solution.
6[f] $(CH_3)_3C$, N–O$\cdot$, $(CH_3)_3C$	Stable to oxygen even above 100°C.

a. C. F. Koelsch, *J. Am. Chem. Soc.* **79**, 4439 (1957).
b. K. Ziegler and B. Schnell, *Justus Liebigs Ann. Chem.* **445**, 266 (1925).
c. G. D. Mendenhall, D. Griller, D. Lindsay, T. T. Tidwell, and K. U. Ingold, *J. Am. Chem. Soc.* **96**, 2441 (1974).
d. M. Ballester, J. Riera, J. Castañer, C. Badia, and J. M. Monsó, *J. Am. Chem. Soc.* **93**, 2215 (1971).
e. K. V. Scherer, Jr., T. Ono, K. Yamanouchi, R. Fernandez, and P. Henderson, *J. Am. Chem. Soc.* **107**, 718 (1985).
f. G. M. Coppinger, *J. Am. Chem. Soc.* **79**, 501 (1957); P. D. Bartlett and T. Funahashi, *J. Am. Chem. Soc.* **84**, 2596 (1962).
g. A. K. Hoffmann and A. T. Henderson, *J. Am. Chem. Soc.* **83**, 4671 (1961).

Although the existence of the stable and persistent free radicals we have discussed is of significance in establishing that free radicals can have extended lifetimes, most free-radical reactions involve highly reactive intermediates that have relatively fleeting lifetimes and can only be studied at very low concentrations. The techniques for study of radicals under these conditions are the subject of the next section.

12.1.3. Direct Detection of Radical Intermediates

The distinguishing characteristic of free radicals is the presence of an unpaired electron. Species with an unpaired electron are said to be *paramagnetic*. The most useful method for detecting and characterizing unstable radical intermediates is *electron paramagnetic resonance* (EPR) spectroscopy. *Electron spin resonance* (ESR) spectroscopy is synonymous. This method of spectroscopy detects the transition of an electron between the energy levels associated with the two possible orientations of electron spin in a magnetic field. An EPR spectrometer records the absorption of energy when an electron is excited from the lower to the higher state. The energy separation is very small on an absolute scale and corresponds to the energy of microwaves. EPR spectroscopy is a highly specific tool for detecting radical species since only molecules with unpaired electrons give rise to EPR spectra. As with other spectroscopic methods, detailed analysis of the absorption spectrum can give rise to structural information. One feature that is determined is the g value, which specifies the separation of the two spin states as a function of the magnetic field strength of the spectrometer.

$$hv = E = g\mu_B H$$

where μ_B is a constant, the Bohr magneton ($=9.274 \times 10^{-21}$ erg/G), and H is the magnetic field in gauss. The measured value of g is a characteristic of the particular type of radical, just as the line positions in IR and NMR spectra are characteristic of the absorbing species.

A second type of structural information can be deduced from the *hyperfine splitting* in EPR spectra. The origin of this line splitting is closely related to the factors that cause spin–spin splitting in proton NMR spectra. Certain nuclei have a magnetic moment. Those which are of particular interest in organic chemistry include ^1H, ^{13}C, ^{14}N, ^{19}F, and ^{31}P. Interaction of the unpaired electron with one or more of these nuclei splits the signal arising from the electron. The number of lines is given by the equation

$$\text{number of lines} = 2nI + 1$$

where I is the nuclear spin quantum number, and n is the number of equivalent interacting nuclei. For ^1H, ^{13}C, ^{19}F and ^{31}P, $I = \frac{1}{2}$. Thus, a single hydrogen splits a signal into a doublet. Interaction with three equivalent hydrogens, as in a methyl group, gives rise to splitting that produces four lines. This splitting is illustrated in Fig. 12.1. Nitrogen (^{14}N), with $I = 1$, splits each energy level into three lines. Neither

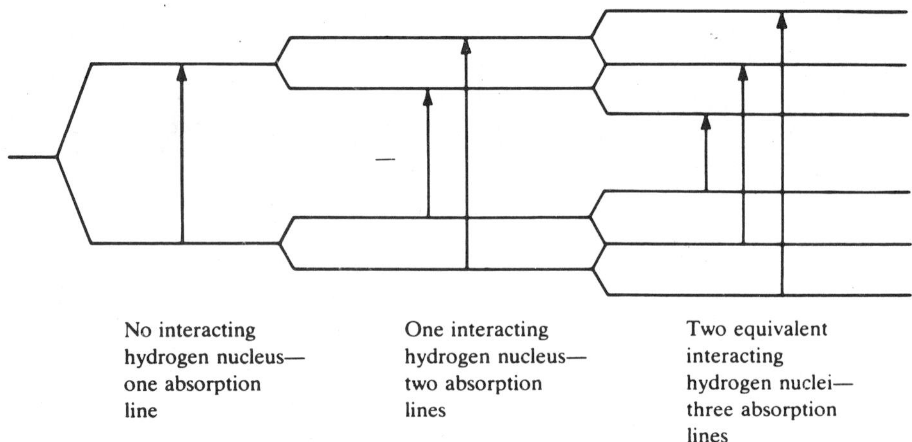

Fig. 12.1. Hyperfine splitting in EPR spectra.

^{12}C nor ^{16}O has a nuclear magnetic moment, and just as they cause no splitting in NMR spectra, they have no effect on the multiplicity in EPR spectra.

A great deal of structural information can be obtained by analysis of the hyperfine splitting pattern of a free radical. If we limit our discussion for the moment to radicals without heteroatoms, the number of lines indicates the number of interacting hydrogens, and the magnitude of the splitting, given by the hyperfine splitting constant a, is a measure of the unpaired electron density in the hydrogen $1s$ orbital. For planar systems in which the unpaired electron resides in a π-orbital system, the relationship between electron spin density and the splitting constant is given by the McConnell equation[6]:

$$a = \rho Q$$

where a is the hyperfine coupling constant for a proton, Q is a proportionality constant (about 23 G), and ρ is the spin density on the carbon to which the hydrogen is attached. For example, taking $Q = 23.0$ G, the hyperfine splitting in the benzene radical anion may be readily calculated by taking $\rho = \frac{1}{6}$, since the one unpaired electron must be distributed equally among the six carbon atoms. The calculated value of $a = 3.83$ is in good agreement with the observed value. The spectrum (Fig. 12.2a) consists of seven lines separated by a coupling constant of 3.75 G.

The EPR spectrum of the ethyl radical presented in Fig. 12.2b is readily interpreted, and the results are of interest with respect to the distribution of unpaired electron density in the molecule. The 12-line spectrum is a triplet of quartets resulting from unequal coupling of the electron spin to the α and β protons. The two coupling constants are $a_\alpha = 22.38$ G and $a_\beta = 26.87$ G and imply extensive delocalization of spin density through the σ bonds.

EPR spectra have been widely used in the study of reactions to detect free-radical intermediates. An interesting example involves the cyclopropylmethyl radical. Much

6. H. M. McConnell, *J. Chem. Phys.* **24**, 764 (1956).

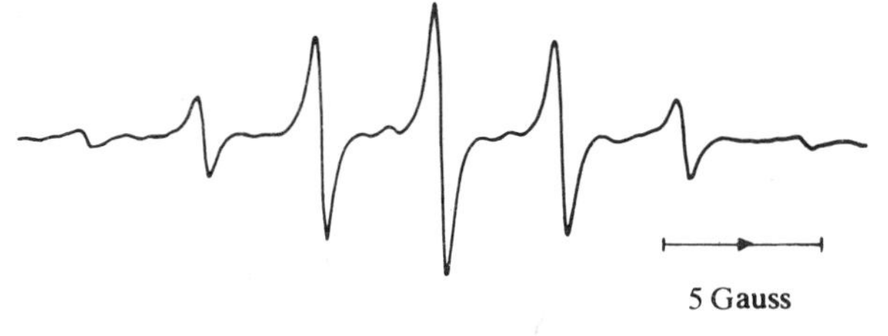

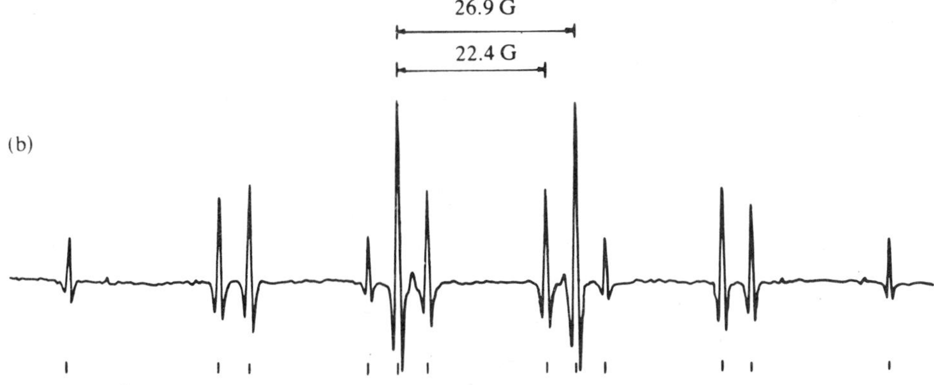

Fig. 12.2. Some EPR spectra of small organic free radicals: (a) Spectrum of the benzene radical anion. [From J. R. Bolton, *Mol. Phys.* **6**, 219 (1963). Reproduced by permission of Taylor and Francis, Ltd.] (b) Spectrum of the ethyl radical. [From R. W. Fessenden and R. H. Schuler, *J. Chem. Phys.* **33**, 935 (1960); **39**, 2147 (1963); Reproduced by permission of the American Institute of Physics.]

chemical experience has indicated that this radical is unstable, giving rise to 3-butenyl radical rapidly after being generated.

$$\triangleright\!\!-CH_2\cdot \longrightarrow \cdot CH_2-CH_2-CH{=}CH_2$$

The radical was generated by photolytic decomposition of di-*t*-butyl peroxide in methylcyclopropane, a process that leads to selective abstraction of a methyl hydrogen from methylcyclopropane.

$$(CH_3)_3COOC(CH_3)_3 \xrightarrow{h\nu} 2\,(CH_3)_3C{-}O\cdot$$

$$(CH_3)_3C{-}O\cdot + \triangleright\!\!-CH_3 \longrightarrow \triangleright\!\!-CH_2\cdot + (CH_3)_3COH$$

Below −140°C, the EPR spectrum observed was that of the cyclopropylmethyl radical. If the photolysis was done above −140°C, however, the spectrum of a second species was seen, and above −100°C, this was the only spectrum observed. This spectrum could be shown to be that of the 3-butenyl radical.[7] This study also established that the 3-butenyl radical did not revert to the cyclopropylmethyl radical on being cooled back to −140°C. The conclusion is that the ring opening of the

7. J. K. Kochi, P. J. Krustic, and D. R. Eaton, *J. Am. Chem. Soc.* **91**, 1877 (1969).

cyclopropyl radical is a very facile process, so that the lifetime of the cyclopropyl radical above $-100°C$ is very short. The reversal of the ring opening can be detected by isotopic labeling experiments, which reveal the occurrence of deuterium migration:

$$CH_2{=}CHCH_2CD_2\cdot \rightleftharpoons \underset{H_2C-CD_2}{\overset{H\;\;CH_2\cdot}{\diagdown\!\!\!\!\triangle}} \rightleftharpoons CH_2{=}CHCD_2CH_2\cdot$$

The rates of both the ring opening ($k = 2 \times 10^8\ s^{-1}$ at 25°C) and the ring closure ($k = 3 \times 10^3\ s^{-1}$) have been measured and show that only a very small amount of the cyclopropylmethyl radical is present at equilibrium, in agreement with the EPR results.[8]

It is important to emphasize that direct studies such as those carried out on the cyclopropylmethyl radical can be done with low *steady-state* concentrations of the radical. In the case of the study of the cyclopropylmethyl radical, removal of the source of irradiation would lead to rapid disappearance of the EPR spectrum, because the radicals would react rapidly and not be replaced by continuing radical formation. Under many conditions, the steady-state concentration of a radical intermediate may be too low to permit direct detection. Failure to observe an EPR signal therefore cannot be taken as conclusive evidence against a radical intermediate.

A technique called *spin trapping* can sometimes be used to study radicals in this circumstance. A diamagnetic molecule that has the property of reacting rapidly with radicals to give a stable paramagnetic species is introduced into the reaction system being studied. As radical intermediates are generated, they are trapped by the reactive molecule to give more stable, detectable radicals. The most useful spin traps are nitroso compounds. They rapidly react with radicals to give stable nitroxide radicals.[9] Analysis of the EPR spectrum of the nitroxide radical product can often provide information about the structure of the original radical.

$$R\cdot + R'N{=}O \longrightarrow \begin{matrix} R \\ & \diagdown \\ & & N{-}\dot{O} \\ & \diagup \\ R' \end{matrix}$$

Another technique that is highly specific for radical processes is known as CIDNP, an abbreviation for *chemically induced dynamic nuclear polarization.*[10] The instrumentation required for such studied is a normal NMR spectrometer. CIDNP is observed as a strong perturbation of the intensity of NMR signals for products formed in certain types of free-radical reactions. The variation in intensity results

8. A. Eiffio, D. Griller, K. U. Ingold, A. L. J. Beckwith, and A. K. Serelis, *J. Am. Chem. Soc.* **102**, 1734 (1980); L. Mathew and J. Warkentin, *J. Am. Chem. Soc.* **108**, 7981 (1986).
9. E. G. Janzen, *Acc. Chem. Res.* **4**, 31 (1971); E. G. Janzen, in *Free Radicals in Biology*, W. A. Pryor (ed.), Vol. 4, Academic Press, New York, 1980, pp. 115–154.
10. H. R. Ward, *Acc. Chem. Res.* **5**, 18 (1972); R. G. Lawler, *Acc. Chem. Res.* **5**, 25 (1972).

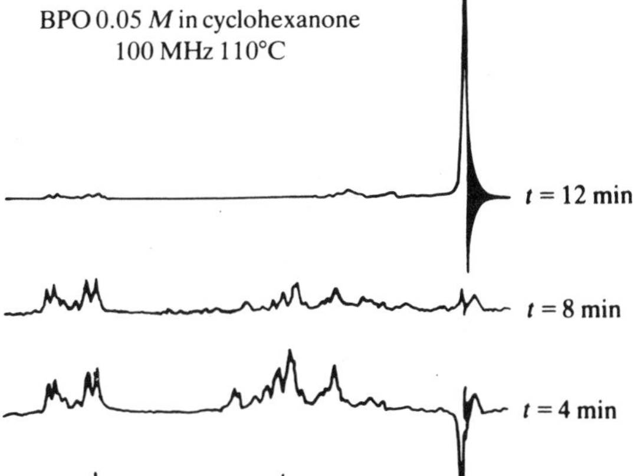

Fig. 12.3. NMR spectra recorded during thermal decomposition of dibenzoyl peroxide. Singlet at high field is due to benzene; other signals are due to dibenzoyl peroxide. [From H. Fischer and J. Bargon, *Acc. Chem. Res.* **2**, 110 (1969). Reproduced by permission of the American Chemical Society.]

when the normal population of nuclear spin states dictated by the Boltzmann distribution is disturbed by the presence of an unpaired electron. The magnetic moment associated with an electron causes a redistribution of the nuclear spin states. Individual nuclei can become overpopulated in either the lower or upper spin state. If the lower state is overpopulated, an enhanced absorption signal is observed. If the upper state is overpopulated, an emission signal is observed. The CIDNP method is not as general as EPR spectroscopy because not all free-radical reactions can be expected to exhibit the phenomenon.[11]

Figure 12.3 shows the observation of CIDNP during the decomposition of dibenzoyl peroxide in cyclohexanone:

$$\mathrm{PhC(=O)OOC(=O)Ph} \longrightarrow 2\,\mathrm{Ph\cdot} + 2\,\mathrm{CO_2}$$

$$\mathrm{Ph\cdot} + \mathrm{S{-}H} \longrightarrow \mathrm{C_6H_6} + \mathrm{S\cdot}$$

The emission signal corresponding to benzene confirms that it is formed by a free-radical process. As in steady-state EPR experiments, the enhanced emission and absorption are observed only as long as the reaction is proceeding. When the

11. For a discussion of the theory of CIDNP and the conditions under which spin polarization occurs, see G. L. Closs, *Adv. Magn. Reson.* **7**, 157 (1974); R. Kaptein, *Adv. Free Radical Chem.* **5**, 318 (1975); G. L. Closs, R. J. Miller, and O. D. Redwine, *Acc. Chem. Res.* **18**, 196 (1985).

reaction is complete or is stopped in some way, the signals rapidly return to their normal intensity, because the equilibrium population of the two spin states is rapidly reached.

One aspect of both EPR and CIDNP studies that should be kept in mind is that either technique is capable of detecting very small amounts of radical intermediates. This aspect makes both techniques quite sensitive but can also present a pitfall. The most prominant features of either EPR or CIDNP spectra may actually be due to radicals that account for only minor products of the total reaction process. An example of this was found in a CIDNP study of the decomposition of trichloroacetyl peroxide in alkenes:

$$Cl_3C\overset{O}{\overset{\|}{C}}OO\overset{O}{\overset{\|}{C}}CCl_3 \longrightarrow 2\,Cl_3C\cdot + 2\,CO_2$$

$$Cl_3C\cdot + CH_2{=}C(CH_3)_2 \longrightarrow Cl_3CH_2\dot{C}(CH_3)_2$$

$$Cl_3C\cdot + Cl_3CCH_2\dot{C}(CH_3)_2 \longrightarrow Cl_3CH + Cl_3CCH_2\underset{CH_3}{\underset{|}{C}}{=}CH_2$$

In addition to the emission signals of $CHCl_3$ and $Cl_3CCH_2C(CH_3){=}CH_2$, which are the major products, a strong emission signal for Cl_3CCHCl_2 was identified. However, this compound is a very minor product of the reaction, and when the signals have returned to their normal intensity, Cl_3CCHCl_2 is present in such a small amount that it cannot be detected.[12]

12.1.4. Sources of Free Radicals

There are several reactions that are quite commonly used as sources of free radicals, both for the study of radical structure and reactivity and also in synthetic processes. Some of the most general methods are outlined here. Examples of many of these will be encountered again when specific reactions are discussed. For the most part, we will defer discussion of the reactions of the radicals until then.

Peroxides are a common source of radical intermediates. An advantage of the generation of radicals from peroxides is that reaction generally occurs at relatively low temperature. The oxygen–oxygen bond in peroxides is weak ($\sim$30 kcal/mol), and activation energies for radical formation are low. Diacyl peroxides are sources of alkyl radicals because the carboxyl radicals that are initially formed lose CO_2 very rapidly.[13] In the case of aroyl peroxides, products may be derived from the carboxyl radical or the radical formed by decarboxylation.[14]

12. H. Y. Loken, R. G. Lawler, and H. R. Ward, *J. Org. Chem.* **38**, 106 (1973).
13. J. C. Martin, J. W. Taylor, and E. H. Drew, *J. Am. Chem. Soc.* **89**, 129 (1967); F. D. Greene, H. P. Stein, C.-C. Chu, and F. M. Vane, *J. Am. Chem. Soc.* **86**, 2080 (1964).
14. D. F. DeTar, R. A. J. Long, J. Rendleman, J. Bradley, and P. Duncan, *J. Am. Chem. Soc.* **89**, 4051 (1967).

$$CH_3\overset{O}{\overset{\|}{C}}OO\overset{O}{\overset{\|}{C}}CH_3 \xrightarrow{80\text{–}100°C} 2\,CH_3\overset{O}{\overset{\|}{C}}O\cdot \longrightarrow 2\,CH_3\cdot + 2\,CO_2$$

$$Ph\overset{O}{\overset{\|}{C}}OO\overset{O}{\overset{\|}{C}}Ph \xrightarrow{80\text{–}100°C} 2\,Ph\overset{O}{\overset{\|}{C}}O\cdot \longrightarrow 2\,Ph\cdot + 2\,CO_2$$

Alkyl hydroperoxides give alkoxy radicals and the hydroxyl radical. *t*-Butyl hydroperoxide is often used as a radical source. Detailed studies have been reported on the mechanism of the decomposition, which is a somewhat more complicated process than simple unimolecular decomposition.[15] Dialkyl peroxides decompose to give two alkoxy radicals.[16]

$$C_2H_5OOC_2H_5 \xrightarrow{80°C} 2C_2H_5O\cdot$$

Peroxyesters are also sources of radicals. The acyloxy portion normally loses carbon dioxide, so peroxyesters yield an alkyl (or aryl) and an alkoxy radical.[17]

$$R\overset{O}{\overset{\|}{C}}OOC(CH_3)_3 \rightarrow R\cdot + CO_2 + \cdot OC(CH_3)_3$$

The decomposition of peroxides, which occurs thermally in the examples cited above, can also be readily accomplished by photochemical excitation. The alkyl hydroperoxides are also sometimes used in conjunction with a transition metal ion. Under these conditions, an alkoxy radical is produced, but the hydroxyl portion appears as hydroxide ion as a result of one-electron reduction by the metal ion.[18]

$$(CH_3)_3COOH + M^{2+} \rightarrow (CH_3)_3CO\cdot + {}^-OH + M^{3+}$$

The thermal decompositions described above are unimolecular reactions that should exhibit first-order kinetics. Under many conditions, peroxides decompose at rates faster than expected for unimolecular thermal decomposition, and with more complicated kinetics. This behavior is known as *induced decomposition* and occurs when part of the peroxide decomposition is the result of bimolecular reactions with radicals present in solution, as illustrated specifically for diethyl peroxide:

$$X\cdot + CH_3CH_2OOCH_2CH_3 \rightarrow CH_3\dot{C}HOOCH_2CH_3 + H{-}X$$
$$CH_3\dot{C}HOOCH_2CH_3 \rightarrow CH_3CH{=}O + \cdot OCH_2CH_3$$

The amount of induced decomposition that occurs depends on the concentration and reactivity of the radical intermediates and the susceptibility of the substrate to radical attack. The racial X· may be formed from the peroxide, but it can also be derived from subsequent reactions with the solvent. For this reason, both the structure

15. R. Hiatt, T. Mill, and F. R. Mayo, *J. Org. Chem.* **33**, 1416 (1968), and accompanying papers.
16. W. A. Pryor, D. M. Huston, T. R. Fiske, T. L. Pickering, and E. Ciuffarin, *J. Am. Chem. Soc.* **86**, 4237 (1964).
17. P. D. Bartlett and R. R. Hiatt, *J. Am. Chem. Soc.* **80**, 1398 (1958).
18. W. H. Richardson, *J. Am. Chem. Soc.* **87**, 247 (1965).

of the peroxide and the nature of the reaction medium are important in determining the extent of induced decomposition relative to unimolecular homolysis.

Another quite general source of free radicals is the decomposition of azo compounds. The products are molecular nitrogen and the radicals derived from the substituent groups.

$$\text{R—N=N—R}' \xrightarrow[h\nu]{\Delta \text{ or}} \text{R}\cdot + \text{N}\equiv\text{N} + \cdot\text{R}'$$

Both symmetrical and unsymmetrical azo compounds can be made so a single radical or two different ones may be generated. The energy for the decomposition can be either thermal or photochemical.[19] In the thermal decomposition, it has been established that the temperature at which decomposition occurs depends on the nature of the substituent groups. Azomethane does not decompose to methyl radicals and nitrogen until temperatures above 400°C are reached. Azo compounds that generate relatively stable radicals decompose at much lower temperatures. Azo compounds derived from allyl groups decompose somewhat above 100°C, for example:

$$CH_3CH_2CH_2N{=}NCH_2CH{=}CH_2 \xrightarrow{130^\circ C} CH_3CH_2CH_2\cdot + N_2 + CH_2{=}CHCH_2\cdot \quad \text{Ref. 20}$$

Unsymmetrical azo compounds must be used to generate phenyl radicals because azobenzene is very stable thermally. Phenylazotriphenylmethane decomposes readily because of the stability of the triphenylmethyl radical.

$$PhN{=}NC(Ph)_3 \xrightarrow{60^\circ C} Ph\cdot + Ph_3C\cdot + N_2 \quad \text{Ref. 21}$$

Many azo compounds also generate radicals when photolyzed. This can occur by a thermal decomposition of the *cis* azo compounds that are formed in the photochemical step.[22] The *cis* isomers are thermally much more labile than the *trans* isomers.

$$\text{R—N=N—R (trans)} \rightarrow \text{R—N=N—R (cis)} \rightarrow \text{R}\cdot \;\; N_2 \;\; \cdot\text{R}$$

N-Nitrosoanilides are an alternative source of aryl radicals. There is a close mechanistic relationship between this route and the decomposition of azo compounds. The *N*-nitrosoanilides rearrange to an intermediate with a nitrogen–nitrogen double bond. This intermediate then decomposes to generate aryl radicals.[23]

$$\text{PhN(N=O)C(=O)R} \rightarrow [\text{PhN=N—O—C(=O)R}] \rightarrow \text{Ph}\cdot + N_2 + RCO_2\cdot$$

19. P. S. Engel, *Chem. Rev.* **80**, 99 (1980).
20. K. Takagi and R. J. Crawford, *J. Am. Chem. Soc.* **93**, 5910 (1971).
21. R. F. Bridger and G. A. Russell, *J. Am. Chem. Soc.* **85**, 3754 (1963).
22. M. Schmittel and C. Rüchardt, *J. Am. Chem. Soc.* **109**, 2750 (1987).
23. C. Rüchardt and B. Freudenberg, *Tetrahedron Lett.*, 3623 (1964); J. I. G. Cadogan, *Acc. Chem. Res.* **4**, 186 (1971).

A technique that is a convenient source of radicals for study by EPR involves photolysis of a mixture of di-*t*-butyl peroxide, triethylsilane, and the alkyl bromide corresponding to the radical to be studied.[24] Photolysis of the peroxide gives *t*-butoxy radicals, which selectively abstract hydrogen from the silane. This reactive silicon radical in turn abstracts bromine, generating the alkyl radical at a steady-state concentration suitable for EPR study.

$$(CH_3)_3COOC(CH_3)_3 \xrightarrow{h\nu} 2\,(CH_3)_3CO\cdot$$

$$(CH_3)_3CO\cdot + (C_2H_5)_3SiH \longrightarrow (CH_3)_3COH + (C_2H_5)_3Si\cdot$$

$$(C_2H_5)_3Si\cdot + RBr \longrightarrow (C_2H_5)_3SiBr + R\cdot$$

The acyl derivatives of *N*-hyroxypyridine-2-thione are a synthetically versatile source of free radicals.[25] These compounds are readily prepared from reactive acylating agents, such as acid chlorides, and a salt of the *N*-hydroxypyridine-2-thione.

$$RC(=O)Cl + {}^{-}O{-}N(\text{pyridine-2-thione}) \longrightarrow RC(=O){-}O{-}N(\text{pyridine-2-thione})$$

Radicals react at the sulfur, and decomposition generating an acyloxy radical ensues. The acyloxy radical undergoes decarboxylation. Usually, the radical then gives product and another radical, which can continue a chain reaction. The process can be illustrated by the reactions with tri-*n*-butylstannane and bromotrichloromethane.

(a) Reductive decarboxylation by reaction with tri-n-butylstannane

$$RC(=O){-}O{-}N(\text{pyridine-2-thione}) + Bu_3Sn\cdot \longrightarrow RC(=O){-}O{-}N(\text{2-}Bu_3SnS\text{-pyridinyl}\cdot) \longrightarrow R{-}CO_2\cdot + 2\text{-}(Bu_3SnS)\text{pyridine}$$

Ref. 26

$$R{-}CO_2\cdot \rightarrow R\cdot + CO_2 \qquad R\cdot + Bu_3Sn{-}H \rightarrow R{-}H + Bu_3Sn\cdot$$

(b) Conversion of aromatic carboxylic acid to aryl bromide by reaction with bromotrichloromethane

$$ArC(=O){-}O{-}N(\text{pyridine-2-thione}) + \cdot CCl_3 \longrightarrow ArCO_2\cdot + 2\text{-}(Cl_3CS)\text{pyridine}$$

Ref. 27

$$ArCO_2\cdot \longrightarrow Ar\cdot + CO_2 \qquad Ar\cdot + BrCCl_3 \rightarrow Ar{-}Br + \cdot CCl_3$$

24. A. Hudson and R. A. Jackson, *J. Chem. Soc., Chem. Commun.*, 1323 (1969); D. J. Edge and J. K. Kochi, *J. Am. Chem. Soc.* **94**, 7695 (1972).
25. D. H. R. Barton, D. Crich, and W. B. Motherwell, *Tetrahedron* **41**, 3901 (1985).
26. D. H. R. Barton, D. Crich, and W. B. Motherwell, *J. Chem. Soc., Chem. Commun.*, 939 (1983).
27. D. H. R. Barton, B. Lacher, and S. Z. Zard, *Tetrahedron Lett.* **26**, 5939 (1985).

12.1.5. Structural and Stereochemical Properties of Radical Intermediates

EPR studies and other physical methods have provided the basis for some insight into the detailed geometry of radical species.[28] Deductions about structure can also be drawn from the study of the stereochemistry of reactions involving radical intermediates. Several structural possibilities must be considered. If discussion is limited to alkyl radicals, the possibilities include a rigid pyramidal structure, rapidly inverting pyramidal structures, or a planar structure.

rigid pyramidal   flexible pyramidal   planar

Precise description of the pyramidal structures would also require that the bond angles be specified. The EPR spectrum of the methyl radical leads to the conclusion that its structure could be either planar or a very shallow pyramid.[29] The IR spectrum of methyl radical has been recorded at very low temperatures in frozen argon.[30] The IR spectrum puts a maximum of ~5° on the deviation from planarity.

The *t*-butyl radical has been studied extensively. While experimental results have been interpreted in terms of both planar and slightly pyramidal structures, theoretical calculations favor a pyramidal structure.[31] It appears that simple alkyl radicals are generally pyramidal, although the barrier to inversion is very small. *Ab initio* molecular orbital calculations suggest that two factors are of principal importance in favoring a pyramidal structure. One is a torsional effect in which the radical center tends to adopt a staggered conformation of the radical substituents. There is also a hyperconjugative interaction between the half-filled orbital and the hydrogen that is aligned with it. This hyperconjugation is stronger in the conformation in which the pyramidalization is such as to minimize eclipsing.[32] The theoretical results also indicate that the barrier to inversion is no more than 1–2 kcal/mol, so rapid inversion will occur.

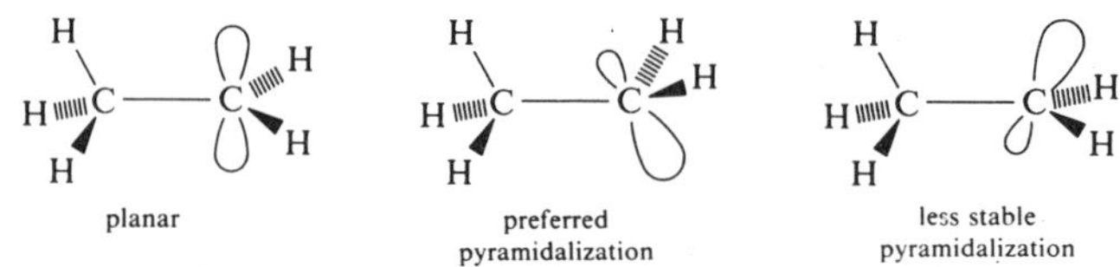

planar   preferred pyramidalization   less stable pyramidalization

28. For a review, see J. K. Kochi, *Adv. Free Radicals Chem.* **5**, 189 (1975).
29. M. Karplus and G. K. Fraenkel, *J. Chem. Phys.* **35**, 1312 (1961).
30. L. Andrews and G. C. Pimentel, *J. Chem. Phys.* **47**, 3637 (1967).
31. L. Bonazolla, N. Leroy, and J. Roncin, *J. Am. Chem. Soc.* **99**, 8348 (1977); D. Griller, K. U. Ingold, P. J. Krusic, and H. Fischer, *J. Am. Chem. Soc.* **100**, 6750 (1978); J. Pacansky and J. S. Chang, *J. Phys. Chem.* **74**, 5539 (1978); B. Schrader, J. Pacansky, and U. Pfeiffer, *J. Phys. Chem.* **88**, 4069 (1984).
32. M. N. Paddon-Row and K. N. Houk, *J. Am. Chem. Soc.* **103**, 5046 (1981); M. N. Paddon-Row and K. N. Houk, *J. Phys. Chem.* **89**, 3771 (1985).

Radical geometry is significantly affected by substituent groups that can act as π donors. Addition of fluorine or oxygen substituents, in particular, favors a pyramidal structure. Analysis of the EPR spectra of the mono-, di-, and trifluoromethyl radicals indicates a progressive distortion from planarity.[33] Both EPR and IR studies of the trifluoromethyl radical show it to be pyramidal.[34] The basis of this structural effect has been probed by molecular orbital calculations and is considered to result from interactions of both the σ and the π type. There is a repulsive interaction between the singly occupied p orbital and the filled orbitals occupied by "lone pair" electrons on the fluorine or oxygen substituents. This repulsive interaction is minimized by adoption of a pyramidal geometry. The tendency for pyramidal geometry is reinforced by an interaction between the p orbital on carbon and the σ^* antibonding orbitals associated with the C—F or C—O bonds. The energy of the p orbital can be lowered by interaction with the σ^* orbital. This interaction increases electron density on the more electronegative fluorine or oxygen atom. This stabilizing p-σ^* interaction is increased with pyramidal geometry.

stabilizing interaction with σ^*

There have been many studies aimed at deducing the geometry of radical sites by examining the stereochemistry of radical reactions. The most direct kind of study involves the generation of a radical at a carbon that is a chiral center. A planar or rapidly inverting radical would lead to racemization, whereas a rigid pyramidal structure should lead to product of retained configuration. Some examples of reactions that have been subjected to this kind of study are shown in Scheme 12.2. In each case, racemic product is formed, indicating that alkyl radicals do not retain the tetrahedral geometry of their precursors.

Cyclic molecules also permit deductions about stereochemistry without the necessity of using resolved chiral compounds. The stereochemistry of a number of reactions of 4-substituted cyclohexyl radicals has been investigated.[35] In general, reactions starting from pure *cis* or *trans* stereoisomers give mixtures of *cis* and *trans* products. This result indicates that the radical intermediates do not retain the stereochemistry of the precursor. Radical reactions involving cyclohexyl radicals are not usually very stereoselective, but some show a preference for formation of the *cis* product. This has been explained in terms of a torsional effect. If the cyclohexyl radical is planar or a shallow pyramid, equatorial attack leading to *trans* product causes the hydrogen at the radical site to become eclipsed with the two

33. P. J. Krusic and R. C. Bingham, *J. Am. Chem. Soc.* **98**, 230 (1976); F. Bernardi, W. Cherry, S. Shaik, and N. D. Epiotis, *J. Am. Chem. Soc.* **100**, 1352 (1978).
34. R. W. Fessenden and R. H. Schuler, *J. Chem. Phys.* **43**, 2704 (1965); G. A. Carlson and G. C. Pimentel, *J. Chem. Phys.* **44**, 4053 (1966).
35. F. R. Jensen, L. H. Gale, and J. E. Rodgers, *J. Am. Chem. Soc.* **90**, 5793 (1968).

Scheme 12.2. Stereochemistry of Radical Reactions at Chiral Carbon Atoms

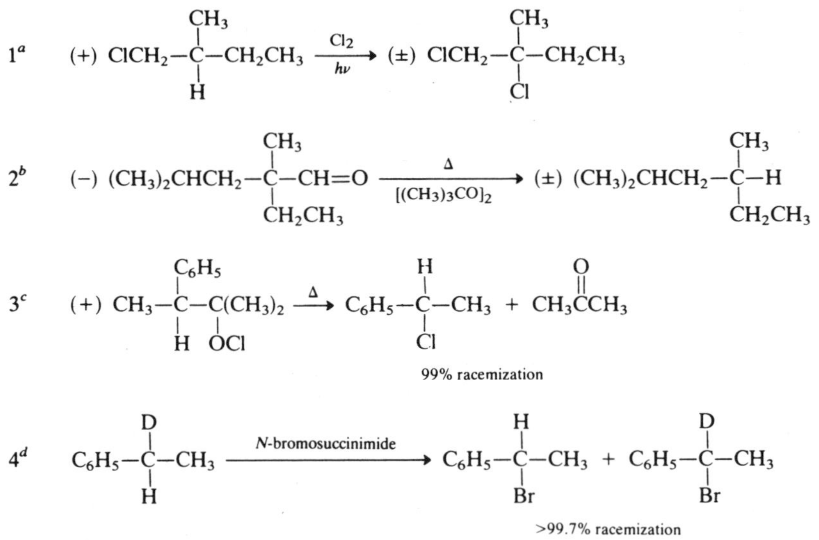

1[a] (+) $ClCH_2-C(CH_3)(H)-CH_2CH_3 \xrightarrow[h\nu]{Cl_2}$ (±) $ClCH_2-C(CH_3)(Cl)-CH_2CH_3$

2[b] (−) $(CH_3)_2CHCH_2-C(CH_3)(CH_2CH_3)-CH{=}O \xrightarrow[[(CH_3)_3CO]_2]{\Delta}$ (±) $(CH_3)_2CHCH_2-C(CH_3)(CH_2CH_3)-H$

3[c] (+) $CH_3-C(C_6H_5)(H)-C(CH_3)_2(OCl) \xrightarrow{\Delta} C_6H_5-C(H)(Cl)-CH_3 + CH_3CCH_3$ (O above C)

99% racemization

4[d] $C_6H_5-C(D)(H)-CH_3 \xrightarrow{N\text{-bromosuccinimide}} C_6H_5-C(H)(Br)-CH_3 + C_6H_5-C(D)(Br)-CH_3$

>99.7% racemization

a. H. C. Brown, M. S. Kharasch, and T. H. Chao, *J. Am. Chem. Soc.* **62**, 3435 (1940).
b. W. von E. Doering, M. Farber, M. Sprecher, and K. B. Wiberg, *J. Am. Chem. Soc.* **74**, 3000 (1952).
c. F. D. Greene, *J. Am. Chem. Soc.* **81**, 2688 (1959); D. B. Denney and W. F. Beach, *J. Org. Chem.* **24**, 108 (1959).
d. H. J. Dauben, Jr., and L. L. McCoy, *J. Am. Chem. Soc.* **81**, 5404 (1959).

neighboring equatorial hydrogens. Axial attack does not suffer from this strain, since the hydrogen at the radical site moves away from the equatorial hydrogens toward the staggered conformation that is present in the chair conformation of the ring. The pyramidalization of the radical would be expected to be in the direction favoring axial attack.

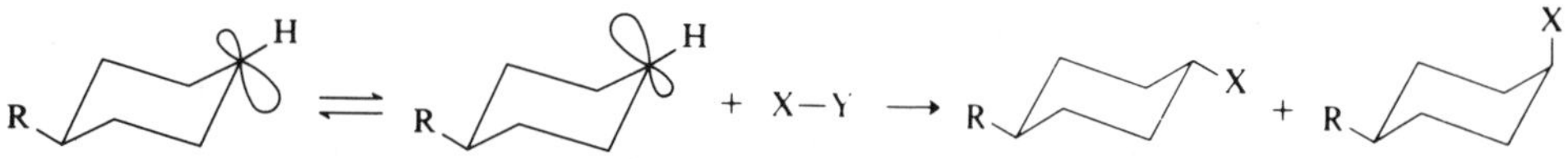

R, H ⇌ R, H + X—Y ⟶ R, X + R, X

Another approach to obtaining information about the geometric requirements of free radicals has been to examine bridgehead systems. It will be recalled that small bicyclic rings strongly resist formation of carbocations at bridgehead centers because the skeletal geometry prevents attainment of the preferred planar geometry by the carbocation. In an early study, the decarbonylation of bridgehead aldehydes by a free-radical reaction was found to proceed without difficulty[36]:

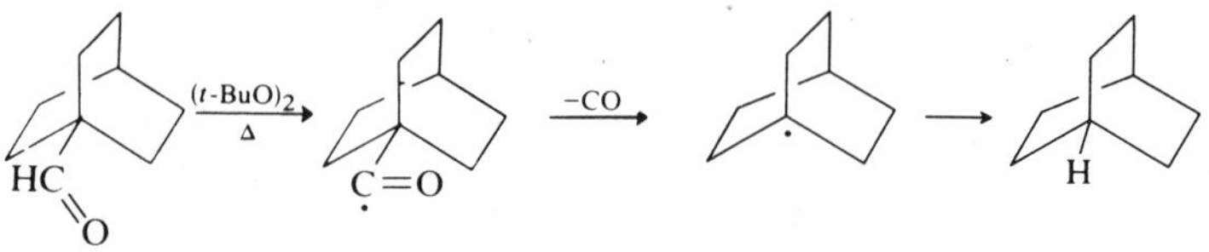

HC=O $\xrightarrow[\Delta]{(t\text{-BuO})_2}$ Ċ=O $\xrightarrow{-CO}$ · ⟶ H

36. W. v. E. Doering, M. Farber, M. Sprecher, and K. B. Wiberg, *J. Am. Chem. Soc.* **74**, 3000 (1952).

Subsequent rate studies have shown that there is significant rate retardation for reactions in which the norbornyl radical is generated in a rate-determining step.[37] Typically, such reactions proceed 500 to 1000 times slower than the corresponding reaction generating the *t*-butyl radical. This is a much smaller rate retardation than that of 10^{14} found in S_N1 solvolysis. Rate retardation is smaller for less strained bicyclic systems. The EPR spectra of the bridgehead radicals **A** and **B** are consistent with pyramidal geometry at the bridgehead carbon atoms.[38]

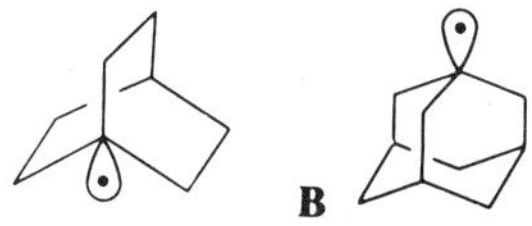

A **B**

The general conclusion of all these studies is that alkyl radicals are shallow pyramids, and the barrier to inversion of the pyramidal structures is low. Radicals also are able to tolerate some geometric distortion associated with strained ring systems.

The allyl radical would be expected to be planar in order to maximize delocalization. Molecular structure parameters have been obtained from EPR, IR, and electron diffraction measurements and confirm that the radical is planar.[39]

There has also been study of the structure of vinyl free radicals.[40] Stereochemical results indicate that radicals formed at trigonal centers rapidly undergo interconversion with the geometric isomer. As a result, reactions proceeding through vinyl radical intermediates usually give rise to the same mixture from both the *cis* and *trans* precursors.

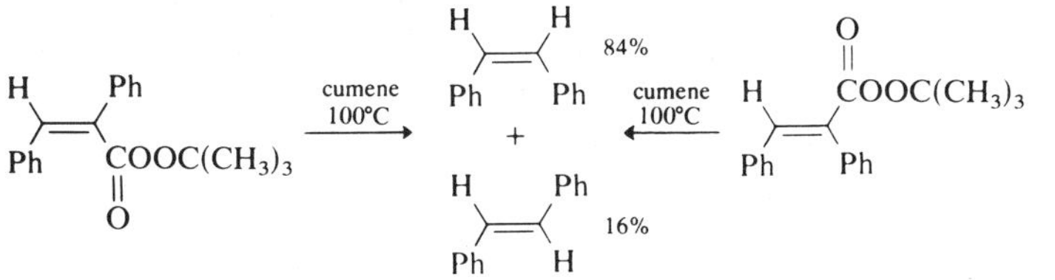

H Ph / Ph $COOC(CH_3)_3$ —cumene, 100°C→ H H / Ph Ph (84%) + H Ph / Ph H (16%) ←cumene, 100°C— H $COOC(CH_3)_3$ / Ph Ph   Ref. 41

In this particular case, there is evidence from EPR spectra that the radical is not linear in its ground state, but is an easily inverted bent species.[42] The barrier to

37. A. Oberlinner and C. Rüchardt, *Tetrahedron Lett.*, 4685 (1969); L. B. Humphrey, B. Hodgson, and R. E. Pincock, *Can. J. Chem.* **46**, 3099 (1968); D. E. Applequist and L. Kaplan, *J. Am. Chem. Soc.* **87**, 2194 (1965).
38. P. J. Krusic, T. A. Rettig, and P. v. R. Schleyer, *J. Am. Chem. Soc.* **94**, 995 (1972).
39. R. W. Fessenden and R. H. Schuler, *J. Chem. Phys.* **39**, 2147 (1963); A. K. Maltsev, V. A. Korolev, and O. M. Nefedov, *Izd. Akad. Nauk SSSR, Ser. Khim.*, 555 (1984); E. Vajda, J. Tremmel, B. Rozandai, I. Hargittai, A. K. Maltsev, N. D. Kagrammanov, and O. M. Nefedov, *J. Am. Chem. Soc.* **108**, 4352 (1986).
40. For reviews of the structure and reactivity of vinyl radicals, see W. G. Bentrude, *Annu. Rev. Phys. Chem.* **18**, 283 (1967); L. A. Singer, in *Selective Organic Transformations*, Vol II, B. S. Thyagarajan (ed.), John Wiley, New York, 1972, p. 239; O. Simamura, *Top. Stereochem.* **4**, 1 (1969).
41. L. A. Singer and N. P. Kong, *J. Am. Chem. Soc.* **88**, 5213 (1966); J. A. Kampmeier and R. M. Fantazier, *J. Am. Chem. Soc.* **88**, 1959 (1966).
42. R. W. Fessenden and R. H. Schuler, *J. Chem. Phys.* **39**, 2147 (1963).

Table 12.1. Oxidation and Reduction Potentials for Some Aromatic Hydrocarbons[a]

Hydrocarbon	$Ar{-}H \xrightarrow{+e^-} Ar{-}H^{\cdot -}$	$Ar{-}H \xrightarrow{-e^-} Ar{-}H^{\cdot +}$
Benzene	−3.42[b]	+2.06
Naphthalene	−2.95	+1.33
Anthracene	−2.36	+0.89
Phenanthrene	−2.87	+1.34
Tetracene	−1.92	+0.57

a. Except where noted otherwise, the data are from C. Madec and J. Courtot-Coupez, *J. Electroanal. Chem. Interfacial Electrochem.* **84**, 177 (1977).
b. J. Mortensen and J. Heinze, *Angew. Chem. Int. Ed. Engl.* **23**, 84 (1984).

inversion is very low (~2 kcal), so that the lifetime of the individual isomers is very short (~10^{-9} s).

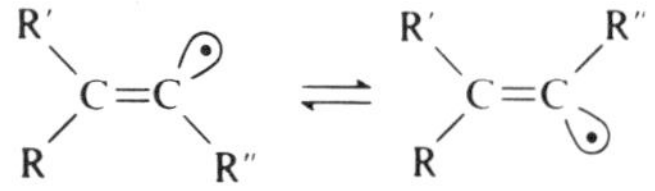

12.1.6. Charged Radical Species

Unpaired electrons can be present in charged species as well as in the neutral systems that have been considered up to this point. There have been many studies of such radical cations and radical anions, and we will consider some representative examples in this section.

Various aromatic and conjugated polyunsaturated hydrocarbons undergo one-electron reduction by alkali metals.[43] Benzene and naphthalene are examples. The EPR spectrum of the benzene radical anion was shown in Fig. 12.2a (p. 657). These reductions must be carried out in aprotic solvents, and ethers are usually used. The ease of formation of the radical anion increases as the number of fused rings increases. The electrochemical reduction potentials of some representitive compounds are given in Table 12.1. The potentials correlate with the energy of the LUMO as calculated by simple Hückel MO theory.[44] A correlation that includes a more extensive series of compounds can be observed with the use of somewhat more sophisticated molecular orbital methods.[45]

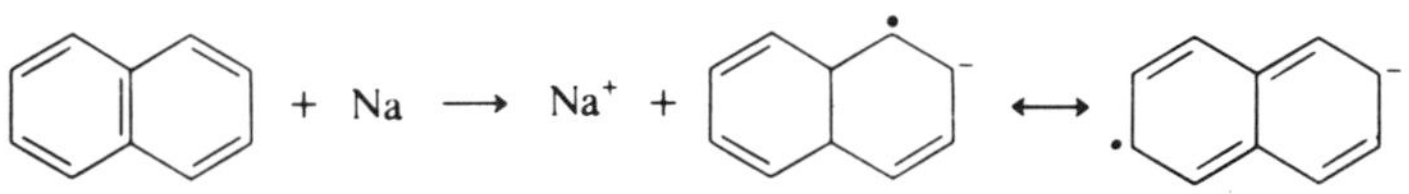

(many resonance structures)

43. D. E. Paul, D. Lipkin, and S. I. Weissman, *J. Am. Chem. Soc.* **78**, 116 (1956); T. R. Tuttle, Jr., and S. I. Weissman, *J. Am. Chem. Soc.* **80**, 5342 (1958).
44. E. S. Pysh amd N. C. Yang, *J. Am. Chem. Soc.* **85**, 2124 (1963); D. Bauer and J. P. Beck, *Bull. Soc. Chim. Fr.*, 1252 (1973); C. Madec and J. Courtot-Coupez, *J. Electroanal. Chem. Interfacial Electrochem.* **84**, 177 (1977).
45. C. F. Wilcox, Jr., K. A. Weber, H. D. Abruna, and C. R. Cabrera, *J. Electroanal. Chem. Interfacial Electrochem.* **198**, 99 (1986).

In the presence of a proton source, the radical anion is protonated and further reduction occurs (the Birch reduction, Part B, Section 5.5). In general, when no proton source is present, it is relatively difficult to add a second electron. Solutions of the radical anions of aromatic hydrocarbons can be maintained for relatively long periods in the absence of oxygen or protons.

Cyclooctatetraene provides a significant contrast to the usual preference of aromatic hydrocarbons for one-electron reduction. It is converted to a diamagnetic dianion by addition of two electrons.[46] It is easy to understand the ease with which the cyclooctatetraene radical accepts a second electron because of the aromaticity of the 10-π-electron system that results.

$$\text{C}_8\text{H}_8 + 2\text{K} \longrightarrow \text{C}_8\text{H}_8^{=} + 2\text{K}^+$$

Radical cations can be derived from aromatic hydrocarbons or alkenes by one-electron oxidation. Antimony trichloride and pentachloride are among the chemical oxidants that have been used.[47] Photodissociation and γ radiation also generate radical cations from aromatic hydrocarbons.[48] Most radical cations derived from hydrocarbons have limited stability, but EPR spectral parameters have permitted structural characterization.[49] The radical cations can be generated electrochemically, and the oxidation potentials are included in Table 12.1. The potentials correlate with the HOMO levels of the hydrocarbons. The higher the HOMO, the more easily oxidized is the hydrocarbon.

Two classes of charged radicals derived from ketones have been well studied. *Ketyls* are radical anions formed by one-electron reduction of carbonyl compounds. The formation of the benzophenone radical anion by reduction with sodium metal is an example. This radical anion is deep blue in color and is very reactive toward both oxygen and protons. Many detailed studies on the structure and spectral properties of this and related radical anions have been carried out.[50] A common chemical reaction of the ketyl radicals is coupling to form a diamagnetic dianion. This occurs reversibly for simple aromatic ketyls. The dimerization is promoted by protonation of one or both of the ketyls since the electrostatic repulsion is then removed. The coupling process leads to reductive dimerization of carbonyl com-

46. T. J. Katz, *J. Am. Chem. Soc.* **82**, 3784 (1960).
47. I. C. Lewis and L. S. Singer, *J. Chem. Phys.* **43**, 2712 (1965); R. M. Dessau, *J. Am. Chem. Soc.* **92**, 6356 (1970).
48. R. Gschwind and E. Haselbach, *Helv. Chim. Acta* **62**, 941 (1979); T. Shida, E. Haselbach, and T. Bally, *Acc. Chem. Res.* **17**, 180 (1984); M. C. R. Symons, *Chem. Soc. Rev.* **13**, 393 (1984).
49. J. L. Courtneidge and A. G. Davies, *Acc. Chem. Res.* **20**, 90 (1987).
50. For a summary, see N. Hirota, in *Radical Ions*, E. T. Kaiser and L. Kevan (eds.), Interscience, New York, 1968, pp. 35–85.

pounds, a reaction that will be discussed in detail in Part B, Section 5.5.

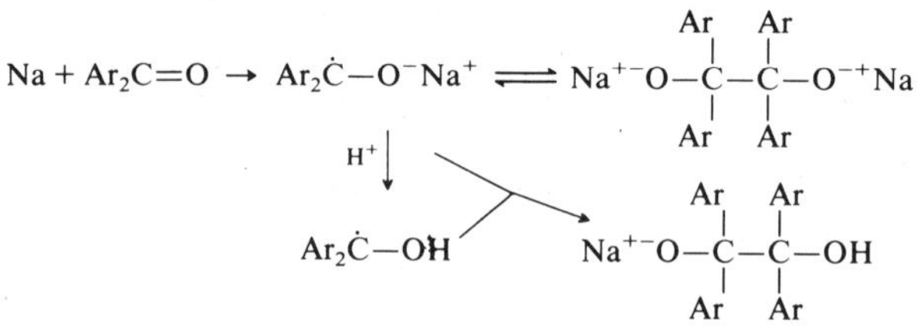

$$Na + Ar_2C{=}O \rightarrow Ar_2\dot{C}{-}O^-Na^+ \rightleftharpoons Na^{+-}O{-}C(Ar)_2{-}C(Ar)_2{-}O^{-+}Na$$

$$Ar_2\dot{C}{-}O^-Na^+ \xrightarrow{H^+} Ar_2\dot{C}{-}OH \longrightarrow Na^{+-}O{-}C(Ar)_2{-}C(Ar)_2{-}OH$$

One-electron reduction of α-dicarbonyl compounds gives radical anions known as *semidiones.*[51] Closely related are the one-electron reduction products of aromatic quinones, the *semiquinones.* Both the semidiones and the semiquinones can be protonated to give neutral radicals which are relatively stable.

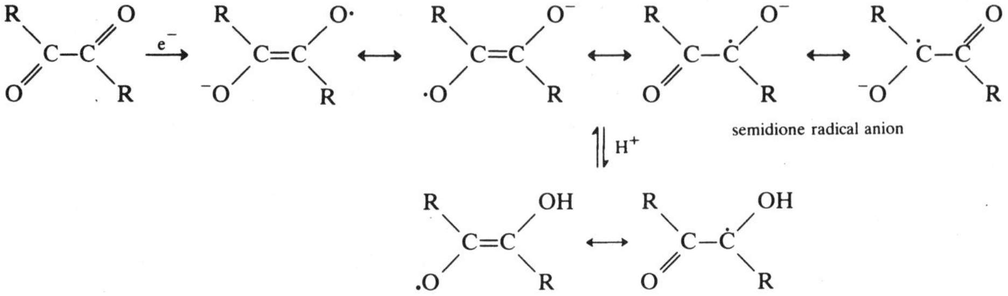

$$RC(=O){-}C(=O)R \xrightarrow{e^-} \cdot O(R)C{=}C(R)O^- \longleftrightarrow \cdot O(R)C{=}C(R)O^- \longleftrightarrow O{=}C(R){-}\dot{C}(R)O^- \longleftrightarrow {}^-O(R)\dot{C}{-}C(R){=}O$$

semidione radical anion

$$\rightleftharpoons H^+$$

$$\cdot O(R)C{=}C(R)OH \longleftrightarrow O{=}C(R){-}\dot{C}(R)OH$$

neutral semidione radical

$$O{=}C_6H_4\cdot{-}O^- \longleftrightarrow {}^-O{-}\cdot C_6H_4{=}O$$

semiquinone radical anion

Reductants such as zinc or sodium dithionite generate the semidione from diketones. Electrolytic reduction can also be used. Enolates can reduce diones to semidiones by electron transfer:

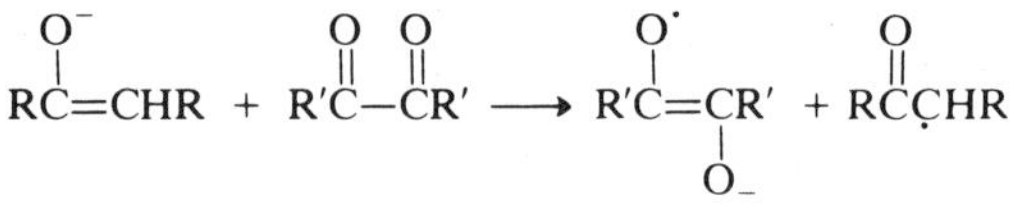

$$RC(O^-){=}CHR + R'C(=O){-}C(=O)R' \longrightarrow R'C(O^\cdot){=}CR'(O^-) + RC(=O)\dot{C}HR$$

The radicals formed from the enolate in this process are rapidly destroyed so that only the stable semidione species remains detectable for EPR study.

Semidiones can also be generated oxidatively from ketones by reaction with oxygen in the presence of base[52].

51. G. A. Russell, in *Radical Ions*, E. T. Kaiser and L. Kevan (eds.), Interscience, New York, 1968, pp. 87–150.
52. G. A. Russell and E. T. Strom, *J. Am. Chem. Soc.* **86**, 744 (1964).

$$\mathrm{RC(=O)CH_2R} \xrightarrow{\text{base}} \mathrm{RC(O^-)=CHR} \xrightarrow{O_2} \mathrm{RC(=O)CH(OOH)R} \xrightarrow{\text{base}} \mathrm{RC(=O)-C(=O)R}$$

$$\mathrm{RC(O^-)=CHR} + \mathrm{RC(=O)-C(=O)R} \rightarrow \mathrm{RC(O^{\cdot})=C(O^-)R} + \mathrm{RC(=O)\dot{C}HR}$$

The diketone is presumably generated oxidatively and then reduced to the semidione via reduction by the enolate derived from the original ketone.

The EPR spectra of semidione radical anions can provide information on the spin density at the individual atoms. The semidione derived from 2,3-butane dione, for example, has a spin density of 0.22 at each oxygen and 0.23 at each carbonyl carbon. The small amount of remaining spin density is associated with the methyl groups. This extensive delocalization is consistent with the resonance picture of the semidione radical anion.

12.2. Characteristics of Reaction Mechanisms Involving Radical Intermediates

12.2.1. Kinetic Characteristics of Chain Reactions

Certain aspects of free-radical reactions are unique in comparison with other reaction types that have been considered to this point. The underlying difference is that many free-radical reactions are chain reactions; that is, the reaction mechanism consists of a cycle of repetitive steps which form many product molecules for each initiation event. The hypothetical mechanism below illustrates a typical chain reaction.

	A–A	$\xrightarrow{k_i}$	2A·	initiation
repeated many times	A· + B–C	$\xrightarrow{k_{p1}}$	A–B + C·	propagation
	C· + A–A	$\xrightarrow{k_{p2}}$	A–C + A·	
	A· + B–C	$\xrightarrow{k_{p1}}$	A–B + C·	
	C· + A–A	$\xrightarrow{k_{p2}}$	A–C + A·	
	2A·	$\xrightarrow{k_{t1}}$	A–A	termination
	2C·	$\xrightarrow{k_{t2}}$	C–C	
	A· + C·	$\xrightarrow{k_{t3}}$	A–C	
overall reaction	A–A + B–C	$\longrightarrow$	A–B + A–C	

The step in which the reactive intermediate, in this case A·, is generated is called the *initiation step.* In the next four equations in the example, a sequence of two reactions is repeated; this is the *propagation phase.* Chain reactions are characterized by a *chain length,* which is the number of propagation steps that take place per initiation step. Finally, there are *termination steps.* These include any reactions that destroy one of the reactive intermediates necessary for the propagation of the chain. Clearly, the greater the frequency of termination steps, the lower the chain length will be.

The overall rate of a chain process is determined by the rates of initiation, propagation, and termination reactions. Analysis of the kinetics of chain reactions normally depends on application of the steady-state approximation to the radical intermediates. Such intermediates are highly reactive, and their concentrations are low and nearly constant throughout the course of the reaction:

$$\frac{d[\mathrm{A}\cdot]}{dt} = \frac{d[\mathrm{C}\cdot]}{dt} = 0$$

The result of the steady-state condition is that the overall rate of initiation must equal the total rate of termination. The application of the steady-state approximation and the resulting equality of the initiation and termination rates permits formulation of a rate law for the reaction mechanism above. The overall stoichiometry of a free-radical chain reaction is independent of the initiating and termination steps because the reactants are consumed and products formed almost entirely in the propagation steps.

$$\mathrm{A_2} + \mathrm{B{-}C} \rightarrow \mathrm{A{-}B} + \mathrm{A{-}C}$$

The overall reaction rate is given by

$$\text{rate} = \frac{d[\mathrm{A{-}B}]}{dt} = \frac{d[\mathrm{A{-}C}]}{dt} = \frac{-d[\mathrm{A_2}]}{dt} = \frac{-d[\mathrm{B{-}C}]}{dt}$$

Setting the rate of initiation equal to the rate of termination and assuming that k_{t2} is the dominant termination process gives

$$k_i[\mathrm{A_2}] = 2k_{t2}[\mathrm{C}\cdot]^2$$

$$[\mathrm{C}\cdot] = \left(\frac{k_i}{2k_{t2}}\right)^{1/2} [\mathrm{A_2}]^{1/2}$$

In general, the rate constants for termination reactions involving coupling of two radicals are very large. Since the concentration of the reactive intermediates is very low, however, the overall rate of termination is low enough that the propagation steps can compete since these steps involve the reactants, which are present at much higher concentration. The rate of the overall reaction is that of either propagation step:

$$\text{rate} = k_{p2}[\mathrm{C}\cdot][\mathrm{A_2}]$$

Both propagation steps must proceed at the same rate or the concentration of A· or C· would build up. By substituting for the concentration of the intermediate C·, we obtain:

$$\text{rate} = k_{p2}\left(\frac{k_i}{2k_t}\right)_2^{1/2} [\mathrm{A_2}]^{3/2} = k_{obs}[\mathrm{A_2}]^{3/2}$$

The observed rate law is then three-halves order in the reagent A_2.

In most real systems, the situation would be complicated by the possibility that more than one termination reaction makes a contribution to the total termination rate. A more complete discussion of the effect of termination steps on the form of the rate law has been given by Huyser.[53]

The overall rates of chain reactions can be greatly modified by changing the rate at which initiation or termination steps occur. The idea of initiation has been touched on in Section 12.1.4, where sources of free radicals were discussed. Many chain reactions of interest in organic chemistry depend on the presence of an *initiator*, which is a source of free radicals to start chain sequences. Peroxides are frequently used as initiators, since they give radicals by thermal decomposition at relatively low temperatures. Azo compounds are another very useful class of initiators, with azoisobutyronitrile, AIBN, being the most commonly used compound. Initiation by irradiation of a photosensitive compound that generates radical products is also a common procedure. Conversely, chain reactions can be greatly retarded by *inhibitors*. A compound can act as an inhibitor if it is sufficiently reactive toward a radical involved in the chain process that it effectively traps the radical, thus terminating the chain. The sensitivity of the rates of free-radical chain reactions to both initiators and inhibitors can be used in mechanistic studies to distinguish radical chain reactions from polar or concerted processes. Certain stable free radicals, for example, galvinoxyl (see entry 6, Scheme 12.1), are used in this way. Since they contain an unpaired electron, they are usually very reactive toward radical intermediates.

Free-radical chain inhibitors are of considerable economic importance. The term *antioxidant* is commonly applied to inhibitors that retard the free-radical chain oxidations that can cause relatively rapid deterioration of many commercial materials derived from organic molecules, including foodstuffs, petroleum products, and many plastics. The chain mechanism for autoxidation of hydrocarbons is

$$\begin{array}{ll} \text{initiation} & \text{In}\cdot + \text{R—H} \rightarrow \text{R}\cdot + \text{In—H} \\ \text{propagation} & \text{R}\cdot + \text{O}_2 \rightarrow \text{ROO}\cdot \\ & \text{ROO}\cdot + \text{R—H} \rightarrow \text{ROOH} + \text{R}\cdot \end{array}$$

The function of an antioxidant is to divert the peroxy radicals and thus prevent a chain process. Other antioxidants function by reacting with potential initiators and thus retard oxidative degradation by preventing the initiation of autoxidation chains. The hydroperoxides generated by autoxidation are themselves potential chain initiators, and autoxidations therefore have the potential of being autocatalytic. Certain antioxidants function by reducing such hydroperoxides and thereby preventing their accumulation. Other antioxidants function by diverting intermediates in the chain process. They function in the same way as free-radical inhibitors. Such antioxidants are frequently called *free-radical scavengers.*

Molecular oxygen is an important participant in chain oxidative degradation. The oxygen molecule, with its two unpaired electrons, is extremely reactive toward most free-radical intermediates. The product that is formed is a reactive peroxyl

53. E. S. Huyser, *Free Radical Chain Reactions*, Wiley-Interscience, New York, 1970, pp. 39–54.

radical, which can propagate a chain reaction.

$$R\cdot + O_2 \rightarrow R{-}O{-}O\cdot$$

The presence of oxygen frequently can modify the course of a free-radical chain reaction. This occurs if a radical intermediate is diverted by reaction with molecular oxygen.

12.2.2. Structure–Reactivity Relationships

Structure-reactivity relationships can be probed by measurements of rates and equilibria, as was discussed in Chapter 4. Direct kinetic measurements have been used relatively less often in the study of radical reactions than for heterolytic reactions. Instead, *competition methods* have frequently been used. The basis of competition methods lies in the rate expression for a reaction, and the results obtained can be just as valid a basis for comparison of relative reactivity as directly measured rates, *provided the two competing processes are of the same kinetic order.* Suppose it is desired to compare the reactivity of two related compounds, B—X and B—Y, in a hypothetical sequence:

$$A{-}A \longrightarrow 2A\cdot$$

$$A\cdot + B{-}X \xrightarrow{k_X} A{-}B + X\cdot$$

$$X\cdot + A{-}A \rightarrow A{-}X + A\cdot$$

and

$$A\cdot + B{-}Y \xrightarrow{k_Y} A{-}B + Y\cdot$$

$$Y\cdot + A{-}A \rightarrow A{-}Y + A\cdot$$

The data required are the relative magnitudes of k_X and k_Y. When both B—X and B—Y are present in the reaction system, they will be consumed at rates that are a function of their reactivity and concentration:

$$\frac{-d[B{-}X]}{dt} = k_X[A\cdot][B{-}X]$$

$$\frac{-d[B{-}Y]}{dt} = k_Y[A\cdot][B{-}Y]$$

$$\frac{k_X}{k_Y} = \frac{d[B{-}X]/[B{-}X]}{d[B{-}Y]/[B{-}Y]}$$

Integration of this expression with the limits $[B{-}X] = [B{-}X]_{in}$ to $[B{-}X]_t$, where t is a point in time during the course of the reaction, gives

$$\frac{k_X}{k_Y} = \frac{\ln([B{-}X]_{in}/[B{-}X]_t)}{\ln([B{-}Y]_{in}/[B{-}Y]_t)}$$

This relationship permits the measurement of the ratio k_X/k_Y. The initial concentrations $[B{-}X]_{in}$ and $[B{-}Y]_{in}$ are known from the conditions of the experiment. The reaction can be stopped at some point when some of both B—X and B—Y remain unreacted, or an excess of B—X and B—Y can be used so that neither is completely consumed when A—A has completely reacted. Determination of $[B{-}X]_t$ and $[B{-}Y]_t$ then provides the information needed to calculate k_X/k_Y. Is it clear why the reactions being compared must be of the same order? If they were not, division of the two rate expressions would leave uncanceled concentration terms.

Another experiment that can be considered to be of the competition type involves the comparison of the reactivity of different atoms in the same molecule. For example, gas phase chlorination of butane can lead to 1- or 2-chlorobutane. The relative reactivity (k_p/k_s) of the primary and secondary hydrogens is the sort of information that helps to characterize the details of the reaction process.

$$Cl\cdot + CH_3CH_2CH_2CH_3 \xrightarrow{k_p} HCl + \cdot CH_2CH_2CH_2CH_3$$

$$Cl_2 + \cdot CH_2CH_2CH_2CH_3 \longrightarrow Cl\cdot + ClCH_2CH_2CH_2CH_3$$

$$Cl\cdot + CH_3CH_2CH_2CH_3 \xrightarrow{k_s} HCl + CH_3\dot{C}HCH_2CH_3$$

$$CH_3\dot{C}HCH_2CH_3 + Cl_2 \longrightarrow Cl\cdot + CH_3\underset{\displaystyle Cl}{\underset{|}{C}}HCH_2CH_3$$

The value of k_p/k_s can be determined by measuring the ratio of the products, 1-chlorobutane:2-chlorobutane, during the course of the reaction. A statistical correction must be made to take account of the fact that the primary hydrogens outnumber the secondary ones by 3:2. This calculation provides the relative reactivity of chlorine atoms toward the primary and secondary hydrogens in butane.

$$\frac{k_p}{k_s} = \frac{2[\text{1-chlorobutane}]_t}{3[\text{2-chlorobutane}]_t}$$

Recent developments of techniques for measuring the rates of very fast reactions have permitted absolute rates to be measured for some fundamental types of free-radical reactions. Some examples where absolute rates and E_a values are available are given in Table 12.2.

The strength of the bond to the reacting hydrogen is a major determinant of the rate at which hydrogen atom abstraction occurs. It is for this reason that trisubstituted stannanes are among the most useful reagents as hydrogen atom donors. As indicated by entries 6–8 in Table 12.2, hydrogen abstractions from stannanes proceed with rates greater than $10^6\ M^{-1}\ s^{-1}$ and have very low activation energies. This high reactivity correlates with the low bond strength of the Sn—H bond (74 kcal/mol, Table 12.4, p. 683). For comparison, entries 1–3 give the rates of hydrogen abstraction from two of the more reactive C-H hydrogen atom donors, tetrahydrofuran and isopropylbenzene. For the directly comparable reactions with the phenyl radical (entries 1 and 8), tri-*n*-butylstannane is about 100 times more reactive than tetrahydrofuran as a hydrogen atom donor.

Table 12.2. Absolute Rates of Some Free-Radical Reactions[a]

	Reaction		Ref.
A.	Hydrogen atom abstraction		
1	Ph· + tetrahydrofuran → Ph—H + tetrahydrofuranyl·	$k = 4.8 \times 10^6 M^{-1} s^{-1}$	b
2	$(CH_3)_3CO\cdot$ + $(CH_3)_2CH$—Ph → $(CH_3)_3COH$ + $(CH_3)_2\dot{C}$—Ph	$k = 8.7 \times 10^5 M^{-1} s^{-1}$	c
3	$(CH_3)_3CO\cdot$ + tetrahydrofuran → $(CH_3)_3COH$ + tetrahydrofuranyl·	$k = 8.3 \times 10^6 M^{-1} s^{-1}$	c
4	Cl· (free) + cyclopentane → H—Cl + cyclopentyl·	$k = 4.7 \times 10^9 M^{-1} s^{-1}$	d
5	Cl· (benzene complex) + cyclopentane → H—Cl + cyclopentyl·	$k = 4.3 \times 10^7 M^{-1} s^{-1}$	d
6	$CH_3\cdot + Bu_3Sn—H \rightarrow CH_4\cdot + Bu_3Sn\cdot$	$k = 1.0 \times 10^7 M^{-1} s^{-1}$, $E_a = 3.2$ kcal	e
7	$(CH_3)_3C\cdot + Bu_3Sn—H \rightarrow (CH_3)_3CH + Bu_3Sn\cdot$	$k = 1.8 \times 10^6 M^{-1} s^{-1}$, $E_a = 2.95$ kcal	e
8	$Ph\cdot + Bu_3Sn—H \rightarrow Ph—H + Bu_3Sn\cdot$	$k = 5.9 \times 10^8 M^{-1} s^{-1}$	f
9	$PhCH_2\cdot + PhSH \rightarrow PhCH_3 + PhS\cdot$	$k = 3.1 \times 10^5 M^{-1} s^{-1}$	g
B.	Additions to alkenes and aromatics		
10	$Ph\cdot + PhCH{=}CH_2 \rightarrow Ph\dot{C}H—CH_2Ph$	$k = 1.1 \times 10^8 M^{-1} s^{-1}$	b
11	Ph· + benzene → (Ph)(H)cyclohexadienyl·	$k = 4.5 \times 10^5 M^{-1} s^{-1}$	b
12	$(CH_3)_3CO\cdot + CH_2{=}CH(CH_2)_5CH_3$ → $(CH_3)_3COCH_2\dot{C}H(CH_2)_5CH_3$	$k = 1.5 \times 10^6 M^{-1} s^{-1}$	c
C.	Cyclization and ring opening		
13	cyclopropylmethyl· → 3-butenyl·	$k = 2.1 \times 10^8 s^{-1}$, $E_a = 7.6$ kcal	h

Table 12.2. continued.

	Reaction		Ref.
14	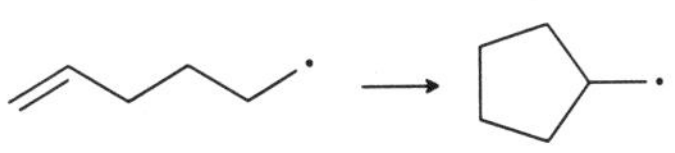 →	$k = 1.0 \times 10^5\ s^{-1}$ $E_a = 6.1$ kcal	i
15	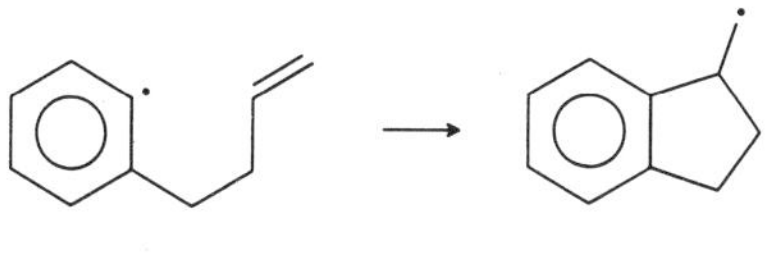 →	$k = 4.0 \times 10^8\ s^{-1}$ $E_a = 3.6$ kcal	f
16	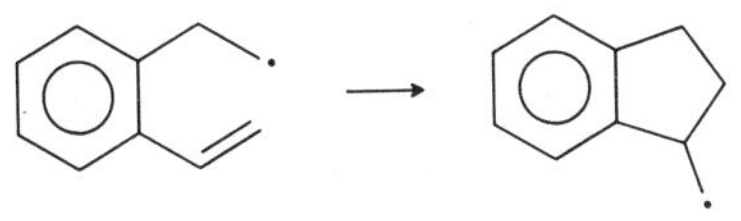 →	$k = 1.5 \times 10^5\ s^{-1}$ $E_a = 7.3$ kcal	j
17	 →	$k = 2 \times 10^{-1}\ s^{-1}$ $E_a = 16.3$ kcal	k
18	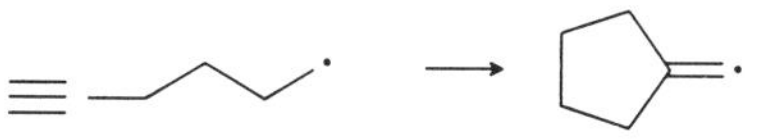 →	$k = 2.8 \times 10^4\ s^{-1}$ $E_a = 8.3$	l
D.	Other reactions		
19	$(CH_3)_3C\cdot + O_2 \rightarrow (CH_3)_3C{-}O{-}O\cdot$	$k = 4.9 \times 10^9 M^{-1}\ s^{-1}$	f
20	$PhCH_2\cdot + O_2 \rightarrow PhCH_2{-}O{-}O\cdot$	$k = 2.4 \times 10^9 M^{-1}\ s^{-1}$	f
21	$(CH_3)_3C{-}C(=O)\cdot \rightarrow (CH_3)_3C\cdot + C{\equiv}O$	$k = 1.2 \times 10^5\ s^{-1}$ $E_a = 9.3$ kcal	i
22	$PhCH_2C(=O)\cdot \rightarrow PhCH_2\cdot + C{\equiv}O$	$k = 5.2 \times 10^7\ s^{-1}$ $E_a = 7.2$ kcal	i

a. Rates quoted are for temperatures near 300 K. The original literature should be consulted for precise temperatures and other conditions.
b. J. C. Scaiano and L. C. Stewart, *J. Am. Chem. Soc.* **105**, 3609 (1983).
c. A. Baignee, J. A. Howard, J. C. Scaiano, and L. C. Stewart, *J. Am. Chem. Soc.* **105** 6120 (1983).
d. N. J. Bunce, K. U. Ingold, J. P. Landers, J. Lustzyk, and J. Scaiano, *J. Am. Chem. Soc.* **107**, 564 (1985).
e. C. Chatgilialoglu, K. U. Ingold, and J. C. Scaiano, *J. Am. Chem. Soc.* **103**, 7739 (1981).
f. L. J. Johnson, J. Lusztyk, D. D. M. Wayner, A. N. Abeywickreyma, A. L. Beckwith, J. C. Scaiano, and K. U. Ingold, *J. Am. Chem. Soc.* **107**, 4594 (1985).
g. J. A. Franz, N. K. Suleman, and M. S. Alnajjar, *J. Org. Chem.* **51**, 19 (1986).
h. L. Mathew and J. Warkentin, *J. Am. Chem. Soc.* **108**, 7981 (1986).
i. D. Griller and K. U. Ingold, *Acc. Chem. Res.* **13**, 317 (1980).
j. J. A. Franz, R. D. Barrows, and D. M. Camaioni, *J. Am. Chem. Soc.* **106**, 3964 (1984).
k. J. A. Franz, M. S. Alnajjar, R. D. Barrows, D. L. Kaisaki, D. M. Camaioni, and N. K. Suleman, *J. Org. Chem.* **51**, 1446 (1986).
l. A. L. J. Beckwith and C. H. Schiesser, *Tetrahedron Lett.* **26**, 373 (1985).
m. B. Maillard, K. U. Ingold, and J. C. Scaiano, *J. Am. Chem. Soc.* **105**, 5095 (1983).

Entries 4 and 5 point to another important aspect of free-radical reactivity. The data given illustrate that the observed reactivity of the chlorine atom is strongly influenced by the presence of benzene. Evidently, a complex is formed that attenuates the reactivity of the chlorine atom. This is probably a general feature of radical chemistry, but there are relatively few data available on solvent effects on either absolute or relative reactivity of radical intermediates.

Section C of Table 12.2 shows some reactions involving cyclization of unsaturated radicals. This type of reaction has become an important application of free-radical chemistry in synthesis and will be discussed more thoroughly in Section 10.3.4 of Part B. Rates of cyclization reactions have also proved useful in mechanistic studies, where they can serve as reference points for comparison with other reaction rates.

The remaining entries in part C of Table 12.2 are examples of ring closures of unsaturated radicals. They all display a feature that is common to such cyclizations, namely, there is a preference for five-membered ring formation over six-membered ring formation.[54] This is observed even though it results in formation of a less stable primary radical. The cause for this preference has been traced to stereoelectronic effects. In order for a bonding interaction to occur, the radical center must interact with the π^* orbital of the alkene. Bond formation takes place as the result of initial interaction with the LUMO, which is the π^* orbital. According to MO calculations, the preferred direction of attack is from an angle of about 70° with respect to the plane of the double bond.[55]

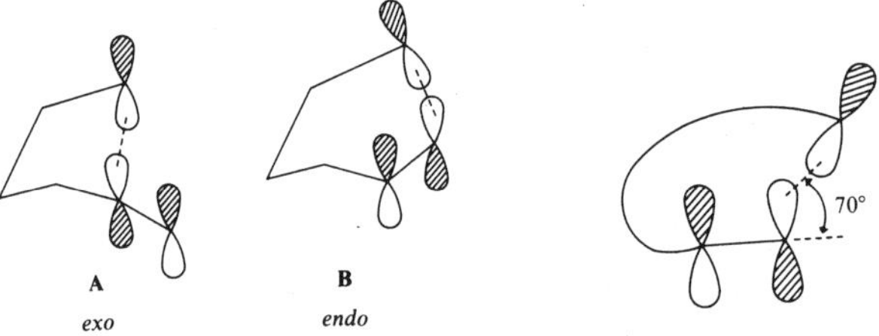

When this stereoelectronic requirement is combined with a calculation of the steric and angle strain imposed on the transition state, as determined by molecular mechanics calculations, preferences for the *exo* versus *endo* modes of cyclization are predicted to be as summarized in Table 12.3.

The observed results show at least the correct qualitative trend. The observed preference for ring formation is $5 > 6$, $6 > 7$, and $8 > 7$, which is in agreement with the calculated preference. The relationship only holds for terminal double bonds. An additional alkyl substituent at either end of the double bond reduces the relative reactivity by a steric effect.

The relatively low rate and high activation energy noted for entry 17 in Table 12.2 also reflects a stereoelectronic effect. The preference for delocalization at the radical center requires coplanarity of the substituents at the radical site. In view of

54. A. L. J. Beckwith, C. J. Eaton, and A. K. Serelis, *J. Chem. Soc., Chem. Commun.*, 482 (1980); A. L. J. Beckwith, T. Lawrence, and A. K. Serelis, *J. Chem. Soc., Chem. Commun.*, 484 (1980); A. L. J. Beckwith, *Tetrahedron* **37**, 3073 (1981).
55. M. J. S. Dewar and S. Olivella, *J. Am. Chem. Soc.* **100**, 5290 (1978); D. C. Spellmeyer and K. N. Houk, *J. Org. Chem.* **52**, 959 (1987); A. L. J. Beckwith and C. H. Schiesser, *Tetrahedron* **41**, 3925 (1985).

Table 12.3. Regioselectivity of Radical Cyclization as a Function of Ring Size[a]

		exo : endo Ratio	
Ring size	n	Calc.	Found
5:6	2	10:1	50:1
6:7	3	>100:1	10:1
7:8	4	1:5.8	<1:100

a. D. C. Spellmeyer and K. N. Houk, *J. Org. Chem.* **52**, 959 (1987).

the restrictions on the mode of approach of the radical to the double bond, significant strain develops at the transition state, and this requires rotation of the benzylic methylene group out of its preferred coplanar alignment.[56]

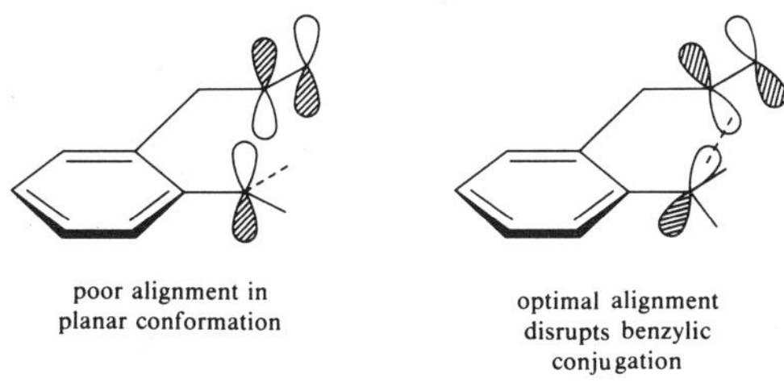

Some general remarks about structure-reactivity relationships in radical reactions can be made at this point. The reactivity of C—H groups toward radicals that abstract hydrogen is usually primary < secondary < tertiary. Vinyl and phenyl substituents at a reaction site increase the reactivity toward radicals. This reactivity order reflects the bond dissociation energies of the various C—H bonds, which are in the order allyl ~ benzyl < tertiary < secondary < primary.[57] The relative reactivity of primary, secondary, and tertiary positions in aliphatic hydrocarbons toward hydrogen abstraction by methyl radicals is 1:4.3:46.[58] The relative reactivity toward the *t*-butoxy radical is 1:10:44.[59] An allylic or benzylic hydrogen is more reactive toward a methyl radical by a factor of about 9, compared to a corresponding unactivated hydrogen.[58] Data for other types of radicals have been obtained and tabulated.[59] The trend of reactivity tertiary > secondary > primary is consistently observed, and the range of reactivity is determined by the nature of the reacting radical. In the gas phase, the bromine atom, for example, is very selective, with relative reactivities of 1:250:6,300 for primary, secondary, and tertiary hydrogens.[60]

The stabilizing effects of vinyl groups (in allylic radicals) and phenyl groups (in benzyl radicals) are very significant and can be satisfactorily rationalized in

56. J. A. Franz, N. K. Suleman, and M. S. Alnajjar, *J. Org. Chem.* **51**, 19 (1986).
57. J. A. Kerr, *Chem. Rev.* **66**, 465 (1966).
58. W. A. Pryor, D. L. Fuller, and J. P. Stanley, *J. Am. Chem. Soc.* **94**, 1632 (1972).
59. C. Walling and B. B. Jacknow, *J. Am. Chem. Soc.* **82**, 6108 (1960).
60. A. F. Trotman-Dickenson, *Adv. Free Radical Chem.* **1**, 1 (1965).

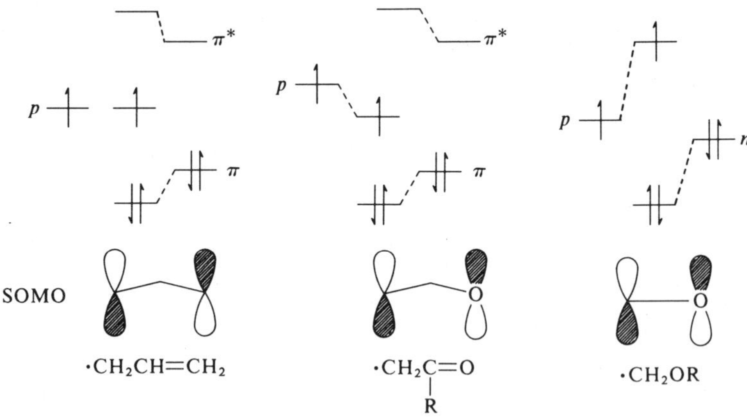

Fig. 12.4. PMO representation of p-orbital interactions with (a) C=C, (b) C=O, and (c) $\ddot{\text{O}}$R substituents. Form of SOMO (singly occupied molecular orbital) is shown.

resonance terminology:

$$CH_2{=}CH{-}CH_2\cdot \leftrightarrow \cdot CH_2{-}CH{=}CH_2$$

$$C_6H_5{-}CH_2\cdot \longleftrightarrow {=}CH_2 \longleftrightarrow {=}CH_2$$

The stabilizing role of other functional groups can also be described in resonance terms. Both electron-attracting groups such as carbonyl and cyano and electron-donating groups such as methoxy and dimethylamino have a stabilizing effect on a radical intermediate at an adjacent carbon. The resonance structures that depict these interactions indicate delocalization of the unpaired electron onto the adjacent substituents:

$$\dot{C}{-}C{=}\ddot{O}: \leftrightarrow C{=}C{-}\dot{\ddot{O}}: \qquad \dot{C}{-}C{\equiv}N \leftrightarrow C{=}C{=}\dot{N}:$$

$$\dot{C}{-}\ddot{O}{-}CH_3 \leftrightarrow {}^{-}C{-}\dot{O}^{+}{-}CH_3 \qquad \dot{C}{-}\ddot{N}(CH_3)_2 \leftrightarrow {}^{-}C{-}\dot{N}^{+}(CH_3)_2$$

A description of the radical-stabilizing effect of both types of substituents can also be presented in MO terms. In this case, the question we ask is how will the unpaired electron in a p orbital interact with the orbitals of the adjacent substituent, such as vinyl, carbonyl, or methoxy? Figure 12.4 presents a qualitative description of the situation.

For the vinyl substituent, we can analyze the stabilization in terms of simple Hückel MO theory. The interaction of a p orbital with an adjacent vinyl group creates the allyl radical. In Hückel calculations, the resulting orbitals have energies of $\alpha + 1.4\beta$, α, and $\alpha - 1.4\beta$. Thus, the interaction of the p orbital with both the π and π^* orbitals leaves it at the same energy level but the π and π^* levels are transformed to ψ_1 and ψ_3 of the allyl radical. There is a net stabilization of 0.4β. The stabilization of the allyl radical has been estimated to be about 14 kcal/mol.[61] Since the stabilizing interaction is maximum in a planar structure, there is a barrier to rotation of either of the terminal methylene groups with respect to the rest of the molecule. The measured barrier is 15.7 kcal/mol.

The basic tenet of PMO theory that the orbitals which are closest in energy will interact most strongly is employed in the analysis of the effect of electron-withdrawing and electron-donating substituents. In the case of an electron-accepting substituent, such as a carbonyl group, the strongest interaction is with the carbonyl LUMO, π^*. This results in a lowering of the energy of the orbital containing the unpaired electron, that is, the radical is stabilized. For an electron-donating substituent, the strongest interaction is between the electron in the p orbital and a nonbonding pair on the electron donor. This interaction results in lowering of the energy of the orbital occupied by the electron pair and raising of the energy of the orbital occupied by the single electron. The net effect is still stabilizing since there are two electrons in the stabilized orbital and only one in the destabilized one.

Radicals are particularly strongly stabilized when both an electron-attracting and an electron-donating substituent are present at the radical site. This has been called "*mero-stabilization*"[62] or "*capto-dative stabilization*."[63] This type of stabilization results from mutual reinforcement of the two substituent effects. Scheme 12.3 gives some information on the stability of this type of radical.

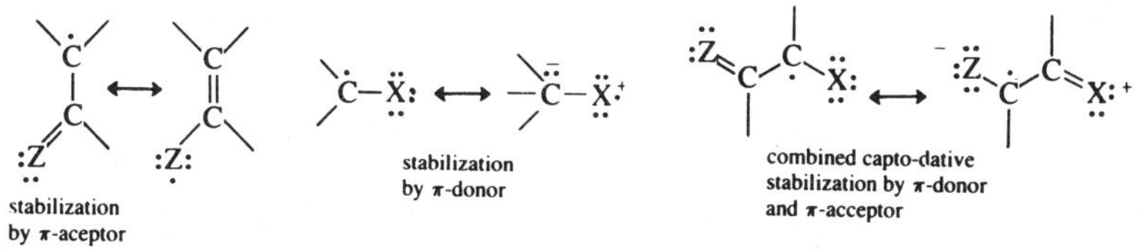

stabilization by π-aceptor

stabilization by π-donor

combined capto-dative stabilization by π-donor and π-acceptor

A comparison of the rotational barriers in allylic radicals **A**–**D** provides evidence for the stabilizing effect of the capto-dative combination:

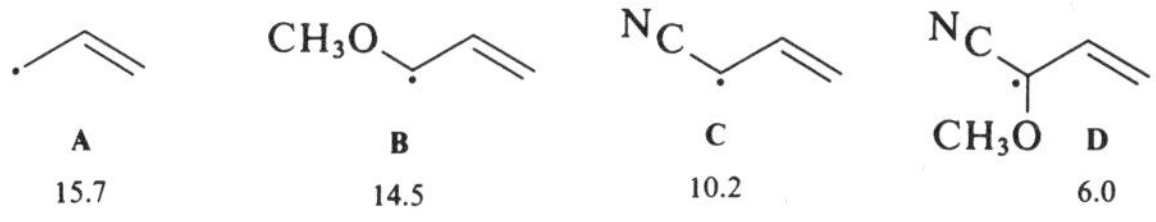

A	B	C	D
15.7	14.5	10.2	6.0

The decreasing barrier at the formal single bond along the series **A** to **D** implies decreasing π-allyl character of this bond. The decrease in the importance of the π bonding in turn reflects a diminished degree of interaction of the radical center with the adjacent double bond. The fact that the decrease from **C** to **D** is greater than that from **A** to **B** indicates a synergistic effect, as implied by the capto-dative

61. H. G. Korth, H. Trill, and R. Sustmann, *J. Am. Chem. Soc.* **103**, 4483 (1981).
62. R. W. Baldock, P. Hudson, A. R. Katritzky, and F. Soti, *J. Chem. Soc., Perkin Trans. 1,* 1422 (1974).
63. H. G. Viehe, R. Merenyi, L. Stella, and Z. Janousek, *Angew Chem. Int. Ed. Engl.* **18**, 917 (1979).

Scheme 12.3. Radicals with Capto-dative Stabilization

1[a]	$(CH_3)_2N^{+}$=C₆H₄•–$N(CH_3)_2$	Wurster's salts. Generated by one-electron oxidation of the diamine. Indefinitely stable.
2[b]	$C_2H_5O_2C$–(pyridinyl•) :N–CH_3	Generated by one-electron reduction of the pyridinium salt. Stable, distillable, and only moderately reactive to oxygen.
3[c]	2-acetyl-N-methylpyridinyl radical (C(=O)–CH_3; N–CH_3)	Stable and distillable. A small amount of dimer is present in equilibrium with the radical.
4[d]	morpholinone radical (O, C=O, •–CH_3, NH, H_3C, CH_3)	Generated by spontaneous dissociation of the dimer. In equilibrium with dimer.
5[e]	morpholinone radical (O, C=O, •–Ph, NH, H_3C, CH_3)	Generated by spontaneous dissociation of the dimer. Stable for several days at room temperature. Oxidized by oxygen.
6[f]	$(CH_3)_2N$–C•(C≡N)$_2$	Generated spontaneously from dimethyl-amino-malononitrile at room temperature. Observed to be persistent over many hours by EPR.

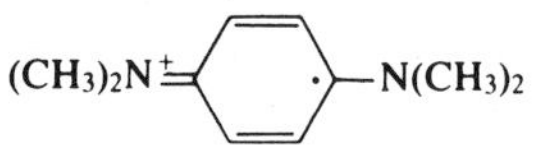

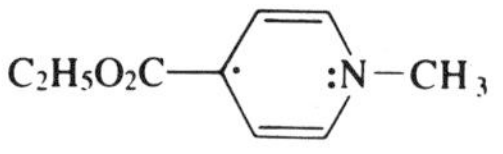

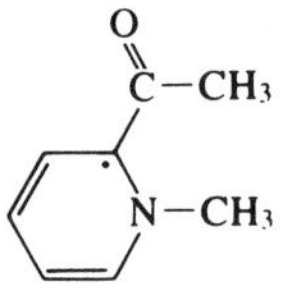

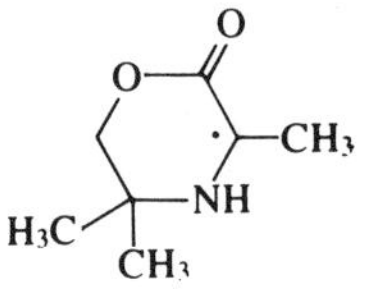

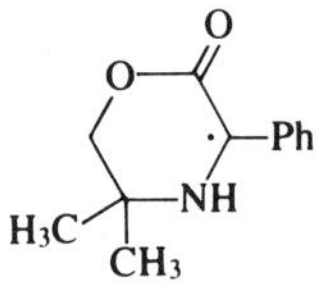

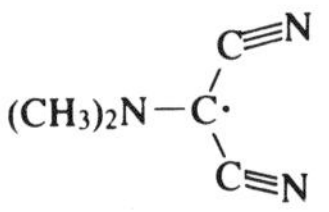

a. A. R. Forrester, J. M. Hay, and R. H. Thompson, *Organic Chemistry of Stable Free Radicals*, Academic Press, New York, 1968, pp. 254–261.
b. J. Hermolin, M. Levin, and E. M. Kosower, *J. Am. Chem. Soc.* **103**, 4808 (1981).
c. J. Hermolin, M. Levin, Y. Ikegami, M. Sawayangai, and E. M. Kosower, *J. Am. Chem. Soc.* **103**, 4795 (1981).
d. T. H. Koch, J. A. Oleson, and J. DeNiro, *J. Am. Chem. Soc.* **97**, 7285 (1975).
e. J. M. Burns, D. L. Wharry, and T. H. Koch, *J. Am. Chem. Soc.* **103**, 849 (1981).
f. L. de Vries, *J. Am. Chem. Soc.* **100**, 926 (1978).

formulation. The methoxy group is more stabilizing when it can interact with the cyano group than as an isolated substituent, as it is in **B**.[64]

The capto-dative effect has also been demonstrated by studying the bond dissociation process in a series of 1,5-dienes substituted at C-3 and C-4.

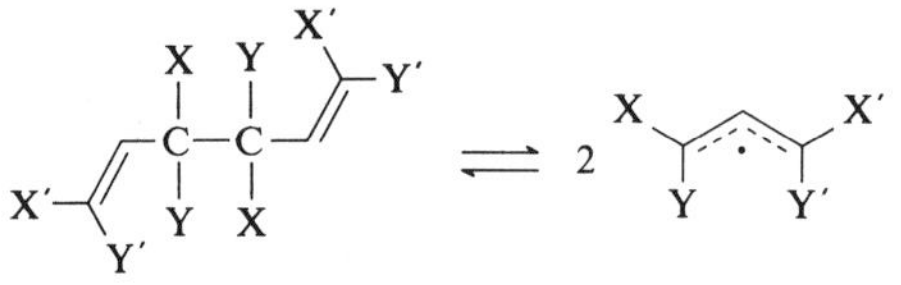

X'(Y')C=CH–C(X)(Y)–C(Y)(X)–CH=C(X')(Y') ⇌ 2 X(Y)C•–CH=C(X')(Y')

X	Y	X′	Y′	H
CO_2R	CO_2R	CO_2R	CO_2R	38.1
CO_2R	CO_2R	CO_2R	OR	28.2
CN	OR	CN	OR	24.5
CN	NR_2	CN	NR_2	8.1

64. H.-G. Korth, P. Lommes, and R. Sustmann, *J. Am. Chem. Soc.* **106**, 663 (1984).

Table 12.4. Bond Dissociation Energies (kcal/mol)[a]

Bond	D.E.	Bond	D.E.
CH_3-H	104	$H-Br$	87.5
CH_3CH_2-H	98	$H-I$	71
$(CH_3)_2CH-H$	94.5	$HOCH_2-H$	92
$(CH_3)_3C-H$	91	$CH_3CH_2OCH(-H)CH_3$	92[b]
$CH_2=CH-H$	104		
cyclopropyl C–H (CH_2–CH_2–$CH-H$ ring)	101	$CH_3C(=O)CH_2-H$	92
$PhCH_2-H$	85	$N\equiv CCH_2-H$	86
$CH_2=CHCH_2-H$	85	$PhS-H$	82[c]
1,4-cyclohexadiene C–H	73[b]	$(CH_3)_3Si-H$	90[d]
		$(CH_3)_3Ge-H$	82[d]
		$(C_4H_9)_3Sn-H$	74[b]
F_3C-H	106	$CH_3C(=O)O-OC(=O)CH_3$	30
Cl_3C-H	96		
C_2H_5-F	106		
C_2H_5-Cl	81	$(CH_3)_3CO-OH$	44
C_2H_5-Br	69	$F-F$	38
C_2H_5-I	53	$Cl-Cl$	58
$H-F$	136	$Br-Br$	46
$H-Cl$	103	$I-I$	36

a. Except where noted otherwise, data are from J. A. Kerr, *Chem. Rev.* **66**, 465 (1966); S. W. Benson, *J. Chem. Educ.* **42**, 502 (1965).
b. T. J. Burkey, M. Majewski, and D. Griller, *J. Am. Chem. Soc.* **108**, 2218 (1986).
c. S. W. Benson, *Chem. Rev.* **78**, 23 (1978).
d. R. A. Jackson, *J. Organomet. Chem.* **166**, 17 (1979).

When one or both of the combinations X, Y and X′, Y′ are of the capto-dative type, as is the case for an alkoxy and an ester group, the enthalpy of bond dissociation is 10–15 kcal lower than when all four groups are electron-attracting. When the captodative combination CN/NR_2 occupies both X, Y and X′, Y′ positions, the enthalpy for dissociation of the C-3—C-4 bond is less than 10 kcal/mol.[65]

The radical stabilization provided by various functional groups results in reduced bond dissociation energies for bonds to the stabilized radical center. Some bond dissociation energy values are given in Table 12.4. As an example of the substituent effect on bond dissociation energies, it can be seen that the primary C—H bonds in acetonitrile (86 kcal/mol) and acetone (92 kcal/mol) are significantly weaker than a primary C—H bond in ethane (98 kcal/mol).

By analysis of heats of formation of compounds incorporating radical fragments and assignment of standard sets of bond energies, it is possible to arrive at energies corresponding to the stabilization of the radical fragment. This energy then reflects

65. M. Van Hoecke, A. Borghese, J. Pennelle, R. Merenyi, and H. G. Viehe, *Tetrahedron Lett.* **27**, 4569 (1986).

Table 12.5. Thermochemical Stabilization Energies for Some Substituted Radicals

	SE^{0a}	SE^{b}
$\dot{C}H_3$	−1.67	−3.9
$CH_3\dot{C}H_2$	2.11	−1.5
$CN\dot{C}H_2$	6.50	
$NH_2\dot{C}H_2$	3.92	
$CH_3NH\dot{C}H_2$	12.18	
$(CH_3)_2N\dot{C}H_2$	14.72	
$HO\dot{C}H_2$	3.13	1.7
$CH_3O\dot{C}H_2$	3.64	
$F\dot{C}H_2$	−1.89	−0.7
$(CH_3)_2\dot{C}H$	2.57	
$(CN)_2\dot{C}H$	5.17	
$(HO)_2\dot{C}H$	−2.05	
$(CN)(NH_2)\dot{C}H$	12.94	
$(CN)(OH)\dot{C}H$	2.15	
$(CN)F\dot{C}H$	−2.79	
$F_2\dot{C}H$	−4.11	1.7
$(CH_3)_3\dot{C}$	4.35	
$CH_3\dot{C}(CN)_2$	3.92	
$CH_3\dot{C}(OH)_2$	0.15	
$CH_3\dot{C}(CN)(OH)$	2.26	
$\dot{C}F_3$	−4.17	
$\dot{C}Cl_3$	−13.79	
$CH_2{=}CH{-}\dot{C}H_2$	13.28	10.5
$C_6H_5{-}\dot{C}H_2$	12.08	13.9
$CH_2{=}\dot{C}H$	−6.16	−10.0
$HC{\equiv}C^{\cdot}$	−15.57	
$(C_6H_5)^{\cdot}$	−10.27	−12.3
$(C_5H_5)^{\cdot}$	19.24	
$CH_3\overset{\cdot}{C}{=}O$ ($CH_3C\cdot$ with C=O)	7.1	
$H\overset{\cdot}{C}{=}O$ ($HC\cdot$ with C=O)		5.8

a. Stabilization energy as defined by C. Leroy, D. Peeters, and C. Wilante, *Theochem* **5**, 217 (1982).
b. Reorganization energy as defined by R. T. Sanderson, *J. Org. Chem.* **47**, 3835 (1982) but given here as a stabilization energy of opposite sign.

the stabilization or destabilization of the radical center in the particular structural environment. Table 12.5 lists values assigned by two such approaches. Although the two sets of values differ by a few kilocalories, the trends are consistent with the qualitative predictions of radical stability based on PMO theory. These data reveal the familiar stabilization of allyl and benzyl radicals, which appears as positive stabilization energies. The trend of increasing stability of alkyl radicals in the order tertiary > secondary > primary > methyl is also apparent. According to this analysis, donor substituents, such as hydroxy, are not as strongly stabilizing as acceptor substituents, such as the carbonyl group. The assignment of significantly negative stabilization energies (i.e., destabilization) implies that the substituent will increase the bond energy of adjacent bonds. The phenyl and vinyl radicals are examples of radicals having negative stabilization energies, implying that there will be an increase in the bond energy to the substituents. These negative values are in agreement with the relative instability of phenyl and vinyl radicals and the difficulty of abstracting a hydrogen atom from aromatic rings or double bonds.

Bond dissociation energies such as those in Table 12.4 are also useful for estimation of the energy balance in individual steps in a free-radical reaction sequence. This is an important factor in assessing the feasibility of chain reaction sequences since only reactions with low activation energies are rapid enough to sustain a chain process. If individual steps are identified as being endothermic by more than a few kilocalories, it is very unlikely that a chain mechanism can operate.

Example 12.1. Calculate the enthalpy for each step in the bromination of ethane by bromine atoms from molecular bromine. Determine the overall enthalpy of the reaction.

Initiation $Br\text{–}Br \longrightarrow 2\,Br\cdot$ Br–Br +46 kcal/mol

propagation:

$Br\cdot + CH_3\text{–}CH_3 \longrightarrow H\text{–}Br + \cdot CH_2\text{–}CH_3$ H–Br −87
H–C +98
+11

$Br\text{–}Br + \cdot CH_2\text{–}CH_3 \longrightarrow BrCH_2\text{–}CH_3 + Br\cdot$ Br–Br +46
Br–C −69
−23

total −12 kcal/mol

The enthalpy of the reaction is given by the sum of the propagation steps and is −12 kcal/mol. Analysis of the enthalpy of the individual steps indicates the first step is somewhat endothermic. This endothermicity is the lower limit of the activation energy for the step. Radical chain processes depend on a series of rapid steps that maintain the reactive intermediates at low concentration. Since termination reactions are usually very fast, the presence of an endothermic step in a chain sequence means that the chains will be short. The value for ethane is borderline and suggests that radical bromination of ethane would exhibit only short chain lengths. Since the enthalpies of the corresponding steps for abstraction of secondary or tertiary hydrogen are less positive, the bromination selects for hydrogens in the order tertiary > secondary > primary in compounds with more than one type of hydrogen. Enthalpy calculations cannot give a direct evaluation of the activation energy of either exothermic or endothermic steps. These will depend on the energy of the transition state. The bond dissociation energies can therefore provide only permissive, not definitive, conclusions.

Radical stability is reflected in a variety of ways in addition to the bond dissociation energy of the corresponding C—H bond. It has already been indicated that radical structure and stability determines the temperature at which azo compounds undergo decomposition with elimination of nitrogen (Section 12.1.4). Similar trends have been established in other radical-forming reactions. Rates of thermal decomposition of *t*-butyl peroxyesters, for example, vary over a wide range, depending on the structure of the carbonyl substituent.[66] This clearly indicates that the bonding changes involved in the rate-determining step are not completely localized in the O—O bond. Radical character must also be developing at the alkyl group by partial cleavage of the alkyl-carbonyl bond.

$$R{-}\overset{O}{\overset{\|}{C}}{-}O{-}O{-}C(CH_3)_3 \longrightarrow R\cdots\overset{O}{\overset{\|}{C}}{\overset{-}{-}}O\cdots OC(CH_3)_3 \longrightarrow R\cdot + CO_2 + \cdot OC(CH_3)_3$$

R	Relative rate at 60°C
CH_3	1
Ph	17
$PhCH_2$	290
$(CH_3)_3C$	1,700
Ph_2CH	19,300
$Ph(CH_3)_2C$	41,500
$PhCHCH{=}CH_2$	125,000

The same is true for decarbonylation of acyl radicals. The rates of decarbonylation have been measured over a very wide range of structural types.[67] There is a very strong dependence of the rate on the stability of the radical that results from decarbonylation. For example, decarbonylations giving tertiary benzylic radicals are on the order of $10^8\ s^{-1}$, whereas the benzoyl radical decarbonylates with a rate on the order of $1\ s^{-1}$.

Free-radical reactions written in the simplest way imply no separation of charge. The case of toluene bromination can be used to illustrate this point:

$$Br\cdot + H{-}CH_2Ph \longrightarrow Br\cdots H\cdots CH_2Ph \longrightarrow Br{-}H + \cdot CH_2Ph$$

$$PhCH_2\cdot + Br{-}Br \longrightarrow PhCH_2\cdots Br\cdots Br \longrightarrow PhCH_2Br + Br\cdot$$

Nevertheless, many free-radical processes respond to introduction of polar substituents, just as do heterolytic processes that involve polar or ionic intermediates. The substituent effects on toluene bromination, for example, are correlated by the Hammett equation, and the correlation gives a ρ value of -1.4, indicating that the benzene ring acts as an electron donor in the transition state.[68] Other radicals, for

66. P. D. Bartlett and R. R. Hiatt, *J. Am. Chem. Soc.* **80**, 1398 (1958).
67. H. Fischer and H. Paul, *Acc. Chem. Res.* **20**, 200 (1987).
68. J. Hradil and V. Chvalovsky, *Collect. Czech. Chem. Commun.* **33**, 2029 (1968); S. S. Kim, S. Y. Choi, and C. H. Kong, *J. Am. Chem. Soc.* **107**, 4234 (1985); G. A. Russell, C. DeBoer, and K. M. Desmond, *J. Am. Chem. Soc.* **85**, 365 (1963); C. Walling, A. L. Rieger, and D. D. Tanner, *J. Am. Chem. Soc.* **85**, 3129 (1963).

example, the *t*-butyl radical, show a positive ρ for hydrogen abstraction reactions involving toluene.[69]

Why do free-radical reactions involving neutral reactants and intermediates respond to substituent changes that modify electron distribution? One explanation has been based on the idea that there would be some polar character in the transition state because of the electronegativity differences of the reacting atoms.[70]:

$$\text{Br}\cdot\cdots\text{H}\cdots\text{CH}_2-\text{C}_6\text{H}_4-\text{X} \longleftrightarrow \overset{-}{\text{Br}}\text{:}\cdots\dot{\text{H}}\cdots\overset{+}{\text{C}}\text{H}_2-\text{C}_6\text{H}_4-\text{X}$$

This idea receives general support from the fact that the most negative ρ values are found for reactions of more electronegative radicals such as Br·, Cl· and Cl_3C·. There is, however no simple correlation with a single property, and this probably reflects the fact that the *selectivity* of the radicals is also different. Furthermore, in hydrogen abstraction reactions, where many of the quantitative measurements have been done, the C—H bond dissociation energy is also subject to a substituent effect.[71] Thus, the extent of bond cleavage and formation at the transition state may be different for different radicals. Successful interpretation of radical reactions therefore requires consideration of factors such as the electronegativity and polarizability of the radicals and the bond energy of the reacting C—H bond. The relative importance of these effects may vary from system to system. As a result, substituent effect trends in radical reactions appear to be more complicated than those for heterolytic reactions, where substituent effects are usually dominated by the electron-withdrawing or electron-donating capacity of the substituent.[72]

Despite their overall electrical neutrality, carbon-centered radicals show pronounced electrophilic or nucleophilic character, depending on the substituents present. This electrophilic or nucleophilic character is then reflected in rates of reaction with nonradical species, for example, in additions to substituted alkenes. Unsubstituted alkyl radicals and α-alkoxyalkyl radicals are nucleophilic in character and react most rapidly with alkenes having electron-attracting substituents.[73] Radicals having electron-attracting groups, such as the radicals derived from malonate esters, react preferentially with double bonds having electron-releasing substituents.[74] Some representative data are given in Table 12.6.

These reactivity trends are in line with what would be expected from a frontier orbital interpretation. As shown in Fig. 12.5, electron-releasing substituents will

69. W. A. Pryor, F. Y. Tang, R. H. Tang, and D. F. Church, *J. Am. Chem. Soc.* **104**, 2885 (1982); R. W. Henderson and R. O. Ward, Jr., *J. Am. Chem. Soc.* **96**, 7556 (1974); W. A. Pryor, D. F. Church, F. Y. Tang, and T. H. Tang, in *Frontiers of Free Radical Chemistry*, W. A. Pryor (ed.), Academic Press, New York, 1980, pp. 355-379.
70. E. S. Huyser, *Free Radical Chain Reactions*, Wiley-Interscience, New York, 1970, Chapter 4; G. A. Russell, in *Free Radicals*, J. Kochi (ed.), Vol. 1, Wiley, New York, 1973, Chapter 7.
71. A. A. Zavitsas and J. A. Pinto, *J. Am. Chem. Soc.* **94**, 7390 (1972).
72. W. H. Davis, Jr., and W. A. Pryor, *J. Am. Chem. Soc.* **99**, 6365 (1977); W. H. Davis, Jr., J. H. Gleason, and W. A. Pryor, *J. Org. Chem.* **42**, 7 (1977); W. A. Pryor, G. Gojon, and D. F. Church, *J. Org. Chem.* **43**, 793 (1978).
73. B. Giese, *Angew. Chem. Int. Ed. Engl.* **22**, 753 (1983).
74. B. Giese, H. Horler, and M. Leising, *Chem. Ber.* **119**, 444 (1986).

Table 12.6. Relative Rates of Radical Addition as a Function of Alkene Substitution[a]

A. Addition to Substituted Ethenes, $CH_2{=}CH{-}X$

X	$\cdot CH_3$	$\cdot C_2H_5$	cyclohexyl radical
CN	2.2	5.1	24
$COCH_3$	2.3		13
CO_2CH_3	1.3	1.9	6.7
Ph	1.0	1.0	1.0
O_2CCH_3		0.05	0.016

B. Addition to Substituted Styrenes, $CH_2{=}C(X){-}Ph$

X	cyclohexyl radical	$\cdot CH(CO_2C_2H_5)_2$
CN	122	
$CO_2C_2H_5$	11.7	0.28
Ph	1.0	1.0
CH_3	0.28	1.06
OCH_3		0.78
$N(CH_3)_2$		6.6

a. Data from B. Giese, H. Horler, and M. Leising, *Chem. Ber.* **119**, 444 (1986); B. Giese, *Angew. Chem. Int. Ed. Engl.* **22**, 753 (1983).

raise the energy of the singly occupied molecular orbital (SOMO) and increase the strength of interaction with the relatively low-lying LUMO of alkenes having electron-withdrawing groups. When the radical site is substituted by an electron-attracting group, the SOMO is lower, and the strongest interaction will be with the HOMO of the alkene. This interaction is strengthened by donor substituents on the alkene.

12.3. Free-Radical Substitution Reactions

12.3.1. Halogenation

Free-radical bromination is an important method of selective functionalization of hydrocarbons.[75] The process is a chain reaction involving the following steps:

initiation
$$Br{-}Br \longrightarrow 2\,Br\cdot$$

propagation
$$Br\cdot + R_3CH \longrightarrow R_3C\cdot + HBr$$
$$R_3C\cdot + Br_2 \longrightarrow R_3CBr + Br\cdot$$

75. W. A. Thaler, *Methods Free Radical Chem.* **2**, 121 (1969); A. Nechvatal, *Adv. Free Radical Chem.* **4**, 175 (1972).

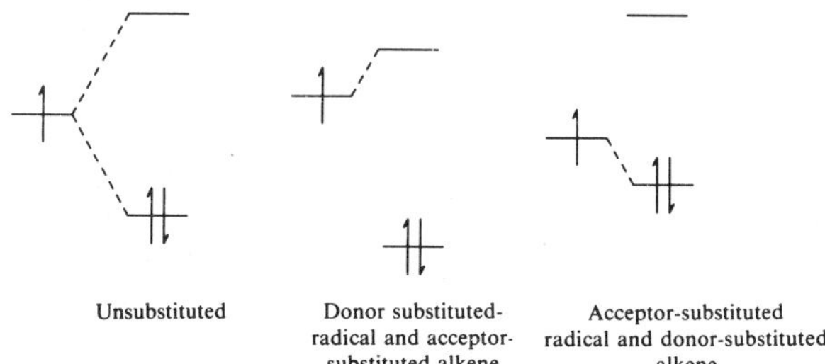

Fig. 12.5. Frontier orbital interactions between different combinations of substituted radicals and alkenes.

The reaction is often initiated by photolysis of bromine. The hydrogen atom abstraction step is rate-limiting, and the product composition is governed by the selectivity of the hydrogen abstraction step. The enthalpy requirements for abstraction of hydrogen from methane, ethane (primary), propane (secondary), and isobutane (tertiary) by bromine atoms are +16.5, +10.5, +7.0, and +3.5 kcal/mol, respectively.[76] These difference are reflected in the activation energies, and there is a substantial kineric preference for hydrogen abstraction in the order tertiary > secondary > primary. Substituents that promote radical stability, such as phenyl, vinyl, or carbonyl groups, also lead to kinetic selectivity in radical brominations. Bromination at benzylic positions is a particularly efficient process, as illustrated by entries 2 and 4 in Scheme 12.4.

There are important differences in the reactions of other halogens relative to bromination. In the case of chlorination, although the same chain mechanism as for bromination is operative, there is a key difference in the *greatly diminished selectivity of the chlorination.* Because of the greater reactivity of the chlorine atom, abstractions of primary, secondary, and tertiary hydrogens are all *exothermic.* As a result of this exothermicity, the stability of the product radical has less influence on the activation energy. In terms of the Hammond postulate, the transition state would be expected to be very *reactant-like.* As an example of the low selectivity, ethylbenzene is chlorinated at both the methyl and the methylene positions, despite the much greater stability of the benzyl radical.[77]

$$C_6H_5CH_2CH_3 \xrightarrow[\text{nitrobenzene}]{Cl_2,\ 40°C} C_6H_5CHClCH_3 + C_6H_5CH_2CH_2Cl$$

Radical chlorination reactions show a substantial polar effect. Positions substituted by electron-withdrawing groups are relatively unreactive toward chlorination, even though the substituents may be potentially capable of stabilizing the

76. E. S. Huyser, *Free Radical Chain Reactions*, Wiley-Interscience, New York, 1970, p. 91.
77. G. A. Russell, A. Ito, and D. G. Hendry, *J. Am. Chem. Soc.* **86**, 2976 (1963).

Scheme 12.4. Radical Halogenations

Molecular bromine

1[a] $(CH_3)_2CHCH_2CH_2CH_3 \xrightarrow[h\nu]{Br_2} (CH_3)_2C(Br)CH_2CH_2CH_3$ 90%

2[b] o-xylene ($C_6H_4(CH_3)_2$) $\xrightarrow[h\nu]{Br_2}$ $C_6H_4(CH_2Br)_2$ 48%–53%

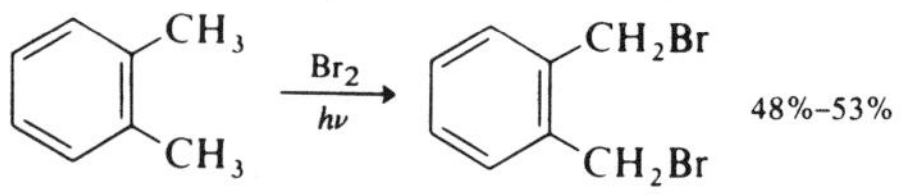

N-Bromosuccinimide

3[c] $CH_3(CH_2)_3CH{=}CHCH_3 \xrightarrow[(PhCO_2)_2]{NBS} CH_3(CH_2)_2CH(Br)CH{=}CHCH_3$ 58%–64%

4[d] 3-methylthiophene (CH_3) $\xrightarrow[(PhCO_2)_2]{NBS}$ 3-thienyl-CH_2Br 79%

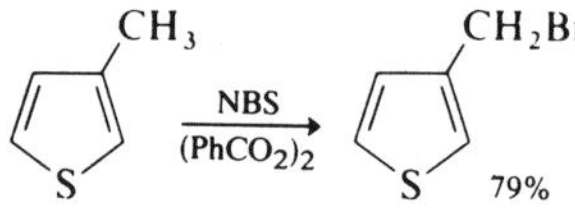

Other halogenating agents

5[e] $C_6H_5CH_2CH_2CH_3 \xrightarrow[40°C]{(CH_3)_3COCl}$ $C_6H_5CH(Cl)CH_2CH_3$ 65% + $C_6H_5CH_2CH(Cl)CH_3$ 25% + $C_6H_5CH_2CH_2CH_2Cl$ 10%

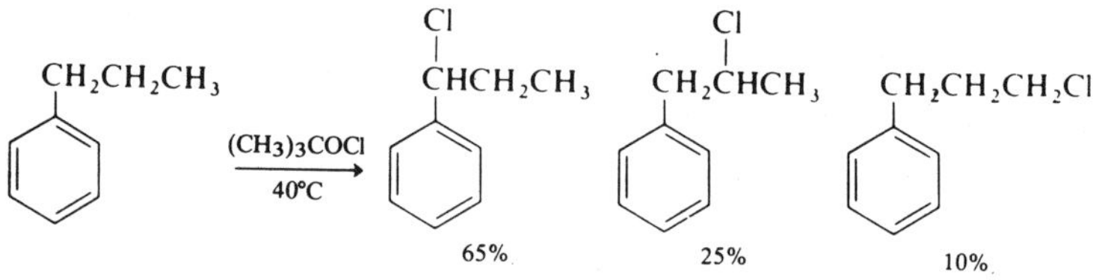

6[f] $CH_2{=}CHCH_2CH_3 + (CH_3)_3COCl \xrightarrow[-78°C]{h\nu} CH_3CH{=}CHCH_2Cl$ 74%
$+ CH_2{=}CHCH(Cl)CH_3$ 26%

7[g] $C_6H_5CH_2CH_2CH_3 \xrightarrow[(PhCO_2)_2]{SO_2Cl_2} C_6H_5CH(Cl)CH_2CH_3$ 50%

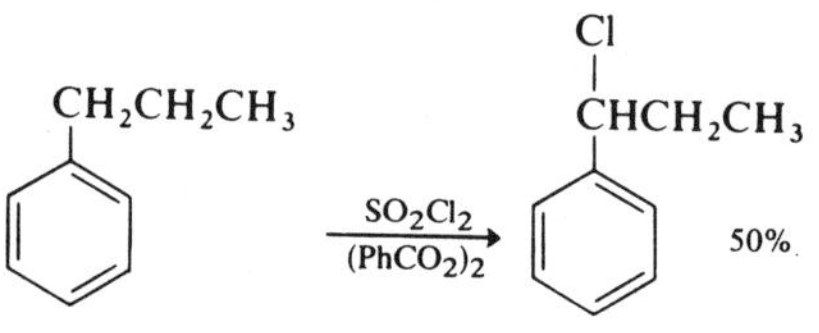

a. G. A. Russell and H. C. Brown, *J. Am. Chem. Soc.* **77**, 4025 (1955).
b. E. F. M. Stephenson, *Org. Synth.* **IV**, 984 (1963).
c. F. L. Greenwood, M. D. Kellert, and J. Sedlak, *Org. Synth.* **IV**, 108 (1963).
d. E. Campaigne and B. F. Tullar, *Org. Synth.* **IV**, 921 (1963).
e. C. Walling and B. B. Jacknow, *J. Am. Chem. Soc.* **82**, 6108 (1960).
f. C. Walling and W. Thaler, *J. Am. Chem. Soc.* **83**, 3877 (1961).
g. H. C. Brown and A. B. Ash, *J. Am. Chem. Soc.* **77**, 4019 (1955).

free-radical intermediate.[78]

$$CH_3CH_2CH_2CN + Cl_2 \longrightarrow \underset{\substack{|\\ Cl}}{CH_3CHCH_2CN}\ (69\%) + ClCH_2CH_2CH_2CN\ (31\%)$$

The polar effect is attributed to the fact that the chlorine atom is an electrophilic species, and the relatively electron-deficient carbon atoms adjacent to the electron-withdrawing groups are avoided. Similarly, carboxylic acid and ester groups tend to direct chlorination to the β and γ positions, because attack at the α position is electronically disfavored. The effect of an electron-withdrawing substituent is to decrease the electron density at the potential radical site, even though there would be a net stabilization of the radical that would be eventually formed. Since the chlorine atom is highly reactive, the reaction would be expected to have a very early transition state. The polar substituent effect dominates the *kinetic selectivity* of the reaction, and the stability of the radical intermediate has relatively little influence.

Radical substitution reactions by iodine are not practical because the abstraction of hydrogen from hydrocarbons by iodine is highly endothermic, even for stable radicals. The enthalpy of the overall reaction is also slightly endothermic. Thus, both because of the kinetic problem excluding a chain reaction and an unfavorable equilibrium constant for substitution, iodination cannot proceed by a radical chain mechanism.

Fluorination presents problems of the other extreme. Both steps in the substitution chain reaction are so exothermic that the reaction is violent if not performed under carefully controlled conditions. Furthermore, fluorine atoms are capable of cleaving carbon–carbon bonds:

$$F\cdot + CH_3CH_3 \rightarrow CH_3F + CH_3\cdot$$

Saturated hydrocarbons such as neopentane, norbornane, and cyclotane have been converted to the corresponding perfluoro derivatives in 10–20% yield by gas phase reaction with fluorine gas diluted with helium at −78°C.[79] Simple ethers can be completely fluorinated under similar conditions.[80] Crown polyethers can be fluorinated by passing a F_2/He stream over a solid mixture of sodium fluoride and the crown ether.[81] Liquid phase fluorination of hydrocarbons has also been observed, but the reaction is believed to be ionic, rather than radical, in character.[82]

Halogenations of organic molecules can also be carried out with several other chemical reagents in addition to the molecular halogens. *N*-Bromosuccinimide (NBS) has been used extensively, especially for allylic and benzylic bromination. Mechanistic investigations have established that molecular bromine is the active

78. A. Bruylants, M. Tits, C. Dieu, and R. Gauthier, *Bull. Soc. Chim. Belg.* **61**, 266 (1952); A. Bruylants, M. Tits, and R. Danby, *Bull. Soc. Chim. Belg.* **58**, 210 (1949); M. S. Kharasch and H. C. Brown, *J. Am. Chem. Soc.* **62**, 925 (1940).
79. N. J. Maraschin, D. B. Catsikis, L. H. Davis, G. Jarvinen, and R. J. Lagow, *J. Am. Chem. Soc.* **97**, 513 (1975).
80. D. F. Persico, H.-N. Huang, R. J. Lagow, Jr., and L. C. Clark, *J. Org. Chem.* **50**, 5156 (1985).
81. W.-H. Lin, W. I. Bailey, Jr., and R. J. Lagow, *J. Chem. Soc., Chem. Commun.*, 1350 (1985).
82. C. Gal and S. Rozen, *Tetrahedron Lett.* **25**, 449 (1984).

halogenating agent under the usual conditions used for NBS bromination.[83] Molecular bromine is maintained at a low concentration throughout the course of the reaction by formation from NBS and hydrogen bromide.

$$Br\cdot + H\text{-}R \longrightarrow HBr + R\cdot$$

HBr + (N-bromosuccinimide) ⟶ (succinimide, N—H) + Br_2

$$Br_2 + R\cdot \longrightarrow RBr + Br\cdot$$

The fact that the bromine concentration remains at very low levels is important to the success of the allylic halogenation process. Substitution at the allylic position in preference to addition is the result of the reversibility of electrophilic attack by bromine at the double bond. With a low concentration of bromine, the rate of bromine addition is low enough to permit hydrogen atom abstraction to compete successfully.

N-Bromosuccinimide can also be used to brominate alkanes. For example, cyclopropane, cyclopentane, and cyclohexane give the corresponding bromides when irradiated with NBS in dichloromethane.[84] Under these conditions, the succinimidyl radical appears to be involved as the hydrogen-abstracting intermediate.

$$Br\cdot + R\text{—}H \longrightarrow R\cdot + HBr$$

R· + (N-bromosuccinimide, N—Br) ⟶ RBr + (succinimidyl radical, N·)

(succinimidyl radical, N·) + R—H ⟶ (succinimide, NH) + R·

Another reagent that effects chlorination by a radical mechanism is *t*-butyl hypochlorite. The hydrogen-abstracting species in the chain mechanism is the *t*-butoxy radical.

$$(CH_3)_3CO\cdot + H\text{-}R \longrightarrow (CH_3)_3COH + R\cdot$$

$$R\cdot + (CH_3)_3COCl \longrightarrow (CH_3)_3CO\cdot + RCl$$

83. R. E. Pearson and J. C. Martin, *J. Am. Chem. Soc.* **85**, 354, 3142 (1963); G. A. Russell, C. DeBoer, and K. M. Desmone, *J. Am. Chem. Soc.* **85**, 365 (1963); J. H. Incremona and J. C. Martin, *J. Am. Chem. Soc.* **92**, 627 (1970); J. C. Day, M. J. Lindstrom, and P. S. Skell, *J. Am. Chem. Soc.* **96**, 5616 (1974).
84. J. G. Traynham and Y.-S. Lee, *J. Am. Chem. Soc.* **96**, 3590 (1974).

Table 12.7. Relative Reactivities of Some Aromatic Hydrocarbons toward Oxygen[a]

$PhCH(CH_3)_2$	1.0	$PhCH_2CH_3$	0.18
$PhCH_2CH{=}CH_2$	0.8	$PhCH_3$	0.015
$(Ph)_2CH_2$	0.35		

a. Data from G. A. Russell, *J. Am. Chem. Soc.* **78**, 1047 (1956).

This radical is intermediate in selectivity between chlorine and bromine atoms. The selectivity is also solvent and temperature dependent. A typical ratio, in chlorobenzene as solvent, is tertiary:secondary:primary = 60:10:1.[85] Scheme 12.4 (p. 690) gives a number of specific halogenation reactions that proceed by radical chain mechanisms.

12.3.2. Oxidation

Free-radical chain oxidation of organic molecules by molecular oxygen is often referred to as *autoxidation*. The general mechanism is outlined below:

initiation $\quad In\cdot + R{-}H \longrightarrow In{-}H + R\cdot$

propagation $\quad R\cdot + O_2 \longrightarrow R{-}O{-}O\cdot$

$$R{-}O{-}O\cdot + R{-}H \longrightarrow R{-}O{-}O{-}H + R\cdot$$

The rate of reaction of oxygen with most radicals is very rapid because of the triplet character of molecular oxygen. The ease of autoxidation is therefore largely governed by the ease of hydrogen abstraction in the second step of the propagation sequence. The alkylperoxy radicals that act as the chain carrier are fairly selective. Substrates that are relatively electron-rich or that provide particularly stable radicals are the most easily oxidized. Benzylic, allylic, and tertiary positions are especially susceptible to oxidation. This selectivity make radical chain oxidation a preparatively useful reaction in some cases.

The reactivities of a series of hydrocarbons toward oxygen measured under a standard set of conditions can give some indication of the susceptibility of various structural units to autoxidation.[86] Table 12.7 gives the results for a series of aromatic hydrocarbons. These data illustrate the activating effect of alkyl, vinyl, and phenyl substituents.

The best preparative results from autoxidation are encountered when only one relatively reactive hydrogen is available for abstraction. The oxidation of isopropylbenzene (cumene) is carried out on an industrial scale, with the ultimate products

85. C. Walling and P. J. Wagner, *J. Am. Chem. Soc.* **86**, 3368 (1964).
86. G. A. Russell, *J. Am. Chem. Soc.* **78**, 1047 (1956).

being acetone and phenol.

$$C_6H_5CH(CH_3)_2 + O_2 \longrightarrow C_6H_5C(CH_3)_2OOH \xrightarrow{H_2SO_4} C_6H_5OH + (CH_3)_2C{=}O$$

The benzylic position in tetralin is selectively oxidized to the hydroperoxide.[87]

$$\text{tetralin} + O_2 \xrightarrow[48\ h]{70°C} \text{1-hydroperoxytetralin (OOH)} \quad 44\%\text{–}57\%$$

Functional groups that stabilize radicals would be expected to increase susceptibility to autoxidation. This is illustrated by two cases that have been relatively well studied. Aldehydes, in which abstraction of the aldehyde hydrogen is facile, are easily autoxidized. The autoxidation initially forms a peroxycarboxylic acid, but usually the corresponding carboxylic acid is isolated because the peroxy acid oxidizes additional aldehyde in a parallel reaction.

$$RCH{=}O + In\cdot \longrightarrow R\overset{O}{\overset{\|}{C}}\cdot + InH$$

$$R\overset{O}{\overset{\|}{C}}\cdot + O_2 \longrightarrow R\overset{O}{\overset{\|}{C}}OO\cdot$$

$$R\overset{O}{\overset{\|}{C}}OO\cdot + R\overset{O}{\overset{\|}{C}}H \longrightarrow R\overset{O}{\overset{\|}{C}}OOH + R\overset{O}{\overset{\|}{C}}\cdot$$

$$R\overset{O}{\overset{\|}{C}}OOH + R\overset{O}{\overset{\|}{C}}H \longrightarrow 2RCO_2H$$

The final step is not a radical reaction but is an example of the Baeyer–Villiger reaction, which will be discussed in Part B, Section 12.5.2.

Similarly, the α position in ethers is autoxidized quite readily to give α-hydroperoxy ethers.

$$\text{tetrahydrofuran} + O_2 \longrightarrow \text{2-hydroperoxytetrahydrofuran (OOH)}$$

This reaction is the basis of a widely recognized laboratory hazard. The peroxides formed from several commonly used ethers such as diethyl ether and tetrahydrofuran are explosive. Appreciable amounts of such peroxides can build up in ether samples that have been exposed to the atmosphere. Since the hydroperoxides are less volatile than the ethers, they are concentrated by evaporation or distillation, and the

87. H. B. Knight and D. Swern, *Org. Synth.* **IV**, 895 (1963).

concentrated peroxide solutions may explode. For this reason, extended storage of ethers that have been exposed to oxygen is extremely hazardous.

12.4. Free-Radical Addition Reactions

12.4.1. Addition of Hydrogen Halides

The anti-Markownikoff addition of hydrogen bromide to alkenes was one of the earliest free-radical reactions to be put on a firm mechanistic basis. In the presence of a suitable initiator, such as peroxides, a radical chain mechanism becomes competitive with the ionic mechanism for addition of hydrogen bromide.

$$\mathrm{In\cdot + HBr \longrightarrow Br\cdot + HIn}$$
$$\mathrm{Br\cdot + RCH{=}CH_2 \longrightarrow R\dot{C}HCH_2Br}$$
$$\mathrm{R\dot{C}HCH_2Br + HBr \longrightarrow RCH_2CH_2Br + Br\cdot}$$

Since the bromine atom adds to the less substituted carbon atom of the double bond, thus generating the more substituted radical intermediate, the regioselectivity of radical chain hydrobromination is opposite to that of ionic addition. The work on the radical mechanism originated in studies of the addition of hydrogen bromide that were undertaken to understand why Markownikoff's rule was violated under certain circumstances. The cause was found to be conditions, such as peroxide impurities or light, which initiated the radical chain process. Some examples of radical chain additions of hydrogen bromide to alkenes are included in Scheme 12.5.

The stereochemistry of radical addition of hydrogen bromide to alkenes has been studied with both acylic and cyclic alkenes.[88] *Anti* addition is favored.[89,90] This is contrary to what would be expected if the sp^2 carbon of the radical were rapidly rotating with respect to the remainder of the molecule:

(RCH=CHR, trans) + Br· ⟶ R(H)Ċ—C(Br)(H)(R) ⇌ H(R)Ċ—C(Br)(H)(R)

The stereospecificity of the radical addition can be explained in terms of a bridged structure similar to that involved in discussion of ionic bromination of alkenes[91]:

(RCH=CHR, trans) + Br· ⟶ bridged :Br· radical between C(R)(H)—C(H)(R) ⟶ H(R)C(H)—C(Br)(H)(R)

88. B. A. Bohm and P. I. Abell, *Chem. Rev.* **62**, 599 (1962).
89. P. S. Skell and P. K. Freeman, *J. Org. Chem.* **29**, 2524 (1964).
90. H. L. Goering and D. W. Larsen, *J. Am. Chem. Soc.* **81**, 5937 (1959).
91. P. S. Skell and J. G. Traynham, *Acc. Chem. Res.* **17**, 160 (1984).

Scheme 12.5. Free-Radical Chain

Hydrogen bromide

1[a] $CH_3CH{=}C(CH_3)_2 \xrightarrow{HBr} CH_3CH(Br)CH(CH_3)_2$ 55%

2[b] $CH_2{=}CHCO_2CH_3 \xrightarrow{HBr} BrCH_2CH_2CO_2CH_3$ 80%–84%

3[c] $CH_2{=}CHCH_2C(CH_3)_2C(=O)CH_3 \xrightarrow[h\nu]{HBr} BrCH_2CH_2CH_2C(CH_3)_2C(=O)CH_3$ 76%

Addition of halomethanes

4[d] $CCl_4 + H_2C{=}CH(CH_2)_5CH_3 \xrightarrow{(PhCO_2)_2} Cl_3CCH_2CH(Cl)(CH_2)_5CH_3$ 75%

5[e] $BrCCl_3 + CH_2{=}C(C_2H_5)_2 \xrightarrow{h\nu} Cl_3CCH_2C(Br)(C_2H_5)_2$ 91%

6[f] norbornene + $CCl_4 \xrightarrow{(PhCO_2)_2}$ exo-CCl_3, endo-Cl product (73%) + exo-CCl_3, exo-Cl product (4%)

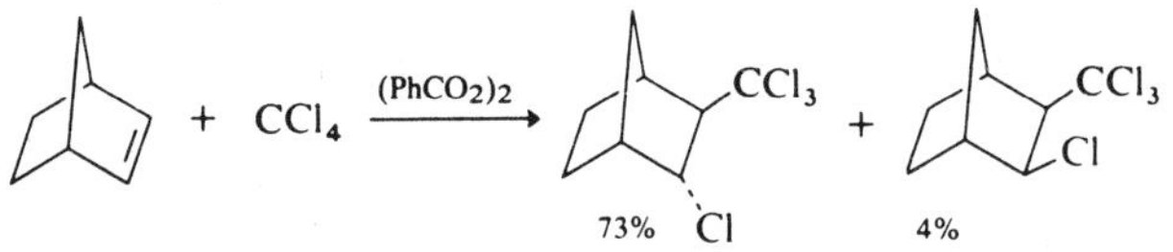

7[g] $CCl_4 + CH_2{=}CHC(OH)(CH_3)_2 \xrightarrow[80°C]{(PhCO_2)_2} Cl_3CCH_2CH(Cl)C(OH)(CH_3)_2$ 70%

a. W. J. Bailey and S. S. Hirsch, *J. Org. Chem.* **28**, 2894 (1963).
b. R. Mozingo and L. A. Patterson, *Org. Synth.* **III**, 576 (1955).
c. H. O. House, C.-Y. Chu, W. V. Phillips, T. S. B. Sayer, and C.-C. Yau, *J. Org. Chem.* **42**, 1709 (1977).
d. M. S. Kharasch, E. V. Jensen, and W. H. Urry, *J. Am. Chem. Soc.* **69**, 1100 (1947).
e. M. S. Kharasch and M. Sage, *J. Org. Chem.* **14**, 537 (1949).
f. C. L. Osborn, T. V. Van Auken, and D. J. Trecker, *J. Am. Chem. Soc.* **90**, 5806 (1968).
g. P. D. Klemmensen, H. Kolind-Andersen, H. B. Madsen, and A. Svendsen, *J. Org. Chem.* **44**, 416 (1979).
h. M. S. Kharasch, W. H. Urry, and B. M. Kuderna, *J. Org. Chem.* **14**, 248 (1949).

Further evidence for a bromine-bridged radical comes from radical substitution of optically active 2-bromobutane. Most of the 2,3-dibromobutane that is formed is racemic, indicating that the chiral center has been involved in the reaction. When the 3-deuterated reagent is used, it can be shown that the hydrogen (or deuterium) that is abstracted is replaced by bromine with *retention of stereochemistry*.[92] These results are consistent with a bridged bromine radical.

92. P. S. Skell, R. R. Pavlis, D. C. Lewis, and K. J. Shea, *J. Am. Chem. Soc.* **95**, 6735 (1973).

Addition of other carbon radicals

8[h] $$CH_3CH{=}O + CH_2{=}CH(CH_2)_5CH_3 \longrightarrow CH_3\overset{O}{\overset{\|}{C}}(CH_2)_7CH_3 \quad 64\%$$

9[i] $$CH_3(CH_2)_5CH{=}O + H_5C_2O_2CCH{=}CHCO_2C_2H_5 \longrightarrow CH_3(CH_2)_5\overset{O}{\overset{\|}{C}}\underset{CH_2CO_2C_2H_5}{\underset{|}{C}}HCO_2C_2H_5$$

10[j] $$CH_2{=}CHCH_2CH_2CO_2CH_3 + H\overset{O}{\overset{\|}{C}}NH_2 \xrightarrow[\text{acetone}]{h\nu} H_2N\overset{O}{\overset{\|}{C}}(CH_2)_4CO_2CH_3 \quad 58\%$$

11[k] $$\text{1,3-dioxolane} + H_3CO_2CCH{=}CHCO_2CH_3 \xrightarrow[\text{acetone}]{h\nu} \text{(1,3-dioxolan-2-yl)}-\underset{CH_2CO_2CH_3}{\underset{|}{C}}HCO_2CH_3$$

12[l] $$\text{norbornene} + CH_3\overset{O}{\overset{\|}{C}}CH_3 \xrightarrow{h\nu} \text{(2-norbornyl)}CH_2\overset{O}{\overset{\|}{C}}CH_3$$

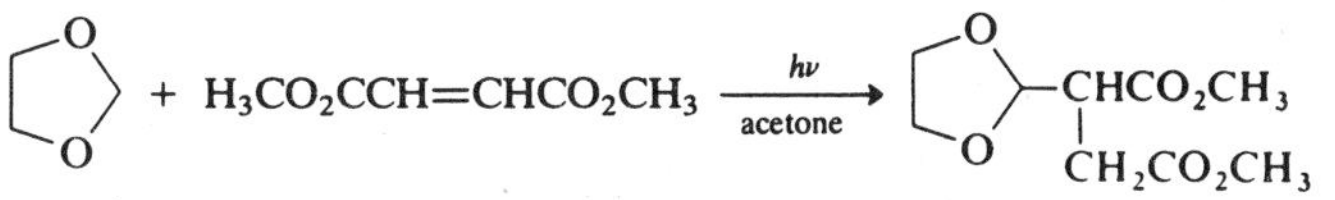

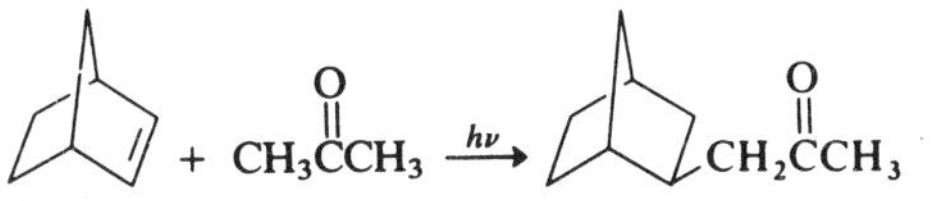

13[m] $$CH_2{=}CH(CH_2)_5CH_3 + CH_2(CO_2C_2H_5)_2 \xrightarrow{\text{ROOR}} CH_3(CH_2)_7CH(CO_2C_2H_5)_2 \quad (67\text{–}85\%)$$

14[n] $$CH_2{=}CH(CH_2)_5CH_3 + N{\equiv}CCH_2CO_2C_2H_5 \xrightarrow{\text{CuO}} C_2H_5O_2C\underset{C\equiv N}{\underset{|}{C}}H(CH_2)_7CH_3 \quad 75\%$$

Additions of thiols and thio acids

15[o] $$CH_3\overset{O}{\overset{\|}{C}}SH + C_6H_5CH{=}CHCH{=}O \longrightarrow CH_3\overset{O}{\overset{\|}{C}}S\overset{C_6H_5}{\overset{|}{C}}HCH_2CH{=}O \quad (90\%)$$

16[p] $$CH_3CH_2CH_2CH_2SH + CH_2{=}CH(CH_2)_4CH_3 \xrightarrow{(PhCO_2)_2} CH_3(CH_2)_3S(CH_2)_6CH_3 \quad 68\%$$

i. T. M. Patrick, Jr., and F. B. Erikson, *Org. Synth.* **IV**, 430 (1963).
j. D. Elad and J. Rokach, *J. Org. Chem.* **29**, 1855 (1964).
k. I. Rosenthal and D. Elad, *J. Org. Chem.* **33**, 805 (1968).
l. W. Reusch, *J. Org. Chem.* **27**, 1882 (1962).
m. J. C. Allen, J. I. G. Cadogan, B. W. Harris and D. H. Hey, *J. Chem. Soc.*, 4468 (1962).
n. A. Hajek and J. Malek, *Synthesis*, 454 (1977).
o. R. Brown, W. E. Jones, and A. R. Pinder, *J. Chem. Soc.*, 2123 (1951).
p. D. W. Grattan, J. M. Locke, and S. R. Wallis, *J. Chem. Soc. Perkin Trans. 1*, 2264 (1973).

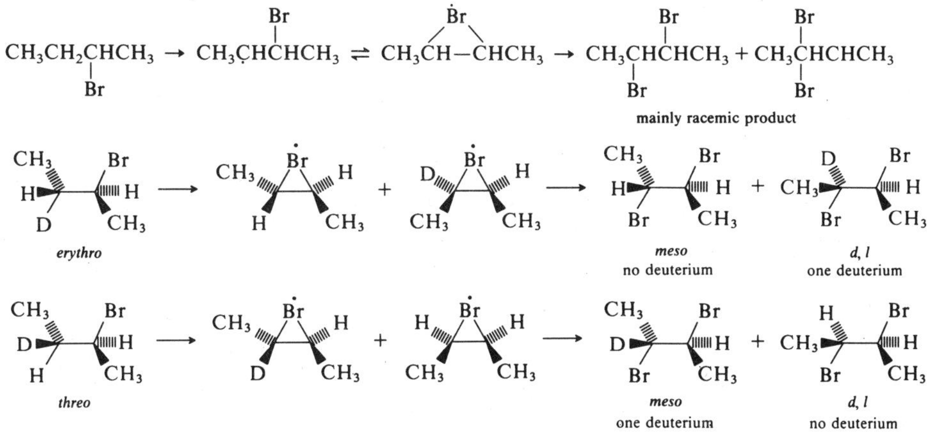

$$CH_3CH_2\underset{Br}{\underset{|}{C}}HCH_3 \rightarrow CH_3\dot{C}H\overset{Br}{\overset{|}{C}}HCH_3 \rightleftharpoons CH_3CH\overset{\dot{Br}}{\frown}CHCH_3 \rightarrow CH_3\underset{Br}{\underset{|}{C}}H\overset{Br}{\overset{|}{C}}HCH_3 + CH_3\underset{Br}{\underset{|}{C}}H\overset{Br}{\overset{|}{C}}HCH_3$$

mainly racemic product

erythro → bridged bromine radicals (CH₃, H / H, CH₃ and D, CH₃ / H, CH₃) → *meso* (no deuterium) + *d, l* (one deuterium)

threo → bridged bromine radicals (CH₃, D / H, CH₃ and H, CH₃ / H, CH₃) → *meso* (one deuterium) + *d, l* (no deuterium)

Other mechanisms must also operate, however, to account for the fact that 5–10% of the product is formed with retained configuration at the chiral center. Isotopic labeling studies have also demonstrated that the 3-bromo-2-butyl radical undergoes reversible loss of bromine atom to give 2-butene at a rate that is competitive with that of the bromination reaction.

$$CH_3\dot{C}HCH(Br)CH_3 \rightleftharpoons CH_3CH{=}CHCH_3 + Br^{\cdot}$$

This process can account for some of the observed loss of optical purity by a mechanism which does not involve the bridged intermediate.[93]

trans-Diaxial addition is the preferred stereochemical mode for addition to cyclohexene and its derivatives[94]:

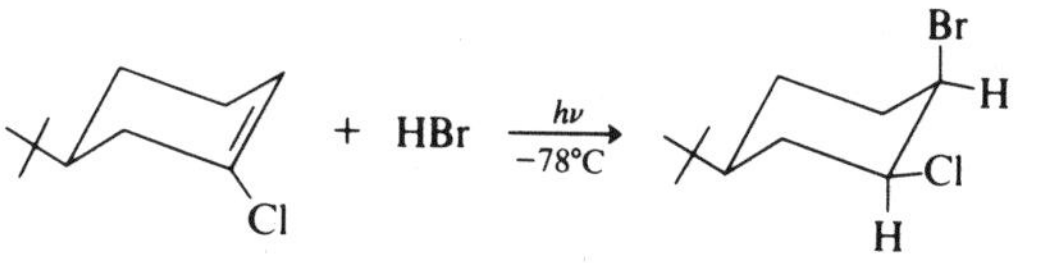

$$\text{4-}t\text{-butyl-1-chlorocyclohexene} + HBr \xrightarrow[-78^\circ C]{h\nu} \text{product (Br, H; H, Cl)}$$

This stereochemistry can be explained in terms of a bromine-bridged intermediate.

Product mixtures from radical chain addition of hydrogen chloride to alkenes are much more complicated than is the case for hydrogen bromide. The problem is that the rate of abstraction of hydrogen from hydrogen chloride is not large relative to addition of the alkyl radical to the alkene. This results in the formation of low-molecular-weight polymers in competition with simple addition.

$$RCH_2CH_2Cl + Cl\cdot$$

$$Cl\cdot + RCH{=}CH_2 \longrightarrow R\dot{C}HCH_2Cl \xrightarrow{RCH=CH_2} RCHCH_2(CH(R)CH_2)_nCl \xrightarrow{HCl} RCH_2CH_2(CH(R)CH_2)_nCl$$

(with $R\dot{C}HCH_2Cl \xrightarrow{HCl} RCH_2CH_2Cl + Cl\cdot$)

Radical chain additions of hydrogen fluoride and hydrogen iodide to alkenes are not observed. In the case of hydrogen iodide, the addition of an iodine atom to an alkene is an endothermic process and is too slow to permit a chain reaction, even though the hydrogen abstraction step would be favorable. In the case of hydrogen fluoride, the abstraction of hydrogen from hydrogen fluoride is energetically prohibitive.

12.4.2. Addition of Halomethanes

One of the older preparative free-radical reactions is the addition of polyhalomethanes to alkenes. Examples of addition of carbon tetrabromide, carbon

93. D. D. Tanner, E. V. Blackburn, Y. Kosugi, and T. C. S. Rao, *J. Am. Chem. Soc.* **99**, 2714 (1977).
94. H. L. Goering and L. L. Sims, *J. Am. Chem. Soc.* **77**, 3465 (1955); N. A. LeBel, R. F. Czaja, and A. DeBoer, *J. Org. Chem.* **34**, 3112 (1969); P. D. Readio and P. S. Skell, *J. Org. Chem.* **31**, 753 (1966); H. L. Goering, P. I. Abell, and B. F. Aycock, *J. Am. Chem. Soc.* **74**, 3588 (1952).

tetrachloride, and bromoform have been recorded.[95] The reactions are chain processes that depend on facile abstraction of halogen or hydrogen from the halomethane.

$$\mathrm{In\cdot} + \mathrm{CBr_4} \longrightarrow \mathrm{InBr} + \mathrm{\cdot CBr_3}$$
$$\mathrm{\cdot CBr_3} + \mathrm{CH_2{=}CHR} \longrightarrow \mathrm{Br_3CCH_2\dot{C}HR}$$
$$\mathrm{Br_3CCH_2\dot{C}HR} + \mathrm{CBr_4} \longrightarrow \mathrm{Br_3CCH_2CH(Br)R} + \mathrm{\cdot CBr_3}$$

or

$$\mathrm{In\cdot} + \mathrm{HCBr_3} \longrightarrow \mathrm{InH} + \mathrm{\cdot CBr_3}$$
$$\mathrm{\cdot CBr_3} + \mathrm{CH_2{=}CHR} \longrightarrow \mathrm{Br_3CCH_2\dot{C}HR}$$
$$\mathrm{Br_3CCH_2\dot{C}HR} + \mathrm{HCBr_3} \longrightarrow \mathrm{Br_3CCH_2CH_2R} + \mathrm{\cdot CBr_3}$$

Bromotrichloromethane can also be used effectively in the addition reaction. Because of the preferential abstraction of bromine, a trichloromethyl unit is added to the less substituted carbon atom of the alkene.

$$\mathrm{BrCCl_3} + \mathrm{CH_2{=}CHR} \xrightarrow[\Delta]{\text{peroxide}} \mathrm{Cl_3CCH_2CH(Br)R}$$

The efficiency of the halomethane addition process depends on the relative rate of halogen atom abstraction versus that of addition to the alkene.

$$\mathrm{X_3CCH_2\dot{C}HR} \xrightarrow{\text{atom abstraction}} \mathrm{X_3CCH_2CH(X)R}$$
$$\mathrm{X_3CCH_2\dot{C}HR} \xrightarrow{\text{addition to } \mathrm{RCH{=}CH_2}} \mathrm{X_3CCH_2CH(R)CH_2\dot{C}HR} \longrightarrow \text{telomer}$$

For a given alkene, the order of reactivity of the halomethanes is $\mathrm{CBr_4} > \mathrm{CBrCl_3} > \mathrm{CCl_4} > \mathrm{CH_2Cl_2} > \mathrm{CHCl_3}$. The efficiency of 1:1 addition for a given alkene depends on the ease with which it undergoes radical chain polymerization, since rapid polymerization will compete with the halogen atom abstraction step in the chain mechanism. Polymerization is usually most rapid for terminal alkenes bearing stabilizing substituents such as a phenyl or ester group. Several specific examples of additions of polyhaloalkanes are included in Scheme 12.5.

The addition of bromotrichloromethane to cyclohexene gives a nearly 1:1 mixture of the two possible stereoisomers.[96]

$\mathrm{BrCCl_3}$ + cyclohexene $\xrightarrow{h\nu}$ 1-bromo-2-(trichloromethyl)cyclohexane (Br, $\mathrm{CCl_3}$) 55% + stereoisomer (Br, $\mathrm{CCl_3}$) 45%

95. E. Sosnovsky, *Free Radical Reactions in Preparative Organic Chemistry*, MacMillan, New York, 1964, Chapter 2.
96. J. G. Traynham, A. G. Lane, and N. S. Bhacca, *J. Org. Chem.* **34**, 1302 (1969).

This result shows that the initially added trichloromethyl group has little influence on the stereochemistry of the subsequent bromine atom abstraction. In contrast, in the case of $\Delta^{2,3}$-octahydronaphthalene, the addition is exclusively *trans*-diaxial:

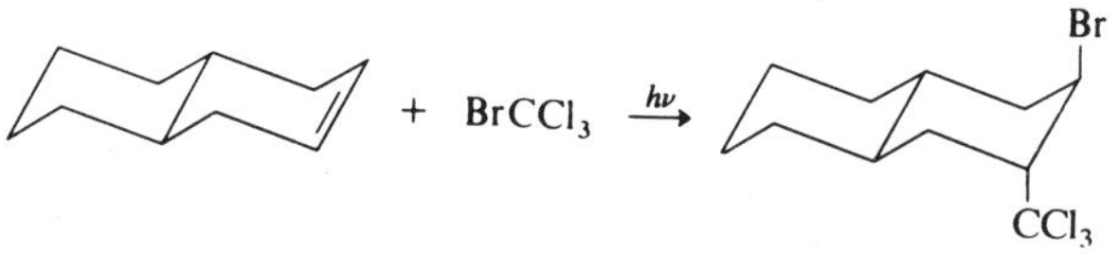

$$+ \ BrCCl_3 \xrightarrow{h\nu}$$

Br

CCl_3

The *trans*-fused decalin system is conformationally rigid, and the stereochemistry of the product indicates that the initial addition of the trichloromethyl radical is from an axial direction. This would be expected on stereoelectronic grounds, because the radical should initially interact with the π^* orbital. The axial trichloromethyl group then shields the adjacent radical position enough to direct the bromine abstraction in the *trans* sense. The results from the addition to cyclohexene indicate that the flexible 2-trichloromethylcyclohexyl radical must undergo conformational relaxation faster than bromine atom abstraction.

Addition of bromotrichloromethane to norbornene is also *anti.*

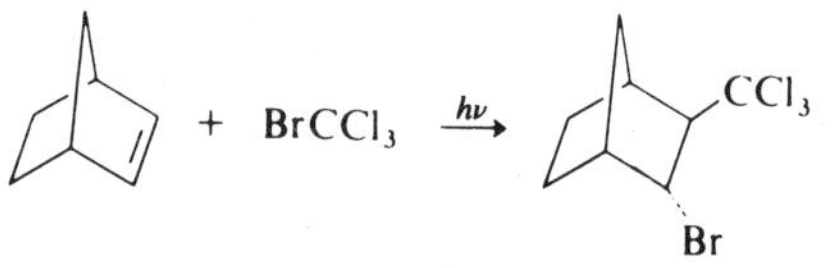

$$+ \ BrCCl_3 \xrightarrow{h\nu}$$

CCl_3

Br

This is again the result of steric shielding by the trichloromethyl group, which causes the bromine atom to be abstracted from the *endo* face of the intermediate radical.

12.4.3. Addition of Other Carbon Radicals

Other functional groups can also provide sufficient stabilization of radicals to permit successful chain additions to alkenes. Acyl radicals are formed by abstraction of the formyl hydrogen from aldehydes. As indicated in Table 12.5, the resulting acyl radicals are somewhat stabilized. The chain process results in formation of a ketone by addition of the aldehyde to an alkene.

$$In\cdot + RCH(=O) \longrightarrow RC(=O)\cdot + InH$$

$$RC(=O)\cdot + CH_2{=}CHR' \longrightarrow RC(=O)CH_2\dot{C}HR'$$

$$RC(=O)CH_2\dot{C}HR' + RCH(=O) \longrightarrow RC(=O)CH_2CH_2R' + RC(=O)\cdot$$

Some specific examples are included in Scheme 12.5.

The chain addition of formamide to alkenes is a closely related reaction. It results in the formation of primary amides.[97] The reaction is carried out with irradiation in acetone. The photoexcited acetone initiates the chain reaction by abstracting hydrogen from formamide.

$$CH_3COCH_3 \xrightarrow{h\nu} CH_3COCH_3^*$$

$$CH_3COCH_3^* + HCONH_2 \longrightarrow \cdot CONH_2 + (CH_3)_2\dot{C}OH$$

$$\cdot CONH_2 + RCH{=}CH_2 \longrightarrow R\underset{\cdot}{C}HCH_2CONH_2$$

$$R\underset{\cdot}{C}HCH_2CONH_2 + HCONH_2 \longrightarrow RCH_2CH_2CONH_2 + \cdot CONH_2$$

12.4.4. Addition of Thiols and Thiocarboxylic Acids

The addition of S—H compounds to alkenes by a radical chain mechanism is a quite general and efficient reaction.[98] The mechanism is analogous to that for hydrogen bromide addition. The energetics of both the hydrogen abstraction and addition steps are favorable. Entries 15 and 16 in Scheme 12.5 are examples.

$$In\cdot + R'SH \longrightarrow In{-}H + R'S\cdot$$

$$R'S\cdot + CH_2{=}CHR \longrightarrow R'SCH_2\underset{\cdot}{C}HR$$

$$R'SCH_2\underset{\cdot}{C}HR + R'SH \longrightarrow R'SCH_2CH_2R + R'S\cdot$$

The preferred stereochemistry of addition to cyclic alkenes is *anti*.[99] The additions are not as highly stereoselective as hydrogen bromide additions, however.

12.5. Intramolecular Free-Radical Reactions

Both substitution and addition reactions can occur intramolecularly. Intramolecular substitution reactions that involve hydrogen abstraction have some important synthetic applications, since they permit functionalization of carbon atoms relatively remote from the initial reaction site. The preference for a six-membered cyclic transition state in the hydrogen abstraction step imparts considerable position selectivity to the process.

[Scheme: C–C–C–C–C–C chain with $\dot{X}$ abstracting H via six-membered transition state → carbon radical C· with X–H → (Y') → C–Y product with X–H]

97. D. Elad and J. Rokach, *J. Org. Chem.* **29**, 1855 (1964).
98. K. Griesbaum, *Angew Chem. Int. Ed. Engl.* **9**, 273 (1970).
99. N. A. LeBel, R. F. Czaja, and A. DeBoer, *J. Org. Chem.* **34**, 3112 (1969); P. D. Readio and P. S. Skell, *K. Org. Chem.* **31**, 759 (1966); F. G. Bordwell, P. S. Landis, and G. S. Whitney, *J. Org. Chem.* **30**, 3764 (1965); E. S. Huyser, H. Benson, and H. J. Sinnige, *J. Org. Chem.* **32**, 622 (1967).

There are several reaction sequences which involve such intramolecular hydrogen abstraction steps. One example is the photolytically initiated decomposition of N-haloamines in acidic solution, which is known as the *Hofmann–Loeffler reaction.*[100] The reaction leads initially to δ-haloamines, but these are usually converted to pyrrolidines by intramolecular nucleophilic substitution.

$$RCH_2CH_2CH_2CH_2\overset{+}{N}(Cl)HCH_3 \xrightarrow{h\nu} RCH_2CH_2CH_2CH_2\overset{+}{\dot{N}}HCH_3 + Cl\cdot$$

$$RCH_2CH_2CH_2CH_2\overset{+}{\dot{N}}HCH_3 \longrightarrow R\dot{C}HCH_2CH_2CH_2\overset{+}{N}H_2CH_3$$

$$R\dot{C}HCH_2CH_2CH_2\overset{+}{N}H_2CH_3 + RCH_2CH_2CH_2CH_2\overset{+}{N}(Cl)HCH_3 \longrightarrow RCH(Cl)CH_2CH_2CH_2\overset{+}{N}H_2CH_3 + RCH_2CH_2CH_2CH_2\overset{+}{\dot{N}}HCH_3$$

$$RCH(Cl)CH_2CH_2CH_2\overset{+}{N}H_2CH_3 \xrightarrow{NaOH} \text{2-R-1-methylpyrrolidine } (N\text{-}CH_3)$$

There are related procedures involving N-haloamides which lead to lactones via iminolactone intermediates.[101]

$$RCH_2(CH_2)_2\overset{O}{\overset{\|}{C}}NH_2 \xrightarrow[CH_2Cl_2]{(CH_3)_3COCl, I_2} RCH_2(CH_2)_2\overset{O}{\overset{\|}{C}}NHI \xrightarrow{h\nu} RCH(I)(CH_2)_2\overset{O}{\overset{\|}{C}}NH_2$$

$$RCH(I)(CH_2)_2\overset{O}{\overset{\|}{C}}NH_2 \longrightarrow \text{(5-R-iminolactone, } =\overset{+}{N}H_2\ I^-) \xrightarrow{H_2O} \text{(5-R-}\gamma\text{-lactone)}$$

A significant point about the final step in these reaction sequences is that the intramolecular nucleophilic attack by the amide group involves the oxygen, not the nitrogen, as the site of nucleophilic reactivity. Amides are generally more nucleophilic at oxygen than at nitrogen. This reflects both the fact that the oxygen is a site of relatively high electron density and also that the O-alkylated product retains a conjugated four-π-electron system.

A procedure for intramolecular functionalization of alcohols has been developed in studies carried out primarily with steroid derivatives.[102] The alcohol is converted to a nitrite ester by reaction with nitrosyl chloride. Photolysis effects

100. M. E. Wolff, *Chem. Rev.* **63**, 55 (1963).
101. D. H. R. Barton, A. L. J. Beckwith, and A. Goosen, *J. Chem. Soc.*, 181 (1965); R. S. Neale, N. L. Marcus, and R. G. Schepers, *J. Am. Chem. Soc.* **88**, 3051 (1966).
102. R. H. Hesse, *Adv. Free Radicals Chem.* **3**, 83 (1969); D. H. R. Barton, J. M. Beaton, L. E. Geller, and M. M. Pechet, *J. Am. Chem. Soc.* **83**, 4076 (1961).

introduction of a nitroso function at an adjacent unsubstituted carbon atom. The nitrosoalkyl group is equivalent to an aldehyde or ketone group, since alkyl nitroso compounds rearrange to oximes. This reaction involves a hydrogen abstraction by photolytically generated alkoxy radicals but is not believed to be a chain process, since the quantum yield is less than unity.[103] Labeling studies using nitrogen-15 have established that the NO group is transferred intermolecularly, rather than in a cage process.[104]

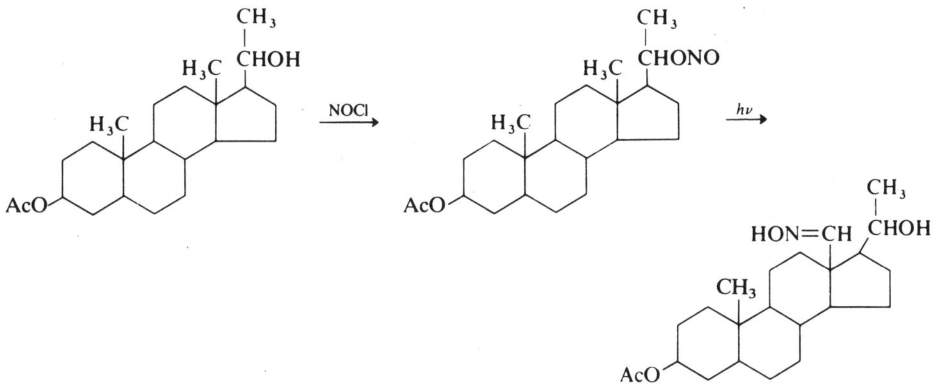

Intramolecular hydrogen abstraction reactions have also been observed in medium-sized rings. The reaction of cyclooctene with carbon tetrachloride and bromotrichloromethane is an interesting case. As shown in the equation below, bromotrichloromethane adds in a completely normal manner, but carbon tetrachloride gives some 4-chloro-1-trichloromethylcyclooctane as well as the expected product.[105]

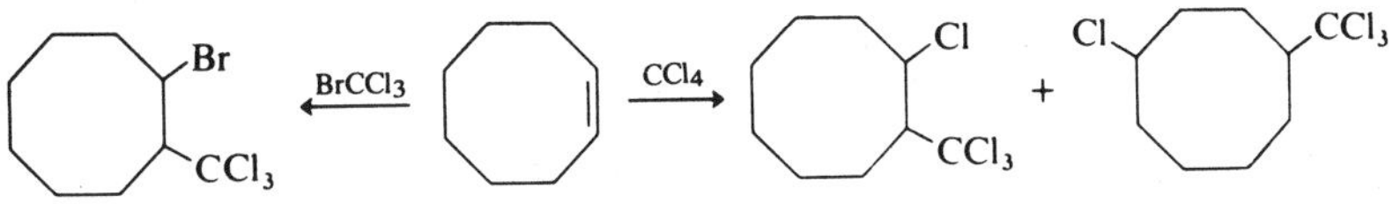

In the case of carbon tetrachloride, the radical intermediate is undergoing two competing reactions; intramolecular hydrogen abstraction is competitive with abstraction of a chlorine atom from carbon tetrachloride.

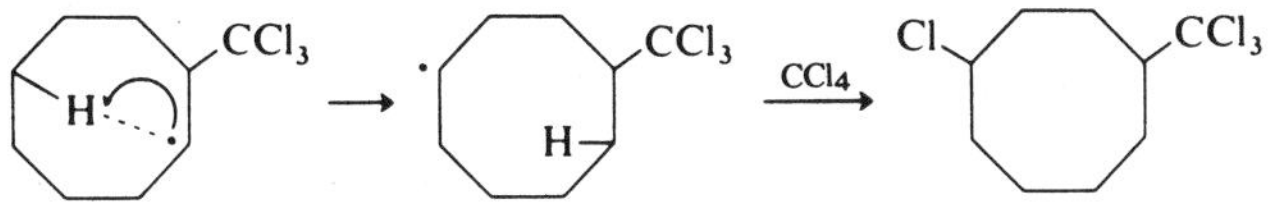

103. P. Kabasakalian and E. R. Townley, *J. Am. Chem. Soc.* **84**, 2711 (1962).
104. M. Akhtar and M. M. Pechet, *J. Am. Chem. Soc.* **86**, 265 (1964).
105. J. G. Traynham, T. M. Couvillon, and N. S. Bhacca, *J. Org. Chem.* **32**, 529 (1967); J. G. Traynham and T. M. Couvillon, *J. Am. Chem. Soc.* **87**, 5806 (1965); J. G. Traynham and T. M. Couvillon, *J. Am. Chem. Soc.* **89**, 3205 (1967).

No product derived from the transannular hydrogen abstraction is observed in the addition of bromotrichloromethane because bromine atom abstraction is sufficiently rapid to prevent effective competition by the intramolecular hydrogen abstraction.

Intramolecular addition reactions are quite common when radicals are generated in molecules with unsaturation in a sterically favorable position.[106] Cyclization reactions based on intramolecular addition of radical intermediates have become synthetically useful, and several specific cases will be considered in Part B, Section 10.3.4.

12.6. Rearrangement and Fragmentation Reactions of Free Radicals

12.6.1. Rearrangement Reactions

Compared with rearrangement of cationic species, rearrangements of radical intermediates are quite rare. However, for specific structural types, free-radical migrations can be expected. The groups that are frequently capable of migration in free-radical intermediates include aryl, vinyl, acyl, and other unsaturated substituents. Migration of saturated groups is very unusual, and there is a simple structural reason for this. In cationic intermediates, migration occurs through a bridged transition state (or intermediate) that involves a three-center two-electron bond:

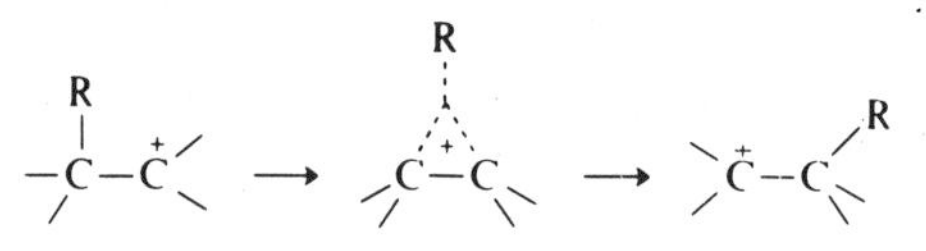

In a free radical, there is a third electron in the system. It cannot occypy the same orbital as the other two electrons. It must instead be in an antibonding level. As a result, the transition state for migration is less favorable than for the corresponding carbocation.

The more facile migration of aryl and other unsaturated groups is the result of formation of bridged intermediates by an addition process. In the case of aryl migration, the intermediate is a cyclohexadienyl radical.

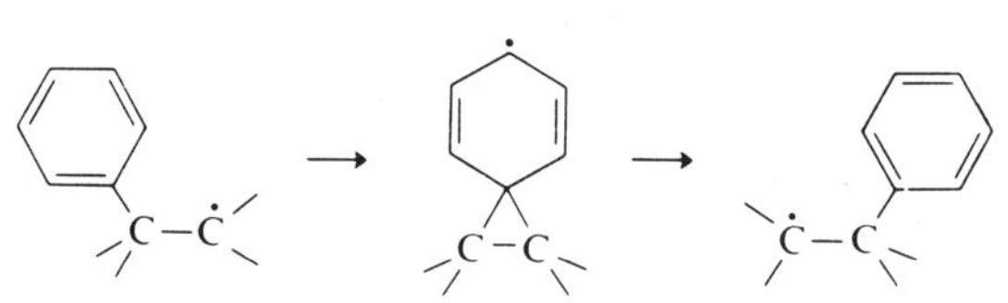

Aryl migrations are promoted by steric crowding in the initial radical. This trend is illustrated by data from the thermal decomposition of a series of diacyl peroxides.

106. A. L. J. Beckwith, *Tetrahedron* **37**, 3073 (1981).

The amount of product derived from rearrangement increases with the size and number of substituents.[107]

$$(PhCR_1R_2CH_2CO_2)_2 \xrightarrow{\Delta} R_1R_2\dot{C}CH_2Ph$$

R_1	R_2	Rearrangement
CH_3	H	39%
Ph	H	63%
Ph	Ph	100%

It has been possible to measure absolute rates and activation energies for migration in a series of 2-substituted 2,2-dimethylethyl radicals. The rates at 25°C and the E_a values for several substituents are indicated below.[108]

$$X{-}C(CH_3)_2{-}CH_2\cdot \rightarrow \cdot C(CH_3)_2{-}CH_2{-}X$$

X	k (s^{-1})	E_a (kcal/mol)
Ph	7.6×10^2	11.8
$CH_2{=}CH$	10^7	5.7
$(CH_3)_3CC{\equiv}C$	9.3	12.8
$(CH_3)_3CC{=}O$	1.7×10^5	7.8
$C{\equiv}N$	9.0	16.4

The rapid rearrangement of vinyl and acyl substituents can be explained as proceeding through intermediate cyclopropyl species.

$$(CH_3)_2C(HC{=}CH_2){-}CH_2\cdot \longrightarrow \text{(cyclopropane with } CH_2\cdot \text{ and } CH_3, CH_3) \longrightarrow (CH_3)_2\dot{C}CH_2CH{=}CH_2$$

$$(CH_3)_2C(CH_3C{=}O){-}CH_2\cdot \longrightarrow \text{(cyclopropane with } O\cdot, CH_3 \text{ and } CH_3, CH_3) \longrightarrow (CH_3)_2\dot{C}CH_2\overset{O}{\overset{\|}{C}}CH_3$$

The relatively slower rearrangement of alkynyl and cyano substituents can be attributed to the reduced stability of the intermediate derived from cyclization of the triply bonded substituents.

$$(CH_3)_2C(C{\equiv}CH){-}CH_2\cdot \longrightarrow \text{(cyclopropane with } {=}CH\cdot \text{ and } CH_3, CH_3) \qquad (CH_3)_2C(C{\equiv}N){-}CH_2\cdot \longrightarrow \text{(cyclopropane with } {=}N\cdot \text{ and } CH_3, CH_3)$$

107. W. Rickatson and T. S. Stevens, *J. Chem. Soc.*, 3960 (1963).
108. D. A. Lindsay, J. Lusztyk, and K. U. Ingold, *J. Am. Chem. Soc.* **106**, 7087 (1984).

Scheme 12.6. Free-Radical Rearrangements

1[a] $PhC(CH_3)_2CH_2CH{=}O \xrightarrow{ROOR} (CH_3)_2CHCH_2Ph + PhC(CH_3)_3$
35% 35%

2[b] 1-phenyl-1-($CH_2CH{=}O$)cyclohexane $\xrightarrow{ROOR}$ 1-phenyl-1-(CH_3)cyclohexane + cyclohexyl–CH_2Ph
5% 42%

3[c] $(CH_3)_3CBr + (CH_3)_3COCl \xrightarrow{h\nu} (CH_3)_2CClCH_2Br$ 92%

4[d] cyclopentenyl–$CH_2COOC(CH_3)_3$ (with C=O) $\xrightarrow{140°C}$ cyclohexene + cyclopentenyl–CH_3
47% 12%

5[e] $(Ph)_3CCH_2Ph \xrightarrow[(PhCO_2)_2]{N\text{-bromosuccinimide}} Ph_2C{=}CPh_2$

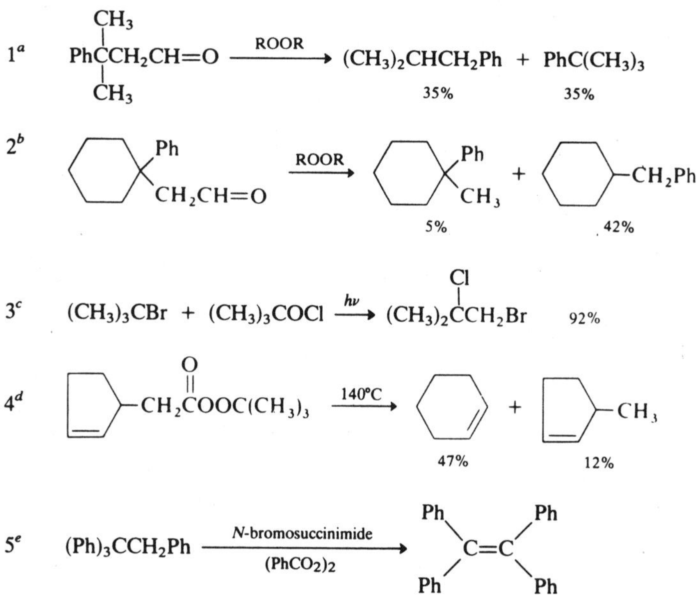

a. S. Winstein and F. H. Seubold, Jr., *J. Am. Chem. Soc.* **69**, 2916 (1947).
b. J. W. Wilt and H. P. Hogan, *J. Org. Chem.* **24**, 441 (1959).
c. P. S. Skell, R. G. Allen, and N. D. Gilmour, *J. Am. Chem. Soc.* **83**, 504 (1961).
d. L. H. Slaugh, *J. Am. Chem. Soc.* **87**, 1522 (1965).
e. H. Meislich, J. Costanza, and J. Strelitz, *J. Org. Chem.* **33**, 3221 (1968).

Scheme 12.6 gives some examples of reactions in which free-radical rearrangements have been observed.

12.6.2. Fragmentation Reactions

In earlier sections, we have already brought forward several examples of radical fragmentation reactions, although the terminology was not explicitly used. The facile decarboxylation of acyloxy radicals was mentioned earlier.

$$RCOOC(CH_3)_3 \longrightarrow RCO\cdot \longrightarrow R\cdot + CO_2$$

For the acetoxy radical, the E_a for decarboxylation is about 6.5 kcal/mol and the rate is about $10^9\ s^{-1}$ at 60°C and $10^6\ s^{-1}$ at −80°C.[109] Thus, only very rapid reactions can compete with decarboxylation. As would be expected because of the lesser stability of aryl radicals, the rates of decarboxylation of aroyloxy radicals are less. The rate for *p*-methoxylbenzoyloxy radical has been determined to be about 3 ×

109. J. Chateauneuf, J. Lusztyk, and K. U. Ingold, *J. Am. Chem. Soc.* **109**, 897 (1987).

$10^5\ s^{-1}$ near room temperature.[109] Hydrogen donation by very reactive hydrogen atom donors such as triethylsilane can compete with decarboxylation at moderate temperatures.

Acyl radicals can fragment with loss of carbon monoxide. Decarbonylation is slower than decarboxylation, but the rate also depends on the stability of the radical that is formed.[110] For example, when reaction of isobutyraldehyde with carbon tetrachloride is initiated by *t*-butyl peroxide, both isopropyl chloride and isobutyryl chloride are formed. Decarbonylation is competitive with the chlorine atom abstraction.

$$(CH_3)_2CHCH{=}O + In\cdot \longrightarrow (CH_3)_2CHC({=}O)\cdot$$

$$(CH_3)_2CHC({=}O)\cdot \xrightarrow{CCl_4} (CH_3)_2CHC({=}O)Cl + \cdot CCl_3$$

$$(CH_3)_2CHC({=}O)\cdot \xrightarrow{-CO} (CH_3)_2CH\cdot + CO \xrightarrow{CCl_4} (CH_3)_2CHCl + \cdot CCl_3$$

Another common fragmentation reaction is the cleavage of an alkoxy radical to an alkyl radical and a carbonyl compound.[111]

$$(CH_3)_3CO\cdot \longrightarrow CH_3\cdot + (CH_3)_2C{=}O$$

This type of fragmentation is involved in the chain decomposition of alkyl hypochlorites.[112]

$$RCH_2OCl \longrightarrow RCH_2O\cdot + Cl\cdot$$

$$RCH_2O\cdot \longrightarrow R\cdot + CH_2{=}O$$

$$R\cdot + RCH_2OCl \longrightarrow RCl + RCH_2O\cdot$$

In this reaction, too, the stability of the radical being eliminated is the major factor in determining the rate of fragmentation.

In cyclic systems, the fragmentation of alkoxy radicals can be a reversible process. The 10-decalyloxy radical can undergo fragmentation of either the 1-9 or the 9-10 bond:

B ⇌ A ⇌ C

110. D. E. Applequist and L. Kaplan, *J. Am. Chem. Soc.* **87**, 2194 (1965); W. H. Urry, D. J. Tecker, and H. D. Hartzler, *J. Org. Chem.* **29**, 1663 (1964); H. Fischer and H. Paul, *Acc. Chem. Res.* **20**, 200 (1987).
111. P. Gray and A. Williams, *Chem. Rev.* **59**, 239 (1959).
112. F. D. Greene, M. L. Savitz, F. D. Osterholtz, H. H. Lau, W. N. Smith, and P. M. Zanet, *J. Org. Chem.* **28**, 55 (1963); C. Walling and A. Padwa, *J. Am. Chem. Soc.* **85**, 1593, 1597 (1963).

By using various trapping reagents, it has been deduced that the transannular fragmentation is rapidly reversible. The cyclization of the fragmented radical **C** is less favorable, and it is trapped at rates which exceed that for recyclization under most circumstances.[113]

Radicals derived from ethers and acetals by hydrogen abstraction are subject to fragmentation, with formation of a ketone or an ester, respectively.

$$R_2\dot{C}OR' \longrightarrow R_2C{=}O + R'\cdot$$

$$R\dot{C}(OR')_2 \longrightarrow R\overset{\overset{\displaystyle O}{\|}}{C}OR' + R'\cdot$$

These fragmentations are sufficiently slow, however, that the initial radicals can undergo reactions, addition to alkenes being one example, at rates that are competitive with that of fragmentation.

A special case of fragmentation is that of 1,4-diradicals where fragmentation can lead to two stable molecules. In the case of 1,4-diradicals without functional group stabilization, reclosure to cyclobutanes is normally competitive with fragmentation to two molecules of alkene.

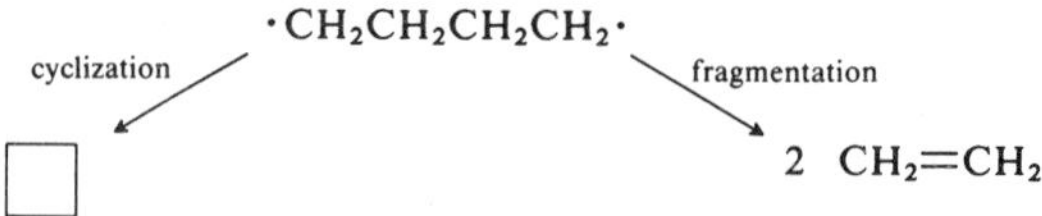

Theoretical calculations on the simplest such radical, 1,4-butadiyl, indicate that both processes are exothermic and can proceed with little if any barrier.[114]

A study of the biradicals **A**, **B**, and **C**, which were generated from the corresponding azo compounds, found that the lifetimes decreased in the order **A** > **B** > **C**. The lifetime of **A** is on the order of 1×10^{-6} s.[115]

A **B** **C**

The major factor identified in controlling the lifetimes of these radicals is the orientation of the singly occupied orbitals with respect to one another. In **A**, they are essentially parallel. This is a poor orientation for interaction. Diradical **B** is more flexible and rapidly reacts to give the coupling and fragmentation products. The geometry of the bicyclic ring system in radical C directs the half-filled orbitals

113. A. L. J. Beckwith, P. Kazlauskas, and M. R. Syner-Lyons, *J. Org. Chem.* **48**, 4718 (1983).
114. C. Doubleday, Jr., R. N. Camp, H. F. King, J. W. McIver, Jr., D. Mullaly, and M. Page, *J. Am. Chem. Soc.* **106**, 447 (1984).
115. W. Adam, K. Hanneman, and R. M. Wilson, *J. Am. Chem. Soc.* **106**, 7646 (1984); W. Adam, K. Hanneman, and R. M. Wilson, *Angew. Chem. Int. Ed. Engl.* **24**, 1071 (1985).

towards one another and its lifetime is less than 1×10^{-10} sec.

$$\cdot C_6H_{12}\cdot \longrightarrow \text{bicyclo[2.2.0]hexane} + CH_2{=}CHCH_2CH_2CH{=}CH_2$$

12.7. Electron Transfer Reactions Involving Transition Metal Ions

Most of the free-radical mechanisms discussed thus far have involved some combination of homolytic bond dissociation, atom abstraction, and addition steps. In this section, we will discuss reactions that include discrete electron transfer steps. Addition to or removal of one electron from a diamagnetic organic molecule generates a radical. Many organic reactions that involve electron transfer steps are mediated by transition metal ions. Many transition metal ions have two or more relatively stable oxidation states differing by one electron. Transition metal ions therefore frequently participate in electron transfer processes.

The decomposition of peroxy esters has been shown to be strongly catalyzed by Cu(I). The process is believed to involve oxidation of the copper to Cu(II):

$$RC(=O)OOR + Cu(I) \longrightarrow RC(=O)O^- + RO\cdot + Cu(II)$$

An example of this reaction is the reaction of cyclohexene with *t*-butyl perbenzoate, which is mediated by Cu(I).[116] The initial step is the reductive cleavage of the peroxy ester. The *t*-butoxy radical then abstracts hydrogen from cyclohexene to give an allylic radical. The radical is oxidized by Cu(II) to the carbocation, which captures benzoate ion.

$$(CH_3)_3C{-}O\cdot + \text{cyclohexene} \longrightarrow (CH_3)_3COH + \text{cyclohexenyl radical} \xrightarrow{Cu(II)} \text{cyclohexenyl cation} \xrightarrow{PhCO_2^-} \text{3-}(O_2CPh)\text{cyclohexene}$$

The reactions of copper salts with diacyl peroxides have been investigated quite thoroughly, and the mechanistic studies indicate that both radicals and carbocations are involved as intermediates. The radicals are oxidized to carbocations by Cu(II), and the final products can be recognized as having arisen from carbocations because characteristic patterns of substitution, elimination, and rearrangement can be

116. K. Pedersen, P. Jakobsen, and S.-O. Lawesson, *Org. Synth.* **V**, 71 (1973).

discerned.[117]

$$[(CH_3)_3CCH_2CO_2]_2 + Cu(I) \longrightarrow (CH_3)_3CCH_2CO_2^- + (CH_3)_3CCH_2\cdot + CO_2 + Cu(II)$$

$$(CH_3)_3CCH_2\cdot + Cu(II) \longrightarrow (CH_3)_3CCH_2^+ + Cu(I)$$

$$(CH_3)_3CCH_2^+ \longrightarrow (CH_3)_2\overset{+}{C}CH_2CH_3 \longrightarrow (CH_3)_2C{=}CHCH_3$$

$$\downarrow$$

$$(CH_3)_2C(O_2CCH_3)CH_2CH_3$$

When the radicals have β hydrogens, alkenes are formed by a process in which carbocations are probably bypassed. Instead, the oxidation and elimination of a proton probably occur in a single step through an alkylcopper species. The oxidation state of copper in such an intermediate is Cu(III).

$$R_2CHCH_2\cdot + Cu(II) \rightarrow R_2CHCH_2Cu(III) \rightarrow R_2C{=}CH_2 + Cu(I) + H^+$$

When halide ions or anions such as thiocyanate or azide are present, these anions are incorporated into the organic radical generated by decomposition of the peroxide. This anion transfer presumably occurs in the same step as the redox interaction with Cu(II), and such reactions have been called *ligand transfer reactions*.[118]

$$(RCO_2)_2 + Cu(I)X \rightarrow RCO_2^- + R\cdot + CO_2 + Cu(II)X$$

$$R\cdot + Cu(II)X \rightarrow R{-}Cu(III)X$$

$$R{-}Cu(III)X \rightarrow R{-}X + Cu(I)$$

These reactions do not appear to involve free carbocations, because they proceed effectively in nucleophilic solvents that would successfully compete with halide or similar anions for free carbocations. Also, rearrangements are unusual under these conditions, although they have been observed in special cases.

A unified concept of these reactions is provided by the proposal that alkylcopper intermediates are involved in each of these reactions at the stage of the oxidation of the radical.[119]

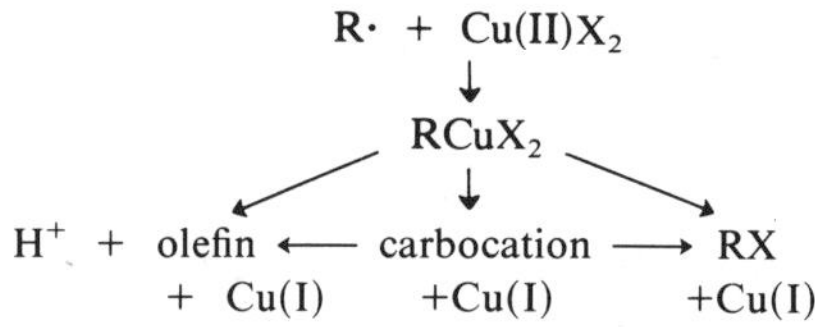

117. J. K. Kochi, *J. Am. Chem. Soc.* **85**, 1958 (1963); J. K. Kochi and A. Bemis, *J. Am. Chem. Soc.* **90**, 4038 (1968).
118. C. L. Jenkins and J. K. Kochi, *J. Am. Chem. Soc.* **94**, 856 (1972).
119. C. L. Jenkins and J. K. Kochi, *J. Am. Chem. Soc.* **94**, 843 (1972).

The organocopper intermediate has three possible fates, and the preferred path is determined by the structure of the group R and the identity of the copper ligand X. If R is potentially a very stable carbocation, the intermediate breaks down to generate the carbocation and the products are derived from it. When X is halide or a pseudohalide such as ^-CN, ^-SCN, or $^-N_3$, the preferred pathway is ligand transfer leading to the alkyl halide or pseudohalide. If the R group is not capable of sustaining formation of a carbocation and no easily transferred anion is present, the organocopper intermediate is converted primarily to alkene by elimination of a proton.

One-electron oxidation of carboxylate ions generates acyloxy radicals, which can undergo the usual decarboxylation. Such electron transfer reactions can be effected by strong one-electron oxidants, such as Mn(III), Ag(II), Ce(IV), and Ph(IV).[120] These metal ions are also capable of oxidizing the radical intermediate, so the products are those expected from carbocations. The oxidative decarboxylation by Pb(IV) in the presence of halide salts leads to alkyl halides.[121] For example, oxidation of pentanoic acid with lead tetraacetate in the presence of lithium chloride gives 1-chlorobutane in 71% yield.

$$CH_3(CH_2)_3CO_2H \xrightarrow[\text{LiCl}]{\text{Pb(OAc)}_4} CH_3(CH_2)_3Cl + CO_2$$

A chain mechanism is proposed. The first step is oxidation of a carboxylate ion coordinated to Pb(IV), with formation of alkyl radical, carbon dioxide, and Pb(III). The alkyl radical then abstracts halogen from a Pb(IV) complex, generating a Pb(III) species that decomposes to Pb(II) with release of an alkyl radical. This alkyl radical can continue the chain process. The step involving abstraction of halide from a complex with a change in metal ion oxidation state is a ligand transfer type reaction.

$$RCO_2-\overset{X}{\underset{|}{\overset{|}{Pb}}}(IV)- \longrightarrow R\cdot + \overset{X}{\underset{|}{\overset{|}{Pb}}}(III)- + CO_2$$

$$R\cdot + X-\underset{|}{\overset{|}{Pb}}(IV)- \longrightarrow R-X + \underset{|}{\overset{|}{Pb}}(III)-$$

$$\underset{|}{\overset{|}{Pb}}(III)- + RCO_2^- \longrightarrow R\cdot + CO_2 + Pb(II)$$

In the absence of halide salts, the principal products may be alkanes, alkenes, or acetate esters.

A classic reaction involving electron transfer and decarboxylation of acyloxy radicals is the Kolbe electrolysis, in which an electron is abstracted from a carboxy-

120. J. M. Anderson and J. K. Kochi, *J. Am. Chem. Soc.* **92**, 2450 (1970); J. M. Anderson and J. K. Kochi, *J. Am. Chem. Soc.* **92**, 1651 (1970); R. A. Sheldon and J. K. Kochi, *J. Am. Chem. Soc.* **90**, 6688 (1968); W. A. Mosher and C. L. Kehr, *J. Am. Chem. Soc.* **75**, 3172 (1953); J. K. Kochi, *J. Am. Chem. Soc.* **87**, 1811 (1965).
121. J. K. Kochi, *J. Org. Chem.* **30**, 3265 (1965); R. A. Sheldon and J. K. Kochi, *Org. React.* **19**, 279 (1972).

late ion at the anode of an electrolysis system. This reaction gives products derived from coupling of the decarboxylated radicals:

$$RCO_2^- \longrightarrow RCO_2\cdot + e^-$$
$$RCO_2\cdot \longrightarrow R\cdot + CO_2$$
$$2\,R\cdot \longrightarrow RR$$

Other transformations of the radicals are also possible. For example, the 5-hexenyl radical partially cyclizes in competition with coupling[122]:

$$H_2C{=}CH(CH_2)_4CO_2^- \xrightarrow{\text{electrolysis}} \text{cyclopentyl}{-}(CH_2)_2{-}\text{cyclopentyl} + \text{cyclopentyl}{-}(CH_2)_5CH{=}CH_2 + CH_2{=}CH(CH_2)_8CH{=}CH_2$$

Carbocations can also be generated during the electrolysis, and they give rise to alcohols and alkenes. The carbocations are presumably formed by an oxidation of the radical at the electrode before it reacts or diffuses into solution. For example, an investigation of the electrolysis of phenylacetic acid in methanol has led to the identification of benzyl methyl ether (30%), toluene (1%), benzaldehyde dimethylacetal (1%), methyl phenylacetate (6%), and benzyl alcohol (5%), in addition to the coupling product bibenzyl (26%).[123]

12.8. $S_{RN}1$ Substitution Processes

Electron transfer processes are also crucially involved in a group of reactions which are designated by the mechanistic description $S_{RN}1$. This refers to a nucleophilic substitution via a radical intermediate which proceeds by unimolecular decomposition of a radical anion derived from the substrate. There are two families of such reactions that have been developed to a stage of solid mechanistic understanding and also synthetic utility. The common mechanistic pattern involves electron transfer to the substrate, generating a radical anion which then expels the leaving group. This becomes a chain process if the radical generated by expulsion of the leaving group then reacts with the nucleophile to give a radical anion capable of sustaining a chain reaction.

initiation $\quad R{-}X + e^- \rightarrow R{-}X^{\dot{-}} \rightarrow R\cdot + X^-$

propagation $\quad R\cdot + Nu^- \rightarrow R{-}Nu^{\dot{-}}$

$$R{-}Nu^{\dot{-}} + R{-}X \rightarrow R{-}Nu + R{-}X^{\dot{-}}$$
$$R{-}X^{\dot{-}} \rightarrow R\cdot + X^-$$

A mechanism of this type permits substitution of certain aromatic and aliphatic nitro compounds by a variety of nucleophiles. These reactions were discovered as the result of efforts to explain the mechanistic basis for high-yield carbon alkylation

122. R. F. Garwood, C. J. Scott, and B. C. L. Weedon, *J. Chem. Soc., Chem. Commun.*, 14 (1965).
123. S. D. Ross and M. Finkelstein, *J. Org. Chem.* **34**, 2923 (1969).

of the 2-nitropropane anion by *p*-nitrobenzyl chloride. The corresponding bromide and iodide and benzyl halides that do not contain a nitro substituent give mainly the unstable oxygen alkylation product with this ambident anion.[124]

$$O_2N-C_6H_4-CH_2Cl + (CH_3)_2\bar{C}NO_2 \longrightarrow O_2N-C_6H_4-CH_2C(CH_3)_2(NO_2)$$

$$C_6H_5-CH_2Cl + (CH_3)_2\bar{C}NO_2 \longrightarrow + \; C_6H_5-CH_2O\overset{+}{N}(O^-)=C(CH_3)_2$$

The mixture of carbon and oxygen alkylation is what would be expected for an S_N2 substitution process. The high preference for carbon alkylation suggested that a new mechanism operates with *p*-nitrobenzyl chloride. This conclusion was further strengthened by the fact that the chloride is more reactive than would be predicted on the basis of the usual $I > Br > Cl$ reactivity trend for leaving groups in S_N2 reactions. The involvement of a free-radical process was indicated by EPR studies and by the observation that typical free-radical inhibitors decrease the rate of the carbon alkylation process. The mechanism proposed is a free-radical chain process initiated by electron transfer from the nitronate anion to the nitroaromatic compound.[125] This process is the principal reaction only for the chloride, because with the better leaving groups bromide and iodide, a direct S_N2 process is more rapid.

$$O_2N-C_6H_4-CH_2Cl + (CH_3)_2\bar{C}NO_2 \longrightarrow {}^{\dot{-}}O_2N-C_6H_4-CH_2Cl + (CH_3)_2\dot{C}NO_2$$

$${}^{\dot{-}}O_2N-C_6H_4-CH_2Cl \longrightarrow O_2N-C_6H_4-CH_2\cdot + Cl^-$$

$$O_2N-C_6H_4-CH_2\cdot + (CH_3)_2\bar{C}NO_2 \longrightarrow O_2N-C_6H_4-CH_2C(CH_3)_2-NO_2^{\dot{-}}$$

$$O_2N-C_6H_4-CH_2C(CH_3)_2-NO_2^{\dot{-}} + O_2N-C_6H_4-CH_2Cl \longrightarrow O_2N-C_6H_4-CH_2C(CH_3)_2-NO_2 + {}^{\dot{-}}O_2N-C_6H_4-CH_2Cl$$

The absolute rate of dissociation of the radical anion of *p*-nitrobenzyl chloride has been measured as $4 \times 10^3\ s^{-1}$. The *m*-nitro isomer does not undergo a corresponding reaction.[126]

The synthetic value of the reaction has been developed from this mechanistic understanding. The reaction has been shown to be capable of providing highly substituted carbon skeletons that would be inaccesible by normal S_N2 processes.

124. N. Kornblum, *Angew Chem. Int. Ed. Engl.* **14**, 734 (1975); N. Korblum, in *The Chemistry of Amino, Nitroso and Nitro Compounds and Their Derivatives*, S. Patai (ed.), Wiley Interscience, New York, 1982, Chapter 10.
125. N. Kornblum, R. E. Michel, and R. C. Kerber, *J. Am. Chem. Soc.* **88**, 5662 (1966); G. A. Russell and W. C. Danen, *J. Am. Chem. Soc.* **88**, 5663 (1966).
126. R. K. Norris, S. D. Baker, and P. Neta, *J. Am. Chem. Soc.* **106**, 3140 (1984).

Scheme 12.7. Carbon Alkylation via Nitroalkane Radical Anions Generated by Electron Transfer

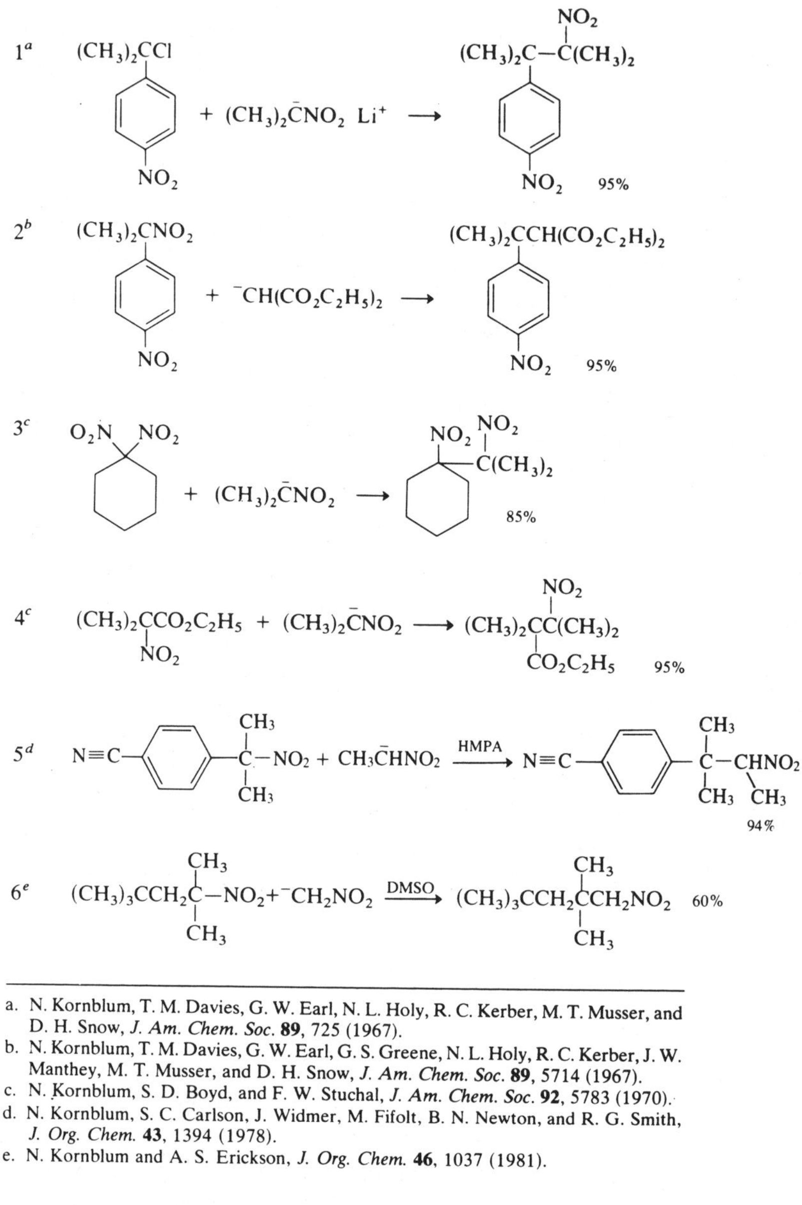

1[a] $p\text{-}O_2NC_6H_4C(CH_3)_2Cl + (CH_3)_2\bar{C}NO_2\ Li^+ \longrightarrow p\text{-}O_2NC_6H_4C(CH_3)_2C(CH_3)_2NO_2$ 95%

2[b] $p\text{-}O_2NC_6H_4C(CH_3)_2NO_2 + {}^-CH(CO_2C_2H_5)_2 \longrightarrow p\text{-}O_2NC_6H_4C(CH_3)_2CH(CO_2C_2H_5)_2$ 95%

3[c] 1,1-dinitrocyclohexane (O_2N, NO_2) $+ (CH_3)_2\bar{C}NO_2 \longrightarrow$ 1-nitro-1-[$C(CH_3)_2NO_2$]cyclohexane 85%

4[c] $(CH_3)_2C(NO_2)CO_2C_2H_5 + (CH_3)_2\bar{C}NO_2 \longrightarrow (CH_3)_2C(CO_2C_2H_5)C(CH_3)_2NO_2$ 95%

5[d] $N{\equiv}C\text{-}C_6H_4\text{-}C(CH_3)_2\text{-}NO_2 + CH_3\bar{C}HNO_2 \xrightarrow{HMPA} N{\equiv}C\text{-}C_6H_4\text{-}C(CH_3)_2\text{-}CH(CH_3)NO_2$ 94%

6[e] $(CH_3)_3CCH_2C(CH_3)_2\text{-}NO_2 + {}^-CH_2NO_2 \xrightarrow{DMSO} (CH_3)_3CCH_2C(CH_3)_2CH_2NO_2$ 60%

a. N. Kornblum, T. M. Davies, G. W. Earl, N. L. Holy, R. C. Kerber, M. T. Musser, and D. H. Snow, *J. Am. Chem. Soc.* **89**, 725 (1967).
b. N. Kornblum, T. M. Davies, G. W. Earl, G. S. Greene, N. L. Holy, R. C. Kerber, J. W. Manthey, M. T. Musser, and D. H. Snow, *J. Am. Chem. Soc.* **89**, 5714 (1967).
c. N. Kornblum, S. D. Boyd, and F. W. Stuchal, *J. Am. Chem. Soc.* **92**, 5783 (1970).
d. N. Kornblum, S. C. Carlson, J. Widmer, M. Fifolt, B. N. Newton, and R. G. Smith, *J. Org. Chem.* **43**, 1394 (1978).
e. N. Kornblum and A. S. Erickson, *J. Org. Chem.* **46**, 1037 (1981).

For example, tertiary *p*-nitrocumyl halides can act as alkylating agents in high yield. The nucleophile need not be a nitroalkane anion, but can be such anions as thiolate, phenolate, or a carbanion such as those derived from malonate esters.[127] The same

127. N. Kornblum, T. M. Davies, G. W. Earl, N. L. Holy, R. C. Kerber, M. T. Musser, and D. H. Snow, *J. Am. Chem. Soc.* **89**, 725 (1967); N. Kornblum, L. Cheng, T. M. Davies, G. W. Earl, N. L. Holy, R. C. Kerber, M. M. Kestner, J. W. Manthey, M. T. Musser, H. W. Pinnick, D. H. Snow, F. W. Stuchal, and R. T. Swiger, *J. Org. Chem.* **52**, 196 (1987).

mechanism operates as for the nitronate anion. Furthermore, the leaving group need not be a halide. Displacement of nitrite ion from α,p-nitrocumene occurs with good efficiency.[128]

$$O_2N-C_6H_4-C(CH_3)_2-NO_2 + (CH_3)_2\bar{C}NO_2 \longrightarrow O_2N-C_6H_4-C(CH_3)_2-C(CH_3)_2-NO_2 \quad 95\%$$

Azido, sulfonyl, and quaternary nitrogen groups can also be displaced by this mechanism.[129]

The $S_{RN}1$ reaction also proceeds with tertiary benzyl nitro compounds lacking a p-nitro substituent. The nitro group at the benzyl position acts as the electron acceptor, and decomposes to the benzyl radical and nitrite anion. The nitronate anion nucleophiles are then alkylated.[130] Entry 5 in Scheme 12.7 provides a specific example.

A similar mechanism has been proposed for the alkylation of amines by p-nitrocumyl chloride[131]:

$$O_2N-C_6H_4-C(CH_3)_2-Cl + \text{quinuclidine (N)} \longrightarrow O_2N-C_6H_4-C(CH_3)_2-\overset{+}{N}(\text{quinuclidinium}) \quad Cl^-$$

Clearly, the tertiary nature of the chloride would make any proposal of an S_N2 mechanism highly suspect. Furthermore, the nitro substituent is essential to the success of these reactions. Cumyl chloride itself undergoes elimination of HCl on reaction with amines.

A related process constitutes a method of carrying out alkylation reactions to give highly branched alkyl chains that could not easily be formed by S_N2 mechanisms. The alkylating agent must contain a nitro group and a second electron-attracting group. These compounds react with nitronate anions to effect displacement of the nitro group.[132]

$$R_2C(X)-NO_2 + (R')_2\bar{C}NO_2 \longrightarrow R_2C(X)-C(R')_2(NO_2)$$

$$X = C{\equiv}N,\ CO_2C_2H_5,\ \overset{O}{\overset{\|}{C}}R,\ NO_2$$

Experiments in which radical scavengers are added indicate that a chain reaction is involved, because the reaction is greatly retarded in the presence of the scavengers.

128. N. Kornblum, T. M. Davies, G. W. Earl, G. S. Greene, N. L. Holy, R. C. Kerber, J. W. Manthey, M. T. Musser, and D. H. Snow, *J. Am. Chem. Soc.* **89**, 5714 (1967).
129. N. Kornblum, P. Ackerman, J. W. Manthey, M. T. Musser, H. W. Pinnick, S. Singaram, and P. A. Wade, *J. Org. Chem.* **53**, 1475 (1988).
130. N. Kornblum, S. G. Carlson, J. Widmer, M. J. Fifolt, B. N. Newton, and R. G. Smith, *J. Org. Chem.* **43**, 1394 (1978).
131. N. Kornblum and F. W. Stuchal, *J. Am. Chem. Soc.* **92**, 1804 (1970).
132. N. Kornblum and S. D. Boyd, *J. Am. Chem. Soc.* **92**, 5784 (1970).

The mechanism shown below indicates that one of the steps in the chain process is an electron transfer and that none of the steps involves atom abstraction. The elimination of nitrite occurs as a unimolecular decomposition of the radical anion intermediate, and the $S_{RN}1$ mechanistic designation would apply.

initiation

$$R_2\underset{\displaystyle NO_2}{\underset{|}{C}}-X + (R')_2\bar{C}NO_2 \longrightarrow R_2\underset{\displaystyle NO_2^{\dot{-}}}{\underset{|}{C}}-X \longrightarrow R_2\dot{C}-X + NO_2^-$$

$$+ (R')_2\dot{C}NO_2$$

propagation

$$R_2\dot{C}-X + (R')_2\bar{C}NO_2 \longrightarrow R_2\underset{\displaystyle X}{\underset{|}{C}}-\underset{\displaystyle NO_2^{\dot{-}}}{\underset{|}{C}}(R')_2 \longrightarrow R_2\underset{\displaystyle X}{\underset{|}{C}}-\underset{\displaystyle NO_2}{\underset{|}{C}}(R')_2 + R_2\underset{\displaystyle NO_2^{\dot{-}}}{\underset{|}{C}}-X$$

This reaction can also be applied to tertiary nitroalkanes lacking any additional functional group. The reactions with nitro compounds lacking additional anion-stabilizing groups are carried out in dimethyl sulfoxide solution.[133]

$$R_3CNO_2 + {}^-CH_2NO_2 \xrightarrow{\text{DMSO}} R_3CCH_2NO_2$$

These reactions also appear to be chain reactions that proceed through the $S_{RN}1$ mechanism. Dimethyl sulfoxide is a particularly favorable solvent for this reaction, probably because its conjugate base acts as an efficient chain initiator by transferring an electron to the nitroalkane.

Although the nitro group plays a crucial role in most of these $S_{RN}1$ reactions, they have synthetic application beyond the area of nitro compounds. The nitromethyl groups can be converted to other functional groups, including aldehydes and carboxylic acids.[134] Nitro groups at tertiary positions can be reductively removed by reaction with the methanethiolate anion.[135] This reaction also appears to be of the electron transfer type, with the methanethiolate anion acting as the electron donor.

$$R_3C-NO_2 + CH_3S^- \rightarrow R_3C-NO_2^{\dot{-}} + CH_3S\cdot$$

$$R_3C-NO_2^{\dot{-}} \rightarrow R_3C\cdot + NO_2^-$$

$$R_3C\cdot + CH_3S^- \rightarrow R_3C-H + H_2\dot{C}S^-$$

$$H_2\dot{C}S^- + R_3C-NO_2 \rightarrow H_2C{=}S + R_3C-NO_2^{\dot{-}}$$

The unique feature of the $S_{RN}1$ reactions of substituted alkyl nitro compounds is the facility with which carbon–carbon bonds between highly branched centers can be formed. This point is illustrated by several of the examples in Scheme 12.7.

A second general reaction which proceeds by an $S_{RN}1$ mechanistic pattern involves aryl halides. Aryl halides undergo substitution be certain nucleophiles by

133. N. Kornblum and A. S. Erickson, *J. Org. Chem.* **46**, 1037 (1981).
134. N. Kornblum, A. S. Erickson, W. J. Kelly, and B. Henggeler, *J. Org. Chem.* **47**, 4534 (1982).
135. N. Kornblum, S. C. Carlson, and R. G. Smith, *J. Am. Chem. Soc.* **101**, 647 (1979).

a chain mechanism of the $S_{RN}1$ class.[136] Many of the reactions are initiated photochemically, and most have been conducted in liquid-ammonia solution.

initiation $ArX + Nu^- \xrightarrow{h\nu} [ArX]^{\cdot -} + Nu\cdot$

propagation $[ArX]^{\cdot -} \longrightarrow Ar\cdot + X^- \xrightarrow{Nu^-} [ArNu]^{\cdot -}$

$[ArNu]^{\cdot -} + ArX \longrightarrow ArNu + [ArX]^{\cdot -}$

The reactions can also be initiated by a strong chemical reductant or electrochemically.[137] There are several lines of evidence which support the operation of a chain mechanism, one of the most general observations being that the reactions are stopped or greatly retarded by radical traps. The reaction is not particularly sensitive to the aromatic ring substituents. Both electron-releasing groups such as methoxy and electron-attracting groups such as benzoyl can be present.[138] Groups which easily undergo one-electron reduction, especially the nitro group, cause the reaction to fail. The nucleophiles which have been used successfully include sulfide and phosphide anions, dialkyl phosphite anions, and certain enolates. Scheme 12.8 illustrates some typical reactions.

Kinetic studies have shown that the enolate and phosphorus nucleophiles all react at about the same rate. This suggests that the only step directly involving the nucleophile (step 2 of the propagation sequence) occurs at essentially the diffusion-controlled rate so that there is little selectivity among the individual nucleophiles.[139] The synthetic potential of the reaction lies in the fact that other substituents which activate the halide to substitution are not required, in contrast to aromatic nucleophilic substitution which proceeds by an addition–elimination mechanism (Section 10.5).

Instances of substitution of hindered alkyl halides by the $S_{RN}1$ mechanism have also been documented.[140] Some examples are shown below.

$$(CH_3)_3CCH_2Br + PhS^- \xrightarrow[NH_3]{h\nu} (CH_3)_3CCH_2SPh \quad 60\%$$ Ref. 141

136. J. F. Bunnett, *Acc. Chem. Res.* **11**, 413 (1978); R. A. Rossi and R. H. deRossi, *Aromatic Substitution by the $S_{RN}1$ Mechanism*, American Chemical Society Monograph No. 178, Washington, D.C., 1983.
137. C. Amatore, J. Chaussard, J. Pinson, J.-M. Saveant, and A. Thiebault, *J. Am. Chem. Soc.* **101**, 6012 (1979).
138. J. F. Bunnett and J. E. Sundberg, *Chem. Pharm. Bull.* **23**, 2620 (1975); R. A. Rossi, R. H. deRossi, and A. F. Lopez, *J. Org. Chem.* **41**, 3371 (1976).
139. C. Galli and J. F. Bunnett, *J. Am. Chem. Soc.* **103**, 7140 (1981); R. G. Scamehorn, J. M. Hardacre, J. M. Lukanich, and L. R. Sharpe, *J. Org. Chem.* **49**, 4881 (1984).
140. S. M. Palacios, A. N. Santiago, and R. A. Rossi, *J. Org. Chem.* **49**, 4609 (1984).
141. A. B. Pierini, A. B. Penenory, and R. A. Rossi, *J. Org. Chem.* **50**, 2739 (1985).

Scheme 12.8. Aromatic Substitution by the $S_{RN}1$ Mechanism

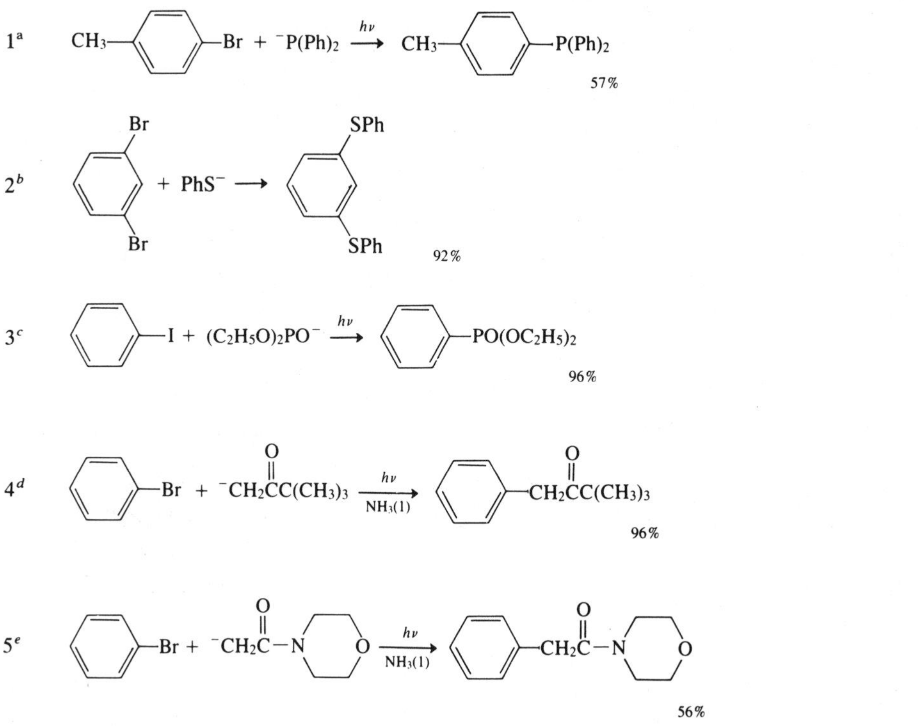

1[a] $CH_3-C_6H_4-Br + {}^{-}P(Ph)_2 \xrightarrow{h\nu} CH_3-C_6H_4-P(Ph)_2$ 57%

2[b] m-dibromobenzene + PhS^- $\longrightarrow$ m-bis(phenylthio)benzene (SPh, SPh) 92%

3[c] $C_6H_5-I + (C_2H_5O)_2PO^- \xrightarrow{h\nu} C_6H_5-PO(OC_2H_5)_2$ 96%

4[d] $C_6H_5-Br + {}^{-}CH_2\overset{O}{\overset{\|}{C}}C(CH_3)_3 \xrightarrow[NH_3(l)]{h\nu} C_6H_5-CH_2\overset{O}{\overset{\|}{C}}C(CH_3)_3$ 96%

5[e] $C_6H_5-Br + {}^{-}CH_2\overset{O}{\overset{\|}{C}}-N\ \ O$ (morpholine) $\xrightarrow[NH_3(l)]{h\nu} C_6H_5-CH_2\overset{O}{\overset{\|}{C}}-N\ \ O$ (morpholine) 56%

a. J. E. Swartz and J. F. Bunnett, *J. Org. Chem.* **44**, 340 (1979).
b. J. F. Bunnett and X. Creary, *J. Org. Chem.* **39**, 3611 (1974).
c. J. F. Bunnett and X. Creary, *J. Org. Chem.* **39**, 3612 (1974).
d. M. F. Semmelhack and T. Bargar, *J. Am. Chem. Soc.* **102**, 7765 (1980).
e. R. A. Rossi and R. A. Alonso, *J. Org. Chem.* **45**, 1239 (1980).

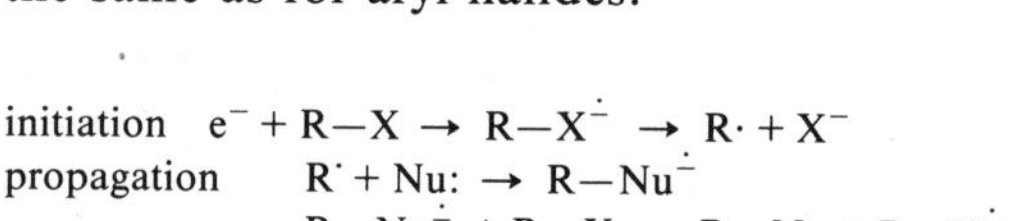

1-adamantyl–Br + $^{-}PPh_2 \xrightarrow[NH_3]{h\nu}$ 1-adamantyl–PPh_2 Ref. 142

bicyclo[4.1.0]heptyl–Br + $^{-}PPh_2 \xrightarrow[NH_3]{h\nu}$ bicyclo[4.1.0]heptyl–PPh_2 87% Ref. 143

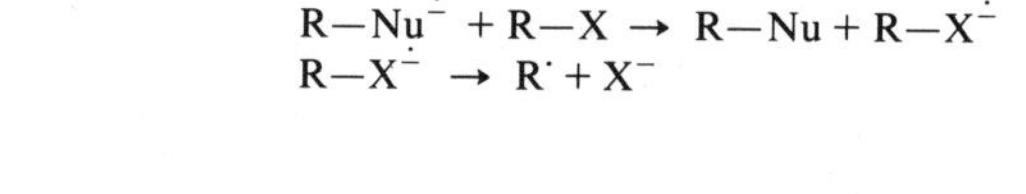

The mechanism is the same as for aryl halides:

$$\text{initiation} \quad e^- + R{-}X \rightarrow R{-}X^{\dot{-}} \rightarrow R\cdot + X^-$$
$$\text{propagation} \quad R^\cdot + Nu{:} \rightarrow R{-}Nu^{\dot{-}}$$
$$R{-}Nu^{\dot{-}} + R{-}X \rightarrow R{-}Nu + R{-}X^{\dot{-}}$$
$$R{-}X^{\dot{-}} \rightarrow R^\cdot + X^-$$

142. R. A. Rossi, S. M. Palacios, and A. N. Santiago, *J. Org. Chem.* **47**, 4654 (1982).
143. R. A. Rossi, A. N. Santiago, and S. M. Palacios, *J. Org. Chem.* **49**, 3387 (1984).

General References

Reactions and Mechanisms

A. L. J. Beckwith and K. U. Ingold, in *Rearrangements in Ground and Excited States*, P. de Mayo (ed.), Academic Press, New York, 1980, Chapter 4.
B. Giese, *Radicals in Organic Synthesis*; *Formation of Carbon-Carbon Bonds*, Pergamon, Oxford, 1986.
E. S. Huyser, *Free Radical Chain Reactions*, Wiley-Interscience, New York, 1970.
W. H. Pryor, *Free Radicals*, McGraw-Hill, New York, 1966.
L. Reich and S. S. Stivala, *Autoxidation of Hydrocarbons and Polyolefins*, Marcel Dekker, New York, 1969.
G. Scott, *Atmospheric Oxidation and Antioxidants*, Elsevier, Amsterdam, 1965.
C. Walling, *Free Radicals in Solution*, Wiley, New York, 1957.

Stable Free Radicals

A. R. Forrester, J. M. Hay, and R. H. Thompson, *Organic Chemistry of Stable Free Radicals*, Academic Press, New York, 1968.

EPR Spectroscopy and CIDNP

M. Bersohn and J. C. Baird, *An Introduction to Electron Paramagnetic Resonance*, W. A. Benjamin, New York, 1966.
N. Hirota and H. Ohya-Nishiguchi, in *Investigation of Rates and Mechanisms of Reaction*, C. Bernasconi (ed.), *Techniques of Chemistry*, Fourth Edition, Vol VI, Part 2, Wiley-Interscience, New York, 1986, Chapter XI.
L. Kevan, in *Methods of Free Radical Chemistry*, Vol. 1, E. Huyser (ed.), Marcel Dekker, New York, 1969, pp. 1-33.
A. G. Lawler and H. R. Ward, in *Determination of Organic Structures by Physical Methods*, Vol. 5, F. C. Nachod and J. J. Zuckerman (eds.), Academic Press, New York, 1973, Chapter 3.

Charged Radicals

E. T. Kaiser and L. Kevan (eds.), *Radical Ions*, Wiley-Interscience, New York, 1968.

Problems

(*References for these problems will be found on page* 784.)

1. Predict the products of the following reactions.

(a) O; $CH_2CH{=}CH_2$; CH_3 $\xrightarrow[HBr]{h\nu}$

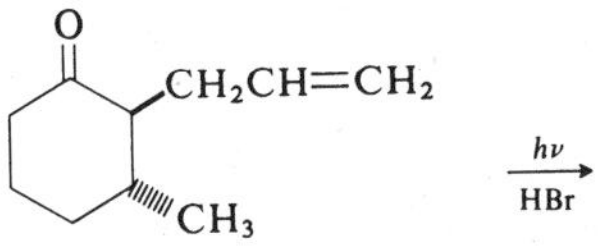

(b) [maleic anhydride] + $(CH_3)_2CHOH \xrightarrow[130°C]{(CH_3)_3COOC(CH_3)_3}$

(c) $CH_3(CH_2)_6CH_2OCl \xrightarrow[1\,M \text{ in } CCl_4]{h\nu}$

(d) $CH_3CH{=}O + H_2C{=}CHCH(OCH_3)_2 \xrightarrow[(PhCO_2)_2]{\Delta}$

(e) [cyclohexene] + $PhCO_2C(CH_3)_3$ (with C=O) $\xrightarrow[80°C]{CuBr}$

(f) [cyclooctadiene] + $HCCl_3 \xrightarrow[65°C]{(PhCO_2)_2}$

(g) $O_2NC(CH_3)_2CH_2CH_2CO_2^- \xrightarrow{\text{electrolysis}}$

(h) $CH_2{=}CHC(CH_3)_2(OH) + CCl_4 \xrightarrow{(PhCO_2)_2}$

(i) [1-chloronaphthalene] + $K^{+-}CH_2CCH_3$ (with C=O) $\xrightarrow[NH_3]{h\nu}$

(j) $C_4H_9SH + CH_3(CH_2)_4CH{=}CH_2 \xrightarrow{(PhCO_2)_2}$

2. Using Table III in Ref. 58 (p. 679), calculate the expected product composition from the gas phase photochemical chlorination and bromination of 3-methylpentane under conditions (excess hydrocarbon) in which only monohalogenation would occur.

3. A careful study of the photoinitiated addition of HBr to 1-hexene established the following facts: (1) The quantum yield is 400; (2) the products are 1-bromohexane, 2-bromohexane, and 3-bromohexane. The amounts of 2- and 3-bromohexane formed are always nearly identical and increase from about 8% each at 4°C to about 22% at 63°C; (3) during the course of the reaction, small amounts of 2-hexene can be detected. Write a mechanism that could accommodate all these facts.

4. The irradiation of 1,3-dioxolane in the presence of alkenes and a photochemically activated initiator at 30°C leads to 2-alkyldioxolanes:

$$\text{1,3-dioxolane} + CH_2{=}CHR \xrightarrow{h\nu} \text{2-}(CH_2CH_2R)\text{-1,3-dioxolane}$$

The reaction is particularly effective with alkenes with electron-attracting substituents such as diethyl maleate. When the reaction is conducted thermally with a peroxide initiator at 160°C, the product mixture is much more complex:

$$\text{1,3-dioxolane} + CH_2{=}CHR \xrightarrow[160°C]{ROOR} \text{2-}(CH_2CH_2R)\text{-1,3-dioxolane} + \text{4-}(CH_2CH_2R)\text{-1,3-dioxolane} + HC(=O)OCH_2CH_2CH_2CH_2R$$

(a) Provide a mechanism for the formation of the observed products.
(b) Why is diethyl maleate an especially good reactant?
(c) Why does the photochemical method lead to a different product distribution than the peroxide-catalyzed reaction?

5. Provide a detailed mechanistic explanation of the following results.

(a) Photochemical bromination of **1**, $\alpha_D + 4.21°$, affords **2**, which is optically active, $\alpha_D - 3.23°$, but **3** under the same conditions gives **4**, which is optically inactive.

$$CH_3CH_2C(H)(CH_3)CH_2X \xrightarrow[h\nu]{Br_2} CH_3CH_2C(Br)(CH_3)CH_2X$$

1: X = Br   **2**: X = Br
3: X = F   **4**: X = F

(b) The stereoisomerization shown below proceeds efficiently, with no other chemical change occurring at a comparable rate, when the compound is warmed with *N*-bromosuccinimide and a radical chain initiator.

$$\text{bicyclo[3.1.0]hex-2-ene-6-}CO_2CH_3 \text{ (endo)} \longrightarrow \text{bicyclo[3.1.0]hex-2-ene-6-}CO_2CH_3 \text{ (exo)}$$

(c) There is a substantial difference in the reactivity of the two stereoisomeric compounds shown below toward abstraction of a hydrogen atom by the *t*-butoxy radical.

2,4,6-trimethyl-1,3-dioxane (2-H axial, CH_3 equatorial) is 7–10 times more reactive than 2,4,6-trimethyl-1,3-dioxane (2-CH_3 axial, H equatorial)

(d) Free-radical chlorination of optically active 1-chloro-2-methylbutane yields six dichloro derivatives of which four are optically active and two are not. Identify the optically active and optically inactive products and provide an explanation for the origin of each product.

(e) Irradiation of a mixture of the hydrocarbon **B** and di-*t*-butyl peroxide generates a free radical which can be identified as the 2-phenylethyl radical by its EPR spectrum. This is the only spectrum which can be observed, even when the photolysis is carried out at low temperature (−173°C).

B

(f) Among the products from heating 1,5-heptadiene with 1-iodoperfluoropropane in the presence of azobis(isobutyronitrile) are two saturated 1:1 adducts. Both adducts gave the same olefin on dehydrohalogenation, and this olefin was shown by spectroscopic means to contain a $CH_2{=}C$ unit. Give the structures of the two adducts and propose a mechanism for their formation.

6. Write mechanisms which satisfactorily account for the following reactions.

(a) $(CH_3)_2C(Ph)CH_2CH_2CH{=}O \xrightarrow[100°C]{ROOR}$ 4,4-dimethyl-1-tetralone (18%) + $(CH_3)_2C(Ph)CH_2CH_3$ 30% + $(CH_3)_2CHCH_2CH_2C(=O)Ph$ 18%

(b) $Ph(CH_2)_5CH{=}CH_2 \xrightarrow{ROOR,\ 140°C}$ 1-phenyl-2-methylcyclohexane (CH_3, Ph)

(c) $N{\equiv}C{-}C_6H_4{-}C(CH_3)_2{-}NO_2 \xrightarrow[(PhCO_2)_2O]{Bu_3SnH} N{\equiv}C{-}C_6H_4{-}CH(CH_3)_2$

(d) $CH_3(CH_2)_5CH{=}CH_2 + ClN(H)CO_2C_2H_5 + Cr(II) \xrightarrow{-78°C} CH_3(CH_2)_5CH(Cl)CH_2NHCO_2CH_3$

(e) $N\langle C_5H_4\rangle{-}C(CH_3)_2{-}NO_2 + N\langle C_5H_4\rangle{=}C(CH_3)_2 \xrightarrow{NH_3} N\langle C_5H_4\rangle{-}C(CH_3)_2{-}C(CH_3)_2{-}\langle C_5H_4\rangle N$

(f) $C_6H_5C(OH)(CH_3)CH(CH_3)_2 \xrightarrow[CH_3C{\equiv}N]{\text{electrolysis},\ R_4N^+BF_4^-} C_6H_5C(=O)CH_3 + CH_3C(=O)NHCH(CH_3)_2$

(g) C_6H_5–cyclopropane with CH_3, CH_3 and $=C(H)H$ $\xrightarrow{100°C}$ C_6H_5–cyclopropane with $=C(CH_3)CH_3$ (H_3C, CH_3)

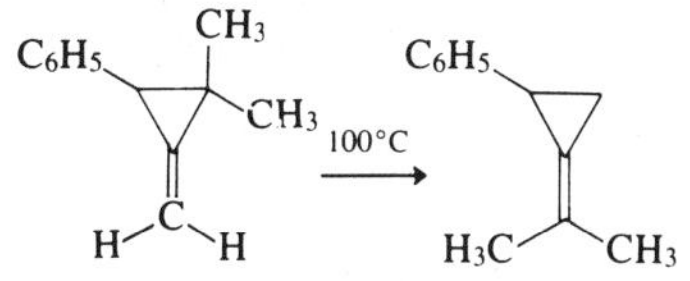

(h) $CH_2{=}CHCH_2CH{=}C(CH_3)_2 \xrightarrow[O_2]{PhSH}$ $PhSCH_2$–(1,2-dioxolane, O—O)∼$C(CH_3)_2$OOH

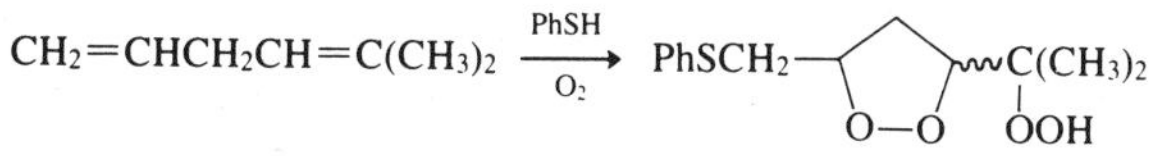

7. The decarbonylation of the two labeled pentenals shown below has been studied. Write a mechanism that could explain the distribution of deuterium label found in the two products.

$$CH_2{=}CHCH_2CD_2CH{=}O \xrightarrow{ROOR} CH_2{=}CHCD_2CH_3 + CH_2{=}CHCH_2CHD_2$$

1 : 1 ratio in dilute solution, increasing to 1 : 1.5 in concentrated solution

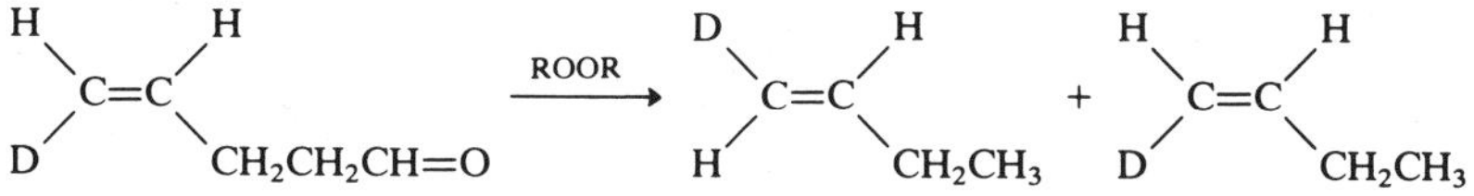

(H)(D)C=C(H)(CH$_2$CH$_2$CH=O) $\xrightarrow{ROOR}$ (D)(H)C=C(H)(CH$_2$CH$_3$) + (H)(D)C=C(H)(CH$_2$CH$_3$)

1 : 1 ratio in dilute solution, increasing to 1 : 1.4 in concentrated solution

8. Decomposition of the *trans*-decalyl perester **A** gives a 9:1 ratio of *trans*: *cis* hydroperoxide product at all oxygen pressures studied. The product ratio from the *cis* isomer is dependent on the oxygen pressure. At 1 atm O_2, it is 9:1 *trans*: *cis*, as with the *trans* substrate, but this ratio decreases and eventually inverts with increasing O_2 pressure. It is 7:3 *cis*: *trans* at 545 atm oxygen pressure. What deduction about the stereochemistry of the decalyl free radical can be made from these data?

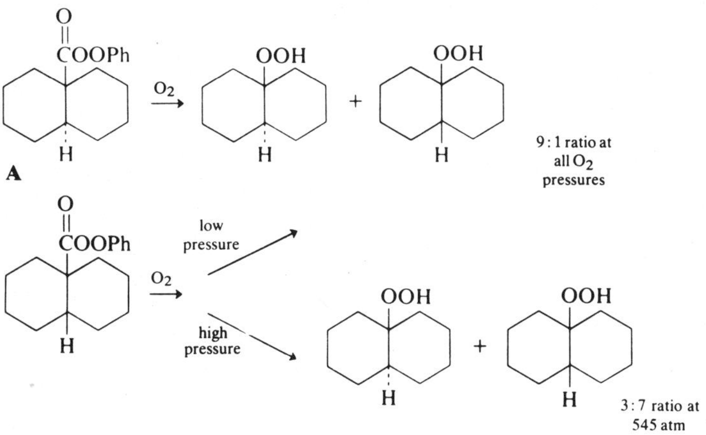

A (*trans*-decalyl, C(=O)OOPh, H) $\xrightarrow{O_2}$ *trans*-decalyl OOH (H) + *cis*-decalyl OOH (H)

9 : 1 ratio at all O_2 pressures

cis-decalyl C(=O)OOPh (H) $\xrightarrow{O_2}$ low pressure ↗ ; high pressure ↘ *trans*-decalyl OOH (H) + *cis*-decalyl OOH (H)

3 : 7 ratio at 545 atm

9. (a) Trichloromethanesulfonyl chloride, Cl_3CSO_2Cl, can chlorinate hydrocarbons as described in the stoichiometric equation below. The reaction is a chain process. Write at least two possible sequences for chain propagation. Suggest some likely termination steps.

$$C_6H_5{-}CH_3 + Cl_3CSO_2Cl \xrightarrow[\Delta]{(PhCO_2)_2} C_6H_5{-}CH_2Cl + HCCl_3 + SO_2$$

(b) Given the following additional information, choose between the chain propagation sequences you have postulated in part (a).

(1) In the reaction

$$R{-}H + BrCCl_3 \xrightarrow{h\nu} RBr + HCCl_3$$

the reactivity of cyclohexane is about one-fifth of toluene.

(2) In the chlorination by trichloromethanesulfonyl chloride, cyclohexane is about three times as reactive as toluene.

10. A highly selective photochemical chlorination of esters, amides, and alcohols can be effected in 70%–90% H_2SO_4 using N-chlorodialkylamines as chlorinating agents. Mechanistic studies indicate that a chain reaction is involved:

$$R_2\overset{+}{N}(Cl)H \xrightarrow{\text{initiation}} R_2\overset{+}{\dot{N}}H + Cl\cdot$$

$$R_2\overset{+}{\dot{N}}H + CH_3CH_2(CH_2)_nX \rightarrow R_2\overset{+}{N}H_2 + CH_3\dot{C}H(CH_2)_nX$$

$$CH_3\dot{C}H(CH_2)_nX + R_2\overset{+}{N}(Cl)H \rightarrow CH_3CH(Cl)(CH_2)_nX + R_2\overset{+}{\dot{N}}H$$

where $X = CO_2CH_3$, CH_2OH, or $CONH_2$. A very interesting feature of the reaction is that the chlorine atom is introduced on the next to terminal carbon atom for reactant molecules with $n = 4$ or 6. In contrast, chlorination of these same compounds with chlorine in nonpolar solvents shows little position selectivity. Rationalize the observed selectivity.

11. Analyze the hyperfine coupling in the spectrum of the butadiene radical anion given in Fig. 12.P11. What is the spin density at each carbon atom according to the McConnell equation?

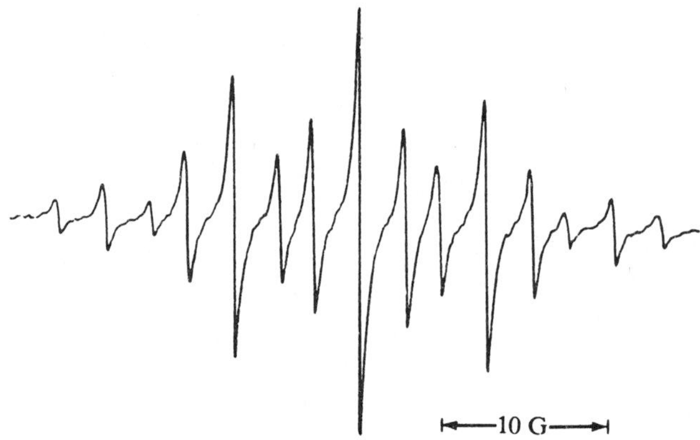

Fig. 12.P11. Spectrum of the butadiene radical anion. [From D. H. Levy and R. J. Meyers, *J. Chem. Phys.* **41**, 1062 (1964).]

(b) A representation of the EPR spectrum of allyl radical is presented below. Interpret the splitting pattern and determine the values of the hyperfine splitting constants.

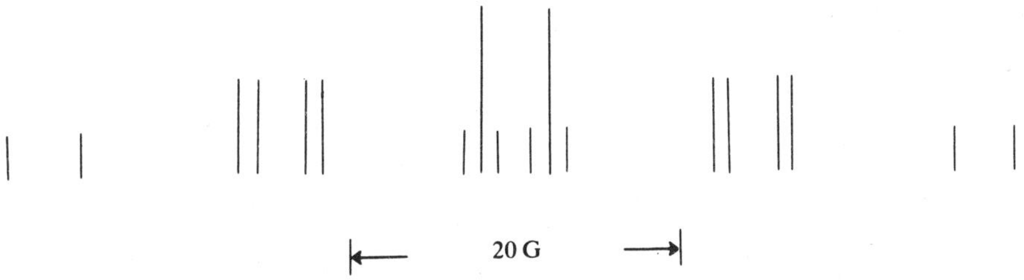

20 G

12. The oxidation of norbornadiene by *t*-butyl perbenzoate and Cu(I) leads to 7-*t*-butoxynorbornadiene. Similarly, oxidation with dibenzoyl peroxide and CuBr leads to 7-benzyloxynorbornadiene. In both cases, when a 2-monodeuterated sample of norbornadiene is used, the deuterium is found distributed at all seven carbons in the product. Provide a mechanism which could account for this result. In what ways does this mechanism differ from the general mechanism discussed on pp. 709–710?

13. A very direct synthesis of certain lactones can be achieved by heating an alkene, a carboxylic acid, and the Mn(III) salt of the acid. Suggest a mechanism by which this reaction might proceed.

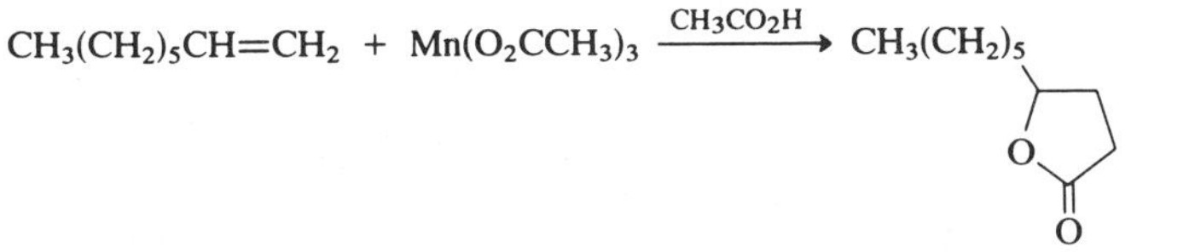

$$CH_3(CH_2)_5CH{=}CH_2 + Mn(O_2CCH_3)_3 \xrightarrow{CH_3CO_2H} CH_3(CH_2)_5\text{-}(\gamma\text{-butyrolactone})$$

14. Indicate mechanisms that would account for each of the products observed in the thermal decomposition of compound **A**:

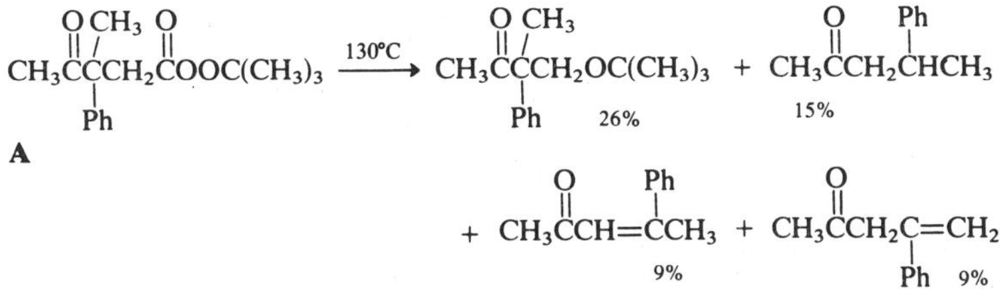

$$\underset{\mathbf{A}}{CH_3C(=O)C(CH_3)(Ph)CH_2C(=O)OOC(CH_3)_3} \xrightarrow{130°C} \underset{26\%}{CH_3C(=O)C(CH_3)(Ph)CH_2OC(CH_3)_3} + \underset{15\%}{CH_3C(=O)CH_2CH(Ph)CH_3}$$

$$+ \underset{9\%}{CH_3C(=O)CH{=}C(Ph)CH_3} + \underset{9\%}{CH_3C(=O)CH_2C(Ph){=}CH_2}$$

15. The *spiro* peroxide **A**, which is readily prepared from cyclohexanone and hydrogen peroxide, decomposes thermally to give substantial amounts of cyclodecane and 11-undecanolactone (**C**). Account for the efficient formation of these macrocyclic compounds.

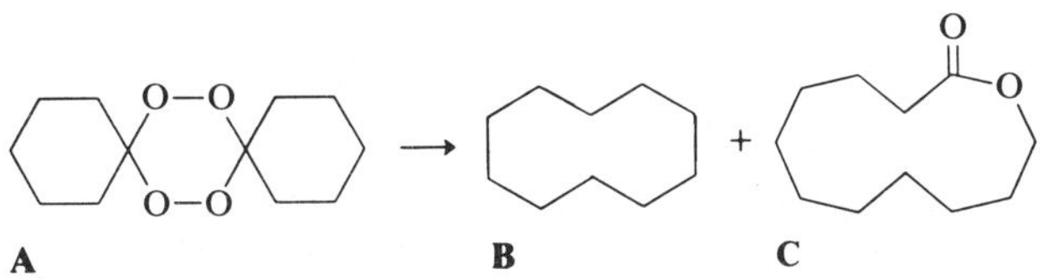

A (O—O, O—O) → **B** + **C**

16. Methylcyclopropane shows strikingly different reactivity toward chlorination and bromination under radical chain conditions. With chlorine, cyclopropyl chloride (56%) is the major product, along with small amounts of 1,3-dichlorobutane (7%). Bromine gives a quantitative yield of 1,3-dibromobutane. Offer an explanation for the difference.

17. Electrolysis of 3,3-diphenylpropanoic acid in acetic acid-acetate solution gives the products shown below. Propose mechanisms for the formation of each of the major products.

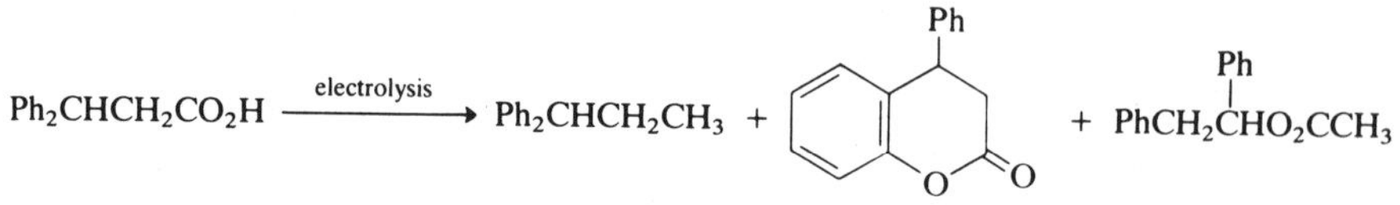

$$Ph_2CHCH_2CO_2H \xrightarrow{\text{electrolysis}} Ph_2CHCH_2CH_3 + \text{(4-phenyl-3,4-dihydrocoumarin)} + PhCH_2CH(Ph)O_2CCH_3$$

18. Write a mechanism to account for the formation of the observed product of each of the following reactions.

(a)

$$CH_3CH(NO_2)CH_2CH_3 \xrightarrow[K_3Fe(CN)_6]{NaOH,\ KCN} CH_3C(CN)(NO_2)CH_2CH_3$$

(b)

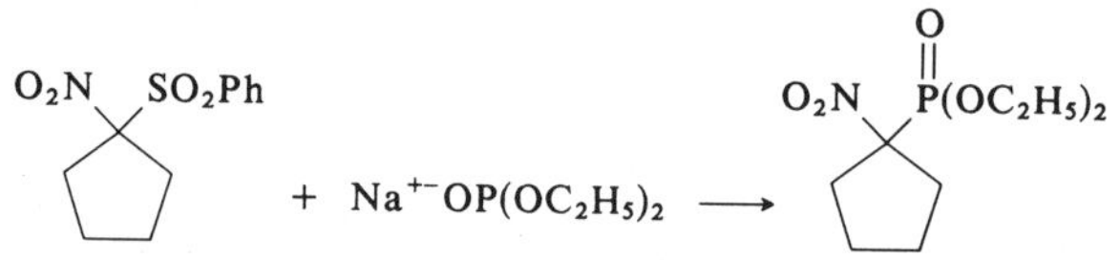

$$\text{1-nitro-1-(phenylsulfonyl)cyclopentane } (O_2N,\ SO_2Ph) + Na^{+-}OP(OC_2H_5)_2 \longrightarrow \text{1-nitro-1-}[P(=O)(OC_2H_5)_2]\text{cyclopentane}$$

(c)

$$\text{C}_6\text{H}_{11}\text{—HgCl} + \text{C}_6\text{H}_{10}\text{=}NO_2^-\,Na^+ \longrightarrow \text{1-nitro-1-cyclohexylcyclohexane } (NO_2)$$

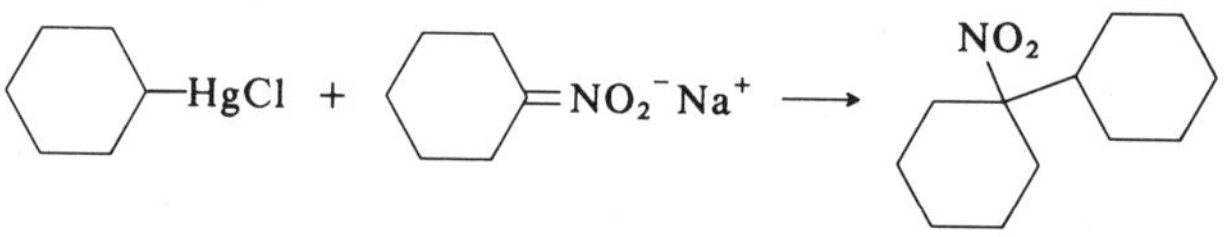

19. The N-benzoyl methyl esters of the amino acids valine, alanine, and glycine have been shown to react with N-bromosuccinimide to give monobromination products containing bromine at the α carbon of the amino acid structure. The order of reactivity is glycine > alanine > valine (23:8:1). Account for the observed trend in reactivity.

20. By measurements in an ion cyclotron resonance spectrometer, the proton affinity (PA) of free radicals can be measured.

$$RH^+ \rightarrow R^\cdot + H^+, \qquad \Delta H_{298} = PA$$

These data can be combined with ionization potential (IP) data according to the scheme below to determine bond dissociation energies (BDE).

$$\begin{array}{ll} RH^+ \rightarrow R^\cdot + H^+ & \Delta H_{298} = PA \\ RH \rightarrow RH^+ + e^- & \Delta H_{298} = IP \\ e^- + H^+ \rightarrow H^\cdot & \Delta H_{298} = -313.6 \text{ kcal/mol} \\ \hline RH \rightarrow R^\cdot + H^\cdot & \Delta H_{298} = BDE \end{array}$$

Data for PA and IP are given for several hydrocarbons of interest.

	IP	PA
$PhCH_2—H$	203	198
—H	190	200
—H	198	199
—H	224	180
$CH_2{=}CHCH_2—H$	224	180
—H	232	187
$CH_2{=}CH—H$	242	183

According to these data, which structural features provide stabilization of radical centers? Determine the level of agreement between these data and the "radical stabilization energies" given in Table 12.5 if the standard C—H bond dissociation energy is taken to be 98.8 kcal/mol. (Compare the calculated and observed bond dissociation energies for the benzyl, allyl, and vinyl systems.)

13

Photochemistry

The photochemical reactions of organic compounds attracted great interest in the 1960s. As a result, many useful and fascinating reactions were uncovered, and photochemistry became an important synthetic tool in organic chemistry. There is also a firm basis for mechanistic discussion of many photochemical reactions. Some of the more general types of photochemical reactions will be discussed in this chapter. In Section 13.2, the relationship of photochemical reactions to the principles of orbital symmetry will be considered. In later sections, characteristic photochemical reactions of alkenes, dienes, carbonyl compounds, and aromatic rings will be considered.

13.1. General Principles

A broad recognition of the fundamental elements of a photochemical reaction is not difficult. The first condition that must be met is that the compound absorb light emitted by the irradiation source. For light to be absorbed, the molecule must have an energy level corresponding in energy to that of the radiation. Most organic photochemical reactions involve excited electronic states. Depending on functionality, organic compounds can have electronic absorption bands from the far ultraviolet to the visible region of the spectrum. Table 13.1 lists the general regions of absorption for the classes of organic molecules that we will discuss in this chapter. A number of light sources can be used. The most common sources for preparative-scale work are mercury vapor lamps, which emit mainly at 254, 313, and 366 nm. The composition of the radiation reaching the sample can be controlled by use of filters. For example, if the system is constructed so that light must pass through borosilicate glass, only wavelengths longer than 300–310 nm will reach the sample, because the glass absorbs below this wavelength. Pure fused quartz, which transmits down to 200 nm, must be used if the 254-nm radiation is desired. Other materials have cutoff

Table 13.1. General Wavelength Ranges for Lowest-Energy Absorption Band of Some Classes of Photochemical Substrates

Substrates	Absorption maxima (nm)
Simple alkenes	190–200
Acyclic dienes	220–250
Cyclic dienes	250–270
Styrenes	270–300
Saturated ketones	270–280
α,β-Unsaturated ketones	310–330
Aromatic ketones and aldehydes	280–300
Aromatic compounds	250–280

points between those of quartz and Pyrex. Filter solutions that absorb in specific wavelength ranges can also be used to control the energy of the light reaching the sample.[1]

When a quantum of light is absorbed, the electronic configuration changes to correspond to an excited state. Two general points about this process should be emphasized:

1. At the instant of excitation, only electrons are reorganized; the heavier nuclei retain their ground state geometry. The statement of this condition is referred to as the *Franck–Condon principle.* A consequence is that the intially generated excited state will have a non-minimal-energy geometry.
2. The electrons do not undergo spin inversion at the instant of excitation. Inversion is forbidden by quantum-mechanical selection rules, which require that there be conservation of spin during the excitation process.

Thus, in the very short time (10^{-15} s) required for excitation, the molecule does not undergo changes in nuclear position nor a spin change of the promoted electron. After the excitation, however, the changes can occur very rapidly. The new minimum-energy geometry associated with the excited state is rapidly achieved by vibrational relaxation. The excited state molecule transfers thermal energy to the solvent in this process. Sometimes, chemical reactions of the excited molecule are fast relative to this vibrational relaxation, but this circumstance is rare in solution. When reaction proceeds faster than vibrational relaxation, the reaction is said to involve a "hot excited state," that is, one with excess vibrational energy. The excited state can also undergo *intersystem crossing.* This process involves inversion of spin of an electron in a half-filled orbital and gives a *triplet state,* in which both unpaired electrons have the same spin. The triplet state will also adopt a new minimum-energy molecular geometry.

1. Detailed information on the emission characteristics of various sources and the transmission properties of glasses and filter solutions can be found in A. J. Gordon and R. A. Ford, *The Chemist's Companion,* Wiley-Interscience, New York, 1972, pp. 348-368, and in S. L. Murov, *Handbook of Photochemistry,* Marcel Dekker, New York, 1973.

The general situation can be represented for a hypothetical molecule by a potential energy diagram (Fig. 13.1). The designations S and T are used for singlet and triplet states, respectively. The excitation is a "vertical transition," that is, it involves no distortion of the molecular geometry. Horizontal displacement on the diagram corresponds to motion of the atoms relative to one another. Since the potential energy curves of the excited states are displaced from that of the ground state, the species formed by excitation is excited both electronically and vibrationally. The energy well for the triplet states also correspond to a different minimum-energy molecular geometry. Vibrational relaxation corresponds to dissipation of the vibrational energy as the molecule moves to the bottom of its energy well. One of the central issues in the description of any photochemical reaction is the question of whether a singlet or triplet excited state is involved. This will depend on the rate of intersystem crossing in comparison with the rate of chemical reaction of the singlet excited state. If intersystem crossing is fast relative to reaction, the triplet state will be reached and reaction will occur through the triplet excited state. If reaction is faster than intersystem crossing, reaction will occur from the singlet state.

Photosensitization is an important alternative to direct excitation of molecules, and usually this method results in reaction occurring via triplet excited states. If a reaction is to be carried out by photosensitization, a substance, the *sensitizer*, is

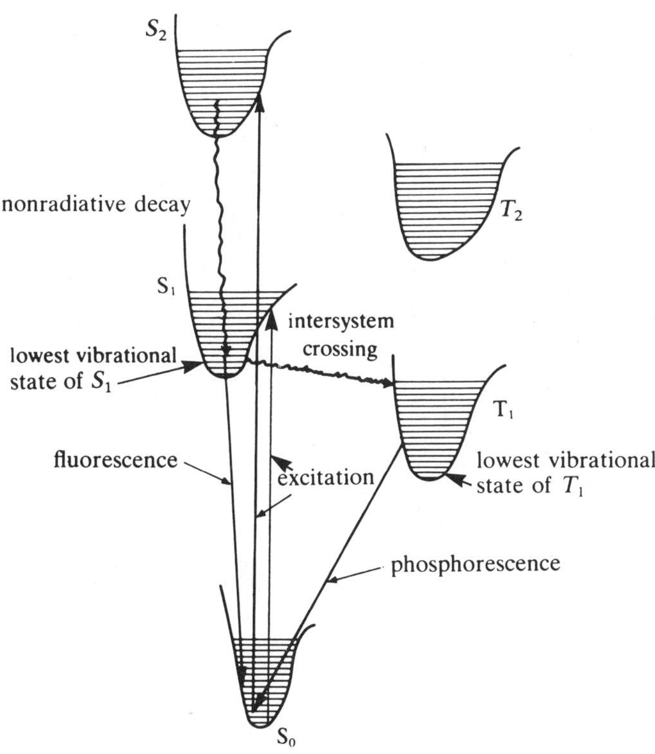

Fig. 13.1. Energy level diagram and summary of photochemical processes.

present in the system. This substance is chosen to meet the following criteria:

1. It must be excited by the irradiation to be used.
2. It must be present in sufficient concentration to absorb more strongly than the other reactants under the conditions of the experiment so that it is the major light absorber.
3. It must be able to transfer energy to the desired reactant.

The usual case and the one that will be emphasized here is triplet sensitization. In this case, the intersystem crossing of the sensitizer must be faster than energy transfer to the reactant or solvent from the singlet excited state.

The transfer of energy from the sensitizer to the reactant must proceed with net conservation of spin. In the usual case, the acceptor molecule is a ground state singlet, and its reaction with the triplet state of the sensitizer will produce the triplet state of the acceptor. The mechanism for triplet photosentization is outlined below;

$\text{Sens} \longrightarrow {}^1\text{Sens}^*$	Sensitizer singlet formed
${}^1\text{Sens}^* \longrightarrow {}^3\text{Sens}^*$	Intersystem crossing of sensitizer
${}^3\text{Sens}^* + \text{React} \longrightarrow \text{Sens} + {}^3\text{React}^*$	Energy transfer to reactant molecule

A corollary of requirement 3 is that the energy of the triplet excited state of the sensitizer be equal to or higher than that of the reactant. If this condition is not met, step 3 of the mechanism above becomes endothermic and loses out in competition with other means for deactivation of ${}^3\text{Sens}^*$.

Once the excited state of the reactant has been formed, either by direct or sensitized energy transfer, the stage is set for a photochemical reaction. There are still, however, competitive processes that can occur and result in the return of unreacted starting material. The excited state can decay to the ground state by emission of light, a *radiative transition.* The rate of emission is very high ($k = 10^5$–$10^9\ \text{s}^{-1}$) for radiative transitions between electronic states of the same multiplicity and somewhat lower ($k = 10^3$–$10^5\ \text{s}^{-1}$) between states of different multiplicity. The two processes are known as *fluorescence* and *phosphorescence*, respectively. Once energy has been emitted as light, the reactant is no longer excited, of course, and a photochemical reaction will not occur.

Excited states can also be *quenched.* Quenching is the same physical process as sensitization, but the word "quenched" is applied when a photoexcited state of the reactant is deactivated by transferring its energy to another molecule in solution. The substance to which energy is transferred is called a *quencher.*

Finally, *nonradiative decay* can occur. This name is given to the process by which the energy of the excited state is transferred to the surrounding molecules as vibrational (thermal) energy without light emission.

The kinds of processes that can occur after photochemical excitation are summarized in Fig. 13.1.

Because of the existence of these competing processes, not every molecule that is excited undergoes a photochemical reaction. The fraction of molecules that react relative to those that are excited is called the *quantum yield.* This yield is a simple measure of the efficiency of the absorption of light in producing reaction product. A quantum yield of one means that each molecule excited (which equals the number of quanta of light absorbed) goes to product. If the quantum yield is 0.01, then only one one-hundredth of the molecules that are excited undergo photochemical reaction. This yield can vary widely, depending on the structure of the reactants and the reaction conditions. For example, quantum yields can be very large in chain reactions, in which a single photoexcitation initiates a repeating series of reactions leading to many molecules of product per initiation step.

Because photochemical processes are very fast, special techniques are required to obtain rate measurements. One method is flash photolysis. The irradiation is effected by a short pulse of light in an apparatus designed to monitor fast spectroscopic changes. The kinetic characteristics of the reactions following irradiation can be determined from these spectroscopic changes.

Another useful technique for measuring the rates of certain reactions involves measuring the quantum yield as a function of quencher concentration. A plot of the inverse of the quantum yield versus quencher concentration is then made (*Stern–Volmer plot*). Since the quantum yield indicates the fraction of excited molecules that go on to product, it is a function of the rates of the processes that result in other fates for the excited molecule. These processes are described by the rate constants k_q (quenching) and k_n (other nonproductive decay to the ground state):

$$\Phi = \frac{k_r}{k_r + k_q[Q] + k_n}$$

A plot of $1/\Phi$ versus $[Q]$ then gives a line with the slope k_q/k_r. It is usually possible to assume that quenching is diffusion-controlled, permitting assignment of a value to k_q. The rate of photoreaction, k_r, for the excited intermediate can then be calculated.

In this chapter, the discussion will center on the reactions of excited states, rather than on the other routes available to them for dissipation of their excess energy. The chemical reactions of photoexcited molecules are of interest primarily for three reasons:

1. Excited states have a great deal of energy and can therefore undergo reactions that would be highly endothermic if initiated from the ground state. For example, we can calculate from the relationship $E = h\nu$ that excitation by 350-nm light corresponds to energy transfer of 82 kcal/mol.
2. The population of an antibonding orbital in the excited state allows the occurrence of chemical transformations that are electronically not available to ground state species.
3. Either the singlet or the triplet state may be involved in a photochemical reaction, while only singlet species are involved in most thermal processes. This permits the formation of intermediates that are unavailable under thermal conditions.

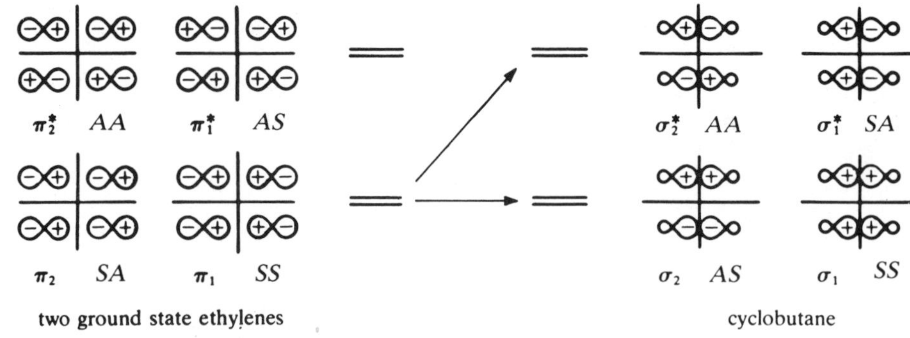

Fig. 13.2. Orbital correlation diagram for two ground state ethylenes and cyclobutane. The symmetry designations apply, respectively, to the horizontal and vertical planes for two ethylene molecules approaching one another in parallel planes.

13.2. Orbital Symmetry Considerations Related to Photochemical Reactions

The complementary relationship between thermal and photochemical reactions can be illustrated by considering some of the same reaction types discussed in Chapter 11 and applying orbital symmetry considerations to the photochemical mode of reaction.

The case of [2 + 2] cycloaddition of two alkenes can serve as an example. This reaction was classified as a forbidden thermal reaction. The correlation diagram (Fig. 13.2) shows that the ground state molecules would lead to an excited state of cyclobutane, and this process would therefore involve a prohibitive energy requirement.

How does the situation change when a photochemical reaction involving one ground state alkene and one excited state alkene is considered? We can assume the same symmetrical approach as in the thermal reaction, so the same array of orbitals is involved. The occupation of the orbitals is different, however; the π_1 (*SS*) orbital is doubly occupied, but π_2 (*SA*) and π_2^* (*AS*) are singly occupied. The reaction is therefore allowed. Although the correlation diagram illustrated in Fig. 13.3 might suggest that the product would initially be formed in an excited state, this in not necessarily the case. The concerted process can involve a transformation of the reactant excited state to the ground state of product. This transformation will be discussed shortly.

Consideration of the HOMO–LUMO interactions also indicates that the [2 + 2] additions would be allowed photochemically. The HOMO in this case is the excited alkene π^* ortibal. The LUMO is the π^* of the ground state alkene, and a bonding interaction is present between the carbons where new bonds must be formed in the cycloaddition reaction:

HOMO

LUMO

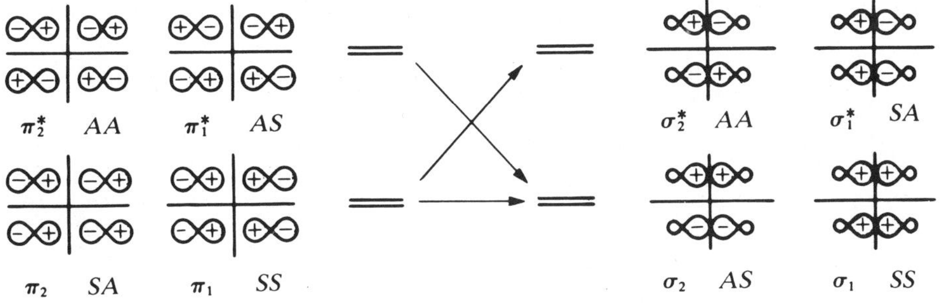

Fig. 13.3. Orbital correlation diagram for one ground state alkene and one excited state alkene. The symmetry designations apply, respectively, to the horizontal and vertical planes for two ethylene molecules approaching one another in parallel planes.

A striking illustration of the relationship between orbital symmetry considerations and the outcome of photochemical reactions can be found in the stereochemistry of electrocyclic reactions. In Chapter 11, the distinction between the conrotatory and the disrotatory mode of reaction as a function of the number of electrons in the system was described. Orbital symmetry considerations predict, and it has been verified experimentally, that photochemical electrocyclic reactions show a reversal of stereochemistry[2]:

Number of π-electrons	Thermal	Photochemical
2	disrotatory	conrotatory
4	conrotatory	disrotatory
6	disrotatory	conrotatory
8	conrotatory	disrotatory

The most fundamental way of making this prediction is to construct an electronic energy state diagram for the reactant and product molecules and observe the correlation between the states.[3] Those reactions will be permitted in which the reacting state correlates with a state of the product that is not appreciably higher in energy.[4]

The states involved in the photochemical butadiene-to-cyclobutene conversion are ψ_1, ψ_2, and ψ_3 of the first excited state of butadiene and σ, π, and π^* for cyclobutene. The appropriate elements of symmetry are the plane of symmetry for the conrotatory process and the axis of symmetry for the disrotatory process. The

2. R. B. Woodward and R. Hoffmann, *J. Am. Chem. Soc.* **87**, 395 (1965).
3. H. C. Longuet-Higgins and E. W. Abrahamson, *J. Am. Chem. Soc.* **87**, 2045 (1965).
4. R. B. Woodward and R. Hoffmann, *The Conservation of Orbital Symmetry*, Academic Press, New York, 1970.

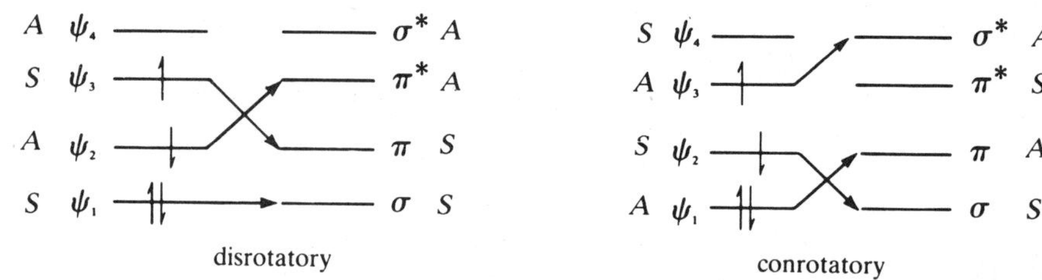

Fig. 13.4. Correlation of energy states involved in the photochemical butadiene-to-cyclobutene conversion.

correlation diagram for this reaction is shown in Fig. 13.4.

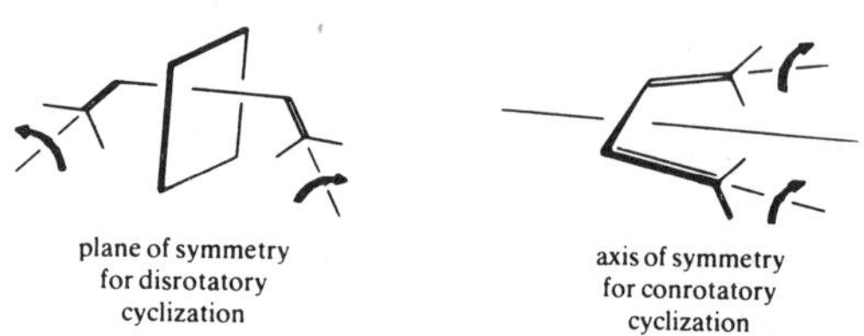

This analysis shows that disrotatory cyclization is allowed, while conrotation would lead to a highly excited $\sigma^1, \pi^2, \sigma^{*1}$ configuration of cyclobutene. The same conclusion is reached if it is assumed that the frontier orbital will govern reaction stereochemistry[2]:

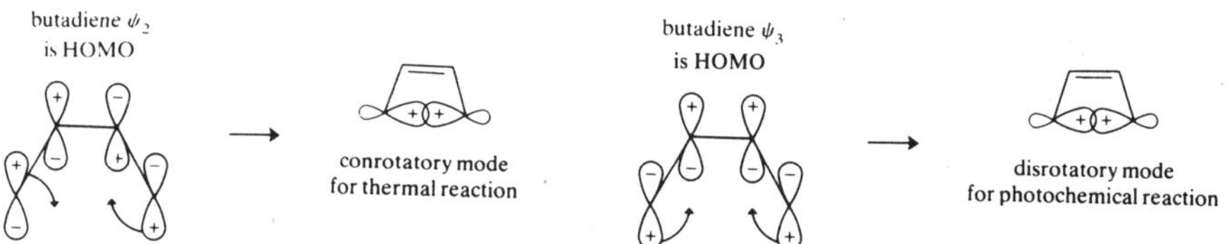

In fact, it is a general result that the Woodward–Hoffmann rules predict that photochemical reactions will be precisely complementary to thermal reactions. What is allowed photochemically is forbidden thermally, and vice versa. The physical basis for this complementary relationship is that the high barrier associated with forbidden thermal reactions provides a point for strong interaction of the ground state and excited state species. This interaction is necessary for efficient photochemical reactions.[5] An energy diagram illustrating this relationship is shown in Fig. 13.5.

This diagram, which is based on quantum-chemical calculations of the butadiene and cyclobutene molecules in the geometries traversed during interconversion, shows

5. H. E. Zimmerman, *J. Am. Chem. Soc.* **88**, 1566 (1966); W. Th. A. M. van der Lugt and L. J. Oosterhoff, *J. Chem. Soc., Chem. Commun.*, 1235 (1968); *J. Am. Chem. Soc.* **91**, 6042 (1969); R. C. Dougherty, *J. Am. Chem. Soc.* **93**, 7187 (1971); J. Michl, *Top. Curr. Chem.* **46**, 1 (1974); *Photochem. Photobiol.* **25**, 141 (1977).

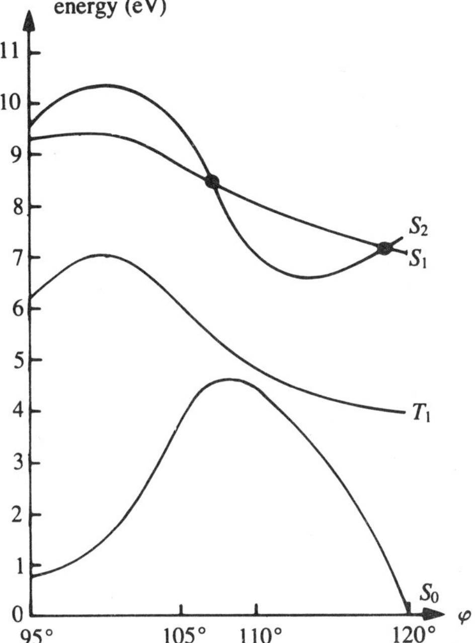

Fig. 13.5. Energy diagram showing potential energy curves for interconversion of ground (S_0), first and second singlet (S_1 and S_2), and first triplet (T_1) excited states. The angle φ is the C—C—C bond angle at C-2 and C-3. [From D. Grimbert, G. Segal, and A. Devaquet, *J. Am. Chem. Soc.* **97**, 6629 (1975).]

a minimum in the S_2 excited state.[6] It also shows that, with a small activation energy (which is calculated to be about 5 kcal/mol), the molecule in the S_1 state can reach points where the S_1 and S_2 energy surfaces cross. The excited molecules, whether generated from cyclobutene or butadiene, would be expected to follow the S_2 surface to the minimum located above the S_0 transition state. By loss of energy to the surrounding medium, the excited state molecule can drop to the S_0 surface and then be transformed to butadiene or cyclobutene. This diagram also provides an explanation of how the excited state shown in Fig. 13.4, which has one singly occupied antisymmetric orbital, returns to ground state cyclobutene, in which only symmetric orbitals are occupied. The S_2 state has the same symmetry as the ground state and can therefore decay directly to the ground state.

The experimental results on the photochemical cyclobutene-to-butadiene ring opening are not as straightforward as for the thermal reaction. For simple alkylcyclobutenes, the photolysis must be done in the vacuum ultraviolet (<200 nm) because of the high energy of the absorption maximum. Cyclobutene ring opening

6. D. Grimbert, G. Segal, and A. Devaquet, *J. Am. Chem. Soc.* **97**, 6629 (1975).

bond is lengthened, and the dipole moment is reduced from 2.34 to 1.56 D.[9] The reduction of the dipole moment results from the transfer of electron density from an orbital localized on oxygen to one that also encompasses the carbon atom. The oxygen atom in this excited state is relatively electron deficient, and its reactivity resembles that of an oxygen-centered radical.

An alternative excited state involves promotion of a bonding π electron to the antibonding π^* orbital. This is called the π–π^* excited state and is most likely to be involved when the carbonyl group is conjugated with an extensive π-bonding system, as is the case for aromatic ketones.

It is not possible to draw unambiguous Lewis structures of excited states of the sort that are so useful in depicting ground state chemistry. Instead, it is common to asterisk the normal carbonyl structure and provide information about the nature and multiplicity of the excited state:

$[H_2C{=}O]^*$	$^3[H_2C{=}O]^*$	$^1[H_2C{=}O]^*$
excited state, multiplicity unspecified	excited-state triplet	excited-state singlet

The MO diagrams for the n–π^* and π–π^* states can be readily depicted:

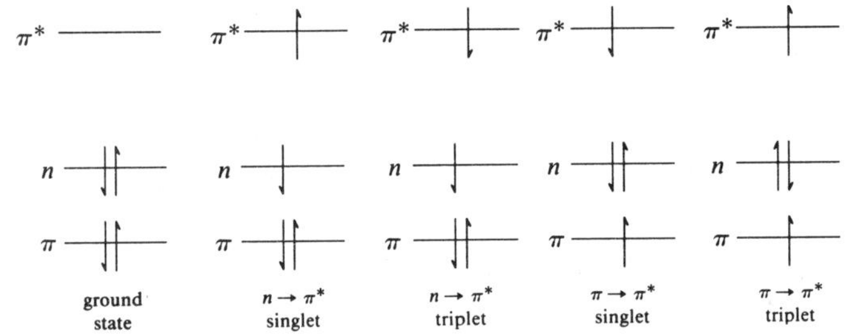

For conjugated carbonyl compounds, such as α,β-enones, the orbital diagram would be similar, except for the recognition that the HOMO of the ground state is ψ_2 of the enone system, rather than the oxygen lone pair orbital. The excited state can sometimes be usefully represented as a dipolar or diradical intermediate:

While single Lewis structures of this type are not adequate descriptions of the structure of the excited states, they do correspond to the MO picture by showing distortion of charge and the presence of reactive polar or radical-like centers. The excited states are much more reactive than the corresponding ground state molecules.

9. J. C. D. Brand and D. G. Williamson, *Adv. Phys. Org. Chem.* **1**, 365 (1963); D. E. Freeman and W. Klemperer, *J. Chem. Phys.* **45**, 52 (1966).

In addition to the increased energy content, this high reactivity is associated with the presence of half-filled orbitals. The atoms which contribute to the two SOMO orbitals in the excited states have enhanced radical, cationic, or anionic character.

One of the most common reactions of photoexcited carbonyl groups is hydrogen atom abstraction from solvent or some other hydrogen donor. A second common reaction is cleavage of the carbon-carbon bond adjacent to the carbonyl group:

$$R_2C{=}O \xrightarrow{h\nu} [R_2C{=}O]^* \begin{cases} \xrightarrow{X-H} R_2\dot{C}{-}OH + \dot{X} \\ \text{or} \\ \longrightarrow R\dot{C}{=}O + R\cdot \end{cases}$$

The hydrogen atom abstraction can be either intramolecular or intermolecular. Many aromatic ketones react by hydrogen atom abstraction, and the stable products are diols formed by coupling of the resulting α-hydroxybenzyl radicals:

$$\underset{\|}{\text{ArCR}}\ (=O^*) + S{-}H \longrightarrow \text{Ar}\dot{C}R(OH) + S\cdot \longrightarrow \text{ArC(R)(OH)}{-}\text{C(R)(OH)Ar}$$

These reactions usually occur via the triplet excited state T_1. The intersystem crossing of the initially formed singlet excited state is so fast ($k \approx 10^{10}\ s^{-1}$) that reactions of the S_1 state are usually not observed. The reaction of benzophenone has been particularly closely studied. Some of the facts that have been established in support of the general mechanism outlined above are as follows:

1. For a given hydrogen donor S—H, replacement by S—D leads to a decreased rate of reduction, relative to nonproductive decay to the ground state.[10] This decreased rate is consistent with a primary isotope effect in the hydrogen abstraction step.
2. The photoreduction can be quenched by known triplet quenchers. The effective quenchers are those which have T_1 states less than 69 kcal/mol above S_0. Quenchers with higher triplet energies are ineffective because the benzophenone π-π^* triplet is then not sufficiently energetic to effect energy transfer.
3. The intermediate diphenylhydroxymethyl radical has been detected after generation by flash photolysis.[11] Photolysis of benzophenone in benzene solution containing potential hydrogen donors results in the formation of two intermediates that are detectable, and their rates of decay have been measured. One intermediate is the $Ph_2\dot{C}OH$ radical. It disappears by combination with another radical in a second-order process. A much shorter-lived species disappears with first-order kinetics in the presence of excess amounts of various hydrogen donors. The pseudo-first-order rate constants vary with

10. W. M. Moore, G. S. Hammond, and R. P. Foss, *J. Am. Chem. Soc.* **83**, 2789 (1961); G. S. Hammond, W. P. Baker, and W. M. Moore, *J. Am. Chem. Soc.* **83**, 2795 (1961).
11. J. A. Bell and H. Linschitz, *J. Am. Chem. Soc.* **85**, 528 (1963).

the structure of the donor; for 2,2-diphenylethanol, for example, $k = 2 \times 10^6\ s^{-1}$. The rate is much less with poorer hydrogen atom donors. The rapidly disappearing intermediate is the triplet excited state of benzophenone.

4. In 2-propanol, the quantum yield for photolytic conversion of benzophenone to the coupled reduction product is 2.0.[12] The reason is that the radical remaining after abstraction of a hydrogen atom from 2-propanol transfers a hydrogen atom to ground state benzophenone in a nonphotochemical reaction. Because of this transfer, two molecules of benzophenone are reduced for each one that is photoexcited:

$$Ph_2C{=}O \xrightarrow{h\nu} [Ph_2C{=}O]^*$$

$$[Ph_2C{=}O]^* + (CH_3)_2CHOH \longrightarrow Ph_2\dot{C}OH + (CH_3)_2\dot{C}OH$$

$$(CH_3)_2\dot{C}OH + Ph_2C{=}O \longrightarrow Ph_2\dot{C}OH + (CH_3)_2C{=}O$$

$$2\ Ph_2\dot{C}OH \longrightarrow Ph_2C(OH){-}C(OH)Ph_2$$

The efficiency of reduction of benzophenone derivatives is greatly diminished when an *ortho* alkyl substituent is present because a new photoreaction, intramolecular hydrogen atom abstraction, then becomes the dominant process. The abstraction takes place from the benzylic position on the adjacent alkyl chain, giving an unstable enol that can revert to the original benzophenone without photoreduction. This process is known as *photoenolization.*[13] Photoenolization can be detected, even though no net transformation of the reactant occurs, by photolysis in deuterated hydroxylic solvents. The proton of the enolic hydroxyl is rapidly exchanged with solvent, so deuterium is introduced at the benzylic position. Deuterium is also introduced if the enol is protonated at the benzylic carbon by solvent.

CH2R / CPh / O → CHR / H / C=O* / Ph → CHR / OH / C / Ph ⇌(R'OD) CHR / OD / C / Ph; R'OD → D / CHR / CPh / O

The dominant photochemical reaction of ketones in the gas phase is cleavage of one of the carbonyl substituents, which is followed by decarbonylation and subsequent reactions of the alkyl free radicals that result.

$$RC(=O)R' \xrightarrow{h\nu} R\dot{C}{=}O + {\bullet}R' \longrightarrow R{\bullet} + CO + {\bullet}R' \longrightarrow R{-}R'$$

12. N. J. Turro, *Molecular Photochemistry*, W. A. Benjamin, New York, 1965, pp. 143, 144.
13. P. G. Sammes, *Tetrahedron*, **32**, 405 (1976).

This reaction is referred to as the *type-I* or *α-cleavage* reaction of carbonyl compounds. This type of reaction is not so common in solution, although some cyclic ketones do undergo decarbonylation:

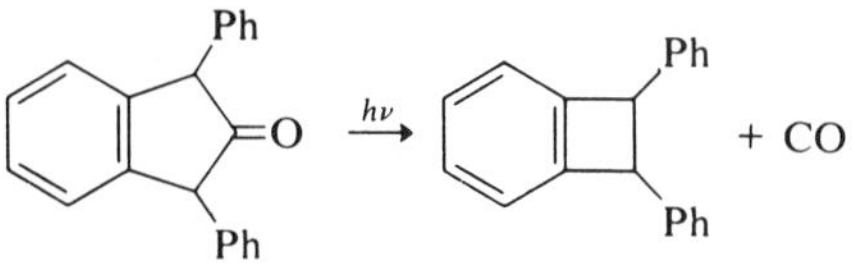

Ref. 14

The facility with which this reaction occurs in solution depends on the stability of the radical fragments that can be ejected. Benzylic ketones, for example, are readily cleaved photolytically.[14] Similarly, *t*-butyl ketones undergo α cleavage quite readily on photolysis in solution.[15]

$$PhCH_2\overset{O}{\overset{\|}{C}}CHPh_2 \xrightarrow{h\nu} PhCH_2CH_2Ph + PhCH_2CHPh_2 + Ph_2CHCHPh_2 \qquad \text{Ref. 14}$$

$$(CH_3)_3C\overset{O}{\overset{\|}{C}}CH_3 \xrightarrow{h\nu} (CH_3)_3CH + (CH_3)_2C{=}CH_2 + CH_3CH{=}O \qquad \text{Ref. 15}$$

Ketones such as 2,2,5,5-tetraphenylcyclopentanone and 2,2,6,6-tetraphenylcyclohexanone decarbonylate rapidly because of the stabilization afforded by the phenyl groups. The products result from recombination, disproportionation, or fragmentation of the diradical intermediate.[16]

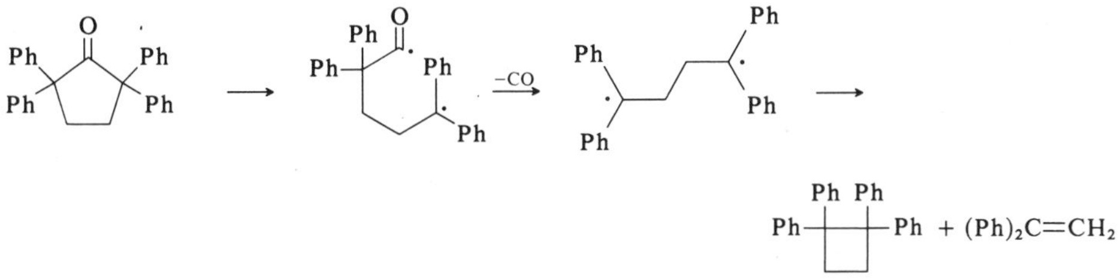

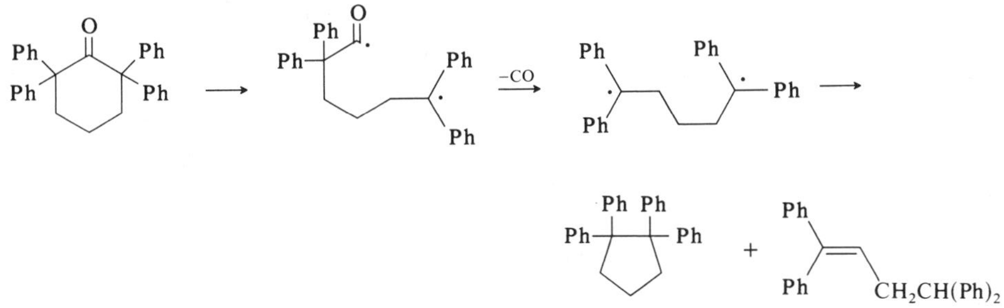

$$\xrightarrow{-CO}$$

$$+ (Ph)_2C{=}CH_2$$

$$\xrightarrow{-CO}$$

$$+ \quad CH_2CH(Ph)_2$$

14. G. Quinkert, K. Opitz, W. W. Wiersdorff, and J. Weinlich, *Tetrahedron Lett.*, 1863 (1963).
15. N. C. Yang and E. D. Feit, *J. Am. Chem. Soc.* **90**, 504 (1968).
16. D. H. R. Barton, B. Charpiot, K. U. Ingold, L. J. Johnston, W. B. Motherwell, J. C. Scaiano, and S. Stanforth, *J. Am. Chem. Soc.* **107**, 3607 (1985).

With cyclic ketones, the α cleavage can also be followed by intramolecular hydrogen abstraction that leads eventually to an unsaturated ring-opened aldehyde.[17]

$$\text{cyclohexanone} \xrightarrow{h\nu} H_2C(\dot{C}{=}O)\text{–}CH_2\text{–}CH_2\text{–}CH_2\text{–}\dot{C}H_2 \longrightarrow H_2C(CH{=}O)\text{–}CH_2\text{–}CH_2\text{–}CH{=}CH_2$$

In ketones having propyl or longer alkyl groups as a carbonyl substituent, intramolecular hydrogen abstraction by the excited carbonyl group can be followed by either cleavage of the bond between the α- and β-carbon atoms or by formation of a cyclobutanol.

$$RC({=}O)CH_2CH_2CH_2R' \xrightarrow{h\nu} R\dot{C}(OH)CH_2CH_2\dot{C}HR' \xrightarrow{\text{ring closure}} \text{1-R-2-R'-cyclobutanol}$$

or

$$\xrightarrow{\text{fragmentation}} H_2C{=}CHR' + RC(OH){=}CH_2 \longrightarrow RCCH_3\ (C{=}O)$$

Cleavage between C_α and C_β is referred to as *type-II* photoelimination to distinguish it from α cleavage. Type-II photoeliminations are observed for both aromatic and aliphatic ketones. Studies aimed at establishing the identity of the reactive excited state indicate that both S_1 and T_1 are involved for aliphatic ketones, but when one of the carbonyl substituents is aryl, intersystem crossing is very fast, and T_1 is the reactive state. Usually, Type II cleavage is the dominant reaction, with cyclobutanol yields being rather low, but there are exceptions. The 1,4-diradical intermediate generated by intramolecular hydrogen abstraction is very short-lived, with a lifetime of probably not more than 10^{-7}–10^{-9} s. The competition between Type II cleavage and cyclization can be generally understood in terms of stability of the various intermediates.[18] Activation energies for hydrogen abstraction from a methylene group are about 4 kcal/mol. The activation energy for carbon–carbon bond cleavage is 8–12 kcal/mol. The nature of substituents on the aryl ring can affect the balance between the two competing reactions, but the overall reactivity pattern remains the same.[19]

$$ArC({=}O)(CH_2)_2CH(H)R \rightarrow ArC({=}O)\cdot + \cdot CH_2CH_2CH_2R$$

$$\downarrow$$

$$Ar\dot{C}(OH)CH_2CH_2\dot{C}HR \rightarrow ArC(OH){=}CH_2 + CH_2{=}CHR \quad \textit{and/or} \quad \text{1-Ar-1-HO-2-R-2-H-cyclobutane}$$

17. W. C. Agosta and W. L. Schreiber, *J. Am. Chem. Soc.* **93**, 3947 (1971); P. J. Wagner and R. W. Spoerke, *J. Am. Chem. Soc.* **91**, 4437 (1969).
18. P. J. Wagner, *Acc. Chem Res.* **4**, 168 (1971).
19. M. V. Encina, E. A. Lissi, E. Lemp, A. Zanocco, and J. C. Scaiano, *J. Am. Chem. Soc.* **105**, 1856 (1983).

Intramolecular hydrogen atom abstraction is also an important process for acyclic α,β-unsaturated ketones.[20] The intermediate diradical then cyclizes to give the enol of a cyclobutyl ketone. Among the by-products of such photolyses are cyclobutanols resulting from alternative modes of cyclization of the diradical intermediate:

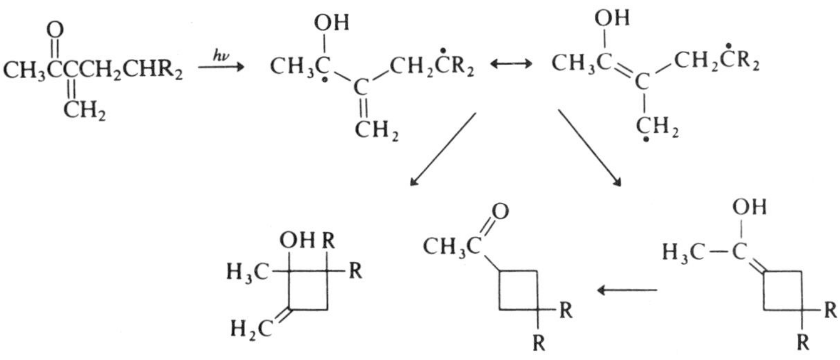

α-β-Unsaturated ketones with γ-hydrogens can undergo hydrogen atom transfer resulting in formation of a dienol. Because the hydrogen atom transfer occurs through a cyclic transition state, the originally formed product has Z-stereochemistry.

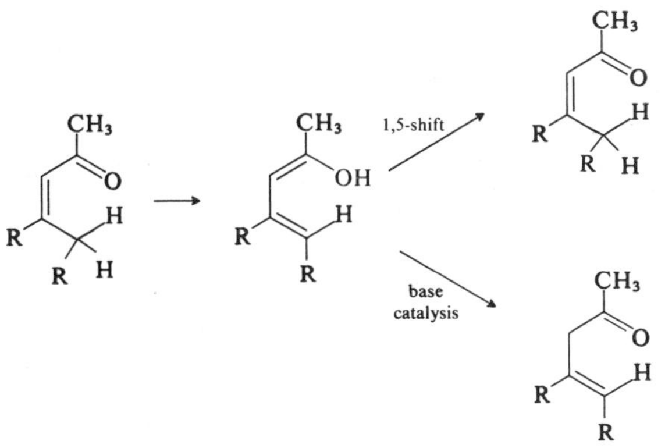

The dienol is unstable, and two separate processes have been identified for reketonization. These are a 1,5-sigmatropic shift of hydrogen leading back to the α,β-enone and a base-catalyzed proton transfer which leds to the β,γ-enone.[21] The deconjugated enone is formed because of the kinetic preference for reprotonation of the dienolate at the α-carbon. Photochemical deconjugation is a synthetically useful way of effecting isomerization of α,β-unsaturated ketones and esters to the β,γ-isomers.

4,4-Dialkylcyclohexenones undergo a photochemical rearrangement which involves the formal shift of the C-4—C-5 bond to C-3 and formation of a new

20. R. A. Cormier, W. L. Schreiber, and W. C. Agosta, *J. Am. Chem. Soc.* **95**, 4873 (1973); R. A. Cormier and W. C. Agosta, *J. Am. Chem. Soc.* **96**, 618 (1974).
21. R. Richard, P. Sauvage, C. S. K. Wan, A. C. Weeden, and D. F. Wong, *J. Org. Chem.* **51**, 62 (1986).

C-2—C-4 bond[22]:

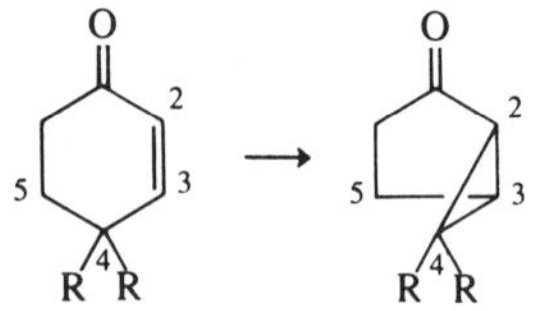

This reaction is quite general and also proceeds in the case of 4-alkyl-4-aryl-cyclohexenones:

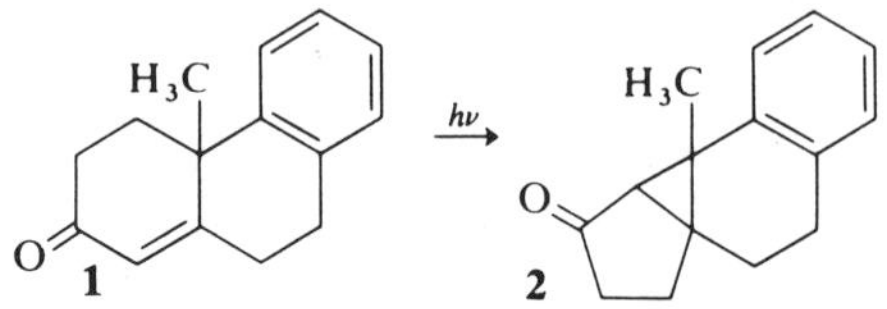

Ref. 23

The reaction is stereospecific and can be described as a $[\pi 2_a + \sigma 2_a]$ cycloaddition. This mechanism requires that inversion of configuration occur at C-4 as the new σ bond is formed at the back lobe of the reacting C-4—C-5 σ bond:

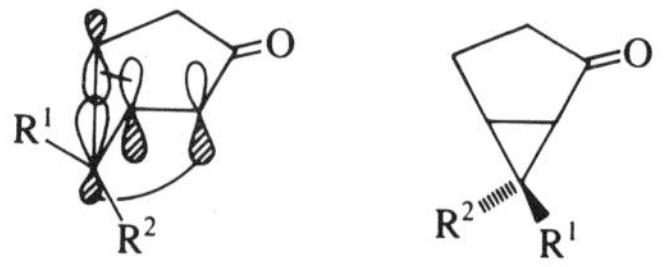

It has been demonstrated in several systems that the reaction is in fact stereospecific with the expected inversion occurring at C-4. The ketone **3** provides a specific example. The stereoisomeric products **4** and **5** are both formed, but in each product inversion has occurred at C-4:

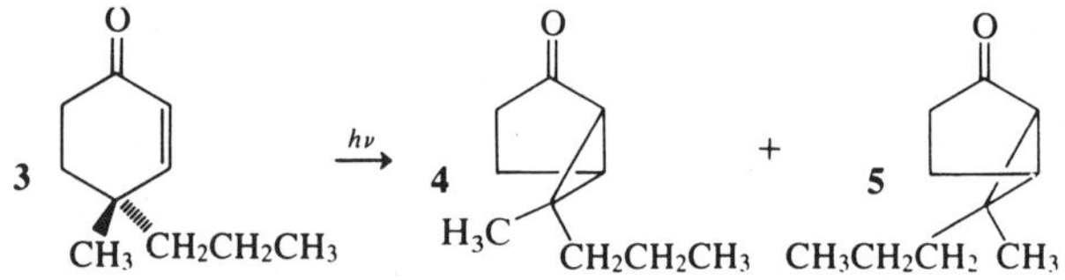

Ref. 24

With 4,4-diarylcyclohexenones, the reaction takes a slightly different course involving aryl migration. In compounds in which the two aryl groups are substituted differently, it is found that substituents that stabilize radical character favor migration. Thus the *p*-cyanophenyl substituent migrates in preference to the phenyl

22. For a review of this reaction, see D. I. Schuster, in *Rearrangements in Ground and Excited States*, Vol. 3, P. de Mayo (ed.), Academic Press, New York, 1980, Chapter 17.
23. O. L. Chapman, J. B. Sieja, and W. J. Welstead, Jr., *J. Am. Chem. Soc.* **88**, 161 (1966).
24. D. I. Schuster and J. M. Rao, *J. Org. Chem.* **46**, 1515 (1981); D. I. Schuster, R. H. Brown, and B. M. Resnick, *J. Am. Chem. Soc.* **100**, 4504 (1978).

substituent in **6**:

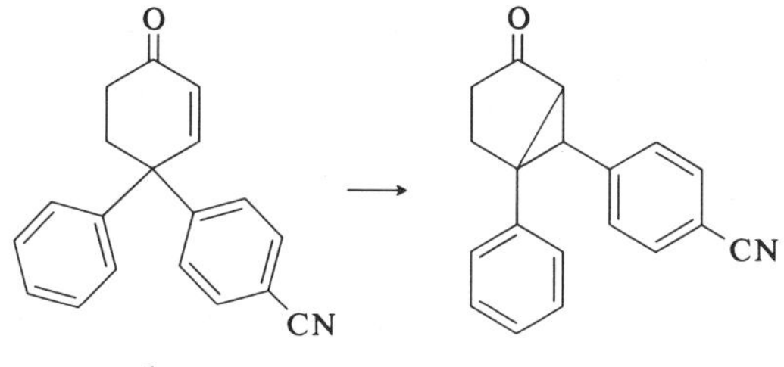

6 Ref. 25

This rearrangement can be considered to occur via a transition state in which C-2—C-4 bridging is accompanied by a 4 → 3 aryl migration[26]:

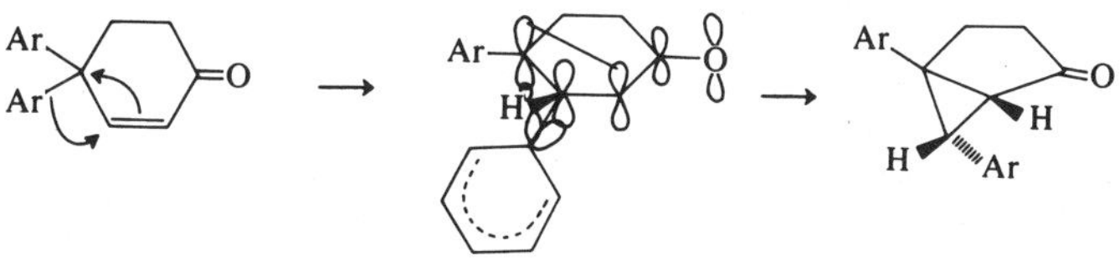

In contrast to the rearrangement described for 4,4-dialkylcyclohexenones, this reaction is not entirely stereospecific and a minor stereoisomer is formed:

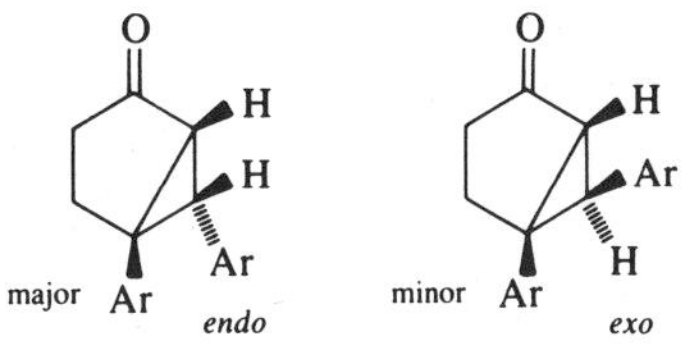

major *endo* minor *exo*

This suggests that the existence of a competitive reaction pathway which is not concerted. Note that the *endo* product is predicted by the concerted mechanism. It is the major product, even though it is sterically more congested than the *exo* isomer.

With other ring sizes, the photochemistry of unsaturated cyclic ketones takes different courses. For cyclopentenones, the principal products result from hydrogen abstraction processes. Irradiation of cyclopentenone in cyclohexane gives a mixture of 2- and 3-cyclohexylcyclopentanone.[27] These products can be formed by intermolecular hydrogen abstraction, followed by recombination of the resulting radicals:

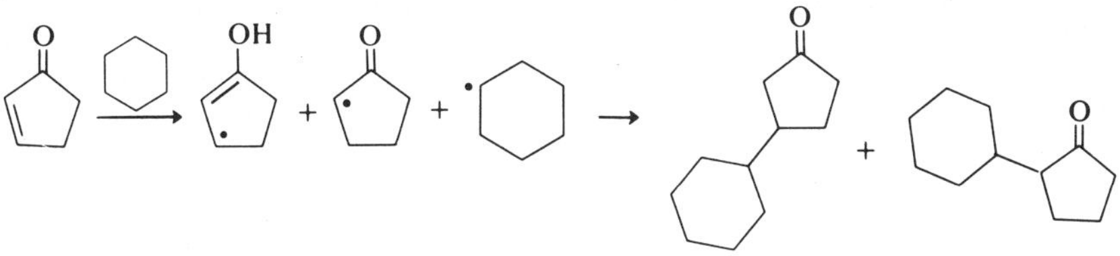

25. H. E. Zimmerman, R. D. Rieke, and J. R. Scheffer, *J. Am. Chem. Soc.* **89**, 2033 (1967).
26. H. E. Zimmerman, *Tetrahedron* **30**, 1617 (1974).
27. S. Wolff, W. L. Schreiber, A. B. Smith II, and W. C. Agosta, *J. Am. Chem. Soc.* **94**, 7797 (1972).

If a substituent chain is present on the cyclopentenone ring, an intramolecular hydrogen abstraction can take place:

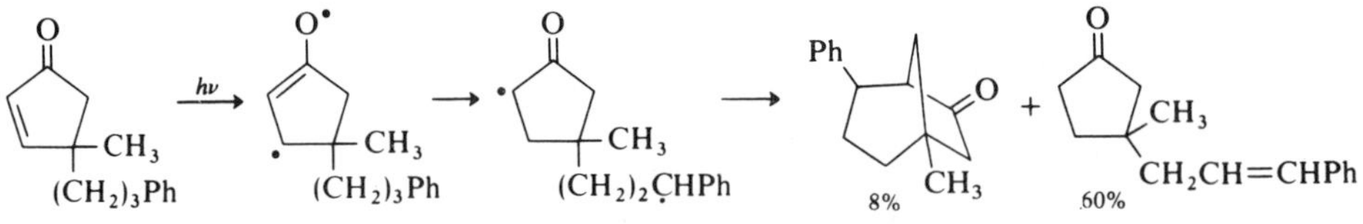

The bicyclic product is formed by coupling of the two radical sites, while the alkene results from an intramolecular hydrogen atom transfer. These reactions can be sensitized by aromatic ketones and quenched by typical triplet quenchers and are therefore believed to proceed via triplet excited states.

In the case of the cycloheptenone and larger rings, the main initial photoproducts are the *trans*-cycloalkenones produced by photoisomerization. In the case of the seven- and eight-membered rings, the *trans* double bonds are sufficiently strained that rapid reactions follow. In non-nucleophilic solvents, dimerization occurs, whereas in nucleophilic solvents addition occurs[28]:

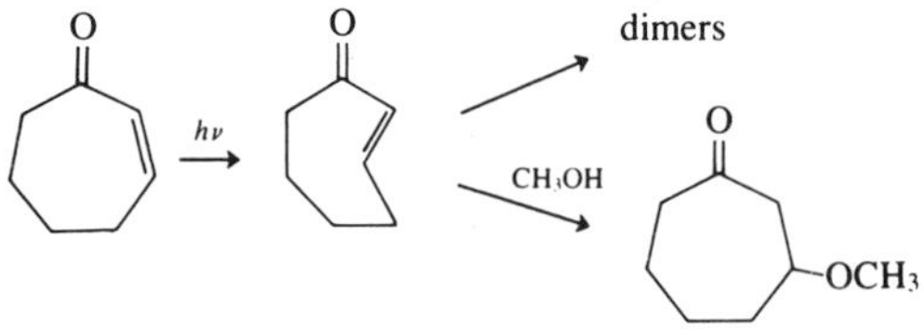

Ketones having a double bond in the β-γ-position are likely candidates for α cleavage because of the stability of the allyl radical that is formed. This is an important process on direct irradiation. Products then arise by recombination of the radicals or by recombination after decarbonylation:

$$\mathrm{RCCH_2CH{=}CHR'} \xrightarrow{h\nu} \mathrm{R\dot{C}{=}O} + \cdot\mathrm{CH_2CH{=}CHR'} \rightarrow \text{products}$$

In cyclic ketones, the diradical intermediates usually recombine, leading to isomerized ketones:

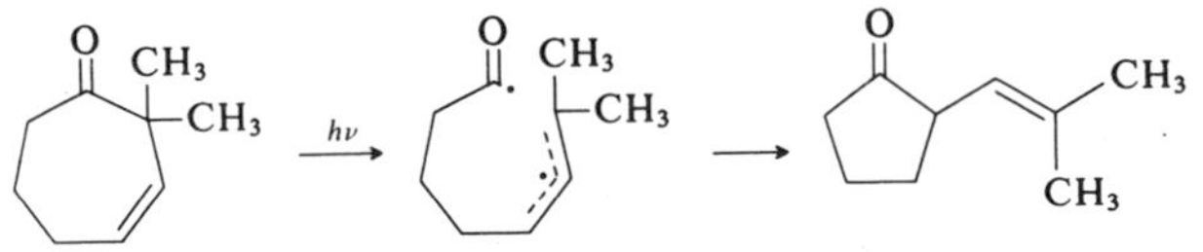

Ref. 29

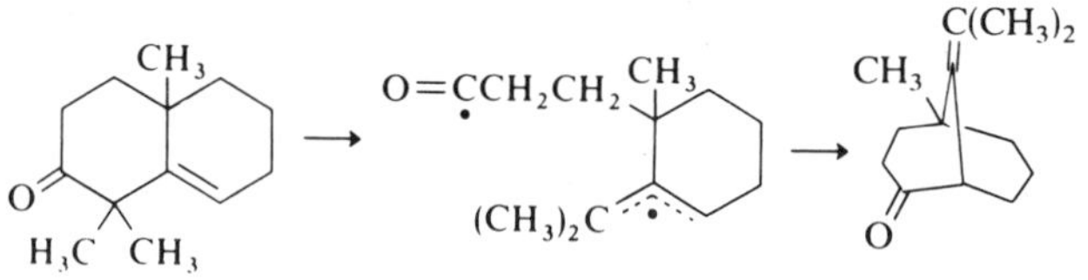

Ref. 30

28. H. Hart, B. Chen, and M. Jeffares, *J. Org. Chem.* **44**, 2722 (1979).
29. L. A. Paquette, R. F. Eizember, and O. Cox, *J. Am. Chem. Soc.* **90**, 5153 (1968).
30. H. Sato, N. Furutachi, and K. Nakanishi, *J. Am. Chem. Soc.* **94**, 2150 (1972).

Excitation of acyclic $\beta\gamma$-unsaturated ketones can give cyclopropyl ketones[31]:

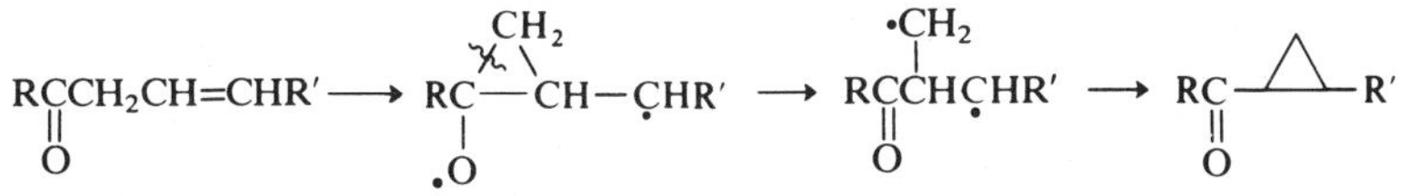

$$\mathrm{RCCH_2CH{=}CHR'} \longrightarrow \mathrm{RC{-}CH{-}\dot{C}HR'}\ (\text{with } \mathrm{CH_2},\ \cdot\mathrm{O}) \longrightarrow \mathrm{RC\dot{C}H\dot{C}HR'}\ (\cdot\mathrm{CH_2}) \longrightarrow \mathrm{RC}(\text{cyclopropyl})\mathrm{R'}$$

Another class of carbonyl compounds that has received much attention is the cyclohexadienones.[32] The primary photolysis product of 4,4-diphenylcyclohexadienone, for example, is **8**[33]:

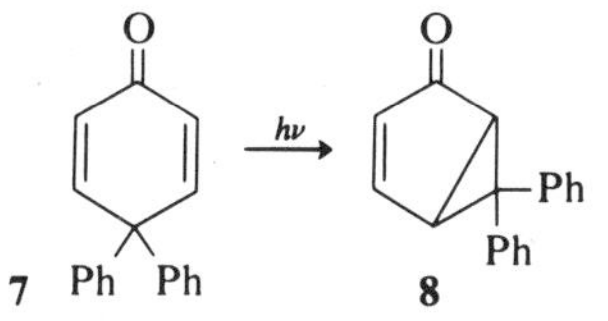

$$\mathbf{7} \xrightarrow{h\nu} \mathbf{8}$$

Quenching and photosensitization experiments on representative examples have indicated that the reaction proceeds through a triplet excited state. A scheme that delineates the bonding changes is outlined below:

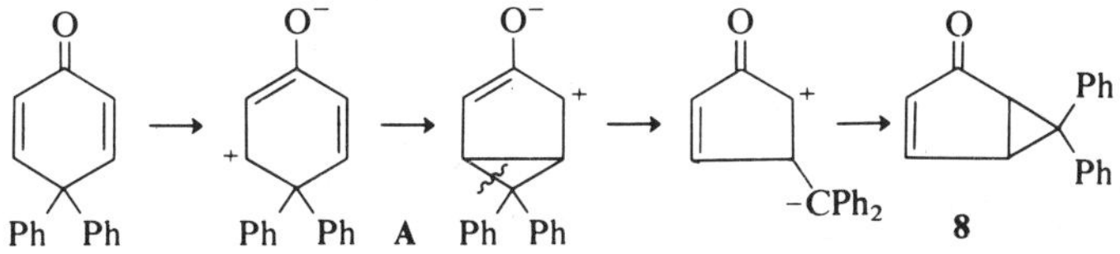

It is believed that a reactive ground state species, the zwitterion **A**, is the initial photoproduct and that it rearranges to the observed product.[34] To test this mechanism, generation of species **A** by nonphotochemical means was undertaken.[35] α-Haloketones, when treated with strong base, ionize to such dipolar intermediates. Thus, the bromoketone **9** is a potential precursor of **A**:

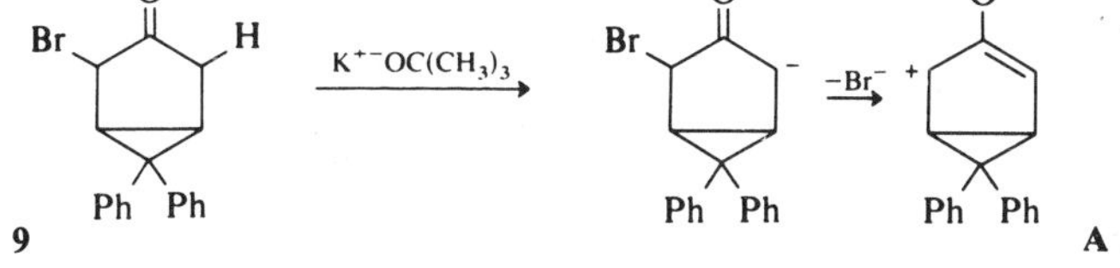

$$\mathbf{9} \xrightarrow{\mathrm{K^{+}\,^{-}OC(CH_3)_3}} \text{carbanion} \xrightarrow{-\mathrm{Br^-}} \mathbf{A}$$

The zwitterion prepared by this route did indeed lead to **8**, as required if it is an intermediate in the photochemical reaction. Further study of this process established

31. W. G. Dauben, M. S. Kellog, J. I. Seeman, and W. A. Spitzer, *J. Am. Chem. Soc.* **92**, 1786 (1970).
32. H. E. Zimmerman, *Angew. Chem. Int. Ed. Engl.* **8**, 1 (1969); K. Schaffner and M. Demuth, in *Rearrangements in Ground and Excited States*, Vol. 3, P. de Mayo (ed.), Academic Press, New York, 1980, Chapter 18; D. I. Schuster, *Acc. Chem. Res.* **11**, 65 (1978).
33. H. E. Zimmerman and D. I. Schuster, *J. Am. Chem. Soc.* **83**, 4486 (1961).
34. H. E. Zimmerman and J. S. Swenton, *J. Am. Chem. Soc.* **89**, 906 (1967).
35. H. E. Zimmerman, D. Döpp, and P. S. Huyffer, *J. Am. Chem. Soc.* **88**, 5352 (1966).

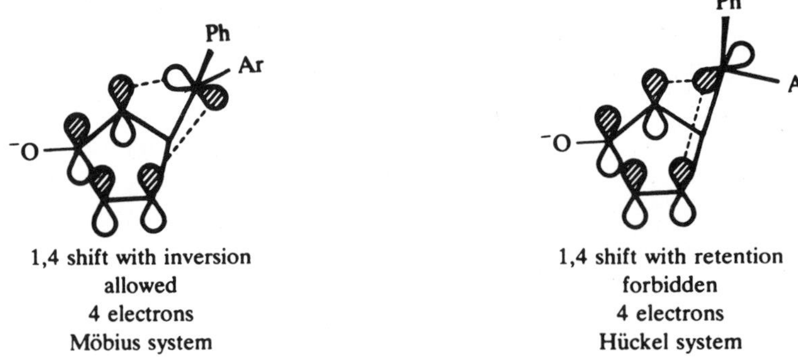

Fig. 13.6. Symmetry properties for [1,4]-sigmatropic shifts with inversion and retention.

another aspect of the reaction mechanism. The product could be formed by a process involving inversion at C-4 or by one involving a pivot about the C-3—C-4 bond:

or

The two mechanisms lead to the formation of stereochemically different products when the aryl groups at C-4 are different. When the experiment was carried out on **10**, only the product corresponding to inversion of configuration at C-4 was observed.[36]

$K^{+-}OC(CH_3)_3$

10 **11**

The rearrangement step itself is a ground state thermal process and may be classified as a [1,4]-sigmatropic shift of carbon across the face of a 2-oxybutenyl cation. The Woodward–Hoffmann rules requires a sigmatropic shift of this type to proceed with inversion of configuration. The orbitals involved in a [1,4]-sigmatropic shift are shown in Fig. 13.6.

As is clear from the preceding examples, there are a variety of overall reactions that can be initiated by photolysis of ketones. The course of photochemical reactions

36. H. E. Zimmerman and D. S. Crumrine, *J. Am. Chem. Soc.* **90**, 5612 (1968).

of ketones is very dependent on the structure of the reactant. Despite the variety of overall processes that can be observed, the number of individual steps involved is limited. For ketones, the most important are inter- and intramolecular hydrogen abstraction, cleavage α to the carbonyl group, and substituent migration to the β-carbon atom of α,β-unsaturated ketones. Reexamination of the mechanisms illustrated in this section will reveal that most of the reactions of carbonyl compounds that have been described involve combinations of these fundamental processes. The final product usually result from rebonding of reactive intermediates generated by these steps.

Some ketones undergo cycloaddition reaction with alkenes to form oxetanes:

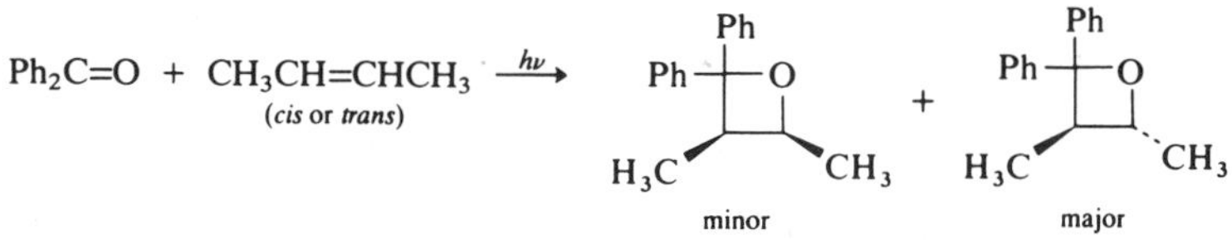

$$Ph_2C{=}O + \underset{(cis\text{ or }trans)}{CH_3CH{=}CHCH_3} \xrightarrow{h\nu} \underset{\text{minor}}{\text{oxetane}} + \underset{\text{major}}{\text{oxetane}}$$

The reaction is ordinarily stereoselective, favoring the more stable adduct, and a long-lived triplet diradical intermediate is implicated[37]:

$$^3[Ph_2C{=}O]^* + CH_3CH{=}CHCH_3 \longrightarrow Ph_2\dot{C}OCH(CH_3)\dot{C}HCH_3 \longrightarrow \text{products}$$

The diradical is believed to be preceded on the reaction path by a complex of the alkene with excited state benzophenone. This reaction, particularly its stereochemistry and regioselectivity, will be considered in more detail in Part B, Chapter 6.

13.4. Photochemistry of Alkenes and Dienes

The photochemistry of alkenes and dienes has already been examined in part, since these compounds are particularly illustrative of the principles of orbital control in electrocyclic processes. The orbital symmetry rules for cycloadditions and electrocyclic processes were covered in Section 13.2. Cycloadditions are also considered, from a synthetic viewpoint, in Part B, Chapter 6. This section will emphasize unimolecular photoreactions of alkenes and dienes.

A characteristic photochemical reaction of alkenes is interconversion of *cis* and *trans* isomers. Usually, the *trans* isomer is thermodynamically more stable, and photolysis establishes a mixture which is richer in the *cis* isomer. Irradiation therefore provides a means of converting a *trans* alkene to the *cis* isomer.

The composition of the photostationary state depends on the absorption spectra of the isomeric alkenes. A hypothetical case is illustrated in Fig. 13.7. Assume that

37. R. A. Caldwell, G. W. Sovocool, and R. P. Gajewski, *J. Am. Chem. Soc.* **95**, 2549 (1973).

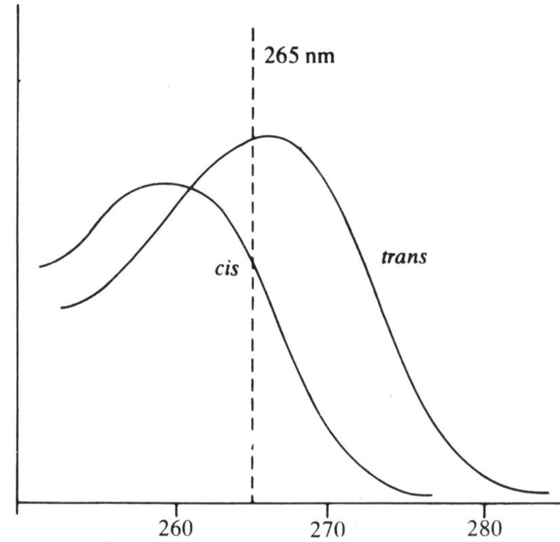

Fig. 13.7. Absorption spectra of a *cis-trans* isomer pair.

the vertical line at 265 nm is the limit of light impinging on the system. This wavelength limit can be controlled by use of appropriate sources and filters. Because of its longer wavelength maximum and higher extinction coefficient, the *trans* isomer will be absorbing substantially more light than the *cis* isomer. The relative amount of light absorbed at any wavelength will be proportional to the extinction coefficients at that wavelength. Assuming that the quantum yield for *cis* → *trans* is approximately equal to that for *trans* → *cis* conversion, the conversion of *trans* alkene to *cis* will occur faster than the converse process when the two isomers are in equal concentration. On continued photolysis, a photostationary state will be achieved at which the rate of *trans* → *cis* is equal to that of *cis* → *trans.* At this point, the concentration of the *cis* isomer will be greater than that of the *trans* isomer. The relationship can be expressed quantitatively for monochromatic light as

$$\frac{[t]_s}{[c]_s} = \left(\frac{\varepsilon_c}{\varepsilon_t}\right)\left(\frac{\phi_{c\to t}}{\phi_{t\to c}}\right)$$

The isomerization of alkenes is believed to take place via an excited state in which the two sp^2 carbons are twisted 90° with respect to one another. This state is referred to as the *p* (perpendicular) geometry. This geometry is believed to be the minimum-energy geometry for both the singlet and triplet excited states.

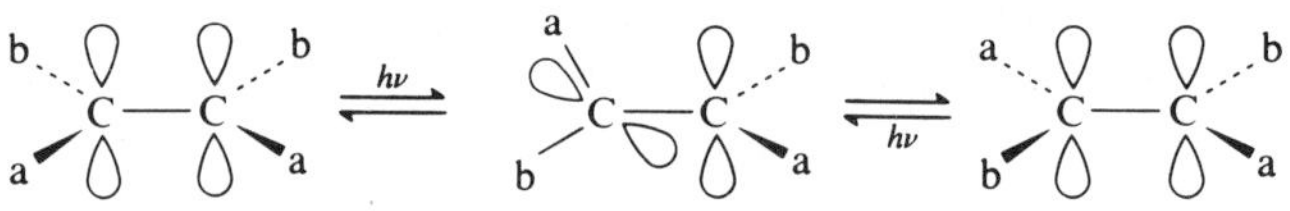

The perpendicular geometry for the excited state permits the possibility for returning to either the *cis* or *trans* configuration of the ground state.

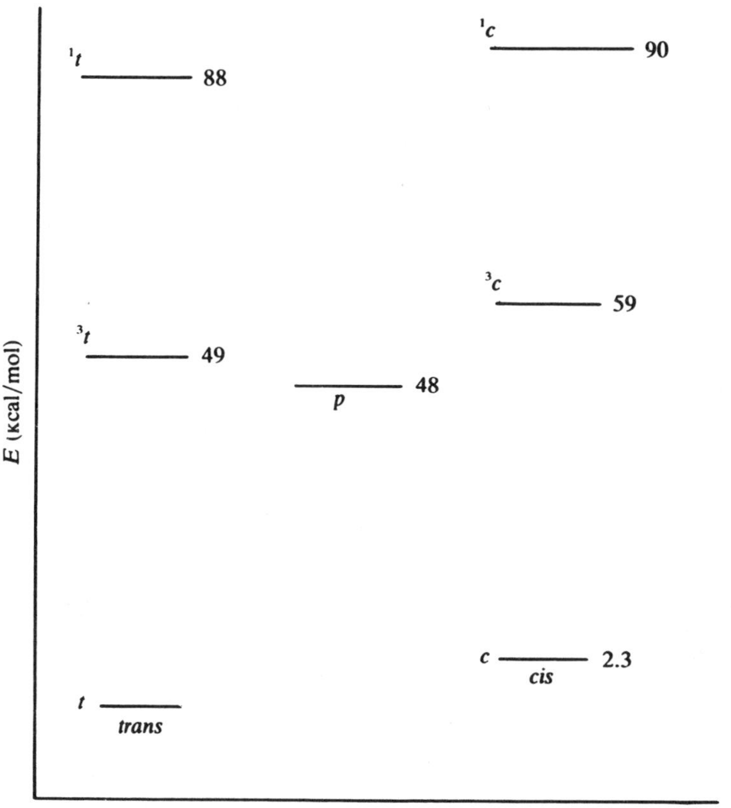

Fig. 13.8. Energy of excited states involved in *cis-trans* isomerization of stilbene. [From J. Saltiel and J. L. Charlton, *Rearrangments in Ground and Excited States*, P. de Mayo (ed.), Academic Press, 1980, Chap. 14.]

Especially detailed study of the mechanism of photochemical configurational isomerism has been done on *cis*- and *trans*-stilbene.[38] Spectroscopic data have established the energies of the singlet and triplet states of both *cis*- and *trans*-stilbene and of the twisted excited states that are formed from both isomers. This information is summarized in Fig. 13.8. It is believed that the geometries of the species ^3t and ^3p are very similar. The state ^3c is believed to vibrationally convert rapidly to ^3p. The excited states derived from both *cis*- and *trans*-stilbene can readily attain the common p states.

Direct irradiation leads to isomerization via singlet state intermediates.[39] The isomerization presumably involves a twisted singlet state that can be reached from either the *cis* or the *trans* isomer. The temperature dependence of the isomerization further reveals that the process of formation of the twisted state involves a small activation energy. This energy is required for conversion of the initial excited state

38. J. Saltiel, J. T. D'Agostino, E. D. Megarity, L. Metts, K. R. Neuberger, M. Wrighton, and O. C. Zafiriow, *Org. Photochem.* **3**, 1 (1973); J. Saltiel and J. L. Charlton, in *Rearrangments in Ground and Excited States*, Vol. 3, P. de Mayo (ed.), Academic Press, New York, 1980, Chapter 14.
39. J. Saltiel, *J. Am. Chem. Soc.* **89**, 1036 (1967); **90**, 6394 (1968).

to the perpendicular geometry associated with the ^1S state. Among the pieces of evidence indicating that a triplet intermediate is not involved for direct irradiation is the fact that azulene, which is known to intercept stilbene triplets, has only a minor effect on the efficiency of the direct photoisomerization.[40]

The photoisomerization can also be carried out by photosensitization. Under these conditions, the composition of the photostationary state depends on the triplet energy of the sensitizer. With sensitizers having triplet energies above 60 kcal/mol, $[c]/[t]$ is slightly more than 1, but a range of sensitizers having triplet energies of 52–58 kcal/mol affords much higher *cis* : *trans* ratios in the photostationary state.[41] The high *cis* : *trans* ratio in this region results from the fact that the energy required for excitation of *trans*-stilbene is less than for excitation of *cis*-stilbene (see Fig. 13.8). Thus, sensitizers in the range 52–58 kcal/mol selectivity excite the *trans* isomer. Since the rate of conversion of *trans* to *cis* is increased, the composition of the photostationary state is the enriched in *cis* isomer.

Direct photochemical excitation of unconjugated alkenes requires light with $\lambda \leq 230$ nm. There have been relatively few studies of direct photolysis of alkenes in solution because of the experimental difficulties imposed by this wavelength restriction. A study of *cis*- and *trans*-2-butene diluted with neopentane demonstrated that *cis–trans* isomerization was competitive with the photochemically allowed $[2+2]$ cycloaddition which is the dominant reaction in pure liquid alkene.[42] The cycloaddition reaction is completely stereospecific for each isomer, which requires that the *excited intermediates involved in cycloaddition must retain a geometry which is characteristic of the reactant isomer.* As the ratio of neopentane to butene is increased, the amount of cycloaddition decreases relative to *cis–trans* isomerization. This effect presumably is the result of the lifetime of the intermediate responsible for cycloaddition being very short. When the alkene is diluted by inert hydrocarbon, the rate of encounter of a second alkene molecule is reduced, and the unimolecular isomerization becomes the dominant reaction.

Aromatic compounds such as toluene, xylene, and phenol can photosensitize *cis–trans* interconversion of simple alkenes. This is a case where the sensitization process must be somewhat endothermic because of the energy relationships between the excited state of the alkene and the sensitizers. The photostationary state obtained under these conditions favors the less strained of the two isomers. The explanation for this effect can be summarized with reference to Fig. 13.9. Isomerization takes place through a twisted triplet state. This state is achieved by a combination of energy transfer from the sensitizer and thermal activation. Because the *cis* isomer is somewhat higher in energy, its requirement for activation to the excited state is somewhat less than for the *trans* isomer. If it is also assumed that the excited state forms the *cis* and *trans* isomers with equal ease, the rate of *cis* $\rightarrow$ *trans* conversion

40. J. Saltiel, E. D. Megarity, and K. G. Kneipp, *J. Am. Chem. Soc.* **88**, 2336 (1966); J. Saltiel and E. D. Megarity, *J. Am. Chem. Soc.* **91**, 1265 (1969).
41. G. S. Hammond, J. Saltiel, A. A. Lamola, N. J. Turro, J. S. Bradshaw, D. O. Cowan, R. C. Counsell, V. Vogt, and C. Dalton, *J. Am. Chem. Soc.* **86**, 3197 (1964); S. Yamauchi and T. Azumi, *J. Am. Chem. Soc.* **95**, 2709 (1973).
42. H. Yamazaki and R. J. Cventanovic, *J. Am. Chem. Soc.* **91**, 520 (1969); H. Yamazaki, R. J. Cventanovic, and R. S. Irwin, *J. Am. Chem. Soc.* **98**, 2198 (1976).

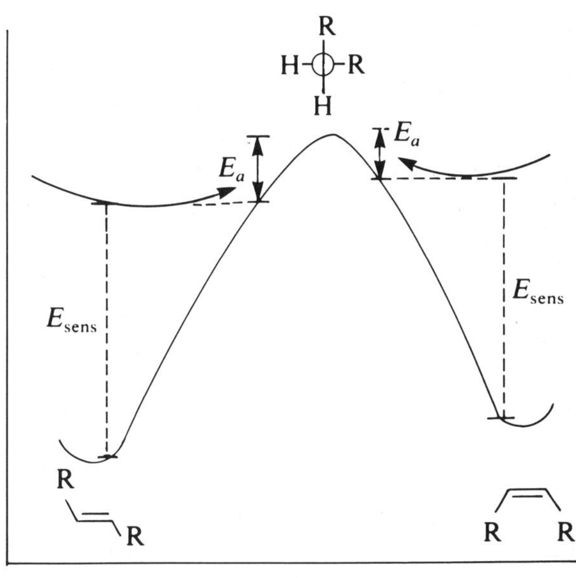

Fig. 13.9. Energy diagram illustrating differential in energy deficit for photosensitized isomerization of *cis* and *trans* isomers.

exceeds that for *trans* → *cis* conversion ($k_{c\rightarrow t} > k_{t\rightarrow c}$), and, at the photostationary state, $[trans]k_{c\rightarrow t} = [cis]k_{t\rightarrow c}$ and therefore $[trans] > [cis]$.[43]

The reaction course taken by photoexcited cycloalkenes in hydroxylic solvents depends very much on ring size. 1-Methylcyclohexene, 1-methylcycloheptene, and 1-methylcyclooctene all add methanol, but neither 1-methylcyclopentene nor norbornene does so. The key intermediate in the addition reactions is believed to be the highly reactive *trans* isomer of the cycloalkene.

It appears that the *trans*-cycloalkenes can be protonated exceptionally easily, because of the enormous relief of strain that accompanies protonation.[44,45] The *trans* isomers of cyclopentene and norbornene are too strained to be formed. Cyclopentene and norbornene give products that result from hydrogen abstraction processes.[46] The reactivity of the excited state is that of a diradical species.

43. J. J. Snyder, F. P. Tise, R. D. Davis, and P. J. Kropp, *J. Org. Chem.* **46**, 3609 (1981).
44. P. J. Kropp, E. J. Reardon, Jr., Z. L. F. Gaibel, K. F. Williard, and J. H. Hattaway, Jr., *J. Am. Chem. Soc.* **95**, 7058 (1973).
45. J. A. Marshall, *Acc. Chem. Res.* **2**, 33 (1969).
46. P. J. Kropp, *J. Am. Chem. Soc.* **91**, 5783 (1969).

As was mentioned in Section 13.2, the [2 + 2] photocycloaddition of alkenes is an allowed reaction according to orbital symmetry considerations. Among the most useful reactions in this category, from a synthetic point of view, are intramolecular [2 + 2] cycloadditions of dienes and intermolecular [2 + 2] cycloadditions of alkenes with cyclic α,β-unsaturated carbonyl compounds. These reactions will be discussed in more detail in Part B, Section 6.4.

Conjugated dienes can undergo a variety of photoreactions, depending on whether excitation is direct or photosensitized. The benzophenone-sensitized excitation of 1,3-pentadiene, for example, results in stereochemical isomerization and dimerization[47]:

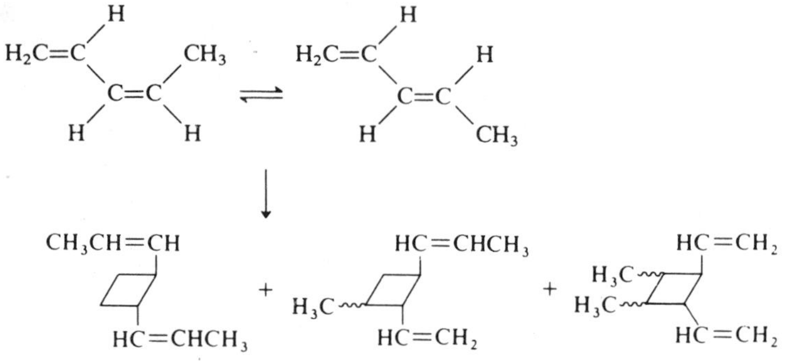

H
H_2C=C CH_3 ⇌ H_2C=C H
C=C C=C
H H H CH_3

↓

CH_3CH=CH + HC=$CHCH_3$ + HC=CH_2
H_3C H_3C
HC=$CHCH_3$ HC=CH_2 H_3C HC=CH_2

Alkyl derivatives of 1,3-butadiene usually undergo photosensitized *cis–trans* isomerization when photosensitizers that can supply at least 60 kcal/mol are used. Two conformers of the diene, the *s-cis* and *s-trans*, exist in equilibrium, so there are two nonidentical ground states from which excitation can occur. Two triplet excited states that do not readily interconvert are derived from the *s-trans* and *s-cis* conformers. Theoretical calculations suggest that the minimum-energy geometry for the excited states of conjugated dienes can be described as an alkyl radical and an orthogonal allyl system. The species can be described as an allylmethylene diradical:

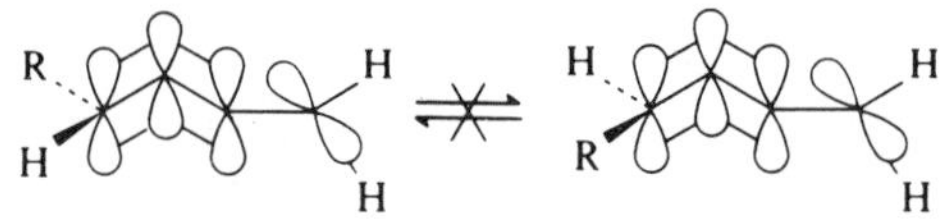

R H ⇌̸ H H
H H R H

Such a structure implies that there would be a barrier to rotation about the C-2—C-3 bond and would explain why the *s-trans* and *s-cis* conformers lead to different excited states.[48]

Another result that can be explained in terms of the two noninterconverting excited states is the dependence of the ratio of [2 + 2] and [2 + 4] addition products on sensitizer energy. The *s-cis* geometry is suitable for cyclohexene formation, but

47. J. Saltiel, D. E. Townsend, and A. Sykes, *J. Am. Chem. Soc.* **95**, 5968 (1973).
48. R. Hoffmann, *Tetrahedron* **22**, 521 (1966); N. C. Baird and R. M. West, *J. Am. Chem. Soc.* **93**, 4427 (1971).

the *s-trans* is not. The excitation energy for the *s-cis* state is slightly lower than that for the *s-trans*. With low-energy sensitizers, therefore, the *s-cis* excited state is formed preferentially, and the ratio of cyclohexene to cyclobutane product increases.[49]

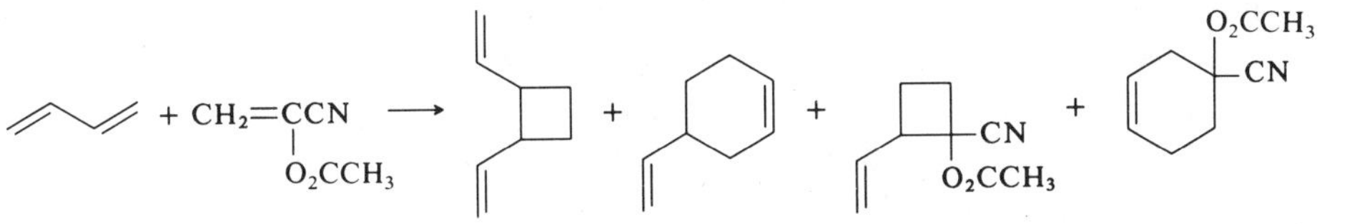

+ $CH_2{=}C(O_2CCH_3)CN$ → + + CN, O_2CCH_3 + O_2CCH_3, CN

Sensitizer	E_T	Butadiene dimers 2+2	Butadiene dimers 2+4	α-Acetoxy acrylonitrile adducts 2+2	α-Acetoxy acrylonitrile adducts 2+4
Acetophenone	73.6	97	3	98	2
Benzophenone	68.5	96	4	98	2
Triphenylene	66.6			97	3
Anthraquinone	62.4			95	5
Biacetyl	54.9	71	29	76	24
Benzil	53.7	66	34	74	26
Pyrene	48.7			69	31
Anthracene	42.5			87	13

The structure of the excited state of 1,3-dienes is also significant with respect to *cis–trans* isomerization. If the excited state is an allylmethylene diradical, only one of the two double bonds would be isomerized in any single excitation event:

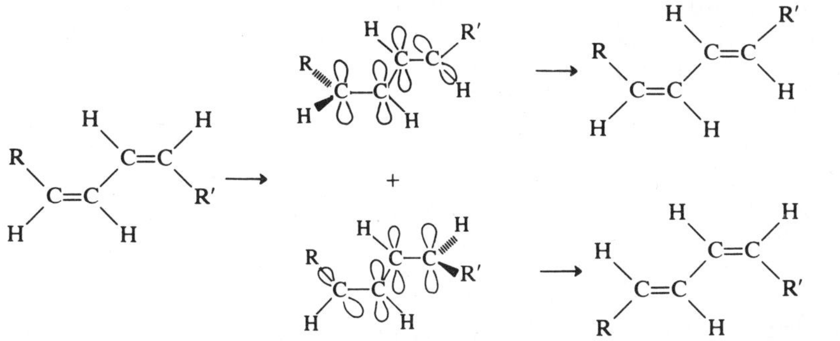

but not

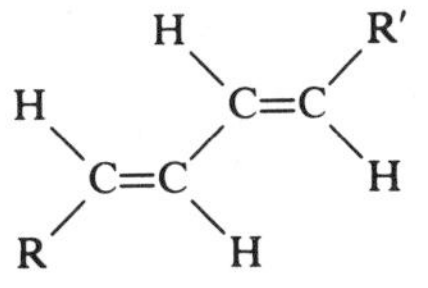

49. R. S. Liu, N. J. Turro, Jr., and G. S. Hammond, *J. Am. Chem. Soc.* **87**, 3406 (1965); W. L. Dilling, R. D. Kroening, and J. C. Little, *J. Am. Chem. Soc.* **92**, 928 (1970).

On the other hand, if the two possible allylmethylene diradicals interconvert rapidly, excitation could lead to isomerization at both double bonds. It is this latter situation that apparently exists in the triplet state. The triplet state has a high bond order between C-2 and C-3 and resists rotation about this bond, but the barrier to rotation at either of the terminal carbons is low.[43,50] Both double bonds can isomerize through this excited state. In contrast, it has been shown that in direct irradiation of 2,4-hexadiene, only one of the double bonds isomerizes on excitation.[51] The singlet state apparently retains a substantial barrier to rotation about the bonds in the allyl system:

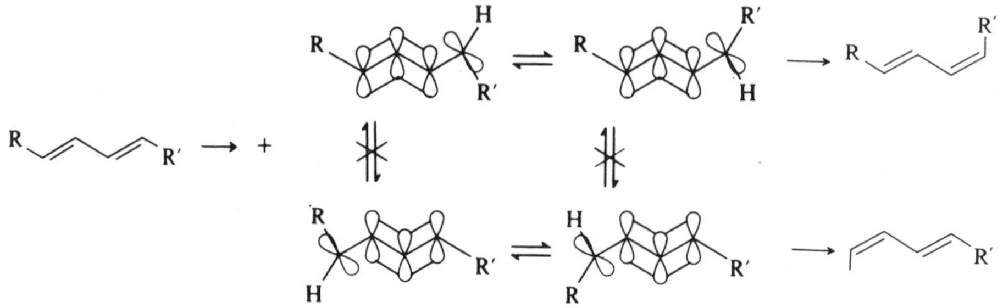

An alternative description of the singlet excited state is a cyclopropylmethyl singlet diradical. Only one of the terminal carbons would be free to rotate in such a structure.

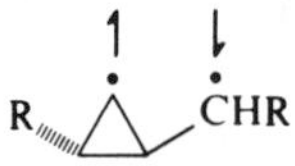

Orbital symmetry control of subsequent ring opening could account for isomerization at only one of the double bonds. Taking ψ_3 as the controlling frontier orbital, it can be seen than a concerted return to ψ_2 leads to rotation at only one terminus of the diene:

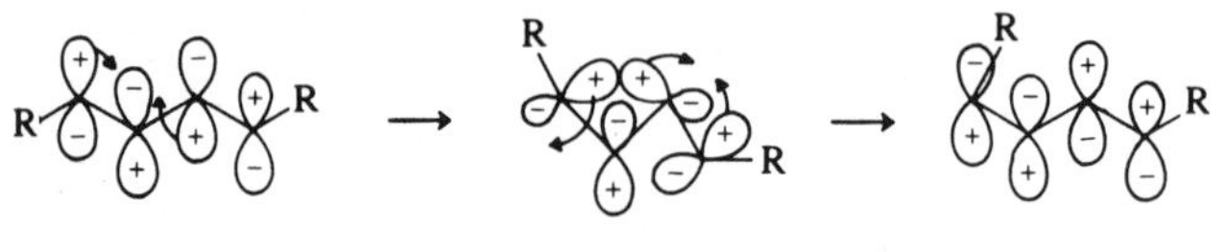

On direct irradiation of 1,3-pentadiene, *cis–trans* isomerization is accompanied by cyclization to 1,3-dimethylcyclopropene and 3-methylcyclobutene:

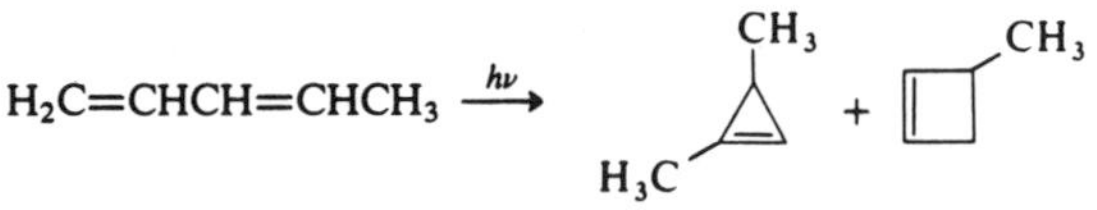

$$H_2C{=}CHCH{=}CHCH_3 \xrightarrow{h\nu}$$

50. J. Saltiel, A. D. Rousseau, and A. Sykes, *J. Am. Chem. Soc.* **94**, 5903 (1972).
51. J. Saltiel, L. Metts, and M. Wrighton, *J. Am. Chem. Soc.* **92**, 3227 (1970).

The reaction to form the latter product is an example of a concerted, photochemically allowed, electrocyclic reaction. A hydrogen atom migration from the cyclopropyldimethyl diradical can account for the cycloprepene formation.[52] This product, then, is suggestive of a ring structure in the excited state:

$H_2C{=}CHCH{=}CHCH_3 \longrightarrow$ [diradical $H_2\dot{C}$–cyclopropyl–CH_3] $\longrightarrow$ [1,3-dimethylcyclopropene]

Theoretical calculations indicate that it becomes progressively easier for a photochemical *cis–trans* isomerization to occur as the conjugation is extended. Furthermore, it is easiest to isomerize the central double bond and hardest to isomerize a terminal double bond.[53] This conclusion is consistent with simple Hückel theory since the stability of the two orthogonal radicals will be greatest when the bond broken is the central π bond.

Cyclohexadienes represent a special case among conjugated dienes since the occurrence of *cis–trans* isomerization is precluded by the ring geometry. Electrocyclic ring opening (see Section 13.2) can occur:

[Structure: tricyclic diene with $CH(CH_3)_2$, H_3C, CO_2H] $\underset{}{\overset{h\nu}{\rightleftharpoons}}$ [bicyclic triene with H_2C, CH_2, $CHCCH(CH_3)_2$, H_3C, CO_2H] Ref. 54

5,5-Diphenylcyclohexadiene shows divergent photochemical behavior, depending on whether the reaction is induced by direct irradiation or by photosensitization. On direct irradiation, the electrocyclic conversion to 1,1-diphenylhexatriene is dominant, whereas a rearrangement involving one of the aromatic rings is the major reaction of the triplet excited state formed by photosensitization[55]:

[Scheme: 5,5-diphenylcyclohexadiene —sensitized→ spiro diradical (Ph) → diradical (Ph, Ph) → bicyclohexene (Ph, Ph); —direct→ $H_2C{=}CHCH{=}CHCH{=}CPh_2$]

52. S. Boué and R. Srinivasan, *J. Am. Chem. Soc.* **94**, 3226 (1970).
53. I. Ohmine and K. Morokuma, *J. Chem. Phys.* **73**, 1907 (1980); M. Said, D. Maynau, and J.-P. Malrieu, *J. Am. Chem. Soc.* **106**, 580 (1984).
54. W. G. Dauben and R. M. Coates, *J. Org. Chem.* **29**, 2761 (1964).
55. H. E. Zimmerman and G. A. Epling, *J. Am. Chem. Soc.* **94**, 8749 (1972); J. S. Swenton, J. A. Hyatt, T. J. Walker, and A. L. Crumrine, *J. Am. Chem. Soc.* **93**, 4808 (1971).

The latter reaction is an example of the *di-π-methane* rearrangement.[56] This rearrangement is very general reaction for 1,4-dienes and other systems that have two π systems separated by an sp^3 carbon atom:

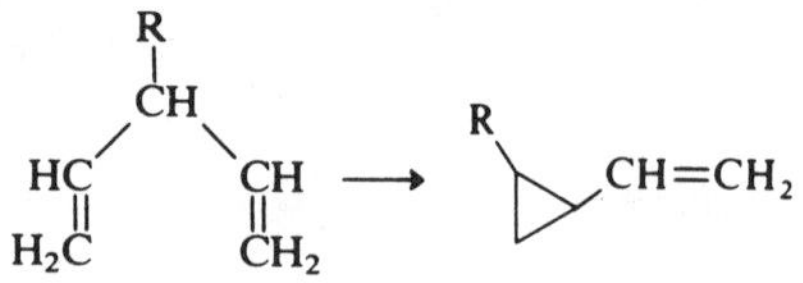

The transformation can be formulated in terms of a diradical species formed by bonding between C-2 and C-4:

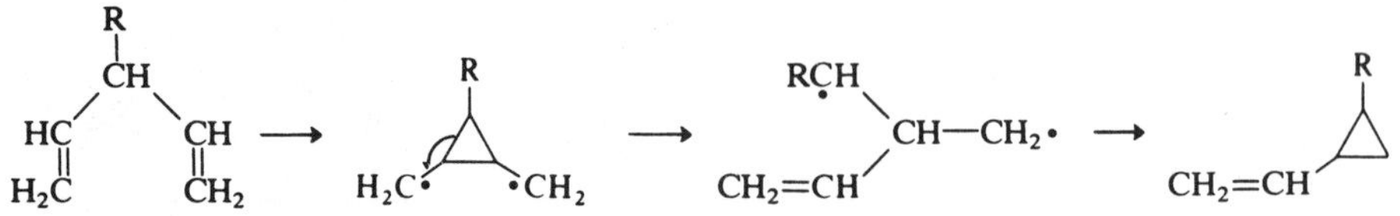

It has been found that the di-π-methane rearrangement can proceed through either a singlet or a triplet excited state.[57] The reaction can be formulated as a concerted process, and this mechanism is followed in the case of some acyclic dienes and for cyclic systems in which a concerted process is sterically feasible.

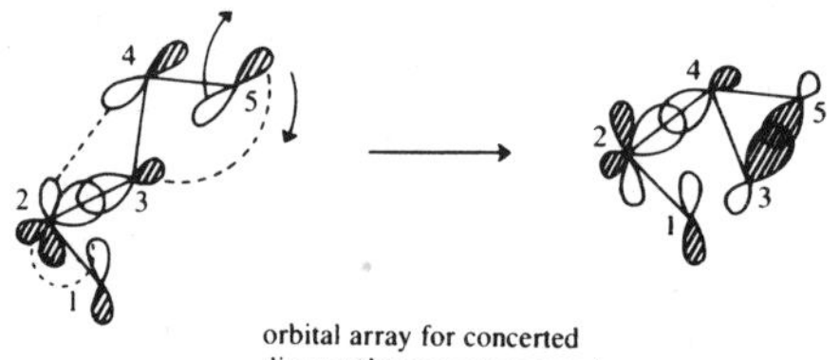

orbital array for concerted di-π-methane rearrangement

Notice that the orbital array is of the Möbius topology with a phase change depicted between the C-1 and C-2 positions. This corresponds to an allowed photochemical process since there are six electrons involved in bonding changes.

The di-π-methane rearrangement has been studied in a sufficient number of cases to elucidate some of the patterns regarding substituent effects. When the central sp^3 carbon is unsubstituted, the di-π-methane mechanism becomes less favorable. The case of 1,1,5,5-tetraphenyl-1,4-pentadiene is illustrative. Although one of the products has the expected structure, labeling with deuterium proves that an alternative mechanism operates.

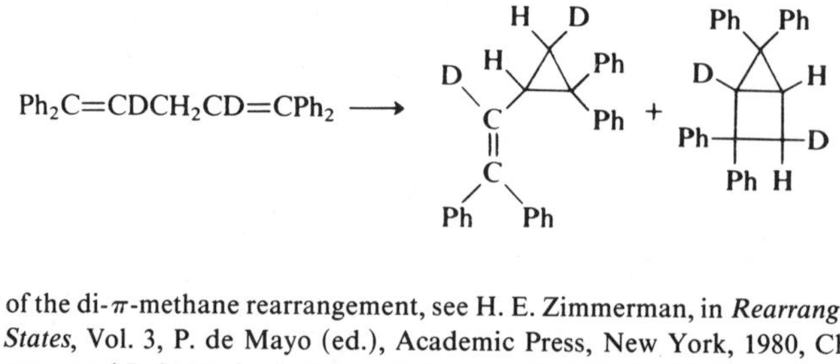

$$Ph_2C{=}CDCH_2CD{=}CPh_2 \longrightarrow$$

56. For a review of the di-π-methane rearrangement, see H. E. Zimmerman, in *Rearrangements in Ground and Excited States*, Vol. 3, P. de Mayo (ed.), Academic Press, New York, 1980, Chapter 16.
57. H. E. Zimmerman and P. S. Mariano, *J. Am. Chem. Soc.* **91**, 1718 (1969); P. S. Mariano, R. B. Steitle, D. G. Watson, M. J. Peters, and E. Bay, *J. Am. Chem. Soc.* **98**, 5899 (1976).

The cyclopropane bridge is formed only after hydrogen atom migration. The driving force for this migration may be the fact that a more stable allylic radical results.

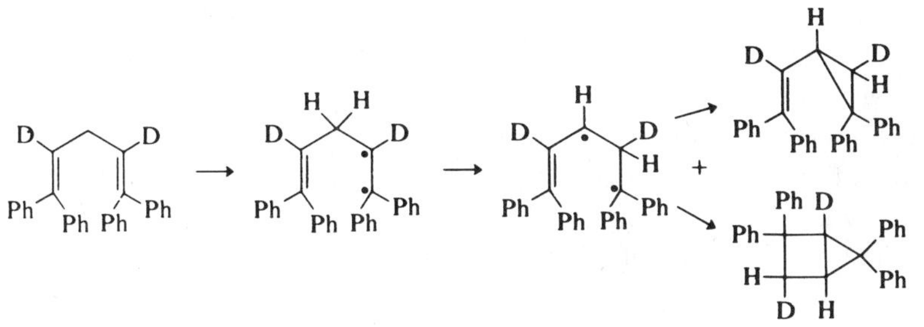

The resistance of the unsubstituted system to the di-π-methane rearrangement probably occurs at the second step of the rearrangement.[58] If the central carbon is unsubstituted, this step results in the formation of a primary radical and would be energetically unfavorable.

The groups at the termini of the 1,4-pentadiene system also affect the efficiency and direction of the reaction. The general trend is that cyclization will tend to occur at the diene terminus that best stabilizes radical character. Thus, a terminus substituted with aryl groups will cyclize in preference to an unsubstituted or alkyl-substituted terminus.

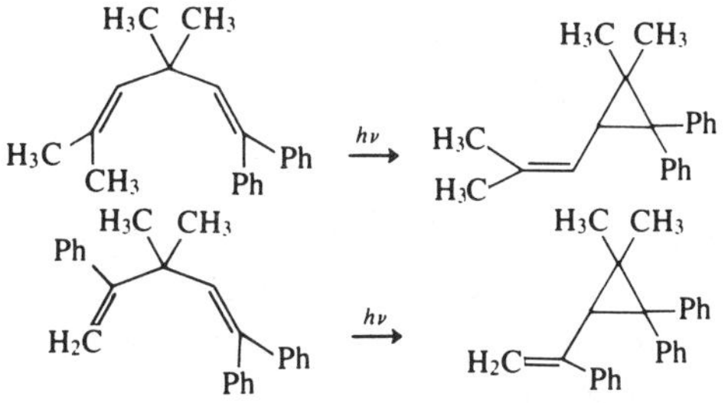

Ref. 59

Ref. 60

This result can be rationalized in terms of a diradical structure by noting that the bond cleavage of the initial intermediate will occur to give the more stable of the two possible 1,3-diradicals.[61] The cyclopropane ring in the final product will then incorporate this terminus:

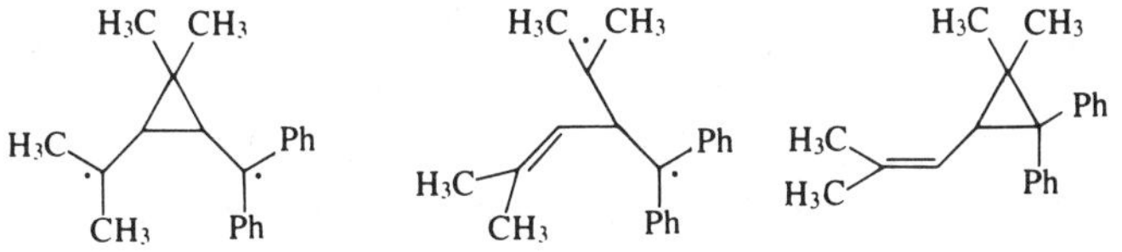

58. H. E. Zimmerman and J. O. Pincock, *J. Am. Chem. Soc.* **95**, 2957 (1973).
59. H. E. Zimmerman and A. C. Pratt, *J. Am. Chem. Soc.* **92**, 1409 (1970).
60. H. E. Zimmerman and A. A. Baum, *J. Am. Chem. Soc.* **93**, 3646 (1971).
61. H. E. Zimmerman and A. C. Pratt, *J. Am. Chem. Soc.* **92**, 6259, 6267 (1970).

This interpretation can be expressed in terms of the concerted mechanism by regarding the "diradical" to be a contributing structure to the transition state of the concerted process.

The di-π-methane rearrangement is a stereospecific reaction. There are several elements of stereochemistry to be considered. It is known that the double bond which remains uncyclized retains the *E*- or *Z*-configuration present in the starting material. This result immediately excludes any intermediate with a freely rotating terminal radical. The concerted transition state implies that C-3 would undergo inversion of configuration since the new C-3—C-5 bond is formed using the back lobe of the C-2—C-3 σ bond. This inversion of configuration has been confirmed.[62]

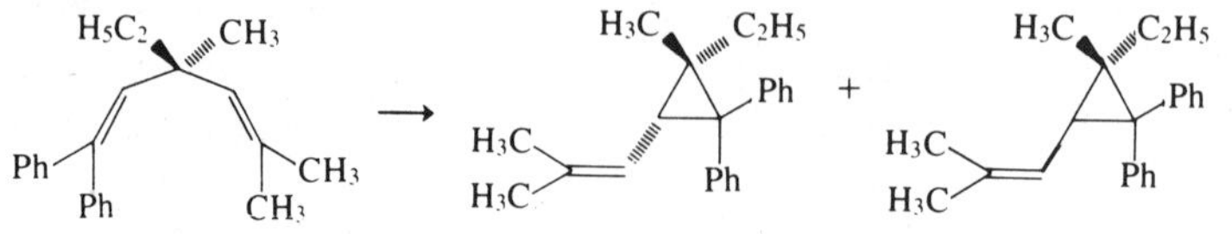

Thus, the transition state depicted for the concerted reaction correctly predicts the stereochemical course of the di-π-methane rearrangement.

An alternative pathway for reaction of 1,4-dienes is intramolecular cycloaddition giving bicyclo[2.1.0]pentanes. This pathway is usually not observed, but 1,5-diphenyl-1,4-pentadiene is an example of a compound that takes this course[63]:

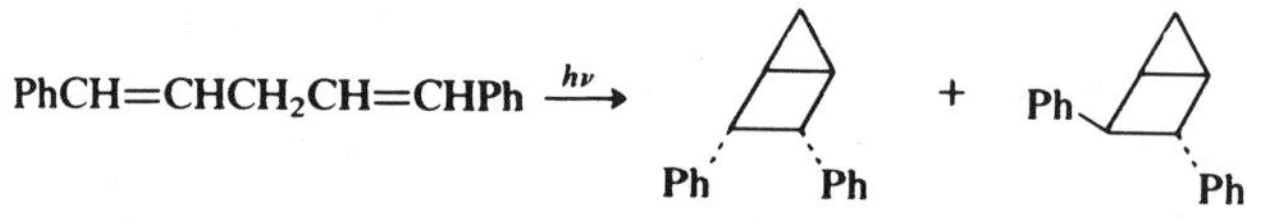

$$PhCH{=}CHCH_2CH{=}CHPh \xrightarrow{h\nu}$$

The structural explanation that has been offered to rationalize this altered reaction course is that the phenyl substituent may stabilize the diradical formed by 2,4-bridging, permitting it to exist long enough for closure to occur in preference to the concerted di-π-methane rearrangement. Intramolecular cycloaddition is also observed for 1,4-cyclooctadiene[64]:

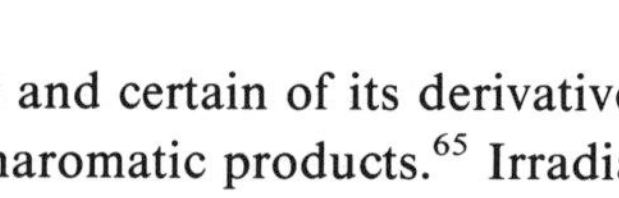

13.5. Photochemistry of Aromatic Compounds

Irradiation of benzene and certain of its derivatives results in bond reorganization and formation of nonaromatic products.[65] Irradiation of liquid benzene with

62. H. E. Zimmerman, J. D. Robbins, R. D. McKelvey, C. J. Samuel, and L. R. Sousa, *J. Am. Chem. Soc.* **96**, 1974, 4630 (1974).
63. E. Block and H. W. Orf, *J. Am. Chem. Soc.* **94**, 8438 (1972).
64. S. Moon and C. R. Ganz, *Tetrahedron Lett.*, 6275 (1968).
65. D. Bryce-Smith and A. Gilbert, *Tetrahedron* **32**, 1309 (1976).

light of 254-nm wavelength results in the accumulation of a very small amount of tricyclo[$3.1.0.0^{2,6}$]hex-3-ene, also known as benzvalene[66]:

$h\nu$

The maximum conversion to this product in liquid benzene is only 0.01%. A higher concentration (~1%) is achieved in solutions diluted with saturated hydrocarbons.

Because of the low photostationary concentration of benzvalene, photolysis is not an efficient way of accumulating this species. The highly reactive molecule can be trapped, however, if it is generated in the presence of other molecules with which it reacts. Irradiation of benzene in acidic hydroxylic solvents gives products formally resulting from 1,3-bonding in the benzene ring and addition of a molecule of solvent:

ROH, H^+ → **12** (OR) + **13** (OR)

These compounds are not direct photoproducts, however. The compounds arise by solvolysis of benzvalene, the initial photoproduct. Products of type **13** are secondary photoproducts derived from **12**.[67]

The photoisomerization of aromatic rings has also been studied with 1,3,5-tri-*t*-butylbenzene. The composition of the photostationary state is shown below[68]:

7.3% ⇌ <0.7% ⇌ 20.6% ⇌ 7.1% ⇌ 64.8%

R = *t*-butyl

These various photoproducts are all valence isomers of the normal benzenoid structure. These alternative bonding patterns are reached from the excited state, but it is difficult to specify a precise mechanism. The presence of the *t*-butyl groups introduces a steric factor that works in favor of the photochemical valence isomerism. Whereas the *t*-butyl groups are coplanar in the aromatic system, the geometry of the bicyclic products results in reduced steric interactions between adjacent *t*-butyl groups.

Irradiation of solutions of alkenes in benzene or substituted benzenes gives primarily 1:1 adducts in which the alkene bridges *meta* positions of the aromatic

66. K. E. Wilzbach, J. S. Ritscher, and L. Kaplan, *J. Am. Chem. Soc.* **89**, 1031 (1967).
67. L. Kaplan, D. L. Rausch, and K. E. Wilzbach, *J. Am. Chem. Soc.* **94**, 8638 (1972).
68. K. E. Wilzbach and L. Kaplan, *J. Am. Chem. Soc.* **87**, 4004 (1965).

ring[69]:

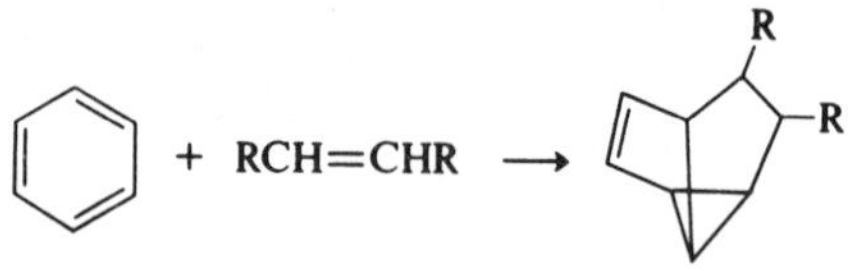

+ RCH=CHR →

These reactions are believed to proceed through a complex of the alkene with the singlet excited state of the aromatic compound (an exciplex). The alkene and aromatic ring are presumed to be oriented in such a manner that the alkene π system reacts with p orbitals on 1,3-carbons of the aromatic:

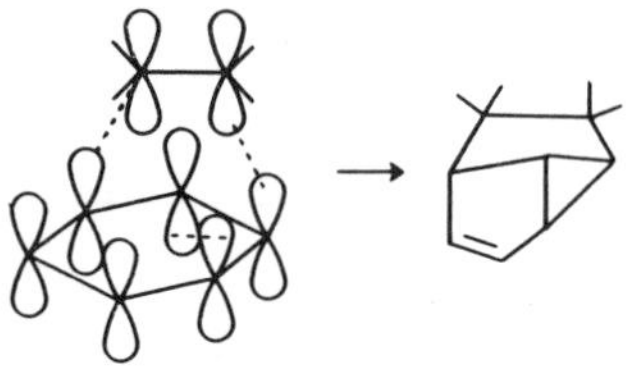

This addition to the aromatic ring is believed to be concerted since the relative geometry of the substituents on the alkene is retained in the product. Lesser amounts of products involving addition to 1,2- or 1,4-positions of the aromatic ring are also formed in such photolyses.[70] This type of addition reaction has also been realized intramolecularly when the distance between the alkene and the phenyl substituent is sufficient to permit interaction.

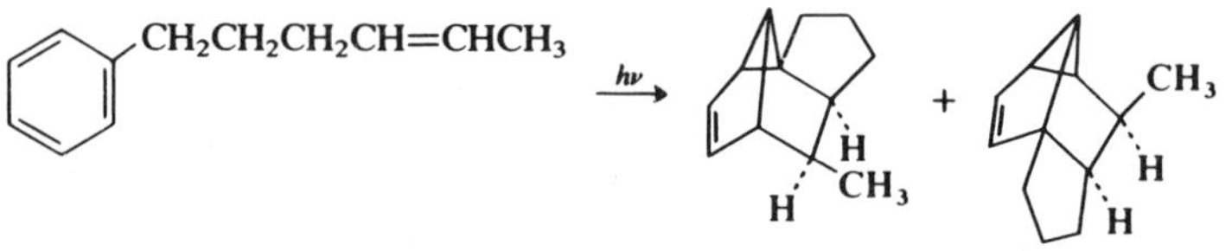

$CH_2CH_2CH_2CH{=}CHCH_3$ $\xrightarrow{h\nu}$ +

General References

D. R. Arnold, *Adv. Photochem.* **6**, 301 (1968).
D. R. Arnold, N. C. Baird, J. R. Bolton, J. C. D. Brand, P. W. M. Jacobs, P. de Mayo, and W. R. Ware, *Photochemistry: An Introduction*, Academic Press, New York, 1974.
D. O. Cowan and R. L. Drisko, *Elements of Organic Photochemistry*, Plenum, New York, 1976.
J. M. Coxon and B. Halton, *Organic Photochemistry*, Cambridge University Press, London, 1974.
J. C. Dalton and N. J. Turro, *Annu. Rev. Phys. Chem.* **21**, 499 (1970).

69. K. E. Wilzbach and L. Kaplan, *J. Am. Chem. Soc.* **88**, 2066 (1966); J. Cornelisse, V. Y. Merritt, and R. Srinivasan, *J. Am. Chem. Soc.* **95**, 6197 (1973); A. Gilbert and P. Yianni, *Tetrahedron* **37**, 3275 (1981); D. Bryce-Smith and A. Gilbert, *Tetrahedron* **33**, 2459 (1977); T. Wagner-Jauregg, *Synthesis* 165 769 (1980).
70. K. E. Wilzbach and L. Kaplan, *J. Am. Chem. Soc.* **93**, 2073 (1971).

P. de Mayo (ed.) *Rearrangements in Ground and Excited States*, Col. 3, Academic Press, New York, 1980.
W. L. Dilling, *Chem. Rev.* **66**, 373 (1966).
D. Neckers, *Mechanistic Organic Photochemistry*, Reinhold, New York, 1967.
N. J. Turro, *Molecular Photochemistry*, W. A. Benjamin, New York, 1967.
N. J. Turro, *Modern Molecular Photochemistry*, Benjamin-Cummings, Menlo Park, California, 1978.
M. A. West, *Flash and Laser Photolysis* in *Investigation of Rates and Mechanisms of Reactions*, C. F. Bernasconi (ed.), *Techniques of Chemistry*, Vol. VI, Part 2, Wiley-Interscience, New York, 1986.
H. E. Zimmerman, *Top. Curr. Chem.* **100**, 45 (1982).

Problems

(*References for these problems will be found on page* 785.)

1. The bridged radical **A** has been suggested as a possible intermediate in the photochemical decarbonylation of 3-phenylpropanal. Suggest an experiment to test this hypothesis.

A $H_2C{-}CH_2$

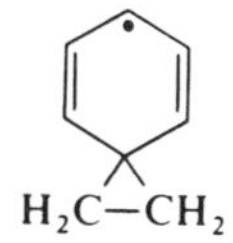

2. Predict the structure, including all aspects of stereochemistry, for the product expected to result from direct irradiation of each compound:

(a) 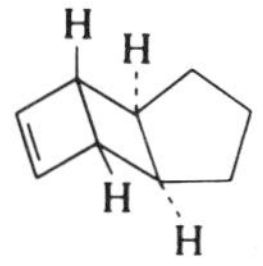 H H H H

(b) 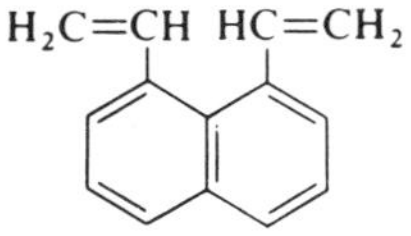 $H_2C{=}CH$ $HC{=}CH_2$

(c) 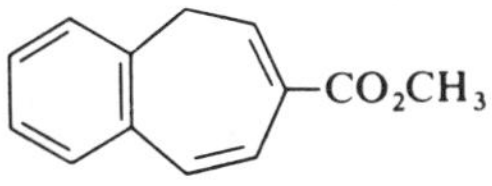 $-CO_2CH_3$

(d) 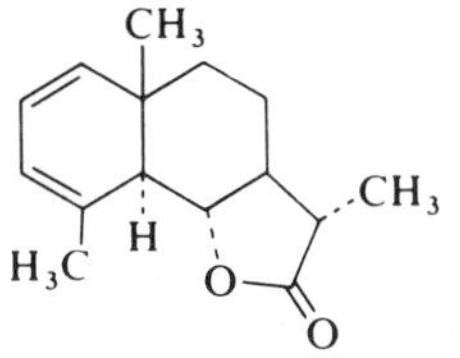 CH_3 CH_3 H H_3C O O

(e)

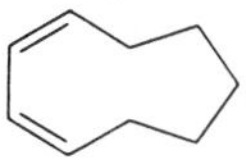

(f)

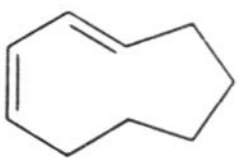

(g)

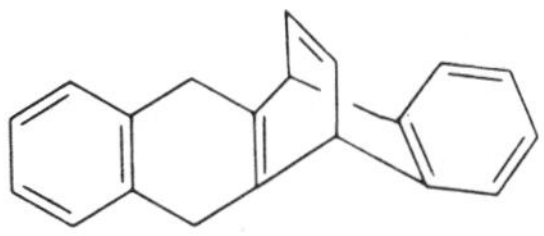

(h) 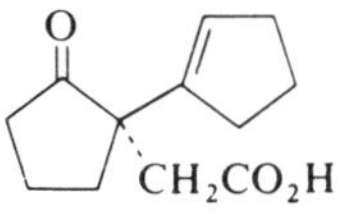 O CH_2CO_2H

3. Suggest reasonable explanations for the following observations:
   (a) Optically active 2,3-pentadiene is racemized rapidly under conditions of toluene-sensitized photolysis.
   (b) Direct photolysis of diene **A** at 254 nm produces a photochemical stationary state containing 40% **A** and 60% triene **B**. When the irradiation is carried out at 300 nm, no **B** is produced, and **C** is the observed product.

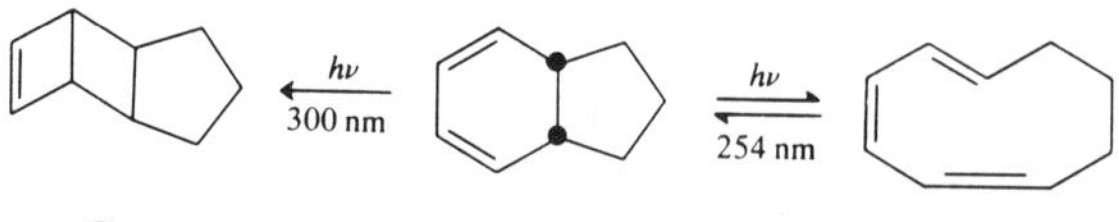
hν
300 nm
hν
254 nm

C **A** **B**

   (c) Photochemical cyclization of the acetal **D** to **E** in benzene is efficiently catalyzed by benzoic acid.

D

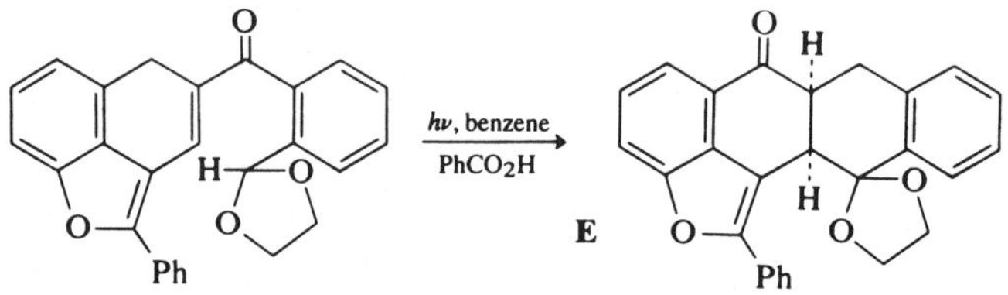
O
H
O
Ph
hν, benzene
PhCO2H
O
H
H
O
O
O
Ph

E

4. Provide a mechanistic rationalization for each of the following reactions.

(a)

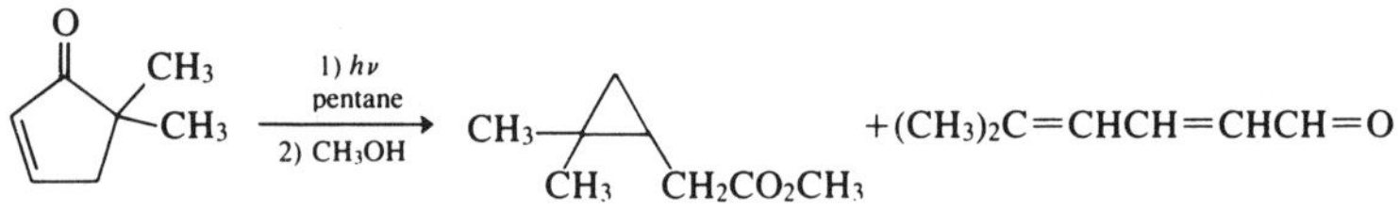
O
CH3
CH3
1) hν
pentane
2) CH3OH
CH3
CH3
CH2CO2CH3
+(CH3)2C=CHCH=CHCH=O

(b)

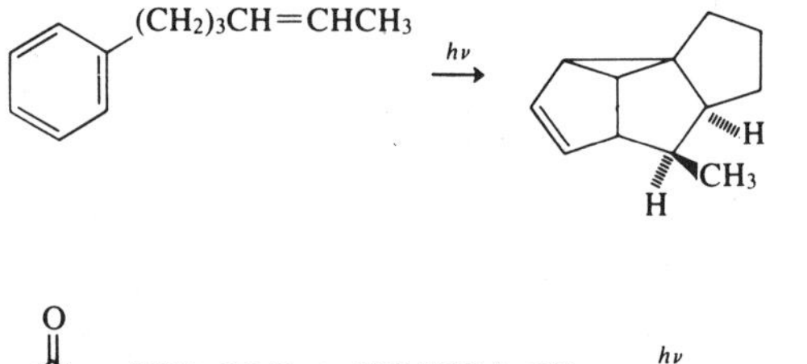
(CH2)3CH=CHCH3
hν
H
CH3
H

(c)

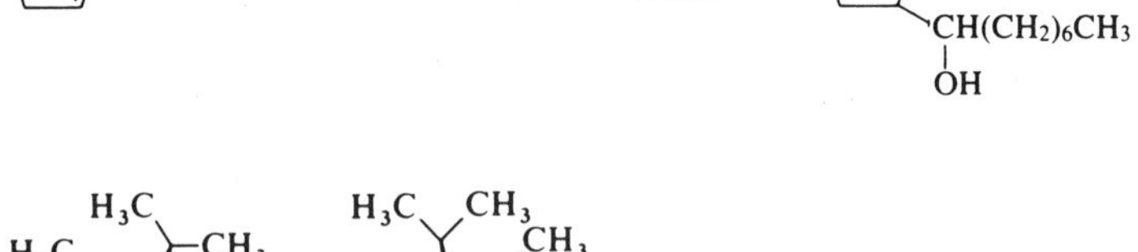
O
(CH2)6CO2H + CH3(CH2)7OH
hν
benzophenone
350 nm
O
(CH2)6CO2H
CH(CH2)6CH3
OH

(d)

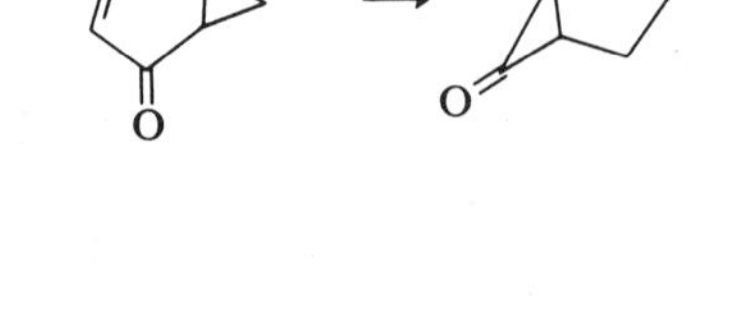
H3C
H3C
CH3
O
hν
H3C
CH3
CH3
O

(e) $\xrightarrow[CH_3OH]{h\nu}$ [cyclobutanone with isopropylidene group: CH_3, CH_3] → [tetrahydrofuran with OCH_3 and $=C(CH_3)CH_3$]

(f) [H_3C-substituted cycloheptadienone with gem-dimethyl CH_3, CH_3] $\xrightarrow{h\nu}$ [bicyclic enone with H_3C, CH_3, H_3C]

(g) $C_6H_4(CH{=}CD_2)_2$ $\xrightarrow{h\nu}$ [benzo-fused bicyclic product labelled H, D, D, H, D, D]

(h) [cyclooctatriene] $\xrightarrow{h\nu}$ [tricyclic product]

(i) [polycyclic diene] $\xrightarrow{h\nu}$ [rearranged polycyclic diene]

(j) [bicyclic ketone] $\xrightarrow[\text{benzophenone}]{h\nu}$ [polycyclic ketone]

(k) Ph(H)C=C(H)–C(H_3C)(CH_3)–$\overset{O}{\|}$CCH$(CH_3)_2$ $\xrightarrow[\text{acetophenone}]{h\nu}$ [cyclopropane bearing $\overset{O}{\|}$CCH$(CH_3)_2$, H, H_3C, H_3C, H, Ph]

(l) [bicyclic ketone with $\overset{O}{\|}$CPh and H] $\xrightarrow{h\nu}$ [cyclopentene with $CH_2\overset{O}{\|}$CPh]

(m) [6,6-dimethylcyclohexadienone, H_3C CH_3] $\xrightarrow[\text{ether–water}]{h\nu}$ $(CH_3)_2C{=}CHCH{=}CHCH_2CO_2H$

(n)

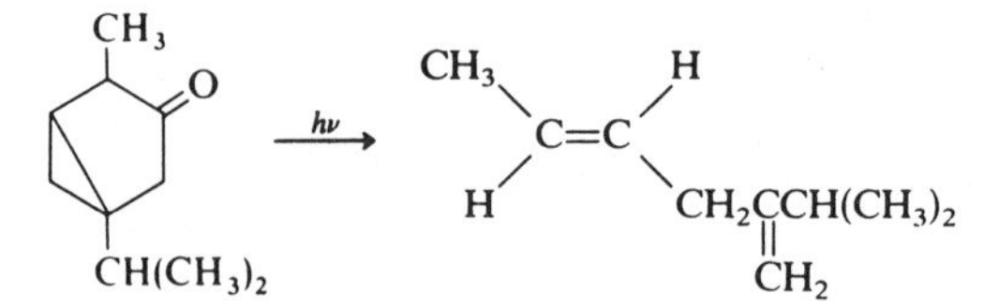

$\xrightarrow{h\nu}$ $CH_3(H)C{=}C(H)CH_2C({=}CH_2)CH(CH_3)_2$

(o)

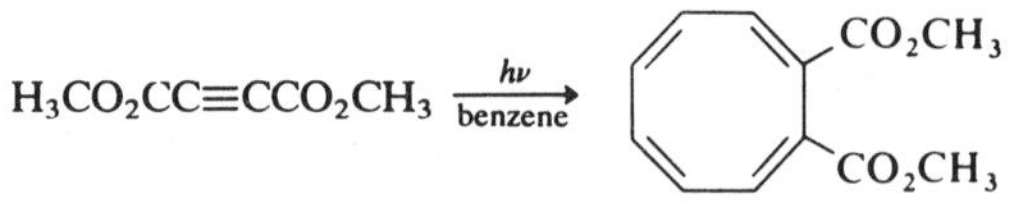

$H_3CO_2CC{\equiv}CCO_2CH_3 \xrightarrow[\text{benzene}]{h\nu}$ (cyclooctatetraene with CO_2CH_3, CO_2CH_3)

(p)

$PhCH{=}O + H_3CC{\equiv}CCH_3 \xrightarrow{h\nu} PhCH{=}C(CH_3)C({=}O)CH_3$

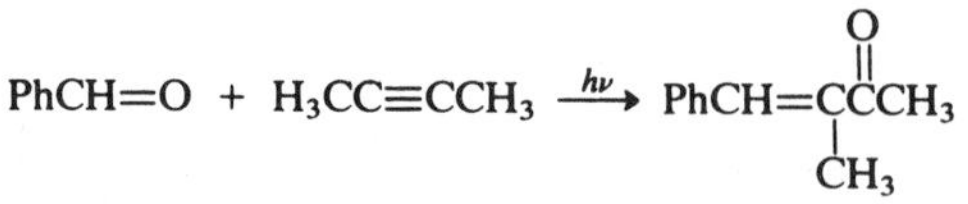

5. Benzene-sensitized photolysis of methyl 3-cyclohexene-1-carboxylate in acetic acid leads to addition of acetic acid to the double bond. Only the *trans* adducts are formed. What factor(s) is (are) responsible for the reaction stereochemistry? Which of the two possible addition products, **A** or **B**, do you expect to be the major product?

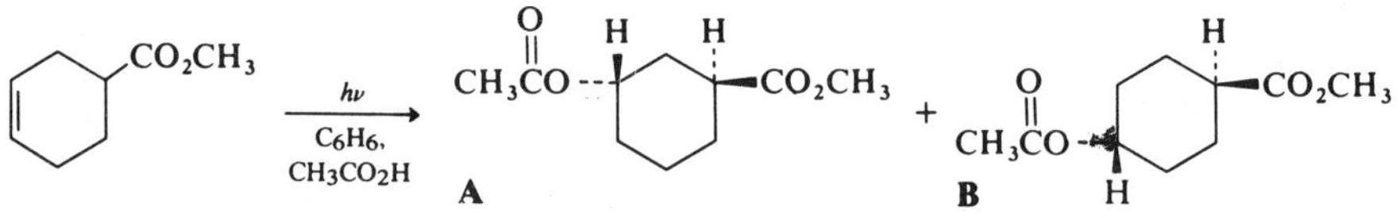

CO_2CH_3 $\xrightarrow[\text{C}_6\text{H}_6,\ \text{CH}_3\text{CO}_2\text{H}]{h\nu}$ **A** + **B**

6. Photolysis of bicyclo[2.2.2]octan-2-one (**A**) gives **B** in good yield. When **A** labeled as shown is used, the aldehyde group carries deuterium to the extent of 51.7%. Write a mechanism to account for the overall transformation. Calculate the isotope effect for the step in which hydrogen atom transfer occurs. What mechanistic conclusion do you draw from the magnitude of the isotope effect?

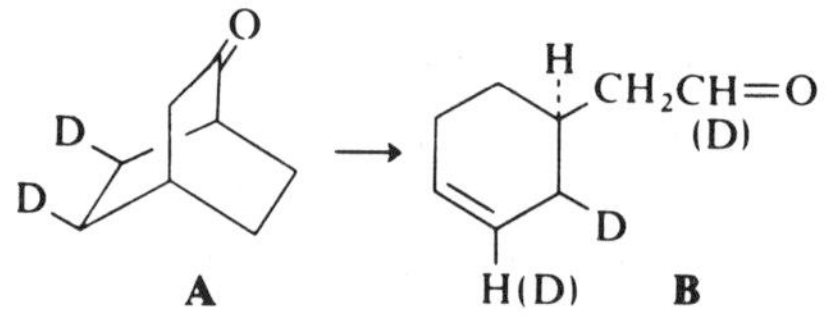

A → **B** ($CH_2CH{=}O$ (D), H(D))

7. The photolysis of benzobarrelene, **A**, has been studied in considerable detail. Direct photolysis gives **C**, but when acetone is used as a photosensitizer, the di-π-methane rearrangement product **B** is formed.

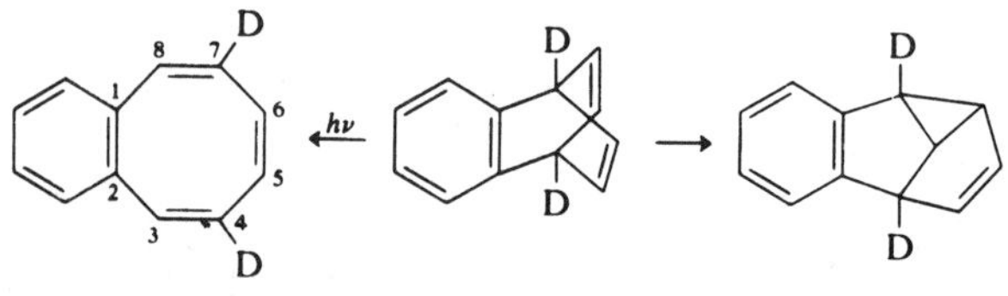

C $\xleftarrow{h\nu}$ **A** → **B**

(A small amount of D is at C-3,8.)

A deuterium labeling study has been performed with the results shown. Discuss the details of the mechanism that are revealed by these results. Is there a feasible mechanism that would have led to **B** having an alternative label distribution?

8. Quantum yield data for several processes that occur on photolysis of *S*-4-methyl-1-phenyl-1-hexanone have been determined. The results are tabulated below for benzene as solvent:

Process	Quantum yield
Type-II elimination	0.23
Cyclobutanol formation	0.03
Racemization	0.78

What information about the mechanism operating under these conditions can be drawn from these data?

9. Show by a diagram why the energy of radiation emitted from an excited electronic state (by fluorescence or phosphorescence) is of lower energy than the exciting radiation. Would you expect the shift to lower energy to be more pronounced for fluorescence or phosphorescence? Explain.

10. *cis*-2-Propyl-4-*t*-butylcyclohexanone is photolyzed to 4-*t*-butylcyclohexanone. The *trans* isomer is converted to the *cis* isomer, which then undergoes cleavage. Offer a rationale for this pronounced stereochemical effect.

11. The quantum yield for formation of 3-methylcyclobutene from *trans*-1,3-pentadiene is 10 times greater than for the cyclization of *cis*-1,3-pentadiene. Can you offer an explanation?

12. The irradiation of 2-methoxytropone (**A**) leads to methyl (4-oxo-2-cyclopentenyl)acetate (**D**). The reaction can be followed by analytical gas chromatography, and two intermediates are observed. These have the structure **B** and **C**. Indicate a mechanism by which each of the three successive reactions might occur. The first two steps are photochemical, while the third is probably an acid-catalyzed reaction which occurs under the photolysis conditions.

CH_3O O **A** → CH_3O O **B** → CH_3O O **C** → O $CH_3O_2CCH_2$ **D**

13. When an aryl substituent is placed at C-5 of a 4-substituted cyclohexenone, a new product type containing a cyclobutanone ring is formed.

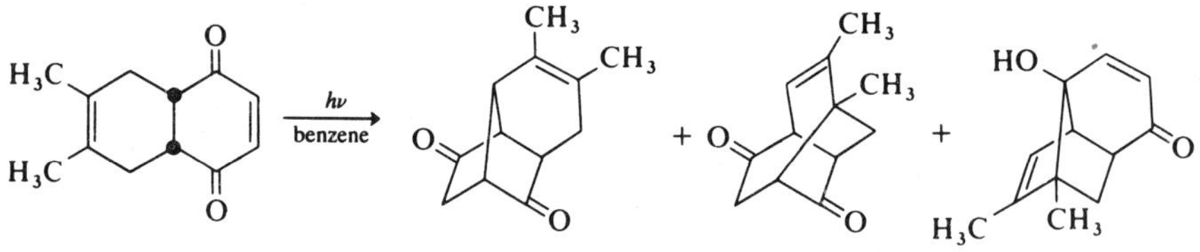

H_3C H_3C O O $\xrightarrow[\text{benzene}]{h\nu}$ CH_3 CH_3 O O + O CH_3 CH_3 O + HO O H_3C CH_3

20. The direct irradiation of **A** gives predominantly **B**, but the photosensitized reaction gives more **C**. Explain.

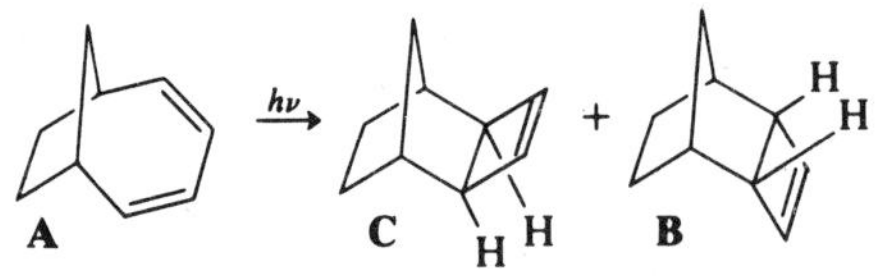

A $\xrightarrow{h\nu}$ **C** H H + **B** H H

	C	B
Direct:	28%	72%
Sensitized:	68%	32%

References for Problems

Chapter 1

2a. H. Prinzbach, *Pure Appl. Chem.* **28**, 281 (1971).

b. E. S. Gould, *Mechanism and Structure in Organic Chemistry*, Holt-Dryden, New York, 1959, pp. 69–70.

c. T. J. Barton, R. W. Roth, and J. G. Verkade, *J. Am. Chem. Soc.* **94**, 8854 (1972).

3a. R. L. Reeves and W. F. Smith, *J. Am. Chem. Soc.* **85**, 724 (1963).

b. R. B. Martin, *J. Chem. Soc., Chem. Commun.*, 793 (1972).

c, d. J. A. Joule and G. F. Smith, *Heterocyclic Chemistry*, Van Nostrand Reinhold, London, 1972, c: pp 194–195; d: p. 64.

4a. H. L. Ammon and G. L. Wheeler, *J. Am. Chem. Soc.* **97**, 2326 (1975).

b. R. Breslow, T. Eicher, A. Krebs, R. A. Peterson, and J. Posner, *J. Am. Chem. Soc.* **87**, 1320 (1965).

c. R. B. Woodward, *J. Am. Chem. Soc.* **63**, 1123 (1941).

5. R. T. Sanderson, *J. Am. Chem. Soc.* **97**, 1367 (1975).

6b. B. B. Ross and N. S. True, *J. Am. Chem. Soc.* **106**, 2451 (1984).

c. H. Slebocka-Tilk and R. S. Brown, *J. Org. Chem.* **52**, 805 (1987); V. Somayaji and R. S. Brown, *J. Org. Chem.* **51**, 2676 (1986).

7a. R. C. Bingham, *J. Am. Chem. Soc.* **98**, 535 (1976).

b. W. L. Jorgensen and L. Salem, *The Organic Chemist's Book of Orbitals*, Academic Press, New York, 1973, pp. 179–184.

c. H. Stafast and H. Bock, *Tetrahedron* **32**, 855 (1976).

d. I. Fleming, *Frontier Orbitals and Organic Chemical Reactions*, Wiley-Interscience, New York, 1976, pp. 51–57.

e. L. Libit and R. Hoffmann, *J. Am. Chem. Soc.* **96**, 1370 (1974).

f. W. J. Hehre, L. Radom, and J. A. Pople, *J. Am. Chem. Soc.* **94**, 1496 (1972).

9. W. L. Jorgensen and L. Salem, *The Organic Chemist's Book of Orbitals*, Academic Press, New York, 1973, pp 93–94.

11a. R. E. Lehr and A. P. Marchand, *Orbital Symmetry: A Problem-Solving Approach*, Academic Press, New York, 1972, p. 37.

b. C. A. Coulson and A. Streitwieser, Jr., *Dictionary of π-Electron Calculations*, W. H. Freeman, San Francisco, 1965, p. 184.

12. D. W. Davis, D. A. Shirley, and T. D. Thomas, *J. Am. Chem. Soc.* **94**, 6565 (1972).

13. C. A. Coulson and A. Streitwieser, Jr., *Dictionary of π-Electron Calculations*, W. H. Freeman, San Francisco, 1965, p. 1.

15. N. L. Allinger and N. A. Pamphilis, *J. Org. Chem.* **38**, 316 (1973).

16. K. B. Wiberg and F. H. Walker, *J. Am. Chem. Soc.* **104**, 5239 (1982).

17. R. B. Turner, B. J. Mallon, M. Tichy, W. v. E. Doering, W. R. Roth, and G. Schröder, *J. Am. Chem. Soc.* **95**, 8605 (1973).

21a. R. Breslow and J. M. Hoffmann, Jr., *J. Am. Chem. Soc.* **94**, 2110 (1972).

b. R. Breslow and P. Dowd, *J. Am. Chem. Soc.* **85**, 2729 (1963).

c. R. Breslow and J. M. Hoffmann, *J. Am. Chem.* **94**, 2110 (1972); M. Saunders, R. Berger, A. Jaffe, M. McBride, J. O'Neill, R. Breslow, J. M. Hoffmann, Jr., C. Perchonock, E. Wasserman, R. S. Hutton, and V. J. Kuck, *J. Am. Chem. Soc.* **95**, 3017 (1973); R. Breslow and S. Mazur, *J. Am. Chem. Soc.* **95**, 584 (1973).

22. W. Kirmse, K. Kund, E. Pitzer, A. Dorigo, and K. N. Houk, *J. Am. Chem. Soc.* **108**, 6045 (1986).

23. K. N. Houk, N. G. Rondan, M. N. Paddon-Row, C. W. Jefford, P. T. Huy, P. D. Burrow, and K. N. Jordan, *J. Am. Chem. Soc.* **105**, 5563 (1983).

24. A. Pross, L. Radom, and N. V. Riggs, *J. Am. Chem. Soc.* **102**, 2253 (1980).

Chapter 2

2. M. Sprecher and D. B. Sprinson, *J. Org. Chem.* **28**, 2490 (1963).

3. K. L. Marsi, *J. Org. Chem.* **39**, 265 (1974).

4a. J. A. Pettus, Jr., and R. E. Moore, *J. Am. Chem. Soc.* **93**, 3087 (1971).

b. K. T. Black and H. Hope, *J. Am. Chem. Soc.* **93**, 3053 (1971).

c. J. Dillon and K. Nakanishi, *J. Am. Chem. Soc.* **96**, 4055 (1974).

d. M. Miyano and C. R. Dorn, *J. Am. Chem. Soc.* **95**, 2664 (1973).

e. M. Koreeda, G. Weiss, and K. Nakanishi, *J. Am. Chem. Soc.* **95**, 239 (1973).

f. P. A. Apgar and M. L. Ludwig, *J. Am. Chem. Soc.* **94**, 964 (1972).

g. M. R. Jones and D. J. Cram, *J. Am. Chem. Soc.* **96**, 2183 (1974).

5a. J. W. deHaan and L. J. M. van den Ven, *Tetrahedron Lett.*, 2703 (1971).

b. T. Sone, S. Terashima, and S. Yamada, *Synthesis*, 725 (1974).

c. B. Witkop and C. M. Foltz, *J. Am. Chem. Soc.* **79**, 197 (1957).

e. E. W. Yankee, B. Spencer, N. E. Howe, and D. J. Cram, *J. Am. Chem. Soc.* **95**, 4220 (1973).

f. R. S. Lenox and J. A. Katzenellenbogen, *J. Am. Chem. Soc.* **95**, 957 (1973).

g. R. Bausch, B. Bogdanovic, H. Dreeskamp, and J. B. Koster, *Justus Liebigs Ann. Chem.*, 1625 (1974).

6. J. A. Marshall, T. R. Konicek, and K. E. Flynn, *J. Am. Chem. Soc.* **102**, 3287 (1980).

7. E. Vogel, W. Tückmantel, K. Schlög, M. Widham, E. Kraka, and D. Cremer, *Tetrahedron Lett.* **25**, 4925 (1984).

8. R. A. Pascal, Jr., and R. B. Grossman, *J. Org. Chem.* **52**, 4616 (1987); A. Rici, R. Danieli, and S. Rossini, *J. Chem. Soc., Perkin Trans. 1,* 1691 (1976).

9. K. Koyama, S. Natori, and Y. Iitaka, *Chem. Pharm. Bull.* **35**, 4049 (1987).

10. K. J. Thottathil, C. Przybyla, M. Malley, and J. Z. Gougoutas, *Tetrahedron Lett.* **27**, 1533 (1986).

12. H. W. Gschwend, *J. Am. Chem. Soc.* **94**, 8430 (1972).

13a, b. M. Kainosho, K. Akjisaka, W. H. Pirkle, and S. D. Beare, *J. Am. Chem. Soc.* **94**, 5924 (1972).

c. M. Rabin and K. Mislow, *Tetrahedron Lett.*, 3961 (1966).

d. R. K. Hill, S. Yan, and S. M. Arfin, *J. Am. Chem. Soc.* **95**, 7857 (1973).

e. V. J. Morlino and R. B. Martin, *J. Am. Chem. Soc.* **89**, 3107 (1967).

f. J. A. Elvidge and R. G. Foster, *J. Chem. Soc.*, 981 (1964); L. S. Rattet and J. H. Goldstein, *Org. Mang. Reson.* **1**, 229 (1969).

14a. G. Helmchen and G. Staiger, *Angew. Chem. Int. Ed. Engl.*, **16**, 116 (1977).

b. M. Farina and C. Morandi, *Tetrahedron* **30**, 1819 (1974).

c. D. T. Longone and M. T. Reetz, *J. Chem. Soc., Chem. Commun.*, 46 (1967).

d. I. T. Jacobson, *Acta Chem. Scand.* **21**, 2235 (1967).

e. R. J. Ternansky, D. W. Balogh, and L. A. Paquette, *J. Am. Chem. Soc.* **104**, 4503 (1982).

f. C. Ganter and K. Wicker, *Helv. Chim. Acta* **53**, 1693 (1970).

g. M. S. Newman and D. Lednicer, *J. Am. Chem. Soc.* **78**, 4765 (1956).

h. T. Otsubo, R. Gray, and V. Boekelheide, *J. Am. Chem. Soc.* **100**, 2449 (1978).

i. J. H. Brewster and R. S. Jones, Jr., *J. Org. Chem.* **34**, 354 (1969).

j. H. Gerlach, *Helv. Chim. Acta* **51**, 1587 (1968).

k, l. H. Gerlach, *Helv. Chim. Acta* **68**, 1815 (1985).

m. R. K. Hill, G. H. Morton, J. R. Peterson, J. A. Walsh, and L. A. Paquette, *J. Org. Chem.* **50**, 5528 (1985).

15. T. C. Bruice and S. Benkovic, *Bioorganic Mechanisms*, Vol. 1, W. A. Benjamin, New York, 1966, p. 305.
16. M. Cohn, J. E. Pearson, E. L. O'Connell, and I. A. Rose, *J. Am. Chem. Soc.* **92**, 4095 (1970).
17. R. K. Hill, S. Yan, and S. M. Arfin, *J. Am. Chem. Soc.* **95**, 7857 (1973).
18. D. J. Cram and F. A. Abd Elhafez, *J. Am. Chem. Soc.* **74**, 5828 (1952).
19. Y.-F. Cheung and C. Walsh, *J. Am. Chem. Soc.* **98**, 3397 (1976).
20. D. M. Jerina, H. Selander, H. Yagi, M. C. Wells, J. F. Davey, V. Magadevan, and D. T. Gibson, *J. Am. Chem. Soc.* **98**, 5988 (1976).
21. Y. Fujimoto, F. Irreverre, J. M. Karle, I. L. Karle, and B. Witkop, *J. Am. Chem. Soc.* **98**, 5988 (1976).
22a. M. Raban, S. K. Lauderback, and D. Kost, *J. Am. Chem. Soc.* **97**, 5178 (1975).
b. P. Finocchiaro, D. Gust, and K. Mislow, *J. Am. Chem. Soc.* **96**, 2165 (1974).
23. F.-C. Huang, L. F. H. Lee, R. S. D. Mittal, P. R. Ravikumar, J. A. Chan, C. J. Sih, E. Capsi, and C. R. Eck, *J. Am. Chem. Soc.* **97**, 4144 (1975).
24. J. Dominguez, J. D. Dunitz, H. Gerlach, and V. Prelog, *Helv. Chim. Acta* **45**, 129 (1962); H. Gerlach and V. Prelog, *Justus Liebigs Ann. Chem.* **669**, 121 (1963); B. T. Kilbourn, J. D. Dunitz, L. A. R. Pioda, and W. Simon, *J. Mol. Biol.* **30**, 559 (1967).

Chapter 3

1a. R. L. Lipnick and E. W. Garbisch, Jr., *J. Am. Chem. Soc.* **95**, 6375 (1973).
b. L. Lunazzi, D. Macciantelli, F. Bernardi, and K. U. Ingold, *J. Am. Chem. Soc.* **99**, 4573 (1977).
c. B. Rickborn and M. T. Wuesthoff, *J. Am. Chem. Soc.* **92**, 6894 (1970).
2. H. C. Brown and W. C. Dickason, *J. Am. Chem. Soc.* **91**, 1226 (1969).
3. E. L. Eliel and S. H. Schroeter, *J. Am. Chem. Soc.* **87**, 5031 (1965).
4. N. L. Allinger and M. T. Tribble, *Tetrahedron Lett.*, 3259 (1971); W. F. Bailey, H. Connor, E. L. Eliel, and K. B. Wiberg, *J. Am. Chem. Soc.* **100**, 2202 (1978).
5a. C. L. Stevens, J. B. Filippi, and K. G. Taylor, *J. Org. Chem.* **31**, 1292 (1966).
b. M. Miyamoto, Y. Kawamatsu, M. Shinohara, Y. Nakadaira, and K. Nakanishi, *Tetrahedron* **22**, 2785 (1966).
c. D. H. Williams and J. R. Kalman, *J. Am. Chem. Soc.* **99**, 2768 (1977).
6a. J. E. Baldwin and J. A. Reiss, *J. Chem. Soc., Chem. Commun.*, 77 (1977).
b. H. C. Brown, J. H. Kawakami, and S. Ikegami, *J. Am. Chem. Soc.* **92**, 6914 (1970).
c. R. E. Lyle, E. W. Southwick, and J. J. Kaminski, *J. Am. Chem. Soc.* **94**, 1413 (1972).
d. E. C. Ashby and S. A. Noding, *J. Am. Chem. Soc.* **98**, 2010 (1976).
e. R. D. G. Cooper, P. V. DeMarco, and D. O. Spry, *J. Am. Chem. Soc.* **91**, 1528 (1969).
f. J. E. Baldwin and M. J. Lusch, *Tetrahedron* **38**, 2939 (1982).
g. S. K. Taylor, G. H. Hockerman, G. L. Karnick, S. B. Lyle, and S. B. Schramm, *J. Org. Chem.* **48**, 2449 (1983).
7a. H. Tanida, S. Yamamoto, and K. Takeda, *J. Org. Chem.* **38**, 2792 (1973).
b. J. E. Baldwin and L. I. Kruse, *J. Chem. Soc., Chem. Commun.*, 233 (1977).
c. N. L. Allinger and J. C. Graham, *J. Org. Chem.* **36**, 1688 (1971).
d. C. Galli, G. Illuminati, L. Mandolini, and P. Tamborra, *J. Am. Chem. Soc.* **99**, 2591 (1977).
e. C. M. Evans, R. Glenn, and A. J. Kirby, *J. Am. Chem. Soc.* **104**, 4706 (1982).
f. J. F. Bunnett, S. Sekiguchi, and L. A. Smith, *J. Am. Chem. Soc.* **103**, 4865 (1981).
g. P. Müller and J.-C. Perlberger, *J. Am. Chem. Soc.* **98**, 8407 (1976).
8a. J. C. Little, Y.-L. C. Tong, and J. P. Heeschen, *J. Am. Chem. Soc.* **91**, 7090 (1969).
b. D. J. Pasto and D. R. Rao, *J. Am. Chem. Soc.* **92**, 5151 (1970).
c. P. E. Pfeffer and S. F. Osman, *J. Am. Chem. Soc.* **37**, 2425 (1972).
d. P. L. Durrette and D. Horton, *Carbohydr. Res.* **18**, 57 (1971).
e. B. Fuchs and A. Ellencweig, *J. Org. Chem.* **44**, 2274 (1979).
f. M. J. Anteunis, D. Tavernier, and F. Borremans, *Heterocyclics* **4**, 293 (1976).
g. P. Deslongchamps, D. D. Rowan, N. Pothier, T. Sauvé, and J. K. Saunders, *Can. J. Chem.* **59**, 1105 (1981).
h. P. v. R. Schleyer and A. J. Kos, *Tetrahedron* **39**, 1141 (1983).
9. K. B. Wiberg, *J. Am. Chem. Soc.* **108**, 5817 (1986).
10. L. Lunazzi, D. Macciantelli, F. Bernardi, and K. U. Ingold, *J. Am. Chem. Soc.* **99**, 4573 (1977).

11a. E. N. Marvell and R. S. Knutson, *J. Org. Chem.* **35**, 388 (1970).
b. R. J. Oulette, J. D. Rawn, and S. N. Jreissaty, *J. Am. Chem. Soc.* **93**, 7117 (1971).
c. V. S. Mastryukov, E. L. Olsina, L. V. Vilkov, and R. L. Hilderbrandt, *J. Am. Chem. Soc.* **99**, 6855 (1977).
d. D. D. Danielson and K. Hedberg, *J. Am. Chem. Soc.* **101**, 3730 (1979).
12. J. B. Lambert, R. R. Clikeman, and E. S. Magyar, *J. Am. Chem. Soc.* **96**, 2265 (1974).
13. J. G. Vintner and H. M. R. Hoffmann, *J. Am. Chem. Soc.* **96**, 5466 (1974).
14. A. Bienvenue, *J. Am. Chem. Soc.* **95**, 7345 (1973).
15. E. Ghera, Y. Gaoni, and S. Shoua, *J. Am. Chem. Soc.* **98**, 3627 (1976).
16. D. K. Dalling and D. M. Grant, *J. Am. Chem. Soc.* **94**, 5318 (1972).
17a. H. C. Brown, J. H. Kawakami, and S. Ikegami, *J. Am. Chem. Soc.* **92**, 6914 (1970).
b. B. Waegell and C. W. Jefford, *Bull. Soc. Chim. Fr.* 844 (1964).
c. H. C. Brown and J. H. Kawakami, *J. Am. Chem. Soc.* **92**, 1990 (1970).
d. L. A. Spurlock and K. P. Clark, *J. Am. Chem. Soc.* **94**, 5349 (1972).
e. K. B. Wiberg and K. A. Saegebarth, *J. Am. Chem. Soc.* **79**, 2822 (1957).

Chapter 4

1. W. Kusters and P. deMayo, *J. Am. Chem. Soc.* **96**, 3502 (1974).
2a. A. Streitwieser, Jr., H. A. Hammond, R. H. Jagow, R. M. Williams, R. G. Jesaitis, C. J. Chang, and R. Wolf, *J. Am. Chem. Soc.* **92**, 5141 (1970).
b. M. T. H. Liu and D. H. T. Chien, *J. Chem. Soc., Perkin Trans. 2,* 937 (1974).
3. W. E. Billups, K. H. Leavell, E. S. Lewis, and S. Vanderpool, *J. Am. Chem. Soc.* **95**, 8096 (1973).
4. O. Exner, in *Correlation Analysis in Chemistry*, N. B. Chapman and B. Shorter (eds.), Plenum Press, New York, 1978, Chapter 10.
5. P. R. Wells, *Linear Free Energy Relationships*, Academic Press, New York, 1968, pp. 12, 13.
6a. O. R. Zaborsky and E. T. Kaiser, *J. Am. Chem. Soc.* **92**, 860 (1970).
b. W. M. Schubert and J. R. Keefe, *J. Am. Chem. Soc.* **94**, 559 (1972).
c. D. H. Rosenblatt, L. A. Hull, D. C. DeLuca, G. T. Davis, R. C. Weglein, and H. K. R. Williams, *J. Am. Chem. Soc.* **89**, 1158 (1967).
7a. W. K. Kwok, W. G. Lee, and S. I. Miller, *J. Am. Chem. Soc.* **91**, 468 (1969).
b. H. C. Brown and E. N. Peters, *J. Am. Chem. Soc.* **95**, 2400 (1973).
c. W. M. Schubert and D. F. Gurka, *J. Am. Chem. Soc.* **91**, 1443 (1969).
d. J. Rocek and A. Riehl, *J. Am. Chem. Soc.* **88**, 4749 (1966).
8. D. E. Applequist and R. D. Gdanski, *J. Org. Chem.* **46**, 2502 (1981).
9a. V. J. Shiner, Jr., and J. O. Stoffer, *J. Am. Chem. Soc.* **92**, 3191 (1970).
b. M. H. Davies, *J. Chem. Soc., Perkin Trans. 2,* 1018 (1974).
c. J. E. Baldwin and J. A. Kapecki, *J. Am. Chem. Soc.* **91**, 3106 (1969).
d. H. G. Bull, K. Koehler, T. C. Pletcher, J. J. Ortiz, and E. H. Cordes, *J. Am. Chem. Soc.* **93**, 3002 (1971).
e. R. B. Timmons, J. deGuzman, and R. E. Varnerin, *J. Am. Chem. Soc.* **90**, 5996 (1968).
f. Y. Pocker and J. H. Exner, *J. Am. Chem. Soc.* **90**, 6764 (1968).
g. K. D. McMichael, *J. Am. Chem. Soc.* **89**, 2943 (1967).
h. R. J. Cvetanovic, F. J. Duncan, W. E. Falconer, and R. S. Irwin, *J. Am. Chem. Soc.* **87**, 1827 (1965).
10. J. J. Eisch and S.-G. Rhee, *J. Am. Chem. Soc.* **96**, 7276 (1974).
11. C. G. Swain, A. L. Powell, W. A. Sheppard, and C. R. Morgan, *J. Am. Chem. Soc.* **101**, 3576 (1979).
12. R. S. Shue, *J. Am. Chem. Soc.* **93**, 7116 (1971).
13. D. N. Kevill and G. M. L. Lin, *J. Am. Chem. Soc.* **101**, 3916 (1979).
14. C. G. Swain, J. E. Sheats, and K. G. Harbison, *J. Am. Chem. Soc.* **97**, 783 (1975).
15. G. E. Hall and J. D. Roberts, *J. Am. Chem. Soc.* **93**, 2203 (1971).
16. R. K. Lustgarten and H. G. Richey, Jr., *J. Am. Chem. Soc.* **96**, 6393 (1974).
17. C. A. Bunton and G. Stedman, *J. Chem. Soc.*, 2420 (1958); C. A. Bunton and E. A. Halevi, *J. Chem. Soc.*, 4917 (1952).
18. P. G. Gassman and K. Sato, *Tetrahedron Lett.* **22**, 1311 (1981).
19. T. B. McMahon and P. Kebarle, *J. Am. Chem. Soc.* **94**, 2222 (1977).
20. J. Bromilow, R. T. C. Brownlee, V. O. Lopez, and R. W. Taft, *J. Org. Chem.* **44**, 4766 (1979).

21. A. Fischer, W. J. Galloway, and J. Vaughan, *J. Chem. Soc.*, 3591 (1964); C. L. Liotta, E. M. Perdue, and H. P. Hopkins, Jr., *J. Am. Chem. Soc.* **96**, 7308 (1974).
22. E. S. Lewis, J. T. Hill, and E. R. Newman, *J. Am. Chem. Soc.* **90**, 662 (1968).
23. E. M. Arnett and D. R. McKelvey, *J. Am. Chem. Soc.* **88**, 2598 (1966).

Chapter 5

1a. H. C. Brown and E. N. Peters, *J. Am. Chem. Soc.* **99**, 1712 (1977).
b. X. Creary, M. E. Mehrsheikh-Mohammadi, and M. D. Eggers, *J. Am. Chem. Soc.* **109**, 2435 (1987).
c. R. J. Blint, T. B. McMahon, and J. L. Beauchamp, *J. Am. Chem. Soc.* **96**, 1269 (1974).
d. J. L. Fry, E. M. Engler, and P. v. R. Schleyer, *J. Am. Chem. Soc.* **94**, 4628 (1972).
e. F. G. Bordwell and W. T. Brannen, Jr., *J. Am. Chem. Soc.* **86**, 4645 (1964).
2a. R. K. Crossland, W. E. Wells, and V. J. Shiner, Jr., *J. Am. Chem. Soc.* **93**, 4217 (1971).
b. R. L. Buckson and S. G. Smith, *J. Org. Chem.* **32**, 634 (1967).
c. C. J. Norton, Ph.D. Thesis, Harvard University, 1955, cited by P. V. R. Schleyer, W. E. Watts, R. C. Fort, Jr., M. B. Comisarow, and G. A. Olah, *J. Am. Chem. Soc.* **86**, 5679 (1964).
d. E. N. Peters and H. C. Brown, *J. Am. Chem. Soc.* **97**, 2892 (1975).
e. B. R. Ree and J. C. Martin, *J. Am. Chem. Soc.* **92**, 1660 (1970).
f. F. G. Bordwell and W. T. Brannen, Jr., *J. Am. Chem. Soc.* **86**, 4645 (1964).
g, h. D. D. Roberts, *J. Org. Chem.* **34**, 285 (1969).
i. K. L. Servis and J. D. Roberts, *J. Am. Chem. Soc.* **87**, 1331 (1965).
j. R. G. Lawton, *J. Am. Chem. Soc.* **83**, 2399 (1961).
k. G. M. Bennett, F. Heathcoat, and A. N. Mosses, *J. Chem. Soc.*, 2567 (1929).
l. S. Kim, S. S. Friedrich, L. J. Andrews, and R. M. Keefer, *J. Am. Chem. Soc.* **92**, 5452 (1970).
3b. D. H. Ball, E. D. Eades, and L. Long, Jr., *J. Am. Chem. Soc.* **86**, 3579 (1964).
c. H. G. Richey, Jr., and D. V. Kinsman, *Tetrahedron Lett.*, 2505 (1969).
d. P. v. R. Schleyer, W. E. Watts, and C. Cupas, *J. Am. Chem. Soc.* **86**, 2722 (1964).
e. P. E. Peterson and J. E. Duddey, *J. Am. Chem. Soc.* **85**, 2865 (1963).
f. A. Colter, E. C. Friedrich, N. J. Holness, and S. Winstein, *J. Am. Chem. Soc.* **87**, 378 (1965).
g, h. M. Cherest, H. Felkin, J. Sicher, F. Sipos, and M. Tichy, *J. Chem. Soc.*, 2513 (1965).
i. J. C. Martin and P. D. Bartlett, *J. Am. Chem. Soc.* **79**, 2533 (1957).
j. S. Archer, T. R. Lewis, M. R. Bell, and J. W. Schulenberg, *J. Am. Chem. Soc.* **83**, 2386 (1961).
k. D. A. Tomalia and J. N. Paige, *J. Org. Chem.* **38**, 422 (1973).
l. P. Wilder, Jr., and W.-C. Hsieh, *J. Org. Chem.* **40**, 717 (1975).
m. C. W. Jefford, J.-C. Rossier, J. A. Zuber, S. C. Suri, and G. Mehta, *Tetrahedron Lett.*, 4081 (1980).
4. H. L. Goering and B. E. Jones, *J. Am. Chem. Soc.* **102**, 1628 (1980).
5. K. Banert and W. Kirmse, *J. Am. Chem. Soc.* **104**, 3766 (1982).
6. Y. Pocker and M. J. Hill, *J. Am. Chem. Soc.* **93**, 691 (1971).
7. R. Baird and S. Winstein, *J. Am. Chem. Soc.* **85**, 567 (1963).
8. J. B. Lambert and S. I. Featherman, *J. Am. Chem. Soc.* **99**, 1542 (1977).
9a. K. Yano, *J. Org. Chem.* **40**, 414 (1975).
b. N. C. Deno, N. Friedman, J. D. Hodge, and J. J. Houser, *J. Am. Chem. Soc.* **85**, 2995 (1963).
c. I. Lillien and L. Handloser, *J. Am. Chem. Soc.* **93**, 1682 (1971).
d. J. C. Barborak and P. v. R. Schleyer, *J. Am. Chem. Soc.* **92**, 3184 (1970); P. Ahlberg, C. Engdahl, and G. Jonsäll, *J. Am. Chem. Soc.* **103**, 1583 (1981).
10. S. Winstein and E. T. Stafford, *J. Am. Chem. Soc.* **79**, 505 (1957).
11a. M. Brookhart, A. Diaz, and S. Winstein, *J. Am. Chem. Soc.* **88**, 3135 (1966).
b. G. A. Olah, J. M. Bollinger, C. A. Cupas, and J. Lukas, *J. Am. Chem. Soc.* **89**, 7627 (1970).
c. G. A. Olah and R. D. Porter, *J. Am. Chem. Soc.* **92**, 7627 (1970).
d. G. A. Olah and G. Liang, *J. Am. Chem. Soc.* **97**, 2236 (1975).
e, f. G. A. Olah and G. Liang, *J. Am. Chem. Soc.* **93**, 6873 (1971).
12. L. A. Paquette, I. R. Dunkin, J. P. Freeman, and P. C. Storm, *J. Am. Chem. Soc.* **94**, 8124 (1972).
13. C. Paradisi and J. F. Bunnett, *J. Am. Chem. Soc.* **103**, 946 (1981).
14. J. P. Richard and W. P. Jencks, *J. Am. Chem. Soc.* **104**, 4689, 4691 (1982).
15a. J. W. Wilt and P. J. Chenier, *J. Am. Chem. Soc.* **90**, 7366 (1968); S. J. Cristol and G. W. Nachtigall, *J. Am. Chem. Soc.* **90**, 7132, 7133 (1968).

b. H. C. Brown and E. N. Peters, *J. Am. Chem. Soc.* **97**, 1927 (1975).
c. P. G. Gassman and W. C. Pike, *J. Am. Chem. Soc.* **97**, 1250 (1975).
d. I. Tabushi, Y. Tamura, Z. Yoshida, and T. Sugimoto, *J. Am. Chem. Soc.* **97**, 2886 (1975).
e. H. Weiner and R. A. Sneen, *J. Am. Chem. Soc.* **87**, 287 (1965).
f. J. R. Mohrig and K. Keegstra, *J. Am. Chem. Soc.* **89**, 5492 (1967).
g. E. N. Peters and H. C. Brown, *J. Am. Chem. Soc.* **96**, 265 (1974).
h. L. R. C. Barclay, H. R. Sonawane, and J. C. Hudson, *Can. J. Chem.* **50**, 2318 (1972).
i. C. D. Poulter, E. C. Friedrich, and S. Winstein, *J. Am. Chem. Soc.* **92**, 4274 (1970).
j. J. M. Harris, J. R. Moffatt, M. G. Case, F. W. Clarke, J. S. Polley, T. K. Morgan, Jr., T. M. Ford, and R. K. Murray, Jr., *J. Org. Chem.* **47**, 2740 (1982).
k. F. David, *J. Org. Chem.* **47**, 2740 (1982).
16. W. Franke, H. Schwarz, and D. Stahl, *J. Org. Chem.* **45**, 3493 (1980); Y. Apeloig, J. B. Collins, D. Cremer, T. Bally, E. Haselbach, J. A. Pople, J. Chandrasekhar, and P. v. R. Schleyer, *J. Org. Chem.* **45**, 3496 (1980).
17. J. S. Haywood-Farmer and R. E. Pincock, *J. Am. Chem. Soc.* **91**, 3020 (1969).
18. G. A. Olah, A. L. Berrier, M. Arvanaghi, and G. K. Surya Prakash, *J. Am. Chem. Soc.* **103**, 1122 (1981).
19. J. J. Tufariello and R. J. Lorence, *J. Am. Chem. Soc.* **91**, 1546 (1969); J. Lhomme, A. Diaz, and S. Winstein, *J. Am. Chem. Soc.* **91**, 1548 (1969).
20. R. Fuchs and L. L. Cole, *J. Am. Chem. Soc.* **95**, 3194 (1973).

Chapter 6

1a. K. Yates, G. H. Schmid, T. W. Regulski, D. G. Garratt, H.-W. Leung, and R. McDonald, *J. Am. Chem. Soc.* **95**, 160 (1973).
b. J. F. King and M. J. Coppen, *Can. J. Chem.* **49**, 3714 (1971).
c. N. C. Deno, F. A. Kish, and H. J. Peterson, *J. Am. Chem. Soc.* **87**, 2157 (1965).
d. M. A. Cooper, C. M. Holden, P. Loftus, and D. Whittaker, *J. Chem. Soc., Perkin Trans. 2,* 665 (1973).
e. R. A. Bartsch and J. F. Bunnett, *J. Am. Chem. Soc.* **91**, 1376 (1969).
f. K. Yates and T. A. Go, *J. Org. Chem.* **45**, 2377 (1980).
g. C. B. Quinn and J. R. Wiseman, *J. Am. Chem. Soc.* **95**, 1342 (1973).
2a. H. Wong, J. Chapuis, and I. Monkovic, *J. Org. Chem.* **39**, 1042 (1974).
b. P. K. Freeman, F. A. Raymond, and M. F. Grostic, *J. Org. Chem.* **32**, 24 (1967).
c. T. I. Crowell, R. T. Kemp, R. E. Lutz, and A. A. Wall, *J. Am. Chem. Soc.* **90**, 4638 (1968).
d. L. C. King, L. A. Subluskey, and E. W. Stern, *J. Org. Chem.* **21**, 1232 (1956).
e. M. L. Poutsma and P. A. Ibarbia, *J. Am. Chem. Soc.* **93**, 440 (1971).
f. S. P. Acharya and H. C. Brown, *J. Chem. Soc., Chem. Commun.*, 305 (1968).
g. A. Lewis and J. Azoro, *J. Org. Chem.* **46**, 1764 (1981).
h. K. Yates and T. A. Go, *J. Org. Chem.* **45**, 2377 (1980).
3. D. Y. Curtin, R. D. Stolow, and W. Maya, *J. Am. Chem. Soc.* **81**, 3330 (1959).
4. W. H. Saunders and A. F. Cockerill, *Mechanisms of Elimination Reactions*, Wiley, New York, 1973, pp. 79–80.
5. M. Anteunis and H. L. Peeters, *J. Org. Chem.* **40**, 307 (1975).
6a, b. D. S. Noyce, D. R. Hartter, and R. M. Pollack, *J. Am. Chem. Soc.* **90**, 3791 (1968).
c. C. L. Wilkins and T. W. Regulski, *J. Am. Chem. Soc.* **94**, 6016 (1972).
7. R. A. Bartsch, *J. Am. Chem. Soc.* **93**, 3683 (1971).
8. K. Oyama and T. T. Tidwell, *J. Am. Chem. Soc.* **98**, 947 (1976).
9a. K. D. Berlin, R. O. Lyerla, D. E. Gibbs, and J. P. Devlin, *J. Chem. Soc., Chem. Commun.*, 1246 (1970).
b. R. J. Abraham and J. R. Monasterios, *J. Chem. Soc., Perkin Trans. 2,* 574 (1975).
c. H. C. Brown and K.-T. Liu, *J. Am. Chem. Soc.* **97**, 2469 (1975).
d. D. J. Pasto and J. A. Gontarz, *J. Am. Chem. Soc.* **93**, 6902 (1971).
10a. J. F. Bunnett and S. Sridharan, *J. Org. Chem.* **44**, 1458 (1979); L. F. Blackwell, A. Fischer, and J. Vaughan, *J. Chem. Soc., B*, 1084 (1967).
b. J. E. Nordlander, P. O. Owuor, and J. E. Haky, *J. Am. Chem. Soc.* **101**, 1288 (1979).

c. G. E. Heasley, T. R. Bower, K. W. Dougharty, J. C. Easdon, V. L. Heasley, S. Arnold, T. L. Carter, D. B. Yaeger, B. T. Gipe, and D. F. Shellhamer, *J. Org. Chem.* **45**, 5150 (1980).
d. P. Cramer and T. T. Tidwell, *J. Org. Chem.* **46**, 2683 (1981).
e. R. C. Fahey and C. A. McPherson, *J. Am. Chem. Soc.* **91**, 3865 (1965); R. C. Fahey, M. W. Monahan, and C. A. McPherson, *J. Am. Chem. Soc.* **92**, 2810 (1970).
f. C. H. DePuy and A. L. Schultz, *J. Org. Chem.* **39**, 878 (1974).
11a. T. T. Coburn and W. M. Jones, *J. Am. Chem. Soc.* **96**, 5218 (1974).
b. R. B. Miller and G. McGarvey, *J. Org. Chem.* **44**, 4623 (1979).
c. P. F. Hudrlik and D. Peterson, *J. Am. Chem. Soc.* **97**, 1464 (1975).
d. G. Bellucci, A. Marsili, E. Mastrorilli, I. Morelli, and V. Scartoni, *J. Chem. Soc., Perkin Trans. 2,* 201 (1974).
e. K. J. Shea, A. C. Greeley, S. Nguyen, P. D. Beauchamp, D. H. Aue, and J. S. Witzeman, *J. Am. Chem. Soc.* **108**, 5901 (1986).
12. G. H. Schmid, A. Modro, and K. Yates, *J. Org. Chem.* **45**, 665 (1980); A. D. Allen, Y. Chiang, A. J. Kresge, and T. T. Tidwell, *J. Org. Chem.* **47**, 775 (1982).
13. G. Belluci, R. Bianchini, and S. Vecchiani, *J. Org. Chem.* **52**, 3355 (1987).
14. T. Hasam, L. B. Sims, and A. Fry, *J. Am. Chem. Soc.* **105**, 3987 (1983).
15. B. Badat, M. Julia, J. N. Mallet, and C. Schmitz, *Tetrahedron Lett.* **24**, 4331 (1983).
16. D. S. Noyce, D. R. Hartter, and F. B. Miles, *J. Am. Chem. Soc.* **90**, 3794 (1968).
17. Y. Pocker, K. D. Stevens, and J. J. Champoux, *J. Am. Chem. Soc.* **91**, 4199 (1969).
18. M. F. Ruasse, A. Argile, and J. E. Dubois, *J. Am. Chem. Soc.* **100**, 7645 (1978).
19. W. K. Chwang, P. Knittel, K. M. Koshy, and T. T. Tidwell, *J. Am. Chem. Soc.* **99**, 3395 (1977).
20. D. J. Pasto and J. F. Gadberry, *J. Am. Chem. Soc.* **100**, 1469 (1978).

Chapter 7

2b. G. L. Closs and L. E. Closs, *J. Am. Chem. Soc.* **85**, 2022 (1963).
f. R. Breslow, *J. Am. Chem. Soc.* **79**, 1762 (1957).
g. K. Ogura and G. Tsuchihashi, *Tetrahedron Lett.*, 3151 (1971).
h. G. L. Closs and R. B. Larabee, *Tetrahedron Lett.*, 287 (1965).
i. R. B. Woodward and C. Wintner, *Tetrahedron Lett.*, 2689 (1969).
j. J. A. Zoltewicz, G. M. Kauffman, and C. L. Smith, *J. Am. Chem. Soc.* **90**, 5939 (1968).
3a. T.-Y. Luh and L. M. Stock, *J. Am. Chem. Soc.* **96**, 3712 (1974).
b. H. W. Amburn, K. C. Kaufman, and H. Shecter, *J. Am. Chem. Soc.* **91**, 530 (1969); F. G. Bordwell, J. C. Branca, C. R. Johnson, and N. R. Vanier, *J. Org. Chem.* **45**, 3884 (1980).
c. F. G. Bordwell, G. E. Drucker, and H. E. Fried, *J. Org. Chem.* **46**, 632 (1981); E. M. Arnett and K. G. Venkatasubramanian, *J. Org. Chem.* **48**, 1569 (1983).
d. T. B. Thompson and W. T. Ford, *J. Am. Chem. Soc.* **101**, 5459 (1979).
4a. G. A. Abad, S. P. Jindal, and T. T. Tidwell, *J. Am. Chem. Soc.* **95**, 6326 (1973).
b. A. Nickon and J. L. Lambert, *J. Am. Chem. Soc.* **84**, 4604 (1962).
6. G. B. Trimitsis and E. M. Van Dam, *J. Chem. Soc., Chem. Commun.*, 610 (1974).
7. A. Streitwieser, Jr., W. B. Hollyhead, A. H. Pudjaatmaka, P. H. Owens, T. L. Kruger, P. A. Rubenstein, R. A. MacQuarrie, M. L. Brokaw, W. K. C. Chu, and H. M. Niemeyer, *J. Am. Chem. Soc.* **93**, 5088 (1971).
8. F. G. Bordwell and F. J. Cornforth, *J. Org. Chem.* **43**, 1763 (1978).
9a. P. T. Lansbury, *J. Am. Chem. Soc.* **83**, 429 (1961).
b. D. W. Griffiths and C. D. Gutsche, *J. Am. Chem. Soc.* **93**, 4788 (1971).
c. T. D. Hoffman and D. J. Cram, *J. Am. Chem. Soc.* **91**, 1000 (1969).
d. G. Büchi, D. M. Foulkes, M. Kurono, G. F. Mitchell, and R. S. Schneider, *J. Am. Chem. Soc.* **89**, 6745 (1967).
e. G. Stork, G. L. Nelson, F. Rouessac, and O. Grigone, *J. Am. Chem. Soc.* **93**, 3091 (1971).
10. S. Danishefsky and R. K. Singh, *J. Am. Chem. Soc.* **97**, 3239 (1975); E. M. Arnett and J. A. Harrelson, Jr., *J. Am. Chem. Soc.* **109**, 809 (1987).
11. A. Streitwieser, Jr., R. G. Lawler, and C. Perrin, *J. Am. Chem. Soc.* **87**, 5383 (1965); J. E. Hofmann, A. Schriesheim, and R. E. Nichols, *Tetrahedron Lett.*, 1745 (1965).
12. E. J. Stamhuis, W. Mass, and H. Wynberg, *J. Org. Chem.* **30**, 2160 (1965).

13. B. G. Cox, *J. Am. Chem. Soc.* **96**, 6823 (1974).
14. E. L. Eliel, A. A. Hartmann, and A. G. Abatjoglou, *J. Am. Chem. Soc.* **96**, 1807 (1974); J.-M. Lehn and G. Wipff, *J. Am. Chem. Soc.* **98**, 7498 (1976).
15a. N. S. Mills, J. Shapiro, and M. Hollingsworth, *J. Am. Chem. Soc.* **103**, 1263 (1981).
b. N. S. Mills, *J. Am. Chem. Soc.* **104**, 5689 (1982).
16. R. B. Woodward and G. Small, Jr., *J. Am. Chem. Soc.* **72**, 1297 (1950).
17a. E. J. Corey, T. H. Hopie, and W. A. Wozniak, *J. Am. Chem. Soc.* **77**, 5415 (1955).
b. E. W. Garbisch, Jr., *J. Org. Chem.* **30**, 2109 (1965).
c. F. Caujolle and D. Q. Quan, *C. R. Acad. Sci., C* **265**, 269 (1967).
d. C. W. P. Crowne, R. M. Evans, G. F. H. Green, and A. G. Long, *J. Chem. Soc.*, 4351 (1956).
e. N. C. Deno and R. Fishbein, *J. Am. Chem. Soc.* **95**, 7445 (1973).
18a. P. L. Stotter and K. A. Hill, *J. Am. Chem. Soc.* **96**, 6524 (1974).
b. B. J. L. Huff, F. N. Tuller, and D. Caine, *J. Org. Chem.* **34**, 3070 (1969).
c. H. O. House and T. M. Bare, *J. Org. Chem.* **33**, 943 (1968).
d. R. S. Matthews, P. K. Hyer, and E. A. Folkers, *J. Chem. Soc., Chem. Commun.*, 38 (1970).
e. C. H. Heathcock, R. A. Badger, and J. W. Patterson, Jr., *J. Am. Chem. Soc.* **89**, 4133 (1967).
f. J. E. McMurry, *J. Am. Chem. Soc.* **90**, 6821 (1968).
19. M. S. Newman, V. DeVries, and R. Darlak, *J. Org. Chem.* **31**, 2171 (1966).
20. H. M. Walborsky and L. M. Turner, *J. Am. Chem. Soc.* **94**, 2273 (1972).
21. Y. Jasor, M. Gaudry, and A. Marquet, *Tetrahedron Lett.*, 53 (1976).

Chapter 8

1. R. P. Bell, *Adv. Phys. Org. Chem.* **4**, 1 (1966).
2a. A. Lapworth and R. H. F. Manske, *J. Chem. Soc.*, 1976 (1930).
b. T. H. Fife, J. E. C. Hutchins, and M. S. Wang, *J. Am. Chem. Soc.* **97**, 5878 (1975).
c. K. Bowden and A. M. Last, *J. Chem. Soc., Chem. Commun.*, 1315 (1970).
d. P. R. Young and W. P. Jencks, *J. Am. Chem. Soc.* **99**, 1206 (1977).
e. H. Dahn and M.-N. Ung-Truong, *Helv. Chim. Acta* **70**, 2130 (1987).
3a, b. M. M. Kreevoy and R. W. Taft, Jr., *J. Am. Chem. Soc.* **77** 5590 (1955).
c, d. M. M. Kreevoy, C. R. Morgan, and R. W. Taft, Jr., *J. Am. Chem. Soc.* **82**, 3064 (1960).
e. T. H. Fife and L. H. Brod, *J. Org. Chem.* **33**, 4136 (1968).
4. R. G. Bergstrom, M. J. Cashen, Y. Chiang, and A. J. Kresge, *J. Org. Chem.* **44**, 1639 (1979).
5a. T. Maugh II and T. C. Bruice, *J. Am. Chem. Soc.* **93**, 3237 (1971).
b. L. E. Eberson and L.-A. Svensson, *J. Am. Chem. Soc.* **93**, 3827 (1971).
c. A. Williams and G. Salvadori, *J. Chem. Soc., Perkin Trans. 2,* 883 (1972).
d. G. A. Rogers and T. C. Bruice, *J. Am. Chem. Soc.* **96**, 2463 (1974).
e. K. Bowden and G. R. Taylor, *J. Chem. Soc., B*, 145, 149 (1971); M. S. Newman and A. L. Leegwater, *J. Am. Chem. Soc.* **90**, 4410 (1968).
f. T. C. Bruice and S. J. Benkovic, *J. Am. Chem. Soc.* **85**, 1 (1963).
6. E. H. Cordes and W. P. Jencks, *J. Am. Chem. Soc.* **85**, 2843 (1963).
7. E. Anderson and T. H. Fife, *J. Am. Chem. Soc.* **95**, 6437 (1973).
8. M. W. Williams and G. T. Young, *J. Chem. Soc.*, 3701 (1964).
9a. V. Somayaji and R. S. Brown, *J. Org. Chem.* **51**, 2676 (1986).
b. A. J. Briggs, C. M. Evans, R. Glenn, and A. J. Kirby, *J. Chem. Soc., Perkin Trans. 2,* **1637** (1983); C. B. Quin and J. R. Wiseman, *J. Am. Chem. Soc.* **95**, 1342 (1973).
10a. N. S. Nudelman, Z. Gatto, and L. Bohe, *J. Org. Chem.* **49**, 1540 (1984).
b. P. Baierwick, D. Hoell, and K. Mueller, *Angew. Chem. Int. Ed. Engl.* **24**, 972 (1985).
c. C. Alvarez-Ibarra, P. Perez-Ossorio, A. Perez-Rubalcaba, M. L. Quiroga, and M. J. Santemases, *J. Chem. Soc., Perkin Trans. 2,* 1645 (1983).
11. H. R. Mahler and E. H. Cordes, *Biological Chemistry*, Harper and Row, New York, 1966, p. 201.
12. J. Hajdu and G. M. Smith, *J. Am. Chem. Soc.* **103**, 6192 (1981).
13. T. Okuyama, H. Shibuya, and T. Fueno, *J. Am. Chem. Soc.* **104**, 730 (1982).
14. R. P. Bell, M. H. Rand, and K. M. A. Wynne-Jones, *Trans. Faraday Soc.* **52**, 1093 (1956).
15. S. T. Purrington and J. H. Pittman, *Tetrahedron Lett.* **28**, 3901 (1987).
16. J. T. Edward and S. C. Wong, *J. Am. Chem. Soc.* **101**, 1807 (1979).

17. R. Breslow and C. McAllister, *J. Am. Chem. Soc.* **93**, 7096 (1971).
18. B. Capon and K. Nimmo, *J. Chem. Soc., Perkin Trans. 2,* 1113 (1975).
19. D. P. Weeks and D. B. Whitney, *J. Am. Chem. Soc.* **103**, 3555 (1981).
20. R. L. Schowen, C. R. Hopper, and C. M. Bazikian, *J. Am. Chem. Soc.* **94**, 3095 (1972).
21. E. Tapuhi and W. P. Jencks, *J. Am. Chem. Soc.* **104**, 5758 (1982).
22. C. M. Evans, R. Glenn, and A. J. Kirby, *J. Am. Chem. Soc.* **104**, 4706 (1982).
23a. T. C. Bruice and I. Oka, *J. Am. Chem. Soc.* **96**, 4500 (1974).
b. R. Hershfield and G. L. Schmir, *J. Am. Chem. Soc.* **95**, 7359 (1973).
c. T. H. Fife and J. E. C. Hutchins, *J. Am. Chem. Soc.* **94**, 2837 (1972).
d. J. Hine, J. C. Craig, Jr., J. G. Underwood II, and F. A. Via, *J. Am. Chem. Soc.* **92**, 5194 (1970).
24. T. S. Davies, P. D. Feil, D. G. Kubler, and D. J. Wells, Jr., *J. Org. Chem.* **40**, 1478 (1975).
25. L. doAmaral, M. P. Baston, H. G. Bull, and E. H. Cordes, *J. Am. Chem. Soc.* **95**, 7369 (1973).
26. A. J. Kirby and P. W. Lancaster, *J. Chem. Soc., Perkin Trans. 2,* 1206 (1972).
27. E. H. Cordes and W. P. Jencks, *J. Am. Chem. Soc.* **85**, 2843 (1963).
28a. M. Hässermann, *Helv. Chim. Acta* **34**, 1482 (1951).
b. A. C. Anderson and J. A. Nelson, *J. Am. Chem. Soc.* **73**, 232 (1954).
c. W. A. Kleschick, C. T. Buse, and C. H. Heathcock, *J. Am. Chem. Soc.* **99**, 247 (1977); C. H. Heathcock, C. T. Buse, W. A. Kleschick, M. C. Pirrung, J. E. Sohn, and J. Lampe, *J. Org. Chem.* **45**, 1066 (1980).
d. A. T. Nielsen and W. J. Houlihan, *Org. React.* **16**, 1 (1968).

Chapter 9

1. P. Reeves, T. Devon, and R. Pettit, *J. Am. Chem. Soc.* **91**, 5890 (1969).
2a. H. L. Ammon and G. L. Wheller, *J. Am. Chem. Soc.* **97**, 2326 (1975).
b. W. v. E. Doering and C. H. DePuy, *J. Am. Chem. Soc.* **75**, 5955 (1953).
c. J. H. M. Hill, *J. Org. Chem.* **32**, 3214 (1967).
d. D. J. Bertelli, *J. Org. Chem.* **30**, 891 (1965).
3a. W. v. E. Doering and F. L. Detert, *J. Am. Chem. Soc.* **73**, 876 (1951).
b. E. F. Jenny and J. D. Roberts, *J. Am. Chem. Soc.* **78**, 2005 (1956).
4a. J. J. Eisch, J. E. Galle, and S. Kozima, *J. Am. Chem. Soc.* **108**, 379 (1986).
b. P. J. Garratt, N. E. Rowland, and F. Sondheimer, *Tetrahedron* **27**, 3157 (1971).
c. R. Concepcion, R. C. Reiter, and G. R. Stevenson, *J. Am. Chem. Soc.* **105**, 1778 (1983).
d. G. Jonsäll and P. Ahlberg, *J. Am. Chem. Soc.* **108**, 3819 (1986).
e. J. M. Herbert, P. D. Woodgate, and W. A. Denny, *J. Med. Chem.* **30**, 2081 (1987).
f. R. F. X. Klein and V. Horak, *J. Org. Chem.* **51**, 4644 (1986).
5. D. Cremer, T. Schmidt, and C. W. Bock, *J. Org. Chem.* **51**, 4644 (1986).
6. B. A. Hess, Jr., and L. J. Schaad, *J. Am. Chem. Soc.* **93**, 305 (1971).
7. D. Bostwick, H. F. Henneike, and H. P. Hopkins, Jr., *J. Am. Chem. Soc.* **97**, 1505 (1975).
8. T. Otsubo, R. Gray, and V. Boekelheide, *J. Am. Chem. Soc.* **100**, 2449 (1978).
9a. Z. Yoshida, M. Shibata, and T. Sugimoto, *Tetrahedron Lett.* **24**, 4585 (1983); *J. Am. Chem. Soc.* **106**, 6383 (1984).
b. K. Nakasuji, K. Yoshida, and I. Murata, *J. Am. Chem. Soc.* **105**, 5136 (1983).
c. K. Takahashi, T. Nozoe, K. Takase, and T. Kudo, *Tetrahedron Lett.* **25**, 77 (1984).
d. A. Minsky, A. Y. Meyers, K. Hafner, and M. Rabinovitz, *J. Am. Chem. Soc.* **105**, 3975 (1983).
10. R. A. Wood, T. R. Welberry, and A. D. Rae, *J. Chem. Soc., Perkin Trans. 2,* 451 (1985); W. E. Rhine, J. H. Davis, and G. Stucky, *J. Organomet. Chem.* **134**, 139 (1977); Y. Cohen, N. H. Roelofs, G. Reinhardt, L. T. Scott, and M. Rabinovitz, *J. Org. Chem.* **52**, 4207 (1987); B. Eliasson and H. Edlund, *J. Chem. Soc., Perkin Trans. 2,* 1837 (1983); B. C. Becker, W. Huber, C. Schneiders, and K. Müller, *Chem. Ber.* **116**, 1573 (1983); A. Minsky, A. Y. Meyers, K. Hafner, and M. Rabinovitz, *J. Am. Chem. Soc.* **105**, 3975 (1983).
11. A. Greenberg, R. P. T. Tomkins, M. Dobrovolny, and J. E. Liebman, *J. Am. Chem. Soc.* **105**, 6855 (1983); F. Gavina, A. M. Costero, P. Gil, and S. V. Luis, *J. Am. Chem. Soc.* **106**, 2077 (1984); P. J. Garratt, *Aromaticity*, Wiley-Interscience, New York, 1986, pp. 173-181.
12. H. Prinzbach, V. Freudenberg, and U. Scheidegger, *Helv. Chim. Acta* **50**, 1087 (1967).

13. J. F. M. Oth, K. Müller, H.-V. Runzheimer, P. Mues, and E. Vogel, *Angew. Chem. Int. Ed. Engl.* **16**, 872 (1977).
14. G. Jonsäll and P. Ahlberg, *J. Am. Chem. Soc.* **108**, 3819 (1986).
15. G. R. Stevenson and B. E. Forch, *J. Am. Chem. Soc.* **102**, 5985 (1980).

Chapter 10

1a. C. K. Ingold and E. H. Ingold, *J. Chem. Soc.*, 2249 (1928).
b. R. J. Albers and E. C. Kooyman, *Rec. Trav. Chim.* **83**, 930 (1964).
c. J. R. Knowles and R. O. C. Norman, *J. Chem. Soc.*, 2938 (1961).
d. A. Gastaminza, T. A. Modro, J. H. Ridd, and J. H. P. Utley, *J. Chem. Soc., B*, 534 (1968).
e. J. R. Knowles, R. O. C. Norman, and G. K. Radda, *J. Chem. Soc.*, 4885 (1960).
f. F. L. Riley and E. Rothstein, *J. Chem. Soc.*, 4885 (1960).
2. T. C. van Hoek, P. E. Verkade, and B. M. Wepster, *Rev. Trav. Chim.* **77**, 559 (1958); A. van Loon, P. E. Verkade, and B. M. Wepster, *Rec. Trav. Chim.* **79**, 977 (1960).
3. G. A. Olah, S. J. Kuhn, S. H. Flood, and J. C. Evans, *J. Am. Chem. Soc.* **84**, 3687 (1962).
4. J. E. Dubois, J. J. Aaron, P. Alcais, J. P. Doucet, F. Rothenberg, and R. Uzan, *J. Am. Chem. Soc.* **94**, 6823 (1972).
5. R. M. Roberts and D. Shiengthong, *J. Am. Chem. Soc.* **86**, 2851 (1964).
6. R. L. Dannley, J. E. Gagen, and K. Zak, *J. Org. Chem.* **38**, 1 (1973); R. L. Dannley and W. R. Knipple, *J. Org. Chem.* **38**, 6 (1973).
7a. C. D. Gutsche and K. H. No, *J. Org. Chem.* **47**, 2708 (1982).
b. G. D. Figuly and J. C. Martin, *J. Org. Chem.* **45**, 3728 (1980).
c. K. Key, C. Eaborn, and D. R. M. Walton, *Organomet. Chem. Synth.* **1**, 151 (1970–1971).
d. S. Winstein and R. Baird, *J. Am. Chem. Soc.* **79**, 756 (1957).
8. D. S. Noyce, P. A. Kittle, and E. H. Bannitt, *J. Org. Chem.* **33**, 1500 (1958).
9. M. Essiz, G. Guillaumet, J.-J. Brunet, and P. Caubere, *J. Org. Chem.* **45**, 240 (1980).
10. W. Nagata, K. Okada, and T. Akoi, *Synthesis*, 365 (1979).
11. W. G. Miller and C. U. Pittman, Jr., *J. Org. Chem.* **39**, 1955 (1974).
12. T. A. Modro and K. Yates, *J. Am. Chem. Soc.* **98**, 4247 (1976).
13. A. V. R. Rao, V. H. Deshpande, and N. L. Reddy, *Tetrahedron Lett.*, 4373 (1982).
14. L. M. Jackman and V. R. Haddon, *J. Am. Chem. Soc.* **96**, 5130 (1974); M. Gates, D. L. Frank, and W. C. von Felten, *J. Am. Chem. Soc.* **96**, 5138 (1974).
15. C. K. Ingold, *Structure and Mechanism in Organic Chemistry*, Second Edition, Cornell University Press, Ithaca, New York, 1969, pp. 340–344; C. G. Swain and D. R. Crist, *J. Am. Chem. Soc.* **94**, 3195 (1972).
16. M. L. Bird and C. K. Ingold, *J. Chem. Soc.*, 918 (1938); J. D. Roberts, J. K. Sanford, F. L. J. Sixma, H. Cerfontain, and R. Zagt, *J. Am. Chem. Soc.* **76**, 4525 (1954).
17. E. Baciocchi, F. Cacace, G. Ciranni, and G. Illuminati, *J. Am. Chem. Soc.* **94**, 7030 (1972).
18. R. B. Moodie and K. Schofield, *Acc. Chem. Res.* **9**, 287 (1976).
19. P. C. Myhre, M. Beug, and L. L. James, *J. Am. Chem. Soc.* **90**, 2105 (1968).
20a. L. R. C. Barclay, B. A. Ginn, and C. E. Milligan, *Can. J. Chem.* **42**, 579 (1964).
b. A. A. Khalaf and R. M. Roberts, *J. Org. Chem.* **34**, 3571 (1969); A. A. Khalaf, *Rev. Chim. (Bucharest)* **19**, 1373 (1974).
21a. G. P. Stahly, *J. Org. Chem.* **50**, 3091 (1985).
b. A. I. Meyers and P. D. Pansegrau, *Tetrahedron Lett.* **24**, 4935 (1983).
c. M. Maksoza and J. Winiarski, *J. Org. Chem.* **49**, 1494 (1984).
d. I. J. Anthony and D. Wege, *Aust. J. Chem.* **37**, 1283 (1984).
e. Y. Konayashi, T. Nagai, I. Kumadaki, M. Takahashi, and T. Yamauchi, *Chem. Pharm. Bull.* **32**, 4382 (1984).

Chapter 11

2a. E. Vogel, *Justus Liebigs Ann. Chem.* **615**, 14 (1958).
b. A. C. Cope, A. C. Haven, Jr., F. L. Ramp, and E. R. Turnbull, *J. Am. Chem. Soc.* **74**, 4867 (1952).

c. R. Pettit, *J. Am. Chem. Soc.* **82**, 1972 (1960).
d. M. Oda, M. Oda, and Y. Kitahara, *Tetrahedron Lett.*, 839 (1976).
e. K. M. Rapp and J. Daub, *Tetrahedron Lett.*, 2011 (1976).
f. E. J. Corey and D. K. Herron, *Tetrahedron Lett.*, 1641 (1971).
3. A. G. Anastassiou, V. Orfanos, and J. H. Gebrian, *Tetrahedron Lett.*, 4491 (1969); P. Radlick and G. Alford, *J. Am. Chem. Soc.* **91**, 6529 (1969).
4a. K. B. Wiberg, V. Z. Williams, Jr., and L. E. Friedrich, *J. Am. Chem. Soc.* **90**, 5338 (1968).
b. P. S. Wharton and R. A. Kretchmer, *J. Org. Chem.* **33**, 4258 (1968).
c. R. L. Danheiser, C. Martinez, and J. M. Morin, *J. Org. Chem.* **45**, 1340 (1980); R. L. Danheiser, C. Martinez-Davilla, R. J. Auchus, and J. T. Kadonaga, *J. Am. Chem. Soc.* **103**, 2443 (1981).
d. R. K. Hill and M. G. Bock, *J. Am. Chem. Soc.* **100**, 637 (1978).
e. M. Newcomb and W. T. Ford, *J. Am. Chem. Soc.* **95**, 7186 (1973).
f. R. K. Hill, C. B. Giberson, and J. V. Silverton, *J. Am. Chem. Soc.* **110**, 497 (1988).
5a. L. A. Paquette and M. Oku, *J. Am. Chem. Soc.* **96**, 1219 (1974).
b. A. E. Hill, G. Greenwood, and H. M. R. Hoffmann, *J. Am. Chem. Soc.* **95**, 1338 (1973).
c. S. W. Staley and T. J. Henry, *J. Am. Chem. Soc.* **93**, 1292 (1971).
d. T. Kauffmann and E. Köppelmann, *Angew. Chem. Int. Ed. Engl.* **11**, 290 (1972).
e. C. W. Jefford, A. F. Boschung, and C. G. Rimbault, *Tetrahedron Lett.*, 3387 (1974).
f. M. F. Semelhack, H. N. Weller, and J. S. Foos, *J. Am. Chem. Soc.* **99**, 292 (1977).
g. R. K. Boeckman, Jr., M. H. Delton, T. Nagasaka, and T. Watanabe, *J. Org. Chem.* **42**, 2946 (1977).
h. I. Hasan and F. W. Fowler, *J. Am. Chem. Soc.* **100**, 6696 (1978).
i. R. Subramanyan, P. D. Bartlett, G. Y. M. Iglesias, W. H. Watson, and J. Galloy, *J. Org. Chem.* **47**, 4491 (1982).
6. A. Anastassiou and R. P. Cellura, *J. Chem. Soc., Chem. Commun.*, 1521 (1969).
7a. L. A. Feiler, R. Huisgen, and P. Koppitz, *J. Am. Chem. Soc.* **96**, 2270 (1974).
b. H. H. Wasserman, J. U. Piper, and E. V. Dehmlow, *J. Org. Chem.* **38**, 1451 (1973).
c. D. A. Evans and A. M. Golob, *J. Am. Chem. Soc.* **97**, 4765 (1975).
d. K. Oshima, H. Takahashi, H. Yamamoto, and H. Nozaki, *J. Am. Chem. Soc.* **95**, 2693 (1973).
e. N. Shimizu, M. Tanaka, and Y. Tsuno, *J. Am. Chem. Soc.* **104**, 1330 (1982).
f. V. Cere, E. Dalcanale, C. Paolucci, S. Pollicino, E. Sandri, L. Lunazzi, and A. Fava, *J. Org. Chem.* **47**, 3540 (1982).
g. L. A. Paquette and M. J. Wyvratt, *J. Am. Chem. Soc.* **96**, 4671 (1974); D. McNeil, B. R. Vogt, J. J. Sudol, S. Theodoropulos, and E. Hedaya, *J. Am. Chem. Soc.* **96**, 4673 (1974).
h. D. Bellus, H.-C. Mez, G. Rihs, and H. Sauter, *J. Am. Chem. Soc.* **96**, 5007 (1974).
i. W. Grimme, *J. Am. Chem. Soc.* **95**, 2381 (1973).
j. W. Weyler, Jr., L. R. Byrd, M. C. Caserio, and H. W. Moore, *J. Am. Chem. Soc.* **94**, 1027 (1972).
k. M. Nakazaki, K. Naemura, H. Harada, and H. Narutaki, *J. Org. Chem.* **47**, 3470 (1982).
8. W. H. Rastetter and T. J. Richard, *J. Am. Chem. Soc.* **101**, 3893 (1979).
9. R. Huisgen and W. E. Konz, *J. Am. Chem. Soc.* **92**, 4102 (1970).
10. K. Maruyama, N. Nagai, and Y. Naruta, *J. Org. Chem.* **51**, 5083 (1986).
11. R. Subramanyam, P. D. Bartlett, G. Y. M. Iglesia, W. H. Watson, and J. Galloy, *J. Org. Chem.* **47**, 4491 (1982).
12a. L. A. Paquette and R. S. Beckley, *J. Am. Chem. Soc.* **97**, 1084 (1975).
b. K. C. Nicolaou, N. A. Petasis, R. E. Zipkin, and J. Uenishi, *J. Am. Chem. Soc.* **104**, 5555 (1982).
c. B. M. Trost and A. J. Bridges, *J. Am. Chem. Soc.* **98**, 5017 (1976).
d. K. C. Nicolaou, N. A. Petasis, R. E. Zipkin, and J. Uenishi, *J. Am. Chem. Soc.* **104**, 5555 (1982).
e. K. J. Shea and R. B. Phillips, *J. Am. Chem. Soc.* **100**, 654 (1978).
13. R. K. Hill, J. W. Morgan, R. V. Shetty, and M. E. Synerholm, *J. Am. Chem. Soc.* **96**, 4201 (1974); H. M. R. Hoffmann, *Angew. Chem. Int. Ed. Engl.* **8**, 556 (1969).
14a. T. J. Brocksom and M. G. Constantino, *J. Org. Chem.* **47**, 3450 (1982).
b. L. E. Overman, G. F. Taylor, K. N. Houk, and L. N. Domelsmith, *J. Am. Chem. Soc.* **100**, 3182 (1978).
c. P. W. Tang and C. A. Maggiulli, *J. Org. Chem.* **46**, 3429 (1981).
d. R. B. Woodward, F. Sondheimer, D. Taub, K. Heusler, and W. M. McLamore, *J. Am. Chem. Soc.* **74**, 4223 (1952).
e. T. Cohen and Z. Kosarych, *J. Org. Chem.* **47**, 4005 (1982).
15. S. V. Ley and L. A. Paquette, *J. Am. Chem. Soc.* **96**, 2887 (1974).
16a. A. K. Cheng, F. A. L. Anet, J. Mioduski, and J. Meinwald, *J. Am. Chem. Soc.* **96**, 2887 (1974).
b. J. S. McKennis, L. Brener, J. S. Ward, and R. Pettit, *J. Am. Chem. Soc.* **93**, 4957 (1971).

c. W. Grimme, H. J. Riebel, and E. Vogel, *Angew. Chem. Int. Ed. Engl.* **7**, 823 (1968).
d. W. Grimme, *J. Am. Chem. Soc.* **94**, 2525 (1972).
J. J. Gajewski, L. K. Hoffman, and C. N. Shih, *J. Am. Chem. Soc.* **96**, 3705 (1974).
f. D. P. Lutz and J. D. Roberts, *J. Am. Chem. Soc.* **83**, 2198 (1961).
17. A. Krantz, *J. Am. Chem. Soc.* **94**, 4020 (1972).
18. H.-D. Martin and E. Eisenmann, *Tetrahedron Lett.*, 661 (1975).
19a. A. Viola and L. Levasseur, *J. Am. Chem. Soc.* **87**, 1150 (1965).
b. S. F. Reed, Jr., *J. Org. Chem.* **30**, 1663 (1965).
c. T. S. Cantrell and H. Shechter, *J. Am. Chem. Soc.* **89**, 5868 (1967).
d. R. B. Woodward, R. E. Lehr, and H. H. Inhoffen, *Justus Liebigs Ann. Chem.* **714**, 57 (1968).
e. R. B. Woodward and T. J. Katz, *Tetrahedron* **5**, 70 (1959).
f. N. J. Turro and W. B. Hammond, *Tetrahedron* **24**, 6029 (1968).
g. J. S. McConaghy, Jr., and J. J. Bloomfield, *Tetrahedron Lett.*, 3719 (1969).
h. W. J. Linn and R. E. Benson, *J. Am. Chem. Soc.* **87**, 3657 (1965).
i. J. K. Crandall and W. H. Machleder, *J. Am. Chem. Soc.* **90**, 7292 (1968).
j. M. Jones, Jr., S. D. Reich, and L. T. Scott, *J. Am. Chem. Soc.* **92**, 3118 (1970).
k. M. Jones, Jr., and B. Fairless, *Tetrahedron Lett.*, 4881 (1968); R. T. Seidner, N. Nakatsuka, and S. Masamune, *Can. J. Chem.* **48**, 187 (1970).
l. P. G. Gassman, J. J. Roos, and S. J. Lee, *J. Org. Chem.* **49**, 717 (1984).
m. Y. N. Gupta, M. J. Don, and K. N. Houk, *J. Am. Chem. Soc.* **104**, 7336 (1982).
n. V. Glock, M. Wette, and F.-G. Klärner, *Tetrahedron Lett.* **26**, 1441 (1985).
o. P. A. Zoretic, R. J. Chambers, G. D. Marbury, and A. A. Riebiro, *J. Org. Chem.* **50**, 2981 (1985).
p. S. Sato, K. Tomita, H. Fujita, and Y. Sabo, *Heterocycles* **22**, 1045 (1984).
q. M. Nakazaki, K. Naemura, H. Harada, and H. Narutaki, *J. Org. Chem.* **47**, 3470 (1982).

Chapter 12

1a. H. O. House, C.-Y. Chu, W. V. Phillips, T. S. B. Sayer, and C.-C. Yau, *J. Org. Chem.* **42**, 1709 (1977).
b. K. Fujunishi, Y. Inoue, Y. Kishimoto, and F. Mashio, *J. Org. Chem.* **40**, 628 (1975).
c. C. Walling and D. Bristol, *J. Org. Chem.* **37**, 3514 (1972).
d. D. H. Miles, P. Loew, W. S. Johnson, A. F. Kluge, and J. Meinwald, *Tetrahedron Lett.*, 3019 (1972).
e. K. Pedersen, P. Jakobsen, and S.-O. Lawesson, *Org. Synth.* **V**, 70 (1973).
f. R. Dowbenko, *Org. Synth.* **V**, 93 (1973).
g. W. H. Sharkey and C. M. Langkammerer, *Org. Synth.* **V**, 445 (1973).
h. P. D. Klemmensen, H. Kolind-Andersen, H. B. Madsen, and A. Svendsen, *J. Org. Chem.* **44**, 416 (1979).
i. R. A. Rossi, R. H. deRossi, and A. F. Lopez, *J. Am. Chem. Soc.* **98**, 1252 (1976).
j. D. W. Grattan, J. M. Locke, and S. R. Wallis, *J. Chem. Soc., Perkin Trans. 1*, 2264 (1973).
3. L. H. Gale, *J. Am. Chem. Soc.* **88**, 4661 (1961).
4. I. Rosenthal and D. Elad, *J. Org. Chem.* **33**, 805 (1968); R. Lalande, B. Maillard, and M. Cazaux, *Tetrahedron Lett.*, 745 (1969).
5a. D. D. Tanner, H. Yabuuchi, and E. V. Blackburn, *J. Am. Chem. Soc.* **93**, 4802 (1971).
b. E. Müller, *Tetrahedron Lett.*, 1835 (1974).
c. A. L. J. Beckwith and C. J. Easton, *J. Am. Chem. Soc.* **103**, 615 (1981).
d. H. C. Brown, M. S. Kharasch, and T. H. Chao, *J. Am. Chem. Soc.* **62**, 3435 (1940).
e. A. Effio, D. Griller, K. U. Ingold, J. C. Scaiano, and S. J. Sheng, *J. Am. Chem. Soc.* **102**, 6063 (1980).
f. N. O. Brace, *J. Am. Chem. Soc.* **86**, 523 (1964).
6a. W. H. Urry, D. J. Trecker, and H. D. Hartzler, *J. Org. Chem.* **29**, 1663 (1964).
b. H. Pines, N. C. Sih, and D. B. Rosenfield, *J. Org. Chem.* **31**, 2255 (1966).
c. D. D. Tanner, E. V. Blackburn, and G. E. Diaz, *J. Am. Chem. Soc.* **103**, 1557 (1981).
d. H. Driquez, J. M. Paton, and J. Lessard, *Can. J. Chem.* **55**, 700 (1977).
e. H. Feuer, J. Doty, and J. P. Lawrence, *J. Org. Chem.* **38**, 417 (1973).
f. E. A. Mayeda, *J. Am. Chem. Soc.* **97**, 4012 (1975).
g. X. Creary, *J. Org. Chem.* **45**, 280 (1980).
h. A. L. J. Beckwith and R. D. Wagner, *J. Am. Chem. Soc.* **101**, 7099 (1979).

7. L. K. Montgomery and J. W. Matt, *J. Am. Chem. Soc.* **93**, 4802 (1971).
8. P. D. Bartlett, R. E. Pinnock, J. H. Rolston, W. G. Schindel, and L. A. Singer, *J. Am. Chem. Soc.* **87**, 2590 (1965).
9. E. S. Huyser and B. Giddings, *J. Org. Chem.* **27**, 3391 (1962).
10. N. C. Deno, W. E. Billups, R. Fishbein, C. Pierson, R. Whalen, and J. C. Wyckoff, *J. Am. Chem. Soc.* **93**, 438 (1971).
11a. D. H. Levy and R. J. Myers, *J. Chem. Phys.* **41**, 1062 (1964).
b. J. K. Kochi and P. J. Krusic, *J. Am. Chem. Soc.* **90**, 7157 (1968).
12. H. Tanida and T. Tsuji, *J. Org. Chem.* **29**, 849 (1964); P. R. Story, *Tetrahedron Lett.*, 401 (1962).
13. E. J. Heiba, R. M. Dessau, and W. J. Koehl, Jr., *J. Am. Chem. Soc.* **90**, 7157 (1968).
14. C. L. Karl, E. J. Maas, and W. Reusch, *J. Org. Chem.* **37**, 2834 (1972).
15. P. R. Story, D. D. Denson, C. E. Bishop, B. C. Clark, Jr., and J.-C. Farine, *J. Am. Chem. Soc.* **90**, 817 (1968).
16. K. J. Shea and P. S. Skell, *J. Am. Chem. Soc.* **95**, 6728 (1973).
17. W. A. Bonner and F. D. Mango, *J. Org. Chem.* **29**, 430 (1964).
18a. N. Kornblum, H. K. Singh, and W. J. Kelly, *J. Org. Chem.* **48**, 332 (1983).
b. G. A. Russell, F. Ros, J. Hershberger, and H. Tashtoush, *J. Org. Chem.* **47**, 1480 (1982).
c. G. A. Russell, J. Hershberger, and K. Owens, *J. Am. Chem. Soc.* **101**, 1312 (1979).
19. S. S. D. Brown, J. J. Colquhoun, W. McFarlane, M. Murray, I. D. Salter, and V. Sik, *J. Chem. Soc., Chem. Commun.*, 53 (1986).
20. D. J. De Frees, R. J. McIver, Jr., and W. J. Hehre, *J. Am. Chem. Soc.* **102**, 3334 (1980).

Chapter 13

1. C. C. Lee and D. Unger, *Can. J. Chem.* **50**, 593 (1972).
2a. E. Vogel, W. Grimme, and E. Dinne, *Tetrahedron Lett.*, 391 (1965).
b. J. Meinwald and J. W. Young, *J. Am. Chem. Soc.* **93**, 725 (1971).
c. D. M. Madigan and J. S. Swenton, *J. Am. Chem. Soc.* **93**, 6316 (1971).
d. E. J. Corey and A. G. Hortmann, *J. Am. Chem. Soc.* **85**, 4033 (1963).
e, f. K. M. Shumate, P. N. Neumann, and G. J. Fonken, *J. Am. Chem. Soc.* **87**, 3996 (1965); K. M. Shumate and G. J. Fonken, *J. Am. Chem. Soc.* **88**, 1073 (1966).
g. H. E. Zimmerman and M.-L. Viriot-Villaume, *J. Am. Chem. Soc.* **95**, 1274 (1973).
h. R. L. Coffin, R. S. Givens, and R. G. Carlson, *J. Am. Chem. Soc.* **96**, 7554 (1974).
3a. O. Rodriguez and H. Morrison, *J. Chem. Soc., Chem. Commun.*, 679 (1971).
b. W. G. Dauben and M. S. Kellogg, *J. Am. Chem. Soc.* **93**, 3805 (1971).
c. D. H. R. Barton, D. L. J. Clive, P. D. Magnus, and G. Smith, *J. Chem. Soc., C*, 2193 (1971).
4a. W. C. Agosta and A. B. Smith III, *J. Am. Chem. Soc.* **93**, 5513 (1971).
b. W. Ferree, Jr., J. B. Grutzner, and H. Morrison, *J. Am. Chem. Soc.* **93**, 5502 (1971).
c. A. Wissner, *J. Org. Chem.* **42**, 356 (1977).
d. G. W. Shaffer and M. Pesaro, *J. Org. Chem.* **39**, 2489 (1974).
e. D. R. Morton and N. J. Turro, *J. Am. Chem. Soc.* **95**, 3947 (1973).
f. H. Hart and A. F. Naples, *J. Am. Chem. Soc.* **94**, 3256 (1972).
g. M. Pomerantz and G. W. Gruber, *J. Am. Chem. Soc.* **93**, 6615 (1971).
h. O. L. Chapman, G. W. Borden, R. W. King, and B. Winkler, *J. Am. Chem. Soc.* **86**, 2660 (1964); A. Padwa, L. Brodsky, and S. Clough, *J. Am. Chem. Soc.* **94**, 6767 (1972).
i. R. K. Russell, R. E. Wingard, Jr., and L. A. Paquette, *J. Am. Chem. Soc.* **96**, 7483 (1974).
j. D. I. Schuster and C. W. Kim, *J. Am. Chem. Soc.* **96**, 7437 (1974).
k. W. G. Dauben, M. S. Kellogg, J. I. Seeman, and W. A. Spitzer, *J. Am. Chem. Soc.* **92**, 1786 (1970).
l. A. Padwa and W. Eisenberg, *J. Am. Chem. Soc.* **92**, 2590 (1970).
m. D. H. R. Barton and G. Quinkert, *J. Chem. Soc.*, 1 (1960).
n. R. S. Cooke and G. D. Lyon, *J. Am. Chem. Soc.* **93**, 3840 (1971).
o. D. Bryce-Smith and J. E. Lodge, *J. Chem. Soc.*, 695 (1963); E. Grovenstein, Jr., and D. V. Rao, *Tetrahedron Lett.*, 148 (1961).
p. L. E. Friedrich and J. D. Bower, *J. Am. Chem. Soc.* **95**, 6869 (1973).
5. T.-Y. Leong, T. Imagawa, K. Kimoto, and M. Kawanisi, *Bull. Chem. Soc. Jpn*, **46**, 596 (1973).
6. W. B. Hammond and T. S. Yeung, *Tetrahedron Lett.*, 1173 (1975).

7. H. E. Zimmerman, R. S. Givens, and R. M. Pagni, *J. Am. Chem. Soc.* **90**, 6096 (1968).
8. P. J. Wagner, P. A. Kelso, and R. G. Zepp, *J. Am. Chem. Soc.* **94**, 7480 (1972).
10. N. J. Turro and D. S. Weiss, *J. Am. Chem. Soc.* **90**, 2185 (1968).
11. S. Boue and R. Srinivasan, *J. Am. Chem. Soc.* **92**, 3226 (1970); J. Saltiel *et al.*, *Org. Photochem.* **3**, 1 (1973).
12. W. G. Dauben, K. Koch, S. L. Smith, and O. L. Chapman, *J. Am. Chem. Soc.* **85**, 2616 (1963).
13. H. E. Zimmerman and R. D. Solomon, *J. Am. Chem. Soc.* **108**, 6276 (1986).
14. H. E. Zimmerman, *Angew. Chem. Int. Ed. Engl.* **8**, 1 (1969).
15. K. R. Huffman, M. Burger, W. A. Henderson, Jr., M. Loy, and E. F. Ullman, *J. Org. Chem.* **34**, 1 (1969).
16. F. D. Lewis, R. W. Johnson, and D. E. Johnson, *J. Am. Chem. Soc.* **96**, 6090 (1974).
17. H. E. Zimmerman, R. J. Boettcher, N. E. Buehler, G. E. Keck, and M. G. Steinmetz, *J. Am. Chem. Soc.* **98**, 7680 (1976).
18. H. E. Zimmerman, R. J. Boettcher, and W. Braig, *J. Am. Chem. Soc.* **95**, 2155 (1973).
19. J. R. Scheffer, K. S. Bhandari, R. E. Gayler, and R. A. Wostradowski, *J. Am. Chem. Soc.* **97**, 2178 (1975).
20. C. W. Jefford and F. Delay, *J. Am. Chem. Soc.* **97**, 2272 (1975).

Index